Numerical Mathematics and Advanced Applications

Karl Kunisch • Günther Of • Olaf Steinbach
Editors

Numerical Mathematics and Advanced Applications

Proceedings of ENUMATH 2007,
the 7th European Conference on Numerical
Mathematics and Advanced Applications,
Graz, Austria, September 2007

Springer

Karl Kunisch
Institut für Mathematik und
Wissenschaftliches Rechnen
Karl Franzens Universität Graz
Heinrichstrasse 36
8010 Graz
Austria
karl.kunisch@uni-graz.at

Günther Of
Olaf Steinbach
Institut für Numerische Mathematik
Technische Universität Graz
Steyrergasse 30
8010 Graz
Austria
of@tugraz.at
o.steinbach@tugraz.at

Cover photo - © by Graz Tourismus

ISBN 978-3-540-69776-3 e-ISBN 978-3-540-69777-0
DOI: 10.1007/978-3-540-69777-0

Library of Congress Control Number: 2008933508

Cover design: WMXDesign GmbH, Heidelberg, Germany

Printed on acid-free paper

9 8 7 6 5 4 3 2 1

springer.com

Preface

The European Conference on Numerical Mathematics and Advanced Applications (ENUMATH) is a series of conferences held every two years to provide a forum for discussion on recent aspects of numerical mathematics and their applications. The first ENUMATH conference was held in Paris (1995), and the series continued by the one in Heidelberg (1997), Jyvaskyla (1999), Ischia (2001), Prague (2003), and Santiago de Compostela (2005).

This volume contains a selection of invited plenary lectures, papers presented in minisymposia, and contributed papers of ENUMATH 2007, held in Graz, Austria, September 10–14, 2007.

We are happy that so many people have shown their interest in this conference. In addition to the ten invited presentations and the public lecture, we had more than 240 talks in nine minisymposia and fifty four sessions of contributed talks, and about 316 participants from all over the world, specially from Europe. A total of 98 contributions appear in these proceedings.

Topics include theoretical aspects of new numerical techniques and algorithms, as well as to applications in engineering and science. The book will be useful for a wide range of readers, giving them an excellent overview of the most modern methods, techniques, algorithms and results in numerical mathematics, scientific computing and their applications.

We would like to thank all the participants for the attendance and for their valuable contributions and discussions during the conference. Special thanks go the minisymposium organizers, who made a large contribution to the conference, the chair persons, and all speakers.

We would like to address our thanks to the invited speakers: A. Arnold (Austria), J. Barrett (United Kingdom), J. Chleboun (Czech Republic), M. Griebel (Germany), P. Joly (France), U. Langer (Austria), D. Marini (Italy), G. Papanicolaou (USA), G. Wanner (Switzerland), J. Xu (USA), and to P. Deuflhard (Germany) who gave the public lecture during ENUMATH 2007.

A big share of the success of this conference series should be given to the members of the Programme Committee: F. Brezzi, M. Feistauer, R. Glowinski, R. Jeltsch, Y. Kuznetsov, J. Periaux, R. Rannacher.

We are greatly indebted to the Scientific Committee (C. Bernardi, H. G. Bock, A. Borzi, C. Canuto, A. Bermudes, G. Haase, R. Hoppe, A. Iserles, F. Kappel, G. Kobelkov, M. Krizek, O. Pironneau, A. Quarteroni, S. Repin, S. Sauter, J. Sanz–Serna, C. Schwab, E. Süli), and to the local Organizing Committee (A. Borzi, B. Carpentieri, G. Haase, M. Hintermüller, F. Kappel, S. Keeling, V. Kovtunenko, G. Peichl, G. Propst, W. Ring, S. Volkwein). Special thanks go to S. Fürtinger, E. Hötzl, A. Kimeswenger, O. Lass, A. Laurain M. Leykauf, B. Pöltl, E. Rath, J. Rubesa, M. Taus, C. Thenius, F. Tschiatschek, G. von Winckel for their help in the administration, computer support, and during the conference.

We gratefully acknowledge the financial support by the Austrian Ministry for Science and Research, the State of Styria, the Association for Applied Mathematics and Mechanics (GAMM), the Start Prize project Interfaces and Free Boundaries, the Karl Franzens University Graz, and Graz University of Technology.

We would like to thank all authors for their contributions to this volume. Moreover, we also like to thank all anonymous referees for their work, their criticism, and their proposals. These hints were very helpful to improve the contributions. Finally, we would like to thank Springer Heidelberg for the continuous support and patience while preparing this volume.

Graz,
July 2008

Karl Kunisch
Günther Of
Olaf Steinbach

Contents

Biomedical Applications

Computational Electromagnetism

Computational Methods, Preconditioners, Solvers

Domain Decomposition Methods

Finance, Stochastic Applications

Fluid Mechanics

Optimal Control Problems

Optimization, Inverse Problems

Ordinary Differential Equations

Plenary Lectures

The Worst Scenario Method: A Red Thread Running Through Various Approaches to Problems with Uncertain Input Data

J. Chleboun

Abstract Three ingredients constitute mathematical models dependent on parameters whose value is uncertain: a compact set $\mathscr{U}_{\mathrm{ad}}$ of admissible parameters a, a state problem $A(a)u = f(a)$ with an a-dependent state $u \equiv u(a)$, and a continuous quantity of interest $\Psi(a) = \Phi(a, u(a))$. In the worst scenario method (WSM), the maximum of Ψ over $\mathscr{U}_{\mathrm{ad}}$ is identified. By mastering the WSM and if an adequate characterization of input uncertainty is available, the analyst can easily step forward to a more complex uncertainty analysis, namely that based on the Dempster-Shafer theory or fuzzy set theory. Elements of the above non-stochastic approaches to uncertainty modeling are presented with the emphasis on uncertain functions appearing in problems driven by differential equations.

1 Introduction

Since uncertainty in input parameters accompanies most, if not all, mathematical and computational models, its impact on model outputs deserves attention. We will focus on the worst scenario method (WSM) that can be applied as a stand-alone method (Subsection 2.1) or used as a fundamental part of other approaches such as the Dempster-Shafer theory (Subsection 2.2) and fuzzy set theory (Subsection 2.3). That is, by mastering the WSM, the analyst can easily step forward to a more complex uncertainty analysis if an adequate characterization of input uncertainty is available. Attention is paid to uncertain functions appearing in problems driven by differential equations (Section 3). The goal of this paper is two-fold: (A) to provide the reader with an insight into non-stochastic uncertainty modeling, and (B) to show the reader how non-stochastic uncertainty in input functions can be treated.

Jan Chleboun
Institute of Mathematics, Academy of Sciences, Žitná 25, 115 67 Praha 1,
and Faculty of Engineering, Czech Technical University, Karlovo nám. 13, 121 35 Praha 2,
Czech Republic, e-mail: chleb@math.cas.cz

Although other sources aiming at (A) can be found in the literature, (B) seems to be a rather uncommon subject.

The assessment of uncertainty in data is, essentially, equivalent to the weighting of data. Consequently, as uncertainty propagates through a model, the model outputs are also weighted and the determination of these weights counts among the analyst's ultimate goals. Different weighting approaches result in different methods or even theories.

Stochastic methods stem from weighting the values of input parameters by the probability of their occurrence. Stochastic methods can yield strong results but the analyst should be aware of the fact that they also assume rather strong input information such as the probability distribution of uncertain input parameters and a possible correlation between them, for example. Such information is not always available or it is itself highly uncertain. If this is the case, other methods of weighting input data can be more appropriate, reliable, and realistic.

2 Non-Stochastic Methods

Three representatives of non-stochastic methods will be introduced. Let us start with the basic mathematical framework that will be shared by all the presented methods:

(a) Let the state problem be represented by $A(a)u = f(a)$, an a-dependent equation where a is an input parameter. The existence and uniqueness of the state solution $u \equiv u(a)$ is assumed for any a considered.

(b) Let the a-dependent solution $u(a)$ be evaluated by $\Phi(a, u(a))$, a real-valued criterion-functional often called the quantity of interest that can directly depend on a. Owing to the uniqueness of $u(a)$, the criterion-functional Φ gives rise to the criterion-functional $\Psi(a) = \Phi(a, u(a))$. It is assumed that both u and Ψ depend continuously on a.

Both (a) and (b) deserve a few comments. State problems are not limited to equations; variational inequalities, for instance, are also possible; see [16]. The parameter a can be a scalar, a vector, a tensor, a function, or an n-tuple of functions.

The criterion-functional can represent quantities such as local temperature, local stress invariants, potential energy, or the distance between u and an *a priori* given function.

To illustrate (a) and (b), let us consider a steady heat flow problem depending on a thermal conductivity coefficient a; see also (11)-(13). The state equation (together with relevant boundary conditions) determines the temperature field $u(a)$ in the problem domain. Let $\Psi(a)$, the a-dependent quantity of interest, be defined as an average temperature in a small fixed subdomain; see (14). A change in a can cause a change in $u(a)$ and $\Psi(a)$.

2.1 Worst Scenario Method

It happens quite often that the parameter a cannot be uniquely determined and that we only know that a belongs to $\mathscr{U}_{\mathrm{ad}}$, a set of admissible values. These can originate from measurements or expert opinions, for instance. In other words, a is uncertain, so are $u(a)$ and $\Psi(a)$.

In the worst scenario method, the input values are not weighted. The significance of $a_1 \in \mathscr{U}_{\mathrm{ad}}$ is equal to the significance of $a_2 \in \mathscr{U}_{\mathrm{ad}}$. Given $\mathscr{U}_{\mathrm{ad}}$, the goal of the method is to find $a^0 \in \mathscr{U}_{\mathrm{ad}}$ such that

$$a^0 = \arg\max_{a \in \mathscr{U}_{\mathrm{ad}}} \Psi(a). \tag{1}$$

Since large values of quantities commonly used in engineering (such as mechanical stress, displacement, temperature) are usually considered dangerous, the maximum values correspond to the worst scenario that can happen among all $\mathscr{U}_{\mathrm{ad}}$-driven scenarios. Problem (1) is also known as *anti-optimization*; see [8, 9].

A slight modification of (1) leads to the best scenario problem: find $a_0 \in \mathscr{U}_{\mathrm{ad}}$ such that

$$a_0 = \arg\min_{a \in \mathscr{U}_{\mathrm{ad}}} \Psi(a). \tag{2}$$

It is not generally guaranteed that such a_0 and a^0 exist. If $\mathscr{U}_{\mathrm{ad}}$ is a compact subset of a Banach space and Ψ is continuous, then a^0 and a_0 exist and, if $\mathscr{U}_{\mathrm{ad}}$ is connected, determine I_Ψ, the range of $\Psi|_{\mathscr{U}_{\mathrm{ad}}}$:

$$I_\Psi = [\Psi(a_0), \Psi(a^0)]. \tag{3}$$

From the computational standpoint, convex $\mathscr{U}_{\mathrm{ad}}$ are preferred.

The above assumptions are fulfilled in many engineering problems; see [16] for examples from heat transfer, elasticity and plasticity theory as well as other fields. A short survey of mostly PDE-oriented applications of the method appeared in [14].

2.2 Dempster-Shafer Theory

Although the range (3) is useful to know when one analyzes the impact of uncertainty in input parameters on the quantity of interest, the plain range is dissatisfactory in many practical problems where some weights can be attributed to the input values even if these weights are not probabilistic. Then the analyst should strive for determining the weights of model outputs.

In the approach stemming from the works of Dempster and Shafer (see [6, 19]), sets are weighted. Details and examples can also be found in [1, 3], for instance.

Let us confine ourselves to the most essential ideas relevant to our purpose. We assume that U_i, where $i = 1, 2, \ldots, k$, are given convex and compact subsets (called focal elements) of a Banach space. Moreover, let each U_i have an assigned weight $m_U(U_i) > 0$ such that $\sum_{i=1}^{k} m_U(U_i) = 1$. These weights represent the information we have about U_i. Some U_i, for instance, can originate from less reliable measurements than the others. This would be indicated by the lower weights of these U_i.

By solving (1) and (2), where $\mathscr{U}_{\mathrm{ad}} = U_i$, we obtain the respective scenarios a_0^i and a_i^0. Consequently, see (3), we arrive at intervals I_Ψ^i that will constitute a new family of focal elements, now in $\mathbb{R}$, the space of real numbers. If it happens that $I_\Psi^i = I_\Psi^j$ for some $i \neq j$, the interval is considered only once; thus a family of $\hat{k}$ intervals $\hat{I}_\Psi^l$ is established, where $l = 1, 2, \ldots, \hat{k}$ and $1 \leq \hat{k} \leq k$.

The extension principle allows for deriving $m_\Psi(\hat{I}_\Psi^l)$, the weight of $\hat{I}_\Psi^l$:

$$m_\Psi(\hat{I}_\Psi^l) = \sum_{\{j \in \{1,2,\ldots,k\}:\ I_\Psi^j = \hat{I}_\Psi^l\}} m_U(U_j), \quad l = 1, 2, \ldots, \hat{k}. \tag{4}$$

The quantity $m_\Psi(\hat{I}_\Psi^l)$ can be interpreted as a measure of the amount of "likelihood" (the weight) that is assigned to $\hat{I}_\Psi^l$; see [17]. This assignment is determined by the criterion-functional Ψ and by the "likelihood" assigned to the sets U_i.

Once $m_\Psi(\hat{I}_\Psi^l)$ is determined for $l \in K = \{1, 2, \ldots, \hat{k}\}$ and $m_\Psi(\emptyset) = 0$ is defined, two mappings from subsets of $\mathbb{R}$ to the interval $[0,1]$ can be introduced. These are *Bel*, *belief*, and *Pl*, *plausibility*:

$$Bel(S) = \sum_{\{l \in K:\ \hat{I}_\Psi^l \subset S\}} m_\Psi(\hat{I}_\Psi^l), \qquad Pl(S) = \sum_{\{l \in K |\ \hat{I}_\Psi^l \cap S \neq \emptyset\}} m_\Psi(\hat{I}_\Psi^l), \qquad S \subset \mathbb{R}. \tag{5}$$

Referring to [17] again, we can interpret $Bel(S)$ as a lower bound on the likelihood of S and $Pl(S)$ as an upper bound on the likelihood of S. According to [1], $Bel(S)$ (and similarly $Pl(S)$) can also be interpreted as a lower (upper) limit on the strength of evidence at hand.

Example 1. Let us consider a loaded cantilever beam with one end fixed and the other supported by a spring whose stiffness a is uncertain and represented by five different intervals U_i with respective weights 0.1, 0.4, 0.1, 0.25, and 0.15. Let Ψ be defined as the displacement of the supported tip of the cantilever. Let $[72,82]$, $[68,74]$, $[73,79]$, $[71,83]$, and $[76,84]$ be the respective displacement intervals I_Ψ^i determined by the worst (best) scenario problems (1)–(2) solved for $a \in U_i$, $i = 1,2\ldots,5$. Then

$$m_\Psi(I_\Psi^1) = m_\Psi([72,82]) = 0.1, \qquad m_\Psi(I_\Psi^2) = m_\Psi([68,74]) = 0.4, \tag{6}$$

$$m_\Psi(I_\Psi^3) = m_\Psi([73,79]) = 0.1, \qquad m_\Psi(I_\Psi^4) = m_\Psi([71,83]) = 0.25, \tag{7}$$

$$m_\Psi(I_\Psi^5) = m_\Psi([76,84]) = 0.15. \tag{8}$$

To analyze the uncertainty in Ψ, let us graph $Bel([x, x+d])$ and $Pl([x, x+d])$, where $d \in \{1,2\}$ is fixed and $x \in [60,90]$. In other words, the intervals $[x, x+d]$ chosen in the space of output data (that is, displacements) will be assessed through the evidence that we have about the input datasets. Fig. 1 shows the results for

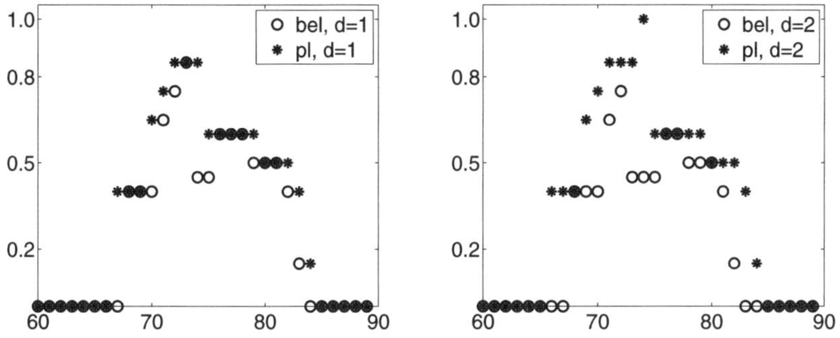

Fig. 1 Example 1; the vertical axis shows $Bel([x,x+d])$ and $Pl([x,x+d])$, the horizontal axis shows x.

$x = 60, 61, \ldots, 90$. Such graphs help the analyst to formulate a conclusion or make a decision. Thinking of the uncertain displacement magnitude in the above example, the analyst would hardly overlook the significance of values around 73, for instance.

Although the sets of scalar values were considered in this example, U_i could be sets of functions as well. Take, for instance, a set of functions representing an uncertain non-constant thickness of the beam.

2.3 Fuzzy Set Theory

In fuzzy set theory, points are weighted by a membership function with values in the interval $[0,1]$; see [1, 3, 7, 20, 21, 22]. For our purposes, a zero membership value will not indicate that the point does not belong to the (fuzzy) set. Indeed, we assume that a compact and convex admissible set $\mathscr{U}_{\mathrm{ad}}$ is given together with a membership function $\mu_{\mathscr{U}_{\mathrm{ad}}} : \mathscr{U}_{\mathrm{ad}} \to [0,1]$. A non-constant membership function indicates that not all members of $\mathscr{U}_{\mathrm{ad}}$ are equally possible. The higher $\mu_{\mathscr{U}_{\mathrm{ad}}}(a)$, the higher the possibility of a. We allow for $\mu_{\mathscr{U}_{\mathrm{ad}}}(a)$ to be equal to zero. Typically, $\mu_{\mathscr{U}_{\mathrm{ad}}}(a) > 0$ if a belongs to the interior of $\mathscr{U}_{\mathrm{ad}}$.

Special nested subsets of $\mathscr{U}_{\mathrm{ad}}$, called α-cuts, are defined as follows:

$$\mathscr{U}_{\mathrm{ad}}^{\alpha} = \{a \in \mathscr{U}_{\mathrm{ad}} : \mu_{\mathscr{U}_{\mathrm{ad}}}(a) \geq \alpha\}, \qquad \alpha \in [0,1]. \tag{9}$$

For any $\alpha \in [0,1]$, let us assume that the set $\mathscr{U}_{\mathrm{ad}}^{\alpha}$ is a convex and compact subset of $\mathscr{U}_{\mathrm{ad}}$; the compactness is guaranteed if, for instance, $\mu_{\mathscr{U}_{\mathrm{ad}}}$ is a continuous map.

By determining the best and the worst scenarios in $\mathscr{U}_{\mathrm{ad}}^{\alpha}$, we infer I_{Ψ}^{α}, the α-dependent intervals; cf. (3). These intervals are nothing else than the α-cuts of I_{Ψ}, the image of $\mathscr{U}_{\mathrm{ad}}$ under the map Ψ. To characterize the fuzziness of I_{Ψ}, the relevant membership function μ_{Ψ} is inferred (the extension principle):

$$\mu_\Psi(y) = \max\{\alpha : y \in I_\Psi^\alpha\}, \qquad y \in I_\Psi. \tag{10}$$

The degree of possibility of $\Psi(a)$, the a-dependent quantity of interest, is given by $\mu_\Psi(\Psi(a))$, $a \in \mathscr{U}_{\mathrm{ad}}$. A computational example will be presented later.

Remark 1. In information-gap decision theory [2], a non-fuzzy approach is introduced (besides other concepts) that also leads to the calculation of α-dependent worst scenarios. It is assumed there that α controls the amount of uncertainty present in an admissible set $\mathscr{U}_{\mathrm{ad}}^\alpha$ (α controls the "size" of the admissible set; the larger the α, the larger the size of $\mathscr{U}_{\mathrm{ad}}^\alpha$). It is also assumed that a value α exists such that $\Psi(a^0)$ determined by the worst scenario in $\mathscr{U}_{\mathrm{ad}}^\alpha$ is less than $q \in \mathbb{R}$, a given maximum acceptable value of the quantity of interest.

The goal is to find the maximum $\alpha_{\max} \in \mathbb{R}$ such that $\Psi(a^0) \le q$, where $a^0 \in \mathscr{U}_{\mathrm{ad}}^{\alpha_{\max}}$ maximizes Ψ over $\mathscr{U}_{\mathrm{ad}}^{\alpha_{\max}}$, that is, the maximum acceptable amount of uncertainty is to be identified.

3 Admissible Sets of Functions

In differential equations and the associated boundary conditions, parameters and right-hand sides often take the form of functions and are burdened with uncertainty. To introduce uncertain functions, we will present an approach stemming from the definition of admissible functions used in shape optimization; see [11].

For illustration, let us consider the following quasilinear PDE defined in Ω, a bounded domain in $\mathbb{R}^2$,

$$-\operatorname{div}(a(u)\operatorname{grad} u) = f(x,u), \tag{11}$$

$$u|_{\partial\Omega} = 0, \tag{12}$$

where a does not directly depend on $x \in \Omega$ but depends on the solution u; the right-hand side f depends both on the spatial variable x and the solution u. This boundary value problem can model a nonlinear thermal conductivity problem; we refer to [15] for a more general setting applied to modeling the temperature field in a transformer.

An admissible set $\mathscr{U}_{\mathrm{ad}}$, typical of many applications, can be defined as follows

$$\mathscr{U}_{\mathrm{ad}} = \left\{a \in \mathscr{U}_{\mathrm{ad}}^0(C_L) : a_{\min}(t) \le a(t) \le a_{\max}(t) \quad \forall t \in \mathbb{R}\right\}, \tag{13}$$

$$\mathscr{U}_{\mathrm{ad}}^0(C_L) = \big\{a \in C^{(0),1}(\mathbb{R}) \text{ (i.e., Lipschitz functions on } \mathbb{R}\text{)}:$$
$$|\mathrm{d}a/\mathrm{d}t| \le C_L \text{ a.e. in } \mathbb{R},\ a(t) = \text{const. for } t \notin [T_0, T_1]\big\},$$

where $a_{\min}, a_{\max} \in \hat{\mathscr{U}}_{\mathrm{ad}} = \left\{a \in \mathscr{U}_{\mathrm{ad}}^0(C_L) : 0 < a_1 \le a(t) \le a_2 < +\infty \quad \forall t \in \mathbb{R}\right\}$ are given functions and C_L, a_1, a_2, T_0, T_1 are given constants such that $C_L > 0$, $a_1 < a_2$, and $-\infty < T_0 < T_1 < +\infty$; see [16, 13, 4].

The criterion-functional

$$\Psi(a) = (\mathrm{meas}_2\, G)^{-1} \int_G u(a)(x)\,\mathrm{d}x \tag{14}$$

represents the a-dependent temperature u averaged over a fixed set $G \subset \Omega$.

It can be proved that the worst scenario problem (1) based on (11)–(14) (where the boundary conditions can be more complex) has at least one solution; see [13, 16]. Two features of the problem are crucial for the proof: (i) $\Psi(a)$ is continuous with respect to $a \in \mathscr{U}_{\mathrm{ad}}$ and the standard norm in $C(\mathbb{R})$, the space of functions continuous on $\mathbb{R}$; and (ii) $\mathscr{U}_{\mathrm{ad}}$ is compact in $C(\mathbb{R})$ (by virtue of the Arzelà-Ascoli theorem).

Generally speaking, variants of both (i) and (ii) appear in the analysis of other worst scenario problems with uncertain functions (13) or similar; see [16] for examples from continuum mechanics (e.g., elasticity or plasticity). In (i), the continuous dependence of $u(a)$ on a is the most substantial but usually also the most demanding part of the proof. The solvability of (2) is also ensured by (i)–(ii).

Remark 2. To ensure the compactness of the admissible set $\mathscr{U}_{\mathrm{ad}}$, rather strict assumptions are employed in (13). These, however, can be too restrictive in problems where other families of input functions have to be considered (discontinuous or oscillating functions, for instance). Consequently, such an admissible set might not be compact in a standard space of functions, and its compactification in a special space is necessary. Such *relaxed* problems appear and are analyzed in optimization-oriented modeling (see [18] and the references therein) and could also be considered in uncertainty modeling.

3.1 Approximation

To solve the state problem $A(a)u = f(a)$ (imagine (11)–(12), for instance, and allow an a-dependent f), one has to resort to a numerical method such as the finite element method (FEM), the finite difference method, the boundary element method, etc. These methods deliver an approximate state solution u_h defined on a mesh characterized by $h > 0$, the discretization parameter. Let us note that the uniqueness of u_h may be an open problem in certain situations even if u is unique; see [13, 15]. Non-unique state solutions u_h can be handled under some assumptions; see [16, Chapter II]. The uniqueness of u_h is assumed henceforth.

The functions from the admissible set $\mathscr{U}_{\mathrm{ad}}$ can be approximated by continuous, piece-wise linear functions controlled by the vertical position of M nodes bound by possible constraints; see C_L, $a_{\min}$, and $a_{\max}$ in (13). These functions constitute the approximate admissible set $\mathscr{U}_{\mathrm{ad}}^M$, which is identifiable with a compact subset of $\mathbb{R}^M$.

The approximate best and worst scenario problems

$$a_{0h}^M = \arg\min_{a \in \mathscr{U}_{\mathrm{ad}}^M} \Phi(a, u_h(a)) \quad \text{and} \quad a_h^{0M} = \arg\max_{a \in \mathscr{U}_{\mathrm{ad}}^M} \Phi(a, u_h(a)) \tag{15}$$

are, in fact, finite dimensional constrained optimization problems.

The typical relationship between a_h^{0M} and a^0 (or a_{0h}^M and a_0) is as follows: If $\{a_h^{0M}\}$ is a sequence of the solutions to (15) controlled by $h \to 0+$ and $M \to \infty$, then a subsequence $\{a_{h_k}^{0M_k}\}$ exists such that, for $k \to \infty$,

$$a_{h_k}^{0M_k} \to a^0, \quad u_{h_k}(a_{h_k}^{0M_k}) \to u(a^0), \quad \text{and} \quad \Phi(a_{h_k}^{0M_k}, u_{h_k}(a_{h_k}^{0M_k})) \to \Phi(a^0, u(a^0)),$$

where the first and second sequences converge in proper spaces and topologies; see [10]. Similar convergence results for various worst scenario problems can be found in [16].

If it happens that more than one admissible set are available for the analyzed problem, say (11)-(12), and that the analyst can assess each $\mathcal{U}_{\mathrm{ad}}^i$ by $m(\mathcal{U}_{\mathrm{ad}}^i)$, the "likelihood" of $\mathcal{U}_{\mathrm{ad}}^i$, then the transition from the WSM to the Dempster-Shafer approach is straightforward.

Indeed, by finding the worst and the best scenarios, one determines the ranges (3) for each $\mathcal{U}_{\mathrm{ad}}^i$. By identifying U^i with $\mathcal{U}_{\mathrm{ad}}^i$ and obtaining m_Ψ (see (4)), the analyst is ready for the assessment of various sets $S \subset \mathbb{R}$ through (5), that is, for the assessment of the bounds of the likelihood that S is related to the uncertain values of Ψ.

Let us pay more attention to the fuzzy set approach.

3.2 Fuzzification of $\mathcal{U}_{\mathrm{ad}}$

Different concepts of fuzziness can be merged with functions see [1, Section 2.4.9]. We will simply retain $\mathcal{U}_{\mathrm{ad}}$ as a set of crisp functions but we will add a membership function to $\mathcal{U}_{\mathrm{ad}}$. In other words, we will weight $a \in \mathcal{U}_{\mathrm{ad}}$. Two forms of weighting will be introduced; see also [5].

The first approach is rather straightforward. It is based on the distance between $a \in \mathcal{U}_{\mathrm{ad}}$ and a given function a_{mid}; the details follow.

For illustration, let us recall (13) and define $a_{\mathrm{mid}}(t) = (a_{\min}(t) + a_{\max}(t))/2$ and $a_{\mathrm{dif}}(t) = (a_{\max}(t) - a_{\min}(t))/2$, where $t \in \mathbb{R}$. It is assumed that a_{dif} is positive on the real axis. For $\alpha \in [0,1]$, we then define

$$\mathcal{U}_{\mathrm{ad}}^{\alpha} = \{a \in \mathcal{U}_{\mathrm{ad}}^0(C_L) : \ |a(t) - a_{\mathrm{mid}}(t)| \le (1-\alpha) a_{\mathrm{dif}}(t) \ \forall t \in \mathbb{R}\}, \tag{16}$$

that is, we define the α-cuts of $\mathcal{U}_{\mathrm{ad}}$. This concept is close to fuzzy functions [1] or to controlling the amount of uncertainty through α; see [2]. Nevertheless, in (16), we still consider crisp functions. If $\alpha = 1$, then $\mathcal{U}_{\mathrm{ad}}^{\alpha} = \{a_{\mathrm{mid}}\}$. If $\alpha = 0$, then $\mathcal{U}_{\mathrm{ad}}^{\alpha} = \mathcal{U}_{\mathrm{ad}}$.

The membership function value (the weight) of $a \in \mathcal{U}_{\mathrm{ad}}$ is defined as

$$\mu(a) = \max\{\alpha \in [0,1] : \ a \in \mathcal{U}_{\mathrm{ad}}^{\alpha}\}. \tag{17}$$

With this μ, definition (9) leads to $\mathcal{U}_{\mathrm{ad}}^{\alpha}$ defined in (16).

If $\mathcal{U}_{\mathrm{ad}}$ is fuzzy, so is $\mathcal{U}_{\mathrm{ad}}^M$. The approximate problems (15) result in optimization problems with simple bounds (determined by $a_{\min}$ and $a_{\max}$) and linear constraints (determined by C_L). The approximate best and worst scenarios in $\mathscr{U}_{\mathrm{ad}}^{M,\alpha}$, an α-cut of $\mathscr{U}_{\mathrm{ad}}^{M}$, are again obtained through solving optimization problems with simple bounds (determined by $a_{\min}$, $a_{\max}$, and α) and linear constraints (determined by C_L). Common optimization software coupled with FEM software can often be applied to solve such problems.

The other approach to weighting $\mathscr{U}_{\mathrm{ad}}$ is motivated by the observation described below. Let $a_1, a_2 \in \mathscr{U}_{\mathrm{ad}}^{\alpha}$ and let the inequality in (16) becomes the equality on the entire set $\mathbb{R}$ if a_1 is considered, and at a single point $t_0 \in \mathbb{R}$ if a_2 is considered. Moreover, let a_2 coincide with a_{mid} except for an interval containing t_0. These a_1 and a_2 share the same α-cuts of $\mathscr{U}_{\mathrm{ad}}$. In many applications, however, the weight of a_2 would be expected greater than the weight of a_1 because a_2 is "closer" to a_{mid}, which has the highest degree of possibility.

We will design a membership function able to separate a_1 from a_2. We first define an auxiliary continuous function $\rho : Q \to [0,1]$, where $Q = \{[t,y] \in \mathbb{R}^2 : t \in [T_0,T_1], y \in [a_{\min}(t), a_{\max}(t)]\}$. It is assumed that $\rho(t,\cdot)$ is a concave function for each $t \in [T_0,T_1]$. The functions $\rho(t,\cdot)$ can be viewed as auxiliary membership functions (weights) assessing the degree of possibility of $a(t)$ if $a \in \mathscr{U}_{\mathrm{ad}}$. The graph of $\rho(t,\cdot)$ is shaped accordingly; it is triangular or trapezoidal, which is common in fuzzy set theory. The function ρ can be derived from measurements, estimates, or expert opinions.

We are ready to define $\mu_\rho : \mathscr{U}_{\mathrm{ad}} \to [0,1]$, the membership function associated with $\mathscr{U}_{\mathrm{ad}}$:

$$\mu_\rho(a) = (T_1 - T_0)^{-1} \int_{T_0}^{T_1} \rho(t, a(t))\, \mathrm{d}t. \tag{18}$$

It is evident that we can obtain $\mu_\rho(a_1) < \mu_\rho(a_2)$ if ρ is properly shaped.

Unlike (16), the identification of all the functions a that comprise a particular α-cut is not straightforward. This difficulty also appears in the search for the approximate best and worst scenarios, where, moreover, (18) gives rise to a nonlinear constraint in the definition of $\mathscr{U}_{\mathrm{ad}}^{\alpha}$. If ρ is nonsmooth, μ_ρ is not differentiable at some a. This partial lack of differentiability is also observed in μ_{ρ^M}, a $\mathscr{U}_{\mathrm{ad}}^{M}$-related approximation of μ_ρ based on a piece-wise linear auxiliary function ρ^M that approximates ρ.

Since the use of nonsmooth (triangular, trapezoidal) $\rho(t,\cdot)$ is common and the piece-wise linearity of $\rho(t,\cdot)$ is advantageous in many respects, nonsmooth optimization seems to be unavoidable in solving (15)-like problems on the α-cuts determined by μ_{ρ^M}.

A closer inspection reveals, however, that the approximate (15)-like problems defined on $\mathscr{U}_{\mathrm{ad}}^{M,\alpha}$, the μ_{ρ^M}-based α-cuts of $\mathscr{U}_{\mathrm{ad}}^{M}$, can be decomposed into a finite sequence of smooth optimization subproblems.

Indeed, $a^M \in \mathscr{U}_{\mathrm{ad}}^{M}$ is uniquely determined by the values $a_i \equiv a(t_i)$ at fixed points t_i, where $i = 1,2,\ldots,M$. Let us assume that ρ^M is piece-wise linear and determined by the continuous functions $\rho(t_i,\cdot)$ that are linear on intervals $\theta_{ij} = [y_{i,j}, y^{i,j}]$, where $i = 1,2,\ldots,M$, $j = 1,2,\ldots,N$, and $y^{i,j} = y_{i,j+1}$ if $j = 1,2,\ldots,N-1$. It is $[y_{i,1}, y^{i,N}] =$

$[a_{\min}(t_i), a_{\max}(t_i)]$. Typically, $N = 2$ ($N = 3$) if $\rho(t,\cdot)$ is triangularly (trapezoidally) shaped.

As long as $a_i \in \Theta_{ij}$ for $i = 1,2,\ldots,M$ and for a *fixed* set $\mathscr{J}$ of indices j, $\mu_{\rho M}$ is differentiable (left- and right-differentiable at the ends of Θ_{ij}) and the related optimization subproblem is smooth. The differentiability is lost at one point when a_i passes from the current interval Θ_{ij} to its neighbor Θ_{ik}, $k \neq j$, but it is again restored if $a_i \in \Theta_{ik}$ and $\mathscr{J}$ is updated. The updated set of indices determines a new smooth optimization subproblem.

The partial derivative of $\mu_{\rho M}$ with respect to a_i, where $i = 1,2,\ldots,M$, can be obtained in a closed form in each of the subproblems. Consequently, the analytic gradient of $\mu_{\rho M}$ exists except for some points and can be employed in the calculation of $\partial\Psi/\partial a_i$, which is important in a gradient-based search for the best and worst scenarios in $\mathscr{U}_{\mathrm{ad}}^{M,\alpha}$.

Example 2. Let u, the a-dependent solution to the boundary value problem

$$-(a(x)u'(x))' = f \text{ on } \Omega = (0,1), \quad u(0) = 0 = u(1),$$

be evaluated through the criterion-functional (quantity of interest)

$$\Psi(a) = \int_{\Omega} (u(x) - \sin(2\pi x))^2 \,\mathrm{d}x.$$

In the state problem, f is chosen in such a way that if $a(x) = 1 + x$, then $u(x) = \sin(2\pi x)$ and, consequently, $\Psi(a) = 0$.

The parameter a belongs to the admissible set $\mathscr{U}_{\mathrm{ad}}$ determined by the quadratic function $g(x) = 1.5 + x^2$ and two constants. In detail,

$$\mathscr{U}_{\mathrm{ad}} = \left\{ a \in C^{(0),1}([0,1]) : \, |a(x) - g(x)| \le 0.5 \text{ and } |a'(x) - g'(x)| \le 0.8 \right\}.$$

The auxiliary function ρ is "triangular", that is, $\rho(x,\cdot)$ is determined by the linear interpolation of the points $[x, g(x) - 0.5, 0]$, $[x, g(x), 1]$, and $[x, g(x) + 0.5, 0]$, where $x \in [0,1]$. The membership function μ_ρ is given by (18), where $T_0 = 0$ and $T_1 = 1$.

The goal is to infer μ_Ψ, the membership function of the quantity of interest; see Subsection 2.3 and (10).

To achieve the goal at least approximately, see Fig. 2, the state equation was solved by the finite element method with piece-wise linear basis functions, and $\mathscr{U}_{\mathrm{ad}}$ was approximated by continuous piece-wise linear functions constituting $\mathscr{U}_{\mathrm{ad}}^{M}$, where $M = 15$. The optimization problems, see (15), were solved on the α-cuts of $\mathscr{U}_{\mathrm{ad}}^{M}$ for $\alpha = 0, 0.05, 0.1, \ldots, 1$.

The gradient of Ψ was calculated via the adjoint equation technique [12]; an explicit formula was obtained for the gradient of $\mu_{\rho M}$ at the points of differentiability. The search for the best and the worst scenarios in the α-cuts was based on the NAG® Foundation (MATLAB®) Toolbox `E04UCF` routine for constrained sequential quadratic programming.

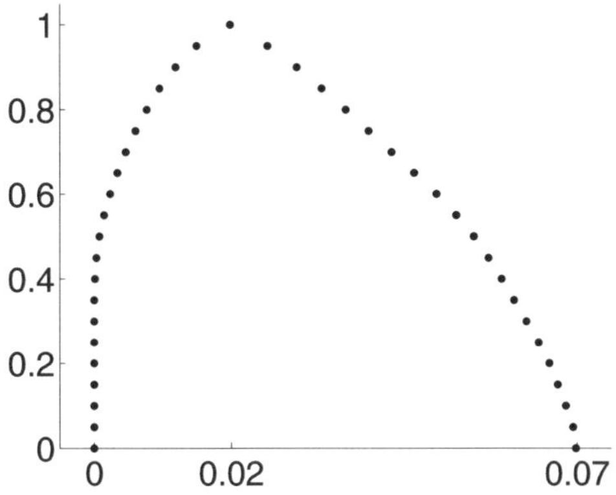

Fig. 2 Example 2. The approximation of μ_Ψ inferred from (10), where $\alpha = 0, 0.05, 0.1, \ldots, 1$. The horizontal axis shows the Ψ values, the vertical axis shows the α values. We observe that $a(x) = 1 + x$ belongs to the α-cuts if $\alpha = 0, 0.05, 0.1, \ldots, 0.35$. Indeed, for these α, the best scenario implies the zero value of Ψ. If $\alpha = 1$, then the α-cut comprises only the function g.

4 Conclusions

The worst scenario method is appropriate if we know only the set of admissible inputs but we do not have information that would enable us to weight the importance (possibility or likelihood) of input data. Since searching for the best scenario is mathematically equivalent to the worst scenario search, the WSM eventually delivers the range of the quantity of interest Ψ induced by the uncertainty in inputs.

If more extensive information on inputs is available (inputs can be weighted in some sense) and if it complies with the Dempster-Shafer or fuzzy set theory assumptions (which are less demanding than the probability theory assumptions), the uncertainty in an output quantity of interest can be weighted too. To achieve this, the WSM has to be repeatedly applied to obtain (3)-like ranges that are pivotal in the other two approaches for obtaining m_Ψ and μ_Ψ; see Subsection 2.2 and (4) as well as Section 2.3 and (10).

From the computational standpoint, solving (15)-like problems is crucial in all the above-mentioned methods. In the case of smooth problems, the gradients of both Ψ and the constraints are available, which can speed up the search for the minimum (maximum) of Ψ.

If (15) leads to a nonsmooth optimization problem, we can (a) try to decompose it to smooth subproblems, (b) use a subgradient-based technique, or (c) apply an evolution strategy that partly or completely avoids the need for the (sub)gradient.

However, it is fair to say that the worst scenario method is computationally challenging because it asks for solving a global optimization problem. Nevertheless, we can benefit from theoretical and software tools that have proved themselves well in optimal design, control theory, parameter identification, and sensitivity analysis.

Acknowledgements The research was supported by the Academy of Sciences of the Czech Republic through Institutional Research Plan No. AV0Z10190503 and grant No. IAA100190803 from the Grant Agency of AS CR.

References

1. Ayyub, B.M., Klir, G.J.: Uncertainty Modeling and Analysis in Engineering and the Sciences. Chapman & Hall/CRC, Taylor & Francis Group, Boca Raton (2006)
2. Ben-Haim, Y.: Information-gap Decision Theory. Academic Press, San Diego (2001)
3. Bernardini, A.: What are the random and fuzzy sets and how to use them for uncertainty modelling in engineering systems? In: I. Elishakoff (ed.) Whys and Hows in Uncertainty Modelling, Probability, Fuzziness and Anti-Optimization, CISM Courses and Lectures No. 388. Springer–Verlag, Wien, New York (1999)
4. Chleboun, J.: On a reliable solution of a quasilinear elliptic equation with uncertain coefficients: sensitivity analysis and numerical examples. Nonlinear Anal. Theory Methods Appl. **44**, 375–388 (2001)
5. Chleboun, J.: On fuzzy input data and the worst scenario method. Appl. Math. (Prague) **48**, 487–496 (2003)
6. Dempster, A.P.: Upper and lower probabilities induced by a multivalued mapping. Ann. Math. Stat. **38**, 325–339 (1967)
7. Dubois, D., Prade, H. (eds.): Fundamentals of Fuzzy Sets. Foreword by Lotfi Zadeh, *The Handbooks of Fuzzy Sets Series*, vol. 7. Kluwer Academic Publishers, Dordrecht (2000)
8. Elishakoff, I.: An idea of the uncertainty triangle. Shock Vib. Dig. **22**, 1 (1990)
9. Elishakoff, I.E., Haftka, R.T., Fang, J.: Structural design under bounded uncertainty – optimization with anti-optimization. Comput. Struct. **53**, 1401–1405 (1994)
10. Harasim, P.: On the worst scenario method: A modified convergence theorem and its application to an uncertain differential equation. Appl. Math. (2008). To appear
11. Haslinger, J., Neittaanmäki, P.: Finite Element Approximation for Optimal Shape, Material and Topology Design. J. Wiley, Chichester (1996)
12. Haug, E.J., Choi, K.K., Komkov, V.: Design Sensitivity Analysis of Structural Systems. Academic Press, Orlando (1986)
13. Hlaváček, I.: Reliable solution of a quasilinear nonpotential elliptic problem of a nonmonotone type with respect to the uncertainty in coefficients. J. Math. Anal. Appl. **212**, 452–466 (1997)
14. Hlaváček, I.: Uncertain input data problems and the worst scenario method. Appl. Math. **52**, 187–196 (2007)
15. Hlaváček, I., Křížek, M., Malý, J.: On Galerkin approximations of a quasilinear nonpotential elliptic problem of a nonmonotone type. J. Math. Anal. Appl. **184**, 168–189 (1994)
16. Hlaváček, I., Chleboun, J., Babuška, I.: Uncertain Input Data Problems and the Worst Scenario Method, *North-Holland Series in Applied Mathematics and Mechanics*, vol. 46. Elsevier, Amsterdam (2004)
17. Oberkampf, W.L., Helton, J.C., Sentz, K.: Mathematical representation of uncertainty. Research Article AIAA 2001-1645, American Institute of Aeronautics and Astronautics, Reston, VA (2001)
18. Roubíček, T.: Relaxation in Optimization Theory and Variational Calculus. De Gruyter Series in Nonlinear Analysis and Applications. Walter de Gruyter, Berlin (1997)
19. Shafer, G.: A Mathematical Theory of Evidence. Princeton University Press, Princeton, NJ (1976)
20. Zadeh, L.A.: Fuzzy sets. Inf. Control **8**, 338–353 (1965)
21. Zadeh, L.A.: Fuzzy sets as a basis for theory of possibility. Fuzzy Sets Syst. **1**, 3–28 (1978)
22. Zimmermann, H.J.: Fuzzy Set Theory — and its Applications, fourth edn. Kluwer Academic Publishers, Boston (2001)

Boundary and Finite Element Domain Decomposition Methods

U. Langer

Abstract Since Boundary Element (BE) Methods and Finite Element (FE) Methods exhibit certain complementary properties, it is sometimes very useful to couple these discretization techniques within a Domain Decomposition (DD) framework. We give a short review of the symmetric coupling technique and of primal, dual and dual-primal iterative substructuring solvers for coupled FE-BE equations. The boundary and interface concentrated FE methods have some features in common with data-sparse BE methods, but their applicability is far wider. We present primal and dual iterative substructuring solvers which exhibit the same complexity as the corresponding data-sparse BE solvers. Finally, we use the BE DD technology in order to construct FE approximations on polygonal and polyhedral meshes.

1 Introduction

Boundary Element Methods (BEM) and Finite Element Methods (FEM) have certain complementary properties. The BEM only requires a triangulation of the boundary of the computational domain, whereas the FEM needs a volume mesh. The handling of unbounded domains is straightforward in the BEM, whereas the FEM needs special modifications like absorbing boundary conditions, perfectly matched layers, or infinite elements. The approximation of singularities is easier in BEM than in FEM. In electromagnetics, the BEM can easily handle large air subdomains and moving parts. However, the application of the BEM requires the knowledge of the fundamental solution of the underlying differential operator. On the other hand, the FEM is more general. It can treat variable coefficients in the Partial Differential Equations (PDE), source terms and non-linearities with the same technology. Thus, it is quite natural to couple both discretization techniques within a Domain

---

Ulrich Langer
Institute for Computational Mathematics, Johannes Kepler University Linz, Altenberger Str. 69, A-4040 Linz, Austria, e-mail: ulanger@numa.uni-linz.ac.at

Decomposition (DD) framework and to benefit from the advantages of these discretization methods in different subdomains depending on the local properties of the problem which we are going to solve. O. C. Zienkiewicz, D. M. Kelly and P. Bettes are among the first authors who successfully started coupling BEM and FEM in engineering applications [27]. In [28], they refer to the BEM-FEM coupling as the *"Marriage a la mode - the best of both worlds"*. At that time the standard BEM was based on the collocation technique, which does not really fit to the Galerkin technique that has been used in FEM from the very beginning. Unsymmetric Galerkin BEM-FEM couplings were first studied by mathematicians at the end of the 1970s and at the beginning of the 1980s [5, 15]. In [8], M. Costabel introduced and analyzed the symmetric Galerkin BEM-FEM coupling, which perfectly fits to the Galerkin FEM. Since then, the symmetric coupling has been used in many applications. G.C. Hsiao and W.L. Wendland first used the symmetric coupling to construct a pure boundary element substructuring method [14]. The first fast solver for symmetrically coupled BE-FE DD equations was proposed and analyzed in [18]. Further contributions to efficient parallel DD solvers for symmetrically coupled BE-FE DD equations have been made in [10, 6]. We also refer to the recent survey articles [25] by E. Stephan and [20] by U. Langer and O. Steinbach where the reader can find an excellent overview on BE-FE coupling and BE-FE DD solvers. A comprehensive presentation of DD (FE) methods can be found in [26].

The standard collocation as well as Galerkin boundary element discretizations lead to dense matrices. This was the major drawback of the BEM in large-scale applications, especially in 3D. In practice, dense matrices represent a complexity barrier that restricts the application of the BEM to small-scale problems. Indeed, the memory demand for dense BE matrices and the cost for the matrix-vector multiplication, which is the basic operation in iterative solvers, grow like $O(h^{-2(d-1)})$ whereas the sparse FE matrices require $O(h^{-d})$ storage units and $O(h^{-d})$ arithmetical operations for the matrix-vector multiplication. Here h denotes the usual discretization parameter such that the number of unknowns (degrees of freedom = DOF) behaves like $O(h^{-(d-1)})$ and $O(h^{-d})$ in BEM and FEM, respectively, where d is the dimension of the domain Ω where the boundary value problem is given. The development of data-sparse representations of the dense BE matrices was very crucial for breaking down this complexity barrier. Data-sparse techniques like the Fast Multipole Method (FMM) [24, 7], the panel clustering method [13], the $\mathcal{H}$-matrix technology [12] and the Adaptive Cross Approximation (ACA) [1, 2], see also the book [23], allow us to reduce the BE complexity from $O(h^{-2(d-1)})$ to $O(h^{-(d-1)}(\log h^{-1})^l)$ with some $l \in \mathbb{N}_0$ (e.g. $l = 2$ for the FMM).

If we now use data-sparse approximation of the BE matrices in the BE subdomains, then the FE parts in a coupled BE-FE solver suddenly dominates the complexity, at least asymptotically. B.N. Khoromskij and J.M. Melenk proposed the boundary-concentrated FEM that allow us to reduce the FE complexity from $O(h^{-d})$ to $O(h^{-(d-1)})$ [16]. The use of the boundary-concentrated FEM in a DD framework, which is called Interface-Concentrated (IC) FEM, and the corresponding primal and dual DD solvers were investigated in [3]. Tearing and interconnecting solvers for coupled data-sparse BE and IC FE DD equations were studied in [19].

The remainder of this paper is organized as follows. Section 2 gives a short review of the symmetric BE-FE-coupling in a DD framework. Section 3 is devoted to DD solvers for interface-concentrated finite element equations. In Section 4, we symmetrically couple these IC FE equations with data-sparse BE equations and look at fast and robust DD solvers. Section 5 deals with a local Trefftz FEM that originates from the symmetric BE DD coupling.

2 A Review of the Symmetric BE-FE-Coupling in a Domain Decomposition Framework

A good starting point for non-overlapping DD methods is the skeleton variational formulation of the Boundary Value Problem (BVP) which we are going to solve. For our model Dirichlet BVP,

$$-\mathrm{div}(a\nabla u) = f \text{ in } \Omega \quad \text{and} \quad u = g \text{ on } \Gamma = \partial\Omega, \tag{1}$$

the skeleton variational problem reads as follows [20]: Given $g \in H^{1/2}(\Gamma)$, find $u \in H_g^{1/2}(\Gamma_S)$ such that

$$\sum_{i=1}^{p} \int_{\Gamma_i} (S_i u)(x) v(x) ds_x = \sum_{i=1}^{p} \int_{\Gamma_i} (N_i f)(x) v(x) ds_x \;\; \forall v \in H_0^{1/2}(\Gamma_S), \tag{2}$$

where Ω is shape-regularly decomposed into p non-overlapping subdomains Ω_i with the local boundaries $\Gamma_i = \partial\Omega_i$, $i = 1, \ldots, p$. The skeleton Γ_S is nothing but the union of all subdomain boundaries Γ_i including the Dirichlet boundary $\Gamma_D = \Gamma$. The pseudo-differential operators S_i and N_i denote the Steklov-Poincaré and the Newton potential operators, respectively. $H_g^{1/2}(\Gamma_S)$ and $H_0^{1/2}(\Gamma_S)$ are given by all functions from skeleton trace space $H^{1/2}(\Gamma_S) := \{v|_{\Gamma_S} : v \in H^1(\Omega)\}$ with the trace g and 0 on Γ, respectively. For simplicity of the description of the discretization, we assume that the domain Ω and subdomains Ω_i are polygons or polyhedra in 2D or 3D, respectively. Furthermore, we assume that $a = a_i = \mathrm{const} > 0$ and $f = 0$ in Ω_i for $i = 1, \ldots, q$, and $0 < \mathrm{const} = \underline{a}_i \leq a(\cdot) \leq \overline{a}_i = \mathrm{const}$ in Ω_i with a small ratio $\overline{a}_i/\underline{a}_i$ and $f \in L_2(\Omega_i)$ for $i = q+1, \ldots, p$.

Now it is clear from the properties of the data that we will use the BEM in the first q and the FEM in the remaining subdomains for discretizing (2). The discretization of (2) starts with a quasi-regular surface triangulation of the skeleton Γ_S followed by a conforming regular volume triangulation of the FE subdomains $\overline{\Omega}_i$, $i = q+1, \ldots, p$. Using, for simplicity, piecewise linear boundary respectively finite elements shape functions for approximating the potential u and piecewise constant BE shape functions for approximating the normal derivatives (tractions) $t = \partial u/\partial n_i$ on the boundaries Γ_i of the BE subdomains, we arrive at the following BE / FE approximations

$$\mathbf{S}_{C,i}^{BE} = \alpha_i \mathbf{D}_{C,i} + \alpha_i \left(\frac{1}{2}\mathbf{M}_{C,i}^{\top} + \mathbf{K}_{C,i}^{\top}\right)\left(\mathbf{V}_{C,i}\right)^{-1}\left(\frac{1}{2}\mathbf{M}_{C,i} + \mathbf{K}_{C,i}\right), \quad i = 1,\ldots,q, \tag{3}$$

$$\mathbf{S}_{C,i}^{FE} = \mathbf{K}_{CC}^{(i)} - \mathbf{K}_{CI}^{(i)}\left(\mathbf{K}_{II}^{(i)}\right)^{-1}\left(\mathbf{K}_{CI}^{(i)}\right)^{\top}, \quad i = q+1,\ldots,p, \tag{4}$$

and

$$\mathbf{N}_{C,i}^{FE} = \left[\mathbf{I}_C \middle| -\mathbf{K}_{CI}^{(i)}\left(\mathbf{K}_{II}^{(i)}\right)^{-1}\right], \quad i = q+1,\ldots,p, \tag{5}$$

to the Steklov-Poincaré operators S_i and to the Newton potential N_i, respectively. The block matrices $\mathbf{V}_{C,i}$, $\mathbf{K}_{C,i}$, $\mathbf{D}_{C,i}$ and $\mathbf{M}_{C,i}$ building the BE Schur complement $\mathbf{S}_{C,i}^{BE}$ arise from the BE Galerkin approximation to the single layer potential operator V_i, double layer potential operator K_i, hypersingular integral operator D_i and the identity operator I_i living on Γ_i, respectively, see, e.g., [20] for the precise definition. The block matrices $\mathbf{K}_{CC}^{(i)}$, $\mathbf{K}_{CI}^{(i)} = (\mathbf{K}_{IC}^{(i)})^{\top}$ and $\mathbf{K}_{II}^{(i)}$ building the FE Schur complement $\mathbf{S}_{C,i}^{FE}$ and the FE Newton potential matrix $\mathbf{N}_{C,i}^{FE}$ coincide with the blocks in the subdomain stiffness matrix

$$\mathbf{K}_i = \begin{pmatrix} \mathbf{K}_{II}^{(i)} & (\mathbf{K}_{CI}^{(i)})^{\top} \\ \mathbf{K}_{CI}^{(i)} & \mathbf{K}_{CC}^{(i)} \end{pmatrix}, \tag{6}$$

which is nothing else than the FE approximation to the Neumann problem in Ω_i whereas $\mathbf{K}_{II}^{(i)}$ approximates the Dirichlet problem. Thus, for our potential equation (1), $\mathbf{K}_i$ is singular whereas $\mathbf{K}_{II}^{(i)}$ is regular. We mention that the indices "C" and "I" are associated with nodes located on the (coupling) boundary Γ_i and in the interior of the subdomain Ω_i, respectively.

Now this BE / FE Galerkin approximation to the skeleton variational formulation (2) immediately leads to the BE-FE Schur-complement system

$$\sum_{i=1}^{q} \mathbf{R}_{i,0}^{\top}\mathbf{S}_{C,i}^{BE}\mathbf{R}_i\mathbf{u}_C + \sum_{i=q+1}^{p} \mathbf{R}_{i,0}^{\top}\mathbf{S}_{C,i}^{FE}\mathbf{R}_i\mathbf{u}_C = \sum_{i=q+1}^{p} \mathbf{R}_{i,0}^{\top}\mathbf{N}_i^{FE}(\mathbf{f}_{C,i}^{\top},\mathbf{f}_{I,i}^{\top})^{\top} \tag{7}$$

subject to the Dirichlet boundary condition $\mathbf{R}_D\mathbf{u}_C = \mathbf{g}_D$, where the vector $\mathbf{u}_C$ contains all nodal parameters living on the skeleton Γ_S including the Dirichlet boundary Γ. The restriction operators $\mathbf{R}_i$, $\mathbf{R}_{i,0}^{\top}$ and $\mathbf{R}_D$ are defined as usual (see, e.g., [19]). Homogenizing the Dirichlet condition $\mathbf{R}_D\mathbf{u}_C = \mathbf{g}_D$ in (7), we arrive at the symmetric and positive definite (SPD) coupled BE-FE Schur-complement system that can be solved by the Schur-complement PCG method with a suitable preconditioner for the assembled BE-FE Schur-complement. Such preconditioners are available, see, e.g., [26] for the pure FE case and [20] for the pure BE and the coupled BE-FE cases. However, Schur-complement PCG solvers require the solution of systems with the matrices $\mathbf{V}_{C,i}$ and $\mathbf{K}_{II}^{(i)}$ in every step of the PCG iteration. Usually, in a pre-iteration step, the LU-factorization of these matrices is performed. Thus, in every iteration step, the multiplication with the Schur-complements only requires the foreward and backward substitutions, which are far less costly than the LU-factorization. To avoid Schur-complements, we can unfold the Schur-complements resulting in a larger system of algebraic equations that is still SPD in the pure FE case and that is a saddle-point system (symmetric, but indefinite) if BE subdomains are present. In this case we speak about inexact iterative substructuring methods. Solvers and preconditioners for the inexact substructuring DD equations have been investigated in [11] for the FE case and in [18, 10, 6] for the pure BE- and the coupled FE-BE-cases. On the other hand, tearing and interconnecting methods have been developed for solving pure FE-, pure BE- and coupled FE-BE DD equations now called FETI, BETI and FETI-BETI methods (see [26] and [20] for an overview and relevant references). In the next section we present such primal and dual iterative substructuring solvers for IC-FE DD equations.

3 Interface-Concentrated FEM in Domain Decomposition

Let us consider the pure FE case where $q = 0$ and let us consider the 2D case $(d = 2)$ for simplicity. Further, let us assume that every subdomain Ω_i is triangulated by a geometric triangular mesh resulting in a conforming triangulation of the whole computational domain Ω as illustrated in Figure 1 (left). It is clear that the geometric mesh only produces $O(H_i/h_i)$ triangles and nodal points in contrast to a uniformly refined FE mesh that has $O((H_i/h_i)^2)$ triangles and nodal points. Now we use piecewise linear finite element shape functions only on the triangles which are close to the subdomain boundaries, and hierarchically increase the polynomial degree of the basis functions from the finer to the coarser levels in the geometric mesh (the larger the distance of an element from the subdomain boundary the higher the degree) as is shown in Figure 1. We refer the reader to [17] for the technical details of the BC FE discretization and the derivation of the corresponding discretization error estimates. We only mention that under appropriate smoothness assumptions the same discretization error estimates are obtained as in the case of a uniformly refined FE

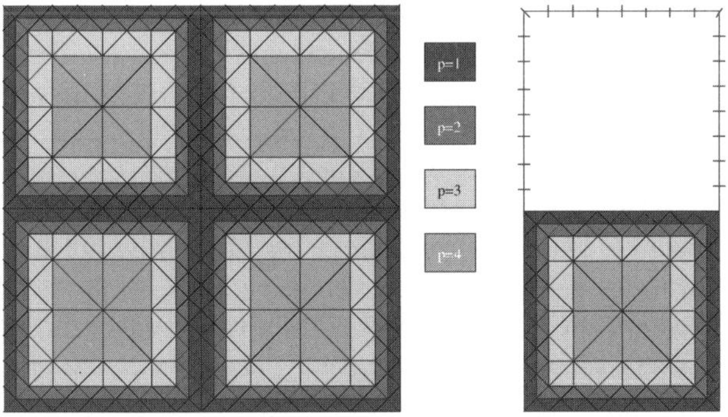

Fig. 1 Geometric (interface-concentrated) mesh with polynomial degree distribution (left) and boundary element and interface-concentrated finite element mesh (right).

mesh whereas the number of unknowns (DOF) per subdomain still behaves like the number of unknowns on the subdomain boundary Γ_i, namely like $O(H_i/h_i)$.

The IC-FE-discretization leads to the SPD FE system

$$\mathbf{K}_h \mathbf{u}_h = \mathbf{f}_h, \tag{8}$$

where the stiffness matrix has the following structure

$$\mathbf{K}_h = \begin{pmatrix} \mathbf{K}_{II} & \mathbf{K}_{IC} \\ \mathbf{K}_{CI} & \mathbf{K}_{CC} \end{pmatrix} \tag{9}$$

after a corresponding reordering of the unknowns $\mathbf{u}_h = \left(\mathbf{u}_I^\top, \mathbf{u}_C^\top\right)^\top$, where the index I refers to all interior unknowns and the index C to all coupling nodes. The interior unknowns are arranged subdomain by subdomain. Hence, $\mathbf{K}_{II} = \mathrm{diag}(\mathbf{K}_{II}^{(i)})$. The load vector $\mathbf{f}_h$ has the same structure as the solution vector $\mathbf{u}_h$. Eliminating the interior unknowns $\mathbf{u}_I$ from (8)–(9), we arrive at the Schur-complement system (7) that is equivalent to (8).

From the factorization

$$\mathbf{K}_h = \begin{pmatrix} \mathbf{I}_{II} & \mathbf{0} \\ \mathbf{K}_{CI}\mathbf{K}_{II}^{-1} & \mathbf{I}_{CC} \end{pmatrix} \begin{pmatrix} \mathbf{K}_{II} & \mathbf{0} \\ \mathbf{0} & \mathbf{S}_{CC} \end{pmatrix} \begin{pmatrix} \mathbf{I}_{II} & \mathbf{K}_{II}^{-1}\mathbf{K}_{IC} \\ \mathbf{0} & \mathbf{I}_{CC} \end{pmatrix} \tag{10}$$

of the global stiffness matrix $\mathbf{K}_h$, we can derive the factorized SPD preconditioner

$$\mathbf{C}_h = \begin{pmatrix} \mathbf{I}_{II} & \mathbf{0} \\ -\mathbf{E}_{IC}^\top & \mathbf{I}_{CC} \end{pmatrix} \begin{pmatrix} \mathbf{C}_{II} & \mathbf{0} \\ \mathbf{0} & \mathbf{C}_{CC} \end{pmatrix} \begin{pmatrix} \mathbf{I}_{II} & -\mathbf{E}_{IC} \\ \mathbf{0} & \mathbf{I}_{CC} \end{pmatrix}, \tag{11}$$

where $\mathbf{C}_{II} = \mathrm{diag}(\mathbf{C}_{II}^{(i)})$ is a preconditioner for $\mathbf{K}_{II} = \mathrm{diag}(\mathbf{K}_{II}^{(i)})$ and $\mathbf{C}_{CC}$ is a preconditioner for the Schur-complement $\mathbf{S}_{CC} = \mathbf{K}_{CC} - \mathbf{K}_{CI}\mathbf{K}_{II}^{-1}\mathbf{K}_{IC}$. The block $\mathbf{E}_{IC}$ is nothing else than the matrix representation of a bounded extension operator from the coupling boundary Γ_C to the interior of the subdomains.

In [3], we proposed block preconditioners $\mathbf{C}_{II}$ and $\mathbf{C}_{CC}$ as well as an extension operator $\mathbf{E}_{IC}$ such that the preconditioner $\mathbf{C}_h$ is spectrally equivalent to the stiffness matrix $\mathbf{K}_h$ with constants which are independent of h and H. The complexity of the preconditioning operation $\mathbf{C}_h^{-1}\mathbf{r}_h$ in the PCG iteration is the same as the matrix-vector multiplication $\mathbf{K}_h\mathbf{v}_h$, namely $O(H/h)$ in a parallel regime. Therefore, the total complexity of the PCG iteration is proportional to $O((H/h)\log(\varepsilon^{-1}))$, where $\varepsilon \in (0,1)$ is the relative accuracy of the usual PCG iteration error.

The FETI methods avoid the preconditioning of the assembled Schur-complement $\mathbf{S}_{CC}$ by tearing the global skeleton potential vector $\mathbf{u}_C$ on all subdomain boundaries Γ_i including the Dirichlet boundary Γ_D. Thus, we introduce the local unknowns $\mathbf{u}_{C,i} = \mathbf{R}_i\mathbf{u}_C$ in every subdomain $\overline{\Omega}_i$ separately, and enforce the global continuity across the interfaces and Dirichlet conditions by the constraints

$$\sum_{i=1}^{p} \mathbf{B}_{C,i}\mathbf{u}_{C,i} = \mathbf{g} \tag{12}$$

where the definition of the matrices $\mathbf{B}_{C,i}$ and the vector $\mathbf{g}$ is straightforward (see [26] and [19]). The incorporation of the Dirichlet conditions into the constraints (12) is called All-Floating (AF) FETI [21] or total FETI [9]. Introducing Lagrange multipliers λ, we can obviously transform the SPD Schur-complement system (7) to the following equivalent saddle point problem

$$\begin{pmatrix} \mathbf{S}_{C,1} & & & \mathbf{B}_{C,1}^\top \\ & \ddots & & \vdots \\ & & \mathbf{S}_{C,p} & \mathbf{B}_{C,p}^\top \\ \mathbf{B}_{C,1} & \dots & \mathbf{B}_{C,p} & \mathbf{0} \end{pmatrix} \begin{pmatrix} \mathbf{u}_{C,1} \\ \vdots \\ \mathbf{u}_{C,p} \\ \lambda \end{pmatrix} = \begin{pmatrix} \mathbf{b}_{C,1} \\ \vdots \\ \mathbf{b}_{C,p} \\ \mathbf{g} \end{pmatrix}, \tag{13}$$

with the singular Schur complements $\mathbf{S}_{C,i} = \mathbf{S}_{C,i}^{FE} = \mathbf{S}_{C,i}^{IC-FE}$ $(\ker(\mathbf{S}_{C,i}) = \text{span}\{\mathbf{1}_{C,i}\})$ and the right-hand sides $\mathbf{b}_{C,i} = \mathbf{f}_{C,i} - \mathbf{K}_{CI,i}\mathbf{K}_{II,i}^{-1}\mathbf{f}_{I,i}$, $i = 1,\dots,p$. System (13) is called AF-FETI-2 system.

Now we can again unfold the Schur complements $\mathbf{S}_{C,i}$ arriving at the larger saddle point problem

$$\begin{pmatrix} \mathbf{K}_1 & & & \mathbf{B}_1^\top \\ & \ddots & & \vdots \\ & & \mathbf{K}_p & \mathbf{B}_p^\top \\ \mathbf{B}_1 & \dots & \mathbf{B}_p & \mathbf{0} \end{pmatrix} \begin{pmatrix} \mathbf{u}_1 \\ \vdots \\ \mathbf{u}_p \\ \lambda \end{pmatrix} = \begin{pmatrix} \mathbf{f}_1 \\ \vdots \\ \mathbf{f}_p \\ \mathbf{g} \end{pmatrix}, \tag{14}$$

with the Neumann matrices $\mathbf{K}_i$ on the main diagonal. System (14) or, more precisely, its regularized version (see [3] for details) is called AF-FETI-3 system.

On the other hand, we can eliminate the primal variables $\mathbf{u}_{C,i}$ from the AF-FETI-2 system (13) and arrive at the AF-FETI-1 system

$$\mathbf{P}^\top \mathbf{F} \mathbf{P} \lambda = \mathbf{P}^\top \mathbf{b} \tag{15}$$

for determining the Lagrange multipliers λ, with the FETI operator $\mathbf{F}$, the FETI projector $\mathbf{P}$, and the corresponding right-hand side $\mathbf{b}$ (see, e.g., [26] and [3] for a detailed description).

The FETI-1 system (15) can be solved by means of a subspace PCG method with the exact Dirichlet FETI preconditioner $\mathbf{C}_F^{-1} = \mathbf{A}\,\text{diag}(\mathbf{S}_{C,i})\,\mathbf{A}^\top$, where $\mathbf{A}$ is an appropriately chosen scaling matrix. It can be shown that the number $I(\varepsilon)$ of PCG iterations only slowly grows like $O((1+\log(H/h))\log(\varepsilon^{-1}))$ and is robust with respect to coefficient jumps [3]. However, in every PCG iteration step, the solution of local Neumann and Dirichlet problems are hidden. In order to avoid the solution of local Neumann and Dirichlet problems, inexact FETI methods can be used. Inexact FETI methods are nothing else but a Krylov subspace solver (e.g., Bramble-Pasciak's PCG for saddle-point problems [4]) for the FETI-3 system (14). In order to construct an appropriate preconditioner, we need block preconditioners $\mathbf{C}_i$ for $\mathbf{K}_i$ and a FETI preconditioner $\mathbf{C}_F$. In [3], we have proved the following theorem.

Theorem 1. *Let us assume that the AF-FETI-3 system (14) is solved by means of the Bramble-Pasciak PCG method with the appropriately scaled local Neumann* *preconditioners* $\mathbf{C}_i$ *constructed in the same way as the factorized preconditioner* $\mathbf{C}_h$ *for* $\mathbf{K}_h$ *(see formulas (11) and (10)) and with the scaled inexact Dirichlet FETI preconditioner* $\mathbf{C}_F^{-1} = \mathbf{A}\, diag(\overline{\mathbf{K}}_{CC}^{(i)} + \mathbf{E}_{IC,i}^\top \mathbf{K}_{II}^{(i)} \mathbf{E}_{IC,i} - \mathbf{K}_{CI}^{(i)} \mathbf{E}_{IC,i} - \mathbf{E}_{IC,i}^\top \mathbf{K}_{IC}^{(i)})\, \mathbf{A}^\top$. *Then not more than* $I(\varepsilon) = O((1+\log(H/h))\log(\varepsilon^{-1}))$ *iterations and* $ops(\varepsilon) = O((H/h)$ $(1+\log(H/h))\log\varepsilon^{-1})$ *arithmetical operations are required in order to reduce the initial error by the factor* $\varepsilon \in (0,1)$ *in a parallel regime, where* $H/h = \max H_i/h_i$. *The number of iterations* $I(\varepsilon)$ *is robust with respect to the jumps in the coefficients.*

Let us discuss some results of our numerical experiments with the IC-FEM. We consider the potential equation (1) modelling a magnetic valve in 2D, where the coefficient $a(.)$ now denotes the reluctivity (which typically has large coefficient jumps across the interfaces !), the current density $f(.)$ is concentrated in the coil, and $g = 0$. The geometry of the computational domain Ω and the IC mesh is shown in Figure 2. Table 1 provides a comparison of the IC-FEM with the standard FEM obtained by uniform mesh refinement. Both FE versions produce approximate solutions of the same accuracy. However, the DOF grow with the factor 2 for the IC-FEM, whereas the DOF quadruplicates for the standard FEM if the discretization parameter h is halved. In both cases we use the one-level FETI-1 solver, where the local Neumann problems in the matrix-vector multiplication and the local Dirichlet problems in the FETI preconditioning step are solved by a direct solver which is of course not optimal with respect to the complexity (see the CPU time in the last two columns). The numbers of PCG iterations are almost the same for both cases (see column 7 and 8) and grow like $O(1+\log(H/h))$ as predicted by the theory. Moreover, the iteration numbers will not change significantly if the jumps of $a(.)$ will artificially be varied across the interfaces.

The numerical features (CPU time in seconds) of a parallel implementation of the IC-FETI-1 algorithm on a 4 $\times$ Dualcore Intel Xeon CPU 3.40 GHz (8 processors in total) with 16 MB processor cache are presented in Table 2.

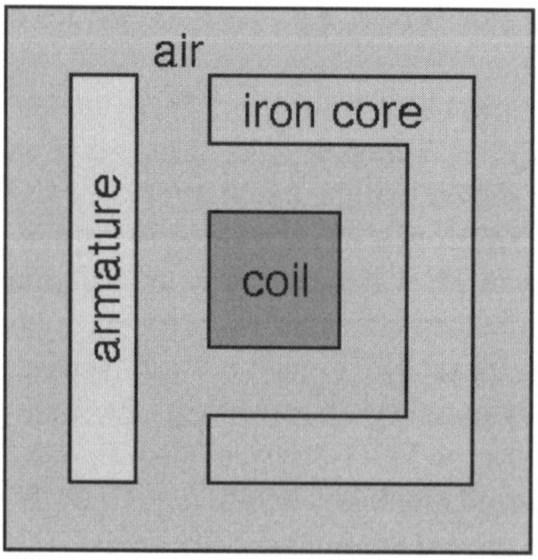

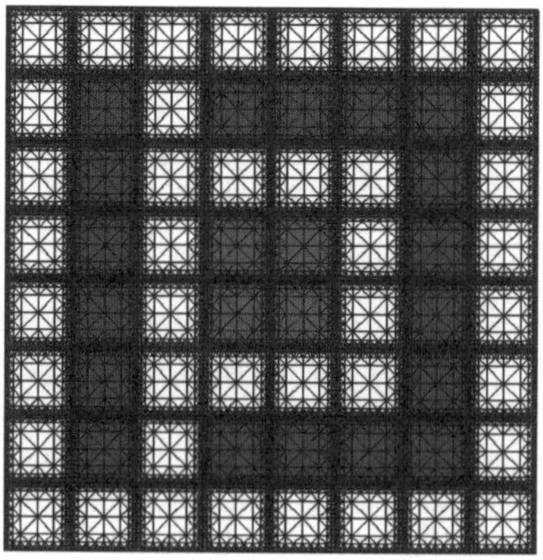

Fig. 2 The magnetic valve (left) and the IC mesh for the valve (right).

Table 1 FETI1: PCG $= I(\varepsilon = 10^{-8})$, CPU $=$ [sec]

FETI global dof	IC-FETI global dof	FETI local dof	IC-FETI local dof	Lagr. dof	H/h	FETI PCG	IC-F PCG	FETI CPU	IC-FETI CPU
289	289	9	9	406	2	**11**	**11**	3.2	3.3
1089	833	25	21	630	4	**13**	**14**	6.9	7.4
4225	2945	81	61	1078	8	**16**	**16**	8.3	8.1
16641	10241	289	189	1974	16	**18**	**18**	9.6	9.3
66049	30977	1089	541	3766	32	**19**	**20**	12.5	11.7
263169	82689	4225	1405	7350	64	**21**	**21**	26.7	16.6
1050625	201473	16641	3373	14518	128	**23**	**23**	95.3	29.5
4198401	460545	66049	7645	28854	256	**25**	**25**	489.7	61.9
16785409	1007361	263169	16637	57526	512	**26**	**26**	2478.7	136.8
67125249	2137857	1050625	35179	114870	1024	**–**	**28**	–	327.2
268468225	4444929	4198401	73073	229558	2048	**–**	**30**	–	823.6
1073807361	9115393	16785401	149597	458934	4096	**–**	**30**	–	2043.3

Table 2 IC-FETI on a parallel computer [CPU-time in seconds]

global DOF	460545	1007361	2137867	4444929	9115393
local DOF	7645	16637	35197	73037	149597
1 processor	60.0	132.6	317.1	956.8	2043.3
2 processors	31.3	69.0	168.1	566.3	1184.4
4 processors	16.2	34.8	82.5	265.5	541.6
8 processors	9.6	19.2	56.8	160.7	342.5

4 Coupling of Data-Sparse BEM with Interface-Concentrated FEM

The construction of coupled data-sparse BE and interface-concentrated FE equations in a non-overlapping DD framework is based on the representation (7), where the BE matrices $\mathbf{V}_{C,i}$, $\mathbf{K}_{C,i}$ and $\mathbf{D}_{C,i}$ occuring in the BE Schur-complement (3) are replaced by data-sparse approximations $\widetilde{\mathbf{V}}_{C,i}$, $\widetilde{\mathbf{K}}_{C,i}$ and $\widetilde{\mathbf{D}}_{C,i}$ without perturbing the spectral properties and discretization error estimates. However, the complexity of the matrix-vector multiplications and the storage demand can be reduced from $O((H/h)^{2(d-1)})$ to $O((H/h)^{(d-1)})$ up to a polylogarithmic factor. In the FE subdomains Ω_i, $i = q+1,\ldots,p$, we use the IC-FEM described in the preceding section, i.e. $\mathbf{S}_{C,i} = \mathbf{S}_{C,i}^{FE} = \mathbf{S}_{C,i}^{IC-FE}$. The coupling is illustrated in Figure 1 (right).

As in the IC-FE case, we can construct primal and dual iterative substructuring solvers. In [19], we study all-floating coupled data-sparse boundary and interface-concentrated finite element tearing and interconnecting methods. In particular, for BETI-FETI-3 solvers, we can show the same results as formulated in Theorem 1 with the only difference that we have to add a polylogarithmic factor to the complexity estimates, e.g. $(1+\log(H/h))^2$ for the Fast Multipole Method (see [19] for details). In the same paper, we present also some numerical results for the valve problem in $\mathbb{R}^2$, where the the exterior domain is included in the domain decomposition as an additional subdomain $\Omega_0 = \mathbb{R}^2 \setminus \overline{\Omega}$.

5 A Boundary-Element-Based FEM

The pure BE case ($q = p$) was first considered by G.C. Hsiao and W.L. Wendland in [14]. They already mentioned that there are two typical situations in DD, namely, h_i tends to 0 for fixed H_i and H_i tends to 0 for fixed H_i/h_i. The former is typical for DD with a fixed number p of subdomains (processors). The latter one means a growing number p of subdomains with a fixed number of DOF per subdomain which is typical in massively parallel computing.

Let us now consider the special case $H_i = h_i$ of the latter case, where $q = p$ is now the number N_e of polygonal (2D) or polyhedral (3D) elements. Similar to (7), we obtain a linear system of the form (8) that can be considered as a sparse system of FE equations, where the global stiffness matrix $\mathbf{K}_h$ and the global load vector $\mathbf{f}_h$ are assembled from the boundary element stiffness matrices $\mathbf{S}_{C,i}^{BE}$ and the boundary element load vectors $\mathbf{f}_{C,i}^{BE}$. The incorporation of the boundary conditions, in particular, of the Dirichlet condition can be done in the same way as in the usual FEM. The element load vectors $\mathbf{f}_{C,i}^{BE}$ are defined by the relation $\mathbf{f}_{C,i}^{BE} = \mathbf{M}_{C,i}^{\top}\left(\mathbf{V}_{C,i}\right)^{-1}\mathbf{f}_{C,i}^{N}$, where the vector $\mathbf{f}_{C,i}^{N}$ is given by the Newton potential identity

$$(\mathbf{f}_{C,i}^{N}, \mathbf{t}_{C,i}) = \int_{\Gamma_i}\int_{\Omega_i} U^*(x,y)f(y)dy\, t_{h,i}(x)ds_x \tag{16}$$

for all vectors $\mathbf{t}_{C,i}$ corresponding to the piecewise constant functions $t_{h,i}$ on the triangulation of Γ_i, where U^* denotes the fundamental solution of the Laplace operator.

In our first numerical experiments we apply this BE-based FEM to the Laplace equation with prescribed Dirichlet conditions $g(x) = \log\|x - x^*\|$ on the boundary $\Gamma = \partial\Omega$, where the singularity $x^* = (1.1, 1.1)^{\top}$ is located outside the computational domain $\Omega = (0,1) \times (0,1)$. It is easy to see that in the case of Courant's triangular element the BE-based FE stiffness matrix coincides with the standard FE stiffness matrix. In the case of the Laplace equation the right-hand sides are also identical. Therefore, the same discretization error estimates are valid. In this case, the generation time for the BE-based FE stiffness matrix is approximately three times higher than for the Courant stiffness matrix (e.g. 13.3 seconds vs. 4.7 seconds for 263169 unknowns). However, the BE-based FEM can obviously be used for general polygonal (2D) and polyhedral (3D) elements Ω_i. We performed two series of experiments. In the first series, we generated random quadrangular meshes where the nodes of a uniform quadrangular mesh were randomly shifted within some neighborhood of the nodes (see left picture of Figure 3), whereas in the second series we generated hexagonal meshes (see right picture of Figure 3). Table 3 shows the numerical results for 4 different grids, where the Algebraic Multigrid (AMG) PCG, implemented in the AMG code PEBBLES [22], was used for solving the system of algebraic equations. L denotes the number of levels which were generated by the AMG code. The generation time for the stiffness matrix and the setup time for the AMG in seconds are given in the columns with the headers K_h and *Setup*, respectively. The CPU times reflect the linear complexity of the generation procedure and the AMG set up. The number $I(\varepsilon)$ of PCG iterations for reducing the initial error by the factor $\varepsilon = 10^{-12}$ and the CPU time per iteration cycle are presented in the columns with the headers *Cycle* and $I(\varepsilon)$, respectively. Finally, we present the discretization error in the $L_2(\Omega)$ - norm that behaves like expected.

Table 3 Numerical features for the random quadrangular (left) and the hexagonal (right) meshes

		AMG		PCG					AMG		PCG		
N_h	L	K_h	Setup	Cycle	$I(\varepsilon)$	$\|u-u_h\|_{0,\Omega}$	N_h	L	K_h	Setup	Cycle	$I(\varepsilon)$	$\|u-u_h\|_{0,\Omega}$
16641	4	0.9	0.6	0.06	16	3.3 E-6	25440	4	1.6	2.1	0.14	12	2.0 E-5
66049	5	3.7	2.8	0.26	17	8.4 E-7	102080	5	6.5	10.7	0.57	10	1.9 E-6
263169	6	15.0	15.8	1.07	19	2.1 E-7	408960	6	26.0	58.4	2.34	10	1.9 E-7

Acknowledgements The author would like to thank his colleagues S. Beuchler, D. Copeland, C. Pechstein and D. Pusch for their contributions to this paper. Last but not least, the support by the Austrian Science Fund (FWF) under the grant P19255 is gratefully acknowledged.

References

1. Bebendorf, M.: Approximation of boundary element matrices. Numerische Mathematik **86**, 565–589 (2000)
2. Bebendorf, M., Rjasanow, S.: Adaptive low–rank approximation of collocation matrices. Computing **70**, 1–24 (2003)
3. Beuchler, S., Eibner, T., Langer, U.: Primal and dual interface concentrated iterative substructuring methods. SIAM J. Numer. Anal. (2008). To appear
4. Bramble, J.H., Pasciak, J.E.: A preconditioning technique for indefinite systems resulting from mixed approximations of elliptic problems. Math. Comp. **50**(181), 1–17 (1988)
5. Brezzi, F., Johnson, C.: On the coupling of boundary integral and finite element methods. Calcolo **16**, 189–201 (1979)
6. Carstensen, C., Kuhn, M., Langer, U.: Fast parallel solvers for symmetric boundary element domain decomposition equations. Numerische Mathematik **79**, 321–347 (1998)
7. Cheng, H., Greengard, L., Rokhlin, V.: A fast adaptive multipole algorithm in three dimensions. J. Comput. Phys. **155**(2), 468–498 (1999)

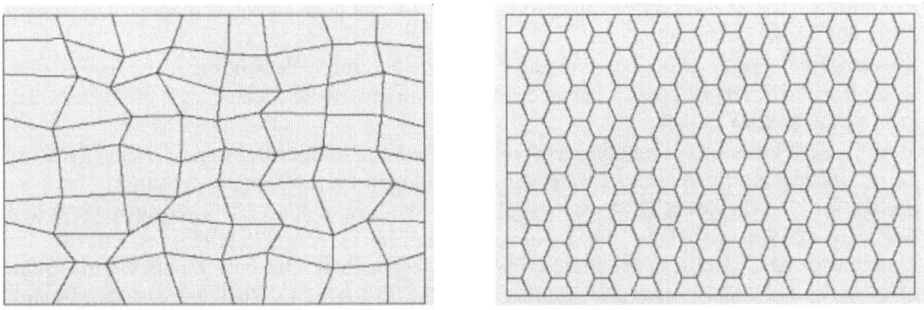

Fig. 3 Random quadrangular mesh (left) and hexagonal mesh (right).

8. Costabel, M.: Symmetric methods for the coupling of finite elements and boundary elements. In: C. Brebbia, W. Wendland, G. Kuhn (eds.) Boundary Elements IX, pp. 411–420. Springer, Berlin, Heidelberg, New York (1987)
9. Dostál, Z., Horák, D., Kučera, R.: Total FETI - an easier implementable variant of the FETI method for numerical solution of elliptic PDE. Comm. Numer. Methods Engrg. **22**(12), 1155–1162 (2006)
10. Haase, G., Heise, B., Kuhn, M., Langer, U.: Adaptive domain decomposition methods for finite and boundary element equations. In: W. Wendland (ed.) Boundary Element Topics, pp. 121 –147. Springer, Berlin, Heidelberg, New York (1997)
11. Haase, G., Langer, U., Meyer, A.: The approximate Dirichlet decomposition method. Part I: An algebraic approach, Part II: Applications to 2nd-order elliptic BVPs. Computing **47**, 137–167 (1991)
12. Hackbusch, W.: A sparse matrix arithmetic based on $\mathscr{H}$-matrices. Part I: Introduction to $\mathscr{H}$-matrices. Computing **62**(2), 89–108 (1999)
13. Hackbusch, W., Nowak, Z.P.: On the fast matrix multiplication in the boundary element method by panel clustering. Numer. Math. **54**(4), 463–491 (1989)
14. Hsiao, G.C., Wendland, W.L.: Domain decomposition in boundary element methods. In: Proceedings of the Fourth Intern. Symposium on Domain Decomposition Methods for Partial Differential Equations (ed. by R. Glowinski and Y.A. Kuznetsov and G. Meurant and J. Périaux and O. B. Widlund), Moscow, May 21–25, 1990, pp. 41–49. SIAM, Philadelphia (1991)
15. Johnson, C., Nédélec, J.C.: On coupling of boundary integral and finite element methods. Math. Comp. **35**, 1063–1079 (1980)
16. Khoromskij, B.N., Melenk, J.M.: An efficient direct solver for the boundary concentrated FEM in 2D. Computing **69**, 91–117 (2002)
17. Khoromskij, B.N., Melenk, J.M.: Boundary concentrated finite element methods. SIAM J. Numer. Anal. **41**(1), 1–36 (2003)
18. Langer, U.: Parallel iterative solution of symmetric coupled FE/BE–equation via domain decomposition. In: A. Quarteroni, J. Periaux, Y. Kuznetsov, O. Widlund (eds.) Sixth Intern. Conf. on Domain Decomposition Methods in Science and Engineering, Como, June 15–19, 1992, *Contempory Mathematics*, vol. 157, pp. 335–344. AMS, Providence, Rhode Island (1994)
19. Langer, U., Pechstein, C.: All-floating coupled data-sparse boundary and interface-concentrated finite element tearing and interconnecting methods. Computing and Visualization in Science (2008). DOI 10.1007/s00791-008-0100-6
20. Langer, U., Steinbach, O.: Coupled finite and boundary element domain decomposition methods. In: M. Schanz, O. Steinbach (eds.) Boundary Element Analysis: Mathematical Aspects and Applications, *LNACM*, vol. 29, pp. 29–59. Springer, Berlin (2007)
21. Of, G.: The all-floating BETI method: Numerical results. In: U. Langer, M. Discacciati, D. Keyes, O. Widlund, W. Zulehner (eds.) Domain Decomposition Methods in Science and Engineering XVII, *LNCSE*, vol. 60, pp. 294–302. Springer, Heidelberg (2008)
22. Reitzinger, S.: Algebraic Multigrid Methods for Large Scale Finite Element Equations. Reihe C - Technik und Naturwissenschaften. Universitätsverlag Rudolf Trauner, Linz (2001)
23. Rjasanow, S., Steinbach, O.: The Fast Solution of Boundary Integral Equations. Springer, New York (2007)
24. Rokhlin, V.: Rapid solution of integral equations of classical potential theory. J. Comput. Phys. **60**(2), 187–207 (1985)
25. Stephan, P.: Coupling of boundary element methods and finite element methods. In: E. Stein, R. de Borst, T.J. Hughes (eds.) Encyclopedia of Computational Mechanics, vol. 1, chap. 12. John Wiley & Sons (2004)
26. Toselli, A., Widlund, O.: Domain Decoposition Methods - Algorithms and Theory, *Springer Series in Computational Mathematics*, vol. 34. Springer, Berlin, Heidelberg (2004)
27. Zienkiewicz, O.C., Kelly, D.M., Bettes, P.: The coupling of the finite element method and boundary solution procedures. Int. J. Numer. Meth. Eng **11**, 355–376 (1977)
28. Zienkiewicz, O.C., Kelly, D.M., Bettes, P.: Marriage a la mode - the best of both worlds (finite elements and boundary integrals). In: R. Glowinski, E. Rodin, O. Zienkiewicz (eds.) Energy Methods in Finite Element Analysis, chap. 5, pp. 81–106. J. Wiley and Son, London (1979)

Discontinuous Galerkin Elements for Reissner-Mindlin Plates

L.D. Marini

Abstract We present an overview of some families of locking-free elements for Reissner-Mindlin plates recently introduced and analyzed in [2] and [1]. They are all based on the ideas of discontinuous Galerkin approach, and they vary in the amount of interelement continuity required.

1 Introduction

The Reissner–Mindlin model for moderately thick clamped plates consists in loocking for the rotation vector $\boldsymbol{\theta}$ and the transverse displacement w which minimize over $\boldsymbol{H}_0^1(\Omega) \times H_0^1(\Omega)$ the (scaled) plate energy

$$J(\boldsymbol{\theta}, w) = \frac{1}{2} \int_\Omega \mathsf{C}\varepsilon(\boldsymbol{\theta}) : \varepsilon(\boldsymbol{\theta}) \, \mathrm{d}x + \frac{1}{2} \lambda t^{-2} \int_\Omega |\nabla w - \boldsymbol{\theta}|^2 \, \mathrm{d}x - \int_\Omega g w \, \mathrm{d}x, \tag{1}$$

where the coefficients C and λ depend on the material properties of the plate, g is the scaled load, and t is the plate thickness. If one minimizes the energy over subspaces consisting of low order finite elements, then the resulting approximation suffers from the problem of *locking*, which can be described as follows. As t tends to 0, the solution of (1) tends to $(\boldsymbol{\theta}_0, w_0)$, where $\boldsymbol{\theta}_0 = \nabla w_0$ which, in general, will not be zero (actually, w_0 will be the solution of the Kirchhoff model). If we discretize the problem directly by seeking $\boldsymbol{\theta}_h \in \boldsymbol{\Theta}_h$ and $w_h \in W_h$ minimizing $J(\boldsymbol{\theta}, w)$ over $\boldsymbol{\Theta}_h \times W_h$, then as t vanishes, $(\boldsymbol{\theta}_h, w_h)$ will converge to some $(\boldsymbol{\theta}_{0,h}, w_{0,h})$ where, again, $\boldsymbol{\theta}_{0,h} = \nabla w_{0,h}$. For low order finite element spaces, this last condition is too restrictive. In particular, if continuous piecewise linear functions are used to approximate both variables, then $\boldsymbol{\theta}_{0,h} \equiv \nabla w_{0,h}$ would be continuous *and* piecewise constant, with zero boundary conditions. Only the choice $\boldsymbol{\theta}_{0,h} = 0$ can satisfy all

L. Donatella Marini
Università di Pavia and IMATI-CNR, Via Ferrata 1, 27100 Pavia(Italy), e-mail: marini@imati.cnr.it

these conditions. For t very small, the quantity $\boldsymbol{\theta}_h - \nabla w_h$, although not necessarily zero, must be very small, and hence $\boldsymbol{\theta}_h$ will be very close to zero, instead of being close to $\boldsymbol{\theta}$ which, in turn, will be close to $\boldsymbol{\theta}_0$. Another way of looking at this problem is from the point of view of approximation: for small t, one cannot find suitable interpolants $\boldsymbol{\theta}^I$ and w^I that are close to $\boldsymbol{\theta}$ and w, respectively, if one requires $\boldsymbol{\theta}^I - \nabla w^I$ to be of the order of t^2.

A number of approaches have been developed to avoid the locking problem. One successful idea has been to introduce an additional finite element space $\boldsymbol{\Gamma}_h$ and a reduction operator $\boldsymbol{P}_h : \boldsymbol{\Theta}_h \to \boldsymbol{\Gamma}_h$, and then look for $\boldsymbol{\theta}_h \in \boldsymbol{\Theta}_h$ and $w_h \in W_h$ minimizing a modified energy functional

$$J_h(\boldsymbol{\theta}, w) = \frac{1}{2}\int_\Omega \mathbb{C}\varepsilon(\boldsymbol{\theta}) : \varepsilon(\boldsymbol{\theta})\,\mathrm{d}x + \frac{1}{2}\lambda t^{-2}\int_\Omega |\nabla w - \boldsymbol{P}_h\boldsymbol{\theta}|^2\,\mathrm{d}x - \int_\Omega gw\,\mathrm{d}x. \tag{2}$$

A crucial assumption is that ∇W_h is a subset of $\boldsymbol{\Gamma}_h$, and in particular of the image of $\boldsymbol{P}_h$. As t tends to 0, the limiting condition will now be the much less demanding

$$\boldsymbol{P}_h\boldsymbol{\theta}_{0,h} = \nabla w_{0,h}. \tag{3}$$

Various locking-free finite elements have been obtained in this way (see, e.g., [3], [5], [8], [11], [12], [9], [13], [10]).

In [2], the techniques of Discontinuous Galerkin (DG) methods were used to develop two families of odd-degree locking-free elements. Since DG solutions are not required to satisfy the standard interelement continuity conditions of conforming finite element methods (that is, continuous elements in the case of the Reissner–Mindlin plate problem), the method allows a greater flexibility.

Starting from the approach of [2], other elements were introduced ([9], [13], [10]) for the functional (2), while in [1] a collection of families of locking-free elements which do not need the reduction operator $\boldsymbol{P}_h$ were developed. The common feature in all the methods considered in [1] is to choose W_h to be piecewise polynomials of degree $\leq k$ (with $k \geq 2$), and $\boldsymbol{\Theta}_h = \boldsymbol{\Gamma}_h$ to be piecewise polynomials of degree $\leq k-1$. The methods vary in the amount of interelement continuity required.

In the present paper we shall give an overview of some DG elements, and we shall report the convergence results, referring for the proofs to the corresponding papers.

2 Discontinuous Galerkin Discretization

Introducing the shear stress $\boldsymbol{\gamma} = \lambda t^{-2}(\nabla w - \boldsymbol{\theta})$ as an auxiliary variable, and writing the Euler equations for the energy functional (1) we may write the Reissner–Mindlin equations as:

$$-\operatorname{div} \mathsf{C}\varepsilon(\boldsymbol{\theta}) - \boldsymbol{\gamma} = 0 \quad \text{in } \Omega, \tag{4}$$
$$-\operatorname{div} \boldsymbol{\gamma} = g \quad \text{in } \Omega, \tag{5}$$
$$\nabla w - \boldsymbol{\theta} - t^2 \boldsymbol{\gamma} = 0 \quad \text{in } \Omega, \tag{6}$$
$$\boldsymbol{\theta} = 0, \; w = 0 \text{ on } \partial\Omega. \tag{7}$$

Equation (6) should actually be $\nabla w - \boldsymbol{\theta} - \lambda^{-1} t^2 \boldsymbol{\gamma} = 0$, where λ is the *shear correction factor*, but we set $\lambda = 1$ to simplify the presentation. By setting

$$a(\boldsymbol{\theta}, \boldsymbol{\eta}) = (\mathsf{C}\varepsilon(\boldsymbol{\theta}), \varepsilon(\boldsymbol{\eta})) \quad \text{for } \boldsymbol{\theta}, \boldsymbol{\eta} \in \boldsymbol{H}^1(\Omega)$$

the variational formulation of equations (4)–(7) is:
Given $g \in L^2(\Omega)$, find $\boldsymbol{\theta} \in \boldsymbol{H}_0^1(\Omega)$, $w \in H_0^1(\Omega)$ and $\boldsymbol{\gamma} \in \boldsymbol{L}^2(\Omega)$ such that

$$a(\boldsymbol{\theta}, \boldsymbol{\eta}) + (\boldsymbol{\gamma}, \nabla v - \boldsymbol{\eta}) = (g, v) \quad \forall (\boldsymbol{\eta}, v) \in \boldsymbol{H}_0^1(\Omega) \times H_0^1(\Omega), \tag{8}$$
$$(\nabla w - \boldsymbol{\theta}, \boldsymbol{\tau}) - t^2 (\boldsymbol{\gamma}, \boldsymbol{\tau}) = 0 \quad \forall \boldsymbol{\tau} \in \boldsymbol{L}^2(\Omega). \tag{9}$$

Before proceeding we need to introduce some notations. We shall use the usual Sobolev spaces such as $H^s(\Omega)$, with the corresponding seminorm and norm denoted by $|\cdot|_s$ and $\|\cdot\|_s$, respectively. By convention, we use boldface type for the vector-valued analogues ($\boldsymbol{H}^s(\Omega) = [H^s(\Omega)]^2$), and calligraphic type for symmetric-tensor-valued analogues ($\mathscr{H}^s(\Omega) = [H^s(\Omega)]^2_{\text{sym}}$); we use parentheses $(\cdot, \cdot)$ to denote the inner product in any of the spaces $L^2(\Omega)$, $\boldsymbol{L}^2(\Omega)$, or $\mathscr{L}^2(\Omega)$.

We recall the following result (see [3], [4] for a more general case). If Ω is a convex polygonal domain, and C is smooth, then problem (8)–(9) has a unique solution that verifies

$$\|\boldsymbol{\theta}\|_2 + \|w\|_2 + \|\boldsymbol{\gamma}\|_0 + t\|\boldsymbol{\gamma}\|_1 \leq C(\|g\|_{-1} + t\|g\|_0), \tag{10}$$

where C is a constant depending only on Ω and on the coefficients in C.

Let now $\mathscr{T}_h$ be a family of shape-regular decompositions of Ω into triangles T and let $\mathscr{E}_h$ denote the set of all the edges in $\mathscr{T}_h$. For piecewise polynomial spaces, we use the notation

$$\mathscr{L}_k^s(\mathscr{T}_h) = \{ v \in H^s(\Omega) : v|_T \in \mathscr{P}_k(T), T \in \mathscr{T}_h \}, \tag{11}$$

where, as usual, $\mathscr{P}_k(T)$ is the set of polynomials of degree at most k on T. Since we will work with discontinuous finite elements not belonging to $H^1(\Omega)$, we define the space

$$H^1(\mathscr{T}_h) := \{ v \in L^2(\Omega) : v|_T \in H^1(T), T \in \mathscr{T}_h \}. \tag{12}$$

Differential operators can be applied to this space only piecewise. We indicate this by a subscript h on the operator. Hence, the space $H^1(\mathscr{T}_h)$ will be equipped with the seminorm $|v|_{1,h} = \|\nabla_h v\|_0$ and the corresponding norm $\|v\|_{1,h}^2 = |v|_{1,h}^2 + \|v\|_0^2$.

Finally, before deriving a DG discretization of (8)–(9) we need to introduce typical tools as *averages* and *jumps* on the edges of $\mathscr{T}_h$. Let e be an internal edge of $\mathscr{T}_h$, shared by two elements T^+ and T^-, and let $\boldsymbol{n}^+$ and $\boldsymbol{n}^-$ denote the unit normals to e, pointing outward from T^+ and T^-, respectively. If φ belongs to $H^1(\mathscr{T}_h)$ (or possibly the vector- or tensor-valued analogue), we define the average $\{\varphi\}$ on e as usual:

$$\{\varphi\} = \frac{\varphi^+ + \varphi^-}{2}.$$

For a scalar function $\varphi \in H^1(\mathscr{T}_h)$ we define its jump on e as

$$[\![\varphi]\!] = \varphi^+ \boldsymbol{n}^+ + \varphi^- \boldsymbol{n}^-,$$

which is a vector normal to e. The jump of a vector $\boldsymbol{\varphi} \in \boldsymbol{H}^1(\mathscr{T}_h)$ is the symmetric matrix-valued function given on e by:

$$[\![\boldsymbol{\varphi}]\!] = \boldsymbol{\varphi}^+ \odot \boldsymbol{n}^+ + \boldsymbol{\varphi}^- \odot \boldsymbol{n}^-,$$

where $\boldsymbol{\varphi} \odot \boldsymbol{n} = (\boldsymbol{\varphi} \otimes \boldsymbol{n} + \boldsymbol{n} \otimes \boldsymbol{\varphi})/2$ is the symmetric part of the tensor product of $\boldsymbol{\varphi}$ and $\boldsymbol{n}$.

On a boundary edge, the average $\{\varphi\}$ is defined simply as the trace of φ, while for a scalar-valued function we define $[\![\varphi]\!]$ to be $\varphi \boldsymbol{n}$ (with $\boldsymbol{n}$ the outward unit normal), and for a vector-valued function we define $[\![\boldsymbol{\varphi}]\!] = \boldsymbol{\varphi} \odot \boldsymbol{n}$.

It is easy to check that, (using the symbol $\langle \cdot , \cdot \rangle$ to denote L^2-inner product of functions or vectors on $\mathscr{E}_h$)

$$\sum_{T \in \mathscr{T}_h} \int_{\partial T} \boldsymbol{\varphi} \cdot \boldsymbol{n}_T v \, \mathrm{d}s = \langle \{\boldsymbol{\varphi}\}, [\![v]\!] \rangle, \quad \boldsymbol{\varphi} \in \boldsymbol{H}^1(\Omega), v \in H^1(\mathscr{T}_h). \tag{13}$$

Similarly,

$$\sum_{T \in \mathscr{T}_h} \int_{\partial T} \mathscr{S} \boldsymbol{n}_T \cdot \boldsymbol{\eta} \, \mathrm{d}s = \langle \{\mathscr{S}\}, [\![\boldsymbol{\eta}]\!] \rangle, \quad \mathscr{S} \in \mathscr{H}^1(\Omega), \boldsymbol{\eta} \in \boldsymbol{H}^1(\mathscr{T}_h). \tag{14}$$

To derive a finite element method for the Reissner–Mindlin system based on discontinuous elements, we test (4) against a test function $\boldsymbol{\eta} \in \boldsymbol{H}^2(\mathscr{T}_h)$ and (5) against a test function $v \in H^1(\mathscr{T}_h)$, integrate by parts, and add. Since $\boldsymbol{\eta}$ and v may be discontinuous across element boundaries, we obtain terms at the interelement boundaries that we manipulate using (13)-(14). We obtain:

$$(\mathsf{C}\varepsilon_h(\boldsymbol{\theta}), \varepsilon_h(\boldsymbol{\eta})) - \langle \{\mathsf{C}\varepsilon_h(\boldsymbol{\theta})\}, [\![\boldsymbol{\eta}]\!] \rangle + (\boldsymbol{\gamma}, \nabla_h v - \boldsymbol{\eta}) - \langle \{\boldsymbol{\gamma}\}, [\![v]\!] \rangle = (g, v), \tag{15}$$
$$(\boldsymbol{\eta}, v) \in \boldsymbol{H}^2(\mathscr{T}_h) \times H^1(\mathscr{T}_h),$$
$$(\nabla_h w - \boldsymbol{\theta}, \boldsymbol{\tau}) - t^2(\boldsymbol{\gamma}, \boldsymbol{\tau}) = 0, \qquad \boldsymbol{\tau} \in \boldsymbol{H}^1(\mathscr{T}_h). \tag{16}$$

The second and fourth terms in (15) involve integrals over the edges and would not be present in conforming methods. They arise from the integration by parts and are necessary to maintain consistency.

We now proceed as is common for DG methods. (For a different point of view on this type of derivation see [6]). First, we add terms to symmetrize this formulation so that it is adjoint-consistent as well. Second, to stabilize the method, we add *interior penalty* terms $p_\Theta(\boldsymbol{\theta}, \boldsymbol{\eta})$ and $p_W(w, v)$ in which the functions p_Θ and p_W will depend only on the jumps of their arguments. Following [2] we set

$$p_\Theta(\boldsymbol{\theta},\boldsymbol{\eta}) = \sum_{e\in\mathscr{E}_h} \frac{\kappa^\Theta}{|e|}\int_e [\![\boldsymbol{\theta}]\!] : [\![\boldsymbol{\eta}]\!]\,\mathrm{d}s, \quad p_W(w,v) = \sum_{e\in\mathscr{E}_h} \frac{\kappa^W}{|e|}\int_e [\![w]\!]\cdot[\![v]\!]\,\mathrm{d}s, \tag{17}$$

so that $p_\Theta(\boldsymbol{\eta},\boldsymbol{\eta})$, $(p_W(v,v)$, respectively) can be viewed as a measure of the deviation of $\boldsymbol{\eta}$ (v, respectively) from being continuous. The parameters κ^Θ and κ^W are positive constants to be chosen; they must be sufficiently large to ensure stability. Since $[\![\boldsymbol{\theta}]\!] = 0$ and $[\![w]\!] = 0$, equations (15)–(16) can then be written as

$$(\mathrm{C}\varepsilon_h(\boldsymbol{\theta}),\varepsilon_h(\boldsymbol{\eta})) - \langle\{\mathrm{C}\varepsilon_h(\boldsymbol{\theta})\},[\![\boldsymbol{\eta}]\!]\rangle - \langle[\![\boldsymbol{\theta}]\!],\{C\varepsilon_h(\boldsymbol{\eta})\}\rangle\rangle + (\boldsymbol{\gamma},\nabla_h v - \boldsymbol{\eta})$$

$$-\langle\{\boldsymbol{\gamma}\},[\![v]\!]\rangle + p_\Theta(\boldsymbol{\theta},\boldsymbol{\eta}) + p_W(w,v) = (g,v), \quad (\boldsymbol{\eta},v)\in \boldsymbol{H}^2(\mathscr{T}_h)\times H^1(\mathscr{T}_h), \tag{18}$$

$$(\nabla_h w - \boldsymbol{\theta},\boldsymbol{\tau}) - \langle[\![w]\!],\{\boldsymbol{\tau}\}\rangle - t^2(\boldsymbol{\gamma},\boldsymbol{\tau}) = 0, \quad \boldsymbol{\tau}\in \boldsymbol{H}^1(\mathscr{T}_h). \tag{19}$$

To obtain a DG discretization, we have to choose finite dimensional subspaces $\boldsymbol{\Theta}_h \subset \boldsymbol{H}^2(\mathscr{T}_h)$, $W_h \subset H^1(\mathscr{T}_h)$, and $\boldsymbol{\Gamma}_h \subset \boldsymbol{H}^1(\mathscr{T}_h)$, and then write the discrete problem:

Find $(\boldsymbol{\theta}_h, w_h)\in\boldsymbol{\Theta}_h\times W_h$ *and* $\boldsymbol{\gamma}_h\in\boldsymbol{\Gamma}_h$ *such that*

$$\begin{aligned}(\mathrm{C}\varepsilon_h(\boldsymbol{\theta}_h),\varepsilon_h(\boldsymbol{\eta})) - \langle\{\mathrm{C}\varepsilon_h(\boldsymbol{\theta}_h)\},[\![\boldsymbol{\eta}]\!]\rangle - \langle[\![\boldsymbol{\theta}_h]\!],\{\mathrm{C}\varepsilon_h(\boldsymbol{\eta})\}\rangle\rangle \\ +(\boldsymbol{\gamma}_h,\nabla_h v - \boldsymbol{\eta}) - \langle\{\boldsymbol{\gamma}_h\},[\![v]\!]\rangle \\ +p_\Theta(\boldsymbol{\theta}_h,\boldsymbol{\eta}) + p_W(w_h,v) = (g,v), \quad (\boldsymbol{\eta},v)\in\boldsymbol{\Theta}_h\times W_h,\end{aligned} \tag{20}$$

$$(\nabla_h w_h - \boldsymbol{\theta}_h,\boldsymbol{\tau}) - \langle[\![w_h]\!],\{\boldsymbol{\tau}\}\rangle - t^2(\boldsymbol{\gamma}_h,\boldsymbol{\tau}) = 0, \quad \boldsymbol{\tau}\in\boldsymbol{\Gamma}_h. \tag{21}$$

For any choice of the finite element spaces $\boldsymbol{\Theta}_h$, W_h, and $\boldsymbol{\Gamma}_h$, and any interior penalty functions p_Θ and p_W depending only on the jumps of their arguments, this gives a consistent finite element method since no reduction operator $\boldsymbol{P}_h$ is used. If instead $\boldsymbol{P}_h$ is needed, there will be a consistency error to be estimated, and equations (20)-(21) will be modified into:

$$\begin{aligned}(\mathrm{C}\varepsilon_h(\boldsymbol{\theta}_h),\varepsilon_h(\boldsymbol{\eta})) - \langle\{\mathrm{C}\varepsilon_h(\boldsymbol{\theta}_h)\},[\![\boldsymbol{\eta}]\!]\rangle - \langle[\![\boldsymbol{\theta}_h]\!],\{\mathrm{C}\varepsilon_h(\boldsymbol{\eta})\}\rangle\rangle \\ +(\boldsymbol{\gamma}_h,\nabla_h v - \boldsymbol{P}_h\boldsymbol{\eta}) - \langle\{\boldsymbol{\gamma}_h\},[\![v]\!]\rangle \\ +p_\Theta(\boldsymbol{\theta}_h,\boldsymbol{\eta}) + p_W(w_h,v) = (g,v), \quad (\boldsymbol{\eta},v)\in\boldsymbol{\Theta}_h\times W_h,\end{aligned} \tag{22}$$

$$(\nabla_h w_h - \boldsymbol{P}_h\boldsymbol{\theta}_h,\boldsymbol{\tau}) - \langle[\![w_h]\!],\{\boldsymbol{\tau}\}\rangle - t^2(\boldsymbol{\gamma}_h,\boldsymbol{\tau}) = 0, \quad \boldsymbol{\tau}\in\boldsymbol{\Gamma}_h. \tag{23}$$

In the next section we shall recall different choices of the finite element spaces.

3 The Finite Elements

We recall in this section some DG-elements/families developed so far. We refer to the original papers for detailed proofs, and we will just recall the resulting error estimates obtained in the DG-norms defined as

$$
\begin{aligned}
|||\boldsymbol{\eta}|||_{\Theta}^2 &:= \|\boldsymbol{\eta}\|_{1,h}^2 + \sum_{e\in\mathscr{E}_h}\left(\frac{1}{|e|}\|[\![\boldsymbol{\eta}]\!]\|_{0,e}^2 + |e|\,\|\{\mathrm{C}\varepsilon_h(\boldsymbol{\eta})\}\|_{0,e}^2\right), \qquad \boldsymbol{\eta}\in \boldsymbol{H}^2(\mathscr{T}_h),\\
|||v|||_{W}^2 &:= |v|_{1,h}^2 + \sum_{e\in\mathscr{E}_h}\frac{1}{|e|}\|[\![v]\!]\|_{0,e}^2, \qquad v\in H^1(\mathscr{T}_h), \qquad\qquad (24)\\
|||\boldsymbol{\tau}|||_{\Gamma}^2 &:= \|\boldsymbol{\tau}\|_{0}^2 + \sum_{e\in\mathscr{E}_h}|e|\,\|\{\boldsymbol{\tau}\}\|_{0,e}^2, \qquad \boldsymbol{\tau}\in \boldsymbol{H}^1(\mathscr{T}_h).
\end{aligned}
$$

3.1 DG-Elements Based on the Use of the Reduction Operator $\boldsymbol{P}_h$

Example 3.1.1 The following family of elements of odd degree $k\geq 1$ was introduced in [2]:

$$
\boldsymbol{\Theta}_h = \mathscr{L}_k^0(\mathscr{T}_h), \quad W_h = \mathscr{L}_k^0(\mathscr{T}_h), \quad \boldsymbol{\Gamma}_h = \mathscr{L}_{k-1}^0(\mathscr{T}_h), \tag{25}
$$

where $\mathscr{L}_k^0(\mathscr{T}_h)$ denotes the space of discontinuous piecewise polynomials of degree $\leq k$ (see (11)). The penalty term $p_\Theta(\boldsymbol{\theta},\boldsymbol{\eta})$ is taken as in (17), while $p_W(w,v)$ is somewhat weaker:

$$
p_W(w,v) = \sum_{e\in\mathscr{E}_h}\frac{\kappa^W}{|e|}\int_e \boldsymbol{Q}_e[\![w]\!]\cdot\boldsymbol{Q}_e[\![v]\!]\,ds, \tag{26}
$$

and $\boldsymbol{Q}_e$ is the projection onto polynomials of degree $k-1$. The error estimates in the norms (24) are:

$$
\begin{aligned}
&|||\boldsymbol{\theta}-\boldsymbol{\theta}_h|||_{\Theta} + |||w-w_h|||_W + t|||\boldsymbol{\gamma}-\boldsymbol{\gamma}_h|||_{\Gamma} \\
&\qquad \leq Ch^k\left(\|\boldsymbol{\theta}\|_{k+1,\Omega} + \|w\|_{k+1,\Omega} + t\|\boldsymbol{\gamma}\|_{k,\Omega} + \|\boldsymbol{\gamma}\|_{k-1,\Omega}\right),
\end{aligned} \tag{27}
$$

which are optimal in terms of order of convergence, and for the case $k=1$ also in terms of regularity (see (10)). The definition of $\boldsymbol{P}_h$ is quite complicated and will not be detailed here. We note however that, for the lowest order case $k=1$, the reduction operator $\boldsymbol{P}_h$ is simply the L^2 projection onto the piecewise constant space $\mathscr{L}_0^0(\mathscr{T}_h)$. The degrees of freedom are shown in Fig. 1.

Example 3.1.2 In the spirit of [2], a linear nonconforming element plus a quadratic nonconforming bubble was first obtained and analyzed in [9]. Then Lovadina in [13] showed that the bubble is actually not needed, and also proved optimal L^2–estimates (see also [10]). Denoting by P_1^{nc} the space of piecewise linear polynomials continuous at the midpoint of each edge of $\mathscr{T}_h$, the choice of spaces is

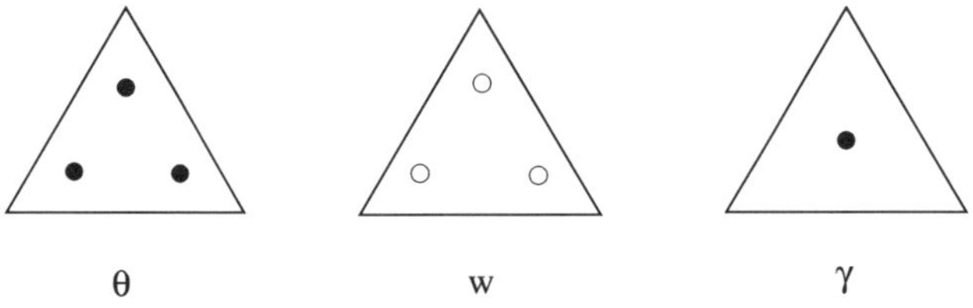

Fig. 1 Totally discontinuous elements: d.o.f. for the lowest order case

$$\boldsymbol{\Theta}_h = P_1^{nc}, \quad W_h = P_1^{nc}, \quad \boldsymbol{\Gamma}_h = \mathscr{L}_{k-1}^0(\mathscr{T}_h), \tag{28}$$

and the degrees of freedom are shown in Fig. 2. For this element, optimal estimates

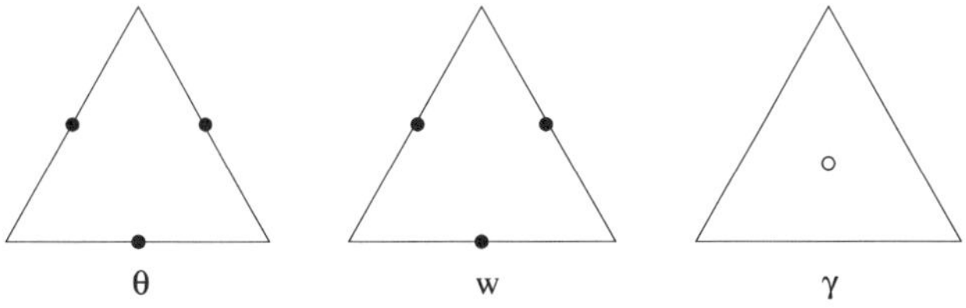

Fig. 2 D.o.f. for the nonconforming element

were proved in [13]

$$||\boldsymbol{\theta} - \boldsymbol{\theta}_h||_{1,h} + ||w - w_h||_{1,h} + ||\boldsymbol{\gamma} - \boldsymbol{\gamma}_h||_{\Gamma} + t||\boldsymbol{\gamma} - \boldsymbol{\gamma}_h||_0 \leq Ch||g||_0, \tag{29}$$

and in [10] for the L^2 error:

$$||\boldsymbol{\theta} - \boldsymbol{\theta}_h||_0 + ||w - w_h||_0 \leq Ch^2||g||_0. \tag{30}$$

3.2 DG-Elements without Reduction Operator $\boldsymbol{P_h}$

Two families of elements for the formulation (20)–(21) were developed in [1]. In all the cases the transverse displacement w is approximated with piecewise polynomials of degree at most k, with $k \geq 2$, while the rotations $\boldsymbol{\theta}$ and the shear stresses $\boldsymbol{\gamma}$ with piecewise polynomials of degree $\leq k-1$, and the methods differ in the amount of continuity required at the interelement boundaries. In all the cases the spaces satisfy

$$\nabla W_h \subseteq \boldsymbol{\Theta}_h = \boldsymbol{\Gamma}_h. \tag{31}$$

Example 3.2.1 In the first family of elements w is approximated by continuous finite elements, so that equations (20)–(21) simplify into:

$$a_h(\boldsymbol{\theta}_h, \boldsymbol{\eta}) + (\boldsymbol{\gamma}_h, \nabla v - \boldsymbol{\eta}) = (g, v), \qquad (\boldsymbol{\eta}, v) \in \boldsymbol{\Theta}_h \times W_h, \tag{32}$$

$$(\nabla w_h - \boldsymbol{\theta}_h, \boldsymbol{\tau}) - t^2(\boldsymbol{\gamma}_h, \boldsymbol{\tau}) = 0, \qquad \boldsymbol{\tau} \in \boldsymbol{\Gamma}_h. \tag{33}$$

Inclusion (31) forbids the use of a space $\boldsymbol{\Theta}_h$ consisting of *continuous* functions. However, since w_h is continuous, it allows choices where the tangential component is continuous (as well as totally discontinuous choices). We recall here the choice that minimizes the number of degrees of freedom. For other possible choices see [1]. We take

$$W_h = \mathscr{L}_k^1, \quad \boldsymbol{\Theta}_h = \boldsymbol{\Gamma}_h = \mathbf{BDM}_{k-1}^R \qquad k \geq 2, \tag{34}$$

where $\mathbf{BDM}_{k-1}^R$ denotes the rotated Brezzi-Douglas-Marini space of degree $k-1$, i.e., the space of all piecewise polynomial vector fields of degree at most $k-1$ with tangential components continuous at the interelements [7]. With this choice, the inclusion (31) is clearly satisfied. The following estimates were proved in the norms (24):

$$|||\boldsymbol{\theta} - \boldsymbol{\theta}_h|||_\Theta + t\|\boldsymbol{\gamma} - \boldsymbol{\gamma}_h\|_0 \leq Ch^{k-1}(\|\boldsymbol{\theta}\|_k + t\|\boldsymbol{\gamma}\|_{k-1}), \tag{35}$$

$$\|\nabla(w - w_h)\|_0 \leq C(h^k + th^{k-1})(\|\boldsymbol{\theta}\|_k + t\|\boldsymbol{\gamma}\|_{k-1}), \tag{36}$$

and in L^2:

$$\|w - w_h\|_0 + \|\boldsymbol{\theta} - \boldsymbol{\theta}_h\|_0 \leq Ch^k(\|\boldsymbol{\theta}\|_k + t\|\boldsymbol{\gamma}\|_{k-1}). \tag{37}$$

Estimates (35)–(36) are optimal with respect to order of convergence (and also with respect to regularity for the case $k = 2$, according to (10)) while (37) is optimal for $\boldsymbol{\theta}$ and suboptimal of one order for w.

Fig. 3 shows the degrees of freedom for the lowest order element of the family:

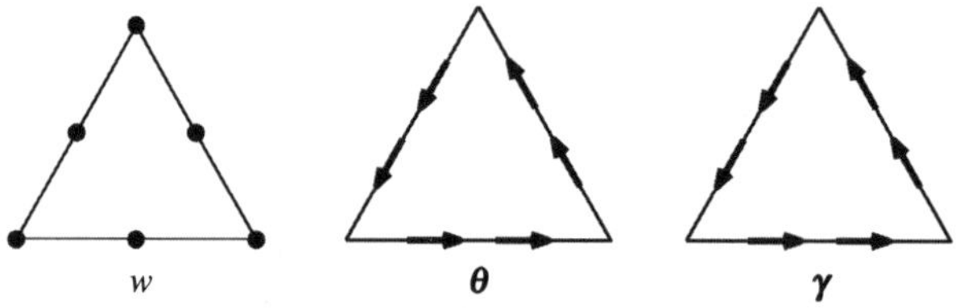

Fig. 3 Continuous w: lowest-order elements without reduction operator

Example 3.2.2 The second family consists of totally discontinuous elements. Thus, the spaces are

$$W_h = \mathscr{L}_k^0, \quad \boldsymbol{\Theta}_h = \boldsymbol{\Gamma}_h = \mathscr{L}_{k-1}^0 \qquad k \geq 2, \tag{38}$$

and the inclusion (31) is obviously verified. For this family the following error estimates were proved in the norms (24):

$$|||\boldsymbol{\theta} - \boldsymbol{\theta}_h|||_\Theta + t\|\boldsymbol{\gamma} - \boldsymbol{\gamma}_h\|_0 + [p_W(w - w_h, w - w_h)]^{1/2} \leq Ch^{k-1}(\|\boldsymbol{\theta}\|_k + \|\boldsymbol{\gamma}\|_{k-1}), \tag{39}$$

$$|||w - w_h|||_W \leq Ch^{k-1}(\|\boldsymbol{\theta}\|_k + \|\boldsymbol{\gamma}\|_{k-1} + \|w\|_k), \tag{40}$$

and in L^2:

$$\|\boldsymbol{\theta}-\boldsymbol{\theta}_h\|_0+\|w-w_h\|_0 \leq Ch^k(\|\boldsymbol{\theta}\|_k+\|\boldsymbol{\gamma}\|_{k-1}). \tag{41}$$

The bad feature of these estimates is the lack of the factor t in the norm $\|\boldsymbol{\gamma}\|_{k-1}$ on the right hand side. Since this norm behaves like $t^{-(k-3/2)}$ as $t \to 0$, the extra factor of t helps to control the size of this term, and for $k=2$ guarantees that it remains bounded. A better estimate in this respect can be obtained by assuming that the Helmholtz decomposition for $\boldsymbol{\gamma}$ holds. In this case we have:

$$\begin{aligned} |||\boldsymbol{\theta}-\boldsymbol{\theta}_h|||_\Theta+t\|\boldsymbol{\gamma}-\boldsymbol{\gamma}_h\|_0+[p_W(w-w_h,w-w_h)]^{1/2} \\ \leq Ch^{k-1}(\|\boldsymbol{\theta}\|_k+t\|\boldsymbol{\gamma}\|_{k-1}+\|\boldsymbol{\gamma}\|_{\boldsymbol{H}^{k-2}(\mathrm{div})}), \\ |||w-w_h|||_W \leq Ch^{k-1}(\|\boldsymbol{\theta}\|_k+t\|\boldsymbol{\gamma}\|_{k-1}+\|\boldsymbol{\gamma}\|_{\boldsymbol{H}^{k-2}(\mathrm{div})}+\|w\|_k), \end{aligned} \tag{42}$$

and in L^2:

$$\|\boldsymbol{\theta}-\boldsymbol{\theta}_h\|_0+\|w-w_h\|_0 \leq Ch^k(\|\boldsymbol{\theta}\|_k+t\|\boldsymbol{\gamma}\|_{k-1}+\|\boldsymbol{\gamma}\|_{\boldsymbol{H}^{k-2}(\mathrm{div})}). \tag{43}$$

We point out that the regularity of $\boldsymbol{\gamma}$ is such that, for the lowest-order case $k=2$, the Helmholtz decomposition holds, and estimates (42)–(43) are optimal with respect to regularity. Indeed, $\|\boldsymbol{\gamma}\|_{\boldsymbol{H}^{k-2}(\mathrm{div})} \equiv \|\operatorname{div}\boldsymbol{\gamma}\|_0 \equiv \|g\|_0$ which does not explode when $t \to 0$. In terms of order of convergence they are optimal for $\boldsymbol{\theta}$, and suboptimal of one order for w. The lowest-order elements are depicted in Fig. 4.

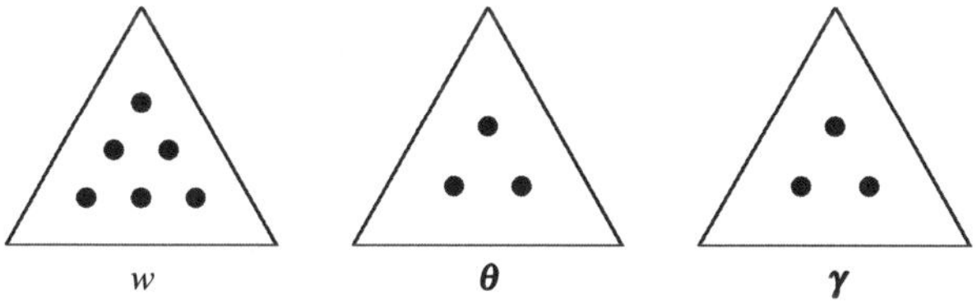

Fig. 4 Totally discontinuous elements without reduction operator: lowest-order elements

4 Conclusions

We presented a quick overview of some locking-free finite elements for Reissner-Mindlin plates, obtained through the use of Discontinuous Galerkin techniques. Since DG solutions are not required to satisfy the interelement continuity conditions of conforming finite elements, DG methods result more flexible and offer possibilities, in terms of degree of the finite elements, which are forbidden with conforming elements. For instance, the simple linear element of Example 3.1.1 would be unthinkable for conforming approximations. Similarly, the nonconforming linear element of Example 3.1.2 would have been hard to derive without the DG techniques.

The counterpart is that discontinuity implies an increasing of the number of unknowns, and efficient techniques to handle the final linear systems might be needed. The low order elements of Subsection 3.1 are very appealing, but their behavior might depend on the choice of the parameters in the penalty terms. By increasing these parameters one increases continuity, and the elements get closer to conforming elements, with the risk of locking. This dependence should be checked in practice, and sound numerical tests should be performed to compare the new elements with existing conforming elements, but this goes beyond the scope of this paper. We refer to [10] for numerical results on various linear nonconforming elements.

References

1. Arnold, D.N., Brezzi, F., Falk, R.S., Marini, L.D.: Locking-free Reissner–Mindlin elements without reduced operator. Comput. Methods Appl. Mech. Engrg. **196**, 3660–3671 (2007)
2. Arnold, D.N., Brezzi, F., Marini, L.D.: A Family of Discontinuous Galerkin Finite Elements for the Reissner–Mindlin plate. J. Scientific Computing **22**, 25–45 (2005)
3. Arnold, D.N., Falk, R.S.: A uniformly accurate finite element method for the Reissner–Mindlin plate. SIAM J. Numer. Anal. **26**, 1276–1290 (1989)
4. Arnold, D.N., Falk, R.S.: The boundary layer for the Reissner–Mindlin plate model. SIAM J. Math. Anal. **21**, 281–312 (1990)
5. Brezzi, F., Bathe, K.J., Fortin, M.: Mixed-interpolated elements for Reissner–Mindlin plates. Internat. J. Numer. Methods Engrg. **28**, 1787–1801 (1989)
6. Brezzi, F., Cockburn, B., Marini, L.D., Süli, E.: Stabilization mechanisms in Discontinuous Galerkin finite element methods. Comput. Methods Appl. Mech. Engrg. **195**, 3293–3310 (2006)
7. Brezzi, F., Douglas, J., Marini, L.D.: Two families of mixed finite elements for second order elliptic problems. Numer. Math. **47**, 217–235 (1985)
8. Brezzi, F., Fortin, M., Stenberg, R.: Error analysis of mixed-interpolated elements for Reissner–Mindlin plates. Math. Models Methods Appl. Sci. **1**, 125–151 (1991)
9. Brezzi, F., Marini, L.D.: A nonconforming element for the Reissner–Mindlin plate. Computers & Structures **81**, 515–522 (2003)
10. Chinosi, C., Lovadina, C., Marini, L.D.: Nonconforming locking-free finite elements for Reissner–Mindlin plates. Comput. Methods Appl. Mech. Engrg. **195**, 3448–3460 (2006)
11. Duran, R., Liberman, E.: On mixed finite-element methods for the Reissner–Mindlin plate model. Math. Comp. **58**, 561–573 (1992)
12. Falk, R.S., Tu, T.: Locking-free finite elements for the Reissner–Mindlin Plate. Math. Comp. **69**, 911–928 (2000)
13. Lovadina, C.: A low-order nonconforming finite element for Reissner–Mindlin plates. SIAM J. Numer. Anal. **42**, 2688–2705 (2005)

Contributed Lectures

A Posteriori Error Estimation and Adaptive Methods

Functional Type A Posteriori Error Estimates for Maxwell's Equations

A. Hannukainen

Abstract In this note, we consider functional type a posteriori error estimation for eddy current and time-harmonic approximations of the Maxwell's equations. Derivation of an upper bound is presented in both cases. The derived bound for the eddy current case is illustrated in numerical examples.

1 Introduction

Estimating the reliability of numerical solutions is crucial in all computational simulations. Two main approaches applied in this context are a priori and a posteriori error estimates. A priori error estimates give information on the behavior of the error, i.e. the difference between exact and approximate solutions, in the asymptotic range. Since the asymptotic range might be impossible to reach with reasonable computational resources, a prior estimates mostly play a theoretical role by guaranteeing convergence of the numerical method. A posteriori error estimates, on the other-hand, can deliver information on the numerical solution at hand, thus having more value in practical computations.

A posteriori error estimates in the context of the finite element method have been studied extensively during the past three decades. However, a posterior error estimation of finite element solutions to different models based on Maxwell's equations is quite recent field of study. The focus is mainly on problems with elliptic structure (e.g., the eddy current problem), which have been studied by several authors [2, 10]. Some work on the time-harmonic problem has been done in [6].

Majority of the existing research has focused on developing residual type error estimates. Such estimates contain mesh dependent constants from Clémént interpolation. Upper bounds for these Clémént interpolation constants are computationally

Antti Hannukainen
Institute of Mathematics, Helsinki University of Technology, P. O. Box 1100, FIN–02015 TKK, Finland, antti.hannukainen@hut.fi

expensive and suffer from large overestimation (see [3]). Hence, the existing residual based a posterior error estimates cannot deliver sharp upper bound for the error and serve mainly as error indicators.

A recent approach for obtaining computable a posterior upper bounds is the functional type error estimation. In the context of Maxwell's equations, an error bound for the eddy current problem can be obtained based on abstract bounds presented in [8]. A different derivation, presented also in this note, can be found from [9]. To the authors knowledge, a functional type bound has not been applied to time-harmonic problems, neither has the eddy-current bound been tested in practice.

In this note, we will derive a functional type error bound for the time-harmonic problem and present numerical examples from the functional type error estimate in the case of the eddy current problem. We will also propose a simple computational technique for choosing the required parameter. The obtained results clearly indicate the potential of functional type a posteriori error estimates for this type of problems.

2 Preliminaries

The model problem is

$$\begin{aligned} \nabla\times\nabla\times\mathbf{u}+\beta\mathbf{u} &= \mathbf{f} \text{ in } \Omega \\ \mathbf{n}\times\mathbf{u} &= 0 \text{ on } \partial\Omega, \end{aligned} \tag{1}$$

in which $\Omega \subset \mathrm{R}^3$ is a simply connected polyhedral domain with Lipschitz continuos boundary (for details, see [7]) and $f \in L^2(\Omega)$. We will study the model problem in two different cases, $\beta \in \mathrm{R}, \beta > 0$ and $\beta \in \mathrm{C}, \Im\beta \geq 0$. The positive parameter value corresponds to the eddy current problem posed in a cavity with positive conductivity ($\sigma > 0$) with PEC (perfect electric conductor) boundaries. The complex-valued parameter corresponds to the time-harmonic approximation of the Maxwell's equation in a cavity with PEC boundaries.

The weak form of the classical problem (1) is : find $\mathbf{u} \in H_0(\Omega, \mathrm{curl})$ such that

$$a(\mathbf{u},\mathbf{v}) = (\mathbf{f},\mathbf{v}) \ \forall\, \mathbf{v} \in H_0(\Omega,\mathrm{curl}), \tag{2}$$

in which $(\cdot,\cdot)$ is the standard $L^2(\Omega)$ innerproduct and

$$a(\mathbf{u},\mathbf{v}) = (\nabla\times\mathbf{u},\nabla\times\mathbf{v}) + \beta\,(\mathbf{u},\mathbf{v}). \tag{3}$$

The space $H_0(\Omega,\mathrm{curl})$ is defined as

$$H_0(\Omega,\mathrm{curl}) = \{\, \mathbf{u} \in L^2(\Omega) \mid \nabla\times\mathbf{u} \in L^2(\Omega) \text{ and } \mathbf{n}\times\mathbf{u} = 0 \text{ on } \partial\Omega \,\},$$

where the boundary condition is to be understood in the sense of boundary traces.

For positive parameter values, $\beta > 0$, the bilinear form $a(\cdot,\cdot)$ is continuous and elliptic in $H_0(\Omega;\text{curl})$, i.e. there exists positive constants α and C such that for every $\mathbf{v},\mathbf{u} \in H_0(\Omega,\text{curl})$

$$|a(\mathbf{u},\mathbf{v})| \leq C\|u\|_{curl}\|v\|_{curl} \quad \text{and} \quad |a(\mathbf{u},\mathbf{u})| > \alpha\|u\|^2_{curl},$$

where the norm $\|\cdot\|_{curl}$ is the defined as $\|\cdot\|_{curl} := \|\cdot\| + \|\nabla\times\cdot\|$ and $\|\cdot\|$ is the standard $L^2(\Omega)$-norm. Under these conditions, a unique solution to problem (2) can be guaranteed by applying the Lax-Millgramm Lemma.

For complex-valued parameters, the sesquilinear form (3) is not elliptic. In this case, the existence of a unique solution to (2) is guaranteed by a proof presented in [7, Section 4].

3 Error Estimate

In this section, we present functional type a posteriori error estimates for the model problem (2) in the case of positive and complex-valued parameters. Estimates will be derived for any admissible approximate solution $\tilde{\mathbf{u}} \in H_0(\Omega,\text{curl})$. The exact solution is denoted by $\mathbf{u}$ and error by $\mathbf{e} = \mathbf{u} - \tilde{\mathbf{u}}$.

The error is a solution to the problem : find a function $\mathbf{e} \in H_0(\Omega,\text{curl})$ such that

$$a(\mathbf{e},\mathbf{v}) = r(\mathbf{v}) \ \forall\, \mathbf{v} \in H_0(\Omega,\text{curl}), \tag{4}$$

where $r(\cdot)$ is the residual defined as $r(\mathbf{v}) = (\mathbf{f},\mathbf{v}) - a(\tilde{\mathbf{u}},\mathbf{v})$.

For both cases, the error estimation will reduce to estimation of the residual $r(\mathbf{e})$. This is similar to other error estimation techniques. For example, in the case of residual based error estimates, a priori information out of the approximation properties of the finite dimensional solution space are introduced into the residual in the form of Clémént interpolant. This approach requires an approximate solution, which posesses the Galergin orthogonality property.

In the functional type a posteriori error estimation approach an arbitrary parameter $\mathbf{y}^*$ is introduced in to the estimate. This approach does not require the Galergin orthogonality and it can be applied to any approximate solution. However, fixing the parameter $\mathbf{y}^*$ poses an additional computational load.

3.1 Real Case $\beta > 0$

In this subsection, an upper bound for the error is derived in the energy norm

$$|||\mathbf{v}||| := \sqrt{a(\mathbf{v},\mathbf{v})},$$

which is well defined for real positive parameter values. The error in the energy norm satisfies $|||\mathbf{e}|||^2 = r(\mathbf{e})$, thus the main task in deriving an upper bound for the error in the energy norm is to estimate the residual.

Theorem 1. *Let* $\mathbf{u} \in H_0(\Omega, curl)$ *be the solution to (2) with* $\beta > 0$. *For any* $\tilde{\mathbf{u}}, \mathbf{y}^* \in H_0(\Omega, curl)$ *there applies*

$$|||\mathbf{e}|||^2 \leq \|\beta^{-1/2}(\mathbf{f} - \beta\tilde{\mathbf{u}} - \nabla \times \mathbf{y}^*)\|^2 + \|(\nabla \times \tilde{\mathbf{u}} - \mathbf{y}^*)\|^2. \tag{5}$$

Proof. We begin by introducing parameter $\mathbf{y}^* \in H_0(\Omega, \mathrm{curl})$ to the residual

$$r(\mathbf{e}) = (\mathbf{f} - \beta\tilde{\mathbf{u}}, \mathbf{e}) - (\nabla \times \tilde{\mathbf{u}} - \mathbf{y}^*, \nabla \times \mathbf{e}) - (\mathbf{y}^*, \nabla \times \mathbf{e})$$

Using integration by parts formula, $(\mathbf{y}^*, \nabla \times \mathbf{e}) = (\nabla \times \mathbf{y}^*, \mathbf{e})$, gives

$$r(\mathbf{e}) = (\mathbf{f} - \beta\tilde{\mathbf{u}} - \nabla \times \mathbf{y}^*, \mathbf{e}) - (\nabla \times \tilde{\mathbf{u}} - \mathbf{y}^*, \nabla \times \mathbf{e}).$$

The above formula can be written as

$$r(\mathbf{e}) = (\beta^{-1/2}(\mathbf{f} - \beta\tilde{\mathbf{u}} - \nabla \times \mathbf{y}^*), \beta^{1/2}\mathbf{e}) + (\nabla \times \tilde{\mathbf{u}} - \mathbf{y}^*, \nabla \times \mathbf{e}).$$

The proof is completed by applying the Cauchy-Schwarz inequality, the property $|||\mathbf{e}|||^2 = r(\mathbf{e})$ and by reorganizing terms. □

Note, that Estimate (5) is sharp in the sense that $\mathbf{y}^* = \nabla \times \mathbf{u}$ leads to equality. In addition, (5) does not contain any unknown constants. One appropriate and a posteriori computable choice of the parameter $\mathbf{y}^*$ is discussed in Section 4.1 below.

3.2 Complex Case $\beta \in \mathrm{C}$ with $\Im\beta \geq 0$

For complex-valued parameter β, the sesquilinear form (3) is not elliptic, hence it does not induce a natural energy norm. In this case, we will follow [6] and use the $L^2(\Omega)$-norm to measure the error. Establishing a connection between the error measured in the $L^2(\Omega)$-norm and the residual requires the application of the dual problem: find $\mathbf{v} \in H_0(\Omega, \mathrm{curl})$ such that

$$a(\mathbf{w}, \mathbf{v}) = (\mathbf{e}, \mathbf{w}) \ \forall \mathbf{w} \in H_0(\Omega, \mathrm{curl}).$$

Similar dual problems are widely used in a posteriori error estimation in terms of linear functionals for the Poisson equation (see, e.g., [1]). If the original weak problem (2) has a unique solution, so does the dual problem. In [7, Section 13.4.1], an a priori error bound for the solution to the adjoint problem is stated as

$$\|\mathbf{v}\|_{curl} \leq C\|\mathbf{e}\|, \tag{6}$$

in which the constant $C > 0$ is independent of $\mathbf{e}$ and depends only on the parameter β and the domain.

A connection between the $L^2(\Omega)$-norm and the sesquilinear form (3) is established by setting $\mathbf{w} = \mathbf{e}$ in the dual problem. This gives

$$\|\mathbf{e}\|^2 = a(\mathbf{e}, \mathbf{v}) = r(\mathbf{v}).$$

Using this identity, error estimation in the time-harmonic case reduces to the estimation of the residual.

Theorem 2. *Let* $\mathbf{u} \in H_0(\Omega, curl)$ *be the solution to (2) with* $\beta \in C, \Im\beta \geq 0$. *For any* $\tilde{\mathbf{u}}, \mathbf{y}^* \in H_0(\Omega, curl)$ *there applies*

$$\|\mathbf{e}\| \leq C(\|\mathbf{f} - \beta\tilde{\mathbf{u}} - \nabla \times \mathbf{y}^*\| + \|\mathbf{y}^* - \nabla \times \tilde{\mathbf{u}}\|) \tag{7}$$

in which the constant $C > 0$ *is dependent only on the parameter* β *and the domain.*

Proof. The upper bound is established by integration by parts, using Cauchy-Schwarz inequality and the stability estimate (6) as

$$\begin{aligned}
\|\mathbf{e}\|^2 &= r(\mathbf{v}) \\
&= (\mathbf{f}, \mathbf{v}) - (\nabla \times \tilde{\mathbf{u}}, \nabla \times \mathbf{v}) - \beta(\tilde{\mathbf{u}}, \mathbf{v}) \\
&= (\mathbf{f} - \beta\tilde{\mathbf{u}} - \nabla \times \mathbf{y}^*, \mathbf{v}) + (\mathbf{y}^* - \nabla \times \tilde{\mathbf{u}}, \nabla \times \mathbf{v}) \\
&\leq \|\mathbf{f} - \beta\tilde{\mathbf{u}} - \nabla \times \mathbf{y}^*\|\|\mathbf{v}\| + \|\mathbf{y}^* - \nabla \times \tilde{\mathbf{u}}\|\|\nabla \times \mathbf{v}\| \\
&\leq C\|\mathbf{e}\|(\|\mathbf{f} - \beta\tilde{\mathbf{u}} - \nabla \times \mathbf{y}^*\| + \|\mathbf{y}^* - \nabla \times \tilde{\mathbf{u}}\|).
\end{aligned}$$

Dividing by $\|\mathbf{e}\|$ gives the desired estimate. □

Unfortunately, we cannot show that the above bound is sharp. Choosing $\mathbf{y}^* = \nabla \times \mathbf{u}$, as in the previous case, does not yield equality. In addition, the guaranteed upper bound contains an unknown constant C arising from the stability estimate (6). However, the constant does not depend in any way on the computed approximation $\tilde{\mathbf{u}}$, which is a clear improvement over residual based error estimate for the same problem (see [6, Section 13.4.1]).

4 Numerical Examples for $\beta > 0$

In this section, we will present two computational examples out of the bound derived in Theorem 1. All computations are performed using the finite element method with lowest-order Nedelec elements (see e.g. [7]) on a tetrahedral mesh (all meshing is done using the TetGen mesh generator, available from http://tetgen.berlios.de/). The resulting finite element space will be denoted as X_h.

The first test demonstrates that the upper bound behaves well when the computational mesh is uniformly refined. In the second test, we will illustrate that the upper bound can be used to drive adaptive solution processes.

4.1 Computation of $\mathbf{y}^*$ and Adaptive Mesh-Refinement

In order to compute a value for the presented upper-bound (5), we need to fix the parameter $\mathbf{y}^* \in H_0(\Omega, \text{curl})$. Our approach is to choose the parameter $\mathbf{y}^*$ from a finite element space $X_h \subset H_0(\Omega, \text{curl})$ (i.e. from the same lowest-order Nedelec space as the approximate solution $\tilde{\mathbf{u}}$), so that it minimizes the value of the upper bound.

The minimizer can be chosen by solving an auxiliary problem: find $\mathbf{y}^* \in X_h$ such that

$$(\beta^{-1}\nabla \times \mathbf{y}^*, \nabla \times \mathbf{v}) + (\mathbf{y}^*, \mathbf{v}) = \left(\beta^{-1/2}(\mathbf{f} - \beta\tilde{\mathbf{u}}), \nabla \times \mathbf{v}\right) + (\nabla \times \tilde{\mathbf{u}}, \mathbf{v}) \ \forall\, \mathbf{v} \in X_h \quad (8)$$

The auxiliary problem is an eddy current problem with different parameters compared to the original one (2), hence it can be solved by applying the same solver which was used to solve the original problem.

If the bound obtained in this way is not sufficiently sharp, a natural way to obtain a sharper bound is to hierarcially refine the computational mesh and compute new $\mathbf{y}^*$. This strategy, and computationally cheaper constructions of $\mathbf{y}^*$ in the context of the Poisson problem, are studied e.g. in [5]. However, we will not study this option in this note.

The adaptive solution process applied in the second numerical example is based on using elementwise contribution of the upper bound as local error indicator. Based on this local error indicator, we mark elements with local error over θ times the maximal local error for refinement. The mesh refinement is done by re-meshing with volume constraints placed on the marked elements.

4.2 Numerical Experiment with Known Smooth Solution

In this test, the computational domain is a unit cube, $\Omega = (0,1)^3$. The coefficients are chosen such that the solution is a smooth function

$$\mathbf{u} = [0,\ 0,\ \sin \pi x \sin \pi y]^T \quad (9)$$

which clearly satisfies the boundary condition, $\mathbf{n} \times \mathbf{u} = 0$, on all boundaries. The problem data corresponding to the above solution is $\beta = 1$ and

$$\mathbf{f} = \left[0,\ 0,\ \left(1 + 2\pi^2\right) \sin \pi x \sin \pi y\right]^T. \quad (10)$$

The problem was solved on a series of uniformly refined meshes and the upper bound was computed for each obtained solution. In this case, an optimal convergence rate, $O(h)$, for the finite element approximation was observed. Exactly the same convergence rate was also observed for the upper bound, which shows that the presented a posteriori error estimate performs in a reasonable manner for uniformly refined meshes. The behavior of the upper bound and the exact error is visualized in the Figure 1 alongside with the optimal finite-element convergence rate.

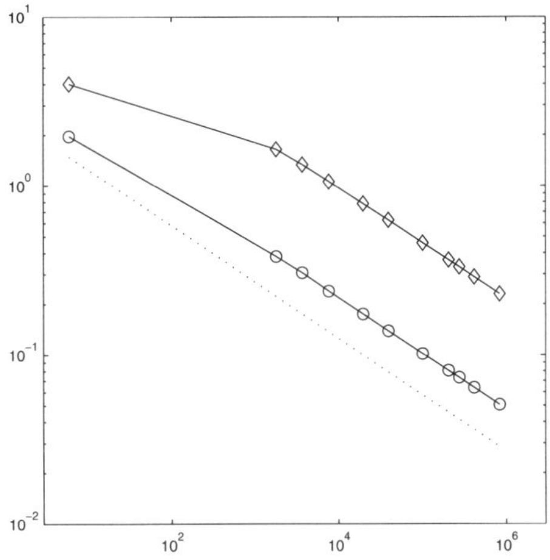

Fig. 1 Behavior of the error for a series of uniformly refined meshes in the test with known smooth solution as a function of degrees of freedom. Diamonds denote the estimated error and circles the exact error. The dotted line is the optimal convergence rate $O(h)$.

4.3 Numerical Experiment with Unknown Solution

In this test, we consider a domain $(-1,1)^3 \setminus [0,1]^3$, which contains a reentrant corner at the origin (0,0,0). The parameter $\beta = 1$ and the right hand side vector is

$$\mathbf{f} = [1,\ 1,\ 1]^T . \tag{11}$$

The problem was solved using a simple adaptive strategy, presented in Section 4.1. The aim was to study, wether the a posteriori error estimator can detect the singular component of the solution which can be expected near the re-entrant corner (see [4]) and improve the convergence rate.

Absolute error decay in the adaptive solution process is plotted in Figure 2 as a function of the mesh size h. The reference error presented in this figure is computed using an overkill mesh with approximately one million degrees of freedom. From the results, one can clearly observe, that the error decay is faster compared to uniformly refined mesh. So, as expected, the singular component is approximated better in adaptive refinement as in uniform refinement.

Acknowledgements The author was supported by the TEKES, The National Technology Agency of Finland (project KOMASI, decision number 210622). I would like to thank Prof. S. Repin for giving an inspiring course on Functional Type error estimates at TKK during fall 2006 and the referee for his excellent comments on the paper.

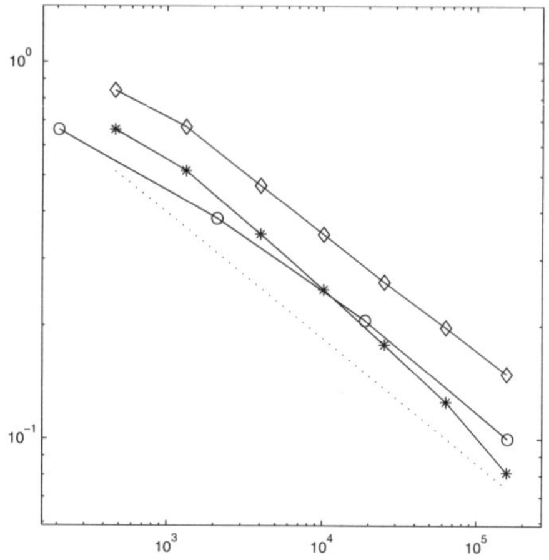

Fig. 2 Behavior of the error in adaptive refinement for the test with unknown solution as a function of degrees of freedom. Diamonds denote the estimated error, stars the reference error and circles error in a solution from uniformly refined mesh. The dotted line is the optimal convergence rate $O(h)$.

References

1. Ainsworth, M., Oden, J.T.: A Posteriori Error Estimation in Finite Element Analysis. John Wiley & Sons Inc., New York (2000)
2. Beck, R., Hiptmair, R., Hoppe, R., Wohlmuth, B.: Residual based a-posteriori error estimators for eddy current computation. Math. Model. Numer. Anal **34**(1), 159–182 (2000)
3. Carstensen, C., Funken, S.A.: Constants in Clément-interpolation error and residual based a posteriori error estimates in finite element methods. East-West J. Numer. Math. **8**, 153–175 (2000)
4. Costabel, M., Dauge, M.: Singularities of electromagnetic fields in polyhedral domains. Arch. Ration. Mech. Anal. **151**(1), 221–276 (2000)
5. Hannukainen, A., Korotov, S.: Computational technologies for reliable control of global and local errors for linear elliptic type boundary value problems, Tech. Rep. A494 (2006), Helsinki University of Technology (to appear in J. Numer. Anal. Indust. Appl. Math.)
6. Monk, P.: A posteriori error indicators for Maxwell's equations. Journal of Computational and Applied Mathematics **100**, 173–190 (1998)
7. Monk, P.: Finite Element Methods for Maxwell's Equations. Calderon Press, Oxford (2003)
8. Neittaanmäki, P., Repin, S.: Reliable Methods for Computer Simulation: Error Control and A Posteriori Error Estimates. Elsevier, Amsterdam (2004)
9. Repin, S.: Functional a posteriori estimates for the maxwell's problem. J. Math. Sci. **1**, 1821–1827 (2007)
10. Schöberl, J.: Commuting quasi-interpolation operators for mixed finite elements. Math. Comp. (submitted) (2006)

Space–Time Adaption for Advection-Diffusion-Reaction Problems on Anisotropic Meshes

S. Micheletti and S. Perotto

Abstract We deal with the approximation of an unsteady advection-diffusion-reaction problem by means of space-time finite elements, continuous affine in space and piecewise constant in time. In particular, we are interested in the advection-dominated framework. To face the trade-off between computational cost and accuracy, we devise a space-time adaptive procedure where both the time step and the spatial grid are adapted throughout the simulation. Two are the key points involved: the derivation of an a posteriori error estimator where the contributions of the spatial and of the temporal discretization are split; a balance of these two contributions via a proper adaptive scheme. The main novelty of the paper is the interest for an anisotropic mesh adaption framework.

1 Introduction

Time dependent advection-dominated problems represent an interesting benchmark for an adaption procedure, due to the (possible) presence of steep internal and/or boundary layers, moving in time. We tackle this issue starting from a theoretically sound space-time adaptive procedure which: i) extends to the time dependent case the anisotropic interpolation error estimates in [7, 8]; ii) generalizes the a posteriori analysis in [13] for a pure diffusive problem to an advective-diffusive-reactive regime. Concerning the pertinent literature, an effective space-time adaptive procedure is proposed in [11] in an optimization framework. In this last case the authors focus on an isotropic goal-oriented analysis for the heat equation. Instead we pursue an anisotropic management of the space adaption procedure. Moreover we control a suitable energy norm of the discretization error as, e.g., in [1, 5]. As far as we know, the only paper dealing with a parabolic problem in an anisotropic framework

Stefano Micheletti and Simona Perotto
MOX, Dipartimento di Matematica "F. Brioschi", Politecnico di Milano, via Bonardi 9, I-20133 Milano, Italy, e-mail: stefano.micheletti, simona.perotto@polimi.it

is [17]. Here the heat equation is considered in an optimal control framework: the time discretization is carried out via the standard backward Euler scheme and no sound time adaption procedure is addressed, in favor of a heuristic approach.

Let us focus on the model parabolic problem for $u = u(\mathbf{x},t)$

$$\begin{cases} Lu = \partial_t u - \nabla\cdot(D\nabla u) + \mathbf{b}\cdot\nabla u + \sigma u = f & \text{in } \Omega\times J,\\ u = 0 & \text{on } \Gamma_{\mathrm{D}}\times J,\\ D\nabla u\cdot\mathbf{n} = g & \text{on } \Gamma_{\mathrm{N}}\times J,\\ u = u_0 & \text{on } \Omega\times\{0\}, \end{cases} \tag{1}$$

where $J = (0,T]$, with $T > 0$, is the considered time span, Ω is a bounded polygonal domain in $\mathbb{R}^2$ with boundary $\partial\Omega$, Γ_{D} and Γ_{N} are nonoverlapping subsets of $\partial\Omega$, each comprising a whole number of sides of $\partial\Omega$ and such that $\partial\Omega = \overline{\Gamma}_{\mathrm{D}} \cup \overline{\Gamma}_{\mathrm{N}}$, and $\mathbf{n}$ is the unit outward normal vector to $\partial\Omega$. Moreover we make the following assumptions on the data: the source $f \in L^2(0,T;L^2(\Omega))$; the Neumann datum $g \in L^2(0,T;H_{00}^{1/2}(\Gamma_{\mathrm{N}})')$; the diffusion tensor $D \in [L^\infty(\Omega)]^{2\times 2}$ and satisfies the standard ellipticity condition; the advective field $\mathbf{b} \in [L^\infty(\Omega)]^2$ with $\nabla\cdot\mathbf{b} \in L^\infty(\Omega)$ and $\mathbf{b}\cdot\mathbf{n} \geq 0$ a.e. on Γ_{N}; the reaction term $\sigma \in L^\infty(\Omega)$ with $\gamma = \sigma - \frac{1}{2}\nabla\cdot\mathbf{b} \geq 0$ a.e. in Ω, while the initial condition $u_0 \in L^2(\Omega)$. Notice that the notation adopted for the function spaces is standard (cf. e.g., [10]). The weak solution to (1) belongs to the space $U = L^2(0,T;H^1_{\Gamma_{\mathrm{D}}}(\Omega)) \cap H^1(0,T;H^1_{\Gamma_{\mathrm{D}}}(\Omega)')$. It is well known that the space U is continuously embedded in $C^0([0,T];L^2(\Omega))$ ([4]).

1.1 Managing the Space-Time

The adopted discrete formulation can be seen as a spatial approximation of a discontinuous in time, dG(0), formulation [18]. Let us first manage the time discretization. We partition the interval J by the time levels $0 = t_0 < t_1 < \ldots < t_{N-1} < t_N = T$, and set $J_n = (t_{n-1},t_n]$, $k_n = t_n - t_{n-1}$. We define the space-time slab $S_n = \Omega\times J_n$, with $n = 1,\ldots,N$. Due to the possible time discontinuity characterizing the dG(0) approximation, for suitable smooth functions $v(\cdot,t)$, we also define the values $v_m^\pm = \lim_{\varepsilon\to 0^+} v(\cdot,t_m \pm \varepsilon)$ and the corresponding temporal jump $[v]_m = v_m^+ - v_m^-$, with $m = 1,\ldots,N-1$. Then we introduce the function space $\mathscr{S}_k = \{v : (0,T] \to H^1_{\Gamma_{\mathrm{D}}}(\Omega) : v(\cdot,t)\big|_{J_n} = \psi(\cdot), \psi \in H^1_{\Gamma_{\mathrm{D}}}(\Omega)\}$, whose elements coincide with polynomials of degree zero in t on each interval J_n, with coefficients in $H^1_{\Gamma_{\mathrm{D}}}(\Omega)$. The functions in $\mathscr{S}_k$ can be discontinuous at each time level, with continuity from the left. Moreover, since $0 \notin J_1$, the value $v(\cdot,0)$ has to be specified separately, $\forall v \in \mathscr{S}_k$.

To discretize the space we resort to a family of conformal decompositions of $\overline{\Omega}$ into triangles, such that there is always a vertex of the triangulation at the interface between $\overline{\Gamma}_{\mathrm{D}}$ and $\overline{\Gamma}_{\mathrm{N}}$ (see, e.g., [3]). The temporal discontinuity allows for the employment of a family $\{\mathscr{T}_{h_n}\}_{h_n}$ of meshes, possibly different on each space-time slab S_n, for $n = 1,\ldots,N$. In particular we define $\mathscr{T}_{h_n} = \{K_n\}$, with K_n triangle of diameter h_{K_n} and $h_n = \max_{K_n} h_{K_n}$, the prism $S_{K_n} = K_n \times J_n$ and its lateral surface $L_{K_n} = \partial K_n \times J_n$. We are now in a position to define the so-called cG(1)-dG(0) space, $\mathscr{S}_{hk} = \{v_{hk} \in \mathscr{S}_k : v_{hk}(\cdot,t)\big|_{J_n} = \psi_h(\cdot), \psi_h \in X^1_{h_n} \cap H^1_{\Gamma_D}(\Omega)\}$, $X^1_{h_n}$ being the space of the finite elements of degree one associated with the mesh $\mathscr{T}_{h_n}$ (see, e.g., [6]). The continuity of the functions $v_{hk} \in \mathscr{S}_{hk}$ is guaranteed with respect to the space, while the discontinuity in time characterizing the space $\mathscr{S}_k$ is maintained.

In view of the cG(1)-dG(0) formulation of (1), we introduce the bilinear and linear forms $B_{\mathrm{DG-GLS}}(\cdot,\cdot)$ and $F_{\mathrm{DG-GLS}}(\cdot)$, given by

$$\begin{aligned}
B_{\mathrm{DG-GLS}}(v,w) &= \sum_{n=1}^{N} \int_{S_n} \{\partial_t v w + D\nabla v \cdot \nabla w + (\mathbf{b}\cdot\nabla v + \sigma v) w\}\, d\mathbf{x}dt \\
&\quad + \int_{\Omega} v_0^+ w_0^+ \, d\mathbf{x} + \sum_{m=1}^{N-1} \int_{\Omega} [v]_m w_m^+ \, d\mathbf{x} + \sum_{n=1}^{N} \sum_{K_n \in \mathscr{T}_{h_n}} \int_{S_{K_n}} \tau_{K_n} Lv Lw \, d\mathbf{x}dt, \\
F_{\mathrm{DG-GLS}}(w) &= \sum_{n=1}^{N} \Big\{ \int_{S_n} f w \, d\mathbf{x}dt + \int_{J_n}\int_{\Gamma_N} g w \, ds dt + \sum_{K_n \in \mathscr{T}_{h_n}} \int_{S_{K_n}} \tau_{K_n} f Lw \, d\mathbf{x}dt \Big\} \\
&\quad + \int_{\Omega} v_0 w_0^+ \, d\mathbf{x},
\end{aligned} \tag{2}$$

respectively, $v_0 = v_0^- \in L^2(\Omega)$ being known. These forms already incorporate a Galerkin Least-Squares (GLS) stabilization [9] to deal with possible numerical instabilities; the $\tau'_{K_n}s$ are suitable anisotropic piecewise constant stabilization coefficients ([16]).

Notice that $[u]_m = 0$, $m = 1,\dots,N-1$, while $u_0^+ = u_0^- = u_0(\cdot)$, as $u \in U$.

The GLS cG(1)-dG(0) discrete formulation of problem (1) is: find $u_{hk} \in \mathscr{S}_{hk}$ such that

$$B_{\mathrm{DG-GLS}}(u_{hk}, v_{hk}) = F_{\mathrm{DG-GLS}}(v_{hk}) \quad \forall v_{hk} \in \mathscr{S}_{hk}, \tag{3}$$

where v_0 in (2) is replaced by $u_h^0 \in X^1_{h_1} \cap H^1_{\Gamma_D}(\Omega)$, i.e., by a proper finite element approximation of the initial data u_0.

It can be proved that the space-time error $e_{hk} = u - u_{hk}$ associated with the approximation u_{hk} satisfies a slabwise Galerkin orthogonality condition with respect to the discrete space $\mathscr{S}_{hk}$.

The bilinear form $B_{\mathrm{DG-GLS}}$ induces the norm

$$\begin{aligned}
|||w|||^2_{\mathrm{DG-GLS}} &= B_{\mathrm{DG-GLS}}(w,w) = \sum_{n=1}^{N} \Big\{ \|D^{1/2}\nabla w\|^2_{[L^2(S_n)]^2} + \|\gamma^{1/2} w\|^2_{L^2(S_n)} \\
&+ \frac{1}{2}\|(\mathbf{b}\cdot\mathbf{n})^{1/2} w\|^2_{L^2(J_n\times\Gamma_N)} + \sum_{K_n \in \mathscr{T}_{h_n}} \|\tau_{K_n}^{1/2} Lw\|^2_{L^2(S_{K_n})} \Big\} + \frac{1}{2}\sum_{m=1}^{N-1} \|w_m^+ - w_m^-\|^2_{L^2(\Omega)} \\
&+ \frac{1}{2}\|w_0^+\|^2_{L^2(\Omega)} + \frac{1}{2}\|w_N^-\|^2_{L^2(\Omega)}
\end{aligned} \tag{4}$$

on the space $U \cup \mathscr{S}_k$ (see, e.g. [9, 14] for further details). This is the energy norm on which we base the a posteriori analysis below.

2 The Anisotropic Framework

With a view to the a posteriori analysis, we recall the basic ideas of the anisotropic setting introduced in [7]. Moreover, we generalize some of the anisotropic interpolation error estimates in [7, 8] to the unsteady case.

Given any slab S_n, $\mathscr{T}_{h_n} = \{K_n\}$ being the associated mesh, we extract the anisotropic information from the invertible affine map $T_{K_n} : \widehat{K} \to K_n$ from the reference triangle $\widehat{K}$ to the general element $K_n \in \mathscr{T}_{h_n}$, such that $K_n = T_{K_n}(\widehat{K}) = M_{K_n}\widehat{K} + \mathbf{t}_{K_n}$, where $M_{K_n} \in \mathbb{R}^{2\times 2}$ and $\mathbf{t}_{K_n} \in \mathbb{R}^2$ denote the Jacobian and the offset associated with T_{K_n}, respectively. Then we introduce the polar decomposition $M_{K_n} = B_{K_n} Z_{K_n}$ of M_{K_n} into a symmetric positive definite matrix $B_{K_n} \in \mathbb{R}^{2\times 2}$ and an orthogonal matrix $Z_{K_n} \in \mathbb{R}^{2\times 2}$, and we further factorize the matrix B_{K_n} in terms of its eigenvectors $\mathbf{r}_{i,K_n}$ and eigenvalues λ_{i,K_n}, for $i = 1,2$, as $B_{K_n} = R^T_{K_n} \Lambda_{K_n} R_{K_n}$, with $\Lambda_{K_n} = \mathrm{diag}(\lambda_{1,K_n}, \lambda_{2,K_n})$ and $R^T_{K_n} = [\mathbf{r}_{1,K_n}, \mathbf{r}_{2,K_n}]$. Notice that Z_{K_n} and $\mathbf{t}_{K_n}$ do not play any role in providing anisotropic information as associated with a rigid rotation and a shift, respectively. We choose $\widehat{K}$ as the equilateral triangle inscribed in the unit circle, with centroid placed at the origin. For this choice, it is possible to completely describe the shape and the orientation of each element K_n through the quantities $\mathbf{r}_{i,K_n}$ and λ_{i,K_n}. The unit circle circumscribed to $\widehat{K}$ is mapped into an ellipse circumscribing K_n: the eigenvectors $\mathbf{r}_{i,K_n}$ and the eigenvalues λ_{i,K_n} provide us with the directions and the length of the semi-axes of such an ellipse, respectively. In particular, we measure the deformation of each element K_n by the so-called stretching factor $s_{K_n} = \lambda_{1,K_n}/\lambda_{2,K_n}$, assuming, without loosing generality, $\lambda_{1,K_n} \geq \lambda_{2,K_n}$, so that $s_{K_n} \geq 1$, the equality holding if and only if K_n is equilateral.

We now state the anisotropic interpolation error estimates used in the a posteriori analysis. We focus on the Lagrange interpolant $\Pi^1_{h_n} : C^0(\overline{\Omega}) \to X^1_{h_n}$. The local interpolant $\Pi^1_{K_n} : \Pi^1_{K_n}(v|_{K_n}) = (\Pi^1_{h_n} v)|_{K_n}$, for any $v \in C^0(\overline{\Omega})$, satisfies the following

Lemma 1. *Let* $v|_{S_{K_n}} \in L^2(J_n; H^2(K_n)) \cap U$; *then it holds*

$$\|v - \Pi^1_{K_n} v\|_{L^2(S_{K_n})} \leq \mathscr{C}_1 \mathscr{L}_{K_n}(v), \quad |v - \Pi^1_{K_n} v|_{H^1(S_{K_n})} \leq \mathscr{C}_2 \mathscr{L}_{K_n}(v),$$
$$|v - \Pi^1_{K_n} v|_{H^2(S_{K_n})} \leq \mathscr{C}_3 \mathscr{L}_{K_n}(v), \quad \|v - \Pi^1_{K_n} v\|_{L^2(L_{K_n})} \leq \mathscr{C}_4 \mathscr{L}_{K_n}(v), \tag{5}$$

where $\mathscr{C}_1 = C_1$, $\mathscr{C}_2 = C_2 \lambda^{-1}_{2,K_n}$, $\mathscr{C}_3 = C_3 \left(\frac{\lambda^2_{1,K_n} + \lambda^2_{2,K_n}}{\lambda^2_{1,K_n} \lambda^2_{2,K_n}} \right)^{1/2}$, $\mathscr{C}_4 = C_4 \left(\frac{\lambda^2_{1,K_n} + \lambda^2_{2,K_n}}{\lambda^3_{2,K_n}} \right)^{1/2}$,

$\mathscr{L}_{K_n}(v) = \left[\sum_{i,j=1}^2 \lambda^2_{i,K_n} \lambda^2_{j,K_n} L^{ij}_{K_n}(v) \right]^{1/2}$, *the constants* C_i, *for* $i = 1, \cdots, 4$ *depending on* $\widehat{K}$ *only. Moreover,* $L^{ij}_{K_n}(v) = \int_{S_{K_n}} (\mathbf{r}^T_{i,K_n} H_{K_n}(v) \mathbf{r}_{j,K_n})^2 \, d\mathbf{x}dt$, *with* $i, j = 1, 2$, *while* $H_{K_n}(v)$ *denotes the Hessian matrix associated with* v.

Estimates (5) generalize the standard (isotropic) results, recovered when $\lambda_{1,K_n} \simeq \lambda_{2,K_n} \simeq h_{K_n}$.

3 The A Posteriori Error Estimate

We provide an a posteriori error estimator, $\eta_{\mathrm{DG-GLS}}$, for the DG-GLS norm (4) of the discretization error e_{hk}. It is essentially a residual-based estimator, weighted by suitable recovered derivatives of the error itself, in the spirit of a Zienkiewicz-Zhu recovery procedure [19, 12]. We define the local residuals, distinguishing between spatial and temporal. For any $K_n \in \mathscr{T}_{h_n}$, with $n = 1, \cdots, N$, let

$$\rho_{K_n} = \left[f - Lu_{hk}\right]\Big|_{S_{K_n}} \quad \text{and} \quad j_{K_n} = \begin{cases} 0 & \text{on } (\partial K_n \cap \Gamma_D) \times J_n, \\ 2(g - D\nabla u_{hk} \cdot \mathbf{n}) & \text{on } (\partial K_n \cap \Gamma_N) \times J_n, \\ -[D\nabla u_{hk} \cdot \mathbf{n}] & \text{on } (\partial K_n \cap \mathscr{E}_h^n) \times J_n, \end{cases}$$

be the interior and boundary residual associated with the cG(1)-dG(0) approximation u_{hk}, respectively, with $\mathscr{E}_h^n$ the skeleton of $\mathscr{T}_{h_n}$ and $[D\nabla u_{hk} \cdot \mathbf{n}] = D\nabla u_{hk} \cdot \mathbf{n}_{K_n} + D\nabla u_{hk} \cdot \mathbf{n}_{K_n'}$ the jump of the diffusive flux across the internal interfaces of K_n, for $(\overline{K_n'} \cap \overline{K_n}) \cap \mathscr{E}_h^n \neq \emptyset$. Then we introduce the temporal and the initial residuals, $\mathscr{J}_n = [-u_{hk}]_n$ and $e_0^- = u_0 - u_h^0$, respectively. The residual $\mathscr{J}_n$ merges the information coming from the different meshes $\mathscr{T}_{h_n}$ and $\mathscr{T}_{h_{n+1}}$. This inevitably entails a careful computation of this term. As a consequence of the dG(0) approximation, it is useful to introduce the time averaged residuals, $\overline{\rho}_{K_n} = k_n^{-1} \int_{J_n} \rho_{K_n}(\cdot,t)\,dt$ and $\overline{j}_{K_n} = k_n^{-1} \int_{J_n} j_{K_n}(\cdot,t)\,dt$, which play an important role in the forthcoming analysis.

We can state the main result of our a posteriori analysis.

Proposition 1. *Let $u \in U$ be the weak solution to (1) and let $u_{hk} \in \mathscr{S}_{hk}$ be the corresponding GLS cG(1)-dG(0) approximation, solution to (3). Then there exists a constant $C = C(\widehat{K})$ such that*

$$|||e_{hk}|||^2_{\mathrm{DG-GLS}} \simeq \eta^2_{\mathrm{DG-GLS}} = C \sum_{n=1}^{N} \sum_{K_n \in \mathscr{T}_{h_n}} \Big(\underbrace{\alpha^{\mathrm{S}}_{K_n} R^{\mathrm{S}}_{K_n} \omega^{\mathrm{S}}_{K_n}}_{\eta^{\mathrm{S}}_{K_n}} + \underbrace{\sum_{i=1}^{4} \alpha^{\mathrm{T}i}_{K_n} R^{\mathrm{T}i}_{K_n} \omega^{\mathrm{T}i}_{K_n}}_{\eta^{\mathrm{T}}_{K_n}} \Big), \quad (6)$$

where $\alpha^{\mathrm{S}}_{K_n} = |\widehat{K}|\,\lambda^2_{1,K_n} \lambda^2_{2,K_n}$, $\alpha^{\mathrm{T}i}_{K_n} = k_n^2$, $i = 1, \cdots, 4$,

$$R^{\mathrm{S}}_{K_n} = |K_n|^{-1/2} \Big\{ \|\overline{\rho}_{K_n}\|_{L^2(S_{K_n})} + (\lambda^2_{1,K_n} + \lambda^2_{2,K_n})^{1/2} \lambda_{2,K_n}^{-3/2} \|\overline{j}_{K_n}\|_{L^2(L_{K_n})}$$
$$+ k_n^{-1/2} (\|\mathscr{J}_{n-1}\|_{L^2(K_n)} + \delta_{1n} \|e_0^-\|_{L^2(K_n)}) + \tau_{K_n} \Big(\lambda^{-1}_{2,K_n} \|\mathbf{b} - \nabla \cdot D\|_{[L^\infty(K_n)]^2}$$
$$+ \|\sigma\|_{L^\infty(K_n)} + (\lambda^2_{1,K_n} + \lambda^2_{2,K_n})^{1/2} (\lambda_{1,K_n} \lambda_{2,K_n})^{-1} \|D\|_{[L^\infty(K_n)]^{2\times 2}} \Big) \|\overline{\rho}_{K_n}\|_{L^2(S_{K_n})} \Big\},$$

$$\omega^{\mathrm{S}}_{K_n} = |K_n|^{-1/2} \Big[s^2_{K_n} L^{11}_{K_n}(e^*_{hk}) + 2 L^{12}_{K_n}(e^*_{hk}) + s^{-2}_{K_n} L^{22}_{K_n}(e^*_{hk}) \Big]^{1/2},$$

$$R_{K_n}^{\mathrm{T1}} = k_n^{-1/2}\Big[\|\rho_{K_n} - \overline{\rho}_{K_n}\|_{L^2(S_{K_n})} + k_n^{-1/2}\Big(\|\mathscr{J}_{n-1}\|_{L^2(K_n)} + \delta_{1n}\|e_0^-\|_{L^2(K_n)}\Big)\Big]$$
$$+\tau_{K_n}\Big(\|\sigma\|_{L^\infty(K_n)}\|\rho_{K_n} - \overline{\rho}_{K_n}\|_{L^2(S_{K_n})} + k_n^{-1}\|\rho_{K_n}\|_{L^2(S_{K_n})}\Big)\Big],$$

$$R_{K_n}^{\mathrm{T2}} = (4k_n)^{-1/2}\|j_{K_n} - \overline{j}_{K_n}\|_{L^2(L_{K_n})}, \quad R_{K_n}^{\mathrm{T4}} = \tau_{K_n} k_n^{-1/2}\|\rho_{K_n} - \overline{\rho}_{K_n}\|_{L^2(S_{K_n})},$$

$$R_{K_n}^{\mathrm{T3}} = R_{K_n}^{\mathrm{T4}}\|\mathbf{b}\|_{[L^\infty(K_n)]^2}, \; \omega_{K_n}^{\mathrm{T1}} = k_n^{-1/2}\|\partial_t e_{hk}^*\|_{L^2(S_{K_n})}, \; \omega_{K_n}^{\mathrm{T2}} = k_n^{-1/2}\|\partial_t e_{hk}^*\|_{L^2(L_{K_n})},$$
$$\omega_{K_n}^{\mathrm{T3}} = k_n^{-1/2}\|\partial_t \nabla e_{hk}^*\|_{[L^2(S_{K_n})]^2}, \quad \omega_{K_n}^{\mathrm{T4}} = k_n^{-1/2}\|\partial_t \nabla\cdot(D\nabla e_{hk}^*)\|_{L^2(S_{K_n})},$$

where δ_{1n} is the Kronecker symbol, and all the terms depending on e_{hk}^* designate suitable space-time recovery quantities that provide computable spatial and temporal derivatives of the discretization error e_{hk}.

Further details concerning the space and time recovery procedures can be found, for instance, in [19, 11, 15], as well as in [14], where the complete proof of (6) is furnished too. We just remark that the quantities $R_{K_n}^{\mathrm{S}}$, $R_{K_n}^{\mathrm{T}i}$, with $i = 1, \cdots, 4$, are scaled (with respect to the size $|K_n|$ of the element and k_n of the time interval, respectively), so that all the spatial and temporal dimensional information is collected into the coefficients $\alpha_{K_n}^{\mathrm{S}}$, $\alpha_{K_n}^{\mathrm{T}i}$, respectively. The weights $\omega_{K_n}^{\mathrm{S}}$ are associated with the anisotropic source, whereas the $\omega_{K_n}^{\mathrm{T}i}$'s drive the time adaption procedure. Finally, $\eta_{K_n}^{\mathrm{S}}$ ($\eta_{K_n}^{\mathrm{T}}$) in (6) represent the local estimators for a pure space-dependent (time-dependent) problem.

3.1 The Adaptive Algorithm

The adaptive algorithm is the same as that introduced in [13]. An equidistribution in space-time of the total error is enforced by splitting a given tolerance τ, equal for each slab, into a space (τ^{S}) and a time (τ^{T}) contribution. The time step and the spatial mesh are successively adapted until both the estimators of the space and time error are within their respective tolerances, i.e., until $\eta_n^{\mathrm{S}} = \sum_{K_n \in \mathscr{T}_{h_n}} \eta_{K_n}^{\mathrm{S}} \simeq \tau^{\mathrm{S}}$, and $\eta_n^{\mathrm{T}} = \sum_{K_n \in \mathscr{T}_{h_n}} \eta_{K_n}^{\mathrm{T}} \simeq \tau^{\mathrm{T}}$. After processing a slab, if the time tolerance is largely satisfied, a new (larger) time step is guessed for the next slab. This algorithm is similar to that in [2] though in our case both space and time adaptivity are carried out via an optimization strategy rather than through a compute-estimate-mark-refine procedure.

3.2 The Rotating Donut

We approximate problem (1) on the cylinder $\Omega \times J = (-1,1)^2 \times (0,10)$, with $D = 10^{-3}I$ (with I the identity tensor), $\mathbf{b} = [-x_2, x_1]^T$, $\sigma = 0$, $\Gamma_N = \emptyset$, and f, u_0 chosen

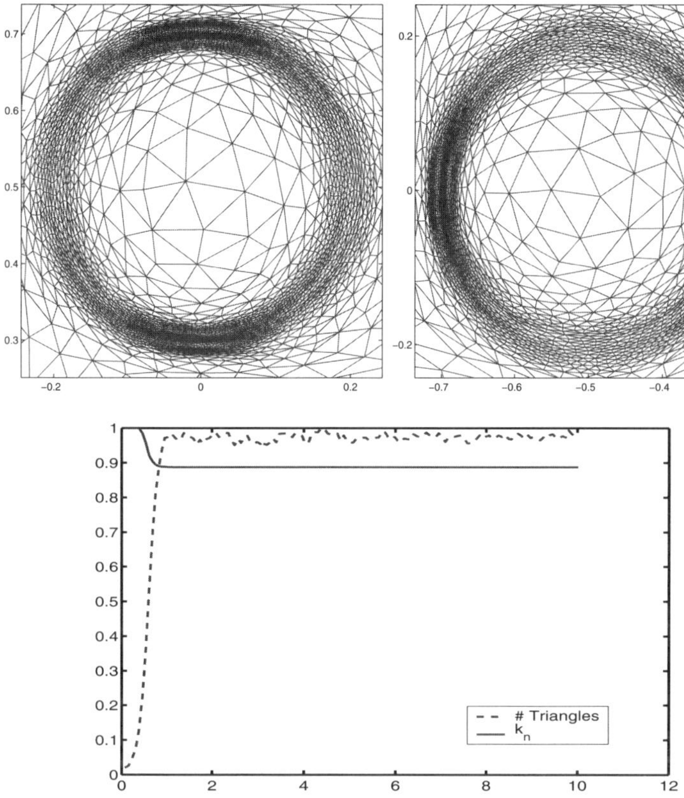

Fig. 1 Details of the adapted meshes at $t \simeq T/4$ (top-left) and at $t \simeq T/2$ (top-right); time evolution of the time step k_n (solid) and of the number of mesh elements (dashed), scaled to their maximum value (bottom)

such that $u = \exp\left(-((r-r_c)/\delta)^2\right)$, where $r = \sqrt{(x_1 - x_{1,R})^2 + (x_2 - x_{2,R})^2}$, $x_{1,R} = R\cos(\omega t), x_{2,R} = R\sin(\omega t)$, with $r_c = 0.2, \delta = 0.01, \omega = 2\pi/10, R = 0.5$. The exact solution is localized in an annular region of thickness $\mathscr{O}(\delta)$ rotating counterclockwise at a constant angular velocity ω. The tolerances for the adaptive algorithm are $\tau^S = \tau^T = 0.01$. Figure 1 shows a detail of the adapted meshes at $t \simeq T/4$ (top-left) and $t \simeq T/2$ (top-right). The mesh is correctly detecting the anisotropic features of the solution. In particular we can appreciate a sort of "wake" that is a clear effect of the donut velocity: this detail would not be spotted in the case of the corresponding stationary problem. The bottom graph in Figure 1 displays the time evolution of the time step k_n (solid) and of the number of mesh elements (dashed), scaled to their maximum value. These time histories show that, after a transient phase, both the time step and the number of triangles level out, as a consequence of the constant angular velocity and of the absence of distortion of the donut. At the final time we obtain the value 1.257 for the effectivity index $EI = \eta_{DG-GLS}/|||e_{hk}|||_{DG-GLS}$, after assuming $C = 1$ in (6).

Concerning the future developments, we are currently extending the above analysis to an optimal control framework.

References

1. Akrivis, G., Makridakis, C., Nochetto, R.: A posteriori error estimates for the Crank-Nicolson method for parabolic equations. Math. Comp. **75**, 511–531 (2006)
2. Cascón, J.M., Ferragut, L., Asensio, M.I.: Space-time adaptive algorithm for the mixed parabolic problem. Numer. Math. **103**, 367–392 (2006)
3. Ciarlet, P.: The Finite Element Method for Elliptic Problems. North-Holland, Amsterdam (1978)
4. Dautray, R., Lions, J.L.: Mathematical Analysis and Numerical Methods for Science and Technology: Evolution Problems I, vol. 5. Springer-Verlag, Berlin (1992)
5. Eriksson, K., Johnson, C.: Adaptive finite element methods for parabolic problems. I: a linear model problem. SIAM J. Numer. Anal. **28**, 43–77 (1991)
6. Eriksson, K., Johnson, C., Thomée, V.: Time discretization of parabolic problems by the discontinuous Galerkin method. RAIRO Modelisation Math. Anal. Numer. **19**, 611–643 (1985)
7. Formaggia, L., Perotto, S.: New anisotropic a priori error estimates. Numer. Math. **89**, 641–667 (2001)
8. Formaggia, L., Perotto, S.: Anisotropic error estimates for elliptic problems. Numer. Math. **94**, 67–92 (2003)
9. Hughes, T.J.R., Franca, L.P., Hulbert, G.M.: A new finite element formulation for computational fluid dynamics: VIII. The Galerkin/least-squares method for advective-diffusive equations. Comput. Methods Appl. Mech. Engrg. **73**, 173–189 (1989)
10. Lions, J.L., Magenes, E.: Non-homogeneous boundary value problems and applications. Vol. I. Springer-Verlag, New York (1972)
11. Meidner, D., Vexler, B.: Adaptive space-time finite element methods for parabolic optimization problems. SIAM J. Control Optim. **46**, 116–142 (2007)
12. Micheletti, S., Perotto, S.: Reliability and efficiency of an anisotropic Zienkiewicz-Zhu error estimator. Comput. Methods Appl. Mech. Engrg. **195**, 799–835 (2006)
13. Micheletti, S., Perotto, S.: Anisotropic mesh adaption for time-dependent problems. To appear in Internat. J. Numer. Methods Fluids (2008). DOI 10.1002/fld.1754
14. Micheletti, S., Perotto, S.: In preparation (2008)
15. Micheletti, S., Perotto, S.: Space–time adaptation for purely diffusion problems in an anisotropic framework. MOX–Report No. 8/2008 (2008)
16. Micheletti, S., Perotto, S., Picasso, M.: Stabilized finite elements on anisotropic meshes: a priori error estimates for the advection-diffusion and the Stokes problems. SIAM J. Numer. Anal. **41**, 1131–1162 (2003)
17. Picasso, M.: Anisotropic a posteriori error estimate for an optimal control problem governed by the heat equation. Numer. Methods Partial Differential Equations **22**, 1314–1336 (2006)
18. Thomée, V.: Galerkin finite element methods for parabolic problems. In: Springer Series in Computational Mathematics, vol. 25. Springer-Verlag, Berlin (2006)
19. Zienkiewicz, O.C., Zhu, J.Z.: The superconvergent patch recovery and a posteriori error estimates. Part I: the recovery technique. Int. J. Numer. Methods Engrg. **33**, 1331–1364 (1992)

A Posteriori Error Analysis for Kirchhoff Plate Elements

J. Niiranen, L. Beirão da Veiga, and R. Stenberg

Abstract We present a posteriori error analysis for two finite element methods for the Kirchhoff plate bending model. The first method is a recently introduced C^0-continuous family, while the second one is the classical nonconforming Morley element.

1 Introduction

In this contribution, we present and compare the main results of a posteriori error analysis for two finite element methods for the Kirchhoff plate model, a stabilized C^0-family of [2, 4], and the classical nonconforming Morley element analyzed in [6, 3, 1].

In the next two sections, we first recall the Kirchhoff plate bending model and then give the finite element formulations of the methods together with a priori error estimates. In Sect. 4, we present a posteriori error indicators for the methods as well as the corresponding reliability and efficiency results. In Sect. 5, we illustrate the robustness of the a posteriori error estimators by benchmark computations.

Jarkko Niiranen
Laboratory of Structural Mechanics, Helsinki University of Technology, P.O. Box 2100, FIN-02015 TKK, Finland, e-mail: jarkko.niiranen@tkk.fi

Lourenço Beirão da Veiga
Dipartimento di Matematica "Federigo Enriques", Università degli Studi di Milano, via Saldini 50, 20133 Milano, Italy, e-mail: beirao@mat.unimi.it

Rolf Stenberg
Institute of Mathematics, Helsinki University of Technology, P.O. Box 1100, FIN-02015 TKK, Finland, e-mail: rolf.stenberg@tkk.fi

2 Kirchhoff Plate Bending Problem

We consider the bending problem of an isotropic linearly elastic plate under the transverse loading g. The midsurface of the undeformed plate is described by a polygonal domain $\Omega \subset \mathbb{R}^2$. The plate is considered to be clamped on the part Γ_C of its boundary $\partial\Omega$, simply supported on the part $\Gamma_\mathrm{S} \subset \partial\Omega$ and free on $\Gamma_\mathrm{F} \subset \partial\Omega$. With $\mathscr{V}$ we indicate the collection of all the corner points in Γ_F corresponding to an angle of the free boundary.

The material constants for the model, the bending stiffness and the shear modulus, respectively, are denoted by

$$\mathsf{D} = \frac{\mathsf{E}t^3}{12(1-\nu^2)} \quad \text{and} \quad \mathsf{G} = \frac{\mathsf{E}}{2(1+\nu)}, \tag{1}$$

with the Young modulus E and the Poisson ratio ν. The thickness of the plate is denoted by t. The physical stress quantities for the problem, the bending moment and shear force, respectively, are defined as

$$\mathbf{M}(\nabla w) = \mathsf{D}\big((1-\nu)\varepsilon(\nabla w) + \nu \operatorname{div} \nabla w \mathbf{I}\big), \tag{2}$$

$$\mathbf{Q}(\nabla w) = -\mathbf{div}\mathbf{M}, \tag{3}$$

where w denotes the deflection and the strain tensor ε is defined as the symmetric tensor gradient. The equilibrium equation $-\operatorname{div}\mathbf{Q} = g$ is now satisfied.

With the notation above, and assuming that the load is sufficiently regular, the Kirchhoff plate bending problem can be written as the well known biharmonic problem:

$$\mathsf{D}\Delta^2 w = g \qquad \text{in } \Omega, \tag{4}$$

with the boundary conditions

$$w = 0, \nabla w \cdot \boldsymbol{n} = 0 \qquad \text{on } \Gamma_\mathrm{C}, \tag{5}$$

$$w = 0,\ \mathbf{Mn}\cdot\mathbf{n} = 0 \qquad \text{on } \Gamma_\mathrm{S}, \tag{6}$$

$$\mathbf{Mn}\cdot\mathbf{n} = 0,\ \frac{\partial}{\partial \mathbf{s}}(\mathbf{Mn}\cdot\mathbf{s}) + (\mathbf{div}\mathbf{M})\cdot\mathbf{n} = 0 \qquad \text{on } \Gamma_\mathrm{F}, \tag{7}$$

$$(\mathbf{Mn}_1\cdot\mathbf{s}_1)(c) = (\mathbf{Mn}_2\cdot\mathbf{s}_2)(c) \qquad \forall c \in \mathscr{V}, \tag{8}$$

where $\mathbf{n}$ and $\mathbf{s}$, respectively, denote the unit outward normal and the unit counterclockwise tangent to the boundary. By the indices 1 and 2 we denote the sides of the boundary angle at a corner point c.

In order to interpret the Kirchhoff model as the limit of the Reissner–Mindlin formulation, it is assumed, as usual, that the loading is scaled as $g = \mathsf{G}t^3 f$ with f fixed. Then the problem (4) becomes independent of the plate thickness:

$$\frac{1}{6(1-\nu)}\Delta^2 w = f \qquad \text{in } \Omega. \tag{9}$$

In the corresponding mixed formulation, the rotation and the scaled shear force, respectively, are taken as new unknowns:

$$\beta = \nabla w \quad \text{and} \quad \mathbf{q} = -\mathbf{div}\,\mathbf{m} = -\mathbf{L}\beta\,, \tag{10}$$

where we have introduced a partial differential operator $\mathbf{L}$ by the scaled moment $\mathbf{m}$. Now the scaled mixed problem reads:

$$-\mathrm{div}\,\mathbf{q} = f\,,\ \mathbf{L}\beta + \mathbf{q} = \mathbf{0}\,,\ \nabla w - \beta = \mathbf{0} \qquad \text{in } \Omega\,, \tag{11}$$

with the boundary conditions $w = 0$, $\beta = \mathbf{0}$ on the clamped boundary Γ_{C}, $w = 0$, $\beta \cdot \mathbf{s} = 0$, $\mathbf{m}(\beta)\mathbf{n} \cdot \mathbf{n} = 0$ on the simply supported boundary Γ_{S}, and $(\nabla w - \beta) \cdot \mathbf{s} = 0$, $\mathbf{m}(\beta)\mathbf{n} \cdot \mathbf{n} = 0$, $\frac{\partial}{\partial \mathbf{s}}(\mathbf{m}(\beta)\mathbf{n} \cdot \mathbf{s}) - \mathbf{q} \cdot \mathbf{n} = 0$ on the free boundary Γ_{F}, $(\mathbf{m}(\beta)\mathbf{n}_1 \cdot \mathbf{s}_1)(c) = (\mathbf{m}(\beta)\mathbf{n}_2 \cdot \mathbf{s}_2)(c)$ for all vertices $c \in \mathscr{V}$.

3 Finite Element Formulations

In what follows, let a a regular family of triangulations $\mathscr{T}_h$ on Ω be given. We will indicate with h_K the diameter of each element $K \in \mathscr{T}_h$, while h will indicate the maximum size of all of the elements in the mesh. Furthermore, E denotes a general edge of the triangulation and h_E is the length of E.

Let the discrete space W_h for the Morley element consist of functions v which are second order piecewise polynomials on $\mathscr{T}_h$ satisfying the following conditions: v is continuous at all the internal vertices and zero at all the vertices on the boundary; the normal derivative $\nabla v \cdot \mathbf{n}_E$ is continuous at all the midpoints of the internal edges and zero at all the midpoints of the boundary edges. Then the finite element approximation of the Kirchhoff problem with the Morley element reads:

Method 1 *Find $w_h \in W_h$ such that*

$$a_h(w_h, v) - (f, v) \quad \forall v \in W_h\,, \tag{12}$$

where the bilinear form a_h is defined as

$$a_h(u,v) = \sum_{K \in \mathscr{T}_h} (\mathbf{E}\varepsilon(\nabla u), \varepsilon(\nabla v))_K \quad \forall u, v \in W_h\,. \tag{13}$$

Let $\mathscr{E}_h$ represent the collection of all the edges of the triangulation, and let $[\![\cdot]\!]$ denote the jump operator which is assumed to be equal to the function value on boundary edges. Introducing now the discrete norm

$$|||v|||_h^2 := \sum_{K \in \mathscr{T}_h} |v|_{2,K}^2 + \sum_{E \in \mathscr{E}_h} h_E^{-3} \| [\![v]\!] \|_{0,E}^2 + \sum_{E \in \mathscr{E}_h} h_E^{-1} \| [\![\frac{\partial v}{\partial \mathbf{n}_E}]\!] \|_{0,E}^2\,, \tag{14}$$

the following a priori error estimate holds for the method [6]:

Proposition 1. *Let $w \in H^3(\Omega)$ and $f \in L^2(\Omega)$. Then there exists a positive constant C such that*

$$|||w - w_h|||_h \le Ch\left(|w|_3 + h\|f\|_0\right). \tag{15}$$

With integer values $k \ge 1$, we next define the discrete spaces for the stabilized C^0-element as

$$W_h = \{v \in H^1(\Omega) \mid v_{|\Gamma_C \cup \Gamma_S} = 0, v_{|K} \in P_{k+1}(K)\ \forall K \in \mathscr{T}_h\}, \tag{16}$$

$$\mathbf{V}_h = \{\eta \in [H^1(\Omega)]^2 \mid \eta_{|\Gamma_C} = \mathbf{0}, \eta \cdot \mathbf{s}_{|\Gamma_S} = 0, \eta_{|K} \in [P_k(K)]^2\ \forall K \in \mathscr{T}_h\}, \tag{17}$$

for the approximations of the deflection and the rotation, respectively. Here $P_k(K)$ denotes the space of polynomials of degree k on K. In addition, let the positive stability constants α and γ be assigned. Now the stabilized C^0-continuous finite element method for the Kirchhoff problem reads:

Method 2 *Find $(w_h, \beta_h) \in W_h \times \mathbf{V}_h$ such that*

$$\mathscr{A}_h(w_h, \beta_h; v, \eta) = (f, v) \quad \forall (v, \eta) \in W_h \times \mathbf{V}_h, \tag{18}$$

where the bilinear form $\mathscr{A}_h$ is defined with $a(\phi, \eta) = (\mathbf{m}(\phi), \varepsilon(\eta))$ as

$$\mathscr{A}_h(z, \phi; v, \eta) = \mathscr{B}_h(z, \phi; v, \eta) + \mathscr{D}_h(z, \phi; v, \eta), \tag{19}$$

$$\mathscr{B}_h(z, \phi; v, \eta) = a(\phi, \eta) - \sum_{K \in \mathscr{T}_h} \alpha h_K^2 (\mathbf{L}\phi, \mathbf{L}\eta)_K \tag{20}$$

$$+ \sum_{K \in \mathscr{T}_h} \frac{1}{\alpha h_K^2} (\nabla z - \phi - \alpha h_K^2 \mathbf{L}\phi, \nabla v - \eta - \alpha h_K^2 \mathbf{L}\eta)_K,$$

$$\mathscr{D}_h(z, \phi; v, \eta) = \sum_{E \in \mathscr{F}_h} \Big((m_{ns}(\phi), (\nabla v - \eta) \cdot \mathbf{s})_E \tag{21}$$

$$+ ((\nabla z - \phi) \cdot \mathbf{s}, m_{ns}(\eta))_E + \frac{\gamma}{h_E} ((\nabla z - \phi) \cdot \mathbf{s}, (\nabla v - \eta) \cdot \mathbf{s})_E \Big)$$

for all (z, ϕ), $(v, \eta) \in W_h \times \mathbf{V}_h$, where $\mathscr{F}_h$ represents the collection of all the boundary edges in Γ_F and $m_{ns} = \mathbf{mn} \cdot \mathbf{s}$.

We next introduce the discrete norm for the deflection and the rotation as

$$|||(v, \eta)|||_h^2 := \sum_{K \in \mathscr{T}_h} |v|_{2,K}^2 + \|v\|_1^2 + \sum_{E \in \mathscr{I}_h} h_E^{-1} \| [\![\frac{\partial v}{\partial \mathbf{n}_E}]\!] \|_{0,E}^2 \tag{22}$$

$$+ \sum_{K \in \mathscr{T}_h} h_K^{-2} \|\nabla v - \eta\|_{0,K}^2 + \|\eta\|_1^2,$$

where $\mathscr{I}_h$ represents the collection of all the internal edges of the triangulation. In addition, for the shear force, we introduce the following notation:

$$\mathbf{V}_* = \{\eta \in [H^1(\Omega)]^2 \mid \eta = \mathbf{0} \text{ on } \Gamma_{\mathrm{C}},\ \eta \cdot \mathbf{s} = 0 \text{ on } \Gamma_{\mathrm{F}} \cup \Gamma_{\mathrm{S}}\}, \tag{23}$$

$$\|\mathbf{r}\|_{-1,*} = \sup_{\eta \in \mathbf{V}_*} \frac{\langle \mathbf{r}, \eta \rangle}{\|\eta\|_1}, \tag{24}$$

$$\mathbf{q}_{h|K} = \frac{1}{\alpha h_K^2}(\nabla w_h - \beta_h - \alpha h_K^2 \mathbf{L}\beta_h)_{|K}. \tag{25}$$

Then the following a priori error estimate holds, cf. Sect. 5 and [2] for more details on the constants C_I and C_I'.

Proposition 2. *Let* $0 < \alpha < C_I/4$, $\gamma > 2/C_I'$, *and* $w \in H^{s+2}(\Omega)$, *with* $1 \le s \le k$. *Then there exists a positive constant* C *such that*

$$|||(w - w_h, \beta - \beta_h)|||_h + \|\mathbf{q} - \mathbf{q}_h\|_{-1,*} \le Ch^s \|w\|_{s+2}. \tag{26}$$

4 A Posteriori Error Estimates

For a posteriori error indicators and estimates below, we indicate the Morley element by $(\mathscr{M})$ and the stabilized C^0-method by $(\mathscr{S})$. The interior error indicators are defined for each element as

$$(\mathscr{M}) \quad \tilde{\eta}_K^2 = h_K^4 \|f\|_{0,K}^2, \tag{27}$$

$$(\mathscr{S}) \quad \tilde{\eta}_K^2 = h_K^4 \|f + \operatorname{div} \mathbf{q}_h\|_{0,K}^2 + h_K^{-2} \|\nabla w_h - \beta_h\|_{0,K}^2, \tag{28}$$

and for inter-element edges as

$$(\mathscr{M}) \quad \eta_E^2 = h_E^{-3} \| [\![w_h]\!] \|_{0,E}^2 + h_E^{-1} \| [\![\frac{\partial w_h}{\partial \mathbf{n}_E}]\!] \|_{0,E}^2, \tag{29}$$

$$(\mathscr{S}) \quad \eta_E^2 = h_E^3 \| [\![\mathbf{q}_h \cdot \mathbf{n}]\!] \|_{0,E}^2 + h_E \| [\![\mathbf{m}(\beta_h)\mathbf{n}]\!] \|_{0,E}^2. \tag{30}$$

Next, for the Morley element, we assume that $\partial\Omega = \Gamma_{\mathrm{C}}$; see [3] and cf. [1] for general boundary conditions. Then for the edges on the clamped boundary Γ_{C} the boundary indicator is defined as

$$(\mathscr{M}) \quad \eta_{E,C}^2 = h_E^{-3} \| [\![w_h]\!] \|_{0,E}^2 + h_E^{-1} \| [\![\frac{\partial w_h}{\partial \mathbf{n}_E}]\!] \|_{0,E}^2. \tag{31}$$

For the stabilized C^0-element, also the simply supported and free boundaries are allowed. Hence, for the edges on Γ_{S} and Γ_{F}, respectively, with $m_{nn} = \mathbf{mn} \cdot \mathbf{n}$, the boundary indicators are defined as

$$(\mathscr{S}) \quad \eta_{E,S}^2 = h_E \|m_{nn}(\beta_h)\|_{0,E}^2, \tag{32}$$

$$(\mathscr{S}) \quad \eta_{E,F}^2 = h_E \|m_{nn}(\beta_h)\|_{0,E}^2 + h_E^3 \| \frac{\partial}{\partial s} m_{ns}(\beta_h) - \mathbf{q}_h \cdot \mathbf{n} \|_{0,E}^2. \tag{33}$$

Finally, let $\mathscr{C}_h$, $\mathscr{S}_h$ and $\mathscr{F}_h$ represent the collections of all the boundary edges in Γ_{C}, Γ_{S} and Γ_{F}, respectively. Then the local and global error indicators are defined as

$$\eta_K = \Big(\tilde{\eta}_K^2 + \frac{1}{2}\sum_{\substack{E\in\mathscr{I}_h\\ E\subset\partial K}} \eta_E^2 + \sum_{\substack{E\in\mathscr{C}_h\\ E\subset\partial K}} \eta_{E,C}^2 + \sum_{\substack{E\in\mathscr{S}_h\\ E\subset\partial K}} \eta_{E,S}^2 + \sum_{\substack{E\in\mathscr{F}_h\\ E\subset\partial K}} \eta_{E,F}^2\Big)^{1/2}, \tag{34}$$

$$\eta = \Big(\sum_{K\in\mathscr{T}_h} \eta_K^2\Big)^{1/2}. \tag{35}$$

We then have the following reliability and efficiency results [3, 2]:

Theorem 1. *There exist positive constants C such that*

$$(\mathscr{M})\quad |||w-w_h|||_h \le C\eta\,, \tag{36}$$

$$(\mathscr{S})\quad |||(w-w_h,\beta-\beta_h)|||_h + \|\mathbf{q}-\mathbf{q}_h\|_{-1,*} \le C\eta\,. \tag{37}$$

Theorem 2. *There exist positive constants C such that*

$$(\mathscr{M})\quad \eta_K \le C\big(|||w-w_h|||_{h,K} + h_K^2\|f-f_h\|_{0,K}\big)\,, \tag{38}$$

$$(\mathscr{S})\quad \eta_K \le C\big(|||(w-w_h,\beta-\beta_h)|||_{h,\omega_K} + h_K^2\|f-f_h\|_{0,\omega_K} + \|\mathbf{q}-\mathbf{q}_h\|_{-1,*,\omega_K}\big)\,, \tag{39}$$

for each element $K\in\mathscr{T}_h$, with some approximation f_h of the load f. Here the domain ω_K denotes the set of all the triangles sharing an edge with K.

5 Numerical Results

In this final section, we present some results on benchmark computations in order to compare the numerical and theoretical results of the two methods proposed. Regarding the stabilized C^0-element, we restrict ourselves to the lowest order element with $k=1$. Hence, the rotation has linear components and the deflection is quadratic, as for the Morley element.

In all of the test cases, the values $\mathsf{E}=1$, $\nu=0.3$, $\alpha=0.1$ and $\gamma=100$ have been used for the material and stability constants. We have implemented the methods in the open-source finite element software Elmer [5] which utilizes local error indicators, and provides complete remeshing with Delaunay triangulations and error balancing strategy for adaptive refinements [4, 5].

The ratio between the estimated and true error is shown as the effectivity index in Fig. 1. With the stabilized C^0-element, we have solved three problems with convex rectangular domains, different types of boundary conditions and known exact solutions. With the Morley element, only a clamped square with uniform loading has been solved. In these test problems, the effectivity index remains on a certain almost constant level uniformly in the mesh size. This indicates that the error estimators can be used as reliable and efficient error measures.

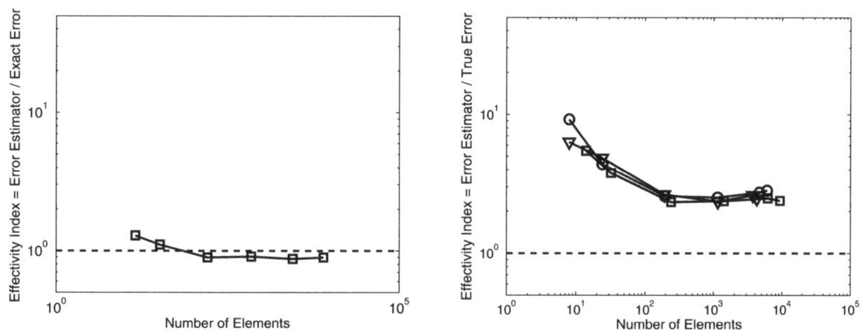

Fig. 1 Effectivity index; *Left*: the Morley element (with C-boundaries); *Right*: the stabilized C^0-element (with C/S/F-boundaries).

In order to compare adaptive and uniform refinements, we consider the stabilized C^0-element and a uniformly loaded L-shaped domain with simply supported boundaries. The convergence graphs for the uniformly (circles) and adaptively (triangles) refined meshes are shown in Fig. 2. The two upper graphs (solid lines) represent the global error estimator, while the lower ones (dashed lines) indicate the maximum local estimator. Moreover, we show in the same figure the convergence rates $\mathcal{O}(h^{1/3})$ and $\mathcal{O}(h)$ (dashed lines) corresponding to the rates with and without the corner singularity of the solution. Now, due to the singular L-corner, $w \in H^{7/3}(\Omega)$.

In Fig. 2, with the uniform refinements (circles), the convergence rate of the error estimator clearly follows the value $\mathcal{O}(h^{1/3})$. Differently, after the first adaptive steps, the method shows its robustness in finding the corner singularity of the solution and refining locally near the L-corner (triangles).

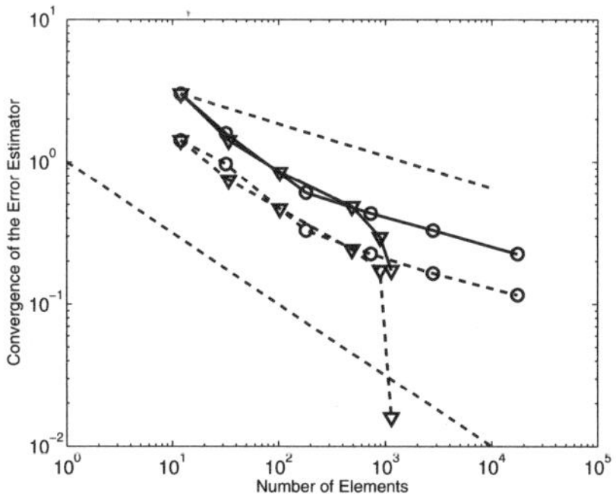

Fig. 2 Simply supported L-domain: Convergence of the global estimator (solid lines) and the maximum local estimator (dashed lines); Circles for the uniform refinements, triangles for the adaptive refinements.

As a conclusion, the Morley element with 6 degrees of freedom is a very simple discontinuous method, while the stabilized C^0-element, with 12 degrees of freedom for $k = 1$, provides additional rotation degrees of freedom and hence exact boundary conditions. In particular, the Morley element possesses nonzero error indicators for clamped edges; however, the corresponding effectivity index seems very uniform. Finally, in Fig. 3, the refinements produced by both of the methods can be compared for an L-shaped clamped domain with uniform loading.

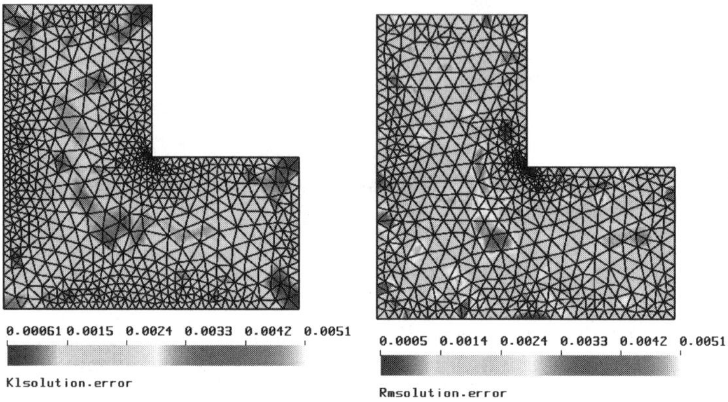

Fig. 3 Distribution of the error estimator after adaptive refinements: *Left*: the Morley element; *Right*: the stabilized C^0-element.

References

1. Beirão da Veiga, L., Niiranen, J., Stenberg, R.: A posteriori error analysis for Morley plate element with general boundary conditions. In preparation
2. Beirão da Veiga, L., Niiranen, J., Stenberg, R.: A family of C^0 finite elements for Kirchhoff plates I: Error analysis. SIAM J. Num. Anal. **45**, 2047–2071 (2007)
3. Beirão da Veiga, L., Niiranen, J., Stenberg, R.: A posteriori error estimates for the Morley plate bending element. Numer. Math. **106**, 165–179 (2007)
4. Beirão da Veiga, L., Niiranen, J., Stenberg, R.: A family of C^0 finite elements for Kirchhoff plates II: Numerical results. Comp. Meths. Appl. Mech. Engrg. **197**, 1850–1864 (2008)
5. Elmer finite element software homepage: `http://www.csc.fi/elmer`
6. Ming, W., Xu, J.: The Morley element for fourth order elliptic equations in any dimensions. Numer. Math. **103**, 155–169 (2006)

Adaptive Finite Element Simulation of Relaxed Models for Liquid-Solid Phase Transition

M. Stiemer

Abstract The purpose of this work is to develop time-space adaptive techniques for phase field models for liquid-solid phase transition. These models are based on coupled systems of the Cahn-Hilliard and the Allen-Cahn equation. The adaptive techniques under consideration allow for mesh adaptation with respect to a functional of interest. They are based on a discontinuous Galerkin approach for time stepping and the exploitation of a parabolic duality argument. For the Bi-Laplace operator arising in linearization of the Cahn-Hilliard equation, a non-conforming Galerkin method is employed.

1 Introduction

If a liquid is carefully cooled under its solidification temperature it may remain in a metastable liquid state. Solidification usually starts from nucleation germs inside the liquid or – more often – from the walls of the basin containing the liquid. The now arising system of two coinciding phases exhibits complicated dynamical phenomena. While on a microscopic level the interface between the phases is subject to molecular dynamic processes of permanent aggregation and disaggregation, it can be modeled as a sharp moving frontier on a phenomenological level. Its evolution is determined by a thermodynamic driving force depending on temperature, concentration gradients and the curvature of the interface. However, sharp interface models are numerically difficult to handle. More convenient is a phase field approach, where a real valued function $\Phi : \Omega \longrightarrow [0,1]$, the so called *phase-field variable* or *order parameter*, is introduced in the domain $\Omega \subset \mathbb{R}^n$ $(n = 2,3)$ of interest, attaining the value 0 in most part of the liquid phase and 1 in most part of the solid phase. Only in a small region about the phase interface, values between 0 and 1 arise such

Marcus Stiemer
Faculty of Mathematics, Technische Universität Dortmund
e-mail: stiemer@math.uni-dortmund.de

that Φ is smooth (see Fig. 1). If the liquid consists of more than one substance, their concentration-gradients have a significant influence on the thermodynamic driving force acting on the interface. Under certain conditions, a solidification process can even take place at approximately constant temperature. In this case it is driven by concentration gradients and the latent heat production during solidification may be neglected. For a binary alloy, a scalar field $c : \Omega \to \mathbb{R}$ representing the concentration of one species suffices to model the influence of concentration gradients. The energy of the isothermal two phase system is represented by the functional[1]

$$\mathscr{F} = \int_\Omega \left[\frac{\varepsilon_\Phi^2}{2}(\nabla\Phi)^2 + \frac{\varepsilon_c^2}{2}(\nabla c)^2 + f(\Phi, c, T) \right] . \tag{1}$$

Here, f denotes a thermodynamic potential to be specified below. The parameter ε_Φ represents the thickness of the interface, i.e. it quantifies the size of the area in which Φ attains values significantly different from 0 or 1. From the point of view of a sharp interface model, ε_Φ takes the role of a regularization parameter. In fact, it has been shown, that under certain assumptions the solutions of phase field models converge to the solution of the corresponding Stefan-problem for $\varepsilon_\Phi \to 0$ (see [10] and the references given therein). Finally, ε_c measures the size of the area, in which mixing of different materials is relevant. Based on the energy-functional (1), equations for the evolution of the fields Φ and c have to be formulated obeying the laws of thermodynamics. In the isothermal situation, the following evolution equations are relevant (see e.g. [2]):

$$\dot{\Phi} = -M_\Phi \left[f_\Phi - \varepsilon_\Phi^2 \nabla^2 \Phi \right] \tag{2}$$

$$\dot{c} = \nabla \left[M_c c(1-c) \nabla \left(f_c - \varepsilon_c^2 \nabla^2 c \right) \right] , \tag{3}$$

where M_Φ, M_c are the mobilities of the phase field and of the concentration field respectively. We assume that M_Φ and M_c are constant. Dots indicate partial derivatives with respect to time, and $f_\Phi = \partial f / \partial \Phi$ and $f_c = \partial f / \partial c$ denote derivatives with respect to Φ and c respectively. The evolution equation (2) for the phase field is known as Cahn-Allen equation, while the equation for the concentrations (3) is a generalized Cahn-Hilliard equation. Next, boundary conditions for each of these equations are required as well as initial conditions. If a region far away from walls is considered, natural boundary conditions are a good choice. The model is completed by specification of the thermodynamic potential f. It typically possesses a double-well shape with respect to Φ, representing the coexistence of a stable and a metastable state (see Fig. 1). Following [2], we assume

$$f_A(\Phi) = f_A^S + W_A g(\Phi) + p(\Phi) \left[f_A^L - f_A^S \right] \tag{4}$$

for a single material with $g(\Phi) = \Phi^2(1-\Phi)^2$ and $p(\Phi) = \Phi^3(6\Phi^2 - 15\Phi + 10)$. Further, W_A represents the height of the energy hump between the solid and the liquid state, and f_A^S and f_A^L denote the ordinary free energy of the material A as

[1] Throughout this presentation, we will suspend volume-elements, area-elements, etc in integrals.

liquid (L) and solid (S) respectively. For a binary alloy, the potentials for the single materials have to be combined and the mixing potential has to be added:

$$f(\Phi,c) = (1-c)f_A(\Phi) + cf_B(\Phi) + R'T[(1-c)\log(1-c) + c\log c] \\ + \{\omega_S[1-p(\Phi)] + \omega_L p(\Phi)\}, \qquad (5)$$

where ω_S and ω_L are regular solution parameters of the liquid and the solid, T denotes the temperature, and R' is the universal gas constant divided by the molar volume of the alloy which is assumed to be constant (cf. [2]). In (5) a typical problem connected to the numerical treatment of the Cahn-Hilliard equation becomes obvious: The thermodynamic potential f is not differentiable with respect to c in $c = 0$ and in $c = 1$. To cope with such problems, a non smooth Newton method has been presented for the numerical solution of the Cahn-Hilliard equation in [11]. In many situations, the logarithmic terms in (5) can sufficiently well be approximated by a polynomial such that f is replaced by a smooth approximating potential h (cf. [4, 13]). In the sequel, we will work with this approximation. However, the here presented adaptive methods can also be combined with a non smooth Newton method to deal with the more realistic case of a non smooth potential. To obtain realistic patterns of solidification, crystalline anisotropy has additionally to be modeled. On a phenomenological level this can be achieved by either making $\varepsilon_\Phi = \varepsilon_\Phi(\nabla\Phi)$ depend on the direction of the interface between solid and liquid phase or by letting the potential h be depending on $\nabla\Phi$. It is known that ε_Φ can be made depending on $\nabla\Phi$ in such a way that the correct sharp interface relations result if ε_Φ tends to zero. Yet, this is not the case for the second approach (see [10]). Finally, thermal noise has to be added to trigger side branching of the solidification front.

Phase field models are extensively studied by physicists to understand phenomena related to phase transitions such as e.g. nucleation [2, 3, 9]. To be able to extend such models to more complex situations, efficient numerical techniques for systems of the above type are required. A main issue is an adaptive numerical treatment, giving consideration to the localized transition zone between different phases. First

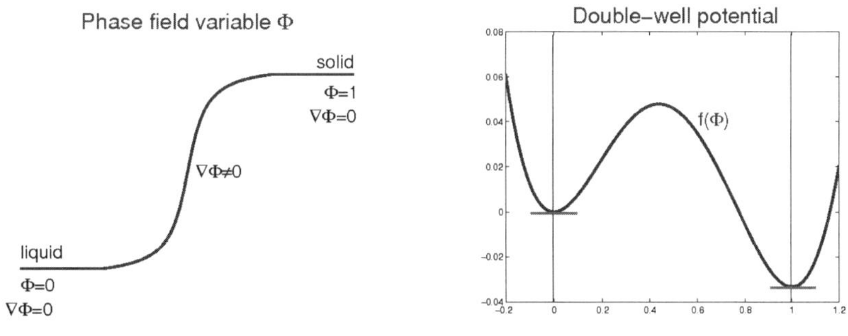

Fig. 1 *Left:* The sharp interface between liquid- and solid phase is approximated by a phase-field (order parameter). *Right:* Double well potential: Here, the solid phase ($\Phi = 1$) is stable and the liquid phase $\Phi = 0$ is metastable.

attempts at adaptive methods have been presented in [5, 12]. The schemes presented are however based on estimators for the energy error or on physically motivated heuristics rather than on a rigorous estimation of the error in more relevant quantities. The purpose of this work is to develop time-space adaptive techniques that allow for mesh adaptation with respect to a functional of interest. These are based on a discontinuous Galerkin approach for time stepping and the exploitation of a parabolic duality argument. For the Bi-Laplace operator arising in linearizations of the Cahn-Hilliard equation, a non-conforming Galerkin method is employed.

2 Discretization

Let $u = (\Phi, c)$. Starting point for the discretization is the weak formulation

$$\begin{aligned}\mathscr{A}(u,v) = &\int_\Omega \dot{\Phi}\tilde{\Phi} + \int_\Omega M_\Phi \varepsilon_\Phi^2 \nabla\Phi\nabla\tilde{\Phi} + \int_\Omega M_\Phi h_\Phi \tilde{\Phi} \\ &+ \int_\Omega \dot{c}\tilde{c} + \int_\Omega M_c \varepsilon_c^2 \nabla^2 c \nabla[c(1-c)\nabla\tilde{c}] + \int_\Omega M_c h_c \nabla[c(1-c)\nabla\tilde{c}] = 0\end{aligned} \tag{6}$$

for all $v = (\tilde{\Phi}, \tilde{c}) \in L^2(H_0^2(\Omega)^2)$. A solution of the system (6) is sought after in the space $L^2(H^2(\Omega)^2)$ of all functions defined on a time interval I of interest with vectors of length 2 as values, each component of which is contained in the Sobolev space $H^2(\Omega)$. See [6] for details on the functional analytic background. All equations have to be understood in the sense of distributions on I.

In the following numerical considerations, we do not incorporate anisotropy or thermal noise. In order to solve (6) numerically, the system needs to be linearized. Fréchet linearization of the operator $A : L^2(H^2(\Omega)^2) \to L^2(H^2(\Omega)^2)$, introduced by the bilinear form $\mathscr{A}$ via $\mathscr{A}(u,v) = (Au, v)_{L^2(H^2(\Omega)^2)}$ with the standard scalar product $(\cdot,\cdot)_{L^2(H^2(\Omega)^2)}$ on $L^2(H^2(\Omega))^2$ yields

$$\begin{aligned}&\int_\Omega \dot{\delta_n^\Phi}\tilde{\Phi} + \int_\Omega M_\Phi \varepsilon_\Phi^2 \nabla\delta_n^\Phi \nabla\tilde{\Phi} + \int_\Omega M_\Phi \left(h_{\Phi\Phi}\delta_n^\Phi + h_{\Phi c}\delta_n^c\right)\tilde{\Phi} \\ &+ \int_\Omega \dot{\delta_n^c}\tilde{c} + \int_\Omega M_c \varepsilon_c^2 \left\{\nabla^2\delta_n^c \nabla[c_n(1-c_n)\nabla\tilde{c}] + \nabla^2 c_n \nabla[(1-2c_n)\delta_n^c\nabla\tilde{c}]\right\} \\ &+ \int_\Omega M_c \left\{\left(h_{c\Phi}\delta_n^\Phi + h_{cc}\delta_n^c\right)\nabla[c_n(1-c_n)\nabla\tilde{c}] + h_c \nabla[(1-2c_n)\delta_n^c\nabla\tilde{c}]\right\} = -\mathscr{A}(u_n, v)\end{aligned} \tag{7}$$

for the Newton update $\delta_n = (\delta_n^\Phi, \delta_n^c) = (\Phi_{n+1} - \Phi_n, c_{n+1} - c_n) \in L^2(H^2(\Omega)^2)$. The equation is required to hold for all $v = (\tilde{\Phi}, \tilde{c}) \in L^2(H^2(\Omega)^2)$. The partial derivatives of h have to be evaluated in $u_n = (\Phi_n, c_n)$. The iteration can be broken down into a separate iteration for each of a finite number of time steps. The result of the Newton iteration for an individual time step is then taken as initial data for the next one. This is convenient in the context of an explicit time stepping scheme.

In the linearized equation, the finite element discretization of the weak Bi-Laplace operator $(\nabla^2 u, \nabla^2 v)_\Omega$ with $(\cdot,\cdot)_\Omega$ being the L^2 standard scalar product on

Ω needs particular care. Aiming at an efficient numerical scheme, we wish to avoid a high order approach such as a H^2-conforming finite element discretization. Although replacing $\Delta^2 c = j$ by the system $\nabla^2 c = \lambda$, $\nabla^2\lambda = j$ leads to a mixed formulation that is inf-sup-stable in $H^1 \times H_0^1$, a convergence result is only known on regular meshes based on superconvergence. It is more adequate to approximate λ in a subspace of $L^2(\Omega)$, while c is approximated in an enlarged space, containing $H^2(\Omega)$. This can be done such that convergence of the Hessian in L^2 is guaranteed [1]. On a triangle-mesh, the resulting scheme is algebraically equivalent to employing the non-conforming Morley-element (with a modified integration of the right hand side) [1]. The corresponding part of the numerical methods destined to solve (7) will now be demonstrated for the simplified version

$$\int_\Omega \varepsilon_c^2 \nabla^2 c \nabla^2 v + \int_\Omega \dot{c} v = \int_\Omega f v, \qquad v \in L^2\left(H^2(\Omega)\right), \tag{8}$$

of equation (7). Although this is a strong simplification, it still allows for discussion of relevant numerical effects. Writing

$$a(c,v) = \int_\Omega \varepsilon_c^2 \nabla^2 c \nabla^2 v, \tag{9}$$

we obtain

$$a(c,v) + (\dot{c},v)_\Omega = (f,v)_\Omega, \qquad v \in L^2\left(H^2(\Omega)\right), \tag{10}$$

which we analyze as a model problem. Equation (10) is the weak form of a classical Cahn-Hilliard equation and hence of interest in itself. A discrete spatial test and trial space is defined based on the non-conforming Morley-element. Let $(\mathscr{T}_h)$ be a family of quasi-uniform triangulations with vertices $\mathscr{V}_h$ and edges $\mathscr{E}_h$. Consider

$$\begin{aligned}\mathscr{S}_h = \Big\{ v \in L^2(\Omega) \ : \ v|_T\,(x,y) = a_0 + a_1 x + a_2 y + a_3 x^2 + a_4 xy + a_5 y^2\,, T \in \mathscr{T}_h\,, \\ v \text{ continuous in } P \in \mathscr{V}_h\,, \nabla v \cdot n_E \text{ continuous on } E \in \mathscr{E}_h \Big\}\,,\end{aligned}$$

with n_E denoting a unit vector normal to the edge $E \subset \partial T$. Note that $\mathscr{S}_h$ is not a subset of $H^2(\Omega)$. Hence, the bilinear form a can not be evaluated for arbitrary test and trial functions. Computing a element-wise and summing up leads to a bilinear form a_h that may attain negative values on $\mathscr{S}_h$. Hence, the stabilized bilinear form

$$\begin{aligned} a_{\sigma,h}(u,v) &= \sigma\varepsilon_c^2 \sum_{T\in\mathscr{T}_h} (\nabla^2 u, \nabla^2 v)_T \\ &\quad + (1-\sigma)\varepsilon_c^2 \int_T 2\frac{\partial^2 u}{\partial x \partial y}\frac{\partial^2 v}{\partial x \partial y} + \frac{\partial^2 u}{\partial x^2}\frac{\partial^2 v}{\partial y^2} + \frac{\partial^2 u}{\partial y^2}\frac{\partial^2 v}{\partial x^2}, \end{aligned} \tag{11}$$

with a $0 < \sigma < 1$ is considered, which is positive definite on $\mathscr{S}_h$. For functions $u, v \in H^2(\Omega)$, $a_{\sigma,h}$ coincides with the bilinear form a (cf. (9)) (see [8]). Consequently, the solution of (10) is not altered when $a(\cdot,\cdot)$ is replaced by $a_{\sigma,h}(\cdot,\cdot)$ for $\sigma, h > 0$. Now let $t_0 < t_1 < \ldots < t_N$ be a partition of the time interval $I = [t_0, t_N]$ under consideration into small intervals $I_n = (t_{n-1}, t_n)$, $n = 1, \ldots, N$. Combining the Morley approach with a discontinuous Galerkin method $DG(q)$, $q \in \mathbb{N}_0$, for time stepping yields the space-time test- and trial space $V_q = \left\{v \in L^2\left(H^1(\Omega)\right) : v|_{I_n} \in V_{q,n})\right\}$ with $V_{q,n} = \left\{v : v = \sum_{j=0}^{q} t^j u_j, u_j \in \mathscr{S}_n\right\}$, where $\mathscr{S}_n$ is the spatial test- and trial-space used in the n-th time interval. An approximation $c^h \in V_q$ is computed via the fully discrete scheme

$$\int_{I_n} \left[a_{\sigma,h}(c^h,v) + (\dot{c}^h,v)_\Omega\right] + \left(\left\lfloor c^h \right\rfloor_{n-1}, v(\cdot,t_{n-1}+)\right)_\Omega = \int_{I_n} (f,v)_\Omega\,,$$

for all $v \in V_q$, $n = 1,\ldots,N$, with $\lfloor c^h \rfloor_n = c^h(\cdot,t_n+) - c^h(\cdot,t_n-)$. It is equivalent to

$$\int_I \left[a_{\sigma,h}(c^h,v) + (\dot{c}^h,v)_\Omega\right] + \sum_{n=1}^{N} \left(\left\lfloor c^h \right\rfloor_{n-1}, v(\cdot,t_{n-1}+)\right)_\Omega = \int_I (f,v)_\Omega\,, \qquad (12)$$

for all $v \in V_q$. In the following, the left hand side of (12) will be abbreviated by $B(c^h,v)$ and the right hand side $L(v)$. Note that with a replaced by $a_{h,\sigma}$, the discrete equation $B(c^h,v) = L(v)$, $v \in V_q$, is formally identical to the equation $B(c,v) = L(v)$, $v \in L^2(H_0^2(\Omega))$, characterizing the solution of the continuous problem (10), except for the different test- and trial-spaces.

3 Adaptivity Based on Time Space Error Control

For non-time depending problems, techniques for a posteriori error control for the Morley-element have been presented in [14]. In the context of time and space adaptivity, we address error control for a given linear functional $J : W \to \mathbb{R}$, where W is a linear space containing both the continuous time-space test and trial space $L^2(H^2(\Omega))$ and its discrete counterpart V_q. To control $J(c) - J(c^h) = J(e)$ with $e = c - c^h$, we consider the primal residual

$$R^p(v) := B(e,v) = L(v) - B(c^h,v)\,, \qquad v \in L^2\left(H_0^2(\Omega)\right) + V_q\,. \qquad (13)$$

It possesses the explicit representation

$$R^p(v) = \int_I \left[(f,v)_\Omega - a_{\sigma,h}(c^h,v) - (\dot{c}^h,v)_\Omega\right] - \sum_{n=1}^{N} \left(\lfloor c^h \rfloor_{n-1}, v(\cdot,t_{n-1}+)\right)_\Omega\,. \qquad (14)$$

The Riesz representation theorem implies the existence of a solution $\zeta \in L^2(H^2(\Omega)) + V_q$ of the dual problem

$$B(v,\zeta) = J(v)\,, \qquad v \in L^2(H_0^2(\Omega))\,. \qquad (15)$$

Equation (15) corresponds to an evolution equation with interchanged initial and final state, which is not yet an inverse problem. Now, let ζ^h be an approximation to the dual problem (15) computed via (12) based on the same discretization as was used to determine c^h. Galerkin orthogonality yields

$$J(e) = B(e,\zeta) = B(e,\zeta-\zeta^h) = R^p(\zeta-\zeta^h). \tag{16}$$

Considering (14), we notice that the error representation given in (16) does not depend on the continuous primal solution c, but only on c^h, ζ^h and on ζ. The first two quantities are known after a finite element computation. If mesh adaptation is the major issue, an approximation of ζ is sufficiently accurately determined from ζ^h by a recovery technique such as e.g. utilized by the ZZ-error estimator [15]. The following local representation of the error can now be derived:

Theorem 1. *With the above notations, we have for $q=0$*

$$\begin{aligned} J(e) = &\sum_{\substack{1\le n\le N\\ T\in\mathscr{T}}} \left[\int_{I_n\times T} f(\zeta-\zeta^h) + \int_T \lfloor c^h\rfloor_{n-1}\left(\zeta-\zeta^h\right)(\cdot,t_{n-1}+)\right] \\ &+(1-\sigma)\varepsilon_c^2 \sum_{\substack{1\le n\le N\\ E\in\mathscr{E}}} \int_{I_n\times E}\left[n_E^\top\lfloor H_{c^h}\rfloor_E\nabla\zeta - n_E^\top\lfloor H_{c^h}\nabla\zeta^h\rfloor_E\right]. \end{aligned} \tag{17}$$

Here, $\lfloor H_{c^h}\rfloor_E$ is the matrix consisting of the jumps of the elements of the Hessian of c^h when passing over the edge E.

Proof. We split (14) with $v=\zeta-\zeta^h$ into a sum of integrals over each single time-space element $I_n\times T$ with $T\in\mathscr{T}$, and $I_n=(t_{n-1},t_n)$. Spatially integrating by parts based on the divergence theorem leads to

$$\begin{aligned} J(e) = &\sum_{\substack{1\le n\le N\\ T\in\mathscr{T}}} \left\{\int_{I_n}\left[\int_T f(\zeta-\zeta^h) + (1-\sigma)\varepsilon_c^2\int_{\partial T} n_E^\top H_{c^h}\nabla(\zeta-\zeta^h)\right]\right. \\ &\left.+\int_T \lfloor c^h\rfloor_{n-1}\left(\zeta-\zeta^h\right)(\cdot,t_{n-1}+)\right\}. \end{aligned} \tag{18}$$

Here H_{c^h} denotes the Hessian of c^h, and n_E a unit outer normal vector. Note that H_{c^h} is element-wise constant but in general not continuous over element edges. All other terms cancel out, since integration by parts results in application of a differential operator of order >2 on the polynomial $c^h|_{I_n\times T}$ which is of second order in space. After summarizing all contributions that belong to a certain edge and eliminating terms due to continuity properties, the statement of the theorem follows. □

Remark 1. Theorem 1 implies also a representation of the error in the stationary situation, represented by (10) without the contribution of $\dot{c}$:

$$J(e) = \sum_{T\in\mathscr{T}}\int_T\left[f(\zeta-\zeta^h)\right] + (1-\sigma)\varepsilon_c^2\sum_{E\in\mathscr{E}}\int_E\left[n_E^\top\lfloor H_{c^h}\rfloor_E\nabla\zeta - n_E^\top\lfloor H_{c^h}\nabla\zeta^h\rfloor_E\right].$$

The representation (17) can – after taking absolute values and employing the Cauchy-Schwarz inequality – be used to indicate the local error contribution of a particular time-space element $I_n \times T$ or $I_n \times E$. The analytical and numerical validation of the resulting error estimator represents work in progress. In a numerical implementation of space- and time-adaptivity, the problem arises that no explicit time stepping is possible anymore, since different cells of the triangulation may have different time steps. However, with an efficient multigrid solver at hand, the complexity of a fully implicit method is principally not larger than that of a problem with explicit time stepping. In a time-space adaptive context for the isothermal solidification problem (6), we would refrain from breaking down the Newton iteration (7) to single time steps, but rather solve the linearized problem on the whole time interval I. For problems which are too small for an efficient solution procedure based on a multigrid solver, ensembles of several time steps could be gathered and solved in a coupled fashion. Further, hanging nodes in time are much less worse than in hyperbolic problems due to energy dissipation. Finally, the non-linear problem can entirely be treated in an adaptive framework, as follows from the results of [7].

References

1. Blum, H., Rannacher, R.: On mixed finite element methods in plate bending analysis. I: The first Herrmann scheme. Comput. Mech. **6**(3), 221–236 (1990)
2. Boettinger, W.J., Warren, J.A., Beckermann, C., Karma, A.: Phase-field simulation of solidification. Annu. Rev. Mater. Res. **32**, 163–194 (2002)
3. Chen, L.Q.: Phase-field models for microstructure evolution. Annu. Rev. Mater. Res. **32**, 113–140 (2002)
4. Copetti, M.I.M., Elliott, C.B.: Numerical analysis of the Cahn-Hilliard equation with a logarithmic free energy. Numer. Math. **63**, 39–65 (1992)
5. Danilov, D., Nestler, B.: Phase-field simulations of solidification in binary and ternary systems using a finite element method (2004). URL http://www.citebase.org/abstract?id=oai:arXiv.org:cond-mat/0407694
6. Dautray, R., Lions, J.L.: Evolution Problems I, *Mathematical Analysis and Numerical Methods for Science and Technology*, vol. 5. Springer-Verlag (1995)
7. Eriksson, K., Johnson, C.: Adaptive finite element methods for parabolic problems. iv: Nonlinear problems. SIAM J. Numer. Anal. **32**(6), 1729–1749 (1995)
8. Goering, H., Roos, H.G., Tobiska, L.: Finite-Elemente-Methode. Akademie-Verlag, Berlin (1993)
9. Gránásy, L., Pustai, T., Warren, J.A.: Modelling polycrystalline solidification using phase field theory. J. Phys.: Condensed Matter **16**, 1205–1235 (2004)
10. Kobayashi, R.: A numerical approach to three-dimensional dendritic solidification. Experimental Mathematics **3** (1994)
11. Kornhuber, R.: Non-smooth Newton methods for the Cahn-Hilliard equation with obstacle potential. MATHEON Workshop on Computational Partial Differential Equations, Humboldt-Universität zu Berlin, Germany, February 02nd, 2006 (2006). With C. Gräser, FU Berlin
12. Provatas, N., Goldenfeld, N., Dantzig, J.: Adaptive mesh refinement computation of solidification microstructures using dynamic data structures. J. Comput. Phys. **148**, 265–290 (1999)
13. Temam, R.: Infinite-Dynamical Systems in Mechanics and Physics, *Applied Mathematical Sciences*, vol. 68. Springer-Verlag (1988)
14. Beirão da Veiga, L., Niiranen, J., Stenberg, R.: A posteriori error estimates for the Morley plate bending element. Numer. Math. **106**(2), 165–179 (2007)
15. Zienkiewicz, O.C., Zhu, J.Z.: A simple error estimator and adaptive procedure for practical engineering analysis. Int. J. Numer. Methods Eng. **24**, 337–357 (1987)

Biomedical Applications

A Numerical Study of the Interaction of Blood Flow and Drug Release from Cardiovascular Stents

C. D'Angelo and P. Zunino

Abstract In this study, we focus on a specific application, the modeling and simulation of drug release from cardiovascular drug eluting stents. In particular, we analyze the interaction between the drug release process and the blood flow, in order to evaluate whether part of the drug released into the blood stream affects the drug deposition into the arterial walls.

1 Introduction

Drug eluting stents (DES) are apparently simple medical implanted devices used to restore blood flow perfusion into stenotic arteries. However, the design of such devices is a very complex task because their performance in widening the arterial lumen and preventing further restenosis is influenced by many factors such as the geometrical design of the stent, the mechanical properties of the materials and the chemical properties of the drug that is released. These properties mutually interact to determine the drug penetration and deposition into the arterial walls.

Since the role of the drug is to prevent restenosis of the artery, most of the computational studies on the efficacy of DES have focused their attention on the transport of the drug into the arterial walls, we refer to [3] and references therein for some examples. In most cases, the blood flow is assumed to have a minor influence on the distribution of the drug into the walls. In particular, it is common to consider that the blood flow acts as a perfect sink with respect to the drug concentration, which is rapidly transported away from the location of the stent. Recently, the analysis pursued in [1] suggested that this assumption is not really justified. Indeed, the drug that is apparently lost in the blood stream significantly affects the drug deposition in the portion of the arterial walls downstream to the stent. However, the computations

Carlo D'Angelo and Paolo Zunino
MOX - Department of Mathematics - Politecnico di Milano, via Bonardi 9, 20133 Milano, Italy, e-mail: carlo.dangelo@polimi.it, paolo.zunino@polimi.it

proposed in [1] are affected by some simplifications of the stent geometry and of the governing equations for the blood flow. In this work we aim to extend the study proposed in [1] to a realistic stent design and arterial geometry in order to accurately evaluate the interaction of the blood flow with the drug release from a stent.

2 Mathematical Model and Numerical Method

To study the interaction between a drug eluting stent and blood flow we consider the Navier-Stokes equations for the fluid dynamics and we describe the drug released in the blood stream as a passive scalar governed by an advection-diffusion equation. In this preliminary study, we neglect the presence of the arterial wall surrounding the lumen but we quantify the drug deposition into the artery by means of the diffusive flux at the interface between the lumen and the wall.

We denote with Ω a portion of a coronary artery where we set up our analysis. This is a cylindric channel deformed by the introduction and the expansion of a stent and truncated from the remaining coronary arterial tree proximally and distally with respect to the stent position. We denote with Γ_{in} and Γ_{out} the proximal and distal sections since they coincide with the inflow and outflow sections of the domain Ω. The remaining part of the boundary of Ω can be subdivided in two regions, the arterial wall and the stent. The former is denoted with Γ_w and the latter with Γ. In conclusion we obtain $\partial\Omega = \Gamma_{in} \cup \Gamma_{out} \cup \Gamma_w \cup \Gamma$.

To analyze the drug release process on a significant time scale we need to consider a time period containing hundreds or thousands of heartbeats. To override this difficulty, we start by considering a stationary flow model and we assume that the blood flow into the arterial lumen is governed by the stationary Navier-Stokes equations,

$$-\nu\Delta\mathbf{u} + (\mathbf{u}\cdot\nabla)\mathbf{u} + \nabla p = \mathbf{0} \text{ and } \nabla\cdot\mathbf{u} = 0 \text{ in } \Omega, \qquad (1)$$

where $\nu = 3\times 10^{-2}$ cm^2/s. This equation must be complemented by suitable boundary conditions, specifying in our case a parabolic inflow profile $\mathbf{u} = \mathbf{u}_{in}$ with a peak of 30 cm/s on Γ_{in}, perfect contact between the blood, the arterial walls and the stent, $\mathbf{u} = 0$ on $\Gamma \cup \Gamma_w$ and zero traction force at the outflow.

The equation governing the interaction of the drug concentration $c(t,\mathbf{x})$ with the blood stream is obtained from the mass conservation principle, taking into account both diffusive and transport fluxes. Contrarily to the assumptions adopted for the fluid dynamics, we consider the time dependent case, because the drug release process in intrinsically transient. By consequence, it is governed by the following equation,

$$\partial_t c + \nabla\cdot(-D\nabla c + \mathbf{u}c) = 0 \text{ in } \Omega, \qquad (2)$$

where $D = 10^{-6}$ cm^2/s is relative to heparin, a common anticoagulant drug. Equation (2) should be supplemented by the initial state of the concentration in the blood stream, $c(t=0) = 0$ in Ω and suitable boundary conditions. On the inflow boundary, Γ_{in}, we prescribe $c = 0$ since the blood does not contain drug proximally to the stent. Assuming that the outflow boundary is far enough to the stent, we neglect the diffusive effects across this section and say $\nabla c \cdot \mathbf{n} = 0$ on Γ_{out}. Furthermore, according to the assumption to neglect the presence of the arterial walls, we prescribe $c = 0$ on Γ_w.

Finally, particular attention should be dedicated to the condition on the interface between the stent and the lumen, because it is primarily responsible to determine the drug release rate. We remind that DES are miniaturized metal structures that are coated with a micro-film containing the drug that will be locally released into the arterial walls for healing purposes. The thickness of this film generally lays within the range of microns. Owing to the fact that the stent coating is extremely thin, we consider the model proposed in [4] where we derived the following formula for the release rate,

$$J(t,\mathbf{x}) = \varphi(t)\big(c_s - c(t,\mathbf{x})\big) \text{ on } \Gamma, \tag{3}$$

c_s being the initial concentration charge of the stent. Given the thickness of the stent coating, $\Delta l = 7\ \mu\text{m}$, and its diffusion parameter, $D_s = 5 \times 10^{-12}$ cm^2/s, the scaling function $\varphi(t)$ is defined as follows,

$$\varphi(t) = \frac{2D_s}{\Delta l} \sum_{n=0}^{\infty} e^{-(n+1/2)^2 kt} \quad \text{with} \quad k = \pi^2 D_s / \Delta l^2.$$

The total drug initially stored into the stent can be expressed as $M_0 = c_s |\Gamma| \Delta l$. Characteristic values of M_0 and c_s are highly variable among different stent models. For this reason, we deal with a normalized concentration with respect to c_s, which is equivalent to set $c_s = 1$ (non-dimensional units). The derivation of (3) is similar to the procedure that leads to the well known Higuchi formula [2], but the former has the advantage to avoid some restrictions that are unnecessary in this context. Owing to (3), the boundary condition on Γ for equation (2) turns out to be the following Robin type condition,

$$-D\nabla c \cdot \mathbf{n} + \varphi(t)(c_s - c) = 0 \text{ on } \Gamma.$$

The initial/boundary value problems relative to equations (1) and (2) are now ready to be approximated by means of suitable numerical methods.

2.1 The Numerical Method

We consider the geometrical model of a coronary artery having length of 9.7 mm and internal diameter of 3.4 mm. The expansion of a realistic model of a Cordis-BX Velocity stent (Johnson & Johnson Interventional Systems, Warren, NY, USA) has been simulated in order to identify the final configuration of the stent and the arterial walls, as described in [3]. Then, the lumen of the artery is subdivided with Gambit (Fluent Inc., Lebanon, NH, USA) into 3.301.271 computational cells, for the approximation of equations (1) and (2) by means of suitable numerical methods. In order to obtain an accurate resolution at reasonable computational cost and memory storage, we applied a nonuniform spacing for the mesh generation. In particular, the central part of the domain has been subdivided by means of variable size tetrahedrons, particularly refined around the stent. Conversely, for the inflow and the outflow part of the artery, where a fine resolution is not needed, we considered a prismatic Cooper mesh, where the height of the prisms increases when approaching the inflow and the outflow boundaries, see figure 1. To solve equations (1) and (2) we applied the commercial solver Fluent (Fluent Inc., Lebanon, NH, USA), which is based on finite volume schemes. We observe that equation (1) is independent of the concentration c, by consequence we computed at first the approximation of the stationary advective field $\mathbf{u}$ and then we solved the problem for the concentration.

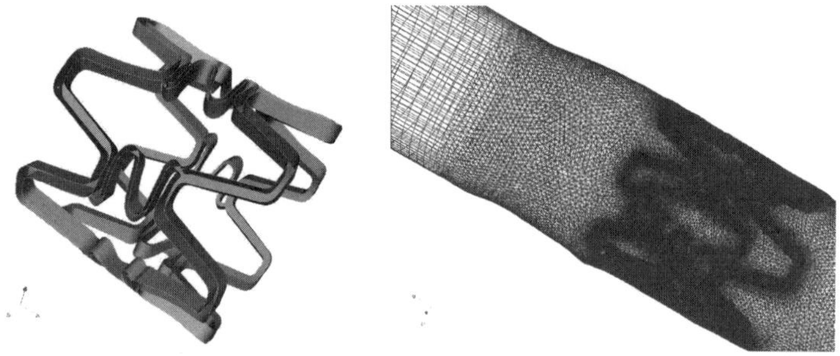

Fig. 1 The geometry of the stent after the expansion (left) with a detail of the computational mesh highlighting the refinement around the stent (bottom-right) and the Cooper mesh approaching the inflow section (top-right).

3 Numerical Results

As already mentioned, we aim to study the interaction of the blood flow with the drug released from the stent. This task is particularly challenging because the complex geometry of the stent highly perturbs the local flow and its pattern significantly influences the drug release into the lumen. We split this analysis in two parts. First of all we focus on the fluid dynamics, trying to put into evidence the main features of the flow around the stent. Secondly, we study how the flow influences the drug release.

3.1 Analysis of the Fluid Dynamics Around the Stent

Looking at the Cordis BX-Velocity stent in figure 1, it is possible to identify two kinds of structures, the struts and the links. The first structure consists of twisted rings that provide the circumferential strength of the stent, whereas the second one is made by tiny connections along the longitudinal axis between subsequent struts. For a preliminary analysis we only focus on the interaction between the struts and the blood flow.

An important feature of the struts is to be twisted in the circumferential direction. For this reason, the blood flow hits the struts with different angles. This suggests that the flow pattern downstream the struts may be substantially different from the well-known flow after a backward facing step that corresponds to the ideal case of a perfectly circular ring that is orthogonal to the flow. This conjecture is confirmed by the fluid dynamics simulations. Indeed, in figure 2 we show the presence of several vortexes downstream the stent. In particular, on the right we highlighted a segment of the strut that is orthogonal to the flow. We observe that in the central part of this segment the vortex is quite stable, but approaching the extrema it is strongly influenced by the change of direction of the strut. The streamlines seem to be guided by the stent and consequently the vortex is stretched and it is absorbed into the main stream. This suggests that the vortex is not only characterized by a planar rotating flow on the plane orthogonal to the axis of the strut, but an out of plane motion is present. This secondary motion is generated by the displacement of the fluid form the center of the vortex to the extrema. Following the streamlines, the fluid is then cast out the vortex into the main stream.

In conclusion, there is evidence that the interaction between the struts and the blood stream generates very complex flow patterns where the recirculation zones downstream the obstacles interact with the main stream. By this way, the fluid that was at some time trapped into a recirculation may join the high speed flow. We will see in the next section that this behavior has important consequences on the drug release process.

3.2 Analysis of the Drug Release in the Blood Stream

The analysis presented in [1] shows that in an idealized geometrical setting, a significant fraction of the drug that is released by the stent into the blood flow resides into the recirculation downstream the struts. This drug can be slowly absorbed into the surrounding arterial walls, with great benefit to the drug deposition into the wall. This interpretation is correct if the recirculation is a planar and stationary vortex, which is the case of 2-dimensional or axial-symmetric flows. However, we showed in the previous section that this is hardly the case for a realistic model of a stent.

The numerical simulation based on equation (2) shows that there is a complex 3-dimensional interaction between the distribution of the drug released into the lumen and the vortexes that appear downstream the struts. The evidence is provided in

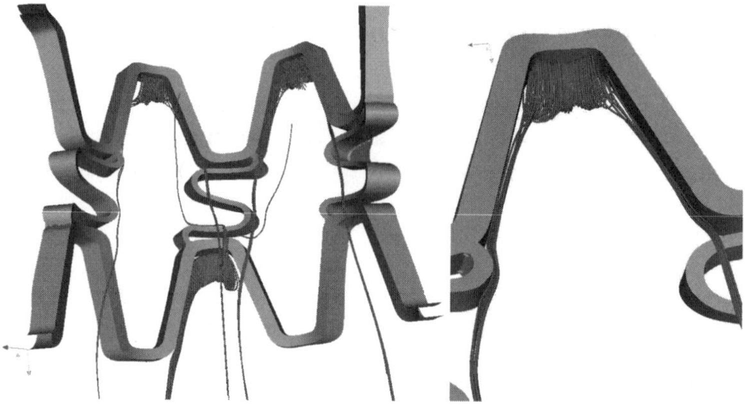

Fig. 2 The interaction between the stent and the blood flow visualized by means of streamlines. The proximal section is located on the top while the distal section in on the bottom.

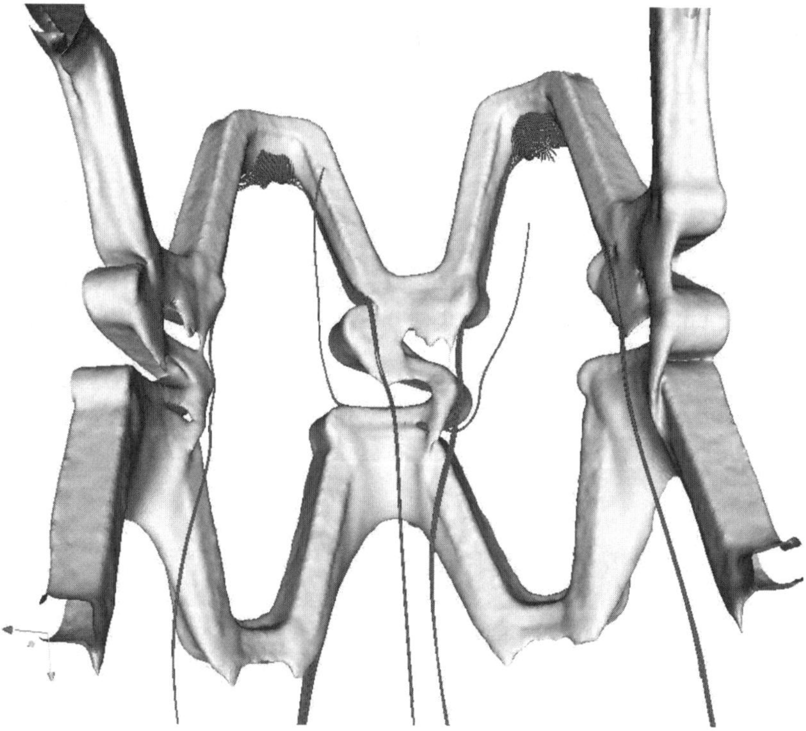

Fig. 3 The interaction between the vortexes downstream the stent and the drug released into the lumen, whose pattern is visualized by means of an isosurface of the drug concentration at 30 minutes after the stent implantation.

figure 3, where we show the streamlines into the recirculations together with an isosurface of the drug concentration in the neighborhood of the stent. The presence of an extended isosurface downstream of the struts shows that part of the drug released into the lumen is trapped into the recirculations. However, the concentration in this region is about 0.01% of the reference concentration c_s, which is much lower than the values provided in [1], where it is estimated that the drug concentration in the blood flow surrounding the stent is up to 40% of c_s. This difference can be explained observing that the drug can easily leave the recirculation regions through the sides of the vortex. This is also confirmed by some details of figure 3 highlighting that the isosurface of the concentration forms a cuff around the streamlines of the vortexes, where the drug leaves the regions of recirculation and it is transported downstream. Furthermore, we observe that the struts located downstream are surrounded by more drug than the ones closer to the proximal section.

In conclusion, the amount of drug trapped into the recirculations is determined as the equilibrium level between the incoming flux released by the stent and the outgoing flux whose major contribution seems to be represented by the transport phenomena at the extrema of the vortex. As a result of this, the concentration at equilibrium and the amount of drug that can be absorbed into the arterial walls from the blood flow are rather low. This interpretation is confirmed by more quantitative results. Starting from the numerical simulation of equation (2), we compute the diffusive flux outgoing the lumen from its interface with the arterial walls and we integrate it over time to compute the corresponding amount of drug, denoted with $M_w(t)$. Then, we compare this quantity with $M(t)$, that is the total amount of drug released into the lumen. Furthermore, instead of analyzing absolute values, it would be easier to study the relative fraction with respect to the total amount of drug initially charged into the stent, M_0. More precisely, we define

$$M(t) = \frac{1}{M_0} \int_0^t \int_\Gamma J(s,\mathbf{x}) \, \mathrm{d}s\mathrm{d}\Gamma, \quad M_w(t) = \frac{1}{M_0} \int_0^t \int_{\Gamma_w} -D\nabla c(s,\mathbf{x}) \cdot \mathbf{n} \, \mathrm{d}s\mathrm{d}\Gamma,$$

where $J(t,\mathbf{x})$ is given in equation (3). In figure 4 we show for a time interval of 8 hours the functions $M(t)$ and $M_w(t)$ together with their ratio $M_w(t)/M(t)$ that represents the fraction of the drug that is absorbed into the arterial walls from the blood flow. We observe that once the equilibrium state is reached, only the 5% of the drug released into the lumen joins the arterial walls while the remaining 95% is transported downstream, outside the computational domain. We finally notice that, although an absorption rate of 5% might be rather small, it is not negligible if compared to the fraction of drug released from the surface of the stent directly in contact with the arterial walls. Indeed, the fraction of drug that is re-absorbed by the artery after being released into the blood corresponds to the 15% of the quantity released by the stent in contact with the arterial walls.

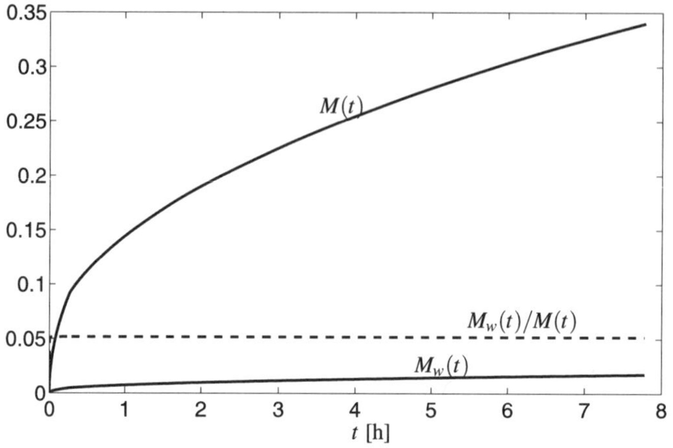

Fig. 4 The fraction of drug released into the lumen and the quantity absorbed into the walls, $M(t)$ and $M_w(t)$ respectively, are plotted together with their ratio $M_w(t)/M(t)$.

4 Conclusions

We analyzed the interactions between the blood flow and the drug release from a stent, showing that a 3-dimensional analysis of the problem accounting for the complex geometry of the stent is mandatory to capture the phenomena into play. In this setting, we studied the contribution of the drug released into the blood flow with respect to the efficacy of drug deposition and penetration into the arterial walls. Even though this effect is not dominant, it seems to give a non negligible contribution to the overall amount of drug released into the artery. In this perspective, further studies addressing the simultaneous drug release in the blood flow and the arterial walls are in order to accurately quantify the drug deposition into the artery.

References

1. Balakrishnan, B., Tzafriri, A., Seifert, P., Groothuis, A., Rogers, C., Edelman, E.: Strut position, blood flow, and drug deposition. implications for single and overlapping drug-eluting stents. Circulation **111**, 2958–2965 (2005)
2. Higuchi, T.: Rate of release of medicaments from ointment bases containing drugs in suspension. J. Pharmac. Sci. **50**, 874–875 (1961)
3. Migliavacca, F., Gervaso, F., Prosi, M., Zunino, P., Minisini, S., Formaggia, L., Dubini, G.: Expansion and drug elution model of a coronary stent. Computer Methods in Biomechanics and Biomedical Engineering **10**, 63–73 (2007)
4. Vergara, C., Zunino, P.: Multiscale modeling and simulation of drug release from cardiovascular stents. Technical Report 15, MOX, Department of Mathematics, Politecnico di Milano (2007). To appear on SIAM Multiscale Model. Simul.

Numerical Modelling of Epidermal Wound Healing

E. Javierre, F.J. Vermolen, C. Vuik, and S. van der Zwaag

Abstract A coupling between wound closure by cell migration and angiogenesis is presented here to model healing of epidermal wounds. The closure of the wound is modelled as a moving interface around which a local grid refinement is applied. The numerical solution combines finite element and finite difference methods to solve the coupled diffusion-reaction equations governing the physiological problem and the hyperbolic equations governing the motion of the interface.

We discuss the accuracy and workload of our numerical model. Furthermore, we illustrate that, under certain circumstances, the healing process may be stopped after initiation.

1 Introduction

Wound healing proceeds by a succession of chemical and mechanical processes: removal of infectious matter by phagocytes, cell mitosis and cell migration to close the wound, extracellular matrix synthesis, reparation of the vascular network (angiogenesis) and, in deeper wounds, reduction of the wound size due to stresses on the tissue (wound contraction). Most of these processes are triggered by the presence or lack of certain growth factors, and are terminated due to a negative feedback mechanism.

E. Javierre and S. van der Zwaag
Fundamentals of Advanced Materials, Faculty of Aerospace Engineering, Delft University of Technology, Kluyverweg 1, 2629 HS Delft, The Netherlands, e-mail: [E.JavierrePerez, S.vanderZwaag]@tudelft.nl

F.J. Vermolen and C. Vuik
Delft Institute of Applied Mathematics, Faculty of Electrical Engineering, Mathematics and Computer Science, Delft University of Technology, Mekelweg 4, 2628 CD Delft, The Netherlands e-mail: [F.J.Vermolen, C.Vuik]@tudelft.nl

Mathematical models for wound healing normally consider only one isolated process. For example, wound closure is studied in [2, 4, 13, 14], angiogenesis in [1, 5, 8] and wound contraction in [9, 10]. However, it is well known that these processes overlap and affect one another. In the present work, we couple wound closure due to cell migration with angiogenesis. In contrast to the discrete models related to cell population dynamics [1, 3], wound closure is modelled here as a closed curve (the wound edge) moving into the wound in the course of time. Hence, the model presented here consists of a number of coupled diffusion-reactions with a moving interface.

The remainder of this paper is organized as follows. The mathematical model is given in Section 2. Subsequently, the computational approach is described in Section 3. Section 4 gives some numerical results that show the potential of the model, and the conclusions are given in Section 5.

2 The Mathematical Model

We combine the models of wound closure due to Arnold and Adam [2] and the model of angiogenesis due to Maggelakis [8] to obtain a new model that couples both processes. In the first model, the closure of the wound is triggered by the production of an epidermic growth factor (EGF) that determines the cell mitosis and motility. Furthermore, the wound edge is identified as the advancing front of cells closing the wound, and the closure rate depends on the curvature of the wound according to a phenomenological relation. In the latter model, the capillaries supply the wound with the necessary oxygen and nutrients needed in the healing process. The lack of oxygen at the wound site stimulates the appearance of macrophages at the wound surface which produce macrophage-derived growth factors (MDGFs) that trigger the regeneration of the vascular system.

Our coupling of both models is based on the following hypotheses [6]: (H1) the production of EGFs only takes place if there is enough oxygen to support it, (H2) the excess of oxygen enhances the production of EGFs, and (H3) the equilibrium capillary density is larger under the wound than anywhere else. Consequently, (H1) delays the actual healing and (H2)-(H3) intend to speed up the healing process after the incubation period.

We consider the distribution of the oxygen concentration u_1, the MDGF concentration u_2, the capillary density u_3 and the EGF concentration u_4 over the computational domain Ω, which is two-dimensional (since the thickness of the epidermis is very small compared to the wound dimensions) and Lipschitz. Moreover, Ω consists of the wounded tissue Ω_w, the active layer Ω_{al} surrounding the wound where the EGFs are being produced, and the outer tissue Ω_{ot}. The wound edge is denoted by Γ. From a mass balance argument we obtain the governing equations describing the transport, production and decay of these quantities:

$$\frac{\partial u_1}{\partial t} = D_1 \Delta u_1 + \lambda_{3,1} u_3 - \lambda_{1,1} u_1, \tag{1}$$

$$\frac{\partial u_2}{\partial t} = D_2 \Delta u_2 + \lambda_{1,2} Q(u_1) - \lambda_{2,2} u_2, \tag{2}$$

$$\frac{\partial u_3}{\partial t} = D_3 \Delta u_3 + \lambda_{3,2} u_2 u_3 \left(1 - \frac{u_3}{u_3^{eq}(1+qH(\phi))} \right), \tag{3}$$

$$\frac{\partial u_4}{\partial t} = D_4 \Delta u_4 - \lambda_{4,4} u_4 + Pf(\mathbf{x},t,u_1,p), \tag{4}$$

where D_i denote the diffusion coefficients, $\lambda_{i,j}$ the production/decay rates, u_i^{eq} the values of the undamaged state (which is an equilibrium state), P the production rate of EGFs and H the heaviside function. Moreover, Q denotes the function describing the production of MDGF when the levels of oxygen are low and f stands for the production of EGF inside the active layer, which, respectively, are defined as

$$Q(u_1) = \begin{cases} 1 - \frac{u_1}{u_1^{\theta_2}}, & \text{if } u_1 < u_1^{\theta_2}, \\ 0, & \text{otherwise}, \end{cases} \tag{5}$$

and

$$f(\mathbf{x},t,u_1,p) = \begin{cases} 1 + p\frac{u_1}{u_1^{\theta_4}}, & \text{if } u_1 \geq u_1^{\theta_4} \text{ and } \mathbf{x} \in \Omega_{al}(t), \\ 0, & \text{otherwise}. \end{cases} \tag{6}$$

Equations (1)-(4) are supplemented with homogeneous Neumann boundary conditions and the following initial conditions:

$$u_i(\cdot,0) = \begin{cases} 0, & \text{in } \Omega_w(0), \\ u_i^{und}, & \text{otherwise}, \end{cases} \quad \text{and} \quad u_j(\cdot,0) = 0 \text{ in } \Omega, \tag{7}$$

where $i = 1,3,\ j = 2,4$ and u_i^{und} denote the (equilibrium) undamaged levels. The parameters p and q are used to describe the influence of oxygen in the production of EGFs and to enhance the capillary regeneration inside the wound respectively, whereas $u_1^{\theta_2}$ and $u_1^{\theta_4}$ denote the oxygen concentration below which MDGFs are produced and EGFs are not produced. Finally, the wound will move towards closure only if the concentration of EGF exceeds a certain threshold value $u_4^{\theta_4}$. Hence, normal velocity v_n of the wound edge is given by

$$v_n(\mathbf{x},t) = \big(\alpha + \beta\kappa(\mathbf{x},t)\big) H\big(u_4(\mathbf{x},t) - u_4^{\theta^4}\big) \quad \text{for } \mathbf{x} \in \Gamma(t), \tag{8}$$

where the normal vector $\mathbf{n}$ points into the wound, α, β are non-negative and κ denotes the local curvature.

Note that each time step we need to find the wound edge in order to properly compute the concentrations u_i, i =1, ..., 4. Hence, we are dealing with a moving boundary problem.

3 The Computational Approach

The solution of equations (1)–(4) is computed using a finite element method with piecewise linear basis functions. The time integration of the governing equations is carried out with an Implicit-Explicit Euler method, where only the nonlinear reaction terms are treated explicitly, and Newton-Cotes integration rules are applied in the calculation of the element matrices and vectors. Further, we have to deal with a moving interface and with a sharp (discontinuous) change of the production of the EGF across the wound edge. In order to track the front position in a fashion that allows us handling changes in the wound geometry easily and to obtain a quick identification of the subparts of the computational domain we use the Level Set Method [11]. Furthermore, we apply an adaptive mesh technique in the vicinity of the wound edge. In this way we recover some of the accuracy lost due to the discontinuous production of EGF and have a higher resolution in the region where the motion of the interface is computed. We refer the interested reader to [7] to find the technical details of the algorithm.

3.1 The Level Set Method for Tracking the Wound Edge

In the level set method, the wound edge is defined as the zero level set of a continuous scalar function ϕ:

$$\mathbf{x} \in \Gamma(t) \iff \phi(\mathbf{x},t) = 0, \quad \forall t \geq 0. \tag{9}$$

Furthermore, the so-called level set function ϕ is initialized and subsequently maintained as a signed distance function ($||\nabla\phi|| = 1$), being positive inside the wound and negative outside. To achieve this, we apply the Fast Marching Method [12] to solve the Eikonal equation $||\nabla\phi|| = 1$ each time step. Hence, if ϕ is a distance function, the domain of computation Ω is parameterized as follows:

$$\Omega_w(t) = \{\mathbf{x} \in \Omega \mid 0 < \phi(\mathbf{x},t)\}, \tag{10}$$

$$\Omega_{al}(t) = \{\mathbf{x} \in \Omega \mid -\delta(\mathbf{x},t) < \phi(\mathbf{x},t) < 0\}, \tag{11}$$

$$\Omega_{ot}(t) = \{\mathbf{x} \in \Omega \mid \phi(\mathbf{x},t) < -\delta(\mathbf{x},t)\}, \tag{12}$$

where δ denotes the thickness of the active layer. The motion of the wound edge is then followed by the advection of the level set function:

$$\frac{\partial\phi}{\partial t} + \mathbf{v}\cdot\nabla\phi = 0, \tag{13}$$

where the advection field $\mathbf{v}$ denotes any continuous extension of the front velocity (8). In this work we advect v_n from the interface position in the normal direction [7].

3.2 The Adaptive Mesh Strategy

We choose as a fixed basis mesh a structured triangulation like the one presented in Fig. 1 (left). At each time step, we refine the elements within a certain distance *dist* from the interface, and the elements adjacent to them in order to preserve mesh consistency. Each edge marked to be refined will be subdivided into equally sized sub-edges that will define the new elements, as depicted in Fig. 1 (center). In order to prevent ill-shaped elements, we will limit ourselves to refinement ratios equal to 2 or 3 (*i.e.* each marked edge will be divided into 2 or 3 sub-edges respectively). The refined mesh inherits the structure of the basis mesh, presenting a refined Cartesian band within the refined region and a coarse Cartesian grid outside, see Fig. 1 (right). We take benefit of this structure as finite difference schemes are applied to the hyperbolic equations (such that the velocity extension and the advection and the reinitialization of the level set function) inherited from the level set formulation, and hence avoid the implementation of stabilization techniques in the finite element approximations.

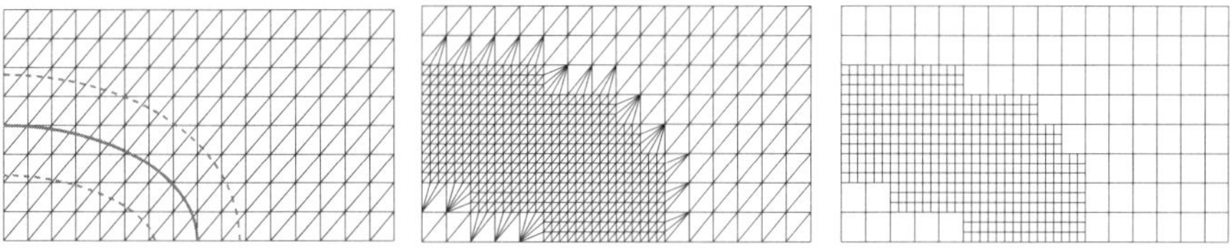

Fig. 1 Left: fixed base FE mesh with the interface position $\phi = 0$ (solid curve) and the contours $\phi = \pm dist$ (dashed curves). The elements within these contours are to be refined. Center: refined FE mesh. Right: the nested Cartesian grids.

4 Numerical Results

4.1 Accuracy and Workload of the Computational Method

The accuracy of the adaptive grid strategy is evaluated here by measuring the relative error of the epidermic growth factor concentration, for the fully uncoupled model, after an incubation period of 45 minutes. The L^2 norm of the relative error in the EGF concentration is presented in the third column of Table 1. The closure of the wound has not been included in this test, and the analytical solution has been obtained for a circular wound after applying separation of variables and expressing the production function f as a series expansion of the eigenfunctions of the homogeneous problem. The numerical results show that, despite the fact that we use linear elements (P_1) in our finite element approximation, the numerical solution is only first order accurate because of the discontinuous production of EGFs. However, it is worth noting that we recover the accuracy for the fixed basis mesh with double number of gridnodes in each dimension if the refinement ratio is set equal to 2 in the coarse grid, and a bit more when it is set equal to 3.

The following columns of Table 1 present the number of elements and the arrangement of the nodal points inside and outside the refined band for several refinement ratios, as well as for the unrefined mesh. The CPU-times employed in the reinitialization of the level set function inside and outside the refined Cartesian band are given in the seventh and eighth columns. The number of elements gives us an estimate of the workload needed to build the system of equations of the discretized problem (*i.e.* updating the right hand side vectors). Since we use the Fast Marching Method, the workload of the reinitialization step is $\mathscr{O}(n \log n)$, where n denotes the number of nodes in the reinitialization region. The numerical results show that the workload sharply increases inside the refined band when the number of nodes increases. Furthermore, the reinitialization of ϕ outside the band cannot be avoided because it is necessary to accurately identify the active layer each time step. Use of a fine fixed basis mesh has been proven to increase the CPU-time per time step. Instead, we suggest to use coarser fixed base grids with higher levels of refinement, since then the workload of the reinitialization steps (inside and outside the refined band) are well balanced and the accuracy is preserved, as can be observed by comparison of the results for N=161 without local refinement and for N=81 with refinement ratio equal to 3.

Table 1 Performance of the computational method. N denotes the number of nodes per Cartesian direction and – indicates that the fixed basis mesh is not refined. The refinement distance *dist* is kept proportional to the mesh width.

N	Ref. ratio	L^2 error	#elements	#nodes inside band	#nodes outside band	Reinitialization inside band	Reinitialization outside band
	-	$3.03 \cdot 10^{-1}$	800	80	361	0.05s	0.27s
21	2	$1.51 \cdot 10^{-1}$	1186	277	361	0.19s	0.26s
	3	$1.31 \cdot 10^{-1}$	1808	592	361	0.45s	0.26s
	-	$1.61 \cdot 10^{-1}$	3200	156	1525	0.11s	1.21s
41	2	$7.55 \cdot 10^{-2}$	3978	549	1525	0.45s	1.19s
	3	$6.09 \cdot 10^{-2}$	5232	1180	1525	1.03s	1.20s
	-	$8.07 \cdot 10^{-2}$	12800	308	6353	0.27s	6.30s
81	2	$3.75 \cdot 10^{-2}$	14362	1093	6253	1.10s	6.28s
	3	$2.87 \cdot 10^{-2}$	16880	2356	6253	2.55s	6.25s
	-	$4.11 \cdot 10^{-2}$	51200	598	25323	0.69s	46.52s
161	2	$2.15 \cdot 10^{-2}$	54294	2157	25323	2.89s	45.75s
	3	$1.52 \cdot 10^{-2}$	59296	4663	25323	6.87s	45.60s

4.2 Healing of an Elliptical Wound

The healing of an elliptical wound is simulated in this section. The solutions u_i ($i = 1,\ldots,4$) after 5% of the wound has healed are plotted in Fig. 2. These plots clearly illustrate the lack of oxygen at the wound site and the resulting high concentration of macrophage-derived growth factors. The plot of the capillary density distinctly shows the role of wound geometry on the healing process. The influence of wound geometry has already been studied in detail for the closure model in [7], and it will be analysed in depth for the coupled model in a future study. Finally, the epidermic growth factor concentration profile shows the location of the active layer. The small wiggles observed in the EGF concentration are believed to be a numerical artifact related to the threshold oxygen level imposed on the production of the EGF.

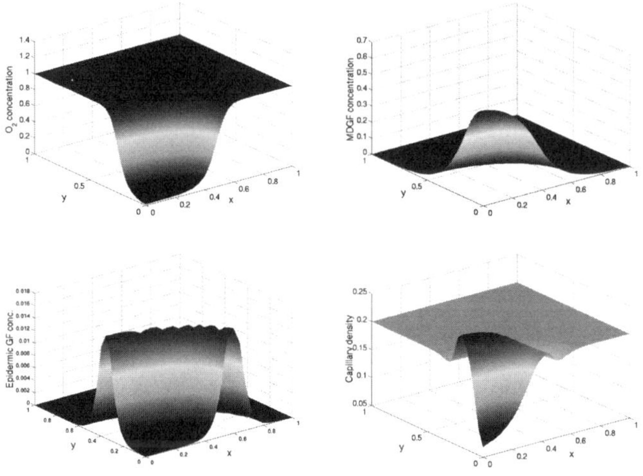

Fig. 2 Oxygen concentration (top left), MDGF concentration (top right), EGF concentration (bottom left) and capillary density (bottom right) for elliptical wound after 5% of it has healed.

4.3 The Revised Critical Size Defect

Arnold and Adam [2] use the EGF model to predict the Critical Size Defect (CSD) of a wound, which is defined as the smallest wound that does not heal during the life time of the animal. Since they do not include the time evolution of the wound, their prediction only allows to elucidate the minimal size of the (circular) wound for which some healing must be expected. However, starting the healing is no guarantee of completing it successfully. Our simulations revealed that there are cases in which the healing process halts prematurely after healing has started. This was observed when the closure rate was excessively slow or the active layer was unable to produce the necessary EGFs.

5 Conclusions

A coupling between wound closure and angiogenesis is presented in this work. The closure of the wound is modelled as a moving interface problem, where the interface is identified with the advancing front of epidermal cells and the Level Set Method is applied to track its position in time. A finite element method is used to solve the governing equations and an adaptive mesh algorithm is implemented because of the discontinuous production of EGFs across the edges of the active layer. The numerical results show that the refinement around the wound edge allows us to recover some of the accuracy lost due aforementioned discontinuity. Furthermore, the width of the fixed basis mesh and the refinement ratio must be chosen in a proper way to bring the workload per time step and the accuracy of the results into balance.

References

1. Anderson, A., Chaplain, M.: Continuous and discrete mathematical models of tumor-induced angiogenesis. Bulletin of Mathematical Biology **60**, 857–900 (1998)
2. Arnold, J.S., Adam, J.A.: A simplified model of wound healing. II. The critical size defect in two dimensions. Math. Comput. Modelling **30**(11-12), 47–60 (1999)
3. Cai, A.Q., Landman, K.A., Hughes, B.D.: Multi-scale modeling of a wound-healing cell migration assay. J. Theoret. Biol. **245**(3), 576–594 (2007)
4. Gaffney, E., Maini, P., Sherratt, J., Tuft, S.: The mathematical modelling of cell kinetics in corneal epithelial wound healing. Journal of Theoretical Biology **197**, 15–40 (1999)
5. Gaffney, E.A., Pugh, K., Maini, P.K., Arnold, F.: Investigating a simple model of cutaneous wound healing angiogenesis. J. Math. Biol. **45**(4), 337–374 (2002)
6. Gordillo, G., Sen, C.: Revisiting the essential role of oxygen in wound healing. The American Journal of Surgery **186**, 259–263 (2003)
7. Javierre, E., Vermolen, F., Vuik, C., Zwaag, S.v.d.: A Mathematical Approach to Epidermal Wound Healing: Model Analysis and Computer Simulations. Tech. Rep. 07-14, Department of Applied Mathematical Analysis, Delft University of Technology (2007)
8. Maggelakis, S.: A mathematical model of tissue replacement during epidermal wound healing. Applied Mathematical Modelling **27**, 189–196 (2003)
9. Murray, J.D.: Mathematical biology. II, *Interdisciplinary Applied Mathematics*, vol. 18, third edn. Springer-Verlag, New York (2003). Spatial models and biomedical applications
10. Olsen, L., Sherrat, J., Maini, P.: A mechanochemical model for adult dermal wound contraction and the permanence of the contracted tissue displacement profile. Journal of Theoretical Biology **177**, 113–128 (1995)
11. Osher, S., Sethian, J.A.: Fronts propagating with curvature-dependent speed: algorithms based on Hamilton-Jacobi formulations. J. Comput. Phys. **79**(1), 12–49 (1988)
12. Sethian, J.A.: Fast marching methods. SIAM Rev. **41**(2), 199–235 (1999)
13. Sherratt, J., Murray, J.: Epidermal wound healing: the clinical implications of a simple mathematical model. Cell Transplantation **1**, 365–371 (1992)
14. Vermolen, F., van Baaren, E., Adam, J.: A simplified model for growth factor induced healing of wounds. Mathematical and Computer Modelling **44**, 887–898 (2006)

Model of Multiple Lexicographical Programming Applied in Cervical Cancer Screening

L. Neamţiu, I. Chiorean, and L. Lupşa

Abstract An important problem for the management of the screening program for cervical cancer is collecting the smears for women who live in remote areas. The issue is to plan the days in which the mobile unit will be used, and its route, such that the total cost and the testing time for all the eligible women are minimal. This paper presents a mathematical model for the Health-Economic problem and a Bellman type theorem for solving this model.

1 The Health-Economics Problem and a Mathematical Model

Screening, in Medicine, is a strategy used to identify diseases in an unsuspecting population. An important problem of the management of the screening program for cervical cancer is to take the smears for women from remote areas. For these women, a mobile unit equipped as a gynecological office is used. The unit goes in every village and the doctor takes the smears from eligible women who have been informed and invited. The unit also transports the smears to the cytological laboratory. The smears are processed and the laboratory provides the results within a maximum of given days (laboratory response time, usually equal to 21 days). From the Health Economics point of view, the problem is to plan the days in which the

Luciana Neamţiu
Oncological Institute "Prof. dr. I. Chiricuţă" of Cluj-Napoca, str. Republicii nr. 34-36, 400015 Cluj-Napoca, Romania, e-mail: luciana@iocn.ro

Ioana Chiorean
Babeş-Bolyai University, Faculty of Mathematics and Computer Science, str. Kogălniceanu nr. 1, 400084 Cluj-Napoca, Romania, e-mail: ioana@cs.ubbcluj.ro

Liana Lupşa
Babeş-Bolyai University, Faculty of Mathematics and Computer Science, str. Kogălniceanu nr. 1, 400084 Cluj-Napoca, Romania, e-mail: llupsa@math.ubbcluj.ro

mobile unit will be used and its route such as the total cost and the testing time for all the eligible women to be, both of them, minimum.

Let m be the number of the villages. For every village $i \in \{1,...m\}$, we denote by n_i, the eligible number of women from the village i which will be tested. We know: the mean time interval for taking a smear, t_r; the maximum time that the mobile unit works every day, t_z; the total number of slides which can be read by laboratory in a day, z_l; the total number of resting slides that should be read by the laboratory besides the smears obtained by the mobile unit, z_l^0; the cost/day for the driver of the mobile unit, c_s; the cost/day for the medical doctor which is on the mobile unit, c_m; overhead/day for mobile unit, c_u; cost of fuel/ km, c_b; mean speed for the mobile unit, v; the laboratory response time, l_r; and the maximum number of the days when all the tests should be done, n_z.

The laboratory receives the slides in the evening of every day and, consequently has to give the answer in $l_r - 1$ days.

We assume that all routes to reach the villages and return to the base O are known. Let p be the number of these routes. For every route j, $j \in \{1,...,p\}$, the length, d_j, of the way and the villages for the mobile unit to pass through are known. In order to identify the affiliation of one village to one route, we introduce the following p vectors $\lambda^j = (\lambda_1^j,...,\lambda_m^j) \in \mathbf{R}^m$, $j \in \{1,...,p\}$, where $\lambda_k^j = 1$, if the route j passes through the village k, and $\lambda_k^j = 0$, if the route j does not pass through the village k.

The problem is to plan the days in which the mobile unit will be used and its route such as the total cost are minimal and, if we have several possibilities, to choose one for which the testing time for all the eligible women is also minimum.

We notice that in the literature, there is not such an approach for this problem. That's why we consider this problem as a dynamic system with finite horizon and vectorial total utility function. The mathematical model permits to obtain an algorithm which solves our problem. The number of steps of the dynamic system is chosen n_z (the maximum days when all the tests should be done). A step corresponds to a day. In each step $h \in \{1,...,n_z\}$, the stage of the system will be described by the vector of state variable $s^h \in \mathbf{N}^{m+1}$: the first component, s_1^h, gives the number of slides existing in the laboratory at the end of day h (this number is equal to the number of slides existing in the evening of the day $h-1$ minus the number of the slides which have been read in day h, plus the number of slides which have been taken in the day h); the following m components, s_i^h, $i \in \{2,...,m+1\}$, contain the number of untested women at the end of day h in the village i, respectively.

Considering that the numbers z_l, z_l^0, n_i, $i \in \{1,...,m\}$, n_z are known, the initial state of the system is described by the vector

$$s^0 = (z_l^0, n_1, ..., n_m). \tag{1}$$

Because s_1^h is the number of existing slides at the end of the day h, the maximum number of the slides which may be taken by laboratory at the end of the day $h+1$ is equal to $\max\{0, (l_r - 1) \cdot z_l - s_1^h\}$. Therefore, in the first day it may be taken only $\max\{0, (l_r - 1) \cdot z_l - z_l^0\}$ slides. In each step $h \in \{1,...,n_z\}$, the decision will be described by the decision vector $x^h \in \mathbf{N}^{m+1}$: the first component, x_1^h, indicates the number of the route done in step h (if this number is 0, in that step no movement exists); the following m components, x_i^h, $i \in \{2,...,m+1\}$, contain the number of women tested in the day h, in village i, respectively. d_j/v is the time necessary to go through the route j. Therefore the decisions set in the stage $h \in \{1,...,n_z\}$, if the system is in the state s^{h-1}, is the set $X_h(s^{h-1})$,

$$X_h(s^{h-1}) = \{0,1,...,p\} \times \tilde{X}_h, \tag{2}$$

where $\tilde{X}_h$ is the set of the solutions of the discrete system

$$\begin{cases} \sum_{k=2}^{m} x_k^h \leq \max\{0, (l_r - 1)\cdot z_l - s_1^{h-1}\} \\ t_r \cdot \sum_{k=2}^{m+1} \lambda_{k-1}^{x_1^h} \cdot x_k^h \leq t_z - \frac{d_{x_1^h}}{v} \\ x_k^h \leq \lambda_{k-1}^{x_1^h} \cdot s_k^{h-1} \cdot sgn(s_1^{h-1}),\ \forall k \in \{2,...,m+1\} \\ x_k^h \in \mathbf{N}, \forall k \in \{1,...,m+1\}. \end{cases} \tag{3}$$

The first inequality indicates that the number of slides taken in day h can not be greater than the number of slides which can be given to the laboratory in the evening. In the second inequality, the l.h.s. term gives the time necessary to take the slides and the r.h.s term gives the available time in a day minus the time spent on the route. In the third inequality, the l.h.s term gives the number of slides planed to be taken from village k in day h, which can not be greater then $\max\{0$, the number of slides remained to be taken in village $k\}$. The relation four indicates that the number of slides has to be a natural number.

The function $f_C : \{1,...,n_z\} \to \mathbf{R}$ describes the cost for each day. Thus the cost of day h is

$$f_C(x^h) = sgnx_1^h \cdot (c_u + c_s + c_m + c_b \cdot d_{x_1^h}), \tag{4}$$

where sgn denotes the function given by $sgnx = 0$, if $x = 0$, $sgnx = 1$, if $x > 0$ and $sgnx = -1$, if $x < 0$.
We remark that the cost is 0, if no movement is done; else it is equal with the sum of the costs.

The function $f_T : \{1,...,n_z\} \to \mathbf{R}$ indicates if in the day h, smears have been taken. Thus

$$f_T(h) = sgnx_1^h. \tag{5}$$

For all $h \in \{1,...,n_z\}$, the dynamic equations are

$$s_1^h = max\{0, s_1^{h-1} - z_l\} + \sum_{i=2}^{m+1} x_i^h, \quad s_i^h = s_i^{h-1} - x_i^h, \forall\, i \in \{2,...,m+1\}. \tag{6}$$

For all $h \in \{1,...,n_z\}$, the stationary equations are

$$s^h \in S_h = \{0,1,...,(l_r-1)\cdot z_l\} \times \{0,1,...,n_1\} \times ... \times \{0,1,...,n_m\}, \tag{7}$$

and

$$x^h \in \{0,1,...,p\} \times \tilde{X}^h. \tag{8}$$

The total utility function is additive, having the value equal to the sum of the values of partial utility effect functions. By denoting this function with F, $F=(F_1,F_2):\{0,1,...,n_z\} \to \mathbf{R}^2$, we have $F(0)=(0,0)$ and $F(h)=F(h-1)+(f_C(h), f_T(h))^T, \forall\, h \in \{1,2,...,n_z\}$.

From practical point of view, our purpose is to obtain a plan of taking the smears such that the function F_1 to be minimum and, if we have possibilities to choose which one assures the minimum for F_2, too. Therefore we obtain the following type of dynamic programming problem:

$$\begin{cases} \left(\sum_{h=1}^{n_z} f_T(x^h),\ \sum_{h=1}^{n_z} f_C(x^h)\right) \to lex-min \\ s_1^h = max\{0, s_1^{h-1} - z_l\} + \sum\limits_{i=2}^{m+1} x_i^h, \forall h \in \{1,...,n_z\}, \\ s_i^h = s_i^{h-1} - x_i^h, \forall\, i \in \{2,...,m+1\}, \forall h \in \{1,...,n_z\}, \\ s^0 \text{ given}, \quad s^h \in S_h, \quad \text{and} \quad x^h \in X_h(s^{h-1}), \ \forall h \in \{1,...,n_z\}, \end{cases} \tag{9}$$

where S_h is given by by (7) and $X_h(s^{h-1})$ by (2) and (3).

By analogy with the definition of lexicographic optimality used in the general context of vectorial programming problem (see [2]) we call this type of problem as *lexicographic dynamic programming problem.*

Remark 1. The subject of dynamic programming problem, when the total utility function is a vectorial function, is discussed in [4]. In [5] fundamental dynamic programming recursive equations are extended to the multi-criteria framework. In that paper, a more detailed procedure for a general recursive solution scheme for the multi-criteria discrete mathematical programming problem is developed. A short note about multi-criteria dynamic programming problem is given in [6]. Recently, multi-criteria dynamic programming is extended for solving variously practical problem. This implies some sort of generalization of Belman's theorem . In [3], an application in Pharmacoeconomics is given. In our paper, we show how the problem (9) can be solved using dynamic programming. But firstly we have to give a generalization of Belman's theorem.

2 Belman's Theorem for Lexicographical Dynamic Programming

Let be a discrete finite stages decision problem, with n stages, with the static equations

$$\begin{cases} s^0 \text{ given}, \\ s^h \in S_h, \ h \in \{1,...,n\}, \\ x^h \in X_h(s^{h-1}), \ h \in H = \{1,...,n\}, \end{cases} \tag{10}$$

and the dynamic equations

$$s^h = g_h(s^{h-1}, x^h), \ h \in H, \tag{11}$$

with s^0 the initial state of the system. S_h denotes the set of the states of system in the stage h and $X_h(s^{h-1})$ denotes the set of the decisions which may be taken in the stage h, if the system is in the state s^{h-1}.

A sequence $(x^1,...,x^n)$, where $x^h \in X_j(s^{h-1})$, for every $h \in \{1,...,n\}$, is called a policy of the system. The set of all the policies of the system will be denoted by *Pol*. In each stage $h \in H$, if we take the decision $x^h \in X_h(s^{h-1})$, the obtained utility is denoted by $f_h(s^{h-1},x^h)$. It is a vector in $\mathbf{R}^p$, where $p \in \mathbf{N}$, $p \geq 1$. The total utility is given by the function $F = (F_1,...,F_p) : Pol \to \mathbf{R}^p$.

Analogously to the classical dynamic programming, for the discrete finite dynamic system with n stages, having the static equation (10) and dynamic equation (11), we build the sets

$$\hat{S}_n := S_n, \quad \hat{S}_{h-1} = \{s \in S_{h-1} \,|\, \exists x \in X_h(s) \text{ such that } g_h(s,x) \in \hat{S}_h\}, \tag{12}$$

for $h = n$, $h = n-1,...,h = 1$. Again, for $h = n$, $h = n-1,...,h = 1$ and for each $s \in \hat{S}_{h-1}$ we build the set

$$\hat{X}_h(s) = \{x \in X_h(s) \,|\, g_h(s,x) \in \hat{S}_h\}. \tag{13}$$

Using the new notations, the lex-min dynamic problem can be rewritten as:

$$(\text{DLP}) \quad \begin{cases} F(x^1,...,x^n) \to lex-\min \\ s^k = g_k(s^{k-1},x^k), \ k \in \{1,...,n\}, \\ s^0 \text{ given}, \\ s^k \in \hat{S}_k, \ k \in \{1,...,n\}, \\ x^k \in \hat{X}_k(s^{k-1}), \ k \in \{1,...,n\}. \end{cases} \tag{14}$$

If $p = 1$, a policy $x \in Pol$ is called optimal, if there is no other policy $y \in Pol$ such that $F(y) < F(x)$. For $p = 1$, an optimal policy can be find using classical Bellman's theorem. For every $h \in \{1,...,n\}$, let's consider the problem

$$(\text{DLPM}_h) \quad \begin{cases} F_h(s^{h-1},x^h,x^{h+1},...,x^n) \to \min \\ s^k = g_k(s^{k-1},x^k), \ k \in \{h,...,n\}, \\ s^{h-1} \text{ given}, \\ s^k \in \hat{S}_k, \ k \in \{h,...,n\}, \\ x^k \in \hat{X}_k(s^{k-1}), \ k \in \{h,...,n\}, \end{cases} \tag{15}$$

where F_h denotes the total utility function if the process begins only at the stage h, the system being in the state s^{h-1}. For all $h \in \{1,...,n-1\}$, let us denote by $Pol_h(s^{h-1})$ the set of the policies of the above problems.

Theorem 1. *(Bellman's theorem [1]). A policy $x = (x^{h-1}, x^h, ..., x^n) \in Pol_{h-1}$ is an optimal policy of the problem $(DLPM_{h-1})$ only if $(x^h, ..., x^n)$ is an optimal policy of the problem $(DLPM_h)$.*

For our problem the classical Bellman's theorem does not work because our function is a vectorial one and not a scalar function. Therefore we have to give a generalization of it.

We say that a policy $x \in Pol$ is *lexicographically minimal* if there is no $y \in Pol$ such that $F(y) <_{lex} F(x)$, where $<_{lex}$ denotes the lexicographical ordering.

We remember that if $u = (u_1, ..., u_p)$ and $v = (v_1, ..., v_p)$ are two points in $\mathbf{R}^p$, then we set:

$$u <_{lex} v, \quad \text{if} \quad \begin{array}{l} \text{there is } i \in \{1,...,p\} \text{ such that } u_i < v_i \text{ and,} \\ \text{if } i > 1, \text{ then } u_j = v_j,\ \forall\, j \in \{1,...,i-1\}. \end{array} \tag{16}$$

We call *lex-min dynamic problem*, the problem of determining a lexicographically minimal policy.

Definition 1. The total utility function is said to be lexicographic prospective increasing separable if there are $n-1$ vectorial functions $\alpha_i : \mathbf{R}^p \times \mathbf{R}^p \to \mathbf{R}^p$, $i \in \{1,...,n-1\}$, such that

$$\begin{aligned} &F(x^1, ..., x^n) \\ &\quad = \alpha_1(f_1(s^0, x^1), \alpha_2(f_2(s^1, x^2), \alpha_3(...\alpha_{n-2}(f_{n-2}(\\ &\qquad s^{n-1}, x^{n-2}), \alpha_{n-1}(f_{n-1}(s^{n-2}, x^{n-1}), f_n(s^{n-1}, x^n)))...))), \end{aligned} \tag{17}$$

for all $(x^1, ..., x^n) \in Pol$, and if for all $i \in \{1,...,n-1\}$, the function α_i is lexicographic increasing in the second argument:

$$\alpha_i(u, v) <_{lex} \alpha_i(u, v'), \text{ for all } (u, v), (u, v') \in \mathbf{R}^p \times \mathbf{R}^p \text{ with } v \leq v'. \tag{18}$$

It is easy to see that if

$$F(x) = \sum_{j=1}^{n} f_h(s^{h-1}, x^h), \text{ for all } x \in Pol, \tag{19}$$

then F is lexicographic prospective increasing separable.

For every $h \in \{1,...,n\}$, by $\varphi_h : (\mathbf{R}^m \times \mathbf{R}^q)^{n+1-h} \to \mathbf{R}^p$ we denote a continuously function which satisfied the condition:

i) if $h \in \{1,...,n-1\}$, then

$$\begin{aligned} &\varphi_h(s^{h-1}, x^h, ..., s^{n-1}, x^n) \\ &= \alpha_h(f_h(s^{h-1}, x^h),\ \alpha_{h+1}(f_{h+1}, (...(\alpha_{n-1}(f_{n-1}(s^{n-2}, x^{n-1}),\ f_n(s^{n-1}, x^n)))...))), \end{aligned} \tag{20}$$

for all $(s^{h-1},x^h,...,s^{n-1},x^n) \in \hat{S}_{h-1} \times \hat{X}_h(s^{h-1}) \times ... \times \hat{S}_{n-1} \times \hat{X}_n(s^{n-1})$;
ii) if $h = n$, then

$$\varphi_n(s^{n-1},x^n) = f_n(s^{n-1},x^n), \text{ for all } (s^{n-1},x^n) \in \hat{S}_{n-1} \times \hat{X}_n(s^{n-1}). \quad (21)$$

We remark that

$$\varphi_h(s^{h-1},x^h,...,s^{n-1},x^n) = \alpha_h(f_h(s^{h-1},x^h), \varphi_{h+1}(s^h,x^{h+1},...,s^{n-1},x^n)), \quad (22)$$

for all $(s^{h-1},x^h,...,s^{n-1},x^n) \in \hat{S}_{h-1} \times \hat{X}_h(s^{h-1}) \times ... \times \hat{S}_{n-1} \times \hat{X}_n(s^{n-1})$. Also, for every $h \in \{1,...,n\}$, we consider the problem

$$\begin{cases} \varphi_h(s^{h-1},x^h,s^h,x^{h+1},...,s^{n-1},x^n) \rightarrow lex - \min \\ s^k = g_k(s^{k-1},x^k), \ k \in \{h,...,n\}, \\ s^{h-1} \text{ given}, \\ s^k \in \hat{S}_k, \ k \in \{h,...,n\}, \\ x^k \in \hat{X}_h(s^{k-1}), \ k \in \{h,...,n\}. \end{cases} \quad (23)$$

This problem could be rewritten as

$$(\text{DLPM}_h) \quad \begin{cases} \alpha_h(f_h(s^{h-1},x^h), \varphi_{h+1}(s^h,x^{h+1},...,s^{n-1},x^n)) \rightarrow lex - \min \\ s^k = g_k(s^{k-1},x^k), \ k \in \{h,...,n\}, \\ s^{h-1} \text{ given}, \\ s^k \in \hat{S}_k, \ k \in \{h,...,n\}, \\ x^k \in \hat{X}_h(s^{k-1}), \ k \in \{h,...,n\}. \end{cases} \quad (24)$$

For all $h \in \{1,...,n-1\}$, let us denote by $Pol_h(s^{h-1})$ the set of the policies of (24).

Theorem 2. *If the total utility function F is lexicographic prospective increasing separable, then the policy $(x^{h-1},x^h,...,x^n) \in Pol_{h-1}$ is a lexicographically minimal policy of the problem $(DLPM_{h-1})$ only if $(x^h,...,x^n)$ is a lexicographically minimal policy of the problem $(DLPM_h)$.*

Proof. Let $(x^{h-1},x^h,...,x^n) \in Pol_{h-1}$ be a lexicographically minimal policy of the problem (DLPM_{h-1}). If we suppose that $(x^h,...,x^n)$ is not a lexicographically minimal policy of the problem (DLPM_h), then there is $(y^h,...,y^n) \in Pol_h$ such that

$$\varphi_h(s^{h-1},y^h,...,s^{n-1}y^n) <_{lex} \varphi_h(s^{h-1},x^h,...,s^{n-1}x^n). \quad (25)$$

As $(y^h,...,y^n) \in Pol_h$ and $(x^{h-1},x^h,...,x^n) \in Pol_{h-1}$, obviously we have

$$(x^{h-1},y^h,...,y^n) \in Pol_{h-1}.$$

The monotony of the function α_{h-1} implies

$$\begin{aligned} &\alpha_{h-1}(f_{h-1}(s^{h-2},x^{h-1}), \varphi_h(s_{h-1},y^h,...,s_{n-1},y^n) <_{lex} \\ &\alpha_{h-1}(f_{h-1}(s^{h-2},x^{h-1}, \varphi_h(s_{h-1},x^h,...,s_{n-1},x^n). \end{aligned} \quad (26)$$

This contradicts the hypotheses that $(x^{h-1}, x^h, \dots, x^n) \in Pol_{h-1}$ is a lexicographically minimal policy of the problem (DLPM$_{h-1}$). □

3 Practical Approach and Conclusions

Let s_i be the number of the routes which connect the base O with a village $i \in \{1,\dots,m\}$, and d^i_j, $j \in \{1,\dots,m\}$, their lengths. If $\min\{d^i_j/v \,|\, j \in \{1,\dots,s_i\}\} + t_r \leq t_z$, then the medical problem has no solution because the time t_z is not enough for the mobile unit to go to village i, to take at least one smears and to come back. In the following we consider that $\min\{d^i_j/v \,|\, j \in \{1,\dots,s_i\}\} + t_r > t_z$, is true for all $i \in \{1,\dots,m\}$.

In the same way that a classical dynamic programming problem can be solved using Bellman's theorem, it is possible to solve the problem (9) using Theorem 2. First we take $G_{n_{z+1}}$ equal to the null function and $\hat{S}_n = \{(s_1^{n_z}, 0, \dots, 0) \,|\, s_1^{n_z} \in \{0, 1, \dots, (l_r - 1) \cdot z_l - z_l^0\}\}$. Then, setting $k = n_z$, $k = n_z - 1, \dots, k = 1$, we solve, for each $s^{k-1} \in \hat{S}_{k-1}$, the problem

$$(P_k)\ \{F_k(s^{k-1}, x^k) + G_{k+1}(g_k(s^{k-1}, x^k)) | x^k \in \hat{X}_k(s^{k-1})\} \to \text{lex-min}, \tag{27}$$

where $F_k(s^{k-1}, x^k) = (\Sigma_{h=k}^{n_z} f_T(x^h),\ \Sigma_{h=k}^{n_z} f_C(x^h))$,

$g_k(s^{k-1}, x^k) = (\max\{0, s_1^{h-1} - z_l\} + \sum_{i=2}^{m+1} x_i^h,\ s_2^{h-1} - x_2^h, \dots, s_{m+1}^{h-1} - x_{m+1}^h)$, and

$G_k(s^{k-1}) = \text{lex-min}\{F_k(s^{k-1}, x^k) + G_{k+1}(g_k(s^{k-1}, x^k)) | x^k \in \hat{X}_k(s^{k-1})\}$. An optimal policy of the problem (9) is $(\hat{u}^1(x^0), \dots, \hat{u}^{n_z}(x^{n_z - 1}))$, where $\hat{u}^k(x^{k-1})$ denotes a lex-min solution of (P_k), $k \in \{1, \dots, k_{n_z}\}$.

Acknowledgements All this investigation comes in connection to the Research Agreement, CEEX No. 125/2006, CanScreen, supported by the Romanian Ministry of Education and Research.

References

1. Bellman, R.: Dynamic Programming. Princeton University Press, Prinseton, New Jersey (1957)
2. Ehrgott, M.: Multicriteria Optimization. Springer, Berlin-Heidelberg (2005)
3. Lupşa, L.: Use of dynamic programming for the supply of a pharmacy. In: E. Popoviciu (ed.) Proc. of the Tiberiu Popoviciu Intinerant Seminar of Functional Equations, Approximation and Convexity, pp. 237–239. SRIMA Publishing House, Cluj-Napoca (2001)
4. Mitten, L.G.: Synthesis of optimal multistage processes. J. Optim. Theory Appl. **12**, 610–619 (1964)
5. Villarreal, B., Karwan, M.H.: Quasimonotonicity, regularity and duality for nonlinear systems of partial differential equations. J. Optim. Theory Appl. **38**(1), 43–69 (1982)
6. Yu, P.L.: Multiple-Criteria Decision Making. Concepts, Techniques, and Extensions. Plenum Press, New York and London (1985)

A Finite Element Model for Bone Ingrowth into a Prosthesis

F.J. Vermolen, E.M. van Aken, J.C. van der Linden, and A. Andreykiv

Abstract We consider a finite element method for a model of bone ingrowth into a prosthesis. Such a model can be used as a tool for a surgeon to investigate the bone ingrowth kinetics when positioning a prosthesis. The overall model consists of two coupled models: the biological part that consists of non-linear diffusion-reaction equations for the various cell densities and the mechanical part that contains the equations for poro-elasticity. The two models are coupled and in this paper the model is presented with some preliminary academic results. The model is used to carry out a parameter sensitivity analysis of ingrowth kinetics with respect to the parameters involved.

1 Introduction

In osteoporosis, fracture risk is high, after a hip fracture a joint that replaces the prosthesis is often the only remedy. In the case of osteoarthritis and rheumatoid arthritis, the cartilage degrades and moving the joints becomes painfull. Ultimately, most patients will receive a prosthesis to restore the function of a diseased joint. Prostheses, which are fixed in the bone by bone ingrowth in a porous layer are usually put in the bone using a screw, to obtain sufficient initial stability. Bone will grow into a porous tantalum layer in the course of time, and hence more stability of the prosthesis is obtained. To investigate the quality and life time of such an artificial joint, one needs to study the effects of the placement of the prosthesis and of the

F.J. Vermolen and E.M. van Aken
Delft Institute of Applied Mathematics, Faculty of Electrical Engineering, Mathematics and Computer Science, Delft University of Technology, Mekelweg 4, 2628 CD Delft, The Netherlands
e-mail: F.J.Vermolen@tudelft.nl,

J.C. van der Linden and A. Andreykiv
Mechanical, Maritime and Materials Engineering, Delft University of Technology, Mekelweg 2, 2628 CD Delft, The Netherlands

materials that are involved in the joint. At present, these effects are often studied using large amounts of data of patients. To predict the life span and performance of artificial joints, numerical simulations are necessary since these simulations give many qualitative insights by means of parameter sensitivity analysis. These insights are hard to obtain by experiments.

Several studies have been done to simulate bone-ingrowth or fracture healing of bones. To list a few of them, we mention the model due to Adam [1], Ament and Hofer [3], Bailon-Plaza *et al.* [5], Huiskes *et al.* [9] and recently by Andreykiv [4]. The model due to Huiskes *et al.* and LaCroix *et al.* [9, 11] will be treated in more detail, since we expect that this model contains most of the biologically relevant processes, such as cell division and differentiation, tissue regeneration, and cell mobilility. Many ideas from modeling fracture healing of bones are used in these models, since bone-ingrowth into a prosthesis resembles the fracture healing process. In the model due to Huiskes, the influence of the mechanical properties on the biological processes are incorporated. Further, we note that Huiskes' model has been compared to animal experiments.

In this paper, we will see a calibrated existing bone ingrowth model (and its numerical solution) in terms of a system of nonlinearly coupled equations from diffusion, reactions and poro-elasticity. This paper concerns a compilation of preliminary results, with some data for a shoulder prosthesis.

2 The Model

Huiskes [9] considers the behavior of mesenchymal cells, that originate from the bone marrow and differentiate into fibroblasts, chondrocytes and osteoblasts. These newly created cell types respectively generate fibrous tissue, cartilage and bone. In Huiskes' model, it is assumed that fibroblasts may differentiate into chondrocytes, chondrocytes may differentiate into osteoblasts. The differentiation processes are assumed to be nonreversible. The differentiation pattern has been sketched in Figure 1. The accumulation at a certain location of all the cell types is determined by

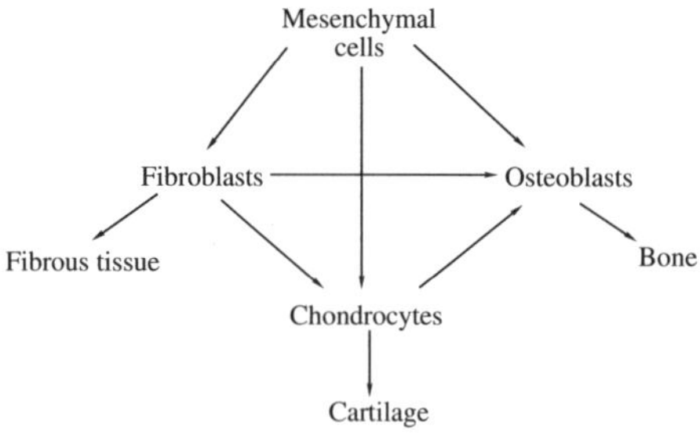

Fig. 1 The scheme of cell differentiation of mesenchymal cells, fibroblasts, chondrocytes and osteoblasts.

cell mobility, cell division and cell differentiation. Let c_m, c_c, c_f and c_b respectively denote the cell density of the mesenchymal cells, chondrocytes, fibroblasts and osteoblasts, in the poro-elastic tantalum of the prosthesis in which bone ingrowth takes place, then, the dynamics of the cell densities are described by

$$\begin{aligned} \frac{\partial c_m}{\partial t} &= \text{div } D_m \text{ grad } c_m + P_m(1-c_{\text{tot}})c_m + \\ &\quad -F_f(1-c_f)c_m - F_c(1-c_c)c_m - F_b(1-c_b)c_m, \\ \frac{\partial c_f}{\partial t} &= \text{div } D_f \text{ grad } c_f + P_f(1-c_{\text{tot}})c_f + \\ &\quad -F_f(1-c_f)c_m - F_c(1-c_c)c_f - F_b(1-c_b)c_f, \end{aligned} \tag{1}$$

where the diffusivities, D_m and D_f, of the mobile cells are determined by the volume fractions of tissues, being denoted by m_c and m_b for cartilage and bone respectively, by

$$\begin{aligned} D_i &= D_i^0(1-m_c-m_b), \\ P_i &= P_i^0(1-m_c-m_b), \end{aligned} \quad i \in \{m, f\}. \tag{2}$$

The chondrocytes and osteoblasts, respectively producing cartilage and bone, are assumed to be immobile. Their reaction processes are modeled by

$$\begin{aligned} \frac{\partial c_c}{\partial t} &= P_c(1-c_{\text{tot}})c_c + F_c(1-c_c)(c_m+c_f) - F_b(1-c_b)c_c, \\ \frac{\partial c_b}{\partial t} &= P_b(1-c_{\text{tot}})c_b + F_b(1-c_b)(c_m+c_f+c_c). \end{aligned} \tag{3}$$

The tissues, fibrous tissue, cartilage and bone are immobile. Let the volume fraction of fibrous tissue be denoted by m_f, then the accumulation of these tissues is modeled by

$$\begin{aligned} \frac{\partial m_f}{\partial t} &= Q_f(1-m_{\text{tot}})c_f - (D_b c_b + D_c c_c)m_f m_{\text{tot}}, \\ \frac{\partial m_c}{\partial t} &= Q_c(1-m_b-m_c)c_c - D_b c_b m_c m_{\text{tot}}, \\ \frac{\partial m_b}{\partial t} &= Q_b(1-m_b)c_b. \end{aligned} \tag{4}$$

The initial concentrations of all tissues and cell types are zero. As boundary conditions, a Dirichlet condition for the mesemchymal cell density at the bone implant and homogeneous Neumann conditions at all other boundaries are applied. In the present paper, the influence of the micromotions is neglected. For the fibroblasts homogeneous Neumann boundary conditions are imposed for all boundary segments. The proliferation, differentiation and diffusion parameters depend on the mechanical stimulus. The mechanical stimulus is given by a linear combination of the maximum shear strain and the fluid velocity relative to the rate of displacement of the solid, that is

$$S = \frac{\gamma}{a} + \frac{v}{\beta}, \tag{5}$$

where γ represents the maximum shear strain and v denotes the relative fluid/solid velocity. Here $\gamma := \frac{1}{2}(\lambda_1 - \lambda_2)$, where $\lambda_{1,2}$ represent the eigenvalues of the strain tensor. The rates of tissue regeneration and differentiation qualitatively depends on the mechanical parameters such that:

- Low strain has a stimulatory effect (in relation to no strain) on the fibroblast proliferation and bone regeneration (if $0 < S < 1$);
- For intermediate values of the strain, cartilage formation is more favorable (if $1 < S < 3$);
- High strains favor the proliferation of fibrous tissue (if $S > 3$).

This gives a coupling of the poro-elasticity model to this biological model. The above set of partial differential equations poses a nonlinearly coupled set of equations. Standard Galerkin Finite Element methods provide a straightforward method to obtain solutions. To get the local strains and stresses in the porous tantalum that are required for the differentiation and mobility characteristics, the equations for poro-elasticity are solved. The model was derived by Biot originally. We will give an explanation for two-dimensional domains. In the poro-elastic domain where $\mathbf{u} = [u\ v]^T$ denote the displacement in the x- and $y-$ direction, we have:

$$
\begin{aligned}
&-\text{div}\,(\mu \text{ grad } u) - \frac{\partial}{\partial x}((\lambda+\mu)\,\text{div}\,\mathbf{u}) + \frac{\partial p}{\partial x} = 0,\\
&-\text{div}\,(\mu \text{ grad } v) - \frac{\partial}{\partial y}((\lambda+\mu)\,\text{div}\,\mathbf{u}) + \frac{\partial p}{\partial y} = 0,\\
&\frac{\partial}{\partial t}(n_f\beta_f p + \text{div}\,\mathbf{u}) - \text{div}\left(\frac{\kappa}{\eta}\text{ grad } p\right) = 0.
\end{aligned}
\tag{6}
$$

Here κ denotes the permeability, η the viscosity, n_f the porosity and finally β_f represents the compressibility. Furthermore, μ and λ are the Lamé parameters that originate from the stiffness and Poisson's ratio of the material. These parameters have to be updated as bone grows into the prosthesis. The Rule of Mixtures is applied to update the mechanical properties (see Lacroix & Prendergast [11]). For more information on the derivation of the above equations, we refer to Bear [6].

Next, we consider a scaled version of equations (6), in which we draw our attention to the third equation. In this scaling argument, we assume that the coefficients in the equations (6) are constant in time and space. Division of this equation by $n_f\beta_f$ (under the assumption that n_f and β_f are constant), and using the dimensionless variables $X,Y := \frac{x,y}{L}$, $\tau := \frac{\kappa}{\eta\beta_f n_f}\frac{t}{L^2}$, and $U,V := \frac{u,v}{L}$, where L is a characteristic length. Then equations (6) change into

$$
\begin{aligned}
&-\overline{\nabla}\cdot(\mu\,\overline{\nabla}\,U) - \frac{\partial}{\partial X}((\lambda+\mu)\,\overline{\nabla}\cdot\mathbf{U}) + \frac{\partial p}{\partial X} = 0,\\
&-\overline{\nabla}\cdot(\mu\,\overline{\nabla}\,V) - \frac{\partial}{\partial Y}((\lambda+\mu)\,\overline{\nabla}\cdot\mathbf{U}) + \frac{\partial p}{\partial Y} = 0,\\
&\frac{\partial}{\partial \tau}(\overline{\nabla}\cdot\mathbf{U}) = n_f\beta_f\left(\overline{\Delta}p - \frac{\partial p}{\partial \tau}\right).
\end{aligned}
\tag{7}
$$

where $\overline{\nabla}(.) := \frac{1}{L}\nabla(.)$, $\overline{\Delta}(.) := \frac{1}{L^2}\Delta(.)$ and $\mathbf{U} := \frac{1}{L}\mathbf{u}$. We see that as $n_f\beta_f \to 0$, then, we reach the incompressible limit, which gives a saddle-point problem where one has to consider LBB condition satisfying elements or a stabilization. The situation becomes analogous to the Stokes' equations.

3 The Method

For a rather recent comprehensive overview of Finite Element methods applied to solid state mechanics, we refer to the book due to Bræss [7]. The above poro-elasticity equations are often solved using non-conforming Finite element methods, such as the Taylor-Hood family: if the pressure is approximated with elements of polynomials of P_n, then, the displacements are approximated using polynomials of P_{n+1}. In the Taylor-Hood elements, one usually uses linear and quadratic basis functions for the pressure and displacements respectively. On the other hand, Crouzeix-Raviart elements, which are often used for Stokes flow problems, are based on a discontuity of the pressure. Since $p \in H^1(\Omega) \subset C(\Omega)$, the Crouzeix-Raviart elements are not suitable here. As long as the compressibility is sufficiently large, one can also make use of linear-linear elements for the pressure and displacement. This was done successfully in the study due to Andreykiv [4]. If $\beta_f = 0$, which is the incompressible case, then the issue of oscillations and the use of appropriate elements or a stabilization becomes more important. For $\beta_f = 0$, the third equation in equation (6) reduces to the version that is solved by Aguilar *et al.* [2].

A Galerkin formulation of the above equation with

$$p = \sum_{j=1}^{m} p_j \psi_j(x,y) \text{ and } \mathbf{u} = \sum_{j=1}^{n} \mathbf{u}_j \phi_j(x,y),$$

is applied to equations (6). For consistency, we require $m \leq 2n$ as $n_f\beta_f \to 0$. This case resembles the classical Stokes' equations. For the classical Taylor-Hood elements, we use $\psi_i \in P_1(\Omega)$ and $\phi_i \in P_2(\Omega)$. Aguilar *et al.* [2] demonstrate for the one-dimensional Terzaghi problem by numerical experiments and the argument that the discretization matrix no longer remains an M-matrix if the time step satisfies $\Delta t < \frac{h}{6}$ that the numerical solution becomes mildly oscillatory. Aguilar *et al.* [2] use a stabilizator term of $\gamma\frac{\partial}{\partial t}\Delta p$ (with $\gamma = \frac{\sigma h^2}{4(\lambda+2\mu)} = O(h^2)$, where $\sigma = 1$) to suppress the spurious oscillations. In our application, the stabilization coefficient is given by $\gamma \approx 1.2 \cdot 10^{-18}$. We, however, think that the incompressible limit is mimiced by equation (7), and here the boundary conditions for the pressure in the problem of Aguilar *et al.* should be removed. Then, the equations can be tackled well with the LBB condition satisfying [8] Taylor-Hood elements.

In this study, we use linear-linear elements to solve equations (6). We verified numerically that these elements gave the same results as the Taylor-Hood elements. A possible reason for this is that for our settings the compressibility term is given by $n_f\beta_f \approx 2.5 \cdot 10^{-16}$, which is larger than the stabilization coefficient γ that was introduced by Aguilar *et al.* [2]. Since this term, and in particular the $\frac{\partial p}{\partial \tau}$-term (also as $\Delta \tau \to 0$), gives an additional contribution to the diagonal entries of the discretization matrix, the *M*-matrix property of the discretization matrix is probably preserved. Hence, the right hand side of equation (7) stabilizes the solution. Note that linear-linear elements are always allowable if the stabilization term due to Aguilar is used. Our approach, which is motivated physically, stabilizes in a similar way as Aguilar's term does. We admit that this issue needs more investigation in mathematical rigor. For the concentrations and densities, linear elements are used too. The diffusion part of the equations for the mesenchymal cells and fibroblasts were solved using an IMEX method, where the diffusivities of the mesenchymal cells and fibroblasts were taken from the previous time step. The reaction parts in all the equations were treated using an IMEX time integration method too. The coupling was treated by the use of information from the previous time step. Until now, no iterative treatment of the coupling has been done in the current preliminary simulations. A state-of-the-art book on several numerical time integrators for stiff problems is the work due to Hundsdorfer & Verwer [10].

To determine the stimulus in equation (5), the strain is computed from the spatial derivatives of the displacements. To determine the strains at the mesh points, we proceed as follows: consider the equation for ε_{xx}, then multiplication by a test-function gives

$$\int_\Omega \varepsilon_{xx} \phi d\Omega = \int_\Omega \frac{\partial u}{\partial x} \phi d\Omega, \text{ for } \phi \in H^1(\Omega), \tag{8}$$

where $\varepsilon_{xx} \in H^1(\Omega)$. Using the set of basis functions as in our finite element solution, gives

$$\sum_{j=1}^{n} \varepsilon_{xx}^{j} \int_\Omega \phi_i \phi_j d\Omega = \sum_{j=1}^{n} u_j \int_\Omega \frac{\partial \phi}{\partial x} \phi_i d\Omega, \qquad \text{for } i \in \{1, \ldots, n\}. \tag{9}$$

This gives a system of n equations with n unknowns. This is applicable for any type of element. For piecewise linear basis functions, the mass matrix is diagonal (lumped) after applying Newton-Cotes' integration rule. Then, the strains and fluid velocities are used for the mechanical stimulus at the mesh points for the ordinary differential equations, which are solved using a time IMEX integrator only.

4 Numerical Experiments

In Figure 4 the distribution of the stimulus, osteoblast density, mesenchymal stem cell density and the bone fraction in the porous tantalum layer after 100 days have been plotted. The prosthesis is assumed to consist of two parts: the top part being the functional part on which an external force is exerted from the outer motion. The botton part is the porous tantalum, in which bone is allowed to grow in from the botton layer. The size of the prosthesis is given by 40×10 mm, in which the prosthesis is divided into the top and botton layer of the same size. The upper force is given by 165.84 N, corresponding to an arm abduction of 30 degrees. In the top part of the prosthesis, the elasticity equations are solved. The prosthesis has been approximated by a two-dimensional geometry, which can be done with the use of cylindrical co-ordinates. The latter has not been done yet.

It can be seen that the osteoblast density is maximal where the stimulus is maximal. This implies that bone develops at the positions where the osteoblast density and stimulus is maximal. This can be seen clearly from the figures. Furthermore, the mesenchymal cell density shows a decrease where the cells differentiate into osteoblasts. The conditions are such that the model only allows the differentiation into osteoblasts and the development of other cell types and tissues is prohibited. To have bone ingrowth in the other parts of the tantalum, it is necessary that the upper arm moves allowing for the stimulus to increase at various positions within the tantalum. This has been observed to take place in preliminary simulations that are not shown in this paper. For arm abductions of 90 degrees, cartilage is also allowed to develop in the tantalum due to a higher outer force that is exerted on the top of the prosthesis. It can be seen that bone develops in the high stimulus domain. Bone remains can only remain at locations where it has been generated. Bone resorption has been disregarded in the model since its effect seems to be of second order only.

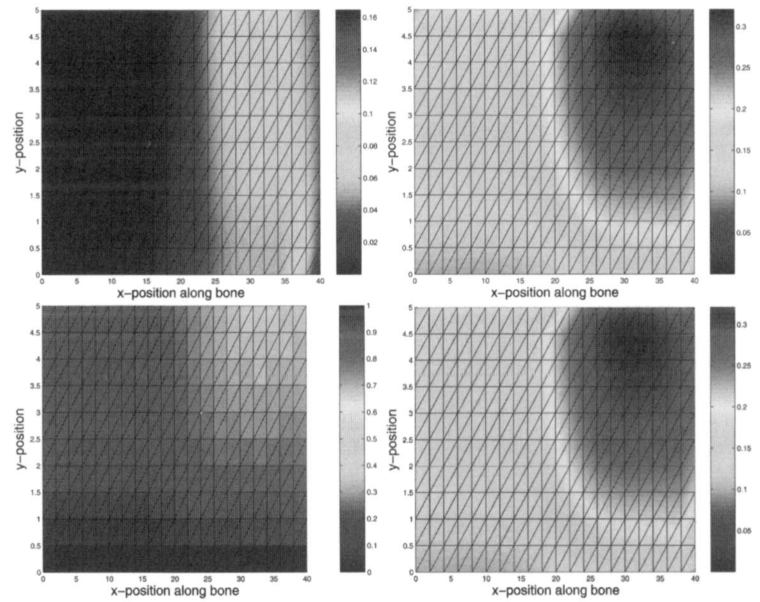

Fig. 2 Some distributions in the porous tantalum after 100 days: Left: The stimulus. Right: The osteoblasts (bone cells). Bottom-Left: The mesenchymal stemcells. Bottom-Right: The bone density.

Some preliminary results reveal that the model is rather insensitive to the diffusion parameters near the current values. There is a high sensitivity with respect to F_b, and Q_b in the present loading regime.

5 Conclusions

A model has been developed for bone-ingrowth into a prosthesis. Parameters that were used were obtained from literature and animal experiments. For small forces exerted, bone develops mainly near the interface and close to the applied force. For large forces, bone develops far away from the interface. For a complete ingrowth, oscillatory forces are to be applied. Linear-linear (displacement-pressure) elements are applicable for this two-dimensional problem.

References

1. Adam, J.: A simplified model of wound healing (with particular reference to the critical size defect). Mathematical and Computer Modelling **30**, 23–32 (1999)
2. Aguilar, G., Gaspar, F., Lisbona, F., Rodrigo, C.: Numerical stabilization of biot's consolidation model by a perturbation on the flow equation. International Journal of Numerical Methods in Engineering, submitted (2007)
3. Ament, C., Hofer, E.: A fuzzy logic model of fracture healing. Journal of Biomechanics **33**, 961–968 (2000)
4. Andreykiv, A.: Simulation of bone ingrowth. Thesis at the Delft University, Faculty of Mechanical Engineering (2006)
5. Bailon-Plaza, A., van der Meulen, M.C.H.: A mathematical framework to study the effect of growth factors that influence fracture healing. Journal of Theoretical Biology **212**, 191–209 (2001)
6. Bear, J.: Dynamics of fluids in porous media. American Elsevier Publishing Inc., New York (1972)
7. Braess, D.: Finite elements: theory, fast solvers, and applications in solid mechanics, 7 edn. Cambridge University Press, Cambridge (2007)
8. Gelhard, T., Lube, G., Olshanskiib, M., Starcke, J.: Stabilized finite element schemes with lbb-stable elements for incompressible flows. Journal of Computational and Applied Mathematics **177**, 243–267 (2005)
9. Huiskes, R., van Driel, W.D., Prendergast, P.J., Søballe, K.: A biomechanical regulatory model for periprosthetic fibrous-tissue differentiation. Journal of Materials Science: Materials in medicine **8**, 785–788 (1997)
10. Hundsdorfer, W., Verwer, J.G.: Numerical solution of time-dependent advection-diffusion-reaction equations. Springer Series in Computational Mathematics, Berlin-Heidelberg (2003)
11. LaCroix, D., Prendergast, P.: A mechano-regulation model for tissue differentiation during fracture healing: analysis of gap size and loading. Journal of Biomechanics **35 (9)**, 1163–1171 (2002)

Computational Electromagnetism

Space and Time Adaptive Calculation of Transient 3D Magnetic Fields

M. Clemens, J. Lang, D. Teleaga, and G. Wimmer

Abstract Transient quasistatic magnetic fields are described by a degenerate parabolic initial-boundary value problems which can be considered as dynamical systems of nonlinear differential-algebraic equations (DAE) of index 1. This DAE structure is retained after the spatial discretization with geometric discretization schemes like the finite element method based on Whitney form functions (WFEM). External transient electric current excitations yield commonly thin layers of eddy currents in electric conductors. Furthermore nonlinear saturation effects have to be taken into account in ferromagnetic materials. A common approach in established simulation tools for solving this problem is the Method of Lines where the space is adaptively discretized at the beginning of the simulation and then kept fixed within the adaptive time integration of the time dependent equation. This approach, however, fails to take into account changes of the solution in the regions of material related to strong local field variation depending on the excitation wave form. This problem is solved by Rothe's method coupling adaptive strategies in space and time.

Markus Clemens
Helmut-Schmidt-University, University of the Federal Armed Forces Hamburg, Holstenhofweg 85, D-22043 Hamburg, Germany, e-mail: m.clemens@hsu-hh.de

Jens Lang
Technische Universität Darmstadt, Fachbereich Mathematik, Schlossgartenstr. 9, D-65789 Darmstadt, Germany, e-mail: lang@mathematik.tu-darmstadt.de

Delia Teleaga
Technische Universität Darmstadt, Fachbereich Mathematik, Schlossgartenstr. 9, D-65789 Darmstadt, Germany, e-mail: dteleaga@mathematik.tu-darmstadt.de, supported by the Deutsche Forschungsgemeinschaft (DFG) under grant LA1372/3-1, LA1372/3-2

Georg Wimmer
Helmut-Schmidt-University, University of the Federal Armed Forces Hamburg, Holstenhofweg 85, D-22043 Hamburg, Germany, e-mail: g.wimmer@hsu-hh.de, supported by the Deutsche Forschungsgemeinschaft (DFG) under grant CL143/3-1, CL143/3-2

1 Introduction

In the past, magnetic field simulation schemes only feature either error control with spatial mesh refinement (e.g. [2], [3], [20]) or with adaptive time step selection with the required implicit time integration schemes (e.g. [5], [8]). The combination of adaptivity in time and space was investigated for nonlinear parabolic equations in [11], [14], [15]. Recently, a combination of adaptive spatial discretizations and variable step-size time discretizations has been considered for transient magnetic field problems. Such an approach allows to detect automatically and discretize appearing and vanishing zones of ferromagnetic saturation and/or regions of eddy currents with sufficient accuracy depending on the variation of the field excitation and a user prescribed error tolerance. In [23] a scheme combining lowest order WFEM and time integration schemes was presented for a linear problem. For the same problem a different technique was presented in [17] using hierarchical bases for the spatial error control combined to a higher order linearly implicit embedded Rosenbrock method. In [7] this approach was extended to nonlinear magnetodynamic problems. In this paper, a different approach is adopted, extending a space and time adaptive solution for 2D planar magnetodynamic problems studied in [21] to linear and nonlinear 3D magnetic field problems. This approach involves the combination of a lowest order WFEM formulation [6] using a suitable Zienkiewicz-Zhu-type gradient recovery method for spatial adaptivity combined with an established higher order embedded SDIRK scheme [5], [8].

2 Magnetic Field Formulation

Magneto-quasistatic (MQS) fields are described by Ampère's law under the MQS assumption of neglecting displacement currents. This is reasonable for low-frequency and high-conductivity applications. The equation can be stated in terms of the magnetic vector potential $\mathbf{A}$ in the form

$$\begin{aligned} \sigma\partial_t\mathbf{A} + \mathrm{curl}\,(\nu(|\mathrm{curl}\mathbf{A}|)\mathrm{curl}\mathbf{A}) &= \mathbf{J}_s \ \text{ in } \Omega\times(0,T] \\ \mathbf{A}\times\mathbf{n} &= 0 \ \text{ on } \partial\Omega\times(0,T] \\ \mathbf{A}(.,0) &= \mathbf{A}_0 \ \text{ in } \Omega\,, \end{aligned} \tag{1}$$

where $\mathbf{J}_s$ is the source current density, σ is the electric conductivity and ν is the magnetic reluctivity. The relevant physical quantities which can be derived from $\mathbf{A}$ are the magnetic flux density $\mathbf{B} = \mathrm{curl}\mathbf{A}$ and the eddy current density $\mathbf{J}_e = -\sigma\partial_t\mathbf{A}$. KARDOS [17] uses a small conductivity in non-concucting regions in order to obtain a consistent linear system after spatial discretization. This non-physical conductivity is avoided in the research code MEQSICO [7] by introducing a current vector potential $\mathbf{T}_s$ for $\mathbf{J}_s$ such that $\mathbf{J}_s = \mathrm{curl}\mathbf{T}_s$ according to [19].

3 Time Discretization

For the discretization of (1) a Rothe approach ("first time, then space") is adopted where the spatial discretization is considered as a perturbation of the implicit time stepping process. Since the resulting equation becomes highly nonlinear and stiff, implicit integrators which avoid time step restrictions forced by stabiltiy requirements are used. The vector potential $\mathbf{A}$ is discretized by values $\mathbf{A}_n \approx \mathbf{A}(\cdot, t_n)$ on a certain time grid

$$0 = t_0 < t_1 < \cdots < t_M = T. \tag{2}$$

Time step methods that need values from several previous time steps are not favorable because the spatial mesh is permanently changing. Hence, one step methods are chosen because only two different meshes are needed in every time step. One mesh on which the starting value is given and a second mesh on which the solution for the new time step is computed. One step time integrators of s stages are considered with the stage variables $\mathbf{A}_{ni}, i = 1, \ldots, s$. The solution at the new time step t_{n+1} has the form

$$\mathbf{A}_{n+1} = \mathbf{A}_n + \sum_{i=1}^{s} m_i \mathbf{A}_{ni} \tag{3}$$

with coefficients $m_i \in \mathbb{R}, 1 \le i \le s$. In sections 3.1 and 3.2 two time integration schemes are introduced: an implicit Runge-Kutta method and a newly designed Rosenbrock method ROS3PL.

3.1 Runge-Kutta Methods

In the i-th stage a nonlinear boundary value problem

$$\frac{\sigma}{\Delta t a_{ii}} \bar{\mathbf{A}}_{ni} + \text{curl}\left(\nu(|\text{curl}\bar{\mathbf{A}}_{ni}|)\text{curl}\bar{\mathbf{A}}_{ni}\right) = \mathbf{F}_{ni} \;\text{ in } \Omega, \quad \bar{\mathbf{A}}_{ni} \times \mathbf{n} = 0 \;\text{ on } \partial\Omega \tag{4}$$

for $\bar{\mathbf{A}}_{ni}$ is solved. The right hand side $\mathbf{F}_{ni}$ is given by

$$\mathbf{F}_{ni} = \mathbf{J}_s(\cdot, t_n + c_i \Delta t) + \frac{\sigma}{\Delta t a_{ii}} \left(\mathbf{A}_n + \tilde{\mathbf{A}}_{ni}\right), \quad \tilde{\mathbf{A}}_{ni} = \sum_{j=1}^{i-1} a_{ij} \mathbf{A}_{nj}. \tag{5}$$

Finally, we obtain

$$\mathbf{A}_{ni} = \left(\bar{\mathbf{A}}_{ni} - \mathbf{A}_n - \tilde{\mathbf{A}}_{ni}\right) / a_{ii}. \tag{6}$$

The step length is given by $\Delta t = t_{n+1} - t_n$. The coefficients a_{ij}, c_i, m_i and details on the error estimation and step size control can be found in [4] (page 51, Pair 2b).

3.2 Rosenbrock Methods

A linear boundary value problem for $\mathbf{A}_{ni}$ is solved in the i-th stage

$$\frac{\sigma}{\Delta t \gamma}\mathbf{A}_{ni} + \mathrm{curl}\,(\mathbf{T}_n \mathrm{curl}\mathbf{A}_{ni}) = \mathbf{R}_{ni} \text{ in } \Omega, \quad \mathbf{A}_{ni} \times \mathbf{n} = 0 \text{ on } \partial\Omega. \tag{7}$$

The right hand side is given by

$$\mathbf{R}_{ni} = -\mathrm{curl}\,(\nu(|\mathrm{curl}\mathbf{A}_i|)\mathrm{curl}\mathbf{A}_i) + \mathbf{J}_s(\cdot,t_i) - \sigma \sum_{j=1}^{i-1} \frac{c_{ij}}{\Delta t}\mathbf{A}_{nj} + \Delta t_n \gamma_i \partial_t \mathbf{J}_s(\cdot,t_n), \tag{8}$$

where $\mathbf{A}_i = \mathbf{A}_n + \sum_{j=1}^{i-1} b_{ij}\mathbf{A}_{nj}$. The operator $\mathbf{T}_n$ is derived from a linearization of the nonlinear operator $\mathrm{curl}\,(\nu(|\mathrm{curl}\mathbf{A}|)\mathrm{curl}\mathbf{A})$ with respect to $\mathbf{A}$ at time t_n

$$\mathbf{T}_n = \nu(|\mathrm{curl}\mathbf{A}_n|)\mathbf{I} + \partial_{|\mathrm{curl}\mathbf{A}|}\nu(|\mathrm{curl}\mathbf{A}|)_{|\mathbf{A}=\mathbf{A}_n}(\mathrm{curl}\mathbf{A}_n)(\mathrm{curl}\mathbf{A}_n)^T/|\mathrm{curl}\mathbf{A}_n|. \tag{9}$$

Here, the identity matrix is denoted by $\mathbf{I}$ and the step length by $\Delta t = t_{n+1} - t_n$. The coefficients γ, γ_i, b_{ij}, c_{ij}, m_i and details on the error estimation and step size control can be found in [7], [18].

4 Space Discretization

The linearization of problem (4) in the context of a Newton method as well as the linear problem (7) can be stated in the form

$$\beta\mathbf{u} + \mathrm{curl}\,(\alpha\mathrm{curl}\mathbf{u}) = \mathbf{f} \text{ in } \Omega, \quad \mathbf{u} \times \mathbf{n} = 0 \text{ on } \partial\Omega. \tag{10}$$

This problem is replaced by the variational formulation in the Hilbert space

$$H_0(\mathrm{curl},\Omega) = \{\mathbf{u} \in L^2(\Omega) | \mathrm{curl}\mathbf{u} \in L^2(\Omega), \mathbf{u} \times \mathbf{n} = 0\}. \tag{11}$$

Find $\mathbf{u} \in H_0(\mathrm{curl},\Omega)$ such that for all $\mathbf{v} \in H_0(\mathrm{curl},\Omega)$

$$\langle \alpha\mathrm{curl}\mathbf{u}, \mathrm{curl}\mathbf{v}\rangle_{L^2(\Omega)} + \langle \beta\mathbf{u}, \mathbf{v}\rangle_{L^2(\Omega)} = \langle \mathbf{f}, \mathbf{u}\rangle_{L^2(\Omega)}. \tag{12}$$

If $\beta \in L^\infty(\Omega)$ and $0 < \underline{\beta} < \beta < \overline{\beta}$ for some $\underline{\beta}, \overline{\beta} > 0$, the Lax-Milgram lemma guarantees existence and uniqueness of the solution. Since the conductivity σ and also β may be zero on sets of positive measure a unique solution can only be found in some quotient space. For the finite element discretization the $H_0(\mathrm{curl},\Omega)$-conforming Whitney finite element space on tetrahedra with lowest order polynomials is used. These functions enforce tangential continuity across interelement boundaries. For the spatial adaptivity, local error estimators based on hierarchical finite elements [1], [7], and also suitable Zienkiewicz-Zhu-type gradient recovery error indicators for edge elements [22] are used.

5 Space-Time Adaptivity

The combination of space and time adaptivity implemented in the code MEQSICO is depicted in Fig. 1. Starting on a coarse grid for each new time step, the spatial dis-

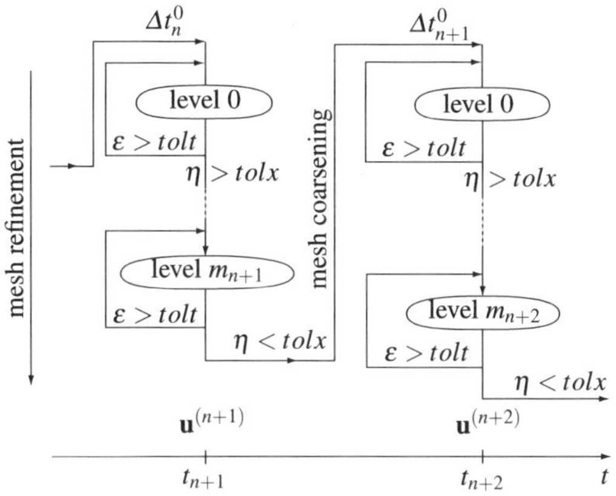

Fig. 1 Flow chart for the space-time adaptive solver MEQSICO.

cretization is adaptively refined using a gradient recovery scheme of Zienkiewicz-Zhu-type for local error detection [8]. For the refinement of the tetrahedral grid a 3D red-green closure strategy is used. The time stepping tolerance *tolt* is tested after every spatial refinement for which the error indicator threshold *tolx* is used. Provided that the new time step solution is accepted, the next step starts again with a coarse grid. The software package KARDOS [10] uses a slightly different strategy where the time step is controlled after a sequence of mesh refinements [16].

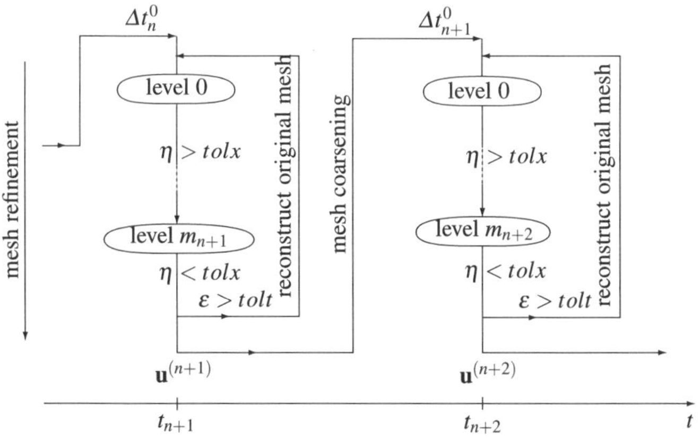

Fig. 2 Flow chart for the space-time adaptive solver KARDOS.

6 Numerical Results

6.1 TEAM 7 Benchmark

The TEAM 7 benchmark problem consists of an aluminium plate with an eccentric hole. A source coil is placed above the plate. Numerical results and experimental data for a time harmonic excitation with a frequency of 50 Hz and a maximum value of 2742 A are given in [13]. In order to test the space-time adaptive algorithm the coil is excited by a ramped sinusoidal current over two periods which reaches the maximum value after 15 ms and the problem is considered as a pure transient eddy current problem. Different combinations of adaptive/fixed time and spatial grids are

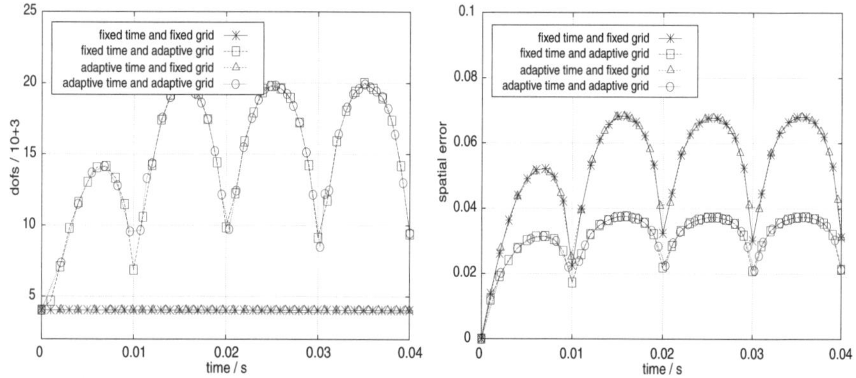

(a) Degrees of freedom (dofs) (b) Spatial error

Fig. 3 TEAM 7: Simulations with different combinations of fixed/adaptive time and spatial grid in MEQSICO with the Runge-Kutta method.

shown in Fig. 3. The evolution of the estimated spatial error in Fig. 3(b) shows that adaptive and fixed time steps produce the same spatial error. Hence time adaptivity alone will not solve the problem sufficiently. Spatial adaptivity decreases the spatial error for account of increasing degrees of freedom (dofs) (see Fig. 3(a)) and in connection with time adaptivity the number of time steps is nearly optimal. Numerical results for KARDOS in [17] show a similar performance. The dofs vary from $0.7 \cdot 10^5$ and $2.1 \cdot 10^5$ and the spatial error is mainly reduced in time intervals with large current excitation.

6.2 Magnetic Write Head Benchmark

This benchmark problem was proposed by the Storage Research Consortium (SRC) in order to test various 3D FEM codes [12]. The model consists of the head which is placed on a rectangular plate with high conductivity and nonlinear material behaviour. The applied magnetomotive force (mmf) in the two coils has a trapezoidal waveform with a frequency of 25 MHz. The $z-$component of the magnetic flux density B_z was measured in front of the small air gap between head and plate. Fig.

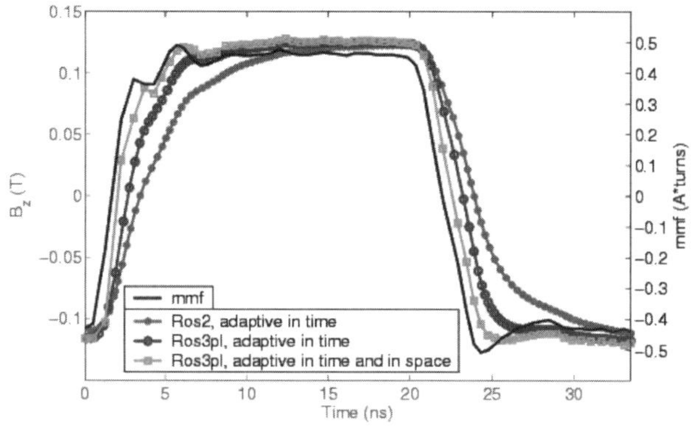

Fig. 4 B_z component of magnetic flux density simulated with KARDOS.

4 shows the calculated results of the head field B_z at the pole tip with KARDOS. Since eddy currents are induced in the conductive material, head and plate, the field is delayed a few nanoseconds with respect to the applied mmf. At the beginning the initial mesh with 109 495 dofs is refined to 357 173 dofs. After successive coarsening approximately 250 000 dofs are sufficient. More details can be found in [7]. The performance of MEQSICO for this benchmark problem is described in [9].

7 Conclusion

A combination of simultaneous use of adaptive space and time discretizations for transient magnetic fields has been presented. This approach features a time evolving mesh where adaptive refinement and coarsening of the spatial resolution take into account appearing and disappearing local transient saturation effects. The numerical results show that accuracy is lost if the mesh or the time steps are kept fixed during the simulation. Different time integration schemes and spatial error indicators have been implemented in the research codes MEQSICO and KARDOS and have been investigated for two benchmark problems.

References

1. Ainsworth, M., Coyle, J.: Hierarchic finite element bases on unstructured tetrahedral meshes. Int. J. Num. Meth. Eng. **58**(14), 2103–2130 (2003)
2. Beck, R., Hiptmair, R., Hoppe, R.H.W., Wohlmuth, B.: Residual based a posteriori error estimators for eddy current computation. Report 112, Universitäten Tübingen und Stuttgart, SFB 382 (1999)
3. Braess, D., Schöberl, J.: Equilibrated residual error estimator for Maxwell's equations. Technical Report 2006-19, Johann Radon Institute for Computational and Applied Mathematics, Austrian Academy of Sciences (2006)
4. Cameron, F.: Low-order Runge-Kutta methods for differential algebraic equations. Ph.D. thesis, Tampere University of Technology (1999)
5. Cameron, F., Piché, R., Forsman, K.: Variable step size time integration methods for transient eddy current problems. IEEE Trans. Magn. **34**(5), 3319–3323 (1998)
6. Castillo, P., Koning, J., Rieben, R., Stowell, M., White, D.: Discrete Differential Forms: A Novel Methodology for Robust Computational Electromagnetics. Lawrence Livermore National Laboratory (2003)
7. Clemens, M., Lang, J., Teleaga, D., Wimmer, G.: Adaptivity in space and time for magnetoquasistatics. JCM Special Issue on Adaptive and Multi-level Methods in Electromagnetics (2008)
8. Clemens, M., Wilke, M., Weiland, T.: 3D transient eddy current simulations using FI^2TD with variable time step size selection schemes. IEEE Trans. Magn. **38**(2), 605–608 (2002)
9. Clemens, M., Wimmer, G., Lang, J., Teleaga, D.: Transient 3d magnetic field simulation with combined space and time adaptivity for lowest order wfem formulations. In: Proc. CEM, Brighton (2008). To appear in IET Research Journal Science, Measurement & Technology.
10. Erdmann, B., Lang, J., Roitzsch, R.: Kardos user's guide. Tech. Rep. ZR-02-42, ZIB (2002)
11. Erikkson, K., Johnson, C.: Adaptive finite element methods for parabolic problems iv: Nonlinear problems. SIAM J. Numer. Anal. **32**(6), 1729–1749 (1995)
12. Fujiwara, K., Ikeda, F., Kameari, A., Kanai, Y., Nakamura, K., Takahashi, N., Tani, K., Yamada, T.: Thin film write head field analysis using a benchmark problem. IEEE Trans. Magn. **36**(4), 1784–1788 (2000)
13. Fujiwara, K., Nakata, T.: Results for benchmark problem 7. COMPEL **9**(3), 137–154 (1990)
14. Lang, J.: Two-dimensional fully adaptive solutions of reaction diffusion equations. Appl. Num. Math. **18**, 223–240 (1995)
15. Lang, J.: Adaptive FEM for reaction-diffusion equations. Appl. Num. Math. **26**, 105–116 (1998)
16. Lang, J.: Adaptive Multilevel Solution of Nonlinear Parabolic PDE Systems: Theory, Algorithm and Application. Springer-Verlag, Berlin, Heidelberg, New York (2001)
17. Lang, J., Teleaga, D.: Towards a fully space-time adaptive FEM for magnetoquasistatics. IEEE Trans. Magn. **44**(6), 1238–1241 (2008)
18. Lang, J., Verwer, J.: Ros3p - an accurate third-order rosenbrock solver designed for parabolic problems. BIT **41**, 730–737 (2001)
19. Ren, Z.: Influence of the r.h.s. on the convergence behaviour of the curl-curl equation. IEEE Trans. Magn. **32**(3), 655–658 (1996)
20. Schöberl, J.: A posteriori error estimates for Maxwell equations. Math. Comp. **77**, 633–649 (2008)
21. Wimmer, G., Steinmetz, T., Clemens, M.: Reuse, recycle, reduce (3r) - strategies for the calculation of transient magnetic fields. Accepted for Publication in a Special Issue of Appl. Numer. Math., devoted to the 11th NUMDIFF Conference, held in Halle, Sept. 2006
22. Zhelezina, E.: Adaptive finite element method for the numerical simulation of electric, magnetic and acoustic fields. Ph.D. thesis, Univ. Erlangen (2005)
23. Zheng, W., Chen, Z., Wang, L.: An adaptive finite element method for the h-ψ formulation of time-dependent eddy current problems. Numer. Math. **103**, 667–689 (2006)

A Boundary Integral Formulation for Nonlocal Electrostatics

C. Fasel, S. Rjasanow, and O. Steinbach

Abstract In the field of protein binding, it is important to gain some knowledge about the electric field surrounding the concerning biomolecules. Most of the interesting action takes place inside cells where the medium is similar to water, which is a nonlocal medium. To handle this nonlocal behavior caused by a network of hydrogen bonds, we formulate an interface problem in terms of a system of coupled partial differential equations involving the Laplace and the Yukawa operator. Furthermore, we deduce for the system a fundamental solution and an associated representation formula. We finally derive a boundary integral formulation to determine the complete Dirichlet and Neumann data.

1 Modelling Nonlocal Electrostatics for Biomolecules in Water

The model given below is a modification of the model as presented in [1, 2]. Before going into detail, we first introduce some notations. Let $\Omega^i \subset \mathbb{R}^3$ be a bounded domain where the molecule is located. Let $\Omega^e = \mathbb{R}^3 \setminus \overline{\Omega}^i$ denote the unbounded domain exterior to Ω^i which is filled with water. $\Gamma = \mathbb{R}^3 \setminus (\Omega^i \cup \Omega^e)$ is the interface between the molecule and water. For $x \in \Gamma$ we define the normal vector $\mathbf{n}(x)$ as the unit vector pointing from Ω^i into Ω^e. The interior and exterior trace operators are denoted by γ_0^{int} and γ_0^{ext}, respectively.

C. Fasel
Saarland University, Department of Mathematics, PF 15 11 50, 66041 Saarbrücken, Germany.
e-mail: fasel@num.uni-sb.de

S. Rjasanow
Saarland University, Department of Mathematics, PF 15 11 50, 66041 Saarbrücken, Germany.
e-mail: rjasanow@num.uni-sb.de

O. Steinbach
Graz University of Technology, Institute of Computational Mathematics, Steyrergasse 30, 8010 Graz, Austria. e-mail: o.steinbach@TUGraz.at

The starting point for our considerations are the Maxwell equations of electrostatics which describe the behavior of the electric field $\mathbf{E}$ and of the displacement field $\mathbf{D}$ by

$$\operatorname{curl} \mathbf{E}(x) = 0, \qquad x \in \Omega^i \cup \Omega^e, \tag{1}$$
$$\operatorname{div} \mathbf{D}(x) = \rho(x), \qquad x \in \Omega^i \cup \Omega^e. \tag{2}$$

The charge distribution is modelled as a sum of partial charges located inside Ω^i,

$$\rho(x) = \sum_{j=1}^{N_c} q_j \delta(x - x_j), \tag{3}$$

where δ denotes the Dirac Delta distribution. Furthermore, it is known that – in absence of surface charges – the tangential component of the electric field and the normal component of the displacement field are continuous through the interface,

$$(\gamma_0^{int}\mathbf{E}(x) - \gamma_0^{ext}\mathbf{E}(x)) \times \mathbf{n}(x) = 0, \qquad x \in \Gamma, \tag{4}$$
$$(\gamma_0^{int}\mathbf{D}(x) - \gamma_0^{ext}\mathbf{D}(x), \mathbf{n}(x)) = 0, \qquad x \in \Gamma. \tag{5}$$

One of the main steps in the modelling is the description of the relation between $\mathbf{E}$ and $\mathbf{D}$. Inside the molecule we assume a local relationship, and therefore,

$$\mathbf{D}(x) = \varepsilon_0 \varepsilon_{\Omega^i} \mathbf{E}(x), \quad x \in \Omega^i, \tag{6}$$

where ε_0 is the permittivity of vacuum, and ε_{Ω^i} is constant inside the molecule. The description in Ω^e is more complicated due to the nonlocal effects of water. There are two different facts that influence the relation between the two fields in Ω^e. Each water molecule can be seen as a dipole and thus would normally align itself with the field and all together would cause a relatively strong shielding effect. On the other side, water molecules build a network of hydrogen bonds. This network has energetic advantages and is strongly dependent on the angle of the water molecules towards each other. The strength of the electric field determines if the water keeps the bonds or aligns with the field. That is the reason why the electric field in the neighbourhood of a point x is important for the displacement field in the point in contrast to the local relationship where only the electric field in the point has to be taken into account. The nonlocal relationship is taken as

$$\mathbf{D}(x) = \varepsilon_0 \left(\varepsilon_\infty \mathbf{E}(x) + \eta^2 \int_{\Omega^e} \frac{e^{-\kappa|x-y|}}{4\pi|x-y|} \mathbf{E}(y)\, dy \right), \quad x \in \Omega^e, \tag{7}$$

where

$$\eta^2 = (\varepsilon_{\Omega^e} - \varepsilon_\infty)\kappa^2, \quad \kappa = \frac{1}{\lambda},$$

λ is the correlation length, and ε_∞ and ε_{Ω^e} are constants describing the dielectricity of the material. The scaling of the system is done with respect to the typical length of an hydrogen bond.

To transfer this system of partial-integro-differential equations into a system of partial differential equations, we make use of a scalar potential for **E**,

$$\mathbf{E}(x) = -\nabla\varphi(x), \quad x \in \Omega^i \cup \Omega^e. \tag{8}$$

Then, the electric field **E** is curl-free and with respect to φ, the material relations can be written as

$$\mathbf{D}(x) = -\varepsilon_{\Omega^i}\nabla\varphi(x), \quad x \in \Omega^i, \tag{9}$$

$$\mathbf{D}(x) = -\varepsilon_\infty\nabla\varphi(x) - \eta^2 \int_{\Omega^e} \frac{e^{-\kappa|x-y|}}{4\pi|x-y|}\nabla\varphi(y)\,dy, \quad x \in \Omega^e. \tag{10}$$

Note that the constant ε_0 has been eliminated during the process of scaling. Applying the divergence to (9) and to (10), it follows by the use of (2) and (8) that

$$-\varepsilon_{\Omega^i}\Delta\varphi(x) = \rho(x), \quad x \in \Omega^i, \tag{11}$$

$$-\varepsilon_\infty\Delta\varphi(x) - \eta^2 \operatorname{div} \int_{\Omega^e} \frac{e^{-\kappa|x-y|}}{4\pi|x-y|}\nabla\varphi(y)\,dy = 0, \quad x \in \Omega^e. \tag{12}$$

In addition, the transmission condition (4) for the electric field takes the form

$$\gamma_0^{int}\varphi(x) = \gamma_0^{ext}\varphi(x), \quad x \in \Gamma. \tag{13}$$

In order to eliminate the volume integral in (12) we introduce an additional unknown function which we denote by **P**. It is defined as

$$\mathbf{P}(x) = -\eta \int_{\Omega^e} \frac{e^{-\kappa|x-y|}}{4\pi|x-y|}\nabla\varphi(y)\,dy, \quad x \in \Omega^i \cup \Omega^e. \tag{14}$$

P describes the polarisation in Ω^e, but has no physical meaning in Ω^i. With this, equation (12) reads

$$-\varepsilon_\infty\Delta\varphi(x) + \eta \operatorname{div}\mathbf{P}(x) = 0, \quad x \in \Omega^e, \tag{15}$$

and the transmission condition (5) for the displacement field becomes

$$\varepsilon_{\Omega^i}\gamma_1^{int}\varphi(x) = \varepsilon_\infty\gamma_1^{ext}\varphi(x) - \eta(\gamma_0^{ext}\mathbf{P}(x), \mathbf{n}(x)), \quad x \in \Gamma, \tag{16}$$

where $\gamma_1^{int/ext}$ denotes the interior and exterior conormal derivative, respectively. To find a second relation between φ and **P**, we use the Yukawa operator $\mathscr{L}_\kappa = \Delta - \kappa^2$, and the fact that the convolution kernel under the integral is its fundamental solution. Thus, we obtain

$$-\mathscr{L}_{\kappa}\mathbf{P}(x)+\eta\nabla\varphi(x)=0\,,\quad x\in\Omega^{e}, \tag{17}$$

$$-\mathscr{L}_{\kappa}\mathbf{P}(x)=0\,,\quad x\in\Omega^{i}. \tag{18}$$

Transmission conditions for the field $\mathbf{P}$ can be deduced from (14) as

$$\gamma_0^{ext}\mathbf{P}(x)=\gamma_0^{int}\mathbf{P}(x)\,,\quad \gamma_1^{ext}\mathbf{P}(x)=\gamma_1^{int}\mathbf{P}(x)\,,\quad x\in\Gamma. \tag{19}$$

To complete the system, we include radiation conditions for φ and $\mathbf{P}$, where the one for $\mathbf{P}$ can be deduced from the one for φ,

$$\varphi(x)\sim\frac{1}{|x|}\,,\quad |\mathbf{P}(x)|\sim\frac{1}{|x|^2}\quad \text{as } |x|\to\infty. \tag{20}$$

The resulting system composed by (11), (13), (15), (16), (17), (18), (19), and (20), is a system of purely partial differential equations that is equivalent to the mixed partial-integro-differential system we started from.

2 Analytical and Fundamental Solution

For the special case of only one charge with strength q, located in the origin of a sphere with radius a, we can transfer all differential operators into spherical coordinates. Then we are able to find an analytical solution, namely

$$\begin{aligned}
\varphi^{int}(x) &= \frac{q}{4\pi\varepsilon_{\Omega_i}|x|}+\frac{q}{4\pi a}\left(\frac{1}{\varepsilon_{\Omega_e}}-\frac{1}{\varepsilon_{\Omega_i}}+B\frac{e^{-\kappa' a}}{4\pi a}\right),\\
\varphi^{ext}(x) &= \frac{q}{4\pi\varepsilon_{\Omega_e}}\left(\frac{1}{|x|}+B\frac{e^{-\kappa'|x|}}{|x|}\right),\\
\mathbf{P}^{int}(x) &= C\frac{q}{4\pi\varepsilon_{\Omega^e}}\frac{(e^{\kappa|x|}(\kappa|x|-1)+e^{-\kappa|x|}(\kappa|x|-1)}{|x|^3}x,\\
\mathbf{P}^{ext}(x) &= \frac{q}{4\pi\varepsilon_{\Omega_e}|x|^3}\left(\frac{1}{\kappa^2}+B\frac{e^{-\kappa'|x|}(1+\kappa'|x|)}{\kappa^2-\kappa'^2}\right)x,
\end{aligned}$$

where

$$\kappa'^2=\kappa^2+\frac{\eta^2}{\varepsilon_\infty},$$

and the constants B and C are given by

$$\begin{aligned}
B &= \frac{\eta(\kappa'^2-\kappa^2)(e^{\kappa a}-e^{-\kappa a})e^{\kappa' a}}{e^{\kappa a}(\kappa^2(\kappa' a+1)+\kappa'^2(\kappa a-1))+e^{-\kappa a}(\kappa'^2(\kappa a+1)-\kappa^2(\kappa' a+1))},\\
C &= \frac{\eta\kappa'^2}{\kappa^2}\frac{1}{e^{\kappa a}(\kappa^2(\kappa' a+1)+\kappa'^2(\kappa a-1))+e^{-\kappa a}(\kappa'^2(\kappa a+1)-\kappa^2(\kappa' a+1))}.
\end{aligned}$$

This example also shows that the model contains the local case. In the local case we have $\lambda \to 0$ which induces $\kappa \to \infty$, and also $\kappa' \to \infty$. For $\kappa' \to 0$ the exponential term in φ^{ext} disappears, and we end up with the solution of the local system, namely,

$$\varphi^{ext}(x) = \frac{q}{4\pi\varepsilon_{\Omega_e}|x|}, \quad \lambda \to 0 .$$

Furthermore, the analytical solution can be used to control future numerical results with respect to their reliability.

In general, the system derived cannot be solved analytically, instead we have to use some numerical approach. Since the domain Ω^i may have a very complicated boundary Γ which is usually given already in a discretized form (see Fig. 1 for the Solvent Excluded Surface of the neurotransmitter Acetylcholinesterase) the use of a boundary integral equation approach seems to be favourable. Moreover, the exterior domain Ω^e is unbound, therefore a boundary element method seems to be more advantageous than a finite element method.

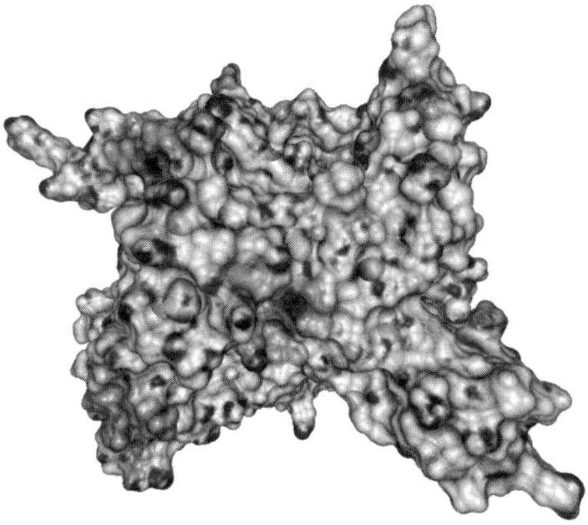

Fig. 1 Solvent excluded surface of Acetylcholinesterase, Source: BALLView [4]

To obtain a boundary integral formulation we first have to consider the interior and the exterior problem separately.

In the interior domain Ω^i we have to deal with the Laplace and with the Yukawa operator, both are well known, for their representation formulae and potentials see e.g., [3, 5, 6]. Thus we will skip this part here and make use of their results later.

To handle the partial differential equation in the exterior domain Ω^e we define the operator

$$\mathscr{A}^{ext} = \begin{pmatrix} -\varepsilon_{\Omega^e}\Delta & \eta\nabla^T \\ \eta\nabla & -\mathscr{L}_\kappa I_3 \end{pmatrix} \tag{21}$$

which turns out to be elliptic. Because of the first order derivatives involved, the operator $\mathscr{A}^{ext}$ is not self-adjoint and, thus, we need to compute the fundamental solution of the adjoint operator to obtain a representation formula for the exterior problem. Since the coefficients of $\mathscr{A}^{ext}$ are constant, we are able to find the fundamental solution by using the Fourier transform and obtain

$$\mathscr{U}^*_{ext}(x,y) = \begin{pmatrix} f_1 & (x-y)^T f_2 \\ (x-y)f_2 & f_3 I_3 + (x-y)(x-y)^T f_4 \end{pmatrix}, \tag{22}$$

where

$$\begin{aligned} f_1 &= \frac{(\kappa'^2 - \kappa^2)e^{-\kappa'|x-y|} + \kappa^2}{4\pi\varepsilon_\infty \kappa'^2 |x-y|}, \\ f_2 &= -\frac{\eta}{4\pi\varepsilon_\infty\kappa'^2} \frac{e^{-\kappa'|x-y|}(\kappa'|x-y|+1)-1}{|x-y|^3}, \\ f_3 &= \frac{e^{-\kappa|x-y|}}{4\pi|x-y|} + \frac{e^{-\kappa|x-y|}(\kappa|x-y|+1)-1}{4\pi\kappa^2|x-y|^3} - \frac{e^{-\kappa'|x-y|}(\kappa'|x-y|+1)-1}{4\pi\kappa'^2|x-y|^3}, \\ f_4 &= \frac{e^{-\kappa'|x-y|}(\kappa'^2|x-y|^2+3\kappa'|x-y|+3)-3}{4\pi\kappa'^2|x-y|^5} - \\ &\quad - \frac{e^{-\kappa|x-y|}(\kappa^2|x-y|^2+3\kappa|x-y|+3)-3}{4\pi\kappa^2|x-y|^5}. \end{aligned}$$

A closer look at the fundamental solution clearly indicates that the only singularities are included in f_1 and in the first part of f_3. Both are first order singularities.

3 Representation Formula and Boundary Integral Formulation

The representation formula is obtained in the usual way by multiplying the partial differential equation with an appropriate test function and integrating over the exterior domain. Afterwards we apply the adjoint operator to the test function, multiply with the solution of the original problem and once again integrate by parts. We end up with the following representation formula for $x \in \Omega^e$

$$\begin{pmatrix} \varphi \\ \mathbf{P} \end{pmatrix}(x) = -\int_\Gamma \gamma_0^{ext} \mathscr{U}^*_{ext}(x,y) \begin{pmatrix} \varepsilon_\infty \gamma_1^{ext}\varphi - \eta(\gamma_0^{ext}\mathbf{P},\mathbf{n}) \\ \gamma_1^{ext}\mathbf{P} \end{pmatrix} ds_y + \int_\Gamma W(x,y) \begin{pmatrix} \gamma_0^{ext}\varphi \\ \gamma_0^{ext}\mathbf{P} \end{pmatrix} ds_y,$$

where W is given as

$$\begin{aligned} W_{11} &= \varepsilon_\infty \gamma_{1,y}^{ext} f_1 + \eta(\gamma_0^{ext}(x-y)f_2, \mathbf{n}(y)), \\ W_{12} &= \gamma_{1,y}^{ext}((x-y)_1 f_2), \\ W_{13} &= \gamma_{1,y}^{ext}((x-y)_2 f_2), \\ W_{14} &= \gamma_{1,y}^{ext}((x-y)_3 f_2), \\ W_{21} &= \varepsilon_\infty \gamma_{1,y}^{ext}(x-y)_1 f_2 + \eta\gamma_0^{ext}(f_3\mathbf{e}_1 + (x-y)_1(x-y)f_4), \mathbf{n}(y)), \end{aligned}$$

$$
\begin{aligned}
W_{22} &= \gamma_{1,y}^{ext}(f_3 + (x-y)_1^2 f_4),\\
W_{23} &= \gamma_{1,y}^{ext}((x-y)_1(x-y)_2 f_4) = W_{32},\\
W_{24} &= \gamma_{1,y}^{ext}((x-y)_1(x-y)_3 f_4) = W_{42},\\
W_{31} &= \varepsilon_\infty \gamma_{1,y}^{ext}(x-y)_2 f_2 + \eta \gamma_0^{ext}(f_3\mathbf{e}_2 + (x-y)_2(x-y)f_4), \mathbf{n}(y)),\\
W_{33} &= \gamma_{1,y}^{ext}(f_3 + (x-y)_2^2 f_4),\\
W_{34} &= \gamma_{1,y}^{ext}((x-y)_2(x-y)_3 f_4) = W_{43},\\
W_{41} &= \varepsilon_\infty \gamma_{1,y}^{ext}(x-y)_3 f_2 + \eta \gamma_0^{ext}(f_3\mathbf{e}_3 + (x-y)_3(x-y)f_4), \mathbf{n}(y)),\\
W_{44} &= \gamma_{1,y}^{ext}(f_3 + (x-y)_3^2 f_4).
\end{aligned}
$$

To come up with a boundary integral equation, we follow the direct BEM approach and apply the exterior trace operator γ_0^{ext} to the above representation formula. For the calculation of the matrix entries in the BEM method, we can separate the integrals in parts that are those of the Laplace operator and some additional parts that can be calculated by numerical integration. Mapping properties can be transferred from the ones for the Laplace operator. We end up with the following boundary integral equation

$$
V_{nl}\begin{pmatrix} \gamma_1^{ext}\varphi - \eta(\gamma_0^{ext}\mathbf{P},\mathbf{n}) \\ \gamma_1^{ext}\mathbf{P} \end{pmatrix} = -\frac{1}{2}\gamma_0^{ext}\begin{pmatrix} \varphi \\ \mathbf{P} \end{pmatrix} + K_{nl}\gamma_0^{ext}\begin{pmatrix} \varphi \\ \mathbf{P} \end{pmatrix}, \tag{23}
$$

where

$$
(V_{nl}u)(x) = \gamma_{0,x}^{ext}\int_\Gamma \mathscr{U}_{ext}^*(x,y)u(y)\,ds_y, \tag{24}
$$

$$
(K_{nl}w)(x) = \lim_{\varepsilon\to 0} \int_{y\in\Gamma:|x-y|\geq\varepsilon} W(x,y)w(y)\,ds_y \tag{25}
$$

are the associated single and double layer potential, respectively. In the interior domain Ω^i we obtain by using the particular solution

$$
\mathbf{u}_p(x) = \left(\sum_{j=1}^{N_c}\frac{q_j}{4\pi|x-x_j|}, 0, 0, 0\right)^\top
$$

the boundary integral equation

$$
V_{loc}\gamma_1^{int}\begin{pmatrix} \varphi \\ \mathbf{P} \end{pmatrix} = \frac{1}{2}\gamma_0^{int}\begin{pmatrix} \varphi \\ \mathbf{P} \end{pmatrix} + K_{loc}\gamma_0^{int}\begin{pmatrix} \varphi \\ \mathbf{P} \end{pmatrix} - \gamma_0^{int}\mathbf{u}_p \tag{26}
$$

where

$$
V_{loc} = \begin{pmatrix} V_0 & 0 & 0 & 0 \\ 0 & V_\kappa & 0 & 0 \\ 0 & 0 & V_\kappa & 0 \\ 0 & 0 & 0 & V_\kappa \end{pmatrix} \text{ and } K_{loc} = \begin{pmatrix} K_0 & 0 & 0 & 0 \\ 0 & K_\kappa & 0 & 0 \\ 0 & 0 & K_\kappa & 0 \\ 0 & 0 & 0 & K_\kappa \end{pmatrix}.
$$

V_0 and V_κ denote the single layer potentials of the Laplace and of the Yukawa operator, while K_0 and K_κ denote their double layer potentials. Using the interface conditions (13), (16), and (19), we finally obtain the desired boundary integral equation

$$\left(\begin{pmatrix} \varepsilon_{\Omega^i} & \\ & I_3 \end{pmatrix} V_{loc}^{-1}\left(\frac{1}{2}I+K_{loc}\right)-V_{nl}^{-1}\left(-\frac{1}{2}I+K_{nl}\right)\right)\gamma_0\mathbf{u}=\begin{pmatrix} \varepsilon_{\Omega^i} & \\ & I_3 \end{pmatrix} V_{loc}^{-1}\gamma_0\mathbf{u}_p\,, \tag{27}$$

where

$$\gamma_0\mathbf{u}=\begin{pmatrix} \gamma_0^{int}\varphi, \gamma_0^{int}\mathbf{P} \end{pmatrix}^\top=\begin{pmatrix} \gamma_0^{ext}\varphi, \gamma_0^{ext}\mathbf{P} \end{pmatrix}^\top .$$

Note that the Laplace and Yukawa single layer potentials are elliptic and therefore invertible, and the same holds true for the nonlocal exterior operator. Solving the boundary integral equation (27) leads to the knowledge of the Dirichlet data $\gamma_0\mathbf{u}$ on Γ. When knowing this function, we can calculate the interior Neumann data by solving an associated Dirichlet boundary value problem. The complete Cauchy data of the interior problem then provides the possibility to calculate the exterior Neumann data by using the corresponding transmission condition. In this way, we receive the complete set of Cauchy data for both problems. By the use of the representation formulae in Ω^i and Ω^e, we can evaluate the solution everywhere in $\mathbb{R}^3$.

4 Outlook

Based on the boundary integral equation (27) we are going to solve the problem by using modern boundary element techniques. A good approximation strategy for the matrices based on fast BEM (cf. [5]) has to be chosen to obtain reasonable results in a realistic time. Another major task for future work is the numerical analysis of the approximation scheme.

References

1. Hildebrandt, A., Blossey, R., Rjasanow, S., Kohlbacher, O., Lenhof, H.P.: Novel formulation of nonlocal electrostatics. Phys. Rev. Lett. **93**(10), 104–108 (2004)
2. Hildebrandt, A., Blossey, R., Rjasanow, S., Kohlbacher, O., Lenhof, H.P.: Electrostatic potentials of proteins in water: a structured continuum approach. Bioinformatics **23**(2), 99–103 (2007)
3. McLean, W.: Strongly elliptic systems and boundary integral equations. Cambridge University Press, Cambridge (2000)
4. Moll, A., Hildebrandt, A., Lenhof, H.P., Kohlbacher, O.: BALLView: an object-oriented molecular visualization and modeling framework. J. Comput. Aided Mol. Des. **19**, 791–800 (2005)
5. Rjasanow, S., Steinbach, O.: The fast solution of boundary integral equations. Mathematical and Analytical Techniques with Applications to Engineering. Springer, New York (2007)
6. Steinbach, O.: Numerical Approximation Methods for Elliptic Boundary Value Problems. Finite and Boundary Elements. Springer, New York (2008)

Computational Methods, Preconditioners, Solvers

Efficient Solution of Algebraic Bernoulli Equations Using $\mathscr{H}$-Matrix Arithmetic

U. Baur and P. Benner

Abstract The algebraic Bernoulli equation (ABE) has several applications in control and system theory, e.g., the stabilization of linear dynamical systems and model reduction of unstable systems arising from the discretization and linearization of parabolic partial differential equations (PDEs). As standard methods for the solution of ABEs are of limited use for large-scale systems, we investigate approaches based on the matrix sign function method. This includes the solution of a linear least-squares (LLS) problem. Due to the large-scale setting we propose to solve this LLS problem via normal equations. To make the whole approach applicable in the large-scale setting, we incorporate structural information from the underlying PDE model into the approach. By using data-sparse matrix approximations, hierarchical matrix formats, and the corresponding formatted arithmetic we obtain an efficient solver having linear-polylogarithmic complexity. The proposed solver computes a low-rank representation of the ABE solution.

1 Introduction

We consider the algebraic Bernoulli equation (ABE)

$$A^T X + XA - XBB^T X = 0, \tag{1}$$

where $A \in \mathbf{R}^{n\times n}$, $B \in \mathbf{R}^{n\times m}$, and $X \in \mathbf{R}^{n\times n}$ is the matrix of unknowns. Recent methods for model order reduction of unstable dynamical systems [17] give the

---

U. Baur
Mathematics in Industry and Technology, Faculty of Mathematics, Chemnitz University of Technology e-mail: baur@mathematik.tu-chemnitz.de

P. Benner
Mathematics in Industry and Technology, Faculty of Mathematics, Chemnitz University of Technology e-mail: benner@mathematik.tu-chemnitz.de

motivation for an efficient numerical solution of large-scale ABEs. Thereby, the general assumption

$$\Lambda(A)\cap\mathbf{C}^- \neq \emptyset, \quad \Lambda(A)\cap\mathbf{C}^+ \neq \emptyset, \quad \Lambda(A)\cap i\mathbf{R} = \emptyset \tag{2}$$

is given, using the notation $\Lambda(A)$ for the spectrum of A, $i=\sqrt{-1}$ and $\mathbf{C}^-$ ($\mathbf{C}^+$) for the open left (right) half complex plane. The major part of the computational complexity of the balancing-related methods for model order reduction stems from the solution of large-scale ABEs and Lyapunov equations. In general, numerical methods for matrix equations have a complexity of $\mathcal{O}(n^3)$ (see, e.g., [7, 16]) and therefore, all these approaches are restricted to problems of moderate size. To overcome this limitation for a special class of practically relevant large-scale systems, recent approaches for the solution of Lyapunov equations [2, 3] combine iterative solvers based on the *sign function method* [15] with the hierarchical matrix format and the corresponding arithmetic. This idea is extended to the numerical solution of large-scale ABEs in the following.

This paper is organized as follows. In Section 2, we describe the sign function iteration for the solution of matrix equations and provide some basic facts of the $\mathcal{H}$-matrix format and the corresponding formatted arithmetic. The $\mathcal{H}$-matrix based sign function method is introduced in Section 3. In Section 3.1 we explain how the ABE solution is computed in low-rank factorized form by solving an LLS problem via normal equations. Symmetrizing this solution is explained in Section 3.2 and the derived method is tested on a numerical example in Section 4.

2 Theoretical Background

Necessary basics of the sign function iteration and of the data-sparse hierarchical matrix format are provided in this section.

2.1 The Matrix Sign Function

It is well known that a solution X of an ABE (which is a homogeneous algebraic Riccati equation) can be derived from the invariant subspace of the associated Hamiltonian matrix Z as

$$\underbrace{\begin{pmatrix} A & BB^T \\ 0 & -A^T \end{pmatrix}}_{=:Z} \begin{pmatrix} I_n \\ -X \end{pmatrix} = \begin{pmatrix} I_n \\ -X \end{pmatrix} (A - BB^T X),$$

see for instance [14]. Thus, if (A,B) is stabilizable and (2) holds, the unique stabilizing solution X of (1) can be computed by the Z-invariant subspace corresponding to the stable eigenvalues, i.e. $\Lambda(A-BB^TX) \subset (\Lambda(Z)\cap\mathbf{C}^-)$. Using spectral projection, the kernel of the projector onto the anti-stable Z-invariant subspace $P_+ := (I_{2n} + \text{sign}(Z))/2$ describes the stable Z-invariant subspace. Thus, the stabilizing solution X can be derived from the LLS problem

$$\frac{1}{2}(I_{2n} + \text{sign}(Z)) \begin{pmatrix} I_n \\ -X \end{pmatrix} = 0. \tag{3}$$

One of the numerical methods to compute the sign of Z is based on the Newton iteration for $Z^2 = I$ [15]. To describe this method, consider a matrix $Z \in \mathbf{R}^{n\times n}$ with no eigenvalues on the imaginary axis. The *matrix sign function* for Z is defined by the real version of the Jordan canonical form

$$\text{sign}(Z) := S^{-1} \begin{bmatrix} I_\ell & 0 \\ 0 & -I_{n-\ell} \end{bmatrix} S, \qquad \text{with} \quad Z = S^{-1} \begin{bmatrix} J_\ell^+ & 0 \\ 0 & J_{n-\ell}^- \end{bmatrix} S,$$

and $S \in \mathbf{R}^{n\times n}$, $\Lambda(J_\ell^+) \subset \mathbb{C}^+$, $\Lambda(J_{n-\ell}^-) \subset \mathbb{C}^-$.

To compute the matrix sign function, we use the Newton iteration applied to $(\text{sign}(Z))^2 = I_n$:

$$Z_0 \leftarrow Z, \qquad Z_{j+1} \leftarrow \frac{1}{2}(Z_j + Z_j^{-1}).$$

This so called sign function iteration converges globally quadratically to the sign of Z and is well-behaved in finite-precision arithmetic. In order to solve the ABE (1) satisfying (2), the sign function is applied to the Hamiltonian Z associated with (1). By the block structure of Z, the iteration splits into two parts

$$\begin{aligned} A_0 &\leftarrow A, \quad A_{j+1} \leftarrow \tfrac{1}{2}(A_j + A_j^{-1}), \\ B_0 &\leftarrow B, \quad B_{j+1} \leftarrow \tfrac{1}{\sqrt{2}} \left[B_j, A_j^{-1} B_j \right], \quad j = 0, 1, 2, \ldots, \end{aligned} \tag{4}$$

with quadratic convergence rate and

$$\text{sign}(Z) = \begin{pmatrix} A_\infty & B_\infty B_\infty^T \\ 0 & -A_\infty^T. \end{pmatrix}, \tag{5}$$

using the notations

$$A_\infty := \lim_{j\to\infty} A_j, \quad B_\infty := \lim_{j\to\infty} B_j.$$

In [5, 6], this iteration scheme is used for solving Lyapunov equations and modified for the direct computation of the Cholesky (or full-rank) factors. We review the iteration scheme in Section 3 and propose further improvements using the hierarchical matrix format as briefly introduced in the next section.

2.2 $\mathscr{H}$-Matrix Arithmetic Introduction

In [11], the sign function method for solving algebraic Riccati equations is combined with a data-sparse matrix representation and a corresponding approximate arithmetic. This initiated the idea to use the $\mathscr{H}$-matrix format for computing low-rank solutions of ABEs. As our approach also makes use of this $\mathscr{H}$-matrix format, we will introduce some of its basic facts in the following.

The $\mathscr{H}$-matrix format is a data-sparse representation for a special class of matrices, which often arise in applications. Matrices that belong to this class result, for instance, from the discretization of partial differential or integral equations. Exploiting the special structure of these matrices in computational methods yields reduced computing time and memory requirements. A detailed description of the $\mathscr{H}$-matrix format can be found, e.g., in [9, 10, 12, 13].

The basic idea of the $\mathscr{H}$-matrix format is to partition a given matrix recursively into submatrices $M_{|r\times s}$ that admit low-rank approximations, $\operatorname{rank}(M_{|r\times s}) \le k$, where k denotes the block-wise rank. The corresponding submatrix is stored in factorized form as

$$M_{|r\times s} = AB^T, \quad A \in \mathbb{R}^{r\times k}, B \in \mathbb{R}^{s\times k},$$

all remaining blocks correspond to submatrices which are stored in the usual dense matrix format.

The set of $\mathscr{H}$-matrices of block-wise rank k is denoted by $\mathscr{M}_{\mathscr{H},k}$. The storage requirements for a matrix $M \in \mathscr{M}_{\mathscr{H},k}$ are

$$\mathscr{N}_{\mathscr{M}_{\mathscr{H},k}St} = \mathscr{O}(n\log(n)k)$$

instead of $\mathscr{O}(n^2)$ for the original (full) matrix. We denote by $M_{\mathscr{H}}$ the hierarchical approximation of a matrix M. The formatted arithmetic $\oplus$ ($\ominus$), $\odot$, $(\cdot)^{-1}_{\mathscr{H}}$ is a means to close the set of $\mathscr{H}$-matrices under addition, multiplication and inversion. These operations in formatted arithmetic are performed block-wise with exact addition or multiplication followed by truncating the resulting block back to rank k using a best Frobenius norm approximation. The truncation operator, denoted by $\mathscr{T}_k$, is realized by a truncated singular value decomposition, see, e.g., [10] for more details. For two matrices $A, B \in \mathscr{M}_{\mathscr{H},k}$ and a vector $v \in \mathbf{R}^n$ we consider the formatted arithmetic operations, which all have linear-polylogarithmic complexity:

$$\begin{array}{lll} v \mapsto Av: & & \mathscr{O}(n\log(n)k), \\ A \oplus B & = \mathscr{T}_{\mathscr{H},k}(A+B): & \mathscr{O}(n\log(n)k^2), \\ A \odot B & = \mathscr{T}_{\mathscr{H},k}(AB): & \mathscr{O}(n\log^2(n)k^2). \end{array} \tag{6}$$

In this work the $\mathscr{H}$-inverse $A^{-1}_{\mathscr{H}}$ of a matrix A is computed using an approximate $\mathscr{H}$-LU factorization $A \approx L_{\mathscr{H}} U_{\mathscr{H}}$ followed by an $\mathscr{H}$- forward and $\mathscr{H}$- backward substitution. The complexity of the $\mathscr{H}$-inversion is $\mathscr{O}(n\log^2(n)k^2)$.

Note that in practice, the blockwise rank is chosen adaptively for each matrix block instead of using a fixed rank k. Thus, the rank in each block $M_{|r\times s}$ is determined so that the formatted operation yields an error less than or equal to a prescribed accuracy ε. We will use the $\mathcal{H}$-matrix structure to compute solution factors of ABEs, which reduces the complexity and the storage requirements of the underlying iteration scheme.

3 Efficient Solution of Large-Scale ABEs by Use of the $\mathcal{H}$-Matrix Arithmetic

It is observed that in usual applications which stem from the discretization of some elliptic partial differential operator the B-iterates in (4) and thus the solution X of the ABE (1) have a small (numerical) rank. Thus, memory requirements are reduced by computing low-rank approximations to the factors directly. Furthermore, the hierarchical matrix format is incorporated in the sign function iteration (7) to reduce the cubic complexity and the quadratic storage requirements:

$$A_{j+1} \leftarrow \frac{1}{2}(A_j \oplus A^{-1}_{\mathcal{H},j}), \tag{7}$$

$$B_{j+1} \leftarrow \frac{1}{\sqrt{2}}\left[B_j, A^{-1}_{\mathcal{H},j}B_j\right], \quad j = 0,1,2,\ldots. \tag{8}$$

for details see [1, 3, 6]. Since the number of columns of B_j in (8) is doubled in each iteration step, it is proposed in [6] to apply a rank-revealing LQ factorization (RRLQ) [8] in order to reveal the expected low numerical rank. We denote the numerical rank determined by a threshold τ by t. Thus, after convergence, $B_\infty \in \mathbb{R}^{n \times t}$. Since the spectral norm of an $\mathcal{H}$-matrix can be computed without much effort, it is advised to choose

$$||A_{j+1} - A_j||_2 \leq \text{tol}\, ||A_{j+1}||_2$$

as stopping criterion for the iteration.

3.1 Solving the LLS Problem (3)

When the sign function iteration has converged, including (5) for the sign of Z in (3), the LLS problem is equivalently given by

$$\underbrace{\begin{pmatrix} B_\infty B_\infty^T \\ I_n - A_\infty^T \end{pmatrix}}_{\tilde{A}} X = \underbrace{\begin{pmatrix} I_n + A_\infty \\ 0_n \end{pmatrix}}_{\tilde{b}}. \tag{9}$$

It admits a unique solution if $\operatorname{rank}(\tilde{A}) = n$. Since $\operatorname{rank}(I_n - A_\infty^T) = n - \ell$, where ℓ is the number of unstable eigenvalues of A, we must have $\operatorname{rank}(B_\infty B_\infty^T) \geq \ell$.

To proceed the computations in low complexity we solve the problem using the normal equations

$$\tilde{A}^T\tilde{A}X = \tilde{A}^T\tilde{b},$$

exploiting that B_∞ is of low rank and A_∞ is stored as $\mathscr{H}$-matrix. The matrices involved are computed in the following way:

$$\begin{aligned}
\underbrace{\tilde{A}^T\tilde{A}}_{\mathscr{H}-matrix} &= \underbrace{B_\infty B_\infty^T B_\infty B_\infty^T}_{low\ rank} + \underbrace{(I_n - A_\infty)}_{\mathscr{H}-matrix}\underbrace{(I_n - A_\infty)^T}_{\mathscr{H}-matrix} \\
\underbrace{\tilde{A}^T\tilde{b}}_{low\ rank} &= \underbrace{B_\infty\,[B_\infty^T + B_\infty^T A_\infty]}_{low\ rank}
\end{aligned}$$

where the notation "low rank" indicates that the matrix is stored as product of two rectangular matrices.

Using the $\mathscr{H}$-Cholesky-decomposition of $\tilde{A}^T\tilde{A} = CC^T$ and $\mathscr{H}$-based forward and backward substitutions, we compute the stabilizing solution X of (1) as low rank matrix, i.e.

$$X = X_1X_2^T, \quad X_1 = C^{-1}B_\infty, \quad X_2 = C^{-1}(A_\infty^T B_\infty + B_\infty) = C^{-1}A_\infty^T B_\infty + X_1. \tag{10}$$

3.2 Symmetrizing the Low-Rank Presentation of ABE Solutions Obtained by Normal Equations

The stabilizing solution X of an ABE is known to be symmetric [4]. This property is not reflected in the representation (10) which is not a problem for certain applications as model order reduction of unstable systems. But in case that symmetry of X is required we give a procedure that achieves this task.

Let $B_\infty \in \mathbb{R}^{n\times t}$. From [4] we know that $\operatorname{rank}(X) = \ell$, thus $t \geq \ell$. As $X = X^T \geq 0$ we have $X_1X_2^T = X_2X_1^T$ and $X_1X_2^T$ is the positive semidefinite square root of X^2,

$$X^2 = (X_1X_2^T)^2 = X_1X_2^TX_2X_1^T. \tag{11}$$

Now let $X_1 = Q_1R_1$ be a thin QR decomposition with $Q_1 \in \mathbb{R}^{n\times t}$, $R_1 \in \mathbb{R}^{t\times t}$ upper triangular and compute a singular value decomposition

$$X_2R_1^T = U\Sigma V^T. \tag{12}$$

We then get from (11)

$$\begin{aligned}
X^2 = Q_1R_1X_2^TX_2R_1^TQ_1^T \quad &= Q_1(U\Sigma V^T)^T(U\Sigma V^T)Q_1^T \\
= (Q_1V\Sigma^T)(Q_1V\Sigma^T)^T &= (Q_1V\hat{\Sigma}^2V^TQ_1^T)^2,
\end{aligned}$$

where $\hat{\Sigma} = \Sigma(1:t,:) \in \mathbb{R}^{t\times t}$.

Hence, $Q_1 V \hat{\Sigma} V^T Q_1^T$ is the positive semidefinite square root of X^2 and $Q_1 V \hat{\Sigma}^{\frac{1}{2}} V^T Q_1^T$ is the positive semidefinite square root of X. Due to uniqueness of semidefinite square roots,

$$X = X_1 X_2^T = Q_1 V \hat{\Sigma} V^T Q_1^T.$$

A rank-t factor of X is thus given by

$$Y = Q_1 V \hat{\Sigma}^{\frac{1}{2}}$$

as $X = YY^T$.

Remark 1. Note that the accumulation of $U \in \mathbb{R}^{n \times t}$ in (12) is not necessary which reduces the cost of the SVD computation from $14nt^2 + 8t^3$ to $4nt^2 + 8t^3$ flops.

4 Numerical Results

In this section we examine the accuracy and complexity of the data-sparse approach for the numerical solution of ABEs. As exemplary system we consider the following reaction-diffusion equation

$$\frac{\partial \mathbf{x}}{\partial t}(t,\xi) = \Delta \mathbf{x}(t,\xi) + c\mathbf{x}(t,\xi) + b(\xi)u(t), \qquad \xi \in (0,1)^2, t \in (0,\infty),$$

which is discretized in space by finite elements, leading to the LTI system

$$\dot{x}(t) = \underbrace{(\tilde{A} + cI_n)}_{:=A} x(t) + Bu(t). \tag{13}$$

For the problem sizes $n = 4096$ and $n = 16,384$, we choose the parameter c such that one eigenvalue of the coefficient matrix A has positive real part: $\lambda \approx 0.25$. We compute the relative residual

$$\frac{\|A^T X + XA - XBB^T X\|_F}{2(\|A\|_F \|X\|_F) + \|X\|_F^2 \|BB^T\|_F}$$

of the ABE (1) obtained by applying (7), (8) and (10) to the unstable system (13). We vary the parameters τ for the numerical rank decision in the RRLQ factorization and ε, the approximation error in the adaptive rank choice of the $\mathcal{H}$-matrix arithmetic. The numerical rank of B_∞, the computational time and the accuracy are depicted in Table 1. We observe in Table 1 high accuracy in the solution factors computed with the algorithm in $\mathcal{H}$-matrix arithmetic by low numerical ranks of B_∞ and in low execution time. The results of the parameter variation show the expected behavior, we have increasing accuracy as ε gets smaller and the relative residual is observed to remain bounded from above for increasing problem size.

n	ε	τ	# it.	rank(B_∞, τ)	time [sec]	rel. residual
4096	1.e-04	1.e-04	26	36	261	7.1e-06
	1.e-06	1.e-04	22	14	391	1.8e-07
	1.e-08	1.e-04	19	14	635	3.8e-08
	1.e-06	1.e-06	22	31	395	1.8e-07
	1.e-08	1.e-06	19	21	636	3.7e-08
	1.e-08	1.e-08	19	39	639	3.7e-08
16,384	1.e-04	1.e-04	27	34	2376	2.3e-05
	1.e-06	1.e-04	26	15	4235	5.2e-07
	1.e-08	1.e-04	22	14	7136	6.1e-09
	1.e-06	1.e-06	26	42	4273	5.2e-07
	1.e-08	1.e-06	22	23	7150	6.0e-09
	1.e-08	1.e-08	22	42	7183	5.9e-09

Table 1 Accuracy and rank rank(B_∞, τ) of the computed ABE solution for different problem sizes and parameter combinations.

References

1. Baur, U.: Control-Oriented Model Reduction for Parabolic Systems. Dissertation, Inst. f. Mathematik, TU Berlin, Str. des 17. Juni 136, D-10623 Berlin, FRG (2008)
2. Baur, U., Benner, P.: Factorized solution of the Lyapunov equation by using the hierarchical matrix arithmetic. Proc. Appl. Math. Mech. **4**(1), 658–659 (2004)
3. Baur, U., Benner, P.: Factorized solution of Lyapunov equations based on hierarchical matrix arithmetic. Computing **78**(3), 211–234 (2006)
4. Benner, P.: Computing Low-Rank Solutions of Algebraic Bernoulli Equations. Tech. rep., Fakultät für Mathematik, TU Chemnitz, 09107 Chemnitz, FRG (2007)
5. Benner, P., Claver, J., Quintana-Ortí, E.: Efficient solution of coupled Lyapunov equations via matrix sign function iteration. In: A.D. et al. (ed.) Proc. 3rd Portuguese Conf. on Automatic Control CONTROLO'98, Coimbra, pp. 205–210 (1998)
6. Benner, P., Quintana-Ortí, E.: Solving stable generalized Lyapunov equations with the matrix sign function. Numer. Algorithms **20**(1), 75–100 (1999)
7. Datta, B.: Numerical Methods for Linear Control Systems. Elsevier Academic Press (2004)
8. Golub, G., Van Loan, C.: Matrix Computations, third edn. Johns Hopkins University Press, Baltimore (1996)
9. Grasedyck, L.: Theorie und Anwendungen Hierarchischer Matrizen. Dissertation, University of Kiel, Kiel, Germany (2001). In German, http://e-diss.uni-kiel.de/diss_454
10. Grasedyck, L., Hackbusch, W.: Construction and arithmetics of $\mathcal{H}$-matrices. Computing **70**(4), 295–334 (2003)
11. Grasedyck, L., Hackbusch, W., Khoromskij, B.N.: Solution of large scale algebraic matrix Riccati equations by use of hierarchical matrices. Computing **70**(2), 121–165 (2003)
12. Hackbusch, W.: A sparse matrix arithmetic based on $\mathcal{H}$-matrices. I. Introduction to $\mathcal{H}$-matrices. Computing **62**(2), 89–108 (1999)
13. Hackbusch, W., Khoromskij, B.N.: A sparse $\mathcal{H}$-matrix arithmetic. II. Application to multi-dimensional problems. Computing **64**(1), 21–47 (2000)
14. Lancaster, P., Rodman, L.: The Algebraic Riccati Equation. Oxford University Press, Oxford (1995)
15. Roberts, J.D.: Linear model reduction and solution of the algebraic Riccati equation by use of the sign function. Internat. J. Control **32**, 677–687 (1980)
16. Sima, V.: Algorithms for Linear-Quadratic Optimization, *Pure and Applied Mathematics*, vol. 200. Marcel Dekker, Inc., New York, NY (1996)
17. Zhou, K., Salomon, G., Wu, E.: Balanced realization and model reduction for unstable systems. Int. J. Robust Nonlinear Control **9**(3), 183–198 (1999)

A Purely Algebraic Approach to Preconditioning Based on Hierarchical *LU* Factorizations

M. Bebendorf and T. Fischer

Abstract The efficiency of hierarchical matrices depends on the quality of the block partition. We describe a nested dissection partitioning of the matrix into blocks that uses only the matrix graph and requires a logarithmic-linear number of operations. This block partition allows to compute a hierarchical *LU* decomposition with small fill-in. Furthermore, the algebraic approach admits, in contrast to the usual geometric partitioning, general grids for finite element discretization of elliptic boundary value problems.

1 Introduction

We consider large-scale finite element matrices $A \in \mathbb{R}^{I \times I}$, where I is an index set. Such matrices are usually treated by iterative solvers, which may converge slowly due to ill-conditioning. In order to accelerate the convergence, the FE system is preconditioned. We propose a preconditioning technique which is based on an approximated *LU* decomposition.

In the last years the structure of hierarchical matrices ($\mathscr{H}$-matrices) [5, 7, 2] has proved to be able to handle approximations of discrete solution operators of elliptic partial differential boundary value problems. Hierarchical matrices rely on low-rank approximations on each block of a partition P of the set of matrix indices $I \times I$. In order to guarantee the existence of such low-rank approximations, each block $b = t \times s \in P$ has to satisfy either the admissibility condition

$$\min\{\operatorname{diam} X_t, \operatorname{diam} X_s\} \leq \eta \operatorname{dist}(X_t, X_s) \tag{1}$$

Mario Bebendorf and Thomas Fischer
Mathematical Institute, Faculty of Mathematics and Computer Science, University Leipzig, Johannisgasse 26, 04103 Leipzig, e-mail: fischer@math.uni-leipzig.de

or $\min\{|t|,|s|\} \le n_{\min}$ for given parameters $\eta > 0$ and $n_{\min} \in \mathbb{N}$. Here, X_t denotes the support of a cluster t, which is the union of the supports of the basis functions corresponding to the indices in t:

$$X_t := \bigcup_{i \in t} X_i.$$

The partition is normally generated by recursive subdivision of $I \times I$. The recursion stops in blocks which satisfy (1) or which are small enough. For a given partition P the set of $\mathscr{H}$-matrices with blockwise rank k is defined by

$$\mathscr{H}(P,k) := \{M \in \mathbb{R}^{I \times I} : \operatorname{rank} M_b \le k \text{ for all } b \in P\}.$$

In [1] it was proved that the LU decomposition of FE matrices of uniformly elliptic operators can be approximated by $\mathscr{H}$-matrices with logarithmic-linear complexity. Up to now it was possible to guarantee logarithmic-linear complexity only for quasi-uniform discretizations and for some special grids (see [6]), since the generated cluster trees had to be balanced with respect to both, geometry and cardinality.

This article treats the set up of the approximated factors L and U in the hierarchical matrix format using only the matrix graph

$$G_A := \{(i,j) \in I \times I : a_{ij} \neq 0\} \tag{2}$$

of A. The construction of the partition is described such that instead of the geometric condition (1) the algebraic admissibility condition

$$\min\{\operatorname{diam} t, \operatorname{diam} s\} \le \eta \operatorname{dist}(t,s). \tag{3}$$

is satisfied on each large enough block, where

$$\operatorname{diam} t := \max_{i,j \in t} d_{ij} \quad \text{and} \quad \operatorname{dist}(t,s) := \min_{i \in t,\, j \in s} d_{ij}.$$

Here, d_{ij} is the shortest path between i and j in the matrix graph.

The power of condition (3) is that it does not involve the geometry of the discretization. Hence, clustering has to account only for the cardinality of the clusters. This directly generalizes the theory of $\mathscr{H}$-matrix approximations to arbitrary grids including adaptively refined ones. Additionally, the algebraic approach allows to minimize the interface in nested dissection reorderings. Since the size of the interface determines the quality of the partition P, one can expect an acceleration of the hierarchical LU factorization algorithm. Condition (3), however, involves the distance $\operatorname{dist}(t,s)$ of two clusters t and s in the matrix graph and their diameters $\operatorname{diam} t$ and $\operatorname{diam} s$. The efficient (i.e., with complexity of order $|t|+|s|$) computation of these quantities is a challenge. One should, however, keep in mind that for matrix partitioning it is not required to know their exact values. In this article we will therefore present efficient multilevel algorithms for the computation of approximations of these quantities.

The structure of this article is as follows. The algebraic construction of the cluster tree is presented in the Section 2.1. In the Sections 2.2 and 2.3 we describe the efficient evaluation of (3). The last section contains numerical results which compare $\mathscr{H}$-matrix *LU* factorizations based on geometric and algebraic matrix partitioning with the direct solver PARDISO [11].

2 Algebraic Matrix Partitioning

To construct a partition one usually generates a *cluster tree* $T_I = (V_{T_I}, E_{T_I})$, which is a graph satisfying the following conditions:

1. the index set I is the root of T_I,
2. $t = \cup_{t' \in S_I(t)} t'$ for all $t \in V_{T_I} \backslash \mathscr{L}(T_I)$ and t' are pairwise disjoint,
3. $|S_I(t)| > 1$ for all $t \in V_{T_I} \backslash \mathscr{L}(T_I)$,

where the elements of the set of sons $S_I(t) := \{t' \in V_{T_I} : (t,t') \in E_{T_I}\}$ are pairwise disjoint and $\mathscr{L}(T_I) := \{t \in V_{T_I} : |S_I(t)| = 0\}$ denotes the set of leafs.

Condition (3) does not contain any information about the geometry of the discretization. Hence, the assumption that the cluster tree T_I is geometrically balanced can be omitted, which allows to treat general grids including adaptively refined ones. Therefore, we use a cardinality balanced cluster tree and assume that the diameter of a generated cluster is equivalent to its cardinality in the sense that there are constants $c_1, c_2 > 0$ such that

$$c_1|t| \leq (\operatorname{diam} t)^d \leq c_2|t| \quad \text{for all } t \in T_I. \tag{4}$$

2.1 Algebraic Construction of the Cluster Tree

In order to reduce *fill-in* during *LU* factorization, I is decomposed using the nested dissection method [4]. Nested dissection is based on the matrix graph $G_A = (V, E)$. In each step it partitions the vertex set V into V_1, V_2, S such that V_1, V_2 are of approximately equal size and S separates V_1, V_2 and additionally satisfies $|S| \ll |V_1|$. The vertex sets V_1 and V_2, corresponding to $t_1, t_2 \subset I$, are recursively partitioned, and we achieve a nested dissection cluster tree (see Fig. 1).

Each nested dissection step can be separated in two phases.

(1) The vertices are divided in two disjoint sets V_1', V_2'. The bipartition can be computed using *spectral bisection* based on the Fiedler vector, which is the eigenvector to the second smallest eigenvalue; see [3]. Since computing eigenvectors of large matrices is computationally expensive, multilevel ideas have been introduced to accelerate the process [9]. For this purpose the graph G_A is coarsened into a sequence $G^{(1)}, \ldots, G^{(\kappa)}$ such that $|V| > |V^{(1)}| > \cdots > |V^{(\kappa)}|$. Spectral bisection can then be applied to the smallest graph $G^{(\kappa)}$. The resulting partition P_κ is projected back to

G_A by going through the intermediate partitions $P_{\kappa-1},\dots,P_1$. The partition P_{i+1} can be improved by refinement heuristics such as the Kernighan-Lin algorithm [10].

Subdividing V in this manner in some sense minimizes the *edge cut* C, i.e., a set of edges $C \subset E$ such that $G' = (V, E \backslash C)$ is no longer connected. The size of the edge cut is in $O(|V|^{1-1/d})$.

(2) The vertex set S, which separates V_1 and V_2, is computed. To this end we consider the boundaries

$$\begin{aligned} \partial V_1' &:= \{u \in V_1' : \exists v \in V_2' \text{ and } (u,v) \in E\}, \\ \partial V_2' &:= \{v \in V_2' : \exists u \in V_1' \text{ and } (v,u) \in E\} \end{aligned}$$

of V_1' and V_2' and the edge set $E_{12} = \{(u,v) \in E, u \in \partial V_1', v \in \partial V_2'\}$ between $\partial V_1'$ and $\partial V_2'$. The bipartite graph

$$B := (\partial V_1' \cup \partial V_2', E_{12})$$

is constructed which takes $O(|V|^{1-1/d})$ operations.

In order to get a small separator S, the *minimal vertex cover algorithm* [8] is applied to B. A minimal vertex cover for bipartite graphs can be calculated with complexity $O(|V|^{3/2\cdot(1-1/d)})$. Finally, the vertices belonging to the minimal vertex cover are moved out of V_1' and V_2' to S to obtain a partition V_1, V_2, S of V.

Since the partitioning algorithm ensures that the cardinality of each cluster from the same level in T_I is of the same order of magnitude, i.e., $|t| \sim |I|2^{-\ell}$ for $t \in T_I^{(\ell)}$, we can guarantee logarithmic depth of T_I.

Our algorithm extends the nested dissection cluster tree to ensure that every level of the cluster tree stores a partition of the index set I. To this end the separator index set s of the ℓ-th level of T_I is copied to the next level of the cluster tree as long as the cardinality of s is smaller than $|I|2^{-\ell'}, \ell' > \ell$. A sequence $s^{(1)}, s^{(2)}, \dots, s^{(\ell'-\ell)}$ of separators is obtained, each containing the same index set but in different levels. We say $s^{(1)}, s^{(2)}, \dots, s^{(\ell'-\ell)}$ have the same *virtual depth* ℓ, the depth of $s^{(1)}$. If $|s| \geq |I|2^{-\ell'}$, the separator s is recursively partitioned as described in phase one of the nested dissection algorithm. In Fig. 2 separators s_0 and s_0' contain the same index sets and are in different levels but have the same virtual depth.

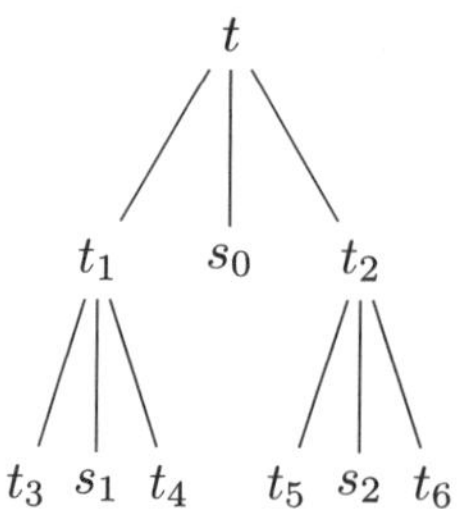

Fig. 1 nested dissection cluster tree

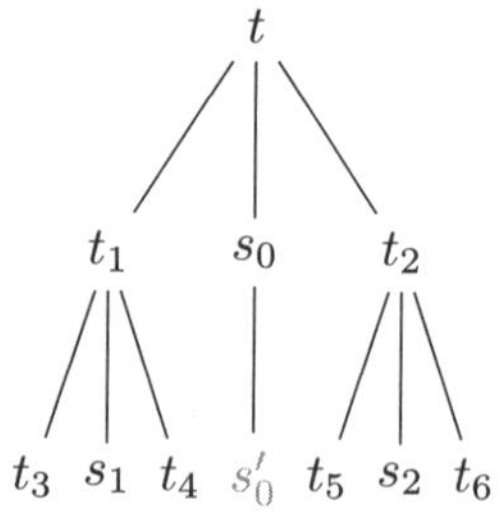

Fig. 2 subtree of T_I rooted at $t \in T_I^{(\ell)}$

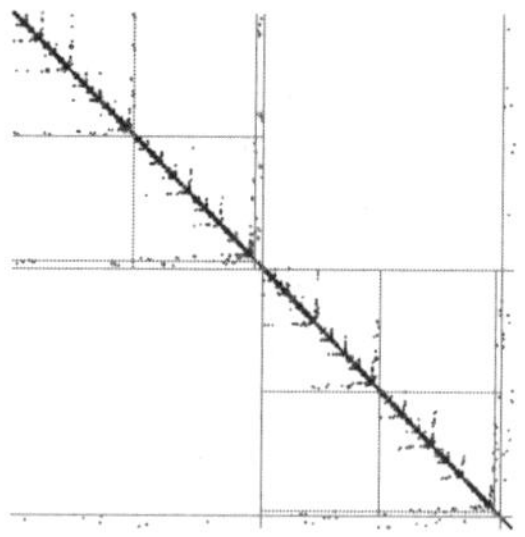

Fig. 3 matrix structure after two nested dissection steps

2.2 The Algebraic Admissibility Condition

In this section we present the evaluation of the admissibility condition (3) based on the cluster tree described in Section 2.1.

Let $t \in T_I^{(\ell-1)}$ be decomposed into $t_1, t_2, s \in T_I^{(\ell)}$, where t_1 and t_2 are separated by s, using the algorithm described in Section 2.1; see Fig. 2. This means that there does not exist any edge between t_1 and t_2. As consequence, $a_{ij} = 0$ and $a_{ji} = 0$ for $i \in t_1, j \in t_2$; see Fig. 3. Most of the information of the matrix is contained in the interface blocks $t_1 \times s$, $t_2 \times s$, $s \times t_1$, $s \times t_2$, and $s \times s$. In order to guarantee logarithmic-linear complexity, these blocks are decomposed into sub-blocks that either can be approximated or are small. The systematical search for pairs of index sets in T_I that form a block which can be approximated by a low rank matrix creates the so called *block cluster tree* $T_{I\times I}$, where the admissible blocks can be found in the leafs $\mathscr{L}(T_{I\times I})$ of $T_{I\times I}$.

In contrast to the usual definition of $T_{I\times I}$, we define the set of sons

$$S_{I\times I}(t\times s) := \begin{cases} \emptyset, & \text{if } S_I(t) = \emptyset \text{ or } S_I(s) = \emptyset, \\ \emptyset, & \text{if } t \neq s \text{ and neither } t \text{ nor } s \text{ are separators,} \\ \emptyset, & \text{if } t \text{ or } s \text{ are separators and satisfy (3),} \\ S_I(t) \times S_I(s), & \text{else.} \end{cases}$$

The block cluster tree is generated by recursively applying $S_{I\times I}$ to the root $I \times I$. The following definition helps to accelerate the admissibility test, i.e., the evaluation of (3).

Definition 1. Two index sets $t_1, t_2 \subset I$ are denoted as *neighbored* if there exists an edge in G_A connecting indices of t and t', i.e., $\exists i \in t_1, \exists j \in t_2$ such that $(i, j) \in E$.

As cluster t is called contiguous if there are two indices $i_{\min}$ and $i_{\max}$ such that

$$t_1 = \{i : i_{\min} \leq i < i_{\max}\}.$$

Note that checking whether two contiguous clusters t_1 and t_2 are neighbored can be be done with $O(\min\{|t_1|, |t_2|\})$ operations.

If $t_1, t_2 \subset I$ are neighbored, it holds that $t_1 \in \mathscr{N}_\eta(t_2)$ and $t_2 \in \mathscr{N}_\eta(t_1)$, where

$$\mathscr{N}_\eta(t) := \{t' \in T_I^{(\ell)} : \operatorname{diam} t > \eta \operatorname{dist}(t, t')\}$$

denotes the *near-field* of t. A block is admissible if $t_2 \notin \mathscr{N}_\eta(t_1)$ or $t_1 \notin \mathscr{N}_\eta(t_2)$. If they are not neighbored, it is necessary to compute the distance $\operatorname{dist}(t_1, t_2)$ between them. The evaluation of $\operatorname{dist}(t_1, t_2)$ involves the computation of $|t_1| \cdot |t_2|$ shortest paths in the matrix graph, each of which takes $O(|I|)$ operations with breadth-first search. Since our aim is to preserve the logarithmic-linear complexity, it is necessary to approximate the distance between t_1 and t_2.

2.3 Approximation of Distance and Diameter

Assume that the father of $t_1 \times t_2 \in T_{I\times I}^{(\ell)}$ is not admissible and t_1, t_2 are not neighbored.

The first step to accelerate the computation of the distance is to calculate all neighbors of the same level in the cluster tree. Assume that the neighbors in the level ℓ of the tree T_I are known. Obviously, for each cluster $t \in T_I^{(\ell)}$ the pairs (t_1, s_0) and (s_0, t_2) (so-called "a-priori neighbors") are neighbored, where $S(t) = \{t_1, s_0, t_2\}$. Since two clusters can be neighbored only if their parents are neighbored, we can restrict the search for neighbors to the set $S_I(t_1) \times S_I(t_2)$, where t_1 and t_2 are neighbored clusters in the ℓ-th level.

Example 1. In Fig. 4 a-priorily known neighbors are symbolized by dashed lines. Computed neighbors are characterized by dotted lines.

In order to compute the approximate distance between $t_1, t_2 \in T_I^{(\ell)}$, we construct a graph G_D. There are predecessors $pre(t_1)$ and $pre(t_2)$ of t_1 and t_2 such that $pre(t_1)$ and $pre(t_2)$ are neighbored. The vertices of G_D consist of the descendants of $pre(t_1)$ and $pre(t_2)$ in the ℓ-th level of T_I. G_D contains a weighted edge between two vertices if and only if the clusters are neighbored. The weight is the difference between ℓ and the virtual depth of the neighbor node. Since $pre(t)$ and $pre(t')$ are neighbored, the graph G_D is connected and it is possible to calculate the approximate distance using Dijkstra's algorithm [12]. In a forthcoming article it is proved that the number of nodes in G_D is bounded from above by a constant.

Example 2. In Fig. 4 assume that $t_7 \times s_0''$ is not admissible. Therefore, the admissibility of the pair (t_{15}, s_0'''), for instance, is checked. Using $pre(t_{15}) = t_3$ and $pre(s_0''') = s_0'$, we obtain the vertex set $\{t_{15}, s_7, t_{16}, s_3', t_{17}, s_8, t_{18}, s_0'''\}$ of G_D, which is depicted in Fig. 5. Dijkstra's algorithm results in an approximate distance between t_{15} and s_0''' of seven.

This approach can be improved by the following iterative refinement procedure. Dijkstra's algorithm not only computes the distance between t_1 and t_2 but also the nodes of the shortest path. We construct a new graph consisting of vertices from level $\ell + m$ for some m rooted at the shortest path nodes. Its edges are defined as in the previous graph G_D. The computation of the shortest path between t_1 and t_2 in this graph will lead to an improved approximation of $\text{dist}(t_1, t_2)$.

Example 3. Assume in Example 2 that Dijkstra's algorithm calculated the shortest path $t_{15}, s_3', t_{18}, s_0'''$. The subtrees rooted at the path nodes are depicted in Fig. 6. We choose vertices of level $\ell + 2$ and determine the edge set considering the neighbors; see Fig. 7. Dijkstra's algorithm is then applied to this refined graph.

It remains to compute an approximation to the diameter of a cluster t. This can be done by a breadth-first search [12]. The result is bounded from below by the radius $r(t) := \min_{i\in t} \max_{j\in t} d_{ij}$ of t and bounded from above by $\operatorname{diam} t$.

In a forthcoming article we prove that it is possible to generate the approximate LU factorization in almost linear time using this matrix.

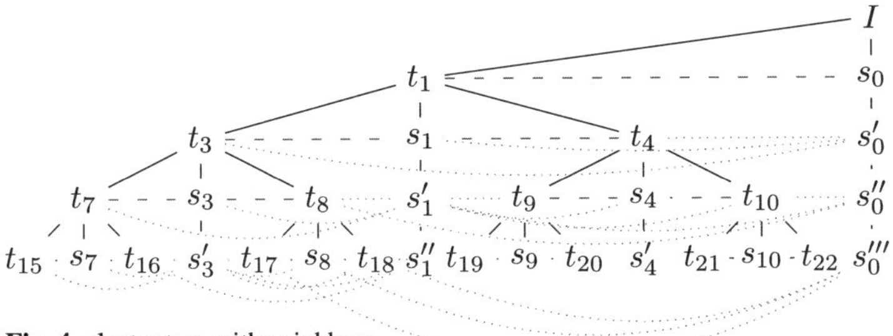

Fig. 4 cluster tree with neighbors

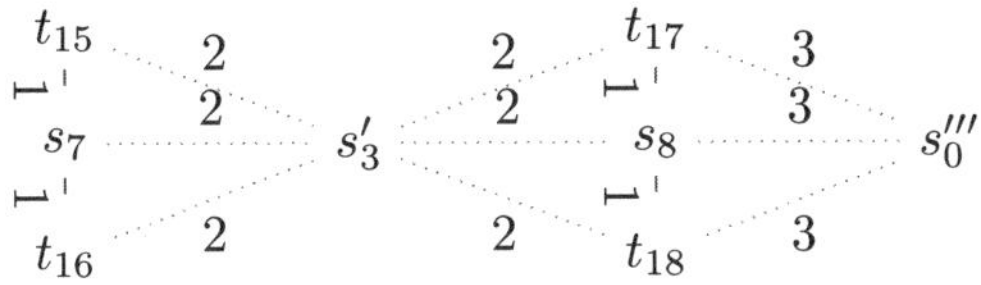

Fig. 5 G_D for computing the approximation for $\text{dist}(t_{15}, s_0''')$

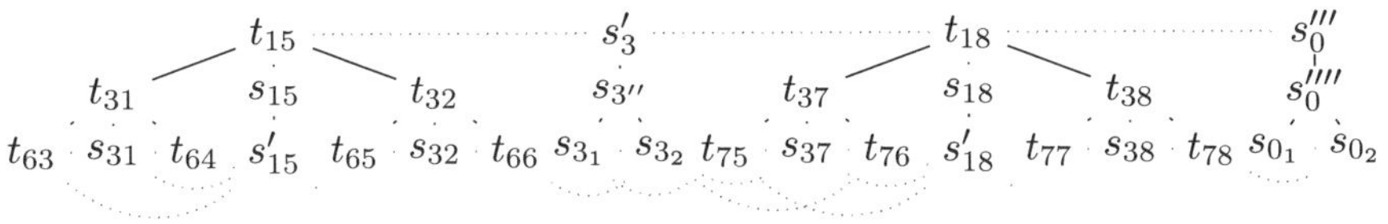

Fig. 6 subtrees and neighbors between clusters

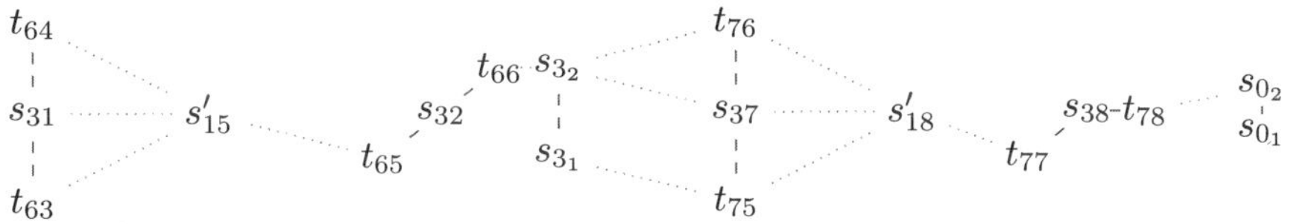

Fig. 7 refined graph

3 Numerical Results

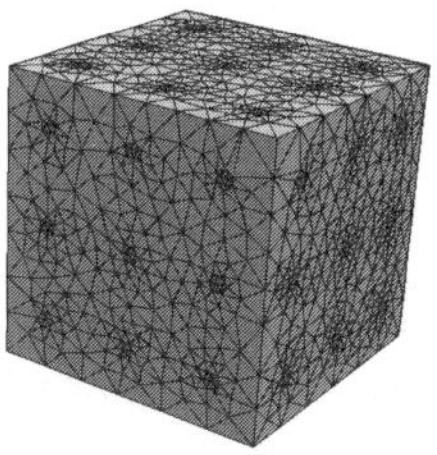

Fig. 8 Computational domain

size	algebraic partitioning	algebraic $\mathscr{H}$-Cholesky		PCG		geometric $\mathscr{H}$-Cholesky		PARDISO Cholesky	
	t in s	t in s	MB	#it	t in s	t in s	MB	t in s	MB
32 429	0.72	0.86	25	18	0.37	1.85	29	0.50	38
101 296	3.06	3.42	77	23	1.62	8.68	105	4.17	198
658 609	25.35	31.38	726	35	19.38	91.61	805	147.81	2659
2 539 954	106.40	158.68	2920	61	142.27	471.54	3507	–	–

Table 1 Algebraic $\mathscr{H}$-Cholesky preconditioner, geometric $\mathscr{H}$-Cholesky preconditioner, PARDISO Cholesky factorization

The results shown in Table 1 were obtained for the Laplacian on the computational domain shown in Fig. 8. The computation were done on an Intel Xeon 3.0 GHz with 16 GB of core memory. The time required to compute the matrix partition based on the algebraic admissibility condition (1) scales almost linearly with the number of degrees of freedom. The accuracy of the approximate LU factorization was chosen to $\delta = 0.5$. Compared with the usual geometric approach to matrix partitioning, the algebraic method leads to a significantly faster computation of the preconditioner. Additionally, the memory consumption of the approximation is reduced.

References

1. Bebendorf, M.: Why finite element discretizations can be factored by triangular hierarchical matrices. SIAM J. Num. Anal. **45**(4), 1472–1494 (2007)
2. Bebendorf, M.: Hierarchical matrices: a means to efficiently solve elliptic boundary value problems, *LNCSE*, vol. 63. Springer (2008)
3. Fiedler, M.: A property of eigenvectors of nonnegative symmetric matrices and its application to graph theory. Czech. Math. J. **25**, 619–633 (1975)
4. George, A.: Nested dissection of a regular finite element mesh. SIAM J. Numer. Anal. **10**(2), 345–363 (1973)
5. Hackbusch, W.: A sparse matrix arithmetic based on $\mathscr{H}$-matrices. Part I: Introduction to $\mathscr{H}$-matrices. Computing **62**(2), 89–108 (1999)
6. Hackbusch, W., Khoromskij, B.N.: $\mathscr{H}$-matrix approximation on graded meshes. In: J.R. Whiteman (ed.) The Mathematics of Finite Elements and Applications X, pp. 307–316. Elsevier (2000)
7. Hackbusch, W., Khoromskij, B.N.: A sparse $\mathscr{H}$-matrix arithmetic. Part II: Application to multi-dimensional problems. Computing **64**(1), 21–47 (2000)
8. Hopcroft, J.E., Karp, R.M.: An $n^{5/2}$ algorithm for maximum matchings in bipartite graphs. SIAM Journal on Computing **2**(4), 225–231 (1973)
9. Karypis, G., Kumar, V.: A fast and high quality multilevel scheme for partitioning irregular graphs. SIAM Journal on Scientific Computing **20**(1), 359–392 (1999)
10. Kernighan, B., Lin, S.: An efficient heuristic procedure for partitioning graphs. The Bell System Technical Journal **29** (1970)
11. Schenk, O., Gärtner, K.: Solving unsymmetric sparse systems of linear equations with PARDISO. Future Gener. Comput. Syst. **20**(3), 475–487 (2004)
12. Sedgewick, R.: Part 5, graph algorithms. In: Algorithms in C, 5, 3 edn. Addison-Wesley (2002)

Additive Schwarz Preconditioners for Degenerate Problems with Isotropic Coefficients

S. Beuchler

Abstract This paper deals with the numerical solution of the degenerate elliptic boundary value problem $-\nabla \cdot x^{\alpha}\nabla u(x,y) = f(x,y)$, $\alpha \geq 0$ in $\Omega = (0,1)^2$. This boundary value problem is discretized by piecewise linear finite elements on regular Cartesian grids. The corresponding linear system of algebraic equations is solved by a preconditioned conjugate gradient method with the Bramble-Pasciak-Xu preconditioner. A uniform bound of the condition number of the preconditioned system matrix is proved for $\alpha \neq 1$. The proof makes use of another additive Schwarz splitting arising from an overlapping domain decomposition of the computational domain Ω. Some numerical experiments show the efficiency of the proposed algorithm also for general polygonal domains.

1 Introduction

In this paper, we investigate a degenerate and isotropic boundary value problem of second order. In the past, degenerate problems have been considered relatively rarely. One reason is the unphysical behavior of the partial differential equation (PDE), which is quite unusual in technical applications. One work focusing on this type of partial differential equation is the book of Kufner and Sändig [5]. Nowadays, problems of this type are becoming more and more popular because there are stochastic PDE's of a similar structure. An example of an isotropic degenerate stochastic PDE is the elliptic part of the Black-Scholes PDE [6]. The discretization of such a boundary value problem using the h-version of the finite element method (FEM) leads to a linear system of algebraic equations. It is well known from the literature that preconditioned conjugate gradient (PCG) methods are among the most efficient iterative solvers for systems of this type.

Sven Beuchler
JKU Linz, Institute of Computational Mathematics, Altenberger Straße 69, 4040 Linz, Austria
e-mail: sven.beuchler@jku.at

In this paper, we analyze the Bramble-Pasciak-Xu (BPX) preconditioner which has been investigated and proposed for uniformly bounded elliptic problems in [4], [7]. We prove the optimality of the BPX-preconditioner for a class of degenerate elliptic problems. Other possible preconditioners are overlapping domain decomposition preconditioners, see [2]. For some other classes of degenerate elliptic problems, multigrid preconditioners or overlapping domain decomposition preconditioners can be used [1], [3].

The outline of this paper is as follows. In section 2, the discretization of the boundary value problem is described. Section 3 is devoted to the definition and analysis of the BPX preconditioner. Section 4 presents some numerical experiments. Section 5 concludes the paper.

Throughout this paper the notation $a \preceq b$ denotes $a \leq cb$ with a constant c which is independent of the discretization parameter. The notation $a \sim b$ means $a \preceq b$ and $a \succeq b$.

2 Setting of the Problem

We investigate the following boundary value problem. Let $\Omega = (0,1)^2$ be the unit square. Find $u \in \mathbb{H}_{\omega,0} := \{u \in L_2(\Omega) : \int_\Omega \omega^2(x)(\nabla u)^\top \nabla u \,\mathrm{d}(x,y) < \infty, u\,|_{\partial\Omega} = 0\}$ such that

$$a(u,v) := \int_\Omega \omega^2(x)(\nabla v)^\top \nabla u \,\mathrm{d}(x,y) = (f,v) \quad \forall v \in \mathbb{H}_{\omega,0} \tag{1}$$

with a weight function $\omega^2(x) = x^\alpha$, $\alpha > 0$. Since $\omega^2(0) = 0$, the bilinear form is not elliptic in $H^1(\Omega)$. Nevertheless, the weight is positive in the open domain Ω. A problem of the type (1) is called a degenerate elliptic problem.

We discretize problem (1) by piecewise linear finite elements on the regular Cartesian grid consisting of congruent, isosceles, right triangles. For this purpose, some notation is introduced. Let k be the level of approximation and $n = 2^k$. Let $x_\mu^k = \left(\frac{\mu_1}{n}, \frac{\mu_2}{n}\right)$ with the multi-index $\mu = (\mu_1, \mu_2) \in \{0,1,\ldots,n\}^2$. The domain Ω is divided into congruent, isosceles, right triangles $\tau_\mu^{s,k}$, where $\mu \in \{0,1,\ldots,n-1\}^2$ and $s = 1,2$, see Figure 1. The triangle $\tau_\mu^{1,k}$ has the three vertices $\left(\frac{\mu_1}{n}, \frac{\mu_2}{n}\right)$, $\left(\frac{\mu_1+1}{n}, \frac{\mu_2+1}{n}\right)$ and $\left(\frac{\mu_1}{n}, \frac{\mu_2+1}{n}\right)$, and $\tau_\mu^{2,k}$ has the three vertices $\left(\frac{\mu_1}{n}, \frac{\mu_2}{n}\right)$, $\left(\frac{\mu_1+1}{n}, \frac{\mu_2+1}{n}\right)$ and $\left(\frac{\mu_1+1}{n}, \frac{\mu_2}{n}\right)$. Piecewise linear finite elements are used on the finite element mesh $\{\tau_\mu^{s,k}\}_{\mu \in \{0,1,\ldots,n-1\}^2, s \in \{1,2\}}$. The subspace of piecewise linear functions ϕ_μ^k with

$$\phi_\mu^k \in H_0^1(\Omega), \ \phi_\mu^k\,|_{\tau_\nu^{s,k}} \in \mathbb{P}_1(\tau_\nu^{s,k})$$

is denoted by $\mathbb{V}_k$ where $\mathbb{P}_1$ is the space of polynomials of total degree ≤ 1. A basis of $\mathbb{V}_k$ is the system of the usual hat-functions $\Phi_k = \{\phi_\mu^k\}_{\mu \in \mathscr{I}_k}$ with $\mathscr{I}_k = \{(\mu_1, \mu_2) \in$

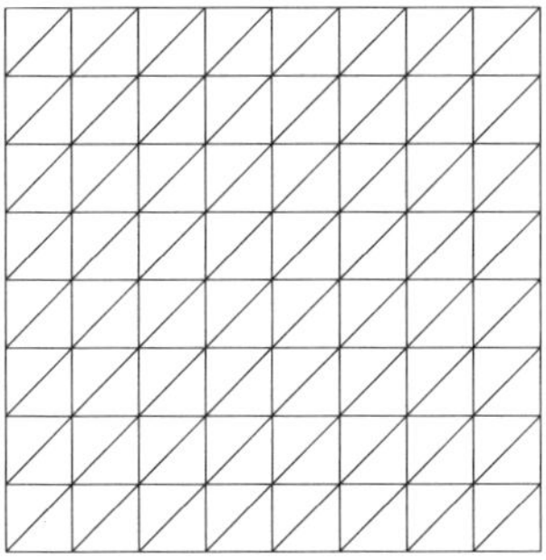

Fig. 1 FE-mesh for (1) and $k = 3$.

$\mathbb{N}^2, \mu_1, \mu_2 \leq 2^k - 1\}$ uniquely defined by

$$\phi_\mu^k(x_\nu^k) = \delta_{\mu\nu}$$

and $\phi_\mu^k \in \mathbb{V}_k$, where $\delta_{\mu\nu}$ is the Kronecker delta for multi-indices. Now, we can formulate the discretized problem. Find $u^k \in \mathbb{V}_k$ such that

$$a(u^k, v^k) = (f, v^k) \quad \forall v^k \in \mathbb{V}_k \tag{2}$$

holds. Problem (2) is equivalent to solving the system of linear algebraic equations

$$K_k \underline{u}_k = \underline{f}_k \tag{3}$$

where $K_k = \left[a(\phi_\mu^k, \phi_\nu^k)\right]_{\mu,\nu \in \mathscr{I}_k}$, $\underline{u}_k = \left[u_\mu\right]_{\mu \in \mathscr{I}_k}$ and $\underline{f}_k = \left[(f, \phi_\nu^k)\right]_{\nu \in \mathscr{I}_k}$. The size of the matrix K_k is $N \times N$, with $N = (n-1)^2$.

3 Formulation of the Main Result

In this section, an additive Schwarz decomposition of the space $\mathbb{V}_k$ is considered. For $l = 0, \ldots, k$, $\mu \in \mathscr{I}_l = \{1, \ldots, 2^l - 1\}^2$, $\mathbb{V}_\mu^l = \operatorname{span}\{\phi_\mu^l\}$ denotes a one-dimensional subspace.

Theorem 1. *Let $a(\cdot,\cdot)$ be defined by* (1) *with $\omega^2(x) = x^\alpha$, $\alpha \geq 0$, $\alpha \neq 1$. Then for any $u^k \in \mathbb{V}_k$ there exists a decomposition $u^k = \sum_{l=0}^{k} \sum_{\mu \in \mathscr{I}_l} u_\mu^l$, with $u_\mu^l \in \mathbb{V}_\mu^l$, such that*

$$\sum_{l=0}^{k} \sum_{\mu \in \mathscr{I}_l} a(u_\mu^l, u_\mu^l) \preceq a(u^k, u^k).$$

Moreover, for all decompositions $u^k = \sum_{l=0}^{k} \sum_{\mu \in \mathscr{I}_l} u_\mu^l$ *with* $u_\mu^l \in \mathbb{V}_\mu^l$*, the estimate*

$$a(u^k, u^k) \preceq \sum_{l=0}^{k} \sum_{\mu \in \mathscr{I}_l} a(u_\mu^l, u_\mu^l)$$

holds.

The proof is given in the end of this section. It requires another additive Schwarz splitting of the bilinear form $a(\cdot,\cdot)$, see [2]. This is based on a decomposition of the domain Ω into $(2^{-j-2}, 2^{-j}) \times (0,1)$, $j = 0, \ldots, k-3$, and $(0, 2^{-j}) \times (0,1)$, $j = k-1, k-2$, see Figure 2. With the index set

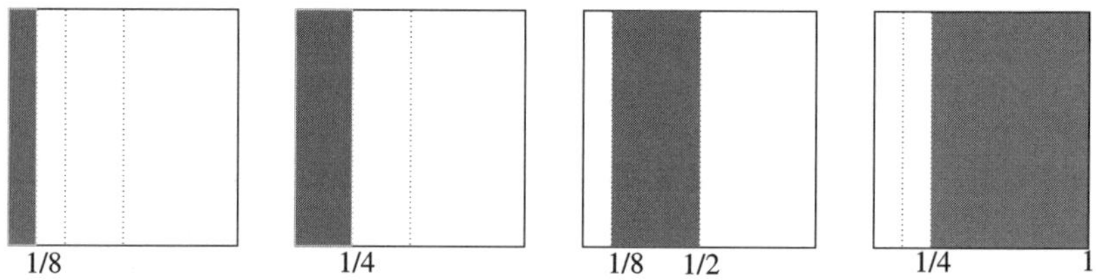

Fig. 2 Decomposition of Ω into stripes for $k = 3$.

$$I_j = \{(\mu_1, \mu_2) \in \mathbb{N}^2 : 2^{k-j-2} < \mu_1 < 2^{k-j}, \mu_2 < 2^k\} \quad \text{for} \quad j = 0, \ldots, k-1,$$

we introduce the finite element spaces

$$\mathbb{V}^{j,k} = \operatorname{span}\left\{\phi_\mu^k\right\}_{\mu \in I_j}.$$

Concerning the space decomposition $\mathbb{V}_k = \oplus_{j=0}^{k-1} \mathbb{V}^{j,k}$, the following result can be shown.

Theorem 2. *Let* $a(\cdot,\cdot)$ *be defined by* (1) *with* $\omega^2(x) = x^\alpha$, $\alpha \geq 0$, $\alpha \neq 1$. *Then for any* $u^k \in \mathbb{V}_k$*, there exists a decomposition* $u^k = \sum_{j=0}^{k-1} u^{j,k}$*, with* $u^{j,k} \in \mathbb{V}^{j,k}$*, such that*

$$\sum_{j=0}^{k-1} a(u^{j,k}, u^{j,k}) \preceq a(u^k, u^k). \tag{4}$$

Moreover, for all decompositions $u^k = \sum_{j=0}^{k-1} u^{j,k}$*, the estimate*

$$a(u^k, u^k) \preceq \sum_{j=0}^{k-1} a(u^{j,k}, u^{j,k})$$

holds.

Proof. The result has been proved in [2].

Now, we are able to prove Theorem 1.

Proof (Theorem 1). Due to the definition of the space $\mathbb{V}^{j,k}$, a function $u \in \mathbb{V}^{j,k}$ has a support which is contained in $[2^{-j-2}, 2^{-j}] \times [0,1]$. Therefore, the weight function lies between $\omega^2(2^{-j-2})$ and $\omega^2(2^{-j})$. This implies the estimate

$$a(u^{j,k}, u^{j,k}) \sim \omega^2\left(2^{-j}\right) \int_\Omega \nabla u^{j,k} \cdot \nabla u^{j,k} \, \mathrm{d}x \, \mathrm{d}y, \quad \forall u^{j,k} \in \mathbb{V}^{j,k} \tag{5}$$

i.e. the bilinear form is spectrally equivalent to the Laplacian multiplied with some factor, where the constants in (5) depend on α but not on j or k. As a consequence of (5), the BPX-like decomposition

$$\mathbb{V}^{j,k} = \bigoplus_{l=0}^{k-j} \bigoplus_{\mu \in I_j} \mathbb{V}^l_\mu$$

is stable, i.e. there exists a decomposition $u = \sum_{l=0}^{k-j} \sum_{\mu \in I_j} u^{j,k}_\mu$, with $u^{j,k}_\mu \in \mathbb{V}^{j,k}_\mu$, such that

$$\sum_{l=0}^{k-j} \sum_{\mu \in I_j} a(u^{j,k}_\mu, u^{j,k}_\mu) \preceq a(u^{j,k}, u^{j,k}) \quad \forall u^{j,k} \in \mathbb{V}^{j,k} \tag{6}$$

(see [7]), where the constants are independent of j and n. Using (4) and (6), one obtains

$$a(u^k, u^k) \succeq \sum_{j=0}^{k-1} \sum_{l=0}^{k-j} \sum_{\mu \in I_j \cap \mathscr{I}_l} a(u^{j,l}_\mu, u^{j,l}_\mu) \quad \forall u^k \in \mathbb{V}^k.$$

Note that $u^{j,l}_\mu$ belongs to the one-dimensional space $\mathbb{V}^l_\mu$. A reordering of the summation gives

$$a(u^k, u^k) \succeq \sum_{l=0}^{k-1} \sum_{\mu \in \mathscr{I}_l} \sum_{j} a(u^{j,l}_\mu, u^{j,l}_\mu) \tag{7}$$

where the last summation runs over all integers j with $\mu \in I_j$. Let

$$\tilde{u}^l_\mu = \sum_{j, \mu \in I_j} u^{j,l}_\mu \in \mathbb{V}^l_\mu. \tag{8}$$

The domain decomposition of Figure 1 implies that there exist not more than two integers j with $\mu \in I_j$ for fixed $\mu \in \mathscr{I}_l$, $0 \le l \le k-1$. These integers are denoted by j_1 and j_2. Using the Cauchy inequality and (8), one obtains

$$a(u^{j_2,l}_\mu, u^{j_2,l}_\mu) + a(u^{j_1,l}_\mu, u^{j_1,l}_\mu) \ge \frac{1}{2} a(u^{j_1,l}_\mu + u^{j_2,l}_\mu, u^{j_1,l}_\mu + u^{j_2,l}_\mu) = \frac{1}{2} a(\tilde{u}^l_\mu, \tilde{u}^l_\mu). \tag{9}$$

Combining (9) and (7), one easily concludes that

$$a(u^k,u^k) \succeq \sum_{l=0}^{k-1} \sum_{\mu\in\mathscr{I}_l} a(\tilde{u}^l_\mu,\tilde{u}^l_\mu) \quad \forall u^k = \sum_{l=0}^{k-1} \sum_{\mu\in\mathscr{I}_l} \tilde{u}^l_\mu \in \mathbb{V}_k.$$

This proves the first estimate, which is the stability of the BPX preconditioner.

The second estimate can be proved in the same way as presented in [7] for uniformly bounded elliptic problems. The techniques presented there depend only on the refinement of the mesh and do not require the ellipticity of the bilinear form.

Remark 1. The results are not valid for the weight function $\omega^2(x) = x$. This is due to the estimate (4). Instead of (4) only the weaker estimate

$$\frac{1}{k^2}\sum_{j=0}^{k-1} a(u^{j,k},u^{j,k}) \preceq a(u^k,u^k) \quad \forall u^k \in \mathbb{V}_k$$

holds, see [2]. This gives a suboptimal condition number estimate for the BPX preconditioner in the case $\omega^2(x) = x$.

4 Numerical Experiments

In this section, we investigate the quality of our preconditioner induced by the space decomposition

$$\mathbb{V}_k = \bigoplus_{l=0}^{k} \bigoplus_{\mu\in\mathscr{I}_l} \mathbb{V}^l_\mu.$$

We investigate three different domains, i.e. a hexagon with the vertices $(1\pm 1,\pm 0.5)$ and $(1,\pm 1)$, the unit square $(0,1)^2$ and a triangle with the vertices $(0,\pm 1)$ and $(2,-0.5)$, and the corresponding singular weight function $\omega^2(x) = x^\alpha$. Note that the boundary with singular diffusion is at $x = -1$.

The coarse finite element meshes are displayed in Figure 3. The finite element mesh on level k is obtained by k refinements of the coarse finite element mesh. The

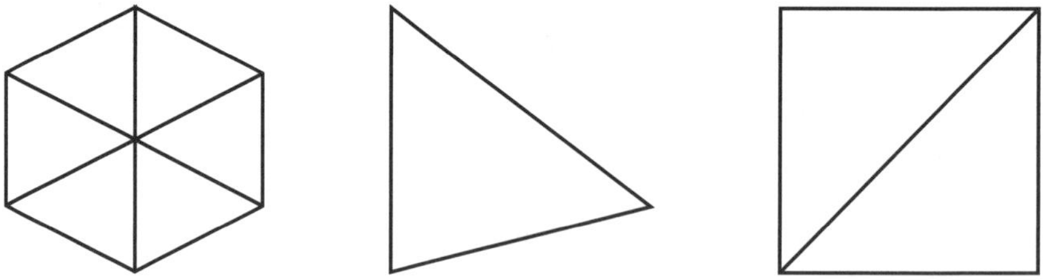

Fig. 3 Finite element coarse meshes: hexagon (left), triangle (middle), square (right).

linear system $K_k\underline{u}_k = \underline{f}_k$ is solved by a PCG method with the BPX preconditioner and a relative accuracy of 10^{-5}. Table 1 shows the PCG-iteration numbers for the fixed weight function $\omega^2(x) = x^2$ and all different meshes. In all experiments, a very small increase of the iteration numbers is observed. In a second experiment, the influence of the weight function is investigated for the hexagonal domain. The PCG iteration numbers are displayed in Table 2. The largest iteration numbers are obtained for $\omega(x)^2 = x$, where the assumptions of Theorem 1 are violated. For $\omega^2(x) = x^\alpha$, $\alpha = 0, 3$, the iteration numbers are clearly bounded.

Level	3	4	5	6	7	8	9
triangle	11	14	17	20	22	24	26
square	14	17	21	23	25	27	29
hexagon	14	16	18	20	22	23	24

Table 1 PCG iteration numbers for (2) using the BPX preconditioner with weight $\omega^2(x) = x^2$.

Level	3	4	5	6	7	8	9
$\omega^2(x) = 1$	12	13	14	14	15	15	15
$\omega^2(x) = x$	13	16	18	21	23	26	29
$\omega^2(x) = x^2$	14	16	18	20	22	23	24
$\omega^2(x) = x^3$	14	16	18	18	19	19	19

Table 2 PCG iteration numbers for (2) using the BPX preconditioner (hexagon).

The paper [2] presented some numerical experiments for the domain decomposition preconditioner induced by the space decomposition which is considered in Theorem 2. The iteration numbers of the BPX preconditioner are lower than the iteration numbers of this overlapping DD preconditioner.

5 Conclusions and the Three-Dimensional Case

In this paper, we have proved the independence of the condition number of the BPX-preconditioned system arising from the discretization of a degenerate elliptic boundary problem in two dimensions from the discretization parameter.

The presented analysis uses two results,

- the condition number estimates of BPX-preconditioned systems for uniformly bounded problems and
- the results about the spectral equivalence of an overlapping DD preconditioner for K_k.

Both results remain true for three-dimensional problems with the weight function $\omega^2(x,y,z) = x^\alpha$, $\alpha \neq 1$. Therefore, PCG methods with BPX preconditioners are also optimal solvers in three spatial dimensions.

Acknowledgements Parts of this work have been supported by the Spezialforschungsbereich (SFB) F013 "Numerical and Symbolic Scientific Computing" of the Austrian Science Fundation (FWF) and the FWF project P20121-N18.

References

1. Beuchler, S.: Multilevel solvers for a finite element discretization of a degenerate problem. SIAM J. Numer. Anal. **42**(3), 1342–1356 (2004)
2. Beuchler, S., Nepomnyaschikh, S.: Overlapping additiv Schwarz preconditioners for isotropic elliptic problems with degenerate coefficients. J. Num. Math. **15**(4), 245–276 (2007)
3. Beuchler, S., Nepomnyaschikh, S.: Overlapping additive schwarz preconditioners for elliptic problems with degenerate locally anisotropic coefficients. SIAM J. Num. Anal. **45**(6), 2321–2344 (2007)
4. Bramble, J., Pasciak, J., Xu, J.: Parallel multilevel preconditioners. Math. Comp. **55**(191), 1–22 (1991)
5. Kufner, A., Sändig, A.: Some applications of weighted Sobolev spaces. B.G.Teubner Verlagsgesellschaft. Leipzig (1987)
6. Pironneau, O., Hecht, F.: Mesh adaption for the Black and Scholes equations. East-West Journal of Numerical Mathematics **8**(1), 25–36 (2000)
7. Zhang, X.: Multilevel Schwarz methods. Numer. Math. **63**, 521–539 (1992)

Performance Analysis of Parallel Algebraic Preconditioners for Solving the RANS Equations Using Fluctuation Splitting Schemes

A. Bonfiglioli, B. Carpentieri, and M. Sosonkina

Abstract We consider iterative solution strategies for solving the Reynolds-Favre averaged Navier-Stokes (RANS) equations on 2D and 3D flow configurations. The novelty of this study is the coupling of an hybrid class of methods for the space discretization, called Fluctuation Splitting (or residual distribution) schemes, and a fully coupled Newton algorithm for solving the RANS equations. This approach is particularly attractive for parallel computations because it gives rise to discretization matrices with a compact stencil resulting in a limited number of nonzero entries. In this paper, we present the solution approach and report on results of numerical experiments with particular emphasis on the design of preconditioners for the inner linear system, which is a critical computational issue of the iterative solution.

1 Introduction

In this study, we consider the iterative solution of Euler and Navier-Stokes equations on unstructured grids. An accurate solution of this problem class is demanded in fluid dynamic simulations, such as those arising in aerodynamic design, oceanography, or turbomachinery. For solving these applications, Finite Element (FE), Finite Volume (FV) and Discontinuous-Galerkin methods are often the methods of choice; the computational domain is decomposed into a finite set of nonoverlapping control volumes and approximate time-dependent values of the conserved variables

Aldo Bonfiglioli
Dip.to di Ingegneria e Fisica dell'Ambiente, University of Basilicata, Potenza, Italy, e-mail: ba001ing@unibas.it

Bruno Carpentieri
Karl-Franzens University, Institute of Mathematics and Scientific Computing, Graz, Austria, e-mail: bruno.carpentieri@uni-graz.at

Masha Sosonkina
Ames Laboratory/DOE, Iowa State University, Ames, USA, e-mail: masha@scl.ameslab.gov

are computed into each control volume. Control volumes are drawn around each gridpoint by joining, in two space dimensions, the centroids of gravity of the surrounding cells with the midpoints of all the edges that connect that gridpoint with its nearest neighbors. In Fig. 1, control volumes are delimited by green lines. The

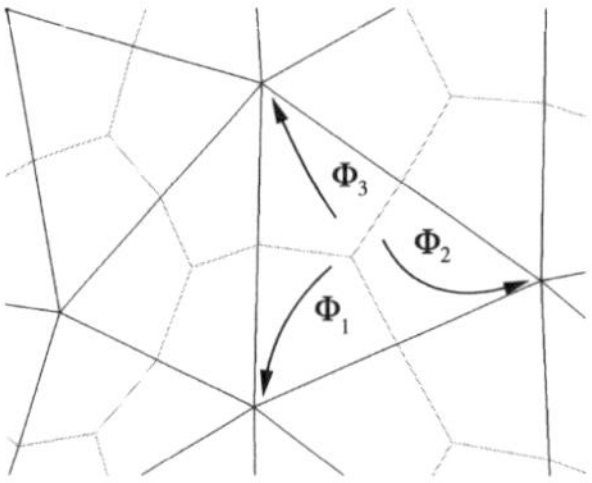

(a) The flux balance of cell T is scattered among its vertices.

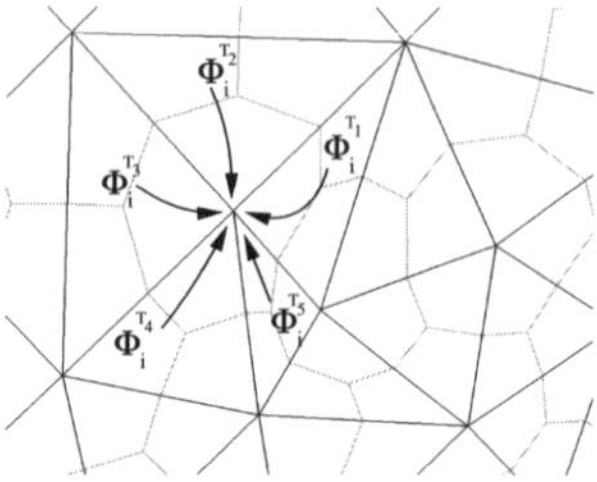

(b) Gridpoint i gathers the fractions of cell residuals from the surrounding cells.

Fig. 1 Residual distribution concept.

space discretization results in a system of nonlinear equations that can be solved using Newton's algorithm. The novelty of this study is to use a hybrid class of methods for the space discretization called Fluctuation Splitting (or residual distribution) schemes in conjunction with a fully coupled Newton algorithm for solving the RANS equations. Introduced in the early eighties by P.L. Roe, and successively developed further by a number of groups worldwide (see e.g. [8, 2]), this class of schemes shares common features with both FE and FV methods. Just as with iso-P1 FE, the dependent variables are stored in the vertices of the mesh, which is made of triangles in two space dimensions (2D) and tetrahedra in three (3D), and assumed to vary linearly in space. We illustrate the discretization approach for the case of a simple advection equation

$$u_t + au_x + bu_y = 0$$

defined on a 2D domain tesselated into a set of triangles $\{T\}$. Fluctuations, i.e. flux balances, are computed over each triangle as for standard linear finite element methods,

$$\Phi^T = \int_T u_t dxdy = \int_T [-(au_x + bu_y)]\, dxdy.$$

Assuming piecewise linear variation of the solution u over each triangle, we may write the fluctuation Φ^T as:

$$\Phi^T = \sum_{i=1}^{3} k_i^T u_i = \sum_{i=1}^{3} \left[\frac{1}{2}(a,b) \cdot \mathbf{n}_i^T \right] u_i,$$

where $\mathbf{n}_i^T$ is a scaled inward normal vector of the edge opposite to node i in the triangle T. The parameters k_i^T are the so-called inflow parameters. Fluctuations are distributed to the nodes of the triangles and then they are accumulated at every node (see Fig. 1(a)). Gridpoint i will then collect fractions Φ_i^T of the flux balances of its surrounding elements, as schematically shown in Fig. 1(b). The method can achieve at most second-order accuracy whatever scheme is used to distribute the fluctuation because of the assumption of linear variation.

2 Governing Equations

Throughout this paper, we use standard notation for the kinematic and thermodynamic variables: $\mathbf{u}$ is the flow velocity, ρ is the density, p is the pressure (divided by the constant density in incompressible, homogeneous flows), T is the temperature, e and h are the specific total energy and enthalpy, respectively, ν is the laminar kinematic viscosity and $\tilde{\nu}$ is a scalar variable related to the turbulent eddy viscosity via a damping function. For compressible flows the sound speed a is a function of temperature while for incompressible flows it is taken constant and equal to the free-stream velocity. Given a control volume C_i, fixed in space and bounded by the control surface ∂C_i with inward normal $\mathbf{n}$, we write the governing conservation laws of mass, momentum, energy and turbulence transport equations as:

$$\int_{C_i} \frac{\partial \mathbf{U}_i}{\partial t} dV = \oint_{\partial C_i} \mathbf{n} \cdot \mathbf{F} dS - \oint_{\partial C_i} \mathbf{n} \cdot \mathbf{G} dS + \int_{C_i} \mathbf{S} dV, \tag{1}$$

where we denote by $\mathbf{U}$ the vector of conserved variables. For compressible flows, we have $\mathbf{U} = (\rho, \rho e, \rho \mathbf{u}, \tilde{\nu})^T$, and for incompressible, constant density flows, $\mathbf{U} = (p, \mathbf{u}, \tilde{\nu})^T$. In (1), the operators $\mathbf{F}$ and $\mathbf{G}$ represent the inviscid and viscous fluxes, respectively. For compressible flows, we have

$$\mathbf{F} = \begin{pmatrix} \rho \mathbf{u} \\ \rho \mathbf{u} h \\ \rho \mathbf{u}\mathbf{u} + p\mathbf{I} \\ \tilde{\nu}\mathbf{u} \end{pmatrix}, \quad \mathbf{G} = \frac{1}{\mathrm{Re}_\infty} \begin{pmatrix} 0 \\ \mathbf{u} \cdot \tau + \nabla q \\ \tau \\ \frac{1}{\sigma}\left[(\nu + \tilde{\nu}) \nabla \tilde{\nu}\right] \end{pmatrix},$$

and for incompressible, constant density flows,

$$\mathbf{F} = \begin{pmatrix} a^2 \mathbf{u} \\ \mathbf{u}\mathbf{u} + p\mathbf{I} \\ \tilde{\nu}\mathbf{u} \end{pmatrix}, \quad \mathbf{G} = \frac{1}{\mathrm{Re}_\infty} \begin{pmatrix} 0 \\ \tau \\ \frac{1}{\sigma}\left[(\nu + \tilde{\nu}) \nabla \tilde{\nu}\right] \end{pmatrix}.$$

Finally, $\mathbf{S}$ is the source term, which has a non-zero entry only in the row corresponding to the turbulence transport equation:

$$\mathbf{S} = \begin{pmatrix} 0 \\ 0 \\ \mathbf{0} \\ c_{b1}\left[1 - f_{t2}\right]\tilde{S}\tilde{\nu} + \frac{1}{\sigma Re}\left[c_{b2}\left(\nabla\tilde{\nu}\right)^2\right] + \\ -\frac{1}{Re}\left[c_{w1}f_w - \frac{c_{b1}}{\kappa^2}f_{t2}\right]\left[\frac{\tilde{\nu}}{d}\right]^2 + Ref_{t1}\Delta U^2 \end{pmatrix}. \tag{2}$$

For a description of the various functions and constants involved in (2) the reader is referred to [6]. In the case of high Reynolds number flows, we account for turbulence effects by the RANS equations that are obtained from the Navier-Stokes (NS) equations by means of a time averaging procedure. The RANS equations have the same structure as the NS equations with an additional term, the Reynolds' stress tensor, that accounts for the effects of the turbulent scales on the mean field. Despite the non-negligible degree of empiricism introduced by the turbulence model, it is recognized that the solution of the RANS equations still remains the only feasible approach to perform computationally affordable simulations of problems of engineering interest on a routine basis. A closure problem arises since the Reynolds' stresses require modeling. Using Boussinesq's hypothesis as the constitutive law for the Reynolds' stresses amounts to link the Reynolds' stress tensor to the mean velocity gradient through a scalar quantity which is called turbulent (or eddy) viscosity. With Boussinesq's approximation, the RANS equations become formally identical to the NS equations, except for an "effective" viscosity (and thermal conductivity), sum of the laminar and eddy viscosities (and similarly for the laminar and turbulent termal conductivity), which appears in the viscous terms of the equations. In the present study, the turbulent viscosity is modeled using the Spalart-Allmaras one-equation model [6].

3 Solution Techniques

Using the fluctuation splitting approach described in Section 1, the integral form of the governing equations (1) is discretized over each control volume C_i evaluating the flux integral over each triangle (or tetrahedron) in the mesh, and then splitting it among its vertices. This approach leads to a space-discretized form of Eq. (1) that reads:

$$\int_{C_i} \frac{\partial \mathbf{U}_i}{\partial t}\, dV = \sum_{T \ni i} \Phi_i^T$$

where

$$\Phi^T = \oint_{\partial T} \mathbf{n} \cdot \mathbf{F}\, dS - \oint_{\partial T} \mathbf{n} \cdot \mathbf{G}\, dS + \int_T \mathbf{S}\, dV$$

is the flux balance evaluated over cell T and Φ_i^T is the fraction of cell residual scattered to vertex i. The properties of the scheme will depend upon the criteria used to distribute the cell residual (see [2]). The discretization of the governing equations in space leads to a system of ordinary differential equations:

$$\mathbf{M}\frac{d\mathbf{U}}{dt} = \mathbf{R}(\mathbf{U}), \tag{3}$$

where t denotes the physical time variable, $\mathbf{M}$ is the mass matrix and $\mathbf{R}(\mathbf{U})$ represents the nodal residual vector of spatial discretization operator, which vanishes at steady state. The residual vector is a (block) array of dimension equal to the number of meshpoints times the number of dependent variables, m; for a one-equation turbulence model, $m = d + 3$ for compressible flows and $m = d + 2$ for incompressible flows, d being the spatial dimension. If the time derivative in equation (3) is approximated using a two-point one-sided finite difference formula we obtain the following implicit scheme:

$$\left(\frac{1}{\Delta t^n}\mathbf{V} - \mathbf{J}\right)\left(\mathbf{U}^{n+1} - \mathbf{U}^n\right) = \mathbf{R}(\mathbf{U}^n), \tag{4}$$

where we denote by $\mathbf{J}$ the Jacobian of the residual $\dfrac{\partial \mathbf{R}}{\partial \mathbf{U}}$. Eq. (4) represents a large nonsymmetric sparse linear system of equations to be solved at each pseudo-time step for the update of the vector of the conserved variables. The nonzero pattern of the sparse coefficient matrix is symmetric, *i.e.* entry (i, j) is nonzero if and only if entry (j, i) is nonzero as well. Due to the compact stencil of the schemes, the sparsity pattern of the Jacobian matrix coincides with the graph of the underlying unstructured mesh, *i.e.* it involves only one-level neighbours; on average, the number of non-zero (block) entries per row equals 7 in 2D and 14 in 3D. The analytical evaluation of the Jacobian matrix, though not impossible, is rather cumbersome [3]. Thus, we currently adopt two alternatives: one is based on an analytically calculated but approximate Jacobian, the other on a numerical approximation of the "true" Jacobian obtained using one-sided finite differences formulae. Calculating the Jacobian matrix requires $(d + 1) \times m$ residual evaluations and pays off only close to the steady state when the quadratic convergence of Newton's method can be exploited. In a practical implementation, a two-step approach is adopted. In the early stages of the calculation, the turbulent transport equation is solved in tandem with the mean flow equations: the mean flow solution is advanced over a single time step using an approximate (Picard) Jacobian while keeping turbulent viscosity frozen, then the turbulent variable is advanced over one or more pseudo-time steps using a FD Jacobian with frozen mean flow variables. This procedure will eventually converge to steady state, but never yields quadratic convergence. Therefore, a true Newton strategy is adopted when the solution has come close to steady state: the mean flow and turbulence transport equation are solved in fully coupled form and the Jacobian is computed by FD.

Although the solution of the RANS equation may require much less computational effort than other simulation techniques such as LES (Large Eddy Simulation) and DNS (Direct Numerical Simulation), severe numerical difficulties may arise when the mean flow and turbulence transport equation are solved in fully coupled form, the Jacobian is computed *exactly* by means of FD and the size of the time-step is rapidly increased to recover Newton's algorithm. Indeed, on 3D unstructured problems, successful experiments are not numerous in the literature (see [9] for a similar study). The crucial computational issue is the design of robust parallel preconditioners for the inner linear systems, which, in the case of turbulent problems defined on anisotropic meshes, can be large unstructured and ill-conditioned. In this study we focus on this component of the numerical solution.

4 Experiments

We initially consider the steady state solution of Euler's equations for the simulation of an incompressible flow around a 2D profile NACA0012 at an incidence angle of 2 degrees. The mesh is formed of 2355 nodes; the pertinent linear system has size 7065 as three unknowns are associated to each meshpoint. In Table 1, we report on results with restarted GMRES [5] and different preconditioners that are ILU with pattern selection strategy based on value and position, Jacobi, block Jacobi, and additive Schwarz (see, *e.g.* [5]). We have tested different accelerators and have observed that QMR-type methods are generally less effective compared to GMRES. Thus we proceed with this solver. In the experiments with block Jacobi and additive Schwarz preconditioners, ILU with zero level of fill-in is used on each block. For all these algorithms, we used the implementations available in the PETSc library [1]. The runs are done on one processor DEC Alpha 2044/233. In the code, preconditioning is implemented from the left and the stopping criterion is the reduction of the original residual by 10^{-5}. In practice, the relative criterion becomes more stringent as the stationary solution is approached, since $||b||_2 = ||\mathbf{R}(\mathbf{U})||_2$ will approach machine zero. The maximum number of iterations allowed for the linear solver is 10^5. The initial guess for the linear solution vector is the approximate solution computed at the previous time step. This choice is reasonable since the solution vector is an increment between successive time levels, and it will thus converge to zero as the steady solution is approached. We see from the results reported in Table 1 that incomplete factorization is very effective to reduce the number of iterations, provided the nonzero pattern is computed from the graph of the matrix. On this problem class dropping strategies based on levels of fill-in are much more robust than strategies based on threshold, such as ILU(t) [5]. Experiments with sparse approximate inverse preconditioners using both static and dynamic pattern strategies, and standard algebraic multigrid methods gave disappointing results.

The three-dimensional test case that we have examined deals with the steady state solution of the RANS equations for the simulation of the internal compressible flow through the so-called Stanitz elbow. The simulation reproduces experiments [7] conducted in early 1950's at the National Advisory Committee for Aeronautics (NACA), presently NASA, to study secondary flows in an accelerating, rectangular elbow with 90° of turning. The chosen flow conditions correspond to a Mach number in the outlet section of 0.68 and Reynolds' number $4.3 \cdot 10^5$. Figure 2(b) shows the geometry along with the computed static pressure contours. The computational mesh consists of 156065 meshpoints and 884736 tetrahedral cells. The simulation

Table 1 Test case *NACA 0012*: comparison of different solvers using Newton linearization.

Linear solver		Nonlinear iterations	Linear iterations (tot)	Execution time (sec)	Memory (Mbytes)
Accel.	Precond.				
GMRES(30)	ILU(0)	11	441	24.781	6.840
”	ILU(1)	11	230	22.368	7.217
”	ILU(2)	11	164	21.433	7.258
”	ILU(t), t=5e-3	10	443	38.744	29.172
”	Jacobi	10	14222	120.369	5.439
”	Block Jacobi, nb=10	11	1991	45.614	7.660
”	ASM nb=10, ov=1	11	927	40.214	10.166
”	ASM nb=10, ov=2	11	696	39.538	12.247

has been run on 16 processors of a Linux Beowulf cluster. Figures 2(a) show the

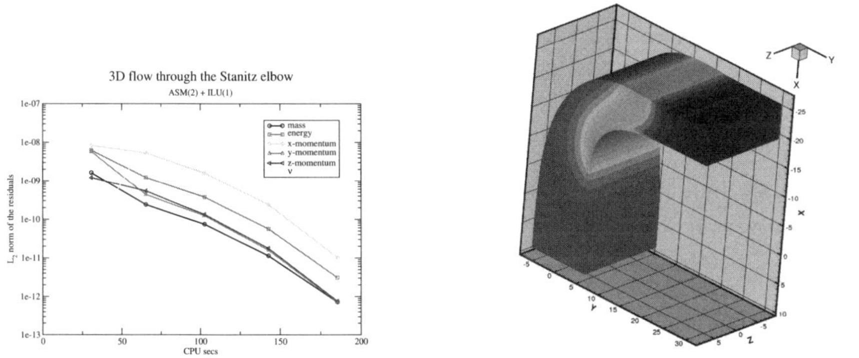

(a) Convergence history for Newton iterations using ASM(2)+ILU(1).

(b) Stanitz elbow geometry and computed static pressure contours.

Fig. 2 Experiments on the Stanitz elbow.

convergence history of the iterative solution; we use an additive Schwarz preconditioner with overlap of 2 for GMRES; the diagonal blocks are approximately inverted using ILU(1). We obtain convergence in only five Newton iterations down to machine precision. In Table 2, we report on comparative results with respect to the number of iterations and solution time (in seconds) with parallel block Jacobi and additive Schwarz methods, using the ILU(1) solver on each subdomain. We see that additive Schwarz delivers the best convergence in terms of number of iterations; we performed tests with different values of overlap observing very similar results.

Table 2 Solution cost for solving the Stanitz problem using GMRES(30)

Newton iter	BJ+ILU(1)		ASM(2)+ILU(1)	
	Iter	CPU time	Iter	CPU time
1	72	10.1	49	8.8
2	121	16.7	76	12.3
3	147	20.0	98	15.8
4	175	23.5	112	17.8
5	200	26.9	127	20.1

5 Concluding Remarks

We have presented iterative solution strategies for the Reynolds-Favre averaged Navier-Stokes equations based on fluctuation splitting schemes and Newton-Krylov solvers. We have analyzed the performance of some algebraic preconditioners for accelerating the solution of the inner linear system arising at each step of the Newton's method. The code is still in a development stage but the numerical results are encouraging. Incomplete factorization with pattern strategy based on level of fill is very effective for solving 2D problems, while additive Schwarz preconditioner is more efficient on 3D applications for its natural parallelism. Perspectives of future research include enhancing the robustness of the code on difficult configurations (e.g. highly anysotropic meshes), and in particular designing multilevel preconditioners based on Schur complement-type incomplete factorization (see *e.g.* [4]) to improve the scalability of the solver on large-scale problems.

References

1. Balay, S., Buschelman, K., Gropp, W., Kaushik, D., Knepley, M., McInnes, L.C., Smith, B., Zhang, H.: PETSc home page (2001). `http://www.mcs.anl.gov/petsc`
2. Bonfiglioli, A.: Fluctuation splitting schemes for the compressible and incompressible Euler and Navier-Stokes equations. IJCFD **14**, 21–39 (2000)
3. Issman, E.: Implicit Solution Strategies for Compressible Flow Equations on Unstructured Meshes. Ph.D. thesis, Université Libre de Bruxelles (1997)
4. Li, Z., Saad, Y., Sosonkina, M.: pARMS: a parallel version of the algebraic recursive multilevel solver. Tech. Rep. umsi-2001-100, Minnesota Supercomputer Institute, University of Minnesota, Minneapolis, MN (2001)
5. Saad, Y.: Iterative Methods for Sparse Linear Systems. SIAM (2003)
6. Spalart, P., Allmaras, S.: A one-equation turbulence model for aerodynamic flows. La Recherche-Aerospatiale **1**, 5–21 (1994)
7. Stanitz, J.D., Osborn, W.M., Mizisin, J.: An experimental investigation of secondary flow in an accelerating, rectangular elbow with 90° of turning. Technical Note 3015, NACA (1953)
8. van der Weide, E., Deconinck, H., Issman, E., Degrez, G.: A parallel, implicit, multidimensional upwind, residual distribution method for the Navier-Stokes equations on unstructured grids. Computational Mechanics **23**, 199–208 (1999)
9. Wong, P., Zingg, D.: Three-dimensional aerodynamic computations on unstructured grids using a Newton-Krylov approach. AIAA Paper (5231) (2005)

On Least-Squares Approximate Inverse-Based Preconditioners

B. Carpentieri

Abstract We discuss approximate inverse preconditioners based on Frobenius-norm minimization. We introduce a novel adaptive algorithm based on truncated Neumann matrix expansions for selecting the sparsity pattern of the preconditioner. The construction of the approximate inverse is based on a dual dropping strategy, namely a threshold to drop small entries and a maximum number of nonzero entries per column. We introduce a post-processing stabilization technique to deflate some of the smallest eigenvalues in the spectrum of the preconditioned matrix which can potentially disturb the convergence. Results of preliminary experiments are reported on a set of linear systems arising from different application fields to illustrate the potential of the proposed algorithm for preconditioning effectively iterative Krylov solvers.

1 Introduction

Approximate inverse methods compute an explicit sparse approximation of a nonsingular matrix A that can be used as preconditioner for solving general linear systems of the form

$$Ax = b,\ A \in \mathbb{C}^{n\times n},\ \ det(A) \neq 0,\ x \in \mathbb{C}^n,\ b \in \mathbb{C}^n. \tag{1}$$

This class of algorithms is computationally attractive for parallelism because applying the preconditioner at each step of an iterative solver simply reduces to carry out one or more sparse matrix-vector products, that are numerically stable and are easy to parallelize. For many problem classes, A^{-1} or its triangular factors are sparse or can be well approximated by a sparse matrix; thus the resulting

Karl-Franzens University, Institute of Mathematics and Scientific Computing, Graz, Austria,
e-mail: bruno.carpentieri@uni-graz.at

preconditioner can reduce the number of iterations significantly. Indeed this class of methods has received much attention in the last years either as stand-alone preconditioners [12, 4, 15], or as local components of multilevel solvers [8], or as smoothers for multigrid algorithms [5, 18]. Library implementations [12] and public-domain codes [1, 7] are available for both sequential and parallel computer architectures. A critical computational issue is the selection of the sparsity pattern of the approximate inverse as, in general, the inverse of an irreducible matrix is structurally dense [11]. The lack of robustness on indefinite problems and the high construction cost are two important drawbacks that can limit the use of approximate inverse methods for solving general problems. In this work we introduce a cost effective adaptive pattern selection strategy based on truncated Neumann matrix series expansions for Frobenius-norm approximate inverse preconditioners and we consider stabilization techniques based on eigenvalue deflation to enhance their robustness on indefinite problems. Comparative experiments with other standard preconditioners are reported on a set of matrix problems arising from various application fields to illustrate the potential of the proposed algorithm.

2 Least-Squares Minimization Preconditioner

We concentrate our attention on Frobenius-norm minimization techniques that compute the matrix M which minimizes the error matrix $||I-MA||_F$ (or $||I-AM||_F$ for right preconditioning) subject to certain sparsity constraints. The Frobenius-norm allows to decouple the constrained minimization problem into n independent linear least-squares problems, one for each column of M (when preconditioning from the right) or row of M (when preconditioning from the left). The independence of these least-squares problems follows immediately from the identity:

$$||I-MA||_F^2 = ||I-AM^T||_F^2 = \sum_{j=1}^{n} ||e_j - Am_j||_2^2 \tag{2}$$

where e_j is the jth unit vector and m_j is the column vector representing the jth row of M. In the case of right preconditioning, the analogous relation

$$||I-AM||_F^2 = \sum_{j=1}^{n} ||e_j - Am_j||_2^2 \tag{3}$$

holds, where m_j is the column vector representing the jth column of M. Clearly, there is considerable scope for parallelism in this approach. Early references to this class of methods can be found in [2, 3]. In the symmetric positive definite (SPD) case, we may preserve symmetry in the preconditioner approximating the inverse of the Cholesky factor L of A by minimizing $||I-GL||_F^2$. The matrix G may be computed without the knowledge of L by solving the normal equations

$$\{GLL^T\}_{ij} = L^T_{ij},\ (i,j) \in \mathscr{S}_L \tag{4}$$

where $\mathscr{S}_L$ is a lower triangular nonzero pattern for G (see e.g. [15]). Equation (4) can be replaced by

$$\{\bar{G}A\}_{ij} = I_{ij},\ (i,j) \in \mathscr{S}_L \tag{5}$$

where $\bar{G} = D^{-1}G$ and D is the diagonal of L; each row of $\bar{G}$ can be computed independently by solving a small linear system. The preconditioned linear system has the form

$$GAG^T = D\bar{G}A\bar{G}^T D.$$

The matrix D is not known and is generally chosen so that the diagonal of GAG^T is all ones.

We construct the preconditioner from a scaled coefficient matrix $\tilde{A} = D_1^{1/2} A D_2^{1/2}$, where D_1 and D_2 are the following diagonal matrices:

$$D_1(i,j) = \begin{cases} \dfrac{1}{\max\limits_i |a_{ij}|}, & \text{if i} = \text{j} \\ 0, & \text{if i} \neq \text{j} \end{cases}, \quad D_2(i,j) = \begin{cases} \dfrac{1}{\max\limits_j |a_{ij}|}, & \text{if i} = \text{j} \\ 0, & \text{if i} \neq \text{j} \end{cases}.$$

Throughout this paper, we denote by A the scaled coefficient matrix $\tilde{A}$. The sparsity structure of column m_j of M, denoted as $struct_j(M)$, is taken as

$$struct_j(M) = \bigcup_{i=1}^{k_j} struct_j\left(A^i\right).$$

The rationale behind this strategy is that the inverse of any nonsingular matrix A can be written in terms of powers of A as

$$A^{-1} = -\frac{1}{\alpha_0} \sum_{j=0}^{m} \alpha_{j+1} A^j,$$

where m is the index sum of the distinct eigenvalues of A and α_j's are the coefficients of the minimal polynomial expansion of A; we may expect that using a low degree polynomial we are able to capture a suitable nonzero pattern of the large entries of m_j. The value of k_j is allowed to vary for each column j, and it is tuned adaptively depending on the user required accuracy. Thus this approach is completely different from other pattern selection strategies for Frobenius-norm minimization methods earlier proposed in the literature (see e.g. [13, 7]). We initially set the value of k_j equal to one, and then we increment it until

$$\left\|Am_j - e_j\right\|_2 < \tau$$

where τ is a fixed tolerance, or until a maximum number of nonzero entries p is computed in column m_j. Note that both parameters τ and p are global, i.e. they apply to all columns of M. The actual entries of vector m_j are calculated by solving a small size least-squares problem using dense QR factorization. We refer to the resulting preconditioner as $FROB(\tau, p)$. In [13], Grote and Huckle provide the following important estimates on the relative distance of the Frobenius-norm minimization preconditioner with respect to the exact inverse matrix and on the norm of the residual matrix:

Theorem 1. *Let $r_j = Am_j - e_j$ be the residual associated with column m_j for $j = 1,2,\ldots,n$, and $q = \max_{1 \le j \le n} \{nnz(r_j)\} \ll n$. Suppose that $\|r_j\|_2 < t$ for $j = 1,2,\ldots,n$, then we have*

$$\begin{array}{ll} \|AM - I\|_F \le \sqrt{n}t, & \left\|M - A^{-1}\right\|_F \le \left\|A^{-1}\right\|_2 \sqrt{n}t, \\ \|AM - I\|_2 \le \sqrt{n}t, & \left\|M - A^{-1}\right\|_2 \le \left\|A^{-1}\right\|_2 \sqrt{n}t, \\ \|AM - I\|_1 \le \sqrt{q}t, & \left\|M - A^{-1}\right\|_1 \le \left\|A^{-1}\right\|_1 \sqrt{q}t. \end{array}$$

Owing to this result, all the eigenvalues of AM lie in the disk centered in 1 and of radius $\sqrt{q}t$; the value of q is not known a priori, though, so that one might enforce the condition $\sqrt{n}t < 1$ to prevent singularity or near-singularity of the preconditioned matrix. To run the algorithm with such a small t is too costly in practice. In fact, for many problems we generally observe a lack of robustness of the approximate inverse due to the presence of clusters of small eigenvalues in the spectrum of the preconditioned matrix (see e.g. [6]). This scenario typically arises when solving indefinite systems since some of the eigenvalues are likely to cluster near zero in their natural trajectory towards point one of the complex plane under the action of the preconditioner. This consideration motivates us to introduce a stabilization step after computing M, which deflates a small group of eigenvalues close to zero in the spectrum of the preconditioned matrix. Let

$$MAx = Mb. \tag{6}$$

be the preconditioned system to solve and

$$MA = V \Lambda V^{-1}, \tag{7}$$

the standard eigenvalue decomposition with $\Lambda = diag(\lambda_i)$, where $|\lambda_1| \le \ldots \le |\lambda_n|$ are the eigenvalues and $V = (v_i)$ the associated right eigenvectors. We denote by V_ε the set of right eigenvectors associated with the set of eigenvalues λ_i with $|\lambda_i| \le \varepsilon$. It is easy to show the following result:

Theorem 2. *Let W be such that $\tilde{A}_c = W^H A V_\varepsilon$ has full rank, $\tilde{M}_c = V_\varepsilon \tilde{A}_c^{-1} W^H$ and $\tilde{M} = M + \tilde{M}_c$. Then $\tilde{M}A$ is similar to a matrix whose eigenvalues are*

$$\begin{cases} \eta_i = \lambda_i & \text{if } |\lambda_i| > \varepsilon, \\ \eta_i = 1 + \lambda_i & \text{if } |\lambda_i| \le \varepsilon. \end{cases}$$

For right preconditioning, that is $AMy = b$, similar results hold (see also [6]). We should point out that, if the symmetry of the preconditioner has to be preserved, an obvious choice exists. For left preconditioning, we can set $W = V_\varepsilon$, but then $\tilde{A}_c$ may not have full rank.

In the SPD case, these results extend as follows.

Theorem 3. *If A and M_1 are SPD, then M_1A is diagonalizable, and $\tilde{A}_c = V_\varepsilon^T A V_\varepsilon$ is SPD. The preconditioner defined by $\tilde{M} = M_1 + \tilde{M}_c$, with $\tilde{M}_c = V_\varepsilon \tilde{A}_c^{-1} V_\varepsilon^T$ is SPD and $\tilde{M}A$ is similar to a matrix whose eigenvalues are*

$$\begin{cases} \eta_i = \lambda_i & \text{if } |\lambda_i| > \varepsilon, \\ \eta_i = 1 + \lambda_i & \text{if } |\lambda_i| \leq \varepsilon. \end{cases}$$

3 Numerical Experiments

We illustrate the numerical behavior of the proposed algorithms on a set of matrix problems arising from various application fields [9, 10]. In Tables 1, for each matrix problem we report on the number of iterations and the elapsed time required by Krylov methods to reduce the initial residual of six orders of magnitude without preconditioner (column 'Unprec') and with $FROB(\tau, p)$ (column 'Iter'), the values used for the parameters τ and p, the number of deflated eigenvalues in the post-processing stabilization step, and the density of the approximate inverse, namely the ratio $nnz(M)/nnz(A)$. The linear system is preconditioned from the right. As iterative solvers, we use restarted GMRES [17] for nonsymmetric problems and the conjugate gradient (CG [14]) for symmetric positive definite problems. We set up the right-hand side so that the exact solution is a vector with all ones, and we start the iterative process from the zero vector. All experiments are run in Fortran on a PC Pentium 4 CPU 3.00 GHz and 2.00 GB of RAM memory.

We observe that for low density values the preconditioner accelerates the convergence of Krylov methods and can reduce significantly the number of iterations required in the unpreconditioned case. The number of nonzeros of M is generally very small for all test problems except for the two most difficult, that are `gre__512` and `epb0`; this suggests that the pattern selection strategy based on Neumann expansions is effective to capture the large entries of the exact inverse, and the resulting preconditioner is reasonably cheap to compute and to apply. The values of the parameters are not tuned for optimal performance as in this study we only intend to illustrate the potential of the preconditioner on problems arising from various fields.

In Tables 3-4 we analyse the numerical and parallel scalability of the approximate inverse for solving large-scale boundary integral equations in electromagnetic scattering applications. The runs are done on eight processors of a Compaq Alpha server (a SMP cluster with four 1.3 GFlops DEC Alpha processors per node). In this case we use $\tau = 1e-2$ and p is the number of one-level neighbors associated with each degree of freedom in the mesh, or equivalently the number of nonzeros per column in the near-field part of the matrix. We see that both the construction and the

Table 1 Numerical experiments with Krylov methods preconditioned by *FROB* on a set of sparse matrix problems.

Problem	Size	Setup FROB τ	p	$\frac{nnz(M)}{nnz(A)}$	Time (sec)	Stabil.	Solver	#None	#FROB
1138_bus	1138	1e-1	30	0.94	0.4	–	cg	1138	56
bcsstk14	1806	1e-1	20	0.37	0.6	–	cg	1806	63
bcsstk27	1224	1e-1	50	0.42	0.5	–	cg	575	65
cavity05	1182	1e-1	20	0.47	1.1	6	gmres(40)	1219	33
cavity16	4562	1e-1	20	0.43	4.7	6	gmres(50)	4733	71
dw2048	2048	1e-1	80	1.12	2.6	6	"	1386	495
e30r0100	9661	1e-1	30	0.45	8.3	6	"	10037	157
epb0	1794	1e-1	40	2.63	1.0	6	"	1821	52
fidap015	6867	1e-1	30	0.95	2.1	6	"	327	107
fidap022	839	1e-1	20	0.52	0.6	6	gmres(30)	865	49
gre__512	512	1e-2	80	2.65	0.3	6	gmres(50)	521	81
orsirr_1	1030	1e-1	20	0.43	0.3	6	gmres(30)	318	78
rdb2048	2048	1e-1	30	0.83	0.5	6	gmres(50)	563	31
rdb3200L	3200	1e-1	30	1.35	1.0	6	"	3329	100

application of the approximate inverse scale very well with respect to the number of processors, and the iterations grows linearly with the problem size. On this application, standard incomplete factorization suffers from ill-conditioning of the triangular factors and exhibits poor performance. In Table 2 we report on results of comparative experiments on a set of indefinite problems using *FROB*, *ILUTP* (the incomplete factorization enhanced with partial pivoting [16]) and *SPAI* (a Frobenius-norm minimization approximate inverse implementing a different adaptive pattern selection strategy [13]), at roughly equal number of nonzeros in the preconditioner. For many problems our method competes and is sometimes more effective than *ILUTP* and *SPAI* mainly thanks to the stabilization step that makes the preconditioner more robust. Finally, we observe that the use of variable powers for different columns of M permits to have a flexible control on the memory storage of the approximate inverse but has an impact also on performance. On the `gre__512` problem we are not able to achieve convergence in more than 1000 iterations using the pattern of A^5 for the approximate inverse, which is $\approx 13\%$ dense, whereas GMRES(50) preconditioned by $FROB(80, 1e-2)$ converges in 81 iterations. We have observed the same behavior on the `epb0` problem.

Table 2 Comparative results of the *FROB* preconditioner with the *ILUTP* and the *SPAI* preconditioners. Notation: $*$ means *nearly singular* preconditioner, – means no convergence in 500 iterations.

Problem	#FROB	#ILUTP no pivoting	#ILUTP piv_thresh=0.05	#ILUTP piv_thresh=0.1	#SPAI
cavity05	33	$*$	109	109	+500
cavity16	71	$*$	$*$	$*$	+500
dw2048	495	144	–	–	+500
e30r0100	157	$*$	$*$	$*$	+500
epb0	52	66	66	109	188
fidap015	107	$*$	$*$	$*$	219
fidap022	49	$*$	219	$*$	116
gre__512	81	$*$	–	–	+500
orsirr_1	78	424	205	205	64
rdb2048	31	668	668	668	227
rdb3200L	100	–	$*$	–	+500

Table 3 Numerical scalability of the preconditioner for solving large-scale boundary integral equations. The symbol • means run on 32 processors. Notation: *m* means minutes, *h* hours.

$dof/freq$	FROB Density	FROB Time	GMRES(∞) Iter	GMRES(∞) Time	GMRES(120) Iter	GMRES(120) Time
104793 / 2.6 Ghz	0.19	6m	234	20m	253	17m
419172 / 5.2 "	0.05	21m	413	2h 44m	571	2h 26m
943137 / 7.8 "	0.02	49m	454	3h 35m•	589	5h 55m

Table 4 Parallel scalability of the approximate inverse for solving large-scale boundary integral equations on a model problem.

n (procs)	Construction time (sec)	Elapsed time precond (sec)
112908 (8)	513	0.39
221952 (16)	497	0.43
451632 (32)	509	0.48
900912 (64)	514	0.60

4 Concluding Remarks

We have discussed approximate inverse preconditioners based on Frobenius-norm minimization methods constructed using a dual dropping strategy. We have introduced a novel adaptive algorithm based on truncated Neumann matrix expansions for selecting the nonzero structure of the preconditioner and we have described a stabilization technique that deflates some of the smallest eigenvalues which can potentially disturb the convergence of Krylov methods. Finally, we have reported on results of preliminary experiments on a set of matrix problems arising from different application fields to illustrate the potential effectiveness of the approximate inverse as alternative to other standard methods of both implicit and explicit type. The numerical experiments show that the proposed method is effective to capture the large entries of the inverse at low computational effort and maintains good scalability. A parallel implementation of the preconditioner for distributed memory computers using the Fortran/MPI standard is envisaged. Perspective of future work also include the design of symmetrization strategies for solving symmetric indefinite systems, blocking strategies for cost reduction and a detailed comparison with other existing implementations.

References

1. Barnard, S.T., Bernardo, L.M., Simon, H.D.: An MPI implementation of the SPAI preconditioner on the T3E. International Journal of High Performance Computing Application **13(2)**, 107–123 (1999)
2. Benson, M.W., Frederickson, P.O.: Iterative solution of large sparse linear systems arising in certain multidimensional approximation problems. Utilitas Mathematica **22**, 127–140 (1982)
3. Benson, M.W., Krettmann, J., Wright, M.: Parallel algorithms for the solution of certain large sparse linear systems. Int J. Comput. Math **16** (1984)
4. Benzi, M., Meyer, C., Tůma, M.: A sparse approximate inverse preconditioner for the conjugate gradient method. SIAM J. Scientific Computing **17**, 1135–1149 (1996)
5. Bröker, O., Grote, M.J.: Sparse approximate inverse smoothers for geometric and algebraic multigrid. Applied Numerical Mathematics: Transactions of IMACS **41**(1), 61–80 (2002). URL http://www.elsevier.com/gej-ng/10/10/28/86/27/31/abstract.html
6. Carpentieri, B., Duff, I.S., Giraud, L.: A class of spectral two-level preconditioners. SIAM J. Scientific Computing **25**(2), 749–765 (2003)
7. Chow, E.: Parallel implementation and performance characteristics of least squares sparse approximate inverse preconditioners. Technical Report UCRL-JC-138883, Lawrence Livermore National Laboratory, Livermore, CA (2000)
8. Chow, E., Saad, Y.: Approximate inverse techniques for block-partitioned matrices. SIAM J. Scientific Computing **18**, 1657–1675 (1997)
9. Matrix Market: a visual web database for numerical matrix data: A collection of Fortran codes for large scale scientific computation (1996). Http://math.nist.gov/MatrixMarket/index.html
10. Davis, T.: Sparse matrix collection (1994). Http://www.cise.ufl.edu/research/sparse/matrices
11. Duff, I.S., Erisman, A.M., Gear, C.W., Reid, J.K.: Sparsity structure and gaussian elimination. SIGNUM Newsletter **23**(2), 2–8 (1988)
12. Gould, N.I.M., Scott, J.A.: Sparse approximate-inverse preconditioners using norm-minimization techniques. SIAM J. Scientific Computing **19(2)**, 605–625 (1998)
13. Grote, M., Huckle, T.: Parallel preconditionings with sparse approximate inverses. SIAM J. Scientific Computing **18**, 838–853 (1997)
14. Hestenes, M.R., Stiefel, E.: Methods of conjugate gradients for solving linear systems. Journal of Research of the National Bureau of Standards **49**(6), 409–436 (1952)
15. Kolotilina, L.Y., Yeremin, A.Y.: Factorized sparse approximate inverse preconditionings. I: Theory. SIAM J. Matrix Analysis and Applications **14**, 45–58 (1993)
16. Saad, Y.: Iterative Methods for Sparse Linear Systems. PWS Publishing, New York (1996)
17. Saad, Y., Schultz, M.H.: GMRES: A generalized minimal residual algorithm for solving nonsymmetric linear systems. SIAM J. Scientific and Statistical Computing **7**, 856–869 (1986)
18. Tang, W.P., Wan, W.: Sparse approximate inverse smoother for multigrid. SIAM J. Matrix Analysis and Applications **21**(4), 1236–1252 (2000)

On the Construction of Stable B-Spline Wavelet Bases

D. Černá and V. Finěk

Abstract The paper is concerned with the construction of wavelet bases on the interval derived from B-splines. The resulting bases generate multiresolution analyses on the unit interval with the desired number of vanishing wavelet moments for primal and dual wavelets. Inner wavelets are translated and dilated versions of well-known wavelets designed by Cohen, Daubechies, Feauveau [4] while boundary wavelets are constructed by combination of methods from [3], [8] and [11]. By this approach, we obtain bases with small Riesz condition numbers. The other important feature of wavelet bases, the sparseness of refinement matrices, is preserved. Finally, Riesz condition numbers of scaling and wavelet bases are computed for some of the constructed bases.

1 Introduction

Wavelets are by now a widely accepted tool in signal and image processing as well as in numerical simulation. In the field of numerical analysis, methods based on wavelets are successfully used especially for preconditioning of large systems arising from discretization of elliptic partial differential equations, sparse representations of some types of operators and adaptive solving of partial differential equations. There are two main approaches for solving partial differential equations by adaptive wavelet methods. The first approach consists in multiresolution adaptive postprocessing, i.e. start from a classical scheme on a uniform grid and use a discrete multiresolution decomposition in order to compress computational time and

Dana Černá
Technical University in Liberec, Studentská 2, 461 17 Liberec, Czech Republic
e-mail: dana.cerna@tul.cz

Václav Finěk
Technical University in Liberec, Studentská 2, 461 17 Liberec, Czech Republic
e-mail: vaclav.finek@tul.cz

memory size, while preserving the accuracy of the initial scheme. This approach is mostly applied to hyperbolic systems such as conservation laws. The second approach is mostly applied to elliptic and parabolic equations and it insists in using wavelets directly as basis and test functions in variational formulation. This leads to methods which are efficient and asymptotically optimal.The quantitative properties of such methods depend on the choice of wavelet basis, in particular on its condition.

Wavelet bases on a bounded domain, which are suitable for solving operator equations, are usually constructed in the following way: Wavelets on the real line are adapted to the interval and then by tensor product technique to n-dimensional cube. Finally by splitting domain into subdomains which are images of $(0,1)^n$ under appropriate parametric mappings one can obtain wavelet bases on a fairly general domain. From the viewpoint of numerical stability, ideal wavelet bases are orthogonal wavelet bases. However, they are usually avoided in numerical treatment of partial differential equations, because they are not accessible analytically, the complementary boundary conditions can not be satisfied and it is not possible to increase the number of vanishing wavelet moments independent from the order of accuracy.

Biorthogonal B-spline wavelet bases on the unit interval were constructed in [8]. Their advantage is that the primal basis is known explicitly. Bases constructed in [3] are well-conditioned, but have globally supported dual basis functions. B-spline bases from [8] have large Riesz condition numbers that cause problems in practical applications. Some modifications which lead to better conditioned bases were proposed in [1, 7, 9]. The recent construction in [11] seems to outperform the previous constructions with respect to Riesz condition numbers, for some numerical experiments see [12] and also [10]. In this paper, we combine approaches from [3, 8, 11] to further improve stability properties of B-spline wavelet bases on the interval.

2 Wavelet Bases

This section provides a short introduction to the concept of wavelet bases. Let V be a separable Hilbert space with inner product $\langle \cdot,\cdot \rangle_V$ and induced norm $\|\cdot\|_V$. Let J be some index set and let each index $\lambda \in J$ takes the form $\lambda = (j,k)$, where $|\lambda| = j \in \mathbb{Z}$ is *scale* or *level*. Assume that J can be decomposed as $J = \cup_{j \geq j_0} J_j$, where $j_0 \in \mathbb{Z}$ is some coarsest level.

Definition 1. Family $\Psi := \{\psi_\lambda \in J\} \subset V$ is called *wavelet basis* of V, if

i) Ψ is a *Riesz basis* for V, that means Ψ generates V, i.e.

$$V = \mathrm{clos}_{\|\cdot\|_V} \mathrm{span} \Psi, \tag{1}$$

and there exist constants $c, C \in (0,\infty)$ such that for all $\mathrm{b} := \{b_\lambda\}_{\lambda \in J} \in l^2(J)$ holds

$$c\,\|\mathrm{b}\|_{l_2(J)} \le \left\| \sum_{\lambda\in J} b_\lambda \psi_\lambda \right\|_V \le C\,\|\mathrm{b}\|_{l_2(J)} \tag{2}$$

Constants c, C are called *Riesz bounds* and C/c is called *Riesz condition number* or the *condition* of Ψ.

ii) Functions are local in the sense that $\operatorname{diam}(\Omega_\lambda) \le C2^{-|\lambda|}$ for all $\lambda \in J$, where Ω_λ is support of ψ_λ.

By the Riesz representation theorem, there exists a unique family of dual functions $\tilde{\Psi} = \{\tilde{\psi}_\lambda, \lambda \in \tilde{J}\} \subset V$, which are biorthogonal to Ψ, i.e. it holds

$$\langle \psi_{i,k}, \tilde{\psi}_{j,l} \rangle_V = \delta_{i,j}\delta_{k,l}, \quad \text{for all} \quad (i,k) \in J, \quad (j,l) \in \tilde{J}. \tag{3}$$

This dual family is also a Riesz basis for V with Riesz bounds C^{-1}, c^{-1}. The pair Ψ, $\tilde{\Psi}$ is often referred to as *biorthogonal system*, Ψ is called *primal* wavelet basis, $\tilde{\Psi}$ is called *dual* wavelet basis. By the above argument, biorthogonality is a necessary for the Riesz basis property (2) to hold. But unfortunately it is not sufficient, see [6].

In many cases, the wavelet system Ψ is constructed with the aid of a multiresolution analysis.

Definition 2. A sequence $S = \{S_j\}_{j\in\mathbb{N}_{j_0}}$ of closed linear subspaces $S_j \subset V$ is called a *multiresolution* or *multiscale analysis*, if the subspaces are nested, i.e.,

$$S_{j_0} \subset S_{j_0+1} \subset \ldots \subset S_j \subset S_{j+1} \subset \ldots V \tag{4}$$

and is dense in V, i.e.,

$$\operatorname{clos}_V \left(\cup_{j\in\mathbb{N}_{j_0}} S_j \right) = V. \tag{5}$$

The nestedness of the multiresolution analysis implies the existence of the *complement* or *wavelet spaces* W_j such that

$$S_{j+1} = S_j \oplus W_j. \tag{6}$$

We now assume that S_j and W_j are spanned by sets of basis functions

$$\Phi_j := \{\phi_{j,k}, k \in I_j\}, \quad \Psi_j := \{\psi_{j,k}, k \in J_j\}, \tag{7}$$

where I_j, J_j are finite or at most countable index sets. We refer to $\phi_{j,k}$ as *scaling functions* and $\psi_{j,k}$ as *wavelets*. The multiscale basis is given by $\Psi_{j_0,s} = \Phi_{j_0} \cup \bigcup_{j=j_0}^{j_0+s-1} \Psi_j$ and the overall wavelet Riesz basis of V is obtained by $\Psi = \Phi_{j_0} \cup \bigcup_{j\ge j_0} \Psi_j$. From the nestedness of S and the Riesz basis property (2), we conclude the existence of bounded linear operators $\mathbf{M}_{j,0} = \left(m_{l,k}^{j,0}\right)_{l\in I_{j+1}, k\in I_j}$ and $\mathbf{M}_{j,1} = \left(m_{l,k}^{j,1}\right)_{l\in I_{j+1}, k\in J_j}$ such that

$$\phi_{j,k} = \sum_{l\in I_{j+1}} m_{l,k}^{j,0} \phi_{j+1,l}, \quad \psi_{j,k} = \sum_{l\in I_{j+1}} m_{l,k}^{j,1} \phi_{j+1,l}. \tag{8}$$

The desired property in applications is the uniform sparseness of $\mathbf{M}_{j,0}$ and $\mathbf{M}_{j,1}$, it means that the number of nonzero entries per row and column remains uniformly bounded in j. The single-scale and the multiscale bases are interrelated by $\mathbf{T}_{j,s} : l^2(I_{j+s}) \to l^2(I_{j+s})$,

$$\Psi_{j,s} = \mathbf{T}_{j,s} \Phi_{j+s}. \tag{9}$$

$\mathbf{T}_{j,s}$ is called the *multiscale* or the *wavelet transform*.

The dual wavelet system $\tilde{\Psi}$ generates a dual multiresolution analysis $\tilde{S}$ with a dual scaling basis $\tilde{\Phi}$ and dual operators $\tilde{\mathbf{M}}_{j,0}$, $\tilde{\mathbf{M}}_{j,1}$.

Polynomial exactness of order $N \in \mathbb{N}$ for primal scaling basis and of order $\tilde{N} \in \mathbb{N}$ for dual scaling basis is another desired property of wavelet bases in $V \subset L^2(\Omega)$, $\Omega \subset \mathbb{R}^n$. It means that

$$\mathbb{P}_{N-1} \subset S_j, \quad \mathbb{P}_{\tilde{N}-1} \subset \tilde{S}_j, \quad j \geq j_0, \tag{10}$$

where $\mathbb{P}_m$ is the space of all algebraic polynomials on Ω of degree less or equal to m.

3 Construction of Stable Wavelet Bases on the Interval

In this section, we assume $V = L^2([0,1])$. The primal scaling bases will be the same as bases designed by Chui and Quak in [3], because they are known to have good condition numbers. A big advantage of this approach is that it readily adapts to the bounded interval by introducing multiple knots at the endpoints. Let N be the desired order of polynomial exactness of primal scaling basis and let $\mathbf{t}^j = \left(t_k^j\right)_{k=-N+1}^{2^j+N-1}$ be a sequence of knots defined by

$$\begin{aligned} t_k &= 0 \quad \text{for} \quad k = -N+1, \ldots, 0, \\ t_k &= \frac{k}{2^j} \quad \text{for} \quad k = 1, \ldots 2^j - 1, \\ t_k &= 1 \quad \text{for} \quad k = 2^j, \ldots, 2^j + N - 1. \end{aligned}$$

The corresponding B-splines of order N are defined by

$$B_{k,N}^j(x) := \left(t_{k+N}^j - t_k^j\right) \left[t_k^j, \ldots, t_{k+N}^j\right]_t (t-x)_+^{N-1}, \quad x \in [0,1], \tag{11}$$

where $(x)_+ := \max\{0, x\}$ and $[t_1, \ldots t_N]_t f$ is the N-th divided difference of f. The set Φ_j of primal scaling functions is then simply defined as

$$\phi_{j,k} = 2^{j/2} B_{k,N}^j, \quad \text{for} \quad k = -N+1, \ldots, 2^j - 1, \quad j \geq 0. \tag{12}$$

Thus there are $2^j - N + 1$ inner scaling functions and $N - 1$ functions on each boundary. Inner functions are translations and dilations of a function ϕ which corresponds to primal scaling function constructed by Cohen, Daubechies, Feauveau in [4]. In the following, we consider ϕ from [4] which is shifted so that its support is $[0,N]$ and we denote $\phi_{j,k}^{\mathbb{R}} = 2^{j/2}\phi\left(2^j\cdot - k\right)$.

The desired property of dual scaling basis $\tilde{\Phi}$ is biorthogonality to Φ and polynomial exactness of order $\tilde{N}$. Let $\tilde{\phi}$ be dual scaling function which was designed in [4] and which is shifted so that its support is $\left[-\tilde{N}+1, N+\tilde{N}-1\right]$. In this case $\tilde{N} \geq N$ and $\tilde{N}+N$ must be an even number. We define basis functions to preserve polynomial exactness in the following way:

$$\theta_{j,k} = 2^j \sum_{l=-N-\tilde{N}+2}^{\tilde{N}-N} \left\langle p_{k+N-1}^{\tilde{N}-1}\left(2^j\cdot\right), \phi_{j,l}^{\mathbb{R}}\right\rangle \tilde{\phi}\left(2^j\cdot - l\right)|_{[0,1]}, \quad k = 1-N, \ldots, \tilde{N}-N,$$

$$\theta_{j,k} = 2^{j/2}\tilde{\phi}\left(2^j\cdot - k\right)|_{[0,1]}, \quad k = \tilde{N}-N+1, \ldots, 2^j-\tilde{N}-1,$$

$$\theta_{j,k} = \theta_{j,2^j-N+1-k}(1-\cdot), \quad k = 2^j-\tilde{N}, \ldots, 2^j-1.$$

Here $p_0^{\tilde{N}-1}, \ldots, p_{\tilde{N}-1}^{\tilde{N}-1}$ is a basis of the space of all algebraic polynomials on $[0,1]$ of degree less or equal to $\tilde{N}-1$. In our case, $p_k^{\tilde{N}-1}$ are Bernstein polynomials defined by

$$p_k^{\tilde{N}-1}(x) := b^{-\tilde{N}+1}\binom{\tilde{N}-1}{k} x^k (b-x)^{\tilde{N}-1-k}, \quad k = 0, \ldots, \tilde{N}-1, \tag{13}$$

because they are known to be well-conditioned on $[0,b]$ relative to the supremum norm. We choose $b=1$, because b does not affect the condition of dual bases after biorthogonalization. Note that there are two types of boundary functions: $\tilde{N}$ functions reproducing Bernstein polynomials and $N-2$ restrictions of $2^{j/2}\tilde{\phi}\left(2^j\cdot - k\right)$. Our next goal is to determine the corresponding wavelet bases. Since the set $\Theta_j := \left\{\theta_{j,k} : k = -N+1, \ldots, 2^j-1\right\}$ is not biorthogonal to Φ_j, we derive a new set $\tilde{\Phi}_j$ from Θ_j by biorthogonalization. Let $\mathbf{A}_j = \left(\left\langle \phi_{j,k}, \theta_{j,l}\right\rangle\right)_{j,l=-N+1}^{2^j-1}$, then viewing $\tilde{\Phi}_j$ and Θ_j as column vectors we define

$$\tilde{\Phi}_j := \mathbf{A}_j^{-T}\Theta_j. \tag{14}$$

For the proof of invertibility of the matrix $\mathbf{A}_j$ for all admissible choices of $N, \tilde{N}$ see [13]. This task is directly connected to the task of determining an appropriate matrices $\mathbf{M}_{j,1}$, $\tilde{\mathbf{M}}_{j,1}$. Thus, the problem has been transferred from functional analysis to linear algebra. We follow a general principle which was proposed in [2].

Definition 3. Any $\mathbf{M}_{j,1} : l_2(J_j) \to l_2(I_{j+1})$ is called a *stable completion* of $\mathbf{M}_{j,0}$, if

$$\left\|\mathbf{M}_j\right\|, \left\|\mathbf{M}_j^{-1}\right\| = O(1), \quad j \to \infty, \tag{15}$$

where $\mathbf{M}_j := \left(\mathbf{M}_{j,0}, \mathbf{M}_{j,1}\right)$.

The idea is to determine first an initial stable completion and then to project it to the desired complement space W_j determined by $\{\tilde{V}_j\}_{j\geq j_0}$. This is summarized in the following theorem.

Theorem 1. *Let Φ_j, $\tilde{\Phi}_j$ be primal and dual scaling basis, respectively. Let $\mathbf{M}_{j,0}$, $\tilde{\mathbf{M}}_{j,0}$ be refinement matrices corresponding to these bases. Suppose that $\check{\mathbf{M}}_{j,1}$ is some stable completion of $\mathbf{M}_{j,0}$ and $\check{\mathbf{G}}_j = \check{\mathbf{M}}_j^{-1}$. Then*

$$\mathbf{M}_{j,1} := \left(\mathbf{I} - \mathbf{M}_{j,0}\tilde{\mathbf{M}}_{j,0}^T\right)\check{\mathbf{M}}_{j,1} \tag{16}$$

is also a stable completion and $\mathbf{G}_j = \mathbf{M}_j^{-1}$ has the form

$$\mathbf{G}_j = \begin{bmatrix} \check{\mathbf{M}}_{j,0}^T \\ \check{\mathbf{G}}_{j,1} \end{bmatrix}. \tag{17}$$

Moreover, the collections

$$\Psi_j := \mathbf{M}_{j,1}^T \Phi_{j+1}, \quad \tilde{\Psi}_j := \check{\mathbf{G}}_{j,1}^T \tilde{\Phi}_{j+1} \tag{18}$$

form biorthogonal systems

$$\left\langle \Psi_j, \tilde{\Psi}_j \right\rangle = \mathbf{I}, \quad \left\langle \Phi_j, \tilde{\Psi}_j \right\rangle = \left\langle \Psi_j, \tilde{\Phi}_j \right\rangle = \mathbf{0}. \tag{19}$$

We found the initial stable completion by the method from [8] with some small changes. Since this construction is quite subtle we don't go into details here.

4 Quantitative Properties of Constructed Bases

In this section, quantitative properties of constructed bases are presented. To further improve the condition of constructed bases we provide a diagonal rescaling in the following way:

$$\phi_{j,k}^N = \frac{\phi_{j,k}}{\sqrt{\left\langle \phi_{j,k}, \phi_{j,k} \right\rangle}}, \quad \tilde{\phi}_{j,k}^N = \tilde{\phi}_{j,k} * \sqrt{\left\langle \phi_{j,k}, \phi_{j,k} \right\rangle}, \quad k \in J_j, \quad j \geq j_0,$$

$$\psi_{j,k}^N = \frac{\psi_{j,k}}{\sqrt{\left\langle \psi_{j,k}, \psi_{j,k} \right\rangle}}, \quad \tilde{\psi}_{j,k}^N = \tilde{\psi}_{j,k} * \sqrt{\left\langle \psi_{j,k}, \psi_{j,k} \right\rangle}, \quad k \in I_j, \quad j \geq j_0.$$

Then the new primal scaling and wavelet bases are normalized with respect to L^2-norm. This can be useful property in adaptive wavelet methods, because it allows to use the same thresholding strategy for all coefficients. The condition of single-scale bases are listed in Table1, for larger scales it does not change significantly. The condition of multiscale bases and wavelet transforms are listed in Table2. The resulting bases for $N = 2$ are the same as those designed in [11], where the primal scaling functions are the same, $\tilde{N}$ dual boundary functions are constructed as in [5] and $N-2$ boundary functions are constructed by finding suitable refinement coefficients. Our bases have comparable or better condition numbers for $N \geq 3$. Due to the compact support of basis functions the sparseness of refinement matrices is preserved. The proof of existence for all admissible choices of N and $\tilde{N}$, the condition of bases adaptated to boundary conditions and other quantitative properties of these bases such as the condition number of stiffness matrices are studied in preparing paper and in [13].

Table 1 The condition of single-scale scaling and wavelet bases

N	$\tilde{N}$	j	Φ_j	$\tilde{\Phi}_j$	Φ_j^N	$\tilde{\Phi}_j^N$	Ψ_j	$\tilde{\Psi}_j$	Ψ_j^N	$\tilde{\Psi}_j^N$
2	2	5	2.00	2.30	1.73	1.97	1.91	2.01	1.91	1.99
2	4	5	2.00	2.09	1.73	1.80	1.99	2.04	1.99	1.99
2	6	5	2.00	2.26	1.73	2.03	1.99	2.29	1.99	2.25
2	8	5	2.00	2.89	1.73	2.78	2.33	3.13	2.22	3.80
3	3	5	3.24	7.36	2.76	5.43	3.98	6.14	3.98	3.95
3	5	5	3.24	4.35	2.76	3.57	4.43	5.93	3.97	3.96
3	7	5	3.24	3.68	2.76	3.16	4.60	5.29	3.97	3.96
3	9	5	3.24	3.75	2.76	3.29	4.40	5.46	3.99	4.06
4	6	6	5.18	19.72	4.42	14.60	13.66	46.66	7.98	10.60
4	8	6	5.18	11.33	4.42	9.25	13.94	29.69	7.97	8.55
4	10	6	5.18	7.80	4.42	6.83	13.80	20.03	7.97	8.04

Table 2 The condition of multi-scale wavelet bases and wavelet transform

N	$\tilde{N}$	j	$\Psi_{j,3}^N$	$\Psi_{j,4}^N$	$\tilde{\Psi}_{j,3}^N$	$\tilde{\Psi}_{j,4}^N$	$\mathbf{T}_{j,1}$	$\mathbf{T}_{j,2}$	$\mathbf{T}_{j,3}$	$\mathbf{T}_{j,4}$
2	2	3	2.50	2.63	2.65	2.73	1.88	2.44	2.83	3.11
2	4	4	2.31	2.33	2.32	2.33	1.58	2.10	2.48	2.72
2	6	4	2.83	2.91	2.87	2.93	1.75	2.19	2.64	2.99
2	8	5	5.34	5.66	5.49	5.73	2.86	3.59	4.46	5.24
3	3	4	6.22	6.82	8.72	8.99	2.79	5.66	8.41	10.62
3	5	4	5.22	5.30	5.33	5.35	2.65	5.22	7.54	8.87
3	7	5	4.60	4.61	4.61	4.61	2.37	4.75	6.53	7.48
3	9	5	4.74	4.75	4.76	4.78	2.29	4.62	6.30	7.17
4	6	5	18.88	19.82	21.41	21.43	5.74	20.89	46.99	64.67
4	8	5	13.93	14.05	14.09	14.09	4.39	16.08	32.15	41.43
4	10	5	11.01	11.03	11.03	11.03	4.22	12.65	22.26	27.60

Acknowledgements The first author was supported by the research center LC06024 of Ministry of Education, Youth and Sports of the Czech Republic. The second author was supported by the research center 1M06047 of Ministry of Education, Youth and Sports of the Czech Republic.

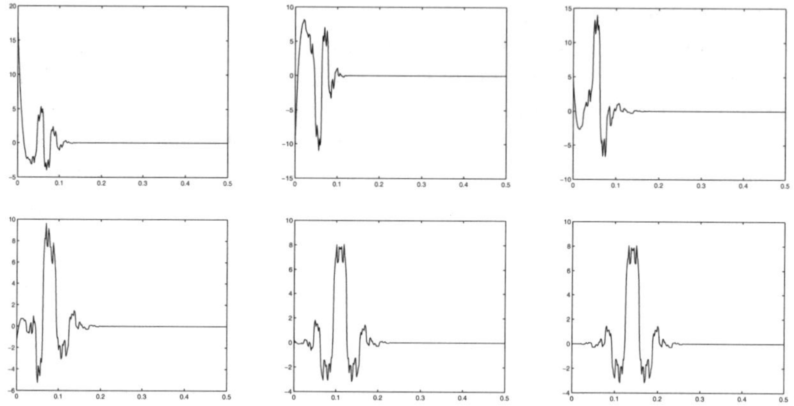

Fig. 1 Plots of dual boundary scaling functions $\tilde{\phi}^{\tilde{N}}_{5,k}$, $k = -2, \ldots 3$ for $N = 3, \tilde{N} = 5$

References

1. Barsch, T.: Adaptive Multiskalenverfahren für elliptische partielle Differentialgleichungen - Realisierung, Umsetzung und numerische Ergebnisse. PhD. thesis, RWTH Aachen, Shaker Verlag (2001)
2. Carnicer, J., Dahmen, W., Peña, J.: Local decompositions of refinable spaces. Appl. Comp. Harm. Anal. **3**, 127–153 (1996)
3. Chui, C., Quak, E.: Wavelets on a bounded interval. In: D. Braess, L. Schumaker (eds.) Numerical Methods of Approximation Theory, International Series of Numerical Mathematics, pp. 53–75. Birkhäser (1992)
4. Cohen, A., Daubechies, I., Feauveau, J.C.: Biorthogonal bases of compactly supported wavelets. Comm. Pure Appl. Math. **45**, 485–560 (1992)
5. Cohen, A., Daubechies, I., Vial, P.: Wavelets on the interval and fast wavelet transforms. Appl. Comp. Harm. Anal. **1**, 54–81 (1993)
6. Dahmen, W.: Multiscale analysis, approximation and interpolation spaces. In: C.K.Chui (ed.) Approximation Theory VIII (2), pp. 47–88 (1995)
7. Dahmen, W., Kunoth, A., Urban, K.: Surface fitting and multiresolution methods **2**. In: A.L. Méhauté, C. Rabut, L. Schumaker (eds.) Vanderbilt University Press, pp. 93–130 (1997)
8. Dahmen, W., Kunoth, A., Urban, K.: Biorthogonal spline-wavelets on the interval - stability and moment conditions. Appl. Comp. Harm. Anal. **6**, 132–196 (1999)
9. Grivet Talocia, S., Tabacco, A.: Wavelets on the interval with optimal localization. Math. Models Meth. Appl. Sci. **10**, 441–462 (2000)
10. Pabel, R.: Wavelet Methods for PDE Constrained Elliptic Control Problems with Dirichlet Boundary Control. Thesis, University Bonn (2005)
11. Primbs, M.: Stabile biorthogonale Spline-Waveletbasen auf dem Intervall. Dissertation, Universität Duisburg-Essen (2006)
12. Raasch, T.: Adaptive Wavelet and Frame Schemes for Elliptic and Parabolic Equations. PhD. thesis, University Marburg (2007)
13. Černá, D.: Biorthogonal Wavelets. PhD. thesis, Charles University, Prague. In preparation. (2008)

The Performance of a Multigrid Algorithm for the Acoustic Single Layer Equation

S. Gemmrich, J. Gopalakrishnan, and N. Nigam

Abstract We study the performance of a multigrid algorithm for an integral equation of the first kind arising in acoustics. The algorithm is adpated from previous works on the single layer potential, and as in the previous works, a key ingredient is the use of a weaker inner product. We report the implementation details and the results of numerical experiments demonstrating the effectiveness of the method.

1 Introduction

Boundary integral equations and their numerical approximations are a popular means to study the scattering of time-harmonic waves from bounded obstacles. A model for these scattering phenomena is given by the Helmholtz equation in the exterior of the scatterer, with appropriate growth conditions on the scattered field. Various reformulations in terms of integral equations on the surface of the scattering object exist. Our focus in this paper lies on integral equations of the first kind, which arise for example in the direct boundary integral method for the Dirichlet problem. The main integral operator involved, i.e. the single layer operator, may be viewed as a pseudo-differential operator of order minus one. Several authors have observed advantages of using integral equations of the first kind (see e.g. [7]), for example when the scattering object is very thin, or indeed reduces to an arc (e.g. [8]).

Due to the non-local behavior of the boundary integral operators, they typically lead to dense linear systems upon discretization. Though one only needs to mesh on a surface of co-dimension one, the fill-in in the matrices corresponding to the integral operators is significant. Without some form of preconditioning or acceleration,

S. Gemmrich and N. Nigam
Dept. of Mathematics and Statistics, McGill University, Montreal, QC,
e-mail: gemmrich@math.mcgill.ca, nigam@math.mcgill.ca

J. Gopalakrishnan
Dept. of Mathematics, University of Florida, Gainesville, Florida, e-mail: jayg@math.ufl.edu

these methods then become prohibitive.

Since the spectra of negative-order pseudodifferential operators link highly oscillatory eigenfunctions to the small magnitude eigenvalues, the use of standard smoothing methods in a multigrid scheme is not appropriate. The key for multigrid methods to work in this context is the use of weaker Sobolev norms in order to modify this spectral behavior, which has first been described in [3] for positive-definite operators. A BPX preconditioner for this problem was analyzed in [4], and the use of Haar basis functions and compression type multilevel algorithms for such equations has been studied in [11]. A slightly different approach has been taken in [10] in order to analyze a multigrid method for large-scale data-sparse approximations to the single layer operator from potential theory. Algebraic multigrid preconditioners for the same problem - based on the smoother in [3] - have been developed in [9].

Discretizations of integral equations for acoustics suffer from a lack of definiteness similar to the discretizations of the associated partial differential equation. A natural tool in the analysis of multigrid methods for such indefinite problems are perturbation type arguments such as in [2] and [6]. The application of such perturbation arguments to the acoustic single layer potential will be carried out in [5].

The purpose of this note is to report the performance of a multigrid algorithm applied to the indefinite acoustic single layer discretization and to clarify a few implementation details. The design of the algorithm heavily relies on the above references. Our aim is to convey the potential of multigrid for this integral equation discretization. Our current codes assemble and multiply matrices in $O(n^2)$ complexity, so the iterative solution by multigrid also costs the same. However, our results show that we can approach optimality in complexity through multigrid once assembly and matrix multiplications are done optimally. The latter is an issue of current research in multipole and hierarchical matrix theories.

2 A Model Problem and a Multigrid Algorithm

2.1 A Model Problem

Let Γ be a simple, polygonal closed curve in the plane with sides Γ_i and let Ω^+ denote its exterior domain. We consider the exterior Helmholtz problem with prescribed Dirichlet data $g \in H^{1/2}(\Gamma)$ and wave-number $\kappa \in \mathbb{R}$:

$$-\Delta u - \kappa^2 u = 0 \quad \text{in } \Omega^+, \qquad u = g \quad \text{on } \Gamma, \qquad \lim_{r\to\infty} r^{\frac{1}{2}}\left(\frac{\partial u}{\partial r} - i\kappa u\right) = 0. \quad (1)$$

Here, r is the usual radial component in polar coordinates, and $H^{1/2}(\Gamma)$ represents the trace space of functions which are locally H^1 in the exterior, i.e. the trace space of $H^1_{loc}(\Omega^+)$.

In order to guarantee unique solvability we assume that κ^2 is not an interior eigenvalue of $-\Delta$. Then there are various ways to solve (1) in terms of boundary integral operators. We use the direct approach. The solution $u(x), x \in \Omega^+$ is given in terms of the integral representation formula [12]:

$$u(x) = \frac{i}{4}\int_\Gamma H_0^1(\kappa|x-y|)\frac{\partial u(y)}{\partial n_y}ds_y - \frac{i}{4}\int_\Gamma \frac{\partial}{\partial n_y}H_0^1(\kappa|x-y|)\,u(y)ds_y. \quad (2)$$

Taking into account the logarithmic singularity of the Hankel function $H_0^1(z)$ at zero, as $x \to \Gamma$, the behaviour of (2) is given in terms of the (weakly singular) single layer operator V, and the (Cauchy singular) integral operator K, defined as

$$V : H^{-1/2}(\Gamma) \longrightarrow H^{1/2}(\Gamma), \quad V\sigma(x) := \frac{i}{4}\int_\Gamma H_0^1(\kappa|x-y|)\,\sigma(y)ds_y \qquad x \in \Gamma,$$

$$K : H^{1/2}(\Gamma) \longrightarrow H^{1/2}(\Gamma), \quad K\mu(x) := \frac{i}{4}\int_\Gamma \frac{\partial}{\partial n_y}H_0^1(\kappa|x-y|)\,\mu(y)ds_y, \quad x \in \Gamma.$$

The Dirichlet scattering problem (1) then becomes the acoustic single layer equation for the unknown Neumann data $\sigma := \partial u/\partial n \in H^{-1/2}(\Gamma)$:

$$V\sigma = f \in H^{1/2}(\Gamma), \quad (3)$$

where the right hand side $f = (-\frac{1}{2}Id + K)g$ depends on the Dirichlet trace g and requires the evaluation of the double layer operator K. The weak form of equation (3) is thus: Given $f \in H^{1/2}(\Gamma)$, find $\sigma \in H^{-1/2}(\Gamma)$ such that

$$\mathscr{V}(\sigma,\mu) = \langle f,\mu\rangle \quad \text{for all } \mu \in H^{-1/2}(\Gamma), \quad (4)$$

where the continuous sesquilinear form $\mathscr{V} : H^{-1/2}(\Gamma) \times H^{-1/2}(\Gamma) \to \mathbb{C}$ is defined by $\mathscr{V}(\sigma,\mu) = \langle V\sigma,\mu\rangle$, and $\langle\cdot,\cdot\rangle$ denotes the duality pairing between $H^{\frac{1}{2}}(\Gamma)$ and $H^{-\frac{1}{2}}(\Gamma)$. When $\kappa = 0$, we denote $\mathscr{V}(\cdot,\cdot)$ as $\Lambda(\cdot,\cdot)$, the positive definite (on suitably scaled regions, [12]), continuous sesquilinear form corresponding to the Laplacian.

2.2 A Multigrid Algorithm

We now adapt the multigrid algorithm from [3], which has been devised for the positive definite operator Λ, for the indefinite equation (4). The main idea is the use of the weaker base inner product. However, this comes at the cost of some computability issues, which we will have to work around later.

Suppose we are given a sequence of finite dimensional approximation spaces $\mathscr{M}_1 \subset \ldots \subset \mathscr{M}_J \subset H^{-1/2}(\Gamma)$, where $\mathscr{M}_i$ is spanned by piecewise constant test functions on meshes which are uniform on every side (patch) Γ_i of Γ. In the following, $(\cdot,\cdot)_{-1,D}$ and $\|\cdot\|_{-1,D}$ denote the the $H^{-1}(D)$-inner product and norm, respectively, on a domain D. When the domain is absent from the notation, we understand it to be Γ. We define discrete operators $V_k : \mathscr{M}_k \to \mathscr{M}_k$ via the relation

$$(V_k\sigma,\mu)_{-1} = \mathscr{V}(\sigma,\mu) \quad \text{for all } \sigma,\mu \in \mathscr{M}_k. \tag{5}$$

Analogously, we choose $f_k \in \mathscr{M}_k$ to satisfy $(f_k,\mu)_{-1} = \langle f,\mu\rangle$ for all $\mu \in \mathscr{M}_k$. Then, on every level k, the equation of interest can be written in operator form as

$$V_k\,\sigma_k = f_k. \tag{6}$$

We define the projections $Q_k : H^{-1}(\Gamma) \to \mathscr{M}_k$ defined by $(Q_k\sigma,\mu)_{-1} := (\sigma,\mu)_{-1}$ for all $\mu \in \mathscr{M}_k$. Since the presence of $(\cdot,\cdot)_{-1}$ introduces computational difficulties, we need additional computable inner products $[\cdot\,,\,\cdot]_k$ on $\mathscr{M}_k$ that are equivalent to the $(\cdot,\cdot)_{-1}$ inner product. We will give an example of such an inner product later.

Once such inner products are available, a simple Richardson smoother suitable for multigrid algorithms (see for example [1]) is given by

$$[R_k\sigma,\theta]_k = \frac{1}{\tilde{\lambda}_k}(\sigma,\theta)_{-1}, \tag{7}$$

where $\tilde{\lambda}_k$ is chosen to be an upper bound for the Ritz quotient involving the definite sesquilinear form $\Lambda(\cdot\,,\,\cdot)$, ie,

$$\tilde{\lambda}_k \geq \lambda_k = \sup_{\theta\in\mathscr{M}_k} \frac{\Lambda(\theta,\theta)}{[\theta,\theta]_k}. \tag{8}$$

Define the operator Λ_k as in (5) but with the sesquilinear form $\mathscr{V}(\cdot,\cdot)$ replaced by $\Lambda(\cdot,\cdot)$. It is then immediate that the smoothers are properly scaled to work for the definite problem, since by the definition of $\tilde{\lambda}_k$ we have $\Lambda(R_k\Lambda_k\theta,\theta) \leq \Lambda(\theta,\theta)$.

Now, given an initial guess $\sigma_0 \in \mathscr{M}_J$, the multigrid iteration computes a sequence of approximate solutions to (4) using an iteration of the form $\sigma_{i+1} = Mg_J(\sigma_i, f_J)$, where $Mg_J(\cdot,\cdot)$ as a mapping of $\mathscr{M}_J \times \mathscr{M}_J$ into $\mathscr{M}_J$ is defined recursively by the following algorithm:

Algorithm 1 *Set* $Mg_1(\sigma,f) = V_1^{-1}f$. *If* $k > 1$ *we define* $Mg_k(\sigma,f)$ *recursively as follows:*

$$\sigma_1 = \sigma + R_k(f - V_k\sigma), \tag{9}$$

$$Mg_k(\sigma,f) = \sigma_1 + Mg_{k-1}(0, Q_{k-1}(f - V_k\sigma_1)). \tag{10}$$

This is a simple variant of a V-cycle multigrid scheme, which only uses pre-smoothing. Equivalently, we can write the iterative scheme as a linear iteration method

$$\sigma_{i+1} = \sigma_i + B_J\,(f_J - V_J\,\sigma_i),$$

with an "approximate inverse" $B_J : \mathscr{M}_J \mapsto \mathscr{M}_J$ defined by

$$B_k f_k = Mg_k(0, f_k) \quad \text{for all } f_k \in \mathscr{M}_k \text{ and } k = 2,\ldots,J.$$

This operator is useful as a preconditioner in preconditioned iterative methods.

3 Numerical Implementation

In this section we give a matrix version of Algorithm 1 in the case of piecewise constant test functions. On every discretization level k assume that $\tau_k = \{\tau_k^i\}$ is a partition of the closed boundary curve Γ, which per side Γ_j consists of $n_k^{(j)}$ uniform elements of length h_k. The considerations here continue to apply even if each Γ_j is discretized by elements of length $h_k^{(j)}$, and we shall present this generalization in [5]. Let ϕ_k^i denote the indicator function of τ_k^i and let $N_k = \sum_j n_k^{(j)}$. Then $\mathcal{M}_k = \mathrm{span}\{\phi_k^i \mid i = 1, \ldots, N_k\}$. The actual implementation of Algorithm 1 hinges on the definition and computability of the **discrete inner products** $[\cdot,\cdot]_k$. To begin with, we would like to evaluate the H^{-1}-inner product of two elements in $\mathcal{M}_k$. This is equivalent to solving a second order boundary value problem on the boundary curve, $\Gamma = \cup_{j=1}^N \Gamma_j$: namely

$$-u'' + u = v \tag{11}$$

with periodic boundary conditions, for given $v \in H^{-1}(\Gamma)$, where the primes denote differentiation with respect to arc-length. Let T be the corresponding solution operator $T : H^{-1}(\Gamma) \longrightarrow H^1(\Gamma)$. This problem is uniquely solvable. Thus, for $v, w \in H^{-1}(\Gamma)$ it is easily verified that $(Tv, w)_\Gamma = (v, Tw)_\Gamma = (v, w)_{-1,\Gamma}$, where $(\cdot,\cdot)_\Gamma$ denotes the complex $L^2(\Gamma)$-inner product. Unfortunately, the use of the exact solution operator T is infeasible. Instead, we discretize (11) using a second-order finite difference method with the discretization stencil $h_k^{-2}\big[-1 \quad (2+h_k^2) \quad 1\big]$. This finite difference method results in an $N_k \times N_k$ linear system $\mathtt{A}_k u_h = v_h$, where $\mathtt{A}_k$ is a tridiagonal, Toeplitz matrix with diagonal entries $(\frac{2}{h_k^2} + 1)$, and super- and sub-diagonal entries $-\frac{1}{h_k^2}$. To enforce periodicity, the $(1, N_k)$ and $(N_k, 1)$ entries are both set to $-\frac{1}{h_k^2}$. The stencil would need to be modified in case different mesh-sizes are used on each Γ_i. The finite difference approximation to u is u_h. The inverse matrix $\mathtt{A}_k^{-1}$ serves as an approximation for the solution operator T. We can now define a computable, discrete inner product on $\mathcal{M}_k$ by

$$[\phi, \psi]_k := \big((\mathtt{A}_k)^{-1}\phi, \psi\big)_\Gamma \quad \text{for all } \phi, \psi \in \mathcal{M}_k. \tag{12}$$

Then, we show in [5] that $\|\cdot\|_{-1,\Gamma}$ and $[\cdot,\cdot]_k$ are equivalent norms, with the equivalence constants independent of the refinement levels $k = 1, 2, \ldots, J$.

The discrete norm $[\cdot,\cdot]_k$ can now be used to define the Richardson smoother.

To present the **matrix version** of the multigrid algorithm suitable for direct implementation, we first introduce some notation to represent Euclidean vectors of inner products and coefficients in the basis expansion of the discrete functions via the following maps:

$$\mathtt{e}_k : \mathscr{M}_k \longrightarrow \mathbb{C}^{N_k}, \qquad [\mathtt{e}_k(g)]_l = \frac{1}{\text{meas}(\tau_k^l)} (g, \phi_k^l)_\Gamma, \tag{13}$$

$$\mathtt{f}_k : \mathscr{M}_k \longrightarrow \mathbb{C}^{N_k}, \qquad [\mathtt{f}_k(g)]_i = (g, \phi_k^i)_{-1}, \tag{14}$$

Since the basis functions ϕ_k^l are the (orthogonal) indicator functions of the segments τ_k^l, the basis expansion for any $g \in \mathscr{M}_k$ is $g = \sum_{l=1}^{N_k} [\mathtt{e}_k(g)]_l \, \phi_k^l$, so the map in (13) gives the vector of coefficients. The other map, namely $\mathtt{f}_k$, gives the vector of $H^{-1}(\Gamma)$-inner products with the basis functions. The $N_k \times N_k$ stiffness matrix of (6) is defined by $[\mathtt{V}_k]_{i,j} = \langle V \phi_k^j \,, \phi_k^i \rangle$. Since $\mathscr{M}_k \subseteq \mathscr{M}_{k+1}$ we can find numbers $c_{i,l}$ such that $\phi_k^i = \sum_{l=1}^{N_{k+1}} c_{i,l} \, \phi_{k+1}^l$. These entries define the $N_k \times N_{k+1}$ restriction matrix $\mathtt{C}_k$ connecting levels k and $k+1$ by $[\mathtt{C}_k]_{i,l} = c_{i,l}$. We denote by $\mathtt{H}_k$ a diagonal matrix whose i^{th} diagonal entry is $\text{meas}(\tau_k^i)$, to store mesh size information. With these definitions, it is straightforward to prove the following identities:

$$\mathtt{f}_k(V_k g) = \mathtt{V}_k \, \mathtt{e}_k(g) \qquad \text{for all } g \in \mathscr{M}_k, \tag{15}$$

$$\mathtt{f}_{k-1}(Q_{k-1} g) = \mathtt{C}_{k-1} \, \mathtt{f}_k(g) \qquad \text{for all } g \in \mathscr{M}_k, \tag{16}$$

$$\mathtt{e}_k(g) = \mathtt{C}_{k-1}^t \mathtt{e}_{k-1}(g) \qquad \text{for all } g \in \mathscr{M}_{k-1}, \tag{17}$$

$$\mathtt{e}_k(R_k g) = \tilde{\lambda}_k^{-1} \, \mathtt{H}_k^{-1} \, \mathtt{A}_k \, \mathtt{f}_k(g) \qquad \text{for all } g \in \mathscr{M}_k, \tag{18}$$

$$\mathtt{e}_1(V_1^{-1} g) = \mathtt{V}_1^{-1} \mathtt{f}_1(g) \qquad \text{for all } g \in \mathscr{M}_1. \tag{19}$$

For example, to prove (18), we use (7) and (13) to get

$$\frac{1}{\tilde{\lambda}_k} [\mathtt{f}_k(g)]_i \equiv \frac{1}{\tilde{\lambda}_k} (g, \phi_k^i)_{-1} = [R_k g, \phi_k^i]_k = [\mathtt{A}_k^{-1} \mathtt{H}_k \, \mathtt{e}_k(R_k g)]_i,$$

and multiply both sides by $\mathtt{H}_k^{-1} \mathtt{A}_k$. Proofs of the other identities are similar.

These identities enable us to state a matrix version of Algorithm 1. For example, applying $\mathtt{e}_k$ to the step (9) of Algorithm 1 and using (18) and (15), we have

$$\begin{aligned} \mathtt{e}_k(\sigma_1) &= \mathtt{e}_k(\sigma) + \mathtt{e}_k(R_k(f - V_k \sigma)) = \mathtt{e}_k(\sigma) + \tilde{\lambda}_k^{-1} \, \mathtt{H}_k^{-1} \, \mathtt{A}_k \, \mathtt{f}_k(f - V_k \sigma) \\ &= \mathtt{e}_k(\sigma) + \tilde{\lambda}_k^{-1} \, \mathtt{H}_k^{-1} \, \mathtt{A}_k \, (\mathtt{f}_k(f) - \mathtt{V}_k \mathtt{e}_k(\sigma)). \end{aligned}$$

Thus, the matrix version of this step is $\mathtt{s}_1 = \mathtt{s} + \tilde{\lambda}_k^{-1} \, \mathtt{H}_k^{-1} \, \mathtt{A}_k \, (\mathtt{b} - \mathtt{V}_k \mathtt{s})$ with $\mathtt{s}_1 = \mathtt{e}_k(\sigma_1)$, $\mathtt{s} = \mathtt{e}_k(\sigma)$, and $\mathtt{b} = \mathtt{f}_k(f)$. Using also the other identities in (15)–(19), we can similarly translate the entire Algorithm 1 and obtain the following matrix version of the algorithm. It defines a procedure $\mathtt{Mg}_J(\mathtt{s}, \mathtt{b})$ that outputs an approximation for the solution of the matrix equation $\mathtt{V}_J \mathtt{u} = \mathtt{b}$, given an input iterate $\mathtt{s}$.

Algorithm 2 *Let* $\mathtt{s}$ *and* $\mathtt{b}$ *be any given vectors in* $\mathbb{C}^{N_k}$. *Define* $\mathtt{Mg}_k(\mathtt{s}, \mathtt{b})$ *recursively as follows. Set* $\mathtt{Mg}_1(\mathtt{s}, \mathtt{b}) = \mathtt{V}_1^{-1} \mathtt{b}$. *If* $k > 1$, *define* $\mathtt{Mg}_k(\mathtt{u}, \mathtt{b})$ *as the vector in* $\mathbb{C}^{N_k}$ *obtained recursively by:*

$$\begin{aligned} \mathtt{s}_1 &= \mathtt{s} + \tilde{\lambda}_k^{-1} \, \mathtt{H}_k^{-1} \, \mathtt{A}_k \, (\mathtt{b} - \mathtt{V}_k \mathtt{s}), \\ \mathtt{Mg}_k(\mathtt{s}, \mathtt{b}) &= \mathtt{s}_1 + \mathtt{C}_{k-1}^t \mathtt{Mg}_{k-1}(0, \mathtt{C}_{k-1}(\mathtt{b} - \mathtt{V}_k \mathtt{s}_1)). \end{aligned}$$

It is important to note that the inverse of $\mathtt{A}_k$ is not needed in the implementation, as we only need to multiply by $\mathtt{A}_k$. Note that a matrix preconditioner $\mathtt{B}_k$ for $\mathtt{V}_k$ is implicit in Algorithm 2 and is defined by $\mathtt{B}_k\mathtt{b} := \mathtt{Mg}_k(0,\mathtt{b})$. A theoretical study of the convergence rate of the algorithm is presented in [5].

4 Numerical Experiments

We now present numerical results on the convergence of the presented algorithm as a linear solver as well as its performance as a preconditioner for GMRES. First, we solve a point-source acoustic scattering problem whose exact solution is known, where the source lies in the interior of Γ, a triangular boundary curve. Second, we investigate the scattering of plane waves from a rectangular obstacle.

Figure 1 refers to three sets of experiments for the point-source problem for triangular Γ. In the left plot in Figure 1, we show the iteration numbers using the method of Algorithm 2 as a linear solver by itself, and as a preconditioner for GMRES. We include iteration counts for unpreconditioned GMRES for comparison. This plot corresponds to low wave-number $\kappa = 2.1$. The middle plot of Figure 1 demonstrates the effect of smoothing on the iteration counts for GMRES with multigrid preconditioning. Again, $\kappa = 2.1$. In one case, we applied a pre-smoothing step alone, and in the other case, used both pre- and post-smoothing. The latter reduced the number of GMRES iterations taken. Finally, in the right plot of Figure 1, we demonstrate the effect of changing wave numbers, with wavenumbers of $\kappa = 2.1$ and $\kappa = 10.2$. It can be seen that the number of iterations taken for the larger wave-number increases, but remains controlled. In all of these experiments, GMRES was run without restart, with a stopping criteria of 10^{-9} relative residual. In all these cases, GMRES with multigrid as preconditioner works well to control iteration numbers.

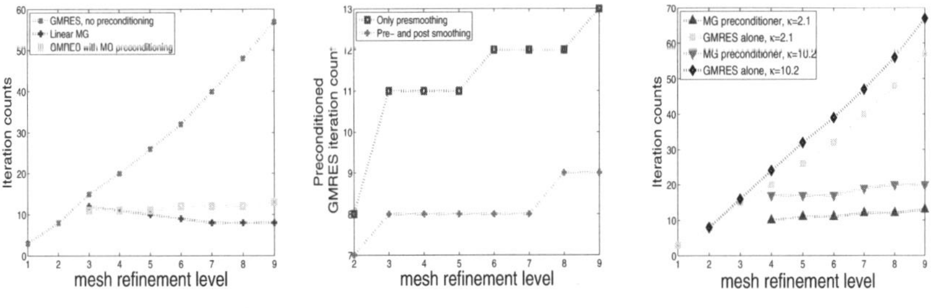

Fig. 1 L-R: MG as solver and preconditioner; effect of smoothing; effect of wave-number κ. x-axis = refinement levels, mesh-size varying from $h = 1/2$ to $h = 1/512$

In Table 1 we present iteration data for scattering of a plane-wave from a rectangle of sides $\frac{2}{9} \times \frac{8}{9}$. We show the effect of the coarse-grid, and the effect of the wave-numbers, on the iteration numbers for a preconditioned GMRES method. Again, the multigrid method presented clearly controls the iteration numbers.

Table 1 Preconditioned GMRES iteration counts for plane-wave scattering from a rectangle for two different wave-numbers κ, mesh sizes **h** and **H** on the finest and coarsest grids

Plane-wave scattering, $\kappa = 2.1$						**Plane-wave scattering, $\kappa = 6$**					
	H				GMRES without		**H**				GMRES without
1/18	10	-	-	-	22	1/18	14	-	-	-	24
1/36	10	10	-	-	29	1/36	14	14	-	-	30
h 1/72	11	10	10	-	36	**h** 1/72	15	14	14	-	38
1/144	11	11	11	10	44	1/144	15	15	14	14	47
1/288	12	11	11	11	55	1/288	15	15	15	15	57
1/576	12	12	12	11	67	1/576	16	16	15	15	70
1/1158	12	12	12	12	81	1/1158	16	16	16	16	85

Acknowledgements JG's research was supported by NSF grants DMS-0713833 and SCREMS-0619080. NN's research was supported by NSERC and FQRNT. We thank Prof. O. Steinbach for helpful discussions.

References

1. Bramble, J.H.: Multigrid methods, *Pitman Research Notes in Mathematics Series*, vol. 294. Longman Scientific & Technical (1993)
2. Bramble, J.H., Kwak, D.Y., Pasciak, J.E.: Uniform convergence of multigrid V-cycle iterations for indefinite and nonsymmetric problems. SIAM J. Numer. Anal. **31**(6), 1746–1763 (1994)
3. Bramble, J.H., Leyk, Z., Pasciak, J.E.: The analysis of multigrid algorithms for pseudodifferential operators of order minus one. Math. Comp. **63**(208), 461–478 (1994)
4. Funken, S.A., Stephan, E.P.: The BPX preconditioner for the single layer potential operator. Appl. Anal. **67**(3-4), 327–340 (1997)
5. Gemmrich, S., Gopalakrishnan, J., Nigam, N.: Convergence analysis of a multigrid algorithm for the acoustic single layer equation (2007). In preparation
6. Gopalakrishnan, J., Pasciak, J.E., Demkowicz, L.F.: Analysis of a multigrid algorithm for time harmonic Maxwell equations. SIAM J. Numer. Anal. **42**(1), 90–108 (electronic) (2004)
7. Hsiao, G., MacCamy, R.C.: Solution of boundary value problems by integral equations of the first kind. SIAM Rev. **15**, 687–705 (1973)
8. Hsiao, G.C., Stephan, E.P., Wendland, W.L.: On the Dirichlet problem in elasticity for a domain exterior to an arc. J. Comput. Appl. Math. **34**(1), 1–19 (1991)
9. Langer, U., Pusch, D., Reitzinger, S.: Efficient preconditioners for boundary element matrices based on grey-box algebraic multigrid methods. Internat. J. Numer. Methods Engrg. **58**(13), 1937–1953 (2003)
10. Langer, U., Pusch, U.: Convergence analysis of geometrical multigrid methods for solving data-sparse boundary element equations. In: Proc. 8th European Multigrid Conference 2005. Springer, Heidelberg (2007)
11. Maischak, M., Mund, P., Stephan, E.P.: Adaptive multilevel BEM for acoustic scattering. Comput. Methods Appl. Mech. Engrg. **150**(1-4), 351–367 (1997)
12. Steinbach, O.: Numerical Approximation Methods for Elliptic Boundary Value Problems. Finite and Boundary Elements. Springer, New York (2008)

Constraints Coefficients in *hp*-FEM

A. Schröder

Abstract Continuity requirements on irregular meshes enforce a proper constraint of the degrees of freedom that correspond to hanging nodes, edges or faces. This is achieved by using so-called constraints coefficients which are obtained from the appropriate coupling of shape functions.

In this note, a general framework for determining the constraints coefficients of tensor product shape functions is presented and its application to shape functions using integrated Legendre or Gauss-Lobatto polynomials. The constraints coefficients in the one-dimensional case are determined via recurrence relations. The constraints coefficients in the multi-dimensional case are obtained as products of these coefficients. The coefficients are available for arbitrary patterns of subdivisions.

1 Introduction

Local refinement processes arising from grid adaption are typically realized either by remeshing or by local refinements of grid elements. In the latter case so-called hanging nodes, edges or faces are unavoidable which result from refining a grid element without the refinement of neighboring elements. Applying conform finite element schemes, one has to ensure the finite element solution to be continuous. If no further local refinements (with possibly complex refinement patterns) are performed to eleminate grid irregularities, one has to constraint the degrees of freedom associated to hanging nodes, edges or faces. This can be done, e.g., by using Lagrange multipliers or static condensation or by incorporating the constraints in the iterative scheme that is used to determine the approximative solution. In all cases, a representation of shape functions in terms of transformed shape functions is needed. Such a representation is given by the so-called constraints coefficients.

Andreas Schröder
Department of Mathematics, Humboldt Universität zu Berlin, Unter den Linden 6, 10099 Berlin, Germany, e-mail: andreas.schroeder@mathematik.hu-berlin.de

In a very general manner, constraints coefficients are defined as follows: Let P_q be a space of polynomials of degree $q \in \mathbb{N}$ on $\mathbb{R}^k$, $k \in \mathbb{N}$, and $\Upsilon : \mathbb{R}^k \to \mathbb{R}^k$ be an affine linear and bijective mapping. Furthermore, let $\xi = \{\xi_i\}_{0 \le i < n} \subset P_q$ be a linear independent set of polynomials. The numbers $\alpha_{ij} \in \mathbb{R}$ with $\xi_i \circ \Upsilon = \sum_{j=0}^{n-1} \alpha_{ij} \xi_j$ are called *constraints coefficients* of ξ for the mapping Υ.

In [3] constraints coefficients of the shape functions

$$\xi_0(x) := \frac{1}{2}(1-x), \quad \xi_1(x) := \frac{1}{2}(1+x), \quad \xi_i(x) := \begin{cases} x^i - 1, & i = 2,4,6,\ldots,q \\ x^i - x, & i = 3,5,7,\ldots,q \end{cases} \tag{1}$$

are determined. Since the functionals $\varphi_0(v) := v(-1)$, $\varphi_1(v) := v(1)$, $\varphi_j(v) := 1/j! d^j v/dx^j(0)$, $j = 2,\ldots,q$ fulfill the duality relation $\varphi_j(\xi_i) = \delta_{ij}$ (where δ_{ij} is the Kronecker delta), one simply obtains $\alpha_{ij} = \varphi_j(\xi_i \circ \Upsilon)$.
In [2] constraints coefficients of the Lagrange shape functions

$$\xi_0(x) := 1-x, \quad \xi_1(x) := x, \quad \xi_i := \frac{x(1-x)}{x_i(1-x_i)} \prod_{\ell=2;\ell \neq i}^{n-1} \frac{x - x_\ell}{x_i - x_\ell}, \quad i = 2,\ldots,q$$

are specified with $x_\ell \in (0,1)$, $\ell = 2,\ldots n-1$. The functionals $\varphi_0(v) := v(0)$, $\varphi_1(v) := v(1)$, $\varphi_j(v) := v(x_j)$, $j = 2,\ldots,n-1$, fulfill the duality relation $\varphi_j(\xi_i) = \delta_{ij}$ only for $i = 2,\ldots,n-1$. We get $\alpha_{i0} = (\xi_i \circ \Upsilon)(0)$ and $\alpha_{i1} = (\xi_i \circ \Upsilon)(1)$ for $i = 0,\ldots,n-1$ and $\alpha_{0j} = \alpha_{1j} = 0$ for $j = 2,\ldots,n-1$. Since $\varphi_j(\xi_i \circ \Upsilon) = \alpha_{i0}\varphi_j(\xi_0) + \alpha_{i1}\varphi_j(\xi_1) + \alpha_{ij}$, the remaining coefficients are determined by $\alpha_{ij} = (\xi_i \circ \Upsilon)(x_j) - \alpha_{i0}(1-x_j) - \alpha_{i1}x_j$.
A widely used family of shape functions are shape functions using integrated Legendre or Gauss-Lobatto polynomials ([7], [8], [9]). These polynomials belong to the family of so-called Gegenbauer polynomials $\{G_i^\rho\}_{i \in \mathbb{N}_0}$ which are defined by

$$(i+1)G_{i+1}^\rho(x) = 2(i+\rho)xG_i^\rho(x) - (i+2\rho-1)G_{i-1}^\rho(x) \tag{2}$$

with $\rho \in \mathbb{R}$, $G_0^\rho(x) := 1$ and $G_1^\rho(x) := 2\rho x$. Theoretical results about equivalent definitions of Gegenbauer polynomials and their special properties can be found, e.g., in [10]. With $\rho := -1/2$, we obtain integrated Legendre ($\beta_i := 1$) and Gauss-Lobatto ($\beta_i := \sqrt{(2i-1)/2}$) shape functions

$$\xi_0(x) := \frac{1}{2}(1-x),\ \xi_1(x) := \frac{1}{2}(1+x),\ \xi_i(x) := \beta_i G_i^{-1/2}(x), \quad i = 2,\ldots,q. \tag{3}$$

Because of the orthogonality relation of the Gegenbauer polynomials (cf. [10]), the functionals $\varphi_0(v) := v(-1)$, $\varphi_1(v) := v(1)$, $\varphi_j(v) := \mu_j \int_{-1}^{1}(1-x^2)^{-1}\xi_j(x)v(x)\,dx$ with $\mu_j := j(j-1)(2j-1)/(2\beta_j^2)$, $j = 2,\ldots,n-1$ fullfill the duality relation $\varphi_j(\xi_i) = \delta_{ij}$ for $i = 2,\ldots,n-1$ and $j = 0,\ldots,n-1$. Similar to the Lagrange shape functions, we obtain $\alpha_{i0} = (\xi_i \circ \Upsilon)(-1)$ and $\alpha_{i1} = (\xi_i \circ \Upsilon)(1)$ for $i = 0,\ldots,n-1$ and $\alpha_{0j} = \alpha_{1j} = 0$ for $j = 2,\ldots,n-1$. Since $\varphi_j(\xi_0) = (-1)^j(2j-1)/(2\beta_j^2)$ and $\varphi_j(\xi_1) = (2j-1)/(2\beta_j^2)$, the remaining coefficients are determined by $\alpha_{ij} = \varphi_j(\xi_i \circ \Upsilon) - (2j-1)/(2\beta_j^2)(\alpha_{i0}(-1)^j + \alpha_{i1})$.

In this note, we present a general framework for constraints coefficients of tensor product polynomials. Furthermore, we present an explicit formula of the constraints coefficients of integrated Legendre and Gauss-Lobatto shape functions without the integral representation given by φ_j. The formula is derived by the use of the recurrence relation (2). At the end of this note, the application of constraints coefficients to irregular grids is briefly discussed. Other areas of applications are *hp*-multigrid schemes (cf. [4], [5]) or grid transfer operations in timedependent problems.

2 Tensor Product Shape Functions

The space of polynomials in one variable of degree q is defined as $S^q := \{v : \mathbb{R} \to \mathbb{R} \mid v(x) = \sum_{0\le i\le q} c_i x^i,\ c_i \in \mathbb{R}\}$, the corresponding tensor product space is denoted by

$$S_k^q := \otimes_{i=0}^{k-1} S^q := \left\{ v : \mathbb{R}^k \to \mathbb{R} \mid v(x_0,\dots,x_{k-1}) = \prod_{i=0}^{k-1} v_i(x_i),\ v_0,\dots,v_{k-1} \in S^q \right\}.$$

Let $\hat{\xi} := \{\hat{\xi}_i\}_{0\le i<m}$ be a subset of S^q and L be an n times k matrix with entries in $\{0,\dots,m-1\}$. Then, we define $\Pi(\hat{\xi},L) := \left\{\prod_{r=0}^{k-1} \hat{\xi}_{L_{ir}}(x_r)\right\}_{0\le i<n} \subset S_k^q$.

For $\Upsilon(x) := \mathrm{diag}(a)x + b$ with $a,b \in \mathbb{R}^k$, it is easy to determine the constraints coefficients of $\Pi(\hat{\xi},L)$: Let $\hat{\alpha}_{ij}(a_r,b_r) \in \mathbb{R}$ be the constraints coefficients of $\hat{\xi}$ for $\Upsilon_r(x_r) := a_r x_r + b_r$. Furthermore, let $\mathscr{L} := \{(L_{i,0},\dots,L_{i,k-1}) \mid 0 \le i < n\}$.

Theorem 1. *Assume that $\Pi(\hat{\xi},L)$ is linear independent and there holds*

$$l \in \{0,\dots,m-1\}^k \backslash \mathscr{L} \Rightarrow \forall 0 \le i < n:\ \exists 0 \le r < k:\ \hat{\alpha}_{L_{ir},l_r} = 0. \tag{4}$$

Then, the constraints coefficients of $\Pi(\hat{\xi},L)$ for Υ are $\alpha_{ij} = \prod_{r=0}^{k-1} \hat{\alpha}_{L_{ir},L_{jr}}(a_r,b_r)$.
Proof: Let $x \in \mathbb{R}^k$. Because of (4), we obtain

$$\Pi(\hat{\xi},L)_i(\Upsilon(x)) = \prod_{r=0}^{k-1} \hat{\xi}_{L_{ir}}(a_r x_r + b_r) = \prod_{r=0}^{k-1}\sum_{l=0}^{m-1} \hat{\alpha}_{L_{ir},l}(a_r,b_r)\hat{\xi}_l(x_r)$$

$$= \sum_{l_0=0}^{m-1} \cdots \sum_{l_{k-1}=0}^{m-1} \left(\prod_{r=0}^{k-1} \hat{\alpha}_{L_{ir},l_r}(a_r,b_r)\right)\left(\prod_{r=0}^{k-1} \hat{\xi}_{l_r}(x_r)\right)$$

$$= \sum_{l\in\mathscr{L}} \left(\prod_{r=0}^{k-1} \hat{\alpha}_{L_{ir},l_r}(a_r,b_r)\right)\left(\prod_{r=0}^{k-1} \hat{\xi}_{l_r}(x_r)\right) = \sum_{j=0}^{n-1}\left(\prod_{r=0}^{k-1} \hat{\alpha}_{L_{ir},L_{jr}}(a_r,b_r)\right)\Pi(\hat{\xi},L)_j(x).$$

Since $\Pi(\hat{\xi},L)$ is assumed to be linear independent, the proof is completed. □

Finite element shape functions are basis polynomials that are defined on a reference element (unit square, cube or simplex). They constitute the global basis functions on the grid elements. In conform approaches shape functions are usually partitioned into nodal modes, edge modes, face modes and inner modes. Nodal modes have the value 1 in exactly one vertex and vanish on the remaining vertices. Edge modes are different from zero on exactly one edge and vanish on the remaining edges and on all non-adjacent faces and all nodes. Face modes are different from zero on exactly one face and vanish on the remaining faces and on all edges and nodes. Inner modes vanish on all nodes, edges and faces, they are only different from zero in the interior. Using the notation $\Pi(\hat{\xi},L)$, the separation is established by splitting the matrix L into submatrices $L^\top := (L^0\ L^1\ \cdots\ L^k)^\top$. The submatrix L^0 generates the nodal modes, L^1 generates the edges modes and so on.

Let $\hat{\xi} = \hat{\xi}^q$ be shape functions in S^q which are partitioned into the nodal modes $\hat{\xi}_0$, $\hat{\xi}_1$ and inner modes $\hat{\xi}_i$, $2 \le i \le q$. With $\alpha(i,j) := i(i+1)/2 + j$, a proper definition of L in the two-dimensional case is, e.g.,

$$
\begin{aligned}
&(L^0)^\top := \begin{pmatrix} 0\,1\,1\,0 \\ 0\,0\,1\,1 \end{pmatrix}^\top, \quad L^1_{i,1} := L^1_{3(q-1)+i,0} := 0, \quad L^1_{q-1+i,0} := L^1_{2(q-1)+i,1} := 1, \\
&L^1_{i,0} := L^1_{q-1+i,1} := L^1_{2(q-1)+i,0} := L^1_{3(q-1)+i,1} := i+2, \quad i = 0,\dots,q-2, \\
&L^2_{\alpha(i,j),0} := j+2, \quad L^2_{\alpha(i,j),1} := i-j+2, \quad i = 0,\dots,q-4+\tau,\ j = 0,\dots,i.
\end{aligned}
\tag{5}
$$

This definition leads to the set of shape functions $\xi = \Pi(\hat{\xi},L)$:

$$
\begin{aligned}
&\xi_0(x_0,x_1) := \hat{\xi}_0(x_0)\hat{\xi}_0(x_1), \quad \xi_1(x_0,x_1) := \hat{\xi}_1(x_0)\hat{\xi}_0(x_1), \\
&\xi_2(x_0,x_1) := \hat{\xi}_1(x_0)\hat{\xi}_1(x_1), \quad \xi_3(x_0,x_1) := \hat{\xi}_0(x_0)\hat{\xi}_1(x_1), \\
&\xi_{4+i}(x_0,x_1) := \hat{\xi}_{i+2}(x_0)\hat{\xi}_0(x_1), \quad \xi_{4+q-1+i}(x_0,x_1) := \hat{\xi}_1(x_0)\hat{\xi}_{i+2}(x_1), \\
&\xi_{4+2(q-1)+i}(x_0,x_1) := \hat{\xi}_{i+2}(x_0)\hat{\xi}_1(x_1), \quad \xi^q_{4+3(q-1)+i}(x_0,x_1) := \hat{\xi}_0(x_0)\hat{\xi}_{i+2}(x_1), \\
&\xi_{4q+\alpha(i,j)}(x_0,x_1) := \hat{\xi}_{j+2}(x_0)\hat{\xi}_{i-j+2}(x_1).
\end{aligned}
$$

For $\tau = 2$ the set $\Pi(\hat{\xi},L)$ is a basis of S^q_2. Assuming that $\hat{\xi}$ is hierarchical (which means that $\hat{\xi}^{\tilde{q}}_i = \hat{\xi}^q_i$ for $0 \le i \le \tilde{q}$ and $\tilde{q} \le q$), the set $\Pi(\hat{\xi},L)$ has some important properties: For $\tau = 0$, we obtain a reduced set of shape functions (also known as Serendipity shape functions) with the same order of approximation (cf., e.g., p. 175 in [1], [7]). Furthermore, the special definition of L implies that the edge modes (edge by edge) and the inner modes are hierarchical as well. This property can be exploited, e.g., for the efficient management of different polynomial degree distributions of neigboring grid elements. One simply omits the edge modes with polynomial degree $p_0 > p_1$, where p_1 is the polynomial degree in the neighboring element. The shape functions $\Pi(\hat{\xi},L)$ with integrated Legendre or Gauss-Lobatto shape functions $\hat{\xi}$ corresponds to the shape functions as proposed in [7] and [9] for *hp*-finite element methods. The use of the recurrence relation (2) admits a stable and fast evaluation of the shape functions and their derivatives. Derivatives of arbitrary order can be easily derived by the relation $\partial^\nu G^\rho_i = 2^\nu(\rho)_\nu G^{\rho+\nu}_{i-\nu}$ with $i,\nu \in \mathbb{N}_0$ and $(\rho)_\nu := \prod_{j=0}^{\nu-1}(\rho+j)$.

3 Constraints Coefficients of Integrated Legendre and Gauss-Lobatto Shape Functions

As a result of Theorem 1, it is sufficient to consider the one-dimensional case to determine the constraints coefficients in the multi-dimensional case.

Theorem 2. *Let $\hat{\xi}$ be a set of hierarchical shape functions and L be defined as in* (5). *Then, the assumption* (4) *is fulfilled for $\tau \in \{0,2\}$.*
Proof. The assumption (4) is obviously fulfilled for $\tau = 2$. Let $q \geq 2$, $\tau = 0$ and $l \in \{0,\dots,q\}^2 \backslash \mathscr{L}$, then $l = (j+2, i-j+2)$ with $i \in \{\max\{q-3,0\}, q-2\}$ and $0 \leq j \leq i$. For the nodal mode ($\kappa = 0$) with index $0 \leq s < 4$ or for the edge mode ($\kappa = 1$) with index $0 \leq s < 4(q-1)$, we obtain $\deg(\hat{\xi}_{L^{\kappa}_{sr}}) = 1$ for at least one $r \in \{0,1\}$. Since $\min\{\deg(\hat{\xi}_{j+2}), \deg(\hat{\xi}_{i-j+2})\} \geq 2$, we have $\hat{\alpha}_{L^{\kappa}_{sr},l_r} = 0$. For $q \geq 4$, the polynomial degree of the inner mode with index $0 \leq s < (q-3)(q-2)/2$ is bounded by $q-2 < \max\{j+2, i-j+2\} = \max\{\deg(\hat{\xi}_{j+2}), \deg(\hat{\xi}_{i-j+2})\}$. Therefore, there exists $r \in \{0,1\}$ such that $\hat{\alpha}_{L^{2}_{sr},l_r} = 0$. □

Theorem 3. *Let $\Upsilon(x) = ax + b$ with $a, b \in \mathbb{R}$ and $i \geq 2$. For integrated Legendre shape functions* (3)*, there holds:*

$$\alpha_{00} = \frac{1+a-b}{2}, \quad \alpha_{10} = \frac{1-a+b}{2}, \quad \alpha_{20} = \frac{1-(a-b)^2}{2},$$

$$\alpha_{i+1,0} = (b-a)\frac{2i-1}{i+1}\alpha_{i,0} - \frac{i-2}{i+1}\alpha_{i-1,0},$$

$$\alpha_{01} = \frac{1-a-b}{2}, \quad \alpha_{11} = \frac{1+a+b}{2}, \quad \alpha_{21} = \frac{1-(a+b)^2}{2},$$

$$\alpha_{i+1,1} = (a+b)\frac{2i-1}{i+1}\alpha_{i,1} - \frac{i-2}{i+1}\alpha_{i-1,1},$$

$$\alpha_{22} = a^2, \quad \alpha_{i+1,2} = \frac{2i-1}{i+1}\left(\frac{a}{5}\alpha_{i,3} + b\alpha_{i,2} + a(\alpha_{i,0} - \alpha_{i,1})\right) - \frac{i-2}{i+1}\alpha_{i-1,2},$$

$$\alpha_{i+1,j} = \frac{2i-1}{i+1}\left(a\frac{j}{2j-3}\alpha_{i,j-1} + a\frac{j-1}{2j+1}\alpha_{i,j+1} + b\alpha_{i,j}\right) - \frac{i-2}{i+1}\alpha_{i-1,j}, \quad j = 3,\dots,i-1,$$

$$\alpha_{i+1,i} = \frac{2i-1}{i+1}\left(a\frac{i}{2i-3}\alpha_{i,i-1} + b\alpha_{ii}\right), \quad i > 2,$$

$$\alpha_{i+1,i+1} = a\alpha_{ii}, \quad \alpha_{i,j} = 0, \quad j > i.$$

Proof. By comparing the coefficients in $\xi_i(ax+b) = \alpha_{i0}\xi_0(x) + \alpha_{i1}\xi_1(x)$, $i = 0,1,2$, we obtain α_{00}, α_{01}, α_{10}, α_{11}, α_{20}, α_{21} and α_{22}. From equation (2) we have:

$$x\xi_j(x) = (2j-1)^{-1}((j+1)\xi_{j+1}(x) + (j-2)\xi_{j-1}(x)), \quad j = 2,3,\dots.$$

Furthermore, we have

$$x\xi_0(x) = \frac{1}{2}x - \frac{1}{2}x^2 = -\frac{1}{2}(1-x) + \frac{1}{2}(1-x^2) = -\xi_0(x) + \xi_2(x),$$
$$x\xi_1(x) = \frac{1}{2}x + \frac{1}{2}x^2 = \frac{1}{2}(1+x) - \frac{1}{2}(1-x^2) = \xi_1(x) - \xi_2(x).$$

This yields

$$\begin{aligned}
&(i+1)\xi_{i+1}(ax+b)\\
&= (2i-1)(ax+b)\xi_i(ax+b) - (i-2)\xi_{i-1}(ax+b)\\
&= b(2i-1)\sum_{j=0}^{i}\alpha_{ij}\xi_j(x) + a(2i-1)x\sum_{j=0}^{i}\alpha_{ij}\xi_j(x) - (i-2)\sum_{j=0}^{i-1}\alpha_{i-1,j}\xi_j(x)\\
&= b(2i-1)\sum_{j=0}^{i}\alpha_{ij}\xi_j(x) + a(2i-1)\sum_{j=2}^{i}\alpha_{ij}\left(\frac{j+1}{2j-1}\xi_{j+1}(x) + \frac{j-2}{2j-1}\xi_{j-1}(x)\right)\\
&\quad + a(2i-1)\left(\alpha_{i,0}(-\xi_0(x)+\xi_2(x)) + \alpha_{i,1}(\xi_1(x)-\xi_2(x))\right)\\
&\quad - (i-2)\sum_{j=0}^{i-1}\alpha_{i-1,j}\xi_j(x)\\
&= a(i+1)\alpha_{ii}\xi_{i+1}(x) + \left(a(2i-1)\frac{i}{2i-3}\alpha_{i,i-1} + b(2i-1)\alpha_{ii}\right)\xi_i(x)\\
&\quad + a(2i-1)\sum_{j=3}^{i-1}\alpha_{i,j-1}\frac{j}{2j-3}\xi_j(x) + a(2i-1)\sum_{j=2}^{i-1}\alpha_{i,j+1}\frac{j-1}{2j+1}\xi_j(x)\\
&\quad + b(2i-1)\sum_{j=0}^{i-1}\alpha_{ij}\xi_j(x) - (i-2)\sum_{j=0}^{i-1}\alpha_{i-1,j}\xi_j(x) + a(2i-1)\left(\alpha_{i,0}-\alpha_{i,1}\right)\xi_2(x)\\
&\quad + a(2i-1)\alpha_{i,1}\xi_1(x) - a(2i-1)\alpha_{i,0}\xi_0(x)\\
&= a(i+1)\alpha_{ii}\xi_{i+1}(x) + \left(a(2i-1)\frac{i}{2i-3}\alpha_{i.i-1} + b(2i-1)\alpha_{ii}\right)\xi_i(x)\\
&\quad + \sum_{j=3}^{i-1}\Bigg(a(2i-1)\frac{j}{2j-3}\alpha_{i,j-1} + a(2i-1)\frac{j-1}{2j+1}\alpha_{i,j+1} + b(2i-1)\alpha_{ij}\\
&\qquad\qquad - (i-2)\alpha_{i-1,j}\Bigg)\xi_j(x)\\
&\quad + \left(a(2i-1)\frac{1}{5}\alpha_{i,3} + b(2i-1)\alpha_{i,2} - (i-2)\alpha_{i-1,2} + a(2i-1)\left(\alpha_{i,0}-\alpha_{i,1}\right)\right)\xi_2(x)\\
&\quad + \left(b(2i-1)\alpha_{i,1} - (i-2)\alpha_{i-1,1} + a(2i-1)\alpha_{i,1}\right)\xi_1(x)\\
&\quad + \left(b(2i-1)\alpha_{i,0} - (i-2)\alpha_{i-1,0} - a(2i-1)\alpha_{i,0}\right)\xi_0(x)
\end{aligned}$$

Division by $i+1$ completes the proof. □

It is easy to see, that the constraints coefficients of Gauss-Lobatto shape functions are $\sqrt{(2i-1)/(2j-1)}\alpha_{ij}$, $i,j \geq 2$. Furthermore, Theorem 3 can be extended to the case of Gegenbauer polynomials or general Jacobi polynomials.

4 Application to Hanging Nodes

Let $\mathscr{T}$ be a subdivision of $\Omega \subset \mathbb{R}^k$ consisting of quadrangles ($k=2$) or hexahedrons ($k=3$) and let $\Psi_T : [-1,1]^k \to T \in \mathscr{T}$ be a bijective and sufficiently smooth mapping. In conform finite element methods, the space of admissable functions is defined as $S^p(\mathscr{T}) := \{v \in C^0(\Omega) \mid \forall T \in \mathscr{T} : v_{|T} \circ \Psi_T \in S_k^{p_T}\}$ with the degree distribution $p = \{p_T\}_{T\in\mathscr{T}}$, $p_T \leq q$. By using so-called connectivity matrices $\pi_T \in \mathbb{R}^{\ell \times n_k}$, a basis $\{\phi_r\}_{0\leq r<\ell}$ of $S^p(\mathscr{T})$ is constructed via

$$\phi_{r|T} := \sum_{s=0}^{n_k-1} \pi_{T,rs}\hat{\phi}_{T,s}$$

with $\hat{\phi}_{T,s} := \Pi(\hat{\xi}, L)_s \circ \Psi_T^{-1}$, $0 \leq s < n_k$, where n_k is the number of shape functions. In particular, the stiffness matrix K and the load vector b are assembled via $K := \sum_{T\in\mathscr{T}} \pi_T K_T \pi_T^\top$ and $b := \sum_{T\in\mathscr{T}} \pi_T b_T$ with local stiffness matrices $K_T \in \mathbb{R}^{n_k \times n_k}$ and local load vectors $b_T \in \mathbb{R}^{n_k}$.

In the presence of hanging nodes, the definition of π_T is the crucial point. The entries are ± 1 (or 0), if the associated shape functions are related to a non-hanging node, edge or face. Otherwise, the entries are given by the constraints coefficients as introduced in the previous sections. Figure 1a shows a typical situation in 3D which is obtained by refining the neighbored grid element of the left hexahedron (denoted by T_L), for example by dividing it into eight small hexahedrons. One of them (denoted by T_R) is examplarly depicted on the right hand side of T_L. The entries of the connectivity matrix of T_L related to the nodes v_0 and v_1, to the edges e_0, e_1, e_2 and to the face f are defined as follows. The entries related to v_0 and e_0 are given by the constraints coefficients α_{ij} of the one-dimensional case: Let $\phi_{\hat{r}}$ be a basis function of $\{\phi_r\}_{0\leq r<\ell}$, that belongs to V_0, V_1 or E. Furthermore, let $\{\hat{\phi}_{T_L,s}\}_{s\in\mathscr{S}_L}$ be the polynomials of $\{\hat{\phi}_{T_L,s}\}_{0\leq s<n_3}$, that belong to V_0, V_1 and E, and let $\{\hat{\phi}_{T_R,s}\}_{s\in\mathscr{S}_R}$ be the polynomials of $\{\hat{\phi}_{T_R,s}\}_{0\leq s<n_3}$, that belong to V_0, v_0 and e_0. Since V_0, V_1 and E are non-hanging, it holds

$$\pm\hat{\phi}_{T_L,\hat{s}|e_0} = \phi_{\hat{r}|e_0} = \sum_{s\in\mathscr{S}_R} \pi_{T,\hat{r}s}\hat{\phi}_{T_R,s|e_0}$$

with $\hat{s} \in \mathscr{S}_L$. Provided that E is subdivided into two subedges with proportions of division z and $1-z$, $z \in (0,1)$, and e_0 is its first subedge, we define a mapping Υ by $\Upsilon(x) := zx + z - 1$ which maps $[-1, (2-z)/z]$ onto $[-1,1]$. If e_0 is the second subedge of E, we set $\Upsilon(x) := (1-z)x + z$ which maps $[(z+1)/(z-1), 1]$ onto $[-1,1]$. Due to the tensor structure of $\Pi(\hat{\xi}, L)$, there exist bijective mappings $\Delta_L : \{0,\ldots,n_1-1\} \to \mathscr{S}_L$, $\Delta_R : \{0,\ldots,n_1-1\} \to \mathscr{S}_R$, and $\Psi_{e_0} : [-1,1] \to e_0$, such that $\hat{\phi}_{T_L,\hat{s}|e_0} \circ \Psi_{e_0} = \hat{\xi}_{\Delta_L^{-1}(\hat{s})} \circ \Upsilon_{|[-1,1]}$ and $\hat{\phi}_{T_R,\Delta_R(j)|e_0} \circ \Psi_{e_0} = \hat{\xi}_j$, $0 \leq j < n_1$. Therefore, we obtain

$$\pm\hat{\xi}_{\Delta_L^{-1}(\hat{s})} \circ \Upsilon = \sum_{j=0}^{n_1-1} \pi_{T_R,\hat{r},\Delta_R(j)}\hat{\xi}_j$$

and, finally, $\pi_{T_R,\hat{r},\Delta_R(j)} = \pm\alpha_{\Delta_L^{-1}(\hat{s}),j}$.

By analogy, the entries related to v_1, e_1, e_2 and f are the constraints coefficients of the two-dimensional case. We consider the polynomials of $\{\hat{\phi}_{T_L,s}\}_{0\le s<n_3}$, that belong to F and its nodes and edges, restricted to F and those of $\{\hat{\phi}_{T_R,s}\}_{0\le s<n_3}$, that belong to v_1, e_1, e_2 and f, restricted to f. For more details, see [6].

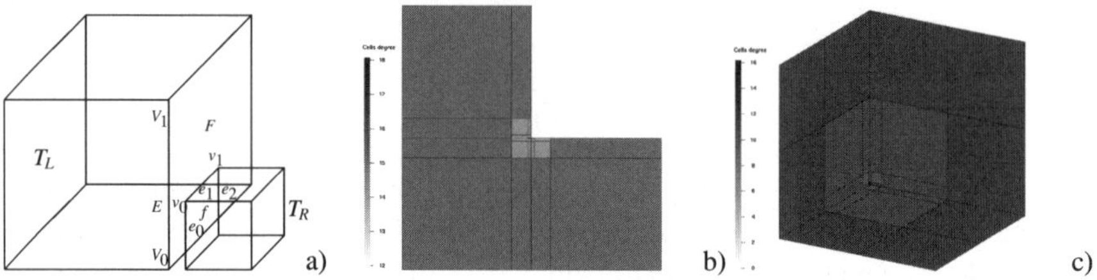

Fig. 1 **a**: Local refinement in 3D. **b-c**: *hp*-adaptive grids with unsymmetric divisions.

References

1. Arnold, D.N., Boffi, D., Falk, R.S.: Approximation by quadrilateral finite elements. Math. Comput. **71**(239), 909–922 (2002)
2. Demkowicz, L., Gerdes, K., Schwab, C., Bajer, A., Walsh, T.: HP90: A general and flexible Fortran 90 *hp*-FE code. Comput. Vis. Sci. **1**(3), 145–163 (1998)
3. Demkowicz, L., Oden, J.T., Rachowicz, W., Hardy, O.: Toward a universal h-p adaptive finite element strategy. i: Constrained approximation and data structure. Comp. Meth. Appl. Mech. Engrg **77**, 79–112 (1989)
4. Paszyński, M., Kurtz, J., Demkowicz, L.: Parallel, fully automatic *hp*-adaptive 2*d* finite element package. Comput. Methods Appl. Mech. Eng. **195**(7-8), 711–741 (2006)
5. Rachowicz, W., Pardo, D., Demkowicz, L.: Fully automatic hp-adaptivity in three dimensions. Comput. Methods Appl. Mech. Eng. **195**(37-40), 4816–4842 (2006)
6. Schröder, A.: Error controlled adaptive *h*- and *hp*- finite elements methods for contact problems with applications in production engineering. (Fehlerkontrollierte adaptive *h*- und *hp*-Finite-Elemente-Methoden für Kontaktprobleme mit Anwendungen in der Fertigungstechnik.). Ph.D. thesis, Dortmund: Univ. Dortmund, Fachbereich Mathematik (Diss.). Bayreuther Math. Schr. 78, xviii, 216 p. (2006)
7. Schwab, C.: $p-$ and $hp-$finite element methods. Theory and applications in solid and fluid mechanics. Numerical Mathematics and Scientific Computation. Clarendon Press, Oxford (1998)
8. Solin, P., Segeth, K., Delezel, I.: Higher-order finite element methods. Studies in Advanced Mathematics. CRC Press, Boca Raton (2004)
9. Szabo, B., Babuska, I.: Finite element analysis. Wiley-Interscience Publication. John Wiley & Sons Ltd., New York (1991)
10. Tricomi, F.G.: Vorlesungen über Orthogonalreihen. Die Grundlehre der mathematischen Wissenschaften. Springer-Verlag, Berlin (1955)

A Boundary Element Algorithm for the Dirichlet Eigenvalue Problem of the Laplace Operator

O. Steinbach and G. Unger

Abstract A novel boundary element method for the solution of the interior Dirichlet eigenvalue problem for the Laplace operator is presented and analyzed. Hereby, the linear eigenvalue problem for the partial differential operator is transformed into a nonlinear eigenvalue problem for an associated boundary integral operator. This nonlinear eigenvalue problem is solved by using a Newton scheme. We discuss the convergence and the boundary element discretization of this algorithm, and give some numerical results.

1 Introduction

We consider the interior Dirichlet eigenvalue problem of the Laplace operator,

$$-\Delta u(x) = \lambda u(x) \quad \text{for } x \in \Omega, \quad u(x) = 0 \quad \text{for } x \in \Gamma = \partial\Omega, \tag{1}$$

where $\Omega \subset \mathbb{R}^3$ is a bounded Lipschitz domain. This eigenvalue problem can be transformed into the equivalent nonlinear eigenvalue problem, see [11],

$$\frac{1}{4\pi}\int_\Gamma \frac{\cos(\kappa|x-y|)}{|x-y|} t(y) ds_y = 0 \quad \text{for x } \in \Gamma \tag{2}$$

where the unknowns $\kappa = \sqrt{\lambda}$ and $t(y) = n_y \cdot \nabla u(y)$ for $y \in \Gamma$, the corresponding normal derivative, have to be found. Associated eigenfunctions of (1) can then be represented by

$$u(x) = \frac{1}{4\pi}\int_\Gamma \frac{\cos(\kappa|x-y|)}{|x-y|} t(y) ds_y \quad \text{for x } \in \Omega.$$

Olaf Steinbach and Gerhard Unger
Institute of Computational Mathematics, Graz University of Technology, Steyrergasse 30, 8010 Graz, Austria, e-mail: {o.steinbach, gerhard.unger}@tugraz.at

In [3, 4] similar boundary element approaches for the eigenvalue problem (1) are suggested. The boundary element approximations in those works lead to polynomial approximations of the nonlinear eigenvalue problem (2).

In this work we consider an iterative solution approach for the nonlinear eigenvalue problem (2) which is an analogon of the inverse iteration for linear and for nonlinear matrix eigenvalue problems, see, e.g., [6, 8, 12]. In fact, we will apply a Newton scheme to solve the nonlinear equation (2) where in addition we introduce an appropriate scaling condition. However, our theoretical approach is restricted to simple eigenvalues only. Further a Galerkin boundary element method to solve the nonlinear eigenvalue problem is formulated and the results of the numerical analysis are presented. Numerical examples given in Section 3 confirm not only the theoretical results, the experiments indicate that our approach also works for multiple eigenvalues. For a detailed numerical analysis we refer to [11].

2 Boundary Element Methods

The nonlinear eigenvalue problem (2) can be written as

$$(V_\kappa t)(x) = \frac{1}{4\pi}\int_\Gamma \frac{\cos(\kappa|x-y|)}{|x-y|} t(y) ds_y = 0 \quad \text{for } x \in \Gamma \tag{3}$$

where for fixed κ the operator $V_\kappa : H^{-1/2}(\Gamma) \to H^{1/2}(\Gamma)$ is linear and bounded, see, e.g., [5]. To normalize the eigensolutions $t \in H^{-1/2}(\Gamma)$ of (3) we introduce a scaling condition by using an equivalent norm in $H^{-1/2}(\Gamma)$,

$$\|t\|_V^2 = \langle Vt, t\rangle_\Gamma = \frac{1}{4\pi}\int_\Gamma t(x)\int_\Gamma \frac{1}{|x-y|} t(y) ds_y ds_x = 1, \tag{4}$$

where $V : H^{-1/2}(\Gamma) \to H^{1/2}(\Gamma)$ is the single layer potential of the Laplace operator. Now we have to find solutions $(t,\kappa) \in H^{-1/2}(\Gamma) \times \mathbb{R}$ of the nonlinear eigenvalue problem

$$F_1(t,\kappa) = (V_\kappa t)(x) = 0 \quad \text{for } x \in \Gamma, \quad F_2(t,\kappa) = \langle Vt, t\rangle_\Gamma - 1 = 0. \tag{5}$$

Hence we define the function $F : H^{-1/2}(\Gamma) \times \mathbb{R} \to H^{1/2}(\Gamma) \times \mathbb{R}$ as

$$F(t,\kappa) = \begin{pmatrix} F_1(t,\kappa) \\ F_2(t,\kappa) \end{pmatrix} = \begin{pmatrix} \displaystyle\frac{1}{4\pi}\int_\Gamma \frac{\cos(\kappa|x-y|)}{|x-y|} t(y) ds_y \\ \langle Vt, t\rangle_\Gamma - 1 \end{pmatrix}.$$

Then, to obtain eigensolutions of the scaled eigenvalue problem (5) we have to find solutions $(t,\kappa) \in H^{-1/2}(\Gamma) \times \mathbb{R}$ of the nonlinear equation

$$F(t,\kappa) = 0 \tag{6}$$

which is to be solved by applying Newton's method. For the Fréchet derivative of $F(t,\kappa)$ we obtain

$$F'(t,\kappa) = \begin{pmatrix} V_\kappa & -A_\kappa t \\ 2\langle Vt,\cdot\rangle_\Gamma & 0 \end{pmatrix} : H^{-1/2}(\Gamma)\times\mathbb{R} \to H^{1/2}(\Gamma)\times\mathbb{R} \tag{7}$$

where

$$(A_\kappa t)(x) = \frac{1}{4\pi}\int_\Gamma \sin(\kappa|x-y|)t(y)ds_y \quad \text{for } x\in\Gamma.$$

When applying a Newton scheme to find solutions $(t_*,\kappa_*)\in H^{-1/2}(\Gamma)\times\mathbb{R}$ of the nonlinear equation (6) the new iterates $(t_{n+1},\kappa_{n+1})\in H^{-1/2}(\Gamma)\times\mathbb{R}$ can be determined by the linear operator equation

$$F'(t_n,\kappa_n)\begin{pmatrix} t_{n+1}-t_n \\ \kappa_{n+1}-\kappa_n \end{pmatrix} + F(t_n,\kappa_n) = 0. \tag{8}$$

In the following theorem we give sufficient conditions that equation (8) is unique solvable and that Newton's method locally converges to an eigensolution (t_*,κ_*).

Theorem 1. *[11] Let (t_*,κ_*) be a solution of $F(t,\kappa)=0$. Assume*

(A1) *κ_* is a simple eigenvalue of $V_\kappa t = 0$,*
(A2) *$A_{\kappa_*}t_* \notin \mathscr{R}(V_{\kappa_*})$.*

Then $F'(t_*,\kappa_*)$ is invertible and Newton's method converges for all initial values in a sufficient small neighborhood $U_\rho(t_*,\kappa_*)$ to (t_*,κ_*), where $\rho>0$ with

$$\|t_*-t_n\|^2_{H^{-1/2}(\Gamma)} + |\kappa_*-\kappa_n|^2 \le \rho^2. \tag{9}$$

Remark 1. For multiple eigenvalues κ_* the Fréchet derivative $F'(t_*,\kappa_*)$ is not invertible, because $F'(t_*,\kappa_*)$ is not injective. Nevertheless Newton's method may also converge [1, 2]. The convergence rate may then be smaller and the convergence domain is not a small neighborhood of the solution but rather a restricted region which avoids the set on which F' is singular. In our case numerical examples show that Newton's method converges also for multiple eigenvalues, see the numerical example in Section 3.

The linearized equation (8) is equivalent to a saddle point problem to find $(t_{n+1},\kappa_{n+1})\in H^{-1/2}(\Gamma)\times\mathbb{R}$ such that

$$\begin{aligned} \langle V_{\kappa_n}t_{n+1},w\rangle_\Gamma - \kappa_{n+1}\langle A_{\kappa_n}t_n,w\rangle_\Gamma &= -\kappa_n\langle A_{\kappa_n}t_n,w\rangle_\Gamma \\ 2\langle Vt_n,t_{n+1}\rangle_\Gamma &= \langle Vt_n,t_n\rangle_\Gamma + 1 \end{aligned} \tag{10}$$

is satisfied for all $w\in H^{-1/2}(\Gamma)$. For a Galerkin discretization of (10) we first define trial spaces $S_h^0(\Gamma)$ of piecewise constant basis functions ψ_k which are defined with respect to a globally quasi–uniform boundary element mesh of mesh size h. Then the Galerkin discretization of (10) reads to find $(t_{n+1,h},\kappa_{n+1,h}) \in S_h^0(\Gamma)\times\mathbb{R}$ such that

$$\begin{aligned}\langle V_{\kappa_n} t_{n+1,h}, w_h\rangle_\Gamma - \kappa_{n+1,h}\langle A_{\kappa_n} t_n, w_h\rangle_\Gamma &= -\kappa_n\langle A_{\kappa_n} t_n, w_h\rangle_\Gamma \\ 2\langle V t_n, t_{n+1,h}\rangle_\Gamma &= \langle V t_n, t_n\rangle_\Gamma + 1.\end{aligned} \tag{11}$$

is satisfied for all $w_h \in S_h^0(\Gamma)$. In the following theorem the solvability of the linear system (11) is discussed and an error estimate for the approximate solution $(t_{n+1,h},\kappa_{n+1,h})$ is given.

Theorem 2. *[11] Let* (t_*,κ_*) *be a solution of* $F(t,\kappa)=0$ *and let the assumptions* (A1) *and* (A2) *be satisfied. Let* $(t_n,\kappa_n)\in U_\rho(t_*,\kappa_*)$ *be satisfied where* ρ *is appropriately chosen as discussed in Theorem* 1. *Then, for a sufficient small mesh size* $h<h_0$, *the Galerkin variational problem* (11) *has a unique solution* $(t_{n+1,h},\kappa_{n+1,h})\in S_h^0(\Gamma)\times\mathbb{R}$ *satisfying the error estimate*

$$\|t_{n+1}-t_{n+1,h}\|^2_{H^{-1/2}(\Gamma)} + |\kappa_{n+1}-\kappa_{n+1,h}|^2 \le c \inf_{w_h\in S_h^0(\Gamma)} \|t_{n+1}-w_h\|^2_{H^{-1/2}(\Gamma)}. \tag{12}$$

In practical computations we have to replace in (11) $(t_n,\kappa_n)\in H^{-1/2}(\Gamma)\times\mathbb{R}$ by previously computed approximations $(\hat{t}_{n,h},\hat{\kappa}_{n,h})\in S_h^0(\Gamma)\times\mathbb{R}$. In particular we have to find $(\hat{t}_{n+1,h},\hat{\kappa}_{n+1,h})\in S_h^0(\Gamma)\times\mathbb{R}$ such that

$$\begin{aligned}\langle V_{\hat{\kappa}_{n,h}} \hat{t}_{n+1,h}, w_h\rangle_\Gamma - \hat{\kappa}_{n+1,h}\langle A_{\hat{\kappa}_{n,h}} \hat{t}_{n,h}, w_h\rangle_\Gamma &= -\hat{\kappa}_{n,h}\langle A_{\hat{\kappa}_{n,h}} \hat{t}_{n,h}, w_h\rangle_\Gamma \\ 2\langle V \hat{t}_{n,h}, \hat{t}_{n+1,h}\rangle_\Gamma &= \langle V \hat{t}_{n,h}, \hat{t}_{n,h}\rangle_\Gamma + 1\end{aligned} \tag{13}$$

is satisfied for all $w_h \in S_h^0(\Gamma)$. To analyze the perturbed variational problem (13) we also need to consider the continuous variational problem to find $(\hat{t}_{n+1},\hat{\kappa}_{n+1})\in H^{-1/2}(\Gamma)\times\mathbb{R}$ such that

$$\begin{aligned}\langle V_{\hat{\kappa}_{n,h}} \hat{t}_{n+1}, w\rangle_\Gamma - \hat{\kappa}_{n+1}\langle A_{\hat{\kappa}_{n,h}} \hat{t}_{n,h}, w\rangle_\Gamma &= -\hat{\kappa}_{n,h}\langle A_{\hat{\kappa}_{n,h}} \hat{t}_{n,h}, w\rangle_\Gamma \\ 2\langle V \hat{t}_{n,h}, \hat{t}_{n+1}\rangle_\Gamma &= \langle V \hat{t}_{n,h}, \hat{t}_{n,h}\rangle_\Gamma + 1\end{aligned} \tag{14}$$

is satisfied for all $w\in H^{-1/2}(\Gamma)$. Note that (13) is the Galerkin discretization of (14). In the next theorem we discuss the solvability of the linear system (14) and give an error estimate for the discrete Newton iterate $(\hat{t}_{n+1,h},\hat{\kappa}_{n+1,h})$ with respect to the continuous Newton iterate (t_{n+1},κ_{n+1}).

Theorem 3. *[11] Let* (t_*,κ_*) *be a solution of* $F(t,\kappa)=0$ *and let the assumptions* (A1) *and* (A2) *be satisfied. Let* $(\hat{t}_{n,h},\hat{\kappa}_{n,h})\in S_h^0(\Gamma)\times\mathbb{R}\cap U_\rho(t_*,\kappa_*)$ *be satisfied where* ρ *is appropriately chosen as discussed in Theorem* 1. *Then, for a sufficient small mesh size* $h<h_0$, *the Galerkin variational problem* (13) *has a unique solution* $(\hat{t}_{n+1,h},\hat{\kappa}_{n+1,h})\in S_h^0(\Gamma)\times\mathbb{R}$ *satisfying the error estimate*

$$\|t_{n+1}-\hat{t}_{n+1,h}\|^2_{H^{-1/2}(\Gamma)} + |\kappa_{n+1} - \hat{\kappa}_{n+1,h}|^2$$
$$\le c\left[\|t_n-\hat{t}_{n,h}\|^2_{H^{-1/2}(\Gamma)} + |\kappa_n - \hat{\kappa}_{n,h}|^2 + \inf_{w_h\in S_h^0(\Gamma)}\|t_{n+1}-w_h\|^2_{H^{-1/2}(\Gamma)}\right] \tag{15}$$

where the constant c depends on (t_*, κ_*), *and on* ρ.

Considering the approximation property of $S_h^0(\Gamma)$ we get from (15) the following error estimate, see [11],

$$\|t_{n+1}-\hat{t}_{n+1,h}\|^2_{H^{-1/2}(\Gamma)} + |\kappa_{n+1} - \hat{\kappa}_{n+1,h}|^2 \le c\left[\rho^4 + h^3\right]$$

when assuming $t_* \in H^1_{\text{pw}}(\Gamma)$. The constant c depends on (t_*, κ_*), n, and on ρ.

When using the Aubin–Nitsche trick, see for example [10], it is possible to derive error estimates in Sobolev spaces with lower Sobolev index. In particular we obtain the error estimate

$$\|t_{n+1}-\hat{t}_{n+1,h}\|^2_{H^{-2}(\Gamma)} + |\kappa_{n+1} - \hat{\kappa}_{n+1,h}|^2 \le c\left[\rho^4 + h^6\right]$$

when assuming $t_* \in H^1_{\text{pw}}(\Gamma)$. Hence we can expect a cubic convergence rate for the eigenvalues,

$$|\kappa_{n+1} - \hat{\kappa}_{n+1,h}| \le c\left[\rho^4 + h^6\right]^{1/2} = \mathcal{O}(h^3).$$

3 Numerical Results

In this section we present some numerical results to investigate the behavior of the nonlinear boundary element approach as presented in this paper. As a model problem we consider the interior Dirichlet eigenvalue problem (1) where the domain $\Omega = (0, \frac{1}{2})^3$ is a cube. Hence the eigenvalues are given by

$$\lambda_k = 4\pi^2\left[k_1^2 + k_2^2 + k_3^2\right]$$

and the associated eigenfunctions are

$$u_k(x) = (\sin 2\pi k_1 x_1)(\sin 2\pi k_2 x_2)(\sin 2\pi k_3 x_3).$$

It turns out that the first eigenvalue $(k_1 = k_2 = k_3 = 1)$

$$\lambda_1 = 12\pi^2, \quad \kappa_1 = 2\sqrt{3}\pi$$

is simple, while the second eigenvalue $(k_1 = 2, k_2 = k_3 = 1)$

$$\lambda_2 = 24\pi^2, \quad \kappa_2 = 2\sqrt{6}\pi$$

is multiple.

For the boundary element discretization the boundary $\Gamma = \partial\Omega$ was decomposed into N uniform triangular boundary elements. The numerical results to approximate the simple eigenvalue $\kappa_1 = \sqrt{\lambda_1}$ are given in Table 1.

Table 1 Approximation of $\kappa_1 = 2\sqrt{3}\pi \approx 10.8828$, simple eigenvalue

N	$\kappa_{1,N}$	$\lvert\kappa_1 - \kappa_{1,N}\rvert$	rate
384	10.8768	6.0e-03	-
1536	10.8821	7.0e-04	8.6
6144	10.8827	8.6e-05	8.1

Note that the convergence rate of approximately 8 corresponds to the cubic convergence as predicted in (2). Next we consider the case of a multiple eigenvalue, the results to approximate $\kappa_2 = \sqrt{\lambda_2}$ are given in Table 2.

Table 2 Approximation of $\kappa_2 = 2\sqrt{6}\pi \approx 15.3906$, multiple eigenvalue

N	$\kappa_{2_1,N}$	$\lvert\kappa_2 - \kappa_{2_1,N}\rvert$	rate	$\kappa_{2_2,N}$	$\lvert\kappa_2 - \kappa_{2_2,N}\rvert$	$\kappa_{2_3,N}$	$\lvert\kappa_2 - \kappa_{2_3,N}\rvert$
384	15.3739	1.7e-02	-	14.7057	0.68	15.8867	0.50
1536	15.3887	1.9e-03	8.9	14.6902	0.70	15.8579	0.47
6144	15.3904	2.3e-04	8.3	14.6839	0.71	15.8499	0.46

As in other boundary element approaches for eigenvalue problems [3, 4, 13] the problem of the so–called spurious eigenvalues occurs close to multiple eigenvalues. In particular, several distinct discrete eigenvalues are obtained to approximate a multiple eigenvalue. This phenomenon also occurs for algebraic eigenvalue problems when an approximation of the matrix is used, see e.g. [9].

The spurious eigenvalues can be filtered out with an a posteriori error control by using the complex valued fundamental solution for an eigensolution (t,κ),

$$\frac{1}{4\pi}\int_\Gamma \frac{e^{i\kappa|x-y|}}{|x-y|}t(y)ds_y = (V_\kappa t)(x) + i\frac{1}{4\pi}\int_\Gamma \frac{\sin\kappa|x-y|}{|x-y|}t(y)ds_y = 0. \tag{16}$$

Then the norm of the residual

$$r(t_h,\kappa_h) = \frac{1}{4\pi}\int_\Gamma \frac{e^{i\kappa_h|x-y|}}{|x-y|}t_h(y)ds_y$$

for actual approximations of eigensolutions (t_h,κ_h) is significant smaller than for spurious eigensolutions, see Table 3.

When an analogous algorithm is used which is based on the complex valued fundamental solution (16) no spurious eigenvalues occur. But then complex arithmetics

Table 3 Residual for true and spurious eigenvalues

N	$\|r(t_{1,N},\kappa_{1,N})\|$	$\|r(t_{2_1,N},\kappa_{2_1,N})\|$	$\|r(t_{2_2,N},\kappa_{2_2,N})\|$	$\|r(t_{2_3,N},\kappa_{2_3,N})\|$
384	6.3e-05	1.9e-04	7.5e-03	2.3e-02
1536	7.7e-06	2.2e-05	3.8e-03	1.2e-02
6144	9.6e-07	2.5e-06	1.9e-03	6.1e-03

has to be used so that the computational complexity is twice expensive as for the real valued version. In Table 4 and 5 the approximations for κ_1 and κ_2 are given.

Table 4 Approximation of $\kappa_1 = 2\sqrt{3}\pi \approx 10.8828$, simple eigenvalue

N	$\kappa_{1,N}$	$\|\kappa_1 - \kappa_{1,N}\|$	rate
384	10.8768-1.0e-06i	6.0e-03	-
1536	10.8821-2.4e-07i	7.0e-04	8.6
6144	10.8827-6.0e-09i	8.6e-05	8.1

Table 5 Approximation of $\kappa_2 = 2\sqrt{6}\pi \approx 15.3906$, multiple eigenvalue

N	$\kappa_{2,N}$	$\|\kappa_2 - \kappa_{2,N}\|$	rate
384	15.3739-5.1e-06i	1.7e-02	-
1536	15.3887-9.4e-07i	1.9e-03	8.9
6144	15.3904-2.1e-08i	2.3e-04	8.3

Note that the real part of the approximations of the complex valued algorithm are the same as of the real valued version.

4 Conclusions

In this paper we have presented and analyzed a boundary element method for the solution of the interior Dirichlet eigenvalue problem for the Laplace operator. Hereby, the linear eigenvalue problem for the partial differential operator is transformed into a nonlinear eigenvalue problem for an associated boundary integral operator which is solved via a Newton iteration. The discretization by using a Galerkin boundary element method gives a cubic order of convergence of the approximated eigenvalues. When using fast boundary element methods [7] an almost optimal computational complexity can be obtained. For this, also efficient preconditioned iterative solution methods to solve the Galerkin equations (13) are mandatory. As already mentioned in Remark 1 a further analysis in the case of multiple eigenvalues is needed.

Finally we mention that the proposed approach can be used to solve the interior Neumann eigenvalue problem for the Laplace operator, and to solve related eigenvalue problems in linear elastostatics.

References

1. Decker, D.W., Kelley, C.T.: Newton's method at singular points. I. SIAM J. Numer. Anal. **17**(1), 66–70 (1980)
2. Decker, D.W., Kelley, C.T.: Convergence acceleration for Newton's method at singular points. SIAM J. Numer. Anal. **19**(1), 219–229 (1982)
3. Kamiya, N., Andoh, E., Nogae, K.: Eigenvalue analysis by the boundary element method: new developments. Engng. Anal. Bound. Elms. **12**, 151–162 (1993)
4. Kirkup, S.M., Amini, S.: Solution of the Helmholtz eigenvalue problem via the boundary element method. Internat. J. Numer. Methods Engrg. **36**, 321–330 (1993)
5. McLean, W.: Strongly elliptic systems and boundary integral equations. Cambridge University Press, Cambridge (2000)
6. Mehrmann, V., Voss, H.: Nonlinear eigenvalue problems: a challenge for modern eigenvalue methods. GAMM Mitt. Ges. Angew. Math. Mech. **27**(2), 121–152 (2005) (2004)
7. Rjasanow, S., Steinbach, O.: The fast solution of boundary integral equations. Mathematical and Analytical Techniques with Applications to Engineering. Springer, New York (2007)
8. Ruhe, A.: Algorithms for the nonlinear eigenvalue problem. SIAM J. Numer. Anal. **10**, 674–689 (1973)
9. Saad, Y.: Numerical methods for large eigenvalue problems. Algorithms and Architectures for Advanced Scientific Computing. Manchester University Press, Manchester (1992)
10. Steinbach, O.: Numerical approximation methods for elliptic boundary value problems. Finite and boundary elements. Springer, New York (2008)
11. Steinbach, O., Unger, G.: A boundary element method for the Dirichlet eigenvalue problem of the Laplace operator. Berichte aus dem Institut für Numerische Mathematik, TU Graz, Bericht 2007/5 (2007)
12. Wilkinson, J.H.: The algebraic eigenvalue problem. Clarendon Press, Oxford (1965)
13. Yeih, W., Chen, J.T., Chen, K.H., Wong, F.: A study on the multiple reciprocity method and complex-valued formulation for the Helmholtz equation. Adv. Engng. Software **29**, 1–6 (1998)

On Efficient Solution of Linear Systems Arising in *hp*-FEM

T. Vejchodský

Abstract This contribution studies the *static condensation of internal degrees of freedom* which allows for efficient solution of linear algebraic systems arising in higher-order finite element methods. On each element, the static condensation eliminates the degrees of freedom corresponding to the internal (or bubble) basis functions. The elimination is local in elements and can be done in parallel. The resulting Schur complement system is considerably smaller and, moreover, it has less nonzero elements and better condition number in comparison with the original system. This paper focuses on the numerical performace of the static condensation and shows its CPU time efficiency.

1 Introduction and Higher-Order Finite Elements

In the standard finite element method (FEM) or more precisely in its h-version (h-FEM), the decrease of the discretization error is achieved by successive refinement of the mesh. The method converges if the size of the elements tends to zero, and the rate of this convergence is proved to be algebraic. In an alternative approach called the p-version (p-FEM), the geometry of the mesh is fixed and the polynomial degrees of the elements vary. The convergence is achieved by increasing the polynomial degrees and the convergence rate is exponential if the exact solution is C^∞-smooth. A combination of these two approaches is known as the hp-version (hp-FEM), see, e.g., [2, 4, 5, 8]. To decrease the discretization error in the hp-FEM we either refine the elements or we increase their polynomial degrees or we both refine the elements and redistribute the polynomial degrees on the subelements in a suitable way. If this hp-refinement is done in a correct way, then the hp-FEM converges exponentially fast even in the presence of singularities.

Tomáš Vejchodský
Institute of Mathematics, Czech Academy of Sciences, Žitná 25, 115 67 Prague 1, Czech Republic
e-mail: vejchod@math.cas.cz

The higher-order FEM leads to linear algebraic systems with a special structure. This special structure can be utilized to design efficient algebraic solvers. In particular, a characteristic feature of the higher-order FEM is the presence of the so-called bubble (or internal) basis functions that are supported in a single element only. The static condensation of internal degrees of freedom (DOFs) eliminates these bubble functions from the whole system by a local (element-by-element) procedure. After this elimination, we obtain a reduced system of linear algebraic equations – the Schur complement system. From this system, we compute the other (non-internal) DOFs which correspond to vertices, edges, and faces of the elements. The number of the internal DOFs grows with the polynomial degree p by an order of magnitude faster than the number of the non-internal DOFs. Thus, for higher values of p, the number of the internal DOFs dominates and their static condensation leads to a significant decrease of the size of the linear algebraic system.

The technique of the static condensation of the internal DOFs is described in Section 2. The core of this paper lies in Section 3, where the performance of the static condensation is tested by various numerical experiments. Brief conclusions are given in Section 4.

2 Static Condensation of Internal Degrees of Freedom

To simplify the exposition, we only consider 2D elliptic problems discretized by triangular finite elements of an arbitrary order. However, the static condensation of the internal DOFs can be used in any dimension, for much wider class of problems, and for various types of higher-order finite elements.

Let $\Omega \subset \mathbb{R}^2$ be a polygon. We consider a problem whose weak formulation reads: find $u \in V$ such that

$$a(u,v) = \mathscr{F}(v) \quad \forall v \in V, \tag{1}$$

where V is a suitable Hilbert space, $a : V \times V \to \mathbb{R}$ is a continuous V-elliptic bilinear form, and $\mathscr{F}$ is a continuous linear functional on V. Problem (1) possesses a unique solution due to the Lax-Milgram lemma. For example, if

$$V = H_0^1(\Omega), \quad a(u,v) = \int_\Omega \nabla u \cdot \nabla v \,\mathrm{d}x, \quad \text{and} \quad \mathscr{F}(v) = \int_\Omega f v \,\mathrm{d}x, \tag{2}$$

then (1) corresponds to the Poisson problem with homogeneous Dirichlet boundary conditions.

We discretize problem (1) by the hp-FEM. Let $\mathscr{T}_{hp}$ be a triangulation of Ω, let p_K stand for the polynomial degree assigned to the element $K \in \mathscr{T}_{hp}$, and let

$$V_{hp} = \{v_{hp} \in V : v_{hp}|_K \in P^{p_K}(K),\ K \in \mathscr{T}_{hp}\}$$

be the finite element space, where $P^{p_K}(K)$ denotes the space of polynomials of degree at most p_K on the triangle K. The hp-FEM solution $u_{hp} \in V_{hp}$ is defined by

$$a(u_{hp}, v_{hp}) = \mathscr{F}(v_{hp}) \quad \forall v_{hp} \in V_{hp}. \tag{3}$$

We consider a standard *hp*-FEM basis $\varphi_1, \varphi_2, \ldots, \varphi_N$ of V_{hp}, where $N = \dim(V_{hp})$, see, e.g., [2, 4, 5, 8]. These basis functions are constructed element by element as

$$\varphi_i|_K = \varphi^K_{\iota_K^{-1}(i)}, \quad i = 1, 2, \ldots, N,$$

where φ^K_m, $m = 1, 2, \ldots, N^K$, denote the *shape* functions that only are supported in the single element K and $\iota_K : \{1, 2, \ldots, N^K\} \to \{1, 2, \ldots, N\}$ is the standard connectivity mapping, see [8, 7] for more details and Fig. 1 for an illustration. Notice that if $i \notin \mathrm{Dom}(\iota_K)$, i.e., if $\iota_K^{-1}(i)$ is not defined, then $\varphi^K_{\iota_K^{-1}(i)}$ is considered to be zero.

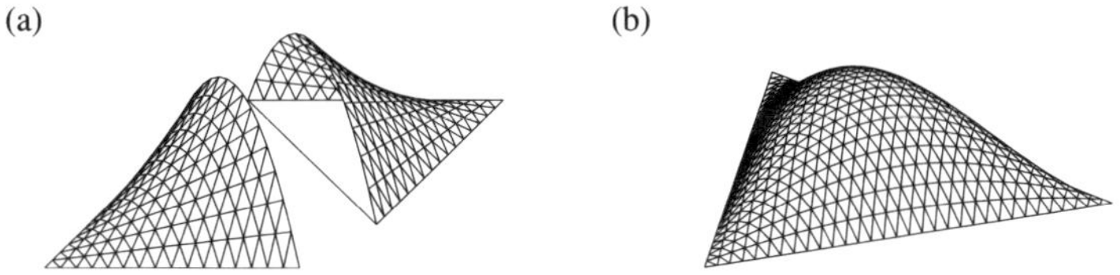

Fig. 1 (a) Two edge shape functions on two neighbouring elements form an edge basis function. (b) The bubble shape function coincides with the bubble basis function.

Problem (3) is equivalent to the system of linear algebraic equations

$$\mathbb{A}\mathbb{Y} = \mathbb{F}, \quad \mathbb{A}_{ij} = a(\varphi_j, \varphi_i), \quad \mathbb{F}_i = \mathscr{F}(\varphi_i), \quad i, j = 1, 2, \ldots, N, \tag{4}$$

where $\mathbb{A} \in \mathbb{R}^{N \times N}$ and $\mathbb{F} \in \mathbb{R}^N$ are the (global) stiffness matrix and the (global) load vector, respectively. The vector $\mathbb{Y} \in \mathbb{R}^N$ contains the expansion coefficients of u_{hp} in the finite element basis.

The global stiffness matrix and the global load vector are assembled from the local stiffness matrices $\mathbb{A}^K \in \mathbb{R}^{N^K \times N^K}$ and from the local load vectors $\mathbb{F}^K \in \mathbb{R}^{N^K}$, $K \in \mathscr{T}_{hp}$. These local matrices and vectors are defined by

$$\mathbb{A}^K_{\ell m} = a_K(\varphi_{\iota_K(m)}, \varphi_{\iota_K(\ell)}) \quad \text{and} \quad \mathbb{F}^K_\ell = \mathscr{F}_K(\varphi_{\iota_K(\ell)}), \quad \ell, m = 1, 2, \ldots, N^K,$$

where the local bilinear form $a_K(\cdot,\cdot)$ and the local linear functional $\mathscr{F}_K$ satisfy

$$a(\varphi_j, \varphi_i) = \sum_{K \in \mathscr{T}_{hp}} a_K(\varphi_j, \varphi_i) \quad \text{and} \quad \mathscr{F}(\varphi_i) = \sum_{K \in \mathscr{T}_{hp}} \mathscr{F}_K(\varphi_i), \quad i, j = 1, 2, \ldots, N.$$

For example, if $a(\cdot,\cdot)$ and $\mathscr{F}$ are given by (2), then the local bilinear form $a_K(\cdot,\cdot)$ and the local linear functional $\mathscr{F}_K$ are defined as

$$a_K(\varphi_j, \varphi_i) = \int_K \nabla \varphi_j \cdot \nabla \varphi_i \,\mathrm{d}x \quad \text{and} \quad \mathscr{F}_K(\varphi_i) = \int_K f \varphi_i \,\mathrm{d}x.$$

With this notation, the standard finite element assembling procedure can be written as

$$\mathbb{A}_{ij} = \sum_{K\in\mathscr{T}_{hp}} \mathbb{A}^K_{\iota_K^{-1}(i),\iota_K^{-1}(j)} \quad \text{and} \quad \mathbb{F}_i = \sum_{K\in\mathscr{T}_{hp}} \mathbb{F}^K_{\iota_K^{-1}(i)}, \quad i,j=1,2,\ldots,N. \tag{5}$$

From now on we will consider a special enumeration of the basis functions. We enumerate the bubbles first, and then the other basis functions. Hence if M denotes the number of the bubble functions, then $\varphi_1,\ldots,\varphi_M$ stand for the bubbles and $\varphi_{M+1},\ldots,\varphi_N$ stand for the other basis functions. Similarly, we enumerate the shape functions in all elements. In each element $K\in\mathscr{T}_{hp}$, the first M^K shape functions are the bubbles and the other N^K-M^K shape functions are the non-bubbles. This enumeration splits the global and local stiffness matrices and the global and local load vectors into natural blocks

$$\mathbb{A} = \begin{pmatrix} A & B^T \\ B & C \end{pmatrix}, \quad \mathbb{A}^K = \begin{pmatrix} A^K & (B^K)^T \\ B^K & C^K \end{pmatrix}, \quad \mathbb{F} = \begin{pmatrix} F \\ G \end{pmatrix}, \quad \mathbb{F}^K = \begin{pmatrix} F^K \\ G^K \end{pmatrix}, \tag{6}$$

where $A\in\mathbb{R}^{M\times M}$, $B\in\mathbb{R}^{(N-M)\times M}$, $A^K\in\mathbb{R}^{M^K\times M^K}$, $B^K\in\mathbb{R}^{(N^K-M^K)\times M^K}$, etc.

Since the bubble functions are supported in a single element, the corresponding matrix A is block diagonal with the diagonal blocks being A^K, i.e., $A=\operatorname{blockdiag}\{A^K, K\in\mathscr{T}_{hp}\}$. Thus, the matrix A is easily invertible and this makes the static condensation of the internal DOFs efficient.

The block structure (6) reshapes the global stiffness system (4) as follows

$$\begin{pmatrix} A & B^T \\ B & C \end{pmatrix}\begin{pmatrix} \mathbf{x} \\ \mathbf{y} \end{pmatrix} = \begin{pmatrix} F \\ G \end{pmatrix}, \qquad \text{where } \begin{pmatrix} \mathbf{x} \\ \mathbf{y} \end{pmatrix} = \mathbb{Y}. \tag{7}$$

The idea of the static condensation is to express $\mathbf{x}\in\mathbb{R}^M$ as $\mathbf{x}=A^{-1}(F-B^T\mathbf{y})$ and substitute this into the second block-row of (7) to obtain the Schur complement system for $\mathbf{y}\in\mathbb{R}^{N-M}$

$$S\mathbf{y}=\widetilde{G}, \quad \text{where} \quad S=C-BA^{-1}B^T \quad \text{and} \quad \widetilde{G}=G-BA^{-1}F. \tag{8}$$

It is shown in [6] that the Schur complement S and the right-hand side $\widetilde{G}$ can be obtained by the standard finite element assembling procedure, cf. (5),

$$S_{ij} = \sum_{K\in\mathscr{T}_{hp}} S^K_{\iota_K^{-1}(M+i),\iota_K^{-1}(M+j)} \quad \text{and} \quad \widetilde{G}_i = \sum_{K\in\mathscr{T}_{hp}} \widetilde{G}^K_{\iota_K^{-1}(M+i)}, \tag{9}$$

$i,j=1,2,\ldots,N-M$, where $S^K=C^K-B^K(A^K)^{-1}(B^K)^T$ are the local Schur complements and $\widetilde{G}^K=G^K-B^K(A^K)^{-1}F^K$ are the corresponding local right-hand sides.

The static condensation of the internal DOFs can also be interpreted as an orthogonalization of the non-bubble basis functions with respect to the bubbles. It can be shown that the static condensation and the partial orthogonalization of the basis are just two interpretations of the same arithmetic procedure. Moreover, if the Schur complement system (8) is solved by the ILU-PCG, then this arithmetic procedure is equivalent to the ILU-PCG applied to the original system (4). However, the usage of ILU-PCG for (4) is less efficient than the static condensation because in ILU-PCG we eliminate the internal DOFs superfluously in every iteration while it suffices to do it once. Furthermore, it can be shown that the sparsity patterns of the Schur complement S and of the original block C are identical. Hence, no fill-in appears during the construction of S. Finally, notice that the Schur complement S only depends on the space of the bubbles and not on the particular basis. All these facts are proven in [7], where more technical details can be found.

Another interesting fact, see [3], is that the conditioning of S cannot be worse than the conditioning of $\mathbb{A}$. In practice, however, the condition number of S is observed to be much smaller than the condition number of $\mathbb{A}$.

3 Numerical Performance

This section presents several numerical experiments to compare the performance of the ILU-PCG with and without the static condensation. More precisely, we compare two approaches. First, we use the static condensation and construct the Schur complement system (8), where we explicitely invert the local blocks A^K. The Schur complement system (8) is then solved by ILU-PCG. In the second approach we directly apply the ILU-PCG to system (4). We show in [7] that these approaches are two different implementations of the same arithmetic procedure and hence the number of ILU-PCG iterations N_{iter} is the same in both cases.

For the following tests, we consider the Possion problem

$$-\Delta u = f \quad \text{in } \Omega = (-1,1)^2, \quad u = 0 \quad \text{on } \partial\Omega.$$

The right-hand side $f = u\pi^2/2$ is chosen in agreement with the exact solution $u = \cos(x\pi/2)\cos(y\pi/2)$.

We stress that the static condensation can easily be implemented with the same memory requirements as the standard approach. The memory columns in Tables 1–4 below show the total number of entries in the local stiffness matrices $\mathbb{A}^K$.

The first two experiments illustrate the standard h- and p-version. In the h-FEM, the most efficient way is to use the same polynomial degrees in all elements. However, we use different polynomial degrees to study the performance of the static condensation in the context of the hp-FEM. Therefore, the initial mesh for the h-version consists of four elements with polynomial degrees 4,5,6,7, see Fig. 2(a). In every refinement step, we split each triangular element into four similar subtriangles with the same polynomial degree as the parent element has. In Table 1 we present: N, the total number of DOFs (the size of $\mathbb{A}$); $M-N$, the number of DOFs after the elimination of the internal DOFs (the size of S); the memory requirements (specified above); the relative discretization error $\|u-u_h\|/\|u_h\|$ measured in the energy norm; the number of ILU-PCG iterations N_{iter}; and the CPU times needed to solve the stiffness system with and without the static condensation.

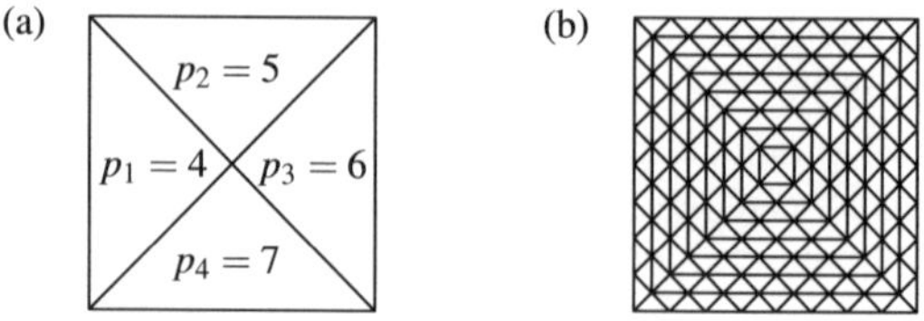

Fig. 2 (a) The initial mesh for the h-FEM. (b) The initial mesh for the p-FEM consits of linear elements ($p = 1$). There are eight elements along each edge of the square.

Similarly, Table 2 shows the same quantities for the p-FEM. Here we start with the first order elements and increase this order by one in every step. The initial mesh was uniform with 256 elements, see Fig. 2(b). We remark that the values of the relative discretization error for $p = 8$ and $p = 9$ are already polluted by the round-off errors and by the precission of the used numerical quadrature because the discretization error is already close to the machine precission.

The results in Tables 1 and 2 show that for the presented range of polynomial degrees the static condensation of the internal DOFs decreases the solver CPU time up to ten times. We remark that the polynomial degrees higher than ten are rarely used in practice.

Notice the exponential decrease of the error for the p-version in Table 2 and in Fig. 3. This is due to the C^∞-smoothness of the exact solution. However, the number of DOFs grows very rapidly with increasing p. The question is whether the error would decrease if we fix the number of DOFs and increase p only. The answer is given in Table 3. Practically, for a given value of p we construct a uniform triangulation of Ω such that the number of DOFs is more-less fixed. Clearly, the number of elements decreases with growing p. In Table 3 we can observe the decrease of the discretization error as well as the speed-up obtained by the static condensation.

Nevertheless, the memory requirements grow with p even if the number of DOFs is fixed. This is due to the fact that the stiffness matrix $\mathbb{A}$ is more dense for higher polynomial degrees. Hence, we can modify the previous experiment in order to keep the memory requirements fixed. For a given p we construct a uniform triangulation of Ω such that the resulting memory requirements are constant. Table 4 summarizes the results. Interestingly, see also Fig. 3, the number of DOFs decreases quite rapidly but the discretization error decreases as well. However, the rate of the error decrease is not as fast as in the previous cases, which is not surprising.

Table 1 The standard h-FEM.

ref.	N	$N-M$	memory	rel. err.		solver CPU time [s]	
step	(size $\mathbb{A}$)	(size S)	[$\times 10^3$]	[%]	N_{iter}	stat. con.	no conden.
0	50	16	2.7	1.2	3	0.004	0.005
1	225	89	11.0	5.4×10^{-2}	5	0.012	0.017
2	953	409	43.9	3.4×10^{-3}	7	0.049	0.130
3	3921	1745	175.7	2.1×10^{-4}	11	0.389	1.665
4	15905	7201	703.0	1.3×10^{-5}	21	4.697	26.10
5	64065	29249	2811.9	8.2×10^{-7}	40	71.10	415.5

Table 2 The standard *p*-FEM.

p	N (size $\mathbb{A}$)	$N-M$ (size S)	memory [$\times 10^3$]	rel. err. [%]	N_{iter}	solver CPU time [s] stat. con.	no conden.
1	113	113	2.3	$1.0\times10^{+1}$	8	—	0.007
2	481	481	9.2	5.1×10^{-1}	10	—	0.049
3	1105	849	25.6	1.7×10^{-2}	12	0.102	0.105
4	1985	1217	57.6	4.1×10^{-4}	13	0.171	0.350
5	3121	1585	112.9	7.7×10^{-6}	14	0.302	0.987
6	4513	1953	200.7	1.3×10^{-7}	14	0.504	2.333
7	6161	2321	331.8	1.7×10^{-9}	15	0.808	4.933
8	8065	2689	518.4	2.3×10^{-11}	16	1.218	9.582
9	10225	3057	774.4	1.1×10^{-11}	16	1.773	17.407

Table 3 The *p*-FEM with fixed number of DOFs.

p	N (size $\mathbb{A}$)	$N-M$ (size S)	memory [$\times 10^3$]	rel. err. [%]	N_{iter}	solver CPU time [s] stat. con.	no conden.
1	28561	28561	518	6.9×10^{-1}	82	—	16.9
2	28561	28561	518	9.3×10^{-3}	83	—	28.1
3	28561	22161	640	1.3×10^{-4}	53	26.7	44.4
4	28561	17761	810	2.1×10^{-6}	41	21.2	59.8
5	28561	14737	1 016	3.2×10^{-8}	34	17.6	74.7
6	28561	12561	1 254	5.3×10^{-10}	29	15.0	89.4
7	28085	11221	1 498	8.9×10^{-12}	26	12.8	101.2
8	28561	9661	1 823	4.2×10^{-12}	24	11.7	118.5
9	27145	9663	2 045	1.3×10^{-11}	22	9.6	121.4

Table 4 The *p*-FEM with fixed memory requirements.

p	N (size $\mathbb{A}$)	$N-M$ (size S)	memory [$\times 10^3$]	rel. err. [%]	N_{iter}	solver CPU time [s] stat. con.	no conden.
1	28561	28561	518	6.9×10^{-1}	65	—	15.0
2	28561	28561	518	9.3×10^{-3}	62	—	26.8
3	23113	17929	518	1.8×10^{-4}	40	17.8	28.5
4	18241	11329	518	5.1×10^{-6}	28	8.8	23.9
5	14281	7345	510	1.8×10^{-7}	22	4.5	18.5
6	12013	5253	530	7.1×10^{-9}	19	2.8	15.6
7	9661	3661	518	3.6×10^{-10}	17	1.7	12.0
8	8065	2689	518	2.3×10^{-11}	16	1.2	9.6
9	7813	2325	593	1.2×10^{-11}	15	1.1	10.3

4 Conclusions

The presented experiments show that the static condensation of the internal DOFs can lead to a considerable speed-up of the solver. Asymptotically, however, if the polynomial degrees tend to the infinity and the number of elements stays fixed, then the algorithm of the static condensation is close to the computation of the inverse of the (almost) fully populated matrix, which is not efficient. On the other hand, high polynomial degrees are rare in practical computations.

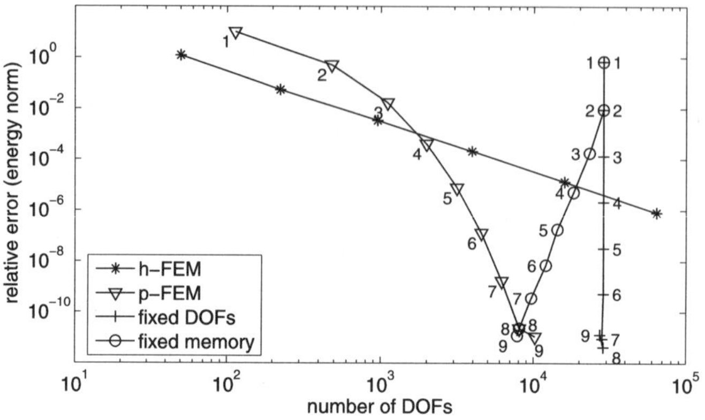

Fig. 3 The error plot in the log-log scale. The numbers indicate the polynomial degrees.

Finally we mention that more elaborate preconditioners than ILU are available for higher-order FEM, see, e.g., [1], where almost optimal preconditioners for the p-FEM are derived. However, even these preconditioners can be implemented either with or without the static condensation. For these preconditioners the static condensation would lead to the same speed-up per iteration as for the ILU preconditioner.

Acknowledgements The author thanks an anonymous referee for pointing out reference [6]. This research has been supproted by the Grant Agency of the Czech Academy of Sciences, project No. IAA100760702, and by the Czech Academy of Sciences, institutional research plan No. AV0Z10190503. This support is gratefully acknowledged.

References

1. Babuška, I., Craig, A., Mandel, J., Pitkäranta, J.: Efficient preconditioning for the p-version finite element method in two dimensions. SIAM J. Numer. Anal. **28**(3), 624–661 (1991)
2. Demkowicz, L.: Computing with hp-adaptive finite elements. Vol. 1. One and two dimensional elliptic and Maxwell problems. Chapman & Hall/CRC, Boca Raton, FL (2007)
3. Mandel, J.: On block diagonal and Schur complement preconditioning. Numer. Math. **58**(1), 79–93 (1990)
4. Melenk, J.: hp-finite element methods for singular perturbations. Springer-Verlag, Berlin (2002)
5. Schwab, C.: p- and hp-finite element methods. Theory and applications in solid and fluid mechanics. The Clarendon Press, Oxford University Press, New York (1998)
6. Smith, B.F., Bjørstad, P.E., Gropp, W.D.: Domain decomposition. Cambridge University Press, Cambridge (1996)
7. Vejchodský, T., Šolín, P.: Static condensation, partial orthogonalization of basis functions, and ILU preconditioning in the hp-FEM. J. Comput. Appl. Math. (2007). DOI 10.1016/j.cam.2007.04.044
8. Šolín, P., Segeth, K., Doležel, I.: Higher-order finite element methods. Chapman & Hall/CRC, Boca Raton, FL (2004)

Convection, Diffusion, Conservation, and Hyperbolic Systems

Coupling Two Scalar Conservation Laws via Dafermos' Self-Similar Regularization

A. Ambroso, B. Boutin, F. Coquel, E. Godlewski, and P.G. LeFloch

Abstract We are interested in the problem of coupling two scalar conservation laws with distinct flux-functions. This problem arises, for instance, in modeling fluid flows in media with discontinuous porosity and has important possible applications in the numerical computation of a singular pressure drop. This problem is also well-known to exhibit several technical difficulties due to the presence of non-conservative terms and to the resonant behavior of the system of equations. We present here a global approach consisting of two scalar problems in a half-space coupled through an algebraic jump relation. We view this problem as a 2×2 system of conservation laws, and introduce a viscous regularization *à la Dafermos*. We establish that this approximation converges as the viscosity tends to zero and we analyze the structure of the entropy solutions constructed in this way.

1 Introduction

We are interested in the coupling of two scalar conservation laws with distinct flux-functions, each one being posed on a half-space:

A. Ambroso
DEN/DANS/DM2S/SFME/LETR CEA-Saclay, 91191 Gif-sur-Yvette, France,
e-mail: annalisa.ambroso@cea.fr

B. Boutin
DEN/DANS/DM2S/SFME/LETR CEA-Saclay, 91191 Gif-sur-Yvette, France,
and Laboratoire Jacques-Louis Lions & Centre National de la Recherche Scientifique, Université Pierre et Marie Curie-Paris6, UMR7598, 75252 Paris, France,
e-mail: Benjamin.Boutin@cea.fr

F. Coquel, E. Godlewski, P.G. LeFloch
Laboratoire Jacques-Louis Lions & Centre National de la Recherche Scientifique, Université Pierre et Marie Curie-Paris6, UMR7598, 75252 Paris, France,
e-mail: coquel@ann.jussieu.fr, godlewski@ann.jussieu.fr, lefloch@ann.jussieu.fr

$$\begin{aligned} \partial_t u + \partial_x f_L(u) &= 0, \quad x < 0, \quad t > 0, \\ \partial_t u + \partial_x f_R(u) &= 0, \quad x > 0, \quad t > 0, \end{aligned} \tag{1}$$

were f_L, f_R are smooth functions. These equations are supplemented with the following coupling condition at the (fixed) interface $x = 0$:

$$\Phi_L\big(u(0^-,t)\big) = \Phi_R\big(u(0^+,t)\big), \quad t > 0 \tag{2}$$

and the following initial Riemann data:

$$u(x,0) = \begin{cases} u_l, \ x < 0, \\ u_r, \ x > 0, \end{cases} \tag{3}$$

where Φ_L, Φ_R are given smooth functions, and u_L, u_R are constant states. These functions allows to treat different coupling conditions, for example with $\Phi_L = f_L$ and $\Phi_R = f_R$ one expects the continuity of the flux at the interface and a conservative coupling, and with $\Phi_{L,R} = \mathrm{Id}$ one expects the continuity of the variable u at the interface, then the coupling is not conservative. For convenience in this presentation, we restrict attention to this case $\Phi_L(u) = \Phi_R(u) = u$ for all u.

The main difficulty in tackling the problem (1–3) is that a resonance phenomenon can occur near the interface, so that (2) cannot be realized in a strong sense. Following Dubois and LeFloch [5] on the weak formulation of the boundary value problem for nonlinear hyperbolic equations and systems, one may consider the problem obtained by sticking together two half-space boundary Riemann problems. This approach has been investigated by Godlewski and Raviart [6] and, more recently, Ambroso *et al.* [1]. Our standpoint in the present paper is different and we construct the entropy solutions directly in the whole space, using the self-similar viscosity method proposed by Dafermos [3]. We follow here earlier work by Joseph and LeFloch [7] on the boundary Riemann problem via self-similar approximation with vanishing viscosity. In this short note, we only present the main ideas in the simpler case of two scalar equations; for further details and a generalization to systems, we refer to a follow-up paper of Boutin, Coquel, and LeFloch [2].

1.1 Global Model

We begin by introducing an (auxilliary) unknown function v, say the Heaviside step function, which allows us to single out the half-space under consideration. A regularization of v should be introduced in order to connect continuously the two equations in the problem, that is, a flux-function extending continuously the given flux-functions f_L and f_R should be introduced. We rewrite the two partial differential equations in the half space (1) into a single system of PDE's on the real line $\mathbb{R}_x$

$$\begin{aligned} &\partial_t u + \big((1-v)f_L'(u) + v f_R'(u)\big)\partial_x u = 0, \\ &\partial_t v = 0, \end{aligned} \tag{4}$$

with Riemann initial conditions

$$u(x,0) = \begin{cases} u_l, & x < 0, \\ u_r, & x > 0, \end{cases} \qquad v(x,0) = \begin{cases} 0, & x < 0, \\ 1, & x > 0. \end{cases} \tag{5}$$

Note that this system contains nonconservative products which are defined rigorously in the sense of Dal Maso, LeFloch, and Murat [4].

1.2 Dafermos' Self-Similar Regularization

To handle the nonconservative products and the resonance phenomena in the regime where system (4) is not strictly hyperbolic, we introduce the following viscous approximation

$$\begin{aligned} \partial_t u_\varepsilon + \lambda(u_\varepsilon, v_\varepsilon)\partial_x u_\varepsilon &= t\varepsilon\ \partial_{xx} u_\varepsilon, \\ \partial_t v_\varepsilon &= t\varepsilon^2 \partial_{xx} v_\varepsilon, \end{aligned} \tag{6}$$

where we have set

$$\lambda(u,v) := (1-v) f_L'(u) + v f_R'(u). \tag{7}$$

The resonance happens precisely when the characteristic speed of the first equation vanishes.

The choice of different viscosity scales for u_ε and v_ε in (6) allows some interaction between two features of the problem: the viscous approximation of solutions and the regularization of the flux-functions. The factor $t\varepsilon$ factor proposed in [3] enables us to look for self-similar solutions depending only on the ratio $\xi = x/t$. Such solutions satisfy the following nonlinear system of ODE's:

$$\begin{aligned} \big(-\xi + \lambda(u_\varepsilon, v_\varepsilon)\big) d_\xi u_\varepsilon &= \varepsilon\ d_{\xi\xi} u_\varepsilon, \\ -\xi d_\xi v_\varepsilon &= \varepsilon^2\ d_{\xi\xi} v_\varepsilon. \end{aligned} \tag{8}$$

Thanks to the property of finite speed of propagation (for hyperbolic equations), we can restrict attention to a bounded domain $[-L, L]$ and, therefore, we consider the boundary conditions

$$\begin{aligned} u_\varepsilon(-L) &= u_l, & v_\varepsilon(-L) &= 0, \\ u_\varepsilon(+L) &= u_r, & v_\varepsilon(+L) &= 1. \end{aligned} \tag{9}$$

Our investigation is now focused on finding the solution to (8)-(9) and analyzing its structural properties, especially when the viscosity ε vanishes.

2 Theoretical Results

A standard fixed point argument allows us to obtain existence for all $\varepsilon > 0$. The solution u_ε to the nonlinear problem (8)–(9) is given implicitly by the following representation formula

$$u_\varepsilon(\xi) = u_L + (u_R - u_L)\frac{\displaystyle\int_{-L}^{\xi} e^{-h_\varepsilon(\zeta)/\varepsilon}\,d\zeta}{\displaystyle\int_{-L}^{L} e^{-h_\varepsilon(\zeta)/\varepsilon}\,d\zeta}, \tag{10}$$

$$h_\varepsilon(\zeta) = \int_\alpha^\zeta (\omega - \lambda(u_\varepsilon(\omega), v_\varepsilon(\omega)))\,d\omega. \tag{11}$$

Theorem 1. *The solutions u_ε converge strongly in the L^1 norm to a limit function u (as ε tends to 0) which satisfies the following properties:*

- *u is an entropy solution in each half space for the Riemann problems (1).*
- *There exists a boundary layer profile U having the same monotonicity as u and satisfying over $\mathbb{R}$*

$$\ddot{U} = \lambda(U,V)\dot{U}, \tag{12}$$

such that at the interface

$$\begin{array}{ll} f_L(u(0^-)) = f_L(U(-\infty)), & q_L(u(0^-)) \geq q_L(U(-\infty)), \\ f_R(U(+\infty)) = f_R(u(0^+)), & q_R(U(+\infty)) \geq q_R(u(0^+)), \end{array} \tag{13}$$

where q_L and q_R are entropy fluxes associated with any convex entropy, and relative to f_L and f_R respectively.

In other words, the "macroscopic" stationary discontinuities $(u(0^-), u(0^+))$ at the interface are associated with a stationary shock on the left-hand side for the (left) Riemann problem beetween $(u(0^-)$ and $U(-\infty))$, and a non-constant "microscopic" boundary layer U connecting $U(-\infty)$ to $U(+\infty)$ and/or a stationary shock on the right for the (right) Riemann problem beetween $(U(+\infty)$ and $u(0^+))$. As expected, this result does not provide uniqueness in the resonant cases, as the example in the following section shows.

3 Example : Burgers-Burgers Coupling

As a model case, consider now the case of two quadratic flux functions with different sonic points, say 0 and c, respectively:

$$f_L(u) = \frac{u^2}{2}, \quad f_R(u) = \frac{(u-c)^2}{2}.$$

3.1 Theoretical Aspects

In Figure 1 and Figure 2, we plot the structure of all possible solutions for each data (u_L, u_R), for $c > 0$ and for $c < 0$, respectively.

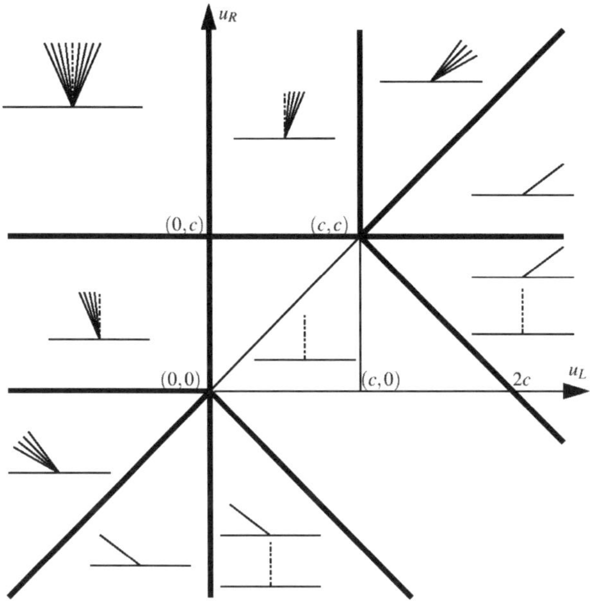

Fig. 1 Diagram of solutions for $c > 0$

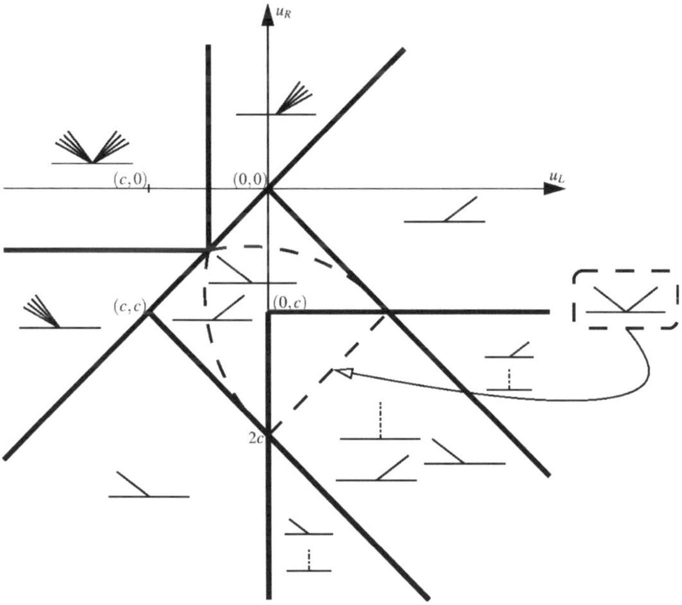

Fig. 2 Diagram of solutions for $c < 0$

These diagrams are obtained by a complete study of all cases covered by Theorem 1. Note that both sonic lines $u_L + u_R = 0$ (for f_L) and $u_L + u_R = 2c$ (for f_R) appear in reversed order according to the sign of c, which explains the complexity of the case $c < 0$. In the resonant domains (when $f_L'(u)f_R'(u) < 0$ for some u beetween u_L and u_R) we do not get uniqueness of the solution since we may obtain up to four distinct solutions. However we observe that the (multivalued) set of solutions is connected in the $L^1_{\rm loc}$ tolopogy.

3.2 Numerical Experiments

We perform some numerics with a scheme based on the Engquist-Osher flux:

$$u_j^{n+1} = u_j^n - \frac{\Delta t^n}{\Delta x}(F_j(u_j^n, u_{j+1}^n) - F_j(u_{j-1}^n, u_j^n)). \tag{14}$$

The time step Δt^n is determined by a standard CFL condition. The numerical fluxes F_j are chosen in a cell of "color" v_j with the relation to the corresponding flux $f_j(u) = (1 - v_j)f_L(u) + v_j f_R(u)$ inside this cell (cf. Figure 3). In this way, each interface involves two different fluxes $F_j(u_j^n, u_{j+1}^n)$ and $F_{j+1}(u_j^n, u_{j+1}^n)$ for example. More precisely, we have

$$F_j(r,s) = \frac{1}{2}\Big(f_j(r) + f_j(s) - \int_r^s |f_j'(w)|\,dw\Big). \tag{15}$$

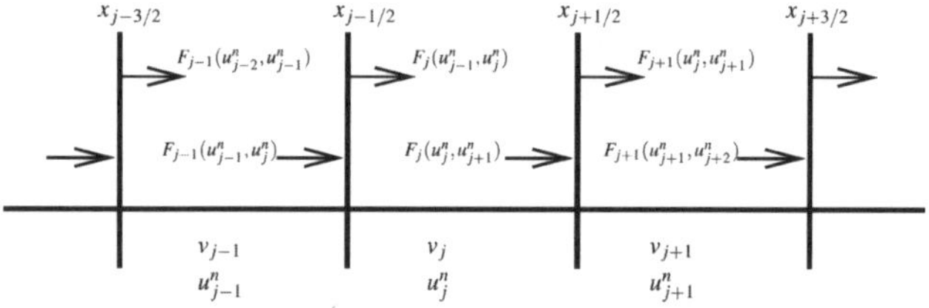

Fig. 3 Two fluxes scheme

The computations are performed with two color variables v: a Heaviside function, and then a regularized one corresponding to a thin and to a thick interface, respectively. In Figure 4, Figure 5 and Figure 6 we plot, in the plane (u_L, u_R) of initial data, the diagram of solutions computed in these two cases for $c > 0$ and for $c < 0$, respectively.

We observe a good agreement with the theoretical results in the non-resonant cases. In the case $c > 0$ the numerical scheme turns out to select the solution that ensures the $L^1_{\rm loc}$ continuity of the solution in terms of (u_L, u_R); this suggests that certain solutions in the non-uniqueness regions may be unstable. However, in the

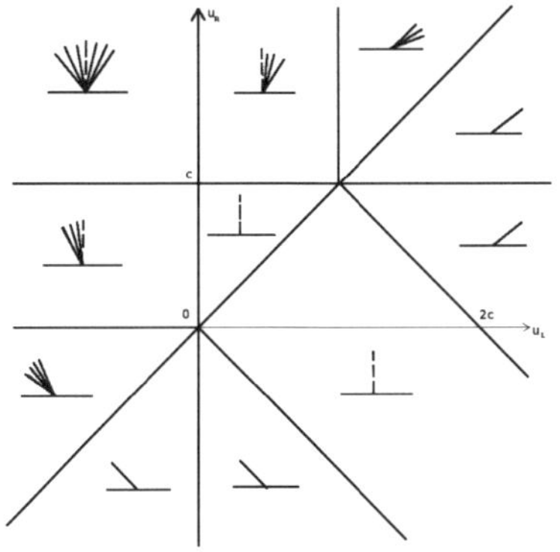

Fig. 4 Numerical solutions, $c > 0$, thin or thick interface

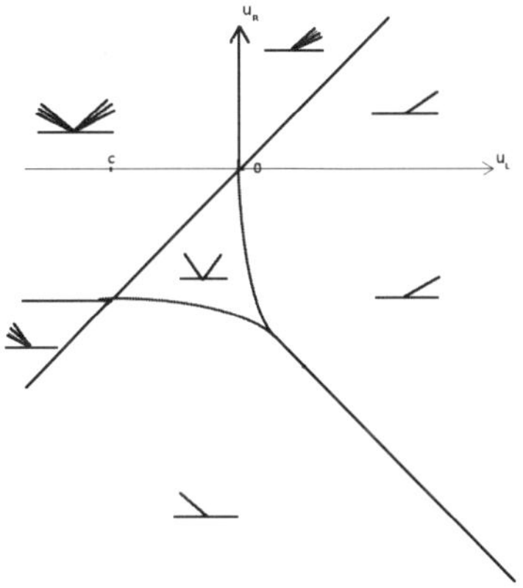

Fig. 5 Numerical solutions, $c < 0$, thin interface

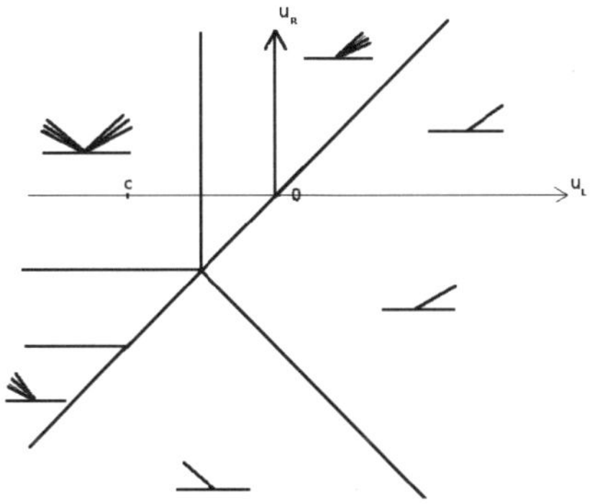

Fig. 6 Numerical solutions, $c < 0$, thick interface

case $c < 0$ it is more difficult to conclude. We observe, as pointed out in [6], that the numerical scheme (due to its own viscosity) has selected some of the expected solutions.

4 Conclusion

We have presented here a new theoretical approach to understand the coupling of conservation laws with distinct flux-functions; this approach is based on the self-similar approximation originally introduced by Dafermos. We have shown that the solutions are standard entropy solutions of some initial problem in each half-space, but are connected by a discontinuity selected by the "microscopic" features taking place in the boundary layer. The numerical experiments allow us to observe, for the considered scheme at least, a selection of certain particular solutions in the resonant situation, due to the viscosity effect of this scheme. The stability of solutions will be addressed in a forthcoming work and could provide an additional argument to select the physically relevant solutions.

Acknowledgements The authors were partially supported by the Centre National de la Recherche Scientifique (CNRS) and the Commissariat à l'Énergie Atomique (CEA - Saclay), thanks to a joint research program on coupling methods for multiphase flows with the Laboratoire J.L. Lions.

References

1. Ambroso, A., Chalons, C., Coquel, F., Godlewski, E., Lagoutière, F., Raviart, P.A., Seguin, N.: Coupling to general Lagrangian systems. Math. Comp. (2007)
2. Boutin, B., Coquel, F., LeFloch, P.G.: Self-similar approximation to the riemann problem for coupled hyperbolic systems. (In preparation)
3. Dafermos, C.M.: Solution of the Riemann problem for a class of hyperbolic systems of conservation laws by the viscosity method. Arch. Ration. Mech. Anal. **52**, 1–9 (1973)
4. Dal Maso G. LeFloch, P.G., Murat, F.: Definition and weak stability of nonconservative products. J. Math. Pures Appl. **74**, 483–548 (1995)
5. Dubois, F., LeFloch, P.G.: Boundary conditions for nonlinear hyperbolic systems of conservation laws. J. Differential Equations **71**, 93–122 (1988)
6. Godlewski, E., Raviart, P.A.: The numerical interface coupling of nonlinear hyperbolic systems of conservation laws, I: The scalar case. Numer. Math. **97**, 81–130 (2004)
7. Joseph, K.T., LeFloch, P.G.: Boundary layers in weak solutions of hyperbolic conservation laws. self-similar vanishing diffusion limits. Comm. Pure Appl. Anal. **1**, 51–76 (2002)

A Sharp Interface and Fully Conservative Scheme for Computing Nonclassical Shocks

B. Boutin, C. Chalons, F. Lagoutière, and P.G. LeFloch

Abstract We present a sharp interface and fully conservative numerical strategy for computing nonclassical solutions of scalar conservation laws. The difficult point is to impose at the discrete level a prescribed kinetic relation along each nonclassical discontinuity. Our method is based on a relevant reconstruction technique operating on each cell which is expected to contain a nonclassical shock. To prove the validity of our approach, we state some important stability properties and numerical tests are proposed. The convergence is also illustrated numerically.

1 Introduction

We are interested in the numerical approximation of *nonclassical* solutions. Our model consists of a scalar conservation law given by

$$\begin{cases} \partial_t u + \partial_x f(u) = 0, \ u(x,t) \in \mathbb{R}, \ (x,t) \in \mathbb{R} \times \mathbb{R}_+ \setminus \{0\}, \\ u(x,0) = u_0(x), \end{cases} \tag{1}$$

B. Boutin
Laboratoire J.L. Lions, Université Pierre et Marie Curie Paris 6, Boîte courrier 187, 75252 Paris Cedex 05, France, and DEN/DANS/DM2S/SFME/LETR CEA-Saclay, 91191 Gif-sur-Yvette, France, e-mail: boutin@ann.jussieu.fr

C. Chalons
Laboratoire J.L. Lions & Université Paris Diderot (Paris 7), Boîte courrier 187, 75252 Paris Cedex 05, France, and DEN/DANS/DM2S/SFME/LETR CEA-Saclay, 91191 Gif-sur-Yvette, France, e-mail: chalons@math.jussieu.fr

F. Lagoutière
Laboratoire J.L. Lions & Université Paris Diderot (Paris 7), Boîte courrier 187, 75252 Paris Cedex 05, France, e-mail: lagoutie@math.jussieu.fr

P.G. LeFloch
Laboratoire J.L. Lions, Centre National de la Recherche Scientifique, Université Pierre et Marie Curie Paris 6, Boîte courrier 187, 75252 Paris Cedex 05, France, e-mail: lefloch@ann.jussieu.fr

and supplemented with the validity of an entropy inequality of the following form

$$\partial_t U(u) + \partial_x F(u) \leq 0. \quad (2)$$

Here t is the time variable, x is the one dimensional space variable, $f : \mathbb{R} \to \mathbb{R}$ is the flux function and (U, F) is a mathematical entropy pair with $U : \mathbb{R} \to \mathbb{R}$ and $F : \mathbb{R} \to \mathbb{R}$. In other words, U is strictly convex and F is such that $F' = U'f'$. Equations (1) and (2) are expected to be valid in the weak sense. Importantly, f is taken to be *nonconvex* in this study.

When f is convex, it is well-known that the entropy condition (2) is sufficient to select a unique *classical* solution of (1). When f fails to be convex, it is necessary to supplement (1)-(2) with an additional selection criterion called *kinetic relation* from [5]. More precisely, in this case the Riemann problem associated with (1)-(2) still admits a one-parameter family of solutions, which may contain shock waves violating Lax shock inequalities: the so-called *nonclassical shocks*. In order to ensure the uniqueness, a *kinetic relation* needs to be added along each nonclassical discontinuity connecting a left state u_- to a right state u_+. It takes the form

$$u_+ = \varphi^\flat(u_-) \text{ or } u_- = \varphi^{-\flat}(u_+) \text{ for each nonclassical shock,} \quad (3)$$

where $\varphi^\flat$ is the so-called *kinetic function* and $\varphi^{-\flat}$ its inverse, which means that the right (respectively left) state is no longer free but depends on the left (respectively right) state. The speed of propagation $\sigma(u_-, u_+)$ of such a discontinuity is given by the Rankine-Hugoniot relation:

$$\sigma(u_-, u_+) = \frac{f(u_+) - f(u_-)}{u_+ - u_-}.$$

For further details on the theory of nonclassical solutions selected by a kinetic relation, we refer the reader to the monograph by LeFloch [5] and to the references cited therein.

The numerical approximation of nonclassical solutions is known to be challenging, see [2]. The main difficulty is the respect of the kinetic relation at the discrete level. In this paper, our objective is to design the first *sharp interface* and *fully conservative* scheme for approximating nonclassical solutions of scalar conservation laws. Our strategy is based on the so-called discontinuous reconstruction schemes proposed by the third author [4], [3]. Aim of these algorithms is to get numerical (classical) shock profiles that are not or little bit diffused. Since a major difficulty when dealing with nonclassical solutions is related to the the small scales and the numerical diffusion (see [5], [1]), these strategies are clearly interesting for our purpose. Our algorithm then follows this approach and is based on a suitable reconstruction strategy involving the kinetic function $\varphi^\flat$. To prove the validity of the method, we show that it enjoys important stability properties like consistency and exact capture of isolated nonclassical discontinuities. Numerical illustrations are also proposed.

2 Numerical Approximation of the Nonclassical Solutions

We put ourselves in the well-known context of finite volume methods. Introducing two constant steps Δx and Δt (for space and time discretizations) and setting $x_{j+1/2} = j\Delta x$, $j \in \mathbb{Z}$, and $t^n = n\Delta t$, $n \in \mathbb{N}$, it consists in computing at each time t^n a piecewise constant approximate solution $x \rightarrow u_{\Delta x,\Delta t}(x,t^n)$ of the exact solution $u(x,t^n)$ on the cell $\mathscr{C}_j = [x_{j-1/2}; x_{j+1/2})$:

$$u_{\Delta x,\Delta t}(x,t^n) = u_j^n \text{ for all } x \in C_j,\ j \in \mathbb{Z},\ n \in \mathbb{N}.$$

At time $t = 0$, the exact solution is given by u_0 and we thus define the sequence $(u_j^0)_{j\in\mathbb{Z}}$:

$$u_j^0 = \frac{1}{\Delta x}\int_{x_{j-1/2}}^{x_{j+1/2}} u_0(x)dx, \text{ for all } j \in \mathbb{Z}.$$

Assuming a given sequence $(u_j^n)_{j\in\mathbb{Z}}$ at time t^n, it is thus a question of defining its evolution towards the next time level t^{n+1}. In the following notations, ν refers to the ratio $\Delta t/\Delta x$. Since we seek for a *conservative* finite volume scheme, we start from the following usual update formula:

$$u_j^{n+1} = u_j^n - \nu(f_{j+1/2}^n - f_{j-1/2}^n), \text{ for all } j \in \mathbb{Z}, \tag{4}$$

where $f_{j+1/2}^n$ has to be defined. The latter represents an approximate value of the flux that passes through the interface $x_{j+1/2}$ between times t^n and t^{n+1}. We assume for simplicity that

$$\text{either } f'(u) \geq 0,\ \forall u \quad \text{or } f'(u) \leq 0,\ \forall u, \tag{5}$$

so that propagation takes place in one direction only.

The main idea is as follows. Usually, u_j^n is seen as an approximate value of the average on the cell $\mathscr{C}_j$ of the exact solution at time t^n. We propose here to understand u_j^n as the projection on constant values of a nonclassical discontinuity which is possibly located in the cell $\mathscr{C}_j$. The left and right states of this discontinuity are noted $u_{j,l}^n$ and $u_{j,r}^n$ (we will assume $u_{j,l}^n \neq u_{j,r}^n$), and $d_j^n \in [0,1]$ is such that $d_j^n\Delta x$ represents the distance between $x_{j-1/2}$ and the reconstructed discontinuity. See Fig. 1 for an illustration. Let us discuss the definition of these quantities. We set $\overline{x}_j = x_{j-1/2} + d_j^n\Delta x$. First of all, one requires the reconstructed discontinuity to satisfy the following conservation property:

$$(\overline{x}_j - x_{j-1/2})u_{j,l}^n + (x_{j+1/2} - \overline{x}_j)u_{j,r}^n = (x_{j+1/2} - x_{j-1/2})u_j^n.$$

It equivalently recasts as

$$\overline{x}_j = x_{j-1/2} + \frac{u_{j,r}^n - u_j^n}{u_{j,r}^n - u_{j,l}^n}\Delta x,$$

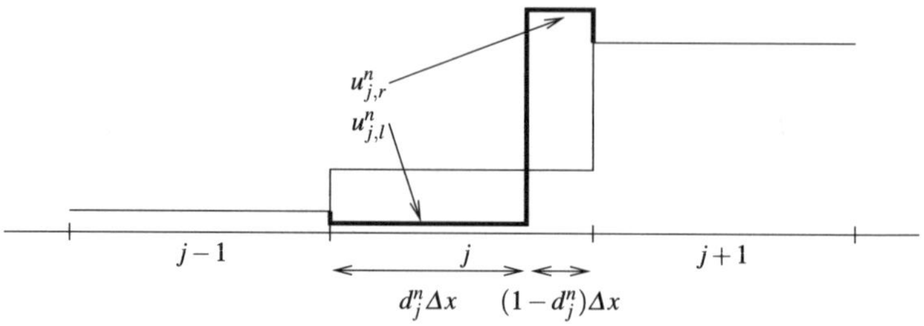

Fig. 1 An example of discontinuous reconstruction

which means that

$$d^n_j = \frac{u^n_{j,r} - u^n_j}{u^n_{j,r} - u^n_{j,l}}.$$

Then, we propose to set

$$u^n_{j,l} = \varphi^{-\flat}(u^n_{j+1}), \quad u^n_{j,r} = \varphi^{\flat}(u^n_{j-1}). \tag{6}$$

This choice is motivated by the particular situation where u^n_{j-1} and u^n_{j+1} can be joined by an admissible nonclassical discontinuity, that is if $u^n_{j+1} = \varphi^{\flat}(u^n_{j-1})$. It is expected in this case that the reconstructed discontinuity coincides with the nonclassical shock. Formulas (6) achieve this goal.

At last, we introduce $\Delta t_{j+1/2}$ (respectively $\Delta t_{j-1/2}$) to be the time needed by the reconstructed discontinuity to reach the interface $x_{j+1/2}$ (resp. $x_{j-1/2}$) if $f'(u) \geq 0\, \forall u$ (resp. $f'(u) \leq 0\, \forall u$). Assume for a moment that $f'(u) \geq 0\, \forall u$. Then, the flux that passes through $x_{j+1/2}$ between times t^n and $t^{n+1} = t^n + \Delta t$ naturally equals $f(u^n_{j,r})$ until $t^n + \Delta t_{j+1/2}$, and $f(u^n_{j,l})$ after (if $\Delta t_{j+1/2} < \Delta t$). Therefore, we naturally set for all $j \in \mathbb{Z}$:
(i) if $f'(u) \geq 0\, \forall u$:

$$\Delta t f^n_{j+1/2} = \begin{cases} \min(\Delta t_{j+1/2}, \Delta t) f(u^n_{j,r}) + \max(\Delta t - \Delta t_{j+1/2}, 0) f(u^n_{j,l}) & \text{if} \quad 0 \leq d^n_j \leq 1, \\ \Delta t f(u^n_j) & \text{otherwise,} \end{cases} \tag{7}$$

with

$$\Delta t_{j+1/2} = \frac{1 - d^n_j}{\sigma(u^n_{j,l}, u^n_{j,r})} \Delta x. \tag{8}$$

Similarly, we set
(ii) if $f'(u) \leq 0\, \forall u$:

$$\Delta t f^n_{j-1/2} = \begin{cases} \min(\Delta t_{j-1/2}, \Delta t) f(u^n_{j,l}) + \max(\Delta t - \Delta t_{j-1/2}, 0) f(u^n_{j,r}) & \text{if} \quad 0 \leq d^n_j \leq 1, \\ \Delta t f(u^n_j) & \text{otherwise,} \end{cases} \tag{9}$$

with

$$\Delta t_{j-1/2} = \frac{d_j^n}{\sigma(u_{j,l}^n, u_{j,r}^n)} \Delta x. \tag{10}$$

Remark 1. The above argumentation does not make sense if d_j^n does not belong to $[0,1]$. In this case, the proposed numerical flux is the usual upwind conservative scheme.

Remark 2. Note that the local time step $\Delta t_{j+1/2}$ (respectively $\Delta t_{j-1/2}$) given by (8) (respectively (10)) is only a prediction of the time needed by the reconstructed discontinuity to reach the interface $x_{j+1/2}$ (respectively $x_{j-1/2}$). This prediction is however exact in the case of an isolated nonclassical discontinuity, that is as soon as u_{j-1}^n and u_{j+1}^n verify $u_{j+1}^n = \varphi^\flat(u_{j-1}^n)$.

We now state some important properties enjoyed by our algorithm. The proof is given in [1].

Theorem 1. *Assume that f is a smooth function satisfying (5). Then, under the CFL restriction*

$$\frac{\Delta t}{\Delta x} \max_u |f'(u)| \leq 1,$$

where the maximum is taken over all the u under consideration, the conservative scheme (4) with $f_{j+1/2}^n$ defined for all $j \in \mathbb{Z}$ by (6)-(7)-(8)–(9)-(10) is consistent with (1)-(2)-(3) in the following sense:

(i) Flux consistency : Assume that $u := u_{j-1}^n = u_j^n = u_{j+1}^n$, then $f_{j+1/2}^n = f(u)$ if $f' \geq 0$ and $f_{j-1/2}^n = f(u)$ if $f' \leq 0$.

(ii) Classical solution (remaining in the region of convexity - or concavity - of f): Let us assume that the sequence $(u_j^n)_j$ belongs to the same region of convexity - or concavity - of f. Then the definition u_j^{n+1} given by the conservative scheme (4)-(7)-(9) coincides with the one given by the usual upwind conservative scheme.

(iii) Isolated nonclassical shock: Let u_l and u_r be two initial states such that $u_r = \varphi^\flat(u_l)$. Assume that $u_j^0 = u_l$ if $j \leq 0$ and $u_j^0 = u_r$ if $j \geq 1$. Then the conservative scheme (4)-(6)-(7)-(8)–(9)-(10)provides an exact numerical solution on each cell $\mathscr{C}_j$ in the sense that

$$u_j^n = \frac{1}{\Delta x} \int_{x_{j-1/2}}^{x_{j+1/2}} u(x,t^n)dx, \ \forall j \in \mathbb{Z}, \ \forall n \in \mathbb{N},$$

where u denotes the exact Riemann solution given by $u(x,t) = u_l$ if $x < \sigma(u_l,u_r)t$ and $u(x,t) = u_r$ otherwise, and is convergent towards u. In particular, the numerical discontinuity is diffused on one cell at most.

Let us briefly comment this result. Property (i) shows that the proposed numerical flux function is consistent in the classical sense of finite volume methods. Properties

(ii) and (iii) are more note-worthy and can be seen as stability/accuracy properties. Indeed, they state that the method is actually *convergent* if the solution either remains in the same region of convexity of f ((ii)), or more importantly consists in an isolated nonclassical discontinuity satisfying the prescribed kinetic relation ((iii)). Up to our knowledge, none of the *conservative* schemes already existing in the literature verifies the latter property. To the authors'mind, this explains the very good numerical results obtained in the next section.

3 Numerical Experiments

This section presents a couple of numerical illustrations. Additional simulations and evaluations of the method can be found in [1].

We consider $f(u) = u^3 + u$ thus f is concave-convex. Concerning the entropy-entropy flux pair (U,F) used in (2), we set

$$U(u) = u^2, \quad F(u) = \frac{3}{2}u^4 + u^2.$$

The kinetic function $\varphi^\flat$ is taken to be

$$\varphi^\flat(u) = -\beta u, \quad \beta = \frac{3}{4}.$$

The first test corresponds to the Riemann problem

$$u_0(x) = \begin{cases} 4, & x < 0, \\ -5, & x > 0, \end{cases}$$

for which the solution is a nonclassical shock followed by a rarefaction wave (see [1]). The computations presented on the left part of Fig. 2 are performed successively with $\Delta x = 0.01$ and $\Delta x = 0.002$. We observe a very good agreement between the exact and numerical solutions. Moreover, we note that the nonclassical shock is sharp, more precisely localised in only one cell. The right part of the figure represents the logarithm of the L1-norm between the exact and the numerical solution versus the logarithm of Δx. The numerical order of convergence is here of about 0.8374. In the second test (Fig. 3), we choose another Riemann initial condition that develops a nonclassical shock followed by a classical shock:

$$u_0(x) = \begin{cases} 4, & x < 0, \\ -2, & x > 0. \end{cases}$$

We have the same observation as previously concerning the nonclassical shock computation that is well captured and arises in a tenuous spatial domain. However note that the classical shock diffuses: in fact our scheme is exactly the Godunov scheme

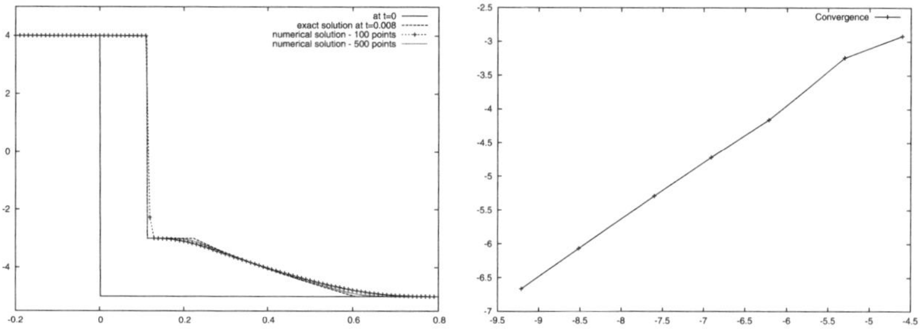

Fig. 2 Test 1 - Nonclassical shock and rarefaction – L1 convergence ($\log(E_{L^1})$ versus $\log(\Delta x)$).

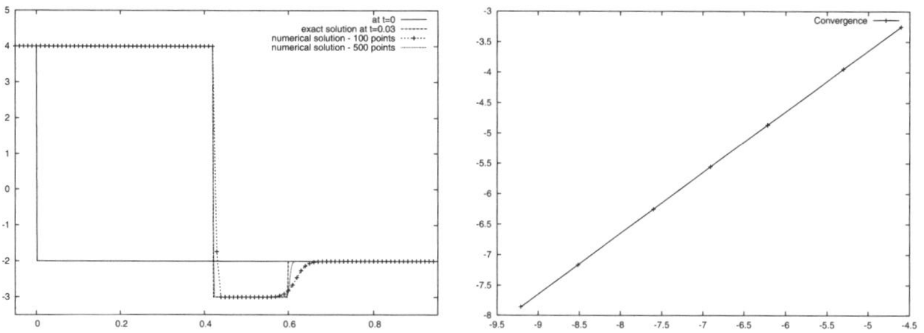

Fig. 3 Test 2 - Nonclassical and classical shocks – L1 convergence ($\log(E_{L^1})$ versus $\log(\Delta x)$).

in domains of same convexity of the flux f. Once again, the L1-error figure ensures numerical convergence with numerical order of about 0.9999.

4 Conclusion

We have presented a new numerical strategy for computing nonclassical solutions of scalar conservation laws. The method is based on a relevant reconstruction technique operating on each cell which is expected to contain a nonclassical shock. Importantly, the whole algorithm remains conservative and exactly propagates any isolated admissible nonclassical discontinuity. The convergence of the method is obtained numerically for several test cases (see also [1]).

To our feeling, the proposed method brings a new idea for the hard task of dealing with nonclassical shocks and kinetic functions. We think its efficiency makes it relevant and it could serve in particular as the basis for future developments and prospects in the field. We may quote among several interesting and open questions:

existence of a total variation bound for the scheme, extension to systems of conservation laws and real applications like phase transitions, and extension to high-order schemes. These points are currently under investigation.

Acknowledgements This work fits into a joint research program on multiphase flows between CEA-Saclay and Laboratoire J.L. Lions.

References

1. Boutin, B., Chalons, C., Lagoutière, F., LeFloch, P.G.: A convergent and conservative scheme for nonclassical solutions based on kinetic relations. Submitted
2. Hayes, B.T., LeFloch, P.G.: Nonclassical shocks and kinetic relations : finite difference schemes. SIAM J. Numer. Anal. **35**, 2169–2194 (1998)
3. Lagoutière, F.: Non-dissipative entropy satisfying discontinuous reconstruction schemes for hyperbolic conservation laws. Submitted
4. Lagoutière, F.: Stability of reconstruction schemes for scalar hyperbolic conservation laws. to appear in Communications in Mathematical Sciences
5. LeFloch, P.G.: Hyperbolic Systems of Conservation Laws: The theory of classical and nonclassical shock waves. E.T.H. Lecture Notes Series, Birkhäuser (2002)

A Numerical Descent Method for an Inverse Problem of a Scalar Conservation Law Modelling Sedimentation

R. Bürger, A. Coronel, and M. Sepúlveda

Abstract This contribution presents a numerical descent method for the identification of parameters in the flux function of a scalar nonlinear conservation law when the solution at a fixed time is known. This problem occurs in a model of batch sedimentation of an ideal suspension. We formulate the identification problem as a minimization problem of a suitable cost function and derive its formal gradient by means of a first-order perturbation of the solution of the direct problem, which yields a linear transport equation with source term and discontinuous coefficients. For the numerical approach, we assume that the direct problem is discretized by the Engquist-Osher scheme and obtain a discrete first order perturbation associated to this scheme. The discrete gradient is used in combination with the conjugate gradient and coordinate descent methods to find numerically the flux parameters.

1 Introduction

There is a large list of authors who have proposed analytical and numerical methods for inverse problems in nonlinear hyperbolic PDEs; see for instance [11, 12, 15] and the references cited in these works. The inverse problem consists in recovering such data from experimental observations that allow to improve the validation of the model by identifying the constitutive laws of the flux. We focus in this paper on the problem of identifying the flux of a scalar conservation law motivated by a model of sedimentation [8, 13].

Raimund Bürger and Mauricio Sepúlveda
Departamento de Ingeniería Matemática, Universidad de Concepción, Casilla 160-C, Concepción, Chile, e-mail: {rburger,mauricio}@ing-mat.udec.cl

Aníbal Coronel
Departamento de Ciencias Básicas, Facultad de Ciencias, Universidad del Bío-Bío, Casilla 447, Campus Fernando May, Chillán, Chile, e-mail: acoronel@roble.fdo-may.ubiobio.cl

From an applicative point of view, batch sedimentation is a classical procedure to separate a suspension of small particles dispersed in a viscous fluid into a concentrated sediment and a clear fluid. This process is used, for example, in mineral processing, food industry, and wastewater treatment. The experimental setup is a vertical column in which the mixture is allowed to settle. For continuous operation, a surface feed at the top and a surface discharge at the bottom may be provided. Under the influence of gravity, the solid particles settle and form a sediment, which is collected at the bottom of the column either batchwise or continuously. We refer to [8, 13] for more complete descriptions and details. Under several constitutive and simplifying assumptions, it turns out that this mixture of two continuous media (fluid and solid flocs) can be described by a single nonlinear hyperbolic equation.

This model is a special case of a more general strongly degenerate parabolic equation. Numerical methods for identification of parameters for the strongly degenerate parabolic equation modelling sedimentation processes can be found in [2, 3, 9], where the authors consider a Lagrangian formulation which provides an associated discrete adjoint state. In the limit and for the continuous case, the well-posedness of the adjoint state is not well known. In the case of a hyperbolic equation, there exists a unique reversible solution of the adjoint state under the assumption of a $W^{1,\infty}$-regularity of the final condition in order to guarantee a one-sided Lipschitz continuity (OSLC) condition and the well-posedness of the solution [4, 5, 12].

Here, we consider a different approach consisting in computing a first-order perturbation associated to the scheme and the hyperbolic equation in order to calculate the gradient of a suitable cost function. The perturbation of the equation gives a linear transport equation with source term and discontinuous coefficients. The discrete gradient is used in combination with the conjugate gradient and coordinate descent methods to numerically find the flux parameters.

The remainder of this paper is organized as follows. In Section 2 we describe the conservation law arising in the sedimentation model, and formulate the inverse problem. Section 3 is devoted to the computation of the gradient for the formal and numerical point of view. Next, in Section 4 we describe the descent algorithm and we give some remarks on the convergence of the method. Finally, a numerical example is computed in Section 5.

2 Description of the Model and the Inverse Problem

We start by describing the direct problem, that is, the underlying model. Batch and continuous sedimentation of ideal suspensions of monosized spheres under the influence of gravity are quite well modeled by Kynch's sedimentation theory. In this framework, the conservation of mass yields the following one-dimensional initial-value boundary-flux problem [8]:

$$\partial_t u + \partial_x f(u,t) = 0, \qquad u(x,0) = u_0(x), \qquad f(u,t)\big|_{x=x_\ell} = \Gamma_\ell(t), \tag{1}$$

where $(x,t) \in Q_T := I \times \mathscr{T} = (0,1) \times (0,T)$, t is the time, x is the spatial coordinate, T is the final time, u is the unknown function (i.e., the solids concentration), $f(u,t) = q(t)u + b(u)$ is the total flux with a control function q and a material-dependent flux function b, Γ_0 and Γ_1 are two functions describing the discharge and feed rate.

The basic assumptions of the kinematic sedimentation model by Kynch [8, 13] on the coefficients of the model and on the initial and boundary data are the following. The function $q(t) \leq 0$ is the volume average velocity of the suspension, while $b(u)$ is a continuous, piecewise smooth function satisfying:

$$b(u) < 0 \text{ for } u \in (0, u_{\max}) \text{ and } b(u) = 0 \text{ for } u \in \mathbb{R} - (0, u_{\max})$$

where $u_{\max} \in (0,1]$ is the maximum concentration value. The function $b(u)$ models the concentration-dependent hindrance of the settling of solid particle due to the presence of other particles. The initial condition $u_0(x)$ is piecewise continuous function such that $0 \leq u_0(x) \leq u_{\max}$. Finally, Γ_ℓ are the boundary fluxes with $\Gamma_0 = q(t)u(0)$ and $\Gamma_1 = q_1 u(1)$, $q_1 \geq 0$. In particular, we have $\Gamma_0 = \Gamma_1 = 0$ for batch sedimentation. The left hand of Figure 1 displays some typical features of the sedimentation model.

Let us now turn to the description of the inverse problem. In modelling practice, the principal assumption made by Kynch [13], namely that the local settling velocity or solid phase velocity is a function of solids concentration u only, implies that the flux function $f(u,t)$ or the material specific function $b(u)$ should be estimated from constitutive relations and experimental data. We consider as observation the concentration profile with respect to the height of the vessel at fixed time, for instance the end time T, denoted by $u^{\text{obs}}(x)$. The flux determination implies the solution of the following inverse problem: given the functions u_0, q, Γ_0, Γ_1 and the observation data $u^{\text{obs}}(x)$ at the end time T, find f such that $u_f(x,T)$ is as close as to $u^{\text{obs}}(x)$, where $u_f(x,T)$ is the solution profile of (1) for some f. A quite natural formulation of this problem is the following optimization problem:

$$\left.\begin{aligned} &\text{minimize } J\big(u_f(\cdot,T)\big) \text{ with respect to } f, \\ &\text{where } J(u) := \frac{1}{2}\int_I \left|u(x,T) - u^{\text{obs}}(x)\right|^2 dx. \end{aligned}\right\} \tag{2}$$

In cases of practical interest, this general situation is reduced to a parameter identification problem, that is, we do not attempt to identify the functions f or b "freely". Rather, assuming that the constitutive relations depends on a finite vector of parameters $\mathbf{k} \in \mathbb{R}^M$, the flux function $b(u) = b(u;\mathbf{k})$ reflects the specific material properties of the suspension, and the functions $u_0, q, \Gamma_0, \Gamma_1$ are control functions which do not depends on the set of parameters as long as no automatic control is prescribed. Thus, the inverse problem (2), in the case of parameter identification, can be rewritten as the following minimization problem in $\mathbb{R}^M$:

$$\left.\begin{aligned} &\text{minimize } \tilde{J}(\mathbf{k}) \text{ with respect to } \mathbf{k} \in \mathbb{R}^M, \\ &\text{where } \tilde{J}(\mathbf{k}) = J(u_f) \text{ and } f(u,t) = q(t)u + b(u;\mathbf{k}). \end{aligned}\right\} \tag{3}$$

3 Gradient of Cost Function

3.1 Formal Continuous Gradient of Cost Function

Let f^ε a family of first-order perturbation of flux, i.e. admits an expansion of the form $f^\varepsilon = f + \varepsilon\delta f$. We denote by $u^\varepsilon(x,t)$ the solution of (1) with flux f^ε. We assume a suitable regularity conditions such that $u^\varepsilon(x,t) = u(x,t) + \varepsilon w(x,t) + o(\varepsilon)$, where $\lim_{\varepsilon\to 0}\varepsilon^{-1}\|o(\varepsilon)\|_{L^1} = 0$ and $w = \delta u$ denotes the first-order perturbation of u and solves the following linear initial-value boundary-flux problem

$$\partial_t w + \partial_x\big(f_u(u,t)w\big) = -\partial_x\delta f(u,t), \quad \delta u(x,0) = 0, \quad f_u w + \delta f\big|_{x=x_\ell} = 0. \tag{4}$$

Therefore, the derivation of cost function (2) can be (formally) computed as

$$J'(u) := \frac{dJ(u)}{du} = \int_I \big(u(x,T) - u^{\mathrm{obs}}(x)\big) w(x,T)\,dx.$$

In the case of the identification problem (3), the gradient is formally given by

$$\nabla J(u) = \int_I \big(u(x,T) - u^{\mathrm{obs}}(x)\big)\mathbf{v}(x,T)\,dx,$$

where $\mathbf{v} = \nabla_{\mathbf{k}} u$ and v_i is a solution of an IBVP like (4) with $\delta f = \partial_{k_i} f$. This gradient can be either used as a first-order necessary condition for a minimum: $\nabla J(u) = 0$, or to employ a gradient scheme. However, the analytic solution for the direct problem it is not available. Consequently, there is not an obvious way to directly discretize the perturbation equation. Thus, the subsequent strategy is to first discretize and then optimize.

3.2 Discretization of the Direct Problem

For the direct problem a homogeneous discretization of space and time is introduced. Let $J,N \in \mathbb{N}$. We recall the standard notation of finite difference schemes. We denote by $\Delta x := L/J$ and $\Delta t := T/N$, the size of space and time steps and denote by $\lambda := \Delta t/\Delta x$ and $\mu := \Delta t/(\Delta x)^2$ the ratios of these quantities. We define the grid points as $(x_j,t^n) := (j\Delta x, n\Delta x)$ and denote by u_j^n the numerical solution on the finite volume cell $Q_j^n :=]x_{j-1/2}, x_{j+1/2}[\times]t^n, t^{t+1}[$, where $x_{j+1/2} = (x_j + x_{j+1})/2$ for $j = 0,\dots,J-1$ with $x_{-1/2} = x_0$ and $x_{J+1/2} = x_J$. Thus, by the standard arguments of finite volume methods, the following scheme (interior and boundary) can be stated:

$$u_j^{n+1} = u_j^n - \lambda\left(g_{j+1/2}^n - g_{j-1/2}^n\right), \tag{5}$$

where the numerical flux is $g_{j+1/2}^n = q(t^n)u_{j+1}^n + g\left(u_j^n, u_{j+1}^n\right)$ for $j = 1,\dots,J-1$, $g_{-1/2}^n = q(t^n)u_0^n + \Gamma_0(t^n)$ and $g_{J+1/2}^n = \Gamma_1(t^n)$. In this work we employ the well-known Engquist-Osher generalized upwind flux [10]:

$$g(u,v) = f_b^{\mathrm{EO}}(u,v) := b(0) + \int_0^u \max\{b'(s),0\}\,ds + \int_0^v \min\{b'(s),0\}\,ds. \quad (6)$$

3.3 Discrete Gradient and First-Order Perturbation Scheme

We define the discrete cost function by discretizing (2) on the interval I:

$$J_\Delta(u_\Delta) = \frac{\Delta x}{2} \sum_{j=0}^{J} |u_j^N - u_j^{\mathrm{obs}}|^2, \quad (7)$$

where u_j^N, for $j = 0,\ldots,J$, is the solution of the direct problem obtained with the scheme (5). Introducing the notation $\mathbf{v}_j^n = \partial_{\mathbf{k}} u_j^n$ and $\tilde{J}_\Delta(\mathbf{k}) = J_\Delta(u_\Delta(\mathbf{k}))$, and following the lines of discrete gradient computation given in [11], we deduce that:

$$\nabla \tilde{J}_\Delta(\mathbf{k}) = \Delta x \sum_{j=0}^{J} (u_j^N - u_j^{\mathrm{obs}})\mathbf{v}_j^N,$$

where $\mathbf{v}_0^N,\ldots,\mathbf{v}_J^N$ is the solution at time step N of the following first-order perturbation scheme, which is analogous to the continuous perturbation equations (4):

$$\mathbf{v}_j^{n+1} = \mathbf{v}_j^n - \lambda\left(\delta_{\mathbf{k}} g_{j+1/2}^n - \delta_{\mathbf{k}} g_{j-1/2}^n\right), \quad \delta_{\mathbf{k}} g_{-1/2}^n = q(t^n), \quad \delta_{\mathbf{k}} g_{J+1/2}^n = 0. \quad (8)$$

The initial condition is $\mathbf{v}_j^0 = 0$, for $j = 0,\ldots,J$, and the derivative operators $\delta_{\mathbf{k}}$ are computed by the relations

$$\delta_{\mathbf{k}} g_{j+1/2}^n = \quad q(t^n)\mathbf{v}_{j+1}^{n+1} + \partial_1 g(u_j^n, u_{j+1}^n)\mathbf{v}_j^n + \partial_2 g(u_j^n, u_{j+1}^n)\mathbf{v}_{j+1}^n + \partial_{\mathbf{k}} g(u_j^n, u_{j+1}^n).$$

The Engquist-Osher numerical flux is differentiated as $\partial_1 g(u,v) = \max\{b'(u),0\}$ and $\partial_2 g(u,v) = \min\{b'(v),0\}$.

4 The Descent Algorithm

The most efficient methods for solve numerically minimization problems are the gradient methods although they have the disadvantage of local convergence. For instance, the conjugate gradient algorithm is given in the following lines:

Step 0. Choose $\mathbf{k}^0$ and set $i = 0$ and $\mathbf{g}^0 = \mathbf{h}^0 = -\nabla \tilde{J}_\Delta(\mathbf{k}^0)$. If $\nabla \tilde{J}_\Delta(\mathbf{k}^0) = 0$, stop.

Step 1. Find λ_i such that $\tilde{J}_\Delta(\mathbf{k}^i + \lambda_i \mathbf{h}^i) = \inf_{\lambda>0} \tilde{J}_\Delta(\mathbf{k}^i + \lambda \mathbf{h}^i)$.

Step 2. Set $\mathbf{k}^{i+1} = \mathbf{k}^i + \lambda_i \mathbf{h}^i$. If $\nabla \tilde{J}_\Delta(\mathbf{k}^{i+1}) = 0$, then stop.

Step 3. Define $\mathbf{g}^{i+1} = -\nabla \tilde{J}_\Delta(\mathbf{k}^{i+1}), \ \ \mathbf{h}^{i+1} = \mathbf{g}^{i+1} + \gamma_i \mathbf{h}^i, \ \ \gamma_i := \dfrac{(\mathbf{g}^{i+1} - \mathbf{g}^i) \cdot \mathbf{g}^{i+1}}{\mathbf{g}^i \cdot \mathbf{g}^i}.$

Step 4. Set $i = i + 1$ and return to step 1.

The purpose of any gradient algorithm is to build a sequence of vectors $\{\mathbf{k}^i\}_{i \in \mathbb{N}}$ which tends to some minimization point $\mathbf{k}^\Delta$ of $\tilde{J}_\Delta$.

The solution of the direct problem is more sensitive to some parameters. Thus, the gradient of $\tilde{J}_\Delta$ is far greater in some directions than in others. This behavior implies that $\lambda_i = 0$ for some $\mathbf{k}^i$ far from the minimum $\mathbf{k}^\Delta$. In this paper we consider the following modification of linear minimization:

Step 1'. Find λ_i such that

$$\tilde{J}_\Delta(\mathbf{k}^i + \lambda_i \mathbf{h}^i) = \min_{\lambda \in \mathbb{R}} \{ \tilde{J}_\Delta(\mathbf{k}^i + \lambda \mathbf{d}^i) \mid \mathbf{d}^i = h^i_k \mathbf{e}^k \text{ for } k = 1, \dots, M \text{ and } \mathbf{d}^{M+1} = \mathbf{h}^i \},$$

where $\mathbf{e}^k$ for $k = 1, \dots, M$ are the vectors of the canonical basis of $\mathbb{R}^M$.

This kind of relaxed determination of the step comes from the coordinate descent method (see [14]).

Remark. The existing analytical results on the convergence of the method are valid only in the case that t $f(\cdot,t)$ is strictly convex (see [12, 15]). The gradient convergence analysis can be split into four essential questions: the well-posedness of the perturbed scheme, the differentiability of the continuous cost function, the convergence of the numerical scheme for the direct problem and the convergence of the perturbed scheme.

The perturbed equation (4) is a transport equation with source term and with discontinuous coefficients. Thus, there is no general well-posedness theory. The main ingredient of the theory of reversible-dual solutions introduced by Bouchut and James, according with references [4] and [5], is the OSLC condition, which can be violated in the case of a non-convex flux function when shocks interact and form a rarefaction wave. However, in the interesting cases of batch sedimentation, when the initial condition is constant or monotonic, we know that this kind of behavior does not appear since the numerical and exact solutions are monotonically decreasing [6, 7].

The low regularity of the cost function is, in some sense, independent of the convexity condition of the flux and is a consequence of the presence of discontinuities in the solution of the direct problem u_f. Thus, we can apply the ideas given in [12], where the authors leave the rigorous definition of the continuous gradient and consider the subgradient characterization.

Concerning to the convergence of the numerical scheme of the direct problem we can apply the recent results given in [6, 7], where, in the setting of strongly degenerate parabolic equations, the convergence to the unique BV entropy solution of the problem, up to satisfaction of one of the boundary conditions, is done.

The convergence of the perturbed scheme needs a monotonicity property of the coefficients and the continuous OSLC condition. The monotonical behavior of the coefficients are guaranteed by construction, but the OSLC condition is not satisfied in the general non-convex case.

To conclude, we can apply, with some careful analysis, the ideas developed in the case of convex flux function to the inverse problem in the sedimentation case, when the initial solution is constant and monotonic and when the rarefaction waves are not originated from shock interaction.

5 A Numerical Example

We consider the analytic (parametric dependent) form of the flux density function proposed by Barton et al. [1]:

$$b(u) = v_0 u(1-u/u_m)^C + v_1 u^2(u_m - u), \qquad u \in (0, u_{\max}), \qquad u_m = u_{\max} = 0.9.$$

The set of parameters to be identified is $\mathbf{k} = (C, v_0, v_1)$.

The observation profile at $T = 12273$ is generated by a numerical simulation with $\mathbf{k}^{obs} = (5, -1.18\times 10^{-4}, -1.0\times 10^{-5})$ and $J = 1000$ and with the initial condition $u_0(x) = 0.6$ for $x \in [0.8, 1]$ and $u_0(x) = 0$ for $x \in [0, 0.8[$.

The vector of parameters for the initial guess is $\mathbf{k}^0 = (10, -1.0\times 10^{-4}, -1.0\times 10^{-6})$. The identified set of parameters are shown in Table 1 and the profiles on the right hand of Figure 1.

Table 1 Identified parameters

J	C	v_0	v_1	Cost value
100	4.54347155	-1.089×10^{-4}	-0.9901×10^{-5}	0.000381
1000	4.48301583	-1.075×10^{-4}	-0.0276×10^{-5}	0.000375

Acknowledgements We acknowledge support by Conicyt through Fondap in Applied Mathematics, Fondecyt 1070694, Fondecyt 1050728 and Fondecyt 11060400.

References

1. Barton, N.G., Li, C., Spencer, S.J.: Control of a surface of discontinuity in continuous thickeners. J. Austral. Math. Soc. **33**(9), 269–280 (1992)
2. Berres, S., Bürger, R., Coronel, A., Sepúlveda, M.: Numerical identification of parameters for a flocculated suspension from concentration measurements during batch centrifugation. Chem. Eng. J. **111**, 91–103 (2005)

3. Berres, S., Bürger, R., Coronel, A., Sepúlveda, M.: Numerical identification of parameters for a strongly degenerate convection-diffusion problem modelling centrifugation of flocculated suspensions. Appl. Numer. Math. **52**(4), 311–337 (2005)
4. Bouchut, F., James, F.: Équations de transport unidimensionnelles à coefficients discontinus. C. R. Acad. Sci. Paris Sér. I Math. **320**(9), 1097–1102 (1995)
5. Bouchut, F., James, F.: One-dimensional transport equations with discontinuous coefficients. Nonlinear Anal. **32**(7), 891–933 (1998)
6. Bürger, R., Coronel A.and Sepúlveda, M.: On an upwind difference scheme for strongly degenerate parabolic equations modelling the settling of suspensions in centrifuges and non-cylindrical vessels. Appl. Numer. Math. **56**(10), 1397–1417 (2006)
7. Bürger, R., Coronel A.and Sepúlveda, M.: A semi-implicit monotone difference scheme for an initial-boundary value problem of a strongly degenerate parabolic equation modeling sedimentation-consolidation processes. Math. Comp. **75**(253), 91–112 (electronic) (2006)
8. Bustos, M.C., Concha, F., Bürger, R., Tory, E.: Sedimentation and Thickening. Kluwer Academic Publishers, Dordrecht, The Netherlands (1999)
9. Coronel, A., James, F., Sepúlveda, M.: Numerical identification of parameters for a model of sedimentation processes. Inverse Problems **19**(4), 951–972 (2003)
10. Engquist, B., Osher, S.: One-sided difference approximations for nonlinear conservation laws. Math. Comp. **36**(154), 321–351 (1981)
11. James, F., Postel, M.: Numerical gradient methods for flux identification in a system of conservation laws. J. Engrg. Math. **60**(3-4), 293–317 (2008)
12. James, F., Sepúlveda, M.: Convergence results for the flux identification in a scalar conservation law. SIAM J. Control Optim. **37**(3), 869–891 (electronic) (1999)
13. Kynch, G.: A theory of sedimentation. Trans. Faraday Soc. **48**, 166–176 (1952)
14. Luo, Z.Q., Tseng, P.: On the convergence of the coordinate descent method for convex differentiable minimization. J. Optim. Theory Appl. **72**(1), 7–35 (1992)
15. Ulbrich, S.: A sensitivity and adjoint calculus for discontinuous solutions of hyperbolic conservation laws with source terms. SIAM J. Control Optim. **41**(3), 740–797 (2002)

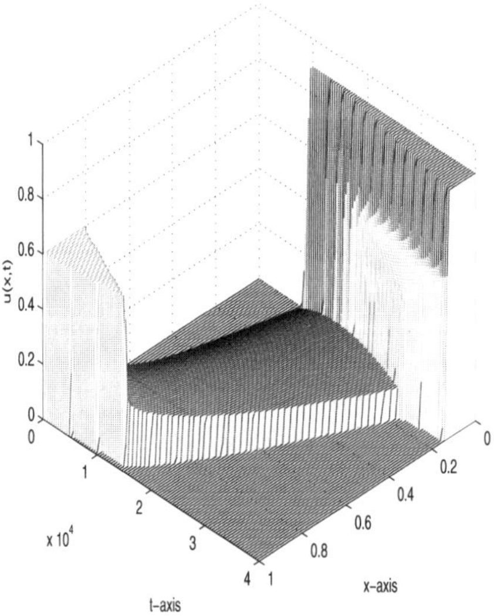

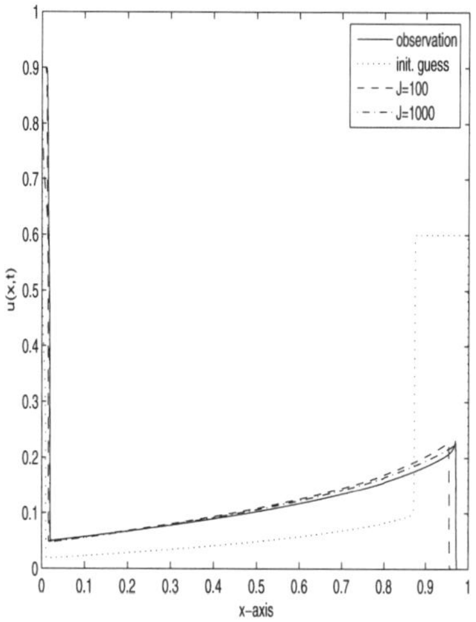

Fig. 1 Numerical solution of (1) with $T = 4 \times 10^4$ and the set of parameters given $\mathbf{k}^{obs}$. Observed and identified profiles

PSI Solution of Convection-Diffusion Equations with Data in L^1

J. Casado-Díaz, T. Chacón Rebollo, V. Girault, M. Gómez Mármol, and F. Murat

Abstract This paper is devoted to the analysis of finite element approximations of convection-diffusion equations with data in L^1. We discretize the convection operator by the PSI (Positive Streamwise Implicit) scheme, and the diffusion operator by the standard Galerkin method, using conforming $\mathbb{P}_1$ finite elements. We give the main idea in the proof of convergence of the approximations to the unique renormalized solution in $W^{1,q}(\Omega)$, $1 \le q < \frac{d}{d-1}$.

1 Introduction

This paper is devoted to the analysis of finite element approximations of convection-diffusion equations with data in L^1. This kind of problem models several phenomena arising in applied sciences. For instance, it is satisfied by the turbulent kinetic energy in turbulence modeling, and by the heat equation in thermoelectric modeling.

We shall study the following problem:

$$\begin{cases} a \cdot \nabla u - \operatorname{div}(A\nabla u) = f & \text{in } \Omega, \\ u = 0 \quad \text{on } \partial\Omega, \end{cases} \tag{1}$$

where Ω is a bounded domain of $\mathbb{R}^d$, where the integer $d \ge 1$ is the dimension, $a \in L^\infty(\Omega)^d$ is a given divergence-free velocity field,

J. Casado-Díaz, T. Chacón Rebollo and M. Gómez Mármol
Dpto. Ecuaciones Diferenciales y Análisis Numérico. Universidad de Sevilla. Spain, e-mail: jcasadod@us.es, chacon@us.es, macarena@us.es

V. Girault and F. Murat
UPMC, Univ Paris 06, UMR 7598, Laboratoire Jacques Louis Lions. F75005, Paris, France, e-mail: girault@ann.jussieu.fr, murat@ann.jussieu.fr

$$\nabla \cdot a = 0, \tag{2}$$

$A \in L^\infty(\Omega)^{d\times d}$ is a uniformly positive definite and bounded matrix field.

$$M|\xi|^2 \geq A(x)\xi\xi \geq \alpha|\xi|^2, \text{ for a.e. } x \in \Omega, \quad \forall \xi \in \mathbb{R}^d, \tag{3}$$

for some $M \geq \alpha > 0$, and f is the source term in $L^1(\Omega)$. In reference [9], it is proved that this problem has a unique *renormalized* solution in $W^{1,q}(\Omega)$, $1 \leq q < \frac{d}{d-1}$. We recall the concept of renormalized solution in Section 2.

In the case of pure diffusion ($a = 0$), the analysis of finite element approximations of problem (1) has been recently carried on in [3] (see also [7]). It is proved that the standard Galerkin piecewise affine finite element approximation converges in $W^{1,q}(\Omega)$, $1 \leq q < \frac{d}{d-1}$ to the renormalized solution for suitable constructions of the triangulations. A discrete version of the Boccardo-Gallouet estimates, cf. [2], is developed. This is essentially based upon the fact that the discrete diffusion matrix is an M-matrix for such triangulations.

Here we treat the case of convection-diffusion equations. This presents new technical difficulties, because we need to find a discretization of the convection term such that the discrete convection-diffusion matrix is an M-matrix. In [4], it is proved that non-linear residual distribution schemes, such as the PSI (Positive Streamwise Implicit) have this property in the standard case $f \in H^{-1}(\Omega)$.

Here, we report the extension of the result of [3] to convection-diffusion equations. We discretize the convection operator by the PSI scheme. We use as technical tools the analysis developed in [3] and [4], and a new comparison result in $W^{1,q}(\Omega)$ for approximate solutions of (1), which is the main technical innovation of this work.

The paper is structured as follows: In Section 2 we recall the definition of renormalized solution of problem (1), and some basic results about its theoretical analysis. In Section 3 we introduce our discretization of problem (1): The convection term is discretized by the PSI scheme, and the diffusion term by the standard Galerkin scheme. We also recall the main result obtained in [3]. We analyze the well-possedness of our discretization in $W^{1,q}(\Omega)$, $1 \leq q < \frac{d}{d-1}$ in Section 4. Finally, in Section 5 we report our main convergence result.

2 Renormalized Solution

Let us start by recalling the definition of the renormalized solution for problem (1):

Definition 1. A function u is a renormalized solution of (1) if it satisfies

- $u \in L^1(\Omega)$
- $\forall k > 0, \quad T_k(u) \in H_0^1(\Omega)$, where $T_k(u) = \begin{cases} u & |u| \leq k, \\ k\frac{u}{|u|} & |u| > k. \end{cases}$
- $\lim\limits_{k\to\infty} \frac{1}{k} \int_{|u|\leq k} |\nabla u|^2 \, dx = 0$

- $\forall S \in \mathscr{C}_c^1(\mathbb{R}) = \{v \in \mathscr{C}^1(\mathbb{R})$ with compact support $\}$, the equation

$$(a \cdot \nabla u)S - (\mathrm{div}(A\nabla u))S = fS$$

is satisfied in the distributions sense.

The above definition of renormalized solutions was introduced by P.L. Lions & F. Murat [9] (see also [5], [10]). Two other definitions of solutions, the entropy solution and the solution obtained as limit of approximations, were introduced at the same time respectively by Bénilan et al. in [1] and by Dall'Aglio in [6]. In the linear case considered in the present work, the three definitions are also equivalent to the solution by transposition introduced in 1969 by Stampacchia.

The main interest of the definition of renormalized solution is the following existence, uniqueness and continuity theorem.

Theorem 1. *Assume that a and A respectively satisfy* (2) *and* (3). *Then there exists a unique renormalized solution of* (1). *Moreover,*

$$u \in W_0^{1,q}(\Omega) \text{ for every } q \text{ with } 1 \leq q < \frac{d}{d-1}.$$

Finally, this unique solution depends continuously on the right-hand side f in the following sense: if f^ε is a sequence which satisfies

$$f^\varepsilon \to f \quad \text{strongly in } L^1(\Omega), \quad \text{as } \varepsilon \to 0,$$

then the sequence u^ε of the renormalized solutions of (1) *for the right-hand sides f^ε satisfies for every q with $1 \leq q < \frac{d}{d-1}$,*

$$\lim_{\varepsilon \to 0} \|u^\varepsilon - u\|_{W_0^{1,q}} = 0,$$

where u is the renormalized solution of (1) *for the right-hand side f.*

The proof of this Theorem can be found in [10].

3 Discrete Problem

To approximate problem (1), let us assume that $\Omega \subset \mathbb{R}^d$ ($d = 2$ or 3) is a polytopic domain. Consider a conforming triangulation $\mathscr{T}_h$ of Ω by triangles if $d = 2$ and tetrahedra if $d = 3$. As usual we assume that h denotes the largest diameter of the elements of $\mathscr{T}_h$. We consider the finite-dimensional space of piecewise affine finite elements built on $\mathscr{T}_h$:

$$V_h = \{v_h \in \mathscr{C}^0(\overline{\Omega}) / \ v_h|_T \in \mathbb{P}_1 \ \forall T \in \mathscr{T}_h, \ v_h = 0 \ \text{ on } \partial\Omega\}. \tag{4}$$

Denote by $\{b_j\}_{j=1}^N$ the nodes of the mesh located on $\overline{\Omega} \setminus \partial\Omega$, i.e. the interior nodes. We consider the nodal basis of V_h, $\{\varphi_j\}_{j=1}^N$ defined by

$$\varphi_i(b_j) = \delta_{ij},\ 1 \le i, j \le N.$$

We also associate with $\mathscr{T}_h$ and an arbitrary function s_h of V_h a discrete space of piecewise constant functions, denoted by $W_h(s_h)$. This space is defined through its *nodal* basis functions $\lambda_1, \lambda_2, \cdots, \lambda_N$ (also depending on s_h), that we assume known for the time being:

$$W_h(s_h) = \text{ span } \{\lambda_1(s_h), \lambda_2(s_h), \ldots, \lambda_N(s_h)\}. \tag{5}$$

The functions λ_j have supports that look for "upwind" information with respect to the velocity field a (See Fig. 1).

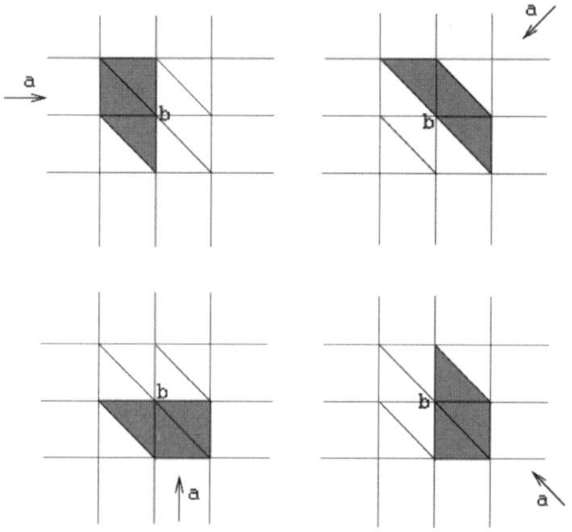

Fig. 1 Typical supports of the basis functions $\lambda_j(s_h)$

The dependence upon s_h is due to the non-linear nature of the PSI method; indeed, the constant values taken by λ_i in each element $T \in \mathscr{T}_h$ depend on s_h: $\lambda_i|_T = \lambda_i|_T(s_h)$.

We construct an associated interpolation operator taking values in W_h, denoted by Π_{s_h}, as follows:

$$\Pi_{s_h} : \mathscr{C}^0(\overline{\Omega}) \cap H_0^1(\Omega) \longrightarrow W_h$$

$$z \longrightarrow \Pi_{s_h} z = \sum_{i=1}^{N} z(b_i)\lambda_i. \tag{6}$$

In particular applying Π_{s_h} to φ_j, $j = 1, \ldots N$, formula (6) gives $\lambda_j = \Pi_{s_h}\varphi_j$. Thus, Π_{s_h} is a bijection from V_h onto W_h.

We shall refer to Π_{s_h} as the Distributed Interpolation Operator generated by s_h. We may characterize each actual Fluctuation Splitting method by its associated Distributed Interpolation Operator, through the definition of the basis functions λ_j. For

PSI and in general for conservative Residual Distribution Schemes, these basis functions satisfy the following property:

For any element $T \in \mathscr{T}_h$,

$$\lambda_{i_T}|_T \geq 0,\ i = 1,\ldots,d+1, \qquad \sum_{i=1}^{d+1} \lambda_{i_T}|_T = 1, \tag{7}$$

where i_T is the global index corresponding to the local index i, $i = 1,\ldots,d+1$ on element T, and $\lambda_{i_T}|_T$ is the restriction of λ_i to element T.

Next, we define the bilinear form $a_h : V_h \times V_h \mapsto \mathbb{R}$ as

$$a_h(u_h, v_h) = \int_\Omega (a \cdot \nabla u_h) \Pi_{u_h} v_h \, dx + \int_\Omega A \nabla u_h \cdot \nabla v_h \, dx. \tag{8}$$

We may now formulate our discrete variational approximation of the convection-diffusion problem (1), as follows:

$$\begin{cases} \text{Find } u_h \in V_h \text{ such that} \\ a_h(u_h, v_h) = \langle f, \Pi_{u_h} v_h \rangle \ \forall v_h \in V_h. \end{cases} \tag{9}$$

The term $\langle f, \Pi_{u_h} v_h \rangle$ is well defined for $f \in L^1(\Omega)$.

Note that problem (9) is a non-linear problem, due to the non-linear nature of the Distributed Interpolation operator.

Consider the convection $C(u_h)$ and diffusion D matrices of the discrete problem (8), with entries

$$C_{ij}(u_h) = \int_\Omega (a \cdot \nabla \varphi_i) \Pi_{u_h} \varphi_j \, dx = \int_\Omega (a \cdot \nabla \varphi_i) \lambda_j(u_h) \, dx \tag{10}$$

and

$$D_{ij} = \int_\Omega A \nabla \varphi_i \nabla \varphi_j \, dx. \tag{11}$$

Also, denote by $Q(u_h)$ the convection-diffusion matrix $C(u_h) + D$.

For the PSI method, $C(s_h)$ is a quasi-M matrix for any $s_h \in V_h$, in the sense that

$$C_{ii}(u_h) \geq 0, \quad C_{ij}(u_h) \leq 0 \quad \text{if} \quad i \neq j.$$

Then, $Q(u_h)$ is an M-matrix whenever D is an M-matrix, which is the case studied in [3]. For general matrices A we do not know conditions that ensure this property. But in the particular case where $A = Id$ (i. e., the diffusion operator is the Laplacian), the next lemma follows from a result proved in [3]:

Lemma 1. *Assume that all angles between sides (when $d=2$) or that all dihedral angles between faces (when $d=3$) of all elements of the grid are bounded by $\pi/2$ degrees. Then*

$$Q(u_h) \text{ is an M-matrix, for any } u_h \in V_h. \tag{12}$$

It is proved in [4] that the PSI scheme is well-balanced up to the second order for the pure convection equation. Hence we use a second-order scheme such as PSI to obtain an M-matrix for $Q(u_h)$. In fact, this property is very close to the maximum principle for discretizing (8). And it is a classical result that no linear scheme which is well-balanced up to the second order accuracy for the pure convection equation can satisfy the maximum principle.

4 Existence of Solutions of the Discrete Problem

In this section we study the existence of solutions of the discrete problem (9). The proof of Theorem 2 is given in [8] and it uses a particular form of Brouwer's Fixed Point Theorem.

Theorem 2. *If $f \in L^p(\Omega)$ with $p \geq 1$ if $d = 2$ and $p = 1$ or $p > 6/5$ if $d = 3$, then problem (9) has at least one solution $u_h \in V_h$ that satisfies the estimate:*

- *For $p > 6/5$ if $d = 3$ and $p > 1$ if $d = 2$*

$$\|u_h\|_{H_0^1} \leq C_1 \|f\|_{L^p}, \tag{13}$$

- *For $p = 1$*

$$\|u_h\|_{H_0^1} \leq C_2(h) \|f\|_{L^1}, \tag{14}$$

where $C_1 > 0$ is a constant that only depend on d, p and Ω and $C_2 > 0$ is a constant that tends to infinity as h goes to 0.

Note that the estimate (14) is not uniform in h, because the unique renormalized solution of the continuous problem does not belong to $H_0^1(\Omega)$: it belongs to $W_0^{1,q}$. The next result gives a sharper a priori estimate for u_h in $W_0^{1,q}$.

Theorem 3. *Assume that a and A respectively satisfy* (2) *and* (3), *the triangulation is regular in the sense of Ciarlet and satisfies the hypothesis of Lemma 1. For every $h > 0$, let u_h be one solution of problem (9), then $\{u_h\}_{h>0}$ is bounded in $W_0^{1,q}$ and there exists a constant $C > 0$ independent of h, such that*

$$\|u_h\|_{W_0^{1,q}} \leq C \|f\|_{L^1}.$$

This result was proved in [3, Theorem 2.1]. The proof is essentially based upon two facts. On one hand, the property that $Q(u_h)$ is an M-matrix (condition (12)) and on the other hand, a piecewise $\mathbb{P}_1$ variant of Boccardo-Gallouet estimates [2].

5 Convergence

We next prove the convergence in $W_0^{1,q}$-norm of each solution of the discrete scheme (9) to the unique renormalized solution of the continuous problem (1).

First, we present a convergence result when the right-hand side $g \in L^p$ with $p > 1$ if $d = 2$ and $p > 6/5$ if $d = 3$.

Theorem 4. *We retain the assumptions of Theorem 3. Let the data g belong to $L^p(\Omega)$ with $p > 1$ if $d = 2$ and $p > 6/5$ if $d = 3$ and let v_h be a solution of:*

$$\int_\Omega (a\cdot\nabla v_h)\Pi_{v_h} w_h\,dx + \int_\Omega A\nabla v_h\cdot\nabla w_h\,dx = \int_\Omega g\Pi_{v_h} w_h\,dx, \qquad \forall w_h \in V_h,$$

and let v be the unique solution of:

$$\int_\Omega (a\cdot\nabla v)w\,dx + \int_\Omega A\nabla v\cdot\nabla w\,dx = \int_\Omega gw\,dx, \qquad \forall w \in H_0^1.$$

Then

$$\lim_{h\to 0}\|v_h - v\|_{H_0^1} = 0.$$

The proof of this result can be found in [4].

Now, we present the original contribution of this work. The following result, comparing two solutions of the discrete problem, allows us to derive convergence in our case.

Theorem 5. *Let $f^\varepsilon = T_{\frac{1}{\varepsilon}} f$ and let u_h be a solution of*

$$\int_\Omega (a\cdot\nabla u_h)\Pi_{u_h} w_h\,dx + \int_\Omega A\nabla u_h\cdot\nabla w_h\,dx = \int_\Omega f\Pi_{u_h} w_h\,dx, \qquad \forall w_h \in V_h,$$

and let u_h^ε be a solution of

$$\int_\Omega (a\cdot\nabla u_h^\varepsilon)\Pi_{u_h^\varepsilon} w_h\,dx + \int_\Omega A\nabla u_h^\varepsilon\cdot\nabla w_h\,dx = \int_\Omega f^\varepsilon\Pi_{u_h^\varepsilon} w_h\,dx, \qquad \forall w_h \in V_h.$$

If the assumptions of Theorem 3 hold, $\{u_h^\varepsilon - u_h\}$ is bounded in $W_0^{1,q}$ and satisfies

$$\|u_h^\varepsilon - u_h\|_{W_0^{1,q}} \le C\left(\|f^\varepsilon - f\|_{L^1} + \frac{h^2}{\varepsilon^2} + h^{2(1+d(1/2-1/q))}\right), \tag{15}$$

where the constant C is independent of ε and h.

This comparison estimate allows to prove our main result,

Theorem 6. *We retain the assumptions of Theorem 3. For every $h > 0$, let u_h be a solution of (9). Then for every q with $1 \le q < \frac{d}{d-1}$,*

$$\lim_{h\to 0}\|u_h - u\|_{W_0^{1,q}} = 0,$$

where u is the unique renormalized solution of (1).

Proof. We retain the notation of Theorem 5. Let u^ε be the unique solution of problem (1) with data f^ε. The dicretization error can be split as follows:

$$\|u-u_h\|_{W_0^{1,q}} \le \|u-u^\varepsilon\|_{W_0^{1,q}} + \|u^\varepsilon - u_h^\varepsilon\|_{W_0^{1,q}} + \|u_h^\varepsilon - u_h\|_{W_0^{1,q}} = I + II + III.$$

Each term (I), (II), (III) converges to 0 when h and ε go to 0. Indeed, the continuity of Theorem 1 implies that (I) tends to zero. Next, by applying Theorem 4 with data f^ε, we see that $\|u^\varepsilon - u_h^\varepsilon\|_{H_0^1}$ converges to 0, when h goes to 0, for any $\varepsilon > 0$. Therefore (II) also tends to zero.

Finally, by choosing $\varepsilon = \varepsilon_0$ small enough in (15), we can make the first term in (15) small and next we can adjust $h_0 = h_0(\varepsilon_0)$ such that for all $h \le h_0$ the two remaining terms in (15) are also small. Therefore the term (III) converges to 0 with h. □

Acknowledgements The research of J. Casado Díaz and F. Murat was partially supported by the Spanish Ministerio de Educación y Ciencia Research Project MTM 2005-04914. The research of T. Chacón Rebollo, V. Girault and M. Gómez Mármol was partially supported by the Spanish Ministerio de Educación y Ciencia MTM Research Project 2006-01275.

References

1. Bénilan, P., Boccardo, L., Gallouet, T., Gariepy, R., Pierre, M., Vázquez, J.L.: An L^1-theory of existence and uniqueness of solutions of nonlinear elliptic equations. Ann. Scuola Norm. Sup. Pisa Cl. Sci. **22**(2), 241–273 (1995)
2. Boccardo, L., Gallouet, T.: Nonlinear elliptic and parabolic equations involving measure data. J. Funct. Anal. **87**, 149–169 (1989)
3. Casado-Díaz, J., Chacón Rebollo, T., Girault, V., Gómez Mármol, M., Murat, F.: Finite elements approximation of second order linear elliptic equations in divergence form with right-hand side in L^1. Numer. Math. **105**(3), 337–374 (2007)
4. Chacón Rebollo, T., Gómez Mármol, M., Narbona Reina, G.: Numerical Analysis of the PSI Solution of Advection-Diffusion Problems through a Petrov-Galerkin Formulation. Math. Models Methods Appl. Sci. **17**(11), 1905–1936 (2007)
5. Dal Maso, G., Murat, F., Orsina, L., Prignet, A.: Renormalized solutions of elliptic equations with general measure data. Ann. Scuola Norm. Sup. Pisa Cl. Sci. **28**(4), 741–808 (1999)
6. Dall'Aglio, A.: Approximated solutions of equations with L^1 data. Application to the H-convergence of quasi-linear parabolic equations. Ann. Mat. Pura Appl. **170**, 207–240 (1996)
7. Gallouet, T., Herbin, R.: Convergence of linear finite elements for diffusion equations with measure data. C.R: Math. Acad. Sci. Paris Sér.I **1**(338), 81–84 (2004)
8. Girault, V., Raviart, P.: Finite Element Methods for Navier-Stokes Equations. Springer-Verlag, Berlin (1986)
9. Lions, P.L., Murat, F.: Solutions renormalisées d'équations elliptic non linéaires. To appear.
10. Murat, F.: Soluciones renormalizadas de edp elípticas no lineales. Publication du Laboratoire d'Analyse Numérique de l'Université Paris VI. (93023) (1993)

High Order Two Dimensional Numerical Schemes for the Coupling of Transport Equations and Shallow Water Equations

M.J. Castro, E.D. Fernández Nieto, A.M. Ferreiro Ferreiro, J.A. García Rodríguez, and C. Parés

Abstract In this article we apply a *second order* 2d scheme for non-conservative hyperbolic systems based on state reconstructions, for the modelization of the *transport* of a pollutant in a fluid. The mathematical model consists in the coupling of a system of shallow water system and a transport equation. That coupling gives rise to a new linearly degenerated field in the system. Therefore, to approximate the evolution of the pollutant, it is necessary to consider numerical methods that can capture accurately those contact discontinuities.

1 Shallow Water Equations with Pollutant Transport

The mathematical model for the pollutant transport problem is obtained coupling a one layer shallow water system with a transport equation. The unknowns are the thickness of the layer of fluid $h(\boldsymbol{x},t)$, the flux $q(\boldsymbol{x},t) = (q_x(\boldsymbol{x},t), q_y(\boldsymbol{x},t))$ and the pollutant concentration $C(\boldsymbol{x},t)$. H is the bottom bathimetry measured from a fixed reference level, $\sigma(\boldsymbol{x},t)$ are the emitting sources (measured in m^2/s) and C_σ is the concentration of the substance at these sources. The obtained system is given by:

M.J. Castro, C. Parés
U. de Málaga, Dpto. de Análisis Matemático, e-mail: {castro,pares}@anamat.cie.uma.es

E.D. Fernández Nieto
U. de Sevilla, Dpto. Matemática Aplicada I, e-mail: edofer@us.es

A.M. Ferreiro Ferreiro, José A. García Rodríguez
U. de A Coruña, Dpto. de Matemáticas, e-mail: {aferreiro, jagrodriguez}@udc.es

$$\begin{cases} \dfrac{\partial h}{\partial t} + \dfrac{\partial q_x}{\partial x} + \dfrac{\partial q_y}{\partial y} = 0, \\ \dfrac{\partial q_x}{\partial t} + \dfrac{\partial}{\partial x}\left(\dfrac{q_x^2}{h} + \dfrac{1}{2}gh^2\right) + \dfrac{\partial}{\partial y}\left(\dfrac{q_x q_y}{h}\right) = gh\dfrac{\partial H}{\partial x}, \\ \dfrac{\partial q_y}{\partial t} + \dfrac{\partial}{\partial x}\left(\dfrac{q_x q_y}{h}\right) + \dfrac{\partial}{\partial y}\left(\dfrac{q_y^2}{h} + \dfrac{1}{2}gh^2\right) = gh\dfrac{\partial H}{\partial y}, \\ \dfrac{\partial hC}{\partial t} + \dfrac{\partial q_x C}{\partial x} + \dfrac{\partial q_y C}{\partial y} = \sigma C_\sigma. \end{cases} \quad (1)$$

The system (1) can be written as a two dimensional non-conservative system (see [5] and [1]),

$$\frac{\partial W}{\partial t} + \mathscr{A}_1(W)\frac{\partial W}{\partial x} + \mathscr{A}_2(W)\frac{\partial W}{\partial y} = 0, \quad (2)$$

where $W(x,t) : \mathscr{O} \times (0,T) \to \Omega \subset \mathbb{R}^N$, $\mathscr{O}$ is a bounded domain of $\mathbb{R}^2$, Ω is a convex subset of $\mathbb{R}^N$, $\mathscr{A}_i : \Omega \to M_{N\times N}$ are regular and locally bounded functions.

Given an unitary vector $\boldsymbol{\eta} = (\eta_x, \eta_y) \in \mathbb{R}^2$ we define the matrix: $\mathscr{A}(W,\boldsymbol{\eta}) = \mathscr{A}_1(W)\eta_x + \mathscr{A}_2(W)\eta_y$. We suppose that the system (2) is hyperbolic, that is, for all $W \in \Omega \subset \mathbb{R}^N$ and $\forall\, \boldsymbol{\eta} \in \mathbb{R}^2$, the matrix $\mathscr{A}(W,\boldsymbol{\eta})$ has N real eigenvalues: $\lambda_1(W,\boldsymbol{\eta}) \leq \ldots \leq \lambda_N(W,\boldsymbol{\eta})$, being $R_j(W,\boldsymbol{\eta})$, $j = 1,\ldots,N$ the associated eigenvectors; and the matrix $\mathscr{A}(W,\boldsymbol{\eta})$ is diagonalizable: $\mathscr{A}(W,\boldsymbol{\eta}) = \mathscr{K}(W,\boldsymbol{\eta})\mathscr{L}(W,\boldsymbol{\eta})\mathscr{K}^{-1}(W,\boldsymbol{\eta})$, where $\mathscr{L}(W,\boldsymbol{\eta})$ is the diagonal matrix which coefficients are the eigenvalues of $\mathscr{A}(W,\boldsymbol{\eta})$ and $\mathscr{K}(W,\boldsymbol{\eta})$ is the matrix whose columns are the eigenvectors $R_j(W,\boldsymbol{\eta})$, $j = 1,\ldots,N$. In this case the eigenvalues of the system are $\lambda_1 = \boldsymbol{u}\cdot\boldsymbol{\eta} - \sqrt{gh}\|\boldsymbol{\eta}\|$, $\lambda_2 = \boldsymbol{u}\cdot\boldsymbol{\eta}$, $\lambda_3 = \boldsymbol{u}\cdot\boldsymbol{\eta}$, $\lambda_4 = \boldsymbol{u}\cdot\boldsymbol{\eta} + \sqrt{gh}\|\boldsymbol{\eta}\|$. Therefore, the coupled system (1) has two linearly degenerated fields. Note that if initially the substance only occupies a portion of the fluid, the boundaries of this portion will propagate following a contact discontinuity (see [6]). Therefore, to approximate with precision the spill evolution it is necessary to use numerical schemes that are able to accurately capture these contact discontinuities (see [5]).

2 High Order Finite Volume Methods Based on State Reconstructions

To discretize the system (2), the computational domain is split into subsets of simple geometry called cells or control volumes, $V_i \subset \mathbb{R}^2$, that will be supposed to be closed polygons. We use the following notation: given such a finite volume V_i, N_i represents its center, $\mathscr{N}_i$ is the set of all the indexes j such that V_j is a neighbor of V_i, E_{ij} is the common edge between two neighbor cells V_i and V_j, being $|E_{ij}|$ its length, $\boldsymbol{\eta}_{ij} = (\eta_{ij,x}, \eta_{ij,y})$ is the unitary outward normal vector to the edge E_{ij} pointing towards cell V_j (see Figure 1). V_{ij} is the triangle defined by the center of the cell, N_i, and the edge E_{ij}. $|V_j|$ and $|V_{ij}|$ represent the areas of V_i and V_{ij}, respectively.

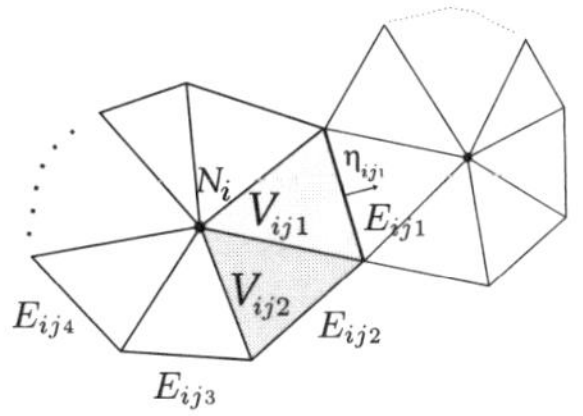

Fig. 1 General finite volume.

The discretization of the system (2) is made using a finite volume scheme. If $W(\boldsymbol{x},t)$ is the exact solution we denote by $\overline{W}_i^n$ the average of the solution at time step t^n,

$$\overline{W}_i^n = \frac{1}{|V_i|}\int_{V_i} W(\boldsymbol{x},t^n)d\boldsymbol{x},$$

W_i^n denotes the approximation of $\overline{W}_i^n$ in t^n, that is, $W_i^n \simeq \overline{W}_i^n$.

Given a volume V_i we denote by P_i the reconstruction operator over the volume. More precisely, P_i depends of a family of values $\{W_j\}_{j\in\mathscr{B}_i}$, where $\mathscr{B}_i$ is a set of indexes of neighbors or control volumes close to V_i (see [1]). When the values of this succession depend on time, we denote the state reconstruction operator by P_i^t.

For a given vector $\boldsymbol{\eta}_{ij}$ pointing towards cell V_j, we denote by $W_{ij}^-(t,\boldsymbol{s})$ and $W_{ij}^+(t,\boldsymbol{s})$ $\forall \boldsymbol{s}\in E_{ij}$, the limit of P_i^t (P_j^t respectively) when $\boldsymbol{x}$ tends to $\boldsymbol{s}$, $\boldsymbol{x}$ inside V_i (inside V_j respectively):

$$\lim_{\substack{\boldsymbol{x}\to\boldsymbol{s}\\ \boldsymbol{x}\cdot\boldsymbol{\eta}_{ij}<k_{ij}}} P_i^t(\boldsymbol{x}) = W_{ij}^-(t,\boldsymbol{s}), \qquad \lim_{\substack{\boldsymbol{x}\to\boldsymbol{s}\\ \boldsymbol{x}\cdot\boldsymbol{\eta}_{ij}>k_{ij}}} P_j^t(\boldsymbol{x}) = W_{ij}^+(t,\boldsymbol{s}). \tag{3}$$

We propose the following high order numerical scheme for non-conservative systems which is the natural extension to 2D domains of the high order numerical introduced in [2]. For more details about its derivation see [1]:

$$\begin{aligned} W_i'(t) = &-\frac{1}{|V_i|}\sum_{j\in\mathscr{N}_i}\int_{E_{ij}} \mathscr{A}_{ij}^-(t,\boldsymbol{s})(W_{ij}^+(t,\boldsymbol{s})-W_{ij}^-(t,\boldsymbol{s}))d\boldsymbol{s} \\ &-\int_{V_i}\left(\mathscr{A}_1(P_i^t(\boldsymbol{x}))\frac{\partial P_i^t}{\partial x}(\boldsymbol{x})+\mathscr{A}_2(P_i^t(\boldsymbol{x}))\frac{\partial P_i^t}{\partial y}(\boldsymbol{x})\right)d\boldsymbol{x}, \end{aligned} \tag{4}$$

where $\mathscr{A}_{ij}(t,\boldsymbol{s}) = \mathscr{A}_{ij}(W_{ij}^+(t,\boldsymbol{s}),W_{ij}^-(t,\boldsymbol{s}),\boldsymbol{\eta}_{ij})$, $\mathscr{A}_{ij}^-(t,\boldsymbol{s}) = \frac{1}{2}(\mathscr{A}_{ij}(t,\boldsymbol{s})-|\mathscr{A}_{ij}(t,\boldsymbol{s})|)$, being $\mathscr{A}_{ij}(W_{ij}^+(t,\boldsymbol{s}),W_{ij}^-(t,\boldsymbol{s}),\boldsymbol{\eta}_{ij})$ the Roe matrix associated to the 1D non-conservative problem over the edge E_{ij} and the states $W_{ij}^+(t,\boldsymbol{s})$ and $W_{ij}^-(t,\boldsymbol{s})$. The non-conservative products makes difficult the definition of weak solution for this kind of systems. Following the theory developed in [3] it is possible to define a non-conservative product as a Borel measure, depending on the selection of a family of paths in the phase space. For this case, we must also chose a family of paths to define the Roe matrix (see [5]). In this work the family of segments has been chosen.

Finally, the integral terms in (4) should be approximated by quadrature formulae. They are chosen in terms of the reconstruction operator used.

In [1] the order of accuracy of the previous numerical scheme has been studied as well as its well-balance properties. In particular, when it is applied to shallow-water systems it exactly preserves the solutions corresponding to water at rest.

2.1 Second Order Reconstruction Operator over 2D Unstructured Meshes

In this section, we propose a MUSCL-type second order reconstruction operator for edge-based finite volume unstructured meshes. The edge type volume can be written as $V_i = V_{i1} \cup V_{i2} \cup V_{i3} \cup V_{i4}$, where V_{ik}, $k = 1,2,3,4$, are triangles (see Figure 2) defined by $\mathbf{C}_i$ (middle point of the edge over which the finite volume of edge type is built) and the four edges of the control volume. By $b_{i,k}$, $k = 1,2,3,4$, we denote the barycenter of this triangles, respectively.

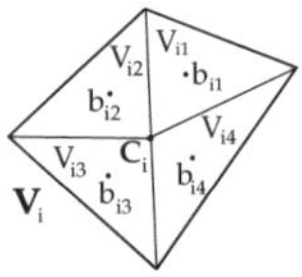

Fig. 2 Triangles that compose a finite volume of edge type, V_i.

We consider the following MUSCL-type state reconstruction operator $P_i(\boldsymbol{x})$ of order two (see [4] for details), defined by

$$P_i(\boldsymbol{x}) = \overline{W}_i + p(\boldsymbol{x}), \quad \text{with} \quad p(\boldsymbol{x}) = \nabla W_i(\boldsymbol{x} - N_i). \tag{5}$$

where ∇W_i is an approximation, at least of order one, of the gradient of the solution $W(\boldsymbol{x})$, and N_i is the point defined by:

$$N_i = \sum_{k=1}^{4} \frac{|V_{ik}|}{|V_i|} b_{i,k}. \tag{6}$$

With the purpose of simplicity, lets us suppose that V_i is a interior finite volume and let us denote by $V_{i,1}, \ldots, V_{i,4}$ its four neighbors. Let us consider the points $N_{i,j}$, $j = 1,\ldots,4$, associated to $V_{i,j}$, $j = 1,\ldots 4$, respectively, given by (6). Let us also consider four triangles $T_1, \ldots, T_4$ defined by the union of the points $N_{i,1}, \ldots, N_{i,4}$ and the point N_i (see Figure 3). Over each triangle T_i, $i = 1,\ldots,4$, we consider a linear approximation of the solution, by the values W_i, $W_{i,1}$, ..., $W_{i,4}$ which are second order approximation of $W(N_i), W(N_{i,1}), \ldots, W(N_{i,4})$, respectively.

At each triangle T_j, $j = 1,\ldots,4$, with vertices $\{N_i, N_{i,j}, N_{i,ip(j)}\}$, where $ip(j)$, $j = 1,\ldots,4$ take the values in the ordered set $\{2,3,4,1\}$, we consider an approximation

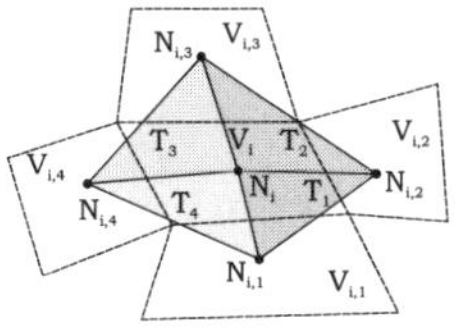

Fig. 3 Triangles T_1, T_2, T_3, T_4, used to approximate the gradient of $W(\boldsymbol{x})$ in V_i.

of the gradient $\nabla W_{|T_j}$ using the linear approximation previously constructed (see [4] for details). Finally, to give an approximation of $\nabla W(\boldsymbol{x})$ in V_i, we consider a weighted averaged of $\nabla W_{|T_j}$, $j = 1, \ldots, 4$:

$$\nabla W(\boldsymbol{x}) \approx \nabla W_i = \sum_{j=1}^{4} |T_j| \nabla W_{|T_j} \Big/ \sum_{j=1}^{4} |T_j|. \tag{7}$$

Frequently, the solution of hyperbolic systems presents discontinuities. To obtain a state reconstruction operator that approximates with order two the solution in regular areas and at same time that captures the regions where $W(\boldsymbol{x})$ is discontinuous, it is necessary to modify the reconstruction operator (5), using a slope limiter function.

In [4] we prove that the reconstruction operator previously described achieves second order accuracy. Finally, numerous numerical experiments have been performed in the framework of the one-layer and two-layer shallow-water systems achieving very good results as well as good well-balancing properties (see [1]).

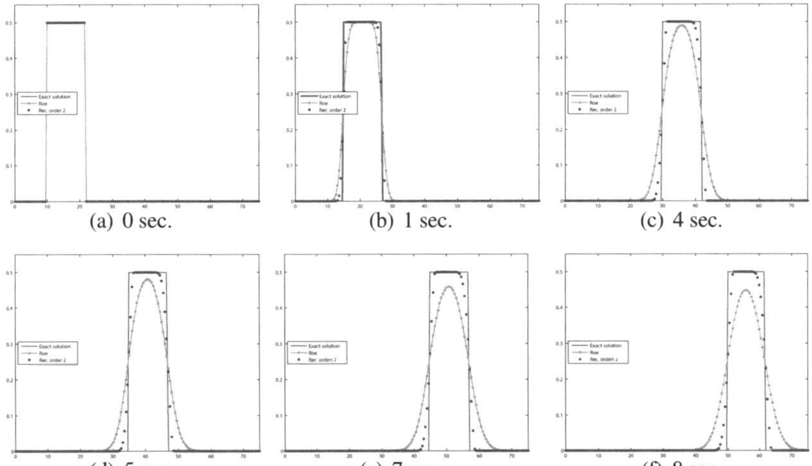

(a) 0 sec. (b) 1 sec. (c) 4 sec.

(d) 5 sec. (e) 7 sec. (f) 8 sec.

Fig. 4 Evolution of the pollutant concentration at different time steps. Longitudinal central section of the channel.

3 Numerical Experiments

3.1 Pollutant Transport in a Rectangular Channel with Planar Bed

In this bidimensional test we study the evolution of a pollutant that initially occupies a circle and that is transported at constant velocity in a rectangular channel of dimensions $75\,m \times 30\,m$. We are going to suppose that do not exist emitting sources and we take like initial condition a circular region with a pollutant concentration equal to 0.5. So the initial conditions are:

$$h(x,y,0) = 2,\; q_x(x,y,0) = 10,\; q_y(x,y,0) = 0;$$

and the initial pollutant concentration is given by the following equation:

$$C(x,y,0) = \begin{cases} 0.5 & \text{if } (x-15)^2 + (y-15)^2 \leq 36, \\ 0 & \text{in other case.} \end{cases}$$

The CFL condition is equal to 0.8. We use an unstructured mesh of 9000 finite volumes. We impose the flux $\boldsymbol{q} = (10,0)$ in the boundary corresponding to $x = 0$ and $x = 75$. In the lateral walls we impose a sliding condition $\boldsymbol{q} \cdot \boldsymbol{\eta} = 0$. With this data the experiment runs until $t = 10$ seconds.

The exact solution is given by a circle that is translated with constant velocity in the direction of the axis X, $v_x = \dfrac{q_x}{h} = 5\; m/s$. In the Figure 4 we present a comparison in the longitudinal section, $y = 15$, between the exact solution, the numerical solution using a method of order 1 (Roe) and the approximated solution obtained using the numerical scheme of the section 2.1. It can be observed that the high order scheme preserves the initial concentration 0.5 and preserves better the boundary of the pollutant. In Figure 5 is shown a comparison using iso-levels between the Roe scheme and the scheme (4) using the second order reconstruction operator defined in section 2.1. Note the artificial diffusion introduced by the Roe scheme and how it is reduced by considering the second order scheme.

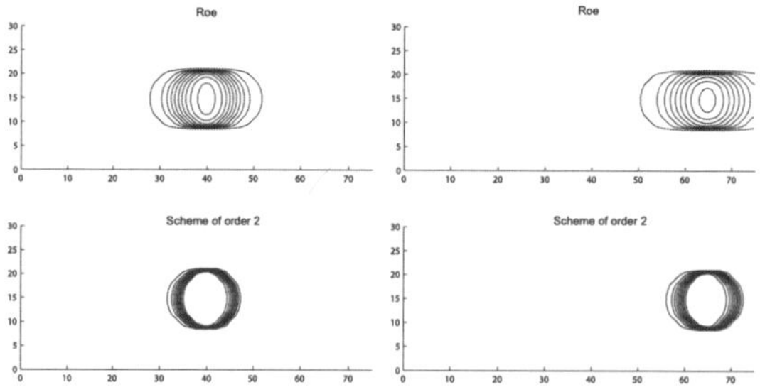

(a) Time 5 s. (b) Time 10s.

Fig. 5 Pollutant evolution (from up to down, Roe vs. Second order reconstruction): iso-levels.

3.2 Pollutant Transport in a Rectangular Channel with Variable Bed

This test is done in a rectangular channel of dimensions $75m \times 30m$. The bed of the channel presents a bump given by the following function:

$$B(x,y) = e^{-0{,}075\cdot(x-37{,}5)^2}. \tag{8}$$

Initially we suppose the pollutant occupies a circle of center $(15,15)$ and radius 6 m. Initially, we suppose a constant flux equal to $\boldsymbol{q} = (1,0)$ and a constant free surface equal to 3 m. of height in its deepest point. As boundary conditions we impose a sliding condition $\boldsymbol{q}\cdot\boldsymbol{n} = 0$ along the boundaries $y = 0$ and $y = 30$, and free conditions in $x = 0$ and $x = 75$. We consider $CFL = 0.8$ and the simulation runs until the time $t = 120$ s.

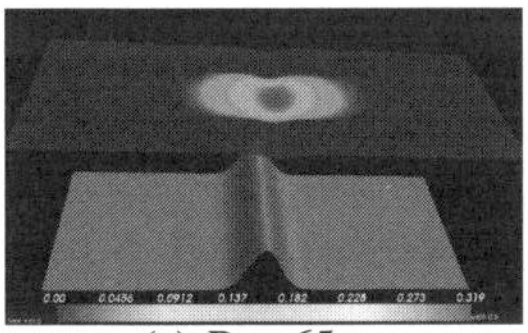
(a) Roe 65 s.

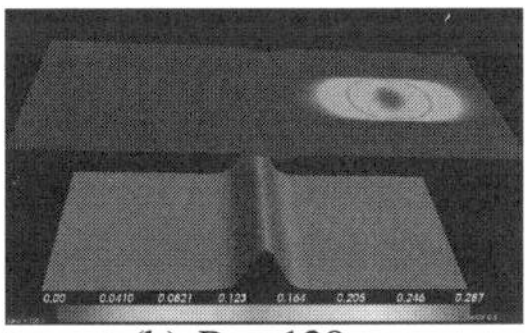
(b) Roe 120 s.

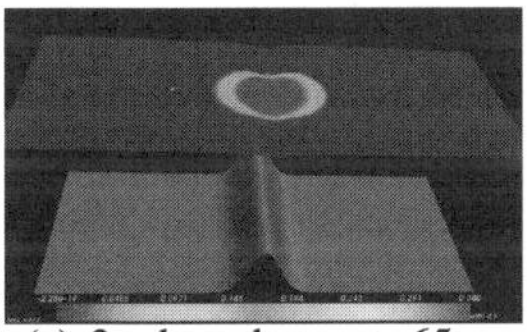
(c) 2order scheme, t=65 s.

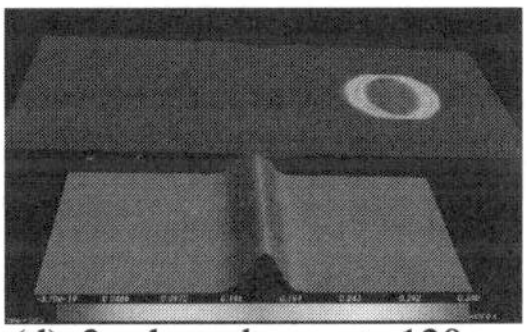
(d) 2order scheme, t=120 s.

Fig. 6 Pollutant concentration evolution at different instants.

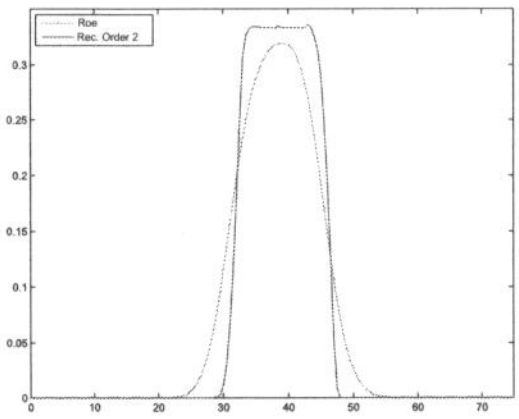

(a) Time 65 s.

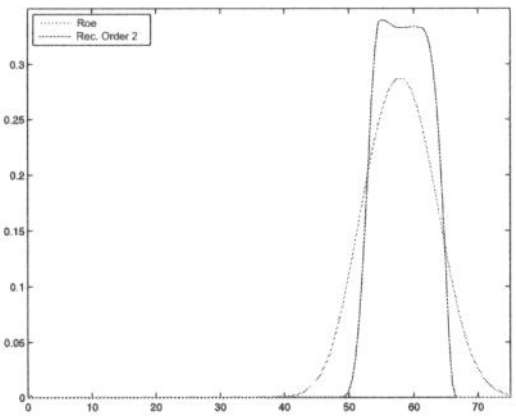

(b) Time 120s.

Fig. 7 Pollutant concentration evolution: central longitudinal section (Roe vs. Second order scheme).

In Figure 6 we present the obtained results using Roe scheme (Figures 6-(a) and 6-(b)) and the scheme of order two (4) (Figures 6-(c) and 6-(d)).

In Figure 7 we show a comparison between Roe scheme and the scheme (4) along a central longitudinal section. The solution obtained with Roe scheme is represented in dotted line, and the continuous line corresponds with the solution obtained with the second order reconstructions scheme.

4 Conclusions

In this work we have solved, in a coupled way, a one layer shallow water system together with a transport equation. To better capture the evolution and concentration of the spill, we consider a second order scheme for non-conservative problems based on a MUSCL-type state reconstruction. Moreover, one of the main advantage of the presented numerical scheme is its well-balance character. In fact, in the framework of shallow-water systems it can be easily proved that it is exactly well-balanced for the solutions corresponding to water at rest. Finally, the accuracy of the second order scheme has been assessed with two numerical experiments.

Acknowledgements This work has been partially supported by the research projects MTM2006-08075 and MTM2006-01275, funded by the Spanish Government.

References

1. Castro, M., Fernández Nieto, E., Ferreiro, A., García, J., Parés, C.: High order extension of roe schemes for two dimensional nonconservative hyperbolic systems. (Submitted)
2. Castro, M., Gallardo, J.M. Parés, C.: Finite volume schemes based on weno reconstruction of states for solving nonconservative hyperbolic systems. applications to shallow water systems. Mathematics of Computation **255**(75), 1103–1134 (2006)
3. Dal Maso, G., LeFloch, P., Murat, F.: Definition and weak stability of nonconservative products. J. Math. Pures Appl. (74), 483–548 (1995)
4. Ferreiro Ferreiro, A.: Desarrollo de técnicas de post-proceso de flujos hidrodinámicos, modelización de problemas de transporte de sedimentos y simulación numérica mediante técnicas de volúmenes finitos. Phd. University of Sevilla (2006)
5. Parés, C., Castro, M.: On the well-balance property of roe's method for nonconservative hyperbolic systems. applications to shallow-water systems. ESAIM: M2AN **5**(38), 821–852 (2004)
6. Toro, E.: Shock-capturing methods for free-surface shallow flows. Wiley (2001)

Well-Balanced High-Order MUSTA Schemes for Non-Conservative Hyperbolic Systems

M.J. Castro, C. Parés, A. Pardo, and E.F. Toro

Abstract We introduce a Multi-Stage (MUSTA) approach for constructing upwind numerical schemes for nonconservative hyperbolic systems. MUSTA schemes for hyperbolic conservation laws were introduced in [8] as an approximate Riemann solver based on a GFORCE scheme and a predictor-corrector procedure. In [2] a path-conservative GFORCE numerical scheme (in the sense introduced in [6]) for nonconservative hyperbolic systems is proposed. Here, we propose a predictor-corrector procedure based on this extension of GFORCE to obtain a generalization of MUSTA schemes. These schemes can be applied to systems of conservation laws with source terms and nonconservative products. In particular, some applications to two-layer shallow-water flows are presented.

1 Introduction

MUSTA schemes for hyperbolic conservation laws were introduced by Toro and Titarev in [8] as a predictor-corrector procedure based on the use of the so-called GFORCE method, whose numerical flux is a convex linear combination of the Lax-Frieridchs and the Lax-Wendroff fluxes. In [2] we present a generalization of GFORCE to nonconservative hyperbolic system:

$$\frac{\partial W}{\partial t} + \mathscr{A}(W)\frac{\partial W}{\partial x} = 0, \quad x \in \mathbb{R},\, t > 0, \tag{1}$$

which is path-conservative in the sense introduced in [6]. A particular case of (1) is the class of systems of conservation laws with source terms and nonconservative

M.J. Castro, C. Parés, A. Pardo
Depto. de Análisis mátematico, U. de Málaga, Spain, e-mail: castro@anamat.cie.uma.es

E.F. Toro
University of Trento. Laboratory of Applied Mathematics. Faculty of Engineering, 38050 Mesiano di Povo, Trento, Italy

products of the form:

$$w_t + F(w)_x + \mathscr{B}(w) \cdot w_x = S(w)\sigma_x, \tag{2}$$

where $\sigma(x)$ is a known function. Following an idea introduced in [4], System (2) can be considered as the particular case of (1) corresponding to the choice:

$$W = \begin{bmatrix} w \\ \sigma \end{bmatrix}, \quad \mathscr{A}(W) = \left[\begin{array}{c|c} \mathscr{J}(w) + \mathscr{B}(w) & -S(w) \\ \hline 0 & 0 \end{array}\right],$$

where

$$\mathscr{J}(w) = \frac{\partial F}{\partial w}(w).$$

The equations governing the flow of a stratified fluid composed by two superposed shallow layers of immiscible liquids can be formulated under the form (2) (see [1]).

In this work, we first present the expression of the generalized GFORCE method. Then, we extend the MUSTA procedure to nonconservative system on the basis of GFORCE and a prediction-correction procedure which is interpreted as a reconstruction of states. These MUSTA schemes are first order accurate but they can be extended to higher order methods by using a reconstruction operator, following the general methodology introduced in [6]. Finally, some applications to the two-latyer shallow-water systems are presented.

2 Some Numerical Schemes for Nonconservative Hyperbolic Systems

In this section we present some numerical schemes for solving Cauchy problems related to the nonconservative hyperbolic system (1), where the unknown $W(x,t)$ takes values on an open convex set Ω. The system is supposed to be strictly hyperbolic and every characteristic field is supposed to be either genuinely nonlinear or linearly degenerate.

For discontinuous solutions W, the nonconservative product $\mathscr{A}(W)W_x$ in (1) may not make sense as a distribution. Nevertheless, after the theory developed by Dal Maso, LeFloch and Murat in [3], it is possible to give a rigorous definition of weak solutions associated to the choice of a family of paths in Ω:

Definition 1. A family of paths in $\Omega \subset \mathbb{R}^N$ is a locally Lipschitz map

$$\Phi\colon [0,1] \times \Omega \times \Omega \mapsto \Omega,$$

such that:

- $\Phi(0; W_L, W_R) = W_L$ and $\Phi(1; W_L, W_R) = W_R$, for any $W_L, W_R \in \Omega$;
- for every arbitrary bounded set $\mathscr{O} \subset \Omega$, there exists a constant k such that

$$\left|\frac{\partial \Phi}{\partial s}(s;W_L,W_R)\right| \leq k|W_R - W_L|,$$

for any $W_L, W_R \in \mathscr{O}$ and almost every $s \in [0,1]$;

- for every bounded set $\mathscr{O} \subset \Omega$, there exists a constant K such that

$$\left|\frac{\partial \Phi}{\partial s}(s;W_L^1,W_R^1) - \frac{\partial \Phi}{\partial s}(s;W_L^2,W_R^2)\right| \leq K(|W_L^1 - W_L^2| + |W_R^1 - W_R^2|),$$

for any $W_L^1, W_R^1, W_L^2, W_R^2 \in \mathscr{O}$ and almost every $s \in [0,1]$.

The numerical schemes to be introduced here are path-conservative in the sense of the following definition introduced in [6], which is a generalization of that of conservative scheme for systems of conservation laws:

Definition 2. Given a family of paths Ψ, a numerical scheme is said to be a Ψ-conservative numerical scheme if it can be written under the form:

$$W_i^{n+1} = W_i^n - \frac{\Delta t}{\Delta x}\left(D_{i-1/2}^+ + D_{i+1/2}^-\right), \tag{3}$$

where

$$D_{i+1/2}^{\pm} = D^{\pm}\left(W_{i-q}^n, \ldots, W_{i+p}^n\right), \tag{4}$$

D^- and D^+ being two continuous functions from Ω^{p+q+1} to Ω satisfying

$$D^{\pm}(W,\ldots,W) = 0, \quad \forall W \in \Omega, \tag{5}$$

and

$$D^-(W_{-q},\ldots,W_p) + D^+(W_{-q},\ldots,W_p) = \int_0^1 \mathscr{A}\left(\Psi(s;W_0,W_1)\right)\frac{\partial \Psi}{\partial s}(s;W_0,W_1)\,ds, \tag{6}$$

for every set $\{W_{-q},\ldots,W_p\} \subset \Omega$.

2.1 Nonconservative Roe Scheme

Roe methods can be generalized to nonconservative systems. First, a Roe linearization $\mathscr{A}(W_L,W_R)$ in the sense defined in [9] has to be chosen:

Definition 3. Given a family of paths Ψ, a function $\mathscr{A}_\Psi : \Omega \times \Omega \mapsto \mathscr{M}_{N\times N}(\mathbb{R})$ is called a Roe linearization if it verifies the following properties:

- for each $W_L, W_R \in \Omega$, $\mathscr{A}_\Psi(W_L,W_R)$ has N distinct real eigenvalues,
- $\mathscr{A}_\Psi(W,W) = \mathscr{A}(W)$, for every $W \in \Omega$,
- for any $W_L, W_R \in \Omega$,

$$\mathscr{A}_\Psi(W_L,W_R)(W_R-W_L)=\int_0^1 \mathscr{A}(\Psi(s;W_L,W_R))\frac{\partial\Psi}{\partial s}(s;W_L,W_R)\,ds. \tag{7}$$

Once a Roe linearization $\mathscr{A}_\Psi$ has been chosen, some straightforward calculations (see [6]) allow to show that a nonconservative Roe scheme can be written under the form:

$$W_i^{n+1}=W_i^n-\frac{\Delta t}{\Delta x}(D^+_{i-1/2}+D^-_{i+1/2}), \tag{8}$$

$$D^-_{i+1/2}=\mathscr{A}^-_{i+1/2}(W^n_{i+1}-W^n_i);D^+_{i+1/2}=\mathscr{A}^+_{i+1/2}(W^n_{i+1}-W^n_i); \tag{9}$$

where $\mathscr{A}_{i+1/2}=\mathscr{A}_\Psi(W_i^n,W_{i+1}^n)$, $\mathscr{L}_{i+1/2}$ is the diagonal matrix composed of the eigenvalues of $\mathscr{A}_{i+1/2}$ ($\lambda_1^{i+1/2}<\lambda_2^{i+1/2}<\cdots<\lambda_N^{i+1/2}$), $\mathscr{L}^\pm_{i+1/2}$ the positive and the negative part of $\mathscr{L}_{i+1/2}$ and $\mathscr{K}_{i+1/2}$ is a $N\times N$ matrix whose columns are associated eigenvectors. Finally, $\mathscr{A}^\pm_{i+1/2}=\mathscr{K}_{i+1/2}\mathscr{L}^\pm_{i+1/2}\mathscr{K}^{-1}_{i+1/2}$.

Roe methods have been applied successfully for nonconservative systems: in general, they are robust and have good well-balance properties (see [7, 6]). Nevertheless, these methods also present some drawbacks. In particular, their implementation requires the explicit knowledge of the eigenvalues and eigenvectors of the intermediate matrices. When their analytic expression is not available, as it is the case for the two-layer shallow water system, the eigenvalues and eigenvectors of the matrix have to be numerically calculated at every interface and at every time step, which is computationally expensive. We look here for numerical schemes that overcome this drawback.

2.2 Nonconservative GFORCE Scheme

Given a Roe linearization $\mathscr{A}_\Psi$ we consider the following family of numerical schemes:

$$W_i^{n+1}=W_i^n-\frac{\Delta t}{\Delta x}(D^+_{i-1/2}+D^-_{i+1/2}), \tag{10}$$

with:

$$\begin{aligned}
D^+_{i+1/2}&=(1-\omega)\left(\frac{1}{2}\mathscr{A}_{i+1/2}(W^n_{i+1}-W^n_i)+\frac{1}{2}\frac{\Delta x}{\Delta t}I(W^n_{i+1}-W^n_i)\right)\\
&\quad+\omega\left(\frac{1}{2}\mathscr{A}_{i+1/2}(W^n_{i+1}-W^n_i)+\frac{1}{2}\frac{\Delta t}{\Delta x}\mathscr{A}^2_{i+1/2}(W^n_{i+1}-W^n_i)\right),\\
D^-_{i+1/2}&=(1-\omega)\left(\frac{1}{2}\mathscr{A}_{i+1/2}(W^n_{i+1}-W^n_i)-\frac{1}{2}\frac{\Delta x}{\Delta t}I(W^n_{i+1}-W^n_i)\right)\\
&\quad+\omega\left(\frac{1}{2}\mathscr{A}_{i+1/2}(W^n_{i+1}-W^n_i)-\frac{1}{2}\frac{\Delta t}{\Delta x}\mathscr{A}^2_{i+1/2}(W^n_{i+1}-W^n_i)\right).
\end{aligned}$$

Here, $\mathscr{A}_{i+1/2}=\mathscr{A}_\Psi(W_i^n,W_{i+1}^n)$ and I is the identity matrix.

The choices $\omega = 0, 1, 1/2$ generalize the usual Lax-Friedrichs, Lax-Wendroff, and the FORCE methods for conservative problems. GFORCE scheme corresponds to the choice $\omega = \dfrac{1}{1+CFL}$, being CFL the usual CFL number. It can be easily shown that, if the problem is conservative, i.e. if $\mathscr{A}(W)$ is the Jacobian of a flux function $F(W)$ then the numerical schemes can be rewritten as a conservative method whose numerical flux is a convex linear combination of the Lax-Friedrichs and Lax-Wendroff fluxes.

This family of numerical schemes do not preserve the well-balanced properties of the Roe scheme corresponding to the chosen linearization. Nevertheless, these properties can be recovered if the identity matrix appearing in the definition of $D^{\pm}_{i+1/2}$ is changed by:

$$\widehat{I}_{i+1/2} = K_{i+1/2} \cdot \widehat{I} \cdot K^{-1}_{i+1/2}, \tag{11}$$

being $\widehat{I}$ the diagonal matrix whose j-th coefficient is 1 if $\lambda_j^{i+1/2} \neq 0$, or 0 if $\lambda_j^{i+1/2} = 0$.

The generalization of the Lax-Wendroff scheme corresponding to the choice $\omega = 1$ is not a second order scheme for general nonconservative problem, but for conservative ones. A second order extension of this method can be found in [2].

The particular expression of the numerical scheme (10)-(11) to system (2) can be found in [2].

2.3 Nonconservative MUSTA Scheme and High-Order Schemes

In this paragraph we extend to nonconservative systems the MUSTA procedure, introduced in [8] for systems of conservation laws. The idea is as follows: let us suppose that the approximations W_i^n at the n-th time level have been obtained. As usual in approximate Riemann solvers, a Riemann problem is associated at every inter-cell $x_{i+1/2}$:

$$\begin{cases} \dfrac{\partial U}{\partial t} + \mathscr{A}(U)\dfrac{\partial U}{\partial x} = 0, \quad x \in \mathbb{R},\ t > 0, \\ U(x,0) = \begin{cases} W_i^n & if\ x < 0, \\ W_{i+1}^n & if\ x > 1, \end{cases} \end{cases} \tag{12}$$

The idea now is to use a first order path-conservative numerical scheme to solve numerically these Riemann problems. To do this, first a local mesh is considered, with space step h and time step k. These steps are chosen so that $\Delta x = Mh$ and $\Delta t = Nk$ for some positive integers M, N (in particular these integers may be equal to one). Let us denote by $\tilde{x}_{j+1/2}$ the inter-cells of the local mesh. We suppose for simplicity that $\tilde{x}_{1/2} = 0$. We perform now N time iterations of a first order path-conservative numerical scheme:

$$U_j^{n+1} = U_j^n - \frac{k}{h}\left(D^+_{1,j-1/2} + D^-_{1,j+1/2}\right), \quad n = 1,\ldots,N. \tag{13}$$

In practice, these iterations are performed in a truncated domain, i.e. j takes values between two integers $l < 0$ and $r > 0$ (possibly $-l = r = M$). Therefore, some transmissive conditions have to be chosen at the boundary cells. In this work, we use the standard technique based on the use of two ghost and the duplication of the states.

At the end of this first stage, two approximations

$$W^-_{i+1/2} = U_0^N, \quad W^+_{i+1/2} = U_1^N. \tag{14}$$

of the limits to the left and to the right of $x = 0$ of the solution of the Riemann problem (12) at time Δt are available. Now, these approximations are applied to calculate W_i^{n+1} by using a new first order path-conservative numerical scheme (3) (possibly the same) and a reconstruction procedure. More precisely, the approximations at time $n+1$ are calculated as follows:

$$W_i^{n+1} = W_i^n - \frac{\Delta t}{\Delta x}\left(E^+_{i-1/2} + E^-_{i+1/2}\right), \quad \forall i, \tag{15}$$

where

$$E^+_{i+1/2} = D^+(W^-_{i+1/2}, W^+_{i+1/2}) + \int_0^1 \mathscr{A}(\varphi(s; W^+_{i+1/2}, W^n_{i+1})\frac{\partial \varphi}{\partial s}(s; W^+_{i+1/2}, W^n_{i+1})\, ds;$$
$$E^-_{i+1/2} = D^-(W^-_{i+1/2}, W^+_{i+1/2}) + \int_0^1 \mathscr{A}(\varphi(s; W^n_i, W^-_{i+1/2})\frac{\partial \varphi}{\partial s}(s; W^n_i, W^-_{i+1/2})\, ds.$$

Here, φ represents the family of segments, i.e.

$$\varphi(s; W_L, W_R) = (1-s)W_L + sW_R.$$

It can be easily shown that this numerical scheme is path-conservative and thus consistent for smooth solutions (see [6]).

Here, we consider a MUSTA procedure in which the two first-order path-conservative numerical schemes are GFORCE, and $N = M = 1$ or 2.

2.4 High Order Extensions

In [6] the construction of high-order discrete numerical scheme for (1) based on a first order path-conservative numerical scheme and a reconstruction operator was presented. We follow here that methodology. In particular, we consider the third order PHM monotone reconstruction operator (piecewise hyperbolic method) introduced in [5].

3 Application to Bilayer Shallow Water Equations with Depth Variations

We consider in this paragraph the system of partial differential equations governing the one-dimensional flow of two superposed immiscible layers of shallow water fluids studied in [1]:

$$\begin{cases} \dfrac{\partial h_1}{\partial t} + \dfrac{\partial q_1}{\partial x} = 0, \\[2ex] \dfrac{\partial q_1}{\partial x} + \dfrac{\partial}{\partial x}\left[\dfrac{q_1^2}{h_1} + \dfrac{g}{2}h_1^2\right] = -gh_1\dfrac{\partial h_2}{\partial x} + gh_1\dfrac{\partial H}{\partial x}, \\[2ex] \dfrac{\partial h_2}{\partial t} + \dfrac{\partial q_2}{\partial x} = 0, \\[2ex] \dfrac{\partial q_2}{\partial x} + \dfrac{\partial}{\partial x}\left[\dfrac{q_2^2}{h_2} + \dfrac{g}{2}h_2^2\right] = -rgh_2\dfrac{\partial h_1}{\partial x} + gh_2\dfrac{\partial H}{\partial x} \end{cases} \tag{16}$$

In these equations, index 1 makes reference to the upper layer and index 2 to the lower one. The fluid is assumed to occupy a straight channel with constant rectangular cross-section and constant width. The coordinate x refers to the axis of the channel, t is the time, and g is gravity. $H(x)$ represents the depth function measured from a fixed level of reference. Each layer is assumed to have a constant density, ρ_i, $i = 1,2$ ($\rho_1 < \rho_2$). The unknowns $q_i(x,t)$ and $h_i(x,t)$ represent respectively the mass-flow and the thickness of the i-th layer at the section of coordinate x at time t.

We have constructed a GFORCE scheme for this system based on the family of segments (see [2] for details).

We consider a numerical test which is designed to assess the long time behavior and the convergence to a steady state including a regular transition and a shock. The axis of the channel is the interval $[0,10]$. The bottom topography is given by the function $H(x) = 1.0 - 0.47e^{-(x-5.0)^2}$. The initial conditions are $q_1(x,0) = q_2(x,0) = 0$, and

$$h_1(x,0) = \begin{cases} 0.5 & \text{if } x < 5, \\ 0.03 & \text{otherwise,} \end{cases} \quad h_2(x,0) = \begin{cases} 0.5 - 0.47e^{-(x-5)^2} & \text{if } x < 5, \\ 0.97 - 0.47e^{-(x-5)^2} & \text{otherwise.} \end{cases}$$

As boundary conditions, the following relations are imposed at both ends $q_1(\cdot,t) = -q_2(\cdot,t)$, and the free surface is fixed to $z = 0$ at $x = 10$, that is $h_1(10,t) + h_2(10,t) - H(x) = 0$. The CFL parameter is set to 0.8. The final time is $t = 300$.

A reference solution is computed by using a mesh of 3200 points.

Figure 1 (left) shows the comparison of the numerical solutions computed with the numerical schemes: ROE, GFORCE and MUSTA, with the reference solution at time $t = 300$. Note that, the numerical solution computed with MUSTA scheme is better than the one computed with GFORCE scheme. Figure 1 (right) shows the comparison of the numerical solutions computed with the third order extension of MUSTA and the reference solution at time $t = 300$.

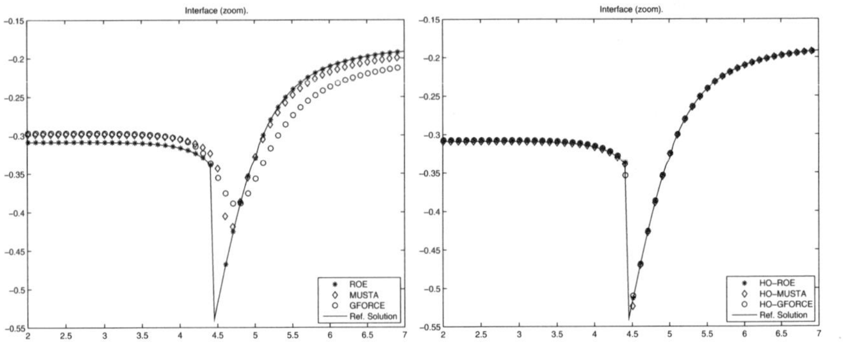

Fig. 1 Test 1. Interface of the two fluids: comparison with the reference solution.

References

1. Castro, M., Macías, J., Parés, C.: A q-scheme for a class of systems of coupled conservation laws with source term. application to a two-layer 1-d shallow water system. Math. Mod. Num. Anal. (2001). DOI 35(1):107–127
2. Castro, M., Pardo, A., Parés, C., Toro, E.: Coefficent-splitting numerical schemes for nonconservative hyperbolic systems and high order order extensions. (to appear)
3. Dal Maso, G., LeFloch, P., Murat, F.: Definition and weak stability of nonconservative products. J. Math. Pures Appl. (1995). DOI 74:483–548
4. LeFloch, P.: Shock waves for nonlinear hyperbolic systems in nonconservative form. Appl., Minneapolis, Preprint 593 (1989)
5. Marquina, A.: Local piecewise hyperbolic reconstructions for nonlinear scalar conservation laws. SIAM J. Sci. Comp (1994)
6. Parés, C.: Numerical methods for nonconservative hyperbolic systems: a theoretical framework. SIAM J. Num. Anal. (2006). DOI 44(1):300–321
7. Parés, C., CastroM.J.: On the well-balance property of roes method for nonconservative hyperbolic systems. applications to shallow-water systems. Math. Mod. Num. Anal. (2004). DOI 38(5):821–852
8. Toro, E., Titarev, V.: Musta fluxes for systems of conservation laws. J. Comput. Phys. (2006). DOI 216(2):403–429
9. Toumi, I.: A weak formulation of roe approximate riemann solver. J. Comp. Phys. (1992). DOI 360–373

Local Time Stepping for Implicit-Explicit Methods on Time Varying Grids

F. Coquel, Q.-L. Nguyen, M. Postel, and Q.-H. Tran

Abstract In the context of nonlinear conservation laws a model for multiphase flows where slow kinematic waves co-exist with fast acoustic waves is discretized with an implicit-explicit time scheme. Space adaptivity of the grid is implemented using multiresolution techniques and local time stepping further enhances the computing time performances. A parametric study is presented to illustrate the robustness of the method.

1 Introduction

In the context of offshore oil production, we are interested in accurate and fast computation of two-phase flows in pipelines. A one dimensional model of nonlinear hyperbolic equations shows that two sets of waves interplay in the phenomenon: a slow transport wave which actually models the propagation of the gas and oil phases and fast acoustic waves which are not interesting for the engineers but require special numerical treatment. The initial system of equations is split into a Lagrange-projection method. The Lagrange step, which handles the fast acoustic wave is treated with

F. Coquel
CNRS, UMR 7598, Laboratoire Jacques-Louis Lions, F-75005, Paris, France,
e-mail: coquel@ann.jussieu.fr

Q.-L. Nguyen
Département Mathématiques Appliquées, Institut Français du Pétrole, 1 et 4 avenue de Bois-Préau, 92852 Rueil-Malmaison Cedex, France, e-mail: Q-Long.NGUYEN@ifp.fr

M. Postel
UPMC Univ Paris 06, UMR 7598, Laboratoire Jacques-Louis Lions, F-75005, Paris, France,
e-mail: postel@ann.jussieu.fr

Q.-H. Tran
Département Mathématiques Appliquées, Institut Français du Pétrole, 1 et 4 avenue de Bois-Préau, 92852 Rueil-Malmaison Cedex, France, e-mail: Q-Huy.Tran@ifp.fr

an implicit numerical scheme, therefore relaxing the stability constraint on the time step. The projection step which deals with the transport phenomenon must be solved as accurately as possible, with an explicit numerical scheme. This method is detailed in [3] along with a derivation of an explicit stability condition for the time step, ensuring positivity of physical quantities such as the density or the gas mass fraction. Since the ratio between the acoustic and kinetic waves speeds is typically more than 10, it is very interesting to be able to treat the acoustic wave implicitly. The time step is monitored by the explicit step requirement and therefore gains a factor of 10. Still the numerical scheme calls for performance improvement, specially since the closure laws entering the equations are in practice very costly to compute. The multiresolution method initially designed for an explicit scheme in the context of scalar conservation laws [2] has been successfully extended to implicit-explicit schemes in [5, 1] and for our specific Lagrange-projection method in [4]. Also in [4], we have extended the local time stepping method designed by Müller and Stiriba in [6] for a scalar hyperbolic equation, to our PDE system in the explicit-explicit and implicit-explicit versions of the Lagrange projection method. We present here a parametric study of the standard multiresolution and local time stepping algorithms in the implicit-explicit case.

2 Modeling of the Physical Problem

The density of the mixture ρ, velocity u and the gas mass fraction Y are solution of the following problem

$$\begin{cases} \partial_t(\rho) \;\; + \partial_x(\rho u) \qquad = 0, \\ \partial_t(\rho Y) + \partial_x(\rho Y u) \quad = 0, \\ \partial_t(\rho u) \; + \partial_x(\rho u^2 + P) = 0, \end{cases} \tag{1}$$

whose space of states is

$$\Omega_{\mathbf{V}} = \left\{ \mathbf{V} = (\rho, \rho Y, \rho u) \in \mathbb{R}^3, \rho > 0, Y \in [0,1], u \in \mathbb{R} \right\}.$$

The thermodynamical closure law $P(\rho, Y)$ entering in (1) can be in real applications very costly to evaluate. Under the assumption $c^2(\mathbf{V}) = \partial_\rho P(\rho, Y)_{|Y} > 0$, the system (1) is hyperbolic with three distinct eigenvalues $u - c < u < u + c$. The intermediate eigenvalue corresponds to the slow transport wave and is linearly degenerate, the remaining ones are much larger and correspond to nonlinear acoustic waves. The main idea consists in decomposing the flux in an acoustic part, associated with the nonlinear waves, and a transport part, associated with the linearly degenerate waves. The Lagrange step where we deal with the acoustic part of the flux is treated implicitly, which enables us to use a larger time step, basically driven by the transport phenomenon, which is treated explicitly for better accuracy.

We denote by $\mathbf{V}_j^n = \rho_j^n(1, Y_j^n, u_j^n)$, for $j = 0, \dots J-1$, the numerical solution on cell Ω_j at time $n\Delta t$ and by $\mathbf{V}_j^{n\sharp} = \rho_j^{n\sharp}(1, Y_j^{n\sharp}, u_j^{n\sharp})$ the numerical solution at the end of the Lagrange step. The cell length is $\Delta x = L/J$, where L is the length of the domain.

The **implicit Lagrange step** consists in

$$\begin{cases} \rho_j^n \dfrac{\tau_j^{n\sharp} - \tau_j^n}{\Delta t} - \dfrac{\widetilde{u}_{j+1/2}^{n\sharp} - \widetilde{u}_{j-1/2}^{n\sharp}}{\Delta x} = 0, \\ \rho_j^n \dfrac{Y_j^{n\sharp} - Y_j^n}{\Delta t} = 0, \\ \rho_j^n \dfrac{u_j^{n\sharp} - u_j^n}{\Delta t} + \dfrac{\widetilde{P}_{j+1/2}^{n\sharp} - \widetilde{P}_{j-1/2}^{n\sharp}}{\Delta x} = 0, \end{cases} \tag{2}$$

where τ denotes the specific volume $1/\rho$. The intermediate states $\widetilde{u}_{j+1/2}^{n\sharp}$ and $\widetilde{P}_{j+1/2}^{n\sharp}$ are given by

$$\begin{cases} \widetilde{u}_{j+1/2}^{n\sharp} = \dfrac{1}{2}(u_j^{n\sharp} + u_{j+1}^{n\sharp}) - \dfrac{1}{2a_n}(P_{j+1}^{n\sharp} - P_j^{n\sharp}), \\ \widetilde{P}_{j+1/2}^{n\sharp} = \dfrac{1}{2}(P_j^{n\sharp} + P_{j+1}^{n\sharp}) - \dfrac{a_n}{2}(u_{j+1}^{n\sharp} - u_j^{n\sharp}), \end{cases} \tag{3}$$

where we have set

$$P_j^{n\sharp} = P_j^n - a_n^2(\tau_j^{n\sharp} - \tau_j^n). \tag{4}$$

In (3) and (4), a_n is a stabilizing coefficient coming from the relaxation formulation of problem (1) as described in [3]. It is set globally for all cells at each time step by the Whitham condition

$$a_n^2 > \max_{j=0,\dots,J-1} -\partial_\tau P(\tau_j^n, Y_j^n).$$

The projection step computes the conservative state at time $(n+1)\Delta t$ according to

$$\mathbf{V}_j^{n+1} = \mathbf{V}_j^n - \frac{\Delta t}{\Delta x}\left(\mathbf{F}_{j+1/2}^{n,\sharp} - \mathbf{F}_{j-1/2}^{n,\sharp}\right),$$

where

$$\mathbf{F}_{j-1/2}^{n,\sharp} = \left(0, 0, \widetilde{P}_{j-1/2}^{n\sharp}\right)^T + (\widetilde{u}_{j-1/2}^{n\sharp})^+ \mathbf{V}_{j-1}^{n\sharp} + (\widetilde{u}_{j-1/2}^{n\sharp})^- \mathbf{V}_j^{n\sharp}.$$

We refer to [3] for a comprehensive study of the boundary conditions and we implement here nonreflecting ones. The stability of the resulting scheme has also been thoroughly studied in [3] and is ensured by a CFL–like condition. This condition is enforced on the time step along with another bound ensuring that the implicit step does not smooth too much the acoustic waves. We sum up these two requirements by

$$\Delta t < \mathrm{CFL}_{\mathrm{exp}}(\mathbf{V})\Delta x, \quad \text{and } \Delta t < \mathrm{CFL}_{\mathrm{imp}}(\mathbf{V})\Delta x, \tag{5}$$

where $\mathrm{CFL_{imp}}$ is set so that acoustic waves can travel at most N_a cells within one time step. N_a is set to 10 in the numerical simulations presented here.

3 Multiresolution

It is well known that since the fast acoustic waves are treated implicitly, they are smoothed out very early in the computation (see [5]). The wave of interest which moves with the slow speed and is computed explicitly, may present on the other hand singularities that we want to compute as precisely as possible. It is of course natural to discretize the solution finely in the region of these singularities and more coarsely elsewhere where it is smooth. In answer to this observation, we have adapted to the semi-implicit scheme the multiresolution techniques established for explicit schemes in [2] (see [4] and references herein). A further enhancement involving adapting the time step to the local grid size is also presented in [4].

Basics of Multiresolution Analysis

We consider a uniform reference mesh with step size Δx, and a hierarchy of $K+1$ discretization levels of step size $2^{K-k}\Delta x$ for $k=0,\dots,K$. The finite volume representation of the solution at a level k can be *encoded* at a coarser level $k-1$ by averaging. Inversely, the solution at a coarse level k can be *decoded* to reconstruct the solution at the finer level $k+1$, using a local polynomial operator. The differences -or details- between the predicted values and the actual values at level $k+1$ measure the local smoothness of the solution. The solution is represented on an adaptive grid designed by locally selecting the level above which the details are negligible up to a given tolerance ε.

This non-uniform grid evolves with time, with a strategy based on the prediction of the displacement and formation of the singularities in the solution. The wavelet basis used to perform the multiscale analysis enables to reconstruct the solution at any time back to the finest level of discretization, within an error tolerance controlled by the threshold parameter ε.

Local Time Stepping

In the first works [5, 2], the time step is dictated by the size of the smallest cell in the adaptive grid. In [4], the local time stepping approach developed by Müller and Stiriba in [6] is adapted to the explicit and semi-implicit case: since the stability of the scheme is controlled by a CFL condition we can design an elementary time step Δt which can be used to update the solution in the small cells of size Δx – corresponding to stiff variation areas– while larger time steps $2^k\Delta t$ are used in larger cells of size $2^k\Delta x$ belonging to coarser levels k, for $k=0,\dots K-1$. While the largest cells of the grid, of size $2^K \Delta x$, are updated in a single step, the smaller cells on finer levels k require 2^k intermediate updates to be synchronized in time. The evolution algorithm consists of a time loop on the macro time step of length $2^K \Delta t$. Within each macro time step a second loop on 2^K intermediate time steps is performed, summarized in the following flow chart

- Loop on intermediate time steps $i = 1, \ldots, 2^K$
  - Loop on levels $k = K \searrow k_i$ (synch. level at time step i)
    Update fluxes where needed
    Update solution on level k using time step $2^k \Delta t$
  - If i even, partial regriding on levels k_i to K.

At a given intermediate time step, only cells finer or in the current *synchronization level* are updated using the adequate local time step. This is a somewhat technical but very powerful notion which is illustrated in Figure 1. The conservativity of the resulting scheme is ensured by the design of transition zones between cells of size $2^k \Delta x$ and $2^{k-1} \Delta x$: some cells on the coarse level $k-1$ are still updated with the small time step $2^k \Delta t$. The same ratio $\lambda = \Delta t / \Delta x$ is valid at all levels, except in the transition zones where half this value is used. At each intermediate time step

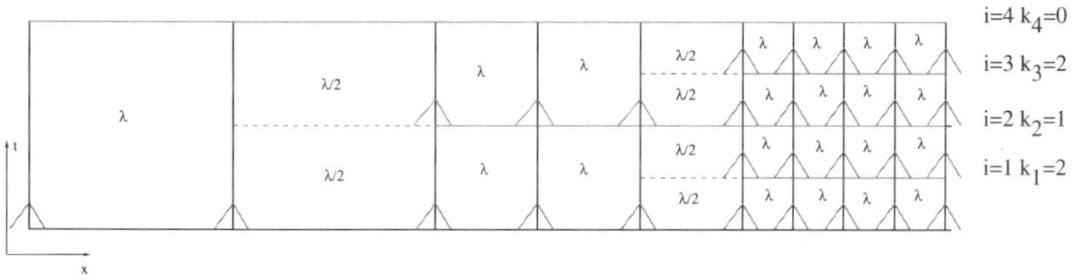

Fig. 1 A three-level adaptive grid with transition zones of width 1 represented with dotted lines. The intermediate step number is denoted by $i = 1, \ldots, 4$, and k_i denotes the synchronization level. The arrows indicate the fluxes that are computed at the corresponding time step

i, the implicit system (2) providing the solution of the Lagrange step is designed. Recall that the Lagrange step deals with fast waves, traveling at a speed much larger than the transport speed. The linear system must therefore include the cells in levels $k = k_i, \ldots, K$, plus all cells in coarser levels within a distance $N_a 2^{k_i} \Delta x$ (see (5)).

4 Numerical Validation

The adaptive algorithms has been extensively tested in [4] in the fully explicit, and the implicit-explicit version. The robustness in terms of precision and computing time performance have been illustrated on Riemann problems. We present here another comparison of the Multiresolution and the Local Time Stepping schemes for an initial value problem with a moving discontinuity separating two smoothly varying regions. This is closer to typical operating conditions and allows a more realistic study of performances. Figure 2 displays the density profile computed on

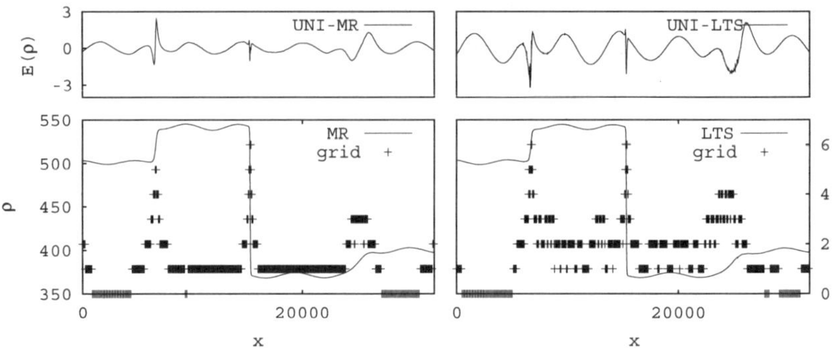

Fig. 2 Density profile at $t = 37$s computed using standard multiresolution (MR) and local time stepping (LTS) schemes. The + denote the level of discretization used locally, to be read on the right vertical axis. The top curves represent the error with the density computed using the reference scheme on the finest uniform grid (UNI).

a grid with cells of 1m at time $t = 36.864$s, with the multiresolution (MR) and the local time stepping (LTS) schemes. Both uniform and MR schemes use the same time step $\Delta t = 0.018$s. After 2048 time steps, the slow wave has moved 750 meters to the left at a speed approximately equal to -20m/s. One acoustic wave has propagated to the right at a speed roughly given by 250m/s while another one has propagated to the left at a mean speed of -255m/s. The curves at the top of the figures display the difference between the density field computed by the finest grid uniform scheme and the two adaptive schemes. The two acoustic waves are slightly distorted by the multiresolution schemes. The transport wave discontinuity, on the other hand, is almost identically resolved by both adaptive schemes. The adaptive grid has 7 levels of discretization numbered from 0 to 6 on the right vertical axis and symbols + indicate which level is used locally. The coarsest level of discretization, with cells of length 64m is used in areas where the solution is very smooth, while near the transport discontinuity and also the left-going acoustic wave, the finest cells are needed.

We have performed a **parametric study** on this test case, using several threshold values ε between 10^{-6} and 10^{-1} and several numbers of levels in the multiscale hierarchy, between 1 and 12. For each set of parameters we compute the relative L^1 error on the gas mass fraction obtained either with the multiresolution or the local time stepping schemes with respect to the solution computed on the fine uniform grid. In Figure 3 the error is displayed versus the computing time gain. This is the ratio between the computed time needed by the uniform scheme and by either the multiresolution or the local time stepping scheme. Three multiscale grids are tested for each adaptive scheme with 1, 6 or 11 levels of coarsening starting from the fine grid. On each curve, the different points correspond to different thresholding

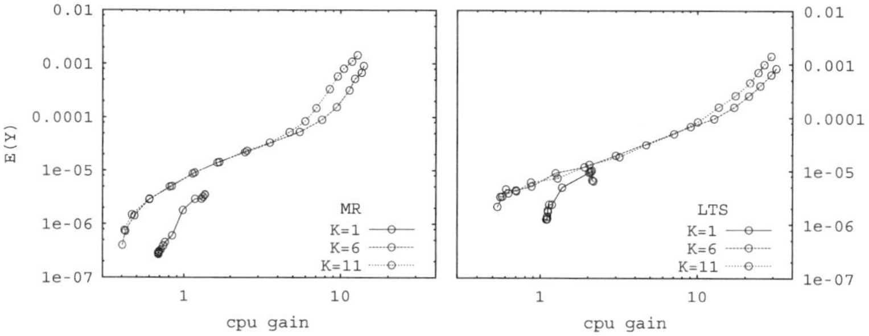

Fig. 3 Relative error on the gas mass fraction versus computing time gain for the standard multiresolution (MR) scheme on the left and the local time stepping (LTS) scheme on the right. Results for grids with 1, 6 and 11 coarsening levels are represented and on each curve the points correspond to simulations with different thresholds ε, varying between 10^{-6} and 10^{-1}.

values. Small values of ε mean that almost all details must be kept therefore ensuring small error, and little coarsening of the grid will be performed, therefore allowing only poor computing time gain. Using only one level of coarsening gives very good results in terms of precision but hardly any benefit in terms of computing time, due to the book-keeping and coding/decoding operations. For large number of levels, the computing time gain is as high as 10 for the multiresolution scheme and three times better using the local time stepping for the same error level.

Since we plan to use this scheme in an engineering context using realistic and very costly state laws in (1), it is also very informative to know the gain in terms of calls to the state laws, which is more or less the same as the gain in grid size. We display in Figure 4 the relative error on the gas mass fraction as a function of the ratio between the number of calls to the state laws in the uniform and adaptive grid simulation. Results for all possible depths of resolution –1 to 12– levels of coarsening and all threshold values from 10^{-6} to 10^{-1} are represented. The best gain is 42 for the standard multiresolution and 650 for the local time stepping scheme, both achieving a relative error of 0.0013 on the gas mass fraction.

Acknowledgements This work was supported by the Ministère de la Recherche under grant ERT-20052274: *Simulation avancée du transport des hydrocarbures* and by the Institut Français du Pétrole.

References

1. Andrianov, N., Coquel, F., Postel, M., Tran, Q.H.: A relaxation multiresolution scheme for accelerating realistic two-phase flows calculations in pipelines. Internat. J. Numer. Methods Fluids **54**(2), 207–236 (2007)
2. Cohen, A., Kaber, M.S., Müller, S., Postel, M.: Fully adaptive multiresolution finite volume schemes for conservation laws. Math. Comp. **72**(241), 183–225 (2003)

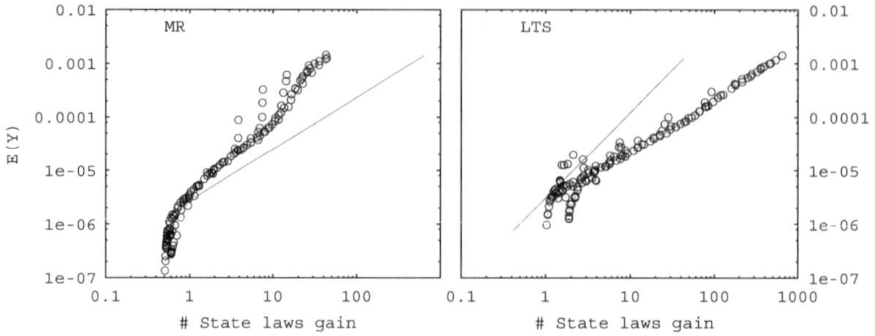

Fig. 4 Relative error on the gas mass fraction versus state law calls ratio for the standard multiresolution (MR) scheme on the left and the local time stepping (LTS) scheme on the right. Each point corresponds to a simulation with a different threshold ε, varying between 10^{-6} and 10^{-1} and a different number of levels in the adaptive grid, varying between 1 and 12. The straight line recalls the behavior of the other scheme.

3. Coquel, F., Nguyen, Q.L., Postel, M., Tran, Q.H.: Entropy-satisfying relaxation method with large time-steps for Euler IBVPs. Submitted, prepublication LJLL R08004
4. Coquel, F., Nguyen, Q.L., Postel, M., Tran, Q.H.: Local time stepping applied to implicit-explicit methods for hyperbolic systems. Submitted, prepublication LJLL R07058
5. Coquel, F., Postel, M., Poussineau, N., Tran, Q.H.: Multiresolution technique and explicit-implicit scheme for multicomponent flows. J. Numer. Math **14**(3), 187–216 (2006)
6. Müller, S., Stiriba, Y.: Fully adaptive multiscale schemes for conservation laws employing locally varying time stepping. J. Sci. Comput. **30**(3), 493–531 (2007)

A ‘TVD-like’ Scheme for Conservation Laws with Source Terms

R. Donat Beneito and A. Martínez Gavara

Abstract The theoretical foundations of high-resolution TVD schemes for homogeneous scalar conservation laws and linear systems of conservation laws have been firmly established through the work of Harten [5], Sweby [11], and Roe [9]. These TVD schemes seek to prevent an increase in the total variation of the numerical solution, and are successfully implemented in the form of flux-limiters or slope limiters for scalar conservation laws and systems. However, their application to conservation laws with source terms is still not fully developed. In this work we analyze the properties of a second order, flux-limited version of the Lax-Wendroff scheme preserving steady states [3]. Our technique is based on a flux limiting procedure applied only to those terms related to the physical flow derivative.

1 Introduction

The theory of numerical schemes for homogeneous scalar conservation laws is well established. Total Variation Diminishing (TVD) schemes have proved to be particularly successful at capturing shock waves and discontinuous solutions. A problem of increasing importance in Computational Fluids Dynamics is the application of numerical methods to inhomogeneous problems such as shallow water equations. In such cases the TVD property is no longer valid. Although certain source terms may preserve the TVD property of the homogeneous part, others will actively increase the variation in the solution. An adapted one-step second-order scheme gives a very good accuracy of the solution in smooth regions although the inevitable presence of spurious overshoots in the proximity of the shock, typical of second order schemes,

Anna Martínez Gavara
Universitat de València, Doctor Moliner,50 46100 Burjassot (Spain),
e-mail: Ana.Martinez-Gavara@uv.es

Rosa Donat Beneito
Universitat de València, Doctor Moliner,50 46100 Burjassot (Spain), e-mail: Rosa.M.Donat@uv.es

has been observed. As in the homogeneous case, the oscillations are not reduced if we use a fine mesh (see Fig. 1). This motivates the use of TVD-like schemes for inhomogeneous problems, however, although care needs to be taken in the inclusion of the source terms.

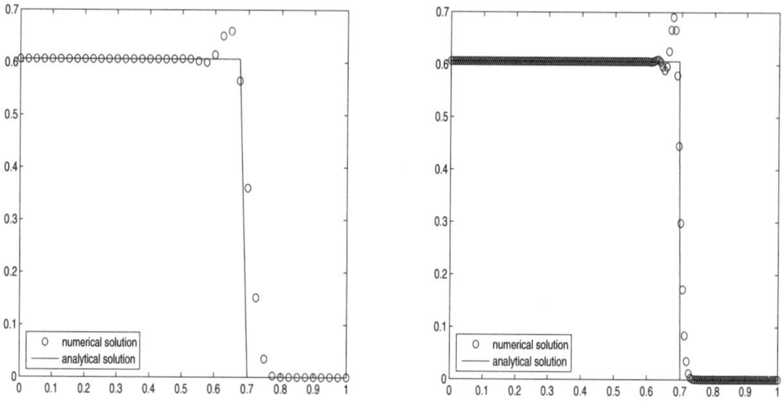

$\triangle x = 0.025$ $\triangle x = 0.00625$

Fig. 1 Second order scheme applied to $u_t + u_x = -u$.

2 Operator-Splitting

A popular method of treating inhomogeneous hyperbolic equations of the form

$$u_t + f(u)_x = s(x,u), \tag{1}$$

is to split the problem, over a time step $\triangle t$, into a homogeneous conservation part

$$u_t + f(u)_x = 0 \tag{2}$$

and an ODE part

$$u_t = s(x,u) \tag{3}$$

and to alternate between the two solutions. The numerical solution for a general scalar problem of the type (1) would be to find the numerical solution, $\bar{u}^{n+1}$, of (2) with initial data $u(x,t^n) = u^n$, a high order TVD scheme would be suitable, then use a numerical ODE solver , like various Runge-Kutta type methods, to obtain u^{n+1} from (3) with initial data $u = \bar{u}^{n+1}$.

The advantages of such an approach are clear since numerical schemes for both (2) and (3) are well developed and can be chosen to optimal effect. This is particularly true for stiff problems where much work has been undertaken using implicit ODE solvers [7]. Despite their advantages in problems with stiff source terms, the situation is by no means ideal, however. The solution of the homogeneous PDE part may cause a large departure from the true solution which will need to be recovered by the ODE solver. If the recovery is not exact, numerical errors will be introduced.

There are some other potential pitfalls in using a fractional-step method to handle source terms. This approach performs very poorly in those situations where u_t is small relative to the other two terms, in particular when steady or quasi-steady solutions are being sought. For such solutions, highly accurate numerical simulations can only be obtained from numerical methods that respect the balance that occurs between the flux gradient and the source term when u_t is small, and it is known ([6]) that this balance is not likely to be respected when using a fractional step approach.

3 An Adapted Second-Order Method

Many numerical methods, like fractional step methods, have difficulties preserving steady states and cannot accurately calculate small perturbations of such states, as we have observed in the previous section.

The source term has to be incorporated into the algorithm, avoiding fractional steps. In general, the source term can be approximated in two ways: A pointwise approach, where the source term approximation is calculated at the nodal points, and an upwind characteristic based approach, where the source term is approximated in a more physical way. Roe [10] put forward the idea of upwinding the source terms in inhomogeneous conservation laws, in a manner similar to that for constructing numerical fluxes for solving homogeneous conservation laws. Further work in this direction was carried out by Bermúdez and Vázquez-Cendón [1], who started by considering the problem

$$u_t + au_x = s(x,u). \tag{4}$$

The solution of this inhomogeneous linear equation with nonlinear source, considering a constant ($a > 0$), at time $t = (n+1)\triangle t$ can be calculated by integrating along the characteristic through (x_i, t_{n+1}) between t_n and t_{n+1} to give

$$u(x_i,t_{n+1}) = u(x_i - a\triangle t, t_n) - \int_{t_n}^{t_{n+1}} b(x_i - a(t_{n+1}-\xi), u(x_i - a(t_{n+1}-\xi), \xi) d\xi. \tag{5}$$

In the above integral b is clearly dependent on data in the upwind domain, indicating a need for an upwind treatment of source term.

In this sense, Gascón and Corberán in [3] presented an extension of the one-step Lax-Wendroff scheme for inhomogeneous conservation laws by rewriting (1) as

$$u_t + g(x,u)_x = 0 \quad \text{where} \quad g(x,u) = f(u) - \int_0^x s(\xi, u(\xi,t))d\xi. \tag{6}$$

A second order method is obtained by the scheme

$$U_i^{n+1} = U_i^n - \frac{\Delta t}{\Delta x}(g_{i+\frac{1}{2}}^{n+\frac{1}{2}} - g_{i-\frac{1}{2}}^{n+\frac{1}{2}}), \tag{7}$$

where the estimation of the new flux, g, at the point mid-way between grid points is obtained by an expression based on Taylor's expansion

$$g_{i+\frac{1}{2}}^{n+\frac{1}{2}} = g_{i+\frac{1}{2}}^{n} + \frac{\Delta t}{2}\frac{\partial g}{\partial t}\bigg|_{i+\frac{1}{2}}^{n}. \tag{8}$$

By introducing the following notation:

$$\alpha_{i+\frac{1}{2}}^n = \frac{\Delta t}{\Delta x}\frac{\partial f}{\partial u}\bigg|_{i+\frac{1}{2}}^n, \quad \beta_{i+\frac{1}{2}}^n = \frac{\Delta t}{2}\frac{\partial s}{\partial u}\bigg|_{i+\frac{1}{2}}^n, \quad b_{ik}^n \approx \int_{x_i}^{x_k} -s(\xi, u(\xi, t_n))d\xi^1,$$

and using simple algebraic manipulations, the scheme admits the expression

$$\begin{aligned} U_i^{n+1} &= U_i^n - \frac{\Delta t}{\Delta x}(f_{i+\frac{1}{2}}^{LW} - f_{i-\frac{1}{2}}^{LW}) - \frac{\Delta t}{\Delta x}(b_{i-\frac{1}{2}i}^n + b_{ii+\frac{1}{2}}^n) \\ &- \frac{\Delta t}{2\Delta x}(\beta_{i+\frac{1}{2}}^n(f_{i+1}^n - f_i^n + b_{ii+1}^n) + \beta_{i-\frac{1}{2}}^n(f_i^n - f_{i-1}^n + b_{i-1i}^n)) \end{aligned} \tag{9}$$

with

$$f_{i+\frac{1}{2}}^{LW} = \frac{1}{2}(f_{i+1}^n + f_i^n - b_{ii+\frac{1}{2}}^n + b_{i+\frac{1}{2}i+1}^n - \alpha_{i+\frac{1}{2}}^n(f_{i+1}^n - f_i^n + b_{ii+1}^n)). \tag{10}$$

4 A Flux Limiter Scheme

The motivation for this work is to analyze the properties of a second order, flux-limited version of the Lax-Wendroff scheme which preserves the TVD property, in the sense that it avoids oscillations around discontinuities, while preserving steady states ([3]).

We consider the Lax-Wendroff method (9) adapted to a balance law (1), this is a second order method that generates spurious oscillations near discontinuities (see Fig. 1). In order to construct a "flux-limiting" method, we consider the numerical flux in (9) of the form

$$F_{i+\frac{1}{2}}^n = F_{i+\frac{1}{2}}^{LO} + \phi_{i+\frac{1}{2}}^n(F_{i+\frac{1}{2}}^{HI} - F_{i+\frac{1}{2}}^{LO}), \tag{11}$$

---

[1] We make sure that b_{ik}^n approximation guarantee that the scheme satisfy the exact C-property, i.e., it is exact when applied to the stationary case.

using (10) as a high order numerical flux($F^{HI}_{i+\frac{1}{2}}$). As a low order numerical flux($F^{LO}_{i+\frac{1}{2}}$), our choice is

$$F^{LO}_{i+\frac{1}{2}} = \frac{1}{2}(f^n_{i+1} + f^n_i - b^n_{ii+\frac{1}{2}} + b^n_{i+\frac{1}{2}i+1} - sign(\alpha^n_{i+\frac{1}{2}})(f^n_{i+1} - f^n_i + b^n_{ii+1})). \quad (12)$$

We called this method the **TVDB** scheme, and we can notice that the numerical flux incorporates information on the source term in its definition, for this reason, we define the variable $r^n_{i+\frac{1}{2}}$, that is always the ratio of the upwind change to the local change, as:

$$r^n_{i+\frac{1}{2}} = \begin{cases} \dfrac{f^n_i - f^n_{i-1} + b^n_{i-1i}}{f^n_{i+1} - f^n_i + b^n_{ii+1}}, & sign(\alpha^n_{i+\frac{1}{2}}) > 0; \\ \\ \dfrac{f^n_{i+2} - f^n_{i+1} + b^n_{i+1i+2}}{f^n_{i+1} - f^n_i + b^n_{ii+1}}, & sign(\alpha^n_{i+\frac{1}{2}}) < 0. \end{cases} \quad (13)$$

As a flux limiter function $\phi^n_{i+\frac{1}{2}} = \phi(r^n_{i+\frac{1}{2}})$ we use the minmod limiter,

$$\phi^n_{i+\frac{1}{2}} = max(0, min(r^n_{i+\frac{1}{2}}, 1)) \quad (14)$$

In Fig 2 left, we display the numerical results obtained after applying this scheme to $u_t + u_x = -u$. A slight oscillation can be observed, whose amplitude decreases with the mesh width, as shown in Fig. 2 right. The oscillatory behavior can be completely avoided by using an implicit scheme (see [8]).

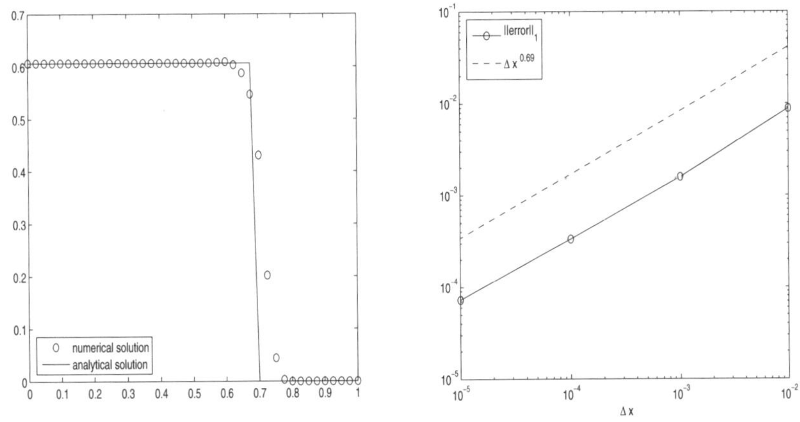

Fig. 2 $u_t + u_x = -u$. Left: TVB method. Right: Error for the TVDB method.

5 Burgers' Equation with Source Terms

In a variety of physical problems one encounters source terms that are balanced by internal forces and this balance supports multiple steady state solutions that are stable. Typical of these are gravity-driven flows such as those described by the shallow water equations over a nonuniform ocean bottom. In this section we show a scalar 1-D approximation of balance laws of this kind.

5.1 The Embid Problem

This problem was presented in [2] as a simple scalar approximation to the 1-D equations that model the flow of a gas through a duct of variable cross-section.

$$\begin{cases} u_t + (\frac{u^2}{2})_x = (6x-3)u, & 0 < x < 1 \\ u(0,t) = 1, u(1,t) = -0.1. \end{cases} \tag{15}$$

There are two entropy satisfying steady solutions for the Embid problem. One is stable in time with a standing shock at $x_1 = 0.18$ and the other with an unstable standing shock at $x_2 = 0.82$. The steady solutions for the Embid problem are

$$u(x) = \begin{cases} 1 + 3x^2 - 3x, & x < x_i; \\ -0.1 + 3x^2 - 3x, & x > x_i. \end{cases} \tag{16}$$

for $i = 1, 2$. We computed the steady profiles by taking initial data with a jump at the stable location, using a CFL number equal to 0.8 and by marching in time until convergence criterion

$$\sum_i |u_i^{n+1} - u_i^n| \leq 10^{-10}$$

was satisfied (Fig. 3 left).

The TVDB numerical solution reproduces the exact steady solution except for one internal shock point. The scheme requires 383 iterations to reach the stationary solution with a residual less than 10^{-10} and using the minmod limiter. Fig. 3 right shows the logarithm of the residual errors with respect to the number of iterations for both schemes.

5.2 Greenberg et al. Tests

In order to test the methods described above, we show the numerical result following the tests in [4]. Let us consider the equation

$$u_t + (\frac{u^2}{2})_x + a_x(x)u = 0. \tag{17}$$

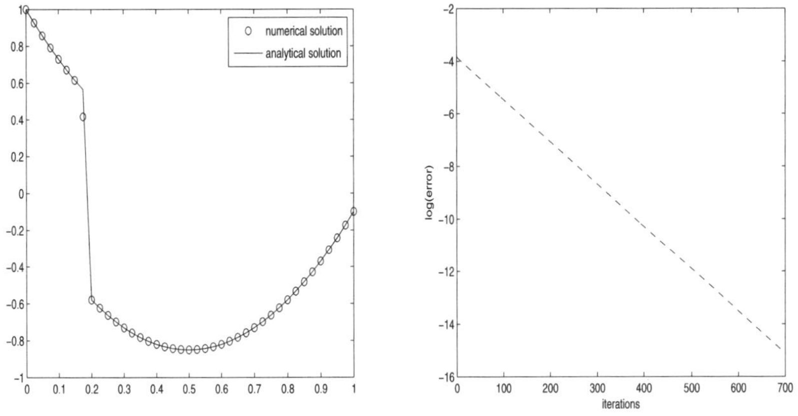

Fig. 3 Embid problem. Left: TVB scheme. Right: Convergence history

where

$$a(x) = 0.9 \begin{cases} 0, & x < 0; \\ (\cos(\pi \frac{x-1}{2}))^{30}, & 0 \le x \le 2; \\ 0, & 2 < x. \end{cases} \tag{18}$$

Fig. 4 left is the numerical solution of (17) with the initial data $u + a = 1$ at time

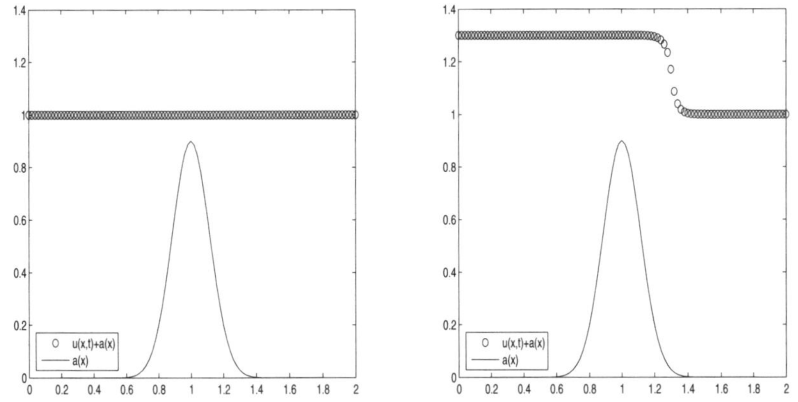

Fig. 4 Left: Experiment 1. Right: Experiment 2.

1 (Experiment 1). The l_1-error is $6.9044 \cdot 10^{-17}$ for the TVDB scheme, thus the C-property is ensured.

On the other hand, the initial condition used to generate Fig. 4 right at time 1.5 (Experiment 2) is

$$u+a=\begin{cases}1.3, & x<0.2;\\ 1, & 0<x.\end{cases} \tag{19}$$

In this case, we cannot observe any spurious oscillations in the numerical solution.

Acknowledgements The authors acknowledge support from Spanish MTM2005-07214.

References

1. Bermúdez, A., Vázquez, M.: Upwind methods for hyperbolic conservation laws with source terms. Computers & Fluids **23**, 1049–1071 (1994)
2. Embid, P., Goodman, J., Majda, A.: Multiple steady states for 1-d transonic flow. SIAM Journal on Scientific and Statistical Computing **5**, 21–41 (1984)
3. Gascón, L., Corberán, J.M.: Construction of second-order tvd schemes for nonhomogeneous hyperbolic conservation laws. J. Comput. Phys. **172**(1), 261–297 (2001). DOI http://dx.doi.org/10.1006/jcph.2001.6823
4. Greenberg, J.M., Leroux, A.Y.: A well-balanced scheme for the numerical processing of source terms in hyperbolic equations. SIAM J. Numer. Anal. **33**(1), 1–16 (1996). DOI http://dx.doi.org/10.1137/0733001
5. Harten, A.: High resolution schemes for hyperbolic conservation laws. Journal of Computational Physics **135(2)**, 260–278 (1997)
6. LeVeque, R.J.: Balancing source terms and flux gradients in high-resolution godunov methods: the quasi-steady wave-propagation algorithm. J. Comput. Phys. **146**(1), 346–365 (1998). DOI http://dx.doi.org/10.1006/jcph.1998.6058
7. LeVeque, R.J., Yee, H.C.: A study of numerical methods for hyperbolic conservation laws with stiff source terms. J. Comput. Phys. **86**(1), 187–210 (1990). DOI http://dx.doi.org/10.1016/0021-9991(90)90097-K
8. Patankar, S.V.: Numerical Heat Transfer and Fluid Flow (1980)
9. Roe, P.: Generalized formulation of tvd lax-wendroff schemes. ICASE **84**, 53 (1984)
10. Roe, P.: Upwind differencing schemes for hyperbolic conservation laws with source terms. In: in proceedings of Nonlinear Hyperbolic problems,edited by C. Carasso, P.Raviart, and D. Serre, Lecture Notes in Mathematics (Springer-Verlag), vol. 1270, p. 41 (1986)
11. Sweby, P.K.: High resolution schemes using flux limiters for hyperbolic conservation laws. SIAM Journal on Numerical Analysis **21**(5), 995–1011 (1984). DOI 10.1137/0721062. URL http://link.aip.org/link/?SNA/21/995/1

Application of the WAF Method to Shallow Water Equations with Pollutant and Non-Constant Bottom

E.D. Fernández-Nieto and G. Narbona-Reina

Abstract In this work we perform the extension of the WAF method [3] to discretize non-homogeneous Shallow Water Equations with pollutant.

We propose a well-balanced extension: the numerical scheme preserves all stationary solutions up to second order, and exactly preserves water at rest. The difficulty lies in the treatment of the pollutant component that includes an extra term related with the approximation of the intermediate wave.

Finally, we perform several numerical tests, by comparing it with the HLLC solver, analytical solutions and reference solutions.

1 The WAF Method

In this section we summarize the WAF method for the homogeneous SWE with pollutant. As a general reference for this section, see [4].

We begin by considering the homogeneous SWE given by the system

$$\begin{cases} \partial_t W + \partial_x F(W) = 0, & x \in [0,L], t \in [0,T], \\ W(x,0) = W_0 & x \in [0,L]; \end{cases} \tag{1}$$

where $W = (h,q,r)$, and $F(W) = \left(q, \dfrac{q^2}{h} + \dfrac{1}{2}gh^2, \dfrac{qr}{h}\right)$.

E.D. Fernández Nieto
Dpto. de Matemática Aplicada I, E.T.S. Arquitectura. Universidad de Sevilla. Avda. Reina Mercedes 2. 41012 Sevilla, Spain, e-mail: edofer@us.es

G. Narbona-Reina
Dpto. de Matemática Aplicada I, E.T.S. Arquitectura. Universidad de Sevilla. Avda. Reina Mercedes 2. 41012 Sevilla, Spain, e-mail: gnarbona@us.es

The unknowns are h, the height of the water column, q, the discharge; if we denote by ψ the pollutant concentration, then $r = h\psi$. W_0 is the initial data, L is the length of the domain, T the final time and g is the constant gravity.

We consider a partition of the domain $\{x_i\}_i = \{i\Delta x\}_i$ where, by simplicity, we take Δx a constant space step, and we denote by $t^n = t^{n-1} + \Delta t$ the time values, with Δt the time step. If we use a finite volume method in conservative form to approximate the solution of this problem, we have

$$W_i^{n+1} = W_i^n - \frac{\Delta t}{\Delta x}(\phi_{i+1/2}^n - \phi_{i-1/2}^n), \tag{2}$$

where we denote by W_i^n an approximation of the mean value of the solution on the control volume $(x_{i-1/2}, x_{i+1/2})$ at time $t = t^n$, and by $\phi_{i+1/2}^n = \phi(W_i^n, W_{i+1}^n)$ the numerical flux function that characterize each method.

To obtain the numerical flux of the WAF method we integrate in the computational grid $[-\Delta x/2, \Delta x/2] \times [0, \Delta t]$ (see Fig. 1), getting:

$$\phi_{i+1/2}^{WAF} = \frac{1}{2}(F_i + F_{i+1}) - \frac{1}{2}\sum_{k=1}^{N} M_k \Delta F_{i+1/2}^{(k)}, \tag{3}$$

where $M_k = sign(S_k)A_k$, A_k is a flux limiter function and S_k the approximation of characteristic velocities. We have denoted by $\Delta F_{i+1/2}^{(k)} = F_{i+1/2}^{(k+1)} - F_{i+1/2}^{(k)}$, with $F_{i+1/2}^{(k)}$ the value of the flux function in the interval k (see Fig. 1).

If we denote by $[\cdot]_j$ the *j-th* component, the usual choice for $F^{(k)}$ definition is the HLLC flux, given by:

$$\begin{aligned}
&[\phi_{i+1/2}^{HLLC}]_j = [\phi_{i+1/2}^{HLL}]_j \quad j = 1,2;\\
&[\phi_{i+1/2}^{HLLC}]_3 = [\phi_{i+1/2}^{HLL}]_1 \psi_* \quad \text{where } \psi_* = \begin{cases} \psi_i & \text{if } S_2 \geq 0 \\ \psi_{i+1} & \text{if } S_2 < 0. \end{cases}
\end{aligned} \tag{4}$$

Where the HLL flux $\phi_{i+1/2}^{HLL}$ is defined as: F_i if $S_1 \geq 0$, F_{i+1} if $0 \geq S_3$ and $F^{HLL} = \frac{S_3 F_i - S_1 F_{i+1} + S_3 S_1 (W_{i+1} - W_i)}{S_3 - S_1}$ when $S_1 \leq 0 \leq S_3$. For S_i, $1 \leq i \leq 3$, for example we can set:

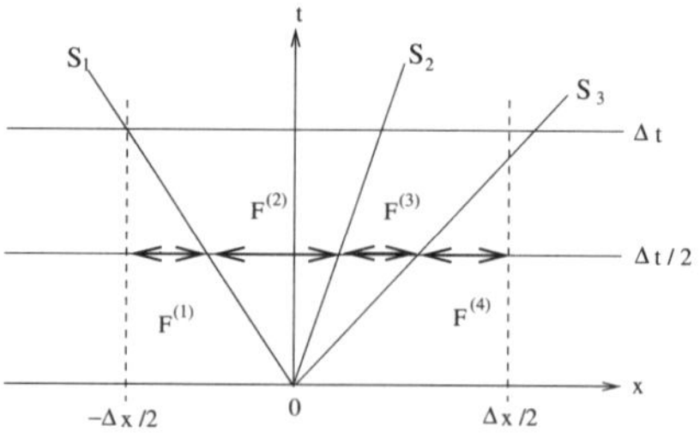

Fig. 1 Computational grid and intermediate waves to compute WAF method.

$$S_1 = u_i - \sqrt{gh_i} \quad S_2 = \frac{S_1 q_{i+1} - S_3 q_i - S_1 S_3 (h_{i+1} - h_i)}{q_{i+1} - q_i - h_{i+1} S_3 + h_i S_1} \quad S_3 = u_{i+1} + \sqrt{gh_{i+1}}. \tag{5}$$

Finally we can write the WAF flux as follows:

$$[\phi^{WAF}]_j = \left[\frac{F_i + F_{i+1}}{2} - \frac{1}{2}\frac{1}{S_3 - S_1}\left((M_3 S_3 - M_1 S_1)(F_{i+1} - F_i) - \right.\right.$$
$$\left.\left. - \; S_3 S_1 (M_3 - M_1)(W_{i+1} - W_i)\right)\right]_j ; \text{ for } j = 1,2. \tag{6}$$

$$[\phi^{WAF}]_3 = \left[\frac{F_i + F_{i+1}}{2}\right]_3 - \frac{1}{2}\frac{1}{S_3 - S_1}\left[(S_3 M_3 \psi_{i+1} - S_1 M_1 \psi_i)(F_{i+1} - F_i) + \right.$$
$$+ \; S_1 S_3 (M_1 \psi_i - M_3 \psi_{i+1})(W_{i+1} - W_i) +$$
$$\left. + \; M_2(\psi_{i+1} - \psi_i)\big(S_3 F_i - S_1 F_{i+1} + S_1 S_3 (W_{i+1} - W_i)\big)\right]_1. \tag{7}$$

2 Extension to Non-Homogeneous System

We consider now the non-homogeneous shallow-water equations with pollutant:

$$\begin{cases} \partial_t W + \partial_x F(W) = G(x,W), \; x \in [0,L], \, t \in [0,T], \\ W(x,0) = W_0 \qquad\qquad\qquad x \in [0,L]. \end{cases} \tag{8}$$

If we take the topography source term and we denote by $z_b(x)$ the height of the topography at point x, we have: $G(x,W) = \big(0, -ghz_b'(x), 0\big)$.

In order to obtain the structure of the numerical scheme in conservative form, we must integrate equation (8) on the control volume $(x_{i-1/2}, x_{i+1/2})$. Thus, we obtain

$$\frac{W_{i+1}^n - W_i^n}{\Delta t} + \frac{F_{i+1/2} - F_{i-1/2}}{\Delta x} = \frac{1}{\Delta x}\int_{x_{i-1/2}}^{x_{i+1/2}} G(x, W(x))dx.$$

Where, $F_{i+1/2}$ denotes an approximation of $F(\tilde{W}(x_{i+1/2}))$ with $\tilde{W}$ the solution of the Riemann problem associated to the non-homogeneous system. The key point is that the Riemann problem depends on the original system, then $F_{i+1/2}$ must depend on the source term. So we use the following structure for the scheme to approximate (8):

$$\frac{W_{i+1}^n - W_i^n}{\Delta t} + \frac{\phi_{G,i+1/2}^n - \phi_{G,i-1/2}^n}{\Delta x} = G_{Ci}^n. \tag{9}$$

We have denoted by ϕ_G the numerical flux depending on the source term G and by G_{Ci} a centered approximation of G at point x_i.

So, the objective of this section is to define ϕ_G, as a generalization of the WAF method depending on G. To do it, we focus on the well-balanced properties.

2.1 Studying the Numerical Viscosity of the Method

The technique introduced in [1] is to study the numerical viscosity of the method to extend it to non-homogeneous system. The objective is that the numerical viscosity term must vanishes for all stationary solutions.

If we can write the numerical flux function in the following form

$$\phi_{i+1/2} = \frac{F_i + F_{i+1}}{2} + \nu_1(W)(W_{i+1} - W_i) + \nu_2(W)(F_{i+1} - F_i), \tag{10}$$

with ν_1 and ν_2 the viscosity coefficients, the numerical scheme defined by (2)-(10) can be viewed as a centered discretization of the equivalent system:

$$\partial_t W + \partial_x F(W) + \Delta x\left[\partial_x\left(\nu_1 \partial_x W\right) + \partial_x\left(\nu_2 \partial_x F(W)\right)\right] = G(x,W). \tag{11}$$

If W is a stationary solution of (8), and if we assume that the jacobian matrix of F, A, is not singular, then it satisfies: $\partial_x F(W) = G(x,W)$ and $\partial_x W = A^{-1}(W)G(x,W)$. So, W is a stationary solution of the equivalent system

$$\begin{aligned}\partial_t W + \partial_x F(W) + \Delta x\big[\partial_x\left(\nu_1(\partial_x W - A^{-1}(W)G(x,W))\right) + \\ +\partial_x\left(\nu_2(\partial_x F(W) - G(x,W))\right)\big] = G(x,W).\end{aligned} \tag{12}$$

Finally, we propose the following definition of ϕ_G,

$$\begin{aligned}\phi_{G,i+1/2} = {} & \frac{F_i + F_{i+1}}{2} + \Delta x\left[\nu_1(W_i, W_{i+1})\left(\frac{W_{i+1} - W_i}{\Delta x} - \tilde{A}^{-1}_{i+1/2} G_{i+1/2}\right) + \right. \\ & \left. + \ \nu_2(W_i, W_{i+1})\left(\frac{F_{i+1} - F_i}{\Delta x} - G_{i+1/2}\right)\right],\end{aligned} \tag{13}$$

where $\tilde{A}^{-1}_{i+1/2}$ is an approximation of $A^{-1}(W_{i+1/2})$.

2.2 Extension to Non-Homogeneous System

In this section we shall see that we can rewrite the WAF flux under the structure (10) for first and second components. The third component includes another term that can be written in terms of S_2. Finally, following Subsect. 2.1 we propose the extension of WAF method to non-homogeneous system.

Firstly we focus on the first two components, given by (6). It can be written under the form (10) by setting the following values for the viscosity coefficients, :

$$\nu_1(W) = -\frac{1}{2}\frac{S_3 S_1}{S_3 - S_1}(M_1 - M_3) \quad \nu_2(W) = -\frac{1}{2}\frac{M_3 S_3 - M_1 S_1}{S_3 - S_1}.$$

So as for the third one, defined by equation (7), we could take:

$$v_1(W) = -\frac{1}{2}\frac{S_3S_1}{S_3-S_1}(M_1\psi_i - M_3\psi_{i+1}) \quad v_2(W) = -\frac{1}{2}\frac{M_3S_3\psi_{i+1} - M_1S_1\psi_i}{S_3-S_1},$$

but the last term in the definition cannot be put in this form. We noted it by rhs_3:

$$rhs_3 = M_2(\psi_{i+1} - \psi_i)\Big[\underbrace{\big(S_3F_i - S_1F_{i+1} + S_1S_3(W_{i+1} - W_i)\big)}_{(*)}\Big]_1. \tag{14}$$

On the other hand the definition of S_2, given by (5), can also be rewritten in function of $(*)$: $S_2 = -\frac{\big[S_3F_i - S_1F_{i+1} + S_1S_3(W_{i+1} - W_i)\big]_1}{q_{i+1} - q_i - h_{i+1}S_3 + h_iS_1}$. Moreover in [2] the following extension of S_2 to non-homogeneous systems is proposed:

$$S_{2G} = \frac{S_1q_{i+1} - S_3q_i - S_1S_3(h_{i+1} - h_i - \Delta x[\tilde{A}^{-1}_{i+1/2}G_{i+1/2}]_1)}{q_{i+1} - q_i - h_{i+1}S_3 + h_iS_1}. \tag{15}$$

This definition verifies for example that S_{2G} is equal to zero for water at rest, whereas S_2 is non zero for water at rest when topography is not flat. We also consider the corresponding M_{2G} associated to S_{2G} (see (3)). For term $(*)$ in equation (14) we consider an analogous extension to the non-homogeneous case as for S_2. Finally, we propose the following extension of the WAF method:

For the first and second components, $j = 1,2$:

$$[\phi_G^{WAF}]_j = \Big[\frac{F_i + F_{i+1}}{2} - \frac{1}{2}\frac{1}{S_3 - S_1}\big((M_3S_3 - M_1S_1)(F_{i+1} - F_i - \Delta x G_{i+1/2}) - S_3S_1(M_3 - M_1)(W_{i+1} - W_i - \Delta x(A^{-1}G)_{i+1/2})\big)\Big]_j. \tag{16}$$

and for the third component:

$$[\phi_G^{WAF}]_3 = \left[\frac{F_i + F_{i+1}}{2}\right]_3 - \frac{1}{2}\frac{1}{S_3 - S_1}\Big[(S_3M_3\psi_{i+1} - S_1M_1\psi_i)(F_{i+1} - F_i - \Delta x G_{i+1/2}) + S_1S_3(M_1\psi_i - M_3\psi_{i+1})(W_{i+1} - W_i - \Delta x(A^{-1}G)_{i+1/2}) + M_{2G}(\psi_{i+1} - \psi_i)\big(S_3F_i - S_1F_{i+1} + S_1S_3(W_{i+1} - W_i - \Delta x(A^{-1}G)_{i+1/2})\big)\Big]_1. \tag{17}$$

2.3 Well-Balance Property

In this section we state the main result of well-balanced property obtained for the proposed extension of WAF method.

We consider the *asymptotically well-balance* property introduced in [1]:

Definition 1. We say that the scheme (9) is asymptotically well-balanced if there is an increasing sequence of compact sets $\{K_n\}_n$ such that:

1. $\mu([0,L]-\cup_n K_n)=0$, being μ the Lebesgue measure in R.
2. For all n there exists a $\delta_n>0$ such that if $0<\Delta x<\delta_n$, then the scheme balances system (8) up to second order in K_n.

Theorem 1. *We consider the scheme (9), then:*

i) The scheme is asymptotically well-balance for all stationary solutions of the SWE.
ii) The scheme preserves exactly the stationary solution of water at rest.

3 Numerical Tests

In this section we present several numerical tests by comparing the WAF method with HLLC and with exact or reference solutions.

Test 1: A transport of pollutant test with not flat bottom. With this test we show that the WAF scheme can produce very different results, for both approximations of the intermediate wave speed, S_2 given by (5) or S_{2G} given by (15) (see [2]). To consider S_{2G} is essential for the good behavior of the pollutant concentration, as we will see below.

We set a domain of 4 meters, we take a constant space step $\Delta x=0.08$. We state the CFL condition as 0.9. The final time is $T=1$, and the initial conditions:

$$h_0=18-z_b;\quad q_0=h_0;\quad r_0=\begin{cases} h_0 & x<2\\ 0 & x\geq 2\end{cases};\ \text{with } z_b(x)=\begin{cases} 4.5 & x<2\\ 0 & x\geq 2\end{cases}.$$

In Fig. 2 we present the solution obtained using S_{2G}. In a) the bottom function, water surface and pollutant concentration are drawn. In Fig. 2b) we compare the pollutant concentration provided by the WAF solver and HLLC. We observe a less numerical diffusion of the WAF method.

If we use S_2, we obtain a peak of the pollutant concentration near the bump as we observe in Fig. 3b). This is not an unstability, this is the effect of a wrong approximation of the intermediate wave speed. In this problem S_2 must be always positive, nevertheless the definition of S_2 produces a negative value just in $x=2$, as it is shown in Fig. 3a).

Test 2: Stationary transcritical flux with a shock and a periodic time-varying pollutant concentration. We consider a classical test for a stationary solution for h and q but including a source of periodic time pollutant as a boundary condition.

The domain length is 20 meters, the space step $\Delta x=0.1$ and CFL $=0.9$.

The initial conditions are: $h_0=0.33-z_b; q_0=0; r_0=0$, and the boundary conditions: $q=0.18$ and $r=r_p(t)$ at x=0; $h=0.33$ at $x=20$. Being the bottom function:

$$z_b(x)=\begin{cases} 0.2-0.05(x-10)^2 & 8<x<12\\ 0 & \mathit{otherwise}\end{cases}.$$

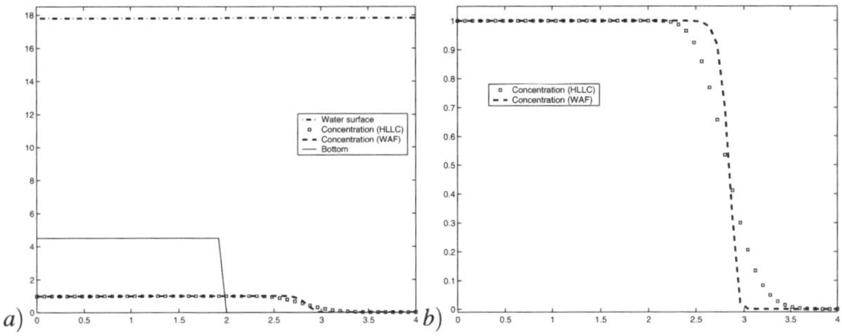

Fig. 2 Test 1: a) Bottom, water surface and pollutant concentration for the HLLC and the WAF method using S_{2G}. b) Pollutant concentrations.

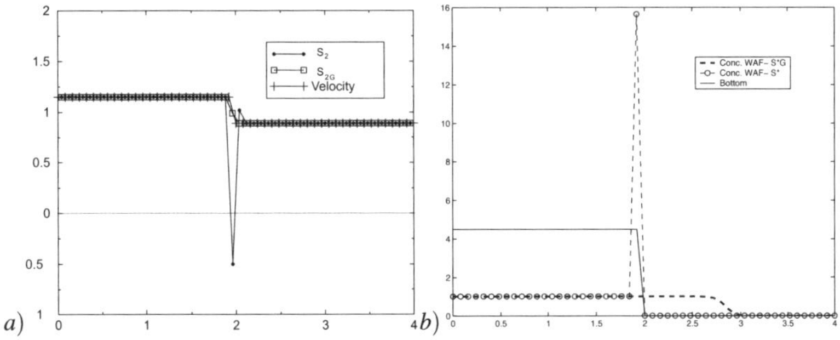

Fig. 3 Test 1: a) Velocity approximations, S_2 and S_{2G}. b) Pollutant concentration for the WAF method taking S_2 or S_{2G}.

The function r_p is a periodical function in time that represents a contribution of pollutant of three seconds, every ten seconds; and it is given by:

$$r_p(t) = \begin{cases} h & t \in [10i, 10i+3] \\ 0 & otherwise \end{cases}, \quad \text{for } i \in \mathbf{N}.$$

In Fig. 4, we compare the exact solution for h and q with the proposed WAF method solution, in both cases we notice an accurate approximation.

In Fig. 5 a) we compare the pollutant concentration for the WAF and HLLC solver at time t = 189 seconds. We observe the great numerical diffusion introduced by HLLC solver in comparison with the WAF method. In Fig. 5b) we compare the WAF method with the approximated solution solving the characteristic curves problem at the same time.

Acknowledgements The research of E.D. Fernández-Nieto and G. Narbona-Reina to carry on this work was partially supported by the Spanish Government Research project MTM2006-01275.

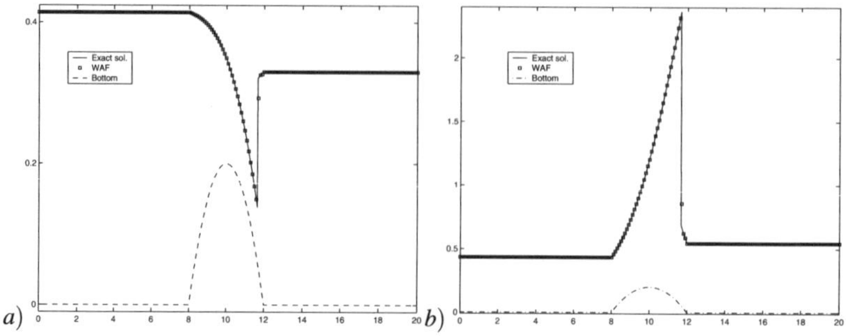

Fig. 4 Test 2: a) Water surface and topography b) Velocity.

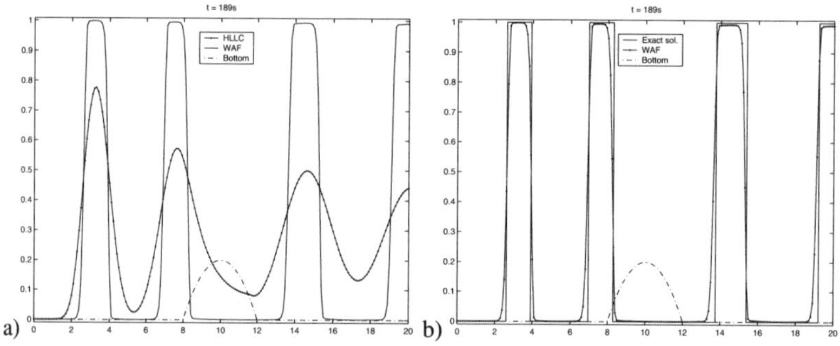

Fig. 5 Test 2: a) Concentration for WAF and HLLC methods. b) WAF and reference solution.

References

1. Chacón Rebollo, T., Domínguez Delgado, A., Fernández Nieto, E.D.: Asymptotically balanced schemes for non-homogeneous hyperbolic systems—application to the shallow water equations. C. R. Math. Acad. Sci. Paris **338**(1), 85–90 (2004)
2. Fernández Nieto, E.D., Bresch, D., Monnier, J.: A consistent intermediate wave speed for a well-balanced hllc solver. Submitted (2007)
3. Toro, E.F.: A weighted average flux method for hyperbolic conservation laws. Proc. Royal Society of London **A 423**, 401–418 (1989)
4. Toro, E.F.: Shock-Capturing Methods for Free-Sufrace Shallow Flows. Wiley and Sons Ltd., Chichester (2001)

A Third Order Method for Convection-Diffusion Equations with a Delay Term

J. Frochte

Abstract The numerical solution of a parabolic convection diffusion equation with delay term is considered. This includes both variants, the initial value problem and the prehistory problem. Equations with a delay or memory term, often called integrodifferential problems, appear in different contexts of heat conduction in materials with memory, viscoelasticity and population models. This work concentrates on the linear convection diffusion case of the prehistory and the initial value problem. One problem concerning delay or memory problems is the data storage. To deal with this problem an adaptivity method of third order in time is developed to save storage data at smooth parts of the solution. Numerical results for higher Péclet numbers are presented.

1 Introduction

Let $T > 0$ and $Q_T = \Omega \times (0,T]$ where Ω is an open bounded region in $\mathbb{R}^n$. First let us consider a parabolic partial integrodifferential equations of the kind

$$\begin{aligned} \frac{d}{dt}u - Au &= f + \int_0^t B(t,s)u\,ds \quad \text{in } \Omega, \\ u &= \hat{u} \quad \text{on } \partial\Omega, \\ u &= u_0 \quad \text{in } \Omega, \text{for } t = 0\,, \end{aligned} \tag{1}$$

with a linear elliptic operator A and a problem depending operator B. Such equations appear for example (see [12], [9], [3], [2] chapter 1) in different contexts of heat conduction in materials with memory, viscoelasticity and population models. These models tend to be nonlinear, but of course first the numerical behaviour of linear

Jörg Frochte
Fachhochschule Südwestfalen, Elektrische Energietechnik Soest, Lübecker Ring 2, 59494 Soest, e-mail: joerg.frochte@uni-due.de

problems has to be studied. Up to now the research in this area has been concentrated on linear and semi-linear cases, see e.g. [11], [5] [7], [6]. We will study a special case of (1).

$$\frac{d}{dt}u - \varepsilon\nabla^2 u + k\cdot\nabla u = f + \int_0^t K(s-t)u\,ds \quad \text{in } \Omega, \tag{2}$$
$$u = \hat{u} \quad \text{on } \partial\Omega,$$
$$u = u_0 \quad \text{in } \Omega, \text{for } t = 0$$

We call this the initial value problem and we also consider the prehistory problem:

$$\frac{d}{dt}u - \varepsilon\nabla^2 u + k\cdot\nabla u = f + \int_{t-d}^t K(s-t+d)u\,ds \quad \text{in } \Omega, \tag{3}$$
$$u = \hat{u} \quad \text{on } \partial\Omega,$$
$$u = u_{history} \quad \text{in } \Omega, \text{for } t \in [-d,0]$$

K is a $C^1(\mathbb{R}\to\mathbb{R})$ function called kernel, $\varepsilon > 0$ is the diffusion coefficient and $k \in C^0(\mathbb{R}^n \to \mathbb{R}^n)$ the convection coefficient. In the prehistory case d is the fix length of the delay.
So additionally to the common problems of partial integrodifferential equations like e.g. memory storage we have to deal with the problems arising in the context of convection diffusion equations. With a rising global Péclet number defined as

$$Pe = \frac{h_\Omega \|k\|_{\max}}{\varepsilon} \tag{4}$$

h_Ω is the characteristic length of the domain Ω. To solve this problem we will use linear finite elements with streamline diffusion stabilisation, see e.g. [8] for details. One of the first approached solver strategies for the initial value problem was published in [10] and uses a left trapez rule to deal with the integral term and a backward Euler scheme for the time discretisation. The result is a scheme of first order for the initial value problem. If we apply this to our initial value problem (2) we achieve:

$$\frac{u^{n+1}-u^n}{\Delta t} - \varepsilon\nabla^2 u^{n+1} + k\cdot\nabla u^{n+1} = f + \Delta t\sum_{j=0}^{n} K(j\Delta t)u^{n-j} \tag{5}$$

To deal with the memory storage problem of the integral term e.g. Thomée advocates integration techniques of higher order so that not every time step has to be stored.

2 Higher Order Integration Scheme

In this paper we propose another approach. First we present a higher order scheme for a fixed time step size using a third order discretisation in time, in our case the

BDF schemes, which is suitable to deal with stiff problems. The higher order in the integral term is performed using hermite interpolation. The resulting integration scheme is independent of the discretisation $\tau = \{t_0, ...t^n\}$ in time so that we can straightforward apply common mechanics for adaptive time step size control to the scheme.

2.1 Integration Using Hermite Interpolation

Let us consider a function $\hat{f}(t) : \mathbb{R} \to \mathbb{R}$ and let $t^j < t^{j+1} \in \mathbb{D}_{\hat{f}}$. So, if we have $f(t^j)$, $\frac{d}{dt}f(t^j)$, $f(t^{j+1})$ and $\frac{d}{dt}f(t^{j+1})$ we can interpolate f on $[t^j, t^{j+1}]$ with a cubic polynomial. Let now the functions in our integral term at a fixed point $\hat{x} \in \Omega$ be denoted as

$$\hat{f}(t) := K(t - t_1) \cdot u(\bar{x}, t) \,.$$

We can now first interpolate $\hat{f}(t)$ using the cubic hermite interpolation over $[t^j, t^{j+1}]$ with $s = (t - t^j)/(t^{j+1} - t^j)$ and

$$u_{t^j} = (1 + 2s)(1 - s)^2, \qquad u_{t^{j+1}} = (3 - 2s)s^2 \tag{6}$$

$$v_{t^j} = s(1 - s)^2, \qquad v_{t_{j+1}} = -s^2(1 - s) \,. \tag{7}$$

Thus we get:

$\hat{f}(t) \approx p(t) = u_{t^j}(t)\hat{f}(t^j) + u_{t^{j+1}}(t)\hat{f}(t^{j+1}) + (t^{j+1} - t^j)\left(v_{t_j}(t)\frac{d\hat{f}(t_j)}{dt} + v_{t_{j+1}}(t)\frac{d\hat{f}(t_{j+1})}{dt}\right)$

This cubic polynomial can be integrated exactly by using simpsons rule.

$$\int_{t_j}^{t^{j+1}} \hat{f}\, dt \approx$$

$$\int_{t_j}^{t^{j+1}} p\, dt = \frac{t^{j+1} - t^j}{2}\left(\hat{f}(t^j) + \hat{f}(t^{j+1})\right) + \frac{(t^{j+1} - t^j)^2}{12}\left(\frac{d\hat{f}(t_j)}{dt} - \frac{d\hat{f}(t_{j+1})}{dt}\right)$$

If we now combine this approach with a time discretisation of third order we receive a method of third order in time. Now there is the question left how to evaluate

$$\frac{\hat{f}(t)}{dt} = u(\bar{x}, t) \cdot \frac{dK(t - t_1)}{dt} + K(t - t_1) \cdot \frac{du(\bar{x}, t)}{dt}$$

of the same order as the time discretisation. While we generally assume that $\frac{dK(t-t_1)}{dt}$ is given as analytical expersion, if not we use the same technique as described for u below, we will have to approximate $\frac{du(\bar{x},t)}{dt}$ anyway. To do this we again choose a BDF approach of the same order as the one used for the time discretisation. Generally it is supposed to compute $\frac{du(\bar{x},t)}{dt}$ once and save it, so that the CPU costs are only spent once. Of course, alternatively it is possible to trade CPU time vs. memory and compute $\frac{du(\bar{x},t)}{dt}$ in every time step.

This approach can be applied in the inner part of the integral. It will not work with the recently added part of the delay integral $[t^n, t^{n+1}]$. Because u^{n+1} is unknown we have to treat it different from the ones before. Many approaches are possible like an extrapolation, a kind of predictor-corrector scheme etc.. In numerical test it turns out that the influence of this part is generally small enough just to use the left trapezium rule for the following exampels.

We also have to consider the beginning of the integral. In this case we have to distinguish between the initial value and the prehistory problem. In both cases it is not clear how to compute an approximation for $\frac{du(\bar{x},0)}{dt}$, respectively $\frac{du(\bar{x},-d)}{dt}$. For in the prehistory case it is quite easy, we can approximate $\frac{du(\bar{x},-d)}{dt}$ using forward instead of backward differences. In the initial value case we can use the same ansatz, but not right from the start. We have to wait a few time steps until enough data has been accumulated. So in the case of the initial value problem until the third time step we will only have an approximation of first or second order for the integral term.

2.1.1 Numerical Results

Let us now consider the following prehistory testproblem:

$$\frac{d}{dt}u - \varepsilon \nabla^2 u + k \cdot \nabla u = f + \int_{t-4}^{t} K(s) u \, ds \quad t \in [0,4]\,, \tag{8}$$

with $K = \exp(s-t)$. f is chosen in a way that

$$u = \frac{g(t)}{2\cosh(a(x-m))\cosh(a(y-n))} \tag{9}$$

is the solution. With $g(t) = (\sin(2\pi t) + 1)/2$ we call this testproblem I.

To verify the order in time we choose the parameters ($a = 1$, $\varepsilon = 1$ and $k = (1,1)^t$) for the testproblem which causes only minor difficulties in space. In Table 1 the results for hermite approach with BDF(3) are displayed. Until the error in space becomes dominant we can see reduction rates of third order. For higher Péclet numbers the stabilisation of the galerkin method is performed by streamline diffusion.

Table 1 Results for the hermite integration and a BDF(3) scheme on testproblem I.1 with $a = 5$. The left table shows the order in time and the right one the behaviour for different Péclet numbers.

Δt	Pe	$\|u-u_h\|_{L_2}$	Quotient
1/8	1	1.322e-2	-
1/16	1	1.794e-3	7.369
1/32	1	2.274e-4	7.889
1/64	1	2.847e-5	7.987
1/128	1	4.112e-6	6.923

Δt	Pe	$\|u-u_h\|_{L_2}$
1/32	10^0	2.274e-4
1/32	10^1	9.301e-4
1/32	10^2	1.286e-3
1/32	10^3	1.329e-3
1/32	10^4	1.335e-3

2.2 Adaptive Step Size Control

One major advantage of the presented hermite integration scheme is the fact that we are able to choose the time step size in every step independent of the ones chosen before. So we are only limited by the used time stepping technique. With a variable step size the coefficients of the BDF scheme have to be recomputed in every time step. For BDF(3) we achieve:

$$\beta_3 = \frac{(t^{j+1}-t^j)(t^{j+1}-t^{j-1})(t^{j+1}-t^j)}{(t^{j-2}-t^{j-1})(t^{j-2}-t^j)(t^{j-2}-t^{j+1})}\,; \quad \beta_2 = \frac{(t^{j+1}-t^j)(t^{j+1}-t^{j-2})(t^{j+1}-t^j)}{(t^{j-1}-t^{j-2})(t^{j-1}-t^j)(t^{j-1}-t^{j+1})}$$

$$\beta_1 = (-1)\frac{(t^{j+1}-t^{j-2})(t^{j+1}-t^{j-1})}{(t^j-t^{j-1})(t^j-t^{j-2})}\,; \quad \beta_0 = 1-\beta_3-\beta_2-\beta_1$$

To construct an adaptive scheme we need an approximation $\bar{u}_h(t^{j+1})$ to compare the computed solution $u_h(t^j)$ with. The computation of this approximation should require low CPU costs, however it should not force the algorithm to unnecessary changes in the time step size. To achieve this we use an extrapolation of third order based on the polynomial of the BDF scheme:

$$\gamma_i = \prod_{j=0,j\neq i}^{k} \frac{t^{n+1}-t^{n+1-j}}{t^{n+1-i}-t^{n+1-j}}\,, \quad \gamma_0 = 1-\sum_{i=1}^{k}\gamma_i \quad \bar{u}^{n+1} = \sum_{i=0}^{k}\gamma_i u^{n-i}$$

Now we set up the error function as followed:

$$\eta = \|\bar{u}^{n+1} - u^{n+1}_{\Delta t^n}\| \le \text{Rtol}\cdot\max\{\|\bar{u}^{n+1}\|, \|u^{n+1}\|\} + \text{Atol} = E\,, \tag{10}$$

with two parameters Rtol and Atol to be chosen and the solution $u^{n+1}_{\Delta t^n}$ computed with the time step size from the last time step. Based on this the next time step size is chosen as followed:

$$h^{n+1} = \alpha h^n \qquad \alpha = \begin{cases} 0.75 & ,\ \text{if } \left(\frac{E}{\eta}\right)^{1/4} < 0.75 \\ \left(\frac{E}{\eta}\right)^{1/4} & ,\ \text{if } 0.75 \le \left(\frac{E}{\eta}\right)^{1/4} \le 1.25 \\ 1.25 & ,\ \text{if } < 1.25 \left(\frac{E}{\eta}\right)^{1/4} \end{cases}$$

For the used variable step size BDF scheme the choice of Δt^{n+1} is restricted by the conditions published by Grigorieff, see [4] and [1]. For the most practical problems the boundaries published by Grigorieff are quite pessimistic so that for our solver we choose $0.75 \le \alpha \le 1.25$ instead of $0.836 \le \alpha \le 1.127$ from [4].

For the initial value problem the adaptivity can be applied strait forward. We have just to add the new value to the history. Considering the prehistory problem there is some additonal work to do. We have to watch carefully that for any new time step size Δt the delay integral is of the given length d. To this we have to consider three different cases illustrated by Fig. 1. Let $\Delta t^0, \Delta t^1, ..., \Delta t^{n-1}$ be the time stepsize (=intervall length) of the last n time steps and $u(t^0), u(t^1), ..., u(t^n)$ the corresponding values of the solution. If $\Delta t^n = \Delta t^0$, case a] in Fig. 1, we are in the same case as for

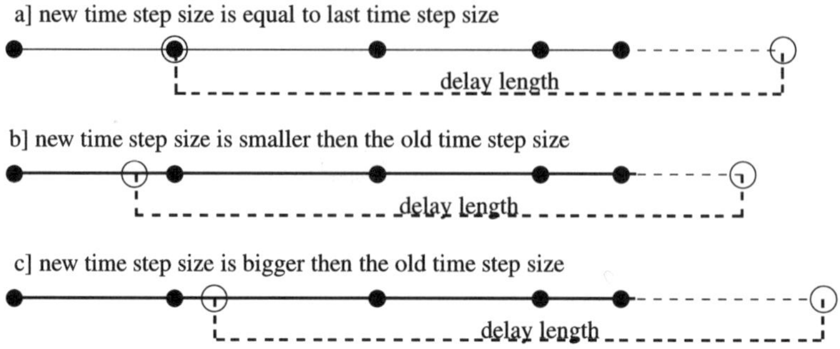

Fig. 1 An outline of the different cases concerning prehistory problems and adaptivity

the non-adaptive technique. We have just to delete Δt^0 and $u(t^0)$ from the history. In the case that $\Delta t^n < \Delta t^0$ (b]) we interpolate u at the required position $t^{n+1} - d$ and set $\Delta t^0 := t^1 - (t^{n+1} - d)$. Finally the case $\Delta t^n > \Delta t^0$ (c]) is left. Here like in case a] we first delete the first value and afterwards continue as in case b].

2.2.1 Numerical Results

If we choose $Atol = Rtol$ =5e-4 in the error indicator (10) we would expect the error control to achieve an accuracy of 1e-3 on the unit square. If we now consider the result for the testproblem I in Fig. 2, we can see that this accuarcy has been achieved. We can see from our test that if we use the presented adaptive approach the error

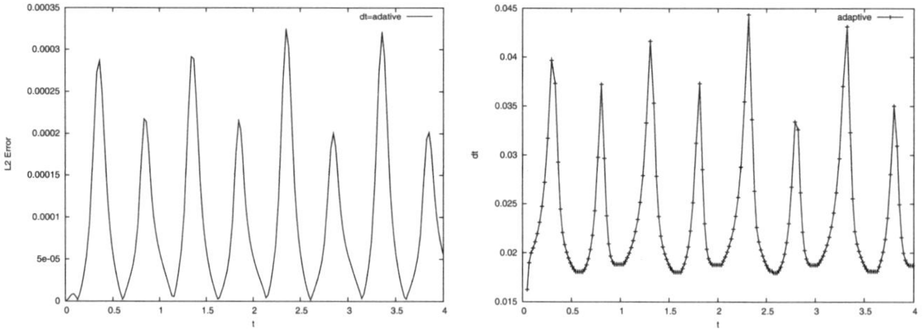

Fig. 2 For the adaptive approach with $Atol = Rtol = 5e-4$ we see the L_2-Error on the left and on the chosen dt the right.

indicators used for parabolic partial differential equations without delay can also be used for problems with a delay. But one should keep in mind that problems with a delay are more sensible than problems without a delay and so smaller tolerance parameters have to be chosen for a required accuracy compared to a problem without delay.

Let us consider now the testproblem II:

$$\frac{d}{dt}u - \varepsilon\nabla^2 u + k\cdot\nabla u = f + \int_{t-2}^{t} K(s)u\,ds \quad t \in [0,4]\,, \tag{11}$$

with $K = \sin(10s)$. f is chosen in a way that

$$u = \frac{g(t)}{2\cosh(a(x-m))\cosh(a(y-n))} \tag{12}$$

with $a = 1$ and $g(t) = \exp(-|t-2|)$ is the solution.

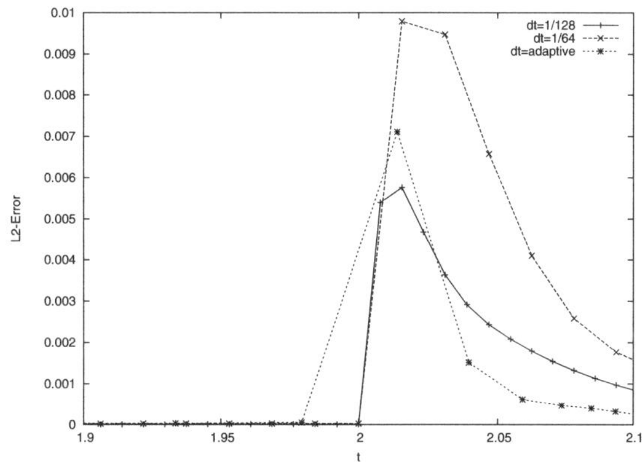

Fig. 3 This figure shows the behaviour of the solution computed with a fixed time step size and the one with an apdaptive chosen time step size in the region around the non-differentiable point $t = 2$.

The solution u is non-differentiable at $t = 2$ and so it contains a special challenge, especially for higher order multistep methods like BDF schemes, compared to onestep methods. Beyond this, we assumed that $u \in C^1[0,4]$ when we used the hermite interpolation. Table 2 displays the results for testproblem II for two fixed time step sizes and the adaptive chosen one. Figure 3 and Table 2 show that this is by nature a problem for this technique but it does not tend to diverge in the case the assumptions are violated.

Table 2 Results for the hermite integration and a BDF(3) scheme on testproblem II

Δt choosen	mean Δt	$\Vert u - u_h \Vert_{L_2}$
fixed	$= 0.015625$	9.794e-3
adaptive	≈ 0.047014	7.108e-3
fixed	$= 0.007812$	5.764e-3

But comparing the result for $dt = 1/64$ and $dt = 1/128$ we see that the reduction rate is far away from the third order in such a case. If we compare the mean timestep size we see that with the adaptive scheme we achieved a higher accuracy with less time steps.

3 Conclusion

A third order algorithm for convection-diffusion problems with a delay was presented. The presented technique is robust for higher Péclet numbers and a violation of the C^1 assumption. The design of the algorithm leads straightforward to adaptivity in time with the potentiality to use techniques common from parabolic PDEs. The adaptivity in time is also one aproach to deal with the memory storage problem. A future prospect will be to develop different storange strategies for kernels with different properties. A mayor goal considering the last results is an adaptive choice of the order of the BDF scheme to deal better with non-differentiable points.

References

1. Calvo, M., Grande, T., Grigorieff, R.D.: On the zero stability of the variable order variable stepsize bdf-formulas. Numerische Mathematik **57**, 39–50 (1990)
2. Chuanmiao, C., Tsimin, S.: Finite Element Methods for Integrodifferential Equations. World Scientific Publishing, Singapore (1998)
3. Dehghan, M.: Solution of a partial integro-differential equation arising from viscoelasticity. International Journal of Computer Mathematics **84**(1), 123–129 (2006)
4. Grigorieff, R.D.: Stability of multistep-methods on variable grids. Numerische Mathematik **42**, 359–377 (1983)
5. Lin, Y., Thomée, V., Wahlbin, L.B.: Ritz-volterra projections to finite-element spaces and applications to integrodifferential and related equations. SIAM J. Num. Ana. **28**, 1047–1070 (1991)
6. Pani, A.K., Peterson, T.E.: Finite element methods with numerical quadrature for parabolic integrodifferential euqations. SIAM Journal on Numerical Analysis **33**(3), 1084–1105 (1996)
7. Pani, A.K., Thomée, V., Wahlbin, L.B.: Numerical methods for hyperbolic and parabolic integro-differential equations. Journal of Integral Equations and Applications **4**(4), 231–252 (1992)
8. Roos, H.G., M., S., L., T.: Numerical Methods for Singulary Perturbed Differential Equations. Springer, Berlin (1996)
9. Ruess, W.: Existence and stability of solutions to partial functional differential equations with delay. Advances in Differential Equations **4**(6), 843–876 (1999)
10. Sloan, H., Thomée, V.: Time discretization of an integro-differential equation of parabolic type. SIAM J. Numer. Anal. **23**(5) (1986)
11. Thomée, V., Zhang, N.V.: Error estimates for semidiscrete finite element methods for parabolic integro-differential equations. Mathematics of Computation **53**(187), 121–139 (1989)
12. Yin, H.M.: On parapolic volterra equations in several space dimensions. SIAM J. Math. Anal. **22**(6), 1723–1737 (1991)

A Third Order WLSQR Scheme on Unstructured Meshes with Curvilinear Boundaries

J. Fürst

Abstract The work deals with the development of a high order finite volume scheme for Euler and Navier–Stokes equations. The accuracy of the scheme is improved by a piecewise quadratic interpolation of cell averaged data. The interpolation procedure uses the weighted least square approach similar to the weighted ENO scheme [2]. The resulting scheme posses extremely good convergence to steady state thanks to single stencil reconstruction with smooth weights. The truncation error for two variants of the simplified scheme for one-dimensional convection-diffusion equation is derived here. The importance of good approximation of the boundary is emphasized and an ENO-like procedure for the approximation of the boundary is described.

1 Introduction

This work deals with the numerical solution of compressible flows described by the system of Navier–Stokes equations

$$U_t + F(U)_x + G(U)_y + H(U)_z = F^v(U)_x + G^v(U)_y + H^v(U)_z + S(U), \qquad (1)$$

where $U = [\rho, \rho u, \rho v, \rho w, e]^T$ is the vector of conservative variables, $F(U)$, $G(U)$ and $H(U)$ are the inviscid fluxes, $F^v(U)$, $G^v(U)$and $H^v(U)$ are the viscous fluxes is the source term, for more details see e.g. [1]. The numerical solution of an IBV problem for this system of equations is obtained with the finite volume method. It is well known, that the basic low-order schemes suffer from excessive numerical diffusion. Therefore a high order method should be used especially for the case of viscous flows.

Jiří Fürst
Czech Technical University in Prague, Karlovo nám. 13, 121 35 Prague, Czech Republic, e-mail: Jiri.Furst@fs.cvut.cz

2 The High Order Finite Volume Scheme

As a base for the numerical method the standard finite volume method with data located in centers of polygonal cells has been chosen. The basic low order semi-discrete method can be written as [5]

$$\frac{dU_i(t)}{dt} = -\frac{1}{|C_i|} \sum_{j \in \mathcal{N}_i} \mathcal{F}(U_i(t), U_j(t), \mathbf{S}_{ij}). \tag{2}$$

Here $U_i(t)$ is the averaged solution over a cell C_i, $\mathcal{N}_i$ denotes the set of indices of neighborhoods of C_i (i.e. if $j \in \mathcal{N}_i$, then cells C_i and C_j share an edge in 2D or a face in 3D), $\mathbf{S}_{ij}$ is the scaled normal vector to the interface between C_i and C_j (oriented to C_j) and $\mathcal{F}$ denotes the so called numerical flux approximating physical flux through the interface between cells C_i and C_j. The AUSMPW+ [6] flux was chosen in this work, nevertheless the other choice of the numerical flux (e.g. Roe's flux etc.) is possible.

A higher order method can be obtained by introducing a cell-wise interpolation $P(\mathbf{x};U) = P_i(\mathbf{x};U)$ for $x \in C_i$ into the basic formula. The higher order method is then

$$\frac{dU_i(t)}{dt} = -\frac{1}{|C_i|} \sum_{j \in \mathcal{N}_i} \sum_q \omega_q \mathcal{F}(P_i(\mathbf{x}_{ij};U), P_j(\mathbf{x}_{ij};U), \mathbf{S}_{ij}), \tag{3}$$

where $\mathbf{x}_{ij}^q$ are the quadrature points at the interface between C_i and C_j and ω_q are the quadrature weights.

The semi-discrete is then solved either by explicit Runge-Kutta method, either by an implicit backward Euler method [5].

3 The Weighted Least Square Reconstruction

The very important part of the above mentioned method is the high order reconstruction (or interpolation). The reconstruction should satisfy following requirements (see e.g. [3], [4]):

1. **Conservativity**, i.e. the mean value of the interpolant $P(x;u)$ over any cell C_i should be equal to cell average of u

$$\int_{C_i} P(\mathbf{x};U)\, d\mathbf{x} = |C_i| U_i. \tag{4}$$

2. **Accuracy**, i.e. for a given smooth function $U(\mathbf{x})$ with cell averages U_i the interpolant $P(\mathbf{x};U)$ should approximate U:

$$\frac{1}{|C_j|} \int_{C_j} P_i(\mathbf{x};U)\, d\mathbf{x} = U_j + \mathcal{O}(h^o),\ \forall j \in \mathcal{M}_i, \tag{5}$$

where h is a characteristical mesh size and o is the order of accuracy.

3. **Non-oscillatory**, i.e. the total variation of the interpolant should be bounded for $h \to 0$.

As soon as the set $\mathscr{M}_i$ contains sufficient number of cell indices, the system becomes overdetermined and it is solved by the means of least square method. The interpolant $P_i(\mathbf{x};U)$ is therefore obtained by minimizing error in (5) for $j \in \mathscr{M}_i$ respect to constraint (4). In order to mimic weighted ENO method the data dependent weights are introduced:

$$P_i(\mathbf{x};U) = \arg\min \sum_{j \in \mathscr{N}_i} \left[w_{ij} \left(\int_{C_j} \tilde{P}(\mathbf{x};U)\, d\mathbf{x} - |C_j| U_j \right) \right]^2, \tag{6}$$

where minimum is take over all linear polynomials $\tilde{P}$ satisfying (4), in other words, P_i is defined as a polynomial satisfying (4) and minimizing errors in (6) in L_2 norm. The data-dependent weights w_{ij} are chosen as (see e.g. [3])

$$w_{ij} = \sqrt{\frac{h^{-r}}{\left| \frac{u_i - u_j}{h} \right|^p + h^q}}, \tag{7}$$

with p, q, and r being constants (e.g. $p = 4$, $q = -2$, $r = 3$).

The stencil $\mathscr{M}_i$ is chosen as

$$\mathscr{M}_i := \mathscr{M}_i^1 = \{ j : \overline{C_i} \cap \overline{C_j} \neq \emptyset \} \tag{8}$$

for the case of piecewise linear interpolations, and

$$\mathscr{M}_i := \mathscr{M}_i^2 = \{ j : \exists k \in \mathscr{M}_i^1 : \overline{C_i} \cap \overline{C_k} \neq \emptyset \}. \tag{9}$$

Note, that for system of equation the polynomial reconstruction is made componentwise.

3.1 The Analysis of Linearized Schemes for Convection-Diffusion Problem

In order to analyze properties of the second and third order schemes we assume scalar linear convection diffusion problem in one space dimension

$$u_t + a u_x = \mu u_{xx}, \; a = const. > 0. \tag{10}$$

The semi-discrete second order finite volume scheme obtained by the least-square interpolation described above taking the weights $w_{ij} = 1$, the upwind flux, and the second order central difference for viscous term is

$$\dot{u}_i = a\frac{-u_{i-2}+5u_{i-1}-3u_i-u_{i+1}}{4h}+\frac{\mu}{h^2}\Delta_i^2 u, \tag{11}$$

where $\Delta_i^2 u = u_{i+1} - 2u_i + u_{i-1}$. The approximation error can be directly computed as

$$u_t + au_x - \mu u_{xx} = \frac{ah^2}{12}u_{xxx} - \frac{\mu h^2}{12}u_{xxxx} + O(h^4). \tag{12}$$

The scheme therefore contains both dissipation and dispersion in the leading term of the approximation error.

If we replace the piecewise-linear interpolation by the piecewise quadratic one, we get a mixed order scheme (i.e. third order for convection combined with second order for diffusion term)

$$\dot{u}_i = a\frac{-41u_{i-3}+18u_{i-2}+508u_{i-1}-457u_i+33u_{i+1}-61u_{i+2}}{510h}+\frac{\mu}{h^2}\Delta_i^2 u. \tag{13}$$

The approximation error is now

$$u_t + au_x - \mu u_{xx} = -\frac{10\mu h^2 + 19ah^3}{120}u_{xxxx} + O(h^4). \tag{14}$$

One can see that the later scheme is is still second order one (due to second order approximation of viscous term). Nevertheless, there is no spurious dispersion up to $O(h^3)$.

4 Curvilinear Boundaries

Another important problem is a good approximation of the boundary. Unfortunately, the high order representation of the boundary shape may be difficult. In order to avoid interface to a CAD software we develop a simple procedure for high order interpolation of 2D boundaries. We assume, that the 2D boundary is given by continuous piecewise smooth curves. Each curve represents a logical part of the boundary (i.e. inlet part, solid wall, etc.). Discretization of each curve gives us an ordered list of points $\mathbf{x}_i$ where $i = 0..N$. $\mathbf{x}$, then the point is in the list. Moreover, we assume that the list contains all endpoints of smooth segments. The interpolation of the boundary is obtained in the following way:

1. compute the approximate arc-lengths $t_i = t_{i-1} + ||\mathbf{x}_i - \mathbf{x}_{i-1}||$, with $t_0 = 0$,
2. then for each segment $\mathbf{x}_i$-$\mathbf{x}_{i+1}$ do:
    a. compute "left" parabola $\mathbf{s}_L$ such that $\mathbf{s}_L(t_{i-1}) = \mathbf{x}_{i-1}$, $\mathbf{s}_L(t_i) = \mathbf{x}_i$, and $\mathbf{s}_L(t_{i+1}) = \mathbf{x}_{i+1}$,
    b. compute "right" parabola $\mathbf{s}^R_{i+1/2}$ such that $\mathbf{s}_R(t_i) = \mathbf{x}_i$, $\mathbf{s}_R(t_{i+1}) = \mathbf{x}_{i+1}$, and $\mathbf{s}_R(t_{i+2}) = \mathbf{x}_{i+2}$,
    c. if $||\mathbf{s}_L''(t_{i+1/2})|| < ||\mathbf{s}_R''(t_{i+1/2})||$, then $\mathbf{s}_{i+1/2} = \mathbf{s}_L$ otherwise $\mathbf{s}_{i+1/2} = \mathbf{s}_R$ (here $t_{i+1/2} = (t_i + t_{i+1})/2$).

The non-polygonal cells are assumed only in the vicinity of boundaries, therefore it has only minor impact on the overall computational time. On the other hand, the numerical experiments show great improvement in the accuracy of solution of an inviscid flow problem compared to simple piecewise linear approximation of the boundary.

5 Numerical Experiments

5.1 Ringleb's Flow

In order to estimate the order of accuracy we solved the so-called Ringleb's flow with known analytical solution. We used the piecewise linear (I1) or quadratic (I2) WLSQR interpolation for the scheme and piecewise linear (B1) or quadratic (B2) representation of the boundary. Namely, we tested the I1B1 combination (formally second order scheme), I2B1 combination (third order scheme with lower order approximation of the shape), and finally I2B2 combination (formally third order scheme). The results were obtained using a coarse unstructured mesh with 1004 triangles, a middle mesh with 3936 triangles and a fine mesh with 15780 triangles. The figure 1 shows the isolines of the Mach number and the distribution of entropy along the left wall. One an see, that the I2B1 variant gives even worst results than the I1B1 scheme.

Comparing numerical results with the exact solution we evaluated the order of accuracy of all three variants. The L_2 order was approximately 2.0 for I1B1 scheme, 2.17 for I2B1 scheme, and 2.97 for I2B2 scheme.

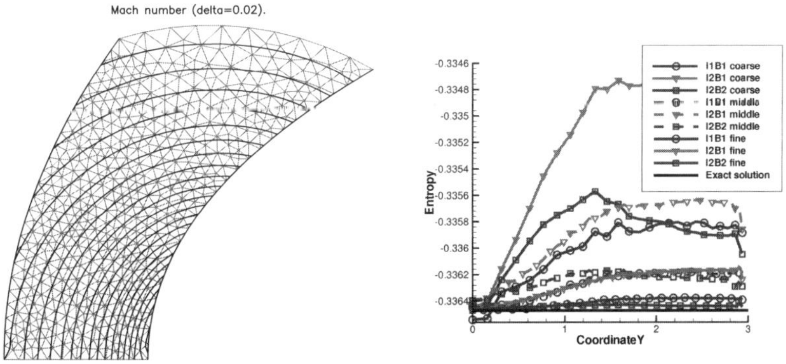

Fig. 1 Solution to Ringleb's flow problem (the coarse mesh with the iso-Mach lines on the left, the distribution of the entropy on the right).

5.2 Inviscid Flow in a Turbine Cascade

Another example shows the solution of the steady inviscid transonic flow in the 2D turbine cascade SE1050. The flow regime is characterized by the inlet angle $\alpha_1 = 19.3°$ and the isentropic outlet Mach number $M_{2i} = 1.198$. An unstructured mesh with 5975 triangles was used in this case. The spatial derivatives were approximated with the above mentioned FV scheme with piecewise linear (I1B1) or piecewise quadratic (I2B2) WLSQR interpolation. The backward Euler method with approximate linearization was chosen for the discretization in time. The resulting system of linear equations was solved with the GMRES method preconditioned by ILU(0).

There is no doubt, that the I2B2 variant is more expensive than I1B1 in terms of CPU time per iteration. On the other hand, our results (see fig. 2) show, that the convergence to steady state can be faster for I2B2 can be faster than for the I1B1.

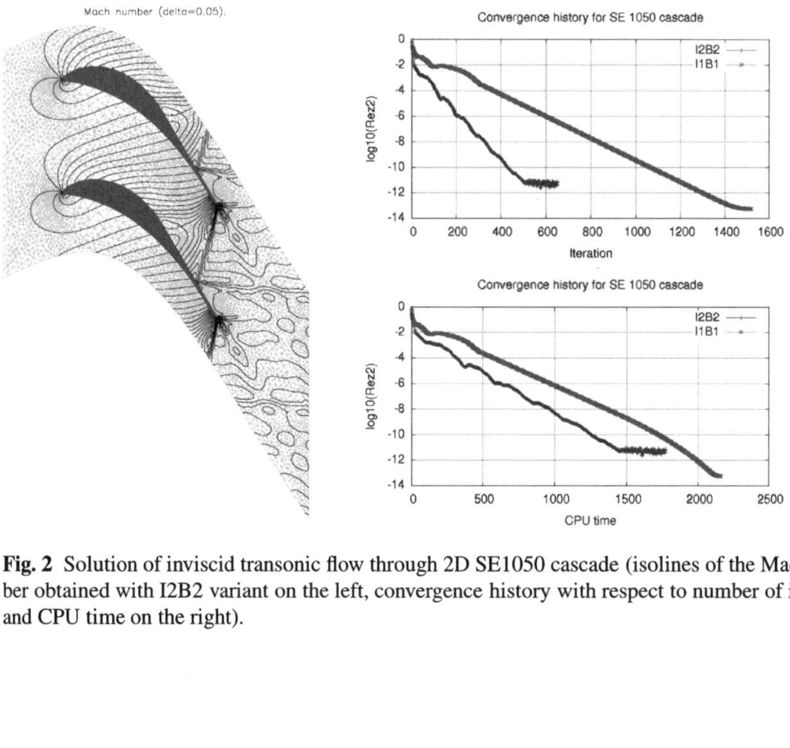

Fig. 2 Solution of inviscid transonic flow through 2D SE1050 cascade (isolines of the Mach number obtained with I2B2 variant on the left, convergence history with respect to number of iteration and CPU time on the right).

5.3 Viscous Turbulent Flow in a Cascade

The last example shows the solution of transonic turbulent flow in the turbine cascade NT24. The flow was modeled by the RANS equations equipped with the two-equations TNT $k-\omega$ of Kok [7]. The flow regime is described by the inlet angle $\alpha_1 = 0°$, outlet isentropic Mach number $M_{2i} = 1.039$ and Reynolds number $Re = 3.6 \cdot 10^5$. The intensity of turbulence at the inlet was chosen 0.5%.

The problem was solved using a hybrid mesh with quadrilaterals near the profile and triangles in the rest of the domain. The mesh with 24632 cells was created with special care near the profile in order to avoid singularities caused by the curvilinear approximation of boundary. The inviscid fluxes were approximated using I1B1, I2B1, or I2B2 variant of WLSQR interpolation combined with the AUSMPW+ flux. Viscous terms were computed with the second order finite volume approach using dual cells. Therefore, the I2B2 and I2B1 schemes correspond to the "mixed order" scheme analyzed in the previous chapter. The solution was compared to the reference solution obtained with the I1B1 scheme on finer mesh with 47452 cells. In both cases, the near wall spacing is chosen in order to fulfill common requirement of having $y_1^+ < 1$.

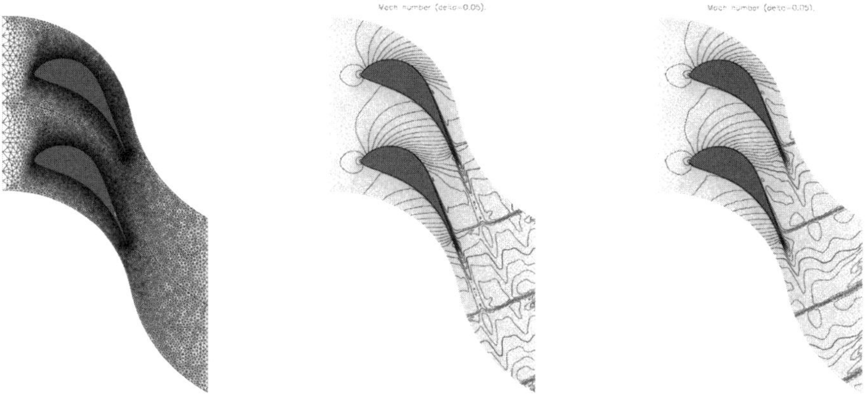

Fig. 3 The mesh and the isolines of the Mach number obtained with I1B1 (center) I2B2 (right) scheme for the flow through NT24 cascade.

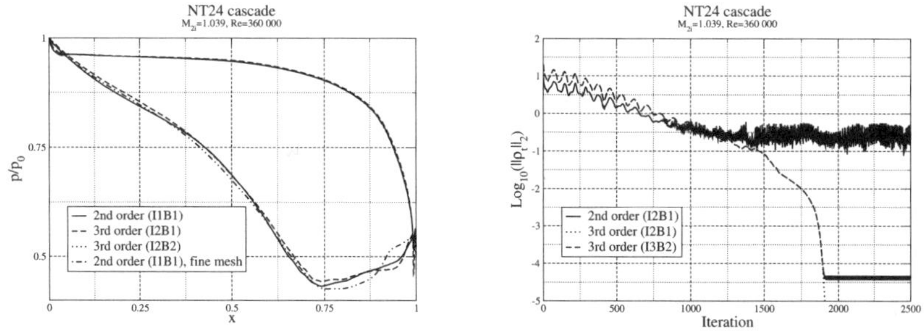

Fig. 4 The distribution of the pressure on the blade and the convergence history for I1B1/I2B2 scheme.

The figures 3 and 4 show, that all I1B1, I2B1 and I2B2 schemes give very similar results. There is no difference between I2B1 and I2B2 results (up to plotting accuracy) and there is only small difference at the suction (upper) side of the blade near the outlet edge between I1B1 and I2B1/I2B2. Therefore, the advantage of better accuracy, which was presented for the case of inviscid flows, is not evident here.[1] On the other hand, the convergence to steady state is again better for I2B2 than for I1B1 scheme.

6 Conclusion

Current progress in the development of the high order FV scheme with WLSQR interpolation show, that it is necessary to take in account the high order representation of the boundary shape especially for the inviscid flows. On the other hand, the necessity of curvilinear representation of the boundary is not evident here and the convergence to steady state can be better for I2B2 and I2B1 variant of the scheme than for the I1B1 one.

Acknowledgements The work has been supported by the project COST OC167, by the Research Plan MSM No. 6840770010, and by the grant no. 201/05/0005 of GACR.

References

1. Feistauer, M., Felcman, J., Straškraba, I.: Mathematical and computational methods for compressible flow. Numerical Mathematics and Scientific Computation. The Clarendon Press Oxford University Press, Oxford (2003)
2. Friedrich, O.: Weighted essentially non-oscillatory schemes for the interpolation of mean values on unstructured grids. J. Comput. Phys. **144**(1), 194–212 (1998)
3. Fürst, J.: A finite volume scheme with weighted least square reconstruction. In: S.R. F. Benkhaldoun D. Ouazar (ed.) Finite Volumes for Complex Applications IV, pp. 345–354. Hermes Science (2005). ISBN 1-905209-48-7
4. Fürst, J.: The weighted least square scheme for multidimensional flows. In: E.O. P. Wesseling, J. Priaux (eds.) Proceedings of "ECCOMAS CFD 2006". TU Delft (2006)
5. Fürst, J., Kozel, K.: Second and third order weighted ENO scheme on unstructured meshes. In: Finite volumes for complex applications, III (Porquerolles, 2002), pp. 723–730. Hermes Sci. Publ., Paris (2002)
6. Kim, K.H., Kim, C., Rho, O.H.: Methods for the accurate computations of hypersonic flows. I. AUSMPW+ scheme. J. Comput. Phys. **174**(1), 38–80 (2001)
7. Kok, J.C.: Resolving the dependence on free stream values for the k-omega turbulence model. Tech. Rep. NLR-TP-99295, NLR (1999)

[1] In order to profit from higher order accuracy, it would be probably necessary to discretize the viscous fluxes and source terms with higher order of accuracy

On the Choice of Parameters in Stabilization Methods for Convection–Diffusion Equations

V. John and P. Knobloch

Abstract A popular finite element approach for the numerical solution of convection–diffusion equations is the streamline upwind/Petrov–Galerkin (SUPG) method. Unfortunately, in the convection–dominated regime, the SUPG solution often contains spurious oscillations along sharp layers. A possible remedy is to introduce an additional artificial diffusion term in the SUPG discretization. We call such approaches spurious oscillations at layers diminishing (SOLD) methods. The properties of the SOLD methods are significantly influenced by the choice of the respective stabilization parameter which determines the amount of the artificial diffusion. The aim of this paper is to discuss various definitions of these stabilization parameters.

1 Introduction

This paper is devoted to the numerical solution of the steady scalar convection–diffusion equation

$$-\varepsilon\Delta u+\mathbf{b}\cdot\nabla u=f \quad \text{in } \Omega, \qquad u=u_b \quad \text{on } \partial\Omega. \tag{1}$$

We assume that Ω is a bounded domain in $\mathbb{R}^2$ with a polygonal boundary $\partial\Omega$, $\varepsilon>0$ is the constant diffusivity, $\mathbf{b}=(b_1,b_2)$, f and u_b are given functions and u is an unknown scalar quantity, e.g., temperature or concentration.

It is well known that the standard Galerkin finite element discretization of (1) loses its stability if convection strongly dominates diffusion. Therefore, various

Volker John
Universität des Saarlandes, Fachbereich 6.1 – Mathematik, Postfach 15 11 50, 66041 Saarbrücken, Germany, e-mail: john@math.uni-sb.de

Petr Knobloch
Charles University, Faculty of Mathematics and Physics, Sokolovská 83, 186 75 Praha 8, Czech Republic, e-mail: knobloch@karlin.mff.cuni.cz

stabilized finite element methods have been developed for the numerical solution of (1). A widely used approach is the streamline upwind/Petrov–Galerkin (SUPG) method proposed in [1]. Denoting by W_h a finite element space approximating the Sobolev space $H^1(\Omega)$, by $u_{bh} \in W_h$ a function whose trace approximates u_b and setting $V_h = W_h \cap H_0^1(\Omega)$, the SUPG method reads:

Find $u_h \in W_h$ such that $u_h - u_{bh} \in V_h$ and

$$\varepsilon\,(\nabla u_h, \nabla v_h) + (\mathbf{b}\cdot\nabla u_h, v_h) + (R_h(u_h), \tau\,\mathbf{b}\cdot\nabla v_h) = (f, v_h) \quad \forall\, v_h \in V_h. \qquad (2)$$

Here $(\cdot,\cdot)$ is the inner product in $L^2(\Omega)$ or $L^2(\Omega)^2$, $R_h(u) = -\varepsilon\,\Delta\,u + \mathbf{b}\cdot\nabla\,u - f$ is the residual (defined elementwise) and τ is a nonnegative stabilization parameter, see Section 2.

Unfortunately, the SUPG method is not monotone and hence a discrete solution satisfying (2) usually contains spurious oscillations along sharp layers. A possible remedy is to add a suitable artificial diffusion term to the left–hand side of the SUPG discretization (2). We call such approaches spurious oscillations at layers diminishing (SOLD) methods, see the review paper [8]. There are three basic types of SOLD terms and they add either isotropic artificial diffusion or crosswind artificial diffusion to the SUPG method (2) or they are based on so–called edge stabilizations. These three types of SOLD terms are respectively defined by

$$(\widetilde{\varepsilon}\,\nabla u_h, \nabla v_h)\,, \qquad (3)$$

$$(\widetilde{\varepsilon}\,\mathbf{b}^\perp\cdot\nabla u_h, \mathbf{b}^\perp\cdot\nabla v_h) \qquad \text{with} \qquad \mathbf{b}^\perp = \frac{(-b_2, b_1)}{|\mathbf{b}|}\,, \qquad (4)$$

$$\sum_{K\in\mathscr{T}_h} \int_{\partial K} \widetilde{\varepsilon}|_K \,\mathrm{sign}\left(\frac{\partial u_h}{\partial \mathbf{t}_{\partial K}}\right) \frac{\partial v_h}{\partial \mathbf{t}_{\partial K}}\, \mathrm{d}\sigma\,, \qquad (5)$$

where $\mathscr{T}_h = \{K\}$ is a triangulation of Ω satisfying the usual compatibility assumptions and $\mathbf{t}_{\partial K}$ is a tangent vector to the boundary ∂K of K. The parameter $\widetilde{\varepsilon}$, which determines the amount of the artificial diffusion added to the SUPG method, is nonnegative and usually depends on u_h. Thus, the resulting methods are nonlinear although the original problem (1) is linear.

Comparative numerical studies of a large number of SOLD methods can be found in, e.g., [6, 7, 8, 9]. It was observed that there are large differences between the SOLD methods. In some cases, many SOLD methods were able to significantly improve the SUPG solution and to provide a discrete solution with negligible spurious oscillations and without an excessive smearing of layers. However, it was not possible to identify a method which could be preferred in all the test cases. The aim of the present paper is to discuss the definitions of the parameters $\widetilde{\varepsilon}$ for those SOLD methods which achieved high rankings in the mentioned numerical studies.

The paper is organized in the following way. In the next section, we present the definitions of the parameter $\widetilde{\varepsilon}$ for several promising SOLD methods. The main part of the paper is Section 3 where we discuss the optimality of these definitions of $\widetilde{\varepsilon}$ for three academic tests problems. We finish the paper by our conclusions in Section 4.

2 Definitions of the Stabilization Parameters

In this section, we present various choices of the parameters in the stabilization terms in (2)–(5). Generally, the parameters should depend on the approximation properties of the finite element space W_h. For simplicity, throughout this paper, we restrict ourselves to spaces

$$W_h = \{v \in H^1(\Omega);\ v|_K \in R(K)\ \ \forall\, K \in \mathscr{T}_h\},$$

where $R(K) = P_1(K)$ if K is a triangle and $R(K) = Q_1(K)$ if K is a rectangle. We assume that the triangulation $\mathscr{T}_h$ consists either of triangles or of rectangles.

The choice of the SUPG parameter τ in (2) may dramatically influence the accuracy of the discrete solution and therefore it has been a subject of an extensive research over the last three decades, see, e.g., the review in [8]. Unfortunately, a general optimal definition of τ is still not known. In our computations, we define τ, on any element $K \in \mathscr{T}_h$, by the formula

$$\tau|_K = \frac{h_K}{2\,|\mathbf{b}|}\left(\coth Pe_K - \frac{1}{Pe_K}\right) \qquad \text{with} \qquad Pe_K = \frac{|\mathbf{b}|\,h_K}{2\,\varepsilon}, \tag{6}$$

where h_K is the element diameter in the direction of the convection vector $\mathbf{b}$, $|\mathbf{b}|$ is the Euclidean norm of $\mathbf{b}$ and Pe_K is the local Péclet number. We refer to [8] for various justifications of this formula. Note that, generally, the parameters h_K, Pe_K and $\tau|_K$ are functions of the points $\mathbf{x} \in K$.

According to the criteria and tests in [6, 7, 8], one of the best choices of $\widetilde{\varepsilon}$ in (3) is to set

$$\widetilde{\varepsilon} = \max\left\{0, \frac{\tau\,|\mathbf{b}|\,|R_h(u_h)|}{|\nabla u_h|} - \tau\,\frac{|R_h(u_h)|^2}{|\nabla u_h|^2}\right\}, \tag{7}$$

as proposed in [4]. Here and in the following, we always assume that $\widetilde{\varepsilon} = 0$ if the denominator of a formula defining $\widetilde{\varepsilon}$ vanishes. In case (4), we suggested in [8] to set, on any $K \in \mathscr{T}_h$,

$$\widetilde{\varepsilon}|_K = \max\left\{0, \eta\,\frac{\operatorname{diam}(K)\,|R_h(u_h)|}{2\,|\nabla u_h|} - \varepsilon\right\}, \tag{8}$$

where $\operatorname{diam}(K)$ is the diameter of K and η is a suitable constant, for which the value $\eta \approx 0.7$ was recommended in [5]. The relation (8) is a slight modification of a formula proposed in [5]. Another promising variant of (4) tested in [6, 7, 8, 9] is defined by

$$\widetilde{\varepsilon} = \frac{\tau\,|\mathbf{b}|\,|R_h(u_h)|}{|\nabla u_h|}\,\frac{|\mathbf{b}|\,|\nabla u_h|}{|\mathbf{b}|\,|\nabla u_h| + |R_h(u_h)|}. \tag{9}$$

This choice of $\widetilde{\varepsilon}$ was proposed in [8] as a simplification of a formula from [2]. For the edge stabilization term (5), acceptable results were computed with

$$\widetilde{\varepsilon}|_K = C\,|K|\,|(R_h(u_h)|_K)| \qquad \forall\, K \in \mathscr{T}_h, \tag{10}$$

where $|K|$ is the area of K and C is a nonnegative constant. Let us mention that, to achieve convergence of the nonlinear iterative process, the sign operator in (5) is regularized by replacing it by the hyperbolic tangent as recommended in [3].

If convection strongly dominates diffusion in Ω and hence the local Péclet numbers Pe_K are very large, the parameter τ defined in (6) satisfies $\tau|_K = h_K/(2|\mathbf{b}|)$ for any $K \in \mathscr{T}_h$. Then we have in (7) and (9)

$$\frac{\tau\,|\mathbf{b}|\,|R_h(u_h)|}{|\nabla u_h|} \approx \frac{h_K\,|R_h(u_h)|}{2\,|\nabla u_h|}\,.$$

Hence, in the definitions of $\widetilde{\varepsilon}$ in (7)–(9), an important role is played by a term of the type $h\,|R_h(u_h)|/|\nabla u_h|$. Moreover, in view of (10), the edge stabilization term (5) can be written in the form

$$\sum_{K\in\mathscr{T}_h} |K| \int_{\partial K} C\, \frac{\left|R_h(u_h)|_K\right|}{\left|\frac{\partial u_h}{\partial \mathbf{t}_{\partial K}}\right|}\, \frac{\partial u_h}{\partial \mathbf{t}_{\partial K}}\, \frac{\partial v_h}{\partial \mathbf{t}_{\partial K}}\, \mathrm{d}\sigma\,,$$

which is an expression of a similar structure as the SOLD terms (3) and (4) with $\widetilde{\varepsilon}$ defined by (7)–(9). Thus, we observe the interesting fact that all three types of SOLD terms with the above described definitions of the parameter $\widetilde{\varepsilon}$ are similar although the formulas for $\widetilde{\varepsilon}$ were derived using completely different arguments.

3 Optimal Choice of Stabilization Parameters for Model Problems

In this section, we shall discuss the optimality of the parameters $\widetilde{\varepsilon}$ introduced in the previous section for three model problems whose solutions possess characteristic features of solutions of (1). We shall confine ourselves to the two types of triangulations depicted in Fig. 1. To characterize these triangulations, we shall use the notion '$N_1 \times N_2$ mesh' where N_1 and N_2 are the numbers of vertices in the horizontal and vertical directions, respectively. The corresponding mesh widths will be denoted by h_1 and h_2, i.e., $h_1 = 1/(N_1-1)$ and $h_2 = 1/(N_2-1)$.

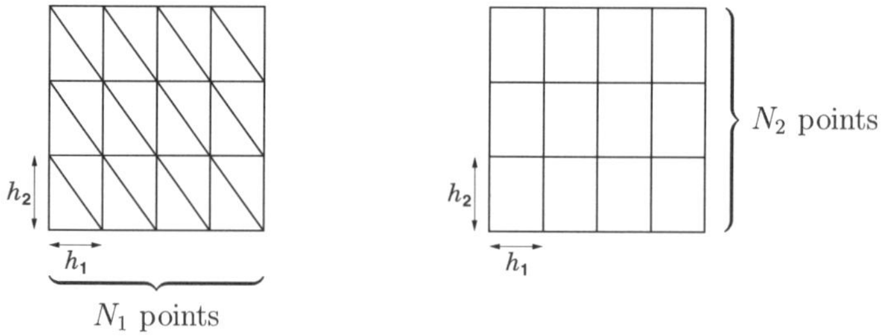

Fig. 1 Triangulations used in Section 3.

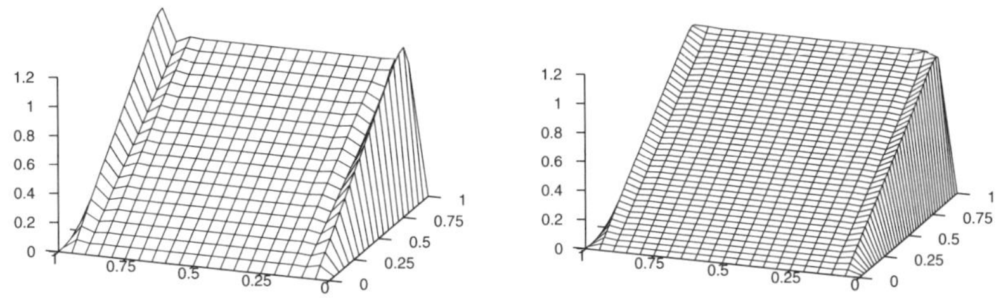

Fig. 2 Example 1, P_1 finite element: SUPG solution on a 21×21 mesh (left) and SOLD solution defined using (4) and (9) on a 41×21 mesh (right).

The analysis below will include the consideration of moderately anisotropic grids. Using such grids might not be reasonable for the considered examples since these grids are not adapted to the layers of the solution. However, convection–diffusion equations are often just a part of a coupled system of equations. For such problems, an adaptation of the grid is performed rather with respect to other equations in the system, for instance with respect to the Navier–Stokes equations in fluid flow applications. Nevertheless, the SOLD methods still should provide satisfactory results.

Example 1 (*Solution with parabolic and exponential boundary layers*). We consider the convection–diffusion equation (1) with $\Omega = (0,1)^2$, $\varepsilon = 10^{-8}$, $\mathbf{b} = (1,0)^T$, $f = 1$, and $u_b = 0$. The solution $u(x,y)$ of this problem possesses an exponential boundary layer at $x = 1$ and parabolic (characteristic) boundary layers at $y = 0$ and $y = 1$. Outside the layers, the solution $u(x,y)$ is very close to x.

For this special example, the stabilization parameter τ given in (6) leads to a nodally exact SUPG solution outside the parabolic layers. However, there are strong oscillations at the parabolic layers, see Fig. 2.

Let us consider a SOLD discretization of (1) with the isotropic SOLD term (3) or the crosswind SOLD term (4) and with $\widetilde{\varepsilon}$ defined by (8). In the triangular case, it is easy to show that η equal to

$$\eta_{opt} = \frac{2h_2}{3\sqrt{h_1^2 + h_2^2}} \tag{11}$$

is optimal for $\varepsilon \to 0$ with respect to the parabolic layers. Indeed, for $\eta = \eta_{opt}$ the discrete solution is nodally exact outside the exponential boundary layer whereas, for $\eta > \eta_{opt}$, the parabolic boundary layers are smeared and, for $\eta < \eta_{opt}$, spurious oscillations along the parabolic boundary layers appear. Moreover, for the nodally exact solution with $\varepsilon \to 0$, the SUPG term $(R_h(u_h), \tau \mathbf{b} \cdot \nabla v_h)$ vanishes outside the exponential boundary layer which shows that the optimal value of η does not depend on the definition of the SUPG stabilization parameter τ. In the quadrilateral case, it is not possible to derive a simple formula for η_{opt} but numerical results suggest that the optimal values of η do not differ much from (11).

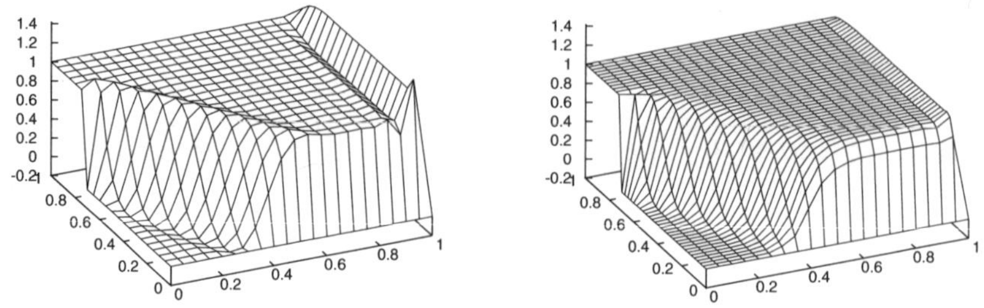

Fig. 3 Example 2, Q_1 finite element: SUPG solution on a 21×21 mesh (left) and SOLD solution defined using (4) and (8) with $\eta = 0.7$ on a 21×41 mesh (right).

Since $\eta_{opt} < 2/3$, spurious oscillations should not appear for the value $\eta \approx 0.7$ recommended in [5]. On the other hand, if we consider $\widetilde{\varepsilon}$ defined by (7) or (9), a comparison of these relations with the formula (8) reveals that spurious oscillations in the discrete solution should be expected for $h_K = h_1 < \eta_{opt}\,\mathrm{diam}(K)$, i.e., for $h_1/h_2 < 2/3$, as it is demonstrated in Fig. 2.

For the edge stabilization term (5) and both the P_1 and Q_1 finite elements, it is easy to derive that the optimal value of C in (10) is $1/6$. However, in practice, the discrete solution slightly differs from the nodally exact solution at the parabolic boundary layers due to the regularization of the sign operator. Moreover, in contrast with the above SOLD methods, the discrete solution is significantly smeared along the exponential boundary layer. A sharp approximation of this layer requires to set $C = 0$ in this region.

The above considerations show that satisfactory numerical results can be obtained generally only using the isotropic or crosswind SOLD term with $\widetilde{\varepsilon}$ defined by (8) or using the edge stabilization (5) with $\widetilde{\varepsilon}$ defined by (10).

Example 2 (*Solution with interior layer and exponential boundary layers*). We consider the convection–diffusion equation (1) with $\Omega = (0,1)^2$, $\varepsilon = 10^{-8}$, $\mathbf{b} = (\cos(-\pi/3), \sin(-\pi/3))^T$, $f = 0$, and

$$u_b(x,y) = \begin{cases} 0 & \text{for } x = 1 \text{ or } y \le 0.7, \\ 1 & \text{else.} \end{cases}$$

The solution possesses an interior (characteristic) layer in the direction of the convection starting at $(0, 0.7)$. On the boundary $x = 1$ and on the right part of the boundary $y = 0$, exponential layers are developed.

We shall assume that $h_1 b_2 + h_2 b_1 < 0$. Then, for both the P_1 and Q_1 finite elements, the SUPG solution of Example 2 contains oscillations along the interior layer and along the boundary layer at $x = 1$. However, there are no oscillations along the boundary layer at $y = 0$ and this layer is not smeared, see Fig. 3.

For a SOLD discretization of (1) with the isotropic SOLD term (3) or the crosswind SOLD term (4) and with $\widetilde{\varepsilon}$ defined by (8), it is easy to derive optimal values of η such that, for $\varepsilon \to 0$, the discrete solution is nodally exact away from the interior

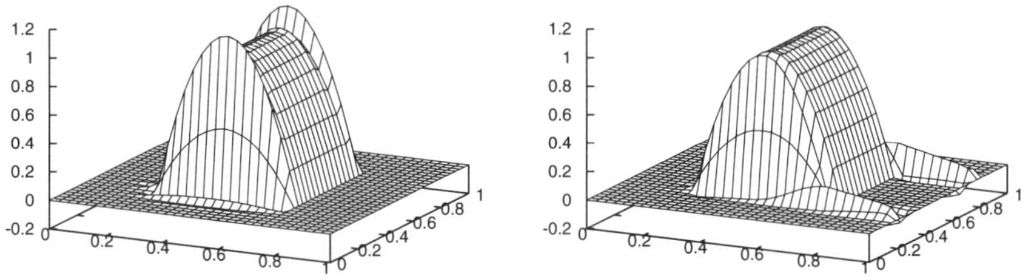

Fig. 4 Example 3, P_1 finite element, 33×33 mesh: SUPG solution (left) and SOLD solution defined using (3) and (7) (right).

layer. First, it is clear that, along the boundary layer at $y = 0$, the optimal choice of η is $\eta = 0$. For the boundary layer at $x = 1$, the optimal values are

$$\eta_{opt}^{isotropic} = \frac{h_1 b_2 + h_2 b_1}{\sqrt{h_1^2 + h_2^2}\, b_2}, \qquad \eta_{opt}^{crosswind} = \frac{(h_1 b_2 + h_2 b_1)|\mathbf{b}|^2}{\sqrt{h_1^2 + h_2^2}\, b_2^3}.$$

These formulas hold for both the P_1 and the Q_1 finite elements. One can see that the optimal choice of η depends not only on the aspect ratio of the elements of the triangulation but also on the direction of the convection vector $\mathbf{b}$. For $\mathbf{b}$ of Example 2 and for $h_1 = 2h_2$, we obtain $\eta_{opt}^{crosswind} \approx 0.85$ and hence we have to expect spurious oscillations for the recommended value $\eta \approx 0.7$. This is really the case as Fig. 3 shows. For $\widetilde{\varepsilon}$ defined by (7) and (9) the oscillations at $x = 1$ are even much larger and, moreover, there are nonnegligible oscillations at the beginning of the interior layer. For the edge stabilization term (5) and both the P_1 and Q_1 finite elements, the optimal value of C in (10) is $C_{opt} = (h_1 b_2 + h_2 b_1)/(4 h_1 b_2)$ along $x = 1$.

The above discussion supports our conclusion to Example 1 and shows that it is in general not sufficient to consider constant values of η and C.

Example 3 (*Solution with two interior layers*). We consider the convection–diffusion equation (1) with $\Omega = (0,1)^2$, $\varepsilon = 10^{-8}$, $\mathbf{b} = (1,0)^T$, $u_b = 0$, and

$$f(x,y) = \begin{cases} 16(1-2x) & \text{for } (x,y) \in [0.25, 0.75]^2, \\ 0 & \text{else.} \end{cases}$$

The solution $u(x,y)$ possesses two interior (characteristic) layers at $(0.25, 0.75) \times \{0.25\}$ and $(0.25, 0.75) \times \{0.75\}$. In $(0.25, 0.75)^2$, it is very close to the quadratic function $(4x-1)(3-4x)$.

As expected, the SUPG solution of Example 3 possesses spurious oscillations along the interior layers, see Fig. 4. Applying any of the SOLD methods discussed above, the spurious oscillations present in the SUPG solution are significantly suppressed, however, the solution is wrong in the region $(0.75, 1) \times (0, 1)$, see Fig. 4. This behaviour is the same for both the P_1 and Q_1 finite elements. Thus, Example 3 represents a problem for which all the SOLD methods described in Section 2 fail.

4 Conclusions

This paper was devoted to the numerical solution of convection–diffusion equations using SOLD methods. It was demonstrated that SOLD methods without user–chosen parameters are in general not able to remove the spurious oscillations of the solution obtained with the SUPG discretization. For the two studied methods involving a parameter, values of the parameter could be given in two examples such that the spurious oscillations were almost removed. The parameter has to be generally non–constant and depends on the mesh and the data of the problem. Therefore, for more complicated problems, it is not clear how suitable parameters can be found. Moreover, an example was presented for which none of the investigated methods provided a qualitatively correct discrete solution. Consequently, we have to conclude that it is in general completely open how to obtain oscillation–free solutions using the considered classes of methods.

Acknowledgements The work of P. Knobloch is a part of the research project MSM 0021620839 financed by MSMT and it was partly supported by the Grant Agency of the Czech Republic under the grant No. 201/08/0012.

References

1. Brooks, A., Hughes, T.: Streamline upwind/Petrov–Galerkin formulations for convection dominated flows with particular emphasis on the incompressible Navier–Stokes equations. Comput. Methods Appl. Mech. Engrg. **32**, 199–259 (1982)
2. Burman, E., Ern, A.: Nonlinear diffusion and discrete maximum principle for stabilized Galerkin approximations of the convection–diffusion–reaction equation. Comput. Methods Appl. Mech. Engrg. **191**, 3833–3855 (2002)
3. Burman, E., Hansbo, P.: Edge stabilization for Galerkin approximations of convection–diffusion–reaction problems. Comput. Methods Appl. Mech. Engrg. **193**, 1437–1453 (2004)
4. do Carmo, E., Galeão, A.: Feedback Petrov–Galerkin methods for convection–dominated problems. Comput. Methods Appl. Mech. Engrg. **88**, 1–16 (1991)
5. Codina, R.: A discontinuity–capturing crosswind–dissipation for the finite element solution of the convection–diffusion equation. Comput. Methods Appl. Mech. Engrg. **110**, 325–342 (1993)
6. John, V., Knobloch, P.: A computational comparison of methods diminishing spurious oscillations in finite element solutions of convection–diffusion equations. In: J. Chleboun, K. Segeth, T. Vejchodský (eds.) Proceedings of the International Conference Programs and Algorithms of Numerical Mathematics 13, pp. 122–136. Academy of Science of the Czech Republic (2006)
7. John, V., Knobloch, P.: On discontinuity–capturing methods for convection–diffusion equations. In: A. Bermúdez de Castro, D. Gómez, P. Quintela, P. Salgado (eds.) Numerical Mathematics and Advanced Applications, Proceedings of ENUMATH 2005, pp. 336–344. Springer–Verlag, Berlin (2006)
8. John, V., Knobloch, P.: On spurious oscillations at layers diminishing (SOLD) methods for convection–diffusion equations: Part I – A review. Comput. Methods Appl. Mech. Engrg. **196**, 2197–2215 (2007)
9. John, V., Knobloch, P.: On the performance of SOLD methods for convection–diffusion problems with interior layers. Int. J. Computing Science and Mathematics **1**, 245–258 (2007)

On Path-Conservative Numerical Schemes for Hyperbolic Systems of Balance Laws

M.L. Muñoz-Ruiz and C. Parés

Abstract This work is concerned with the numerical approximation of Cauchy problems for one-dimensional hyperbolic systems of conservation laws with source terms or balance laws. These systems can be studied as a particular case of nonconservative hyperbolic systems [3, 4, 5]. The theory developed by Dal Maso, LeFloch and Murat [2] is used to define a concept of weak solutions of nonconservative systems based on the choice of a family of paths in the phase space. The notion of path-conservative numerical scheme introduced in [6], which generalizes that of conservative scheme for conservative systems, is also related to the choice of a family of paths. In this work we present an appropriate choice of paths in order to define the concept of weak solution (see [1, 8]) in the particular case of balance laws, together with the notion of path-conservative numerical scheme for this particular case and some properties. We also consider the well-balance property of these schemes and the consistency with the definition of weak solutions, with a result pointing in the direction of a Lax-Wendroff type convergence result.

1 Introduction

This work is concerned with the numerical approximation of Cauchy problems for one-dimensional hyperbolic systems of conservation laws with source terms or balance laws

$$W_t + F(W)_x = S(W)\sigma_x, \quad x \in \mathbf{R},\ t > 0\,, \tag{1}$$

M.L. Muñoz-Ruiz
Dept. Matemática Aplicada, Universidad de Málaga, 29071-Málaga, Spain,
e-mail: munoz@anamat.cie.uma.es

C. Parés
Dept. Análisis Matemático, Universidad de Málaga, 29071-Málaga, Spain,
e-mail: pares@anamat.cie.uma.es

where $W(x,t) \in \Omega$, an open convex set of $\mathbf{R}^N$, $\sigma(x)$ is a known function from $\mathbf{R}$ to $\mathbf{R}$, F is a regular function from Ω to $\mathbf{R}^N$ and S is also a function from Ω to $\mathbf{R}^N$.

System (1) can be studied as a particular case of the nonconservative system

$$W_t + \mathscr{A}(W)W_x = 0, \quad x \in \mathbf{R},\ t > 0\,, \tag{2}$$

where $W(x,t)$ belongs to an open convex set Ω and $\mathscr{A}$ is a smooth locally bounded matrix function. The main difficulty related with system (2) from both the mathematical and the numerical point of view is the presence of nonconservative products, which do not make sense as distributions when the solution is discontinuous.

In Sect. 2 we briefly recall some results concerning systems (2) and in Sect. 3 we study the particular case of systems of balance laws (1).

2 Preliminaries on Hyperbolic Nonconservative Systems

In order to study system (2) we suppose that it is strictly hyperbolic and that the characteristic fields are either genuinely nonlinear or linearly degenerate.

In this work we assume the definition of nonconservative products as Borel measures given by Dal Maso, LeFloch and Murat [2], which is associated to the choice of a family of paths in the phase space Ω. A family of paths in Ω is a locally Lipschitz map $\Phi\colon [0,1] \times \Omega \times \Omega \to \Omega$ such that

$$\Phi(0;W_L,W_R) = W_L \text{ and } \Phi(1;W_L,W_R) = W_R, \text{ for any } W_L, W_R \in \Omega\,,$$

together with certain smoothness hypotheses. Once a family of paths Φ has been chosen, the nonconservative product $\mathscr{A}(W)W_x$ can be interpreted as a Borel measure for $W \in (L^\infty(\mathbf{R}\times\mathbf{R}^+) \cap BV(\mathbf{R}\times\mathbf{R}^+))^N$, denoted by $[\mathscr{A}(W)W_x]_\Phi$. If such W is piecewise $\mathscr{C}^1$ it is said to be a weak solution of (2) if

$$W_t + [\mathscr{A}(W)W_x]_\Phi = 0\,.$$

When no confusion arises, the dependency on Φ will be dropped.

Across a discontinuity, a weak solution must satisfy the generalized Rankine-Hugoniot condition:

$$\int_0^1 \left(\xi\mathscr{I} - \mathscr{A}(\Phi(s;W^-,W^+))\right)\Phi_s(s;W^-,W^+)\,ds = 0\,, \tag{3}$$

where ξ is the speed of propagation of the discontinuity, $\mathscr{I}$ is the identity matrix, and W^- and W^+ are the left and right limits of the solution at the discontinuity.

In the particular case of a system of conservation laws, that is, when $\mathscr{A}(W)$ is the Jacobian matrix of some flux function $F(W)$, the definition of the nonconservative product as a Borel measure does not depend on the choice of paths, and the generalized Rankine-Hugoniot condition reduces to the usual one.

As it occurs in the conservative case, we must assume a concept of entropy solution, as the one due to Lax or one related to an entropy pair.

Once a notion of entropy is chosen the theory of simple waves of hyperbolic systems of conservation laws and the results concerning the solutions of Riemann problems can be extended to systems (2).

The choice of the family of paths is important because it determines the speed of propagation of discontinuities. Although it should be based on the physical aspects of the problem, it is natural from the mathematical point of view to require this family to satisfy some hypotheses concerning the relationship of the paths with the integral curves of the characteristic fields and the solutions of Riemann problems [1, 8]. This is done in Sect. 3 for the particular case of balance laws.

For the discretization of system (2) with initial condition $W(x,0) = W_0(x)$, $x \in \mathbf{R}$, we use computing cells $I_i = [x_{i-1/2}, x_{i+1/2}]$ with constant size Δx. Define $x_{i+\frac{1}{2}} = i\Delta x$ and $x_i = (i-1/2)\Delta x$, the center of the cell I_i. Let Δt be the constant time step and define $t^n = n\Delta t$. We denote by W_i^n the approximation of the cell averages of the exact solution provided by the numerical scheme and we consider the notion of path-conservative numerical scheme proposed in [6]:

Given a family of paths Ψ, a Ψ-conservative numerical scheme is a scheme

$$W_i^{n+1} = W_i^n - \frac{\Delta t}{\Delta x}\left(D_{i-1/2}^{n,+} + D_{i+1/2}^{n,-}\right), \tag{4}$$

where $D_{i+1/2}^{n,\pm} = D^{\pm}\left(W_{i-q}^n, \ldots, W_{i+p}^n\right)$ and D^- and D^+ are two continuous functions from Ω^{p+q+1} to Ω satisfying

$$D^{\pm}(W, \ldots, W) = 0 \quad \forall W \in \Omega, \tag{5}$$

and

$$D^-(W_{-q}, \ldots, W_p) + D^+(W_{-q}, \ldots, W_p) = \int_0^1 \mathscr{A}\left(\Psi(s; W_0, W_1)\right)\Psi_s(s; W_0, W_1)\, ds, \tag{6}$$

for every $W_i \in \Omega$, $i = -q, \ldots, p$.

This definition generalizes that of conservative scheme for conservative problems: in the particular case of a system of conservation laws, a numerical scheme is conservative if and only if it is Ψ-conservative for any family of paths Ψ.

The two properties satisfied by a Ψ-conservative scheme imply a consistency requirement for regular solutions.

The best choice of the family of paths Ψ is obviously $\Psi = \Phi$, being Φ the family selected for the definition of weak solutions. Unfortunately, in practical applications the construction of Φ-conservative schemes can be difficult or very costly and simpler families of paths Ψ have to be chosen, as the family of segments. In that case, discontinuities can be incorrectly treated.

3 The Particular Case of Balance Laws

The goal of this work is the study of systems (1). Adding the trivially satisfied equation $\sigma_t = 0$, the system can be rewritten under the quasilinear form

$$\widetilde{W}_t + \widetilde{\mathscr{A}}(\widetilde{W})\widetilde{W}_x = 0\,, \tag{7}$$

where

$$\widetilde{W} = \begin{bmatrix} W \\ \sigma \end{bmatrix}, \quad \widetilde{\mathscr{A}}(\widetilde{W}) = \begin{bmatrix} A(W) & -S(W) \\ 0 & 0 \end{bmatrix} \quad \text{and} \quad A(W) = \frac{\partial F}{\partial W}(W)\,.$$

Notice that (1) reduces to a conservative system with flux function F if σ is constant.

If we suppose that $A(W)$ has N real distinct eigenvalues $\lambda_1(W) < \ldots < \lambda_N(W)$ which do not vanish, and associated eigenvectors $R_1(W),\ldots,R_N(W)$, then system (7) is strictly hyperbolic, since $\widetilde{\mathscr{A}}(\widetilde{W})$ has $N+1$ real distinct eigenvalues

$$\widetilde{\lambda}_i(\widetilde{W}) = \lambda_i(W), \quad i = 1,\ldots,N; \qquad \widetilde{\lambda}_{N+1}(\widetilde{W}) = 0$$

and eigenvectors

$$\widetilde{R}_i(\widetilde{W}) = \begin{bmatrix} R_i(W) \\ 0 \end{bmatrix}, \quad i = 1,\ldots,N; \qquad \widetilde{R}_{N+1}(\widetilde{W}) = \begin{bmatrix} A^{-1}(W)S(W) \\ 1 \end{bmatrix}.$$

The $(N+1)$-th characteristic field is linearly degenerate and we will suppose, for the sake of simplicity, that it is the only one.

In order to define the weak solutions of (7) a family of paths $\widetilde{\Phi}$ in $\Omega \times \mathbf{R}$ has to be chosen. We will use the notation

$$\widetilde{W}_L = \begin{bmatrix} W_L \\ \sigma_L \end{bmatrix}, \quad \widetilde{W}_R = \begin{bmatrix} W_R \\ \sigma_R \end{bmatrix} \quad \text{and} \quad \widetilde{\Phi}(s;\widetilde{W}_L,\widetilde{W}_R) = \begin{bmatrix} \Phi(s;\widetilde{W}_L,\widetilde{W}_R) \\ \Phi_{N+1}(s;\widetilde{W}_L,\widetilde{W}_R) \end{bmatrix},$$

where $\Phi(s;\widetilde{W}_L,\widetilde{W}_R)$ takes values in $\Omega \subset \mathbf{R}^N$.

The generalized Rankine-Hugoniot condition can be written as

$$\begin{cases} \xi(W_R - W_L) = F(W_R) - F(W_L) - S_{\widetilde{\Phi}}(\widetilde{W}_L,\widetilde{W}_R)\,, \\ \xi(\sigma_R - \sigma_L) = 0\,, \end{cases} \tag{8}$$

where

$$S_{\widetilde{\Phi}}(\widetilde{W}_L,\widetilde{W}_R) = \int_0^1 S(\Phi(s;\widetilde{W}_L,\widetilde{W}_R))\,(\Phi_{N+1})_s\,(s;\widetilde{W}_L,\widetilde{W}_R))\,ds\,. \tag{9}$$

Therefore, discontinuities appearing in weak solutions have to be either stationary ($\xi = 0$) or develop in regions where σ is continuous.

Following [1, 8], the family of paths has to be chosen in such a way that the corresponding weak solutions satisfy some natural requirements:

1. If $\widetilde{W}_L$ and $\widetilde{W}_R$ are such that $\sigma_L = \sigma_R = \bar{\sigma}$, the path $\widetilde{\Phi}(\cdot;\widetilde{W}_L,\widetilde{W}_R)$ satisfies

$$\Phi_{N+1}(s;\widetilde{W}_L,\widetilde{W}_R)=\bar{\sigma} \quad \forall s\in[0,1]\,. \tag{10}$$

2. Given $\widetilde{W}_L$ and $\widetilde{W}_R$ in the same integral curve γ of the linearly degenerate field, the path $\widetilde{\Phi}(\cdot;\widetilde{W}_L,\widetilde{W}_R)$ is a parametrization of the arc of γ linking $\widetilde{W}_L$ and $\widetilde{W}_R$.
3. Let us denote by $\widetilde{\mathscr{RP}}\subset(\Omega\times\mathbf{R})\times(\Omega\times\mathbf{R})$ the set of pairs $(\widetilde{W}_L,\widetilde{W}_R)$ such that the Riemann problem

$$\begin{cases}\widetilde{W}_t+\widetilde{\mathscr{A}}(\widetilde{W})\widetilde{W}_x=0\,,\\ \widetilde{W}(x,0)=\begin{cases}\widetilde{W}_L \text{ if } x<0\,,\\ \widetilde{W}_R \text{ if } x>0\,,\end{cases}\end{cases} \tag{11}$$

has a unique self-similar weak solution composed by at most $N+1$ simple waves connecting $N+2$ intermediate constant states $\widetilde{W}_j$, $j=0,\dots,N+1$. Then, given $(\widetilde{W}_L,\widetilde{W}_R)\in\widetilde{\mathscr{RP}}$, the curve described by the path $\widetilde{\Phi}(\cdot;\widetilde{W}_L,\widetilde{W}_R)$ is equal to the union of those corresponding to the paths $\widetilde{\Phi}(\cdot;\widetilde{W}_{j-1},\widetilde{W}_j)$, $j=1,\dots,N+1$.

These conditions allow to prove the following proposition:

Proposition 1. *If we assume that the concept of weak solutions of (7) is defined on the basis of a family of paths satisfying the previous hypotheses, then:*

1. *If*

$$\widetilde{W}=\begin{bmatrix}W\\ \sigma\end{bmatrix}$$

is a weak solution of (7) with σ constant, $\sigma(x)=\bar{\sigma}$, then W is a weak solution of the conservative problem

$$W_t+F(W)_x=0\,. \tag{12}$$

2. *Given two states $\widetilde{W}_L$ and $\widetilde{W}_R$ belonging to the same integral curve of the linearly degenerate field, the contact discontinuity given by*

$$\widetilde{W}(x,t)=\begin{cases}\widetilde{W}_L \text{ if } x<\xi t\,,\\ \widetilde{W}_R \text{ if } x>\xi t\,,\end{cases}$$

where ξ is the (constant) value of the corresponding eigenvalue through the integral curve, is an entropy weak solution of (7).

3. *Let $(\widetilde{W}_L,\widetilde{W}_R)$ be a pair belonging to $\widetilde{\mathscr{RP}}$ and let $\widetilde{W}$ be the solution of the Riemann problem (11). The following equality holds for every $t>0$:*

$$\left\langle\widetilde{\mathscr{A}}(\widetilde{W}(\cdot,t))\widetilde{W}_x(\cdot,t),1\right\rangle=\int_0^1\widetilde{\mathscr{A}}(\widetilde{\Phi}(s;\widetilde{W}_L,\widetilde{W}_R))\widetilde{\Phi}_s(s;\widetilde{W}_L,\widetilde{W}_R)\,ds\,.$$

Consequently, the total mass of the Borel measure $\left[\widetilde{\mathscr{A}}(\widetilde{W}(\cdot,t))\widetilde{W}_x(\cdot,t)\right]_{\widetilde{\Phi}}$ does not depend on t.

4. Let $(\widetilde{W}_L, \widetilde{W}_R)$ be a pair belonging to $\widetilde{\mathscr{R}\mathscr{P}}$ and $\widetilde{W}_j$ any of the intermediate states involved by the solution of the Riemann problem (11). Then:

$$\int_0^1 \widetilde{\mathscr{A}}(\widetilde{\Phi}(s;\widetilde{W}_L,\widetilde{W}_R))\widetilde{\Phi}_s(s;\widetilde{W}_L,\widetilde{W}_R)\,ds$$
$$= \int_0^1 \widetilde{\mathscr{A}}(\widetilde{\Phi}(s;\widetilde{W}_L,\widetilde{W}_j))\widetilde{\Phi}_s(s;\widetilde{W}_L,\widetilde{W}_j)\,ds$$
$$+ \int_0^1 \widetilde{\mathscr{A}}(\widetilde{\Phi}(s;\widetilde{W}_j,\widetilde{W}_R))\widetilde{\Phi}_s(s;\widetilde{W}_j,\widetilde{W}_R)\,ds\,.$$

The previous hypotheses on the family of paths completely determine the choice of the path linking the states of a pair $(\widetilde{W}_L, \widetilde{W}_R)$ in $\widetilde{\mathscr{R}\mathscr{P}}$.

Let us study now path-conservative schemes in the particular case of balance laws. We consider problem (1) with initial condition $W(x,0) = W^0(x)$ or, equivalently, problem (7) with initial condition $\widetilde{W}(x,0) = \widetilde{W}^0(x)$, where

$$\widetilde{W}^0 = \begin{bmatrix} W^0 \\ \sigma \end{bmatrix}. \tag{13}$$

We will use the notation

$$\widetilde{W}_i^n = \begin{bmatrix} W_i^n \\ \sigma_i \end{bmatrix}, \quad \sigma_i = \frac{1}{\Delta x}\int_{x_{i-1/2}}^{x_{i+1/2}} \sigma(x)\,dx \text{ and } \widetilde{\Psi}(s;\widetilde{W}_L,\widetilde{W}_R) = \begin{bmatrix} \Psi(s;\widetilde{W}_L,\widetilde{W}_R) \\ \Psi_{N+1}(s;\widetilde{W}_L,\widetilde{W}_R) \end{bmatrix},$$

where $\Psi(s;\widetilde{W}_L,\widetilde{W}_R)$ takes values in $\Omega \subset \mathbf{R}^N$, for any given family of paths.

A $\widetilde{\Psi}$-conservative numerical scheme for system (7) with initial condition (13) can be written as

$$W_i^{n+1} = W_i^n - \frac{\Delta t}{\Delta x}\left(D_{i-1/2}^{n,+} + D_{i+1/2}^{n,-}\right), \tag{14}$$

where $D_{i+1/2}^{n,\pm} = D^\pm\left(\widetilde{W}_{i-q}^n, \ldots, \widetilde{W}_{i+p}^n\right)$, being D^- and D^+ two continuous functions such that

$$D^\pm(\widetilde{W},\ldots,\widetilde{W}) = 0 \quad \forall \widetilde{W} \in \Omega \times \mathbf{R} \tag{15}$$

and

$$D^-(\widetilde{W}_{-q},\ldots,\widetilde{W}_p) + D^+(\widetilde{W}_{-q},\ldots,\widetilde{W}_p) = F(W_1) - F(W_0) - S_{\widetilde{\Psi}}(\widetilde{W}_0,\widetilde{W}_1), \tag{16}$$

for $\widetilde{W}_i \in \Omega \times \mathbf{R}$, $i = -q,\ldots,p$, where $S_{\widetilde{\Psi}}(\widetilde{W}_0,\widetilde{W}_1)$ is defined as (9) substituting $\widetilde{\Phi}$ with $\widetilde{\Psi}$ and $\widetilde{W}_L$, $\widetilde{W}_R$ with $\widetilde{W}_0$, $\widetilde{W}_1$.

If we suppose that $\widetilde{\mathscr{R}\mathscr{P}} = (\Omega \times \mathbf{R}) \times (\Omega \times \mathbf{R})$, then Riemann problems always have a unique solution and the following result can be proved:

Proposition 2. *Let $\widetilde{\Phi}$, the family of paths used to define the weak solutions of (7), be such that it satisfies the previously established hypotheses. Then:*

1. A path-conservative numerical scheme is $\widetilde{\Phi}$-conservative if and only if

$$D^-(\widetilde{W}_{-q},\ldots,\widetilde{W}_p)+D^+(\widetilde{W}_{-q},\ldots,\widetilde{W}_p)=F(W_1)-F(W^+_{1/2})+F(W^-_{1/2})-F(W_0)\,, \tag{17}$$

where

$$\widetilde{W}^-_{1/2}=\begin{bmatrix} W^-_{1/2} \\ \sigma_0 \end{bmatrix} \quad \text{and} \quad \widetilde{W}^+_{1/2}=\begin{bmatrix} W^+_{1/2} \\ \sigma_1 \end{bmatrix}$$

are, respectively, the left and right limits at $x=0$ of the solution of the Riemann problem that has $\widetilde{W}_0$ and $\widetilde{W}_1$ as initial condition.

2. *A $\widetilde{\Phi}$-conservative numerical scheme reduces to a conservative numerical scheme in regions where σ is constant.*

In what follows we will suppose that $q=0$ and $p=1$ and use the notation $\overline{W}$ for values $\widetilde{W}$ such that $\sigma=\bar{\sigma}$, being $\bar{\sigma}$ a fixed value, that is

$$\overline{W}=\begin{bmatrix} W \\ \bar{\sigma} \end{bmatrix}.$$

We can prove the following results:

Proposition 3. *A $\widetilde{\Psi}$-conservative numerical scheme reduces to a conservative scheme when σ is constant, $\sigma(x)=\bar{\sigma}$, if and only if*

$$D^{\pm}_{\bar{\sigma}}(\overline{W}_0,\overline{W}_1)=0 \tag{18}$$

and

$$S_{\widetilde{\Psi}}(\overline{W}_0,\overline{W}_1)=0\,, \tag{19}$$

which occurs when

$$\Psi_{N+1}(s;\overline{W}_0,\overline{W}_1)=\bar{\sigma} \quad \forall s\in[0,1]\,,$$

(in particular, when $\widetilde{\Psi}=\widetilde{\Phi}$).

Proposition 4. *A $\widetilde{\Psi}$-conservative numerical scheme which reduces to a conservative scheme when σ is constant, $\sigma(x)=\bar{\sigma}$, can be written in the form*

$$W_i^{n+1}=W_i^n-\frac{\Delta t}{\Delta x}\left(G^n_{i+1/2}-G^n_{i-1/2}\right)+\frac{\Delta t}{\Delta x}\left(H^{n,+}_{i-1/2}+H^{n,-}_{i+1/2}\right)\,, \tag{20}$$

where

$$G^n_{i+1/2}=G\left(W_i^n,W_{i+1}^n\right)\,,$$

being G a continuous function such that

$$G(W,W)=F(W) \tag{21}$$

and

$$H^{n,\pm}_{i+1/2}=H^{\pm}\left(\widetilde{W}_i^n,\widetilde{W}_{i+1}^n\right)\,,$$

being H^- and H^+ continuous functions such that

$$H^{\pm}(\widetilde{W},\widetilde{W}) = 0\,, \tag{22}$$

$$H^{-}(\widetilde{W}_0,\widetilde{W}_1) + H^{+}(\widetilde{W}_0,\widetilde{W}_1) = S_{\widetilde{\Psi}}(\widetilde{W}_0,\widetilde{W}_1) \tag{23}$$

and

$$H^{\pm}(\overline{W}_0,\overline{W}_1) = 0\,. \tag{24}$$

Concerning well-balancing, that is related to the numerical approximation of equilibria, i.e., steady state solutions (see [7] for more details), it can be proved that a necessary (but not sufficient) condition for a path-conservative scheme (14) to be well-balanced is that condition (17) is satisfied for every pair of states $\widetilde{W}_0$ and $\widetilde{W}_1$ in an integral curve of the linearly degenerate field.

The previous results allow us to prove a Lax-Wendroff convergence result for schemes of type (20). In fact, we prove that, when the approximations provided by such a scheme are bounded in $L^{\infty}_{\text{loc}}(\mathbf{R}\times[0,\infty))^N$ and converge in $L^{1}_{\text{loc}}(\mathbf{R}\times[0,\infty))^N$ to some W, then W is a weak solution of the Cauchy problem associated to the system of balance laws.

Acknowledgements This research has been partially supported by the Spanish Government Research project MTM2006-08075.

References

1. Castro, M.J., Gallardo, J.M., Muñoz, M.L., Parés, C.: On a general definition of the Godunov method for nonconservative hyperbolic systems. Application to linear balance laws. In: Numerical Mathematics and Advanced Applications, pp. 662–670. Springer, Berlin (2006)
2. Dal Maso, G., LeFloch, P.G., Murat, F.: Definition and weak stability of nonconservative products. J. Math. Pures Appl. **74**, 483–548 (1995)
3. Gosse, L.: A well-balanced scheme usig non-conservative products designed for hyperbolic systems of conservation laws with source terms. Math. Models Methods Appl. Sci. **11**, 339–365 (2001)
4. Greenberg, J.M., LeRoux, A.Y.: A well balanced scheme for the numerical processing of source terms in hyperbolic equations. SIAM J. Numer. Anal. **33**, 1–16 (1996)
5. LeFloch, P.G.: Shock waves for nonlinear hyperbolic systems in nonconservative form. Institute Math. Appl. Minneapolis, Preprint **593** (1989)
6. Parés, C.: Numerical methods for nonconservative hyperbolic systems: a theoretical framework. SIAM J. Numer. Anal. **44**, 300–321 (2006)
7. Parés, C., Castro, M.J.: On the well-balance property of Roe's method for nonconservative hyperbolic systems. Applications to shallow water systems. M2AN Math. Model. Numer. Anal. **38**, 821–852 (2004)
8. Muñoz Ruiz, M.L., Parés, C.: Godunov method for nonconservative hyperbolic systems. M2AN Math. Model. Numer. Anal. **41**, 169–185 (2007)

Discontinuous Galerkin Methods

An Augmented DG Scheme for Porous Media Equations

T.P. Barrios and R. Bustinza

Abstract We present an augmented local discontinuous Galerkin scheme for Darcy flow, that is obtained adding suitable Galerkin least squares terms arising from constitutive and equilibrium equations. The well-posedness of the scheme is proved applying Lax Milgram's theorem. Finally, we present an a posteriori error estimator, and include one numerical experiment showing that the estimator is reliable and efficient.

1 Introduction

Discontinuous Galerkin (DG) methods have been applied to solve a large class of second order elliptic equations in divergence form related to problems from physics and engineering. We refer to [1] (and the references therein) for an overview of DG methods for elliptic problems. Concerning the Darcy equation, in [7] a mixed DG formulation is analysed, while a stabilized DG method is proposed in [11], by adding a suitable Galerkin least squares term (related to the constitutive equation) to the mixed DG formulation. This is intended to avoid the introduction of lifting operators to prove the existence and uniqueness of the solution, which restringes the choice of the spaces of approximation. As expected, the resulting method is stable and convergent for any combination of velocity and pressure approximations (since it is not required to introduce lifting operators), first-order and higher, taking into account the L^2-norm for velocity. In this direction, we are concerned on the

Tomás P. Barrios
Departamento de Matemática y Física Aplicadas, Universidad Católica de la Santísima Concepción, Casilla 297, Concepción, Chile, e-mail: tomas@ucsc.cl

Rommel Bustinza (corresponding author)
Departamento de Ingeniería Matemática, Universidad de Concepción, Casilla 160-C, Concepción, Chile, e-mail: rbustinz@ing-mat.udec.cl

validity of this good properties, measuring the error in velocity in the piecewise $H(\mathrm{div})$ norm.

On the other hand, if the solution of the boundary value problem is not smooth enough, then the numerical method would need certain knowledge about the singularities of the solution. In order to overcome this difficulty, an adaptive strategy that automatically generates efficiently refined meshes would be most attractive.

These facts motivate us to propose in [3] an augmented DG formulation for the Poisson equation, considering piecewise Raviart-Thomas elements to approximate the vector (gradient) solution on a broken $H(\mathrm{div})$ space, while in [4] we present a generalization of the previous analysis, and develop an a posteriori error analysis for such a problem. One of our aims is therefore to extend/apply the a posteriori error technique for Darcy flow problems.

In this note we report on the main results derived in [2], which extends the applicability of the approach described in [4] to Darcy flow. Further, we obtain optimal rates of convergence (under suitable additional regularity of the exact solution), in the h-version context. In addition, we introduce an a posteriori error estimator, which is reliable and efficient. Finally, we present one numerical example confirming our theoretical results.

2 The Model Problem

We begin by introducing Ω as a simply connected and bounded domain in $\mathbb{R}^2$ with polygonal boundary Γ. Then, given the source terms $\mathbf{f} \in [L^2(\Omega)]^2$, $\omega \in L^2(\Omega)$ and $g \in L^2(\Gamma)$, we look for $(\mathbf{u}, p)$ such that

$$\mathbf{u} + \mathscr{K}\nabla p = \mathbf{f} \quad \text{in} \quad \Omega, \quad \mathrm{div}\,\mathbf{u} = \omega \quad \text{in} \quad \Omega, \quad \text{and} \quad \mathbf{u}\cdot\nu = g \quad \text{on} \quad \Gamma, \tag{1}$$

where $\mathscr{K} \in C(\overline{\Omega})$ is a symmetric and uniformly positive definite tensor representing the permeability of the porous media divided by the viscosity, and ν denotes the unit outward normal to $\partial\Omega$. The source term $\mathbf{f}$ is usually related to the gravity force, ω is the volumetric flow rate, and g is the normal component of the velocity field $\mathbf{u}$ on the boundary Γ. We assume that the data ω and g satisfy the compatibility relation $\int_\Omega \omega - \int_\Gamma g = 0$, and for uniqueness purposes, we suppose that $p \in L^2_0(\Omega) := \{q \in L^2(\Omega) : \int_\Omega q = 0\}$.

Now, in order to apply DG methods, we reformulate the problem (1) to find $(\mathbf{u}, p)$ in appropriate spaces such that, in the distributional sense,

$$\mathscr{K}^{-1}\mathbf{u} + \nabla p = \tilde{\mathbf{f}} \quad \text{in} \quad \Omega, \quad \mathrm{div}\,\mathbf{u} = \omega \quad \text{in} \quad \Omega, \quad \text{and} \quad \mathbf{u}\cdot\nu = g \quad \text{on} \quad \Gamma, \tag{2}$$

where $\tilde{\mathbf{f}} := \mathscr{K}^{-1}\mathbf{f}$. We remark that problem (1) (or (2)), has been already analysed in [7] and [11] using the local discontinuous Galerkin (LDG) method.

3 An Augmented Local DG Method

In this section, we present an augmented local discontinuous Galerkin method for the problem (2) and prove its well-posedness, using the recent results in [3] (see also [4]). We begin with the assumptions on meshes and some basic notations such as average and jumps.

We let $\{\mathscr{T}_h\}_{h>0}$ be a family of shape-regular triangulations of $\bar{\Omega}$ (with possible hanging nodes) made up of straight-side triangles T with diameter h_T and unit outward normal to ∂T given by ν_T. As usual, the index h also denotes $h := \max_{T\in\mathscr{T}_h} h_T$. Then, given $\mathscr{T}_h$, its edges are defined as follows. An *interior edge* of $\mathscr{T}_h$ is the (nonempty) interior of $\partial T \cap \partial T'$, where T and T' are two adjacent elements of $\mathscr{T}_h$, not necessarily matching. Similarly, a *boundary edge* of $\mathscr{T}_h$ is the (nonempty) interior of $\partial T \cap \partial\Omega$, where T is a boundary element of $\mathscr{T}_h$. We denote by $\mathscr{E}_I$ the list of all interior edges of (counted only once) on Ω, and by $\mathscr{E}_\Gamma$ the lists of all boundary edges, and put $\mathscr{E} := \mathscr{E}_I \cup \mathscr{E}_\Gamma$ the interior grid generated by the triangulation $\mathscr{T}_h$. Further, for each $e \in \mathscr{E}$, h_e represents its length. Also, in what follows we assume that $\mathscr{T}_h$ is of *bounded variation*, which means that there exists a constant $l > 1$, independent of the meshsize h, such that $l^{-1} \leq \frac{h_T}{h_{T'}} \leq l$ for each pair $T, T' \in \mathscr{T}_h$ sharing an interior edge.

Next, to define average and jump operators, we let T and T' be two adjacent elements of $\mathscr{T}_h$ and $\mathbf{x}$ be an arbitrary point on the interior edge $e = \partial T \cap \partial T' \in \mathscr{E}_I$. In addition, let q and $\mathbf{v}$ be scalar- and vector-valued functions, respectively, that are smooth inside each element $T \in \mathscr{T}_h$. We denote by $(q_{T,e}, \mathbf{v}_{T,e})$ the restriction of $(q_T, \mathbf{v}_T)$ to e. Then, we define the averages at $\mathbf{x} \in e$ by:

$$\{q\} := \frac{1}{2}\left(q_{T,e} + q_{T',e}\right), \quad \{\mathbf{v}\} := \frac{1}{2}\left(\mathbf{v}_{T,e} + \mathbf{v}_{T',e}\right).$$

Similarly, the jumps at $\mathbf{x} \in e$ are given by

$$[\![q]\!] := q_{T,e}\,\nu_T + q_{T',e}\,\nu_{T'}, \quad [\![\mathbf{v}]\!] := \mathbf{v}_{T,e}\cdot\nu_T + \mathbf{v}_{T',e}\cdot\nu_{T'}.$$

On boundary edges e, we set $\{q\} := q$, $\{\mathbf{v}\} := \mathbf{v}$, as well as $[\![q]\!] := q\,\nu$ and $[\![\mathbf{v}]\!] := \mathbf{v}\cdot\nu$. Hereafter, as usual div_h and ∇_h denote the piecewise divergence and gradient operators, respectively.

3.1 The Discrete Augmented LDG Formulation

Given a mesh $\mathscr{T}_h$, we proceed as in [3] (or [4]) and multiply each one of the equations of (2) by suitable test functions. We wish to approximate the exact solution $(\mathbf{u}, p)$ of (2) by discrete functions $(\mathbf{u}_h, p_h) \in \boldsymbol{\Sigma}_h \times V_h$, where

$$\boldsymbol{\Sigma}_h := \left\{\mathbf{v}_h \in [L^2(\Omega)]^2 : \mathbf{v}_h\big|_T \in \mathbf{RT}_r(T) \quad \forall T \in \mathscr{T}_h\right\},$$

$$V_h := \left\{ q_h \in L^2(\Omega) \,:\, q_h\big|_T \in P_k(T) \quad \forall T \in \mathscr{T}_h \right\}, \qquad \mathscr{V}_h := V_h \cap L^2_0(\Omega),$$

with $k \geq 1$ and $r \geq 0$. Hereafter, given $T \in \mathscr{T}_h$ and an integer $\kappa \geq 0$ we denote by $P_\kappa(T)$ the space of polynomials of degree at most κ on T, while $\mathbf{RT}_\kappa(T)$ denotes the Raviart-Thomas space of order κ on T.

Next, defining the so-called *numerical fluxes* as in [3] (see also [4]), that is, $\widehat{\mathbf{u}}$ and $\widehat{p}$ for each edge of any $T \in \mathscr{T}_h$, are given by

$$\widehat{\mathbf{u}}_{T,e} := \begin{cases} \{\mathbf{u}_h\} + \beta[\mathbf{u}_h] + \alpha[p_h] & \text{if } e \in \mathscr{E}_I \\ g\,\nu & \text{if } e \in \mathscr{E}_\Gamma \end{cases} \text{ and } \widehat{p}_{T,e} := \begin{cases} \{p_h\} - \beta \cdot [p_h] & \text{if } e \in \mathscr{E}_I \\ p_h & \text{if } e \in \mathscr{E}_\Gamma \end{cases} \tag{3}$$

and after adding suitable Galerkin least squares terms arising from constitutive equation $\mathscr{K}^{-1}\mathbf{u} + \nabla p = \tilde{\mathbf{f}}$ in Ω and equilibrium equation $\operatorname{div}\mathbf{u} = \omega$ in Ω, we derive the global discrete augmented LDG formulation: *Find* $(\mathbf{u}_h, p_h) \in \mathbf{\Sigma}_h \times \mathscr{V}_h$ *such that*

$$A^{stab}_{DG}((\mathbf{u}_h, p_h), (\mathbf{v}, q)) = F^{stab}_{DG}(\mathbf{v}, q) \quad \forall (\mathbf{v}, q) \in \mathbf{\Sigma}_h \times \mathscr{V}_h, \tag{4}$$

where the related bilinear form $A^{stab}_{DG} : \Big(\big(H(\operatorname{div};\mathscr{T}_h) \cap [H^\varepsilon(\mathscr{T}_h)]^2\big) \times H^1(\mathscr{T}_h) \Big) \times \Big(\big(H(\operatorname{div};\mathscr{T}_h) \cap [H^\varepsilon(\mathscr{T}_h)]^2\big) \times H^1(\mathscr{T}_h) \Big) \to \mathbb{R}$ and the corresponding linear functional $F^{stab}_{DG} : \Big(\big(H(\operatorname{div};\mathscr{T}_h) \cap [H^\varepsilon(\mathscr{T}_h)]^2\big) \times H^1(\mathscr{T}_h) \Big) \to \mathbb{R}$, are defined by

$$\begin{aligned} A^{stab}_{DG}((\mathbf{u}, r), (\mathbf{v}, q)) &:= \int_\Omega \mathscr{K}^{-1}\mathbf{u} \cdot \mathbf{v} - \int_\Omega r \operatorname{div}_h \mathbf{v} + \int_{\mathscr{E}_I} [\mathbf{v}](\{r\} - \beta \cdot [r]) \\ &+ \int_{\mathscr{E}_\Gamma} r\mathbf{v} \cdot \nu + \int_\Omega q \operatorname{div}_h \mathbf{u} - \int_{\mathscr{E}_I} [\mathbf{u}](\{q\} - \beta \cdot [q]) - \int_{\mathscr{E}_\Gamma} q\mathbf{u} \cdot \nu \\ &+ \int_{\mathscr{E}_I} \alpha[r] \cdot [q] + \frac{1}{2} \int_\Omega (\mathscr{K}^{-1}\mathbf{u} + \nabla_h r) \cdot (\mathscr{K}\nabla_h q - \mathbf{v}) + \int_\Omega \operatorname{div}_h \mathbf{u} \operatorname{div}_h \mathbf{v}, \end{aligned}$$

and

$$F^{stab}_{DG}(\mathbf{v}, q) := \int_\Omega \tilde{\mathbf{f}} \cdot \mathbf{v} + \int_\Omega \omega(q + \operatorname{div}_h \mathbf{v}) - \int_\Gamma g q + \frac{1}{2} \int_\Omega \tilde{\mathbf{f}} \cdot (\mathscr{K}\nabla_h q - \mathbf{v}),$$

for all $(\mathbf{u}, r), (\mathbf{v}, q) \in \big(H(\operatorname{div};\mathscr{T}_h) \cap [H^\varepsilon(\mathscr{T}_h)]^2\big) \times H^1(\mathscr{T}_h)$, with $\varepsilon > 1/2$.

The stabilization parameters α and β, which are needed to define the *numerical fluxes*, are chosen appropriately so that the solvability of the discrete augmented LDG formulation is guaranteed, as well as the optimal rates of convergence. In our case, we know that $\alpha \in \mathscr{O}(1/h)$, while $\beta = \mathscr{O}(1)$ (cf. [4, 2]). Moreover, we remark that the purpose of the term $\operatorname{div}_h \mathbf{u} \operatorname{div}_h \mathbf{v}$ is to give a control on $||\operatorname{div}_h \mathbf{v}||^2_{L^2(\Omega)}$, while the presence of the parameter β could help to prove superconvergence of the method (cf. [10, 9]).

3.2 Well-Posedness, A Priori and A Posteriori Error Estimates

In this subsection we prove the unique solvability of (4), by applying the well-known Lax-Milgram's theorem (as in [4] for Poisson's equation). To this end, we provide the space $\mathbf{\Sigma}_h$ with the usual norm of $\mathbf{\Sigma} := H(\mathrm{div}\,;\mathscr{T}_h)$, which is denoted by $\|\cdot\|_{\mathbf{\Sigma}}$, that is

$$\|\mathbf{v}\|_{\mathbf{\Sigma}}^2 := \|\mathbf{v}\|_{[L^2(\Omega)]^2}^2 + \|\mathrm{div}_h\mathbf{v}\|_{L^2(\Omega)}^2 \quad \forall\,\mathbf{v}\in\mathbf{\Sigma},$$

while for V_h we introduce its seminorm $|||\cdot|||_h : H^1(\mathscr{T}_h)\to\mathbb{R}$ as

$$|||q|||_h^2 := \|\nabla_h q\|_{[L^2(\Omega)]^2}^2 + \|\alpha^{1/2}[q]\|_{[L^2(\mathscr{E}_I)]^2}^2 \quad \forall\,q\in H^1(\mathscr{T}_h).$$

We remark that $|||\cdot|||_h$ is the *so-called* energy norm in $\mathscr{V}_h$ (see [6]). In addition, we define $\|(\cdot,\cdot)\|_{DG} : \mathbf{\Sigma}\times H^1(\mathscr{T}_h)\to\mathbb{R}$ by

$$\|(\mathbf{v},q)\|_{DG}^2 := \|\mathbf{v}\|_{\mathbf{\Sigma}}^2 + |||q|||_h^2 \quad \forall\,(\mathbf{v},q)\in\mathbf{\Sigma}\times H^1(\mathscr{T}_h),$$

which is a norm in the space $\mathbf{\Sigma}\times(H^1(\mathscr{T}_h)\cap L_0^2(\Omega))$.

The following result establishes the well-posedness of problem (4), as well as the optimal rate of convergence of the method.

Theorem 1. *Problem* (4) *has a unique solution* $(\mathbf{u}_h,p_h)\in\mathbf{\Sigma}_h\times\mathscr{V}_h$. *Moreover, assuming that* $(\mathbf{u},p)\in\big(H(\mathrm{div};\Omega)\cap[H^\varepsilon(\mathscr{T}_h)]^2\big)\times\big(H^{1+\varepsilon}(\Omega)\cap L_0^2(\Omega)\big)$, *for some* $\varepsilon>1/2$, *there exists* $C>0$, *independent of the meshsize, such that*

$$\|(\mathbf{u}-\mathbf{u}_h,p-p_h)\|_{DG} \le C \inf_{(\mathbf{v}_h,q_h)\in\mathbf{\Sigma}_h\times\mathscr{V}_h} \|(\mathbf{u}-\mathbf{v}_h,p-q_h)\|_{DG}. \tag{5}$$

In addition, assuming that $\mathbf{u}|_T\in[H^t(T)]^2$, $\mathrm{div}\,\mathbf{u}|_T\in H^t(T)$ *and* $p|_T\in H^{1+t}(T)$ *with* $t>1/2$, *for all* $T\in\mathscr{T}_h$, *there exists* $C_{\mathrm{err}}>0$, *independent of the meshsize, such that*

$$\|(\mathbf{u}-\mathbf{u}_h,p-p_h)\|_{DG}^2$$
$$\le C_{\mathrm{err}} \sum_{T\in\mathscr{T}_h} h_T^{2\min\{t,k,r\}} \Big\{\|\mathbf{u}\|_{[H^t(T)]^2}^2 + \|p\|_{H^{t+1}(T)}^2 + \|\mathrm{div}\,\mathbf{u}\|_{H^t(T)}^2\Big\}. \tag{6}$$

Proof. It is not difficult to see that the bilinear form A_{DG}^{stab} is bounded and strongly coercive on $\Sigma_h\times\mathscr{V}_h$, while the linear functional F_{DG}^{stab} is bounded on $\Sigma_h\times\mathscr{V}_h$, too. Then existence and uniqueness of the solution of problem (4) is a consequence of Lax-Milgram's theorem. Next, a Strang-type estimate is derived, and taking into account that the exact solution is smooth enough, one can check that the consistency term vanishes, yielding the Céa estimate (5). Finally, (6) is obtained from (5), by introducing suitable approximation operators of the exact solution onto its respective space of approximation, and applying then Lemmas 3.1 and 3.2 in [4]. □

We point out that Theorem 1 is valid for any combination of velocity and pressure approximations. However, we consider totally discontinuous Raviart-Thomas elements for the velocity because of its suitability to deal with divergence operator as well as the corresponding approximation theory (see [4] for a discussion for Poisson's problem).

We end this section with the following result, which presents an a posteriori error estimator that results to be reliable and efficient.

Theorem 2. *Let* $(\mathbf{u},p) \in H(\mathrm{div}\,,\Omega) \times (H^1(\Omega) \cap L_0^2(\Omega))$ *be the exact solution of (2) and let* $(\mathbf{u}_h,p_h) \in \mathbf{\Sigma}_h \times \mathscr{V}_h$ *be the unique solution of (4). Then there exist* $C_1, C_2 > 0$*, independent of meshsize, such that*

$$C_1\,\eta \;\le\; ||(\mathbf{u},p)-(\mathbf{u}_h,p_h)||_{DG} \le C_2\,\eta\,, \tag{7}$$

where $\eta^2 := \sum_{T\in\mathscr{T}_h}\eta_T^2$ *with* η_T^2 *given, on each* $T \in \mathscr{T}_h$*, by*

$$\begin{aligned}\eta_T^2 &:= \|\mathbf{f}-\mathbf{u}_h-\mathscr{K}\nabla p_h\|^2_{[L^2(T)]^2} + h_T^2\,\|\omega-\mathrm{div}\,\mathbf{u}_h\|^2_{L^2(T)}\\ &+||\alpha^{1/2}[p_h]||^2_{L^2(\partial T\cap\mathscr{E}_I)} + h_T\,||\mathbf{u}_h\cdot\nu_T-\widehat{\mathbf{u}}\cdot\nu_T||^2_{L^2(\partial T)}\,,\end{aligned} \tag{8}$$

with $\widehat{\mathbf{u}}$ *being the* numerical flux *associated to* $\mathbf{u}$ *(cf. (3)).*

Proof. The proof follows the ideas given in [5] (see also [4] and [8]), and is based on a suitable Helmholtz decomposition of the error in $\mathbf{u}$ (and p). The proof of the efficiency is based on Verfürth's ideas (see [12]), and needs the introduction of local bubble functions, on any edge and on any element of the triangulation. We omit further details. □

4 Numerical Results

In this section we present one numerical result illustrating the performance of the augmented mixed finite element scheme (4) and the a posteriori error estimator η, given in Theorem 2. To this end, we first note that for implementation purposes, the null media condition required by the elements of $\mathscr{V}_h$ can be imposed as a Lagrange multiplier. In other words, we consider the following modified discrete scheme: Find $(\mathbf{u}_h,p_h,\varphi_h) \in \mathbf{\Sigma}_h \times V_h \times \mathbb{R}$ such that

$$\begin{aligned}A_{DG}^{stab}((\mathbf{u}_h,p_h),(\mathbf{v},q)) + \varphi_h\int_\Omega q\,dx &= F_{DG}^{stab}(\mathbf{v},q) \quad \forall(\mathbf{v},q)\in\mathbf{\Sigma}_h\times V_h,\\ \psi\int_\Omega p_h\,dx &= 0 \quad \forall\psi\in\mathbb{R}.\end{aligned} \tag{9}$$

Next, to illustrate the properties of the estimator η and the augmented scheme (9), we take Ω as the square $(0,1)^2$, and choose the data $\mathbf{f}$, ω and g, as well as the matrix $\mathscr{K}$ so that the exact solution for the velocity is $\mathbf{u} = \nabla p$ in Ω with $p(x,y) = \sin(2\pi x)\sin(2\pi y)$ in Ω. The individual and total errors are denoted by $\mathbf{e}(\mathbf{u}) := \|\mathbf{u}-\mathbf{u}_h\|_{\mathbf{\Sigma}}$, $\mathbf{e}(p) := |||p-p_h|||_h$, $\mathbf{e}_0(p) := \|p-p_h\|_{L^2(\Omega)}$, $\mathbf{e} := \left\{\,[\mathbf{e}(\mathbf{u})]^2 + [\mathbf{e}(p)]^2\right\}^{1/2}$, respectively, whereas the effectivity index with respect to η is defined by $\mathbf{e}/\eta$. Now, given two consecutive triangulations with degrees of freedom N and N' and corresponding total errors $\mathbf{e}$ and $\mathbf{e}'$, the experimental rate of convergence is defined by

$$r(\mathbf{e}) := -2\,\frac{\log(\mathbf{e}/\mathbf{e}')}{\log(N/N')}\,.$$

The definition of $r(p)$, $r(\mathbf{u})$ and $r_0(p)$ is done in analogous way.

N	$\mathbf{e}(p)$	$r(p)$	$\mathbf{e}(\mathbf{u})$	$r(\mathbf{u})$	$\mathbf{e}$	$r(\mathbf{e})$	$\mathbf{e}_0(p)$	$r_0(p)$	$\mathbf{e}/\eta$
25	3.4703	—	40.3240	—	40.4731	—	0.3624	—	0.9799
97	2.6140	0.4180	32.2136	0.3312	32.3195	0.3318	0.1453	1.3483	0.9736
385	1.9425	0.4308	14.2623	1.1821	14.3940	1.1735	0.0590	1.3066	0.9526
1537	0.9965	0.9644	7.2996	0.9677	7.3673	0.9676	0.0148	1.9942	0.9327
6145	0.5012	0.9918	3.6716	0.9918	3.7057	0.9918	0.0036	2.0560	0.9016
24577	0.2515	0.9950	1.8386	0.9979	1.8557	0.9978	0.0009	2.0362	0.8019
98305	0.1260	0.9970	0.9197	0.9995	0.9283	0.9994	0.0002	2.0119	0.6916

Table 1 Example 1 with $P_1 - \mathbf{RT}_0$ approximation: uniform refinement.

N	$\mathbf{e}(p)$	$r_h(p)$	$\mathbf{e}(\mathbf{u})$	$r(\mathbf{u})$	$\mathbf{e}$	$r(\mathbf{e})$	$\mathbf{e}_0(p)$	$r_0(p)$	$\mathbf{e}/\eta$
25	3.4703	—	40.3240	—	40.4731	—	0.3624	—	0.9799
97	2.6140	0.4180	32.2136	0.3312	32.3195	0.3318	0.1453	1.3483	0.9736
385	1.9425	0.4308	14.2623	1.1821	14.3940	1.1735	0.0590	1.3066	0.9526
1393	1.1641	0.7964	7.7819	0.9422	7.8685	0.9393	0.0282	1.1476	0.9353
2365	0.9719	0.6816	6.2210	0.8459	6.2965	0.8421	0.0140	2.6528	0.9361
6271	0.5654	1.1111	3.7052	1.0628	3.7481	1.0639	0.0060	1.7203	0.9215
11671	0.4746	0.5637	2.9603	0.7226	2.9981	0.7188	0.0033	1.9667	0.9305
26449	0.3051	1.0797	1.8760	1.1152	1.9006	1.1143	0.0018	1.5298	0.9168
48679	0.2476	0.6847	1.4985	0.7365	1.5188	0.7351	0.0010	1.7315	0.9302
109267	0.1592	1.0936	0.9500	1.1273	0.9633	1.1264	0.0005	1.8985	0.9151

Table 2 Example 1 with $P_1 - \mathbf{RT}_0$ approximation: Adaptive refinement with hanging nodes.

In Tables 1 and 2 we provide the individual and total errors, the experimental rates of convergence, the a posteriori error estimator, and the effectivity index for the uniform and adaptive refinements, respectively, as applied to this example. In this case, uniform refinement means that, given a uniform initial triangulation, each subsequent mesh is obtained from the previous one by dividing each triangle into the four ones arising when connecting the midpoints of its sides. We apply red-refinement technique (with hanging nodes) for the adaptive one (see [12]). The errors are computed on each triangle using a 7-point Gaussian quadrature rule. We observe in Table 1 and 2 that the effectivity index are bounded from above and below, which confirms the reliability as well as the efficiency of the corresponding a posteriori error estimator (cf. Theorem 2). In fact, we notice that the effectivity index related to adaptive refinement, is close to one. We also remark that due to the smoothness of the exact solution, we obtain experimental rates of convergence $O(h)$ for the global error $\mathbf{e}$, in agreement with theoretical results (cf. Theorem 1), for both refinements. Moreover, we notice a quadratic convergence for the error $\mathbf{e}_0(p)$, whose theoretical proof should be deduced from usual duality arguments.

Finally, we point out that a more generalized analysis than the presented in this note, as well as more numerical examples, will be reported in [2].

Acknowledgements This work has been partially supported by CONICIYT-Chile through FONDECYT Grants No. 11060014, 1050842 and 1080168, by the Dirección de Investigación y Postgrado of the Universidad Católica de la Santísima Concepción, and by the Dirección de Investigación of the Universidad de Concepción. Part of this research was initiated while T.P. Barrios visited the Department of Mathematical Sciences, Chalmers University of Technology, Sweden, during May-July, 2007. T.P. Barrios expresses his gratitude to the colleagues and students of that department, in particular to Prof. Peter Hansbo for the hospitality and valuable discussions and suggestions. Finally, but not less important, we would like to thank the anonymous reviewer for constructive suggestions to the first version of this note.

References

1. Arnold, D.N., Brezzi, F., Cockburn, B., Marini, L.D.: Unified analysis of discontinuous Galerkin methods for elliptic problems. SIAM J. Numer. Anal. **39**(5), 1749–1779 (electronic) (2001/02)
2. Barrios, T.P., Bustinza, R.: An a posteriori error analysis of an augmented discontinuous Galerkin formulation for Darcy flow. Pre-print (in preparation), Departamento de Ingeniería Matemática, Universidad de Concepción
3. Barrios, T.P., Bustinza, R.: An augmented discontinuous Galerkin method for elliptic problems. C. R. Math. Acad. Sci. Paris **344**(1), 53–58 (2007)
4. Barrios, T.P., Bustinza, R.: A priori and a posteriori error analyses of an augmented Galerkin discontinuous formulation. Pre-Print 2007-02, Departamento de Ingeniería Matemática, Universidad de Concepción (2007). To appear in IMA J. Numer. Anal.
5. Becker, R., Hansbo, P., Larson, M.G.: Energy norm a posteriori error estimation for discontinuous Galerkin methods. Comput. Methods Appl. Mech. Engrg. **192**(5-6), 723–733 (2003)
6. Brenner, S.C.: Poincaré-Friedrichs inequalities for piecewise H^1 functions. SIAM J. Numer. Anal. **41**(1), 306–324 (electronic) (2003)
7. Brezzi, F., Hughes, T.J.R., Marini, L.D., Masud, A.: Mixed discontinuous Galerkin methods for Darcy flow. J. Sci. Comput. **22/23**, 119–145 (2005)
8. Bustinza, R., Gatica, G.N., Cockburn, B.: An a posteriori error estimate for the local discontinuous Galerkin method applied to linear and nonlinear diffusion problems. J. Sci. Comput. **22/23**, 147–185 (2005)
9. Castillo, P.: A superconvergence result for discontinuous Galerkin methods applied to elliptic problems. Comput. Methods Appl. Mech. Engrg. **192**(41-42), 4675–4685 (2003)
10. Cockburn, B., Kanschat, G., Perugia, I., Schötzau, D.: Superconvergence of the local discontinuous galerkin method for elliptic problems on cartesian grids. SIAM Journal on Numerical Analysis **39**(1), 264–285 (2001)
11. Hughes, T.J.R., Masud, A., Wan, J.: A stabilized mixed discontinuous Galerkin method for Darcy flow. Comput. Methods Appl. Mech. Engrg. **195**(25-28), 3347–3381 (2006)
12. Verfürth, R.: A Review of A Posteriori Error Estimation and Adaptive Mesh-Refinement Techniques. Wiley-Teubner, Chichester (1996)

A Remark to the DGFEM for Nonlinear Convection-Diffusion Problems Applied on Nonconforming Meshes

M. Feistauer

Abstract This paper is concerned with error estimates in $L^2(H^1)$- and $L^\infty(L^2)$-norm of the discontinuous Galerkin finite element method applied to the space semidiscretization of nonlinear nonstationary convection-diffusion problems. We discuss the discontinuos Galerkin method on shape regular meshes, which can be either conforming or nonconforming with hanging nodes. The main goal is to show that the results obtained under restrictive assumptions on the nonconformity of the meshes can be improved by using computational grids with less limiting properties.

1 Continuous Problem

In a number of applications we meet the necessity to solve nonstationary nonlinear convection-diffusion problems. A typical example is the Navier-Stokes system describing compressible viscous flow. One of promising, efficient methods for the solution of compressible flow is the discontinuous Galerkin finite element method (DGFEM) using piecewise polynomial approximation of a sought solution without any requirement on the continuity between neighbouring elements. In this paper we shall be concerned with the analysis of the DGFEM for the solution of a nonlinear nonstationary convection-diffusion equation, which is a simple prototype of the compressible Navier-Stokes system.

Let us consider the problem to find $u : Q_T = \Omega \times (0,T) \to I\!R$ such that

$$\text{a)}\ \frac{\partial u}{\partial t} + \sum_{s=1}^{d} \frac{\partial f_s(u)}{\partial x_s} = \varepsilon \Delta u + g \quad \text{in } Q_T, \tag{1}$$

$$\text{b)}\ u\big|_{\Gamma_D \times (0,T)} = u_D, \quad \text{c)}\ u(x,0) = u^0(x),\ x \in \Omega.$$

Miloslav Feistauer
Charles University Prague, Faculty of Mathematics and Physics, Sokolovská 83, 186 75 Praha 8, Czech Republic, e-mail: feist@karlin.mff.cuni.cz

We assume that $\Omega \subset I\!R^d$, $d = 2,3$, is a bounded polygonal (if $d = 2$) or polyhedral (if $d = 3$) domain with Lipschitz-continuous boundary $\partial\Omega$ and $T > 0$. The diffusion coefficient $\varepsilon > 0$ is a given constant, $g : Q_T \to I\!R$, $u_D : \partial\Omega \times (0,T) \to I\!R$, and $u^0 : \Omega \to I\!R$ are given functions, $f_s \in C^1(I\!R)$, $s = 1,\ldots,d$, are prescribed fluxes.

2 Discrete Problem

Let $\mathscr{T}_h = \{K_i\}_{i\in I}$ ($h > 0, I \subset \{0,1,2,\ldots\}$) denote a partition of the closure $\overline{\Omega}$ of the domain Ω into a finite number of closed triangles (if $d = 2$) or tetrahedra (if $d = 3$) K_i with mutually disjoint interiors. We do not require the usual conforming properties from the finite element method (cf. [1]). We set $h_K = \mathrm{diam}(K)$, $h = \max_{K\in\mathscr{T}_h} h_K$. By ρ_K we denote the radius of the largest ball inscribed into K. If two elements $K_i, K_j \in \mathscr{T}_h$ contain a nonempty open part of their faces, we call them *neighbours*. In this case we put $\Gamma_{ij} = \Gamma_{ji} = \partial K_i \cap \partial K_j$. For $i \in I$ we set $s(i) = \{j \in I; K_j$ is a neighbour of $K_i\}$. We shall also use the symbol Γ_{ij} for sides of K_i which are parts of $\partial\Omega$ and set $\gamma(i) = \{j;\ \Gamma_{ij} \subset \partial K_i \cap \partial\Omega\}$. (We assume that $s(i) \cap \gamma(i) = \emptyset$ for all $i \in I$.) Now, writing $S(i) = s(i) \cup \gamma(i)$, we have $\partial K_i = \bigcup_{j\in S(i)} \Gamma_{ij}$, $\partial K_i \cap \partial\Omega = \bigcup_{j\in\gamma(i)} \Gamma_{ij}$. Furthermore, we use the following notation: $\mathbf{n}_{ij} = ((n_{ij})_1, \ldots, (n_{ij})_d)$ is the unit outer normal to ∂K_i on the face Γ_{ij} and $d(\Gamma_{ij}) = \mathrm{diam}(\Gamma_{ij})$.

Over the triangulation $\mathscr{T}_h$ we introduce the *broken Sobolev space* $H^k(\Omega, \mathscr{T}_h) = \{v;\ v|_K \in H^k(K)\ \forall K \in \mathscr{T}_h\}$ with seminorm

$$|v|_{H^k(\Omega,\mathscr{T}_h)} = \left(\sum_{K\in\mathscr{T}_h} |v|^2_{H^k(K)}\right)^{1/2}, \quad v \in H^k(\Omega, \mathscr{T}_h). \tag{2}$$

For $v \in H^1(\Omega, \mathscr{T}_h)$, $i \in I$, $j \in s(i)$ we use the notation $v|_{\Gamma_{ij}} =$ trace of $v|_{K_i}$ on Γ_{ij}, $v|_{\Gamma_{ji}} =$ trace of $v|_{K_j}$ on Γ_{ji}, $\langle v\rangle_{\Gamma_{ij}} = \frac{1}{2}\left(v|_{\Gamma_{ij}} + v|_{\Gamma_{ji}}\right)$, $[v]_{\Gamma_{ij}} = v|_{\Gamma_{ij}} - v|_{\Gamma_{ji}}$.

The approximate solution is sought in the space of discontinuous piecewise polynomial functions $S_h = S^{p,-1}(\Omega, \mathscr{T}_h) = \{v; v|_K \in P^p(K)\ \forall K \in \mathscr{T}_h\}$, where $P^p(K)$ is the space of all polynomials on K of degree $\leq p$.

In order to introduce the DG space semidiscretization of problem (1), for $u, \varphi \in H^2(\Omega, \mathscr{T}_h)$ we define the forms

$$\begin{aligned} a_h(u,\varphi) &= \sum_{i\in I} \int_{K_i} \varepsilon \nabla u \cdot \nabla\varphi \,\mathrm{d}x \qquad (3)\\ &- \sum_{i\in I} \sum_{\substack{j\in s(i)\\ j<i}} \int_{\Gamma_{ij}} \varepsilon \langle\nabla u\rangle \cdot \mathbf{n}_{ij} [\varphi] \,\mathrm{d}S - \Theta \sum_{i\in I} \sum_{\substack{j\in s(i)\\ j<i}} \int_{\Gamma_{ij}} \varepsilon \langle\nabla\varphi\rangle \cdot \mathbf{n}_{ij} [u] \,\mathrm{d}S\\ &- \sum_{i\in I} \sum_{j\in\gamma(i)} \int_{\Gamma_{ij}} \varepsilon \nabla u \cdot \mathbf{n}_{ij}\, \varphi \,\mathrm{d}S - \Theta \sum_{i\in I} \sum_{j\in\gamma(i)} \int_{\Gamma_{ij}} \varepsilon \nabla \varphi \cdot \mathbf{n}_{ij}\, u \,\mathrm{d}S, \end{aligned}$$

$$l_h(\varphi)(t) = \int_\Omega g(t)\varphi\,dx \tag{4}$$
$$-\Theta\sum_{i\in I}\sum_{j\in\gamma(i)}\int_{\Gamma_{ij}} \varepsilon\nabla\varphi\cdot\mathbf{n}_{ij}u_D(t)\,dS + \sum_{i\in I}\sum_{j\in\gamma(i)}\int_{\Gamma_{ij}}\sigma u_D(t)\varphi\,dS,$$

$$J_h(u,\varphi) = \sum_{i\in I}\sum_{\substack{j\in s(i)\\ j<i}}\int_{\Gamma_{ij}}\sigma[u][\varphi]\,dS + \sum_{i\in I}\sum_{j\in\gamma(i)}\int_{\Gamma_{ij}}\sigma u\varphi\,dS \tag{5}$$

$$b_h(u,\varphi) = -\sum_{i\in I}\int_{K_i}\sum_{s=1}^{2} f_s(u)\frac{\partial\varphi}{\partial x_s}\,dx \tag{6}$$

$$+\sum_{i\in I}\sum_{j\in S(i)}\int_{\Gamma_{ij}} H(u|_{\Gamma_{ij}}, u|_{\Gamma_{ji}}, \mathbf{n}_{ij})\varphi|_{\Gamma_{ij}}\,dS. \tag{7}$$

Taking $\Theta = 1$, 0 and -1, we obtain the symmetric (SIPG), incomplete (IIPG) and nonsymmetric (NIPG) variants of the approximation of the diffusion terms. The weight σ and the numerical flux H will be specified later. If $j\in\gamma(i)$, then in H we set $u|_{\Gamma_{ji}} := u|_{\Gamma_{ij}}$. By $(\cdot,\cdot)$ we denote the $L^2(\Omega)$-scalar product.

Now we can introduce the discrete problem (space semidiscretization with continuous time, also called the method of lines). We define an approximate solution of problem (1) as a function $u_h \in C^1([0,T];S_h)$ satisfying the conditions $u_h(0) = u_h^0 =$ S_h-approximation of u^0, and

$$\frac{d}{dt}(u_h(t),\varphi_h) + b_h(u_h(t),\varphi_h) + \varepsilon J_h(u_h(t),\varphi_h) + a_h(u_h(t),\varphi_h) \tag{8}$$
$$= l_h(\varphi_h)(t), \quad \forall\varphi_h\in S_h,\ \forall t\in(0,T).$$

3 Error Analysis

3.1 Assumptions

Assumptions (H):

1. $H(u,v,\mathbf{n})$ is defined in $\mathbb{R}^2\times B_1$, where $B_1 = \{\mathbf{n}\in\mathbb{R}^d; |\mathbf{n}| = 1\}$, and *Lipschitz-continuous* with respect to u, v:
   $|H(u,v,\mathbf{n}) - H(u^*,v^*,\mathbf{n})| \le C_L(|u-u^*| + |v-v^*|)$, $u, v, u^*, v^*\in\mathbb{R}$, $\mathbf{n}\in B_1$.
2. $H(u,v,\mathbf{n})$ is *consistent*: $H(u,u,\mathbf{n}) = \sum_{s=1}^d f_s(u)\,n_s$, $u\in\mathbb{R}$, $\mathbf{n} = (n_1,\dots,n_d)\in B_1$.
3. $H(u,v,\mathbf{n})$ is *conservative*: $H(u,v,\mathbf{n}) = -H(v,u,-\mathbf{n})$, $u, v\in\mathbb{R}$, $\mathbf{n}\in B_1$.

We shall assume that the weak solution u of problem (1) is regular, namely

$$\frac{\partial u}{\partial t}\in L^2(0,T;H^{p+1}(\Omega)). \tag{9}$$

It is possible to show that the regular solution satisfies the identity

$$\frac{d}{dt}(u(t),\varphi_h)+b_h(u(t),\varphi_h)+\varepsilon J_h(u(t),\varphi_h)+a_h(u(t),\varphi_h) \tag{10}$$
$$=l_h(\varphi_h)(t), \quad \forall\varphi_h\in S_h, \text{ for a.e. } t\in(0,T).$$

Let us consider a regular system $\{\mathscr{T}_h\}_{h\in(0,h_0)}$, $h_0>0$, of partitions of the domain Ω. This means that there exists a constant $C_T>0$ such that

$$\frac{h_K}{\rho_K}\le C_T \quad \forall K\in\mathscr{T}_h \quad \forall h\in(0,h_0). \tag{11}$$

3.2 Some Auxiliary Results

In the analysis of the DGFEM we use the following important tools (see, e.g. [2]).

Multiplicative trace inequality: There exists a constant $C_M>0$ independent of v, h and K such that

$$\|v\|^2_{L^2(\partial K)}\le C_M\left(\|v\|_{L^2(K)}\,|v|_{H^1(K)}+h_K^{-1}\|v\|^2_{L^2(K)}\right), \tag{12}$$
$$K\in\mathscr{T}_h,\ v\in H^1(K),\ h\in(0,h_0).$$

Inverse inequality: There exists a constant $C_I>0$ independent of v, h, and K such that

$$|v|_{H^1(K)}\le C_I h_K^{-1}\|v\|_{L^2(K)}, \quad v\in P^p(K),\ K\in\mathscr{T}_h,\ h\in(0,h_0). \tag{13}$$

Now, for $v\in L^2(\Omega)$ we denote by $\Pi_h v$ the $L^2(\Omega)$-projection of v on S_h. It is possible to show (cf., e.g. [5, Lemma 4.1]) that the operator Π_h has the following properties: There exists a constant $C_A>0$ independent of h,K,v such that

$$\|\Pi_h v-v\|_{L^2(K)}\le C_A h_K^{k+1}|v|_{H^{k+1}(K)}, \tag{14}$$
$$|\Pi_h v-v|_{H^1(K)}\le C_A h_K^{k}|v|_{H^{k+1}(K)},$$
$$|\Pi_h v-v|_{H^2(K)}\le C_A h_K^{k-1}|v|_{H^{k+1}(K)},$$

for all $v\in H^{k+1}(K)$, $K\in\mathscr{T}_h$ and $h\in(0,h_0)$, where $k\in[1,p]$ is an integer.

3.3 Coercivity and Important Estimates

An important step in the analysis of error estimates is the *coercivity* of the form $A_h(u,v)=a_h(u,v)+\varepsilon J_h(u,v)$, which reads

$$A_h(\varphi_h,\varphi_h)\ge\frac{\varepsilon}{2}\left(|\varphi_h|^2_{H^1(\Omega,\mathscr{T}_h)}+J_h(\varphi_h,\varphi_h)\right), \quad \varphi\in S_h,\ h\in(0,h_0). \tag{15}$$

We shall discuss the validity of estimate (15) in various situations.

(I) Conforming Mesh $\mathscr{T}_h$

We assume that the mesh $\mathscr{T}_h$ has the standard properties from the finite element method (cf., e. g. [1]): if $K_i, K_j \in \mathscr{T}_h$, $i \neq j$, then $K_i \cap K_j = \emptyset$ or $K_i \cap K_j$ is a common vertex or $K_i \cap K_j$ is a common edge (or $K_i \cap K_j$ is a common face in the case $d = 3$) of K_i and K_j. In this case we set

$$\sigma|_{\Gamma_{ij}} = \frac{C_W}{d(\Gamma_{ij})}, \quad i \in I, \ j \in S(i). \tag{16}$$

Then (15) holds under the following choice of the constant C_W:

$$C_W > 0 \text{ (e. g. } C_W = 1) \quad \text{for NIPG version,} \tag{17}$$
$$C_W \geq 4C_M(1 + C_I) \quad \text{for SIPG version,} \tag{18}$$
$$C_W \geq 2C_M(1 + C_I) \quad \text{for IIPG version,} \tag{19}$$

where C_M and C_I are constants from (12) and (13), respectively.

(II) Nonconforming Mesh $\mathscr{T}_h$

In this case $\mathscr{T}_h$ is formed by closed triangles with mutually disjoint interiors with hanging nodes in general. Then the coercivity inequality (15) is guaranteed under conditions (17) – (19). However, in this case it is necessary to assume that

$$h_{K_i} \leq C_D d(\Gamma_{ij}), \quad i \in I, \ j \in S(i), \tag{20}$$

with a constant $C_D > 0$, in order to prove the estimate

$$J_h(\eta, \eta) \leq Ch^p |u|_{H^{p+1}(\Omega)}. \tag{21}$$

For the analysis of the cases (I) and (II) we can refer to [2].

(III) Nonconforming Mesh $\mathscr{T}_h$ *without Assumption (20)*

Condition (20) is a subject of a criticism, because it is obviously rather restrictive in some cases. In order to avoid it, we change the definition of the weight σ:

$$\sigma|_{\Gamma_{ij}} = \frac{2C_W}{h_{K_i} + h_{K_j}}, \quad i \in I, \ j \in s(i), \tag{22}$$
$$\sigma|_{\Gamma_{ij}} = \frac{C_W}{h_{K_i}}, \quad i \in I, \ j \in \gamma(i).$$

In theoretical analysis, this definition is used under the assumption

$$h_{K_i} \le C_N h_{K_j}, \quad i \in I,\ j \in s(i). \tag{23}$$

(Hence, $C_N \ge 1$.) Then (15) holds under the following choice of C_W:

$$C_W > 0 \text{ (e. g. } C_W = 1) \qquad \text{for NIPG version,} \tag{24}$$

$$C_W \ge 2C_M(1+C_I)(1+C_N) \qquad \text{for SIPG version,} \tag{25}$$

$$C_W \ge C_M(1+C_I)(1+C_N) \qquad \text{for IIPG version.} \tag{26}$$

Proof. Let us prove, for example, the coercivity inequality (15) in the case (III) for SIPG version.

Using the definition of the forms a_h and J_h and the Cauchy and Young's inequalities, we find that for any $\delta > 0$ we have

$$a_h(\varphi_h, \varphi_h) \ge \varepsilon |\varphi_h|^2_{H^1(\Omega,\mathscr{T}_h)} - \varepsilon\omega - \varepsilon \frac{\delta}{C_W} J_h(\varphi_h, \varphi_h),$$

where

$$\omega = \frac{1}{\delta} \sum_{i\in I} \left(\sum_{\substack{j\in s(i)\\ j<i}} \int_{\Gamma_{ij}} \frac{h_{K_i} + h_{K_j}}{2} |\langle \nabla\varphi_h \rangle|^2 \,\mathrm{d}s + \sum_{j\in\gamma(i)} \int_{\Gamma_{ij}} h_{K_i} |\nabla\varphi_h|^2 \mathrm{d}S \right).$$

In view of (23),

$$\omega \le \frac{1}{\delta} \frac{1+C_N}{2} \sum_{i\in I} h_{K_i} \int_{\partial K_i} |\nabla\varphi_h|^2 \mathrm{d}S.$$

Now, the application of (12) and (13) yields the estimate

$$\omega \le \frac{1}{2\delta} C_M(1+C_I)(1+C_N) |\varphi_h|^2_{H^1(\Omega,\mathscr{T}_h)}.$$

If we set $\delta := C_M(1+C_I)(1+C_N)$ and use assumption (25), we immediately arrive at (15).

In the IIPG case we can proceed similarly.

Moreover, it is possible to show that estimates from [2], obtained in the cases (I) and (II) can be proved also in the case (III). In particular, we mention the following estimate. If u and u_h denote the exact and approximate solutions, we set $\eta(t) = \Pi_h u(t) - u(t), \xi(t) = u_h(t) - \Pi_h u(t)$ for a.e. $t \in (0,T)$. Then, under the above assumptions, we have the estimate

$$|b_h(u,\xi) - b_h(u_h,\xi)| \le C \left(|\xi|^2_{H^1(\Omega,\mathscr{T}_h)} + J_h(\xi,\xi) \right)^{1/2} \left(h^{p+1} |u|_{H^{p+1}(\Omega)} + \|\xi\|_{L^2(\Omega)} \right). \tag{27}$$

As we see, this estimate contains the term h^{p+1}, which is important particularly for the derivation of an optimal $L^\infty(L^2)$-error estimate treated in Section 4.

3.4 Error Estimates

The detailed analysis representing the adaptation of the technique from [2] to the case (III) allows us to summarize the obtained error estimate in the following way.

Theorem 1. *Let assumptions* (H) *and* (11) *be satisfied and let* σ, $d(\Gamma_{ij})$ *and* C_W *satisfy assumptions from the cases (I) or (II) or (III). Let* u *be the exact solution of problem* (1) *satisfying the regularity condition* (9) *and let* u_h *be the approximate solution defined by* (8) *with* $u_h^0 = \Pi_h u^0$. *Then the error* $e_h = u - u_h$ *satisfies the estimate*

$$\max_{t\in[0,T]}\|e_h(t)\|^2_{L^2(\Omega)} + \frac{\varepsilon}{2}\int_0^t \left(|e_h(\vartheta)|^2_{H^1(\Omega,\mathscr{T}_h)} + J_h(e_h(\vartheta),e_h(\vartheta))\right) d\vartheta \le Ch^{2p}, \tag{28}$$

with a constant $C > 0$ *independent of* h.

4 Optimal Error Estimates

The error estimate (28) is optimal in the $L^2(H^1)$-seminorm, but suboptimal in the $L^\infty(L^2)$-norm. In [3], we carried out the analysis of the $L^\infty(L^2)$-optimal error estimate under the following assumptions.

Assumptions (B):

- the discrete diffusion form a_h is symmetric (i.e. we consider the SIPG version of the discrete problem), σ is given by (16) and C_W satisfies (18).
- the polygonal domain Ω is convex,
- the meshes $\mathscr{T}_h$, $h \in (0,h_0)$, are conforming with standard properties from the finite element method (i.e. without hanging nodes),
- the exact solution u of problem (1) satisfies the regularity condition (9),
- conditions (H) and (11) are satisfied,
- $u_h^0 = \Pi_h u^0$.

The derivation of the $L^\infty(L^2)$-optimal error estimate was carried out with the aid of the Aubin-Nitsche technique based on the use of the elliptic dual problem considered for each $z \in L^2(\Omega)$:

$$-\Delta\psi = z \quad \text{in } \Omega, \qquad \psi|_{\partial\Omega} = 0. \tag{29}$$

Then the weak solution $\psi \in H^2(\Omega)$ and there exists a constant $C > 0$, independent of z, such that

$$\|\psi\|_{H^2(\Omega)} \le C\|z\|_{L^2(\Omega)}. \tag{30}$$

For each $h \in (0,h_0)$ and $t \in [0,T]$ we define the function $u^*(t)$ $(= u_h^*(t))$ as the "A_h-projection" of $u(t)$ on S_h, i. e. a function satifying the conditions

$$u^*(t) \in S_h, \qquad A_h(u^*(t), \varphi_h) = A_h(u(t), \varphi_h) \quad \forall \varphi_h \in S_h, \tag{31}$$

and set $\chi = u - u^*$. Using the elliptic dual problem (29), we proved the existence of a constant $C > 0$ such that

$$\|\chi\|_{L^2(\Omega)} \leq Ch^{p+1}|u|_{H^{p+1}(\Omega)}, \quad \|\chi_t\|_{L^2(\Omega)} \leq Ch^{p+1}|u_t|_{H^{p+1}(\Omega)}, \ h \in (0, h_0), \tag{32}$$

which together with (27), multiple application of Young's inequality and Gronwall's lemma represent important tools for obtaining the $L^\infty(L^2)$-error estimate formulated as follows.

Theorem 2. *Let assumptions* (B) *be fulfilled. Then the error* $e_h = u - u_h$ *satisfies the estimate*

$$\|e_h\|_{L^\infty(0,T;L^2(\Omega))} \leq Ch^{p+1}, \tag{33}$$

with a constant $C > 0$ *independent of* h*.*

The assumption that the triangulations $\mathscr{T}_h$, $h \in (0, h_0)$, are conforming is limiting. Using a more sophisticated technique, we have proved estimates (32) and thus, (33) in the case of nonconforming meshes with hanging nodes. The detailed analysis will appear in [4].

Acknowledgements This work is a part of the research project MSM 0021620839 financed by the Ministry of Education of the Czech Republic.

References

1. Ciarlet, P.: The Finite Element Method for Elliptic Problems. North-Holland, Amsterdam (1979)
2. Dolejší, V., Feistauer, M.: Error estimates of the discontinuous Galerkin method for nonlinear nonstationary convection-diffusion problems. Numer. Funct. Anal. Optimiz. **26**, 349–383 (2005)
3. Dolejší, V., Feistauer, M., Kučera, V., Sobotíková, V.: An optimal $L^\infty(L^2)$-error estimate for the discontinuous Galerkin approximation of a nonlinear nonstationary convection-diffusion problem. IMA J. Numer. Anal. (2005). DOI 10.1093/imanum/dri017
4. Feistauer, M., Dolejší, V., Kučera, V., Sobotíková, V.: Optimal error estimates for the DGFEM applied to nonstationary nonlinear convection-diffusion problems on nonconforming meshes. In preparation
5. Feistauer, M., Švadlenka, K.: Discontinuous Galerkin method of lines for solving nonstationary singularly perturbed linear problems. J. Numer. Math. **12**, 98–118 (2004)

BDF-DGFE Method for the Compressible Navier-Stokes Equations

J. Hozman and V. Dolejší

Abstract We deal with a numerical solution of the compressible Navier-Stokes equations. We employ a combination of the discontinuous Galerkin finite element (DGFE) method for the space semi-discretization and the backward difference formulae (BDF) for the time discretization. Moreover, using a linearization of inviscid as well as viscous fluxes and applying a suitable explicit extrapolation to nonlinear terms, we obtain a numerical scheme which is almost unconditionally stable, has a higher degree of approximation with respect to the space and time coordinates and at each time step requires a solution of a linear algebraic problem. We present this approach and compare several variants of the DGFE techniques applied to a steady flow around the NACA0012 profile.

1 Introduction

A specific wide class of problems of fluid mechanics is formed of viscous compressible flow, which is described by the system of the compressible Navier–Stokes equations. Our goal is to develop an efficient, robust and accurate numerical scheme for a solution of the system of the Navier-Stokes equations. In last years the *discontinuous Galerkin method* (DGM) was employed in many papers for the discretization of compressible fluid flow problems, see, e.g., [2], [3], [6], [10], [12], [13] and the references cited therein. DGM is based on a piecewise polynomial, but discontinuous approximation, for a survey see, e.g., [4]. There are several variants of DGM for the solution of a viscous flow, see, e.g., [1]. Within this paper, we consider

J. Hozman
Department of Numerical Mathematics, Faculty of Mathematics and Physics, Charles University Prague, Sokolovská 83, Prague, 186 75, Czech Republic, e-mail: jhozmi@volny.cz

V. Dolejší
Department of Numerical Mathematics, Faculty of Mathematics and Physics, Charles University Prague, Sokolovská 83, Prague, 186 75, Czech Republic, e-mail: dolejsi@karlin.mff.cuni.cz

the so-called *nonsymmetric interior penalty Galerkin* (NIPG), *symmetric interior penalty Galerkin* (SIPG) and *incomplete interior penalty Galerkin* (IIPG) variants of the *discontinuous Galerkin finite element* (DGFE) method.

For the time discretization, the *method of lines* is used. In order to obtain a sufficiently stable scheme with respect to the size of the time step and in order to avoid a solution of nonlinear algebraic problems at each time step, we employ a *semi-implicit* method, which is based on a linearization of the inviscid and viscous fluxes. We especially focus on linearization of the viscous fluxes and the main aim of this paper is study two different approaches called (S1) and (S2).

Technique (S2) represents a direct generalization of the explicit time discretization presented in [6]. Although approach (S2) was presented in [8] for the solution of the Blasius problem, numerical computations presented in the last section show that this (S2) approach does not give satisfactory results for steady-state flow around NACA. Therefore we deal with the approach (S1) which is motivated by papers [12], [3]. However, we presented this approach in [7], a numerical comparison of both approaches is still missing. This is a subject of this contribution where numerical experiments show that technique (S2) does not achieve a satisfactory small steady-state residuum.

2 Compressible Navier-Stokes Equations

We consider the compressible Navier-Stokes equations in an open domain $Q_T = \Omega \times (0,T)$, where $T > 0$ is the final time and $\Omega \subset \mathrm{R}^2$ is the flow domain. We denote the boundary of Ω by $\partial\Omega$, it consists of several disjoint parts — inlet Γ_I, outlet Γ_O and impermeable walls Γ_W. Using this notation, the compressible Navier-Stokes equations can be written in conservative variables $\mathbf{w} = (\rho, \rho v_1, \rho v_2, E)^T$ in dimensionless form

$$\frac{\partial \mathbf{w}}{\partial t} + \sum_{s=1}^{2} \frac{\partial \mathbf{f}_s(\mathbf{w})}{\partial x_s} = \sum_{s=1}^{2} \frac{\partial \mathbf{R}_s(\mathbf{w}, \nabla \mathbf{w})}{\partial x_s} \quad \text{in } Q_T\,, \tag{1}$$

where

$$\mathbf{f}_s(\mathbf{w}) = (\rho v_s,\ \rho v_s v_1 + \delta_{s1} p,\ \rho v_s v_2 + \delta_{s2} p,\ (E+p) v_s)^T,\ s = 1,2\,, \tag{2}$$

are the so-called *inviscid (Euler) fluxes* and

$$\mathbf{R}_s(\mathbf{w}, \nabla \mathbf{w}) = \left(0,\ \tau_{s1}, \tau_{s2},\ \sum_{k=1}^{2} \tau_{sk} v_k + \frac{\gamma}{Re\,Pr} \frac{\partial \vartheta}{\partial x_s}\right)^T,\ s = 1,2\,, \tag{3}$$

are the so-called *viscous fluxes*. We consider the Newtonian type of fluid, i. e., the viscous part of the stress tensor has the form

$$\tau_{sk} = \frac{1}{Re}\left[\left(\frac{\partial v_s}{\partial x_k} + \frac{\partial v_k}{\partial x_s}\right) - \frac{2}{3}\mathrm{div}(\mathbf{v})\,\delta_{sk}\right], \; s,k = 1,2. \tag{4}$$

The following notation is used: ρ – density, $\mathbf{v} = (v_1, v_2)^T$ – velocity field, p – pressure, E – total energy, ϑ – temperature, γ – Poisson adiabatic constant, Re – Reynolds number and Pr – Prandtl number.

In order to close the system (1) – (4) we consider the *state equation for perfect gas* and the *relation for total energy*

$$p = (\gamma - 1)(E - \rho|\mathbf{v}|^2/2), \quad E = c_V \rho\vartheta + \rho|\mathbf{v}|^2/2, \tag{5}$$

where c_V is the specific heat at constant volume which is equal to one in the dimensionless case.

The system (1) – (5) is is equipped with the initial condition

$$\mathbf{w}(x,0) = \mathbf{w}^0(x), \quad x \in \Omega, \tag{6}$$

and the following set of boundary conditions on appropriate parts of boundary:

$$\begin{aligned}
&a)\; \rho = \rho_D,\; \mathbf{v} = \mathbf{v}_D,\; \sum_{k=1}^{2}\left(\sum_{l=1}^{2}\tau_{lk}n_l\right)v_k + \frac{\gamma}{Re\,Pr}\frac{\partial\vartheta}{\partial\mathbf{n}} = 0 \quad \text{on } \Gamma_I,\\
&b)\; \sum_{k=1}^{2}\tau_{sk}n_k = 0,\; s = 1,2,\; \frac{\partial\vartheta}{\partial\mathbf{n}} = 0 \quad \text{on } \Gamma_O,\\
&c)\; \mathbf{v} = 0,\; \frac{\partial\vartheta}{\partial\mathbf{n}} = 0 \quad \text{on } \Gamma_W,
\end{aligned} \tag{7}$$

where ρ_D and $\mathbf{v}_D$ are given functions and $\mathbf{n} = (n_1, n_2)^T$ is the unit outer normal to $\partial\Omega$.

We mention some properties of the inviscid and viscous fluxes (2) and (3), respectively, which are the base of their linearization and the consequent use of a semi-implicit approach.

The Euler fluxes $\mathbf{f}_s$, $s = 1,2$ satisfy

$$\mathbf{f}_s(\mathbf{w}) = \mathsf{A}_s(\mathbf{w})\mathbf{w}, \quad s = 1,2, \tag{8}$$

where $\mathsf{A}_s(\cdot)$, $s = 1,2$ are the Jacobi matrices of $\mathbf{f}_s(\cdot)$, see [11, Lemma 3.1].

In order to linearized the viscous terms, we define the forms $\mathbf{Q}_s(\cdot,\cdot,\cdot,\cdot) : \mathsf{R}^4 \times \mathsf{R}^8 \times \mathsf{R}^4 \times \mathsf{R}^8 \to \mathsf{R}^4$, $s = 1,2$, which formally denote two possible choices of the stabilization terms:

$$\mathbf{Q}_s(\mathbf{w}, \nabla\mathbf{w}, \varphi, \nabla\varphi) = \begin{cases} \sum_{k=1}^{2}\mathsf{K}_{s,k}(\mathbf{w})\frac{\partial\varphi}{\partial x_k}, & \text{(S1)}\\ \mathsf{D}_{s,0}(\mathbf{w},\nabla\mathbf{w})\varphi + \sum_{k=1}^{2}\mathsf{D}_{s,k}(\mathbf{w})\frac{\partial\varphi}{\partial x_k}, & \text{(S2)}\end{cases}, s = 1,2, \tag{9}$$

where $\mathbf{w}, \varphi \in \mathsf{R}^4$, $\mathsf{K}_{s,k} \in \mathsf{R}^{4\times4}$, $k = 1,2, s = 1,2$ are matrices (see, e.g., [11, Section 4.3.1]) and the matrices $\mathsf{D}_{s,k} \in \mathsf{R}^{4\times4}$, $k = 0,1,2, s = 1,2$ were defined in [7].

It is possible to show that the forms $\mathbf{Q}_s$ (for approaches (S1) as well as (S2)) are consistent with viscous fluxes $\mathbf{R}_s$ in the sense that

$$\mathbf{Q}_s(\mathbf{w},\nabla\mathbf{w},\mathbf{w},\nabla\mathbf{w}) = \mathbf{R}_s(\mathbf{w},\nabla\mathbf{w}) \quad \forall \mathbf{w},\ s=1,2. \tag{10}$$

Moreover, the forms $\mathbf{Q}_s(\mathbf{w},\nabla\mathbf{w},\varphi,\nabla\varphi),\ s=1,2$ are linear with respect φ and $\nabla\varphi$ and they are independent of $\nabla\varphi_1$.

3 Discontinuous Finite Element Spaces

Let $\mathscr{T}_h$ $(h>0)$ represents a triangulation of the closure $\overline{\Omega}$ of the domain $\Omega \subset \mathrm{R}^2$ into a finite number of closed elements (triangles or quadrilaterals) K with mutually disjoint interiors. We set $h = \max_{K\in\mathscr{T}_h} \mathrm{diam}(K)$.

By $\mathscr{F}_h$ we denote the smallest possible set of all open edges of all elements $K \in \mathscr{T}_h$. Further, we denote by $\mathscr{F}_h^I$ the set of all $\Gamma \in \mathscr{F}_h$ that are contained in Ω (inner edges) and by $\mathscr{F}_h^w$, $\mathscr{F}_h^i$ and $\mathscr{F}_h^o$ the set of all $\Gamma \in \mathscr{F}_h$ such that $\Gamma \subset \Gamma_W$, $\Gamma \subset \Gamma_I$ and $\Gamma \subset \Gamma_O$, respectively. Moreover, we use a following notation: $\mathscr{F}_h^D$ the set of all $\Gamma \in \mathscr{F}_h$ where the Dirichlet type of boundary conditions is prescribed at least for one component of w (i.e., $\mathscr{F}_h^D = \mathscr{F}_h^w \cup \mathscr{F}_h^i$) and $\mathscr{F}_h^N$ the set of all $\Gamma \in \mathscr{F}_h$ where the Neumann type of boundary conditions is prescribed for all components of w (i.e., $\mathscr{F}_h^N = \mathscr{F}_h^o$). For a shorter notation we put $\mathscr{F}_h^{ID} = \mathscr{F}_h^I \cup \mathscr{F}_h^D$.

Finally, for each $\Gamma \in \mathscr{F}_h$, we define a unit normal vector $\mathbf{n}_\Gamma$. We assume that $\mathbf{n}_\Gamma$, $\Gamma \subset \partial\Omega$ has the same orientation as the outer normal of $\partial\Omega$. For $\mathbf{n}_\Gamma$, $\Gamma \in \mathscr{F}_I$ the orientation is arbitrary but fixed for each edge.

Over the triangulation $\mathscr{T}_h$ we define the *broken Sobolev space*

$$H^k(\Omega,\mathscr{T}_h) = \{v; v|_K \in H^k(K)\ \forall K \in \mathscr{T}_h\}, \tag{11}$$

where $H^k(K) = W^{k,2}(K)$ means the (classical) Sobolev space on element K. Furthermore, we define the space of discontinuous piecewise polynomial functions

$$S_{hp} = \{v; v|_K \in P_p(K)\ \forall K \in \mathscr{T}_h\} \tag{12}$$

where $P_p(K)$ represents the space of all polynomials on K of degree $\leq p$, $K \in \mathscr{T}_h$. We seek the approximate solution (1) – (7) in the space of vector-valued functions $\mathbf{S}_{hp} = \left[S_{hp}\right]^4$.

For each $\Gamma \in \mathscr{F}_h^I$ there exist two elements $K_p, K_n \in \mathscr{T}_h$ such that $\Gamma \subset \overline{K_p} \cap \overline{K_n}$. We use a convention that K_n lies in the direction of $\mathbf{n}_\Gamma$ and for $v \in H^1(\Omega,\mathscr{T}_h)$, by $v|_\Gamma^{(p)} =$ trace of $v|_{K_p}$ on Γ, $v|_\Gamma^{(n)} =$ trace of $v|_{K_n}$ on Γ we denote the *traces* of v on edge Γ, which are different in general. Moreover, $[v]_\Gamma = v|_\Gamma^{(p)} - v|_\Gamma^{(n)}$ and $\langle v\rangle_\Gamma = \left(v|_\Gamma^{(p)} + v|_\Gamma^{(n)}\right)/2$ denotes the *jump* and *mean value* of function v over the edge Γ, respectively.

For $\Gamma \in \partial\Omega$ there exists an element $K_p \in \mathscr{T}_h$ such that $\Gamma \subset \overline{K_p} \cap \partial\Omega$. Then for $v \in H^1(\Omega, \mathscr{T}_h)$, we introduce the following notation: $v|_\Gamma^{(p)} =$ trace of $v|_{K_p}$ on Γ, $\langle v \rangle_\Gamma = [v]_\Gamma = v|_\Gamma^{(p)}$. In case that $\mathbf{n}_\Gamma$, $[\cdot]_\Gamma$ and $\langle\cdot\rangle_\Gamma$ are arguments of $\int_\Gamma \ldots \mathrm{d}S$, $\Gamma \in \mathscr{F}_h$, we omit the subscript Γ and write simply $\mathbf{n}$, $[\cdot]$ and $\langle\cdot\rangle$, respectively.

4 BDF–DGFE Formulation

Let $\bar{\mathbf{w}}, \mathbf{w}, \varphi \in \left[H^2(\Omega, \mathscr{T}_h)\right]^4$. Then, following the approach from [7], we define the diffusive form

$$
\begin{aligned}
\mathbf{a}_h(\bar{\mathbf{w}}, \mathbf{w}, \varphi) = & \sum_{K \in \mathscr{T}_h} \int_K \sum_{s=1}^2 \mathbf{Q}_s(\bar{\mathbf{w}}, \nabla\bar{\mathbf{w}}, \mathbf{w}, \nabla\mathbf{w}) \cdot \frac{\partial \varphi}{\partial x_s} \,\mathrm{d}x \qquad (13)\\
& - \sum_{\Gamma \in \mathscr{F}_h^{ID}} \int_\Gamma \left\langle \sum_{s=1}^2 \mathbf{Q}_s(\bar{\mathbf{w}}, \nabla\bar{\mathbf{w}}, \mathbf{w}, \nabla\mathbf{w}) \right\rangle n_s \cdot [\varphi] \,\mathrm{d}S \\
& - \eta \sum_{\Gamma \in \mathscr{F}_h^{I}} \int_\Gamma \left\langle \sum_{s=1}^2 \mathbf{Q}_s^*(\bar{\mathbf{w}}, \nabla\bar{\mathbf{w}}, \varphi, \nabla\varphi) \right\rangle n_s \cdot [\mathbf{w}] \,\mathrm{d}S \\
& - \eta \sum_{\Gamma \in \mathscr{F}_h^{D}} \int_\Gamma \sum_{s=1}^2 \mathbf{Q}_s^*(\bar{\mathbf{w}}, \nabla\bar{\mathbf{w}}, \varphi, \nabla\varphi)\, n_s \cdot (\mathbf{w} - \mathbf{w}_B) \,\mathrm{d}S,
\end{aligned}
$$

where $\mathbf{n} = (n_1, n_2)$ and forms $\mathbf{Q}_s^*$ are defined (based on the used discretization of viscous fluxes (9)) by

$$
\mathbf{Q}_s^*(\mathbf{w}, \nabla\mathbf{w}, \varphi, \nabla\varphi) = \begin{cases} \sum_{k=1}^2 \mathbb{K}_{s,k}^T(\mathbf{w}) \frac{\partial \varphi}{\partial x_k}, & \text{for (S1)} \\ \mathbf{Q}_s(\mathbf{w}, \nabla\mathbf{w}, \varphi, \nabla\varphi), & \text{for (S2)} \end{cases}, s = 1, 2 \qquad (14)
$$

According to value of η we speak of NIPG ($\eta = -1$), IIPG ($\eta = 0$) or SIPG ($\eta = 1$) variants of the DGFE method. The state vector $\mathbf{w}_B$ is prescribed on $\Gamma_I \cup \Gamma_W$ and given by the boundary conditions (7), namely

$$
\begin{aligned}
\mathbf{w}_B &= (\rho|_{\Gamma_W}, 0, 0, \rho|_{\Gamma_W} \vartheta|_{\Gamma_W}) \text{ on } \Gamma_W, \qquad (15)\\
\mathbf{w}_B &= (\rho_D, \rho_D(\mathbf{v}_D)_1, \rho_D(\mathbf{v}_D)_2, \rho|_{\Gamma_I} \vartheta|_{\Gamma_I} + \frac{1}{2}\rho_D|\mathbf{v}_D|^2) \text{ on } \Gamma_I,
\end{aligned}
$$

where ρ_D and $\mathbf{v}_D$ are given functions from the boundary conditions (7) and $\rho|_\Gamma$ and $\vartheta|_\Gamma$ are the values of density and temperature extrapolated from the interior of Ω on the appropriate boundary part Γ, respectively.

Moreover, we define the convective form

$$\mathbf{b}_h(\bar{\mathbf{w}}_h, \mathbf{w}_h, \varphi_h) = \sum_{K\in\mathscr{T}_h} \int_K \sum_{s=1}^{2} \mathsf{A}_s(\bar{\mathbf{w}}_h(x))\mathbf{w}_h(x) \cdot \frac{\partial \varphi_h(x)}{\partial x_s}\,\mathrm{d}x + \qquad (16)$$

$$+ \sum_{\Gamma\in\mathscr{F}_h} \int_\Gamma \left(\mathsf{P}^+\left(\langle \bar{\mathbf{w}}_h\rangle, \mathbf{n}\right) \mathbf{w}_h|_\Gamma^{(p)} + \mathsf{P}^-\left(\langle \mathbf{w}_h\rangle, \mathbf{n}\right) \mathbf{w}_h|_\Gamma^{(n)}\right) \cdot [\varphi_h]\mathrm{d}S$$

where $\mathsf{A}_s(\cdot) = 1,2$ are the Jacobi matrices of the mappings $\mathbf{f}_s$, $s = 1,2$, $\mathsf{P}^\pm(\cdot,\cdot)$ are the positive and negative parts of the matrix $\mathsf{P}(\mathbf{w},\mathbf{n}) = \sum_{s=1}^{2} \mathsf{A}_s(\mathbf{w})n_s$. A special attention should be paid to the definition of $\mathbf{w}_h|_\Gamma^{(n)}$ for $\Gamma \in \mathscr{F}_h^N \cup \mathscr{F}_h^{ID}$ where the "inviscid boundary conditions" are taken into account, for details see [9].

Finally, we define the interior and boundary penalty form by

$$\mathbf{J}_h^\sigma(\mathbf{w},\varphi) = \sum_{\Gamma\in\mathscr{F}_h^I} \int_\Gamma \sigma\,[\mathbf{w}]\cdot[\varphi]\,\mathrm{d}S + \sum_{\Gamma\in\mathscr{F}_h^D} \int_\Gamma \sigma(\mathbf{w} - \mathbf{w}_B)\cdot\varphi\,\mathrm{d}S\,, \qquad (17)$$

where the penalty parameter σ is defined by $\sigma|_\Gamma = C_W(|\Gamma|Re)^{-1}$, where $C_W \geq 0$ is a suitable constant depending on the used variant of the DGFE scheme and on the degree of polynomial approximation.

In order to simplify a notation we put

$$\mathbf{A}_h(\bar{\mathbf{w}}_h, \mathbf{w}_h, \varphi_h) = \mathbf{a}_h(\bar{\mathbf{w}}_h, \mathbf{w}_h, \varphi_h) + \mathbf{b}_h(\bar{\mathbf{w}}_h, \mathbf{w}_h, \varphi_h) + \mathbf{J}_h^\sigma(\mathbf{w},\varphi), \qquad (18)$$

for $\bar{\mathbf{w}}, \mathbf{w}, \varphi \in \left[H^k(\Omega, \mathscr{T}_h)\right]^4$. It follows from (13) – (18), that form $\mathbf{A}_h(\cdot,\cdot,\cdot)$ is linear with respect to its second and third arguments.

We discretize the problem (1) – (7) by the *semi–implicit* technique, where the linear parts of form $\mathbf{A}_h$ are treated implicitly and the nonlinear ones explicitly. In order to obtain a sufficiently stable and accurate approximation with respect to the time coordinate, we use a multistep *backward difference formulae* (BDF) for the approximation of the time derivative. Moreover, a suitable explicit higher order extrapolation is used in the nonlinear parts of $\mathbf{A}_h$. Since $S_{hp} \subset H^2(\Omega, \mathscr{T}_h)$, the form $\mathbf{A}(\cdot,\cdot,\cdot)$ makes sense for arguments from $\mathbf{S}_{hp}$. Then we are ready to introduce the *discrete problem*:

Let $0 = t_0 < t_1 < \cdots < t_r = T$ be a partition of $(0,T)$ and $\tau_k = t_{k+1} - t_k$, $k = 0,1,\ldots,r-1$ then we seek functions $\mathbf{w}_h^k \in \mathbf{S}_{hp}$, $k = 1,\ldots,r$, satisfying

$$\frac{1}{\tau_k}\left(\sum_{l=0}^{n} \alpha_l \mathbf{w}_h^{k+1-l}, \varphi_h\right) + \mathbf{A}_h\left(\sum_{l=1}^{n} \beta_l \mathbf{w}_h^{k+1-l}, \mathbf{w}_h^{k+1}, \varphi_h\right) = 0 \qquad (19)$$

for all $\varphi_h \in \mathbf{S}_{hp}$ and $k = n-1,\ldots,r-1$, where $n \geq 1$ is the degree of the BDF scheme, the coefficients α_l, $l = 0,\ldots,n$ and β_l, $l = 1,\ldots,n$. The function $\mathbf{w}_h^0$ is given by an initial condition and functions $\mathbf{w}_h^k$, $k = 1,\ldots,n-1$ should be given by a suitable one-step method.

The problem (19) (called BDF–DGFE scheme) represents a system of linear algebraic equations for each time step which should be solved by a suitable solver. We employ the restarted GMRES method with a block-diagonal preconditioning.

5 Numerical Example

We consider a subsonic laminar flow around the profile NACA0012 at the free stream Mach number $M = 0.5$, with zero angle of attack and Reynolds number $Re = 5000$. The walls of the profile are adiabatic. A characteristic feature of this flow problem is the laminar separation of the flow at the trailing edge.

We carried out computations for all combinations of stabilizations (S1) and (S2) with IPG techniques on a fixed relatively coarse triangular grid having 2600 elements, which was adaptive refined along the profile. We used a piecewise quadratic approximation in space and set $C_W = 625$ in the penalty parameter σ introduced in (17). This value guarantees the stability of all interior penalty variants of the DGFE method (NIPG, IIPG and SIPG) and both stabilization (S1) and (S2), see [7].

Table 1 shows achieved steady-state residua and the values of the drag (c_D) coefficients and their pressure ($c_{D,p}$) and viscous ($c_{D,v}$) parts computed by each combination of (S1) and (S2) with NIPG, IIPG and SIPG techniques and comparison with reference values from [2] and [5].

We observe that the use of the stabilization (S2) does not allow to achieve a satisfactory small steady-state residuum. Since the Reynolds number is near to the upper limit for steady laminar flow, the non-convergence of the method (S2) may be explained as a transition to the time-dependent flow regime which does not appear for method (S1) since this scheme contains too much of numerical viscosity in comparison with (S2). However, methods (S1) and (S2) have the same behaviour also for lower Reynolds number flow regimes where the steady state solution exists without any doubt. The values of the drag coefficient c_D rather oscillates for (S2) stabilization and on the other hand, the stabilization of type (S1) gives reasonable similar values of the drag coefficients for the SIPG, NIPG and IIPG variants, from this point of view the stabilization of type (S1) can be considered to be better method. Furthermore, Figure 1 shows details of the isolines of the Mach number around leading and trailing edges obtained by the (S1) stabilization (all IPG variants give in fact identical isolines).

Table 1 Steady-state residua (res) and the computed values of drag (c_D) coefficients and its pressure ($c_{D,p}$) and viscous ($c_{D,v}$) parts for the stabilizations (S1) and (S2) and the SIPG, NIPG and IIPG variants of the BDF-DGFE scheme in comparison with [2] and [5]

IPG	stab (9)	res	c_D	$c_{D,p}$	$c_{D,v}$
SIPG	(S1)	7.7117E-08	0.05322	0.02093	0.03229
NIPG	(S1)	6.0673E-08	0.05327	0.02099	0.03228
IIPG	(S1)	9.5915E-08	0.05325	0.02097	0.03228
SIPG	(S2)	4.0652E-05	0.05780	0.02223	0.03557
NIPG	(S2)	4.3020E-05	0.04618	0.02254	0.02364
IIPG	(S2)	4.3786E-05	0.04422	0.02275	0.02147
ref.	value	[2] – P_2	0.05352	0.01991	0.03361
ref.	value	[5] – P_0	0.05527	0.02281	0.03246

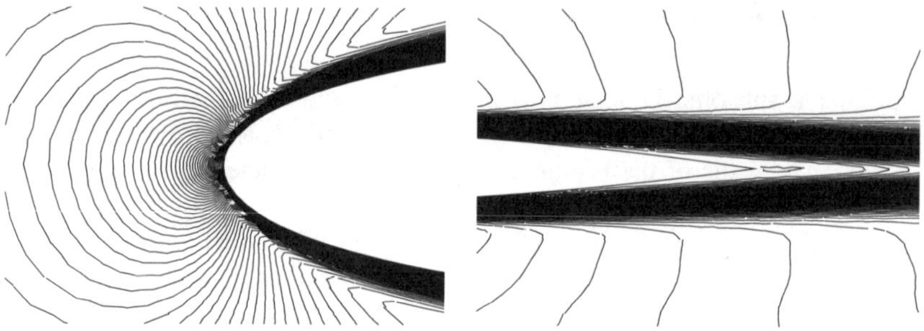

Fig. 1 Mach number isolines with details around the leading (left) and trailing edges (right)

Acknowledgements This work is a part of the research project MSM 0021620839 financed by the Ministry of Education of the Czech Republic and it was partly supported by the Grant No. 316/2006/B-MAT/MFF of the Grant Agency of the Charles University Prague.

References

1. Arnold, D.N., Brezzi, F., Cockburn, B., Marini, L.D.: Unified analysis of discontinuous Galerkin methods for elliptic problems. SIAM J. Numer. Anal. **39**(5), 1749–1779 (2002)
2. Bassi, F., Rebay, S.: A high-order accurate discontinuous finite element method for the numerical solution of the compressible Navier–Stokes equations. J. Comput. Phys. **131**, 267–279 (1997)
3. Baumann, C.E., Oden, J.T.: A discontinuous *hp* finite element method for the Euler and Navier-Stokes equations. Int. J. Numer. Methods Fluids **31**(1), 79–95 (1999)
4. Cockburn, B., Karniadakis, G.E., Shu, C.W. (eds.): Discontinuous Galerkin Methods. Springer, Berlin (2000)
5. Corre, C., Khalfallah, K., Lerat, A.: An efficient relaxation method for a centred Navier-Stokes solver. In: K. Morton, M. Baines (eds.) Numerical Methods for Fluid Dynamics V. Oxford Science Publications (1995)
6. Dolejší, V.: On the discontinuous Galerkin method for the numerical solution of the Navier–Stokes equations. Int. J. Numer. Methods Fluids **45**, 1083–1106 (2004)
7. Dolejší, V.: Semi-implicit interior penalty discontinuous Galerkin methods for viscous compressible flows. Commun. Comput. Phys. **4**(2), 231–274 (2008)
8. Dolejší, V., Hozman, J.: Semi-implicit discontinuous Galerkin method for the solution of the compressible Navier-Stokes equations. In: P. Wesseling, E. Onate, J. Périaux (eds.) European Conference on Computational Fluid Dynamics, Egmond aan Zee, The Netherlands, September 5-8, 2006, pp. 1061–1080. TU Delft, The Netherlands (2006)
9. Dolejší, V., Feistauer, M.: Semi-implicit discontinuous Galerkin finite element method for the numerical solution of inviscid compressible flow. J. Comput. Phys. **198**(2), 727–746 (2004)
10. Dumbser, M., Munz, C.D.: Building blocks for arbitrary high-order discontinuous Galerkin methods. J. Sci. Comput. **27**, 215–230 (2006)
11. Feistauer, M., Felcman, J., Straškraba, I.: Mathematical and Computational Methods for Compressible Flow. Oxford University Press, Oxford (2003)
12. Hartmann, R., Houston, P.: Adaptive discontinuous Galerkin finite element methods for the compressible Euler equations. J. Comput. Phys. **183**(2), 508–532 (2002)
13. Klaij, C.M., van der Vegt, J., der Ven, H.V.: Pseudo-time stepping for space-time discontinuous Galerkin discretizations of the compressible Navier-Stokes equations. J. Comput. Phys. **219**(2), 622–643 (2006)

Discontinuous Galerkin Method for the Numerical Solution of Inviscid and Viscous Compressible Flow

V. Kučera

Abstract In this work we are concerned with the numerical solution of a viscous compressible gas flow (compressible Navier-Stokes equations) with the aid of the discontinuous Galerkin finite element method (DGFEM). Our goal is to incorporate viscous terms into existing semi-implicit DGFEM scheme for the Euler equations, which is capable of solving flows with a wide range of Mach numbers [2, 4]. The nonsymmetric (NIPG), symmetric (SIPG) and incomplete interior penalty Galerkin method (IIPG) are generalized using the unified framework of [1] – derived for the Poisson equation – to the Navier-Stokes viscous terms. The resulting nonlinearities are linearized in a similar manner as nonlinear convective terms in the original scheme, thus enabling semi-implicit time stepping. The resulting scheme has very good stability properties and requires the solution of one sparse linear system per time level.

1 Continuous Problem

We shall discretize the Navier-Stokes equations - the model of viscous compressible two-dimensional flow. Let $T > 0$, $\Omega \subset \mathbb{R}^2$ be a domain with boundary $\partial\Omega = \Gamma_I \cup \Gamma_O \cup \Gamma_W$, the *inlet, outlet* and *wall* boundaries. The model equations are treated in the form of a conservation law for the *state vector* $\mathbf{w}(x,t)$:

$$\frac{\partial \mathbf{w}}{\partial t} + \sum_{s=1}^{2} \frac{\partial \mathbf{f}_s(\mathbf{w})}{\partial x_s} = \sum_{s=1}^{2} \frac{\partial \mathbf{R}_s(\mathbf{w}, \nabla \mathbf{w})}{\partial x_s} \quad \text{in } \Omega \times (0,T), \tag{1}$$

where

Václav Kučera
Charles University in Prague, Faculty of Mathematics and Physics, Sokolovská 83, Praha 8, 186 75, Czech Republic, e-mail: vaclav.kucera@email.cz

$$
\begin{aligned}
&\mathbf{w} = (\rho, \rho v_1, \rho v_2, e)^{\mathrm{T}} \in \mathbb{R}^4, \\
&\mathbf{f}_i(w) = (\rho v_i, \rho v_1 v_i + \delta_{1i} p, \rho v_2 v_i + \delta_{2i} p, (e+p) v_i)^{\mathrm{T}}. \\
&\mathbf{R}_i(\mathbf{w}, \nabla \mathbf{w}) = (0, \tau_{i1}, \tau_{i2}, \tau_{i1} v_1 + \tau_{i2} v_2 + k \partial \theta / \partial x_i)^{\mathrm{T}}, \\
&\tau_{ij} = \lambda \delta_{ij} \mathrm{div} \mathbf{v} + 2\mu d_{ij}(\mathbf{v}),\ d_{ij}(\mathbf{v}) = \frac{1}{2}\left(\frac{\partial v_i}{\partial x_j} + \frac{\partial v_j}{\partial x_i}\right).
\end{aligned} \tag{2}
$$

Here we use the notation ρ-density, $(\rho v_1, \rho v_2)$-momentum and e-internal energy. To system (1) we add thermodynamical relations for pressure p and temperature θ:

$$
p = (\gamma - 1)(e - \rho |\mathbf{v}|^2/2), \quad \theta = \left(\frac{e}{\rho} - \frac{1}{2}|\mathbf{v}|^2\right)/c_v. \tag{3}
$$

We use the following notation: θ – absolute temperature, c_v – specific heat at constant volume, μ, λ – viscosity coefficients, k – heat conduction coefficient. We assume $\mu, k > 0$, $2\mu + 3\lambda \geq 0$. Usually we set $\lambda = -2/(3\mu)$. Finally $\gamma > 1$ is the Poisson adiabatic constant. For example, for air $\gamma = 1.4$.

In the following, we will need the following relation for the fluxes $\mathbf{f}_s$:

$$
\mathbf{f}_s(\mathbf{w}) = \mathbb{A}_s(\mathbf{w})\mathbf{w}, \quad \text{where } \mathbb{A}_s(\mathbf{w}) = \frac{D\mathbf{f}_s(\mathbf{w})}{D\mathbf{w}},\ s = 1,2. \tag{4}
$$

The viscous fluxes $\mathbf{R}_i(\mathbf{w}, \nabla\mathbf{w})$ have a similar property to (4). The term $\mathbf{R}_i(\mathbf{w}, \nabla\mathbf{w})$ can be expressed in the form

$$
\mathbf{R}_i(\mathbf{w}, \nabla \mathbf{w}) = \sum_{j=1}^{2} \mathbb{K}_{ij}(\mathbf{w}) \frac{\partial \mathbf{w}}{\partial x_j}, \tag{5}
$$

where $\mathbb{K}_{ij}$ are 4×4 matrices dependent on $\mathbf{w}$ and independent of $\nabla\mathbf{w}$. Explicit formulae for $\mathbb{A}_s$ and $\mathbb{K}_{ij}$ can be found e.g. in [3].

2 Discretization

Let $\mathscr{T}_h$ be a partition of $\overline{\Omega}$ into a finite number of triangles, whose interiors are mutually disjoint. Let I be a numbering of triangles in $\mathscr{T}_h$. If two elements $K_i, K_j \in \mathscr{T}_h$ share a common face we set $\Gamma_{ij} = \partial K_i \cap \partial K_j$. For $i \in I$ we define $s(i) = \{j \in I; K_j \text{ is a neighbour of } K_i\}$. We denote all boundary faces by S_j, where $j \in I_b \subset \mathbb{Z}^- = \{-1, -2, \ldots\}$ and set $\gamma(i) = \{j \in I_b; S_j \text{ is a face of } K_i\}$. Furthermore we define $S(i) = s(i) \cup \gamma(i)$ and $\gamma_D(i) = \{j \in I_b;$ Dirichlet boundary condition prescribed on $\Gamma_{ij}\}$.

By $\mathbf{n}_{ij}$ we denote the unit outer normal to ∂K_i on the face Γ_{ij}. We set $v_{ij} = v|_{\Gamma_{ij}} =$ *trace* of $v|_{K_i}$ on Γ_{ij}, $\langle v \rangle_{\Gamma_{ij}} = \frac{1}{2}(v_{ij} + v_{ji})$, $[v]_{\Gamma_{ij}} = v_{ij} - v_{ji}$ the average and the jump on an edge, respectively, and

$$
S_h = S^{p,-1}(\Omega, \mathscr{T}_h) = \{v; v|_K \in P_p(K)\ \forall K \in \mathscr{T}_h\}, \tag{6}
$$

where $P_p(K)$ is the space of all polynomials on K of degree $\leq p$. In the current implementation, P_0, P_1 and P_2 approximations are used along with 5th order Gaussian quadrature rules on elements and edges.

In order to derive a discrete formulation of the Navier-Stokes equations, we derive an appropriate weak formulation of (1) with respect to the discontinuous Galerkin finite element space S_h. The discretization of convective terms is rather straightforward and can be carried out as in [2]. The governing equation is multiplied by a test function $\varphi \in [S_h]^4$ and integrated over $K_i \in \mathscr{T}_h$. We apply Green's theorem, sum over all $i \in I$ and incorporate a numerical flux in boundary terms. The treatment of second order elliptic terms is more delicate and in the case of viscous terms in the Navier-Stokes equations, this is further complicated by the fact that we treat a system and the viscous terms are nonlinear with respect to the unknown variable $\mathbf{w}$.

3 Discretization of Viscous Terms

In order to derive an appropriate formulation, we apply the methodology used in [1], which gives a unified framework for the discretization of the Poisson equation. We present a possible generalization of the *nonsymmetric* (NIPG), *symmetric* (SIPG) and *incomplete interior penalty* (IIPG) schemes to nonlinear systems.

The methodology used in [1] is based on introducing an auxiliary variable, which approximates the gradient of the sought solution in a weak sense. This equation is coupled with the weak formulation for the Poisson equation, which results in a system of the first order equations. After discretizing this system with the discontinuous Galerkin method with a special choice of the numerical flux, one can eliminate the auxiliary variable to obtain the so called primal formulation. For an appropriate choice of the numerical flux for the auxiliary equation, one obtains e.g. the NIPG or SIPG methods.

Since we are interested mainly in the discretization of viscous terms, we treat a simplified equation consisting only of the viscous terms contained in the Navier-Stokes equations equipped with a homogeneous Dirichlet boundary condition:

$$\begin{aligned} -\sum_{s=1}^{2} \frac{\partial \mathbf{R}_s(\mathbf{w}, \nabla \mathbf{w})}{\partial x_s} &= 0 \quad \text{in } \Omega, \\ \mathbf{w} &= 0 \quad \text{on } \partial\Omega \end{aligned} \tag{7}$$

In order to derive a discontinuous Galerkin formulation, we shall introduce an auxiliary variable $\sigma \approx \nabla \mathbf{w}$, under the notation $\sigma^{(k)} \approx \frac{\partial \mathbf{w}}{\partial x_k}$, for $k = 1, 2$. We assume that $\mathbf{w}$ is a sufficiently regular classical solution of (7) and write an equivalent formulation of (7) (for simplicity we omit the boundary condition) for unknowns $\mathbf{w}, \sigma^1, \sigma^2$:

$$-\sum_{s=1}^{2}\frac{\partial \mathbf{R}_s(\mathbf{w},\sigma)}{\partial x_s}=0 \quad \text{in } \Omega,$$
$$\sigma^{(k)}=\frac{\partial \mathbf{w}}{\partial x_k} \quad \text{for } k=1,2. \tag{8}$$

To derive a suitable weak formulation, we multiply the first equation by a test function $\varphi \in [S_h]^4$ and the second equation by the test function $\tau \in \Sigma_h$, where Σ_h is an appropriate function space – this need not be specified, since the auxiliary variable σ will be eventually eliminated from the formulation. We integrate over an element $K_i \in \mathscr{T}_h$, apply Green's theorem and sum over all elements:

$$\sum_{i\in I}\int_{K_i}\sum_{s=1}^{2}\mathbf{R}_s(\mathbf{w},\sigma)\cdot\frac{\partial\varphi}{\partial x_s}\,\mathrm{d}x-\sum_{i\in I}\int_{\partial K_i}\sum_{s=1}^{2}\mathbf{R}_s(\mathbf{w},\sigma)n_{K_i}^{(s)}\cdot\varphi\, dS=0, \quad \forall\varphi\in[S_h]^4,$$
$$\sum_{i\in I}\int_{K_i}\sigma^{(k)}\cdot\tau\,\mathrm{d}x=-\sum_{i\in I}\int_{K_i}\mathbf{w}\cdot\frac{\partial\tau}{\partial x_k}\,\mathrm{d}x+\sum_{i\in I}\int_{\partial K_i}\mathbf{w}n_{K_i}^{(k)}\cdot\tau\, dS, \quad \forall\tau\in\Sigma_h. \tag{9}$$

To derive a discontinuous Galerkin formulation of (9), we proceed as in the case of convective terms. We introduce numerical fluxes H_{ij}^{w} and H_{ij}^{σ} into the boundary integrals in each equation:

$$\int_{\Gamma_{ij}}\sum_{s=1}^{2}\mathbf{R}_s(\mathbf{w},\sigma)n_{ij}^{(s)}\cdot\varphi\, dS\approx\int_{\Gamma_{ij}}H_{ij}^{\mathbf{w}}\cdot\varphi\, dS,$$
$$\int_{\Gamma_{ij}}\mathbf{w}n_{ij}^{(k)}\cdot\tau\, dS\approx\int_{\Gamma_{ij}}H_{ij}^{\sigma}n_{ij}^{(k)}\cdot\tau\, dS. \tag{10}$$

The choice of different numerical fluxes H_{ij}^{w} and H_{ij}^{σ} leads to different numerical schemes. For instance, according to [1], for the Poisson equation, the following choices lead to the *nonsymmetric interior penalty* method:

$$H_{ij}^{\mathbf{w}}:=\begin{cases}\sum_{s=1}^{2}\mathbf{R}_s(\mathbf{w}_{ij},\nabla\mathbf{w}_{ij})n_{ij}^{(s)} & \text{for } \Gamma_{ij}\subset\partial\Omega,\\ \sum_{s=1}^{2}\langle\mathbf{R}_s(\mathbf{w},\nabla\mathbf{w})\rangle n_{ij}^{(s)} & \text{otherwise.}\end{cases}$$
$$H_{ij}^{\sigma}:=\begin{cases}2\mathbf{w}_{ij} & \text{for } \Gamma_{ij}\subset\partial\Omega,\\ \langle\mathbf{w}\rangle+[\mathbf{w}] & \text{otherwise.}\end{cases} \tag{11}$$

To obtain the *symmetric* variant, we use a different numerical flux definition. Namely, $H_{ij}^{\mathbf{w}}$ is the same as in (11) and H_{ij}^{σ} is defined as

$$H_{ij}^{\sigma}:=\begin{cases}0 & \text{for } \Gamma_{ij}\subset\partial\Omega,\\ \langle\mathbf{w}\rangle & \text{otherwise.}\end{cases} \tag{12}$$

By replacing boundary terms in (10) by the numerical fluxes defined in (11), we obtain after some manipulation the following discrete formulation of system (8), which leads to the NIPG scheme:

$$\begin{aligned}
&\sum_{i\in I}\int_{K_i}\sum_{s=1}^{2}\mathbf{R}_s(\mathbf{w},\sigma)\cdot\frac{\partial\varphi}{\partial x_s}\,\mathrm{d}x-\sum_{i\in I}\sum_{\substack{j\in s(i)\\ j<i}}\int_{\Gamma_{ij}}\sum_{s=1}^{2}\langle\mathbf{R}_s(\mathbf{w},\nabla\mathbf{w})\rangle n_{ij}^{(s)}\cdot[\varphi]\,dS\\
&\qquad-\sum_{i\in I}\sum_{j\in\gamma(i)}\int_{\Gamma_{ij}}\sum_{s=1}^{2}\mathbf{R}_s(\mathbf{w}_{ij},\nabla\mathbf{w}_{ij})n_{ij}^{(s)}\cdot\varphi_{ij}\,dS=0,\quad\forall\varphi\in[S_h]^4,\\
&\sum_{i\in I}\int_{K_i}\sigma^{(k)}\cdot\tau\,\mathrm{d}x=-\sum_{i\in I}\int_{K_i}\mathbf{w}\cdot\frac{\partial\tau}{\partial x_k}\,\mathrm{d}x+\sum_{i\in I}\sum_{\substack{j\in s(i)\\ j<i}}\int_{\Gamma_{ij}}(\langle\mathbf{w}\rangle+[\mathbf{w}])n_{ij}^{(k)}\cdot[\tau]\,dS\\
&\qquad+\sum_{i\in I}\sum_{j\in\gamma(i)}\int_{\Gamma_{ij}}2\mathbf{w}_{ij}n_{ij}^{(k)}\cdot\tau_{ij}\,dS,\quad k=1,2,\ \forall\tau\in\Sigma_h.
\end{aligned}\tag{13}$$

Similarly as in [1] for the Poisson equation, the choice of numerical fluxes (11) and (12) enables the elimination of the auxiliary variable σ from system (9). For lack of space we omit this technical procedure, which leads to the so-called *primal formulation*. This will, in our case, give a generalization of the NIPG and SIPG schemes from the Poisson equation.

We introduce the notation similar to (5):

$$\widetilde{\mathbf{R}}_i(\mathbf{w},\nabla\varphi):=\sum_{j=1}^{2}\mathbb{K}_{ji}^{T}(\mathbf{w})\frac{\partial\varphi}{\partial x_j}.\tag{14}$$

Now we can write the discrete formulation of equation (1). Find $\mathbf{w}_h(t)\in[S_h]^4$:

$$\frac{d}{dt}(\mathbf{w}_h(t),\varphi)+b_h(\mathbf{w}_h(t),\varphi)+a_h(\mathbf{w}_h(t),\varphi)+J_h(\mathbf{w}_h(t),\varphi)=0,\quad\forall\varphi\in[S_h]^4,\tag{15}$$

where

$$\begin{aligned}
a_h(\mathbf{w},\varphi)&=\sum_{i\in I}\int_{K_i}\sum_{s=1}^{2}\mathbf{R}_s(\mathbf{w},\nabla\mathbf{w})\cdot\frac{\partial\varphi}{\partial x_s}\,\mathrm{d}x\\
&-\sum_{i\in I}\sum_{\substack{j\in s(i)\\ j<i}}\int_{\Gamma_{ij}}\sum_{s=1}^{2}\langle\mathbf{R}_s(\mathbf{w},\nabla\mathbf{w})\rangle n_{ij}^{(s)}\cdot[\varphi]\,dS-\sum_{i\in I}\sum_{j\in\gamma_D(i)}\int_{\Gamma_{ij}}\sum_{s=1}^{2}\mathbf{R}_s(\mathbf{w},\nabla\mathbf{w})n_{ij}^{(s)}\cdot\varphi\,dS\\
&+\Theta\sum_{i\in I}\sum_{\substack{j\in s(i)\\ j<i}}\int_{\Gamma_{ij}}\sum_{s=1}^{2}\langle\widetilde{\mathbf{R}}_s(\mathbf{w},\nabla\varphi)\rangle n_{ij}^{(s)}\cdot[\mathbf{w}]\,dS\\
&+\Theta\sum_{i\in I}\sum_{j\in\gamma_D(i)}\int_{\Gamma_{ij}}\sum_{s=1}^{2}\widetilde{\mathbf{R}}_s(\mathbf{w},\nabla\varphi)n_{ij}^{(s)}\cdot\mathbf{w}\,dS,
\end{aligned}$$

$$b_h(\mathbf{w},\varphi) = -\sum_{K_i\in\mathscr{T}_h}\int_{K_i}\sum_{s=1}^{2}\mathbf{f}_s(\mathbf{w})\cdot\frac{\partial\varphi}{\partial x_s}\,\mathrm{d}x + \sum_{i\in I}\sum_{j\in S(i)}\int_{\Gamma_{ij}}\mathbf{H}(\mathbf{w}|_{\Gamma_{ij}},\mathbf{w}|_{\Gamma_{ji}},\mathbf{n}_{ij})\cdot\varphi\,dS,$$

$$J_h(\mathbf{w},\varphi) = \sum_{i\in I}\sum_{\substack{j\in s(i)\\ j<i}}\int_{\Gamma_{ij}}\sigma[\mathbf{w}]\cdot[\varphi]\,dS + \sum_{i\in I}\sum_{j\in\gamma_D(i)}\int_{\Gamma_{ij}}\sigma\mathbf{w}\cdot\varphi\,dS,$$

$$l_h(\mathbf{w},\varphi) = \sum_{i\in I}\sum_{j\in\gamma_D(i)}\int_{\Gamma_{ij}}\sum_{s=1}^{2}\sigma\mathbf{w}_B\cdot\varphi\,dS + \Theta\sum_{i\in I}\sum_{j\in\gamma_D(i)}\int_{\Gamma_{ij}}\sum_{s=1}^{2}\widetilde{\mathbf{R}}_s(\mathbf{w},\nabla\varphi)n_{ij}^{(s)}\cdot\mathbf{w}_B\,dS.$$

These are the convective form, viscous form, interior and boundary penalty jump terms and right-hand side form, respectively. In the second term of $b_h(\mathbf{w},\varphi)$, we have incorporated an approximation using a numerical flux $\mathbf{H}$, as known from the finite volume method. The approximate solution is defined as $\mathbf{w}_h \in [S_h]^4$ such that (15) holds for all $\varphi_h \in [S_h]^4$. Depending on the value of Θ, we obtain the *nonsymmetric* (NIPG, $\Theta = 1$) *symmetric* (SIPG, $\mathscr{T}_h eta = -1$) and *incomplete interior penalty* (IIPG, $\Theta = 0$) variants of the viscous terms.

4 Time Discretization

Scheme (15) represents a system of ordinary differential equations, which we must discretize with respect to time. Explicit time discretization is however undesirable due to a *CFL*-like condition, which limits the time step proportionally to the Mach number. A fully implicit scheme presents us with the task of solving a large nonlinear system on each time level. We therefore use the method presented in [3]. A forward Euler method is used and the nonlinear terms in the scheme are linearized. The resulting systems are solved using block-Jacobi preconditioned GMRES or direct algorithms (UMFPACK).

The time derivative is discretized as

$$\frac{d}{dt}(\mathbf{w}_h(t_{k+1}),\varphi) \approx \frac{\mathbf{w}_h^{k+1}-\mathbf{w}_h^k}{\tau_k}. \tag{16}$$

The convective form in (15) is linearized using homogeneity of the Euler fluxes and due to the choice of the Vijayasundaram numerical flux, which has appropriate form for linearization:

$$\begin{aligned} b_h(\mathbf{w}_h^{k+1},\varphi) \approx \tilde{b}_h(\mathbf{w}_h^k,\mathbf{w}^{k+1},\varphi) &= -\sum_{i\in I}\int_{K_i}\sum_{s=1}^{2}\frac{D\mathbf{f}_s(\mathbf{w}_h^k)}{D\mathbf{w}}\mathbf{w}_h^{k+1}\cdot\frac{\partial\varphi_h}{\partial x_s}\,\mathrm{d}x \\ &+\sum_{i\in I}\sum_{j\in S(i)}\int_{\Gamma_{ij}}\mathbb{P}^+\left(\langle\mathbf{w}_h^k\rangle,\mathbf{n}_{ij}\right)\mathbf{w}_{ij}^{k+1}+\mathbb{P}^-\left(\langle\mathbf{w}_h^k\rangle,\mathbf{n}_{ij}\right)\mathbf{w}_{ji}^{k+1}\cdot\varphi\,dS. \end{aligned} \tag{17}$$

The matrices $\mathbb{P}^+(\mathbf{w},\mathbf{n})$ and $\mathbb{P}^-(\mathbf{w},\mathbf{n})$ are defined as the positive and negative parts of the matrix $\sum_{s=1}^{2} \mathbb{A}_s(\mathbf{w})n_s$, where $\mathbb{A}_s(\mathbf{w})$ is defined in (4).

The diffusive form is linearized in a similar fashion using the fact that the viscous terms $\mathbf{R}_i(\mathbf{w},\nabla\mathbf{w})$ (and similarly $\widetilde{\mathbf{R}}_i$) can be expressed in the form

$$\mathbf{R}_i(\mathbf{w},\nabla\mathbf{w}) = \sum_{j=1}^{2} \mathbb{K}_{ij}(\mathbf{w})\frac{\partial \mathbf{w}}{\partial x_j}, \tag{18}$$

where $\mathbb{K}_{ij}$ are 4×4 matrices. Explicit formulae for $\mathbb{K}_{ij}$ can be found e.g. in [1]. Thus we can linearize the nonlinearities in $a_h(\mathbf{w}_h^{k+1},\varphi)$ in the following fashion:

$$\mathbf{R}_s(\mathbf{w}_h^{k+1},\nabla\mathbf{w}_h^{k+1}) \approx \sum_{j=1}^{2} \mathbb{K}_{ij}(\mathbf{w}_h^{k})\frac{\partial \mathbf{w}_h^{k+1}}{\partial x_j}. \tag{19}$$

In such a way we obtain a numerical scheme which requires the solution of only one large sparse linear system per time level.

5 Numerical experiments

We treat the compressible flow around a NACA0012 profile with a large angle of attack (25°). The flow is nonstationary with vortex formation and shedding at the upper wall of the profile. The far-field flow has Mach number $M = 0.5$, angle of attack $\alpha = 25°$ and Reynolds number $Re = 5000$. The computational mesh has 2898 elements and is adaptively refined near the profile. Due to the nonstationary character of the flow, the following figures illustrate the flow situation at time $t = 8.5$. Figure 1 shows a detail of the Mach number isolines. The boundary layer and complicated flow structure behind the airfoil are visible. In Figure 2 a detail of streamlines with the vortex structure at $t = 8.5$ is shown. Finally in we plot the entropy, which should be produced only in the boundary layer and convected by the flow field. In Figure 3 we see the entropy isolines are accumulated in the boundary layer and convected by the vortex structures, while outside this region the entropy is constant (no isolines are present).

Acknowledgements The research was supported under the Grant No. 257486 (Finite Element Method for the Analysis of Fluid and Structure Interaction) of the Grant Agency of the Charles University and the Nečas Center for Mathematical Modelling, project LC06052, financed by MSMT.

References

1. Arnold, D.N., Brezzi, F., Cockburn, B., Marini, L.D.: Unified analysis of discontinuous galerkin methods for elliptic problems. SIAM J. Numer. Anal. **39**(5), 1749–1779 (2001)

2. Dolejší, V., Feistauer, M.: A semi-implicit discontinuous galerkin finite element method for the numerical solution of inviscid compressible flow. J. Comput. Phys. **198**(2), 727–746 (2004)
3. Feistauer, M., Felcman, J., Straškraba, I.: Mathematical and Computational Methods for Compressible Flow. Oxford University Press, Oxford (2003)
4. Feistauer, M., Kučera, V.: On a robust discontinuous galerkin technique for the solution of compressible flow. J. Comput. Phys. **224**(1), 208–221 (2007)

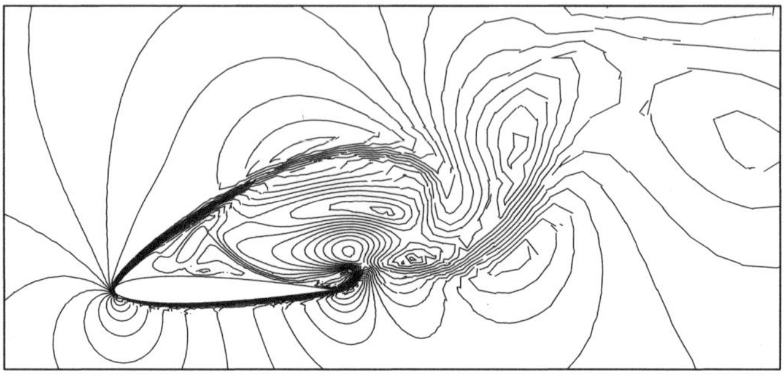

Fig. 1 NACA0012 $\alpha = 25°$ viscous flow, Mach number isolines.

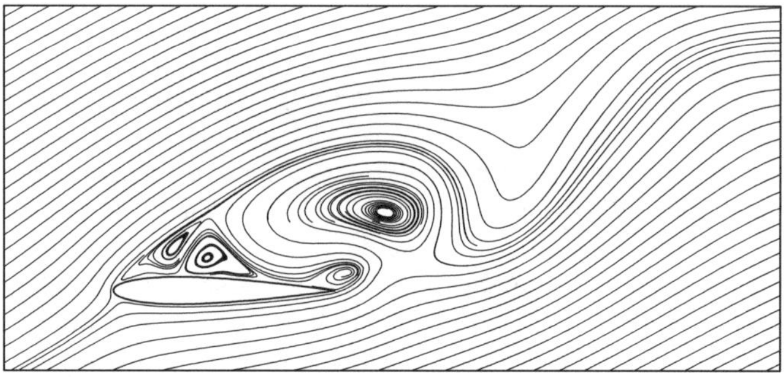

Fig. 2 NACA0012 $\alpha = 25°$ viscous flow, streamlines.

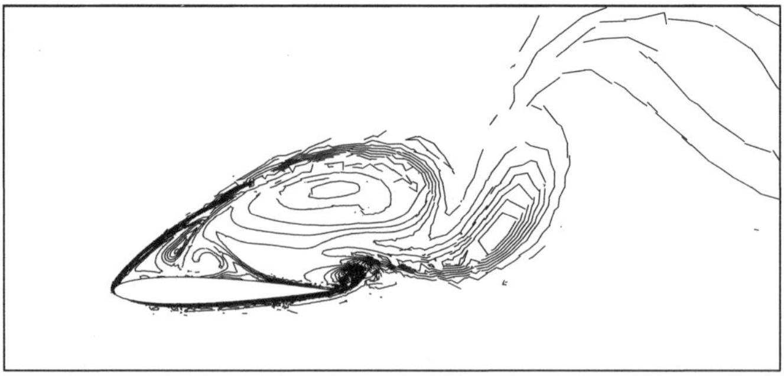

Fig. 3 NACA0012 $\alpha = 25°$ viscous flow, entropy isolines.

Numerical Integration in the Discontinuous Galerkin Method for Nonlinear Convection-Diffusion Problems in 3D

V. Sobotíková

Abstract In this paper the discontinuous Galerkin finite element method is used for the space-semidiscretization of a nonlinear nonstationary convection-diffusion problem in three dimensions. As in practical computations integrals appearing in the forms defining the approximate solution are evaluated with the use of quadrature formulae, the effect of numerical integration in the method is studied. An estimate of the error caused by the numerical integration is presented and it is shown which quadrature formulae guarantee preservation of the accuracy of the method with exact integration.

1 Introduction

We are concerned with the following nonstationary nonlinear convection-diffusion problem: Let $\Omega \subset \mathbb{R}^3$ be a bounded polyhedral domain with Lipschitz-continuous boundary $\partial\Omega = \Gamma_D \cup \Gamma_N$ and $T > 0$. Find a function $u: Q_T = \Omega \times (0,T) \to \mathbb{R}$ such that

$$\frac{\partial u}{\partial t} + \sum_{\ell=1}^{3} \frac{\partial f_\ell(u)}{\partial x_\ell} = \varepsilon \Delta u + g \qquad \text{in } Q_T, \tag{1}$$

$$u\big|_{\Gamma_D \times (0,T)} = u_D, \qquad \varepsilon \frac{\partial u}{\partial n}\Big|_{\Gamma_N \times (0,T)} = g_N, \qquad u(.,0) = u^0.$$

The diffusion coefficient $\varepsilon > 0$ is a given constant, f_ℓ ($\ell = 1,2,3$) are prescribed convective fluxes and g, u_D, g_N and u^0 are given functions.

For the space-semidiscretization of the problem we employ the discontinuous Galerkin finite element method (see e.g. [2] or [3]). We consider its symmetric variant combined with an interior and boundary penalty and we approximate the

Veronika Sobotíková
Czech Technical University Prague, Faculty of Electrical Engineering, Technická 2, 166 27 Praha 6, Czech Republic, e-mail: veronika@math.feld.cvut.cz

nonlinear convective fluxes with the aid of a numerical flux. We evaluate integrals appearing in the forms defining the approximate solution with the use of quadrature formulae.

In [6] an $L^2(H^1)$-optimal error estimate of the error of the method was proved. However, this estimate was only suboptimal in the $L^\infty(L^2)$-norm. The aim of this paper is to derive a general estimate of the error caused by the use of numerical integration and, applying this estimate, to obtain an $L^\infty(L^2)$-optimal error estimate.

2 Discrete Problem

2.1 Space Semidiscretization

Let $\{\mathscr{T}_h\}_{h\in(0,h_0)}$ be a system of partitions of $\overline{\Omega}$ into a finite number of closed tetrahedra K with mutually disjoint interiors. We call $\mathscr{T}_h$ triangulations of Ω, but *do not* require the usual conforming properties from the FEM (cf. [1]).

We set $h_K = \mathrm{diam}(K)$ and by $|K|$ and ρ_K we denote the volume of K and the radius of the largest ball inscribed into K, respectively. All elements of $\mathscr{T}_h$ will be numbered in such a way that $\mathscr{T}_h = \{K_i\}_{i\in I}$, where $I \subset \mathbb{Z}^+$. If the two-dimensional measure of the intersection of boundaries of two elements $K_i, K_j \in \mathscr{T}_h$ is nonzero, we put $\Gamma_{ij} = \Gamma_{ji} = \partial K_i \cap \partial K_j$. For $i \in I$ we set $s(i) = \{j \in I; K_j \text{ is a neighbour of } K_i\}$. The boundary $\partial\Omega$ is formed by a finite number of parts of faces of elements K_i adjacent to $\partial\Omega$. We denote all these "subfaces" by S_j, where $j \in I_b \subset \mathbb{Z}^-$. We set $\Gamma_{ij} = S_j$ for $K_i \in \mathscr{T}_h$ such that $S_j \subset \partial K_i$, $j \in I_b$, and $\gamma(i) = \{j \in I_b; S_j \text{ is a part of a face of } K_i\}$. Obviously, $s(i) \cap \gamma(i) = \emptyset$ for all $i \in I$ and, writing $S(i) = s(i) \cup \gamma(i)$, we have $\partial K_i = \bigcup_{j\in S(i)} \Gamma_{ij}$, $\partial K_i \cap \partial\Omega = \bigcup_{j\in\gamma(i)} \Gamma_{ij}$. (In what follows, we shall often call Γ_{ij} briefly a *face*.) For $i \in I$, by $\gamma_D(i)$ and $\gamma_N(i)$ we denote the subsets of $\gamma(i)$ formed by such indexes j that the faces Γ_{ij} form the parts Γ_D and Γ_N of $\partial\Omega$, respectively. Thus, we suppose that $\gamma(i) = \gamma_D(i) \cup \gamma_N(i)$, $\gamma_D(i) \cap \gamma_N(i) = \emptyset$. Furthermore, we denote by $\mathbf{n}_{ij} = ((n_{ij})_1, (n_{ij})_2, (n_{ij})_3)$ the unit outer normal to ∂K_i on the face Γ_{ij}, by $|\Gamma_{ij}|$ the two-dimensional measure of Γ_{ij}, and we set $s_h = \{\Gamma_{ij}; j \in S(i), i \in I\}$.

We suppose that the system $\{\mathscr{T}_h\}_{h\in(0,h_0)}$ is *regular*, i.e.

$$h_K/\rho_K \le C_R \qquad \forall K \in \mathscr{T}_h,\ h \in (0,h_0),$$

and that

$$h_{K_i} \le C_N h_{K_j} \qquad \forall i \in I,\ j \in s(i),\ h \in (0,h_0),$$

where $C_R > 0$ and $C_N > 0$ are suitable constants.

Remark 1. Let us mention that in two dimensions all faces are always line segments. In three dimensions, if the mesh is nonconforming, the geometry of faces is more complicated. The inner faces, formed by intersections of two triangles, are convex polygons with three up to six vertices. The boundary faces may even be nonconvex. Moreover, regardless of the above assumptions, some faces may be very small and narrow. Nevertheless, it was shown in [6] that if the system $\{\mathscr{T}_h\}_{h\in(0,h_0)}$ satisfies the previous two inequalities, then there exists a constant C_S such that $\mathrm{card}\big(S(i)\big) \le C_S$ for all $i \in I,\ h \in (0, h_0)$.

Over the triangulations $\mathscr{T}_h$ we define for $k \in \mathbb{N},\ k \ge 1$, the *broken Sobolev spaces*

$$H^k(\Omega, \mathscr{T}_h) = \{v; v|_K \in H^k(K)\ \forall K \in \mathscr{T}_h\},$$

equipped with seminorms $|v|_{H^k(\Omega,\mathscr{T}_h)} = \big(\sum_{K\in\mathscr{T}_h} |v|^2_{H^k(K)}\big)^{1/2}$. For $v \in H^1(\Omega, \mathscr{T}_h)$, $i \in I,\ j \in s(i)$, we denote the *traces, average* and *jump of the traces* of v on $\Gamma_{ij} = \Gamma_{ji}$ by $v|_{\Gamma_{ij}} =$ trace of $v|_{K_i}$ on Γ_{ij}, $v|_{\Gamma_{ji}} =$ trace of $v|_{K_j}$ on Γ_{ji}, $\langle v\rangle_{\Gamma_{ij}} = \frac{1}{2}\left(v|_{\Gamma_{ij}} + v|_{\Gamma_{ji}}\right)$ and $[v]_{\Gamma_{ij}} = v|_{\Gamma_{ij}} - v|_{\Gamma_{ji}}$. For $i \in I,\ j \in \gamma(i)$ we put $v|_{\Gamma_{ji}} = v|_{\Gamma_{ij}} =$ trace of $v|_{K_i}$ on Γ_{ij}. (In what follows, we shall write $\langle v\rangle$ and $[v]$ instead of $\langle v\rangle_{\Gamma_{ij}}$ and $[v]_{\Gamma_{ij}}$, respectively.)

The approximate solution of problem (1) is sought in the space S_h of discontinuous piecewise polynomial functions defined by

$$S_h = S^{p,-1}(\Omega, \mathscr{T}_h) = \{v; v|_K \in P^p(K)\ \forall K \in \mathscr{T}_h\},$$

where $P^p(K)$ $(p \ge 1)$ denotes the space of all polynomials on K of degree $\le p$.

In order to introduce the space semidiscretization of problem (1) over mesh $\mathscr{T}_h$ by the DGFEM, we define the following forms for functions $u, \varphi \in H^2(\Omega, \mathscr{T}_h)$:

$$\tilde{a}_h(u,\varphi) = \sum_{i\in I}\int_{K_i} \varepsilon \nabla u \cdot \nabla\varphi \,\mathrm{d}x - \varepsilon \sum_{i\in I}\sum_{\substack{j\in s(i)\\ j<i}} \int_{\Gamma_{ij}} \Big(\langle\nabla u\rangle \cdot \mathbf{n}_{ij}[\varphi]\,\mathrm{d}S + \langle\nabla\varphi\rangle\cdot\mathbf{n}_{ij}[u]\Big)\mathrm{d}S$$

$$- \varepsilon \sum_{i\in I}\sum_{j\in\gamma_D(i)} \int_{\Gamma_{ij}} \Big(\nabla u\cdot\mathbf{n}_{ij}\,\varphi\,\mathrm{d}S + \nabla\varphi\cdot\mathbf{n}_{ij}\,u\Big)\mathrm{d}S,$$

$$\tilde{J}_h^\sigma(u,\varphi) = \sum_{i\in I}\sum_{\substack{j\in s(i)\\ j<i}} \int_{\Gamma_{ij}} \sigma[u]\,[\varphi]\,\mathrm{d}S + \sum_{i\in I}\sum_{j\in\gamma_D(i)}\int_{\Gamma_{ij}} \sigma\, u\,\varphi\,\mathrm{d}S,$$

$$\tilde{\ell}_h(\varphi)(t) = \int_\Omega g(t)\,\varphi\,\mathrm{d}x + \sum_{i\in I}\sum_{j\in\gamma_N(i)}\int_{\Gamma_{ij}} g_N(t)\,\varphi\,\mathrm{d}S$$

$$- \sum_{i\in I}\sum_{j\in\gamma_D(i)}\int_{\Gamma_{ij}} \varepsilon\nabla\varphi\cdot\mathbf{n}_{ij}\,u_D(t)\,\mathrm{d}S + \varepsilon\sum_{i\in I}\sum_{j\in\gamma_D(i)}\int_{\Gamma_{ij}} \sigma\, u_D(t)\,\varphi\,\mathrm{d}S,$$

$$\tilde{b}_h(u,\varphi) = -\sum_{i\in I}\int_K \sum_{\ell=1}^{3} f_\ell(u)\frac{\partial\varphi}{\partial x_\ell}\,\mathrm{d}x + \sum_{i\in I}\sum_{j\in S(i)}\int_{\Gamma_{ij}} H\left(u|_{\Gamma_{ij}}, u|_{\Gamma_{ji}}, \mathbf{n}_{ij}\right)\varphi|_{\Gamma_{ij}}\,\mathrm{d}S.$$

The weight σ in the forms $\tilde{J}_h^\sigma$ and $\tilde{\ell}_h$ is defined by

$$\sigma|_{\Gamma_{ij}} = \frac{2C_W}{h_{K_i} + h_{K_j}} \quad \text{for } j \in s(i),\ i \in I, \qquad \sigma|_{\Gamma_{ij}} = \frac{C_W}{h_{K_i}} \quad \text{for } j \in \gamma(i),\ i\in I, \tag{2}$$

where $C_W > 0$ is a sufficiently large constant (see [6]). The numerical flux $H(u, v, \mathbf{n})$ from the definition of the form $\tilde{b}_h$ we assume to be Lipschitz-continuous with respect to u, v, consistent and conservative (for definitions of these properties see [7]).

Now we can define an *approximate DGFE solution* of problem (1) as a function $\tilde{u}_h \in C^1([0,T]; S_h)$ satisfying the following conditions:

$$\begin{aligned} &\text{a)}\ \left(\frac{\partial \tilde{u}_h(t)}{\partial t}, \varphi_h\right) + \tilde{b}_h(\tilde{u}_h(t), \varphi_h) + \tilde{a}_h(\tilde{u}_h(t), \varphi_h) + \varepsilon \tilde{J}_h^\sigma(\tilde{u}_h(t), \varphi_h) = \tilde{\ell}_h(\varphi_h)(t) \\ &\qquad\qquad\qquad\qquad\qquad\qquad \forall \varphi_h \in S_h \ \forall t \in (0,T), \\ &\text{b)}\ \tilde{u}_h(0) = \tilde{u}_h^0. \end{aligned} \tag{3}$$

By $(\cdot,\cdot)$ we denote the scalar product in the space $L^2(\Omega)$, and $\tilde{u}_h^0$ is the $L^2(\Omega)$-projection $\Pi_h u^0$ of the initial condition u^0 on S_h.

2.2 Numerical Integration

In practical computations, integrals appearing in the forms defining $\tilde{u}_h$ are evaluated with the aid of numerical integration. This means that to integrate a function $\Phi \in C(K)$ over an element $K \in \mathscr{T}_h$, we use an approximation

$$\int_K \Phi \,\mathrm{d}x \approx |K| \sum_{\alpha=1}^{n_K} \omega_\alpha^K \Phi(x_\alpha^K), \tag{4}$$

where $\omega_\alpha^K \in \mathbb{R}$ are integration weights and $x_\alpha^K \in K$ are integration points. To evaluate an integral over a face $\Gamma \in s_h$, we first divide Γ into k_Γ triangles Γ^l with mutually disjoint interiors, $\Gamma = \bigcup_{l=1}^{k_\Gamma} \Gamma^l$. Then for $\Psi \in C(\Gamma)$ we use approximations

$$\int_\Gamma \Psi \,\mathrm{d}S = \sum_{l=1}^{k_\Gamma} \int_{\Gamma^l} \Psi \,\mathrm{d}S, \qquad \int_{\Gamma^l} \Psi \,\mathrm{d}S \approx |\Gamma^l| \sum_{\alpha=1}^{m_{\Gamma,l}} \beta_{\alpha,l}^\Gamma \Psi(x_{\alpha,l}^\Gamma) \tag{5}$$

with integration weights $\beta_{\alpha,l}^\Gamma \in \mathbb{R}$ and integration points $x_{\alpha,l}^\Gamma \in \Gamma^l$. We assume that for each $i \in I$, $j \in s(i)$ and $l \in \{1,\dots,k_{\Gamma_{ij}}\}$, the integration points on $\Gamma_{ij} = \Gamma_{ji}$ satisfy the condition $x_{\alpha,l}^{\Gamma_{ij}} = x_{\alpha,l}^{\Gamma_{ji}}$, and that there exist constants ω, $\beta > 0$ such that $\sum_{\alpha=1}^{n_K} |\omega_\alpha^K| \leq \omega$, $\sum_{l=1}^{k_\Gamma} \sum_{\alpha=1}^{m_{\Gamma,l}} |\beta_{\alpha,l}^\Gamma| \leq \beta$ for all $K \in \mathscr{T}_h$, $\Gamma \in s_h$, $h \in (0,h_0)$.

Using quadrature formulae (4) and (5), we obtain approximations $(\cdot,\cdot)_h$, a_h, J_h^σ, b_h, ℓ_h of the forms $(\cdot,\cdot)$, $\tilde{a}_h$, $\tilde{J}_h^\sigma$, $\tilde{b}_h$, $\tilde{\ell}_h$ and we can define the *discrete problem with numerical integration*: Find $u_h \in C^1([0,T];S_h)$ such that

$$\begin{aligned} &\text{a)}\ \left(\frac{\partial u_h(t)}{\partial t}, \varphi_h\right)_h + b_h(u_h(t), \varphi_h) + a_h(u_h(t), \varphi_h) + \varepsilon J_h^\sigma(u_h(t), \varphi_h) = \ell_h(\varphi_h)(t) \\ &\qquad\qquad\qquad\qquad\qquad\qquad \forall \varphi_h \in S_h \ \forall t \in (0,T), \\ &\text{b)}\ u_h(0) = u_h^0. \end{aligned} \tag{6}$$

By u_h^0 we denote a suitable S_h-approximation of the initial condition u^0. In what follows we shall assume that it is defined either by $(u_h^0 - u^0, \varphi_h)_h = 0 \ \forall \varphi_h \in S_h$ or by the Lagrange interpolation of u^0 (applied elementwise).

3 Error Estimates for the Method of Lines with Numerical Integration

In this section we shall suppose that the exact solution u satisfies the regularity conditions $(\ell = 1,2,3)$

$$u \in L^2(0,T;H^{p+1}(\Omega)),\ \frac{\partial u}{\partial t} \in L^2(0,T;H^p(\Omega)),\ f_\ell(u) \in L^2(0,T;W^{z+1,\infty}(\Omega)), \quad (7)$$

where $z \geq p$ is an integer. We denote the error $u - u_h$ of the method by e_h, and we write $e_h = \tilde{e}_h + e_I$, where $\tilde{e}_h = u - \tilde{u}_h$ and $e_I = \tilde{u}_h - u_h$, respectively. First we show that knowing an estimate of the error $\tilde{e}_h$, we can estimate also the error e_I. We have:

Theorem 1. *Let* $g \in L^2(0,T;H^{\tilde{z}}(\Omega)), g_N \in L^2(0,T;H^{z+1}(\Gamma_N)),\ u^0 \in H^{\tilde{z}}(\Omega),\ u_D \in L^2(0,T;H^{z+2}(\Gamma_D))$, *where* $\tilde{z} = \max\{z,2\}$. *Let the quadrature formulae* (4) *be exact for polynomials of degree* $\leq \max\{\tilde{z}+p-1, 2p\}$ *and let the quadrature formulae* (5) *be exact for polynomials of degree* $\leq z+p$. *Then*

$$\begin{aligned}
&\max_{t\in[0,T]} \|e_I(t)\|^2_{L^2(\Omega)} + \frac{\varepsilon}{2}\int_0^T \left(|e_I(\vartheta)|^2_{H^1(\Omega,\mathscr{T}_h)} + \tilde{J}_h^\sigma(e_I(\vartheta),e_I(\vartheta))\right) \mathrm{d}\vartheta \\
&\quad \leq C\Big\{\frac{1}{\varepsilon}\|\tilde{e}_h\|^2_{L^2(0,T;L^2(\Omega))} + \frac{1}{\varepsilon}h^{2(p+1)}|u|^2_{L^2(0,T;H^{p+1}(\Omega))} \\
&\quad +h^{2z}\Big(|g|^2_{L^2(0,T;H^{\tilde{z}}(\Omega))} + |g_N|^2_{L^2(0,T;H^{z+1}(\Gamma_N))} + (\varepsilon^2+\varepsilon)\|u_D\|^2_{L^2(0,T;H^{z+2}(\Gamma_D))} \\
&\qquad + |u^0|^2_{H^{\tilde{z}}(\Omega)} + \max_{\ell=1,2,3}\{|f_\ell(u)|^2_{L^2(0,T;W^{z+1,\infty}(\Omega))}\}\Big)\Big\} \exp\left(\frac{CT}{\varepsilon}\right)
\end{aligned}$$

with a constant C *independent of* u, g, g_N, u_D, u^0, T, ε *and* $h \in (0,h_0)$.

Proof. From the assumption on the used quadrature formulae and from the fact that $u_h(t) \in S_h = S^{p,-1}(\Omega,\mathscr{T}_h)$ we conclude that for $\varphi_h \in S_h$

$$\left(\frac{\partial u_h}{\partial t},\varphi_h\right) = \left(\frac{\partial u_h}{\partial t},\varphi_h\right)_h,\ a_h(u_h,\varphi_h) = \tilde{a}_h(u_h,\varphi_h),\ J_h^\sigma(u_h,\varphi_h) = \tilde{J}_h^\sigma(u_h,\varphi_h). \quad (8)$$

Setting $\varphi_h := e_I$ in (8), (3) and (6), we obtain after some manipulations

$$\begin{aligned}
&\frac{1}{2}\frac{\mathrm{d}}{\mathrm{d}t}\|e_I\|^2_{L^2(\Omega)} + \tilde{a}_h(e_I,e_I) + \varepsilon\tilde{J}_h^\sigma(e_I,e_I) \\
&\qquad = (\tilde{\ell}_h(e_I) - \ell_h(e_I)) - (\tilde{b}_h(\tilde{u}_h,e_I) - b_h(u_h,e_I)).
\end{aligned} \quad (9)$$

By [4], Corollary 3.10, we have

$$\tilde{a}_h(e_I,e_I) + \tilde{J}_h^\sigma(e_I,e_I) \geq \frac{\varepsilon}{2}\left(|e_I|^2_{H^1(\Omega,\mathscr{T}_h)} + \tilde{J}_h^\sigma(e_I,e_I)\right). \quad (10)$$

The first term on the right-hand side of (9) may be estimated with the aid of Lemma 9 from [6]. We obtain

$$|\tilde{\ell}_h(e_I)-\ell_h(e_I)| \le C\varepsilon h^{z+1/2}\|u_D\|_{H^{z+2}(\Gamma_D)}|e_I|_{H^1(\Omega,\mathscr{T}_h)}$$
$$+C\left(h^{\tilde{z}}|g|_{H^{\tilde{z}}(\Omega)}+h^{z+1/2}|g_N|_{H^{z+1}(\Gamma_N)}+\varepsilon h^{z+1/2}\|u_D\|_{H^{z+2}(\Gamma_D)}\right)\|e_I\|_{L^2(\Omega)}.$$

In order to estimate the second term, we first write

$$\begin{aligned}\tilde{b}_h(\tilde{u}_h,e_I)-b_h(u_h,e_I) &= \left(\tilde{b}_h(\tilde{u}_h,e_I)-\tilde{b}_h(u,e_I)\right)+\left(\tilde{b}_h(u,e_I)-b_h(u,e_I)\right)\\ &+\left(b_h(u,e_I)-b_h(\Pi_h u,e_I)\right)+\left(b_h(\Pi_h u,e_I)-b_h(u_h,e_I)\right)=\vartheta_1+\vartheta_2+\vartheta_3+\vartheta_4,\end{aligned} \tag{11}$$

where $\Pi_h u$ denotes the $L^2(\Omega)$-projection of the exact solution u onto the space S_h.

According to the inequalities $\|\tilde{u}_h-\Pi_h u\|_{L^2(\Omega)} \le \|\tilde{u}_h-u\|_{L^2(\Omega)}+\|u-\Pi_h u\|_{L^2(\Omega)}$, and $\|\Pi_h u-u_h\|_{L^2(\Omega)} \le \|\Pi_h u - u\|_{L^2(\Omega)} + \|u-\tilde{u}_h\|_{L^2(\Omega)} + \|\tilde{u}_h-u_h\|_{L^2(\Omega)}$, the approximation properties of the operator Π_h, (see e.g. [1]), and [6], Lemmas 5–8,

$$|\vartheta_1| \le C\left(|e_I|_{H^1(\Omega,\mathscr{T}_h)}+\tilde{J}_h^\sigma(e_I,e_I)^{1/2}\right)\left(h^{p+1}|u|_{H^{p+1}(\Omega)}+\|\tilde{e}_h\|_{L^2(\Omega)}\right),$$
$$|\vartheta_2| \le C\|e_I\|_{L^2(\Omega)}h^z\max_{\ell=1,2,3}\left\{|f_\ell(u)|_{W^{z+1,\infty}(\Omega)}\right\},$$
$$|\vartheta_3| \le C\left(|e_I|_{H^1(\Omega,\mathscr{T}_h)}+\tilde{J}_h^\sigma(e_I,e_I)^{1/2}\right)h^{p+1}|u|_{H^{p+1}(\Omega)},$$
$$|\vartheta_4| \le C\left(|e_I|_{H^1(\Omega,\mathscr{T}_h)}+\tilde{J}_h^\sigma(e_I,e_I)^{1/2}\right)\left(h^{p+1}|u|_{H^{p+1}(\Omega)}+\|\tilde{e}_h\|_{L^2(\Omega)}+\|e_I\|_{L^2(\Omega)}\right).$$

These estimates and relations (9), (10), (11) imply that

$$\begin{aligned}&\frac{\mathrm{d}}{\mathrm{d}t}\|e_I\|^2_{L^2(\Omega)}+\varepsilon\left(|e_I|^2_{H^1(\Omega,\mathscr{T}_h)}+\tilde{J}_h^\sigma(e_I,e_I)\right)\\ &\le C\left(h^{\tilde{z}}|g|_{H^{\tilde{z}}(\Omega)}+h^{z+1/2}|g_N|_{H^{z+1}(\Gamma_N)}+\varepsilon h^{z+1/2}\|u_D\|_{H^{z+2}(\Gamma_D)}\right)\|e_I\|_{L^2(\Omega)}\\ &+C\varepsilon h^{z+1/2}\|u_D\|_{H^{z+2}(\Gamma_D)}|e_I|_{H^1(\Omega,\mathscr{T}_h)}+Ch^z\max_{\ell=1,2,3}\left\{|f_\ell(u)|_{W^{z+1,\infty}(\Omega)}\right\}\|e_I\|_{L^2(\Omega)}\\ &+C\left(|e_I|_{H^1(\Omega,\mathscr{T}_h)}+\tilde{J}_h^\sigma(e_I,e_I)^{1/2}\right)\left(h^{p+1}|u|_{H^{p+1}(\Omega)}+\|\tilde{e}_h\|_{L^2(\Omega)}+\|e_I\|_{L^2(\Omega)}\right).\end{aligned}$$

With the aid of Young's inequality and integration with respect to time from 0 to $t\in[0,T]$, using the relation $e_I(0)=e_I^0=\tilde{u}_h^0-u_h^0$, which implies that $\|e_I(0)\|_{L^2(\Omega)} \le Ch^{\tilde{z}}|u^0|_{H^{\tilde{z}}(\Omega)}$, and the inequality $z\le\tilde{z}$, we obtain

$$\begin{aligned}&\|e_I(t)\|^2_{L^2(\Omega)}+\frac{\varepsilon}{2}\int_0^t\left(|e_I(\vartheta)|^2_{H^1(\Omega,\mathscr{T}_h)}+\tilde{J}_h^\sigma(e_I(\vartheta),e_I(\vartheta))\right)\mathrm{d}\vartheta\\ &\le Ch^{2z}\Big(|g|^2_{L^2(0,T;H^{\tilde{z}}(\Omega))}+|g_N|^2_{L^2(0,T;H^{z+1}(\Gamma_N))}+(\varepsilon^2+\varepsilon)\|u_D\|^2_{L^2(0,T;H^{z+2}(\Gamma_D))}\\ &+|u^0|^2_{H^{\tilde{z}}(\Omega)}+\max_{\ell=1,2,3}\left\{|f_\ell(u)|^2_{L^2(0,T;W^{z+1,\infty}(\Omega))}\right\}\Big)+\frac{C}{\varepsilon}\|\tilde{e}_h\|^2_{L^2(0,T;L^2(\Omega))}\\ &+\frac{C}{\varepsilon}h^{2(p+1)}|u|^2_{L^2(0,T;H^{p+1}(\Omega))}+C\left(1+\frac{1}{\varepsilon}\right)\int_0^t\|e_I(\vartheta)\|^2_{L^2(\Omega)}\mathrm{d}\vartheta.\end{aligned}$$

Finally, the use of Gronwall's lemma leads to the desired estimate. □

It was shown in [4] that there exists a positive constant $\overline{C} > 0$ (depending on $u, g, g_N, u_D, u^0, T, \varepsilon$ and h_0, but independent of h) such that

$$\max_{t\in[0,T]} \|\tilde{e}_h(t)\|^2_{L^2(\Omega)} + \frac{\varepsilon}{2}\int_0^T \Big(|\tilde{e}_h(\vartheta)|^2_{H^1(\Omega,\mathscr{T}_h)} + \tilde{J}_h^\sigma(\tilde{e}_h(\vartheta),\tilde{e}_h(\vartheta))\Big)\,\mathrm{d}\vartheta \le \overline{C}h^{2p}, \quad (12)$$

which also means that

$$\|\tilde{e}_h(t)\|_{L^2(0,T;L^2(\Omega))} \le \sqrt{T\overline{C}}\, h^p. \quad (13)$$

Hence, as $e_h = \tilde{e}_h + e_I$, combining the estimate from Theorem 1 with relations (12) and (13), we immediately obtain the following estimate of the error of the method of lines with numerical integration:

Theorem 2. *Let us assume that the exact solution u of problem* (1) *satisfies the regularity conditions* (7) *with $z = p$. Moreover, let the quadrature formulae* (4) *and* (5) *be exact for polynomials of degree $\le 2p$. Let $g \in L^2(0,T;H^{\tilde{p}}(\Omega))$, $g_N \in L^2(0,T;H^{p+1}(\Gamma_N))$, $u_D \in L^2(0,T;H^{p+2}(\Gamma_D))$, $u^0 \in H^{\tilde{p}}(\Omega)$, where $\tilde{p} = \max\{p,2\}$. Then for the error e_h we have the estimate*

$$\max_{t\in[0,T]} \|e_h(t)\|^2_{L^2(\Omega)} + \frac{\varepsilon}{2}\int_0^T \Big(|e_h(\vartheta)|^2_{H^1(\Omega,\mathscr{T}_h)} + \tilde{J}_h^\sigma(e_h(\vartheta),e_h(\vartheta))\Big)\,\mathrm{d}\vartheta \le C_1 h^{2p}$$

with a constant C_1 depending on $u, g, g_N, u_D, u^0, T, \varepsilon$ and h_0, but independent of $h \in (0,h_0)$.

For the proof of this theorem without the use of the general error estimate from Theorem 1 see [6], Theorem 2.

The error estimate from Theorem 2 is optimal in the $L^2(H^1)$-norm. However, it is only suboptimal in the $L^\infty(L^2)$-norm. In order to obtain an estimate optimal also in $L^\infty(L^2)$-norm, let us suppose that the domain Ω is convex and that the meshes $\mathscr{T}_h$ are conforming, i.e. two different elements $K, K' \in \mathscr{T}_h$ are either disjoint or $K \cap K'$ is either their common vertex or edge or (whole triangular) face. Further, let us consider the following dual problem: Given $z \in L^2(\Omega)$, find ψ such that

$$-\Delta\psi = z \quad \text{in } \Omega, \qquad \psi|_{\Gamma_D} = 0, \qquad \frac{\partial\psi}{\partial n}\Big|_{\Gamma_N} = 0. \quad (14)$$

It was shown in [5] that if $\psi \in H^2(\Omega)$ and if there exists a constant $C_D > 0$, independent of z, such that

$$\|\psi\|_{H^2(\Omega)} \le C_D\|z\|_{L^2(\Omega)}, \quad (15)$$

then the error $\tilde{e}_h$ satisfies the estimate

$$\|\tilde{e}_h\|_{L^\infty(0,T;L^2(\Omega))} \le \hat{C}h^{p+1}, \quad (16)$$

with a constant $\hat{C} > 0$ independent of h, which also implies that

$$\|\tilde{e}_h\|_{L^2(0,T;L^2(\Omega))} \leq \sqrt{T}\hat{C}h^{p+1}. \tag{17}$$

Now, under the above assumptions, relations (16), (17) and Theorem 1 give the following $L^\infty(L^2)$-optimal estimate of the method with numerical integration:

Theorem 3. *Let us assume that the exact solution u of problem* (1) *satisfies the regularity conditions* (7) *with $z = p+1$ and let the solution of the dual problem* (14) *satisfy* (15). *Moreover, let $g \in L^2(0,T;H^{p+1}(\Omega))$, $g_N \in L^2(0,T;H^{p+2}(\Gamma_N))$, $u_D \in L^2(0,T;H^{p+3}(\Gamma_D))$ and $u^0 \in H^{p+1}(\Omega)$. Let the quadrature formulae* (4) *and* (5) *be exact for polynomials of degree $\leq 2p$ and $\leq 2p+1$, respectively. Then the error e_h satisfies the estimate*

$$\|e_h\|_{L^\infty(0,T;L^2(\Omega))} \leq C_2 h^{p+1},$$

where the constant $C_2 > 0$ depends on u, g, g_N, u_D, u^0, T, ε and h_0, but is independent of $h \in (0, h_0)$.

Remark 2. Let us mention two open problems. One is the weakening of the rather strong assumption $f_\ell(u) \in L^2(0,T;W^{p+1,\infty}(\Omega))$, $\ell = 1,2,3$. The other is whether the assumption of conformity of the meshes $\mathscr{T}_h$ is necessary for deriving the estimate (16), and thus for obtaining the assertion of Theorem 3 as well.

Acknowledgements The research was a part of the research project MSM 6840770010 financed by the Ministry of Education of the Czech Republic.

References

1. Ciarlet, P.G.: The finite element method for elliptic problems. North-Holland, Amsterdam (1979)
2. Cockburn, B.: Discontinuous Galerkin methods for convection dominated problems. In: T.J. Barth, H. Deconinck (eds.) High–Order Methods for Computational Physics, pp. 69–224. Springer, Berlin (1999)
3. Cockburn, B., Karniadakis, G.E., Shu, C.W. (eds.): Discontinuous Galerkin methods. Lecture Notes in Computational Science and Engineering 11. Springer, Berlin (2000)
4. Dolejší, V., Feistauer, M.: Error estimates of the discontinuous Galerkin method for nonlinear nonstationary convection-diffusion problems. Numer. Funct. Anal. Optim. **26**(3), 349–383 (2005)
5. Dolejší, V., Feistauer, M., Kučera, V., Sobotíková, V.: An optimal $L^\infty(L^2)$-error estimate for the discontinuous Galerkim approximation of a nonlinear non-stationary convection-diffusion problem. IMA J. Numer. Anal. (accepted)
6. Sobotíková, V.: Numerical integration in the DGFEM for 3D nonlinear convection-diffusion problems on nonconforming meshes. Numer. Funct. Anal. Optim. (submitted)
7. Sobotíková, V., Feistauer, M.: On the effect of numerical integration in the DGFEM for nonlinear convection-diffusion problems. Numer. Methods Partial Differ. Equ. **23**(6), 1368–1395 (2007)

Implicit-Explicit Runge-Kutta Discontinuous Galerkin Finite Element Method for Convection-Diffusion Problems

M. Vlasák and V. Dolejší

Abstract We deal with a numerical solution of a scalar nonstationary convection-diffusion equation with a nonlinear convection and a linear diffusion terms. We carry out the space semi-discretization with the aid of the nonsymmetric interior penalty Galerkin (NIPG) method and the time discretization by a combination of implicit-explicit Runge-Kutta method. The resulting scheme is unconditionally stable, has a high order of accuracy with respect to space and time coordinates and requires solutions of linear algebraic problems at each time step. We derive a priori error estimates in the L^2-norm.

1 Introduction

We numerically solve a nonstationary nonlinear convection-diffusion equation, which represents a model problem for the system of the compressible Navier-Stokes equations. The class of *discontinuous Galerkin* (DG) methods seems to be one of the most promising candidates to construct high order accurate schemes for solving of convection-diffusion problems. For a survey about DG methods, see [1] or [2]. An analysis of DG methods was presented in many papers, see, e.g., [4], [5], [7], [8].

In [5] we carried out the space semi-discretization of the scalar convection-diffusion equation with the aid of the *discontinuous Galerkin finite element* method and derived a priori error estimates. Within this contribution, we deal with the time discretization of the resulting system of ordinary differential equations. In contrary

Miloslav Vlasák
Charles University Prague, Faculty of Mathematics and Physics, Sokolovská 83, Prague, Czech Republic, e-mail: vlasakmm@volny.cz

Vít Dolejší
Charles University Prague, Faculty of Mathematics and Physics, Sokolovská 83, Prague, Czech Republic, e-mail: dolejsi@karlin.mff.cuni.cz

to [6], where we used the so-called backward difference formulae (BDF) approach, here we employ a combination of implicit-explicit (IMEX) Runge-Kutta methods. We present a formulation of the second order IMEX scheme and derive a priori error estimates.

2 Continuous Problem

Let $\Omega \subset R^d$ ($d = 2$ or 3) be a bounded polyhedral domain and $T > 0$. We set $Q_T = \Omega \times (0,T)$. By $\overline{\Omega}$ and $\partial\Omega$ we denote the closure and boundary of Ω, respectively. Let us consider the following *initial-boundary value problem*: Find $u : Q_T \to R$ such that

$$\frac{\partial u}{\partial t} + \nabla \cdot \mathbf{f}(u) = \varepsilon \Delta u + g \quad \text{in } Q_T, \tag{1}$$

$$u\big|_{\partial\Omega \times (0,T)} = u_D, \tag{2}$$

$$u(x,0) = u^0(x), \quad x \in \Omega. \tag{3}$$

In (1) – (3), $\mathbf{f} = (f_1, \dots, f_d)$, $f_s \in C^2(R)$, $f_s(0) = 0$, $s = 1, \dots, d$ represents convective terms, $\varepsilon > 0$ plays a role of viscosity, $g \in C([0,T]; L^2(\Omega))$ represents volume sources. The Dirichlet boundary condition is given over $\partial\Omega \times (0,T)$ by u_D, which is the trace of some $u^* \in C([0,T]; H^1(\Omega)) \cap L^\infty(Q_T)$ on $\partial\Omega \times (0,T)$ and $u^0 \in L^2(\Omega)$ is an initial condition. We use the standard notation for Lebesgue, Sobolev and Bochner function spaces (see, e. g. [9]).

In order to introduce the concept of a weak solution, we define the forms

$$\begin{aligned}
(u,w) &= \int_\Omega uw \, \mathrm{d}x, \quad u, w \in L^2(\Omega), \\
a(u,w) &= \varepsilon \int_\Omega \nabla u \cdot \nabla w \, \mathrm{d}x, \quad u, w \in H^1(\Omega), \\
b(u,w) &= \int_\Omega \nabla \cdot \mathbf{f}(u) w \, \mathrm{d}x, \quad u \in H^1(\Omega) \cap L^\infty(\Omega), \ w \in L^2(\Omega),
\end{aligned}$$

Definition 1. We say that a function u is a *weak solution* of (1) – (3) if the following conditions are satisfied

$$\begin{aligned}
&\text{a)} \quad u - u^* \in L^2(0,T; H_0^1(\Omega)), \quad u \in L^\infty(Q_T), \\
&\text{b)} \quad \frac{\mathrm{d}}{\mathrm{d}t}(u(t),w) + b(u(t),w) + a(u(t),w) = (g(t),w) \\
&\qquad\quad \text{for all } w \in H_0^1(\Omega) \text{ in the sense of distributions on } (0,T), \\
&\text{c)} \quad u(0) = u^0 \quad \text{in } \Omega.
\end{aligned} \tag{4}$$

By $u(t)$ we denote the function on Ω such that $u(t)(x) = u(x,t)$, $x \in \Omega$.

With the aid of techniques from [10] and [11], it is possible to prove that there exists a unique weak solution. We shall assume that the weak solution u is sufficiently regular, namely,

$$u \in W^{1,\infty}(0,T;H^{p+1}(\Omega)) \cap W^{2,\infty}(0,T;H^1(\Omega)), \quad u_{ttt} \in L^\infty(0,T;L^2(\Omega)).$$

where $u_{ttt} = \partial^3 u/\partial t^3$, an integer $p \geq 1$ will denote a given degree of polynomial approximations. Such a solution satisfies problem (1) – (3) pointwise.

3 Space Semi-Discretization

We discretize problem (4) in space with the aid of the *discontinuous Galerkin finite element method* with *nonsymmetric treatment of stabilization terms* and *interior and boundary penalties*. This approach is called the NIPG variant of the DGFE method, see [1]. We derived the space discretization of (1) – (3) by the NIPG variant of the DGFE method in [5], hence here we present only the final expressions.

Let $\mathscr{T}_h$ $(h > 0)$ be a partition of the domain Ω into a finite number of closed d-dimensional mutually disjoint star–shaped elements K i.e., $\overline{\Omega} = \bigcup_{K \in \mathscr{T}_h} K$. By ∂K we denote the boundary of element $K \in \mathscr{T}_h$ and set $h_K = \mathrm{diam}(K)$, $h = \max_{K \in \mathscr{T}_h} h_K$.

By $\mathscr{F}_h$, we denote the smallest possible set of all open $(d-1)$-dimensional faces of all elements $K \in \mathscr{T}_h$. Further, we denote by $\mathscr{F}_h^I$ the set of all $\Gamma \in \mathscr{F}_h$ that are contained in Ω (inner faces) and by $\mathscr{F}_h^D$ the set of all $\Gamma \in \mathscr{F}_h$ that $\Gamma \subset \partial\Omega$.

Moreover, for each $\Gamma \in \mathscr{F}_h$, we define a unit normal vector $\mathbf{n}_\Gamma$. We assume that $\mathbf{n}_\Gamma$, $\Gamma \in \mathscr{F}_h^D$ has the same orientation as the outer normal of $\partial\Omega$. For $\mathbf{n}_\Gamma$, $\Gamma \in \mathscr{F}_h^I$ the orientation is arbitrary but fixed for each edge.

Let s and p be positive integers denoting Sobolev index and polynomial degree of approximation. Over the triangulation $\mathscr{T}_h$ we define the so-called *broken Sobolev space* corresponding to s

$$H^s(\Omega, \mathscr{T}_h) \equiv \{w; w|_K \in H^s(K)\ \forall K \in \mathscr{T}_h\}. \tag{5}$$

Furthermore, we define the space of discontinuous piecewise polynomial functions associated with p by

$$S_{hp} \equiv \{w \in L^2(\Omega);\ w|_K \in P_p(K)\ \forall K \in \mathscr{T}_h\}, \tag{6}$$

where $P_p(K)$ denotes the space of all polynomials on K of degree $\leq p$.

For each $\Gamma \in \mathscr{F}_h^I$ there exist two elements $K_p, K_n \in \mathscr{T}_h$ such that $\Gamma \subset \overline{K}_p \cap \overline{K}_n$. We use a convention that K_n lies in the direction of $\mathbf{n}_\Gamma$ and K_p in the opposite direction of $\mathbf{n}_\Gamma$. Then for $v \in H^1(\Omega, \mathscr{T}_h)$, we introduce the following notation:

$$\langle w \rangle_\Gamma \equiv \frac{1}{2}\left(w|_\Gamma^{(p)} + w|_\Gamma^{(n)}\right), \quad [w]_\Gamma \equiv w|_\Gamma^{(p)} - w|_\Gamma^{(n)}, \tag{7}$$

where $w|_\Gamma^{(p)}$ and $w|_\Gamma^{(n)}$ denotes the trace of $w|_{K_p}$ and $w|_{K_n}$ on Γ, respectively.

For $\Gamma \in \mathscr{F}_h^D$ there exists an element $K_p \in \mathscr{T}_h$ such that $\Gamma \subset \overline{K}_p \cap \partial\Omega$. Then for $w \in H^1(\Omega, \mathscr{T}_h)$, we introduce the following notation:

$$\langle w \rangle_\Gamma \equiv [w]_\Gamma \equiv w|_\Gamma^{(p)}, \tag{8}$$

where $w|_\Gamma^{(p)}$ denotes the trace of $w|_{K_p}$ on Γ.

In case that $[\cdot]_\Gamma$ and $\langle \cdot \rangle_\Gamma$ are arguments of $\int_\Gamma \ldots \mathrm{dS}$, $\Gamma \in \mathscr{F}_h$ we omit the subscript Γ and write simply $[\cdot]$ and $\langle \cdot \rangle$, respectively.

For $u, v \in H^2(\Omega, \mathscr{T}_h)$, we define the forms (for more details see [5])

$$a_h(u,w) = \varepsilon \sum_{K \in \mathscr{T}_h} \int_K \nabla u \cdot \nabla w \, \mathrm{dx} - \varepsilon \sum_{\Gamma \in \mathscr{F}_h} \int_\Gamma (\langle \nabla \mathrm{u} \cdot \mathbf{n} \rangle [\mathrm{w}] - \langle \nabla \mathrm{w} \cdot \mathbf{n} \rangle [\mathrm{u}]) \, \mathrm{dS}, \tag{9}$$

$$b_h(u,w) = - \sum_{K \in \mathscr{T}_h} \int_K \mathbf{f}(u) \cdot \nabla w \, \mathrm{dx} + \sum_{\Gamma \in \mathscr{F}_h} \int_\Gamma \mathrm{H}\left(\mathrm{u}|_\Gamma^{(p)}, \mathrm{u}|_\Gamma^{(n)}, \mathbf{n}_\Gamma\right) [\mathrm{w}] \, \mathrm{dS}, \tag{10}$$

$$J_h^\sigma(u,w) = \sum_{\Gamma \in \mathscr{F}_h} \int_\Gamma \sigma [u] \, [w] \, \mathrm{dS}, \tag{11}$$

$$\ell_h(w)(t) = \int_\Omega g(t) w \, \mathrm{dx} + \sum_{\Gamma \in \mathscr{F}_h} \int_\Gamma (\nabla \mathrm{w} \cdot \mathbf{n} \mathrm{u_D} + \sigma \mathrm{u_D}(\mathrm{t}) \, \mathrm{w}) \, \mathrm{dS}. \tag{12}$$

The penalty parameter function σ in (11) and (12) is defined by $\sigma|_\Gamma = \varepsilon / \mathrm{diam}(\Gamma)$, $\Gamma \in \mathscr{F}_h$. The function $H(\cdot,\cdot,\cdot)$ in the face integrals in (10) is called the *numerical flux*, well-known from the finite volume method and it approximates the terms $\mathbf{f}(u) \cdot \mathbf{n}_\Gamma$ on $\Gamma \in \mathscr{F}_h$. For simplicity, we put $A_h(u,w) \equiv a_h(u,w) + J_h^\sigma(u,w)$ $\forall u, w \in H^2(\Omega, \mathscr{T}_h)$. Now, we introduce the *discrete problem*.

Definition 2. Let $u_h^0 \in S_{hp}$ be the $L^2(\Omega)$-projection of the initial condition u^0 into S_{hp}. We say that u_h is a DGFE solution of (1) – (3), if

$$\begin{aligned} &\text{a) } u_h \in C^1([0,T]; S_{hp}), \\ &\text{b) } \left(\frac{\partial u_h(t)}{\partial t}, w_h \right) + b_h(u_h(t), w_h) + A_h(u_h(t), w_h) = \ell_h(w_h)\,(t) \\ &\qquad\qquad\qquad\qquad\qquad\qquad \forall w_h \in S_{hp}, \ \forall t \in (0,T), \\ &\text{c) } u_h(0) = u_h^0. \end{aligned} \tag{13}$$

The (semi)-discrete problem (13) represents a system of ordinary differential equations (ODEs) which is solved by a suitable solver in the next section.

4 Time Discretization

Since problem (13) is stiff, it is necessary to solve it with a method having a large stability domain. In [6] we employ the well known backward difference formulae (BDF). Within this contribution, we developed a new approach based on a combination of an implicit and an explicit Runge-Kutta methods, where the linear terms are treated implicitly and the nonlinear ones explicitly. Therefore, the implicit-explicit (IMEX) Runge-Kutta scheme leads to a sufficiently stable method which requires a solution of linear algebraic problems at each time step.

Let us consider the following system of ODEs

$$\frac{\mathrm{d}}{\mathrm{dt}}y(t) = F(t, y(t)), \tag{14}$$

where $y(t) : [0,T] \to R^m$, $F : [0,T] \times R^m \to R^m$, $m \geq 1$ is an integer. Let $t_s = s\tau$, $s = 0,\dots,r$ represent a equidistant partition of $[0,T]$ with time step τ. We denote by y^s an approximation of $y(t_s)$, $s = 0,\dots,r$. Then the q–stages Runge–Kutta method is given by

$$y^{s+1} - y^s = \tau \sum_{v=1}^{q} \omega_v K_v, \tag{15}$$

where

$$K_v = F(t_s + \alpha_v \tau, y^s + \tau \sum_{j=1}^{q} \beta_{v,j} K_j),$$

and ω_v, α_v and $\beta_{v,j}$ are suitable coefficients. For example, when we set $q = 1$, $\omega_1 = 1$, $\alpha_1 = 1$, $\beta_{1,1} = 1$, we obtain the backward Euler method. Since our aim is to have an implicit and an explicit methods that could be easily combined together, we need methods having the same coefficients q, ω_v and α_v. As an example for methods of the first order with such properties, we can get forward and backward Euler method.

Within this contribution, we consider the following two stages scheme.

Definition 3. We say that the set of functions U^s, $s = 1,\dots,r$ is an approximate solution of problem (13) obtained by the two-stage IMEX Runge-Kutta DGFE scheme if

$$(i)\ \tilde{U}^{s+1},\ U^{s+1} \in S_{hp}, \tag{16}$$

$$(ii)\ (\tilde{U}^{s+1} - U^s, w) + \tau A_h(\tilde{U}^{s+1}, w) + \tau b_h(U^s, w) = \tau \ell_h(w)(t_{s+1})\ \forall w \in S_{hp}, \tag{17}$$

$$\begin{aligned}(U^{s+1} - U^s, w) &+ \frac{\tau}{2} A_h(U^{s+1} + U^s, w) + \frac{\tau}{2} b_h(\tilde{U}^{s+1}, w) + \frac{\tau}{2} b_h(U^s, w) \\ &= \frac{\tau}{2}\left(\ell_h(w)(t_{s+1}) + \ell_h(w)(t_s)\right) \quad \forall w \in S_{hp},\end{aligned} \tag{18}$$

$$(iii)\ U^0 = u_h(0), \tag{19}$$

where U^s denotes an approximation of $u_h(t_s)$, $s = 0,\dots,r$.

5 Error Estimates

Our goal is to analyse the error estimates of the approximate solution U^s, $s = 1,\dots,r$ obtained by the method (16) – (19). In the sequel we use the notation $u^s = u(t_s)$,

$\xi^s = U^s - \Pi u^s$, $\eta^s = \Pi u^s - u^s$ and $e^s = U^s - u^s = \xi^s + \eta^s$, where Π be the S_{hp} interpolation described in [12].

Let $\varepsilon|||w|||^2 := A_h(w,w)\ \forall w \in H^2(\Omega, \mathscr{T}_h)$ and $\|\cdot\| := \|\cdot\|_{L^2(\Omega)}$.

Lemma 1. *Let u be sufficiently regular. Then*

$$\begin{aligned}
\left|(u^{s+1} - u^s - \frac{\tau}{2}u_t^{s+1} - \frac{\tau}{2}u_t^s, w)\right| &\le C\tau^3\|w\| \quad \forall w \in S_{hp},\\
\|u^{s+1} - u^s - \tau u_t^{s+1}\| &\le C\tau^2,\\
\|u^{s+1} - u^s\| &\le C\tau,\\
|||u^{s+1} - u^s||| &\le C\tau,\\
|||u^{s+1} - \tau u_t^{s+1} - u^s||| &\le C\tau^2.
\end{aligned}$$

Lemma 2. *Let u be sufficiently regular. Then there exists projection $\Pi : L^2(\Omega) \to S_h$, such that*

$$\begin{aligned}
\|\eta^s\| &\le Ch^{p+1},\\
\left|(\eta^{s+1} - \eta^s, w)\right| &\le C\tau h^{p+1}\|w\| \quad \forall w \in S_{hp},\\
\left|(\tau\eta_t^{s+1}, w)\right| &\le C\tau h^{p+1}\|w\| \quad \forall w \in S_{hp}.
\end{aligned}$$

Lemma 3. *Let u be sufficiently regular. Then it holds that*

$$\begin{aligned}
|A_h(\eta^s, w)| &\le Ch^p\varepsilon|||w||| \quad \forall w \in S_{hp},\\
|A_h(v, w)| &\le C\varepsilon|||v|||\,|||w||| \quad \forall v, w \in S_{hp}.
\end{aligned}$$

Lemma 4. *Let u be sufficiently regular. Then it holds that*

$$\begin{aligned}
|b_h(v,w) - b_h(\bar{v},w)| &\le C(\|v-\bar{v}\| + |||v-\bar{v}|||)\,|||w||| \quad \forall v,\bar{v} \in L^2(\Omega),\ w \in S_{hp},\\
|b_h(v,w) - b_h(\bar{v},w)| &\le C(\|v-\bar{v}\| + |||v-\bar{v}|||)\,\|w\|\\
&\qquad \forall v,\bar{v} \in H^1(\Omega) \cap L^\infty(\Omega),\ w \in S_{hp},\\
|b_h(u,w) - b_h(\Pi u,w)| &\le Ch^{p+1}|||w||| \quad \forall w \in S_{hp},\\
|b_h(\Pi u^s,w) - b_h(U^s,v)| &\le C\|\xi^s\|\,|||w||| \quad \forall w \in S_{hp}.
\end{aligned}$$

Proof. The proof of Lemma 1 can be done similarly as in [6]. The proof of Lemma 2 can be found in [12, 3, 5]. The proof of Lemmas 3 can be found in [5, 6]. The second estimate in Lemma 4 results from $b(.,.) = b_h(.,.)$ under sufficient regularity of arguments and direct estimation. The rest of the estimates in Lemma 4 can be found in [5].

Theorem 1. *Let u be the exact solution of problem (4) satisfying (5). Let the meshes be regular with star–shaped elements and the numerical fluxes H be Lipschitz continuous, conservative and consistent. Let U^s, $s = 0,\dots,r$ be the approximate solution defined by (16) – (19). Then*

$$\max_{s=0,\dots,r} \|U^s - u(t_s)\|^2 \le O(h^{2p} + \tau^4)e^{TC(1+1/\varepsilon+\tau/\varepsilon^2)} \tag{20}$$

Proof. Since $U^s - u(t_s) = \xi^s + \eta^s$, in virtue of Lemma 2, it is sufficient to estimate $\|\xi^s\|$ only. Let us multiply (13) by $\frac{\tau}{2}$ for $t = t_s$ and $t = t_{s+1}$ and substract these equations from (18). Then we have

$$\begin{aligned}
&(\xi^{s+1} - \xi^s, w) + \frac{\tau}{2} A_h(\xi^{s+1} + \xi^s, w) \\
&= -(u^{s+1} - u^s - \frac{\tau}{2} u_t^{s+1} - \frac{\tau}{2} u_t^s, w) - (\eta^{s+1} - \eta^s, w) - \frac{\tau}{2} A_h(\eta^{s+1} + \eta^s, w) \\
&+ \frac{\tau}{2} \left(b_h(u^{s+1}, w) + b_h(u^s, w) - b_h(\tilde{U}^{s+1}, w) - b_h(U^s, w) \right).
\end{aligned} \tag{21}$$

Applying Lemmas 1–4 and under the notation $\tilde{u}^{s+1} := u^s + \tau u_t^{s+1}$, $\tilde{\xi}^{s+1} := \tilde{U}^{s+1} - \Pi \tilde{u}^{s+1}$, we get

$$\begin{aligned}
&(\xi^{s+1} - \xi^s, w) + \frac{\tau}{2} A_h(\xi^{s+1} + \xi^s, w) \\
&\leq \frac{\tau}{2} \|w\|^2 + \tau \frac{C}{\varepsilon} \left(\|\tilde{\xi}^{s+1}\|^2 + \|\xi^s\|^2 \right) + \frac{\tau}{2} A_h(w, w) + \tau q,
\end{aligned} \tag{22}$$

where $q = O(h^{2p} + \tau^4)$. Putting $w := \xi^{s+1} + \xi^s$ in (22), we have

$$\|\xi^{s+1}\|^2 - \|\xi^s\|^2 \leq \tau \|\xi^{s+1}\|^2 + \tau \|\xi^s\|^2 + \tau \frac{C}{\varepsilon} \left(\|\tilde{\xi}^{s+1}\|^2 + \|\xi^s\|^2 \right) + \tau q. \tag{23}$$

Obviously, $\tilde{u}^{s+1}$ satisfies

$$(\tilde{u}^{s+1} - u^s, w) + \tau A_h(u^{s+1}, w) + \tau b_h(u^{s+1}, w) = \tau l(w)(t_{s+1}) \quad \forall w \in S_{hp}. \tag{24}$$

Substracting (24) from (17) and after some manipulation, we obtain

$$\begin{aligned}
&(\tilde{\xi}^{s+1} - \xi^s, w) + \tau A_h(\tilde{U}^{s+1} - u^{s+1}, w) \\
&= \tau (b_h(u^{s+1}, w) - b_h(U^s, w)) - (\tilde{\eta}^{s+1} - \eta^s, w).
\end{aligned} \tag{25}$$

After another manipulation we obtain

$$\begin{aligned}
&(\tilde{\xi}^{s+1} - \xi^s, w) + \tau A_h(\tilde{\xi}^{s+1}, w) \\
&= \tau (b_h(u^{s+1}, w) - b_h(U^s, w)) - (\tau \eta_t^{s+1}, w) - \tau A_h(\tilde{\eta}^{s+1}, w) - \tau A_h(\tilde{u}^{s+1} - u^{s+1}, w).
\end{aligned} \tag{26}$$

Applying Lemmas 1–4, we have

$$(\tilde{\xi}^{s+1} - \xi^s, w) + \tau A_h(\tilde{\xi}^{s+1}, w) \leq \frac{1}{8} \|w\|^2 + \frac{\tau}{2} |||w|||^2 + \tau \frac{C}{\varepsilon} \|\xi^s\|^2 + \tau \tilde{q}, \tag{27}$$

where $\tilde{q} = O(h^{2p} + \tau^4)$. We put $w = 2\tilde{\xi}^{s+1}$ in (27) and immediately obtain

$$\|\tilde{\xi}^{s+1}\|^2 - \|\xi^s\|^2 \leq \frac{1}{2} \|\tilde{\xi}^{s+1}\|^2 + \tau \frac{C}{\varepsilon} \|\xi^s\|^2 + \tau \tilde{q}. \tag{28}$$

From (23) and (28), we get

$$\|\xi^{s+1}\|^2 - \|\xi^s\|^2 \le \tau\|\xi^{s+1}\|^2 + \tau\left(1 + \frac{C}{\varepsilon} + \frac{\tau C}{\varepsilon^2}\right)\|\xi^s\|^2 + \tau q. \tag{29}$$

Summing these inequalities and applying Gronwall's lemma we obtain

$$\|\xi^m\|^2 \le C\left(\|\xi^0\|^2 + Tq\right) e^{TC(1+1/\varepsilon+\tau/\varepsilon^2)}, \tag{30}$$

which proves our theorem, since $\|\xi^0\| \le \|\Pi u^0 - u^0\| = \|\eta^0\| \le Ch^{p+1}$. □

Remark 1. The estimate (20) cannot be used for $\varepsilon \to 0+$, because it blows up exponentially. The nonlinearity of the convective terms represents a serious obstacle.

Acknowledgements This work is a part of the research project MSM 0021620839 financed by the Ministry of Education of the Czech Republic, and it was partly supported by the Grant No. 316/2006/B-MAT/MFF of the Grant Agency of the Charles University Prague. The research of M. Vlasák is also supported by the project LC06052 financed by MSMT (Necas Center for Mathematical Modeling).

References

1. Arnold, D.N., Brezzi, F., Cockburn, B., Marini, L.D.: Unified analysis of discontinuous Galerkin methods for elliptic problems. SIAM J. Numer. Anal. **39**(5), 1749–1779 (2002)
2. Cockburn, B., Karniadakis, G.E., Shu, C.W. (eds.): Discontinuous Galerkin Methods. Springer, Berlin (2000)
3. Dolejší, V., Feistauer, M., Hozman, J.: Analysis of semi-implicit DGFEM for nonlinear convection-diffusion problems. Comput. Methods Appl. Mech. Engrg. **196**, 2813–2827 (2007)
4. Dolejší, V., Feistauer, M., Kučera, V., Sobotíková, V.: An optimal $L^\infty(L^2)$-error estimate of the discontinuous galerkin method for a nonlinear nonstationary convection-diffusion problem. IMA J. Numer. Anal. (in press)
5. Dolejší, V., Feistauer, M., Sobotíková, V.: A discontinuous Galerkin method for nonlinear convection–diffusion problems. Comput. Methods Appl. Mech. Engrg. **194**, 2709–2733 (2005)
6. Dolejší, V., Vlasák, M.: Analysis of a BDF – DGFE scheme for nonlinear convection-diffusion problems. Numer. Math. (2007 (submitted)). Preprint No. MATH-knm-2007/4, Charles University Prague, School of Mathematics, 2007, www.karlin.mff.cuni.cz/ms-preprints
7. Houston, P., Robson, J., Süli, E.: Discontinuous Galerkin finite element approximation of quasilinear elliptic boundary value problems I: The scalar case. IMA J. Numer. Anal. **25**, 726–749 (2005)
8. Houston, P., Schwab, C., Süli, E.: Discontinuous *hp*-finite element methods for advection-diffusion problems. SIAM J. Numer. Anal. **39**(6), 2133–2163 (2002)
9. Kufner, A., John, O., k, S.F.: Function Spaces. Academia, Prague (1977)
10. Lions, P.L.: Mathematical Topics in Fluid Mechanics. Oxford Science Publications (1996)
11. Rektorys, K.: The Method of Discretization in Time and Partial Differential Equations. Reidel, Dodrecht (1982)
12. Verfürth, R.: A note on polynomial approximation in Sobolev spaces. M^{2}AN **33**, 715–719 (1999)

Domain Decomposition Methods

A Domain Decomposition Method Derived from the Primal Hybrid Formulation for 2nd Order Elliptic Problems

C. Bernardi, T. Chacón Rebollo, and E. Chacón Vera

Abstract We introduce a framework for FETI methods using ideas from the decomposition via Lagrange multipliers of $H_0^1(\Omega)$ derived by Raviart-Thomas [17] and complemented with the detailed work on polygonal domains developed by Grisvard [11]. We compute the action of the Lagrange multipliers using the natural $H_{00}^{1/2}$ scalar product. Our analysis allows to deal with cross points and floating subdomains in a natural manner. We obtain that the condition number for the iteration matrix is independent of the mesh size and there is no need for preconditioning. This result improves the standard asymptotic bound for this condition number given by $(1+\log(H/h))^2$ shown by Mandel-Tezaur in [14]. Numerical results that confirm our theoretical analysis are presented in [2] or [4].

1 Introduction

The Lagrange multiplier formulation for elliptic Dirichlet boundary value problems is a classical technique to handle many difficulties such as high-order equations, the divergence-free constraint or non standard boundary conditions. We are interested here in its applications to domain decomposition methods, more precisely to the Finite Element Tearing and Interconnecting (FETI) method hinted by Dihn, Glowinsky and Periaux [10] in 1983, Dorr [6] in 1988, Roux [18]-[19] in 1989 and further developed by Farhat-Roux and collaborators [7]-[9]-[13]-[14]-[15]. This

C. Bernardi
Laboratoire Jacques-Louis Lions, C.N.R.S. et université Pierre et Marie Curie, b.c. 187, 4 place Jussieu 75252 Paris Cedex 05, France, e-mail: bernardi@ann.jussieu.fr,

T. Chacón Rebollo
Dpto. Ecuaciones Diferenciales y Análisis Numérico, Facultad de Matemáticas, Universidad de Sevilla, Tarfia sn. 41012 Sevilla, Spain, e-mail: chacon@us.es

E. Chacón Vera
Dpto. Ecuaciones Diferenciales y Análisis Numérico, Facultad de Matemáticas, Universidad de Sevilla, Tarfia sn. 41012 Sevilla, Spain, e-mail: eliseo@us.es

method has been implemented for large scale engineering problems with excellent results, see for instance [3]-[8]-[20].

In this work we introduce a framework for FETI methods using ideas from the decomposition via Lagrange multipliers of $H_0^1(\Omega)$ derived by Raviart-Thomas [17] and complemented with the detailed work on polygonal domains developed by Grisvard [11]. As a consequence, we obtain a characterization of $H_0^1(\Omega)$ more precise than the one in [17]. Our main ingredient next is the direct computation of the duality $H_{00}^{-1/2} - H_{00}^{1/2}$ using the natural $H_{00}^{1/2}$ scalar product; therefore no consistency error appears. Our analysis allows to deal with cross points and floating subdomains in a natural manner: cross points are dealt with implicitly and the ellipticity on floating subdomains holds naturally because we restrict our work to a subspace that contains the solution and where this ellipticity is satisfied. As a byproduct, we obtain that the condition number for the iteration matrix is independent of the mesh size and does not need any preconditioning. This result improves the standard asymptotic bound for this condition number given by $(1+\log(H/h))^2$, where H and h are the characteristic subdomain size and element size respectively, shown by Mandel-Tezaur in [14]. We refer to [2] or [4] for proofs and the numerical tests.

Standard notation, see Girault and Raviart [12] is used.

2 Motivation of the Method

Our model problem is: Given $f \in L^2(\Omega)$ we look for $u \in H_0^1(\Omega)$ such that

$$(\nabla u, \nabla v)_\Omega + (u,v)_\Omega = (f,v)_\Omega \quad \forall v \in H_0^1(\Omega) \tag{1}$$

where $(\varphi, \psi)_\Omega = \int_\Omega \varphi(x)\,\psi(x)\,dx$ is the scalar product in $L^2(\Omega)$.

Assume now that Ω is a polygonal bounded domain in $\mathbb{R}^2$ that admits a decomposition without overlapping in polygonal subdomains

$$\overline{\Omega} = \cup_{r=1}^{R} \overline{\Omega}_r \qquad \text{and} \qquad \Omega_r \cap \Omega_{r'} = \emptyset, \quad 1 \le r < r' \le R. \tag{2}$$

Then, the solution u of (1) also satisfies that

$$\sum_{r=1}^{R} \{(\nabla u_r, \nabla v_r)_{\Omega_r} + (u_r, v_r)_{\Omega_r}\} = \sum_{r=1}^{R} (f_r, v_r)_{\Omega_r} \quad \forall v \in H_0^1(\Omega) \tag{3}$$

where the subindex r denotes restriction to Ω_r, i.e., for instance $u_r = u_{|\Omega_r}$. Next we assume that the partition (2) of Ω is geometrically conforming in the sense that all interiors interfaces $\Gamma_{r,s} = \overline{\Omega}_r \cap \overline{\Omega}_s \subset \Omega$ are either a common vertex, a common edge or empty. For simplicity, when $\Gamma_{r,s}$ is an internal common edge we will assume that is a straight open segment without corners. A general case on this situation, i.e., $\Gamma_{r,s}$ with corners, could also be handled in the same way but the description would become more cumbersome. We set $\Gamma_{r,0} = \partial\Omega_r \cap \partial\Omega$ and we may allow $\Gamma_{r,0}$

polygonal because we impose zero boundary data on $\partial\Omega$. Now we describe $\partial\Omega_r$ in terms of its edges via $\partial\Omega_r = \Gamma_{r,0} \cup \Gamma_{r,1} \cup \ldots \cup \Gamma_{r,J_r}$, where J_r is a positive integer and $\Gamma_{r,0}$, which might be empty, satisfies $\partial\Omega = \cup_{r=1}^{R} \Gamma_{r,0}$. We call **skeleton** of $\overline{\Omega}$, and denote it by $\mathscr{E}$, the set of all interfaces in $\overline{\Omega}$, and by $\mathscr{E}_0$ the **skeleton** of Ω, i.e., the set of all internal interfaces:

$$\mathscr{E} = \cup_{i=1}^{I} \Gamma_i, \quad \mathscr{E}_0 = \mathscr{E} \cap \Omega = \cup_{i=I_0+1}^{I} \Gamma_i.$$

Here $\Gamma_i = \Gamma_{i,0}$ for $i = 1,..,I_0 \leq R$ describe the boundary $\partial\Omega$, and for $i \geq I_0 + 1$ $\Gamma_i = \Gamma_{r,j}$, for some $r, j \geq 1$, are all the internal interfaces. Then, on each Ω_r we consider the restriction of $H_0^1(\Omega)$ to Ω_r, i.e., the Hilbert space

$$H_b^1(\Omega_r) = \{v_r \in H^1(\Omega_r); v_r = 0 \text{ on } \partial\Omega_r \cap \partial\Omega\},$$

with the classical scalar product $(u_r, v_r)_{1,\Omega_r} = (u_r, v_r)_{\Omega_r} + (\nabla u_r, \nabla v_r)_{\Omega_r}$ and on Ω the Hilbert space X given by

$$X = \{v \in L^2(\Omega); v_r = v_{|\Omega_r} \in H_b^1(\Omega_r), r \leq R, \ [v]_{\Gamma_i} \in H_{00}^{1/2}(\Gamma_i), \quad \forall \Gamma_i \in \mathscr{E}_0\}$$

where $[v]_{\Gamma_i}$ is the jump across $\Gamma_i \in \mathscr{E}_0$ given by $[v]_{\Gamma_i} = v_r - v_s$ when $\Gamma_i = \partial\Omega_r \cap \partial\Omega_s \subset \Omega$. The scalar product on X is given by

$$(u,v)_X = \sum_{r=1}^{R} (u_r, v_r)_{1,\Omega_r} + \sum_{i=I_0+1}^{I} ([u]_{\Gamma_i}, [v]_{\Gamma_i})_{1/2,00,\Gamma_i}, \ \forall\, u, v \in X$$

see Grisvard [11] for instance for the definition of the scalar product in $H^{1/2}(\Gamma)$. The norm on X also measures the jumps across the internal interfaces and thanks to the trace theorems we have the inequality

$$\|v\|_X^2 \leq C \sum_{r=1}^{R} \|v_r\|_{1,\Omega_r}^2, \ \forall\, v \in X \tag{4}$$

that will guarantee the ellipticity of the problems that will be posed later on.

We can identify the space $H_0^1(\Omega)$ with the subspace V of elements of X such that their jumps are zero on the interfaces. Then, the unique solution u of our variational problem (1) also solves the problem: Find $u \in V$ such that for all $v \in V$

$$\sum_{r=1}^{R} \{(\nabla u_r, \nabla v_r)_{\Omega_r} + (u_r, v_r)_{\Omega_r}\} = \sum_{r=1}^{R} (f_{|\Omega_r}, v_r)_{\Omega}. \tag{5}$$

Our purpose now is to get rid of the constrains on the jumps and set (5) on X. This will be achieved by adding the restriction on the jumps via Lagrangian multipliers to (5). Therefore we must characterize $H_0^1(\Omega)$ in X. To achieve this description we follow the idea introduced by Raviart-Thomas [17] and study the linear forms on X that vanish on $H_0^1(\Omega)$. We must guarantee that all the jumps across internal interfaces vanish and we do this via Lagrange multipliers.

A key ingredient is the Green formula on polygonal domains and the localization of the boundary integrals on each element $\Gamma_i \in \mathscr{E}_0$ so as to act on the jumps. It is in this point where we improve the arguments in [17]. As a reward, our analysis will say that cross points do not matter in the computation of the solution and the characterization in [17] will be improved.

Let $\mathscr{O}$ be a polygonal domain in $\mathbb{R}^2$ with edges Γ_j, $1 \le j \le J$. The domain of the divergence operator on $\mathscr{O}$ is

$$H(div;\mathscr{O}) = \{\mathbf{q} \in L^2(\mathscr{O})^2;\ div(\mathbf{q}) \in L^2(\mathscr{O})\};$$

for each j, we also introduce the space

$$H^1_{(j)}(\mathscr{O}) = \{v \in H^1(\mathscr{O});\ v = 0 \,\mathrm{on}\, \partial\mathscr{O} \setminus \overline{\Gamma}_j\}.$$

Let $<\cdot,\cdot>_{-1/2,00,\Gamma_j}$ denote the duality $H_{00}^{-1/2}(\Gamma_j) - H_{00}^{1/2}(\Gamma_j)$ and let E be the space

$$E = \{v \in H^1(\mathscr{O});\ \Delta v \in L^2(\mathscr{O})\}$$

then we have the following integrations by parts

Lemma 1. *When $\mathscr{O} \subset \mathbb{R}^2$ is a polygonal domain and $\partial\mathscr{O} = \cup_{j=1}^J \Gamma_j$, then for each $u \in E$ and any $\mathbf{q} \in H(div;\mathscr{O})$ we have*

$$(\Delta u, v)_{\mathscr{O}} + (\nabla u, \nabla v)_{\mathscr{O}} = \sum_{j=1}^{J} < \partial_{\mathbf{n}_j} u, v >_{-1/2,00,\Gamma_j} \tag{6}$$

$$(\mathbf{q}, \nabla v)_{\mathscr{O}} + (div(\mathbf{q}), v)_{\mathscr{O}} = \sum_{j=1}^{J} < \mathbf{n}_j \cdot \mathbf{q}, v >_{-1/2,00,\Gamma_j} \tag{7}$$

for any $v \in H^1(\mathscr{O})$ with $v_{|\Gamma_j} \in H_{00}^{1/2}(\Gamma_j)$ for $j = 1,2,\dots,J$.

As each Ω_r is a polygonal domain, and as a consequence of the above results, we need to consider the dense subspace W_r of $H^1_b(\Omega_r)$ given by, observe that we consider only internal interfaces,

$$W_r = \{u \in H^1_b(\Omega_r);\ u_{|\Gamma_{r,j}} \in H_{00}^{1/2}(\Gamma_{r,j}),\ j = 1,\dots,J_r\},$$

and the dense subspace in X given by

$$X_0 = \{v \in L^2(\Omega); v_r = v_{|\Omega_r} \in W_r, r = 1,\dots,R\}.$$

The use of X_0 is a key tool in our analysis because the Green's formula on polygonal subdomains can be applied on each Ω_r. Now, as we are only interested in what happens on the internal interfaces, we consider

$$M = \{\mu \in \prod_{r=1}^{R}\prod_{j=1}^{J_r} H_{00}^{-1/2}(\Gamma_{r,j}); \mu_{r,j} = \mathbf{n}_{r,j}\cdot\mathbf{q},\ \text{for some}\ \mathbf{q} \in H(div;\Omega).\}$$

The elements of M will be denoted the Lagrange multipliers on the internal interfaces $\Gamma_i \in \mathscr{E}_0$. We have the characterization of $H_0^1(\Omega)$ as a subspace of X given by

Lemma 2. *Let $b : M \times X \mapsto \mathbb{R}$ be defined for $v \in X$ and $\lambda \in M$ by*

$$b(\lambda, v) = \sum_{i=I_0+1}^{I} < \lambda_i, [v]_{\Gamma_i} >_{-1/2,00,\Gamma_i} . \tag{8}$$

Then $H_0^1(\Omega) = \{v \in X;\ b(\lambda, v) = 0, \quad \forall \lambda \in M\}$.

Define now the bilinear form $a : X \times X \mapsto \mathbb{R}$ given by

$$a(u,v) = (u,v)_X = \sum_{r=1}^{R} (u_r, v_r)_{1,\Omega_r} = \sum_{r=1}^{R} \int_{\Omega_r} \{\nabla u_r \cdot \nabla v_r + u_r v_r\}\, dx. \tag{9}$$

and use the bilinear form $b(\lambda, v)$ given in (8). Then, our Dirichlet problem (1) consists in looking for a pair $(u, \lambda) \in X \times M$ such that

$$a(u,v) + b(\lambda, v) = \sum_{r=1}^{R} (f, v_r)_{\Omega_r}, \ \forall v \in X \tag{10}$$

$$b(\mu, u) = 0, \quad \forall \mu \in M. \tag{11}$$

This formulation is also known as a **Lagrange formulation or primal hybrid formulation** because it mixes the primal variable u with the Lagrange multipliers which weakly enforce continuity. We have the equivalence result

Theorem 1. *$u \in H_0^1(\Omega)$ solves the Dirichlet problem (1) if and only if there exists a unique $\lambda \in M$ such that $(u, \lambda) \in X \times M$ solves problem (10)-(11). Moreover, in this case and for $i = I_0 + 1, \ldots, I$,*

$$\lambda_i = -\partial_{\mathbf{n}_i} u \in H_{00}^{-1/2}(\Gamma_i). \tag{12}$$

The cornerstone now is how to compute the dualities that act on the jumps. This question has been considered from several points of view already. In our approach we use Riesz representation and work with the $H_{00}^{1/2}$ scalar product that is explicitly computed.

3 Domain Decomposition Method

Our saddle point problem (10)-(11) can be written as equations in $X^\star \times M$: find $(u, \lambda) \in X \times M$ such that

$$R^{-1} u + B^\star \lambda = F \quad \text{on } X^\star, \tag{13}$$

$$Bu = 0 \quad \text{on } M, \tag{14}$$

where $R : X^\star \mapsto X$ is the Riesz isomorphism associated with the bilinear form $a(\cdot,\cdot)$ and defined by

$$< R^{-1}u, v >= a(u,v), \quad \forall u, v \in X,$$

B is the continuous mapping $B : X \mapsto M$ defined by $Bv = ([v]_{\Gamma_i})_{i=I_0+1}^{I}$, i.e., Bv gives the jumps across the internal interfaces $\Gamma_i \in \mathscr{E}_0$ of v and $B^\star$ is the transpose operator to B. Then

$$b(\mu,v) = \sum_{i=I_0+1}^{I} (\mu_i, [v]_{\Gamma_i})_{1/2,00,\Gamma_i} = (\mu, Bv)_M \quad \forall v \in X.$$

Finally, we take $F : X \mapsto \mathbb{R}$ given by $< F, v >= \sum_{r=1}^{2} (f, v_r)_{\Omega_r} = (f, v)_\Omega$. We find that

$$u = R(F - B^\star\lambda) \Rightarrow Bu = BRF - BRB^\star\lambda \tag{15}$$

and using $Bu = 0$ from here we have the **dual problem associated to the saddle point problem**

$$(BRB^\star)\lambda = BRF \;\; \text{on}\, M. \tag{16}$$

Thanks to the inf-sup condition, on the infinite dimensional or finite dimensional setting (using the finite element extension theorems, see for instance Bernardi-Maday-Rapetti [5]) **the operator $BRB^\star$ is symmetric positive definite** with eigenvalues in the interval $[\beta^2, \alpha^2]$ where $\beta^2, \alpha^2 > 0$ are independent of the discretization parameter h; it also holds that β^2, α^2 are eigenvalues of $BRB^\star$, see for instance Bacuta [1]. As a consequence, the condition number of the operator $BRB^\star$ is bounded independently of the discretization parameter,

$$\kappa = \kappa(BRB^\star) \le \frac{\alpha^2}{\beta^2}. \tag{17}$$

This result improves the estimate given by Mandel-Tezaur [14] where the estimate on the condition number is expressed asymptotically by

$$C(1 + \log(H/h))^2$$

where H and h are the characteristic subdomain size and element size respectively.

Now the resolution of (16) via the **Conjugate Gradient Method** is possible. This is the basics of the standard FETI methods. We also observe that Conjugate Gradient method does not need any preconditioning.

Recall that $(\cdot,\cdot)_{1/2,00,\Gamma}$ is the scalar product on M. Then, the Conjugate gradient method is:

Take $r_0 = d_0 = BRF - (BRB^\star)\lambda_0$, for $m \ge 0$ set $p_m := (BRB^\star)d_m$ and repeat

$$\alpha_m = \frac{(d_m, r_m)_{1/2,00,\Gamma}}{(d_m, p_m)_{1/2,00,\Gamma}}, \tag{18}$$

$$\lambda_{m+1} = \lambda_m + \alpha_m\, d_m, \qquad \text{on } \Gamma \tag{19}$$

$$r_{m+1} = r_m - \alpha_m p_m, \qquad \text{on } \Gamma \tag{20}$$

$$\beta_m = \frac{(p_m, r_{m+1})_{1/2,00,\Gamma}}{(p_m, d_m)_{1/2,00,\Gamma}}, \tag{21}$$

$$d_{m+1} = r_{m+1} - \beta_m\, d_m, \qquad \text{on } \Gamma. \tag{22}$$

Using (15) the computation of the residual r_0 is made via the computation of u_0

$$u_0 = RF - (RB^\star)\lambda_0 \Rightarrow r_0 = B\, u_0 \tag{23}$$

and for the computation of $p_m := (BRB^\star)d_m$ we set $p_m = B\, w_m$ where w_m solve the auxiliar problem:

$$R^{-1}\, w_m = B^\star\, d_m \qquad \text{on } X^\star. \tag{24}$$

The resolution of (24) is made on each subdomain independently. Following standard convergence results, see for instance Quarteroni-Sacco-Saleri [16], we have **geometric convergence** in a finite number of steps (under exact arithmetic) for this iterative process. Suppose that N is the size of the matrix $BRB^\star$, which amounts to say that N is the number of degrees of freedom on the interfaces, then

Theorem 2. *The method (18)-(22) converges geometrically in at most N steps (under exact arithmetic). For any $m < N$ the error $e_m = \lambda_m - \lambda$ is orthogonal to the direction d_j for $j = 0, 1, 2, \ldots, m-1$ and we have the estimate*

$$\|\lambda_m - \lambda\|_{1/2,00,\Gamma} \leq 2\sqrt{\kappa}\frac{c^m}{1+c^{2m}}\|\lambda_0 - \lambda\|_{1/2,00,\Gamma}$$

where $c = (\sqrt{\kappa}-1)/(\sqrt{\kappa}+1) < 1$ and $\kappa = \alpha^2/\beta^2$ is the spectral condition number of $BRB^\star$ that is independent of the discretization parameter.

Acknowledgements Research partially funded by Spanish government MEC Research Project MTM2006-02175.

References

1. C. Bacuta, *A unified approach for Uzawa algorithm* SIAM J. Numer. Anal., Vol. 44, No. 6, pp. 2633-2649, 2006.
2. C. Bernardi, T . Chacón Rebollo, T . Chacón Vera, *A FETI method with a mesh independent condition number for the iteration matrix* internal report R07031 Laboratoire Jaques Louis-Lions.

3. M. Bhardwaj, D. Day, C. Farhat, M. Lesoinne, K. Pierson, D. Rixen, *Application of the FETI method to ASCI problems–scalability results on 1000 processors and discussion of highly heterogeneous problems* Internat. J. Numer. Methods Engrg., 47, 513-535 (2000).
4. C. Bernardi, T . Chacón Rebollo, T . Chacón Vera, *A FETI method with a mesh independent condition number for the iteration matrix* to appear Computer Methods in Applied Mechanics and Engineering.
5. C. Bernardi, Y. Maday and F. Rapetti, *Discrétisations variationnelles de problèmes aux limites elliptiques.* Mathématiques & Applications, 45. Springer-Verlag, New-York, 2002.
6. M. R. Dorr, *Domain decomposition via Lagrange multipliers* UCRL-98532, Lawrence Livermore National Laboratory, Livermore, CA, 1988.
7. C. Farhat, *A Lagrange multiplier based divide and conquer finite element algorithm* J. Comput. Systems Engrg., 2, 149-156 (1991).
8. C. Farhat, J. Mandel, F.-X. Roux, *Optimal convergence properties of the FETI domain decomposition method* Comput. Methods Appl. Mech. Engrg., 115, 364-385 (1994).
9. C. Farhat, F.-X. Roux, *A method of finite element tearing and interconnecting and its parallel solution algorithm*, Int. J. Numer. Methods Engrg. 32 (1991) 1205-1227.
10. R. Glowinski, Q.V. Dinh and J. Periaux, *Domain decomposition methods for nonlinear problems in fluid dynamics* Comput. Methods Appl. Mech. Engrg., 40, 27-109 (1983).
11. P. Grisvard, *Singularities in Boundary value problems.* Recherches en Mathematiques Appliques 22, Masson, 1992.
12. V. Girault and P.-A. Raviart, *Finite Element Methods for Navier-Stokes Equations. Theory and Algorithms.* Springer Series in Computational Mathematics, 5. Springer-Verlag, Berlin, 1986.
13. J. Mandel, R. Tezaur, On the convergence of a dual-primal substructuring method, Numer. Math. 88 (2001) 543-558.
14. J. Mandel, R. Tezaur: Convergence of a substructuring method with Lagrange multipliers. Numer. Math., 73, 473-487 (1996).
15. J. Mandel, R. Tezaur, C. Farhat, *A scalable substructuring method by Lagrange multipliers for plate bending problems* SIAM J. Numer. Anal., 36, 1370-1391 (1999).
16. A. Quarteroni, R. Sacco and F. Saleri, *Numerical Mathematics* Texts in Applied Mathematics, vol. 37, Springer-Verlag, New-York, 2nd edition, 2006.
17. P.A. Raviart and J.-M. Thomas, *Primal Hybrid Finite Element Methods for second order elliptic equations* Math Comp, Vol 31, number 138, pp. 391-413, 1977.
18. F.-X. Roux,*Méthode de décomposition de domaine à l'aide de multiplicateurs de Lagrange et application à la résolution en parallèle des équations de l'élasticité linéaire*, Thesis, Université Pierre et Marie Curie (1989).
19. F.-X. Roux,*Acceleration of the outer conjugate gradient by reorthogonalization for a domain decomposition method for structural analysis problems.* C.R. Acad. Sci. Paris Sér. I Math. 308, no. 6, 193-198, (1989).
20. D. J. Rixen, C. Farhat, R. Tezaur, J. Mandel *Theoretical comparison of the FETI and algebraically partitioned FETI methods, and performance comparisons with a direct sparse solver* Internat. J. Numer. Methods Engrg., 46, 501-533 (1999).

A Posteriori Error Analysis of Penalty Domain Decomposition Methods for Linear Elliptic Problems

C. Bernardi, T. Chacón Rebollo, E. Chacón Vera, and D. Franco Coronil

Abstract In this work we introduce a new version of the non-overlapping domain decomposition method (DDM) method proposed by Chacón and Chacón in [5]. In the new method a $H_{00}^{1/2}(\Gamma)$ penalty term replaces the $L^2(\Gamma)$ one in the original method. We develop a posteriori error analysis, aimed to design strategies for optimizing the combined choice of the penalty parameter and the adaptation of the grid, to reduce the computational cost. We shall discuss the computational benefits of using $H_{00}^{1/2}(\Gamma)$ penalty versus $L^2(\Gamma)$ penalty.

1 The Penalty Problem

Let $\Omega \subset \mathbb{R}^d$ $(d = 2,3)$ be a simply connected and bounded domain with a Lipschitz-continuous boundary $\partial\Omega$. We consider a simple decomposition of Ω into two non-overlapping subdomains Ω_1 and Ω_2 and set $\Gamma = \partial\Omega_1 \cap \partial\Omega_2$, $\Gamma_i = \partial\Omega_i \cap \partial\Omega$. We assume that all of these boundaries are Lipschitz-continuous $(d-1)$-dimensional manifolds with positive $(d-1)$-dimensional measure.

We shall consider the Poisson problem in Ω as a test problem:
Given $f \in L^2(\Omega)$, find $u \in H_0^1(\Omega)$ such that

$$(\nabla u, \nabla v)_\Omega = (f, v)_\Omega, \quad \forall\, v \in H_0^1(\Omega). \tag{1}$$

Here we introduce a new version of the method studied by T. Chacon and E. Chacon in [6] that use a $H_{00}^{1/2}(\Gamma)$ penalty term, then we recall that $H_{00}^{1/2}(\Gamma)$ is the

C. Bernardi
Laboratoire Jacques-Louis Lions, C.N.R.S. et Université Pierre et Marie Curie. Boite courrier 187, 4 Place Jussieu, 75252 Paris Cedex 05, France. e-mail: bernardi@ann.jussieu.fr

T. Chacón Rebollo, E. Chacón Vera and D. Franco Coronil
Departamento de Ecuaciones Diferenciales y Análisis Numérico, Universidad de Sevilla. C/ Tarfia, s/n. 41080 Sevilla, Spain. e-mail: chacon@us.es, eliseo@us.es, franco@us.es

subspace of $H^{1/2}(\Gamma)$ whose extension by zero to $\partial\Omega_1$ (for instance, it could be also to $\partial\Omega_2$) belongs to $H^{1/2}(\partial\Omega_1)$. An intrinsic scalar product on $H^{1/2}_{00}(\Gamma)$ is defined as

$$[[w,v]]_\Gamma = \int_\Gamma w(x)\,v(x)\,dx + \int_\Gamma\int_\Gamma \frac{(w(x)-w(y))\,(v(x)-v(y))}{|x-y|^d}\,dxdy \qquad (2)$$
$$+ \int_\Gamma \frac{w(x)\,v(x)}{d(x,\partial\Gamma)}\,dx,$$

where the first two summands define the $H^{1/2}(\Gamma)$ scalar product (Cf. [1]).

For brevity, we also denote by $[[\cdot,\cdot]]_\Gamma$ the $L^2(\Gamma)$ scalar product, and study both the $L^2(\Gamma)$ and the $H^{1/2}_{00}(\Gamma)$ penalties at the same time with the same notation. We shall distinguish the two cases when this is necessary.

Consider the Sobolev spaces

$$X_i = H^1(\Omega_i;\Gamma_i) = \{v \in H^1(\Omega_i) \text{ such that } v_{|\Gamma_i} = 0\}, \quad i=1,2; \quad \mathbf{X} = X_1 \times X_2.$$

We define for $\mathbf{u} = (u_1,u_2), \mathbf{v} = (v_1,v_2) \in \mathbf{X}$ the scalar product and norm on $\mathbf{X}$

$$((\mathbf{u},\mathbf{v}))_X = \sum_{i=1}^{2} (\nabla u_i, \nabla v_i)_{\Omega_i}, \quad \|\mathbf{u}\|_X^2 = ((\mathbf{u},\mathbf{u}))_X.$$

We introduce our penalty problem with $H^{1/2}_{00}(\Gamma)$ or $L^2(\Gamma)$ penalties, as

$$(P_\varepsilon) \qquad \begin{cases} \text{Find } \mathbf{u}^\varepsilon \in \mathbf{X} \text{ such that} \\ ((\mathbf{u}^\varepsilon,\mathbf{v}))_X + \dfrac{1}{\varepsilon}\,[[u_1^\varepsilon - u_2^\varepsilon, v_1 - v_2]]_\Gamma = F(\mathbf{v}), \ \forall\, \mathbf{v} = (v_1,v_2) \in \mathbf{X}, \end{cases}$$

with $F(\mathbf{v}) = \sum_{i=1}^{2} (f,v_i)_{\Omega_i}$, where $\varepsilon > 0$ is a parameter destined to tend to zero. This problem has a unique solution for any $\varepsilon > 0$ due to Lax–Milgram Lemma.

Remark 1. The original method introduced in [6] only considers a $L^2(\Gamma)$ penalty term to enhance the continuity of $\mathbf{u}^\varepsilon = (u_1^\varepsilon, u_2^\varepsilon)$ across Γ. We also consider here a $H^{1/2}_{00}(\Gamma)$ penalty term to enhance this continuity in a stronger sense. In both cases, the method may be interpreted as the variational formulation of a coupled system of PDEs with the structure

$$\begin{cases} -\Delta u_1^\varepsilon = f & \text{in } \Omega_1, \\ u_1^\varepsilon = 0 & \text{on } \Gamma_1, \\ \partial_{\mathbf{n}_{12}} u_1^\varepsilon = \dfrac{1}{\varepsilon} b(u_1^\varepsilon - u_2^\varepsilon) & \text{on } \Gamma, \end{cases} \qquad \begin{cases} -\Delta u_2^\varepsilon = f & \text{in } \Omega_2, \\ u_2^\varepsilon = 0 & \text{on } \Gamma_2, \\ \partial_{\mathbf{n}_{21}} u_2^\varepsilon = \dfrac{1}{\varepsilon} b(u_2^\varepsilon - u_1^\varepsilon) & \text{on } \Gamma, \end{cases} \qquad (3)$$

where b is an injective linear bounded boundary operator on Γ, which reduces to the identity for $L^2(\Gamma)$ penalty. So, the method ensures the continuity of the normal fluxes through Γ and forces by penalty the continuity of $\mathbf{u}^\varepsilon$ through Γ.

2 Penalty Error Analysis

The introduction of the $H_{00}^{1/2}(\Gamma)$ penalty instead of the $L^2(\Gamma)$ penalty is suggested by the following a priori penalty error analysis (Cf. [2], [6]):

Theorem 1. *For each $\varepsilon > 0$, we have the following estimates for the penalty problem (P_ε):*

a) Assume that $H_{00}^{1/2}(\Gamma)$ penalty is used and that f belongs to $L^2(\Omega)$, then

$$e_{H_{00}^{1/2}}(\varepsilon) = \sum_{i=1}^{2} |u - u_i^\varepsilon|_{1,\Omega_i} \leq C \|\partial_{\mathbf{n}} u\|_{*,\Gamma}\, \varepsilon, \tag{4}$$

b) Assume that $L^2(\Gamma)$ penalty is used and the solution u of (1) satisfies $\partial_{\mathbf{n}} u \in L^2(\Gamma)$, then

$$e_{L^2}(\varepsilon) = \sum_{i=1}^{2} |u - u_i^\varepsilon|_{1,\Omega_i} \leq C' \, \|\partial_{\mathbf{n}} u\|_{*,\Gamma} \sqrt{\varepsilon}. \tag{5}$$

where C and C' are constants independent of ε and $\|\cdot\|_{*,\Gamma}$ is the dual norm of the scalar product $[[\cdot,\cdot]]_\Gamma$.

Here, we denote by $\mathbf{n}$ the outward normal vector to Ω_1 on Γ pointing into Ω_2.

Remark 2. In [2] we show that the quantity $\eta^P = \|u_1^\varepsilon - u_2^\varepsilon\|_{H_{00}^{1/2}(\Gamma)}$ is an optimal indicator of the penalty error for (P_ε), as it satisfies that, there exist two constants $C > 0$ and $C' > 0$ independent of ε such that

$$C\eta^P \leq \sum_{i=1}^{2} |u - u_i^\varepsilon|_{1,\Omega_i} \leq C' \eta^P, \; i = 1,2. \tag{6}$$

The motivation of introducing the $H_{00}^{1/2}(\Gamma)$ penalty was to observe this estimate when analyzing the error of $L^2(\Gamma)$ penalty.

3 Discretization of the Penalty Problem

To discretize the problem (P_ε), we consider a regular family of triangulations $\{\mathscr{T}_{ih}\}_{\{h>0\}}$ of each $\overline{\Omega}_i$ such that Γ is the union of whole faces or sides of elements K in each $\mathscr{T}_{ih}$. Denote $\mathscr{T}_h = \mathscr{T}_{1h} \cup \mathscr{T}_{2h}$.

We assume that $\mathscr{T}_{1h}$ and $\mathscr{T}_{2h}$ have the same trace sets on Γ, that we denote by $\mathscr{E}_h^\Gamma$. Then, $\{\mathscr{T}_h\}_{\{h>0\}}$ is a regular family of triangulations of Ω.

Next we consider a family of finite element subspaces X_{ih} of X_i $(i = 1,2)$, built on the grids $\mathscr{T}_{ih}$. We set $\mathbf{X}_h = X_{1h} \times X_{2h}$.

Then, our **penalty discrete problem** is the standard finite element Galerkin approximation $\mathbf{u}_h^\varepsilon = (u_{1h}^\varepsilon, u_{2h}^\varepsilon) \in \mathbf{X}_h$ of $\mathbf{u}^\varepsilon$, solution of

$$(P_{\varepsilon,h}) \begin{cases} \text{Find } \mathbf{u}_h^\varepsilon \in \mathbf{X}_h \text{ such that} \\ ((\mathbf{u}_h^\varepsilon, \mathbf{v}_h))_X + \dfrac{1}{\varepsilon}[[u_{1h}^\varepsilon - u_{2h}^\varepsilon, v_{1h} - v_{2h}]]_\Gamma = F(\mathbf{v}_h),\ \forall\ \mathbf{v}_h = (v_{1h}, v_{2h}) \in \mathbf{X}_h. \end{cases}$$

This problem admits a unique solution for any $\varepsilon > 0$.

To solve $(P_{\varepsilon,h})$ numerically, we consider the following **parallel technique** by successive approximations introduced in [6]:

For $n = 0, 1, 2, \ldots$, $u_1^{n+1} = u_{1h}^{n+1,\varepsilon} \in X_{1h}$ and $u_2^{n+1} = u_{2h}^{n+1,\varepsilon} \in X_{2h}$ are computed from u_{1h}^n and u_{1h}^n by solving (we drop the exponent ε and index h for simplicity),

$$(P_{\varepsilon,h}^n) \begin{cases} (\nabla u_1^{n+1}, \nabla v_{1h})_{\Omega_1} + \dfrac{1}{\varepsilon}[[u_1^{n+1} - u_2^n, v_{1h}]]_\Gamma = (f, v_{1h})_{\Omega_1},\ \forall\, v_{1h} \in X_{1h}, \\ \\ (\nabla u_2^{n+1}, \nabla v_{2h})_{\Omega_2} + \dfrac{1}{\varepsilon}[[u_2^{n+1} - u_1^n, v_{2h}]]_\Gamma = (f, v_{2h})_{\Omega_2},\ \forall\, v_{2h} \in X_{2h}. \end{cases} \tag{7}$$

For the analysis that we are going to do in the next sections, we assume that each of the spaces X_{ih} contains the space X_{ih}^* defined

$$X_{ih}^* = \{w_h \in H^1(\Omega_i);\, w_{h_{|K}} \in P_{k_i}(K),\ \forall K \in \mathscr{T}_{ih},\ w_{h_{|\Gamma_i}} = 0\}, \tag{8}$$

where $P_{k_i}(K)$ denotes the space of restrictions to K of polynomials with d variables and total degree $\leq k_i$ for positive integers k_i.

Furthermore we assume that its trace space on Γ is

$$W_{ih} = \{v_h : \overline{\Gamma} \to \mathbb{R} \text{ continuous} :\ v_{h|\partial\Gamma} = 0,\ v_{h|_e} \in P_{k_i}(e), \forall e \in \mathscr{E}_h^\Gamma\}. \tag{9}$$

4 A Posteriori Error Analysis

In this section we develop an a posteriori error analysis of both methods, with the purpose of study the optimality of independent error indicators for both penalty and discretization errors that allows to develop strategies to provide relevant error reductions. Otto and Lube develop in [7] a similar approach to estimate free parameters in a domain decomposition techniques. These error indicators are the following:

$$\textbf{A discretization error indicator :}\quad \eta_h^D = \left(\sum_{i=1}^{2} \sum_{K\in\mathscr{T}_{ih}} \left(\eta_i^K\right)^2\right)^{1/2}; \tag{10}$$

with $\eta_i^K = h_K \|f_h + \Delta u_{ih}^\varepsilon\|_{L^2(K)} + \sum_{e\in\mathscr{E}_K} h_e^{1/2} \|[\partial_{n_K} u_{ih}^\varepsilon]\|_{L^2(e)}$, for each $i = 1, 2$ and $K \in \mathscr{T}_{ih}$, where h_K and h_e stand for for the diameters of K and e, respectively; n_K stands for the unit outward normal to ∂K;

$\mathscr{E}_K$ is the set of edges ($d = 2$) or faces ($d = 3$) of K that are not contained in Γ_i; $[\partial_{n_K} u^{\varepsilon}_{ih}]$ stands for the jump of $\partial_{n_K} u^{\varepsilon}_{ih}$ across e if e is not included in Γ, and for $\partial_{n_K} u^{\varepsilon}_{1h} - \partial_{n_K} u^{\varepsilon}_{2h}$ if e is included in Γ.

A penalty error indicator: $$\eta^P_h = \|u^{\varepsilon}_{1h} - u^{\varepsilon}_{2h}\|_{H^{1/2}_{00}(\Gamma)}. \qquad (11)$$

Our main a posteriori error estimate result, that shows the quasi-optimality of our error indicators, is the following (Cf. [2]):

Theorem 2. *Assume that the trace spaces W_{1h} and W_{2h} defined in (9) coincide. Then, the solution of the discrete penalty problem $(P_{h,\varepsilon})$ satisfies the following a posteriori error estimate*

$$C\,\eta_{h,\varepsilon} \leq \sum_{i=1}^{2} |u - u^{\varepsilon}_{ih}|_{1,\Omega_i} \leq C_h\,\eta_{h,\varepsilon},$$

with

$$\eta_{h,\varepsilon} = \eta^P_h + \left(\sum_{i=1}^{2} \sum_{K \in \mathscr{T}_{ih}} \left(\eta^K_i\right)^2 + h^2_K \|f - f_h\|^2_{L^2(K)} \right)^{1/2},$$

where C is a constant independent of h and ε, and $C_h \simeq h^{-1/2}$ is independent of ε.

Remark 3. This theorem prove the quasi-optimality of our error indicators η^P_h and η^D_h. The full optimality is obtained if $C_h = O(1)$, but we have found serious technical difficulties to prove it. However, as we can see in the next Section, the numerical results confirm this conjecture.

5 Numerical Experiments

In this section we analyze the efficiency of our error indicators, and build strategies to determine both optimal values for the penalty parameter and optimal grids to minimize the computational effort.

We work with the Poisson problem as model problem, in dimension $d = 2$ and in the case where the discrete spaces are made of piecewise affine functions ($k_1 = k_2 = 1$). All experiments are performed on the finite element code FreeFEM++.

To validate each numerical experiment we consider the L-shaped domain, $\Omega =]0,1[^2 \backslash [\frac{1}{2},1]^2$. It is divided into two sub-domains symmetric with respect to the straight line $x = y$. Furthermore, on this domain, we consider a Poisson problem with homogeneous Dirichlet boundary conditions with a known analytical solution, non-symmetric with respect to $x = y$.

Finally, as the scalar product of $H^{1/2}_{00}(\Gamma)$ defined in 2 is so difficult to compute, in the numerical experiments we have used instead, a discrete analogue, that it has been built using quadrature formulas (Cf. [4] for details).

The Efficiency of the Indicators

To test this efficiency we compare the error indicators η_h^P and η_h^D, with the errors $E^{\varepsilon}_{h,H^1} = \sum_{i=1}^{2} |\mathscr{I}_h u - u^{\varepsilon}_{ih}|_{H^1(\Omega_i)}$ and $E^{\varepsilon}_{h,L^2} = \sum_{i=1}^{2} \|\mathscr{I}_h u - u^{\varepsilon}_{ih}\|_{L^2(\Omega_i)}$, where $\mathscr{I}_h u$ is the $\mathbb{P}_1$-Lagrange interpolated of the exact solution u.

To verify first the efficiency of the penalty error indicator η_h^P we have realized some test fixing the mesh, and decreasing ε from 15 to 5×10^{-3}.

Figure 1 shows the test with $h = 1/64$, in the case of $H^{1/2}_{00}(\Gamma)$ penalty. We note that the error indicator η_h^P (in plain line) decreases with ε until the penalization error (which behaves like $c\,\varepsilon$) is of the same order as the finite element discretization error, for a critical value $\varepsilon_c \simeq 0.2$. For ε sufficiently larger than ε_c, the error indicator curve η_h^P is parallel to the curves of the errors (E^{ε}_{h,L^2} in dotted line and E^{ε}_{h,H^1} in dotted dashed line). In contrast, the quantity η_h^D (in dashed line) turns out to be fully independent of ε.

To test now the efficiency of the discrete error indicator η_h^D, we have fixed the penalty parameter ε and we have taken quasi-uniform meshes with grid size h decreasing from 0.25 to 1.5×10^{-2}.

Figure 2 show the test for $\varepsilon = 0.01$ also in the case of $H^{1/2}_{00}(\Gamma)$ penalty. We note the same qualitative behavior that in Figure 1, interchanging the roles of η_h^D and η_h^P.

Let us also remark that the behavior of indicators corresponding to L^2 penalty is similar.

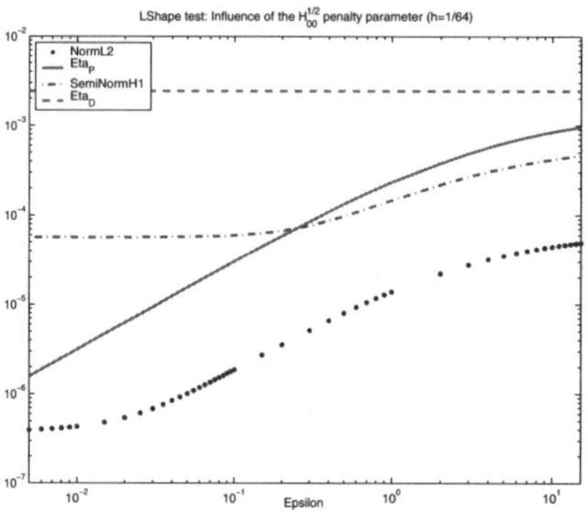

Fig. 1 Efficiency of the penalty error indicator.

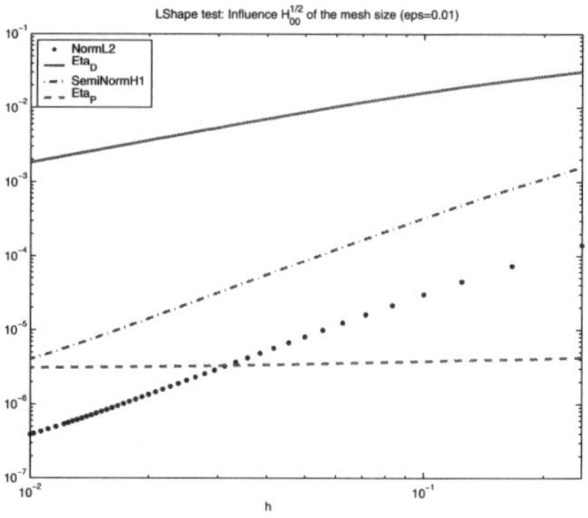

Fig. 2 Efficiency of the discrete error indicator.

Adaptivity Strategy and Penalty Parameter Optimization

Our purpose now is to develop a strategy to jointly determine an optimal value of the penalty parameter and an optimal adapted grid, to balance the penalty and the discretization errors and to set the global error below a given tolerance. These strategies are based upon the error indicators η_h^P and η_h^D and was introduced by Bernardi et al. in [3] in a complete different framework.

This strategy is made in three steps. First, we solve the full discrete problem $(P^n_{\varepsilon,h})$ with a current penalty error ε and a quasi-uniform mesh, and compute the error indicators η_h^P and η_h^D.

Step 1: Adaptation of the penalty parameter. If $\eta_h^P > \eta_h^D$, we divide ε by a constant number of times the η_h^D/η_h^P and go to **Step 2**.
If NOT, we go directly to **Step 2**.
Step 2: Adaptation of the mesh. We compute the all the η_i^K and their mean value $\overline{\eta}_h$. For all K such that η_i^K is larger than $\overline{\eta}_h$, we divide K into smaller triangles such that the diameters of these new elements behave like h_i^K multiplied by the ratio $\overline{\eta}_h/\eta_i^K$.
Step 3: Solution of $(P^n_{\varepsilon,h})$. We iterate the Schwarz procedure to obtain a refined solution of $(P^n_{\varepsilon,h})$.
Stop test: If $\max\{\eta_h^P, \eta_h^D\} \leq \eta$, stop.

We have developed some numerical test, where a targeted tolerance η is given and this strategy is used. Then, we have always achieved values of the errors indicators below this tolerance, for an optimal penalty parameter value and a final adapted mesh. We note that, with this strategy, the necessary CPU time to get this tolerance is drastically reduced with respect to the CPU time we need to give the computation, beginning from the final mesh, the optimal value of ε and an initial solution $u_h^{0,\varepsilon} = 0$.

Concretely, to get values of the errors indicators below $\eta = 0.003$, the CPU time is divided by approximately 4 to $H_{00}^{1/2}(\Gamma)$ penalty and 6 to $L^2(\Gamma)$ penalty. Then, the adaptivity strategy provides a large saving of computational effort for both $H_{00}^{1/2}(\Gamma)$ and $L^2(\Gamma)$ penalties.

Remark 4. If we compare this strategy for the $H_{00}^{1/2}(\Gamma)$ and $L^2(\Gamma)$ penalties, to the same value of η, we observe that the values of the estimators η_h^P and η_h^D and the number of triangles of the final grids are quite close for both penalties. Also, that the optimal ε for $L^2(\Gamma)$ penalty is close to the square of the optimal ε for $H_{00}^{1/2}(\Gamma)$ penalty. However, we obtain a large saving of CPU time for $L^2(\Gamma)$ penalty. This is due to two effects: The faster convergence of the $L^2(\Gamma)$ penalty, that requires much less iterations, and the cost due to the computation of the (even discrete) $H_{00}^{1/2}(\Gamma)$ norm.

Comparison with Other Domain Decomposition Method

We compare our penalty method with the domain decomposition method introduced by Agoshkov and Lebedev in 1985. In the case of two subdomains, the iteration procedure is the following (Cf. [8]): Starting from $u_1^0 = u_2^0 = 0$, we solve

$$\begin{cases} -\Delta u_1^{n+1/2} & = f & \text{in } \Omega_1, \\ u_1^{n+1/2} & = 0 & \text{on } \Gamma_1, \\ \frac{\partial u_1^{n+1/2}}{\partial \mathbf{n}} + p_n u_1^{n+1/2} & = RHS_1 & \text{on } \Gamma, \end{cases} \quad \begin{cases} -\Delta u_2^{n+1/2} & = f & \text{in } \Omega_2, \\ u_2^{n+1/2} & = 0 & \text{on } \Gamma_2, \\ -q_n \frac{\partial u_2^{n+1/2}}{\partial \mathbf{n}} + u_2^{n+1/2} & = RHS_2 & \text{on } \Gamma, \end{cases}$$

$$u_1^{n+1} = u_1^n + \alpha_{n+1}(u_1^{n+1/2} - u_1^n) \text{ in } \Omega_1; \; u_2^{n+1} = u_2^n + \beta_{n+1}(u_2^{n+1/2} - u_2^n) \text{ in } \Omega_2;$$

for $n = 0,1,2,3,...$, with $RHS_1 = \frac{\partial u_2^n}{\partial \mathbf{n}} + p_n u_2^n$ and $RHS_2 = -q_n \frac{\partial u_1^{n+1}}{\partial \mathbf{n}} + u_1^{n+1}$. Here $p_n \geq 0$, $q_n \geq 0$, α_{n+1} and β_{n+1} are free parameters. This algorithm is a generalization, in a wide sense, of the penalty L^2 version of our method to the parameters $\varepsilon = q_n = 1/p_n$. To compare the both methods, we have realized some test with the L-shaped domain fixing the mesh and comparing the results obtained to $\varepsilon = q_n = 1/p_n$ and the optimal choice of α_{n+1} and β_{n+1}. We have obtained that, the Agoshkov-Lebedev method convergence rate is independent of the mesh size with a small number of iterations, while our penalty method is mesh dependent and we need a larger number of iterations. However, for the same tests and measures, numerical experiments show that the discretization error obtained for our method is much better than that obtained for the Agoshkov-Lebedev method.

Acknowledgements This work has been partially funded by Spanish DIG grant MTM2006-01275.

References

1. Adams, R.A.: Sobolev Spaces. Pure and Applied Mathematics, Vol. 65., Academic Press, New York-London (1975)
2. Bernardi, C., Chacón Rebollo, T., Chacón Vera, E., Franco Coronil, D.: A posteriori error analysis for two non-overlapping domain decomposition techniques. Submitted to Appl. Numer. Math.
3. Bernardi, C., Girault, V., Hecht, F.: A posteriori analysis of a penalty method and application to the stokes problem. Math. Models Methods Appl. Sci. **13**(11), 1599–1628 (2003)
4. Casas, E., Raymond, J.P.: The stability in $w^{s,p}(\gamma)$ spaces of l^2-projections on some convex sets. Numer. Funct. Anal. Optim. **27**(2), 117–137 (2006)
5. Chacón Rebollo, T., Chacón Vera, E.: A non-overlapping domain decomposition method for the stokes equations via a penalty term on the interface. C. R. Math. Acad. Sci. Paris **334**(3), 221–226 (2002)
6. Chacón Rebollo, T., Chacón Vera, E.: Study of a non-overlapping domain decomposition method: Poisson and stokes problems. Appl. Numer. Math **48**(2), 169–194 (2004)
7. Otto, F.C., Lube, G.: A posteriori estimates for a non-overlapping domain decomposition method. Computing **62**(1), 27–43 (1999)
8. Quarteroni, A., Valli, A.: Domain decomposition methods for partial differential equations. Numerical Mathematics and Scientific Computation, Oxford Science Publications, New York-London (1999)

BETI-DP Methods in Unbounded Domains

C. Pechstein

Abstract This contribution deals with dual-primal boundary element tearing and interconnecting methods for Poisson-type problems in unbounded domains. To the best of our knowledge, no rigorous analysis of the corresponding preconditioners has been done yet. In the present paper we fill this theoretical gap and generalize the method to unbounded domains. Furthermore, we discuss implementational issues.

1 Introduction

Finite element tearing and interconnecting (FETI) methods are special non-overlapping domain decomposition methods designed for the parallel solution of large-scale systems of finite element equations. The classical one-level FETI method was introduced by Farhat and Roux [5]. Here, the finite element subspaces are given on each subdomain separately and the global continuity of the solution is enforced by Lagrange multipliers. Introducing a special projection, the original problem can be solved by a projected (deflated) preconditioned conjugate gradient (PCG) method in the Lagrange multiplier space. In each step of this PCG iteration one has to solve Dirichlet and Neumann problems locally on the subdomains. Klawonn and Widlund [6] proved that the condition number of the preconditioned system is bounded by $C(1+\log(H/h))^2$, where H is the subdomain diameter, h is the mesh size, and the constant C is independent of H, h and coefficient jumps. For a detailed description and for further references we refer to the monograph by Toselli and Widlund [19].

The principle of the FETI methods can be carried over to boundary-element approximations. The so-called boundary element tearing and interconnecting (BETI) methods were introduced and analyzed by Langer and Steinbach [10]. The same

Clemens Pechstein
Spezialforschungsbereich F013, Johannes Kepler Universität, Altenberger Straße 69, 4040 Linz, Austria, e-mail: clemens.pechstein@numa.uni-linz.ac.at

authors used the tearing and interconnecting principle to couple finite elements and boundary elements in hybrid domain decomposition methods [11].

One issue of one-level FETI methods is the special projection owing to the fact that the local Neumann problems for the Laplace operator on floating subdomains are not uniquely solvable. In linear elasticity, the null spaces of the local Neumann problems which depend on the boundary conditions are often difficult to characterize which complicates the FETI projection. That was one of the main reasons for the development of the dual-primal FETI (FETI-DP) methods introduced by Farhat et. al. [3] and analyzed in two dimensions by Mandel and Tezaur [12]. Algorithms for the three-dimensional case were contributed by Farhat, Lesoinne, and Pierson [4], see also [15]. Finally, a rigorous analysis was given by Klawonn, Widlund, and Dryja [7]. See also Brenner and He [2].

The dual-primal BETI (BETI-DP) methods were briefly introduced by Langer, Pohoaţă, and Steinbach [9]. To our knowledge, however, neither any analysis has been given, nor any efficient implementation has been discussed until now. The aim of the present contribution is to fill this gap and to generalize the BETI-DP methods to the case of unbounded domains. In particular, we prove a condition number bound which is similar to the existing ones for FETI-DP methods. It coincides with the best bound that we could obtain for one-level BETI methods in unbounded domains, see our earlier work [14]. In contrast, our BETI-DP bound is completely independent of the geometric constellation of the subdomains and the Dirichlet boundary.

Let $\Omega \subset \mathbb{R}^d$ (with $d = 2$ or 3) be an open unbounded domain with its complement $\mathbb{R}^d \setminus \overline{\Omega}$ being bounded. We state more specific assumptions on Ω in Section 2. The boundary $\Gamma = \partial\Omega$ is composed into a Dirichlet boundary Γ_D and a Neumann boundary Γ_N, with $\Gamma = \Gamma_D \cup \Gamma_N$ and $\Gamma_D \cap \Gamma_N = \emptyset$. Our model problem reads as follows: Find $u \in H^1_{\mathrm{loc}}(\Omega)$ such that

$$-\nabla \cdot [\alpha \nabla u] = 0 \quad \text{in } \Omega\,, \qquad u = g_D \quad \text{on } \Gamma_D\,, \qquad \alpha \frac{\partial u}{\partial n} = g_N \quad \text{on } \Gamma_N\,, \tag{1}$$

where $\alpha(x)$ is a piecewise constant coefficient and $n(x)$ is the unit normal vector on Γ pointing outside of Ω.

In Section 2 we introduce basic notation, domain decomposition, discretization, and the fundamentals of boundary element methods. Section 3 is devoted to the formulation of BETI-DP methods. In Section 4 we finally discuss implementational issues and give the condition number estimate for the BETI-DP preconditioner.

2 Preliminaries

Basic notation. Generically, we denote the dual of a Banach space V by V^*, and the duality pairing between V^* and V by $\langle \cdot, \cdot \rangle$. We denote the adjoint of a linear operator $T : V \to W^*$ by $T^\top : W \to V^*$. For a Lipschitz domain D, the Sobolev spaces $H^1(D)$, $H^1_{\mathrm{loc}}(D)$, $H^{1/2}(\partial D)$, and $H^{-1/2}(\partial D)$ are defined as usual, see, e. g., [13, 14].

Domain decomposition. The domain Ω is decomposed into $p+1$ open, non-overlapping subdomains Ω_i such that $\overline{\Omega} = \bigcup_{i=0}^{p} \overline{\Omega_i}$ and $\Omega_i \cap \Omega_j = \emptyset$ for $i \neq j$. We assume that the subdomain Ω_0 is unbounded and its complement $\mathbb{R}^d \setminus \overline{\Omega_0}$ is bounded and contractible. The remaining subdomains are assumed to be bounded and contractible. We define the local boundaries $\Gamma_i := \partial\Omega_i$ and the unit normal vectors n_i to Γ_i which point to the outside of Ω_i. In particular, n_0 points to the interior of $\mathbb{R}^d \setminus \overline{\Omega_0}$. Furthermore, we set $H_0 := \operatorname{diam}(\mathbb{R}^d \setminus \Omega_0)$ and $H_i := \operatorname{diam}\Omega_i$ for $i = 1,\dots,p$. According to [19] we define the *skeleton* $\Gamma_S := \bigcup_{i=0}^{p} \Gamma_i$, the local interfaces $\Gamma_{ij} := \Gamma_i \cap \Gamma_j$, and the *interface* $\Gamma_I := \bigcup_{i,j=0}^{p} \Gamma_{ij}$. The interface Γ_I is the union of (topological) subdomain vertices $V \in \mathscr{V}$, edges $E \in \mathscr{E}$ and faces $F \in \mathscr{F}$.

Finally, we assume that the coefficient $\alpha(x)$ is piecewise constant on the subdomains with $\alpha_{|\Omega_i} = \alpha_i = \text{const} > 0$.

Discretization. We consider a triangulation $\mathscr{T}(\Gamma_S)$ of the skeleton Γ_S into simplical elements $T \in \mathscr{T}(\Gamma_S)$ and denote its restriction on Γ_i by $\mathscr{T}(\Gamma_i)$. The sets of nodes of these triangulations are denoted by $\Gamma_{S,h}$ and $\Gamma_{i,h}$. On Γ_S we define the following spaces of piecewise linear functions,

$$V^h(\Gamma_S) := \{v \in \mathscr{C}(\Gamma_S) : v_{|T} \in \mathscr{P}^1(T) \text{ for all } T \in \mathscr{T}(\Gamma_S)\},$$
$$V_D^h(\Gamma_S) := \{v \in V^h(\Gamma_S) : v_{|\Gamma_D} = 0\}.$$

Throughout the paper, we assume that we can find an extension $\widetilde{g}_D \in V^h(\Gamma_S)$ with $\widetilde{g}_D|_{\Gamma_D} = g_D$, and that $\widetilde{g}_N \in H^{-1/2}$ defined by extending g_N by zero fulfills the condition $g_N \in V_D^h(\Gamma_S)^*$. In the same manner we define the spaces $V^h(\Gamma_i)$ and $V_D^h(\Gamma_i)$.

Boundary integral operators. The fundamental solution of the Laplace equation is given by $\mathscr{U}^*(x,y) = -\frac{1}{2\pi}\log|x-y|$ for $d=2$ and $\mathscr{U}^*(x,y) = \frac{1}{2\pi}|x-y|^{-1}$ for $d=3$. On each subdomain boundary Γ_i, for $i = 0,\dots,p$, we define the single layer potential operator $\mathscr{V}_i$, the double layer potential operator $\mathscr{K}_i$, and the hypersingular integral operator $\mathscr{D}_i$ by

$$\begin{aligned}
\mathscr{V}_i : H^{-1/2}(\Gamma_i) \to H^{1/2}(\Gamma_i) : \quad & (\mathscr{V}_i t)(x) := \alpha_i \int_{\Gamma_i} \mathscr{U}^*(x,y)\, t(y)\, ds_y, \\
\mathscr{K}_i : H^{1/2}(\Gamma_i) \to H^{1/2}(\Gamma_i) : \quad & (\mathscr{K}_i u)(x) := \alpha_i \int_{\Gamma_i} \frac{\partial}{\partial n_{i,y}} \mathscr{U}^*(x,y)\, u(y)\, ds_y, \\
\mathscr{D}_i : H^{1/2}(\Gamma_i) \to H^{-1/2}(\Gamma_i) : \quad & (\mathscr{D}_i u)(x) := -\alpha_i \frac{\partial}{\partial n_{i,x}} \int_{\Gamma_i} \frac{\partial}{\partial n_{i,y}} \mathscr{U}^*(x,y)\, u(y)\, ds_y,
\end{aligned} \tag{2}$$

where $x \in \Gamma_i$. Throughout this work, we assume that in two dimensions $H_i < 1$ for all $i = 0,\dots,p$, which can always be obtained by a simple coordinate scaling of the domain Ω. This assumption assures that all the single layer potential operators $\mathscr{V}_i$ are elliptic and therefore invertible. For details see, e. g., [18].

Introducing the space of piecewise constant functions

$$Z^h(\Gamma_i) := \{t \in H^{-1/2}(\Gamma_i) : t_{|T} = \text{const for all } T \in \mathscr{T}(\Gamma_i)\},$$

we define the following discrete boundary integral operators:

$$\begin{aligned}
V_i &: Z^h(\Gamma_i) \to Z^h(\Gamma_i)^* : \langle V_i t, s\rangle = \langle \mathscr{V}_i t, s\rangle \qquad \forall t, s \in Z^h(\Gamma_i),\\
K_i &: V^h(\Gamma_i) \to Z^h(\Gamma_i)^* : \langle t, K_i v\rangle = \langle t, \mathscr{K}_i v\rangle \qquad \forall v \in V^h(\Gamma_i)\,\forall t \in Z^h(\Gamma_i),\\
D_i &: V^h(\Gamma_i) \to V^h(\Gamma_i)^* : \langle D_i v, w\rangle = \langle \mathscr{D}_i v, w\rangle \qquad \forall v, w \in V^h(\Gamma_i),\\
M_i &: V^h(\Gamma_i) \to Z^h(\Gamma_i)^* : \langle t, M_i v\rangle = \langle t, v\rangle \qquad \forall v \in V^h(\Gamma_i)\,\forall t \in Z^h(\Gamma_i).
\end{aligned} \tag{3}$$

Note, that M_i is the identity as an operator. However, its matrix representation is not diagonal but sparse. The remaining matrices can be represented in data-sparse form by the usual boundary element compression techniques, such as the fast multipole method, ACA, HCA, wavelets, etc., see, e. g., [18] and the references therein.

Finally, we define the Steklov-Poincaré operators

$$\mathscr{S}_0 := \mathscr{V}_0^{-1}(\tfrac{\alpha_0}{2}I - K_0), \qquad \mathscr{S}_i := \mathscr{V}_i^{-1}(\tfrac{\alpha_i}{2}I + K_i) \quad \text{for } i = 1,\ldots,p, \tag{4}$$

which are the Dirichlet-to-Neumann maps for the Laplace problem on the subdomains Ω_i, and their discrete symmetric approximations

$$S_i := D_i + \overline{K}_i^{\top} V_i^{-1} \overline{K}_i \qquad \text{for } i = 0,\ldots,p, \tag{5}$$

where $\overline{K}_0 := \frac{\alpha_0}{2}M_0 - K_0$ and $\overline{K}_i := \frac{\alpha_i}{2}M_i + K_i$ for $i = 1,\ldots,p$, cf. [10, 17]. Here, the operators $\mathscr{S}_0$ and S_0 correspond to the exterior problem in Ω_0.

Skeleton variational formulation. As for many non-overlapping domain decomposition methods we use the following formulation for the model problem (1):

$$\text{Find } u_h^{(0)} \in V_D^h(\Gamma_S): \quad \sum_{i=0}^{p} \langle S_i R_i u_h^{(0)}, R_i v_h\rangle = \sum_{i=0}^{p} \langle f_i, R_i v_h\rangle \quad \forall v_h \in V_D^h(\Gamma_S), \tag{6}$$

with $f_i = R_i^* g_N - S_i R_i g_D$, where $R_i : V_D^h(\Gamma_S) \to V_D^h(\Gamma_i)$ and $R_i^* : V_D^h(\Gamma_S)^* \to V_D^h(\Gamma_i)^*$ are suitable restriction operators. Once the solution $u_h = \widetilde{g}_D + u_h^{(0)}$ of (6) is known, the approximate solution of (1) is characterized using the representation formula involving $\mathscr{U}^*(x, y)$ on each subdomain seperately. For details see [10, 17].

3 Formulation of BETI-DP Methods

We start with the minimization problem

$$\min_{u_h^{(0)} \in V_D^h(\Gamma_S)} \sum_{i=0}^{p} \Big[\tfrac{1}{2}\langle S_i R_i u_h^{(0)}, R_i u_h\rangle - \langle f_i, R_i u_h\rangle\Big], \tag{7}$$

which is equivalent to (6). According to the tearing and interconnecting principle we introduce new unknowns u_i for $R_i u_h^{(0)}$ and write $u = [u_i]_{i=0,\ldots,p}$. We define the spaces

$W_i := V_D^h(\Gamma_i)$ and $W := \prod_{i=0}^{p} W_i$ and regard the local Steklov-Poincaré operators S_i as operators mapping W_i to W_i^*. Furthermore, we define $S := \operatorname{diag}(S_i)$, such that $S : W \to W^*$. In general, the functions in W are discontinuous across the subdomain interfaces.

As in all dual-primal methods, we work with subspaces $\widetilde{W} \subset W$ for which sufficiently many continuity constraints are enforced such that the block operator S is SPD on $\widetilde{W}$. Such spaces are constructed as follows. We choose a *primal space* $\widehat{W}_\Pi \subset V_D^h(\Gamma_S)$ and a *dual subspace* $\widetilde{W}_\Delta \subset W$ such that $\widetilde{W} = \widehat{W}_\Pi \oplus \widetilde{W}_\Delta$. Note, that for simplicity we identify continuous functions from $V^h(\Gamma_S)$ with the corresponding ones in the product space W. We denote the i-th component of the product space $\widetilde{W}_\Delta$ by $\widetilde{W}_{\Delta,i}$. Due to space limits, we refer to [19, Algorithms B, C] for particular choices of $\widetilde{W}_\Delta$ and $\widehat{W}_\Pi$. We only mention that for three-dimensional problems one has to add at least edge average constraints to $\widehat{W}_\Pi$ to obtain robustness in h.

We are now ready to formulate the BETI-DP algorithms. Depending on the choice of $\widetilde{W}_\Delta$ and $\widehat{W}_\Pi$, we define the Schur complement $\widetilde{S} : \widetilde{W}_\Delta \to \widetilde{W}_\Delta^*$ by

$$\widetilde{S} := S_\Delta - S_{\Delta\Pi} S_\Pi^{-1} S_{\Pi\Delta}\,, \tag{8}$$

where the block operators $S_\Delta : \widetilde{W}_\Delta \to \widetilde{W}_\Delta^*$, $S_{\Pi\Delta} : \widehat{W}_\Pi \to \widetilde{W}_\Delta^*$, $S_{\Delta\Pi} : \widetilde{W}_\Delta \to \widehat{W}_\Pi^*$ and $S_\Pi : \widehat{W}_\Pi \to \widehat{W}_\Pi^*$ are the Galerkin projections of S to the corresponding spaces. Note, that S_Π is SPD and that $\widetilde{S}$ fulfills the minimizing property

$$\langle \widetilde{S} w_\Delta, w_\Delta \rangle = \min_{w_\Pi \in \widehat{W}_\Pi} \langle S(w_\Delta + w_\Pi), w_\Delta + w_\Pi \rangle$$

Furthermore, we define the reduced right hand side $\widetilde{f} := f_\Delta - S_{\Delta\Pi} S_\Pi^{-1} f_\Pi \in \widetilde{W}_\Delta^*$, where $f_\Delta \in \widetilde{W}_\Delta$ and $f_\Pi \in \widehat{W}_\Pi$ are the corresponding projections of f.

In the following, we incorporate the continuity constraints on the interface but—in contrast to one-level methods—only for the degrees of freedom (dofs) in $\widetilde{W}_\Delta$. These constraints can be summarized by the equation

$$B_\Delta u_\Delta = 0\,,$$

with the jump operator $B_\Delta : \widetilde{W}_\Delta \to \mathbb{R}^M$, where each line corresponds to a constraint of the form $u_i(x) - u_j(x) = 0$ for some $x \in \Gamma_{i,h} \cap \Gamma_{j,h}$, at least in the case of nodal dofs. According to [19], we restrict ourselves to fully redundant constraints and for the sake of simplicity we set $U_\Delta = \operatorname{range} B_\Delta$ which implies $\ker B_\Delta^\top = \{0\}$. In that part of our analysis we identify the space $(\mathbb{R}^M)^*$ with $\mathbb{R}^M$ and U_Δ^* with U_Δ.

With these definitions, we arrive at the minimization problem

$$\min_{u_\Delta \in \widetilde{W}_\Delta\,,\, B_\Delta u_\Delta = 0} \quad \sum_{i=0}^{p} \left[\tfrac{1}{2} \langle \widetilde{S} u_\Delta, u_\Delta \rangle - \langle \widetilde{f}, u_\Delta \rangle \right], \tag{9}$$

which is equivalent to (7). Suppose we have the solution u_Δ of (9), then the overall solution u is given by $u = u_\Delta + u_\Pi$ with $u_\Pi = S_\Pi^{-1}(f_\Pi - S_{\Pi\Delta} u_\Delta)$. Above minimization problem is equivalent to the following saddle point problem:

$$\text{Find } u_\Delta \in \widetilde{W}_\Delta,\, \lambda \in U_\Delta : \quad \begin{pmatrix} \widetilde{S} & B_\Delta^\top \\ B_\Delta & 0 \end{pmatrix} \begin{pmatrix} u_\Delta \\ \lambda \end{pmatrix} = \begin{pmatrix} \widetilde{f} \\ 0 \end{pmatrix}.$$

As an important observation $\widetilde{S}$ is SPD on $\widetilde{W}_\Delta$, and so the inverse $\widetilde{S}^{-1}$ exists. With the definitions $F := B_\Delta \widetilde{S}^{-1} B_\Delta^\top : U_\Delta \to U_\Delta$ and $d := B_\Delta \widetilde{S}^{-1} \widetilde{f} \in U_\Delta$, above saddle point problem reduces to

$$\text{find } \lambda \in U_\Delta : \qquad F\lambda = d. \tag{10}$$

We see that F is SPD on U_Δ. Hence, λ can be computed using a PCG method. The preconditioner $M^{-1} : U_\Delta \to U_\Delta$ is chosen as

$$M^{-1} := B_{D,\Delta}\, S_\Delta\, B_{D,\Delta}^\top, \tag{11}$$

where $B_{D,\Delta} : W_\Delta^* \to U_\Delta$ is a scaled jump operator which takes heterogeneous coefficients into account, cf. [14, 19]. Instead of S_Δ, one can choose the spectrally equivalent block hypersingular operator $D_\Delta := \operatorname{diag}(D_{i,\Delta})$.

4 Implementation and Analysis

In the following we discuss some implementational issues.

- For the implementation of three-dimensional methods one needs to handle the edge and possibly face average constraints. One way is computing with the usual basis and additional constraints, cf. [16]. In connection with hierarchical matrices, it might be possible to construct the matrix representations of V_i, K_i and D_i directly for a new basis with the edge/face averages incorporated.
- The realization of $S_{i,\Delta}$ which is the restriction of S_i to $W_{i,\Delta}$ is performed according to definition (5) of S_i by application of a few boundary element matrices and solving an equation with the single layer potential operator V_i.
- Realization of S_Π^{-1}: The matrix representing S_Π is sparse and can be assembled and factorized once in memory.
- For computing $v_{i,\Delta} = S_{i,\Delta}^{-1} g_{i,\Delta}$ we solve instead the local saddle point problem

$$\begin{pmatrix} D_{i,\Delta} & \overline{K}_{i,\Delta}^\top \\ \overline{K}_{i,\Delta} & -V_i \end{pmatrix} \begin{pmatrix} v_{i,\Delta} \\ t \end{pmatrix} = \begin{pmatrix} w_{g,\Delta} \\ 0 \end{pmatrix}. \tag{12}$$

- Realization of $\widetilde{S}^{-1}$: Computing $v_\Delta = \widetilde{S}^{-1} g_\Delta$ is equivalent to solving

$$\begin{pmatrix} S_\Delta & S_{\Delta\Pi} \\ S_{\Pi\Delta} & S_\Pi \end{pmatrix} \begin{pmatrix} v_\Delta \\ v_\Pi \end{pmatrix} = \begin{pmatrix} g_\Delta \\ 0 \end{pmatrix}. \tag{13}$$

We use the block factorization

$$\begin{pmatrix} S_\Delta & S_{\Delta\Pi} \\ S_{\Pi\Delta} & S_\Pi \end{pmatrix}^{-1} = \begin{pmatrix} I_\Delta & -S_\Delta^{-1}S_{\Delta\Pi} \\ 0 & I_\Pi \end{pmatrix} \begin{pmatrix} S_\Delta^{-1} & 0 \\ 0 & \widetilde{S}_\Pi^{-1} \end{pmatrix} \begin{pmatrix} I_\Delta & 0 \\ -S_{\Pi\Delta}S_\Delta^{-1} & I_\Pi \end{pmatrix}, \quad (14)$$

where $\widetilde{S}_\Pi = S_\Pi - S_{\Pi\Delta}S_\Delta^{-1}S_{\Delta\Pi}$. If $\widetilde{S}_\Pi$ is assembled (by solving as many local problems per subdomain as there are primal dofs) and factorized once in memory, the application of S^{-1} reduces essentially to the application of S_Δ^{-1}, see above.
- Computing u from λ: We solve (13) with $g_\Delta = \widetilde{f} - B_\Delta^\top \lambda$ and set $u = [v_\Delta, v_\Pi]$.

Lemma 1. *If $\mathscr{H}$-matrix techniques are used to approximate the boundary integral operators V_i, K_i, and D_i, and if the $\mathscr{H}$-LU factorization is used to obtain a fast approximate application of $\mathscr{V}_i^{-1}$ and the system matrix in (12), the application of $S_{i,\Delta}$ and $S_{i,\Delta}^{-1}$ can be performed in quasi-optimal complexity. This implies that for each PCG step the number of arithmetical operations is $\mathscr{O}(C_{\mathscr{H}}\,(H_i/h_i)^{d-1}\log(H_i/h_i))$ in a parallel scheme, where $C_{\mathscr{H}}$ depends on the rank in the $\mathscr{H}$-matrix approximation.*

Remark 1. The algorithm will work well if the number $N_{\Pi,0}$ of primal dofs on the exterior subdomain boundary Γ_0 stays small. If $N_{\Pi,0}$ grows, alone assembling the matrix $\widetilde{S}_\Pi$ involves $N_{\Pi,0}$ local solves on Γ_0. This exactly happens when the number of neighboring subdomains of Ω_0 becomes large, and it usually means that also the number of nodal degrees of freedom on Γ_0 is large compared to those on the remaining subdomains Γ_i, which besides effects the sparsity of S_Π and $\widetilde{S}_\Pi$. In that case a sub-parallelization of the operators acting on Γ_0 might be the only way to regain load balancing in a parallel scheme. We believe, however, that this is only possible using inexact BETI-DP methods in the spirit of [8], where the solution of (12) in each PCG step is avoided.

The following theorem states the quasi-optimality of the BETI-DP preconditioner and relies on a typical subdomain regularity assumption [19, Assumption 4.3].

Theorem 1. *First, assume that the partition $\{\Omega_i\}_{i=1}^p$ and the Dirichlet boundary Γ_D fulfill [19, Assumption 4.3]. Secondly, assume that the interior of Γ_0 is the image of a shape regular polygon under a sufficiently smooth map. Finally, let the spaces $\widehat{W}_\Pi$ and $\widetilde{W}_\Delta$ be defined according to [19, Algorithm B or C]. Then the BETI-DP preconditioner (11) fulfills the following condition number estimate*

$$\kappa(M^{-1}F) \le C \max_{i=1}^{p} (1 + \log(H_i/h_i))^2 ,$$

where the constant C is independent of H_i, h_i and the values α_i. We point out that $H_0 = diam\,(int(\Gamma_0))$ does not enter the estimate.

Proof. The detailed proof can be found in [14]. It follows the line of the proof given in [7, 19]. In contrast to the theory of one-level BETI methods in unbounded domains [14] we do not need anything like an extension indicator. The needed tools are spectral equivalence relations between the boundary element approximations S_i, the Steklov-Poincaré operators $\mathscr{S}_i$, and the finite element approximations of them.

The method described in this contribution can easily be generalized to hybrid FETI/BETI-DP methods in unbounded domains, cf. [14] and coupled with interface concentrated FEM approximations, cf. [1].

Acknowledgements The Austrian Science Funds (FWF) supported the author under grant F1306.

References

1. Beuchler, S., Eibner, T., Langer, U.: Primal and dual interface concentrated iterative substructuring methods. SIAM J. Num. Anal. To appear, preprint available at http://www.ricam.oeaw.ac.at/ publications/reports/07/rep07-07.pdf
2. Brenner, S.C., He, Q.: Lower bounds for three-dimensional nonoverlapping domain decomposition algorithms. Numer. Math. **93**(3), 445–470 (2003)
3. Farhat, C., Lesoinne, M., Le Tallec, P., Pierson, K., Rixen, D.: FETI-DP: A dual-primal unified FETI method I: A faster alternative to the two-level FETI method. Internat. J. Numer. Methods Engrg. **50**, 1523–1544 (2001)
4. Farhat, C., Lesoinne, M., Pierson, K.: A scalable dual-primal domain decomposition method. Numer. Linear Algebra Appl. **7**, 687–714 (2000)
5. Farhat, C., Roux, F.X.: A method of finite element tearing and interconnecting and its parallel solution algorithm. Int. J. Numer. Meth. Engrg. **32**, 1205–1227 (1991)
6. Klawonn, A., Widlund, O.: FETI and Neumann-Neumann iterative substructuring methods: connections and new results. Comm. Pure Appl. Math. **54**(1), 57–90 (2001)
7. Klawonn, A., Widlund, O., Dryja, M.: Dual-primal FETI methods for three-dimensional elliptic problems with heterogeneous coefficients. SIAM J. Numer. Anal. **40**(1), 159–179 (2002)
8. Langer, U., Of, G., Steinbach, O., Zulehner, W.: Inexact data-sparse boundary element tearing and interconnecting methods. SIAM J. Sci. Comp. **29**, 290–314 (2007)
9. Langer, U., Pohoaţă, A., Steinbach, O.: Dual-primal boundary element tearing and interconnecting methods. Technical report 2005/6, Institute for Computational Mathematics, University of Technology, Graz (2005)
10. Langer, U., Steinbach, O.: Boundary element tearing and interconnecting method. Computing **71**(3), 205–228 (2003)
11. Langer, U., Steinbach, O.: Coupled boundary and finite element tearing and interconnecting methods. In: Lecture Notes in Computational Sciences and Engineering, vol. 40, pp. 83–97. Springer, Heidelberg (2004). Proceedings of the DD15 conference, Berlin, July 2003
12. Mandel, J., Tezaur, R.: On the convergence of a dual-primal substructuring method. Numer. Math. **88**, 543–558 (2001)
13. McLean, W.: Strongly Elliptic Systems and Boundary Integral Equations. Cambridge University Press, Cambridge, UK (2000)
14. Pechstein, C.: Analysis of dual and dual-primal tearing and interconnecting methods in unbounded domains. SFB-Report 2007-15, Johannes Kepler University, Linz (2007). http://www.sfb013.uni-linz.ac.at/ reports/2007/pdf-files/ rep_07-15_pechstein.pdf
15. Pierson, K.H.: A family of domain decomposition methods for the massively parallel solution of computational mechanics problems. Ph.D. thesis, Aerospace Engineering, University of Colorado at Boulder, Boulder, CO (2000)
16. Rheinbach, O.: Parallel scalable iterative substructuring: Robust exact and inexact FETI-DP methods with applications to elasticity. Ph.D. thesis, Universität Essen-Duisburg (2006)
17. Steinbach, O.: Stability estimates for hybrid coupled domain decomposition methods, *Lecture Notes in Mathematics*, vol. 1809. Springer, Heidelberg (2003)
18. Steinbach, O.: Numerical Approximation Methods for Elliptic Boundary Value Problems. Finite and Boundary Elements. Springer, New York (2008)
19. Toselli, A., Widlund, O.: Domain Decoposition Methods – Algorithms and Theory, *Springer Series in Computational Mathematics*, vol. 34. Springer, Berlin, Heidelberg (2005)

Domain Decomposition and Model Reduction of Systems with Local Nonlinearities

K. Sun, R. Glowinski, M. Heinkenschloss, and D.C. Sorensen

Abstract The goal of this paper is to combine balanced truncation model reduction and domain decomposition to derive reduced order models with guaranteed error bounds for systems of discretized partial differential equations (PDEs) with a spatially localized nonlinearities. Domain decomposition techniques are used to divide the problem into linear subproblems and small nonlinear subproblems. Balanced truncation is applied to the linear subproblems with inputs and outputs determined by the original in- and outputs as well as the interface conditions between the subproblems. The potential of this approach is demonstrated for a model problem.

1 Introduction

Model reduction seeks to replace a large-scale system of differential equations by a system of substantially lower dimension that has nearly the same response characteristics. This paper is concerned with model reduction of systems of discretized partial differential equations (PDEs) with spatially localized nonlinearities. In particular, we are interested in constructing reduced order models for which the error between the input-to-output map of the original system and the input-to-output map of the reduced order model can be controlled.

Balanced truncation is a particular model reduction technique due to [16], which for linear time invariant systems leads to reduced order models which approximate the original input-to-output map with a user controlled error [1, 6]. Although extensions of balanced truncation to nonlinear systems have been proposed, see, e.g.,

Kai Sun, Matthias Heinkenschloss, Danny C. Sorensen
Department of Computational and Applied Mathematics, MS-134, Rice University, 6100 Main Street, Houston, Texas 77005-1892, USA. e-mail: kleinsun/heinken/sorensen@rice.edu

Roland Glowinski
University of Houston, Department of Mathematics, 651 P. G. Hoffman Hall, Houston, Texas 77204-3008, USA. e-mail: roland@math.uh.edu

[9, 14], there are no bounds available for the error between the input-to-output map of the original system and that of the reduced order model. Proper Orthogonal Decomposition (POD) is often used for model reduction of nonlinear systems. Error bounds are available for the error between the so-called snapshots and the reduced order model, see, e.g., [11, 13], but no bounds for the error between the input-to-output map of the original system and that of the reduced order model, unless the so-called snapshot set reflects all possible inputs.

Our approach uses domain decomposition techniques to divide the problem into linear subproblems and small nonlinear subproblems. Balanced truncation is applied only to the linear subproblems with inputs and outputs determined by the original in- and outputs as well as the interface conditions between the subproblems. We expect that this combination of domain decomposition and balanced truncation leads to a substantial reduction of the original problem if the nonlinearities are localized, i.e., the nonlinear subproblems are small relative to the other subdomains, and if the interfaces between the subproblems are relatively small.

To keep our presentation brief, we consider a model problem which couples the 1D Burgers equation to two heat equations. This is motivated by problems in which one is primarily interested in a nonlinear PDE which is posed on a subdomain and which is coupled to linear PDEs on surrounding, larger subdomains. The linear PDE solution on the surrounding subdomains needs to be computed accurately enough to provide acceptable boundary conditions for the nonlinear problem on the 'inner' subdomain. Such situations arise, e.g., in regional air quality models.

Our work is also related to [4], which is an example paper which discusses the coupling of linear and nonlinear PDEs, but no dimension reduction is applied. Domain decomposition and POD model reduction for flow problems with moving shocks are discussed in [15]. POD model reduction is applied on the subdomains away from the shock. The paper [18] discusses a different model reduction technique for second order dynamical systems with localized nonlinearities. The papers [2, 5] and [20] discuss different model reduction and substructuring techniques for second order dynamical systems and model reduction of interconnect systems respectively.

2 The Model Problem

Let $\overline{\Omega} = \bigcup_{k=1}^{3} \overline{\Omega}_k$, where $\Omega_1 = (-10,-1)$, $\Omega_2 = (-1,1)$ and $\Omega_3 = (1,10)$ and let $T > 0$ be given. Our model problem is given by

$$\rho_k \frac{\partial y_k}{\partial t}(x,t) - \mu_k \frac{\partial^2 y_k}{\partial x^2}(x,t) = S_k(x,t), \qquad (x,t) \in \Omega_k \times (0,T), \tag{1a}$$

$$y_k(x,0) = y_{k0}(x), \qquad x \in \Omega_k,\ k = 1,3, \tag{1b}$$

$$\frac{\partial y_1}{\partial x}(-10,t) = 0, \quad \frac{\partial y_3}{\partial x}(10,t) = 0 \qquad t \in (0,T), \tag{1c}$$

$$\rho_2 \frac{\partial y_2}{\partial t}(x,t) - \mu_2 \frac{\partial^2 y_2}{\partial x^2} + y_2 \frac{\partial y_2}{\partial x}(x,t) = 0, \qquad (x,t) \in \Omega_2 \times (0,T), \tag{1d}$$

$$y_2(x,0) = y_{20}(x), \qquad x \in \Omega_2, \tag{1e}$$

with the following interface conditions

$$y_1(-1,t) = y_2(-1,t), \qquad y_2(1,t) = y_3(1,t), \qquad t \in (0,T), \tag{2a}$$

$$\mu_1 \frac{\partial y_1}{\partial x}(-1,t) = \mu_2 \frac{\partial y_2}{\partial x}(-1,t), \quad \mu_2 \frac{\partial y_2}{\partial x}(1,t) = \mu_3 \frac{\partial y_3}{\partial x}(1,t), \quad t \in (0,T). \tag{2b}$$

We assume that the forcing functions S_1, S_3 are given by

$$S_k = \sum_{i=1}^{n_s} b_{ik}(x) u_{ik}(t), \quad k = 1,3. \tag{3}$$

To obtain the weak form of (1) and (2), we multiply the differential equations (1a, d) by test functions $v_i \in H^1(\Omega_i)$, $i = 1,2,3$, respectively, integrate over Ω_i, and apply integration by parts. Using the boundary conditions (1c, h) this leads to

$$\rho_k \frac{d}{dt} \int_{\Omega_k} y_k v_k dx + \mu_k \int_{\Omega_k} \frac{\partial y_k}{\partial x} \frac{\partial v_k}{\partial x} dx - \mu_k \frac{\partial y_k}{\partial x} v_k \Big|_{\partial \Omega_k} = \int_{\Omega_k} S_k v_k dx, \; k = 1,3, \tag{4a}$$

$$\rho_2 \frac{d}{dt} \int_{\Omega_2} y_2 v_2 dx + \mu_2 \int_{\Omega_2} \frac{\partial y_2}{\partial x} \frac{\partial v_2}{\partial x} dx + \int_{\Omega_2} \frac{\partial y_2}{\partial x} y_2 v_2 dx - \mu_2 \frac{\partial y_2}{\partial x} v_2 \Big|_{-1}^{1} = 0. \tag{4b}$$

If $v_k \in H^1(\Omega_k)$, $k = 1,3$, satisfy $v_1(-1) = 1$, $v_3(1) = 1$, then (1c), (4a) imply

$$\mu_1 \frac{\partial y_1(-1)}{\partial x} = -\int_{\Omega_1} S_1 v_1 dx + \rho_1 \frac{\partial}{\partial t} \int_{\Omega_1} y_1 v_1 dx + \mu_1 \int_{\Omega_1} \frac{\partial y_1}{\partial x} \frac{\partial v_1}{\partial x} dx, \tag{5a}$$

$$\mu_3 \frac{\partial y_3(1)}{\partial x} = \int_{\Omega_3} S_3 v_3 dx - \rho_3 \frac{\partial}{\partial t} \int_{\Omega_3} y_3 v_3 dx - \mu_3 \int_{\Omega_3} \frac{\partial y_3}{\partial x} \frac{\partial v_3}{\partial x} dx. \tag{5b}$$

If $v_2 \in H^1(\Omega_2)$ satisfies $v_2(-1) = 1$ and $v_2(1) = 0$, then (4b) implies

$$\mu_2 \frac{\partial y_2(-1)}{\partial x} = -\rho_2 \frac{\partial}{\partial t} \int_{\Omega_2} y_2 v_2 dx - \mu_2 \int_{\Omega_2} \frac{\partial y_2}{\partial x} \frac{\partial v_2}{\partial x} dx - \int_{\Omega_2} \frac{\partial y_2}{\partial x} y_2 v_2 dx. \tag{5c}$$

Finally, if $v_2 \in H^1(\Omega_2)$ satisfies $v_2(-1) = 0$ and $v_2(1) = 1$, then (4b) implies

$$\mu_2 \frac{\partial y_2(1)}{\partial x} = \rho_2 \frac{\partial}{\partial t} \int_{\Omega_2} y_2 v_2 dx + \mu_2 \int_{\Omega_2} \frac{\partial y_2}{\partial x} \frac{\partial v_2}{\partial x} dx + \int_{\Omega_2} \frac{\partial y_2}{\partial x} y_2 v_2 dx. \tag{5d}$$

The identities (5) are used to enforce the interface conditions (2).

We discretize the differential equations in space using piecewise linear functions. We subdivide Ω_j, $j = 1,2,3$, into subintervals. Let x_i denote the subinterval endpoints and let v_i be the piecewise linear basis function with $v_i(x_i) = 1$ and $v_i(x_j) = 0$ or all $j \neq i$. We define the following index sets

$$I_1^I = \{i : x_i \in [-10,-1)\}, \quad I_2^I = \{i : x_i \in (-1,1)\}, \quad I_3^I = \{i : x_i \in (1,10]\},$$
$$I_{12}^\Gamma = \{i : x_i = -1\}, \qquad I_{23}^\Gamma = \{i : x_i = 1\}.$$

Given y_i for $i \in I_{12}^\Gamma \cup I_{23}^\Gamma$, we compute functions

$$y_k(t,x) = \sum_{i \in I_k^I} y_i(t) v_i(x) + \sum_{i \in I^\Gamma} y_i(t) v_i(x), \qquad k = 1,3, \tag{6a}$$

$$y_2(t,x) = \sum_{i \in I_1^I} y_i(t) v_i(x) + \sum_{i \in I_{12}^\Gamma} y_i(t) v_i(x) + \sum_{i \in I_{23}^\Gamma} y_i(t) v_i(x), \tag{6b}$$

where in (6a) we use $I^\Gamma = I_{12}^\Gamma$ if $k = 1$ and $I^\Gamma = I_{23}^\Gamma$ if $k = 3$, as solutions of

$$\rho_k \frac{d}{dt} \int_{\Omega_k} y_k v_i dx + \mu_k \int_{\Omega_k} \frac{\partial}{\partial x} y_k \frac{d}{dx} v_i dx = \int_{\Omega_k} S_k v_i dx, i \in I_k^I, \qquad k = 1,3,$$

$$\rho_2 \frac{d}{dt} \int_{\Omega_2} y_2 v_i dx + \mu_2 \int_{\Omega_2} \frac{\partial}{\partial x} y_2 \frac{d}{dx} v_i dx + \int_{\Omega_2} \frac{\partial}{\partial x} y_2 y_2 v_i dx = 0, i \in I_2^I.$$

If we set $\mathbf{y}_k^I = (y_i)_{i \in I_k^I}$, $k = 1,2,3$, $\mathbf{y}_{jk}^\Gamma = (y_i)_{i \in I_{jk}^\Gamma}$, $jk \in \{12, 23\}$, $\mathbf{y}^\Gamma = (\mathbf{y}_{12}^\Gamma, \mathbf{y}_{23}^\Gamma)^T$, and $\mathbf{u}_k = (u_i)_{i=1,\dots,n_s}$, $k = 1,3$ (cf. (3)), the previous identities can be written as

$$\mathbf{M}_1^{II} \frac{d}{dt} \mathbf{y}_1^I + \mathbf{A}_1^{II} \mathbf{y}_1^I + \mathbf{M}_1^{I\Gamma} \frac{d}{dt} \mathbf{y}_{12}^\Gamma + \mathbf{A}_1^{I\Gamma} \mathbf{y}_{12}^\Gamma = \mathbf{B}_1^I \mathbf{u}_1, \tag{7a}$$

$$\mathbf{M}_2^{II} \frac{d}{dt} \mathbf{y}_2^I + \mathbf{A}_2^{II} \mathbf{y}_2^I + \mathbf{M}_2^{I\Gamma} \frac{d}{dt} \mathbf{y}^\Gamma + \mathbf{A}_2^{I\Gamma} \mathbf{y}^\Gamma + \mathbf{N}^I(\mathbf{y}_2^I, \mathbf{y}^\Gamma) = \mathbf{0}, \tag{7b}$$

$$\mathbf{M}_3^{II} \frac{d}{dt} \mathbf{y}_3^I + \mathbf{A}_3^{II} \mathbf{y}_3^I + \mathbf{M}_3^{I\Gamma} \frac{d}{dt} \mathbf{y}_{23}^\Gamma + \mathbf{A}_3^{I\Gamma} \mathbf{y}_{23}^\Gamma = \mathbf{B}_3^I \mathbf{u}_3. \tag{7c}$$

By construction, the functions y_j, $j = 1,2,3$, in (6) satisfy (2a). To enforce (2b) we insert the identities (5), (6) into (2b). The resulting conditions can be written as

$$\mathbf{M}_1^{\Gamma I} \frac{d}{dt} \mathbf{y}_1^I + \mathbf{A}_1^{\Gamma I} \mathbf{y}_1^I + (\mathbf{M}_1^{\Gamma\Gamma} + \mathbf{M}_{12}^{\Gamma\Gamma}) \frac{d}{dt} \mathbf{y}_{12}^\Gamma + (\mathbf{A}_1^{\Gamma\Gamma} + \mathbf{A}_{12}^{\Gamma\Gamma}) \mathbf{y}_{12}^\Gamma \tag{8a}$$

$$+ \mathbf{M}_{12}^{\Gamma I} \frac{d}{dt} \mathbf{y}_2^I + \mathbf{A}_{12}^{\Gamma I} \mathbf{y}_2^I + \mathbf{N}_{12}^\Gamma(\mathbf{y}_2^I, \mathbf{y}_{12}^\Gamma) = \mathbf{B}_1^\Gamma \mathbf{u}_1, \tag{8b}$$

$$\mathbf{M}_3^{\Gamma I} \frac{d}{dt} \mathbf{y}_3^I + \mathbf{A}_3^{\Gamma I} \mathbf{y}_3^I + (\mathbf{M}_3^{\Gamma\Gamma} + \mathbf{M}_{23}^{\Gamma\Gamma}) \frac{d}{dt} \mathbf{y}_{23}^\Gamma + (\mathbf{A}_3^{\Gamma\Gamma} + \mathbf{A}_{23}^{\Gamma\Gamma}) \mathbf{y}_{23}^\Gamma \tag{8c}$$

$$+ \mathbf{M}_{23}^{\Gamma I} \frac{d}{dt} \mathbf{y}_2^I + \mathbf{A}_{23}^{\Gamma I} \mathbf{y}_2^I + \mathbf{N}_{23}^\Gamma(\mathbf{y}_2^I, \mathbf{y}_{23}^\Gamma) = \mathbf{B}_3^\Gamma \mathbf{u}_3. \tag{8d}$$

To summarize, our discretization of (1) and (2) is given by (7) and (8).

As outputs we are interested in the solution of the PDE at the spatial locations $\xi_1 = -5, \xi_2 = 0, \xi_3 = 5$. Thus the output equations are $y_k(t, \xi_k) = \sum_{i \in I_k^I} y_i(t) v_i(\xi_k)$, $k = 1,2,3$, which can be written as

$$\mathbf{z}_k^I(t) = \mathbf{C}_j^I \mathbf{y}_k^I(t), \quad \text{where } \mathbf{C}_k^I \in \mathbb{R}^{1 \times |I_k^I|}, \qquad k = 1,2,3.$$

3 Balanced Truncation Model Reduction

Given $\mathscr{E} \in \mathbb{R}^{n\times n}$ symmetric positive definite, $\mathscr{A} \in \mathbb{R}^{n\times n}$, $\mathscr{B} \in \mathbb{R}^{n\times m}$, $\mathscr{C} \in \mathbb{R}^{q\times n}$, and $\mathscr{D} \in \mathbb{R}^{q\times m}$, we consider linear time invariant systems in state space form

$$\mathscr{E}\frac{d}{dt}\mathbf{y}(t) = \mathscr{A}\mathbf{y}(t) + \mathscr{B}\mathbf{u}(t), \ \ t \in (0,T), \qquad \mathbf{y}(0) = \mathbf{y}_0, \tag{9a}$$

$$\mathbf{z}(t) = \mathscr{C}\mathbf{y}(t) + \mathscr{D}\mathbf{u}(t), \ \ t \in (0,T). \tag{9b}$$

Projection methods for model reduction generally produce $n \times r$ matrices $\mathscr{V}, \mathscr{W}$ with $r \ll n$ and with $\mathscr{W}^T\mathscr{E}\mathscr{V} = I_r$. One obtains a reduced form of equations (9) by setting $\mathbf{y} = \mathscr{V}\widehat{\mathbf{y}}$ and projecting (imposing a Galerkin condition) so that

$$\mathscr{W}^T[\mathscr{E}\mathscr{V}\frac{d}{dt}\widehat{\mathbf{y}}(t) - \mathscr{A}\mathscr{V}\widehat{\mathbf{y}}(t) - \mathscr{B}\mathbf{u}(t)] = 0, \ \ t \in (0,T).$$

This leads to a reduced system of order r with matrices $\widehat{\mathscr{E}} = \mathscr{W}^T\mathscr{E}\mathscr{V} = I_k$, $\widehat{\mathscr{A}} = \mathscr{W}^T\mathscr{A}\mathscr{V}$, $\widehat{\mathscr{B}} = \mathscr{W}^T\mathscr{B}$, $\widehat{\mathscr{C}} = \mathscr{C}\mathscr{V}$, and $\widehat{\mathscr{D}} = \mathscr{D}$.

Balanced reduction is a particular techniqe for constructing the projecting matrices $\mathscr{V}$ and $\mathscr{W}$, see, e.g., [1, 16]. One first solves the controllability and the observability Lyapunov equation $\mathscr{A}\mathscr{P}\mathscr{E} + \mathscr{E}\mathscr{P}\mathscr{A}^T + \mathscr{B}\mathscr{B}^T = 0$ and $\mathscr{A}^T\mathscr{Q}\mathscr{E} + \mathscr{E}\mathscr{Q}\mathscr{A} + \mathscr{C}^T\mathscr{C}^T = 0$, respectively. Under the assumptions of stability, controllability and observability, the matrices $\mathscr{P}, \mathscr{Q}$ are both symmetric and positive definite. There exist methods to compute (approximations of) $\mathscr{P} = \mathbf{U}\mathbf{U}^T$ and $\mathscr{Q} = \mathbf{L}\mathbf{L}^T$ in factored form. In the large scale setting the factorization is typically a low rank approximation. See, e.g., [8, 17].

The balancing transformation is constructed by computing the singular value decomposition $\mathbf{U}^T\mathscr{E}\mathbf{L} = \mathbf{Z}\mathbf{S}\mathbf{Y}^T$ and then setting $\mathscr{W} = \mathbf{U}\mathbf{Z}_r$, $\mathscr{V} = \mathbf{L}\mathbf{Y}_r$, where $\mathbf{S}_r = diag(\sigma_1, \sigma_2, \ldots, \sigma_r)$ is the $r \times r$ submatrix of $\mathbf{S} = \mathbf{S}_n$. The singular values σ_j are in decreasing order and r is selected to be the smallest positive integer such that $\sigma_{r+1} < \tau\sigma_1$ where $\tau > 0$ is a prespecified constant. The matrices $\mathbf{Z}_r, \mathbf{Y}_r$ consist of the corresponding leading k columns of $\mathbf{Z}, \mathbf{Y}$.

It is well known [6] that $\widehat{\mathscr{A}}$ must be stable and that for any given input $\mathbf{u}$ we have

$$\|\mathbf{z} - \widehat{\mathbf{z}}\|_{\mathscr{L}_2} \leq 2\|\mathbf{u}\|_{\mathscr{L}_2}(\sigma_{r+1} + \ldots + \sigma_n), \tag{10}$$

where $\widehat{\mathbf{z}}$ is the output (response) of the reduced model. Model reduction techniques for infinite dimensional systems are reviewed in, e.g., [3].

We want to apply balanced truncation model to the linear subsystems 1 and 3 in (7) and (8). We need to identify the input-output relations for these subsystems in the context of the coupled system to ensure that balancing techniques applied to these subsystems leads to a reduced model for the coupled system with error bounds.

To identify the appropriate input-output relations, we focus on subsystem 1. Examination of (7a,b) and (8a) shows that $\mathbf{M}_1^{I\Gamma}\frac{d}{dt}\mathbf{y}_{12}^{\Gamma}$, $\mathbf{A}_1^{I\Gamma}\mathbf{y}_{12}^{\Gamma}$ and $\mathbf{B}_1^I\mathbf{u}_1$ are the inputs into system 1 and $\mathbf{C}_1^{\Gamma I}\mathbf{y}_1^I$, $\mathbf{M}_1^{\Gamma I}\frac{d}{dt}\mathbf{y}_1^I + \mathbf{A}_1^{\Gamma I}\mathbf{y}_1^I$ are the outputs. Hence, if

$$\mathbf{M}_1^{I\Gamma} = \mathbf{0} \quad \text{and} \quad \mathbf{M}_1^{\Gamma I} = \mathbf{0}, \tag{11}$$

then we need to apply model reduction to

$$\mathbf{M}_1^{II} \frac{d}{dt} \mathbf{y}_1^I = -\mathbf{A}_1^{II} \mathbf{y}_1^I - \mathbf{A}_1^{I\Gamma} \mathbf{y}_{12}^\Gamma + \mathbf{B}_1^I \mathbf{u}_1 \tag{12a}$$

$$\mathbf{z}_1^I = \mathbf{C}_1^I \mathbf{y}_1^I, \quad \mathbf{z}_1^\Gamma = \mathbf{A}_1^{\Gamma I} \mathbf{y}_1^I. \tag{12b}$$

The system (12) is exactly of the form (9) and we can apply balanced truncation model reduction to obtain

$$\widehat{\mathbf{M}}_1^{II} \frac{d}{dt} \widehat{\mathbf{y}}_1^I = -\widehat{\mathbf{A}}_1^{II} \widehat{\mathbf{y}}_1^I - \widehat{\mathbf{A}}_1^{I\Gamma} \mathbf{y}_{12}^\Gamma + \widehat{\mathbf{B}}_1^I \mathbf{u}_1 \tag{13a}$$

$$\widehat{\mathbf{z}}_1^I = \widehat{\mathbf{C}}_1^I \widehat{\mathbf{y}}_1^I, \quad \widehat{\mathbf{z}}_1^\Gamma = \widehat{\mathbf{A}}_1^{\Gamma I} \widehat{\mathbf{y}}_1^I. \tag{13b}$$

Subsystem 3 can be reduced analogously. The reduced model for the coupled nonlinear system (7) and (8) is now obtained by replacing the subsystem matrices for subsystems 1 and 3 by their reduced matrices. Wether the balanced truncation error bound (10) can be used to derive an error bound between the original coupled problem (7) and (8) and its reduced model is under investigation.

In our finite element discretization we use mass lumping to obtain (11). However other discretizations, such as spectral elements or discontinuous Galerkin methods satisfy (11) directly, see [10, 12].

4 Numerical Results

We subdivide Ω_j into equidistant subintervals of length $h_k = 1/N_k$, $k = 1,2,3$, and we use piecewise linear basis functions.The size of the system (7), (8) is $9(N_1 + N_3) + 2N_2 + 1$. The parameters in the PDE are $\rho_k = 1$, $k = 1,2,3$, and $\mu_1 = 0.05$, $\mu_2 = 0.1$, $\mu_1 = 0.2$. For subsystem 1 and 3 we compute low-rank approximate solutions of the controllability and observability Lyapunov equations using the method described in [8]. We truncate such that $\sigma_{r+1} < \tau\sigma_1$, where $\tau = 10^{-4}$.

The sizes of the full and of the reduced order models for various discretization parameters are shown in Table 1. The subsystems 1 and 3 reduce substantially and the size of the subsystem 2 limits the amount of reduction achieved overall. For example, for $N_1 = N_3 = 20$ the subsystems 1 and 3 are each reduced in size from 180 to 11. The size of the coupled system is reduced from $361 + 2N_2$ to $23 + 2N_2$.

Next, we compare the system output given forcing functions $S_1 = u_1(t)$, $S_3 = u_3(t)$ (cf., (3)) with $u_1(t) = \frac{1}{2}\sin(3t)(1 - 0.8t/T)$, $u_3(t) = \sin(2t)(0.3 + 0.7t/T)$ on $(0,T) = (0,15)$. The full order model (7), (8) and the corresponding reduced order model are solved using the modified θ-scheme [7, 19] with (macro) time step $\Delta t = T/200$. Figure 1 shows the outputs, i.e., the approximate solution of the PDE at $\xi_1 = -5$, $\xi_2 = 0$, $\xi_3 = 5$. The left plot in Figure 2 shows the solution of the reduced order discretized PDE. The solution of the discretized PDE is visually indistinguishable

Table 1 Dimension of the full and of the reduced order models for various discretization parameters N_1, N_2, N_3 and $\tau = 10^{-4}$.

$N_1 = N_3$	N_2	size of full order model	size of reduced order model
10	10	201	41
20	20	401	63
40	40	801	107
20	10	381	43
40	20	761	67

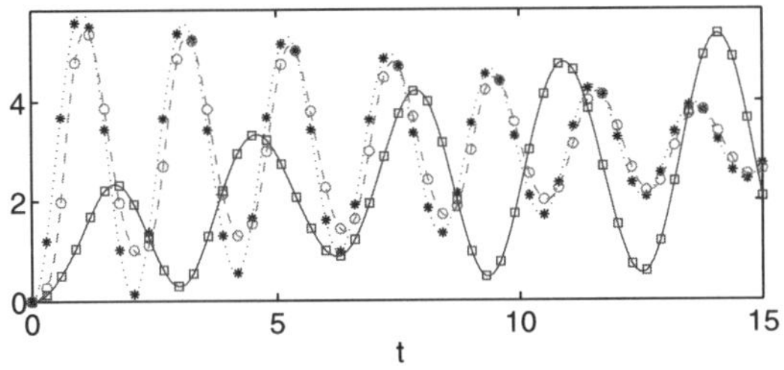

Fig. 1 Outputs 1, 2, 3 of the full order system corresponding to the discretization $N_1 = N_2 = N_3 = 10$ are given by $*$, $\circ$ and $\square$, respectively. Outputs 1, 2, 3 of the reduced order system are given by dotted, dashed and solid lines, respectively.

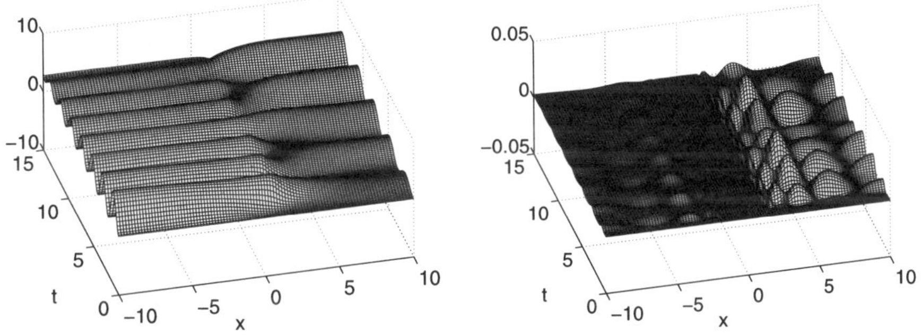

Fig. 2 Solution of the reduced order discretized PDE (left) and error between the solution of the discretized PDE and the reduced order system (right) for discretization $N_1 = N_2 = N_3 = 10$.

from the solution of the reduced order discretized PDE, as indicated by the size of the error shown in the right plot in Figure 2. The error is larger in the right subdomain because the PDE solution is positive and the advection term in (1d) advects the solution to the right.

Our numerical results indicate that the coupling of balanced truncation reduction for linear time variant subsystems with spatially localized nonlinear models leads to a coupled reduced order model with an error in the input-to-output map that is comparable to the error due to balanced truncation model reduction applied to the linear subsystems alone. The efficiency of the approach depends on the size of the interface and on the size of the localized nonlinearity. Investigations for higher dimensional problems are underway to explore the overall gains in efficiency.

Acknowledgements This research was supported in part by NSF grant ACI-0121360 and AFOSR grant FA9550-06-1-0245.

References

1. Antoulas, A.C.: Approximation of Large-Scale Systems. SIAM, Philadelphia (2005)
2. Bai, Z., Liao, B.S.: Towards an optimal substructuring method for model reduction. In: J. Dongarra, K. Madsen, J. Wasniewski (eds.) Applied Parallel Computing, pp. 276–285. Springer Lecture Notes in Computer Science, Vol. 3732 (2006)
3. Curtain, R.F.: Model reduction for control design for distributed parameter systems. In: Research directions in distributed parameter systems (Raleigh, NC, 2000), *Frontiers Appl. Math.*, vol. 27, pp. 95–121. SIAM, Philadelphia, PA (2003)
4. Discacciati, M., Quarteroni, A.: Convergence analysis of a subdomain iterative method for the finite element approximation of the coupling of Stokes and Darcy equations. Comput. Vis. Sci. **6**(2-3), 93–103 (2004)
5. Givoli, D., Barbone, P.E., Patlashenko, I.: Which are the important modes of a subsystem? Internat. J. Numer. Methods Engrg. **59**(12), 1657–1678 (2004)
6. Glover, K.: All optimal Hankel-norm approximations of linear multivariable systems and their L^∞-error bounds. Internat. J. Control **39**(6), 1115–1193 (1984)
7. Glowinski, R.: Finite element methods for incompressible viscous flow. Handbook of Numerical Analysis, Vol. IX. North-Holland, Amsterdam (2003)
8. Gugercin, S., Sorensen, D.C., Antoulas, A.C.: A modified low-rank Smith method for large-scale Lyapunov equations. Numer. Algorithms **32**(1), 27–55 (2003)
9. Hahn, J., Edgar, T.F.: An improved method for nonlinear model reduction using balancing of empirical Gramians. Computers and Chemical Engineering **26**, 1379–1397 (2002)
10. Hesthaven, J.S., Warburton, T.: Nodal Discontinuous Galerkin Methods: Analysis, Algorithms, and Applications. Springer-Verlag, Berlin (2008)
11. Hinze, M., Volkwein, S.: Proper orthogonal decomposition surrogate models for nonlinear dynamical systems: Error estimates and suboptimal control. In: P. Benner, V. Mehrmann, D.C. Sorensen (eds.) Dimension Reduction of Large-Scale Systems, Lecture Notes in Computational Science and Engineering, Vol. 45, pp. 261–306. Springer-Verlag, Heidelberg (2005)
12. Karniadakis, G.E., Sherwin, S.J.: Spectral/*hp* Element Methods for Computational Fluid Dynamics, second edn. Oxford University Press (2005)
13. Kunisch, K., Volkwein, S.: Galerkin proper orthogonal decomposition methods for a general equation in fluid dynamics. SIAM J. Numer. Anal. **40**(2), 492–515 (electronic) (2002)
14. Lall, S., Marsden, J.E., Glavaški, S.: A subspace approach to balanced truncation for model reduction of nonlinear control systems. Int. J. Robust Nonlinear Control **12**(6), 519–535 (2002)
15. Lucia, D.J., King, P.I., Beran, P.S.: Reduced order modeling of a two-dimensional flow with moving shocks. Computer and Fluids **32**, 917–938 (2003)
16. Moore, B.C.: Principal component analysis in linear systems: controllability, observability, and model reduction. IEEE Trans. Automat. Control **26**(1), 17–32 (1981)
17. Penzl, T.: Algorithms for model reduction of large dynamical systems. Linear Algebra Appl. **415**(2-3), 322–343 (2006)
18. Segalman, D.J.: Model reduction of systems with localized nonlinearities. Journal of Computational and Nonlinear Dynamics **2**(3), 249–266 (2007)
19. Turek, S., Rivkind, L., Hron, J., Glowinski, R.: Numerical study of a modified time-stepping θ-scheme for incompressible flow simulations. J. Sci. Comput. **28**(2-3), 533–547 (2006)
20. Vandendorpe, A., Dooren, P.V.: On model reduction of interconnected systems. In: W. Schilders, H. van der Vorst (eds.) Model Order Reduction: Theory, Research Aspects and Applications, Mathematics in Industry, pp. 160–175. Springer Verlag, Heidelberg (2007)

An Adaptive Discontinuous Galerkin Scheme for Second Order Problems with an Interface

P. Zunino

Abstract We discuss the derivation of an a-posteriori local error indicator for a discontinuous Galerkin (DG) method based on weighted interior penalties applied to advection-diffusion-reaction equations featuring a diffusivity parameter that may be discontinuous along a planar interface on a two-dimensional domain. We demonstrate how the weights incorporate into the scheme some a-priori knowledge of the exact solution that improves the efficacy of the local error estimator and of the corresponding adapted mesh. All the theoretical results are illustrated and discussed by means of numerical experiments.

1 Introduction and Problem Setting

We aim to approximate u, the solution of the following boundary value problem,

$$\nabla\cdot(-\varepsilon\nabla u+\beta u)+\mu u=f \text{ in } \Omega\subset\mathbb{R}^2, \quad u=0 \text{ on } \partial\Omega, \tag{1}$$

where Ω is a convex polygonal domain, $\mu\in L^\infty(\Omega)$ is a positive function and $\beta\in[W^{1,\infty}(\Omega)]^2$ is a vector function such that $\mu+\frac{1}{2}(\nabla\cdot\beta)\geq\mu_0>0$, $f\in L^2(\Omega)$. Let Γ be a single planar interface subdividing Ω in two subregions Ω_i, $i=1,2$. For simplicity, the coefficient ε is defined on each subregion by a positive constant.

Given $V:=H_0^1(\Omega)$, the weak formulation of problem (1) corresponds to find $u\in V$ such that

$$a(u,v):=\int_\Omega\Big(\varepsilon\nabla u\cdot\nabla v-\beta u\cdot\nabla v+\mu uv\Big)=F(v):=\int_\Omega fv,\ \forall v\in V. \tag{2}$$

Paolo Zunino
MOX - Department of Mathematics - Politecnico di Milano, via Bonardi 9, 20133 Milano, Italy, e-mail: paolo.zunino@polimi.it

A transmission problem for Poisson equation has already been addressed by means of Nitsche type mortaring in [7, 1], in [5] encompassing a more general geometrical setting, and in [2] for advection-diffusion-reaction equations. Following [3], we address here a discontinuous Galerkin method that automatically accounts for the presence of the interface, provided that it is conforming with the computational mesh. This approach simplifies the implementation of the scheme with respect to a multi-domain mortar method, but sensibly increases the number of degrees of freedom of the discrete problem. In this setting, we develop an a-posteriori local error indicator that suitably exploits the information on the jump of the coefficient ε at the interface, leading to an effective refinement of the mesh.

2 Numerical Approximation

For the numerical approximation of problem (2) we consider a shape regular triangulation T_h of the domain Ω and we define a totally discontinuous approximation space,

$$V_h^p := \{v_h \in L^2(\Omega);\ \forall K \in T_h, v_h|_K \in \mathbb{P}^p\},\ \text{with } p > 0.$$

Let e be an edge of the element $K \in T_h$, which is a triangle in Ω. Let h_e be the size of an edge and h_K be the one of an element. We denote with F_h^i and $F_h^{\partial\Omega}$ the collections of all the internal edges and of all the edges on $\partial\Omega$ respectively. For any interior edge of the mesh we denote with n_e its unit normal vector, and with n the unit normal vector with respect to $\partial\Omega$. For any function v that is discontinuous on the inter-element interface e, we define $v(x)|_e^{\pm} := \lim_{\delta\to 0^+} v(x \pm \delta n_e)$ for a.e. $x \in e$ and we will use the abridged notation $v^{\pm}$. The jump over edges is defined as $[\![v]\!]_e := v^- - v^+$, while we denote with $\{v\}$ and $\{\{v\}\}$ the arithmetic and the harmonic means of v^- and v^+. We also introduce the weighted averages,

$$\{v\}_w := w_e^- v^- + w_e^+ v^+, \quad \{v\}^w := w_e^+ v^- + w_e^- v^+, \quad \forall e \in F_h^i,$$

where the weights necessarily satisfy $w_e^- + w_e^+ = 1$. In order to make the numerical method to be robust for problems featuring a discontinuous and locally vanishing diffusivity, it is convenient to choose the weights to be dependent on the coefficients of the problem, see [3]. To this purpose, we introduce the heterogeneity factor $\lambda_h := \frac{1}{2}\frac{[\![\varepsilon]\!]_e}{\{\varepsilon\}}$ and we set $w_e^{\pm} := \frac{1}{2}(1 \pm \lambda_h)$. It can be easily seen that this leads to the identity $\{\varepsilon\}_w = \{\{\varepsilon\}\}$. Then, we introduce the following bilinear form,

$$
\begin{aligned}
a_h(u_h,v_h) := & \int_{T_h} \Big[(\varepsilon \nabla u_h - \beta u_h) \cdot \nabla v_h + \mu u_h v_h \Big] \\
& + \int_{F_h^i} \Big[\{\beta u_h\} \cdot n_e [\![v_h]\!]_e - \{\varepsilon \nabla u_h\}_w \cdot n_e [\![v_h]\!]_e - \{\varepsilon \nabla v_h\}_w \cdot n_e [\![u_h]\!]_e \\
& + \big(\tfrac{1}{2}|\beta \cdot n_e| + \xi \{\{\varepsilon\}\}\{\{h_e\}\}^{-1}\big) [\![u_h]\!]_e [\![v_h]\!]_e \Big] \\
& + \int_{F_h^{\partial\Omega}} \Big[\big(\tfrac{1}{2}\beta u_h - \varepsilon \nabla u_h\big) \cdot n v_h - \varepsilon \nabla v_h \cdot n u_h + \big(\tfrac{1}{2}|\beta \cdot n| + \varepsilon \xi h_e^{-1}\big) u_h v_h \Big],
\end{aligned}
$$

where we have applied the abridged notation $\int_{T_h} := \sum_{K\in T_h} \int_K$ etc. We notice that the penalty term into $a_h(\cdot,\cdot)$ is suitably adapted to treat nonconforming meshes where h_e may be different on each side of an inter-element interface. The weighted interior penalty method reads as follows: find $u_h \in V_h^p$ such that,

$$
a_h(u_h,v_h) = F(v_h), \ \forall v_h \in V_h^p. \tag{3}
$$

The well posedness of the numerical method is a direct consequence of the positivity of the bilinear form $a_h(\cdot,\cdot)$, we refer to [3] for a proof. Let $H^s(T_h)$ be the broken Sobolev space of order $s>0$ on T_h. We introduce $W := V \cap H^s(T_h)$ with $s > \frac{3}{2}$, such that for any function $u,v \in W$ the bilinear form $a_h(u,v)$ is well defined. Then, we set $W(h) := W \oplus V_h^p$ and we state the following result, remanding to [3] for a proof.

Lemma 1. *Assume that u, the solution of (2), satisfies $u \in W$. Then, $a_h(u,v) = F(v)$ for all $v \in W(h)$ and $a_h(u-u_h,v_h) = 0$ for all $v_h \in V_h^p$.*

We observe that the regularity of W is compatible with the solution of transmission problems on convex polygons and a single planar interface. Furthermore, the results presented in [5] suggest that it is possible to generalize Lemma 1 with less restrictive regularity requirements on the space W. In particular, exploiting theorems 1.5.3.10 and 1.5.3.11 of [4], Lemma 1 could be extended to any $u \in \{v \in H_0^1(\Omega); (\nabla\cdot(-\varepsilon\nabla v + \beta v) + \mu v) \in L^2(K), \forall K \in T_h\}$. In this perspective, the present work could be generalized to any polygonal domain and any interface that is resolved by the computational mesh.

3 Duality Based A-Posteriori Error Analysis

In this section we aim to put into evidence some peculiar advantages of the weighted interior penalty method in the derivation of a local error estimator. For this preliminary study, we opt for the duality based approach because it straightforwardly preserves the robustness of the weighted interior penalty method with respect to locally vanishing diffusivity and it can be also easily adapted to the case of nonconforming elements, see for instance [6].

We start defining the dual problem with respect to (2): find $z \in V$ such that,

$$
a(\varphi,z) = J(\varphi), \quad \forall \varphi \in V, \tag{4}
$$

where J is a linear functional $J(\cdot): V \to \mathbb{R}$ for which we aim to control the error. Then, we consider the discrete dual problem that consists in finding $z_h \in V_h^q$ such that,

$$a_h(\varphi_h, z_h) = J(\varphi_h), \quad \forall\, \varphi_h \in V_h^q, \tag{5}$$

where the discrete dual space V_h^q is generally richer than V_h^p, i.e. $q > p$. Finally, mimicking Lemma 1, we obtain the following result.

Lemma 2. *Assume that z, the solution of* (4), *satisfies $z \in W$. Then, $a_h(\varphi, z) = J(\varphi)$ for all $\varphi \in W(h)$.*

Now, let $e := u - u_h$ be the error relative to our numerical method, where $u \in W$ is the solution of (2) and $u_h \in V_h^p$ satisfies (3). We easily conclude that $e \in W(h)$. Lemma 2 allows us to rewrite the error on the output functional, $J(e) = J(u) - J(u_h)$, in terms of the residuals of the numerical method, more precisely we obtain the following error representation formula in terms of the local residuals.

Lemma 3. *Assume that z, the solution of* (4), *satisfies $z \in W$ and let $u_h \in V_h^p$ be the solution of* (3). *Then, for any $\zeta := (z - v_h) \in W(h)$,*

$$\begin{aligned} J(e) = &- \int_{T_h} R_0(u_h) W_0(\zeta) - \int_{F_h^{\partial\Omega}} \Big(R_4(u_h) W_4(\zeta) + R_5(u_h) W_5(\zeta) \Big) \\ &- \int_{F_h^i} \Big(R_1(u_h) W_1(\zeta) + R_2(u_h) W_2(\zeta) + R_3(u_h) W_3(\zeta) \Big), \end{aligned} \tag{6}$$

where $R_i(\cdot)$ and $W_i(\cdot)$ are defined as follows,

$$\begin{aligned}
&R_0(u_h) = \nabla \cdot (-\varepsilon \nabla u_h + \beta u_h) + \mu u_h - f, && W_0(\zeta) = \zeta, \\
&R_1(u_h) = [\![\varepsilon \nabla u_h - \beta u_h]\!]_e \cdot n_e, && W_1(\zeta) = \{\zeta\}^w, \\
&R_2(u_h) = \{\{\varepsilon\}\} [\![u_h]\!]_e, && W_2(\zeta) = \xi \{\{h_e\}\}^{-1} [\![\zeta]\!]_e - \{\nabla \zeta\} \cdot n_e, \\
&R_3(u_h) = \tfrac{1}{2} \big(|\beta \cdot n_e| - \beta \cdot n_e (w_e^- - w_e^+) \big) [\![u_h]\!]_e, && W_3(\zeta) = [\![\zeta]\!]_e, \\
&R_4(u_h) = \varepsilon u_h, && W_4(\zeta) = h_e^{-1} \zeta - \nabla \zeta \cdot n, \\
&R_5(u_h) = \tfrac{1}{2} \big(|\beta \cdot n| - \beta \cdot n \big) u_h, && W_5(\zeta) = \zeta,
\end{aligned}$$

Proof. Since $e \in W(h)$, owing to Lemma 2 with $\varphi = e$ and the Galerkin orthogonality of the primal problem, we get,

$$J(e) = a_h(e, z) = a_h(e, \zeta).$$

Now, starting from the expression of $a_h(e, \zeta)$, by means of integration by parts and suitable manipulations exploiting the identity $[\![ab]\!] = \{a\}_w [\![b]\!] + \{b\}^w [\![a]\!]$, we exactly obtain (6).□

To develop a local error estimator we start form (6), together with a suitable strategy for the repartition on each element $K \in T_h$ of the error indicators lying on the mesh skeleton F_h^i, namely $R_i(u_h) W_i(\zeta)$ with $i = 1, 2, 3$. This issue is particularly relevant in the case of problems with large variations of the diffusivity across the interface, where the residual $R_1(u_h) = [\![\varepsilon\nabla u_h - \beta u_h]\!]_e \cdot n_e$ is likely to be one of the leading contributions to determine the error. We will see that different strategies to split $R_1(u_h)W_1(\zeta)$ over the neighboring elements lead to remarkably different local error indicators and adapted meshes.

In general, the most common and natural strategy to break up the residuals on each edge is to equally divide them into the neighboring elements $K^{\pm}$. This seems to be the only possibility to treat $R_2(u_h)W_2(\zeta)$ and $R_3(u_h)W_3(\zeta)$, since $W_2(\zeta)$ and $W_3(\zeta)$ are symmetric with respect to $K^{\pm}$ and no information on the dual solution z is a-priori available. This is not the case for $\int_e R_1(u_h)W_1(\zeta)$. Indeed, $W_1(\zeta)$ is the only weighing function that depends on the coefficients of the problem, more precisely $W_1(\zeta) = \{\zeta\}^w = w_e^- \zeta^+ + w_e^+ \zeta^-$. By consequence, exploiting the information into $w_e^{\pm}$, we can conceive different options to separate $W_1(\zeta)$ on $K^{\pm}$. Let us denote with $\int_e R_1(u_h)W_1^*(\zeta)$ the contribution of the error indicator that falls on K^-. We consider the following alternative splitting strategies,

$$W_1^*(\zeta) := \tfrac{1}{2}W_1(\zeta)\ (a); \quad w_e^- \zeta^+\ (b); \quad w_e^+ \zeta^-\ (c); \tag{7}$$

where now $\pm$ refers to the normal vector of each element, namely n_K. Then, we propose the corresponding local estimators,

$$\eta_K^*(u_h,\zeta) = \int_{\partial K\setminus\partial\Omega} \Big(|R_1(u_h)W_1^*(\zeta)| + \tfrac{1}{2}|R_2(u_h)W_2(\zeta)| + \tfrac{1}{2}|R_3(u_h)W_3(\zeta)|\Big)$$
$$+ \int_{\partial K\cap\partial\Omega} \Big(|R_4(u_h)W_4(\zeta)| + |R_5(u_h)W_5(\zeta)|\Big) + \int_K |R_0(u_h)W_0(\zeta)|.$$

We observe that the global error estimator, namely $\sum_{K\in T_h} \eta_K^* \geq |J(e)|$, is the same for all the cases (a), (b), (c). What differs from case to case is the distribution of the error.

In order to build up the local error indicators η_K^* we have to provide a precise definition for the output functional $J(e)$. In the particular case of problems with discontinuous diffusivity, we aim to control the error along those edges where ε is discontinuous. For this reason we consider $J(\varphi) := \int_{F_h^i} |\lambda_h| \varphi$.

Then, we approximate the exact solution of the dual problem (4) by means of its discretization through the scheme (5). Because of the arbitrariness of v_h in the definition of ζ, it is clear that we have to compute z_h with $q > p$ into (6). Let ζ_h be any function of the form $\zeta_h := z_h - v_h$ with $z_h \in V_h^q$ and $v_h \in V_h^p$. To obtain an accurate estimator we choose $p = 1$ and $q = 3$. Then, our approximate local error indicator is given by $\eta_K^*(u_h,\zeta_h)$ on each element $K \in T_h$.

For the set up of a mesh adaptation strategy, we rescale the indicators η_K^* with respect to the estimated global error. By this way, we obtain a piecewise constant function that quantifies to which extent the error on each element contributes to the global one. More precisely, we introduce the *relative* local error indicators, defined by $\rho_K^* := \eta_K^* \big(\sum_{K\in T_h} \eta_K^*\big)^{-1}$. Then, to set up a mesh refinement algorithm we adopt a fixed error reduction strategy. More precisely, the adaptively refined meshes are obtained by bisection of the elements whose estimator ρ_K^* is greater than 40% of the maximum. Then, a new local error estimator is computed on the refined mesh and this procedure is repeated iteratively until a maximal number of iterations is achieved or a stopping test is satisfied.

4 Numerical Results

To compare the three options (a), (b), (c), we consider a one dimensional problem in order to reduce the technical difficulties arising in the mesh adaptation strategy. The numerical method and the a-posteriori error analysis are straightforwardly applied to this simple case. We split the domain Ω into two subregions, $\Omega_1 = (0, \frac{1}{2})$, $\Omega_2 = (\frac{1}{2}, 1)$. The diffusivity $\varepsilon(x)$ is a discontinuous function across the interface $x = \frac{1}{2}$. Precisely, we consider a constant $\varepsilon(x)$ in each subregion with $\varepsilon_{h,1} = 5 \cdot 10^{-3}$ in Ω_1 and $\varepsilon_{h,2} = 1.0$ in Ω_2. In the case $\beta = [1,0]$, $\mu = 0$, $f = 0$ and the boundary conditions $u_1(x=0) = 1$, $u_2(x=1) = 0$, the exact solution of the problem on each subregion Ω_1, Ω_2 can be expressed as an exponential function with respect to x. The global solution $u(x)$ is characterized by means of the value at the interface, $u(x = \frac{1}{2})$, that enforces the continuity of the normal fluxes. We refer to [3] for an explicit formula of $u(x)$.

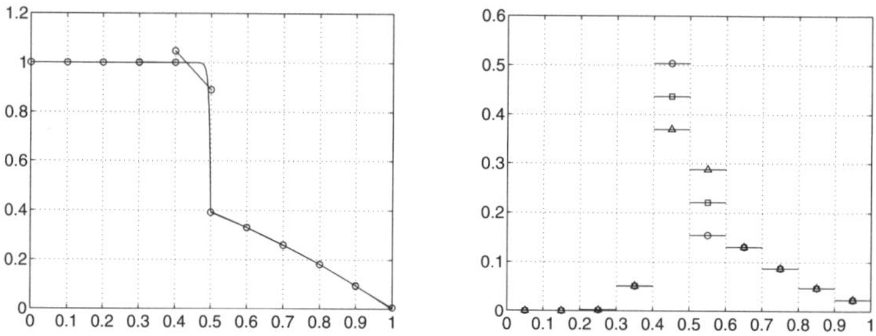

Fig. 1 The exact and the computed solution are reported on the left with solid and dotted lines, respectively. The relative local error estimators ρ_K^* are on the right and the cases (a), (b) and (c), are identified with $\square$, $\circ$, $\triangle$, respectively.

From the comparison between the exact and the computed solution, reported in figure 1 (left), we immediately notice that the element that mostly contribute to the error is the one on the left of the interface where ε is discontinuous, because the exact solution of the problem at hand features a very sharp internal layer in this region. The relative local error indicators ρ_K^* are reported in figure 1 (right) and we observe that they are significantly different on those elements where the heterogeneity factor is not equal to zero and thus the averaging weights $w_e^{\pm}$ differ from $\frac{1}{2}$. More precisely, strategy (b) favorably clusters the error on the elements that lay upwind with respect to the interface, where the internal layer is located, while strategy (a) and in particular strategy (c) promote the dispersion of the local error on both sides. Owing to (7), this is a direct consequence of the definition of the weights, $w_e^{\pm} = 2\varepsilon^{\mp}/\{\varepsilon\}$, while the discrete dual solution does not contribute, because it is almost continuous across the interface.

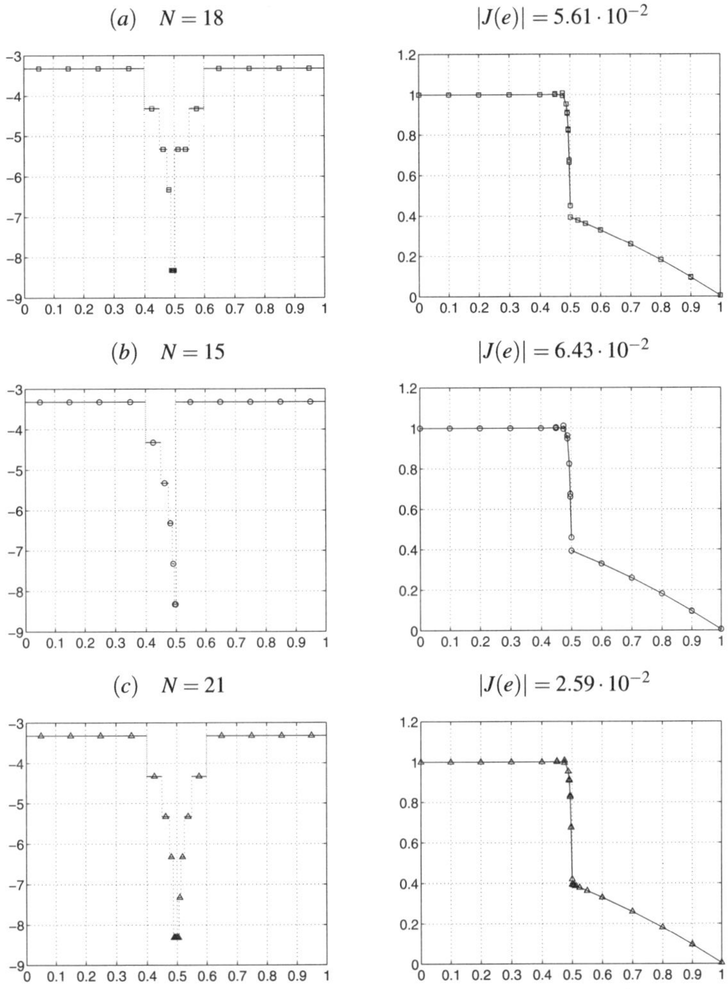

Fig. 2 The adapted grids after 5 refinement cycles (left) and the corresponding solutions (right) for the cases (a), (b) and (c) from top to bottom. The mesh size is represented on the left by $\log_2(h_K)$.

The adapted meshes corresponding to the estimators of type (a), (b) and (c) are reported in figure 2 together with the approximate solutions u_h. As expected, the strategy (b) leads to a strongly upwinded refinement that allows us to capture the internal layer induced by the discontinuity of ε without increasing the number of degrees of freedom on the region where the diffusion dominates. This is not the case for the meshes (a) and (c), where the refined region is located on both sides of the interface $x = \frac{1}{2}$. Figure 2(b) also shows that the weighted interior penalty method is robust with respect to large variations of the mesh size between neighboring elements. We finally analyze the error functional for the three cases (a), (b) and (c) and we notice that the strategy (b) provides an accuracy on $|J(e)|$ that is comparable to (a) and (c), but it exploits less degrees of freedom. This benefit would increase if it were applied to an internal layer in the multi-dimensional case.

5 Conclusions

We have considered a DG method that extends the standard interior penalty schemes by means of weighted averages. It is particularly suited for the approximation of advection-diffusion-reaction problems with a diffusivity coefficient that may be discontinuous and locally vanishing. In particular, we have considered the a-posteriori error analysis of the scheme, putting into evidence that the introduction of weighted interior penalties also helps to improve the efficacy of a local error estimator, leading to a more selective refinement. Furthermore, this technique is not restricted to the case of duality based error estimators, but it could be applied more in general, for instance also to residual based error estimates.

References

1. Becker, R., Hansbo, P., Stenberg, R.: A finite element method for domain decomposition with non-matching grids. M2AN Math. Model. Numer. Anal. **37**(2), 209–225 (2003)
2. Burman, E., Zunino, P.: A domain decomposition method based on weighted interior penalties for advection-diffusion-reaction problems. SIAM J. Numer. Anal. **44**(4), 1612–1638 (electronic) (2006)
3. Ern, A., Stephansen, A., Zunino, P.: A Discontinuous Galerkin method with weighted averages for advection-diffusion equations with locally small and anisotropic diffusivity. Technical Report 07, MOX, Department of Mathematics, Politecnico di Milano (2007). To appear on IMA J. Numer. Anal.
4. Grisvard, P.: Elliptic problems in nonsmooth domains, *Monographs and Studies in Mathematics*, vol. 24. Pitman (Advanced Publishing Program), Boston, MA (1985)
5. Heinrich, B., Nicaise, S.: The Nitsche mortar finite-element method for transmission problems with singularities. IMA J. Numer. Anal. **23**(2), 331–358 (2003)
6. Kanschat, G., Rannacher, R.: Local error analysis of the interior penalty discontinuous Galerkin method for second order elliptic problems. J. Numer. Math. **10**(4), 249–274 (2002)
7. Stenberg, R.: Mortaring by a method of J. A. Nitsche. In: Computational mechanics (Buenos Aires, 1998), pp. CD–ROM file. Centro Internac. Métodos Numér. Ing., Barcelona (1998)

Finance, Stochastic Applications

Abstract Sensitivity Analysis for Nonlinear Equations and Applications

A. Chernov

Abstract Let $u \in Y$ be a unique solution of a nonlinear equation $J(\alpha, u) = 0$ for $\alpha \in X$, where X and Y are suitable Banach spaces. We investigate sensitivity of u with respect to small perturbations of $\alpha = \alpha_0 + r$. The Implicit Function Theorem postulates that the Fréchet differentiability of $u = S(\alpha)$ w.r.t. α is determined by the differentiability properties of J. Then $S(\alpha_0 + r)$ is approximated by its truncated Taylor series at α_0. We derive a sequence of problems, which characterizes the Fréchet derivative $\mathrm{d}^k S(\alpha_0)$ via the Fréchet derivatives of lower order. This gives a recursive procedure for computing the Taylor series of $S(\alpha_0 + r)$. We illustrate the above approach on an elliptic PDE, an elliptic integral equation and on an abstract strongly elliptic problem with small randomly perturbed parameter.

1 Introduction

Many important problems in computational science can be written in an abstract way as problems of finding $u \in Y$ solving an implicit possibly nonlinear equation

$$J(\alpha, u) = 0 \quad \text{in } Z. \tag{1}$$

Here the operator $J : X \times Y \to Z$ represents the system behavior, $\alpha \in X$ stands for the problem parameters, e.g. boundary conditions, operator coefficients, shape of the domain (cf. examples in Section 4), and X, Y, Z are suitable Banach spaces. Suppose (1) is uniquely solvable for $\alpha \in U$, where U is an open subset of X, i.e. there exists a solution operator $S : U \to Y$, such that

$$J(\alpha, S(\alpha)) = 0 \quad \forall \alpha \in U. \tag{2}$$

Alexey Chernov,
Seminar for Applied Mathematics, ETH, 8092 Zürich, Switzerland,
e-mail: achernov@math.ethz.ch

Frequently in practical applications the value of α is not known exactly and computations are performed for some average $\alpha_0 \in X$. The question arises: How sensitive is the solution $u = S(\alpha)$ of (1) w.r.t. small (in certain sense) perturbations of the parameters $\alpha = \alpha_0 + r$?

If $S(\alpha)$ is sufficiently smooth in $U \subset X$, $S(\alpha_0 + r)$ might be approximated by its truncated Taylor series at α_0. We face the following questions:

(a) Existence and of the truncated Taylor series at α_0 and the approximation error
(b) Computation of the Fréchet derivatives $d^k S(\alpha_0)$

It turns out that the classical Implicit Function Theorem handles question (a). After a brief introduction into the differential calculus in Banach spaces in section 2 we investigate (b) and obtain a recursive series of linear problems for computing $\mathrm{d}^k S(\alpha_0)$. In section 4 we apply the above abstract analysis to an elliptic PDE, an elliptic integral equation and to an abstract strongly elliptic problem with small randomly perturbed parameter.

2 Differential Calculus in Banach Spaces

In this section we introduce the main definitions and recall the classical results of differential calculus in Banach spaces, cf. e.g. [1, 2].

Suppose X,Y,Z are Banach spaces and $U \subset X$ is an open set. We denote by $C(U,Y)$ the space of continuous (w.r.t. convergence in norm) maps $U \to Y$, and by $\mathscr{L}_n(X,Y)$, $n \in \mathbb{N}$ the space of n-linear continuous maps $\prod_{i=1}^n X \to Y$.

Definition 1. Let U be an open subset of X. The mapping $F \in C(U,Y)$ is called Fréchet differentiable at $\alpha \in U$ if there exists a linear operator $A \in \mathscr{L}_1(X,Y)$:

$$\|F(\alpha + r) - F(\alpha) - A(r)\|_Y = o(\|r\|_X). \tag{3}$$

The operator $A = \mathrm{d}F(\alpha)$ is uniquely determined and is called the Fréchet derivative of F at α. F is called Fréchet differentiable in U, if F is differentiable at all $\alpha \in U$.

Suppose $J(\cdot,u) \in C(U,Z)$ for some fixed $u \in Y$. The partial Fréchet derivative $\mathrm{d}_\alpha J(\alpha,u) \in \mathscr{L}_1(X,Z)$ at (α,u) with respect to α is naturally defined by

$$\|J(\alpha + r, u) - J(\alpha,u) - \mathrm{d}_\alpha J(\alpha,u)[r]\|_Z = o(\|r\|_X). \tag{4}$$

Classical arguments give the following differentiation rule for composite maps.

Lemma 1. *(Chain rule) Suppose $F : U \to Y$ is differentiable at $\alpha \in U$ and $G : V \to Z$ is differentiable at $u = F(\alpha) \in V$, where U, V are open subsets of X, Y respectively and $F(U) \subset V$. Then $G \circ F : U \to Z$ is differentiable at α and there holds*

$$\mathrm{d}(G \circ F)(\alpha)[r] = \mathrm{d}G(u)[\mathrm{d}F(\alpha)[r]], \quad \textit{with } u = F(\alpha). \tag{5}$$

Higher derivatives are defined by induction. For $F : X \to Y$ we set

$$\mathrm{d}^{n+1}F(\alpha)[r_1,\dots,r_{n+1}] := \mathrm{d}\left(\mathrm{d}^n F(\alpha)[r_1,\dots,r_n]\right)[r_{n+1}] \in \mathscr{L}_1(X,\mathscr{L}_n(X,Y)). \quad (6)$$

Note that $\mathscr{L}_1(X,\mathscr{L}_n(X,Y))$ is isometrically isomorphic to $\mathscr{L}_{n+1}(X,Y)$. In the case $r_i = r \in X$ for $i = 1,\dots,n$, we abbreviate $\mathrm{d}^n F(\alpha)[r]^n := \mathrm{d}^n F(\alpha)[r,\dots,r]$.

We write $F \in C^n(U,Y)$ if $F : U \to Y$ is n times Fréchet differentiable in the open set $U \subset X$, and $\mathrm{d}^n F(\alpha) \in C(U,\mathscr{L}_n(X,Y))$.

Theorem 1. *(Taylor's Theorem) Suppose $F \in C^n(U,Y)$ for $n \geq 1$ and an open set $U \subset X$. Let $\alpha \in U$ and $r \in X$ such that $[\alpha, \alpha + r] \subset U$. Then*

$$F(\alpha + r) = T_n(F,\alpha,r) + R_n(F,\alpha,r), \quad (7)$$

where

$$T_n(F,\alpha,r) = \sum_{k=0}^{n} \frac{1}{k!}\mathrm{d}^k F(\alpha)[r]^k, \qquad \|R_n(F,\alpha,r)\|_Y = o(\|r\|_X^n). \quad (8)$$

Suppose X,Y are Banach spaces. We set

$$\mathrm{Inv}(X,Y) := \{A \in \mathscr{L}_1(X,Y) : A \text{ is invertible}\}. \quad (9)$$

Theorem 2. *(Implicit Function Theorem) Let W be an open subset of $X \times Y$ and $J \in C^n(W,Z)$, $n \geq 1$. Suppose $(\alpha_0,u_0) \in W$, $J(\alpha_0,u_0) = 0$ and $\mathrm{d}_u J(\alpha_0,u_0) \in \mathrm{Inv}(Y,Z)$. Then there exist open neighborhoods $U \subset X$ of α_0 and $V \subset Y$ of u_0 and a unique mapping $S \in C^n(U,Y)$ such that*

i. $J(\alpha,S(\alpha)) = 0$, $\forall \alpha \in U$,
ii. $J(\alpha,u) = 0$, $(\alpha,u) \in U \times V$ *yields* $u = S(\alpha)$,
iii. $\mathrm{d}S(\alpha) = -(\mathrm{d}_u J(\alpha,S(\alpha)))^{-1} \circ \mathrm{d}_\alpha J(\alpha,S(\alpha))$, *for* $\alpha \in U$.

3 Nonlinear Sensitivity Analysis

In this section we answer questions (a) and (b) from the introduction.

We return back to the notations of section 1. Suppose X,Y,Z are Banach spaces and consider an abstract possibly nonlinear equation (1). Assume $J \in C^n(W,Z)$ for some $W \subset X \times Y$ and $n \geq 1$. Suppose $(\alpha_0,u_0) \in W$ satisfies (1) and $\mathrm{d}_u J(\alpha_0,u_0) \in \mathrm{Inv}(Y,Z)$. Then the Implicit Function Theorem ensures existence and uniqueness of the solution operator $S \in C^n(U,Y)$ in some sufficiently small neighborhood U of α_0, such that *i.–iii.* in Theorem 2 hold true. Hence, the truncated Taylor series $T_n(S,\alpha_0,r)$ exists and

$$\|S(\alpha + r) - T(S,\alpha_0,r)\|_Y = o(\|r\|_X^n), \quad (10)$$

which answers the question (a).

In what follows we suggest a method for computation of terms in $T(S,\alpha_0,r)$ and answer (b). Define a composite map

$$F(\alpha) := J(\alpha,S(\alpha)) \quad \forall \alpha \in U. \tag{11}$$

Then $F(\alpha) \equiv 0$ in U and hence all derivatives of F exist in U and vanish

$$\mathrm{d}^k F(\alpha_0)[r]^k = 0 \quad \text{if } [\alpha_0,\alpha_0+r] \subset U. \tag{12}$$

By assumption $J \in C^n(W,Z)$ in $W \subset X \times Y$ yielding $S \in C^n(U,Y)$ and we may use the chain rule, (11) and (12) to determine $\mathrm{d}^k S(\alpha_0)$:

$$\begin{aligned} \mathrm{d}^k F(\alpha_0)[r]^k &= (\mathrm{d}_\alpha(\cdot)[r] + \mathrm{d}_u(\cdot)[\mathrm{d}S(\alpha_0)[r]])^k J(\alpha_0,u_0)[r]^k \\ &= \mathrm{d}_u J(\alpha_0,u_0)[\mathrm{d}^k S(\alpha_0)[r]^k] + Q^k J(\alpha_0,u_0)[r]^k, \end{aligned} \tag{13}$$

where $u_0 = S(\alpha_0)$ and the differential operator Q^k is a linear combination of the mixed derivatives $d^i_{\alpha^j,u^{i-j}} J(\alpha_0,u_0)[r]^j[v]^{i-j}$, $j = 0,\dots i$ of order $i \le k$ and v has the form $d^m S(\alpha_0)[r]^m$ with $m = 1,\dots,k-1$. In particular

- $k = 1$

$$Q^1 J(\alpha_0,u_0)[r] = \mathrm{d}_\alpha J(\alpha_0,u_0)[r] \tag{14}$$

- $k = 2$

$$\begin{aligned} Q^2 J(\alpha_0,u_0)[r]^2 &= \mathrm{d}^2_{\alpha^2} J(\alpha_0,u_0)[r]^2 \\ &\quad + 2\mathrm{d}^2_{\alpha,u} J(\alpha_0,u_0)[r][dS(\alpha_0)[r]] \\ &\quad + \mathrm{d}^2_{u^2} J(\alpha_0,u_0)[dS(\alpha_0)[r]]^2. \end{aligned} \tag{15}$$

Remark 1. Note that $Q^k J(\alpha_0,u_0)[r]^k$ does not depend on $d^m S(\alpha_0)[r]^m$ for $m \ge k$. Thus, $d^k S(\alpha_0)[r]^k$ allows an explicit characterization under assumptions of the Implicit Function Theorem.

Theorem 3. *Suppose the assumptions of the Implicit Function Theorem are satisfied. Then the kth derivative, $k = 1,\dots,n$, of $u = S(\alpha)$ at α_0 exists, is continuous and is given by*

$$\mathrm{d}^k S(\alpha_0)[r]^k = -(\mathrm{d}_u J(\alpha_0,u_0))^{-1} \circ Q^k J(\alpha_0,u_0)[r]^k, \tag{16}$$

where $Q^k J(\alpha_0,u_0)[r]^k$ is defined in (13).

Remark 2. Note that $\mathrm{d}_u J(\alpha_0,u_0) \in \mathscr{L}_1(Y,Z)$ is independent of k, thus computation of $\mathrm{d}^k S(\alpha_0)[r]^k$ requires solving a sequence of *linear* problems (16) with *the same* operator and a varying right-hand side.

4 Applications

In this section we illustrate the abstract analysis from section 3 on an elliptic PDE, an elliptic integral equation and an abstract strongly elliptic problem with small randomly perturbed parameter.

4.1 Diffusion Equation with Perturbed Diffusion Coefficient

Let $D \subset \mathbb{R}^d$, $d = 2,3$ be an open Lipschitz domain and consider a problem of finding $u \in H_0^1(D)$ such that

$$-\mathrm{div}(\alpha \nabla u) = f \quad \text{in } H^{-1}(D). \tag{17}$$

The problem (17) is uniquely solvable if $f \in H^{-1}(D)$ and $\alpha \in L^\infty(D)$ is uniformly positive. Note that u depends nonlinearly on the diffusion coefficient α. According to the notations of section 3 we set

$$\begin{aligned} X &:= \{\alpha \in L^\infty(D) : 0 < \alpha_- \leq \alpha(\mathbf{x}) \leq \alpha_+ < \infty\}, \\ Y &:= H_0^1(D), \quad Z := H^{-1}(D), \quad J(\alpha, u) := -\mathrm{div}(\alpha \nabla u) - f \end{aligned} \tag{18}$$

and $u = S(\alpha)$, where $S : X \to Y$ is the solution operator. The mapping $J : X \times Y \to Z$ is bilinear, hence $J \in C^\infty(X \times Y, Z)$ with the only nonvanishing derivatives

$$\begin{aligned} &\mathrm{d}_\alpha J(\alpha_0, u_0)[r] = -\mathrm{div}(r \nabla u_0), \quad \mathrm{d}_u J(\alpha_0, u_0)[v] = -\mathrm{div}(\alpha_0 \nabla v), \\ &\mathrm{d}^2_{\alpha,u} J(\alpha_0, u_0)[r, v] = -\mathrm{div}(r \nabla v). \end{aligned} \tag{19}$$

Under above assumptions we have $\mathrm{d}_u J(\alpha_0, u_0) \in \mathrm{Inv}(Y, Z)$ and Theorem 3 is applicable. Thus, $u = S(\alpha) \in C^\infty(X, Y)$ and

$$Q^k J(\alpha_0, u_0)[r]^k = k\, \mathrm{d}^2_{\alpha,u} J(\alpha_0, u_0)[r, \mathrm{d}^{k-1} S(\alpha_0)[r]^{k-1}]. \tag{20}$$

The derivatives $S^{(k)} := \mathrm{d}^k S(\alpha_0)[r]^k \in H_0^1(D)$, $k \geq 1$ satisfy the recursive relation

$$-\mathrm{div}(\alpha_0 \nabla S^{(k)}) = k\, \mathrm{div}(r \nabla S^{(k-1)}) \quad \text{in } H^{-1}(D). \tag{21}$$

4.2 Integral Equation on a Perturbed Curve

Suppose $D \Subset \mathbb{R}^2$ is an open, simply connected, bounded domain with smooth boundary $\Gamma := \partial D \in C^\infty$ and an outer normal $\mathbf{n}$. Let B_Γ be an open tubular neighborhood of Γ. Denote by $\gamma_\Gamma : H^{s+1}_{loc}(B_\Gamma) \to H^{s+1/2}(\Gamma)$ the Dirichlet trace operator

$$\gamma_\Gamma u(\mathbf{x}) = \lim_{\mathbf{y}\to\mathbf{x}} u(\mathbf{y}), \quad \mathbf{x}\in\Gamma. \tag{22}$$

Given $g \in H^{s+1}(B_\Gamma)$ for $s \geq 0$, consider the problem of finding $u \in H^{-1/2}(\Gamma)$ satisfying Symm's integral equation

$$\mathscr{V}_\Gamma u = \gamma_\Gamma g \quad \text{in } H^{1/2}(\Gamma) \tag{23}$$

for the single layer operator $\mathscr{V}_\Gamma$

$$\mathscr{V}_\Gamma u(\mathbf{x}) := \int_\Gamma G(\mathbf{x},\mathbf{y})u(\mathbf{y})\,\mathrm{d}s_\mathbf{y}, \quad \mathbf{x}\in\Gamma. \tag{24}$$

Here the kernel $G(\mathbf{x},\mathbf{y}) = -\frac{1}{2\pi}\log\|\mathbf{x}-\mathbf{y}\|$ is the fundamental solution of the Laplace operator and $\mathrm{d}s_\mathbf{y}$ denotes the curve integration on Γ w.r.t. $\mathbf{y}$. In case $\Gamma \in C^\infty$ and $\mathrm{diam}(D) < 1$ the operator $\mathscr{V}_\Gamma : H^{s-1/2}(\Gamma) \to H^{s+1/2}(\Gamma)$ is continuous and bijective for any real s, cf. [6].

Let $\alpha \in C^\infty(\Gamma)$ be a smooth scalar perturbation field. Define a family of closed curves

$$\Gamma_\alpha := \{(Id + T_\alpha)\mathbf{x} : \mathbf{x}\in\Gamma\}, \quad T_\alpha(\mathbf{x}) := \alpha(\mathbf{x})\mathbf{n}(\mathbf{x}), \quad \mathbf{x}\in\Gamma \tag{25}$$

and identify Γ with Γ_0. Note that $\Gamma_\alpha \in C^\infty$ if α is sufficiently small e.g. in the sense $\|\alpha\|_{C^4(\Gamma)} \ll 1$. The perturbed version of Symm's integral equation (23) reads

$$\mathscr{V}_{\Gamma_\alpha} u = \gamma_{\Gamma_\alpha} g \quad \text{in } H^{1/2}(\Gamma_\alpha). \tag{26}$$

In the setting of section 3 we define for $s > 1$

$$X := C^4(\Gamma), \quad Y := H^{s-1/2}(\Gamma), \quad Z := H^{s+1/2}(\Gamma), \quad J(\alpha,u) := \mathscr{V}_{\Gamma_\alpha} u - \gamma_{\Gamma_\alpha} g. \tag{27}$$

We set $u = S(\alpha)$, since (26) is uniquely solvable. Our aim is to show $S \in C^1(X,Y)$ and to compute $dS(\alpha_0)$ at $\alpha_0 = 0$. Theorem 3 requires $J \in C^1(W,Z)$ for $W \subset X\times Y$ and $d_u J(0,u_0) \in \mathrm{Inv}(Y,Z)$. The last is trivial, since $J(\alpha,\cdot)$ is linear and

$$\mathrm{d}_u J(0,u_0) = \mathscr{V}_\Gamma \in \mathrm{Inv}(Y,Z). \tag{28}$$

We sketch the proof of $J(\cdot,u) \in C^1(X,Z)$, cf. [3] for more details. Regularity assumptions on g yield $\mathrm{d}_\alpha(\gamma_{\Gamma_\alpha} g)[r] = \frac{\partial g}{\partial \mathbf{n}} r \in C(X,Z)$. Differentiability of $\mathscr{V}_{\Gamma_\alpha} u$ w.r.t. α follows from differentiability of its kernel, cf. [8, Theorem 3.5]. The smooth change of variables $\mathbf{x}_\alpha = (Id + T_\alpha)\mathbf{x}$, $\mathbf{y}_\alpha = (Id + T_\alpha)\mathbf{y}$, $\mathrm{d}s_{\mathbf{y}_\alpha} = F_\alpha(\mathbf{y}_\alpha)\mathrm{d}s_\mathbf{y}$ yields

$$\mathrm{d}_\alpha(\mathscr{V}_{\Gamma_0} u)[r] = \int_\Gamma \mathrm{d}_\alpha K(\mathbf{x},\mathbf{y},0)[r]\,\mathrm{d}s_\mathbf{y}, \quad K(\mathbf{x},\mathbf{y},\alpha) := G(\mathbf{x}_\alpha,\mathbf{y}_\alpha)u(\mathbf{y})F_\alpha(\mathbf{y}_\alpha) \tag{29}$$

Further, $\mathrm{d}_\alpha G(\mathbf{x},\mathbf{y})[r] = \frac{\partial G}{\partial \mathbf{n}_\mathbf{x}}(\mathbf{x},\mathbf{y})r(\mathbf{x}) + \frac{\partial G}{\partial \mathbf{n}_\mathbf{y}}(\mathbf{x},\mathbf{y})r(\mathbf{y})$, $\mathrm{d}_\alpha F_0(\mathbf{y})[r] = r(\mathbf{y})\mathrm{div}_\Gamma \mathbf{n}(\mathbf{y})$, cf. [10, Lemma 2.49, 2.63] for the last summand. Collecting the terms we obtain

$$\mathrm{d}_\alpha(\mathscr{V}_{\Gamma_0} u)[r] = (\mathscr{K}'_\Gamma u)r + \mathscr{K}_\Gamma(ur) + \mathscr{V}_\Gamma(ur\,\mathrm{div}_\Gamma \mathbf{n}), \tag{30}$$

where $\mathscr{K}_\Gamma$ and $\mathscr{K}_\Gamma'$ are the double layer operator and its adjoint

$$\mathscr{K}_\Gamma u(\mathbf{x}) := \int_\Gamma \frac{\partial G}{\partial \mathbf{n_y}}(\mathbf{x},\mathbf{y})u(\mathbf{y})\,\mathrm{d}s_\mathbf{y}, \qquad \mathscr{K}_\Gamma' u(\mathbf{x}) := \int_\Gamma \frac{\partial G}{\partial \mathbf{n_x}}(\mathbf{x},\mathbf{y})u(\mathbf{y})\,\mathrm{d}s_\mathbf{y}. \tag{31}$$

Herewith we have shown that assumptions of Theorem 3 are satisfied and in particular the following corollary holds.

Corollary 1. *Let $\Gamma \in C^\infty$, $s > 1$ and $\Gamma_r = (Id + T_r)\Gamma$ for $r \in C^4(\Gamma)$: $\|r\|_{C^4(\Gamma)} \ll 1$. Suppose $g \in H^{s+1}(B_\Gamma)$ and $u = S(r) \in H^{s-1/2}(\Gamma_r)$ is a unique solution of (26). Then $S \in C^1(C^4(\Gamma), H^{s-1/2}(\Gamma))$ and its first derivative at $\alpha_0 = 0$ satisfies*

$$\mathscr{V}_\Gamma dS(0)[r] = \left(\frac{\partial g}{\partial \mathbf{n}} - \mathscr{K}_\Gamma' S(0)\right) r - \mathscr{K}_\Gamma(S(0)r) - \mathscr{V}_\Gamma(S(0)\, r\, \mathrm{div}_\Gamma \mathbf{n}). \tag{32}$$

4.3 Random Perturbations with Small Amplitude

Let us recall again the setting of section 3 and consider (1) with randomly perturbed $\alpha = \alpha_0 + r$ with an *almost sure* small perturbation amplitude $r(\omega)$, cf. [3, 5]. In this case the solution $u(\omega) = S(r(\omega))$ of (1) and its Fréchet derivatives are random fields. In what follows we describe, how equation (16) can be used for approximate computation of the statistical moments of $\mathrm{d}S(\alpha_0)[r]$. We consider a class of problems, where X and Y are separable Hilbert spaces, $Z := Y'$ is the dual of Y, and the derivative $\mathrm{d}_u J(\alpha, u)$ is a strongly elliptic operator.

Let (Ω, Σ, P) be a probability space over X consisting of the space of "events" Ω, σ-algebra of its subsets Σ and the probability measure P on Σ. Then for every fixed $\omega \in \Omega$ the functions $r(\omega)$ and $u(\omega)$ belong to X and Y respectively. For an integer $k \geq 1$ and a separable Hilbert space X we define a Bochner space $L^k(\Omega, X)$ of functions $r : \Omega \to X$ endowed with the norm

$$\|r\|_{L^k(\Omega,X)} := \left(\int_\Omega \|r(\omega)\|_X^k \, dP(\omega)\right)^{1/k}.$$

Further, we define the k-fold tensor product space $X^{(k)} := \bigotimes_{i=1}^k X$ with the induced norm $\|\cdot\|_{X^{(k)}}$. Note that $r \in L^k(\Omega, X)$ yields $(\bigotimes_{i=1}^k r) \in L^1(\Omega, X^{(k)})$, cf. [7].

Definition 2. For $r \in L^1(\Omega, X)$ its mean field is defined by

$$\mathbb{E}[r] := \int_\Omega r(\omega)\, dP(\omega) \in X. \tag{33}$$

Moreover, if $r \in L^k(\Omega, X)$, $k \geq 2$, its kth moment is defined by

$$\mathscr{M}^k[r] := \int_\Omega \left(\bigotimes_{i=1}^k r(\omega)\right) dP(\omega) \in X^{(k)}. \tag{34}$$

Note that $\mathscr{M}^k[r]$ is well defined, since $\|\mathscr{M}^k[r]\|_{X^{(k)}} \leq \|r\|^k_{L^k(\Omega,X)}$. Define the tensor product operators $\mathscr{A}^{(k)} : Y^{(k)} \to Z^{(k)}$, $\mathscr{B}^{(k)} : X^{(k)} \to Z^{(k)}$

$$\mathscr{A}^{(k)} := \bigotimes_{i=1}^{k} (-\mathrm{d}_u J(\alpha_0, u_0)), \quad \mathscr{B}^{(k)} := \bigotimes_{i=1}^{k} \mathrm{d}_\alpha J(\alpha_0, u_0). \tag{35}$$

We tensorize the left- and the right-hand side of (16) and take the mean of the tensorized equation, yielding

$$\mathscr{A}^{(k)} \mathscr{M}^{(k)} [\mathrm{d}S(\alpha_0)[r(\omega)]] = \mathscr{B}^{(k)} \mathscr{M}^{(k)} [r(\omega)]. \tag{36}$$

In [7] it is proven that $\mathscr{A}^{(k)} : Y^{(k)} \to Z^{(k)}$ is an invertible operator, thus (36) has a unique solution.

In [4, 7, 9] the above tensorization technique was applied to linear problems. In [5] the described method was applied to the Poisson problem in a randomly perturbed domain. In this paper a general procedure is given, which is applicable to an abstract nonlinear equation (1) with strongly elliptic invertible derivative $d_u J$, cf. also [3].

Acknowledgements The author would like to thank Prof. Christoph Schwab for many helpful discussions.

References

1. Ambrosetti, A., Prodi, G.: A primer of nonlinear analysis, *Cambridge Studies in Advanced Mathematics*, vol. 34. Cambridge University Press, Cambridge (1993)
2. Berger, M.S.: Nonlinearity and functional analysis. Academic Press [Harcourt Brace Jovanovich Publishers], New York (1977). Lectures on nonlinear problems in mathematical analysis, Pure and Applied Mathematics
3. Chernov, A., Schwab, C.: Nonlinear problems under small random perturbations (2008). In preparation
4. Chernov, A., Schwab, C.: Sparse p-version bem for first kind boundary integral equations with random loading. Appl. Numer. Math. (2008). In press
5. Harbrecht, H., Schneider, R., Schwab, C.: Sparse second moment analysis for elliptic problems in stochastic domains. Numer. Math. **109**(3), 385–414 (2008)
6. Hsiao, G.C., Wendland, W.L.: A finite element method for some integral equations of the first kind. J. Math. Anal. Appl. **58**(3), 449–481 (1977)
7. von Petersdorff, T., Schwab, C.: Sparse finite element methods for operator equations with stochastic data. Appl. Math. **51**(2), 145–180 (2006)
8. Potthast, R.: Fréchet Differenzierbarkeit von Randintegraloperatoren und Randwertproblemen zur Helmholtzgleichung und den zeitharmonischen Maxwellgleichungen. Ph.D. thesis, Georg-August-Universität Göttingen (1994)
9. Schwab, C., Todor, R.A.: Sparse finite elements for stochastic elliptic problems—higher order moments. Computing **71**(1), 43–63 (2003)
10. Sokołowski, J., Zolésio, J.P.: Introduction to shape optimization, *Springer Series in Computational Mathematics*, vol. 16. Springer-Verlag, Berlin (1992). Shape sensitivity analysis

Multiscale Analysis for Jump Processes in Finance

N. Reich

Abstract In this work we illustrate how Finite Element methods can be used for asset pricing in generic multidimensional models with jumps. We describe the corresponding partial integrodifferential equations, discuss the numerical challenges, and briefly illustrate possible remedies such as sparse tensor products and wavelet compression.

1 Jump Processes for Asset Pricing

Consider arbitrage-free values $u(x,T)$ of contingent claims on baskets of $s \in \mathbb{N}$ assets. The log-returns of the underlying assets are modeled by a Lévy or, more generally, a Feller process X with state space $\mathbb{R}^d$, $s \leq d$, and $X_0 = x$. For example, the wavelet techniques that we sketch in this work can be applied when X is a Lévy copula process (then $d = s \geq 2$, cf. [9]) or the price process of a Barndorff-Nielsen-Shephard (BNS) stochastic volatility model (then $s = 1$, $d = 2$, cf. [1]).

By the fundamental theorem of asset pricing (see [7]), the arbitrage free price u of an European contingent claim with payoff $g(\cdot)$ is given by the conditional expectation

$$u(x,t) = \mathbb{E}\left(g(X_t) \mid X_0 = x\right),$$

under an a-priori chosen martingale measure equivalent to the historical measure (see e.g. [6, 8] for measure selection criteria).

Deterministic methods to compute $u(x,T)$ are based on the solution of the corresponding backward Kolmogorov equation

$$u_t + \mathscr{A} u = 0, \qquad u|_{t=T} = g. \tag{1}$$

Nils Reich
ETH Zurich, Seminar for Applied Mathematics, Raemistrasse 101, 8092 Zurich, Switzerland, e-mail: reich@math.ethz.ch

Here $\mathscr{A}$ denotes the infinitesimal generator of X with domain $\mathscr{D}(\mathscr{A})$. For the Galerkin-based Finite Element implementation, equation (1) is converted into variational form. Formally, the resulting problem reads: Find u such that

$$\langle \frac{d}{dt}u, v\rangle + \langle \mathscr{A}u, v\rangle = 0, \quad \text{for all } v \in \mathscr{D}(\mathscr{A}). \tag{2}$$

In the classical setting of Black-Scholes, X is a geometric Brownian Motion and $\mathscr{A}$ is a diffusion operator so that a closed form solution of (1) and (2) for plain vanilla contracts is possible in certain cases. For more general Lévy or Feller price processes X, $\mathscr{A}$ is in general a pseudodifferential operator with symbol ψ^X, i.e.

$$(\mathscr{A}u)(x) = (\psi^X(x,D)u)(x) = -\int_{\mathbb{R}^d} e^{i\langle \xi, x\rangle}\psi^X(x,\xi)\widehat{u}(\xi)d\xi. \tag{3}$$

2 Partial Integrodifferential Pricing Equation in High Dimensions

For the numerical solution of the variational problem (2) we employ Finite Element methods as developed in [4, 9, 12, 15, 16, 17]. In this section we further illustrate this approach in case the underlying process X is a Lévy process.

The considerations of this section are based on [9, 21]. Suppose X is a Lévy process with state space $\mathbb{R}^d$ and characteristic exponent

$$\psi^X(\xi) = -i\langle \gamma, \xi\rangle + \frac{1}{2}\langle \xi, Q\xi\rangle + \int_{\mathbb{R}^d\setminus\{0\}} \left(1 - e^{i\langle \xi, y\rangle} + \frac{i\langle \xi, y\rangle}{1+|y|^2}\right)\nu(dy), \tag{4}$$

where $\gamma \in \mathbb{R}^d$ is the drift vector, $Q \in \mathbb{R}^{d\times d}$ is the covariance matrix and $\nu(dy)$ is the Lévy measure. For a detailed overview of Lévy processes we refer to the monographs [2, 22].

Assume the risk-neutral dynamics of $s = d > 1$ assets are given by

$$S_t^i = S_0^i e^{rt + X_t^i}, \quad i = 1, \dots, d,$$

under a risk-neutral measure such that e^{X^i} is a martingale with respect to the canonical filtration $\mathscr{F}_t^0 := \sigma(X_s, s \le t)$, $t \ge 0$, of the multivariate process X. Here $r \ge 0$ denotes a fixed, deterministic interest rate.

Consider an European option with maturity $T < \infty$ and payoff $g(S)$ which is assumed to be Lipschitz. The price $u(t, S_t)$ of this option is given by

$$u(t,S) = \mathbb{E}\left(e^{-r(T-t)}g(S_T)|S_t = S\right), \tag{5}$$

and, sufficient smoothness provided, it can be computed as the solution of a partial integrodifferential equation (PIDE).

Theorem 1. *Assume that the function $u(t,S)$ in (5) satisfies*

$$u(t,S) \in C^{1,2}\left((0,T)\times\mathbb{R}^d_{>0}\right)\cap C^0\left([0,T]\times\mathbb{R}^d_{\geq 0}\right).$$

Then, $u(t,S)$ is the solution of the following PIDE:

$$\frac{\partial u}{\partial t}(t,S)+\frac{1}{2}\sum_{i,j=1}^{d} S_iS_jQ_{ij}\frac{\partial^2 u}{\partial S_i\partial S_j}+r\sum_{i=1}^{d} S_i\frac{\partial u}{\partial S_i}(t,S)-ru(t,S) \tag{6}$$
$$+\int_{\mathbb{R}^d}\left(u(t,Se^z)-u(t,S)-\sum_{i=1}^{d} S_i(e^{z_i}-1)\frac{\partial u}{\partial S_i}(t,S)\right)\nu(dz)=0,$$

in $(0,T)\times\mathbb{R}^d_{\geq 0}$ where $u(t,Se^z):=u(t,S_1e^{z_1},\ldots,S_de^{z_d})$, and the terminal condition is given by

$$u(T,S)=g(S)\quad \forall S\in\mathbb{R}^d_{\geq 0}. \tag{7}$$

Proof. The proof of the PIDE is based on the Itô formulae for multidimensional Lévy processes and semimartingales [14, Theorem 4.57]. For sake of brevity, we refer to [21, Theorem 4.2] for full details.

If the marginal Lévy measures ν_i, $i=1,\ldots,d$, of ν, are absolutely continuous and admit densities $\nu_i(dz)=k_i(z)dz$ with constants $c>0$, $G_i>0$ and $M_i>0$, $i=1,\ldots,d$, such that

$$k_i(z)\leq\begin{cases} ce^{G_iz}, & \text{for all } z<1,\\ ce^{-M_iz}, & \text{for all } z>1,\end{cases} \tag{8}$$

then the PIDE (6) can be transformed into a simpler form.

Corollary 1. *Suppose the marginal Lévy measures ν_i, $i=1,\ldots,d$, satisfy* (8) *with $M_i>1$, $G_i>0$, $i=1,\ldots,d$. Furthermore, let*

$$u(\tau,x)=e^{r\tau}V\left(T-\tau,e^{x_1+(\gamma_1-r)\tau},\ldots,e^{x_d+(\gamma_d-r)\tau}\right),$$

where

$$\gamma_i=\frac{Q_{ii}}{2}+\int_{\mathbb{R}}(e^{z_i}-1-z_i)\,\nu_i(dz_i).$$

Then, u satisfies the PIDE

$$\frac{\partial u}{\partial\tau}+\mathscr{A}_{BS}[u]+\mathscr{A}_J[u]=0, \tag{9}$$

in $(0,T)\times\mathbb{R}^d$ with initial condition $u(0,x):=u_0$. The differential operator is defined for $\varphi\in C_0^2(\mathbb{R}^d)$ by

$$\mathscr{A}_{BS}[\varphi]=-\frac{1}{2}\sum_{i,j=1}^{d} Q_{ij}\frac{\partial^2\varphi}{\partial x_i\partial x_j}, \tag{10}$$

and the integrodifferential operator by

$$\mathscr{A}_J[\varphi] = -\int_{\mathbb{R}^d}\left(\varphi(x+z)-\varphi(x)-\sum_{i=1}^{d} z_i\frac{\partial\varphi}{\partial x_i}(x)\right)\nu(dz)\,. \tag{11}$$

The initial condition is given by $u_0 = g(e^x) := g(e^{x_1},\dots,e^{x_d})$.

Proof. See [21, Corollary 4.3].

For $u,v\in C_0^\infty(\mathbb{R}^d)$ we associate with $\mathscr{A}_{BS}$ the bilinear form

$$\mathscr{E}_{BS}(u,v) = \frac{1}{2}\sum_{i,j=1}^{d} Q_{ij}\int_{\mathbb{R}^d}\frac{\partial u}{\partial x_i}\frac{\partial v}{\partial x_j}dx\,. \tag{12}$$

To the jump part $\mathscr{A}_J$ we associate the so-called canonical bilinear jump form

$$\mathscr{E}_J(u,v) = -\int_{\mathbb{R}^d}\int_{\mathbb{R}^d}\left(u(x+z)-u(x)-\sum_{i=1}^{d} z_i\frac{\partial u}{\partial x_i}(x)\right)v(x)dx\,\nu(dz)\,, \tag{13}$$

and set

$$\mathscr{E}(u,v) = \mathscr{E}_{BS}(u,v)+\mathscr{E}_J(u,v)\,.$$

Herewith, we can now formulate the realization of the abstract problem (2):

$$\begin{aligned}
&\text{Find } u\in L^2((0,T);\mathscr{D}(\mathscr{E}))\cap H^1((0,T);\mathscr{D}(\mathscr{E})^*) \text{ such that}\\
&\langle\frac{\partial u}{\partial\tau},v\rangle_{\mathscr{D}(\mathscr{E})^*,\mathscr{D}(\mathscr{E})}+\mathscr{E}(u,v)=0,\ \tau\in(0,T),\forall v\in\mathscr{D}(\mathscr{E}),\\
&u(0)=u_0\,.
\end{aligned} \tag{14}$$

Here $\mathscr{D}(\mathscr{E})$ denotes the domain of the Dirichlet form $\mathscr{E}$ of X. For the well-posedness of (14) we refer to [9, 21].

To cast (14) into an implementable form it needs to be localized to a bounded domain and $\mathscr{D}(\mathscr{E})$ needs to be replaced by finite dimensional subspaces. For the localization we find, that in Finance the truncation of the original x-domain $\mathbb{R}^d$ to $\Omega_R := [-R,R]^d$, $R>0$, corresponds to approximating the solution u of (6) by the price u_R of a barrier option on Ω_R. In log-price u_R is given by

$$u_R(t,x) = \mathbb{E}\left(g(e^{X_T})1_{\{T<\tau_{\Omega_R,t}\}}|X_t = x\right),$$

where $\tau_{\Omega_R,t} = \inf\{s\ge t|X_s\notin\Omega_R\}$ denotes the first exit time of X_t from Ω_R after time t. In case of semiheavy tails (8), the solution u_R of the localized problem converges pointwise exponentially to the solution u of the original problem (cf. [21]).

Finally, for any function u with support in Ω_R we denote by $\overline{u}$ its extension by zero to the whole of $\mathbb{R}^d$ and define

$$\mathscr{E}_R(u,v) = \mathscr{E}(\overline{u},\overline{v}),$$

with

$$\mathcal{D}(\mathcal{E}_R) = \overline{\{\overline{u} \mid u \in C_0^\infty(\Omega_R)\}},$$

where the closure is taken with respect to the natural norm of $\mathcal{D}(\mathcal{E})$. Thus, we can restate the variational form (14) on the bounded domain Ω_R and existence and uniqueness results for (14) remain valid.

$$\begin{aligned}
&\text{Find } u_R \in L^2((0,T);\mathcal{D}(\mathcal{E}_R)) \cap H^1((0,T);\mathcal{D}(\mathcal{E}_R)^*) \text{ such that} \\
&\langle \frac{\partial u_R}{\partial \tau}, v\rangle + \mathcal{E}_R(u_R,v) = 0, \quad \forall \tau \in (0,T), \forall v \in \mathcal{D}(\mathcal{E}_R), \\
&u_R(0) = u_0|_{\Omega_R}.
\end{aligned} \tag{15}$$

3 Jump Processes from a Wavelet Compression Point of View

Due to the possible non-locality of the integral operator $\mathcal{A}_J$ in (11), Finite Element (FE) discretization of (15) in general leads to linear systems with densely populated matrices of substantial size.

Even on the tensor product domain $[-R,R]^d$, the straightforward application of standard numerical schemes fails due to the "curse of dimension": The number of degrees of freedom on a tensor product Finite Element (FE) mesh of width h in dimension d grows like $\mathcal{O}(h^{-d})$ as $h \to 0$. The non-locality of the underlying operator thus implies that the FE stiffness matrix consists of $\mathcal{O}(h^{-2d})$ non-zero entries.

In this Section we give a very brief illustration of how to construct a sparse tensor product-based wavelet compression scheme for infinitesimal generators of certain multivariate Lévy processes. The scheme preserves the convergence rate and stability while *asymptotically* reducing the Finite Element complexity of a multidimensional non-local operator to that of a one-dimensional local one. For a complete and rigorous treatment we refer to [20, Chapters 3 & 4].

To overcome the "curse of dimension" we choose sparse tensor product Finite Element spaces $\widehat{V}_J$, $J > 0$, for the discretization of $\mathcal{D}(\mathcal{E}_R)$ – assuming that $\widehat{V}_J \subset \mathcal{D}(\mathcal{E}_R)$, which for instance is shown in [9] for Lévy copula processes. For sake of brevity we omit details here but refer to [3, 9, 10, 18, 19] and the references therein for thorough analysis. This approach yields a stiffness matrix consisting of only $\mathcal{O}(h^{-2}|\log h|^{2(d-1)})$ entries as $h \to 0$ while at the same time (essentially) preserving the approximation rate. As shown in e.g. [10], these results require greater smoothness of the approximated function u than the original discretization. As basis functions in the spaces $\widehat{V}_J$ we shall employ piecewise polynomial spline wavelets (cf. [5, 11]).

To further reduce the complexity of the still densely populated sparse tensor product matrix we apply the so-called non-radial wavelet compression of [20, Chapter 4]. To define the requirements of the non-radial compression scheme, we need to introduce some notation: Denote the axes in $\mathbb{R}^d$ by $\Lambda := \{x \in \mathbb{R}^d : x_i = 0 \text{ for some } i \in$

$\{1,\ldots,n\}\}$. Herewith we can define a suitable class of anisotropic symbols and corresponding operators.

Definition 1. A function $p:\mathbb{R}^d\to\mathbb{R}$ is called a symbol in class $\Gamma^{\underline{\alpha}}(\mathbb{R}^d)$, $\underline{\alpha}\in\mathbb{R}^d$, if $p\in C^\infty(\mathbb{R}^d\backslash\Lambda)\cap C(\mathbb{R}^d)$ such that for any $\underline{\tau}\in\mathbb{N}_0^d$ there exists some constant $c_{\underline{\tau}}\geq 0$ such that

$$\left|\partial_\xi^{\underline{\tau}}p(\xi)\right|\leq c_{\underline{\tau}}\cdot\prod_{i\in\mathscr{I}_{\underline{\tau}}}|\xi_i|^{\alpha_i-\tau_i}\cdot\prod_{k\notin\mathscr{I}_{\underline{\tau}}}(1+|\xi_k|^2)^{\frac{\alpha_k}{2}},\qquad\text{for all }\xi\in\mathbb{R}^d,\tag{16}$$

where we set $\mathscr{I}_{\underline{\tau}}:=\{i:\tau_i>0\}$. The multiindex $\underline{\alpha}$ is called the (anisotropic) order of the symbol p and the operator $p(D)u(x)=-\int_{\mathbb{R}^d}e^{i\langle x,\xi\rangle}p(\xi)\widehat{u}(\xi)d\xi$.

Furthermore, one obtains an integral kernel representation of $\mathscr{A}_J$ by writing for any $u\in\mathscr{S}(\mathbb{R}^d)$,

$$\mathscr{A}_Ju(x)=-\int_{\mathbb{R}^d}\int_{\mathbb{R}^d}e^{i\langle x-y,\xi\rangle}\psi^X(x,\xi)u(y)dyd\xi=\int_{\mathbb{R}^d}\kappa(x,y)u(y)dy,\tag{17}$$

with

$$\kappa(x,y):=\int_{\mathbb{R}^d}e^{i\langle x-y,\xi\rangle}\psi^X(x,\xi)d\xi,\tag{18}$$

the inverse Fourier transform (in the sense of oscillatory integrals) of ψ^X at $x-y$.

Remark 1. If X is a Lévy process with absolutely continuous Lévy measure then the following relation holds between $\kappa(\cdot,\cdot)$ and the density $k(\cdot)$ of the Lévy measure of X:

$$\kappa(x,y)=\int_{\mathbb{R}^d\backslash\{0\}}\int_{\mathbb{R}^d}\left(e^{i\langle x-y-z,\xi\rangle}-e^{i\langle x-y,\xi\rangle}+\frac{i\langle z,\xi\rangle}{1+|z|^2}e^{i\langle x-y,\xi\rangle}\right)k(z)d\xi dz,\tag{19}$$

in the sense of distributions. By [13, Lemma 2.8], for any $\xi\in\mathbb{R}^d$, $z\in\mathbb{R}^d$ there holds

$$\left|e^{-i\langle z,\xi\rangle}-1+\frac{i\langle z,\xi\rangle}{1+|z|^2}\right|\leq 7\cdot\frac{|z|^2}{1+|z|^2}\cdot(1+|\xi|^2).$$

Hence, the distributional kernel $\kappa(\cdot,\cdot)$ in (19) is indeed well defined, since $k(\cdot)$ is a Lévy kernel that satisfies

$$\int_{\mathbb{R}^d}(|z|^2\wedge 1)k(z)dz<\infty.$$

Finally, suppose the characteristic exponent ψ^X in (4) of the infinitesimal generator $\mathscr{A}_J=\psi^X(D)$ in (11) satisfies $\psi^X\in\Gamma^{\underline{\alpha}}(\mathbb{R}^d)$ for some $\underline{\alpha}\in\mathbb{R}^d_{\geq 0}$. Then, with the representation (17) of $\mathscr{A}_J$, one may apply the a-priori compression scheme of [20, Section 4.4] to reduce the complexity of the sparse tensor product stiffness matrix without perturbing the sparse tensor product convergence rate. Under the assumption that $Q>0$ in (4) this approach yields essentially optimal $\mathscr{O}(h^{-1}|\log h|^{2(d-1)})$ non-zero matrix entries.

For sake of brevity, we refer to [20, Chapter 4] for the explicit description of the compression scheme as well as consistency and complexity proofs. We conclude by giving some numerical examples:

To analyze the accuracy of the compression scheme, we consider the following model problem: Find the numerical solution of the integrodifferential equation

$$\mathscr{A}u = 0, \quad \text{on } [0,1]^2,$$

where $\mathscr{A}$ denotes the infinitesimal generator of a bivariate Lévy copula process with tempered stable margins and Clayton-type Lévy copula F_θ as defined in [9].

On level $J = 5$, Figure 1 presents the accuracy of the compression scheme. In practice, only the black entries on the right hand side of Figure 1 need to be computed.

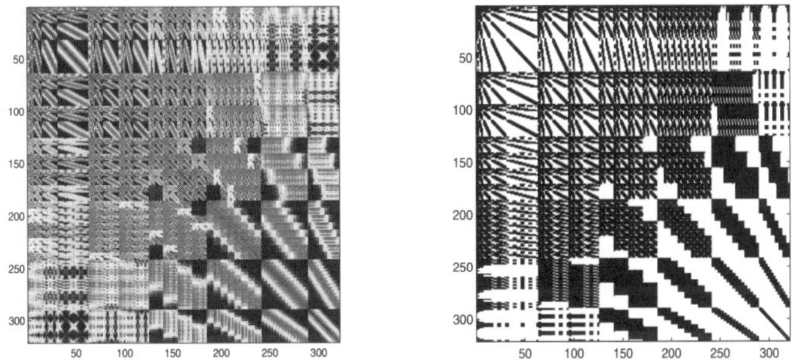

Fig. 1 Structure prediction by the compression scheme. Left: Actual sparsified stiffness matrix of $\mathscr{A}$, 320^2 non-zero entries. Right: A-priori structure prediction by the compression scheme.

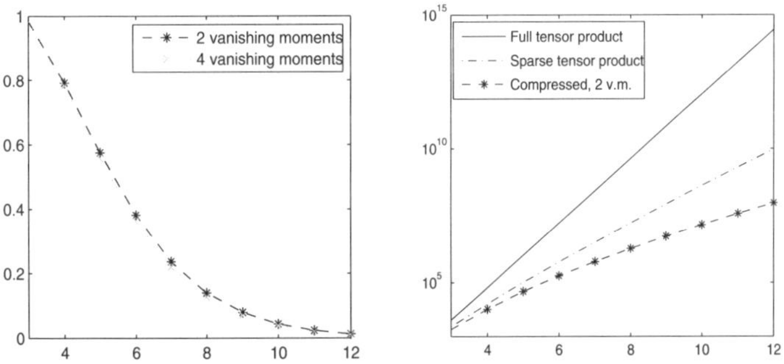

Fig. 2 Left: Percentage of the remaining complexity of the sparse tensor product matrix after compression on levels $J = 3, \ldots, 12$. The dashed line corresponds to wavelet basis functions with 2 vanishing moments, the dotted line describes 4 vanishing moments. Right: Non-zero matrix entries on each level using the piecewise linear spline wavelets described in [9].

The uncompressed sparse tensor product stiffness matrix in Figure 1 was taken from [23]. We also refer to this source for further numerical results.

References

1. Barndorff-Nielsen, O., Shephard, N.: Non-gaussian Ornstein-Uhlenbeck-based models and some of their uses in financial economics. Journal of the Royal Statistical Society **63**, 167–241 (2001)
2. Bertoin, J.: Lévy processes. Cambridge University Press, Cambridge (1996)
3. Bungartz, H.J., Griebel, M.: A note on the complexity of solving Poisson's equation for spaces of bounded mixed derivative. J. Complexity **15**, 167–199 (1999)
4. Cont, R., Voltchkova, E.: A finite difference scheme for option pricing in jump diffusion and exponential Lévy models. SIAM J. Numer. Anal. **43**(4), 1596–1626 (2005)
5. Dahmen, W., Kunoth, A., Urban, K.: Biorthogonal spline wavelets on the interval - stability and moment conditions. Appl. Comp. Harm. Anal. **6**, 259–302 (1999)
6. Delbaen, F., Grandits, P., Rheinländer, T., Samperi, D., Schweizer, M., Stricker, C.: Exponential hedging and entropic penalties. Mathematical Finance **12**, 99–123 (2002)
7. Delbaen, F., Schachermayer, W.: A general version of the fundamental theorem of asset pricing. Math. Ann. **300**, 463–520 (1994)
8. Delbaen, F., Schachermayer, W.: The variance-optimal martingale measure for continuous processes. Bernoulli **2**, 81–105 (1996)
9. Farkas, W., Reich, N., Schwab, C.: Anisotropic stable Lévy copula processes - analytical and numerical aspects. Math. Models and Methods in Appl. Sciences **17**, 1405–1443 (2007)
10. Griebel, M., Oswald, P., Schiekofer, T.: Sparse grids for boundary integral equations. Numer. Math. **83**, 279–312 (1999)
11. Harbrecht, H., Schneider, R.: Biorthogonal wavelet bases for the boundary element method. Math. Nachr. **269-270**, 167–188 (2004)
12. Hilber, N., Matache, A.M., Schwab, C.: Sparse wavelet methods for option pricing under stochastic volatility. Journal of Computational Finance **8**, 1–42 (2005)
13. Hoh, W.: Pseudo Differential Operators generating Markov Processes. Habilitationsschrift, University of Bielefeld (1998)
14. Jacod, J., Shiryaev, A.: Limit theorems for stochastic processes, *Grundlehren der Mathematischen Wissenschaften [Fundamental Principles of Mathematical Sciences]*, vol. 288, second edn. Springer-Verlag, Berlin (2003)
15. Matache, A.M., Nitsche, P., Schwab, C.: Wavelet Galerkin pricing of American contracts on Lévy driven assets. Quantitative Finance **5**, 403–424 (2005)
16. Matache, A.M., von Petersdorff, T., Schwab, C.: Fast deterministic pricing of options on Lévy driven assets. Mathematical Modelling and Numerical Analysis **38**, 37–71 (2004)
17. Matache, A.M., Schwab, C., Wihler, T.: Linear complexity solution of parabolic integrodifferential equations. Numer. Math. **104**, 69–102 (2006)
18. Oswald, P.: On N-term approximation by Haar functions in H^s-norms. In: S. Nikolskij, B. Kashin, A. Izaak (eds.) Metric Function Theory and Related Topics in Analysis, pp. 137–163. AFC, Moscow (1999)
19. von Petersdorff, T., Schwab, C.: Numerical solution of parabolic equations in high dimensions. ESAIM: Math. Model. and Num. Ana. **38**, 93–127 (2004)
20. Reich, N.: Wavelet Compression for Anisotropic Operators in Sparse Tensor Product Spaces. PhD Thesis, ETH Zürich (2008). http://e-collection.ethbib.ethz.ch/view/eth:30174
21. Reich, N., Schwab, C., Winter, C.: On Kolmogorov Equations for Anisotropic Multivariate Lévy Processes. Preprint, Seminar for Applied Mathematics, ETH Zürich (2008)
22. Sato, K.I.: Lévy Processes and Infinitely Divisible Distributions. Cambridge University Press, Cambridge (1999)
23. Winter, C.: Pricing of multidimensional derivatives using Lévy copulas *(tentative title)*. PhD Thesis, ETH Zürich, in preparation

Fluid Mechanics

A Hybrid Numerical Scheme for Aerosol Dynamics

H. Babovsky

Abstract Aerosol coagulation, i.e. the merging of aerosol particles to larger clusters, is commonly described by the Smoluchowski equation. Of special interest is the phenomenon of "gelation", i.e. the formation of "macroparticles". A useful tool for the numerical simulation is the application of Monte Carlo schemes. However, stochastic effects may change qualitative properties, e.g. they may cause the transition from stable states to metastable ones, for example in combination with spatial diffusion.

In [5], a deterministic scheme for the space homogeneous system was formulated and analyzed. It was shown, that the scheme converges monotonically with decreasing discretization parameter. This gives rise to an efficient error control. Moreover, the scheme is capable of monitoring the gelation process with high precision. It turns out that the most efficient system is a hybrid code with a stochastic component for the simulation of higher regions of the state space.

The paper introduces the numerical scheme, discusses its properties in comparison to stochastic schemes and presents numerical examples of the gelation process in diffusive environments.

1 Aerosol Dynamics

Aerosols are suspended particles in fluids. The basic features of aerosol interaction are *coagulation*, i.e. the effect that two particles with masses m and n merge to form one particle with mass $m+n$, and *fragmentation*, the opposite effect, where one particle with mass n may break into two particles with masses i and $n-i$. This latter effect will be suppressed in the present paper. Assuming that all particles have an integer multiple of a unit mass, the (space homogeneous) coagulation dynamics is described by the discrete Smoluchowski equation for the *number density vector*

Hans Babovsky
Technical University Ilmenau, D-98684 Ilmenau, Germany, e-mail: hans.babovsky@tu-ilmenau.de

$f = (f_n)_{n \in \mathbb{N}}$

$$\partial_t f_n = \frac{1}{2} \sum_{i=1}^{n-1} K(i, n-i) f_i f_{n-i} - f_n \sum_{i=1}^{\infty} K(n,i) f_i \tag{1}$$

where $K(i, j)$ is a nonnegative symmetric interaction kernel. An equivalent formulation for the *mass density vector* $g = (g_n) = (n f_n)_{n \in \mathbb{N}}$ is easily derived [2] as

$$\partial_t g_n = \sum_{i=1}^{n-1} \frac{K(i, n-i)}{n-i} g_i g_{n-i} - g_n \sum_{i=1}^{\infty} \frac{K(n,i)}{i} g_i \tag{2}$$

We will make use of this formulation. for the sake of simplicity and clarity we restrict in this paper to the special kernel

$$K(i, j) = ij \tag{3}$$

In this case, the Smoluchowski equation for the mass density reads

$$\partial_t g_n = \sum_{i=1}^{n-1} i g_i g_{n-i} - n g_n \sum_{i=1}^{\infty} g_i. \tag{4}$$

This model has been studied intensely in literature from a theoretical point of view. For a review of different coagulation and fragmentation models, see [1, 7].

As long as $u(t) = \sum_{n=1}^{\infty} n g_n(t)$ (first moment of g) is finite, the right hand side of (4) is absolutely summable, and a straightforward calculation shows that the total mass (zeroth moment) remains constant, i.e.

$$\rho(t) = \sum_{n=1}^{\infty} g_n(t) = \text{const} \tag{5}$$

However, establishing the differential equation for u and rearranging the sums on the right hand side shows that u satisfies

$$\partial_t u = u^2 \tag{6}$$

i.e. u "explodes" after a finite time t_{gel} which is called *gelation time*. At this time, $\rho(t)$ starts to decrease monotonically. The reason for this is *gelation* which means that mass is leaving the state space $\mathbb{N}$ at ∞. Let us define the *gelled mass* g_∞ by $g_\infty(0) = 0$, and

$$\rho(t) + g_\infty(t) = \rho(0) \tag{7}$$

In the case of the special kernel (3) there are two ways to continue the Smoluchowski equation over the gelation time. In the case of of a *passive* gel, the gelled mass is simply removed from the system, and the evolution follows again equation (4) which reads

$$\partial_t g_n = \sum_{i=1}^{n-1} i g_i g_{n-i} - n g_n \overline{\rho} =: S_n[g,g] \tag{8}$$

with $\overline{\rho} = \rho$. In the case of an *active* gel, we have $\overline{\rho} = \rho + g_\infty = \rho(0)$. In the following we will restrict to the case of an active gel.

The aim of this paper is to describe a numerical scheme which is capable of simulating the phase transition to gelation in a space dependent setting, e.g. for the solution of the partial differential equation for $g = g(t,x)$,

$$\partial_t g_n = c_n \Delta_x g_n + S_n[g,g] \tag{9}$$

describing the aerosol evolution in a diffusive environment.

Major difficulties on the way to this end are near at hand and can be studied even in the space homogeneous situation. First, equation (8) describes an *infinite* ODE system. It seems not useful to restrict to a finite system, since transition to gelation means that the higher part of state space plays in some sense a dominant role. Second, (8) is *nonlinear*, of course. Third, as *severe* difficulty, (8) is stiff, as can be seen most clearly from the factor n in the loss term of the equation.

In the next section, we will introduce modifications, which cope with the second and the third aspect, and we will discuss convergence properties of the modified system. Section 3 will deal with the first aspect.

2 Modifications

As a *first modification*, we introduce a small time step $\Delta t > 0$ and replace in the time interval $[k\Delta t, (k+1)\Delta t]$ equation (8) with the linearized system

$$\partial_t g_n = \sum_{i=1}^{n-1} i g_i h_{n-i} - n g_n \overline{\eta} \tag{10}$$

where $h(t) = g(k\Delta t)$ and $\overline{\eta} = \rho(k\Delta t) + g_\infty(k\Delta t)$ ($= \rho(0)$ in the space homogeneous case). This is the simplest possible linearization. It is only of first order in Δt and could be replaced with a more elaborate linear system. However, it turned out in [5] that this discretization allows for a very clear and concise treatment of convergence, since it guarantees a certain monotonicity property (see end of section 2).

Let's consider the evolution of the first moment $u = \sum_n n g_n$ under this modification. Assuming $u(k\Delta t)$ finite, we find $\overline{\eta} = \rho(k\Delta t)$; a rearrangment of the first sum on the right hand side of (10) yields for $t \in [k\Delta t, (k+1)\Delta t]$

$$\partial_t u = u(t) \cdot u(k\Delta t) \tag{11}$$

Thus u stays finite in the whole interval, in contrast to the solution of (6). By induction, u stays finite for all $t > 0$ if $u(0) < \infty$. In particular, since there is no gelation, it follows that $\overline{\eta} = \rho(k\Delta t)$.

Stiffness is an intrinsic property of all systems allowing for gelation (in contrast e.g. to the constant kernel $K(i,j)=1$). On the other hand, it is an unfavorable property of numerical schemes. For convenience, we introduce here a *second modification* reducing stiffness modestly.

Equation (10) is linear; therefore its solution can be represented in terms of fundamental solutions as

$$g(t)=\sum_{r=1}^{\infty} g_r(k\Delta t)\cdot \psi^{(r)}(t) \tag{12}$$

with $\psi^{(r)}$ being the solution of the IVP

$$\partial_t \psi_n^{(r)}=\sum_{i=1}^{n-1} i\psi_i^{(r)}h_{n-i}-n\psi_n^{(r)}\overline{\eta},\quad \psi_n^{(r)}(k\Delta t)=\delta_{n-r} \tag{13}$$

Since $\psi_n^{(r)}(k\Delta t)=0$ for $n<r$, the structure of the differential equation yields $\psi_n^{(r)}(t)=0$ for all $t\geq k\Delta t$. Thus we reduce (but do not axe) stiffness, if we replace in (13) the factors i and n with r. From now on we consider (12) with the modified sequences $\psi^{(r)}$ given by

$$\partial_t \psi_n^{(r)}=r\left(\sum_{i=1}^{n-1} \psi_i^{(r)}h_{n-i}-\psi_n^{(r)}\overline{\eta}\right),\quad \psi_n^{(r)}(k\Delta t)=\delta_{n-r} \tag{14}$$

This modification influences the evolution of the first moment $u(.)$ in a significant way. Since $\sum_n \psi_n^{(r)}(t)=1$ for all t, a straightforward calculation yields

$$u((k+1)\Delta t)=u(k\Delta t)+\Delta t\cdot(u(k\Delta t))^2 \tag{15}$$

which corresponds to the Euler discretization of (6).

From now on we denote for given Δt as $g^{mod}(t)$ the continuous sequence-valued functions given piecewise by (12) under the modification (14) – in contrast to $g(t)$, which is the *exact* solution of the Smoluchowski equation (4). According to this we call the corresponding first moments u^{mod} and u.

Denote ℓ^1 as the space of absolutely summable real-valued sequences and as ℓ^1_+ its positive cone. On ℓ^1_+ we define the partial ordering "$\preceq$" by

$$g\preceq h \quad\Leftrightarrow\quad \sum_{n=1}^{N} g_n \leq \sum_{n=1}^{N} h_n \quad \text{for all } N\in\mathbb{N} \tag{16}$$

For sequences $g,h\in\ell^1_+$ with equal mass $\|g\|_1=\|h\|_1$, the relation $g\preceq h$ implies intuitively that g is concentrated on higher parts of the state space than h. This ordering is well-suited to the Smoluchowski system as well as to the above modified system. Since the coagulation dynamics is directed (mass is flowing from lower to upper state space only), one verifies quickly, that solutions are decreasing in time,

$$g(t_2) \preceq g(t_1), \quad g^{mod}(t_2) \preceq g^{mod}(t_1) \quad \text{for} \quad t_2 > t_1 \tag{17}$$

Moreover, in [5] it was proven that for all $t \geq 0$

$$g(t) \preceq g^{mod}(t) \tag{18}$$

i.e. as a consequence of the modifications the coagulation process is slowed down. On the other hand, because of (6), (15) and the convergence of the Euler scheme, we find for all $t < t_{gel}$

$$u^{mod}(t) \nearrow u(t) \quad \text{as} \quad \Delta t \searrow 0 \tag{19}$$

Exploiting the monotonicity properties and the structure of the Smoluchowski system, one can elaborate in detail the convergence properties. Here, we collect some of the main statements proven in [5].

Theorem 1. *(a) For $T > 0$ there exist $c_n > 0$ such that*

$$\sup_{t\in[0,T]} \left| g_n^{mod}(t) - g_n(t) \right| \leq c_n \Delta t \tag{20}$$

(b) For $t < t_{gel}$,

$$\sum_{n=1}^{\infty} g_n(t) = \lim_{N\to\infty} \lim_{\Delta t \searrow 0} \sum_{n=1}^{N} g_n^{mod}(t) = \lim_{\Delta t \searrow 0} \lim_{N\to\infty} \sum_{n=1}^{N} g_n^{mod}(t) \tag{21}$$

and for $t > t_{gel}$,

$$\sum_{n=1}^{\infty} g_n(t) = \lim_{N\to\infty} \lim_{\Delta t \searrow 0} \sum_{n=1}^{N} g_n^{mod}(t) < \lim_{\Delta t \searrow 0} \lim_{N\to\infty} \sum_{n=1}^{N} g_n^{mod}(t) = \sum_{n=1}^{\infty} g_n(0) \tag{22}$$

We want to point out that our main subject is to model the phase transition to gelation. This requires some thresholding technique, since due to formula (15) there is no blow-up of the first moment in finite time and thus no mass transfer to ∞ (see (22)). We will take this aspect into account in section 3.

3 Numerical Methods

Since the phase space is infinite dimensional, the idea is near at hand to use *Monte Carlo methods* for the above modified system. Such a scheme for g^{mod} was first introduced in [3]. Prior to this there have been different attempts to simulate the space homogeneous Smoluchowski equation via Monte Carlo systems (see [2, 6] and the literature cited there). As it turns out, such methods indeed may produce reliable results. However, there is an important restriction to take into account – in particular with respect to space inhomogeneous simulations. As was observed in [4,

3], in the presence of sinks and sources for the aerosol system, random fluctuations may change stable states into metastable ones with the gelated state as a trap. Thus we have to find ways to reduche stochastic effects.

Let us have a short look on the *truncated Smoluchowski system*. In the case of an active gel, equation (8) for g_n depends on $(g_1, \dots, g_n)$ only. Thus for $N \in \mathbb{N}$ and $g^N := (g_1, \dots, g_N)$, we may restrict to the finite ODE system

$$\partial_t g_n = S_n[g,g] = \sum_{i=1}^{n-1} i g_i g_{n-i} - n g_n \overline{\rho} \quad n = 1, \dots, N, \quad \overline{\rho} = \text{const} \tag{23}$$

Performing again the two modifications of section 2, we end up with

$$g^{N,mod}(t) = \sum_{r=1}^{N} g_r(k\Delta t) \cdot \psi^{(r,N)}(t) \tag{24}$$

where $\psi^{(r,N)}$ is the truncated version of $\psi^{(r)}$ given in (14). In such truncated systems, the only chance to investigate the passage to the gel phase is to introduce a threshold level N and to consider all mass passing this threshold as "gelled" mass. This is the usual way it is done in Monte Carlo simulations like [3]. Here, however, we run into troubles. A low value of N does not meet high demands on accuracy, as was demonstrated in [5]. Increasing this value rapidly blows up the calculational effort – due to the stiffness of the problem.

A promising way out is a hybrid approach coupling the truncated deterministic system (25) resp. (26) with the Monte Carlo scheme [3] for the upper part of state space. To this end we introduce two levels $N_1 < N_2$. The portion $\{1, \dots, N_1\}$ of the state space ("small" particles) are treated by the above truncated system, while the upper part $\{N_1 + 1, \dots, N_2\}$ ("large" particles) is modelled by the stochastic process described in [3]. The particles whose masses pass the level N_1 are treated as a source which feeds the stochastic system. Those which cross the level N_2 represent the gelled mass and drop out of the system. (In the case of an active gel, they still appear as g_∞ in the parameter $\overline{\rho}$ of (8).) The lower part up to N_1 is described deterministically and contains the main portion of the non-gelated mass, while the upper part models the transition to gelation. Particles in this section are rapidly transported to infinity; that's why random fluctuations do not play a crucial role. As was analyzed in [5], this modification is numerically efficient and describes the first approach to the gelation phase with high precision. Thus this version may be taken as the method of choice for the simulation also of space dependent problems.

4 An Application

The applicability of the above numerical scheme is not restricted to the special case treated in this paper. In fact, both the Monte Carlo scheme [3] and the truncated system are readily generalized to the situation of a passive gel as well as to arbitrary interaction kernels $K(i,j)$. (The notion of an active gel is useful only for the particular kernel $K(i,j) = ij$ treated above.) In the following example we are going to present numerical results on the interaction of spatial diffusion models with coagulation models and their transition to the gel phase. In these calculations, we again chose the kernel $K(i,j) = ij$, but compared active and passive gels in combination with different space evolution models. As specific test case we considered the diffusion coagulation system in $[-1,1] \times [-1,1]$ with periodic boundary conditions and with a monomer point source at $x = (0,0)$ given by

$$\partial_t f_n = S_n[f,f] + D_n \Delta_x f_n + s \cdot \delta_{n=1,x=0}. \tag{25}$$

Since we wanted to study the influence of the diffusion model on the gelation process, we calculated two cases with different sets of diffusion coefficient D_n,

Case 1: $D_n = 1/n$
Case 2: $D_n = 1$.

A quick argument suggests the following for the initial phase. In case 1, large particles have less mobility and thus should locally accelerate the gelation process. Since only (the low amount of) small particles move quickly, the spreading of the gelation area should be restricted compared to case 2.

In our simulations we restricted to a spatial 50×50 grid and replaced the Laplace operator by the convenient central difference approximation. Results of the simulation runs are illustrated in Figs. 1 and 2. Fig. 1 shows the first moments of g at

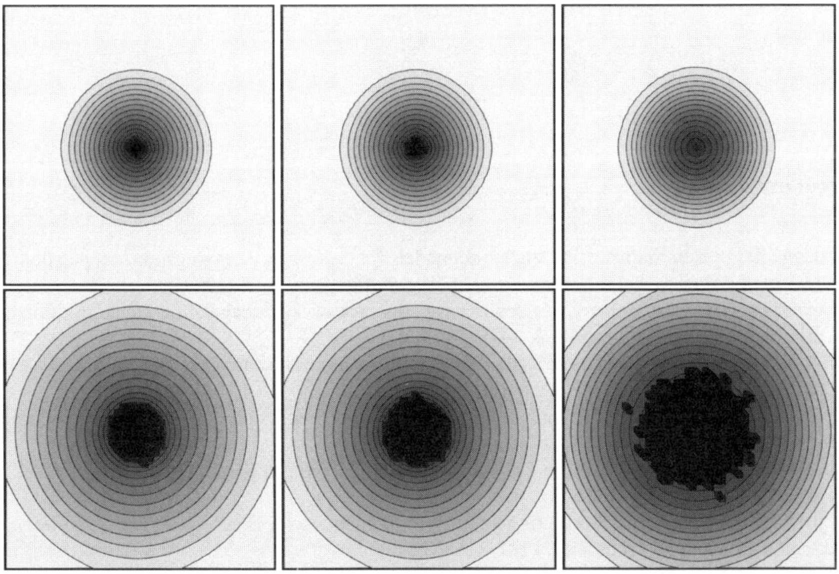

Fig. 1 Development of first moment after short time t_0 (upper row) and after time $3t_0$ (lower row). Left column: active gel, $D_n = 1/n$; middle column: passive gel, $D_n = 1/n$, right column: passive gel, $D_n = 1$.

time t_0 (78 time steps, upper row) and $t_1 = 3.6t_0$ (278 time steps, lower row). The dark regions in the center indicate the areas where gelation has initiated. The results confirm the above arguments. Gelation starts earlier in case 1. (At time t_0, the difference between active and passive gel is not yet relevant.) The lower row illustrates the growth of the gelation area up to time t_1. Its circular boundary is clearly visible, though slightly perturbed from random fluctuations. We recognize that the gelation area spreads much faster in case 2. For active gel, the evolution of gelation is slowest, since the active non-gelled mass is likely to be swallowed by the gel in the area around the source. This is confirmed by Fig. 2, which presents the total non-gelled mass at time t_1.

As a conclusion, the method is numerically efficient and capable of describing (at least qualitatively) the spreading of gelation in spatially inhomogeneous situations.

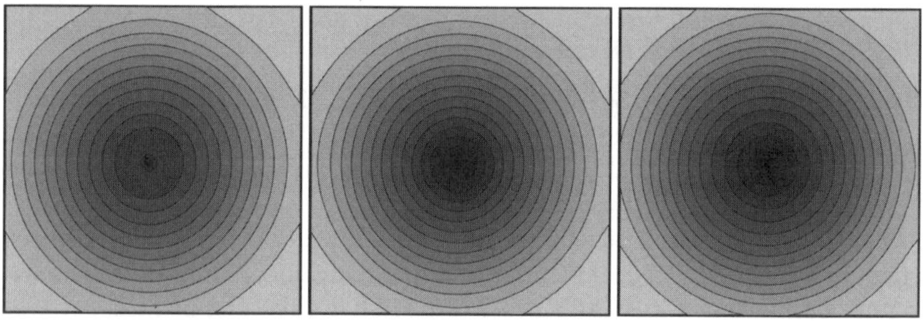

Fig. 2 Development of non-gelled mass (zeroth moment) after time $3t_0$.
Left column: active gel, $D_n = 1/n$; middle column: passive gel, $D_n = 1/n$, right column: passive gel, $D_n = 1$.

References

1. Aldous, D. J.: Deterministic and stochastic models for coalescence (aggregation, coagulation): a review of the mean-field theory for probabilists. Bernoulli **5**, 3–48 (1999)
2. Babovsky, H.: On a Monte Carlo scheme for Smoluchowski's coagulation equation. Monte Carlo Methods and Applications **5**, 1–18 (1999)
3. Babovsky, H.: Gelation of stochastic diffusion-coagulation systems. Physica D **222**, 54–62 (2006)
4. Babovsky, H.: The impact of random fluctuations on the gelation process. Bull. Inst. of Math. Academia Sinica **2**, 329–348 (2007)
5. Babovsky, H.: Approximations to the gelation phase of an aerosol. Preprint 11/06, Inst. f. Math., TU Ilmenau, submitted (2006)
6. Eibeck, A., Wagner, W.: Stochastic particle approximations for Smoluchowski's coagulation equation. Ann. Appl. Probab. **11**, 1137–1165 (2001)
7. Wattis, J. A. D.: An introduction to mathematical models of coagulation-fragmentation processes: A discrete deterministic mean-field approach. Physica D **222**, 1–20 (2006)

Numerical Study of Mixed Finite Element and Multi Point Flux Approximation of Flow in Porous Media

M. Bause and J. Hoffmann

Abstract In this paper the numerical performance properties of some locally mass conservative numerical schemes for simulating flow in a porous medium are studied. In particular, the accuracy of approximations of flows with discontinuous permeability tensors and on irregular grids is analysed. We consider the mixed finite element approach with the lowest order Raviart-Thomas ($\mathbf{RT}_0$) and Brezzi-Douglas-Marini ($\mathbf{BDM}_1$) element and a multi point flux approximation (**MPFA**) control volume method. The $\mathbf{BDM}_1$ method yields a second order accurate approximation of the flux whereas the $\mathbf{RT}_0$ and **MPFA** approach are of first order accuracy only. **MPFA** methods offer explicit discrete fluxes which is not possible to get from mixed finite element methods and allows a wider class of applications.

1 Introduction

Accurate and reliable simulations of moisture fluxes through porous media are desirable in many areas, in particular, in hydrological and environmental engineering. In subsurface flow simulation, the geology which includes composite soil formations, that may be intermittently saturated and drained, and non-orthogonal domains in the media are a major challenge. From the point of view of physical realism, this results in a need to use non-orthogonal grids with a full permeability tensor, which may be discontinuous, in discretizations of the model equations describing the flow.

The mixed finite element method and the multi point flux approximation method have shown to be suitable for the numerical simulation of flow in such a porous medium. Both schemes are locally mass conservative and provide a flux approximation as part of the formulation. In environmental studies, for instance, the reliable prediction of the water flow field is usually of greater importance than the

M. Bause, J. Hoffmann
Department Mathematics, University of Erlangen-Nuremberg, Martensstr. 3, D-91058 Erlangen, Germany e-mail: {bause, hoffm}@am.uni-erlangen.de

approximation of the pressure head, since the flux is responsible for the transport and availability of chemical species in accompanying contaminant transport processes. In this work we investigate the application of the mixed finite element method with lowest order Raviart-Thomas ($\mathbf{RT}_0$) and Brezzi-Douglas-Marini ($\mathbf{BDM}_1$) elements and of the multi point flux approximation (**MPFA**) method to saturated and unsaturated flow in a composite and heterogeneous porous medium and compare the accuracy of the pressure and flux approximation in various norms.

A model for flow in a porous medium, taking into account saturated and unsaturated regions, is given by the assumption of mass conservation and Darcy's law,

$$\partial_t \Theta(\psi) + \nabla \cdot \mathbf{q} = f\,, \quad \mathbf{q} = -\mathbf{K}_s k_{rs}(\psi) \nabla(\psi + z) \quad \text{in } (0,T) \times \Omega\,. \tag{1}$$

In (1), ψ is the pressure head, Θ is the water content, f is a source or sink, $\mathbf{q}$ is the water flux, $\mathbf{K}_s$ is the permeability in the saturated zone, k_{rs} is the relative permeability of water to air in the unsaturated regime, z is the gravity head and Ω is a bounded domain. Functional forms of $\Theta(\cdot)$ and $k_{rs}(\cdot)$ have been derived in the literature. Eq. (1) has to be equipped with an inital condition and boundary conditions. In the saturated zone, where $\Theta(\psi)$ is a constant, Eq. (1) degenerates to an elliptic equation describing single phase Darcy flow. For lack of space, we consider fully saturated and fully unsaturated flow only. For a study of mixed approximations of (1) when the parabolic-elliptic degenerate case is applicable we refer to [3].

2 Numerical Methods

In this section we briefly describe the numerical schemes to be considered. For brevity, this is done for single phase Darcy flow as a prototype for the pressure equation in a simulation of flow in a porous medium. Let p be the pressure, $\mathbf{K}$ be the permeability and $\mathbf{u} = -\mathbf{K}\nabla p$ be the Darcy velocity. For Darcy flow it holds that

$$-\nabla \cdot (\mathbf{K}(\mathbf{x}) \nabla p) = f \ \text{ in } \Omega\,, \quad p = g \ \text{ on } \partial\Omega\,. \tag{2}$$

In terms of p and $\mathbf{u}$ problem (2) can be rewritten as

$$\nabla \cdot \mathbf{u} = f \ \text{ in } \Omega\,, \quad \mathbf{u} = -\mathbf{K}(\mathbf{x}) \nabla p \ \text{ in } \Omega\,, \quad p = g \ \text{ on } \partial\Omega\,. \tag{3}$$

Let $\mathscr{T}_h = \{T\}$ be a finite element decomposition of mesh size h of $\overline{\Omega} \subset \mathbb{R}^2$ into closed triangles T. The decompositions are assumed to be face to face. In the mixed finite element approach (cf. [4]) we form finite dimensional subspaces W_h of $L^2(\Omega)$ and $\mathbf{V}_h$ of $\mathbf{H}(\text{div};\Omega)$ and consider solving the discrete variational form of (3):

Find $\{p_h, \mathbf{u}_h\} \in W_h \times \mathbf{V}_h$ *such that*

$$\langle \nabla \cdot \mathbf{u}_h, w_h \rangle = \langle f, w_h \rangle\,, \quad \langle \mathbf{K}^{-1} \mathbf{u}_h, \mathbf{v}_h \rangle - \langle p_h, \nabla \cdot \mathbf{v}_h \rangle = -\langle g, \mathbf{v}_h \cdot \nu \rangle_{\partial\Omega} \tag{4}$$

holds for all $\{w_h, \mathbf{v}_h\} \in W_h \times \mathbf{V}_h$.

Here, ν denotes outer normal vector to $\partial\Omega$. Let $P_i(T)$, with $i \in \mathbb{N}_0$, denote the set of all polynomials on T of degree less or equal than i. In the BDM_1 mixed finite element approach the spaces W_h and $\mathbf{V}_h$ in the discrete problem (4) are chosen as

$$\begin{aligned} W_h &= \{w_h \in L^2(\Omega) \mid w_{h|T} \in P_0(T)\ \forall T \in \mathscr{T}_h\}, \\ \mathbf{V}_h &= \{\mathbf{v}_h \in \mathbf{H}(\mathrm{div};\Omega) \mid \mathbf{v}_{h|T} \in \mathbf{BDM}_1(T)\ \forall T \in \mathscr{T}_h\}, \end{aligned} \tag{5}$$

where $\mathbf{BDM}_1(T) = (P_1(T))^2$ denotes the lowest order Brezzi–Douglas–Marini space; cf. [4]. A RT_0 mixed approximation of (2) is defined by solving (4) with

$$\mathbf{V}_h = \{\mathbf{v}_h \in \mathbf{H}(\mathrm{div};\Omega) \mid \mathbf{v}_{h|T} \in \mathbf{RT}_0(T)\ \forall T \in \mathscr{T}_h\}$$

and the same W_h as in (5), where $\mathbf{RT}_0(T) = P_0(T)^2 + \mathbf{x}P_0(T)$. Problem (4) leads to an indefinite algebraic system of equations. To solve these equations, we use a hybridization technique. For further details and an application of the technique to nonlinear problems we refer to [2, 3].

Multi point flux approximation control volume methods (cf., e.g., [1]) are discretization techniques that were developed for reservoir simulation and applications of the oil industry. In this approach more than two pressure values are used in the flux approximation. From the class of multi point flux approximation schemes (cf., e.g., [1]) we use the so-called MPFA O-variant in this work. This is done on triangular grids. Its basic idea is to divide each cell of a given triangulation into three subcells; cf. Fig. 1. In each subcell E_i half edge fluxes are then determined by Darcy's law and the assumption of linear pressure variation in E_i. The method is defined by assuming continuous fluxes across each half edge of a subcell and a continuity condition at one point for the pressure across each half edge. The shape of the polylines connecting the involved grid points of the flux molecule form a stylized O (cf. Fig. 1), which explains the name O-method. Multi point flux approximations offer explicit discrete fluxes which is useful, for instance, in simulating multi phase flow.

In detail, using the notation of Fig. 1 and assuming a linear pressure approximation in the subcell E_i, we find by Darcy's law $\mathbf{u} = -\mathbf{K}\nabla p$ that the normal components $U_{[\mathbf{x}_2^i;\mathbf{x}_3^i]}$ and $U_{[\mathbf{x}_3^i;\mathbf{x}_4^i]}$ of the fluxes through the half-edges $[\mathbf{x}_2^i;\mathbf{x}_3^i]$ and $[\mathbf{x}_3^i;\mathbf{x}_4^i]$, multiplied with the respective lengths of the half-edges and orientated as the normal vectors $\nu_{[\mathbf{x}_2^i;\mathbf{x}_3^i]}$ and $\nu_{[\mathbf{x}_3^i;\mathbf{x}_4^i]}$ (cf. Fig. 1), are given by

$$\begin{pmatrix} U_{[\mathbf{x}_2^i;\mathbf{x}_3^i]} \\ U_{[\mathbf{x}_3^i;\mathbf{x}_4^i]} \end{pmatrix} = \frac{1}{2F} \begin{pmatrix} \nu_{[\mathbf{x}_2^i;\mathbf{x}_3^i]} \cdot \mathbf{K}_T \nu_{[\mathbf{x}_4^i;\mathbf{x}_1^i]} & \nu_{[\mathbf{x}_2^i;\mathbf{x}_3^i]} \cdot \mathbf{K}_T \nu_{[\mathbf{x}_1^i;\mathbf{x}_2^i]} \\ \nu_{[\mathbf{x}_3^i;\mathbf{x}_4^i]} \cdot \mathbf{K}_T \nu_{[\mathbf{x}_4^i;\mathbf{x}_1^i]} & \nu_{[\mathbf{x}_3^i;\mathbf{x}_4^i]} \cdot \mathbf{K}_T \nu_{[\mathbf{x}_1^i;\mathbf{x}_2^i]} \end{pmatrix} \begin{pmatrix} \lambda_{\mathbf{x}_2^i} - p_{\mathbf{x}_1^i} \\ \lambda_{\mathbf{x}_4^i} - p_{\mathbf{x}_1^i} \end{pmatrix}. \tag{6}$$

In (6), it is tacitly assumed that the permeability tensor $\mathbf{K}(\mathbf{x})$ is constant in each triangle T of the decompositon $\mathscr{T}_h$ and admits the value $\mathbf{K}_T \in \mathbb{R}^{2,2}$. Further, $p_{\mathbf{x}_1^i}$ is the unknown pressure value in the barycenter $\mathbf{x}_1^i$ of T, $\lambda_{\mathbf{x}_2^i}$ and $\lambda_{\mathbf{x}_4^i}$ are the unkown pressure values in the midpoints $\mathbf{x}_2^i$ and $\mathbf{x}_4^i$ of the edges, and F is the volume of the

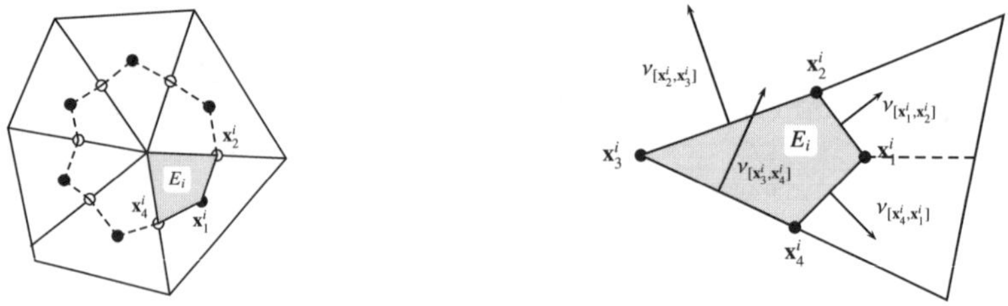

Fig. 1 Interaction region (*left*) and decomposition of a triangle into subcells with notation (*right*).

triangle with vertices $\mathbf{x}_1^i, \mathbf{x}_2^i, \mathbf{x}_4^i$. Moreover, $\nu_{[\mathbf{x}_1^i;\mathbf{x}_2^i]}, \nu_{[\mathbf{x}_2^i;\mathbf{x}_3^i]}, \nu_{[\mathbf{x}_3^i;\mathbf{x}_4^i]}$ and $\nu_{[\mathbf{x}_4^i;\mathbf{x}_1^i]}$ are the normal vectors to the corresponding edges $[\mathbf{x}_k^i;\mathbf{x}_l^i]$ with length $\|\mathbf{x}_k^i - \mathbf{x}_l^i\|_2$; cf. Fig. 1.

By some simple calculations we find that

$$\mathscr{K}^{E_i}\begin{pmatrix} U_{[\mathbf{x}_2^i;\mathbf{x}_3^i]} \\ U_{[\mathbf{x}_3^i;\mathbf{x}_4^i]} \end{pmatrix} = \begin{pmatrix} p_{\mathbf{x}_1^i} - \lambda_{\mathbf{x}_2^i} \\ p_{\mathbf{x}_1^i} - \lambda_{\mathbf{x}_4^i} \end{pmatrix},$$

where $\mathscr{K}_{11}^{E_i} = \frac{2}{|T|}(\mathbf{x}_2^i - \mathbf{x}_1^i)\cdot\mathbf{K}_T^{-1}(\mathbf{x}_3^i - \mathbf{x}_4^i)$, $\mathscr{K}_{12}^{E_i} = -\frac{2}{|T|}(\mathbf{x}_2^i - \mathbf{x}_1^i)\cdot\mathbf{K}_T^{-1}(\mathbf{x}_3^i - \mathbf{x}_2^i)$, $\mathscr{K}_{21}^{E_i} = \frac{2}{|T|}(\mathbf{x}_4^i - \mathbf{x}_1^i)\cdot\mathbf{K}_T^{-1}(\mathbf{x}_3^i - \mathbf{x}_4^i)$ and $\mathscr{K}_{22}^{E_i} = -\frac{2}{|T|}(\mathbf{x}_4^i - \mathbf{x}_1^i)\cdot\mathbf{K}_T^{-1}(\mathbf{x}_3^i - \mathbf{x}_2^i)$. The MPFA O-method on a triangular grid is then defined by the set of equations

$$\begin{aligned} &\mathscr{K}^{E_i}\begin{pmatrix} U_{[\mathbf{x}_2^i;\mathbf{x}_3^i]} \\ U_{[\mathbf{x}_3^i;\mathbf{x}_4^i]} \end{pmatrix} = \begin{pmatrix} p_{\mathbf{x}_1^i} - \lambda_{\mathbf{x}_2^i} \\ p_{\mathbf{x}_1^i} - \lambda_{\mathbf{x}_4^i} \end{pmatrix}, && \forall T \in \mathscr{T}_h,\ E_i \subset T, \\ &\sum_{E_i \subset T}\left(U_{[\mathbf{x}_2^i;\mathbf{x}_3^i]} - U_{[\mathbf{x}_3^i;\mathbf{x}_4^i]}\right) = \int_T f\,d\mathbf{x}, && \forall T \in \mathscr{T}_h, \end{aligned} \tag{7}$$

and by assuming continuity of the half-edge normal fluxes $U_{[\mathbf{x}_2^i;\mathbf{x}_3^i]}, U_{[\mathbf{x}_3^i;\mathbf{x}_4^i]}$ and pressure values $\lambda_{\mathbf{x}_2^i}\ \lambda_{\mathbf{x}_4^i}$ between two adjacent triangles. Alternatively, the MPFA method can be defined by using broken Raviart-Thomas spaces (cf. [6] for quadrilateral elements) which is advantageous for deriving error estimates. Even though this has not been done yet for triangular elements and is our work for the future, we did not observe any convergence problems of the MPFA approach (7) on triangular grids.

For our numerical studies we still introduce the mesh dependent error norms

$$\|p-\lambda_h\|_h^2 = \sum_{T\in\mathscr{T}_h}\sum_{E\subset T}\frac{|T|}{3}(p(\mathbf{x}_E)-\lambda_E)^2, \quad \|p-p_h\|_h^2 = \sum_{T\in\mathscr{T}_h}|T|(p(\mathbf{x}_T)-p_T)^2$$

$$\|\mathbf{u}-\mathbf{u}_h\|_h^2 = \sum_{T\in\mathscr{T}_h}\sum_{E\subset T}\frac{|T|}{3}(\mathbf{u}(\mathbf{x}_E)\cdot\nu_{TE} - u_{TE})^2,$$

where λ_h is the nonconforming Crouzeix-Raviart approximation of p whose degrees of freedom are the Lagrange multipliers of the hybrid mixed finite element approach (cf. [3]) or the edge midpoint pressures of the MPFA method, respectively. In both

	BDM_1							
Cells	$\|p-\lambda_h\|_h$	Red.	$\|p-p_h\|_h$	Red.	$\|\mathbf{u}-\mathbf{u}_h\|_h$	Red.	$\|\mathbf{u}-\mathbf{u}_h\|_{\mathbf{L}^2(\Omega)}$	Red.
8	9.7229211e-02	-	4.9882312e-02	-	6.1752249e-01	-	9.4285564e-01	-
32	3.1180040e-02	3.12	1.5683875e-02	3.18	1.9802025e-01	3.12	2.8323114e-01	3.33
128	8.4702128e-03	3.68	4.2995411e-03	3.65	5.5833886e-02	3.55	7.8642384e-02	3.60
512	2.1725776e-03	3.90	1.1118825e-03	3.87	1.5079887e-02	3.70	2.1172670e-02	3.71
2048	5.4732469e-04	3.97	2.8112307e-04	3.96	4.0007441e-03	3.77	5.6155866e-03	3.77
8192	1.3714569e-04	3.99	7.0547673e-05	3.98	1.0515889e-03	3.80	1.4765975e-03	3.80
32768	3.4370923e-05	3.99	1.7777322e-05	3.97	2.7486533e-04	3.83	3.8614412e-04	3.82

	RT_0							
Cells	$\|p-\lambda_h\|_h$	Red.	$\|p-p_h\|_h$	Red.	$\|\mathbf{u}-\mathbf{u}_h\|_h$	Red.	$\|\mathbf{u}-\mathbf{u}_h\|_{\mathbf{L}^2(\Omega)}$	Red.
8	1.4418195e-01	-	1.1496645e-01	-	1.0212233e+00	-	2.4434378e+00	-
32	6.0908525e-02	2.37	4.0418895e-02	2.84	3.9802766e-01	2.57	1.5110879e+00	1.62
128	1.9139274e-02	3.18	1.2412605e-02	3.26	1.3359828e-01	2.98	8.2889599e-01	1.82
512	5.2366397e-03	3.65	3.4003141e-03	3.65	4.0460505e-02	3.30	4.2920917e-01	1.93
2048	1.3509598e-03	3.87	8.7856169e-04	3.87	1.1608379e-02	3.49	2.1716785e-01	1.98
8192	3.4124114e-04	3.96	2.2208125e-04	3.96	3.2313820e-03	3.59	1.0899513e-01	1.99
32768	8.5575847e-05	3.99	5.5700150e-05	3.99	8.8243636e-04	3.66	5.4560412e-02	2.00

	MPFA							
Cells	$\|p-\lambda_h\|_h$	Red.	$\|p-p_h\|_h$	Red.	$\|\mathbf{u}-\mathbf{u}_h\|_h$	Red.	$\|\mathbf{u}-\mathbf{u}_h\|_{\mathbf{L}^2(\Omega)}$	Red.
8	1.4418195e-01	-	1.1496645e-01	-	1.0212233e+00	-	2.4434378e+00	-
32	6.0908525e-02	2.37	4.0418895e-02	2.84	3.9802766e-01	2.57	1.5110879e+00	1.62
128	1.9139273e-02	3.18	1.2412604e-02	3.26	1.3359828e-01	2.98	8.2889599e-01	1.82
512	5.2366397e-03	3.65	3.4003131e-03	3.65	4.0460507e-02	3.30	4.2920917e-01	1.93
2048	1.3509552e-03	3.88	8.7855455e-04	3.87	1.1608380e-02	3.49	2.1716785e-01	1.98
8192	3.4121886e-04	3.96	2.2204723e-04	3.96	3.2313821e-03	3.59	1.0899513e-01	1.99
32768	8.5573544e-05	3.99	5.5696299e-05	3.99	8.8243620e-04	3.66	5.4560412e-02	2.00

Table 1 Calculated errors and reduction factors (Red.) for BDM_1, RT_0 and MPFA approximation of fully saturated flow with homogeneous right-hand side term on uniformly refined meshes.

cases, mixed finite element and MPFA, these pressure values in the edge midpoints are denoted by λ_E. Further, p_T is the discrete pressure value in the barycenter of triangle T and u_{TE} is the outer normal flux component on edge E with respect to element T. The vector $\mathbf{x}_E$ denotes the midpoint of edge E, $\mathbf{x}_T$ is the barycenter of the triangle T and ν_{TE} is the outer normal to edge E of T. Moreover, L^2-norms of the errors are calculated which is done by a Gaussian quadrature rule of order 7.

3 Numerical Example: Saturated Flow

In our first study of the numerical approximation schemes we prescribe an analytical solution of (2) with $f \equiv 0$, $\Omega = (0,1)^2$ and $\mathbf{K}$ equal to the identity matrix that is given by $u(x,y) = \cos(\pi x)\cosh(\pi(y-0.5))$. Table 1 summarizes the calculated

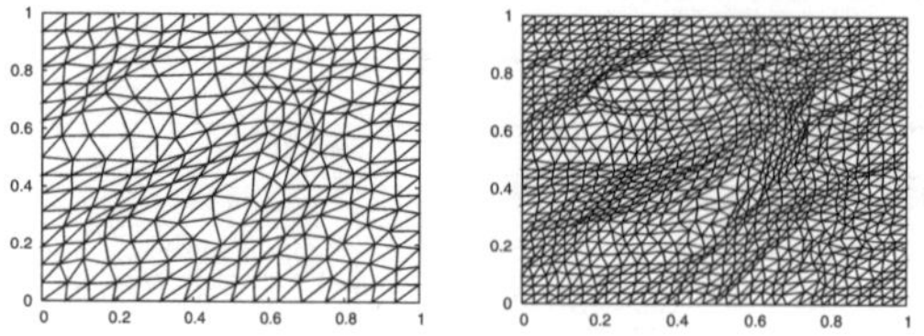

cells	8	32	128	512
h	0.789	0.467	0.261	0.140
red.	–	1.69	1.79	1.86
cells	2048	8192	32768	
h	0.076	0.039	0.021	
red.	1.85	1.97	1.81	

Fig. 2 Stochastically distorted meshes of convergence study and corresponding mesh sizes.

errors and reduction factors of the BDM_1, RT_0 and MPFA approach for a sequence of uniformly refined grids starting with a 8 cell Cartesian grid, i.e. in each mesh refinement step each triangle is devided into four congruent triangles. As far as the RT_0 and MPFA approach is concerned, we observe a superconvergence behavior of $\|p - p_h\|_h$ and $\|\mathbf{u} - \mathbf{u}_h\|_h$. Moreover, the methods lead to identical errors in the significant digits which seems to be suprising. For problem (2) with homogeneous right-hand side f and a piecewise constant permeability $\mathbf{K}$ this observation can be verified theoretically; cf. [5]. For the BDM_1 method a flux approximation of second order accuracy in the $\mathbf{L}^2$-norm is obtained which makes the scheme superior.

In our second study the previous calculations are repeated on a sequence of successively refined and stochastically distorted meshes (cf. Fig. 2) which is done for the sake of physical realism. Now, the mesh size $h = \max_{T \in \mathscr{T}_h} \operatorname{diam}(T)$ is not exactly halved in each refinement step anymore which is due to the stochastic displacement of the grid nodes after each refinement step. As Table 2 shows, superconvergence of $\|\mathbf{u} - \mathbf{u}_h\|_h$ is no longer observed. Identical errors are obtained for the RT_0 and MPFA approach again, and the BDM_1 flux approximation is of higher order accuracy.

4 Numerical Example: Unsaturated Flow

Our next study is devoted to fully unsaturated flow in a layered porous medium. In (1), we put $\Omega = (0,1)^2$, $\Theta(\psi) = 2\psi$ and $k_{rs}(\psi) \equiv 1$. The discontinuous permeability tensor $\mathbf{K} \in \mathbb{R}^{2,2}$ is defined by $\mathbf{K}(\mathbf{x}) = \mathbf{I}$ for $0.25 \le x \le 0.75$, $0 \le y \le 1$ and $\mathbf{K}(\mathbf{x}) = 10 \cdot \mathbf{I}$ else, where $\mathbf{I}$ denotes the identity matrix. We prescribe the solution

$$\psi(x,y,t) = \begin{cases} -(x-t)(1-t-x)(y-0.25)(0.75-y)\,, 0.25 \le x \le 0.75\,, 0 \le y \le 1\,, \\ -(x-t)(1-t-x)(y-0.25)(0.75-y)/10\,, \text{ else}\,. \end{cases}$$

The corresponding right-hand side function f is calculated by means of (1). For the temporal discretization the backward Euler method is used. The time step size is chosen sufficiently small ($\Delta t = 0.0001$) such that the spatial discretization error dominates the error of the discretization in time. The calculations are done on a sequence of uniformly refined grids. The layers of the porous medium are resolved by the grid, i.e., the lines of discontinuity of $\mathbf{K}$ match interelement edges, which is not limiting for our applications in reservoir simulation. The calculated errors are

	BDM$_1$							
Cells	$\|p-\lambda_h\|_h$	Red.	$\|p-p_h\|_h$	Red.	$\|\mathbf{u}-\mathbf{u}_h\|_h$	Red.	$\|\mathbf{u}-\mathbf{u}_h\|_{\mathbf{L}^2(\Omega)}$	Red.
8	9.7027931e-02	-	5.0214314e-02	-	6.1540000e-01	-	9.5579906e-01	-
32	3.2397820e-02	2.99	1.6739760e-02	3.00	1.9992180e-01	3.08	2.9584509e+00	3.23
128	9.2226924e-03	3.51	4.8394301e-03	3.46	5.7660679e-02	3.47	8.4074061e-02	3.52
512	2.4571537e-03	3.75	1.2884092e-03	3.76	1.5719272e-02	3.67	2.2950373e-02	3.66
2048	6.6136415e-03	3.72	3.5390073e-04	3.64	4.2040585e-03	3.74	6.2116904e-03	3.69
8192	1.7448747e-04	3.79	9.4533654e-05	3.74	1.1154094e-03	3.77	1.6578407e-03	3.75
32768	4.6281280e-05	3.77	2.5317323e-05	3.73	2.9380437e-04	3.80	4.4071117e-04	3.76

	RT$_0$							
Cells	$\|p-\lambda_h\|_h$	Red.	$\|p-p_h\|_h$	Red.	$\|\mathbf{u}-\mathbf{u}_h\|_h$	Red.	$\|\mathbf{u}-\mathbf{u}_h\|_{\mathbf{L}^2(\Omega)}$	Red.
8	1.3997074e-01	-	1.1493782e-01	-	1.0369437e+00	-	2.4402589e+00	-
32	6.2410025e-02	2.24	4.1202480e-02	2.79	4.1859052e-01	2.48	1.5294330e+00	1.60
128	2.0945396e-02	2.98	1.3430586e-02	3.07	1.4966002e-01	2.80	8.5665154e-01	1.78
512	6.1447028e-03	3.41	4.0827003e-03	3.29	5.4396147e-02	2.75	4.5243222e-01	1.89
2048	1.7140196e-03	3.59	1.1518377e-03	3.54	2.3357060e-02	2.33	2.3533276e-01	1.92
8192	4.6865718e-04	3.66	3.1810376e-04	3.62	1.0766859e-02	2.17	1.2117518e-01	1.94
32768	1.2797010e-04	3.66	8.7631481e-05	3.63	5.2498907e-03	2.05	6.2388483e-02	1.94

	MPFA							
Cells	$\|p-\lambda_h\|_h$	Red.	$\|p-p_h\|_h$	Red.	$\|\mathbf{u}-\mathbf{u}_h\|_h$	Red.	$\|\mathbf{u}-\mathbf{u}_h\|_{\mathbf{L}^2(\Omega)}$	Red.
8	1.3997074e-01	-	1.1493782e-01	-	1.0369437e+00	-	2.4402589e+00	-
32	6.2410026e-02	2.24	4.1202480e-02	2.79	4.1859052e-01	2.48	1.5294330e+00	1.59
128	2.0945396e-02	2.98	1.3430586e-02	3.07	1.4966002e-01	2.80	8.5665154e-01	1.79
512	6.1447028e-03	3.41	4.0827003e-03	3.29	5.4396148e-02	2.75	4.5243222e-01	1.89
2048	1.7140188e-03	3.59	1.1518362e-03	3.54	2.3357062e-02	2.33	2.3533276e-01	1.92
8192	4.6865629e-04	3.66	3.1810205e-04	3.62	1.0766860e-02	2.17	1.2117518e-01	1.94
32768	1.2796714e-04	3.66	8.7626674e-05	3.63	5.2498917e-03	2.05	6.2388482e-02	1.94

Table 2 Calculated errors and reduction factors (Red.) for BDM$_1$, RT$_0$ and MPFA approximation of fully saturated flow with homogeneous right-hand side term on stochastically distorted meshes.

summarized in Table 3. The errors of the RT$_0$ and MPFA approach are no longer identical (cf. Sec. 3), but they are of similar size. This observation was also made in further studies of the schemes. Even though the solution ψ itself is nonsmooth, the BDM$_1$ method leads to smaller flux discretization errors.

5 Conclusion

In our numerical studies we did not observe any significant differences between the RT$_0$ and MPFA method regarding the accuracy of discrete solutions. Nevertheless, MPFA methods offer explicit discrete fluxes which is not possible to get from mixed finite element methods and is advantageous, for instance, in computing multi phase flow. As long as the flux is smooth, the second order accurate BDM$_1$ approach leads

	BDM$_1$ $\Delta t = 0.0001$					
Cells	$\|\psi - \lambda_h\|_h$	Red.	$\|\psi - \psi_h\|_{L^2(\Omega)}$	Red.	$\|\mathbf{q} - \mathbf{q}_h\|_{\mathbf{L}^2(\Omega)}$	Red.
32	7.27e-4	-	1.61e-3	-	8.78e-3	-
128	1.98e-4	3.67	8.56e-4	1.89	2.29e-3	3.84
512	5.05e-5	3.92	4.34e-4	1.97	5.80e-4	3.94
2048	1.26e-5	4.00	2.17e-4	1.99	1.46e-4	3.98
8192	3.09e-6	4.16	1.09e-4	2.00	3.62e-5	4.02
32768	7.00e-7	4.41	5.44e-5	2.00	8.79e-6	4.12

	RT$_0$ $\Delta t = 0.0001$					
Cells	$\|\psi - \lambda_h\|_h$	Red.	$\|\psi - \psi_h\|_{L^2(\Omega)}$	Red.	$\|\mathbf{q} - \mathbf{q}_h\|_{\mathbf{L}^2(\Omega)}$	Red.
32	3.11e-4	-	1.64e-3	-	3.78e-2	-
128	1.09e-4	2.87	8.59e-4	1.90	1.96e-2	1.93
512	3.01e-5	3.61	4.34e-4	1.98	9.92e-3	1.98
2048	7.76e-6	3.88	2.18e-4	2.00	4.98e-3	1.99
8192	1.97e-6	3.93	1.09e-4	2.00	2.49e-3	2.00
32768	5.20e-7	3.79	5.44e-5	2.00	1.24e-3	2.00

	MPFA $\Delta t = 0.0001$					
Cells	$\|\psi - \lambda_h\|_h$	Red.	$\|\psi - \psi_h\|_{L^2(\Omega)}$	Red.	$\|\mathbf{q} - \mathbf{q}_h\|_{\mathbf{L}^2(\Omega)}$	Red.
32	2.88e-04	-	1.70e-03	-	3.75e-02	-
128	1.078e-04	2.68	8.69e-04	1.96	1.99e-02	1.89
512	3.01e-05	3.58	4.35e-04	2.00	1.01e-02	1.97
2048	7.81e-06	3.86	2.18e-04	2.00	5.07e-03	1.99
8192	2.00e-06	3.90	1.09e-04	2.00	2.54e-03	2.00
32768	5.40e-07	3.71	5.44e-05	2.00	1.27e-03	2.00

Table 3 Calculated errors and reduction factors (Red.) for BDM$_1$, RT$_0$ and MPFA approximation of unsaturated flow with discontinuous permeability tensor.

to smaller discretization errors of the flux variable. This holds even for discontinuous permeabilities, as long as the jumps are resolved by the mesh; cf. [3].

References

1. Aavatsmark, I., et al.: Discretization on unstructured grids for inhomogeneous, anisotropic media. Part I and II. SIAM J. Sci. Comput. **19**, 1700–1736 (1998)
2. Bause, M., Knabner, P.: Computation of variably saturated subsurface flow by adaptive mixed hybrid finite element methods. Adv. Water Resour. **27**, 565–581 (2004)
3. Bause, M.: Higher and lowest order mixed finite element approximation of subsurface flow problems with solutions of low regularity. Adv. Water Resour. **31**, 370–382 (2008)
4. Brezzi, F., Fortin, M.: Mixed and Hybrid Finite Element Methods. Springer, New York (1991)
5. Hoffmann, J.: Equivalence of the lowest-order Raviart-Thomas mixed finite element method and the multi point flux approximation scheme on triangular grids, in preparation (2007)
6. Klausen, R.A., Winther, R.: Robust convergence of multi point flux approximation on rough grids. Numer. Math. **105**, 317–337 (2006)

Local Projection Stabilization for the Oseen System on Anisotropic Cartesian Meshes

M. Braack

Abstract Classical residual-based stabilization techniques, as for instance streamline upwind Petrov-Galerkin, as well as the local projection method are optimal on isotropic meshes. Here we extend the local projection stabilization for the Navier-Stokes system to anisotropic Cartesian meshes. We describe the new method and give an a priori error estimate for the two-dimensional case. The method leads on anisotropic meshes to qualitatively better convergence behavior than other isotropic stabilization methods. The capability of the method is illustrated by means of a numerical test problem.

1 Introduction

The solution of partial differential equations on anisotropic meshes is of substantial importance for efficient solutions of problems with interior layers or boundary layers. For instance, in fluid dynamics at higher Reynolds number anisotropic meshes are usually used in order to resolve sharp gradients of velocity and pressure perpendicular to the boundary. Stabilized finite elements are well established in computational fluid dynamics , e.g. streamline upwind Petrov-Galerkin (SUPG), as introduced by Brooks and Hughes [7], or pressure stabilized Petrov-Galerkin (PSPG), see [14]. For the isotropic case, there is a wide range of methods and their analysis, as e.g. Lube & Tobiska [20], Hansbo & Szepessy [13], Franca & Frey [12], or the interior penalty method by Burman, Fernandez and Hansbo [8]. Codina [9] introduced a weighted global projection method for stabilization in the isotropic case.

It is well-known that stabilized finite element schemes must be modified on anisotropic meshes. Usually, the anisotropic version differs from the isotropic version in the way to compute a characteristic mesh size parameter h_K on each element

M. Braack
Christan-Albrechts-Universität zu Kiel, Mathematisches Seminar, Ludewig-Meyn-Str. 4, D-24098 Kiel, e-mail: braack@math.uni-kiel.de

K which enters in the formulation as an important parameter. Taking this as the cell diameter is not the optimal one as shown by Apel and Lube [16]. Micheletti at al. [19] propose for convection-diffusion problems and for the Stokes system to take instead the minimal eigenvalue $h_K := \lambda_{K,min}$ of the affine mapping of the reference triangle to the physical one, see [11]. Linss [15] derived a particular choice of the stabilization constant for a scalar convection diffusion problems on anisotropic meshes by comparing with the residual free bubble approach. A numerical study shows a considerable reduction of the discretization error. Lube et al. give in [17] an error analysis for SUPG/PSPG on hybrid meshes with a presumed relation of the mesh sizes in the different coordinate directions based on the diffusion constant μ.

In this work we propose a numerical scheme for fluid dynamics based on local projection stabilization (LPS) for anisotropic meshes. The first step of formulating LPS on anisotropic quadrilateral meshes is published in [4] by considering the Stokes system. In the present work we address the by far more relevant and more difficult Oseen system where also the convective terms should be stabilized in such a way that the a priori error analysis remains optimal even when strongly stretched elements are used. In particular, we give an a priori estimate in the case of the Oseen system. LPS techniques are already applied with large success to different fields of computational fluid dynamics, e.g., in 3D incompressible flows [5], compressible flows, reactive flows [6], parameter estimation [1]. An isotropic extension to arbitrary order and more general projections is derived by Matthies et al. [18].

2 A Priori Estimates for the Oseen System

The Oseen system in the domain $\Omega \subset \mathbb{R}^d$ for velocity v and pressure p consists of momentum and continuity equation together with appropriate boundary conditions for v on $\partial\Omega$. We embrace the two variables together in the variable $u := \{v, p\}$. For the analysis we restrict ourself to the case of homogeneous Dirichlet conditions for the velocities. The natural function space is $X := V \times Q$ with the Sobolev space of generalized functions with square-integrable derivatives and vanishing traces, $V := [H_0^1(\Omega)]^d$, and the Hilbert space $Q := L_{2,0}(\Omega)$ consisting of $L_2(\Omega)$ functions with zero mean on Ω, $\int_\Omega p(x)\,dx = 0$. Test functions are denoted by Greek letters, for instance $\phi \in V$ as test function for the momentum equation and $\xi \in Q$ as test function for the continuity equation. Using the notation $(\cdot,\cdot)$ for the $L_2(\Omega)$-scalar product the corresponding Galerkin formulation reads for $f \in L_2(\Omega)$

$$u \in X: \qquad a(u;\phi,\xi) = (f,\phi) \qquad \forall\{\phi,\xi\} \in X, \tag{1}$$

with the bilinear form ($\sigma \geq 0,\ \mu > 0$)

$$a(u;\phi,\xi) := (\sigma v,\phi) + ((\beta\cdot\nabla)v,\phi) + (\mu\nabla v,\nabla\phi) - (p,\operatorname{div}\phi) + (\operatorname{div} v,\xi)\,.$$

The convection field β may vary in space but is solenoidal, $\operatorname{div}\beta = 0$. The viscosity may also vary in space. However, for ease of presentation we assume spatial constant

μ and σ. We take in mind that the result in this work takes over to the case of varying coefficients with minor modifications.

Let $Q_{1,h}$ be the space of d-linear finite elements. The discrete pressure space Q_h is its subspace with zero mean and the discrete velocity space V_h is the corresponding vector-valued space with vanishing traces:

$$Q_h := \{\xi \in Q_{1,h} : (\xi, 1) = 0\}, \quad V_h := \{\phi \in [Q_{1,h}]^d : \phi|_{\partial\Omega} = 0\}.$$

The bilinear form $a(u;\phi,\xi)$ is known to be unstable for such an equal-order interpolation of V and Q due to the violation of the discrete "inf-sup" condition [14]. A further instability is due to the dominant advective term. Stabilization is a standard tool to overcome this short-comings.

Residual-based methods, as for instance SUPG / PSPG, as well as projection-based schemes allow for equal-order finite elements of polynomial order r. It is well-known that their convergence order in terms of the mesh size h in the convection-dominant case is $h^{r+1/2}$ on quasi-uniform isotropic meshes:

$$|||u - u_h|| \leq Ch^{r+1/2} \|u\|_{H^{r+1}}, \tag{2}$$

for sufficiently smooth velocities and pressure (H^{r+1}-regularity). The constant C may depend on the parameters σ, β and μ of the problem. The triplenorm $|||\cdot|||$ in (2) depends on the particular finite element scheme but usually involves the energy norm of the velocities $\mu^{1/2}\|\nabla v\|$, and the L_2-term $\sigma^{1/2}\|v\|$.

On anisotropic meshes, (2) is suboptimal, because $h^{r+1/2}$ must be replaced by the mesh size with respect to the coordinate direction of maximal elongation. For instance in the case of a boundary layer with large second derivatives in y-direction, the mesh should be much finer in y-direction, i.e. $h_y << h_x$, when h_x and h_y are the mesh sizes in x- and y-direction of a Cartesian mesh, respectively. More specifically, the optimal mesh sizes in the two coordinate directions should be

$$(h_x/h_y)^2 \approx \|\partial_{yy}^2 v\| / \|\partial_{xx}^2 v\|. \tag{3}$$

Now, in the estimate (2) the mesh size h must be replaced by the maximal one, i.e. by h_x, so that it becomes for (bi-) linear elements:

$$|||u - u_h||| \leq Ch_x^{3/2}(\|\partial_x \nabla u\| + \|\partial_y \nabla u\|). \tag{4}$$

Obviously, the second derivatives of u in (4) are not well balanced. In particular, the term $h_x^{3/2}\|\partial_y \nabla u\|$ would be the dominant one when the curvature in y-direction is much smaller than the one in x-direction and the mesh sizes are designed in the optimal way (3). Therefore, much more suitable would be an estimate where the partial mesh sizes h_x and h_y are multiplied by the corresponding spatial derivatives. In this work, we will formulate an anisotropic modification of LPS so that we obtain an a priori estimate which satisfies this property. In comparison to the isotropic version (4) we derive an estimate which is better by a factor of $(h_x/h_y)^{3/2}$. For large aspect ratios, this is an enormous gain.

3 Local Projection Stabilization

The mesh $\mathscr{T}_h$ is supposed to be constructed by patches. A coarser mesh $\mathscr{T}_{2h}$ is obtained by one global coarsening of $\mathscr{T}_h$. The correspondence between these two meshes is as follows: Each cell $P \in \mathscr{T}_{2h}$ is cut into 2^d new cells (dividing all lengths of edges of P by 2) in order to obtain the fine partition $\mathscr{T}_h$. Due to this construction we can associate to each cell $K \in \mathscr{T}_h$ an corresponding patch $P = P(K) \in \mathscr{T}_{2h}$ with $K \subset P$. For the formulation of this projection method, hanging nodes are allowed. However, in the analysis of this work, we restrict to the case of anisotropic Cartesian meshes without such irregular nodes. The space Q_{2h}^{disc} consists of patch-wise constants, but discontinuous across patches $P \in \mathscr{T}_{2h}$:

$$Q_{2h}^{disc} := \{\xi \in L^2(\Omega) : \xi\big|_P \equiv \text{const.} \forall P \in \mathscr{T}_{2h}\}.$$

The projection $\pi_h : L^2(\Omega) \to Q_{2h}^{disc}$ is defined as the patch-wise mean:

$$\pi_h q\big|_P := \frac{1}{|P|}\int_P q(x)\,dx \qquad \forall P \in \mathscr{T}_{2h}.$$

The idea of LPS, consists of adding inconsistent stabilization terms $s_h(u_h;\phi,\xi)$ to the Galerkin form involving the difference between the identity I and π_h: $\kappa_h := I - \pi_h$. The concrete form is specified in the following subsections. The discrete system becomes: Find $u_h = \{v_h, p_h\} \in X_h := V_h \times Q_h$ such that

$$a(u_h;\phi,\xi) + s_h(u_h;\phi,\xi) = (f,\phi) \qquad \forall\{\phi,\xi\} \in X_h. \tag{5}$$

LPS for Oseen for anisotropic meshes aligned with the coordinate axes becomes:

$$s_h(u;\phi,\xi) := \sum_{i=1}^{d}\left((\kappa_h(\partial_{x_i}p_h), \alpha_i\partial_{x_i}\xi) + (\kappa_h(\partial_{x_i}v_h), \delta_i\partial_{x_i}\phi)\right), \tag{6}$$

with patch-wise constant parameters $\alpha_{x_i}, \delta_{x_i}$. For the corresponding formulation on rotated meshes, we refer to [2]. It turns out that the optimal choice of the stabilization parameters depends on the minimal Peclet number:

$$\mathrm{Pe}_{min}(K) := \min_{i=1,\ldots,d}(h_{K,i})\|\beta\|_{K,\infty}/\mu.$$

For the isotropic case, this characteristic number is the usual local Peclet number. The optimal choice of the stabilization parameters will be given by:

$$\alpha_i = h_{K,i}^2\mu^{-1}\min(1,\mathrm{Pe}_{min}^{-1}(K)), \quad \delta_i = \|\beta\|_{K,\infty}^2\alpha_i. \tag{7}$$

Remark 1: For residual type stabilization schemes (SUPG, PSPG), the difference between the isotropic and anisotropic case consists in a different choice of the stabilization parameters, see e.g. Formaggia et al. [10], while the additional terms itself remain isotropic. For LPS, the proposed stabilization (6) is anisotropic. In the case of isotropic meshes, the stabilization terms become $\alpha_i \sim \alpha_j$ and $\delta_i \sim \delta_j \sim \|\beta\|^2_{K,\infty}\alpha_i$.

Remark 2: In contrast to residual-based stabilization methods the parameters α and β are independent of the absorption coefficient σ which can also be interpreted as the inverse of a time step, i.e. $\sigma \sim 1/\Delta t$. This is due to the fact that the stabilization does not act on the zero-order term σv in the Oseen system.

4 A Priori Error Estimate

The a priori estimate (2) was shown in [3] for piecewise bilinear($r = 1$) and quadratic ($r = 2$) elements for the triple norm:

$$|||u|||_{lps} := \left(\sigma\|v\|^2 + \mu\|\nabla v\|^2 + s_h(u;u)\right)^{1/2}. \tag{8}$$

The stability follows immediate after integration by parts keeping in mind that β is solenoidal. For the analysis we restrict to the two-dimensional case. The mesh is assumed to be finer in y-direction:

$$h_y < h_x .$$

For ease of presentation, we assume in this section quasi-uniform meshes with respect to each coordinate direction. That means, that h_x represents all mesh sizes in x-directions, and h_y the mesh sizes in y-directions:

$$h_x \sim h_{K,x} \quad \text{and} \quad h_y \sim h_{K,y} \quad \forall K \in \mathscr{T}_h .$$

However, all results can also be understood locally. For this we only need the important *interior angle condition* for neighbor cells $K, L \in \mathscr{T}_h$:

$$h_{K,x} \sim h_{L,x} \quad \text{and} \quad h_{K,y} \sim h_{L,y} .$$

This condition implies that the mesh sizes with respect to the x-direction of neighbor cells are of the same order. The same should hold for the y-directions. Although we allow for varying cells sizes in x-direction (and in y-direction), this change must be moderate from one cell to the next one. In the following, we use the notation "$a \lesssim b$" which stands for an upper bound with a constant C, i.e. $a \leq Cb$. This constant is independent of h and the parameters μ, σ, β.

Theorem 1. *For the choice (7) of the stabilization parameters we obtain for $d = 2$:*

$$\begin{aligned} |||u - u_h|||_{lps} &\lesssim \alpha_x^{1/2} h_x \|\partial_x \nabla p\| + \alpha_y^{1/2} h_y \|\partial_y \nabla p\| \\ &\quad + (\sigma^{1/2} + (\mu^{1/2} + \delta_y^{1/2}) h_y^{-1}) \cdot (h_x^2 \|\partial_x \nabla v\| + h_y^2 \|\partial_y \nabla v\|) . \end{aligned} \tag{9}$$

The proof can be found in [2]. An easy consequence for the isotropic case is:

Theorem 2. *The a priori estimate becomes in the isotropic case:*

$$|||u-u_h|||_{lps} \lesssim \alpha^{1/2}h\|\nabla^2 p\| + \left(\sigma^{1/2}h + \mu^{1/2} + \|\beta\|_\infty \alpha^{1/2}\right) \cdot h\|\nabla^2 v\|, \quad (10)$$

with $\alpha = h\min(\mu^{-1}h, \|\beta\|_\infty^{-1})$.

In particular, for Peclet number larger than one, i.e. the convection dominated case, the right hand side becomes (formally) independent of the viscosity:

$$|||u-u_h|||_{lps} \lesssim h^{3/2}\left(\|\beta\|_\infty^{-1/2}\|\nabla^2 p\| + \left(\sigma^{1/2}h^{1/2} + \|\beta\|_\infty^{1/2}\right)\cdot \|\nabla^2 v\|\right).$$

5 Numerical Validation: Tube Flow with Boundary Layer

In order to validate the proposed discrete scheme for the computation of convection dominated flows on anisotropic meshes we consider the Navier-Stokes equations:

$$(v\cdot\nabla v) - \mu\Delta v + \nabla p = f, \quad \operatorname{div} v = 0. \quad (11)$$

The model problem consists of a flow in a tube length $L \geq 1$ and height $H = 1$, $\Omega = (0,L)\times(0,H)$. We have slip conditions for the velocities $v = v_0$ on Γ_{dir}, at the Dirichlet boundary part $\Gamma_{dir} \subset \partial\Omega$ consisting of the horizonal and the left vertical part of $\partial\Omega$. The remaining boundary (right vertical boundary part) $\Gamma_{out} := \{L\}\times(0,1)$ is the natural "outflow" boundary. The boundary values v_0 and the forcing term f of the momentum equation (11) are taken in such a way that we obtain the following analytical solution:

$$v_1(x,y) = \frac{e^\gamma - e^{\gamma(1-y)}}{e^\gamma - 1}, \quad v_2(x,y) = \varepsilon(1-x/L)^2, \quad p(x,y) = (L-x)xy,$$

with $\varepsilon = 0.01$ and $\gamma = \mu^{-1/2}$. The velocity component in x-direction, v_1, exhibits a boundary layer at the lower boundary $y = 0$ of thickness $\gamma^{-1} = \sqrt{\mu}$. The vertical velocity is chosen small but non-zero so that the velocity stabilization in vertical direction does not vanish for the exact solution and so that v_2 is not in the discrete space $Q_{1,h}$.

The diffusion coefficient is taken between $\mu = 10^{-2}$ and $\mu = 10^{-5}$. For $\mu = 10^{-2}$, it turns out that up to a certain degree of anisotropy the classical isotropic local projection stabilization is still convergent. Therefore, we can compare with the anisotropic version.

In order to see the impact of anisotropy we take the length as $L = 50$ and $L = 100$. The case $L = 1$ needs not be performed because both methods become identical on isotropic meshes. The number of grid points are equal for both coordinate directions and equidistant in x-direction as well as in y-direction. Hence the anisotropy is uniform in space and equal to $a = L/H = L$. Since the scaling of the two velocity components v_1 and v_2 as well as their gradients are very different, the error in each velocity component is listed in the following tables separately.

For aspect ratio $a = 50$, see Table 1 and 2, the isotropic version shows it deficits: the linear solver does not convergence on the grid with $h_y = 1/8$. In contrast to this, the anisotropic version does a very good job and is of order h^2 for the L_2-error of each individual component of the solution. The error in the gradient is still of first order. If we compare this with the isotropic version (at least for the convergent cases), we observe that the error in the pressure is comparable. Surprisingly, the pressure is even slightly better with the isotropic version by a very small factor. A possible explanation is that the pressure does not has a boundary layer. However, the difference is in the range of 4% and hence is not relevant. The error of the velocity in x-direction, v_1, is improved in the anisotropic version by a factor of up to 4. The largest improvement is observed for the accuracy of v_2: Here we gain a factor of more than 100 in the L_2-norm and about 20 in H^1 (except the very first mesh).

For aspect ratio $a = 100$ and $\mu = 10^{-5}$ convergence is obtained only with anisotropic LPS, see Table 3. It was not possible to obtain results with the classical (isotropic) LPS. But with the anisotropic version, we still get second order behavior of the error in L_2 and better than first order in the gradients.

Table 1 Convergence history with **iso**tropic stabilization for $\mu = 10^{-2}$ and aspect ratio $a = 50$.

h_y	$\|p-p_h\|$	$\|\nabla(p-p_h)\|$	$\|v_1-v_{1,h}\|$	$\|\nabla(v_1-v_{1,h})\|$	$\|v_2-v_{2,h}\|$	$\|\nabla(v_2-v_{2,h})\|$
1/4	455	472	4.47	29.8	35.9	173
1/8	no	convergence	no	convergence	no	convergence
1/16	13.6	21.6	0.528	6.00	3.35	12.2
1/32	2.46	7.22	0.196	2.78	0.660	2.64
1/64	0.602	3.24	0.0444	1.30	0.125	0.699
1/128	0.163	1.702	7.05e-03	0.627	0.0223	0.219
1/256	0.0447	0.897	9.82e-04	0.300	2.63e-03	0.0498

Table 2 Convergence history with **aniso**tropic stabilization for $\mu = 10^{-2}$ and aspect ratio $a = 50$.

h_y	$\|p-p_h\|$	$\|\nabla(p-p_h)\|$	$\|v_1-v_{1,h}\|$	$\|\nabla(v_1-v_{1,h})\|$	$\|v_2-v_{2,h}\|$	$\|\nabla(v_2-v_{2,h})\|$
1/4	278.	376.	1.62	12.1	16.9	68.4
1/8	48.1	90.6	1.09	9.91	1.44	6.96
1/16	11.9	25.5	0.306	4.96	0.0982	0.649
1/32	2.97	8.99	0.0490	2.45	6.50e-03	0.0619
1/64	0.743	3.90	0.0112	1.16	4.81e-04	0.0199
1/128	0.186	1.87	2.30e-03	0.508	8.71e-05	9.39e-03
1/256	0.0464	0.924	4.30e-04	0.212	2.16e-05	2.54e-03

Table 3 Convergence history with **aniso**tropic stabilization for $\mu = 10^{-5}$ and aspect ratio $a = 100$.

h_y	$\|p-p_h\|$	$\|v_1-v_{1,h}\|$	$\|\nabla(v_1-v_{1,h})\|$	$\|v_2-v_{2,h}\|$	$\|\nabla(v_2-v_{2,h})\|$
1/4	197	9.80e-02	1.04	0.162	0.639
1/8	47.7	0.399	3.09	1.38e-02	6.84e-02
1/16	11.9	1.99e-02	0.425	9.58e-04	6.43e-03
1/32	2.97	2.29e-03	0.139	7.87e-05	7.41e-04
1/64	0.743	5.02e-04	6.66e-02	1.073e-05	1.31e-04
1/128	0.186	1.25e-04	3.32e-02	2.29e-06	3.41e-05
order	h^2	h^2	h	h^2	h

References

1. Becker, R., Braack, M., Vexler, B.: Parameter identification for chemical models in combustion problems. Appl. Numer. Math. **54**(3–4), 519–536 (2005)
2. Braack, M.: A stabilized finite element scheme for the Navier-Stokes equations on quadrilateral anisotropic meshes. Modél. Math. Anal. Numér. (2007). Submitted
3. Braack, M., Burman, E.: Local projection stabilization for the Oseen problem and its interpretation as a variational multiscale method. SIAM J. Numer. Anal. **43**(6), 2544–2566 (2006)
4. Braack, M., Richter, T.: Local projection stabilization for the Stokes system on anisotropic quadrilateral meshes. In: Bermudez de Castro et al. (ed.) Numerical Mathematics and Advanced Applications, Enumath Proc. 2005, pp. 770–778. Springer (2006)
5. Braack, M., Richter, T.: Solutions of 3D Navier-Stokes benchmark problems with adaptive finite elements. Computers and Fluids **35**(4), 372–392 (2006)
6. Braack, M., Richter, T.: Stabilized finite elements for 3D reactive flow. Int. J. Numer. Methods Fluids **51**(9–10), 981–999 (2006)
7. Brooks, A., Hughes, T.: Streamline upwind Petrov-Galerkin formulation for convection dominated flows with particular emphasis on the incompressible Navier-Stokes equations. Comput. Methods Appl. Mech. Engrg. **32**, 199–259 (1982)
8. Burman, E., Fernandez, M., Hansbo, P.: Edge stabilization for the incompressible Navier–Stokes equations: a continuous interior penalty finite element method. SIAM J. Numer. Anal. **44**(3), 1248–1274 (2006)
9. Codina, R.: Stabilization of incompressibility and convection through orthogonal subscales in finite element methods. Comput. Methods Appl. Mech. Engrg. **190**(13/14), 1579–1599 (2000)
10. Formaggia, L., Micheletti, S., Perotto, S.: Anisotropic mesh adaptation in computational fluid dynamics: Application to the advection-diffusion-reaction and the Stokes problems. Appl. Numer. Math. **51**, 511–533 (2004)
11. Formaggia, L., Perotto, S.: Anisotropic error estimates for elliptic problems. Numer. Math. **94**, 67–92 (2003)
12. Franca, L., Frey, S.: Stabilized finite element methods: II. The incompressible Navier-Stokes equations. Comput. Methods Appl. Mech. Engrg. **99**, 209–233 (1992)
13. Hansbo, P., Szepessy, A.: A velocity-pressure streamline diffusion finite element method for the incompressible Navier-Stokes equations. Comput. Methods Appl. Mech. Engrg. **84**, 175–192 (1990)
14. Hughes, T., Franca, L., Balestra, M.: A new finite element formulation for computational fluid dynamics: V. circumvent the Babuska-Brezzi condition: A stable Petrov-Galerkin formulation for the Stokes problem accommodating equal order interpolation. Comput. Methods Appl. Mech. Engrg. **59**, 89–99 (1986)
15. Linss, T.: Anisotropic meshes and streamline-diffusion stabilization for convection-diffusion problems. Commun. Numer. Meth. in Engng. **21**(10), 515–525 (2005)
16. Lube, G., Apel, T.: Anisotropic mesh refinement in stabilized Galerkin methods. Numer. Math. **74**, 261–282 (1996)
17. Lube, G., Knopp, T., Gritzki, R.: Stabilized FEM with anisotropic mesh refinement for the Oseen problem. In: Proceedings ENUMATH 2005, pp. 799–806. Springer (2006)
18. Matthies, G., Skrzypacz, P., Tobiska, L.: A unifird convergence analysis for local projection stabilisations applied ro the Oseen problem. Tech. rep., Universität Magdeburg, Fakultät für Mathematik, Preprint 44 (2006)
19. Micheletti, S., Perotto, S., Picasso, M.: Stabilized finite elements on anisotropic meshes: A a priori estimate for the advection-diffusion and the Stokes problem. SIAM J. Numer. Anal. **41**, 1131–1162 (2003)
20. Tobiska, L., Lube, G.: A modified streamline diffusion method for solving the stationary Navier-Stokes equations. Numer. Math. **59**, 13–29 (1991)

Simulations of 3D Dynamics of Microdroplets: A Comparison of Rectangular and Cylindrical Channels

C.-H. Bruneau, T. Colin, C. Galusinski, S. Tancogne, and P. Vigneaux

Abstract In this paper, several numerical simulations of diphasic flows in microchannels are presented. The flow in both cylindrical and rectangular channels is considered. The aim is to compute the shape of the droplets and the velocity fields inside and outside the droplets and to quantify the influence of the geometry. The Level Set method is used to follow the interface between the fluids.

1 Introduction

Diphasic flows in microchannels are governed by the pressure gradient and the surface tension at the interface between the fluids. In experimental configurations, jets of one fluid into another are usually not stable. This is due to Rayleigh instability. The jets can break off and therefore can lead to the creation of droplets. The study of the flow inside these droplets is a difficult task that can be achieved only with a numerical approach.

From the experimental point of view, once they are created, droplets propagate at their own speed while the flow between two droplets is essentially a Poiseuille flow,

Charles-Henri Bruneau
e-mail: bruneau@math.u-bordeaux1.fr

Thierry Colin
e-mail: colin@math.u-bordeaux1.fr

Sandra Tancogne
e-mail: tancogne@math.u-bordeaux1.fr

Paul Vigneaux
Institut de Mathématiques de Bordeaux, Inria Bordeaux MC2 - 351 Cours de la Libération - F-33405 Talence Cedex - France, e-mail: Paul.Vigneaux@math.cnrs.fr

Cédric Galusinski
CPT-Imath, Université du Sud - Toulon Var - Avenue de l'Université - BP20132 83957 La Garde Cedex - France, e-mail: galusins@univ-tln.fr

as soon as the droplets are "sufficiently" separated. This implies that some recirculations occurs inside the droplets in order to ensure a non-slip boundary condition on the boundary of the channels. These movements are not only created by the flow itself but also by the surface tension forces that are responsible of the shape of the droplets. From the point of view of the applications these microdroplets can be used as micromixers and as microreactors in order to achieve reactions with very small volumes of products. One of the characteristics of microflows is that they are constrained by the confinement. Typical size of a microchannel is a section of $10^{-8}m^2$ with a length of a few centimeters. Typical velocity of the flows is $1cm/s$. The flow rates are around $3000\mu l$ per hour. This confinement determines strongly the stability of the jets or of the co-flows as well as the shape of the droplets in case of break-up. The geometry of the channels are also very important for the dynamics of the droplets. One can build and use cylindrical channels as well as rectangular ones. For very thin jets, localized in the middle of the channel, the shape of the section of the channel is not determinant and flows in cylindrical and rectangular channels have the same behavior. The explanation is that surface tension forces are important at these scales and the velocity field at the interface is closed to be invariant under rotations. This is observed on the numerical simulations of flows without confinement whatever the shape of the channel is. For larger structures, the flow undergoes the effects of confinement and circular and rectangular channels give rise to different kinds of behavior.

The aim of this note is to present a preliminary comparison of droplets and their associate flows in both cylindrical and rectangular channels. It is organized as follows. In the second part, we recall the model and the numerical method used. In the third part, we try to emphasize the main features of 3D flows in cylindrical and rectangular channels.

2 Modelling and Simulation of Bifluid Microflows

2.1 Governing Equations

We want to modelize incompressible, viscous and Newtonian bifluid flows with surface tension in microchannels. This requires to follow carefully the moving interfaces. We adopt a Level Set approach to capture the interface Γ between fluid 1 and fluid 2 (see [7] and [6]). In this context, the interface is given by the zero level set of a function ϕ :

$$\Gamma(t) = \{x \in \Omega \ / \ \phi(x,t) = 0\}, \forall t \geq 0 \quad (1)$$

where Ω is the 3D bounded computational domain occupied by the fluids. Moreover ϕ satisfies

$$\forall t \geq 0, \begin{cases} \phi(x,t) < 0 \ \forall x \in \text{ fluid 1} \\ \phi(x,t) > 0 \ \forall x \in \text{ fluid 2} \\ \phi(x,t) = 0 \quad \forall x \in \Gamma(t) \end{cases} \tag{2}$$

In the context of microfluidics, the Reynolds number is small and we can neglect the inertial effects. The velocity is then a solution of the Stokes equation

$$-\nabla.(2\eta D\mathbf{u}) + \nabla p = \sigma\kappa\delta(\phi)\mathbf{n} \quad \forall(t,\mathbf{x}) \in \mathbb{R}^+ \times \Omega, \tag{3}$$

$$\nabla.\mathbf{u} = 0 \quad \forall(t,\mathbf{x}) \in \mathbb{R}^+ \times \Omega, \tag{4}$$

where η is the viscosity, p is the pressure and $D\mathbf{u} = (\nabla\mathbf{u} + \nabla^T\mathbf{u})/2$ is the deformation rate tensor, $\sigma \in \mathbb{R}$ is the constant surface tension coefficient, $\mathbf{n}$ is the unit normal to the interface, κ is the curvature of the interface and δ is the Dirac function. The normal and the curvature are computed thanks to the Level Set function :

$$\mathbf{n} = \left. \frac{\nabla\phi}{|\nabla\phi|} \right|_{\phi=0}, \tag{5}$$

$$\kappa = \nabla. \left(\frac{\nabla\phi}{|\nabla\phi|} \right) \Big|_{\phi=0}. \tag{6}$$

In addition, the viscosity is given by

$$\eta = \eta_1 + (\eta_2 - \eta_1)H(\phi), \tag{7}$$

where η_1 (resp. η_2) is the viscosity of fluid 1 (resp. 2) and H is the Heaviside function:

$$H(\phi) = \begin{cases} 0 & \text{if } \phi \leq 0, \\ 1 & \text{if } \phi > 0. \end{cases} \tag{8}$$

The interface moves at the velocity of the fluid and the function ϕ is then defined [8] as the solution of the advection equation :

$$\frac{\partial\phi}{\partial t} + \mathbf{u}.\nabla\phi = 0 \quad \forall(t,\mathbf{x}) \in \mathbb{R}^+ \times \Omega. \tag{9}$$

Finally, two types of geometry are considered, namely rectangular channels and cylindrical channels, as shown on Figure 1.

2.2 Numerical Resolution Procedure

The numerical algorithm is the following one :

1. Compute an initial value for the Level Set function ϕ and related η.
2. Compute the unit normal $\mathbf{n}$ and the curvature κ.

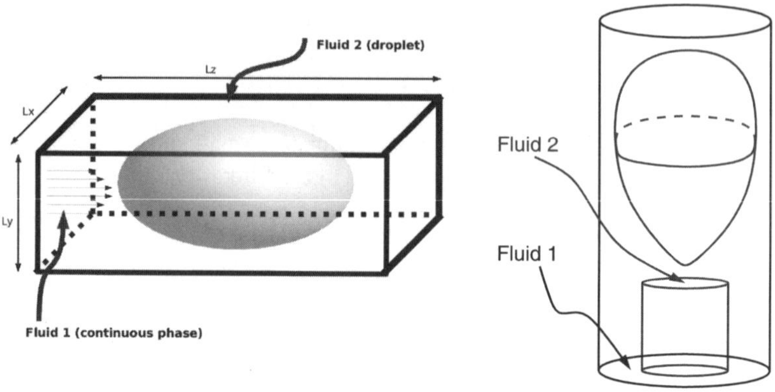

Fig. 1 Geometry of rectangular (on the left) and cylindrical (on the right) channels.

3. Solve the Stokes equation (3)-(4) for $(\mathbf{u}, p)$ with κ and $\mathbf{n}$ obtained by step 2.
4. Update ϕ by solving (9).
5. If needed, apply a redistanciation procedure on ϕ in order to ensure $|\nabla\phi| = 1$ (see [6]).
6. Iterate step 2-5 for each time step.

The Stokes system (3)-(4) is discretized on a staggered mesh using a finite volume scheme. The divergence free condition is ensured by an augmented Lagrangian algorithm. The transport equation (9) is discretized by a fifth order WENO scheme [5].

Note that in microfluidics, surface tension is preponderant and a specific stability condition derived in [2] is used :

$$\Delta t = min\left(c_1 \frac{\Delta x}{\|\mathbf{u}\|_\infty}, c_2 \frac{\eta}{\sigma} \Delta x\right) \tag{10}$$

where Δt is the time step and Δx is the space step. The constant c_1 is linked to the classical CFL condition and depends only on the scheme used to discretize the advection equation. The constant c_2 is associated to the constraint induced by the surface tension term that is discretized explicitly. In our microfluidics applications, $c_2 = 4$ leads to stable computations (see [3]).

We use a penalization method [1] to take into account the spatial structure of the coaxial cylindrical channels shown on the right of Figure 1. Numerical approach for this axisymmetric framework is presented in [9].

3 Numerical Results

3.1 Results in 3D Axisymmetric Channels

The first test case (Fig. 2) concerns the simulation of droplets creation by injecting a fluid into another, thanks to the "injector" geometry of figure 1 (right) where the radius of the external capillary is $R = 300\mu m$. The internal capillary of length R, thickness $50\mu m$ and centered at $r = 75\mu m$ is modelled by a penalization term. Two fluids have a surface tension $\sigma = 33.10^{-3}\ N/m$. In the internal jet, the viscosity is $\eta_2 = 30.10^{-2}\ Pa.s$ and in the external capillary, the viscosity is $\eta_1 = 55.10^{-3}\ Pa.s$. Parabolic profiles are used for the injection velocity at the inlet, with a maximum of $u_2 = 0.07\ m/s$ in the internal tube and $u_1 = 0.01\ m/s$ in external tube. The section R of the computational domain is discretized with 30 cells. The jet breaks up because of Rayleigh instability.

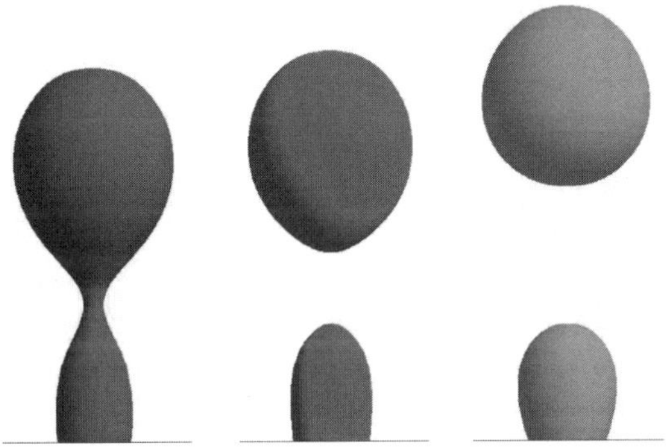

Fig. 2 Droplet creation with an axisymmetric jet. Time increases from left to right.

In the second case (Fig. 3), we then present two examples of flows that exist inside a microdroplet by showing the velocity field in the droplet frame of reference. We use an external capillary which is of radius $R = 60\mu m$. The fluid inside the droplet has a viscosity $\eta_2 = 2.10^{-2}\ Pa.s$ and the viscosity of the continuous phase is $\eta_1 = 4.10^{-2}\ Pa.s$. The injection speed considered is $0.2\ m/s$ and $0.1\ m/s$ respectively for the two numerical simulations. It can be observed that when the velocity is small, the droplet shape is more spherical and a central recirculation region develops towards the front of the droplet. This is due to the increasing influence of the surface tension compared to the driving flow.

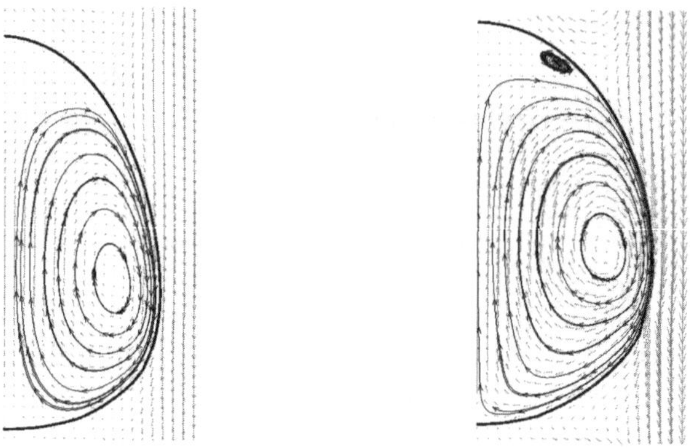

Fig. 3 Hydrodynamics in a cylindrical channel : the droplet is propelled with an injection speed of 0.2 m/s on the left and 0.1 m/s on the right.

3.2 Results in 3D Rectangular Channels

We now present comparative simulations in the 3D cartesian geometry of the left of Figure 1 with $L_x = L_y = 150\mu m$. Again, the viscosity is $\eta_2 = 2.10^{-2}$ $Pa.s$ in the droplet and $\eta_1 = 4.10^{-2}$ $Pa.s$ elsewhere and the surface tension between them is $\sigma = 33.10^{-3}$ N/m. Figures 4 and 5 show a droplet with an injection speed of 0.2 m/s and 0.1 m/s, respectively. A bigger droplet is also shown on Figure 6. The influence of the injection speed is the same as in the axisymmetric case if the global evolution of the shape and the recirculation zones are taken into account (see Fig. 3 and 4, 5). But differences definitely appear when it comes to compare droplets shapes and induced streamlines. First, when looking at a cross section (with respect to the direction of the flow), droplets in a square channel are not spherical – contrary to the cylindrical case – as it can be seen on the back of droplets of Figures 4 and 5 and even more clearly on the slices numbered 3 and 4 on Figure 7 which are at the back of the droplet of Figure 6. On the right of Figure 6, the velocity field in a (x,y) section shows a typical fully 3D behaviour, which is only seen in rectangular configuration : eight vortexes are present inside the droplet near its boundary that correspond to the fluid that focuses at the center. It is clearly not an axisymmetric phenomenon. At the boundary of the channel, the flow is deviated in the direction of the corners. The conservation of the flow rates implies that the liquid has to escape in the longitudinal direction through the four corners. This fully 3D effect is due to the rectangular confinement.

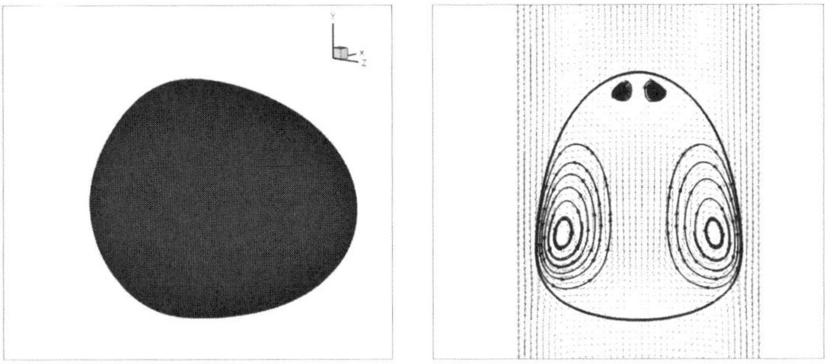

Fig. 4 Droplet hydrodynamics in a 3D cartesian configuration with an injection speed of $0.2\ m/s$.

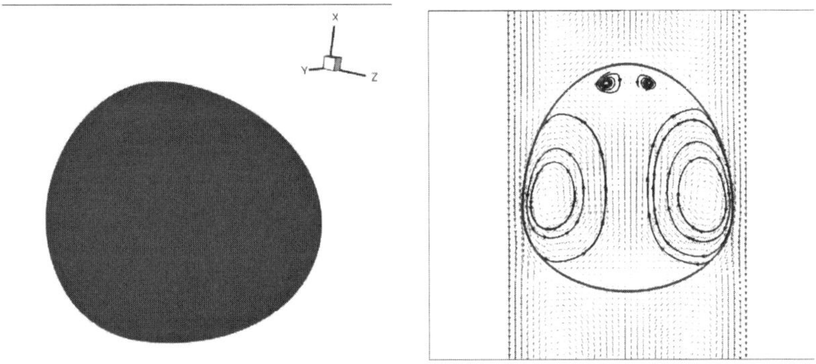

Fig. 5 Droplet hydrodynamics in a 3D cartesian configuration with an injection speed of $0.1\ m/s$.

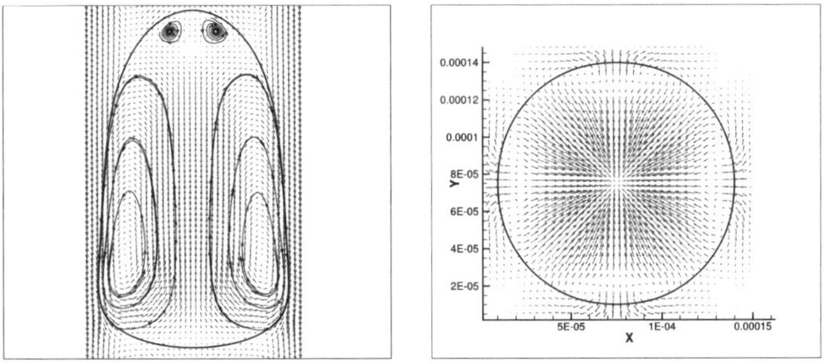

Fig. 6 A bigger droplet with an injection speed of $0.2\ m/s$: velocity field respectively in the droplet's frame of reference (slice in plane (x,z) on the left) and in the global one (slice in (x,y) on the right).

4 Conclusions

In this note, we present various 3D dynamics in microdroplets thanks to a numerical method designed to handle flows driven by surface tension and pressure gradient

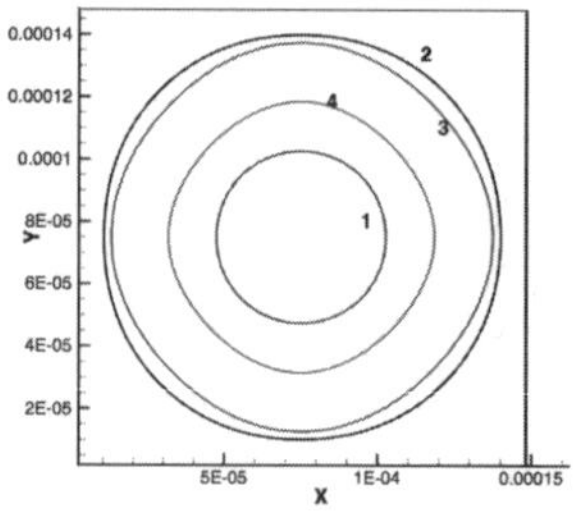

Fig. 7 Slices numbered from 1 to 4 (from the nose of the droplet to the back) in plane (x,y).

in a microfluidic framework. It appears that cylindrical and rectangular microchannels used in practical applications induced clearly different hydrodynamics mainly due to effect of rectangular confinement. This will be further studied by comparing jet stabilities in these two kinds of geometries, as well as mixing in droplets, and compared with equivalent physical experiments in microchannels [4].

References

1. Angot, P., Bruneau, C.H., Fabrie, P.: A penalization method to take into account obstacles in incompressible viscous flows. Numerische Mathematik **81**(4), 497–520 (1999)
2. Galusinski, C., Vigneaux, P.: Level-Set method and stability condition for curvature-driven flows. C. R. Acad. Sci. Paris, Ser. I **344**(11), 703–708 (2007)
3. Galusinski, C., Vigneaux, P.: On stability condition for bifluid flows with surface tension: application to microfluidics. Journal of Computational Physics **227**(12), 6140–6164 (2008). doi:10.1016/j.jcp.2008.02.023
4. Guillot, P.: Ecoulements de fluides immiscibles dans un canal submillimétrique: Stabilité et application à la rhéologie. Ph.D. thesis, Université Bordeaux 1 (2006)
5. Jiang, G.S., Peng, D.: Weighted ENO schemes for Hamilton-Jacobi equations. SIAM Journal of Scientific Computing **21**(No. 6), p. 2126–2143 (2000)
6. Osher, S., Fedkiw, R.: Level Set Methods and Dynamic Implicit Surfaces, *Applied Mathematical Sciences*, vol. 153. Springer (2003)
7. Sethian, J.A.: Level Set Methods and Fast Marching Methods - Evolving interfaces in computational geometry, fluid mechanics, computer vision and materials science, *Cambridge Monographs on Applied and Computational Mathematics*, vol. 3, second edn. Cambridge University Press (1999)
8. Sussman, M., Smereka, P., Osher, S.: A level set approach for computing solutions to incompressible two-phase flow. J. Comput. Phys **114**, p. 146–159 (1994)
9. Vigneaux, P.: An axisymmetric Level-Set method for microdroplets hydrodynamics. under review (2007)

Incomplete Interior Penalty Galerkin Method for a Nonlinear Convection-Diffusion Equation

V. Dolejší

Abstract We deal with a numerical solution of convection-diffusion problems with a nonlinear convection and a quasilinear diffusion. We employ the so-called incomplete interior penalty Galerkin (IIPG) method which is suitable for a discretization of quasilinear diffusive terms. We recall a priori hp error estimates in the L^2-norm and the H^1-seminorm. We present sets of numerical experiments where the experimental orders of convergence in dependence on the polynomial degree of approximation and the regularity of the exact solution are evaluated and discussed.

1 Introduction

We consider the following nonstationary nonlinear convection-diffusion equation, which represents a model problem for the solution of the system of the compressible Navier-Stokes equations. Let $\Omega \subset \mathbb{R}^d$, $d = 2,3$, be a bounded open polygonal domain with Lipschitz-continuous boundary $\partial\Omega = \partial\Omega_D \cup \partial\Omega_N$, $\partial\Omega_D \cap \partial\Omega_N = \emptyset$, and $T > 0$. We seek a function $u : Q_T = \Omega \times (0,T) \to \mathbb{R}$ such that

$$\begin{aligned} &\text{a)}\ \ \frac{\partial u}{\partial t} + \nabla \cdot \mathbf{f}(u) = \nabla \cdot \mathbf{R}(u, \nabla u) + g \quad \text{in } Q_T, \\ &\text{b)}\ \ u\big|_{\partial\Omega_D \times (0,T)} = u_D, \\ &\text{c)}\ \ \mathbf{R}(u,\nabla u)\cdot \mathbf{n}\big|_{\partial\Omega_N \times (0,T)} = g_N, \\ &\text{d)}\ \ u(x,0) = u^0(x), \quad x \in \Omega, \end{aligned} \tag{1}$$

where $g : Q_T \to \mathbb{R}$, $u_D : \partial\Omega_D \times (0,T) \to \mathbb{R}$, $g_N : \partial\Omega_N \times (0,T) \to \mathbb{R}$ and $u^0 : \Omega \to \mathbb{R}$ are given functions, $\mathbf{n} = (n_1, \dots, n_d)$ is a unit outer normal to $\partial\Omega$, $\mathbf{f} = (f_1, \dots, f_d) : \mathbb{R} \to \mathbb{R}^d$ and $\mathbf{R} = (R_1, \dots, R_d) : \mathbb{R}^{d+1} \to \mathbb{R}^d$ (i.e., $R_s = R_s(\zeta) =$

V. Dolejší
Charles University Prague, Faculty of Mathematics and Physics, Sokolovská 83, Prague, Czech Republic, e-mail: dolejsi@karlin.mff.cuni.cz

$R_s(\zeta_0, \zeta_1, \ldots, \zeta_d)$, $s = 1, \ldots, d$) are prescribed functions representing convective and diffusive fluxes, respectively.

The class of *discontinuous Galerkin* (DG) methods seems to be one of the most promising candidates to construct high order accurate schemes for solving partial differential equations of type (1). For a survey about DG methods, see [1] or [4]. An application of a DG method to problems with a linear or a non-linear diffusion was studied in many papers, see, e.g., [7], [8], [10], [11], [12].

The main obstacle in the discetization of the quasilinear diffusive term in (1) is a nonlinear dependence of $\mathbf{R}$ on ∇u. Then, the very popular DG approaches, SIPG (symmetric interior penalty Galerkin) and NIPG (nonsymmetric interior penalty Galerkin), can not be directly applied. Therefore, we use the so-called *incomplete interior penalty Galerkin* (IIPG) method studied in [5], [13]. We analysed the application of the IIPG technique to problem (1) quite recently in [6], where we derived a priori hp-error estimates. See also [3] where more general quasilinear elliptic problem was considered.

In this contribution we recall these results and then we present a set of numerical experiments, where we evaluate the discrete L^2-norm and the discrete H^1-seminorm. We set the experimental orders of convergence in dependence on the polynomial degree of approximation and the regularity of the exact solution. This numerical study can give some hints how to treat with a hp-adaptation technique based on an estimation of the regularity of the exact solution.

We assume that the data of problem (1) satisfy the assumptions:

(A1) function $\mathbf{f}(u)$ is Lipschitz-continuous, i.e., $|\mathbf{f}(v) - \mathbf{f}(w)| \le C_f |v - w|$, $C_f > 0$,
(A2) there exists a constant $c_1 > 0$ such that

$$|R_s(\zeta_0, \ldots, \zeta_d)| \le c_1 \left(1 + \sum_{k=0}^{d} |\zeta_k|\right) \qquad \forall \zeta \in I\!\!R^{d+1},\ s = 1, \ldots, d, \tag{2}$$

(A3) there exists a constant $c_2 > 0$ such that

$$\left|\frac{\partial R_s(\zeta)}{\partial \zeta_k}\right| \le c_2 \qquad \forall \zeta \in I\!\!R^{d+1},\ s = 1, \ldots, d,\ k = 0, \ldots, d, \tag{3}$$

(A4) there exists a constant $c_3 > 0$ such that

$$\sum_{\substack{s=1\\k=0}}^{d} \frac{\partial R_s(\zeta)}{\partial \zeta_k} \psi_s \psi_k \ge c_3 \sum_{k=1}^{d} \psi_k^2 \quad \forall \zeta, \psi \in I\!\!R^{d+1}. \tag{4}$$

In what follows, we shall assume that problem (1) has a unique sufficiently regular solution u. Its regularity, necessary for the theoretical error estimates, will be specified later in (19).

2 Dicretization

2.1 Triangulations

Let $\mathscr{T}_h$ $(h > 0)$ be a family of partitions of the domain Ω into a finite number of closed d-dimensional mutually disjoint simplexes and/or parallelograms K i.e., $\overline{\Omega} = \bigcup_{K\in\mathscr{T}_h} K$. By ∂K we denote the boundary of element $K \in \mathscr{T}_h$ and set $h_K = \mathrm{diam}(K)$, $h = \max_{K\in\mathscr{T}_h} h_K$. By ρ_K we denote the radius of the largest d-dimensional ball inscribed into K and by $|K|$ we denote the d-dimensional Lebesgue measure of K. By $\mathscr{F}_h$ we denote the smallest possible set of all open $(d-1)$-dimensional faces of all elements $K \in \mathscr{T}_h$. Further, we denote by $\mathscr{F}_h^I$ the set of all $\Gamma \in \mathscr{F}_h$ that are contained in Ω (inner faces), by $\mathscr{F}_h^D$ the set of all $\Gamma \in \mathscr{F}_h$ that $\Gamma \subset \partial\Omega_D$ and by $\mathscr{F}_h^N$ the set of all $\Gamma \in \mathscr{F}_h$ that $\Gamma \subset \partial\Omega_N$. For a shorter notation we put $\mathscr{F}_h^{ID} \equiv \mathscr{F}_h^I \cup \mathscr{F}_h^D$ and $\mathscr{F}_h^{DN} \equiv \mathscr{F}_h^D \cup \mathscr{F}_h^N$.

Moreover, for each $\Gamma \in \mathscr{F}_h$ we define a unit normal vector $\mathbf{n}_\Gamma$. We assume that $\mathbf{n}_\Gamma$, $\Gamma \in \mathscr{F}_h^D \cup \mathscr{F}_h^N$ has the same orientation as the outer normal of $\partial\Omega$. For $\mathbf{n}_\Gamma$, $\Gamma \in \mathscr{F}_h^I$ the orientation is arbitrary but fixed for each edge.

We assume that the triangulation is *locally quasi-uniform*, i.e., there exists a constant $C_Q > 0$ such that $h_{K_i} \le C_Q h_{K_j}$ $\forall K_i, K_j \in \mathscr{T}_h$ sharing face $\Gamma_{ij} \in \mathscr{F}_h^I$ and *shape-regular*, i.e., there exists a constant $C_S > 0$ such that $h_K \le C_S \rho_K$ $\forall K \in \mathscr{T}_h$.

2.2 Discontinuous Finite Element Spaces

To each $K \in \mathscr{T}_h$, we assign a positive integer s_K (local Sobolev index) and positive integer p_K (local polynomial degree) . Then we define the vectors

$$\mathsf{s} \equiv \{s_K, K \in \mathscr{T}_h\}, \qquad \mathsf{p} \equiv \{p_K, K \in \mathscr{T}_h\}. \tag{5}$$

Over the triangulation $\mathscr{T}_h$ we define the so-called *broken Sobolev space* corresponding to the vector s

$$H^{\mathsf{s}}(\Omega, \mathscr{T}_h) \equiv \{v; v|_K \in H^{s_K}(K)\ \forall K \in \mathscr{T}_h\}. \tag{6}$$

If $s_K = q$ $\forall K \in \mathscr{T}_h$, $q \in \mathbb{N}$ then we use the notation $H^q(\Omega, \mathscr{T}_h) = H^{\mathsf{s}}(\Omega, \mathscr{T}_h)$.

Furthermore, we define the space of discontinuous piecewise polynomial functions associated with the vector p by

$$S_{h\mathsf{p}} \equiv \{v;\ v \in L^2(\Omega),\ v|_K \in P_{p_K}(K)\ \forall K \in \mathscr{T}_h\}, \tag{7}$$

where $P_{p_K}(K)$ denotes the space of all polynomials on K of degree $\le p_K$, $K \in \mathscr{T}_h$. It means that we use the same polynomial spaces for simplicial as well as parallelogram elements. In order to derive a priori hp error estimates we assume that there exists a constant $C_P \ge 1$ such that $p_K/p_{K'} \le C_P$ $\forall K, K' \in \mathscr{T}_h$ sharing a common face. Moreover, to each $K \in \mathscr{T}_h$ we define the parameter $d(K) \equiv h_K/p_K^2$, $K \in \mathscr{T}_h$.

For each $\Gamma \in \mathscr{F}_h^I$ there exist two elements $K_p, K_n \in \mathscr{T}_h$ such that $\Gamma \subset \overline{K}_p \cap \overline{K}_n$. We use a convention that K_n lies in the direction of $\mathbf{n}_\Gamma$ and K_p in the opposite direction of $\mathbf{n}_\Gamma$. Then for $v \in H^1(\Omega, \mathscr{T}_h)$, we introduce the following notation:

$$\langle v \rangle_\Gamma \equiv \frac{1}{2}\left(v|_\Gamma^{(p)} + v|_\Gamma^{(n)}\right), \quad [v]_\Gamma \equiv v|_\Gamma^{(p)} - v|_\Gamma^{(n)}, \tag{8}$$

where $v|_\Gamma^{(p)}$ and $v|_\Gamma^{(n)}$ denotes the trace of $v|_{K_p}$ and $v|_{K_n}$ on Γ, respectively. Further, we put

$$d(\Gamma) \equiv \min(d(K_p), d(K_n)), \quad \Gamma \in \mathscr{F}_h^I. \tag{9}$$

For $\Gamma \in \mathscr{F}_h^{DN}$ there exists element $K_p \in \mathscr{T}_h$ such that $\Gamma \subset \overline{K}_p \cap \partial\Omega$. Then for $v \in H^1(\Omega, \mathscr{T}_h)$, we introduce the notation $\langle v \rangle_\Gamma \equiv [v]_\Gamma \equiv v|_\Gamma^{(p)}$, where $v|_\Gamma^{(p)}$ denotes the trace of $v|_{K_p}$ on Γ. In virtue of (9), we put

$$d(\Gamma) \equiv d(K_p), \quad \Gamma \in \mathscr{F}_h^{DN}. \tag{10}$$

In case that $[\cdot]_\Gamma$ and $\langle \cdot \rangle_\Gamma$ are arguments of $\int_\Gamma \ldots \,\mathrm{d}S$, $\Gamma \in \mathscr{F}_h$ we omit the subscript Γ and write simply $[\cdot]$ and $\langle \cdot \rangle$, respectively.

2.3 Discrete Problem

The *incomplete interior penalty Galerkin* (IIPG) discretization of the problem (1) reads (for more details see [6])

$$\left(\frac{\partial u}{\partial t}, v\right) + a_h(u,v) + b_h(u,v) + J_h(u,v) = \ell_h(v), \tag{11}$$

where

$$a_h(u,v) = \sum_{K \in \mathscr{T}_h} \int_K \mathbf{R}(u, \nabla u) \cdot \nabla v \,\mathrm{d}x - \sum_{\Gamma \in \mathscr{F}_h^{ID}} \int_\Gamma \langle \mathbf{R}(u, \nabla u) \cdot \mathbf{n} \rangle [v] \,\mathrm{d}S, \tag{12}$$

$$b_h(u,v) = -\sum_{K \in \mathscr{T}_h} \int_K \mathbf{f}(u) \cdot \nabla v \,\mathrm{d}x + \sum_{\Gamma \in \mathscr{F}_h} \int_\Gamma H\left(u|_\Gamma^{(p)}, u|_\Gamma^{(n)}, \mathbf{n}_\Gamma\right) [v] \,\mathrm{d}S, \tag{13}$$

$$J_h^\sigma(u,v) = \sum_{\Gamma \in \mathscr{F}_h^{ID}} \int_\Gamma \sigma [u]\,[v] \,\mathrm{d}S, \tag{14}$$

$$\ell_h(v)(t) = \int_\Omega g(t)\, v \,\mathrm{d}x + \sum_{\Gamma \in \mathscr{F}_h^N} \int_\Gamma g_N(t)\, v \,\mathrm{d}S + \sum_{\Gamma \in \mathscr{F}_h^{ID}} \int_\Gamma \sigma\, u_D(t)\, v \,\mathrm{d}S. \tag{15}$$

The penalty parameter function σ in (14) and (15) is defined by

$$\sigma|_\Gamma = \frac{C_W}{d(\Gamma)}, \quad \Gamma \in \mathscr{F}_h^{ID}, \tag{16}$$

where $d(\Gamma)$ is given either by (9) or (10) and $C_W > 0$ is a suitable constant. We use the value

$$C_W \equiv \frac{1}{4}\frac{C_C c_2(d+1)}{(c_3 - C_m)(1 - C_m)}, \tag{17}$$

where $C_m \equiv \min(1, c_3)/2$, constants c_2 and c_3 were introduced in assumptions (3) – (4) and $C_C \equiv C_M(1 + C_I)$, where constants C_M and C_I appear in the *multiplicative trace inequality* and the *inverse inequality*, respectively, for more details see [6]. The function $H(\cdot,\cdot,\cdot)$ in the face integrals in (13) is called the *numerical flux* well-known from the finite volume method and it approximates the terms $\mathbf{f}(u)\cdot\mathbf{n}_\Gamma$ on $\Gamma \in \mathscr{F}_h$. Now, we introduce the *discrete problem*.

Definition 1. Let $u_h^0 \in S_{hp}$ be the $L^2(\Omega)$-projection of the initial condition u^0 into S_{hp}. We say that u_h is a DGFE solution of (1), if

a) $u_h \in C^1([0,T]; S_{hp})$,

b) $$\left(\frac{\partial u_h(t)}{\partial t}, v_h\right) + b_h(u_h(t), v_h) + a_h(u_h(t), v_h) + J_h^\sigma(u_h(t), v_h) = \ell_h(v_h)(t) \tag{18}$$
$$\forall v_h \in S_{hp},\ \forall t \in (0,T),$$

c) $u_h(0) = u_h^0$.

3 A Priori Error Estimates

For $v \in H^{\mathsf{s}}(\Omega, \mathscr{T}_h)$ we introduce the norm $|||v|||^2 \equiv |v|^2_{H^1(\Omega,\mathscr{T}_h)} + J_h^\sigma(v,v)$. We assume that the exact solution $u = u(x,t)$ of (1) satisfies

$$\left.\frac{\partial u}{\partial t}\right|_K \in L^2(0,T; H^{s_K}(K)), \quad u|_K \in L^\infty(0,T; H^{s_K}(K)) \quad \forall K \in \mathscr{T}_h. \tag{19}$$

It can be ensured, e.g., by $\partial u/\partial t \in L^2(0,T; H^{\bar{s}}(\Omega))$, $u \in L^\infty(0,T; H^{\bar{s}}(\Omega))$, where $\bar{s} = \max\{s_K,\ s_K \in \mathsf{s}\}$. Then it makes sense to define the "element-norm"

$$\|u\|_K^2 \equiv \|u\|^2_{L^2(0,T;H^{s_K}(K))} + \|\partial u/\partial t\|^2_{L^2(0,T;H^{s_K}(K))} + \|u\|^2_{L^\infty(0,T;H^{s_K}(K))},\ K \in \mathscr{T}_h. \tag{20}$$

Theorem 1. *Let u be the exact solution of (1) satisfying (19) and $\mathscr{T}_h$ a partition of the computational domain. Let u_h be the approximate solution given by (18), where the penalty parameter σ satisfies (16) with (17). Let assumptions* **(A1) – (A4)** *be valid. Then the discetization error $e_h \equiv u_h - u$ satisfies the estimate*

$$\max_{t\in[0,T]} \|e_h(t)\|_{L^2(\Omega)} + C_m \int_0^T |||e_h(\vartheta)|||^2 \,\mathrm{d}\vartheta \le Q(T) \sum_{K\in\mathscr{T}_h} \frac{h_K^{2\mu_K-2}}{p_K^{2s_K-3}} \|u\|_K^2, \tag{21}$$

where $\mu_K = \min(p_K + 1, s_K)$, $K \in \mathscr{T}_h$ and $Q(T)$ is a function of T independent of $h_K, p_K, s_K, K \in \mathscr{T}_h$.

Proof. See [6, Theorem 18]. □

Remark 1. The term $Q(t)$ diverges for $C_m \to 0$, hence the estimate (21) cannot be used. The case $C_m \to 0$ corresponds to a vanishing diffusion term $\mathbf{R}(u, \nabla u)$. The blow up of the estimate is caused by the presence of the nonlinear convective term and the use of Gronwall's lemma.

Remark 2. We observe that the error estimate (21) is
i) h-suboptimal in the $L^\infty(0,T,L^2(\Omega))$-norm, namely $O(h^p)$,
ii) h-optimal in the $L^2(0,T,H^1(\Omega))$-seminorm, namely $O(h^p)$,
iii) p-suboptimal in the $L^\infty(0,T,L^2(\Omega))$-norm and the $L^2(0,T,H^1(\Omega))$-seminorm, namely $O(p^{-(s-3/2)})$.

The suboptimality of the h-estimate is caused by the nonsymmetry of the discrete problem which prevent us to use the well-known Aubin-Nitsche theorem (usually employed in order to obtain the optimal L^2-error estimates). The suboptimaly of the p-estimate is a consequence of the used multiplicative trace inequality, similar results was obtained, e.g., in [10].

4 Numerical Examples

In this section, we numerically verify the error estimates (21). We consider the nonlinear convection–diffusion equation

$$\frac{\partial u}{\partial t} + \sum_{s=1}^{2} u \frac{\partial u}{\partial x_s} = \sum_{s=1}^{2} \frac{\partial}{\partial x_s}(\nu(|\nabla u|)\nabla u) + g \quad \text{in } \Omega \times (0,T), \tag{22}$$

where $\nu(w) : (0,\infty) \to I\!R$ is chosen in the form

$$\nu(w) = \nu_\infty + (\nu_0 - \nu_\infty)(1+w)^{-\gamma}, \quad \gamma > 0. \tag{23}$$

We set $\nu_0 = 0.15$, $\nu_\infty = 0.1$, $\gamma = 1/2$, $\Omega = (0,1)^2$, $T = 80$, and define the function g, the initial and boundary conditions in such a way that the exact solution has the form

$$u(x_1,x_2,t) = \left(1 - \mathrm{e}^{-10t}\right)\hat{u}(x_1,x_2), \tag{24}$$

where

$$\hat{u}(x_1,x_2) = 2r^\alpha x_1 x_2 (1-x_1)(1-x_2) = r^{\alpha+2}\sin(2\varphi)(1-x_1)(1-x_2),$$

where (r,φ) are the polar coordinates and $\alpha \in I\!R$ is a constant. For $t = T = 80$ the solution u differs very little from the "steady state" solution $\hat{u}$. The function $\hat{u}$ is equal to zero on $\partial\Omega$ and its regularity depends on the value of α, namely (cf. [2])

$$\hat{u} \in H^\beta(\Omega) \quad \forall \beta \in (0, \alpha+3), \tag{25}$$

where $H^\beta(\Omega)$ denotes (in general) the Sobolev-Slobodetskii space of functions with "non-integer derivatives". In the presented numerical tests we use the value $\alpha =$

$-1/2$, which gives $\hat{u} \in H^{\beta}(\Omega)$, $\beta < 5/2$. The inequality (21) implies the error estimates in $\|\cdot\|_{L^2(\Omega)}$ and $|||\cdot|||$ norms of order

$$O(h^q) \ \text{ with } q \leq \min(3/2, p), \tag{26}$$

where p denotes the degree of polynomial approximation.

We solved the problem (22) – (23) by method (18) using a piecewise linear (P^1), quadratic (P^2), and cubic (P^3) approximations on 6 triangular meshes having 128, 288, 512, 1152, 2048 and 4608 elements. The temporal discretization was carried out by a three-step backward difference formula with a small time step which completely avoid any influence of the time discretization. Since solution $\hat{u}$ has a singularity at the origin ($x_1 = x_2 = 0$), we evaluate the computational errors not only over Ω but also over $\Omega^{\text{reg}} \equiv \Omega \setminus (0, 0.125)^2$.

Table 1 shows computational errors in the $L^2(\cdot)$-norm and in $H^1(\cdot)$-seminorm and the corresponding experimental orders of convergence (EOC) at $T = 20$. We observe the following:

i) EOC in the $|\cdot|_{H^1(\Omega)}$-seminorm perfectly correspond with (26). On the other hand, EOC in the $\|\cdot\|_{L^2(\Omega)}$-norm is better than (26), namely $O(h^q)$ with $q \leq \min(5/2, p+1)$. This precisely corresponds to the result from [9], where for any $\beta \in (1, 5/2)$ we get

$$\|v - I_h v\|_{L^2(\Omega)} \leq C(\beta) h^{\min(\beta, p+1)} \|v\|_{H^{\beta}(\Omega)}, \quad v \in H^{\beta}(\Omega), \tag{27}$$

where $I_h v$ is a piecewise polynomial Lagrange interpolation to v of degree $\leq p$ and $C(\beta)$ is a constant independent of h and v.

ii) Since $\hat{u}$ is regular over Ω^{reg} it follows from (21) that the theoretical error estimates in the $|\cdot|_{H^1(\Omega^{\text{reg}})}$-seminorm and $\|\cdot\|_{L^2(\Omega^{\text{reg}})}$-norm are of order $O(h^p)$. We observe a very good agreement of EOC in the $|\cdot|_{H^1(\Omega^{\text{reg}})}$-seminorm. On the other hand, errors in the $\|\cdot\|_{L^2(\Omega^{\text{reg}})}$-norm do not achieved the expected EOC of order $O(h^{p+1})$ following from (27) but are in agreement with the theoretical results.

Acknowledgements This work is a part of the research project MSM 0021620839 financed by the Ministry of Education of the Czech Republic and it was partly supported by the Grant No. 316/2006/B-MAT/MFF of the Grant Agency of the Charles University Prague.

References

1. Arnold, D.N., Brezzi, F., Cockburn, B., Marini, L.D.: Unified analysis of discontinuous Galerkin methods for elliptic problems. SIAM J. Numer. Anal. **39**(5), 1749–1779 (2002)
2. Babuška, I., Suri, M.: The p- and h-p versions of the finite element method. an overview. Comput. Methods Appl. Mech. Eng. **80**, 5–26 (1990)
3. Böhmer, K., Dolejší, V.: Numerical Methods for Nonlinear Elliptic Differential Equations, chap. DCGMs for Nonlinear Elliptic Differential Equations. A synopsis (in preparation)

Table 1 Computational errors and the corresponding experimental orders of convergence (EOC) in the L^2-norm and the H^1-seminorm

P_1 – approximation

mesh	h	$\lVert e_h\rVert_{L^2(\Omega)}$	EOC	$\lVert e_h\rVert_{L^2(\Omega^{\mathrm{reg}})}$	EOC	$\lvert e_h\rvert_{H^1(\Omega)}$	EOC	$\lvert e_h\rvert_{H^1(\Omega^{\mathrm{reg}})}$	EOC
1	0.884E-01	3.316E-03	–	3.308E-03	–	7.055E-02	–	6.826E-02	–
2	0.589E-01	1.581E-03	1.827	1.578E-03	1.824	4.813E-02	0.943	4.691E-02	0.925
3	0.442E-01	9.193E-04	1.884	9.103E-04	1.913	3.655E-02	0.957	3.471E-02	1.047
4	0.295E-01	4.222E-04	1.919	4.154E-04	1.935	2.468E-02	0.968	2.327E-02	0.986
5	0.221E-01	2.415E-04	1.942	2.367E-04	1.956	1.864E-02	0.976	1.751E-02	0.990
6	0.147E-01	1.092E-04	1.958	1.065E-04	1.969	1.252E-02	0.982	1.171E-02	0.993

P_2 – approximation

mesh	h	$\lVert e_h\rVert_{L^2(\Omega)}$	EOC	$\lVert e_h\rVert_{L^2(\Omega^{\mathrm{reg}})}$	EOC	$\lvert e_h\rvert_{H^1(\Omega)}$	EOC	$\lvert e_h\rvert_{H^1(\Omega^{\mathrm{reg}})}$	EOC
1	0.884E-01	8.137E-05	–	6.170E-05	–	5.719E-03	–	4.230E-03	–
2	0.589E-01	2.834E-05	2.601	2.158E-05	2.591	2.966E-03	1.619	2.200E-03	1.612
3	0.442E-01	1.363E-05	2.543	9.383E-06	2.895	1.872E-03	1.601	1.152E-03	2.248
4	0.295E-01	4.975E-06	2.487	3.294E-06	2.581	9.846E-04	1.584	5.243E-04	1.942
5	0.221E-01	2.463E-06	2.444	1.591E-06	2.530	6.271E-04	1.569	2.981E-04	1.963
6	0.147E-01	9.240E-07	2.418	5.755E-07	2.508	3.338E-04	1.555	1.337E-04	1.978

P_3 – approximation

mesh	h	$\lVert e_h\rVert_{L^2(\Omega)}$	EOC	$\lVert e_h\rVert_{L^2(\Omega^{\mathrm{reg}})}$	EOC	$\lvert e_h\rvert_{H^1(\Omega)}$	EOC	$\lvert e_h\rvert_{H^1(\Omega^{\mathrm{reg}})}$	EOC
1	0.884E-01	1.881E-05	–	1.441E-05	–	1.274E-03	–	5.349E-04	–
2	0.589E-01	6.943E-06	2.458	5.489E-06	2.380	6.856E-04	1.528	2.973E-04	1.449
3	0.442E-01	3.404E-06	2.477	2.057E-06	3.412	4.435E-04	1.515	9.097E-05	4.116
4	0.295E-01	1.239E-06	2.493	5.682E-07	3.172	2.406E-04	1.508	2.728E-05	2.970
5	0.221E-01	6.030E-07	2.503	2.248E-07	3.223	1.560E-04	1.505	1.149E-05	3.005
6	0.147E-01	2.182E-07	2.507	6.026E-08	3.247	8.484E-05	1.503	3.394E-06	3.008

4. Cockburn, B., Karniadakis, G.E., Shu, C.W. (eds.): Discontinuous Galerkin Methods. Springer, Berlin (2000)
5. Dawson, C.N., Sun, S., Wheeler, M.F.: Compatible algorithms for coupled flow and transport. Comput. Meth. Appl. Mech. Engng. **193**, 2565–2580. (2004)
6. Dolejší, V.: Analysis and application of IIPG method to quasilinear nonstationary convection-diffusion problems. J. Comp. Appl. Math. (published online doi:10.1016/j.cam.2007.10.055, 2007)
7. Dolejší, V., Feistauer, M., Kučera, V., Sobotíková, V.: An optimal $L^\infty(L^2)$-error estimate of the discontinuous galerkin method for a nonlinear nonstationary convection-diffusion problem. IMA J. Numer. Anal. (published online 10.1093/imanum/drm023, 2007)
8. Dolejší, V., Feistauer, M., Sobotíková, V.: A discontinuous Galerkin method for nonlinear convection–diffusion problems. Comput. Methods Appl. Mech. Engrg. **194**, 2709–2733 (2005)
9. Feistauer, M.: On the finite element approximation of functions with noninteger derivatives. Numer. Funct. Anal. and Optimiz. **10**(91-110) (1989)
10. Houston, P., Robson, J., Süli, E.: Discontinuous Galerkin finite element approximation of quasilinear elliptic boundary value problems I: The scalar case. IMA J. Numer. Anal. **25**, 726–749 (2005)
11. Houston, P., Schwab, C., Süli, E.: Discontinuous *hp*-finite element methods for advection-diffusion problems. SIAM J. Numer. Anal. **39**(6), 2133–2163 (2002)
12. Rivière, B., Wheeler, M.F., Girault, V.: Improved energy estimates for interior penalty, constrained and discontinuous Galerkin methods for elliptic problems. I. Comput. Geosci. **3**(3-4), 337–360 (1999)
13. Sun, S., Wheeler, M.F.: Symmetric and nonsymmetric discontinuous Galerkin methods for reactive transport in porous media. SIAM J. Numer. Anal. **43**(1), 195–219 (2005)

Numerical Simulations of Incompressible Laminar Flow for Newtonian and Non-Newtonian Fluids

R. Keslerová and K. Kozel

Abstract This paper deals with numerical solution of two dimensional and three dimensional laminar incompressible flows for Newtonian and non-Newtonian fluids through a branching channel. One could describe these problems using Navier-Stokes equations and continuity equation as a mathematical model using two different viscosities. The unsteady system of Navier-Stokes equations modified by unsteady term in continuity equation (artificial compressibility method) is solved by multistage Runge-Kutta finite volume method. Steady state solution is achieved for $t \to \infty$ and convergence is followed by steady residual behaviour. For unsteady solution high compressibility coefficient β^2 is considered. The numerical results for two and three dimensional cases of flows in the branching channel for Newtonian and non-Newtonian fluids are presented and compared.

1 Introduction

In the human body, plasma and bloody cells form a non-Newtonian fluid whose flow properties are uniquely adapted to the architecture of the blood vessels. Therefore study of blood flow in large and medium vessels is a very complex task. In this work blood flow in cardiovascular system is simply simulated by non-Newtonian model flow and it is compared with corresponding Newtonian fluids flow.

R. Keslerová and K. Kozel
Department of Technical Mathematics, Faculty of Mechanical Engineering, Czech Technical University, Karlovo nám. 13, 121 35 Praha 2, Czech Republic, e-mail: keslerov@marian.fsik.cvut.cz, kozelk@fsik.cvut.cz

2 Mathematical Model

2.1 Non-Newtonian Fluids

Firstly, model of non-Newtonian fluids are considered. A system of generalized Navier-Stokes equations and continuity equation is the mathematical model. In conservative form one can read

$$\tilde{R}W_t + F_x + G_y = \frac{\tilde{R}}{\text{Re}}(R_x + S_y) \qquad \tilde{R} = \text{diag}(0,1,1), \qquad \text{Re} = \frac{q_\infty l}{\nu}, \tag{1}$$

where

$$W = (p,u,v)^T, \quad F = (u, u^2+p, uv)^T, \quad G = (v, uv, v^2+p)^T, \tag{2}$$
$$R = (0, g_{11}, g_{21})^T, \quad S = (0, g_{12}, g_{22})^T.$$

The expression of the right hand side is one of the simplest non-Newtonian models

$$g_{ij} = 2\mid e \mid^r e_{ij}, \qquad e_{ij} = \frac{1}{2}\left(\frac{\partial u_i}{\partial x_j} + \frac{\partial u_j}{\partial x_i}\right), \quad r \in [0,1] \tag{3}$$

and

$$\mid e \mid = \sqrt{e_{11}^2 + e_{12}^2 + e_{21}^2 + e_{22}^2} = \sqrt{u_x^2 + \frac{1}{2}(u_y+v_x)^2 + v_y^2}. \tag{4}$$

The definition of the Reynolds number in two or three dimensional case are given by the relations

$$Re = \frac{q_\infty l}{\nu}, \quad Re = \frac{q_\infty 4S}{\nu O} \tag{5}$$

where q_∞ is the reference velocity, l is the reference length, S and O represent volume and circumference of the entrance's cut, ν is the kinematic viscosity.

2.2 Newtonian Fluids

Now we consider Newtonian fluids. It's the special form of non-Newtonian fluids with $r = 0$ (for more details see [1]). The conservative form of the system of Navier-Stokes equation is

$$\tilde{R}W_t + F_x + G_y = \frac{\tilde{R}}{\text{Re}}\Delta W, \tag{6}$$

the definitions of matrix $\tilde{R}$ and Reynolds number Re are the same as above. The wector W and fluxes F, G are given by relations in (2).

2.3 Boundary Conditions

The Dirichlet boundary conditions for velocity vector $(u,v)^T$ are used at the inlet and other variables are computed by extrapolation from the domain. At the outlet a pressure value is given and velocity components are computed by the extrapolation from the domain. The zero Dirichlet boundary conditions for the velocity are used on the wall. For pressure we used extrapolation from computed domain to boundary where normal derivative of pressure is zero - Prandtl's boundary layer relation.

3 Numerical Solution by Finite Volume Method

In this part a steady state solution is considered. In such a case the artificial compressibility method can be applied, i.e. the equation of continuity is completed with a term $\frac{1}{\beta^2}p_t$ [4]. Therefore the system of 2D Navier-Stokes equations for Newtonian fluids (6) has the form

$$W_t + \tilde{F}_x + \tilde{G}_y = 0 \tag{7}$$

where

$$W = \left(\frac{p}{\beta^2}, u, v\right)^T, \quad \beta \in \Re, \quad \tilde{F} = F - \frac{1}{\mathrm{Re}}F^v, \quad \tilde{G} = G - \frac{1}{\mathrm{Re}}G^v \tag{8}$$

and F, G are inviscid fluxes defined by equation (2), F^v, G^v are viscous fluxes [2] $F^v = (0, u_x, v_x)^T, \quad G^v = (0, u_y, v_y)^T$.

For steady state solution β is equal to 1. This system of equations is solved by finite volume method in the form of multistage Runge-Kutta method

$$\begin{aligned} W_{ij}^n &= W_{ij}^{(0)} \\ W_{ij}^{(r)} &= W_{ij}^{(0)} - \alpha_r \Delta t \overline{R}W_{ij}^{(r-1)} \\ W_{ij}^{n+1} &= W_{ij}^{(m)} \qquad\qquad r = 1, \ldots, m, \end{aligned} \tag{9}$$

where $m = 3$, $\alpha_1 = \alpha_2 = 0.5, \alpha_3 = 1.0, \overline{R}W_{ij}^n = RW_{ij}^n - DW_{ij}^n$, and the steady residual RW_{ij} is defined by

$$RW_{ij} = \frac{1}{\mu_{ij}} \sum_{k=1}^{4} \left[\left(F_k - \frac{1}{\mathrm{Re}} F_k^v \right) \Delta y_k - \left(G_k - \frac{1}{\mathrm{Re}} G_k^v \right) \Delta x_k \right]. \tag{10}$$

The term DW_{ij} presents the artificial viscosity of Jameson's type and it's used for higher Reynolds numbers (for more details see, e.g. [5]).

For the satisfaction of the stability condition the time step is computed from the formula (see e.g. [1])

$$\Delta t = \min_{i,j,k} \frac{\text{CFL } \mu_{ij}}{\rho_A \Delta y_k + \rho_B \Delta x_k + \frac{2}{\text{Re}} \left(\frac{(\Delta x_k)^2 + (\Delta y_k)^2}{\mu_{ij}} \right)}, \tag{11}$$

$$\rho_A = |\, \hat{u} \,| + \sqrt{\hat{u}^2 + 1} \qquad \rho_B = |\, \hat{v} \,| + \sqrt{\hat{v}^2 + 1},$$

and $|\, \hat{u} \,|, |\, \hat{v} \,|$ are the maximal values of the components of velocity inside the computational domain, the definition of $\Delta x_k, \Delta y_k$ is shown in Fig. (1). The CFL number for three-stage Runge-Kutta is equal to 2.

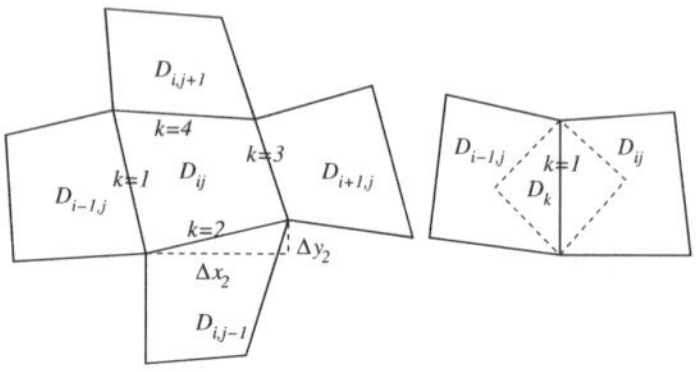

Fig. 1 Finite volume cell D_{ij} and dual volume cell D_k

The computation is performed until the value of the L^2-norm of residual satisfy Rez $W_{ij}^n \leq \varepsilon_{ERR}$ with ε_{ERR} small enough (MN denotes the number of grid cells in the computational domain), where

$$\text{Rez } W_{ij}^n = \sqrt{\frac{1}{MN} \sum_{ij} \left(\frac{W_{ij}^{n+1} - W_{ij}^n}{\Delta t} \right)^2}. \tag{12}$$

3.1 Unsteady Flow

For nonstationary case the artificial compressibility constant β is equal to 10 or sequences $\beta_k \longrightarrow \infty$ (β_k higher) is used. The boundary conditions are the same as for steady computation except one of the outlet pressure value. Here we prescribe pressure by the function

$$p_{out} = \frac{1}{4} \left(1 + \frac{1}{2} sin(\omega t) \right) \tag{13}$$

where ω is the angular velocity $\omega = 2\pi f$ and f is the frequency.

4 Numerical Results

In this section the numerical results of 2D and 3D case for Newtonian and non-Newtonian fluids flow are presented. The branching channel has one entrance and two exit parts. First results for non-Newtonian fluids were presented in [3].

2D results in Fig. 2 and 3 show the velocity magnitude distribution and convergence history for the Newtonian and non-Newtonian fluids. The same shape of geometry with the different angle between axis x and the branch is presented.

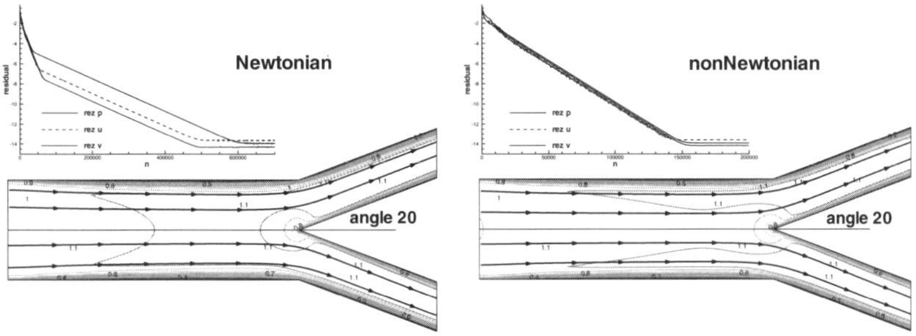

Fig. 2 Velocity isolines for Newtonian fluids (Re=500) and non-Newtonian fluids (Re=2000) with the angle 20 degrees

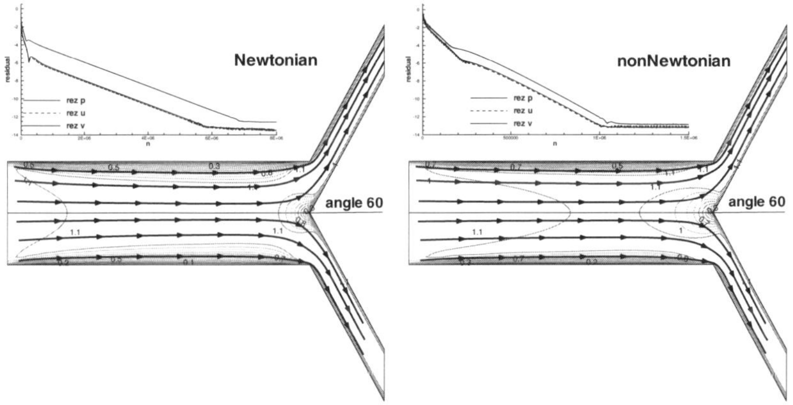

Fig. 3 Velocity isolines for Newtonian fluids (Re=100) and non-Newtonian fluids (Re=1000) with the angle 60 degrees

In the Fig. 4 and Fig. 5 the velocity isolines and history of convergence are shown. The same geometry with two angles is used for the corresponding results for Newtonian and non-Newtonian fluids.

The three dimensional results are presented in this part of this section. The geometry of these channels is similar to 2D geometry. As above the velocity magnitude distributions for Newtonian and non-Newtonian fluids are presented. The history of convergence for residuals is shown for all cases. Fig. 6 and Fig. 8 show the Newtonian fluids and Fig. 7 and Fig. 9 show the corresponding results for the non-Newtonian fluids.

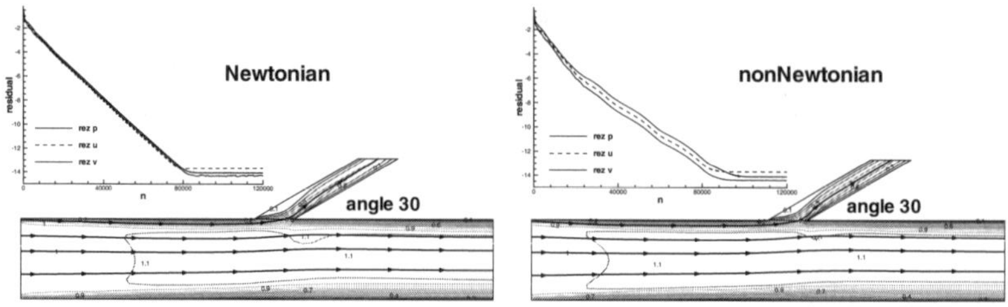

Fig. 4 Velocity isolines for Newtonian fluids (Re=800) and non-Newtonian fluids (Re=2000) with the angle 30 degrees

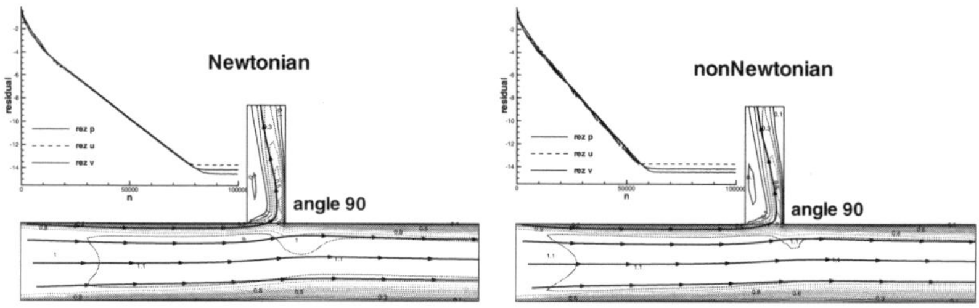

Fig. 5 Velocity isolines for Newtonian fluids (Re=500) and non-Newtonian fluids (Re=1300) with the angle 90 degrees

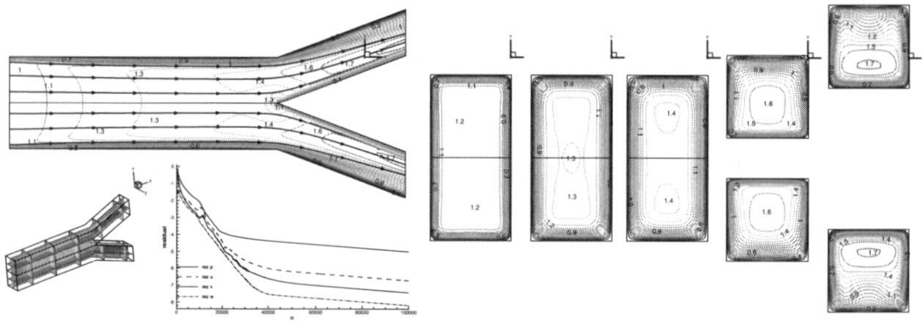

Fig. 6 Velocity magnitude distribution for Newtonian fluids (Re=500), cross-section of plane *xy* and cuts through the channel

First results for unsteady case are shown in this work. We set pressure outlet value as a function of time see 3.1, the frequency f is equal to 2Hz. The results are displayed during one period Fig. 10 and Fig. 11. Corresponding stationary solutions were shown in the Fig. 2.

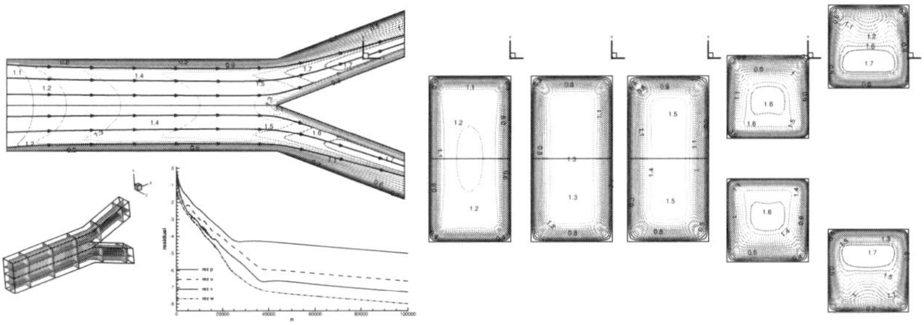

Fig. 7 Velocity magnitude distribution for non-Newtonian fluids (Re=1000), cross-section of plane *xy* and cuts through the channel

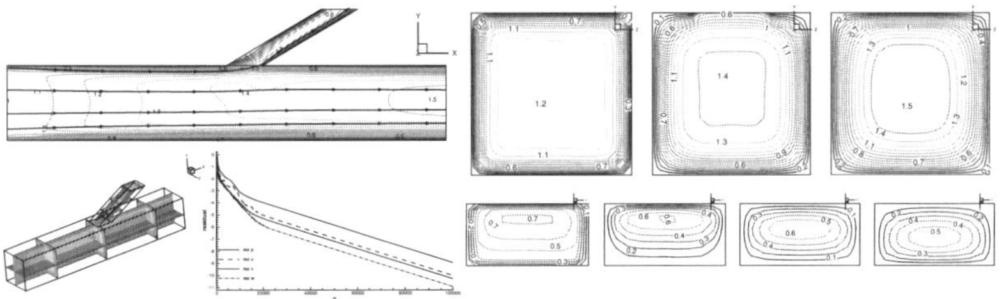

Fig. 8 Velocity magnitude distribution for Newtonian fluids (Re=500), cross-section of plane *xy* and cuts through the main part and small branch of the channel

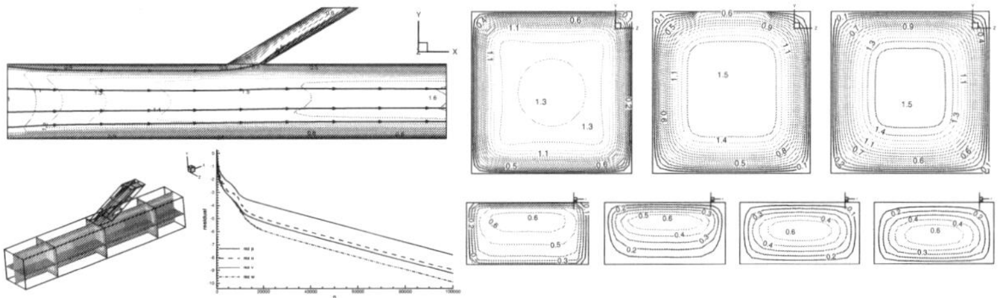

Fig. 9 Velocity magnitude distribution for non-Newtonian fluids (Re=1000), cross-section of plane *xy* and cuts through the main part and small branch of the channel

5 Conclusion

Numerical model for simulation of Newtonian and non-Newtonian fluid flow in a branching channel for two dimensional and three dimensional cases was developed. The method was applied for several different types of channel configurations. For non-Newtonian case the simple model of non-Newtonian flows was used. The presented results should be useful as a blood flow approximation. Two dimensional and three dimensional results for Newtonian fluids were compared with non-Newtonian corresponding results.

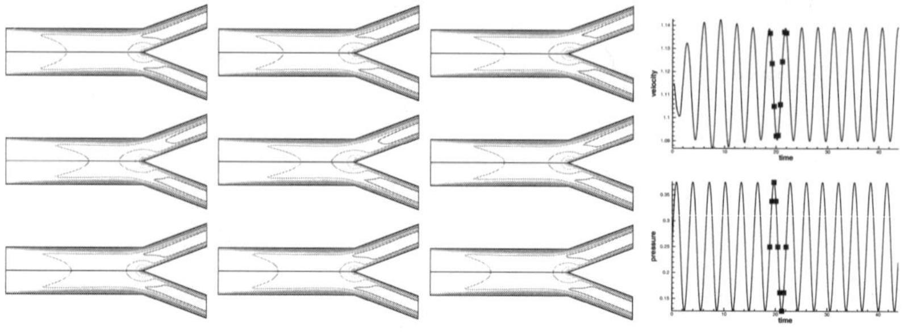

Fig. 10 Velocity isolines for Newtonian fluids

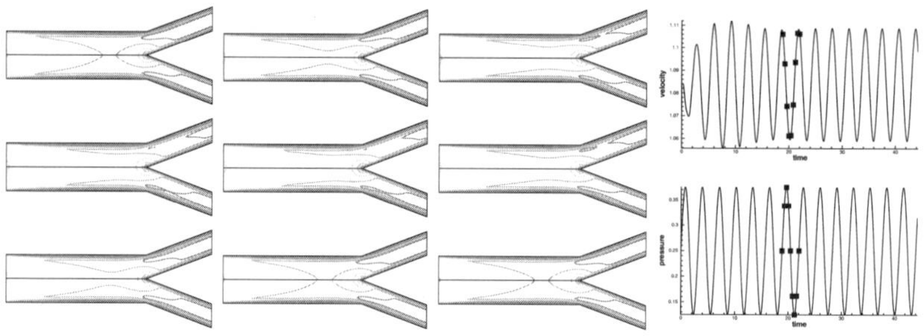

Fig. 11 Velocity isolines for non-Newtonian fluids

Acknowledgements This work was partly supported by grant GAAVCR No. IAA100190804, grant GACR 201/08/0012 and Research Plan MSM 6840770003.

References

1. Dvořák, R., Kozel, K.: Mathematical Modelling in Aerodynamics (in czech). CTU, Prague, Czech Republic (1996)
2. Keslerová, R., Kozel, K.: Numerical simulation of 2d and 3d incompressible laminar flows. In: Seminar of Applied Mathematics, Conference Proceedings pp. 120–127 (2005)
3. Keslerová, R., Kozel, K.: Numerical solution of incompressible laminar flow. In: CMFF'06 Conference Proceedings pp. 857–863 (2006)
4. Keslerová, R., Kozel, K., Prokop, V.: Numerical simulation of newtonian and non-newtonian fluids. In: Topical Problems of Fluid Mechanics 2007, Conference Proceedings pp. 85–88 (2007)
5. Swanson, R., Turkel, E.: Multistage schemes with multigrid for euler and navier-stokes equations. NASA Technical Paper **3631** (1997)

Numerical Solution of 2D and 3D Unsteady Viscous Flows

K. Kozel, P. Louda, and J. Příhoda

Abstract In this work, an artificial compressibility method is used to solve steady and unsteady flows of viscous incompressible fluid. The method is based on implicit higher order upwind discretisation of Navier-Stokes equations. The extension for unsteady simulation is considered by increasing artificial compressibility parameter or by using dual time stepping. Turbulence is modelled by the eddy-viscosity SST (Shear Stress Transport) model. Results for unsteady laminar flows around a circular cylinder and for unsteady turbulent free synthetic jet are presented.

1 Introduction

The work deals with numerical solution of 2D and 3D incompressible viscous (laminar and turbulent) unsteady flows. Solving viscous incompressible flows modelled by the system of Navier-Stokes equations one can find firstly velocity vector $(u,v,w)^{n+1}$ using the numerical solution of momentum equations with $p = p^n$ and then solving a corresponding Poisson equation for p^{n+1} using new velocity $(u,v,w)^{n+1}$ on the right hand side of the Poisson equation. Methods using the Poisson equation for pressure instead of continuity equation are generally limited in time accuracy, especially if they are implicit. On the other hand an implicit discretisation is needed for application of advanced turbulence models which are integrated up to the wall and require refined mesh in the boundary layer. The other problem of this approach is the solution of the Poisson equation itself. In this work, authors use

Karel Kozel
Czech Technical University in Prague, e-mail: kozelk@fsik.cvut.cz

Petr Louda
Czech Technical University in Prague, e-mail: petr.louda@fs.cvut.cz

Jaromír Příhoda
Institute of Thermomechanics, Czech Academy of Sciences, Prague, e-mail: prihoda@it.cas.cz

structured multiblock non-orthogonal non-regular grids. For such grids the convergence of a Poisson solver could deteriorate compared with orthogonal one block grid. That is the reason to apply the approach based on the artificial compressibility in this work. The idea, proposed by Chorin [2], is to complete continuity equation by a pressure time derivative $\frac{1}{\beta^2}\frac{\partial p}{\partial t}$ and then, because the whole inviscid system has the form of the system of compressible Euler equations (non-linear hyperbolic system), one can use a suitable numerical scheme for compressible flow computation. Using steady boundary conditions and time dependent method to solve the system of Euler or Navier-Stokes equations the steady solution of the model will be achieved. An extension for unsteady simulations can be achieved by increasing artificial compressibility parameter or by introducing dual time.

2 Mathematical Model

In the simplest form of the artificial compressibility method [2], only the continuity equation is modified by adding pressure time derivative

$$\frac{1}{\beta^2}\frac{\partial(p/\rho)}{\partial t}+\frac{\partial u_1}{\partial x_1}+\frac{\partial u_2}{\partial x_2}+\frac{\partial u_3}{\partial x_3}=0,\ \rho=const \tag{1}$$

where β is a positive artificial compressibility parameter. The inviscid part of modified Navier-Stokes equations is now hyperbolic and can be solved by standard methods for hyperbolic conservation laws. The system including the modified continuity equation and the momentum equations can be written for a domain of solution Ω in the following form

$$\Gamma\frac{\partial W}{\partial t}+Rez(W)=0,\quad \Gamma=diag[\beta^{-2},1,1,1],\ W=col[p/\rho,u_1,u_2,u_3], \tag{2}$$
$$(x_1,x_2,x_3)\in\Omega,\quad t\in(0,\infty)$$

where W is vector of unknown kinematic pressure and velocity components, and steady residual $Rez(W)$ contains all inviscid and viscous terms. However, the divergence free velocity field is not achieved before steady state, $\partial p/\partial t=0$. In unsteady case, the velocity divergence error may have negligible impact on relevant flow parameters if the β^2 is large enough. Other possibility of obtaining unsteady solution of the system

$$R\frac{\partial W}{\partial t}+Rez(W)=0,\quad R=diag[0,1,1,1] \tag{3}$$

is to use the iterative solution in artificial (dual, iterative) time τ between two time grid levels (t_n,t_{n+1}) described by the following system

$$\Gamma\frac{\partial W}{\partial \tau}+Rez^{uns}(W)=0,\;\; Rez^{uns}(W)=R\frac{\partial W}{\partial t}+Rez(W),\; R=diag[0,1,1,1], \quad (4)$$
$$(x_1,x_2,x_3)\in\Omega,\quad t\in<t_n,t_{n+1}>,\quad \tau\in(0,\infty)$$

for $\tau \in (0,\infty)$. The numerical solution in τ is realised iteratively for $\tau = \tau_0 = t_n,\ \tau_1,\ \tau_2,\ldots$ and the iterative process in τ is considered in such a way that for $\tau \to \infty$ we achieve $W^n \to W^{n+1}$. The advantage is that the unsteady residual $Rez^{uns}(W)$ can be computed in each grid point $t = t_n$ and the convergence to unsteady solution can be observed in similar way as in steady solution realised by time dependent method.

In order to simulate turbulent flows the Reynolds averaging procedure is used. For unsteady simulation it can be understood as a phase averaging since simulated flows have periodic "steady" state. The Reynolds averaged equations formally differ from the Navier-Stokes (NS) equations by additional momentum transport expressed by Reynolds stress tensor. In this work, the Reynolds stress is modelled using the two-equation SST (Shear Stress Transport) turbulence model [4].

3 Numerical Method

The stability limitation of an explicit scheme for 1D inviscid flow requires $\Delta t \sim \Delta x/(|u|+\sqrt{|u|^2+\beta^2})$, where u is flow velocity vector and Δx grid spacing. It becomes more severe as the β increases. The choice of β is problem dependent, there is no general optimum procedure. For robustness it is recommended [7] the β be constant for the whole solution domain, although in general the choice should depend on local grid and local convergence rate. In authors experience with constant β, for orthogonal grids the convergence e.g. in L_2 norm for pressure is as fast as the one for other unknowns (velocity components) and the steady residual level easily is machine zero even for highly refined grids. For non-orthogonal grids the pressure residual settles on higher steady value than the other residuals. This, however, does not seem to affect the accuracy of velocity field. The authors choose the β in order of magnitude of maximum flow velocity in the solution domain for steady simulations.

The time discretisation schemes considered here are implicit both for physical and artificial time. In the single time method for unsteady simulations, a second order accurate three-layer scheme is used in the form

$$\Gamma\frac{3W^{n+1}_{i,j,k}-4W^{n}_{i,j,k}+W^{n-1}_{i,j,k}}{2\Delta t}+Rez(W)^{n+1}_{i,j,k}=0. \quad (5)$$

For the dual time method, the scheme is combination of the backward Euler method for artificial time (superscript ν) and three-layer scheme for physical time (superscript n)

$$\Gamma\frac{W^{\nu+1}_{i,j,k}-W^{\nu}_{i,j,k}}{\Delta\tau}+R\frac{3W^{\nu+1}_{i,j,k}-4W^{n}_{i,j,k}+W^{n-1}_{i,j,k}}{2\Delta t}+Rez(W)^{\nu+1}_{i,j,k}=0. \quad (6)$$

where physical time step Δt is chosen according to the solved problem. The $\Delta\tau$ for explicit scheme is a suitable function of Δt and spatial grid steps which fulfils the condition that iterative process in τ is stable. In our case the scheme in τ is implicit and $\Delta\tau$ is not limited. It is chosen to be very large $\approx 10^7\Delta t$ in order to achieve W^{n+1} quickly in 8-10 iteration steps.

The steady residuals are computed by a cell-centered finite volume method with quadrilateral or hexahedral finite volumes in 2D or 3D case respectively. The discretisation of convective terms uses third order accurate upwind interpolation. Pressure gradient is computed by central approximation. Viscous terms are approximated using the second order central scheme, with cell face derivatives computed on a dual grid of finite volumes constructed over each face of primary grid. The discrete expression for residual is linearized with respect to unknown vector W in each finite volume from the stencil, by means of the Newton method. The derivatives are computed analytically. Only five (2D) or seven (3D) finite volumes are retained in the implicit discrete operator. The resulting linear system is thus five or seven block diagonal. The solution is found by a line relaxation method with direct method for block tri-diagonal matrix inversion. The two equations forming the turbulence model are solved decoupled from NS equations and the discretisation is same as for the NS equations. Only the negative part of source terms is linearized and included into the implicit operator.

4 Numerical Results

4.1 Flows Around Circular Cylinder

Here, the laminar 2D flow around a circular cylinder is simulated. First we consider the flow around a cylinder placed slightly eccentrically inside a channel. The Reynolds number $Re = UD/\nu = 100$, where $U = 2U_m/3$ is the bulk inlet velocity, U_m maximum inlet velocity and D diameter of the cylinder. At this Reynolds number, unsteady periodic flow evolves due to the vortex shedding on the cylinder. The compilation [5] reports results for this case achieved by several numerical methods. The results of present dual time artificial compressibility method with $\beta = U$ in the form of drag and lift coefficients are shown in Fig. 1 and agree well with reference [5]. The instantaneous flow-field for dual time stepping scheme is shown in Fig. 2. The single time method however gave unsatisfactory results, see [3].

Next we consider a cylinder in the free stream of constant velocity U_m. The cylinder is placed in the middle of computational domain $40D$(streamwise) $\times$ $100D$. The results are achieved with dual time method with $\beta = U_m, \Delta t = 0.06D/U_m$.

Fig. 3 shows dependency of the Strouhal number (frequency) of vortex shedding for different Reynolds numbers, in comparison with the empirical corelation. The vortex shedding starts at $Re = 47.5 \pm 0.7$ [8]. In our computation, the flow is steady for $Re = 30$, unsteady but non-periodic for $Re = 40$ and periodic at $Re = 47$ (shown

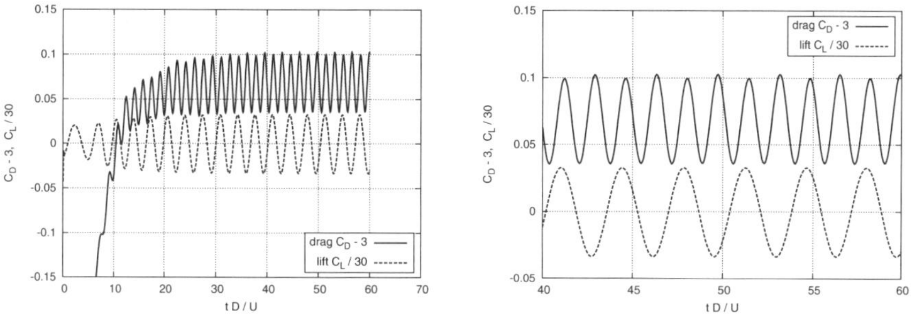

Fig. 1 Laminar flow around cylinder, dual time stepping. Left: evolution from initial state, right: zoom of periodic flow

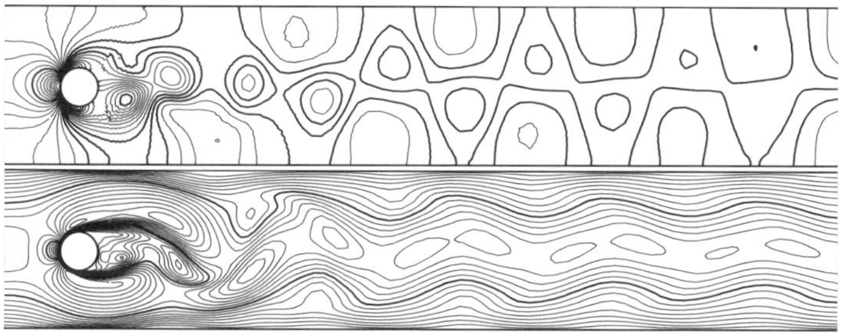

Fig. 2 Instantaneous laminar flow around cylinder, dual time stepping. Isolines of pressure (above) and velocity (below)

in Fig. 3). However for lower *Re* the computed shedding frequency is higher than an empirical correlation $St = 0.266 - 1.016/\sqrt{Re}$ [8] suggests. It should be noted that result of $Re = 140$ was achieved on a finer grid than other results. Original grid has 116 finite volumes on the cylinder surface, total approx. 16800 finite volumes, whereas the finer grid has 133 volumes on the cylinder surface, in total approx. 21500 volumes. The thickness of first finite volume on the cylinder surface was in both cases $2 \cdot 10^{-3}D$. On the finer grid, the Strouhal number decreased to 0.1821 from 0.1839 on the original grid. The results are summarized in Tab. 1. The Strouhal number is more reliable (less sensitive) parameter than force coefficients. The evolution of steady (i.e. in time) and unsteady (i.e. between two time levels, typical for all time steps) residual is shown in Fig. 4 ($Re = 100$).

Table 1 Strouhal number, drag and lift coefficients for cylinder in free space

Re	47	50	60	80	100	120	140
St	0.1336	0.1362	0.1438	0.1547	0.1642	0.1733	0.1821
C_D (mean)	1.55	1.53	1.50	1.43	1.36	1.34	1.34
C_L (ampl.)	0.15	0.18	0.22	0.22	0.23	0.34	0.45

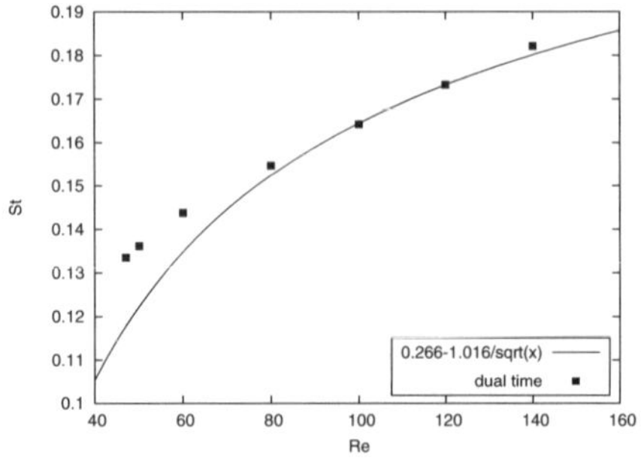

Fig. 3 Variation of the Strouhal number with the Reynolds number

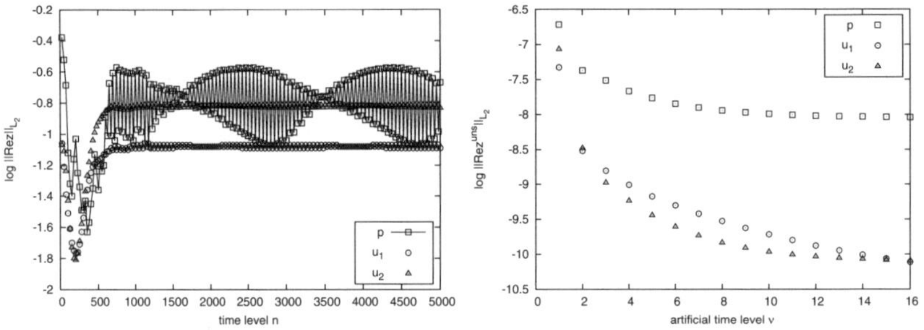

Fig. 4 Convergence history for dual time method (left: steady residuals, right: unsteady residuals)

4.2 Turbulent Synthetic Free Jet

In this part, authors consider the synthetic jet generated by periodical inflow/outflow with zero time average value in the circular nozzle [6]. The nozzle diameter $D = 8$ mm and Reynolds number $Re = U_{max}D/\nu = 13\,325$. The computational domain of conical shape is shown in Fig. 5. The length of the domain is $60D$. The problem is solved in Cartesian coordinates. The grid is split to several structured blocks in order to include paraxial zone, see Fig. 5. The boundary conditions are:

- nozzle ($x = 0$, $\sqrt{(y^2 + z^2)} \le D/2$):

$$u = U\sin(2\pi f t),\; v = 0.1 v_{Nmn},\; w = 0.1 w_{Nmn},$$
$$k = \begin{cases} \frac{3}{2}(uTu)^2 & \text{for} \quad u > 0, \\ k_{Nmn} & \text{otherwise} \end{cases} \qquad \omega = \begin{cases} k/(5\nu) & \text{for} \quad u > 0, \\ \omega_{Nmn} & \text{otherwise} \end{cases}$$

  where frequency $f = 75$Hz, streamwise amplitude $U = U_{max}[1 - (2r/D)^{26}]$, $U_{max} = 27.3$ m/s, where $0 \le r \le D/2$ is the radial position. The v_{Nmn}, w_{Nmn} are values as obtained using the homogenous Neumann condition and 0.1 is ad hoc parameter. The turbulence intensity in the nozzle was set to $Tu = 0.1$.

- wall ($x = 0,\ 0.5D \leq \sqrt{y^2+z^2} \leq 4.95D$): $u = v = w = 0,\ k = 0,\ \omega = \omega_w$, where ω_w is value of ω on the wall. A constant value of $2 \cdot 10^3 U_{max}/D$ was chosen.
- free boundary: This is the rest of domain boundary. Zero Neumann boundary condition was used for velocity, k and ω at $x = 60D$ and zero derivative in the radial direction for these variables on the conical part.

The dual time stepping method with $\beta = U_{max}$, $\Delta t = 1/72$ of forcing period and $\Delta\tau = 10^7 \Delta t$ was used. The Fig. 7 shows time averaged velocity on jet axis. The computational result agrees well with experiment [6], except for distance from the nozzle smaller than $\approx 0.4D$. The computational results achieved using the FLUENT code with an axisymmetrical formulation [6] are also shown. Inlet boundary condition in FLUENT differs from the present one and so does the velocity near nozzle. In FLUENT, the velocity on the axis decreases too fast, which suggests a higher spreading rate of the jet than in experiment. The present computation seems satisfactory here, which means that the turbulence model is acceptable. Next Fig. 6 shows velocity on the axis for several phase angles of inlet excitation. The computed instantaneous velocity corresponds to phase averaged velocity of the experiment [6]. In both cases the unsteadiness reaches up to ≈ 18 nozzle diameters far away. For larger distances, the flow-field corresponds to steady free jet. The Fig. 8 shows self-similarity of velocity profiles, where $r_{0.5}$ denotes the radial distance from jet axis, where the velocity reaches half of the axial velocity. The self-similarity is present except for region near the nozzle. Velocity profiles also agree well with the empirical correlation $U/U_{max} = \exp[-\ln(2)(r/r_{0.5})^2]$ according to [1].

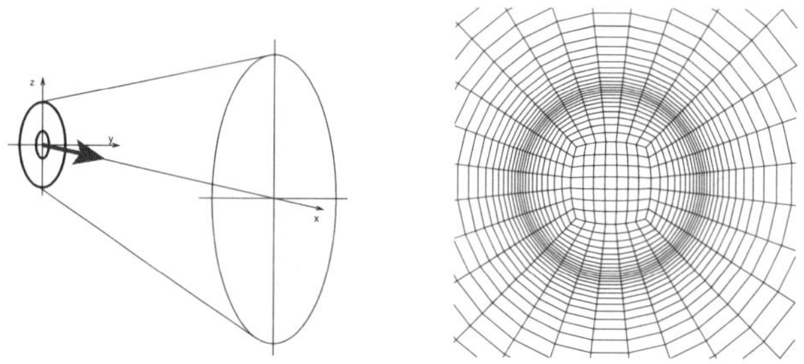

Fig. 5 Computational domain for synthetic jet flow

5 Conclusions

The work presents two variants of an implicit artificial compressibility method for unsteady simulations. The methods were tested using laminar flows around circular cylinder and turbulent synthetic free jet flow. The single time method has unsatisfactory accuracy. The dual time stepping method is found more reliable and of sufficient accuracy.

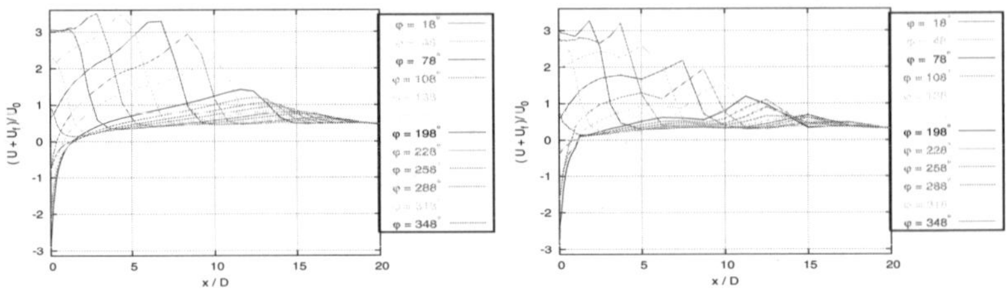

Fig. 6 Phase averaged velocity on jet axis. Left: computation, right: measurement

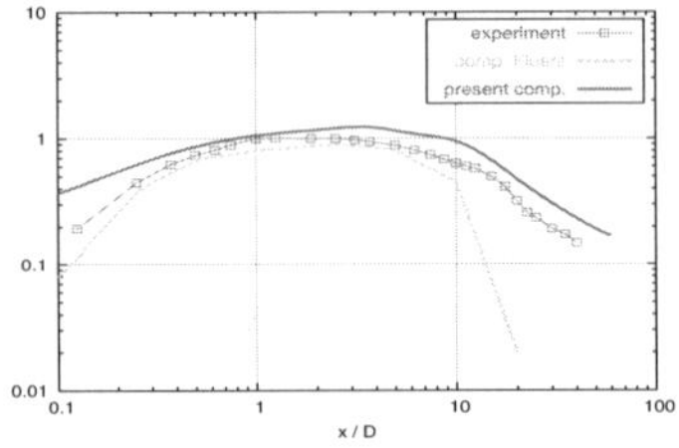

Fig. 7 Time averaged velocity on the jet axis

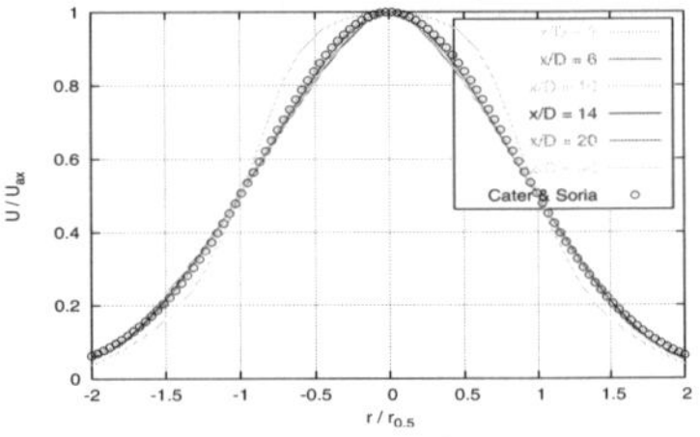

Fig. 8 Time averaged velocity profiles

Acknowledgements The work was supported by grant No. 201/08/0012 of the CSF, the Research Plan MSM 6840770010 and grant IAA200760801 of GA AS CR.

References

1. Cater, J.E., Soria, J.: The evolution of round zero-net-mass-flux jets. J. Fluid. Mech. **472**, 167–200 (2002)
2. Chorin, A.J.: A numerical method for solving incompressible viscous flow problems. J. of Computational Physics **2**(1), 12–26 (1967)
3. Louda, P., Kozel, K., Příhoda, J.: Numerical solution of 3D viscous incompressible steady and unsteady flows using artificial compressibility method. In: Numerical methods for fluid dynamics. Reading (2007). CD-ROM
4. Menter, F.R.: Two-equation eddy-viscosity turbulence models for engineering applications. AIAA Journal **32**(8), 1598–1605 (1994)
5. Schäfer, M., Turek, S.: Benchmark computations of laminar flow around a cylinder. In: NNFM 52 "Flow simulation on high performance computers II", pp. 547–566. Vieweg, Braunschweig (1996)
6. Trávníček, Z., Vogel, J., Vít, T., Maršík, F.: Flow field and mass transfer experimental and numerical studies of a synthetic impinging jet. In: HEFAT2005, 4th International Conference on Heat Transfer, Fluid Mechanics, and Thermodynamics. Cairo, Egypt (2005)
7. Turkel, E.: Algorithms for the Euler and Navier-Stokes equations for supercomputers. Tech. Rep. ICASE Report No. 85-11, NASA (1985)
8. Wang, A.B., Trávníček, Z., Chia, K.C.: On the relationship of effective Reynolds number and Strouhal number for the laminar vortex shedding of a heated circular cylinder. Physics of Fluids **12**(6), 1401–1410 (2000)

Local Projection Stabilization of Finite Element Methods for Incompressible Flows

G. Lube, G. Rapin, and J. Löwe

Abstract A unified analysis for finite element discretizations of linearized incompressible flows using the local projection method with equal-order or inf-sup stable velocity-pressure pairs together with a critical comparison is given.

1 Introduction

A standard numerical approach to the incompressible Navier-Stokes model

$$\partial_t \boldsymbol{u} - \nu \triangle \boldsymbol{u} + (\boldsymbol{u} \cdot \nabla)\boldsymbol{u} + \nabla p = \tilde{\boldsymbol{f}}, \qquad \nabla \cdot \boldsymbol{u} = 0 \qquad \text{in } \Omega \times (0,T) \tag{1}$$

for velocity $\boldsymbol{u}$ and pressure p is to semi-discretize in time first with an A-stable implicit scheme and to apply a fixed-point or Newton-type iteration in each time step. This leads to auxiliary Oseen problems (with $\sigma \geq 0$ from time discretization)

$$-\nu \triangle \boldsymbol{u} + (\boldsymbol{b} \cdot \nabla)\boldsymbol{u} + \sigma \boldsymbol{u} + \nabla p = \boldsymbol{f}, \qquad \nabla \cdot \boldsymbol{u} = 0 \qquad \text{in } \Omega. \tag{2}$$

Residual based stabilization (RBS) methods are the traditional way to cope with spurious solutions of the Galerkin finite element (FE) approximation of (2) caused by violation of the discrete inf-sup stability condition and/or dominating advection. RBS methods are robust and easy to implement, but have severe drawbacks mainly stemming from the strong velocity-pressure coupling in the stabilisation terms.

The key idea of the variational multiscale (VMS) methods [1] is a separation into large, small and unresolved scales. The influence of the unresolved scales has to be modeled. Almost all stabilization methods can be interpreted as VMS methods. In local projection stabilization (LPS) methods [1, 5] the influence of the unresolved scales is modeled by an artificial fine-scale diffusion term.

G. Lube, G. Rapin, and J. Löwe
Math. Department, Georg-August-University Göttingen, Lotzestrasse 16-18, D-37083 Göttingen, Germany, e-mail: lube/grapin@math.uni-goettingen.de

Mostly, an equal-order interpolation of velocity-pressure is applied. Other authors prefer discrete inf-sup stable pairs as the natural choice from regularity point of view for fixed $\nu > 0$. Here, a unified theory of LPS methods for equal-order and inf-sup stable pairs for problem (2) is presented. For full proofs, see [4]. Finally, a comparison of both variants is given.

2 Variational Formulation and LPS-Discretization

Standard notations for Lebesgue and Sobolev spaces are used. The $L^2(G)$ inner product in $G \subset \Omega$ is denoted by $(\cdot,\cdot)_G$ with $(\cdot,\cdot) = (\cdot,\cdot)_\Omega$. The notation $a \lesssim b$ is used if there exists a constant $C > 0$ independent of all relevant quantities s.t. $a \leq Cb$.

The weak formulation for the Oseen problem (2) with homogeneous Dirichlet data reads: Find $U = (\boldsymbol{u}, p) \in \boldsymbol{V} \times Q := [H_0^1(\Omega)]^d \times L_0^2(\Omega)$, s. t. $\forall V = (\boldsymbol{v}, q) \in \boldsymbol{V} \times Q$:

$$A(U,V) = (\nu\nabla\boldsymbol{u},\nabla\boldsymbol{v}) + ((\boldsymbol{b}\cdot\nabla)\boldsymbol{u} + \sigma\boldsymbol{u},\boldsymbol{v}) - (p,\nabla\cdot\boldsymbol{v}) + (q,\nabla\cdot\boldsymbol{u}) = (\boldsymbol{f},\boldsymbol{v}). \quad (3)$$

Let $\Omega \subset \mathbb{R}^d, d \in \{2,3\}$ be a bounded, polyhedral domain and $\nu \in L^\infty(\Omega)$ with $\nu > 0$, $\boldsymbol{f} \in [L^2(\Omega)]^d$, $\boldsymbol{b} \in [L^\infty(\Omega) \cap H^1(\Omega)]^d$ with $\nabla \cdot \boldsymbol{b} = 0$ and $\sigma \in \mathbb{R}^+$. Usually, $\boldsymbol{b}$ is a FE solution of (2) with $(\nabla \cdot b, q_h) = 0$ for some q_h and $\nabla \cdot b$ does not vanish. A remedy is to write the advective term in skew-symmetric form. The analysis can be extended to problems resulting from Newton iteration including the term $(\boldsymbol{u}\cdot\nabla)\boldsymbol{b}$. Sufficiently small time steps ensure coercivity of $A(\cdot,\cdot)$.

Let $\mathcal{T}_h$ be a shape-regular, admissible decomposition of Ω into d-dimensional simplices or quadrilaterals for $d = 2$ or hexahedra for $d = 3$. h_T is the diameter of a cell $T \in \mathcal{T}_h$ and $h = \max h_T$. Let $\hat{T}$ be a reference element of $\mathcal{T}_h$ and $F_T : \hat{T} \to T$ the standard (affine or bi-/trilinear) reference mapping.

Set $P_{k,\mathcal{T}_h} := \{v_h \in L^2(\Omega) \mid v_h|_T \circ F_T \in \mathbb{P}_k(\hat{T}), T \in \mathcal{T}_h\}$ with the set $\mathbb{P}_k(\hat{T})$ of complete polynomials of degree k on $\hat{T}$ and $Q_{k,\mathcal{T}_h} := \{v_h \in L^2(\Omega) \mid v_h|_T \circ F_T \in \mathbb{Q}_k(\hat{T}), T \in \mathcal{T}_h\}$ with the set $\mathbb{Q}_k(\hat{T})$ of all polynomials on $\hat{T}$ with maximal degree k in each coordinate direction. The FE space of the velocity is given by $\mathbf{V}_{h,k_u} = [Q_{k_u,\mathcal{T}_h}]^d \cap \boldsymbol{V}$ or $\mathbf{V}_{h,k_u} = [P_{k_u,\mathcal{T}_h}]^d \cap \boldsymbol{V}$ with scalar components Y_{h,k_u} of $\mathbf{V}_{h,k_u}$. For simplicity, the analysis is restricted to continuous discrete pressure spaces $Q_{h,k_p} = Q_{k_p,\mathcal{T}_h} \cap C(\overline{\Omega})$ resp. $Q_{h,k_p} = P_{k_p,\mathcal{T}_h} \cap C(\overline{\Omega})$. An extension to discontinuous spaces Q_{h,k_p} is given in [6].

The analysis below takes advantage of the inverse inequalities

$$\exists\mu_{inv} \mid \quad |\boldsymbol{v}|_{1,T} \leq \mu_{inv} k_u^2 h_T^{-1} \|\boldsymbol{v}\|_{0,T}, \quad \forall T \in \mathcal{T}_h, \forall \boldsymbol{v}_h \in \boldsymbol{V}_{h,k_u}. \quad (4)$$

The Scott-Zhang quasi-interpolant obeys the interpolation properties

$$\exists C > 0 \mid \; \|v - I_{h,k_u}^u v\|_{m,T} \leq C h_T^{l-m} k_u^{-(r-m)} \|v\|_{r,\omega_T}, \quad 0 \leq m \leq l = \min(k_u+1, r) \quad (5)$$

for $v \in H_0^1(\Omega) \cap H^t(\Omega)$, $t > \frac{1}{2}$ with $v|_{\omega_T} \in H^r(\omega_T)$, $r \geq t$, on the patches $\omega_T := \bigcup_{\overline{T'} \cap \overline{T} \neq \emptyset} T'$. This property can be extended to the vector-valued case with $\boldsymbol{I}^u_{h,k_u} : \boldsymbol{V} \to \boldsymbol{V}_h$. A similar interpolation operator I^p_{h,k_p} satisfying (5) is defined for the pressure.

In LPS-methods the discrete function spaces are split into small and large scales. Stabilization terms of diffusion-type acting only on the small scales are added.

A first variant is to find the large scales on a coarse non-overlapping, shape-regular mesh $\mathscr{M}_h = \{M_i\}_{i \in I}$. $\mathscr{M}_h$ is constructed by coarsening $\mathscr{T}_h$ s. t. each $M \in \mathscr{M}_h$ with diameter h_M consists of one or more neighboring cells $T \in \mathscr{T}_h$. Moreover, suppose that there exists $C \geq 1$ s. t. $h_M \leq C h_T$ for all $T \in \mathscr{T}_h$ with $T \subseteq M \in \mathscr{M}_h$.

Following [5] we define the discrete velocity space D_h^u as a discontinuous FE space on $\mathscr{M}_h$. The restriction to $M \in \mathscr{M}_h$ is denoted by $D_h^u(M) = \{v_h|_M \mid v_h \in D_h^u\}$ The local projection $\pi_M^u : L^2(M) \to D_h^u(M)$ defines the global projection $\pi_h^u : L^2(\Omega) \to D_h^u$ by $(\pi_h^u v)|_M := \pi_M^u(v|_M)$ for all $M \in \mathscr{M}_h$. Denoting the identity on $L^2(\Omega)$ by id, the associated fluctuation operator $\kappa_h^u : L^2(\Omega) \to L^2(\Omega)$ is defined by $\kappa_h^u := id - \pi_h^u$. These operators are applied to vector-valued functions in a component-wise manner.

A discrete space D_h^p and a fluctuation operator κ_h^p are defined similarly.

The second choice consists in choosing lower order discontinuous FE discretizations $D_h^u \times D_h^p$ on $\mathscr{T}_h$ or by enriching $\boldsymbol{V}_{h,k_u} \times Q_{h,k_p}$. The same framework as in the first approach can be used by setting $\mathscr{M}_h = \mathscr{T}_h$.

The LPS scheme reads: find $U_h = (\boldsymbol{u}_h, p_h) \in \boldsymbol{V}_{h,k_u} \times Q_{h,k_p}$ s.t.

$$A(U_h, V_h) + S_h(U_h, V_h) = (\boldsymbol{f}, \boldsymbol{v}_h), \qquad \forall V_h = (v_h, q_h) \in \boldsymbol{V}_{h,k_u} \times Q_{h,k_p}, \tag{6}$$

where the additional stabilization term is given by

$$\begin{aligned} S_h(U_h, V_h) := \sum_{M \in \mathscr{M}_h} \Big[& \tau_M(\boldsymbol{\kappa}_h^u((\boldsymbol{b}\cdot\nabla)\boldsymbol{u}_h), \boldsymbol{\kappa}_h^u((\boldsymbol{b}\cdot\nabla)\boldsymbol{v}_h))_M \\ & + \mu_M(\kappa_h^p(\nabla\cdot\boldsymbol{u}_h), \kappa_h^p(\nabla\cdot\boldsymbol{v}_h))_M + \alpha_M(\boldsymbol{\kappa}_h^u(\nabla p_h), \boldsymbol{\kappa}_h^u(\nabla q_h))_M \Big]. \end{aligned} \tag{7}$$

An alternative is to replace the first two terms of S_h by the projection of $\nabla \boldsymbol{u}_h$.

The constants τ_M, μ_M and α_M will be determined in Section 3 based on an a priori estimate. Please note that the stabilization S_h acts solely on the fine scales.

In order to control the consistency error of the κ_h^u-dependent stabilization terms, the space D_h^u has to be large enough for the approximation property:

Assumption 1 *The fluctuation operator κ_h^u admits for $0 \leq l \leq k_u$, the property:*

$$\exists C_\kappa > 0 \mid \qquad \|\kappa_h^u q\|_{0,M} \leq C_\kappa h_M^l k_u^{-l} |q|_{l,M}, \quad \forall q \in H^l(M),\ \forall M \in \mathscr{M}_h. \tag{8}$$

Assumption 1 is valid for the L^2-projection π_h^u. Due to the consistency of the κ_h^p-dependent term in S_h, thus involving D_h^p, such condition is not needed for D_h^p.

The following property of the symmetric and non-negative term $S_h(\cdot,\cdot)$ is valid for all $U \in \boldsymbol{V} \times Q$, see [4], Lemma 2.1:

$$S_h(U,U) \leq C_S |\boldsymbol{u}|_1^2 + C_\kappa^2 (\max_M \alpha_M) |p|_1^2, \ \ C_S = C_\kappa^2 \max_M \left[\tau_M \|\boldsymbol{b}\|^2_{(L^\infty(M))^d} + \mu_M \right]. \tag{9}$$

Following [5], a special interpolant $j_h^u : H^1(\Omega) \to Y_h$ for the velocity is constructed s.t. the error $v - j_h^u v$ is L^2-orthogonal to D_h^u for all $v \in H_0^1(\Omega)$. A corresponding result can be proved for the pressure too. In order to conserve the standard approximation properties, we additionally assume

Assumption 2 *Let $Y_h(M) := \{v_h|_M \mid v_h \in Y_h, v_h|_{\Omega\setminus M} = 0\}$. There exists β_u, β_p s. t.*

$$\inf_{q_h \in D_h^u} \sup_{v_h \in Y_h(M)} \frac{(v_h, q_h)_M}{\|v_h\|_{0,M}\|q_h\|_{0,M}} \geq \beta_u > 0. \tag{10}$$

$$\inf_{q_h \in D_h^p} \sup_{v_h \in Q_{h,k_p}} \frac{(v_h, q_h)_M}{\|v_h\|_{0,M}\|q_h\|_{0,M}} \geq \beta_p > 0 \tag{11}$$

Remark 1. The space D_h^u must not be too rich w.r.t. (10) but rich enough w.r.t. (8).

Lemma 1. *([4], Lemmata 2, 3) Set $\omega_M := \bigcup_{T\subset M} \omega_T$ for $M \in \mathcal{M}_h$. Under Assumption 2 there are interpolants $j_h^u : \boldsymbol{V} \to \boldsymbol{V}_{h,k_u}$ s.t. for all $v \in [H^l(\Omega)]^d \cap V$:*

$$(v - j_h^u v, q_h) = 0 \qquad \forall q_h \in D_h^u, \tag{12}$$

$$\|v - j_h^u v\|_{0,M} + \frac{h_M}{k_u^2}|v - j_h^u v|_{1,M} \lesssim \left(1 + \frac{1}{\beta_u}\right)\frac{h_M^l}{k^l}\|v\|_{l,\omega_M} \tag{13}$$

and an interpolant $j_h^p : Q \to Q_{h,k_p}$ s.t. for all $v \in Q \cap H^l(\Omega)$:

$$(v - j_h^p v, q_h) = 0, \qquad \forall q_h \in D_h^p, \tag{14}$$

$$\|v - j_h^p v\|_{0,M} + \frac{h_M}{k_p^2}|v - j_h^p v|_{1,M} \lesssim \left(1 + \frac{1}{\beta_p}\right)\frac{h_M^l}{k_p^l}\|v\|_{l,\omega_M}. \tag{15}$$

Remark 2. (13), (15) are optimal w.r.t. h_M but sub-optimal w.r.t. k_u, k_p in $|\cdot|_{1,M}$.

3 A Priori Analysis

The stability of the LPS scheme is given for the mesh-dependent norm

$$|[V]|^2 := \|\sqrt{\nu}\nabla \boldsymbol{v}\|_0^2 + \|\sqrt{\sigma}\boldsymbol{v}\|_0^2 + S_h(V,V), \qquad V = (\boldsymbol{v}, q) \in \boldsymbol{V} \times Q.$$

Then, a "post-processing" argument for the pressure is applied.

Lemma 2. *([4], Lemmata 4 and 5) The following a-priori estimate is valid*

$$\|\sqrt{\nu}\nabla \boldsymbol{u}_h\|_0^2 + \|\sqrt{\sigma}\boldsymbol{u}_h\|_0^2 \leq |[U_h]|^2 = (A + S_h)(U_h, U_h) \leq (\boldsymbol{f}, \boldsymbol{u}_h). \tag{16}$$

There exists a h-independent constant $\gamma > 0$ (depending on the continuous inf-sup constant β and on degree k_u) s. t. (with C_S as in (9) and Poincare constant C_P)

$$\|p_h\|_0 \le \gamma\Big(\sqrt{\nu_\infty}+\sqrt{C_P\sigma}+\min\big(\frac{C_P}{\sqrt{\nu_0}};\frac{1}{\sqrt{\sigma}}\big)\boldsymbol{b}_\infty+\sqrt{C_S}+\max_M\frac{h_M}{\sqrt{\alpha_M}}\Big)|[U_h]|+\frac{\|\boldsymbol{f}\|_{-1}}{\beta}$$

with $\nu_\infty := \|\nu\|_{L^\infty(\Omega)}$, $\nu_0 :=$ *ess inf*$_\Omega \nu(x)$, $\boldsymbol{b}_\infty := \|\boldsymbol{b}\|_{(L^\infty(\Omega))^d}$. *This implies uniqueness and existence of* $(\boldsymbol{u}_h, p_h) \in \boldsymbol{V}_{h,k_u} \times Q_{h,k_p}$ *in (6).*

In LPS methods the Galerkin orthogonality is not fulfilled and a careful analysis of the consistency error has to be done. Subtracting (6) from (3) yields

Lemma 3. *([4], Lemma 6) Let* $U \in \boldsymbol{V} \times Q$ *and* $U_h \in \boldsymbol{V}_{h,k_u} \times Q_{h,k_p}$ *be the solutions of (3) and of (6), respectively. Then, there holds*

$$(A+S_h)(U-U_h,V_h) = S_h(U,V_h), \qquad \forall V_h \in \boldsymbol{V}_{h,k_u} \times Q_{h,k_p}. \tag{17}$$

The consistency error can be estimated using the properties of $S_h(\cdot,\cdot)$.

Lemma 4. *([4], Lemma 7) Let Assumption 1 be fulfilled and* $(u,p) \in \boldsymbol{V} \times Q$ *with* $(\boldsymbol{b}\cdot\nabla)\boldsymbol{u} \in (H^{l_u}(M))^d$, $\nabla\cdot\boldsymbol{u} = 0$, $p \in H^{l_p+1}(M)$ *for all* $M \in \mathcal{M}_h$. *Then, we obtain for* $0 \le l_u, l_p \le k_u$

$$|S_h(U,V_h)| \lesssim \Big(\sum_{M\in\mathcal{M}_h} \tau_M \frac{h_M^{2l_u}}{k_u^{2l_u}}|(\boldsymbol{b}\cdot\nabla)\boldsymbol{u}|^2_{l_u,M} + \alpha_M \frac{h_M^{2l_p}}{k_p^{2l_p}}|p|^2_{l_p+1,M}\Big)^{\frac{1}{2}} |[V_h]|. \tag{18}$$

A combination of the stability and consistency results yields an a-priori estimate.

Theorem 1. *([4], Thm. 1) Let* $U = (\boldsymbol{u},p) \in \boldsymbol{V} \times Q$ *and* $U_h = (\boldsymbol{u}_h,p_h) \in \boldsymbol{V}_{h,k_u} \times Q_{h,k_p}$ *be the solutions of (3) and of (6). Assume that* $U = (\boldsymbol{u},p) \in \boldsymbol{V} \times Q$ *is sufficiently regular, i.e.* $p \in H^{l_p+1}(\Omega)$ *and* $\boldsymbol{u} \in [H^{l_u+1}(\Omega)]^d$, $(\boldsymbol{b}\cdot\nabla)\boldsymbol{u} \in [H^{l_u}(\Omega)]^d$. *Furthermore let the Assumptions 1 and 2 for the coarse velocity space* D_h^u *be satisfied. For the space* D_h^p *we assume that (11) is satisfied. Then, there holds*

$$\begin{aligned}|[U-U_h]|^2 \lesssim \sum_{M\in\mathcal{M}_h} \Big(&\tau_M\big(\frac{h_M}{k_u}\big)^{2l_u}\|(\boldsymbol{b}\cdot\nabla)\boldsymbol{u}\|^2_{l_u,\omega_M} \\ &+(1+\frac{1}{\beta_u})^2 k_u^2\big(\frac{h_M}{k_u}\big)^{2l_u} C_M^u\|\boldsymbol{u}\|^2_{l_u+1,\omega_M} + (1+\frac{1}{\beta_p})^2 k_p^2\big(\frac{h_M}{k_p}\big)^{2l_p} C_M^p\|p\|^2_{l_p+1,\omega_M}\Big)\end{aligned} \tag{19}$$

for $1 \le l_u \le k_u$ *and* $1 \le l_p \le \min\{k_p,k_u\}$ *with*

$$C_M^u := \|\nu\|_{L^\infty(M)} + \frac{h_M^2}{k_u^4}(\sigma+\frac{1}{\tau_M}+\frac{1}{\alpha_M}) + \mu_M + \|\boldsymbol{b}\|^2_{[L^\infty(M)]^d}\tau_M, \; C_M^p := \alpha_M + \frac{1}{\mu_M}\frac{h_M^2}{k_p^4}.$$

Under the notation of Lemma 2 we obtain

$$\|p-p_h\|_0 \lesssim \gamma\Big(\sqrt{\nu_\infty}+\sqrt{C_P\sigma}+\min\big(\frac{C_P}{\sqrt{\nu_0}};\frac{1}{\sqrt{\sigma}}\big)\boldsymbol{b}_\infty+\frac{\sqrt{C_S}}{\beta}+\max_M\frac{h_M}{\sqrt{\alpha_M}}\Big)|[U-U_h]|.$$

Now we calibrate the stabilization parameters α_M, τ_M and μ_M w.r.t. h_M, k_u, k_p and problem data by balancing the terms of the right hand side of error estimate (19).

First, equilibrating the τ_M-dependent terms in C_M^u yields $\tau_M \sim h_M/(\|\boldsymbol{b}\|_{(L^\infty(M))^d} k_u^2)$. Similarly, equilibration of the terms in C_M^u and C_M^p involving μ_M and α_M yields $\mu_M \sim h_M^{l_p-l_u+1}/k^{l_p-l_u+2}$, $\alpha_M \sim h_M^{l_u-l_p+1}/k^{l_u-l_p+2}$ where we used $k \sim k_u \sim k_p$.

Corollary 1. *([4], Corollary 2) Let the assumptions of Theorem 1 be valid. For equal-order interpolation $k = k_u = k_p \geq 1$, let $l = l_u = l_p \leq k$ and set $\mu_M = \mu_0 h_M/k^2$, $\alpha_M = \alpha_0 h_M/k^2$, $\tau_M = \tau_0 h_M/(\|\boldsymbol{b}\|_{(L^\infty(M))^d} k^2)$. Then we obtain*

$$|[U-U_h]|^2 \lesssim \sum_{M\in\mathscr{M}} \Big((1+\frac{1}{\beta_p})^2 \frac{h_M^{2l+1}}{k^{2l}} \|p\|^2_{l+1,\omega_M} + \frac{h_M^{2l+1}}{k^{2l+2}} \|(\frac{\boldsymbol{b}}{\|\boldsymbol{b}\|_{(L^\infty(M))^d}} \cdot \nabla)\boldsymbol{u}\|^2_{l,\omega_M}$$
$$+ (1+\frac{1}{\beta_u})^2 \Big[\|\nu\|_{L^\infty(M)} + \sigma \frac{h_M^2}{k^4} + \|\boldsymbol{b}\|_{(L^\infty(M))^d} \frac{h_M}{k^2} \Big] \frac{h_M^{2l}}{k^{2l-2}} \|\boldsymbol{u}\|^2_{l+1,\omega_M} \Big).$$

For inf-sup stable interpolation with $k_u = k_p + 1$, we assume $l_u = l_p + 1 = k_u$ and set $\alpha_M = \alpha_0 h_M^2/k_u^3$, $\mu_M = \mu_0/k_u$, $\tau_M = \tau_0 h_M/(\|\boldsymbol{b}\|_{(L^\infty(M))^d} k_u^2)$. Then we obtain

$$|[U-U_h]|^2 \lesssim \sum_{M\in\mathscr{M}} \Big((1+\frac{1}{\beta_p})^2 \frac{h_M^{2l_u}}{k_u^{2l_u+1}} \|p\|^2_{l_u,\omega_M} + \frac{h_M^{2l_u+1}}{k_u^{2l_u+2}} \|(\frac{\boldsymbol{b}}{\|\boldsymbol{b}\|_{L^\infty(M)}} \cdot \nabla)\boldsymbol{u}\|^2_{l,\omega_M}$$
$$+ (1+\frac{1}{\beta_u})^2 \Big[\|\nu\|_{L^\infty(M)} + \sigma \frac{h_M^2}{k_u^4} + \|\boldsymbol{b}\|_{[L^\infty(M)]^d} \frac{h_M}{k_u^2} + \frac{1}{k_u} \Big] \frac{h_M^{2l_u}}{k_u^{2l_u-2}} \|\boldsymbol{u}\|^2_{l+1,\omega_M} \Big).$$

- For equal-order pairs $V_{h,k} \times Q_{h,k}$ and Taylor-Hood pairs $V_{h,k+1} \times Q_{h,k}$, we obtain the optimal convergence rates $k + \frac{1}{2}$ and $k+1$, respectively, w.r.t. h_M.
- The estimates are not optimal w.r.t. k_u, see Remark 2. Assume that in Lemma 1 there holds $|v - j_h^u v|_{1,M} \lesssim \left(1 + \frac{1}{\beta_u}\right) \left(\frac{h_M}{k}\right)^{l-1} \|v\|_{l,\omega_M}$ and a similar result for the pressure too. A careful check of the proofs leads to:
  - Equal-order pairs with $k = k_u = k_p$: $\mu_M \sim \alpha_M \sim h_M/k$, $\tau_M \sim \frac{h_M}{\|\boldsymbol{b}\|_{[L^\infty(M)]^d} k_u}$
  - Inf-sup stable pairs with $k_u = k_p + 1$: $\alpha_M \sim \frac{h_M^2}{k_u^2}$, $\mu_M \sim 1$, $\tau_M \sim \frac{h_M}{\|\boldsymbol{b}\|_{(L^\infty(M))^d} k_u}$.

  Then the estimate (19) would be optimal w.r.t. k_u and k_p too with possible exception of the factors depending on β_u, β_p.

Different variants for the choice of the discrete spaces $V_{h,k_u} \times Q_{h,k_p}$ and $D_h^u \times D_h^p$ using simplicial and hexahedral elements are presented in [5] for two variants: a two-level variant with $\mathscr{M}_h = \mathscr{T}_{2h}$ and a one-level variant with $\mathscr{M}_h = \mathscr{T}_h$, thus $h_M = h_K$, with a proper enrichment of $P_{k_u,\mathscr{T}_h}$ by using bubble functions.

Assumption 1 is valid if the local L^2-projection $\pi_M^u : L^2(M) \to D_h^u(M)$ for the velocity and similarly for the pressure is applied, see [5]. In the two-level variant, the constants $\beta_{u/p}$ in Assumption 2 scale like $\mathscr{O}(1/\sqrt{k_{u/p}})$ for simplicial elements and like $\mathscr{O}(1)$ for quadrilateral elements in the affine linear case, see [6].

Please note that the present analysis covers only the case of continuous pressure approximation. An extension to discontinuous discrete pressure approximation, in particular to the case of the case of $Q_k/P_{-(k-1)}$-elements, can be found in [6].

4 Some Numerical Results

A calibration of the LPS parameters requires careful numerical experiments. Some papers validate the design and the convergence rates for the Oseen problem (2) in $\Omega = (0,1)^2$ with the smooth solution $\boldsymbol{u}(x_1,x_2) = \big(\sin(\pi x_1), -\pi x_2 \cos(\pi x_1)\big)$, $p(x_1,x_2) = \sin(\pi x_1)\cos(\pi x_2)$ and data $\boldsymbol{b} = \boldsymbol{u}$, $\sigma = 1$. A study of the one-level variant for equal-order pairs with enrichment of the discontinuous velocity space is given in [7]. The two-level variant is considered in [3] for equal-order and inf-sup stable pairs, see also [6]. Summarizing, all these experiments confirm the calibration of the stabilization parameters w.r.t. h_M and the theoretical a-priori convergence rates.

Here we present some typical results using either Q_2/Q_2 and Q_2/Q_1 pairs for velocity/pressure on unstructured, quasi-uniform meshes for the advection-dominated case $\nu = 10^{-6}$. The coarse spaces of the two-level variant are defined as $D_h^{u/p} := \{v \in [L^2(\Omega)]^d \mid v|_M \in P_{1/1}(M)\}$ and $D_h^{u/p} := \{v \in [L^2(\Omega)]^d \mid v|_M \in P_{1/0}(M)\}$. Table 1 shows comparable results for the best variants of the inf-sup stable Q_2/Q_1

Table 1 Comparison of different variants of stabilization for problem (2) with $\nu = 10^{-6}, h = 1/64$

Pair	τ_0	μ_0	α_0	$\lvert\boldsymbol{u}-\boldsymbol{u}_h\rvert_1$	$\lVert\boldsymbol{u}-\boldsymbol{u}_h\rVert_0$	$\lVert\nabla\cdot\boldsymbol{u}_h\rVert_0$	$\lVert p-p_h\rVert_0$
Q_2/Q_1	0.0000	0.0000	0.0000	2.56E-1	5.42E-4	2.02E-1	2.31E-4
Q_2/Q_1	0.0562	0.5623	0.0000	1.91E-3	6.20E-6	1.66E-4	8.06E-5
Q_2/Q_1	0.0000	0.5623	0.0000	2.61E-3	7.42E-6	1.72E-4	8.05E-5
Q_2/Q_1	3.1623	0.0000	0.0000	1.87E-2	7.50E-5	1.56E-2	1.08E-4
Q_2/Q_2	0.0000	0.0000	0.0178	1.65E-2	3.48E-5	9.37E-3	6.96E-6
Q_2/Q_2	0.0562	1.0000	0.0178	9.30E-4	2.85E-6	2.14E-4	4.31E-6
Q_2/Q_2	0.0562	0.0000	0.0178	1.77E-3	4.18E-6	1.46E-3	3.25E-6
Q_2/Q_2	0.0000	5.6234	0.0178	3.26E-3	7.20E-6	2.00E-4	7.56E-6

and the equal-order Q_2/Q_2 pairs with the exception of the pressure error. Nevertheless, the importance of the stabilization terms is different. The fine-scale SUPG- and PSPG-type terms are necessary for the equal-order case but not for the inf-sup stable pair. On the other hand, the divergence-stabilization gives clear improvement for the inf-sup stable case and some improvement for the other case. Moreover, the PSPG-type term can be omitted for the inf-sup stable case.

Finally, we apply the LPS stabilization to the lid-driven cavity Navier-Stokes flow (1) with $\boldsymbol{f} = \boldsymbol{0}$. No-slip data are prescribed with the exception of the upper part of the cavity where $\boldsymbol{u} = (1,0)^T$ is given. A quasi-uniform mesh is used together with the Q_2/Q_1 and Q_2/Q_2 pairs using the two-level LPS variant with scaling parameter τ_0 and μ_0 according to the Oseen case and $\alpha_0 = 0$.

Fig. 1 shows typical velocity profiles for $Re = 5{,}000$. The results for $h = \frac{1}{64}$ for both variants are in excellent agreement with [2] with well resolved boundary layers. Moreover, the solution for a coarse grid with $h = \frac{1}{16}$ is in good agreement with [2] away from the boundary layers. Similar results are obtained up to $Re = 7.500$ [3]. The results for this nonlinear problem confirm the previous remarks for the

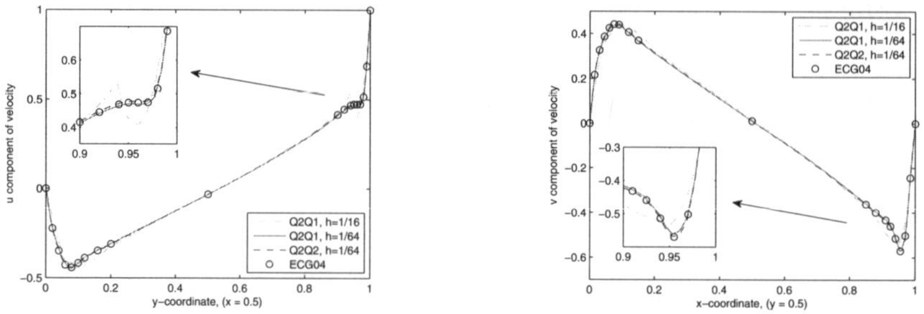

Fig. 1 Lid driven-cavity problem with $Re = 5{,}000$: Cross-sections of the discrete solutions for Q_2/Q_1 pair with $\tau_0 = \alpha_0 = 0$ and $\mu_0 = 1$ and Q_2/Q_2 pair with $\tau_0 = \alpha_0 = \mu_0 = 1$

linear Oseen problem. For the Q_2/Q_1 element, only the divergence stabilization is necessary whereas for the Q_2/Q_2 pair all stabilization terms are relevant.

5 Summary

A unified a-priori analysis of local projection stabilization (LPS) methods is given for equal-order and inf-sup stable velocity-pressure pairs on isotropic meshes. Numerical results confirm the numerical analysis. Compared to residual-based methods, the error estimates are comparable, but the parameter design is much simpler. A major difference between equal-order and inf-sup stable pairs is that LPS-stabilization is always necessary for equal-order pairs. For inf-sup stable pairs, the necessity of stabilization is much less pronounced. In particular, the grad-div stabilization is much more important than the fine-scale SUPG and PSPG stabilization.

References

1. Braack, M., Burman, E.: Local projection stabilization for the Oseen problem and its interpretation as a variational multiscale method. SIAM J. Numer. Anal. **43**(6), 2544–2566 (2006)
2. Erturk, E., Corke, T.C., Gokcol, C.: Numerical Solutions of 2-D Steady Incompressible Driven Cavity Flow at High Reynolds Numbers. ArXiv Computer Science e-prints (2004)
3. Löwe, J.: Local projection stabilization for incompressible flow problems (in German) (2008). Master Thesis, Georg-August University Göttingen
4. Lube, G., Rapin, G., Löwe, J.: Local projection stabilizations for incompressible flows: Equal-order vs. inf-sup stable interpolation (2007). Submitted to ETNA
5. Matthies, G., Skrzypacz, P., Tobiska, L.: A unified convergence analysis for local projection stabilizations applied to the Oseen problem. M^2AN **41**(4), 713–742 (2007)
6. Rapin, G., Löwe, J.: Local projection stabilizations for inf-sup stable finite elements applied to the Oseen problem (2007). Submitted to CMAME
7. Schmaljohann, S.: Local projection stabilization for the Oseen problem (in German) (2007). Master Thesis, Ruhr-Universität Bochum

A Numerical Study of Local Projection Stabilisations Applied to Oseen Problems

G. Matthies and L. Tobiska

Abstract The local projection method has been proven to be a successful stabilisation technique for the Oseen problem. We consider the enrichment approach of the local projection method. After describing the method, known convergence results are recalled. Numerical results which confirm the theoretical predictions are given. Furthermore, the dependence of the results on the choice of the stabilisation parameters is discussed.

1 Introduction

While discretising the Oseen equations with finite element methods, one is usually faced with two difficulties. First, the discrete inf-sup (or Babuška–Brezzi) condition might be violated. This leads to artificial oscillations in the pressure. Second, the dominating convection causes spurious oscillations in the velocity field. The local projection method [1, 2, 3, 5, 7] has been proven to be a successful stabilisation technique for flow problems which overcomes the two mentioned problems.

Compared to residual based stabilisations like the streamline-upwind Petrov–Galerkin method (SUPG) combined with the pressure-stabilised Petrov–Galerkin technique (PSPG), the local projection stabilisation needs much less additional terms to be assembled, in particular for higher order discretisations.

The local projection method was designed for equal-order interpolations. It has been introduced for the Stokes problem in [1], extended to the transport equation in [2], and analysed for the Oseen equations in [3, 7]. The stabilising term of the

Gunar Matthies
Ruhr-Universität Bochum, Fakultät für Mathematik, Universitätsstraße 150, 44780 Bochum, Germany, e-mail: Gunar.Matthies@ruhr-uni-bochum.de

Lutz Tobiska
Otto-von-Guericke-Universität Magdeburg, Institut für Analysis und Numerik, PF 4120, 39016 Magdeburg, Germany, e-mail: Lutz.Tobiska@mathematik.uni-magdeburg.de

local projection method is based on a projection $\pi_h : Y_h \to D_h$ of the finite element space Y_h approximating the pressure and each velocity component into a discontinuous space D_h. The stabilising effects result from weighted L^2-control over fluctuations $(\mathrm{id} - \pi_h)$ of the gradients of velocity and pressure.

Originally, the local projection technique was proposed as a two-level method where the projection space D_h is defined on a coarser grid. The drawback of this approach is an increased discretisation stencil. The general approach given in [5, 7] allows to construct local projection methods such that the discretisation stencil is not increased since approximation space Y_h and projection space D_h are defined on the same mesh. In this case, the approximation space Y_h is enriched compared to standard finite element spaces. We will concentrate in this paper on the enrichment approach of the local projection method.

This paper is organised as follows. Sect. 2 describes the local projection method applied to the Oseen equations. The convergence results are recalled in Sect. 3. Finally, Sect. 4 presents some numerical results. In particular, the choice of the stabilisation parameters is discussed.

Notation. Throughout this paper, C will denote a generic positive constant which is independent of the mesh. We will write shortly $\alpha \sim \beta$ if there are positive constants $\underline{C}$ and $\overline{C}$ such that $\underline{C}\beta \leq \alpha \leq \overline{C}\beta$ holds. The Oseen problem will be considered in the domain $\Omega \subset \mathbb{R}^d$, $d = 2,3$, which is assumed to be a polygonal or polyhedral domain with boundary $\partial\Omega$. For a measurable d-dimensional subset G of Ω, the usual Sobolev spaces $H^m(G)$ with norm $\|\cdot\|_{m,G}$ and semi-norm $|\cdot|_{m,G}$ are used. The L^2-inner product on G is denoted by $(\cdot,\cdot)_G$. Note that the index G will be omitted for $G = \Omega$. This notation of norms, semi-norms, and inner products is also used for the vector-valued and tensor-valued case.

2 Local Projection Stabilisation

2.1 Weak Formulation of the Oseen Equations

We consider the Oseen problem

$$-\nu\triangle u + (b\cdot\nabla)u + \sigma u + \nabla p = f \text{ in } \Omega\,, \quad \operatorname{div} u = 0 \text{ in } \Omega\,, \quad u = 0 \text{ on } \partial\Omega\,, \quad (1)$$

where ν is a small positive parameter and $\sigma > 0$. Furthermore, we assume that $b \in \left(W^{1,\infty}(\Omega)\right)^d$ with $\operatorname{div} b = 0$.

Let $V := \left(H_0^1(\Omega)\right)^d$ and $Q := L_0^2(\Omega) := \{q \in L^2(\Omega) : (q,1) = 0\}$. We introduce on the product space $V \times Q$ the bilinear form A as

$$A\big((v,q);(w,r)\big) := \nu(\nabla v, \nabla w) + \big((b\cdot\nabla)v, w\big) + \sigma(v,w) - (q, \operatorname{div} w) + (r, \operatorname{div} v)\,.$$

A weak formulation of (1) reads:

Find $(u,p) \in V \times Q$ such that for all $(v,q) \in V \times Q$:

$$A\big((u,p);(v,q)\big) = (f,v)\,. \tag{2}$$

Due to $\big((b\cdot\nabla)v,v\big) = 0$ for all $v \in V$, the Lax–Milgram lemma applied in the subspace of divergence-free functions guarantees that (2) has a unique velocity solution u. The existence and uniqueness of the pressure solution p follows from the Babuška –Brezzi condition for the pair (V,Q).

2.2 Discrete Problem and Stabilised Formulation

We are given a family $\{\mathscr{T}_h\}$ of shape-regular decompositions of Ω into d-simplices. The diameter of a cell K is denoted by h_K. The mesh parameter h describes the maximum diameter of all cells $K \in \mathscr{T}_h$.

Let Y_h be a scalar finite element space of continuous, piecewise polynomials functions over $\mathscr{T}_h$. The spaces for approximating the velocity and the pressure are given by $V_h := Y_h^d \cap V$ and $Q_h := Y_h \cap Q$, respectively.

The standard Galerkin discretisation of (1) reads:

Find $(u_h,p_h) \in V_h \times Q_h$ such that for all $(v_h,q_h) \in V_h \times Q_h$:

$$A\big((u_h,p_h);(v_h,q_h)\big) = (f_h,v_h)\,.$$

This discrete problem suffers in general from two shortcomings, the discrete version of the Babuška–Brezzi condition

$$\exists \beta_0 > 0 : \ \forall h > 0 : \ \inf_{q_h\in Q_h} \sup_{v_h\in V_h} \frac{(q_h, \operatorname{div} v_h)}{\|q_h\|_0\,|v_h|_1} \geq \beta_0$$

is violated and the convection dominates in the case $\nu \ll 1$. We will use the local projection stabilisation to handle both difficulties.

Let $D_h(K)$ denote a finite dimensional space on $K \in \mathscr{T}_h$ and $\pi_K : L^2(K) \to D_h(K)$ the L^2-projection into $D_h(K)$. Furthermore, we set

$$D_h := \bigoplus_{K\in\mathscr{T}_h} D_h(K)$$

and define $\pi_h : L^2(\Omega) \to D_h$ by $(\pi_h q)|_K = \pi_K(q|_K)$. The fluctuation operator $\kappa_h : L^2(\Omega) \to L^2(\Omega)$ is given as $\kappa_h := \mathrm{id} - \pi_h$ where $\mathrm{id} : L^2(\Omega) \to L^2(\Omega)$ is the identity mapping on $L^2(\Omega)$. Note that all operators are applied component-wise to vectors and tensors. We introduce the stabilisation terms

$$S_h^a(v_h,w_h) := \sum_{K\in\mathscr{T}_h} \Big[\tau_K\big(\kappa_h(b\cdot\nabla)v_h, \kappa_h(b\cdot\nabla)w_h\big)_K + \mu_K\big(\kappa_h(\operatorname{div} v_h), \kappa_h(\operatorname{div} w_h)\big)_K\Big]$$

and

$$S_h^b(v_h, w_h) := \sum_{K\in\mathcal{T}_h} \delta_K \big(\kappa_h(\nabla v_h), \kappa_h(\nabla w_h)\big)_K$$

where τ_K, μ_K, and δ_K are user-chosen parameters. The stabilised discrete problem reads:

Find $(u_h, p_h) \in V_h \times Q_h$ such that for all $(v_h, q_h) \in V_h \times Q_h$:

$$A\big((u_h, p_h); (v_h, q_h)\big) + S_h(u_h, v_h) + \sum_{K\in\mathcal{T}_h} \alpha_K \big(\kappa_h(\nabla p_h), \kappa_h(\nabla q_h)\big)_K = (f, v_h) \quad (3)$$

where S_h is either S_h^a or S_h^b and α_K are further user-chosen parameters. Note that S_h^b consists only of one term while S_h^a contains two terms. The divergence term in S_h^a introduces a coupling between different velocity components. Numerically, both stabilisation terms give similar results. We will use in our analysis the mesh-dependent norm

$$|||(v,q)||| := \left(\nu|v|_1^2 + \sigma\|v\|_0^2 + S_h(v,v) + \sum_{K\in\mathcal{T}_h} \alpha_K \|\nabla q\|_{0,K}^2 \right)^{1/2}.$$

We will make in the following two assumptions.

Assumption 1. There exist interpolation operators $J_h : H^1(\Omega) \to Y_h$ and $j_h : H^{r+1}(\Omega) \to Y_h$ which satisfy $J_h v \in H_0^1(\Omega)$ for all $v \in H_0^1(\Omega)$ and $j_h v \in H_0^1(\Omega)$ for all $v \in H_0^1(\Omega) \cap H^{r+1}(\Omega)$. They provide the orthogonality properties

$$(v - J_h v, q_h) = 0, \quad (w - j_h w, q_h) = 0 \qquad \forall q_h \in D_h, \forall v \in H^1(\Omega), \forall w \in H^{r+1}(\Omega)$$

and the approximation properties

$$\|v - J_h v\|_{0,K} + h_K |v - J_h v|_{1,K} \le C h_K \|v\|_{1,\omega(K)} \qquad \forall v \in H^1(\omega(K))$$
$$\|w - j_h w\|_{0,K} + h_K |w - j_h w|_{1,K} \le C h_K^{r+1} \|w\|_{r+1,K} \qquad \forall w \in H^{r+1}(K),$$

where $\omega(K)$ denotes a certain neighbourhood of K which appears in the definition of interpolation operators for non-smooth functions, see [4, 8].

In [7], it was shown that interpolation operators fulfilling Assumption 1 can be constructed from interpolation operators with standard approximation properties provided an inf-sup condition between the approximation space Y_h and the projection space D_h holds.

Assumption 2. The fluctuation operator κ_h fulfils

$$\|\kappa_h q\|_{0,K} \le C h^\ell |q|_{\ell,K} \qquad \forall q \in H^\ell(K)\,,\, K \in \mathcal{T}_h\,,\, 0 \le \ell \le r\,.$$

Note that Assumption 2 is satisfied if $P_{r-1}(K) \subset D_h(K)$ holds true for all $K \in \mathcal{T}_h$.

2.3 Approximation and Projection Spaces

Approximation and projection spaces satisfying Assumption 1 and Assumption 2 are given in [7]. Here, we recall simplicial elements.

Let $\widehat{K}$ denote the reference d-simplex with barycentric coordinates $\hat{\lambda}_0, \ldots, \hat{\lambda}_d$. We define

$$\hat{b}(\hat{x}) := (d+1)^{d+1} \prod_{i=0}^{d} \hat{\lambda}_i(\hat{x})$$

as bubble function on $\widehat{K}$ which takes the value 1 in the barycentre of $\widehat{K}$. Furthermore, let

$$P_r^{\text{bubble}}(\widehat{K}) := P_r(\widehat{K}) + \hat{b} \cdot P_{r-1}(\widehat{K}) \,.$$

The pair $Y_h/D_h = P_{r,h}^{\text{bubble}}/P_{r-1,h}^{\text{disc}}$ with

$$\begin{aligned} P_{r,h}^{\text{bubble}} &:= \{v \in H^1(\Omega) \,:\, v|_K \circ F_K \in P_r^{\text{bubble}}(\widehat{K}) \quad \forall K \in \mathscr{T}_h\}\,, \\ P_{r-1,h}^{\text{disc}} &:= \{v \in L^2(\Omega) \,:\, v|_K \circ F_K \in P_{r-1}(\widehat{K}) \qquad \forall K \in \mathscr{T}_h\} \end{aligned}$$

fulfils both assumptions, see [7, Sect. 4.1]. Note that $F_K : \widehat{K} \to K$ is the affine reference mapping.

3 Convergence Results

The properties of the interpolation operator J_h in Assumption 1 together with $\max(\nu, \sigma, \tau_K, \mu_K, \delta_K, h_K^2/\alpha_K) \leq C$ guarantees that the stabilised discrete problem (3) has a unique solution, see [7] where Lemma 2.6 deals with $S_h = S_h^a$ and Corollary 2.14 handles $S_h = S_h^b$.

Theorem 1. *Let Assumption 1 and Assumption 2 be fulfilled. Let* $(u,p) \in \big(H_0^1(\Omega) \cap H^{r+1}(\Omega)\big)^d \times \big(L_0^2(\Omega) \cap H^{r+1}(\Omega)\big)$ *be the solution of* (2) *and* $(u_h, p_h) \in V_h \times Q_h$ *be the solution of the stabilised discrete problem* (3). *We choose either* $S_h = S_h^a$ *with* $\tau_K \sim h_K$, $\mu_K \sim h_K$, $\alpha_K \sim h_K$ *or* $S_h = S_h^b$ *with* $\delta_K \sim h_K$, $\alpha_K \sim h_K$. *Then, there exists a positive constant* C *independent of* ν *and* h *such that*

$$|||(u-u_h, p-p_h)||| \leq C\left(\sum_{K \in \mathscr{T}_h} (\nu + h_K)\, h_K^{2r} \left(\|u\|_{r+1,K}^2 + \|p\|_{r+1,K}^2\right)\right)^{1/2}$$

holds true.

Proof. Follow the proof of Theorem 2.12 and Corollary 2.14 in [7] and use the properties of the interpolation operator j_h in Assumption 1. □

4 Numerical Results

This section presents some numerical results which were obtained by applying the enrichment approach of the local projection method to the Oseen problem. All calculations were performed with the code MooNMD [6] on a Linux PC (Pentium IV, 2.8 GHz).

Let $\Omega = (0,1)^2$. We consider the Oseen equations (1) where the right-hand side f and the Dirichlet boundary conditions are chosen such that

$$u(x,y) = \left(\sin(\pi x), -\pi y \cos(\pi x)\right)^T, \qquad p(x,y) = \sin(\pi x)\cos(\pi y)$$

is the solution for the case $\nu = 10^{-8}$, $\sigma = 1/100$, and $b = u$.

First, we consider calculations which were carried out for the enriched linear pair $Y_h/D_h = P_{1,h}^{\text{bubble}}/P_{0,h}^{\text{disc}}$ on triangles. Fig. 1 shows the convergence rates for the

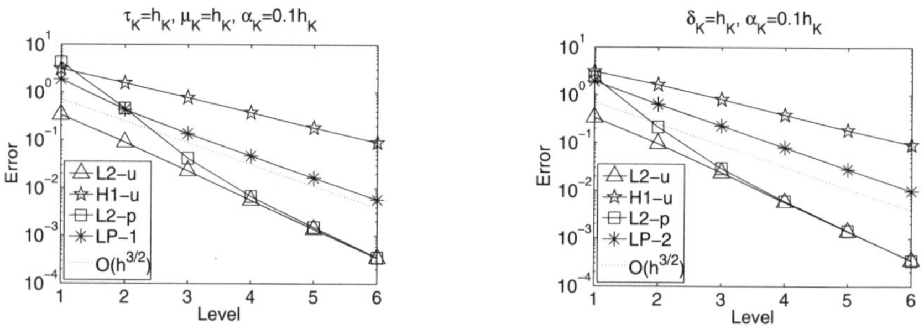

Fig. 1 Convergence order in different norm for the stabilisation terms S_h^a (left) and S_h^b (right).

solution of the stabilised discrete problem (3) where either S_h^a (left) or S_h^b (right) was used as stabilising term S_h. We have chosen the following stabilisation parameters

$$\alpha_K = 0.1\, h_K\,, \quad \mu_K = h_K\,, \quad \tau_K = h_K\,, \quad \delta_K = h_K\,.$$

The convergence rate $\mathcal{O}(h^{3/2})$ in the local projection norm, as predicted by Theorem 1 for both stabilising terms, can be clearly seen in Fig. 1. Furthermore, the results for both stabilising terms differ only slightly. The L^2-norms of pressure and velocity converge with second order while the H^1-semi norm of the velocity shows first order convergence. These convergence rates are optimal with respect to the interpolation error.

The theory gives only the asymptotic choice for the stabilisation parameters, their optimal size is not known. Fig. 2 shows for the stabilisation term S_h^b the dependence of the L^2-norm of velocity (upper left), the H^1-semi norm of velocity (upper right), the L^2-norm of pressure (lower left), and the local projection norm (lower right) on the stabilisation parameters. Since δ_K and α_K should be chosen proportional to h_K, compare Theorem 1, we have performed calculations with $\alpha_K = \alpha_0\, h_K$ and

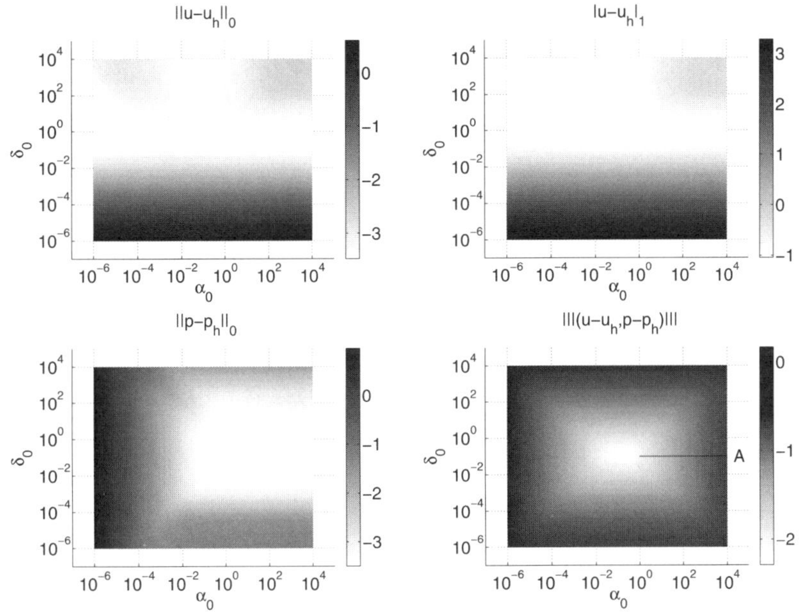

Fig. 2 Dependence of different norms on the stabilisation parameters $\alpha_K = \alpha_0 h_K$, $\delta_K = \delta_0 h_K$ for the stabilisation term S_h^b applied to enriched first order elements.

$\delta_K = \delta_0 h_K$ where the constants δ_0 and α_0 are varied in the range 10^{-6} up to 10^4. Note that the scales for α_0 and δ_0 are logarithmic. Instead of the errors theirselves, their logarithms are shown. It is clearly to see from Fig. 2 that vanishing pressure stabilisation results in large pressure errors while vanishing velocity stabilisation gives large velocity errors. Furthermore, there seems to be no interaction between the velocity stabilisation due to S_h^b and the pressure stabilisation since the pressure error is almost independent of the size of the velocity stabilisation and the velocity error is not influenced by the size of the pressure stabilisation. The behaviour of the local projection norm is different. If the size of the stabilising parameters is too large then the error in the local projection norm increases since the stabilisation terms which depend on the parameters are included in the local projection norm. To be precise, the increase of the error in the local projection norm along the line A in the lower right picture of Fig. 2 is caused only by the increased stabilisation parameter α_0 since the other parts of the local projection norm are almost constant in that region.

Finally, we present the results of our calculations for the enriched third order pair $Y_h/D_h = P_{3,h}^{\text{bubble}}/P_{2,h}^{\text{disc}}$ on triangles. Fig. 3 shows the dependence of the different norms on the coefficients α_0 and δ_0 of the stabilisation parameters $\alpha_K = \alpha_0 h_K$ and $\delta_K = \delta_0 h_K$. We see from Fig. 3 that stabilisation is needed. Compared to the enriched linear elements, the situation is more complex since there seems to be an interaction between pressure and velocity stabilisation terms.

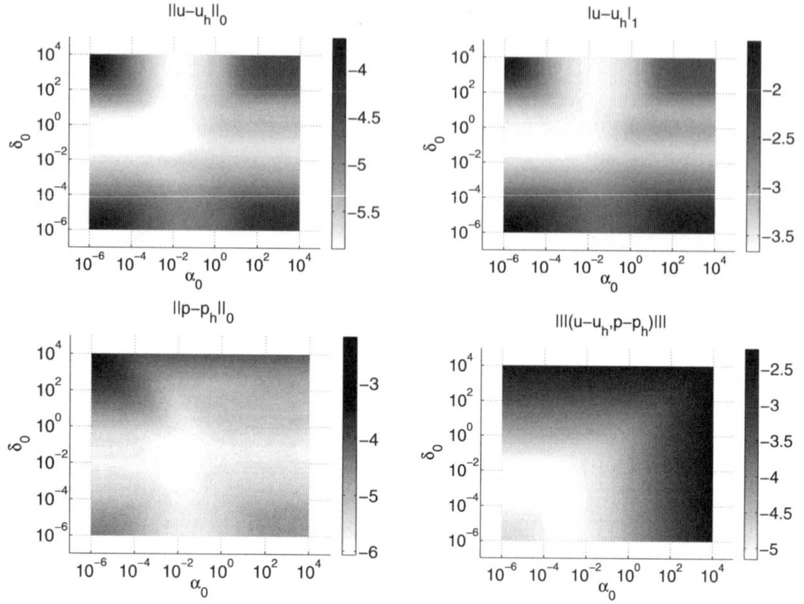

Fig. 3 Dependence of different norms on the stabilisation parameters $\alpha_K = \alpha_0 h_K$, $\delta_K = \delta_0 h_K$ for the stabilisation term S_h^b applied to enriched third order elements.

Acknowledgements The authors like to thank Simon Schmaljohann for performing the numerical calculations.

References

1. Becker, R., Braack, M.: A finite element pressure gradient stabilization for the Stokes equations based on local projections. Calcolo **38**(4), 173–199 (2001)
2. Becker, R., Braack, M.: A two-level stabilization scheme for the Navier-Stokes equations. In: M. Feistauer, V. Dolejší, P. Knobloch, K. Najzar (eds.) Numerical mathematics and advanced applications, pp. 123–130. Springer-Verlag, Berlin (2004)
3. Braack, M., Burman, E.: Local projection stabilization for the Oseen problem and its interpretation as a variational multiscale method. SIAM J. Numer. Anal. **43**(6), 2544–2566 (2006)
4. Clément, P.: Approximation by finite element functions using local regularization. RAIRO Anal. Numer. **9**, 77–84 (1975)
5. Ganesan, S., Matthies, G., Tobiska, L.: Local projection stabilization of equal order interpolation applied to the Stokes problem. Preprint 09/2007, Fakultät für Mathematik, Otto-von-Guericke-Universität Magdeburg (2007)
6. John, V., Matthies, G.: MooNMD - a program package based on mapped finite element methods. Comput. Vis. Sci. **6**(2–3), 163–170 (2004)
7. Matthies, G., Skrzypacz, P., Tobiska, L.: A unified convergence analysis for local projection stabilisations applied to the Oseen problem. M2AN Math. Model. Numer. Anal. **41**(4), 713–742 (2007)
8. Scott, L.R., Zhang, S.: Finite element interpolation of nonsmooth functions satisfying boundary conditions. Math. Comp. **54**(190), 483–493 (1990)

Involutive Completion to Avoid LBB Condition

B. Mohammadi and J. Tuomela

Abstract We propose to use the involutive form of a system of PDEs in numerical computations. We illustrate our approach by applying it to the Stokes system. As in the case of the solution of differential algebraic equations our approach takes explicitly into account the integrability conditions of the system which are only implicitly present in the original formulation. The extra calculation cost is negligible while the discrete form becomes much simpler to handle. One interesting consequence is that the discrete formulation needs not to satisfy the classical LBB compatibility condition. The approach is very general and can be useful for a wide variety of systems not as well known as fluid flow equations.

1 Introduction

Since various systems of PDEs have very different properties, it may seem hopeless to try to say something meaningful about arbitrary systems of PDEs. However, the *formal theory of PDEs* [4, 16, 18] provides tools for analysing general systems of PDEs. The modern formulation of the theory is differential geometric, and hence quite abstract, yet some of the consequences of the theory lead to constructive methods which can be used to analyse a given system of PDEs.

One consequence of the formal theory is the emergence of an important concept, the *involutive form* of the given system. It is important because it turns out that determining the properties of a given system is in general possible only if the system

Bijan Mohammadi
Mathematics and Modeling Institute, Montpellier University, France,
e-mail: bijan.mohammadi@universite-montpellier2.eu

Jukka Tuomela
Department of Mathematics, University of Joensuu, P.O. Box 111, FIN-80101 Joensuu, Finland
e-mail: jukka.tuomela@joensuu.fi

is involutive. For example some systems may not be elliptic/parabolic initially, but their completed forms are elliptic/parabolic [9, 11].

In this article we will argue that the involutive form is also important from the point of view of numerical computations. In fact the framework of formal theory has already been used in the numerical solution of ODE and DAE systems [19, 20, 21]. The approach by formal theory is helpful especially in situations where the physical models have constraints or conserved quantities which make the system essentially overdetermined. However, usually in numerical computations one uses square models (as many equations as unknowns). But then if one "forces" the system to be square by dropping some relevant equations/constraints one may encounter great difficulties in designing appropriate numerical methods.

One of the main issues we would like to insist on is that our numerical model will be solvable with simpler and more generic numerical methods than the original system. Indeed, complex numerical methods are often required to recover what is missed by not considering the right continuous system. Hence our approach allows the use of generic commercial software tools now widely available.

2 Preliminaries

2.1 Involutive Systems

The important concept of involutivity is unfortunately quite difficult to define precisely so we refer to [4, 16, 18] for a rigorous definition. However, for our purposes it is sufficient to explain the idea in concrete terms and indicate how to work with this notion constructively. We will be rather brief because our previous article [14] already discusses this approach.

Let us consider the system $\nabla \times y + y = 0$. Taking the divergence we see that if y is a solution, then it must also satisfy $\nabla \cdot y = 0$. This new equation is called a *differential consequence* or *integrability condition* of the initial system. Hence we have two systems:

$$\mathscr{S} \ :\{\ \nabla \times y + y = 0\}, \qquad \mathscr{S}' \ :\{\ \nabla \times y + y = 0,\ \nabla \cdot y = 0\}\,. \tag{1}$$

We say that $\mathscr{S}'$ is the involutive form of $\mathscr{S}$ because no more new first order differential consequences can be found. So informally we may define involutivity as follows:

> *A system is involutive, if it contains all its differential consequences (up to given order).*

There are many tricky issues involved when one actually tries to compute the involutive form for a given system. Anyway, the important point is that these constructions can be in fact carried out, hence the approach we are proposing here is potentially useful for solving quite general systems of PDE.

Note that in general it is necessary to find the involutive form of the given system before one can analyse any properties of the system. Our simple example above (1) illustrates this point: $\mathscr{S}$ is not elliptic while $\mathscr{S}'$ is elliptic.

Now, it turns out that for the purposes of numerical computation it is sometimes convenient to use not the "full" involutive form of the system, but to add just some of the integrability conditions to the original system. Hence we will use the term *completed system* to indicate that we may not use the full involutive system. We will argue below that it will be useful to use involutive or completed form in numerical computations.

2.2 Augmented Systems

Let us consider our problem in a general form

$$A_0 y = f \,, \tag{2}$$

and let us suppose that A_0 is already in completed form. For definiteness let us also suppose that A_0 is an elliptic operator. We refer to [4, 9, 10] for more information on overdetermined elliptic operators as well as relevant boundary conditions for them.

Now, since A_0 is in general overdetermined, there are typically no solutions for arbitrary f; hence there are some *compatibility conditions* for f. These conditions are given by an operator A_1 such that $A_1 A_0 = 0$ and (2) has a solution only if $A_1 f = 0$. Such an operator A_1 is called the *compatibility operator*; for more technical definition we refer to [4].

Let us now introduce some function spaces V_i such that $A_i : V_i \to V_{i+1}$. It is convenient to represent these spaces and operators with the help of some diagrams. Let us consider a sequence of such operators:

$$\cdots \longrightarrow V_i \xrightarrow{A_i} V_{i+1} \xrightarrow{A_{i+1}} V_{i+2} \longrightarrow \cdots \,.$$

Such a sequence is a *complex*, if $A_{i+1} A_i = 0$ for all i. The complex is *exact at* V_{i+1}, if $\operatorname{image}\big(A_i\big) = \ker\big(A_{i+1}\big)$. It is *exact*, if it is exact at all V_i. For example, the exactness of the complexes

$$0 \longrightarrow V_A \xrightarrow{A} W_A \,, \quad V_B \xrightarrow{B} W_B \longrightarrow 0 \,,$$

means that A is injective and B is surjective.

Let us now suppose that the following complex is exact:

$$0 \longrightarrow V_0 \xrightarrow{A_0} V_1 \xrightarrow{A_1} V_2 \longrightarrow 0 \,.$$

This suggests that we can decompose V_1 as follows:

$$\operatorname{image}(A_0) \oplus \operatorname{image}(A_1^T) \simeq V_1 \ ,$$

where A_1^T is the formal transpose of A_1. Of course, to be able to write equality instead of $\simeq$ we should specify carefully the relevant vector spaces. However, this decomposition is obviously valid if V_i are finite dimensional vector spaces and A_i are linear maps. Anyway, proceeding formally, this decomposition suggests that it is indeed possible to find some functional framework such that the combined operator (A_0, A_1^T) would be bijective or Fredholm. Hence, reasonable discretizations of these operators should yield a well-posed numerical problem.

So instead of trying to solve the original system (2) in some least square sense, we introduce an auxiliary variable z and solve

$$A_0 y + A_1^T z = f \ . \tag{3}$$

We call this system the *augmented system*. This formulation is reasonable because the augmented system is square, hence standard software tools are readily available. Also all the relevant information about the original system is contained in the completed operator A_0 which means that the results will be reliable. The drawback is that we have introduced an extra variable z which increases the computational cost. However, we can use z in error control as shown below; hence the work done for computing z will not be in vain.

In case of our example (1) the augmented system is

$$\mathscr{S}'' \ : \{\nabla \times y + y - \nabla z = 0, \ \ \nabla \cdot y - z = 0\} \ .$$

This system is elliptic, and could be solved in a straightforward manner. On the other hand, a proper discretization of the original system $\mathscr{S}$ would be difficult because the principal part of the operator has an infinite dimensional kernel. Hence, in the numerical solution there may appear components which are approximately in this kernel; these are called spurious solutions [8].

Remark 1. For the augmented Stokes system (7) below space V_0 (resp. V_1 and V_2) is the function space for u and p (resp. f and z). In the same way, for the completed Stokes system (7) in discrete form with any finite element discretization for variables (u, p, z) the space V_0 (resp. V_1 and V_2) will be generated by the corresponding finite element basis for u and p (resp. f and z) on the chosen mesh.

3 Systems from Fluid Mechanics

There is an enormous literature on the numerical solution of Stokes system [6, 15, 17]. We will illustrate our method using these classical models, but our approach can also be applied to the solution of flows in porous media, several biological fluids and microfluids [13].

3.1 Stokes System

Let us consider the Stokes problem:

$$\alpha u - \Delta u + \nabla p = f, \ \ \nabla \cdot u = 0\,, \tag{4}$$

where u is the velocity field, p is the pressure and $\alpha \geq 0$. Usually we have two boundary conditions for u and no condition for p as no natural condition on p is available. By taking the divergence of the first equation we obtain $-\Delta p = -\nabla \cdot f$. Putting $y = (u, p)$ we can write the whole system as

$$A_0 y = \begin{pmatrix} \alpha u - \Delta u + \nabla p \\ -\Delta p \\ -\nabla \cdot u \end{pmatrix} = \begin{pmatrix} f \\ -\nabla \cdot f \\ 0 \end{pmatrix}.$$

The compatibility operator is now given by $A_1 = \big(\nabla\cdot, 1, \alpha I - \Delta\big)$. Hence, the augmented system can be written as

$$A_0 y + A_1^T z = \begin{pmatrix} \alpha u - \Delta u + \nabla p - \nabla z \\ -\Delta p + z \\ -\nabla \cdot u + (\alpha I - \Delta) \end{pmatrix} = \begin{pmatrix} f \\ -\nabla \cdot f \\ 0 \end{pmatrix}. \tag{5}$$

This is elliptic and we need 4 boundary conditions. For u we use of course the original boundary conditions, and for z a natural choice is $z = 0$ on the boundary. However, we still need something for p.

The difficulty with boundary conditions for the pressure also arises when solving stabilized form of the Stokes system [15, 17]:

$$\begin{cases} -\Delta u + \nabla p = f, \\ -\varepsilon \Delta p + \nabla \cdot u = 0 \quad \text{where} \quad \varepsilon \sim h^2. \end{cases} \tag{6}$$

where implicitly some Neumann boundary condition is assumed for p.

In what follows we consider for simplicity only the stationary Stokes equation, and hence take $\alpha = 0$. In instationary problems α is the inverse of the time step. However, everything that follows is also applicable to the case $\alpha > 0$. Proceeding as for the stabilized system (6) we write our system as

$$\begin{cases} -\Delta u + \nabla p - \nabla z = f, \\ -\varepsilon \Delta p = \Delta p - z - \nabla \cdot f, \quad \varepsilon \sim 10^{-10}, \\ -\nabla \cdot u - \Delta z = 0. \end{cases} \tag{7}$$

with the boundary conditions

$$u \quad \text{as in the original system (4)}, \ \frac{\partial p}{\partial n} = 0, \ z = 0\,. \tag{8}$$

We solved the system (5) with finite elements for all variables on triangular meshes. Now one of the issues in the numerical solution of the Stokes problem is that the relevant finite element spaces should satisfy the *inf-sup* or *LBB condition* [1, 2, 7, 5, 12, 15, 17]. In particular it is not possible to use equal order discretization for u and p when solving the system (4). This is unnecessary when solving (6) or better (7). In fact, with (7) one can consider even P^1 elements for velocity and P^2 elements for the pressure (see Fig. 1).

3.2 Cavity Flow

To illustrate the behavior of the new system, we consider the well-known flow in a $(0,1) \times (0,1)$ square cavity. The boundary condition for the velocity $u = (u^1, u^2)$ is given by $u^1 = 4x(1-x)$, $u^2 = 0$ on the upper boundary and u^1 and u^2 are zero on all other boundaries. The numerical results in Fig. 1 show that using our formulation (5) we get the same results with P^1/P^1, P^2/P^1 and P^1/P^2 discretizations for velocity and pressure. z was always discretized with P^1 elements; trials with other finite elements gave essentially same results.

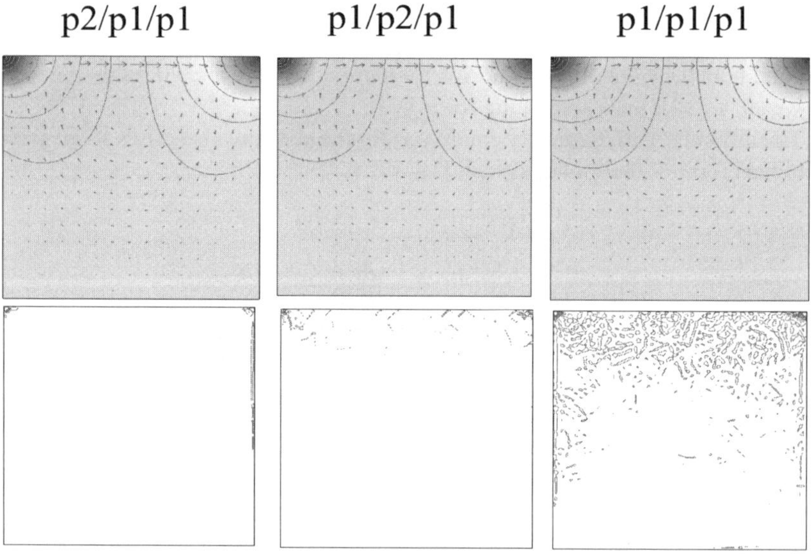

Fig. 1 Stokes flow in a square cavity. Upper row: velocity and pressure for the augmented Stokes system (5) with different discretizations for u, p and z. Lower row: corresponding z iso-contours. The maximum absolute value of z is less than 10^{-5} for all calculations. The same 5000 elements triangular finite element mesh has been used for all cases.

3.3 Flow under Body Forces

Another interesting example is a steady flow subject to a body force. This force may result from a coupling to other systems as in magnetohydrodynamics (MHD). When this force is the gradient of a potential and if *no slip boundary conditions* are applied, the fluid should remain at rest. This is not realized when equal order finite elements are used for the velocity and pressure in a stabilized approach such as (6) or even when finite elements obey the LBB condition. With the original system, one also needs the divergence of any discrete velocity function to belong to the pressure space. We show below the case of a flow submitted to $f = \nabla\phi$ with $\phi = 500(x^2 - y^2)$ solved using (6) and (5) both with equal order discretizations for velocity and pressure on the same mesh. The augmented system better realizes the flow at rest solution, see Fig. 2. In fact, the norm of the velocity in the augmented system is of 2 orders of magnitude smaller than in the standard formulation.

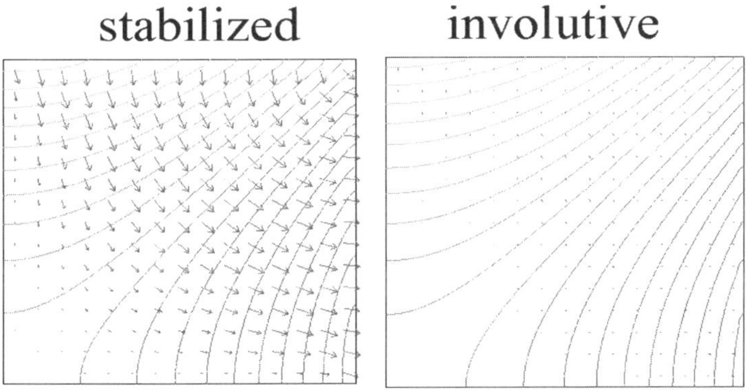

Fig. 2 Stokes flow under potential body force. Left: velocity and pressure for the stabilized Stokes system (6) using equal order linear finite elements for u and p. Right: same but solving augmented Stokes system (5).

4 Concluding Remarks

We have shown how to use involutivity, or more precisely the completed and augmented systems in numerical computations. However, we have not actually discussed how to find the involutive form of the system or how to construct the compatibility operator. There exist in fact tools of computational algebra which can be used for this purpose. However, this is a large topic by itself so it is not possible to present an account of it here. Let us simply mention that *Gröbner bases* play an important role in these constructions. We refer to [3] for a general introduction to Gröbner bases and in [18] one can find more details and references about applying symbolic methods to PDE systems. An important fact about our formulation is that it is more robust than the original system; indeed we did not need to worry about the LBB condition in the solution of Stokes system. This suggests that the involution approach is especially interesting when one would like to solve a new constrained PDE system where, unlike for the Stokes system, few things are known on the adequate numerics.

References

1. Babuska, I.: The finite element method with Lagrangian multipliers. Numer. Math. **20**, 179–192 (1973)
2. Brezzi, F.: On the existence uniqueness and approximation of saddle-point problems arising from Lagrangian multipliers. Revue Françcaise d'Autom. et de rech. opér. **2-129** (1974)
3. Cox, D., Little, J., O'Shea, D.: Ideals, varities and algorithms. Undergraduate Texts in Mathematics. Springer-Verlag (1992)
4. Dudnikov, P., Samborski, S.: Linear overdetermined systems of partial differential equations. In: M. Shubin (ed.) PDE VIII, Encyclopaedia of Mathematical Sciences 65, pp. 1–86. Springer-Verlag, Berlin/Heidelberg (1996)
5. Girault, V., Raviart, P.A.: Finite Element Methods for Navier-Stokes Equations. Springer Verlag (1986)
6. Glowinski, R.: Finite element methods for incompressible viscous flow. In: Handbook of numerical analysis, Vol. IX, pp. 3–1176. North-Holland, Amsterdam (2003)
7. Gunzburger, M.D.: Finite Element Methods for Viscous Compressible Flows. Academic Press (1986)
8. Jiang, B., Wu, J., Povinelli, L.: The origin of spurious solutions in computational electromagnetics. J. Comput. Phys. **7**, 104–123 (1996)
9. Krupchyk, K., Seiler, W., Tuomela, J.: Overdetermined elliptic PDEs. Found. Comp. Math. **6**(3), 309–351 (2006)
10. Krupchyk, K., Tuomela, J.: The Shapiro-Lopatinskij condition for elliptic boundary value problems. LMS J. Comput. Math. **9**, 287–329 (2006)
11. Krupchyk, K., Tuomela, J.: Completion of overdetermined parabolic PDEs. J. Symb. Comput. **43**(3), 153–167 (2008)
12. Ladyshenskaya, O.A.: The Mathematical Theory of Viscous Incompressible Flow. Gordon and Breach, New York (1969)
13. Mohammadi, B., Tuomela, J.: Involutive formulation for electroneutral microfluids. Submitted
14. Mohammadi, B., Tuomela, J.: Simplifying numerical solution of constrained PDE systems through involutive completion. M2AN **39**(5), 909–929 (2005)
15. Pironneau, O.: Finite element methods for fluids. John Wiley & Sons Ltd. (1989)
16. Pommaret, J.F.: Systems of Partial Differential Equations and Lie Pseudogroups, *Mathematics and its applications*, vol. 14. Gordon and Breach Science Publishers (1978)
17. Quarteroni, A., Valli, A.: Numerical approximation of partial differential equations, *Springer Series in Computational Mathematics*, vol. 23. Springer-Verlag, Berlin (1994)
18. Seiler, W.: Involution — the formal theory of differential equations. Habilitation thesis, Universität Mannheim (2001)
19. Tuomela, J., Arponen, T.: On the numerical solution of involutive ordinary differential systems. IMA J. Num. Anal. **20**, 561–599 (2000)
20. Tuomela, J., Arponen, T.: On the numerical solution of involutive ordinary differential systems: Higher order methods. BIT **41**, 599–628 (2001)
21. Tuomela, J., Arponen, T., Normi, V.: On the numerical solution of involutive ordinary differential systems: Enhanced linear algebra. IMA J. Num. Anal. **26**, 811–846 (2006)

Numerical Computation of Unsteady Compressible Flows with Very Low Mach Numbers

P. Punčochářová, K. Kozel, J. Horáček, and J. Fürst

Abstract This study deals with the numerical solution of 2D unsteady flows of a compressible viscous fluid in two types of channels (unsymmetric, symmetric) for a low inlet airflow velocity. The unsteadiness of the flow is caused by a prescribed periodic motion of a part of the channel wall with large amplitudes. The numerical solution is realized by a finite volume method and an explicit predictor-corrector MacCormack scheme with Jameson artificial viscosity using a grid of quadrilateral cells. The moved grid of quadrilateral cells is considered in the form of conservation laws using an Arbitrary Lagrangian-Eulerian method. Numerical results of the unsteady flows in the channels are presented for inlet Mach number $M_\infty \approx 10^{-2}$, Reynolds number Re $\in (5 \times 10^3,\ 1.1 \times 10^4)$ and for a frequency of the wall motion 20 Hz and 100 Hz.

1 Mathematical Model

The 2D system of Navier-Stokes equations in conservative non-dimensional form has been used as mathematical model to describe the unsteady laminar flow of the

Petra Punčochářová
Czech Technical University in Prague, Karlovo náměstí 13, 121 35, Prague 2, Czech Republic, e-mail: puncocha@marian.fsik.cvut.cz

Karel Kozel
Institute of Thermomechanics Academy of Sciences, Dolejškova 5, Prague 8, Czech Republic e-mail: kozelk@fsik.cvut.cz

Jaromír Horáček
Institute of Thermomechanics Academy of Sciences, Dolejškova 5, Prague 8, Czech Republic e-mail: jaromirh@it.cas.cz

Jiří Fürst
Czech Technical University in Prague, Karlovo náměstí 13, 121 35, Prague 2, Czech Republic, e-mail: jiri.furst@fs.cvut.cz

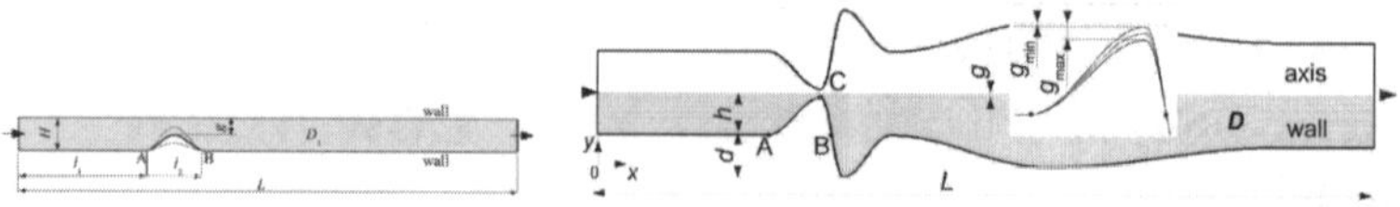

(a) Domain D_1, the unsymmetric channel. (b) Domain D_2, the symmetric channel.

Fig. 1 Computational domains D_1 and D_2.

compressible viscous fluid in a domain [1]:

$$\frac{\partial \mathbf{W}}{\partial t} + \frac{\partial \mathbf{F}}{\partial x} + \frac{\partial \mathbf{G}}{\partial y} = \frac{1}{\mathrm{Re}} \left(\frac{\partial \mathbf{R}}{\partial x} + \frac{\partial \mathbf{S}}{\partial y} \right), \tag{1}$$

where $\mathbf{W} = [\rho, \rho u, \rho v, e]^T$ is the vector of conservative variables, $\mathbf{F}$ and $\mathbf{G}$ are the vectors of inviscid fluxes, $\mathbf{R}$ and $\mathbf{S}$ are the vectors of viscous fluxes, $\mathrm{Re} = (H' \rho'_\infty u'_\infty)/\eta'_\infty$ is the Reynolds number given by the inflow variables marked by infinity subscript (dimensional variables are marked by the prime), ρ denotes the density, u and v are the components of the velocity vector and e is the total energy per unit volume, see [1]. The static pressure is expressed by the equation of state:

$$p = (\kappa - 1) \left[e - \frac{1}{2} \rho \left(u^2 + v^2 \right) \right]. \tag{2}$$

2 Mathematical Formulation

Fig. 1 shows two bounded computational domains which are used for the numerical solution of the system (1).

The domain D_1 in Fig. 1(a) is an unsymmetric channel where the upper boundary is a straight rigid wall, and the lower boundary is a curved wall, oscillating between the points A and B. The shape of this part is described by a simple harmonic function.

The computational domain D_2 is shown in Fig. 1(b). It is a scaled model of the symmetric channel whose shape is inspired by a shape of the vocal folds and supraglottal spaces. The computational domain is only the lower half of the symmetric channel. The upper boundary is the axis of symmetry, the lower boundary is the channel wall, where the part between the points A and B is changing the shape according to a given harmonic function of time and axial coordinates $w(x,t)$. The gap g between the point C and the channel axis is $g = (d+h) - w(x_C, t)$, $g \in (0.01,\ 0.07)$. The points are given by $A = [1.75,\ 0.4]$, $B = [2.4,\ 0.4]$, $C = [2.3,\ w(x_C, t)]$, and further non-dimensional parameters of the channel geometry are: $L = 8$, $h = 0.4$, $d = 0.4$. The boundary conditions are considered in the following formulation:

1. Upstream conditions: $u_\infty = M_\infty \cos(\alpha)$, $v_\infty = M_\infty \sin(\alpha)$, $\rho_\infty = \text{const.}$, p_∞ is extrapolated from the domain D and α is the angle of the incoming flow.
2. Downstream conditions: $p_2 = \text{const.}$, $(\rho, \rho u, \rho v)$ are extrapolated from D.
3. Flow on the wall: $(u,v) = (0, v_{wall})$ and $\frac{\partial T}{\partial \mathbf{n}} = 0$.
4. Flow on the axis of symmetry: $(u,v) \cdot \mathbf{n} = 0$.

3 Numerical Solution

The numerical solution is based on a finite volume method (FVM) in cell centred form on a grid of quadrilateral cells.

The bounded domain D is divided into mutually disjoint sub-domains $D_{i,j}$ (e.g. quadrilateral cells). The system (1) is then integrated over the sub-domains $D_{i,j}$ and reformulated by using Greens formula and the mean value theorem. Due to the unsteady domain the integral form of FVM is derived by using the Arbitrary Lagrangian-Eulerian (ALE) formulation. The ALE method defines a homeomorphic mapping of the reference domain $D_{t=0}$ at the initial time $t = 0$ to a domain D_t at $t > 0$ [2].

An explicit predictor-corrector MacCormack (MC) scheme (3) in the domain with a moving grid of quadrilateral cells is used for the numerical solution of the system (1). The scheme is of 2nd order accuracy in time and space [1]:

$$
\begin{aligned}
\mathbf{W}_{i,j}^{n+1/2} &= \frac{\mu_{i,j}^n}{\mu_{i,j}^{n+1}} \mathbf{W}_{i,j}^n - \frac{\Delta t}{\mu_{i,j}^{n+1}} \sum_{k=1}^{4} \left[\left(\tilde{\mathbf{F}}_k^n - s_{1k} \mathbf{W}_k^n - \frac{1}{Re} \tilde{\mathbf{R}}_k^n \right) \Delta y_k \right. \\
&\quad \left. - \left(\tilde{\mathbf{G}}_k^n - s_{2k} \mathbf{W}_k^n - \frac{1}{Re} \tilde{\mathbf{S}}_k^n \right) \Delta x_k \right], \\
\overline{\mathbf{W}}_{i,j}^{n+1} &= \frac{\mu_{i,j}^n}{\mu_{i,j}^{n+1}} \frac{1}{2} \left(\mathbf{W}_{i,j}^n + \mathbf{W}_{i,j}^{n+1/2} \right) \\
&\quad - \frac{\Delta t}{2\mu_{i,j}^{n+1}} \sum_{k=1}^{4} \left[\left(\tilde{\mathbf{F}}_k^{n+1/2} - s_{1k} \mathbf{W}_k^{n+1/2} - \frac{1}{Re} \tilde{\mathbf{R}}_k^{n+1/2} \right) \Delta y_k \right. \\
&\quad \left. - \left(\tilde{\mathbf{G}}_k^{n+1/2} - s_{2k} \mathbf{W}_k^{n+1/2} - \frac{1}{Re} \tilde{\mathbf{S}}_k^{n+1/2} \right) \Delta x_k \right],
\end{aligned}
\tag{3}
$$

where $\Delta t = t^{n+1} - t^n$ is the time step, $\mu_{i,j} = \int\int_{D_{i,j}} dxdy$ is the volume of the cell $D_{i,j}$, Δx and Δy are the mesh sizes of the grid in x and y directions, and the vector $\mathbf{s}_k = (s_1,\ s_2)_k$ represents the speed of the edge k (see Fig. 2). The physical fluxes $\mathbf{F}$, $\mathbf{G}$, $\mathbf{R}$, $\mathbf{S}$ on the edge k of the cell $D_{i,j}$ are replaced by the numerical fluxes (marked with tilde) $\tilde{\mathbf{F}}$, $\tilde{\mathbf{G}}$, $\tilde{\mathbf{R}}$, $\tilde{\mathbf{S}}$ which are approximations of the physical fluxes.

The approximations of the convective terms $\mathbf{sW}_k$ and of the numerical viscous fluxes $\tilde{\mathbf{R}}_k$, $\tilde{\mathbf{S}}_k$ on the edge k are central. The higher order partial derivatives of the velocity and of the temperature in $\tilde{\mathbf{R}}_k$, $\tilde{\mathbf{S}}_k$ are approximated by using dual volumes V_k' (see [1]) as shown in Fig. 2. The inviscid numerical fluxes are approximated by

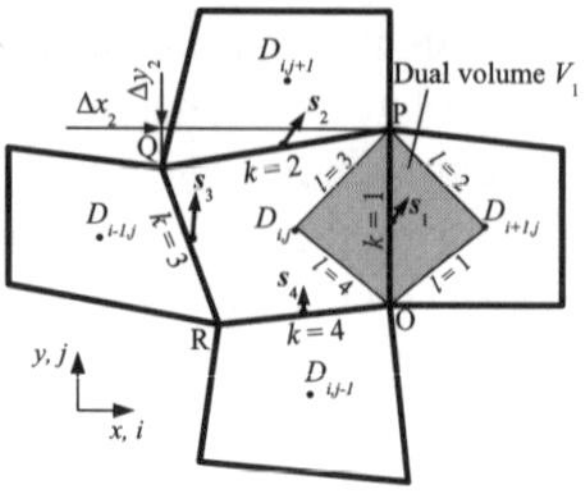

Fig. 2 Finite volume $D_{i,j}$ and the dual volume V_k'.

the physical fluxes as follows:

$$\tilde{\mathbf{F}}_1^n = \mathbf{F}_{i,j}^n,\ \tilde{\mathbf{F}}_1^{n+1/2} = \mathbf{F}_{i+1,j}^{n+1/2},\ \tilde{\mathbf{F}}_3^n = \mathbf{F}_{i-1,j}^n, \tilde{\mathbf{F}}_3^{n+1/2} = \mathbf{F}_{i,j}^{n+1/2},$$
$$\tilde{\mathbf{G}}_2^n = \mathbf{G}_{i,j}^n,\ \tilde{\mathbf{G}}_2^{n+1/2} = \mathbf{G}_{i,j+1}^{n+1/2},\ \tilde{\mathbf{G}}_4^n = \mathbf{G}_{i,j-1}^n, \tilde{\mathbf{G}}_4^{n+1/2} = \mathbf{F}_{i,j}^{n+1/2}, \text{ etc.} \tag{4}$$

The last term used in the MC scheme is the Jameson artificial dissipation [3, 4]:

$$AD(W_{i,j})^n = C_1\gamma_1\left(\mathbf{W}_{i+1,j}^n - 2\mathbf{W}_{i,j}^n + \mathbf{W}_{i-1,j}^n\right) + C_2\gamma_2\left(\mathbf{W}_{i,j+1}^n - 2\mathbf{W}_{i,j}^n + \mathbf{W}_{i,j-1}^n\right), \tag{5}$$

$C_1, C_2 \in \mathbb{R}$ are constants, in our case $C_1 = 1.7,\ C_2 = 1.5$, and the variables $\gamma_1,\ \gamma_2$ have the form

$$\gamma_1 = \frac{|p_{i+1,j}^n - 2p_{i,j}^n + p_{i-1,j}^n|}{|p_{i+1,j}^n| + 2|p_{i,j}^n| + |p_{i-1,j}^n|}, \quad \gamma_2 = \frac{|p_{i,j+1}^n - 2p_{i,j}^n + p_{i,j-1}^n|}{|p_{i,j+1}^n| + 2|p_{i,j}^n| + |p_{i,j-1}^n|}. \tag{6}$$

The term of artificial dissipation has third order accuracy and therefore the original scheme is of second order accuracy. Then the vector of the conservative variables $\mathbf{W}$ can be computed at a new time level t^{n+1}:

$$\mathbf{W}_{i,j}^{n+1} = \overline{\mathbf{W}}_{i,j}^{n+1} + AD(W_{i,j})^n. \tag{7}$$

The stability condition of the scheme (on a regular orthogonal grid) limits the time step reads

$$\Delta t \le CFL\left[\frac{|u_{max}| + c}{\Delta x_{min}} + \frac{|v_{max}| + c}{\Delta y_{min}} + \frac{2}{Re}\left(\frac{1}{\Delta x_{min}^2} + \frac{1}{\Delta y_{min}^2}\right)\right]^{-1}, \tag{8}$$

where c denotes the local speed of sound, u_{max} and v_{max} are the maximum velocities in the domain, and $0 < CFL < 1$ for the non-linear equations. Note that the grid of the channels is successively refined near the wall. The minimum cell size in y direction is $\Delta y_{min} \approx 1/\sqrt{Re}$ to resolve boundary layer effects.

Remark 1. We prefer to use a model of compressible Navier-Stokes equations of parabolic type since there is no need to use preconditioning for low Mach number as for a inviscid hyperbolic (Euler) system. The maximal velocity is limited in the narrow part of the channel, and for higher M_∞ the flow is chocked. These properties are not valid when considering the incompressible Navier-Stokes equations.

4 Numerical Results

The computation of the unsteady solution was carried out in two stages. First, the steady solution is realized, when the channel between the points A and B has a rigid wall in a middle position of the gap. Then, the steady solution is used as initial condition for the unsteady simulations.

4.1 Airflow in the Unsymmetric Channel

The numerical results were obtained for the following input data: Mach number $M_\infty = 0.02$ ($u'_\infty = 6.86\ ms^{-1}$), Re=10900, atmospheric pressure $p_2 = 1/\kappa$ ($p'_2 = 102942$ Pa) at the outlet, and frequency of the wall oscillation f'=20 Hz. The computational domain contains 400×50 cells for L=8 and H=0.5.

Fig. 3(a) shows the steady numerical solution. The maximum of the Mach number computed in the domain is M_{max}=0.050. Fig. 3(b) shows the convergence to the steady state solution computed by using the L_2 norm of momentum residuals (ρu). The convergence seems to be good.

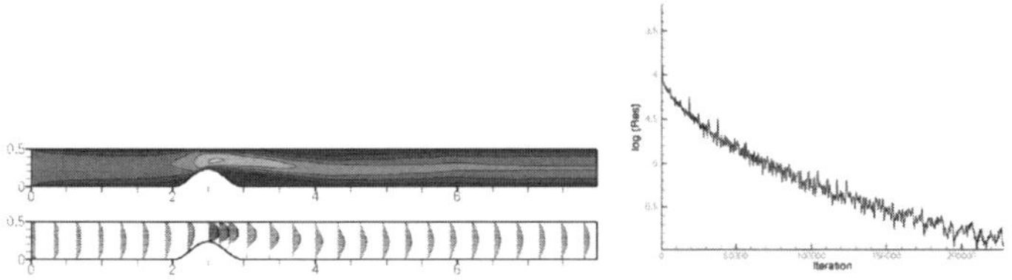

(a) $M_{max} = 0.050,\ g = 0.26.$ (b) Convergence to the steady state solution.

Fig. 3 The steady numerical solution in D_1 - $M_\infty = 0.02$, Re=10900, $p_2 = 1/\kappa$, 400×50 cells.

The unsteady solution in the third period of the wall oscillation at several time layers is shown in Fig. 4. The highest maximum Mach number was achieved at $x = 2.58$ in instant, when the gap is minimal ($t = 9/2\pi$). The flow becomes periodical after the second period of the oscillations.

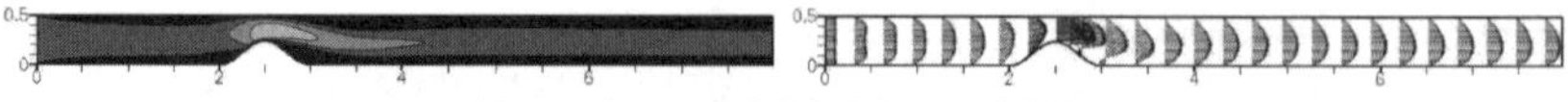

(a) $t = 4\pi,\ g = 0.26,\ M_{max} = 0.048$.

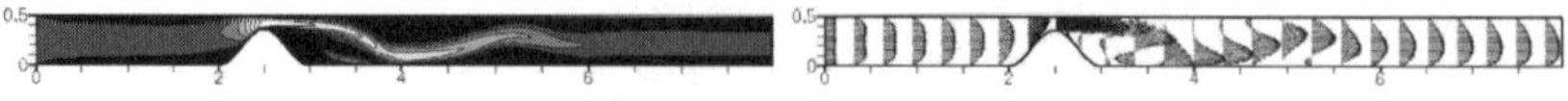

(b) $t = 9/2\pi,\ g = 0.14,\ M_{max} = 0.088$.

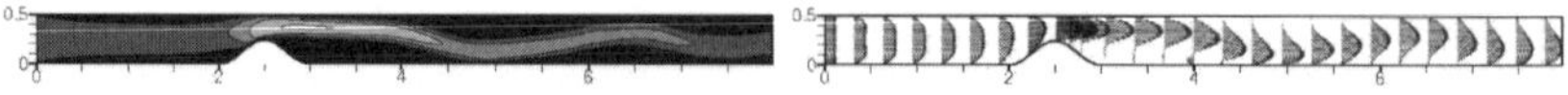

(c) $t = 5\pi,\ g = 0.26,\ M_{max} = 0.050$.

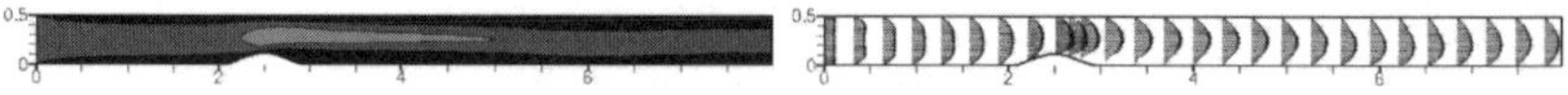

(d) $t = 11/2\pi,\ g = 0.38,\ M_{max} = 0.033$.

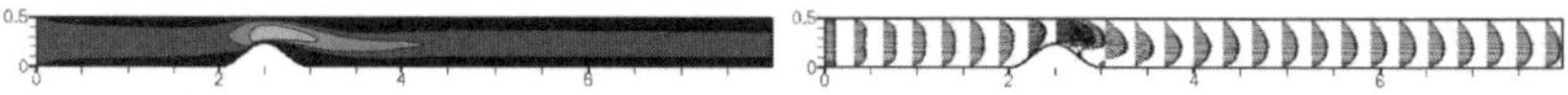

(e) $t = 6\pi,\ g = 0.26,\ M_{max} = 0.048$.

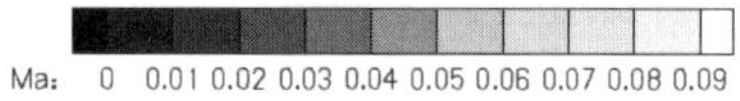

Fig. 4 The unsteady numerical solution of airflow in D_1 for wall motion: f'=20 Hz, $M_\infty = 0.02$, Re=10900, $p_2 = 1/\kappa$, 400×50 cells. Results are mapped by iso-lines of Mach number and by velocity vectors.

4.2 Airflow in the Symmetric Channel

The numerical results were obtained for the following input data: Mach number $M_\infty = 0.012$ ($u'_\infty = 4.11\ ms^{-1}$), Re=5237, atmospheric pressure $p_2 = 1/\kappa$ ($p'_2 = 102942$ Pa) at the outlet, and frequency of the wall oscillation f'=100 Hz. The computational domain contains 450×50 cells for h=0.4 and L=8, $H = 2h$.

Fig. 5(a) shows the steady numerical solution. The maximum of Mach number computed in the domain is $M_{max} = 0.173$ at $x = 2.317$ on the axis. Fig. 5(b) shows the convergence to the steady state solution computed by using the L_2 norm of momentum residuals (ρu). The convergence seems to be satisfactory for this case.

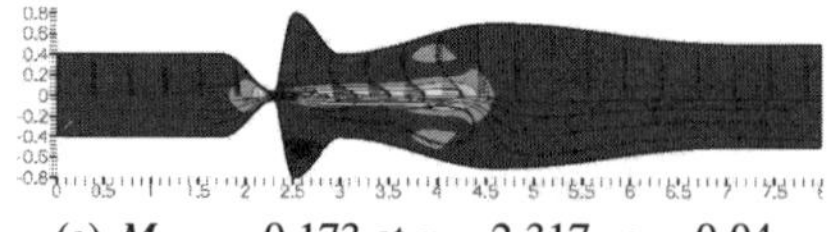

(a) $M_{max} = 0.173$ at $x = 2.317,\ g = 0.04$.

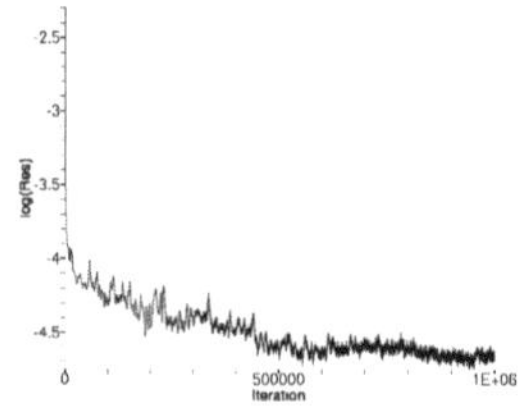

(b) Convergence to the steady state solution.

Fig. 5 The steady numerical solution in D_2 - $M_\infty = 0.012$, Re=5237, $p_2 = 1/\kappa$, 450×50 cells.

The unsteady solution in the fourth period of the wall oscillation is shown in Fig. 6 at several time layers. The highest maximum of Mach number $M_{max} = 0.535$ was achieved in instant when the glottal width is opening after the minimum of the gap is exceeded (see Fig. 7(a)) in time $t = 6\pi + 0.84\pi$.

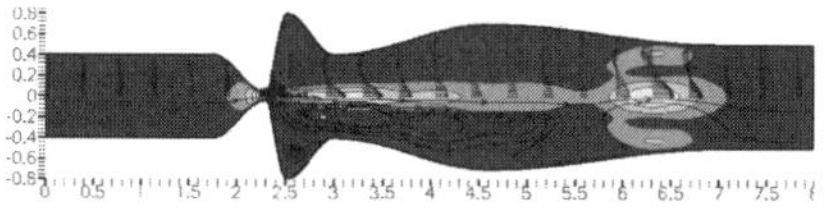

(a) $t = 6\pi$, $g = 0.04$, $M_{max} = 0.154$ at $x = 2.309$.

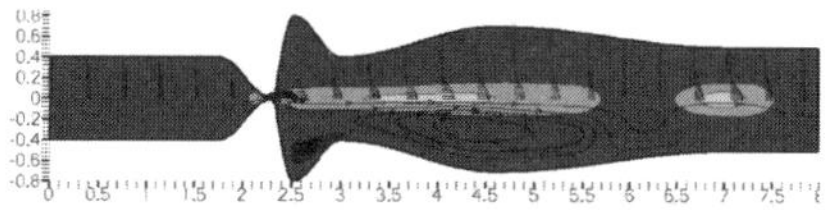

(b) $t = 13/2\pi$, $g = 0.01$, $M_{max} = 0.238$ at $x = 2.294$.

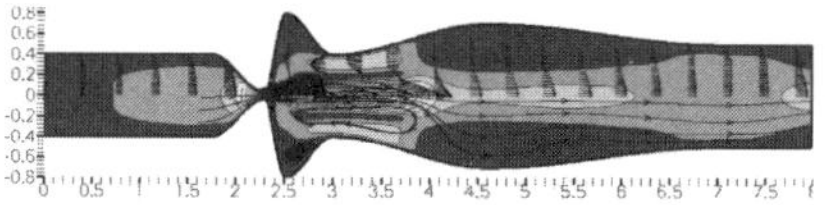

(c) $t = 7\pi$, $g = 0.04$, $M_{max} = 0.361$ at $x = 2.309$.

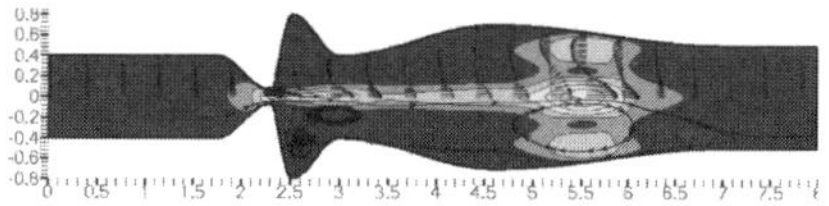

(d) $t = 15/2\pi$, $g = 0.07$, $M_{max} = 0.119$ at $x = 5.516$.

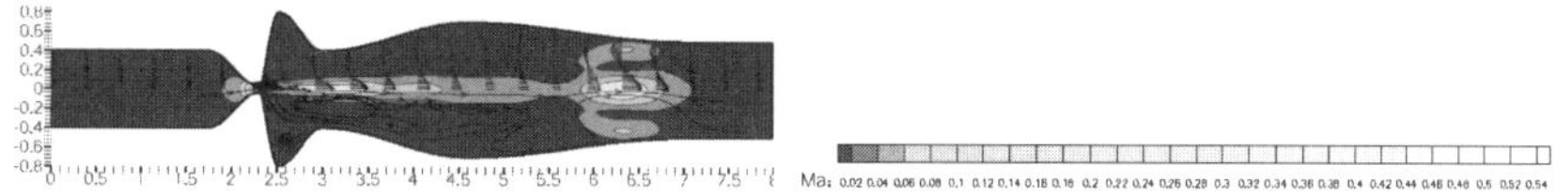

(e) $t = 8\pi$, $g = 0.04$, $M_{max} = 0.154$ at $x = 2.309$.

Fig. 6 The unsteady numerical solution of airflow in D_2 for wall motion: f'=100 Hz, $M_\infty = 0.012$, Re=5237, $p_2 = 1/\kappa$, 450×50 cells. Results are mapped by iso-lines of Mach number, by streamlines (lower part of the channel), and by velocity vectors (upper part of the channel).

Fig. 7(a) shows the Mach number along the axis of symmetry of the channel in several time instants during the oscillation period. Behind the narrowest channel cross-section ($x = x_C$) a second peak of the Mach number is forming which travels as a dying wave to the outlet. Fig. 7(b) shows the changes of the gap g, Mach number and the pressure in real time at the distance $x = 2.3$ on the channel axis. The phase shifts between the minimum glottal gap g and the maximum of Mach number, and pressure fluctuations are about 1.7×10^{-3} s and 7.8×10^{-4} s, respectively. It can be also seen, that the flow becomes periodical after the first period of the oscillations.

5 Summary

A numerical method and a special program code for solving the 2D unsteady Navier-Stokes equations for viscous compressible fluids has been developed. The method has been used for the numerical solution of the airflow in a unsymmetric channel and in a simplified model of the human vocal tract geometry. It is used for the description of the flow field behaviour in these cases because experimental data are not known.

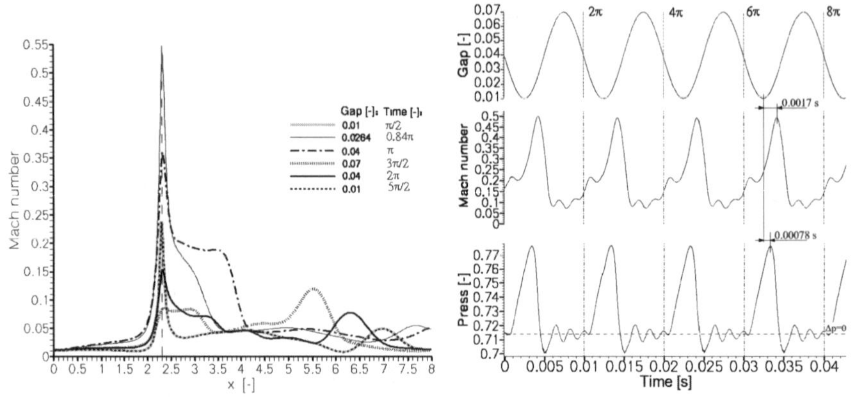

(a) Mach number along the channel axis in several time instants.

(b) Dimensionless gap g, Mach number and pressure at $x_C = 2.3$ on the channel axis in real time.

Fig. 7 Data computed during the fourth oscillation period- $f' = 100$ Hz, $M_\infty = 0.012$, Re=5237, $p_2 = 1/\kappa$, 450×50 cells.

Even if no complete closure of the glottis is modelled (in reality, the airflow coming from the lungs causes the vocal folds self-oscillations, and the glottis is completely closing in normal phonation regimes generating acoustic pressure fluctuations), the numerical simulation of the airflow field in the glottis is complex, and relatively close to reality. When the glottis is closing, the airflow velocity is becoming much higher in the narrowest part of the airways, where also the viscous forces play an important role. Therefore, for a correct modelling of a real flow in the glottis, the compressible, viscous and unsteady fluid-flow model should be considered.

Acknowledgements This work was partially supported by Research Plan MSM 6840770010 and GA ASCR No. IAA200760613.

References

1. Fürst, J., Janda, M., Kozel, K.: Finite volume solution of 2d and 3d Euler and Navier-Stokes equations. In: J. Neustupa, P. Penel (eds.) Mathematical Fluid Mechanics, pp. 173–194. Berlin (2001)
2. Honzátko, R., Horáček, J., Kozel, K.: Solution of inviscid incompressible flow over a vibrating profile. COE Lecture notes **3**, 26–32 (2006). M. Beneš, M. Kimura, T. Nataki (eds.), Kyushu University
3. Jameson, A., Schmidt, W., Turkel, E.: Numerical solution of the Euler equations by the finite volume methods using Runge-Kutta time-stepping schemes. AIAA pp. 81–125 (1981)
4. Punčochářová, P., Kozel, K., Fürst, J.: An unsteady numerical solution of viscous compressible flows in a channel. In: Program and Algorithms of Numerical Mathematics 13, pp. 220–228. Math. Institute ASCR (2006)

A Mixed Hybrid Finite Element Discretization Scheme for Reactive Transport in Porous Media

F.A. Radu, M. Bause, A. Prechtel, and S. Attinger

Abstract We present a model to describe the simultaneous reactive transport in porous media of an arbitrary number of mobile and immobile species. The model includes the effects of advection, dispersion, sorption and degradation catalysed by microbial populations. The locally mass-conservative mixed hybrid finite element method (MHFEM) to discretize this system of coupled convection-diffusion-reaction equations and the algorithmic solution of the resulting nonlinear algebraic equations are described in detail. Further, new ideas regarding the discretization of the convective term are discussed. Finally a comparative numerical study (MHFEM versus conforming FEM) is presented.

1 Introduction

Environmental pollution of soils and groundwater e.g. by organic compounds is nowadays a serious and widespread problem not only in industrial countries. As an alternative to conventional remediation, which is often not feasible, natural attenuation has been recognized as a promising approach. However, the decision to rely on intrinsic bioremediation at a specific site depends essentially on the trustworthy prediction of the fate of the contaminant over long time periods by mathematical models using experimental data. The complexity of the models makes the numerical approach as important as challenging. "Things should be made as simple as possible, but not simpler"- a citation from Albert Einstein, that perfectly applies to

F.A. Radu, S. Attinger
UFZ-Helmholtz Center for Environmental Research, Permoserstr. 15, D-04318 Leipzig, Germany
e-mail: {florin.radu, sabine.attinger}@ufz.de and
University of Jena, Wöllnitzerstr. 7, D-07749, Jena, Germany

M. Bause, A. Prechtel
Mathematics Department, University of Erlangen-Nuremberg, Martensstr. 3, D-91058 Erlangen, Germany e-mail: {bause, prechtel}@am.uni-erlangen.de

mathematical modeling. The interaction of the relevant components (species) via sensitive, highly nonlinear reactive processes stresses the need for very accurate numerical schemes. Artificial numerical diffusion leads to erroneous reactive processes and thus to false predictions of the contaminant fate by overestimating the availability of the reactants. This has been noted by several authors; cf., e.g., [4]. The need of an adequate approximation of the fluid flow by Mixed Finite Element Methods (MFEM) has been recognized in the water resources literature since several decades, see, e.g., [7, 1, 9, 10]. These methods offer the advantage of local mass conservation and continuous flux approximations over the element faces. However, for associated solute transport problems normally conventional methods are applied, e.g. conforming Finite Element Method (FEM) or Finite Volume (FVM).

As in our opinion in particular in the context of reactive nonlinear multicomponent models, which are highly sensitive to numerical errors, the need for high accuracy is evident, we apply the Mixed Hybrid FEM (MHFEM) to the multicomponent transport equations. Such a formulation has to our knowledge not been presented for that type of problem up to date. Thus we introduce this novel formulation of a coupled reactive multicomponent transport model together with a description of water flow in the vadose and saturated zone by the Richards equation in the mixed hybrid finite element setting.

2 Mathematical Model

In the following a mathematical model for subsurface flow and simultaneous reactive multicomponent transport including biodegradation is presented. Let Ω be a bounded domain in $\mathbb{R}^d$, $d = 1,2$, or 3, with sufficiently smooth boundary $\partial\Omega$ and let $J = (0,T]$ be some finite interval with T denoting the final time.

The groundwater flow, taking into account the saturated and the vadose zone is described by the mass conservation principle and Darcy's law:

$$\partial_t \Theta(\psi) + \nabla \cdot \mathbf{q} = 0, \quad \mathbf{q} = -K(\psi)\nabla(\psi + z) \quad \text{in } J \times \Omega . \tag{1}$$

In (1), ψ is the pressure head, Θ is the water content, $\mathbf{q}$ is the water flux, K is the hydraulic conductivity, and z is the gravity head. For the soil-water retention curve $\Theta(\psi)$ and the unsaturated hydraulic conductivity function $K(\psi)$ different functional forms have been derived in the literature (cf. e.g., [8]).

The transport and reaction of N_S species including the effects of advection, dispersion, sorption and degradation is described by the set of equations

$$\partial_t(\Theta c_i) + \rho_b \partial_t s_i - \nabla \cdot (\mathbf{D_i} \nabla c_i - \mathbf{q} c_i) = -\Theta R_i \quad \text{in } J \times \Omega , \tag{2}$$

with $i \in \{1, \dots, N_S\}$, where c_i, s_i denote the concentration of the dissolved and the adsorbed species, respectively, $\mathbf{D_i}$ is the diffusion-dispersion coefficient, ρ_b [ML^{-3}] is the bulk density and R_i denotes the reaction rate of the ith species. For the sake of simplicity we consider only homogeneous Dirichlet boundary conditions in the sequel. For other types of boundary conditions we refer to [8].

Additionally we formulate a mass balance for the ith microbial or other immobile species, $i \in \{N_S+1,\ldots,N_S+M\}$, that are only subject to reactions:

$$\partial_t c_i + k_{di} c_i = R_i \,. \tag{3}$$

Here k_{di} denotes the decay rate of species i. Equations (2) and (3) have to be supplemented with initial conditions $c_i(x,0) = c_{i0}(x)$ for $x \in \Omega$, $i \in \{1,\ldots,N_S+M\}$.

In this work, sorption is considered to be of kinetic type

$$\partial_t s_i = k_i(\phi(c_i) - s_i)\,, \tag{4}$$

with ϕ being an arbitrary sorption isotherm and k_i a rate parameter. The isotherm ϕ can be linear, of Freundlich, Langmuir or Freundlich-Langmuir type or a form free function resulting from an experimental study or a parameter identification process.

3 Discretization

To discretize the flow problem (1) and transport system (2)–(4) in time, we use the backward Euler method. For the spatial discretization of the equations we use the lowest order Raviart-Thomas mixed finite element method. The approximation of the Richards equation is described elsewhere (cf. [3, 8]) and is not considered here. In the sequel we only present the mixed approximation of (2)–(4). In each time step, we then solve first the discretized Richards equation by a damped version of Newton's method. This yields approximations of the water flux and content. Subsequently, we solve the fully coupled species equations again by Newton's method.

To describe the mixed approximation of (2)–(4), let $\mathcal{T}_h$ be a regular decomposition of Ω into closed d-simplices, $d = 1,2$ or 3, with h denoting the mesh-size. Let $0 = t_0 < t_1 < \ldots < t_N = T$ be a partition of the time interval $[0,T]$ and $\Delta t^n = t_n - t_{n-1}$, $n \in \{1,\ldots,N\}$. Let $W = L^2(\Omega)$ be the space of all measurable and square integrable functions on Ω and $V = H(\mathrm{div};\Omega)$ the space of all d-dimensional vector functions with all components and divergence in $L^2(\Omega)$. We denote by $\mathcal{S}_h = \mathcal{S}_h^I \cup \mathcal{S}_h^D$ the set of all faces of $\mathcal{T}_h$, where $\mathcal{S}_h^I$ are interior faces and $\mathcal{S}_h^D$ are faces on the boundary. The discrete subspaces $W_h \subset W$ and $V_h \subset V$ are defined as $W_h = \{p \in W|\ p \text{ is constant on each element } T \in \mathcal{T}_h\}$ and $V_h = \{\mathbf{q} \in V | \mathbf{q}_{|T} = \mathbf{a} + b\mathbf{x}, \mathbf{a} \in \mathbb{R}^d, b \in \mathbb{R} \text{ for all } T \in \mathcal{T}_h\}$. W_h denotes the space of all piecewise constant functions, whereas V_h is refered to as the RT_0 space (cf. [5]).

First, a mixed formulation of the transport equations (2) is obtained by introducing the flux variables $\mathbf{q_i}$, $i \in \{1,\ldots,N_S\}$ as additional unknowns:

$$\partial_t(\Theta c_i) + \rho_b \partial_t s_i + \nabla \cdot \mathbf{q_i} = -R_i, \quad \mathbf{q_i} = -\mathbf{D_i}\nabla c_i + \mathbf{q} c_i \quad \text{in } J \times \Omega\,. \tag{5}$$

The fully discrete mixed approximation of (5) then reads as follows: *For all* $n = 1,\dots,N$ *and* $i \in \{1,\dots,P\}$ *find* $(c^n_{i,h}, \mathbf{q^n_{i,h}}) \in W_h \times V_h$ *such that there holds*

$$\begin{aligned}\langle \Theta^n_h c^n_{i,h} - \Theta^{n-1}_h c^{n-1}_{i,h}, w_h\rangle + \rho_b \langle s^n_{i,h} - s^{n-1}_{i,h}, w_h\rangle + \Delta t^n \langle \nabla \cdot \mathbf{q^n_{i,h}}, w_h\rangle &= -\Delta t^n \langle \Theta^n_h R^n_i, w_h\rangle ,\\ \langle \mathbf{D_i}^{-1}\mathbf{q^n_{i,h}}, \mathbf{v}_h\rangle - \langle \mathbf{D_i}^{-1}\mathbf{q^n_h} c^n_{i,h}, \mathbf{v}_h\rangle - \langle c^n_{i,h}, \nabla\cdot \mathbf{v}_h\rangle &= 0\end{aligned} \tag{6}$$

for all $w_h \in W_h$ and $\mathbf{v_h} \in V_h$.

Unfortunately, the resulting nonlinear algebraic system (6) leads after linearization to a linear system of equations with an indefinite matrix such that standard iterative solvers cannot be applied. To overcome this difficulty, we use a hybridization technique; cf. [5]. Its basic idea is to relax firstly the continuity constraint of the normal components of the fluxes over interelement faces that is implied by $\mathbf{v} \in H(\mathrm{div};\Omega)$; cf. [5]. The continuity constraint is then ensured by means of an additional variational equation involving Lagrange multipliers. Precisely, the space V_h is replaced by $\tilde{V}_h := \{\mathbf{q} \in L^2(\Omega) | \mathbf{q}_{|T} = \mathbf{a} + b\mathbf{x}, \mathbf{a} \in \mathbb{R}^d, b \in \mathbb{R} \text{ for all } T \in \mathcal{T}_h\}$. The discrete space for the Lagrange multiplier is defined by $\Lambda_{h,0} = \{\lambda \in L^2(\mathcal{S}_h) \mid \lambda_{|E} \in P^0(E)\ \forall E \in \mathcal{S}_h \text{ and } \lambda_{|E} = 0\ \forall E \in \mathcal{S}^D_h\}$. The fully-discrete mixed hybrid variational formulation of the overall system (2)–(4) then reads as follows: *For* $n = 1,\dots N$, $i \in \{1,\dots,N_S\}$ *and* $j \in \{N_S+1,\dots,N_S+M\}$ *find* $(c^n_{i,h}, s^n_{i,h}, \lambda^n_{i,h}, \mathbf{q^n_{i,h}}) \in W_h \times W_h \times \Lambda_{h,0} \times \tilde{V}_h$ *and* $c^n_{j,h} \in W_h$ *such that there holds*

$$\begin{aligned}\langle \Theta^n_h c^n_{i,h} - \Theta^{n-1}_h c^{n-1}_{i,h}, w_h\rangle + \rho_b \langle s^n_{i,h} - s^{n-1}_{i,h}, w_h\rangle + \Delta t^n \langle \nabla \cdot \mathbf{q^n_{i,h}}, w_h\rangle &= -\Delta t^n \langle \Theta^n_h R^n_i, w_h\rangle ,\\ \langle \mathbf{D_i}^{-1}\mathbf{q^n_{i,h}}, \mathbf{v}_h\rangle - \langle \mathbf{D_i}^{-1}\mathbf{q^n_h} c^n_{i,h}, \mathbf{v}_h\rangle - \langle c^n_{i,h}, \nabla\cdot \mathbf{v}_h\rangle + \sum_{T\in\mathcal{T}_h} \langle \lambda^n_{i,h}, \mathbf{v}_h \cdot \mathbf{n}\rangle_{\partial T} &= 0, \\ \sum_{T\in\mathcal{T}_h} \langle \mu_h, \mathbf{q^n_{i,h}} \cdot \mathbf{n}\rangle_{\partial T} &= 0\end{aligned} \tag{7}$$

for all $w_h \in W_h$, $\mathbf{v}_h \in \tilde{V}_h$, $\mu_h \in \Lambda_{h,0}$ and

$$\langle s^n_{i,h} - s^{n-1}_{i,h}, w_h\rangle = \Delta t^n \langle k_i(\phi(c^n_{i,h}) - s^n_{i,h}), w_h\rangle , \tag{8}$$

$$\langle c^n_{j,h} - c^{n-1}_{j,h}, w_h\rangle + \Delta t^n \langle k_{dj} c^n_{i,h}, w_h\rangle = \Delta t^n \langle R^n_j, w_h\rangle \tag{9}$$

for all $w_h \in W_h$. In (7), $\mathbf{n}$ denotes the outer normal.

The hybridization increases the complexity of the nonlinear systems. But we can eliminate now internal degrees of freedom. Moreover, the Lagrange multipliers can be used to construct a second order accurate approximation of the primary variables; cf. [5]. This approximation can then also be used to obtain a reliable a posteriori error indicator without almost any computational extra costs; cf. [3].

Let now Θ^n_T denote the water content on the element T, $c^n_{i,T}$ denote the (constant) component of the concentration of species i on the element T, $\{q^n_{i,TS}\}_{S\subset T}$, $\{Q^n_{TS}\}_{S\subset T}$ the components of the flux of species i and water, respectively, in the local Raviart-Thomas space basis $\{\mathbf{w}_{TS}\}_{S\subset T}$ (cf. [5, 3, 8]), $B_{i,TSS'} := \int_T (\mathbf{D_i}^{-1}\mathbf{w}_{TS'}) \cdot$

$\mathbf{w}_{TS}\,d\mathbf{x}$, and $\lambda_{i,S}^n$ be the constant Lagrange multiplier on the face S. From (8) we get that

$$s_{i,T}^n - s_{i,T}^{n-1} = \frac{\Delta t^n k_i}{1+\Delta t^n k_i}\phi(c_{i,T}^n) - \frac{\Delta t^n k_i}{1+\Delta t^n k_i} s_{i,T}^{n-1}. \tag{10}$$

Using (10), we obtain from (7)–(9) the following system of nonlinear equations:

Mass conservation equation for the mobile species $i \in \{1,\ldots,N_S\}$:

$$\begin{aligned} \Theta_T^n c_{i,T}^n - \Theta_T^{n-1} c_{i,T}^{n-1} + \rho_b \frac{\Delta t^n k_i}{1+\Delta t^n k_i}\phi(c_{i,T}^n) - \rho_b \frac{\Delta t^n k_i}{1+\Delta t^n k_i} s_{i,T}^{n-1} & \\ + \frac{\Delta t^n}{|T|}\sum_{S\subset T} q_{i,TS}^n = -\Delta t^n \Theta_T^n R_i^n \qquad \forall\, T \in \mathcal{T}_h, & \qquad (11) \\ s_{i,h}^n - s_{i,h}^{n-1} = \frac{\Delta t^n k_i}{1+\Delta t^n k_i}\phi(c_{i,h}^n) - \frac{\Delta t^n k_i}{1+\Delta t^n k_i} s_{i,h}^{n-1} \qquad \forall\, T \in \mathcal{T}_h. & \end{aligned}$$

Equations for the flux of the mobile species $i \in \{1,\ldots,N_S\}$:

$$\sum_{S'\subset T} B_{i,TSS'} q_{i,TS'}^n = \sum_{S'\subset T} B_{i,TSS'} Q_{TS'}^n c_{i,T}^n + c_{i,T}^n - \lambda_{i,S}^n \qquad \forall\, T \in \mathcal{T}_h, S \subset T. \tag{12}$$

Continuity of the fluxes of the mobile species $i \in \{1,\ldots,N_S\}$ **over faces**:

$$\sum_{T\supset S} q_{i,TS}^n = 0 \quad \forall\, S \in \mathcal{S}_h^I. \tag{13}$$

The local equations for the immobile species, completing the systems, read as

$$c_{i,T}^n + \Delta t^n k_{di} c_{i,T}^n = \Delta t^n R_i^n + c_{i,T}^{n-1} \quad \forall i \in \{N_S+1,\ldots,N_S+M\}. \tag{14}$$

Thus we have to solve locally $N_S + M$ coupled equations for the concentrations, $(d+1)N_S$ for the fluxes and N_S global equations corresponding to the continuity of the fluxes. Therefore we eliminate internal degrees of freedom, also known as static condensation; cf. [8]. Briefly, within each time step, we use (12) to locally eliminate the fluxes. Then we solve on each element the nonlinear system (11), (14) for the concentrations by Newton's method. Afterwards, we compute the fluxes and solve the global system of fully coupled, nonlinear equations (13) for the Lagrange multipliers by Newton's method. By the thus determined multipliers we again solve the local system (11), (14) for the concentrations. We iterate this procedure until a given tolerance is reached. In detail we get from (12) that

$$q_{i,TS}^n = Q_{TS}^n c_{i,T}^n + \sum_{S'\subset T} B_{i,TSS'}^{-1}(c_{i,T}^n - \lambda_{i,S'}^n) \qquad \forall T \in \mathcal{T}_h, S \subset T, \tag{15}$$

where $\mathbf{B_i^{-1}} = \{B_{i,TSS'}^{-1}\}_{S,S'\subset T}$ is the inverse matrix of $\mathbf{B_i} := \{B_{i,TSS'}\}_{S,S'\subset T}$. Let now $S_{i,T} := \sum_{S,S'\subset T} B_{i,TSS'}^{-1}$ and $b_{i,S} = \sum_{S'\subset T} B_{i,TS'S}^{-1}$. Eliminating the flux variable yields on each element T for each species concentration $c_{i,T}^n$, $i \in \{1,\ldots,N_S\}$

$$\Theta_T^n c_{i,T}^n + \rho_b \frac{\Delta t^n k_i}{1+\Delta t^n k_i}\phi(c_{i,T}^n) + \frac{\Delta t^n}{|T|}\sum_{S\subset T} Q_{TS}^n c_{i,T}^n + \frac{\Delta t^n}{|T|} S_{i,T} c_{i,T}^n$$
$$+\Delta t^n \Theta_T^n R_i^n = \frac{\Delta t^n}{|T|}\sum_{S\subset T} b_{i,S}\lambda_{i,S}^n + \Theta_T^{n-1} c_{i,T}^{n-1} + \rho_b \frac{\Delta t^n k_i}{1+\Delta t^n k_i} s_{i,T}^{n-1}. \quad (16)$$

This nonlinear system of $N_S + M$ equations given by (14) and (16) for the $N_S + M$ unknowns $\{c_{i,T}^n\}_{i\in\{1,\dots,N_S+M\}}$ is solved by Newton's method. The residual and the local Jacobian matrix at iteration step k are computed by means of

$$\text{residual}^{k+1}[i] = \Theta_T^n c_{i,T}^{n,k} + \rho_b \frac{\Delta t^n k_i}{1+\Delta t^n k_i}\phi(c_{i,T}^{n,k}) + \frac{\Delta t^n}{|T|}\sum_{S\subset T} Q_{TS}^n c_{i,T}^{n,k}$$
$$+\frac{\Delta t^n}{|T|} S_{i,T} c_{i,T}^{n,k} + \Delta t^n \Theta_T^n R_i^{n,k} - \frac{\Delta t^n}{|T|}\sum_{S\subset T} b_{i,S}\lambda_{i,S}^n - \Theta_T^{n-1} c_{i,T}^{n-1} - \rho_b \frac{\Delta t^n k_i}{1+\Delta t^n k_i} s_{i,T}^{n-1},$$
$$\text{residual}^{k+1}[j] = (1+\Delta t^n k_{dj}) c_{j,T}^{n,k} - \Delta t^n R_j^{n,k} - c_{j,T}^{n-1}$$

for all $i \in \{1,\dots,N_S\}$, $j \in \{N_S+1,\dots,N_S+M\}$ and

$$\text{jac\_loc}^{k+1}[i][j] = \left(\Theta_T^n + \rho_b \frac{\Delta t^n k_i}{1+\Delta t^n k_i}\phi'(c_{i,T}^{n,k}) + \frac{\Delta t^n}{|T|}\sum_{S\subset T} Q_{TS}^n + \frac{\Delta t^n}{|T|} S_{i,T}\right)\delta_{ij}$$
$$+\Delta t^n \Theta_T^n \frac{\partial R_i^{n,k}}{\partial c_j} \qquad \forall (i,j) \in \{1,\dots,N_S\} \times \{1,\dots,N_S+M\},$$
$$\text{jac\_loc}^{k+1}[i][j] = \left[(1+\Delta t^n k_{di}) + \Delta t^n \frac{R_i^{n,k}}{c_{i_{\max}}}\right]\delta_{ij} - \Delta t^n \left(1 - \frac{c_{i,T}^{n,k}}{c_{i_{\max}}}\right)\frac{\partial R_i^{n,k}}{\partial c_j}$$
$$\forall (i,j) \in \{N_S+1,\dots,N_S+M\} \times \{1,\dots,N_S+M\}.$$

Here δ_{ij} denotes the Kronecker symbol. Before calculating the overall global system we still have to recompute the concentration of the adsorbed substances $s_{i,T}^n$ by using (10). Finally, the global system reads

$$\sum_{T\supset S}\{Q_{TS}c_{i,T}^n + \sum_{S'\subset T} B_{i,TSS'}^{-1}(c_{i,T}^n - \lambda_{i,S'}^n)\} = 0 \quad \forall S \in \mathcal{S}_h^I, \quad i \in \{1,\dots,N_S\}. \quad (17)$$

The system (17) is solved by Newton's method. The difficulty here is that we have to determine the derivatives of the concentrations with respect to the Lagrange multipliers. We use (14) to obtain that $\partial c_{i,T}^n/\partial\lambda_{j,S} = (\Delta t^n/|T|)\mathbf{inv\_loc}[i][j] b_{j,S}$, where $\mathbf{inv\_loc} \in \mathbb{R}^{N_S+M,N_S+M}$ denotes the inverse of the local Jacobian matrix **jac_loc** on element T. The local contributions to the global Jacobian matrix are then given by

$$\text{jac}_{i,j}^{S,S'} = \sum_{T\supset S\cap T\supset S'}\left(Q_{TS}\frac{\partial c_{i,T}^n}{\partial\lambda_{j,S'}} + \sum_{S''\subset T} B_{i,TSS''}\left(\frac{\partial c_{i,T}^n}{\partial\lambda_{j,S'}} - \frac{\partial\lambda_{i,S''}}{\partial\lambda_{j,S'}}\right)\right)$$
$$= \sum_{T\supset S\cap T\supset S'}\left(Q_{TS}\frac{\partial c_{i,T}^n}{\partial\lambda_{j,S'}} - B_{i,TSS'}^{-1}\delta_{ij} + \sum_{S''\subset T} B_{i,TSS''}^{-1}\frac{\partial c_{i,T}^n}{\partial\lambda_{j,S'}}\right) \quad (18)$$

with $i,j \in \{1,\dots,N_S\}$ and $S,S' \in \{1,\dots,d+1\}$.

The scheme was implemented in the *UG* toolbox; cf. [2]. The linear system of the Newton iteration is solved by a multigrid method.

Remark 1. A new mixed hybrid FE scheme is obtained by using the Lagrange multipliers, instead of the piecewise constant concentrations, for discretizing the convective term in (1). Instead of (15), this yields

$$\sum_{S' \subset T} B_{i,TSS'} q^n_{i,TS'} = \sum_{S' \subset T} B_{i,TSS'} Q^n_{TS'} \lambda^n_{i,S} + c^n_{i,T} - \lambda^n_{i,S} \qquad \forall T \in \mathcal{T}_h, S \subset T. \qquad (19)$$

The analysis of this scheme is an ongoing work. Numerical tests that are not shown here have indicated an increase in stability for convection dominated problems.

Remark 2. To obtain a discrete maximum principle, the mass matrix should be computed by a quadrature formula (cf., e.g. [6]), which has also been implemented.

4 Numerical Example

The described mixed finite element approximation of the reactive system (2)–(4) is compared numerically to standard linear and quadratic conforming finite element approximations. This is done for an artificial example of two mixing, reacting substances; cf. [4]. In particular, we analyse whether accuracy is gained by the algorithmically more complex but locally mass conservative mixed approach.

Let $\Omega = (0,2) \times (0,3)$, $\mathbf{q} = (0,-1)^\top$, $\Theta = 1.0$, $\mathbf{D}_D = \mathbf{D}_A = 0.1 \cdot \mathbf{I}$ with the identity matrix $\mathbf{I}$, $\alpha_D = 1.0$, $\alpha_A = 2.0$. The final fime is $T = 1$. We solve the model system

$$\partial_t(\Theta c_i) - \nabla \cdot (\mathbf{D_i} \nabla c_i - \mathbf{q} c_i) = -\Theta R_i + f_i\,, \quad R_i = \alpha_i C_D C_A^2\,, \quad i = D,A\,. \qquad (20)$$

The rate R_i implies that degradation of the species C_D and C_A only occurs where both species are available. Numerical diffusion may lead to an artifical mixing of the substances and, thereby, to an overestimation of the degradation process; cf. [4]. For our comparative study of the finite element approaches we prescribe the analytical solution $c_D(x,y,t) = x(2.0-x)y^3\exp(-0.1t)/27$ and $c_A(x,y,t) = (x-1.0)^2y^2\exp(-0.1t)/9$. We determine the corresponding right-hand side functions f_D and f_A by (20), then we compute the finite element approximations of c_D and c_A for the thus obtained right-hand sides and, finally, we compare the computed approximations to the analytical solution.

Table 1 contains the calculated L^2-errors of the finite element approximations of c_D and c_A for a sequence of successively refined meshes and the time step size $\Delta t = 0.001$. Here, C_i^{MFEM} is the nonconforming Crouzeix-Raviart approximation of c_i whose degrees of freedom are the Lagrange multipliers of the hybrid mixed finite element approach; cf. [5]. The scheme based on (19) is used. In the quadratic conforming case the temporal discretization error dominates the spatial error after the first refinement step such that no further error reduction is observed. For smaller

h	$\|c_D - C_D^{\mathrm{MFEM}}\|$	$\|c_D - C_D^{P_1}\|$	$\|c_D - C_D^{P_2}\|$	$\|c_A - C_A^{\mathrm{MFEM}}\|$	$\|c_A - C_A^{P_1}\|$	$\|c_A - C_A^{P_2}\|$
2.00e-1	3.30e-3	2.70e-3	4.46e-5	2.67e-3	2.90e-3	6.48e-5
1.00e-1	8.36e-4	6.61e-4	4.25e-6	6.85e-4	7.09e-4	5.46e-6
5.00e-2	1.90e-4	1.63e-5	1.99e-6	1.56e-4	1.76e-4	6.95e-7
2.50e-2	4.73e-5	3.96e-5	1.95e-6	3.96e-5	4.42e-5	6.79e-7
1.25e-2	1.11e-5	8.85e-6	1.95e-6	1.01e-5	1.14e-5	6.85e-7

Table 1 Calculated L^2-errors for the mixed (C_i^{MFEM}), linear conforming ($C_i^{P_1}$) and quadratic conforming ($C_i^{P_2}$) finite element approximation of c_i, $i = D, A$ with C_i^{MFEM} denoting the nonconforming Crouzeix-Raviart approximation with the Lagrange multipliers as degrees of freedom.

time step sizes convergence of order three is in fact obtained which is not shown here. In Table 1 we do not observe any significant differences between the mixed and the linear conforming approach. The calculated errors are of almost equal size. Thus, in the considered test problem of two mixing reactive species loss of mass and numerical diffusion, the weak points of the linear conforming approach, is no object. The properties of the mixed approach in comparison to conforming approaches will be further analysed in forthcoming studies and for more refined and sophisticated test problems.

References

1. Arbogast, T., Wheeler, M. F., Zhang, N. Y.: A nonlinear mixed finite element method for a degenerate parabolic equation arising in flow in porous media. SIAM J. Numer. Anal. **33**, 1669–1687 (1996)
2. Bastian, P., Birken, K., Johanssen, K., Lang, S., Neuß, N., Rentz-Reichert, H., Wieners, C.: UG–a flexible toolbox for solving partial differential equations. Comput. Visualiz. Sci. **1**,27–40 (1997)
3. Bause, M., Knabner, P.: Computation of variably saturated subsurface flow by adaptive mixed hybrid finite element methods. Adv. Water Resour. **27**, 565–581 (2004)
4. Bause, M., Knabner, P.: Numerical simulation of contaminant biodegradation by higher order methods and adaptive time stepping. Comp. Visualiz. Sci. **7**, 61–78 (2004)
5. Brezzi, F., Fortin, M.: Mixed and Hybrid Finite Element Methods, Springer Verlag, New York (1991)
6. Micheletti, S., Sacco, R., Saleri, F.: On Some Mixed Finite Element Methods with Numerical Integration. SIAM J. Sci. Comput. **23**(1), 245–270 (2001)
7. Mose, R., Siegel, P., Ackerer, P., Chavent,G.: Application of the mixed hybrid finite element approximation in a groundwater flow model: Luxury or necessity?. Water Resources Research **30**, 3001-3012 (1994)
8. Radu, F. A.: Mixed finite element discretization of Richards' equation: error analysis and application to realistic infiltration problems. PhD Thesis, University of Erlangen–Nürnberg (2004)
9. Radu, F. A., Pop, I. S., Knabner, P.: Order of convergence estimates for an Euler implicit, mixed finite element discretization of Richards' equation. SIAM Journal of Numerical Analysis **42**(4), 1452–1478 (2004)
10. Schneid, E., Knabner. P., Radu, F. A.: A priori error estimates for a mixed finite element discretization of the Richards' equation. Num. Math. **98**(2), 353–370 (2004)

Applying Local Projection Stabilization to inf-sup Stable Elements

G. Rapin, G. Lube, and J. Löwe

Abstract In this paper a priori estimates for finite element discretizations of Oseen type problems using local projection stabilization are presented. In contrast to existing papers [5, 1] inf-sup stable velocity-pressure pairs are used. Asymptotic parameter choices are derived and verified by numerical experiments.

1 Introduction

The computation of the non-stationary, incompressible Navier-Stokes equations

$$\partial_t \mathbf{u} - \nu \triangle \mathbf{u} + (\mathbf{u} \cdot \nabla)\mathbf{u} + \nabla p = \tilde{\mathbf{f}}, \qquad \nabla \cdot \mathbf{u} = 0 \qquad \text{in } \Omega$$

for velocity $\mathbf{u}$ and pressure p is still a challenge. Here, $\Omega \subset \mathbb{R}^d$, $d = 2,3$, is a bounded polyhedral domain and $\tilde{\mathbf{f}}$ is a given source term. A common approach is to semi-discretize in time first using any A-stable implicit scheme, cf. [3]. Then, the resulting stationary problems can be solved by a fixed-point or Newton-type scheme in each time step. We end up with auxiliary Oseen problems of the following type:

$$-\nu \triangle \mathbf{u} + (\mathbf{b} \cdot \nabla)\mathbf{u} + \mathscr{C}\mathbf{u} + \nabla p = \mathbf{f}, \qquad \nabla \cdot \mathbf{u} = 0 \qquad \text{in } \Omega \tag{1}$$

with new right hand side $\mathbf{f} \in [L^2(\Omega)]^d$ and coefficients $\mathscr{C} \in [L^2(\Omega)]^{d\times d}$, $\mathbf{b} \in [L^\infty(\Omega) \cap H^1(\Omega)]^d$ with $\nabla \cdot \mathbf{b} = 0$. Assuming sufficient small time steps we can assume that the term $\mathscr{C}\mathbf{u}$ is positive, i.e. there exists a $c_0 > 0$ such that $y^t \mathscr{C} y \geq c_0 y^t y$ for all $y \in \mathbb{R}^d$. Moreover, let $\nu \in L^\infty(\Omega)$ be strictly positive, i.e. $\nu_0 := \inf_{x\in\Omega} \nu(x) > 0$. In contrast to most papers about the Oseen problem we consider the case of a matrix-valued coefficient $\mathscr{C}$ since the Newton linearization contains terms like $(\mathbf{u} \cdot \nabla)\mathbf{b}$.

G. Rapin, G. Lube, and J. Löwe
Math. Dep., University of Göttingen, Germany; e-mail: grapin/lube@math.uni-goettingen.de

It is well known that basic Galerkin approximations of (1) may suffer from two problems: violation of the discrete inf-sup (or Babuška-Brezzi) stability condition and dominating advection, i.e. $\nu \ll \|\mathbf{b}\|_{[L^\infty(\Omega)]^d}$. The traditional way to deal with both aspects in a common framework is the combination of the *Streamline-Upwind/Petrov-Galerkin method (SUPG)* and the *Pressure-Stabilization/ Petrov-Galerkin method (PSPG)*, cf. [4]. The residual based methods have been quite popular in the last twenty years. Unfortunately they have several drawbacks:

- Due to the additional stabilization terms the assembling is expensive. Especially, the discretization of the Laplacian is costly for higher order elements.
- The additional coupling of velocity and pressure destroys the saddle point structure. Thus, standard saddle point preconditioners cannot be applied.

Therefore, other stabilization techniques have become popular. Almost all approaches can be included in the framework of variational multiscale (VMS) methods [2]. The key idea of VMS methods consists in a separating of scales. Mainly, the underlying function spaces are split into different scales: large scales, small scales and unresolved scales. The influence of unresolved scales on other scales has to be modeled. In local projection methods [1, 5] the influence of the unresolved scales on the fine scales is modeled by an artificial diffusion term for the fine scales.

Here, we extend the papers [5, 1] to inf-sup stable elements. In Section 2 the stabilized scheme is introduced. Then, we outline the analysis of [6] and describe the choice of the stabilization parameters (Section 3). Finally, the theoretical results are validated by numerical experiments and some conclusions are given.

2 Stabilization by Local Projection

Throughout this paper standard notations for Lebesgue and Sobolev spaces are used. The L^2 inner product resp. norm in a domain G is denoted by $(\cdot,\cdot)_G$ resp. $\|\cdot\|_G$. If $G=\Omega$ we simply write $(\cdot,\cdot)$ resp. $\|\cdot\|$.

The weak formulation for the Oseen problem (1) with homogeneous Dirichlet data is given as follows: Find $U=(\mathbf{u},p)\in\mathbf{V}\times Q:=[H_0^1(\Omega)]^d\times L_0^2(\Omega)$, such that

$$A(U,W):=(\nu\nabla\mathbf{u},\nabla\mathbf{w})+((\mathbf{b}\cdot\nabla)\mathbf{u}+\mathscr{C}\mathbf{u},\mathbf{w})-(\nabla\cdot\mathbf{w},p)+(\nabla\cdot\mathbf{u},q)=(\mathbf{f},\mathbf{w}) \quad (2)$$

for all $W=(\mathbf{w},q)\in\mathbf{V}\times Q$. Let $\mathscr{T}_h$ denote a shape regular mesh of simplices or hexahedra. For each $T\in\mathscr{T}_h$ and reference element $\hat{T}$ there exists an affine mapping $F_T:\hat{T}\to T$. The finite element spaces $\mathbf{V}_{h,k_u}\subset\mathbf{V}$ and $Q_{h,k_p}\subset Q$ are based on

$$P_{k,\mathscr{T}_h}:=\{v_h\in L^2(\Omega)\mid v_h|_T\circ F_T\in\mathbb{P}_k(\hat{T})\ ,T\in\mathscr{T}_h\},$$
$$Q_{k,\mathscr{T}_h}:=\{v_h\in L^2(\Omega)\mid v_h|_T\circ F_T\in\mathbb{Q}_k(\hat{T})\ ,T\in\mathscr{T}_h\}.$$

$\mathbb{P}_k(\hat{T})$ is the space of complete polynomials of degree k on $\hat{T}$ and $\mathbb{Q}_k(\hat{T})$ is the space of polynomials on $\hat{T}$ whose degree does not exceed k in each coordinate direction.

Now, we define the finite element space of the velocity by $\mathbf{V}_{h,k_u} = [Q_{k_u,\mathscr{T}_h}]^d \cap \mathbf{V}$ or $\mathbf{V}_{h,k_u} = [P_{k_u,\mathscr{T}_h}]^d \cap \mathbf{V}$. The scalar components of $\mathbf{V}_{h,k_u}$ will be denoted as Y_{h,k_u}. The discrete space Q_{h,k_p} for the pressure is defined as $Q_{h,k_p} = Q_{k_p,\mathscr{T}_h}$ resp. $Q_{h,k_p} = P_{k_p,\mathscr{T}_h}$. Then, the Scott-Zhang quasi-interpolant operator I^u_{h,k_u} satisfies for $v \in H^1_0(\Omega) \cap H^t(\Omega)$, $t > \frac{1}{2}$ with $v|_{\omega_T} \in H^r(\omega_T)$, $r \geq t$, on the patches $\omega_T := \bigcup_{\overline{T'} \cap \overline{T} \neq \emptyset} T'$

$$\|v - I^u_{h,k_u} v\|_{m,T} \lesssim h_T^{l-m} k_u^{m-l} \|v\|_{l,\omega_T}, \quad 0 \leq m \leq l = \min(k_u + 1, r). \tag{3}$$

$a \lesssim b$ means $a \leq Cb$ with constant C independent of important sizes like mesh size or polynomial order. We assume the discrete Babuška-Brezzi condition

$$\exists \beta_0 > 0 \mid \qquad \inf_{q_h \in Q_{h,k_p}} \sup_{\mathbf{v}_h \in \mathbf{V}_{h,k_u}} \frac{(q_h, \nabla \cdot \mathbf{v}_h)}{\|q_h\|_0 |\mathbf{v}_h|_1} \geq \beta_0. \tag{A1}$$

The idea of LPS-methods is to split the discrete function spaces into small and large scales. Then, a diffusion-type term is added acting solely on the small scales. Here, the large scales are defined on a coarse mesh $\mathscr{M}_h = \{M_i\}_{i \in I}$.

The mesh $\mathscr{M}_h = \{M_i\}_{i \in I}$ is constructed by coarsening $\mathscr{T}_h$ in the sense of [5]. Each macro element $M \in \mathscr{M}_h$ is the union of one or more neighboring cells $T \in \mathscr{T}_h$. We assume that the interior cells are of the same size than the macro cells:

$$\exists C > 0 \mid \quad h_M \leq C h_T, \quad \forall T \in \mathscr{T}_h, M \in \mathscr{M}_h \text{ with } T \subset M. \tag{4}$$

Following the approach in [5] we define the discrete space D^u_h for the velocity as a discontinuous finite element space defined on the macro partition $\mathscr{M}_h$. The restriction on a macro-element $M \in \mathscr{M}_h$ is denoted by $D^u_h(M) := \{v_h|_M \mid v_h \in D^u_h\}$.

The next ingredient is a local projection $\pi^u_M : L^2(M) \to D^u_h(M)$ which defines the global projection $\pi^u_h : L^2(\Omega) \to D^u_h$ by $(\pi_h w)|_M := \pi_M(w|_M)$ for all $M \in \mathscr{M}_h$. The associated *fluctuation operator* $\kappa^u_h : L^2(\Omega) \to L^2(\Omega)$ is defined by $\kappa^u_h := id - \pi^u_h$.

These operators are applied to vector-valued functions in a component-wise manner using the same notation, e.g. $\boldsymbol{\pi}^u_h : [L^2(\Omega)]^d \to [D^u_h]^d$ and $\boldsymbol{\kappa}^u_h : [L^2(\Omega)]^d \to [L^2(\Omega)]^d$. Please note that the theory covers the case $\mathscr{M}_h = \mathscr{T}_h$. Analogously a discrete space D^p_h and a fluctuation operator κ^p_h can be defined.

Now, we define the stabilized formulation: find $U_h = (\mathbf{u}_h, p_h) \in \mathbf{V}_{h,k_u} \times Q_{h,k_p}$ s.t.

$$A(U_h, V_h) + S_h(U_h, V_h) = (\mathbf{f}, \mathbf{v}_h), \qquad \forall V_h = (\mathbf{v}_h, q_h) \in \mathbf{V}_{h,k_u} \times Q_{h,k_p} \tag{5}$$

where the additional stabilization term is given by

$$S_h(U_h, V_h) := \sum_{M \in \mathscr{M}_h} \Big[\tau_M(\boldsymbol{\kappa}^u_h((\mathbf{b} \cdot \nabla)\mathbf{u}_h), \boldsymbol{\kappa}^u_h((\mathbf{b} \cdot \nabla)\mathbf{v}_h))_M$$
$$+ \mu_M(\kappa^p_h(\nabla \cdot \mathbf{u}_h), \kappa^p_h(\nabla \cdot \mathbf{v}_h))_M + \alpha_M(\boldsymbol{\kappa}^u_h(\nabla p_h), \boldsymbol{\kappa}^u_h(\nabla q_h))_M \Big]. \tag{6}$$

τ_M, μ_M and α_M will be determined later with the help of an a priori estimation. The stabilization $S_h(\cdot,\cdot)$ acts solely on fine scales. Other variants can be found in [5, 1].

Since the additional stabilization terms are not consistent, i.e.

$$A(U-U_h,V_h)=S_h(U_h,V_h), \qquad \forall V_h \in \mathbf{V}_h \times Q_h \tag{7}$$

we have to ensure that the consistency error can be controlled. Therefore, we have to require that the space D_h^u is large enough. Precisely, we assume for $0 \leq l \leq k_u$

$$\exists C_\kappa > 0 \mid \qquad \|\kappa_h^u q\| \leq C_\kappa h_M^l k_u^{-l} |q|_{l,M}, \quad \forall q \in H^l(M),\ \forall M \in \mathscr{M}_h. \tag{A2}$$

Due to the consistency of the stabilization term $\mu_M(\kappa_h^p(\nabla\cdot\mathbf{u}_h), \kappa_h^p(\nabla\cdot\mathbf{v}_h))_M$ a condition like (A2) is not needed for D_h^p.

3 Analysis and Choice of the Parameters

In this section we outline the analysis of [6]. We show stability of the scheme and give an a priori estimation which is used to determine the stabilization parameters.

We start with the stability of the scheme. We define the mesh-dependent norm

$$|||V||| := \left(|[V]|^2 + \delta\|q\|_0^2\right)^{\frac{1}{2}}, \quad |[V]|^2 := \|\nu^{\frac{1}{2}}\nabla\mathbf{v}\|_0^2 + c_0\|\mathbf{v}\|_0^2 + S_h(V,V)$$

for $V=(\mathbf{v},q) \in \mathbf{V}\times Q$ and suitable $\delta > 0$. The following stability result is valid for arbitrary $\mu_M, \tau_M, \alpha_M \geq 0$ and shows the existence and uniqueness of solutions.

Lemma 1. *([6], Lemma 2) There holds*

$$\inf_{V_h=(\mathbf{v}_h,q_h)\in\mathbf{V}_{h,k_u}\times Q_{h,k_p}} \ \sup_{W_h=(\mathbf{w}_h,r_h)\in\mathbf{V}_{h,k_u}\times Q_{h,k_p}} \frac{(A+S_h)(V_h,W_h)}{|||V_h|||\,|||W_h|||} \geq \gamma > 0. \tag{8}$$

Defining $\nu_{max} := \|\nu\|_{L^\infty(\Omega)}$ the constants are given by

$$\gamma \sim 1/(1+\frac{1}{\beta_0}(\nu_{max}+c_0+1)^{\frac{1}{2}}), \qquad \delta \sim \beta_0^2/(\min\{\frac{1}{c_0},\frac{1}{\nu_0}\}+\nu_{max}+1).$$

The constants γ, δ are independent of the mesh size, but we loose control of the pressure in the case of small coefficients c_0 and ν_0.

The a priori estimate is based on a special interpolant $j_h^u : H^1(\Omega) \to Y_{h,k_u}$ for the velocity, introduced in [5], such that the error $w - j_h^u w$ is L^2-orthogonal to D_h^u for all $w \in H_0^1(\Omega)$. In order to conserve the standard approximation properties, we assume for $Y_h(M) := \{v_h|_M \mid v_h \in Y_{h,k_u}, v_h = 0 \text{ on } \Omega\setminus M\}$ the existence of a constant β_u s.t.

$$\inf_{q_h\in D_h^u} \sup_{v_h\in Y_h(M)} \frac{(v_h,q_h)_M}{\|v_h\|_{0,M}\|q_h\|_{0,M}} \geq \beta_u > 0. \tag{A3}$$

Later on, we will present several function spaces D_h^u satisfying (A3).

Using the inverse inequality

$$\exists \mu_{inv} \mid \quad |\mathbf{v}|_{1,T} \leq \mu_{inv} k_u^2 h_T^{-1} \|\mathbf{v}\|_{0,T}, \quad \forall T \in \mathscr{T}_h, \forall \mathbf{v}_h \in \mathbf{V}_{h,k_u} \tag{9}$$

one can prove ([6], Lemma 3)

Lemma 2. *Assume assumption (A3). There exists an operator* $\mathbf{j}_h^u : \mathbf{V} \to \mathbf{V}_{h,k_u}$ *with*

$$(\mathbf{v} - \mathbf{j}_h^u \mathbf{v}, \mathbf{q}_h) = 0, \qquad \forall \mathbf{q}_h \in [D_h^u]^d, \forall \mathbf{v} \in \mathbf{V}, \tag{10}$$

$$\|\mathbf{v} - \mathbf{j}_h^u \mathbf{v}\|_{0,M} + \frac{h_M}{k_u^2} |\mathbf{v} - \mathbf{j}_h^u \mathbf{v}|_{1,M} \lesssim (1 + \frac{1}{\beta_u}) \frac{h_M^l}{k_u^l} \|\mathbf{v}\|_{l,\omega_M}, \forall \mathbf{v} \in [H^l(\Omega)]^d \cap \mathbf{V} \tag{11}$$

for all $T \in \mathscr{T}_h$ *and* $1 \leq l \leq k_u + 1$. $\omega_M := \bigcup_{T \subset M} \omega_T$ *is a neighborhood of* $M \in \mathscr{M}_h$.

The estimate (11) is optimal with respect to the mesh size h_M, but sub-optimal with respect to the polynomial order. This is caused by the inverse inequality (9).

Using the approximated Galerkin orthogonality (7) and splitting the error into

$$U - U_h = (\mathbf{u} - \mathbf{u}_h, p - p_h) = (\mathbf{u} - \mathbf{j}_h^u \mathbf{u}, p - j_h^p p) + (\mathbf{j}_h^u \mathbf{u} - \mathbf{u}_h, j_h^p p - p_h)$$

the following a priori estimation can be proven ([6], Theorem 1)

Theorem 1. *Let* $U = (\mathbf{u}, p) \in \mathbf{V} \times Q$ *be solution of (2) and* $U_h = (\mathbf{u}_h, p_h) \in \mathbf{V}_{h,k_u} \times Q_{h,k_p}$ *of (5). Let be* $p \in H^{l_p+1}(\Omega)$ *and* $\mathbf{u} \in [H^{l_k+1}(\Omega)]^d, (\mathbf{b} \cdot \nabla)\mathbf{u} \in [H^{l_u}(\Omega)]^d$. *Moreover, assume (A2), (A3) for the coarse space* D_h^u *and*

$$\inf_{q_h \in D_h^p} \sup_{v_h \in Q_{h,k_p}} \frac{(v_h, q_h)_M}{\|v_h\|_{0,M} \|q_h\|_{0,M}} \geq \beta_p \tag{A4}$$

for D_h^p. *Then, there holds for* $1 \leq l_u \leq k_u$ *and* $1 \leq l_p \leq \min\{k_p, k_u\}$

$$\begin{aligned} |||U - U_h|||^2 \lesssim \frac{1}{\gamma} \sum_{M \in \mathscr{M}_h} \Bigg(& \tau_M \frac{h_M^{2l_u}}{k_u^{2l_u}} \|(\mathbf{b} \cdot \nabla)\mathbf{u}\|_{l_u,\omega_M}^2 \\ & + \left(1 + \frac{1}{\beta_u}\right)^2 \frac{h_M^{2l_u}}{k_u^{2l_u-2}} C_M^u \|\mathbf{u}\|_{l_u+1,\omega_M}^2 + \left(1 + \frac{1}{\beta_p}\right)^2 \frac{h_M^{2l_p}}{k_p^{2l_p-2}} C_M^p \|p\|_{l_p+1,\omega_M}^2 \Bigg). \end{aligned} \tag{12}$$

with $c_{max} := \sup_{\mathbf{v} \in (L^2(\Omega))^d} \frac{\|\mathscr{C}\mathbf{v}\|}{\|\mathbf{v}\|}$, $C_M^p := \alpha_M + (\delta + \mu_M^{-1}) h_M^2 / k_p^4$ *and*

$$C_M^u := \nu_{max} + \frac{h_M^2}{k_u^4} (c_0 + \frac{c_{max}^2}{c_0} + \tau_M^{-1}) + \mu_M + \|\mathbf{b}\|_{[L^\infty(M)]^d}^2 \tau_M + \delta^{-1}.$$

Remark 1. In the case of continuous pressure we can improve the constant C_M^u:

$$C_M^u = \nu_{max} + \frac{h_M^2}{k_u^4} (c_0 + \frac{c_{max}^2}{c_0} + \tau_M^{-1}) + \mu_M + \|\mathbf{b}\|_{[L^\infty(M)]^d}^2 \tau_M + \min\{\alpha_M^{-1} \frac{h_M^2}{k_u^4}, \delta^{-1}\}.$$

Thus, the error can even be controlled for small viscosity ν and $\mathscr{C} = 0$.

Let us briefly discuss several choices of the discrete spaces $V_{h,k_u} \times Q_{h,k_p}$ and $D_h^u \times D_h^p$ using simplicial and hexahedral elements. The popular choices Q_k/Q_{k-1}, P_k/P_{k-1} and $Q_k/P_{-(k-1)}$ satisfy the assumptions (A2), (A3), (A4) of Theorem 1, if one chooses $D_h^{u/p} = Q_{k_{u/p}-1,\mathcal{M}_h}/P_{k_{u/p}-1,\mathcal{M}_h}$. The constants $\beta_{u/p}$ scales like $\mathcal{O}(\frac{1}{\sqrt{k_{u/p}}})$ for simplicial elements and like $\mathcal{O}(1)$ for quadriliteral elements, cf. [6].

Now we calibrate the stabilization parameters α_M, τ_M and μ_M with respect to the local mesh size h_M, the polynomial degrees k_u and k_p and problem data. We minimize and balance the terms of the right hand side of the a priori error estimation. First, equilibrating the τ_M-dependent terms in C_M^u yields $\tau_M \sim h_M/(\|\mathbf{b}\|_{[L^\infty(M)]^d} k_u^2)$. Similarly, equilibration of the terms in C_M^u and C_M^p yields for continuous pressure spaces $\mu_M \sim h_M^{l_p-l_u+1}/(k^{l_p-l_u+2})$, $\alpha_M \sim h_M^{l_u-l_p+1}/(k^{l_u-l_p+2})$ where we used $k \sim k_u \sim k_p$. For Taylor-Hood pairs with $k_u = k_p+1$, assume $l_u = l_p+1 = k_u$ and set

$$\alpha_M = \alpha_0 h_M^2/k_u^3, \quad \mu_M = \mu_0/k_u, \quad \tau_M = \tau_0 h_M/(\|\mathbf{b}\|_{[L^\infty(M)]^d} k_u^2). \tag{13}$$

We obtain under the assumptions of Theorem 1

$$|[U-U_h]|^2 \lesssim \sum_{M\in\mathcal{M}} \Bigg(\left(1+\frac{1}{\beta_p}\right)^2 \frac{h_M^{2l_u}}{k_u^{2l+1}} \|p\|_{l_u,\omega_M}^2 + \frac{h_M^{2l_u+1}}{k_u^{2l+2}\|\mathbf{b}\|_{[L^\infty(M)]^d}} \|(\mathbf{b}\cdot\nabla)\mathbf{u}\|_{l,\omega_M}^2$$
$$+\left(1+\frac{1}{\beta_u}\right)^2 \left[\|\nu\|_{L^\infty(M)} + (c_0 + \frac{c_{max}^2}{c_0})\frac{h_M^2}{k_u^4} + \|\mathbf{b}\|_{[L^\infty(M)]^d}\frac{h_M}{k_u^2} + \frac{1}{k_u}\right] \frac{h_M^{2l_u}}{k_u^{2l_u-2}} \|\mathbf{u}\|_{l+1,\omega_M}^2 \Bigg).$$

For Taylor-Hood pairs $\mathbf{V}_{h,k_u} = [Q_{k+1,\mathcal{T}_h}]^d \cap \mathbf{V}$, $Q_{h,k_p} = Q_{k,\mathcal{T}_h} \cap C(\Omega) \cap Q$ we get optimal convergence rates $\mathcal{O}(h_M^{k+1})$ w.r.t. h_M. Due to non-optimal interpolation operators $\mathbf{j}_h^u, j_h^p$ these estimates are presumably not optimal w.r.t. k_u.

4 Numerical Results

In this Section we validate convergence rates of the a priori estimation and control the parameter design. As test problem we consider the unit square $\Omega = (0,1)^2$ with Dirirchlet boundary data. The data is chosen such that $p(x,y) = \sin(\pi x)\cos(\pi y)$, $\mathbf{u}(x,y) = (\sin(\pi x), -\pi y\cos(\pi x))^t$ is the solution of (1) with $\mathbf{b} = \mathbf{u}$ and $\mathscr{C}\mathbf{u} = \mathbf{u}$. We use Taylor-Hood elements Q_k/Q_{k-1} on unstructured, isotropic meshes $\mathcal{T}_h$. The coarse spaces are defined by $D_h^u = Q_{k-1,\mathcal{M}_h}$, $D_h^p = \{0\}$.

First, the parameter design is considered. Because of Remark 1 stabilization with $\alpha_T > 0$ is only necessary for small viscosity and small reaction. And indeed in Figure 1 (left) it can be observed for a typical configuration with a small diffusion coefficient ν that the scheme is very robust with respect to the choice of $\alpha_M = \alpha_0 h_M^2$. Therefore, for all further computations the pressure stabilization is neglected.

Next, we consider the remaining parameters $\tau_M = \tau_0 h_M$ and $\mu_M = \mu_0$. In Figure 1 (right) four contour plots are presented for the error w.r.t. the parameters τ_0 and μ_0.

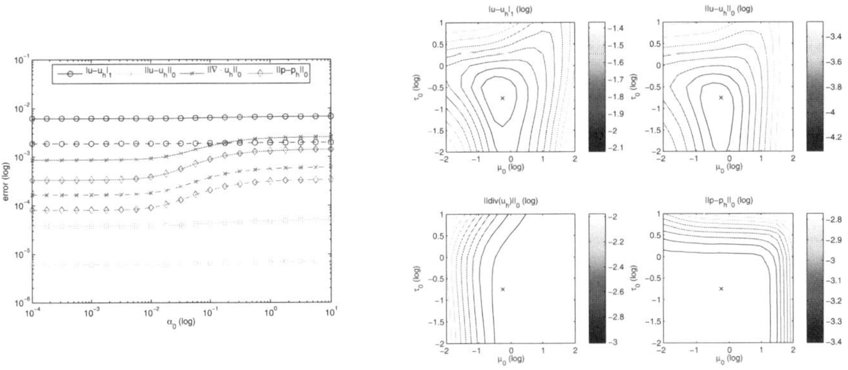

Fig. 1 Parameter choice for Q_2/Q_1 elements: left: Error w.r.t α_0 for $\alpha_M = \alpha_0 h_M^2$, $\nu = 10^{-6}$, $\mu_M = 10^{-1/4}$, $\tau_M = 10^{-3/4} h_M$, $h = \frac{1}{32}, \frac{1}{69}$, right: Contour plot for τ_0, μ_0 ($h = 1/32$, $\nu = 10^{-6}$).

In the upper plots for the velocity errors there exists an optimal value for the parameter pair (τ_0, μ_0). As an approximation we use the cross at $(10^{-1/4}, 10^{-3/4})$.

Our (not presented) numerical experiments show that the scheme is quite robust w.r.t. the choice of τ_M. Therefore we concentrate on the parameter μ_M. In Figure 2 (left) the parameter design w.r.t. the mesh size h is validated ($h \approx \frac{1}{16}, \frac{1}{32}, \frac{1}{69}$). Obviously, the predicted dependency of μ_M on the mesh size is correct.

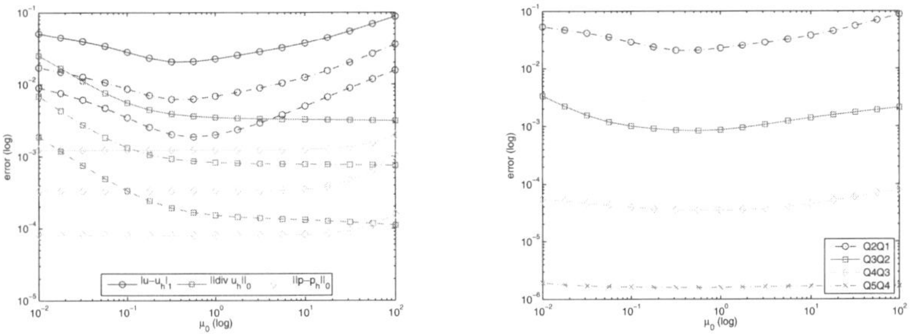

Fig. 2 Errors w.r.t μ_0 for $\nu = 10^{-6}$: left: Q_2/Q_1 ($h = \frac{1}{32}, \frac{1}{69}$), right: Q_k/Q_{k-1} ($h = \frac{1}{16}$).

Next, the dependency of μ_M on the polynomial degree k is considered. In Figure 2 (right) the error of the velocity component is plotted for Q_k/Q_{k-1}-elements, $k = 2, \ldots, 5$. One observes the dependency of μ_0 on the polynomial order k.

The predicted convergence orders can be seen for the advection dominated case in Figure 3. As expected the convergence order of the scheme is 2 for the stabilized norm. The same results hold also for higher order elements, cf. Figure 3 and Table 1. The results are in agreement with the expected convergence rates in Table 1.

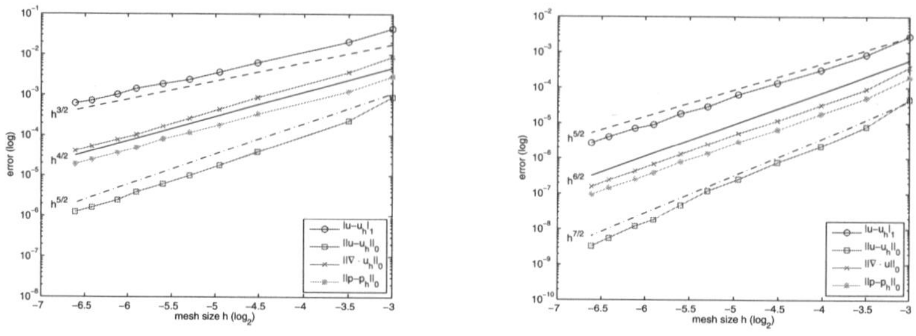

Fig. 3 Convergence for Q_2/Q_1-elements (left) and Q_3/Q_2-elements (right) ($\nu = 10^{-6}$).

ν	1			1e−6		
Element	$\|\|u-u_h\|\|_0$	$\|\|\nabla\cdot u_h\|\|_0$	$\|\|p-p_h\|\|_0$	$\|\|u-u_h\|\|_0$	$\|\|\nabla\cdot u_h\|\|_0$	$\|\|p-p_h\|\|_0$
Q2Q1	3.0201	2.0994	1.9736	2.5645	2.1369	1.9777
Q3Q2	4.1158	3.0887	3.0044	3.6300	3.0271	2.9874
Q4Q3	4.9377	4.0820	4.0830	4.5310	4.0212	3.9910

Table 1 Convergence orders for different Q_k/Q_{k-1}-pairs.

5 Conclusion

In this paper it is shown how the approach for equal-order elements can be extended to inf-sup stable finite elements. A complete stability and a priori analysis is presented. Moreover, various stabilization parameters are specified.

The numerical experiments show that the analysis is sharp with respect to the mesh size. The a priori estimate is sub-optimal with respect to the polynomial order. This is caused by non-optimal interpolation properties of the interpolation operator, which were used for the a priori estimate.

References

1. Braack, M., Burman, E.: Local projection stabilization for the Oseen problem and its interpretation as a variational multiscale method. SIAM J. Numer. Anal. **43**(6), 2544–2566 (2006)
2. Hughes, T., Mazzei, L., Janssen, K.: Large eddy simulation and the variational multiscale method. Comput. Vis. Sci. **3**, 47–59 (2000)
3. John, V., Matthies, G., Rang, J.: A comparison of time-discretization/linearization approaches for the incompressible Navier-Stokes equations. Comput. Methods Appl. Mech. Engrg. **195**, 5995–6010 (2006)
4. Lube, G., Rapin, G.: Residual-based stabilized higher-order FEM for a generalized Oseen problem. Math. Models Methods Appl. Sci. **16**(7), 949–966 (2006)
5. Matthies, G., Skrzypacz, P., Tobiska, L.: A unified convergence analysis for local projection stabilizations applied to the Oseen problem. M2AN **41**, 1293–1316 (2007)
6. Rapin, G., Löwe, J.: Local Projection Stabilizations for inf-sup stable Finite Elements applied to the Oseen Problem (2007). Submitted

Calibration of Model and Discretization Parameters for Turbulent Channel Flow

X.Q. Zhang, T. Knopp, and G. Lube

Abstract The simulation of turbulent incompressible flow in a plane channel is addressed. For $Re_\tau = 395$, discretization and model parameters of LES and DES models are calibrated using a DNS data basis. For higher Re_τ, a non-zonal hybrid method combines the calibrated LES model with wall functions as a near-wall model.

1 Basic Mathematical Model and Discretization

Consider the non-stationary, incompressible Navier-Stokes model

$$\partial_t \mathbf{u} - \nabla \cdot (2\nu \mathbb{S}(\mathbf{u})) + \nabla \cdot (\mathbf{u} \otimes \mathbf{u}) + \nabla p = \mathbf{f} \qquad \text{in } \Omega \times (0,T] \tag{1}$$

$$\nabla \cdot \mathbf{u} = 0 \qquad \text{in } \Omega \times (0,T] \tag{2}$$

for velocity $\mathbf{u}$ and pressure p in a bounded, polyhedral domain $\Omega \subset \mathbb{R}^3$ together with boundary and initial conditions. $\mathbb{S}(\mathbf{u}) = \frac{1}{2}(\nabla \mathbf{u} + \nabla \mathbf{u}^T)$ is the rate of strain tensor.

For the numerical simulation of (1)-(2), the DLR Theta code is used. The spatial discretization is based on a finite volume scheme on unstructured collocated grids. Different upwind schemes (linear upwind scheme (LUDS), quadratic upwind scheme (QUDS)) and the central differencing scheme (CDS) are implemented for the approximation of the convective fluxes. Diffusive fluxes are discretized with the CDS. The interpolation scheme by Rhie and Chow [8] is applied in order to avoid spurious pressure oscillations. The time discretization is performed using the A-stable BDF(2) scheme. The incremental variant of the projection method is used to

X.Q. Zhang and G. Lube
Math. Department, Georg-August-University Göttingen, Lotzestrasse 16-18, D-37083 Göttingen, Germany, e-mail: xiazhang/lube@math.uni-goettingen.de

T. Knopp
Institute of Aerodynamics and Flow Technology, German Aerospace Center (DLR) Göttingen, Bunsenstrasse 10, D-37073 Göttingen, Germany, e-mail: Tobias.Knopp@dlr.de

split the calculation of velocity and pressure within each time step. For a review of semidiscrete error estimates for the time-dependent Stokes problem see [3].

Of special interest here is the wall treatment. In the code, the wall node is shifted to the center of the control volume adjacent to the wall. Denote Γ_w the wall and Γ_δ an artificial inner boundary containing the shifted nodes at wall distance y_δ. Then, as a boundary condition on Γ_w, the wall-shear stress τ_w is prescribed instead of no-slip

$$\mathbf{u}\cdot\mathbf{n}=0, \quad (\mathbb{I}-\mathbf{n}\otimes\mathbf{n})2\nu\mathbb{S}(\mathbf{u})\mathbf{n}=-\tau_w\mathbf{u}_{t,\delta} \quad \text{on } \Gamma_w\,. \tag{3}$$

with $\mathbb{I}-\mathbf{n}\otimes\mathbf{n}$ being the projection operator onto the tangential space of Γ_w, unit velocity vector in wall-parallel direction $\mathbf{u}_{t,\delta}=\mathbf{v}_{t,\delta}/|\mathbf{v}_{t,\delta}|$ and

$$\tau_w=\nu\nabla u_\delta\cdot\mathbf{n}\,, \quad \text{where} \quad u_\delta = |\mathbf{v}_{t,\delta}|\,, \quad \mathbf{v}_{t,\delta}=(\mathbb{I}-\mathbf{n}\otimes\mathbf{n})\mathbf{u}|_{\Gamma_\delta}\,. \tag{4}$$

2 Turbulence Modeling Using LES Type Models

In LES, a scale separation operator subdivides the scales into filtered scales and unresolved scales. Only the filtered scales are solved and the unresolved scales are modeled by a sub-grid stress term of the so-called eddy-viscosity ν_t.

Smagorinsky model: In this classical LES model, the eddy-viscosity is given by $\nu_t=(C_S\Delta)^2|\mathbb{S}|$ with $|\mathbb{S}|=(2\mathbb{S}:\mathbb{S})^{1/2}$. The model constant to be calibrated is C_S. The filter width is $\Delta=nh_c$, $n=1,2,\dots$, with $h_c=(\Delta x\Delta y\Delta z)^{1/3}$, where $\Delta x,\Delta y,\Delta z$ denote the grid spacing in x-, y-, and z-direction respectively.

Near solid walls, the turbulent viscosity ν_t is multiplied with the van Driest damping function $D(y^+)$. For $\mathbf{x}\in\Omega$, denote $\mathbf{x}_w=\mathbf{x}_w(\mathbf{x})\in\Gamma_w$ the corresponding nearest wall point with distance d from $\mathbf{x}$. Then $D(y^+)=(1-\exp(-y^+/A^+))^2$ with $A^+=26$ where $y^+=yu_\tau/\nu$ is the wall-distance of $\mathbf{x}$ from $\mathbf{x}_w$ in viscous units with $y=\mathrm{dist}(\mathbf{x},\mathbf{x}_w(\mathbf{x}))\equiv d$ and $u_\tau=u_\tau|_{\mathbf{x}_w}=\sqrt{\tau_w}$.

Due to its non-local character the van Driest damping is not very suitable for unstructured methods or if parallelization is used A modified definition of Δ by [11] uses $\Delta=\min(\max(C_wd,C_w\Delta_{\max},\Delta_{wn}),\Delta_{\max})$ where $\Delta_{\max}=\max\{\Delta x,\Delta y,\Delta z\}$ with Δ_{wn} denoting the spacing in wall-normal direction. C_w is a calibration parameter.

Detached-eddy simulation model: Detached-eddy simulation (DES) is a single non-zonal hybrid RANS-LES method [10] based on the one-equation RANS model by Spalart & Allmaras [9] which computes the eddy viscosity $\nu_t=f_{\nu1}\tilde{\nu}$ from the auxiliary viscosity $\tilde{\nu}$ using a near-wall damping function $f_{\nu1}=\chi^3/(\chi^3+c_{\nu1}^3)$ with $\chi=\tilde{\nu}/\nu$ which involves only local variables. Here $\tilde{\nu}$ solves the transport equation

$$\partial_t\tilde{\nu}+\mathbf{u}\cdot\nabla\tilde{\nu}-\nabla\cdot\Big(\frac{\nu+\tilde{\nu}}{\sigma}\nabla\tilde{\nu}\Big)-\frac{c_{b2}}{\sigma}(\nabla\tilde{\nu})^2 = c_{b1}\tilde{S}\tilde{\nu}-c_{w1}f_w\Big(\frac{\tilde{\nu}}{d}\Big)^2$$

with $\tilde{S}=|\Omega|+\tilde{\nu}/(\kappa^2d^2)f_{\nu2}$, $|\Omega|=(2\Omega(\mathbf{u}):\Omega(\mathbf{u}))^{1/2}$, and $\Omega(\mathbf{u})=(\nabla\mathbf{u}-(\nabla\mathbf{u})^\top)/2$. The functions f_w and $f_{\nu2}$ and constants σ, c_{b2}, c_{b1}, c_{w1} are given in [9].

In the SA-DES model, d is replaced with $\tilde{d} = \min(d \, , C_{DES}\Delta_{\max})$. The model constant to be calibrated is C_{DES}.

Near-wall treatment for LES: Wall-functions are used to bridge the near-wall region at high Reynolds numbers. The wall shear stress τ_w can be computed from (4) only if $y_\delta^+ < 3$. For larger y_δ^+, $\tau_w = u_\tau^2$ is computed from friction velocity u_τ: The universal velocity profile of RANS-type by Reichardt is matched at the shifted node y_δ with the instantaneous LES solution u_δ

$$\frac{u_\delta}{u_\tau} = F\left(\frac{y_\delta u_\tau}{\nu}\right) , \quad F(y^+) \equiv \frac{\ln(1+0.4y^+)}{\kappa} + 7.8\left(1 - e^{-\frac{y^+}{11.0}} - \frac{y^+}{11.0} e^{-\frac{y^+}{3.0}}\right). \tag{5}$$

Equation (5) is solved for u_τ with Newton's method.

We remark that (5) is an approximative solution of the boundary layer equation in wall-normal direction neglecting convective term and pressure gradient: For each $\mathbf{x}_w \in \Gamma_w$ and given u_δ seek the wall-parallel velocity $u^{\text{RANS}}(y)$ such that

$$\partial_y\left((\nu + \nu_t^{\text{RANS}})\partial_y u^{\text{RANS}}\right) = 0 \quad \text{in} \quad \{\, \mathbf{x}_w - y\mathbf{n} \mid y \in (0, y_\delta)\} \tag{6}$$

$$u^{\text{RANS}}(0) = 0 \, , \qquad u^{\text{RANS}}(y_\delta) = u_\delta \, . \tag{7}$$

3 Calibration for Decaying Isotropic Turbulence

Framework: It is desirable to treat the calibration problem of basic turbulence models within the framework of optimization problems. Consider the abstract equation

$$A(q,u) = f \quad \text{in} \quad \Omega. \tag{8}$$

(here: quasi-stationary turbulent Navier-Stokes model) for the state variable u (here: velocity/pressure) in a Hilbert space $V \subseteq [H^1(\Omega)]^3 \times L^2(\Omega)$ with the parameter vector q (here: model and grid parameter) in the control space $Q := \mathbb{R}^{n_p}$. Let $C : V \to Z$ be a linear observation operator mapping u into the space of measurements $Z := \mathbb{R}^{n_m}$ with $n_m \geq n_p$. Then q is calculated from the constrained optimization problem

$$\text{Minimize} \qquad J(q,u) := \|C(u) - \hat{C}\|_Z^2/2 \tag{9}$$

with the cost functional $J : Q \times V \to \mathbb{R}$ under constraint (8) and using measurements $\hat{C} \in Z$. Assume the existence of a unique solution to (8)-(9) and of an open set $Q_0 \subset Q$ containing the optimal solution. Using the solution operator $S : Q_0 \to V$, one defines via $u = S(q)$ the reduced cost functional $j : Q_0 \to \mathbb{R}$ by $j(q) = J(q, S(q))$. The reduced observation operator $c(q) := C(S(q))$ leads to an unconstrained problem

$$\text{Minimize} \quad j(q) = \|c(q) - \hat{C}\|_Z^2/2, \qquad q \in Q_0. \tag{10}$$

An efficient framework to the solution of the necessary optimality condition $j'(q) = 0$ of (10) provides the adjoint approach, see [4] for a review. The approach can be generalized to time-dependent problems. This makes the optimization problem and solution techniques much more expensive, although sophisticated tools such as a-posteriori based optimization can reduce the costs, e.g. [1].

Seemingly, this approach has not been applied to parameter identification for turbulent flows yet. Main problems occur from the nonlinearity of turbulence models and the simulation over long time intervals to reach a statistically steady solution. Hence, a simpler approach to (10) is applied. As a basic step, a series of numerical simulations for a given flow provide look-up tables for the cost functional depending on relevant parameters as a basis for further systematic considering. In some cases, a Newton type method is feasible to determine optimized parameters.

Application to DIT: The problem of decaying isotropic turbulence (DIT) mimics the experiment by [2] at Taylor microscale Reynolds number $Re_\lambda \sim 150$. We choose a cubic box domain $\Omega = (0, 2\pi)^3$ and an equidistant mesh with N^3 nodes. As initial condition, we use a divergence-free velocity field with energy spectrum $E(k)|_{t=0}$ ($k = |\mathbf{k}|$, $1 \leq k \leq M$, $M = N/2 - 1$) given by data in [2] which can be computed as

$$\mathbf{u}(\mathbf{x})|_{t=0} = \sum_{k_1=0}^{M} \sum_{\substack{k_2,k_3=-M \\ |\mathbf{k}| \leq k_{\max}}}^{M} \left(\frac{E(k)|_{t=0}}{S_k} \right)^{1/2} 2 \left(\mathbb{I} - \frac{\mathbf{k} \otimes \mathbf{k}}{|\mathbf{k}|^2} \right) \gamma(\mathbf{k}) \cos(\mathbf{k} \cdot \mathbf{x} + \Theta(\mathbf{k})). \quad (11)$$

The components of $\gamma(\mathbf{k})$ are real random numbers with Gaussian distribution in $[0, 1]$, S_k is the number of wave-vectors $\mathbf{k}$ with $k - 1/2 \leq |\mathbf{k}| \leq k + 1/2$ and $\Theta(\mathbf{k})$ is a random phase with uniform distribution in $0 \leq \Theta \leq 2\pi$.

The second-order statistics of interest is the energy spectrum

$$E(k,t) = \sum_{k-1/2<|\mathbf{q}|\leq k+1/2} \frac{1}{2} \hat{\mathbf{u}}(\mathbf{q},t) \cdot \hat{\mathbf{u}}^*(\mathbf{q},t), \qquad k = 1, 2, \ldots, M, \qquad (12)$$

where $\hat{\mathbf{u}}^*$ is the complex conjugated of $\hat{\mathbf{u}}$. $\hat{\mathbf{u}}$ is the discrete Fourier transform of $\mathbf{u}$

$$\hat{\mathbf{u}}(\mathbf{k}) = \frac{1}{N^3} \Big(\sum_{x_1,x_2,x_3=0}^{N-1} \mathbf{u}(\mathbf{x}) \cos(-\mathbf{k} \cdot \mathbf{x}) + i \sum_{x_1,x_2,x_3=0}^{N-1} \mathbf{u}(\mathbf{x}) \sin(-\mathbf{k} \cdot \mathbf{x}) \Big). \qquad (13)$$

Then we consider the error functional

$$J(C) = \Big(\sum_{i=1}^{M} \Big[\big(E(k_i,C) - E_{\exp}(k_i)\big)^2_{t=0.87} + \big(E(k_i,C) - E_{\exp}(k_i)\big)^2_{t=2.0} \Big] \Big)^{1/2}.$$

The results in [12] for the spatial discretizations show that CDS is suitable to resolve the large wave-number part of the spectrum, whereas the upwind schemes produce excessive damping at high wave-numbers. Combining QUDS with a skew-symmetric formulation (QUDS_sk) for the convective fluxes gives some improvement. Fig. 1 (left) shows the dependence of the cost functional on the constant C_S for the Smagorinsky model (SMG) and $N = 64$. A Newton-type method (based on numerical differentiation) delivers a minimum with $C_S = 0.094$ for CDS and

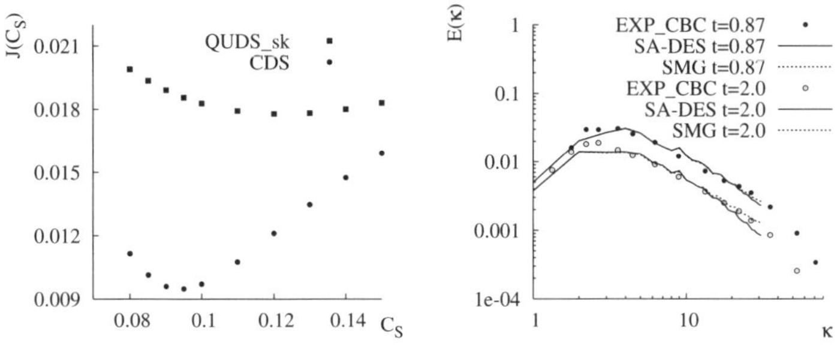

Fig. 1 Left: Calibration of Smagorinsky constant C_S for DIT. Right: Energy spectrum with optimized model constants of Smagorinsky model and of SA-DES model for CDS scheme.

$C_S = 0.123$ for QUDS_sk. For the SA-DES model, a similar Newton-type approach yields a minimum of $J(C)$ for $C_{\mathrm{DES}} = 0.67$. In Fig. 1 (right), the corresponding energy spectra for CDS with optimized constants for SMG and SA-DES are shown.

4 Parameter Calibration for Channel Flow

Consider now the benchmark problem of fully developed turbulent channel flow in the domain $\Omega = (0,2\pi) \times (0,2) \times (0,\pi)$. Periodic boundary conditions in streamwise x-direction, a no-slip condition for the walls in y-direction and symmetry planes in the spanwise z-direction are imposed. We consider a moderate Reynolds number $Re_\tau = u_\tau H/\nu = 395$ with channel half width $H = 1$, for which DNS data are available [6]. In order to achieve a constant mass flux, the streamwise forcing term is adjusted dynamically by taking into account the time step size δt_n and the bulk velocity from the DNS data and the bulk velocity at the present time t_n

$$\mathbf{f} \;=\; \tau_{\mathrm{w}}\mathbf{e}_x \;+\; (\delta t_n)^{-1}(U_{\mathrm{bulk,DNS}} - U_{\mathrm{bulk}}(t_n))\mathbf{e}_x\,, \quad U_{\mathrm{bulk}} = H^{-1}\int_0^H u(y)\mathrm{d}y \qquad (14)$$

where $\mathbf{e}_x$ denotes the unit-vector in x-direction. As initial condition we use a randomly perturbed velocity field $\mathbf{u}|_{t=0} = u_\tau F(yu_\tau/\nu)\mathbf{e}_x + 0.1U_{\mathrm{bulk}}\psi$ where F is given by (5) and each component of ψ is a random number in $(-1,1)$. The spatial discretization uses $N_x \times N_y \times N_z = 64 \times 64 \times 64$ nodes. The equidistant spacing in x- and z direction corresponds to $\Delta x^+ = \Delta x u_\tau/\nu = 38.8$ and $\Delta z^+ = \Delta z u_\tau/\nu = 19.4$ respectively. The grid in wall-normal direction is stretched using a hyperbolic tangent function $y(j)/H \;=\; \tanh[\gamma(2j/N_y - 1)]/\tanh(\gamma) + \;1.0,\; j = 0,1,\ldots,N_y - 1$ where $y(j)$ is the coordinate of the jth grid point in y direction providing thus an anisotropic, layer-adapted mesh, see [5]. The parameter γ allows to move the position $y^+(1)$ of the shifted wall node. The time step is chosen as $\delta t^+ \equiv \delta t u_\tau^2/\nu = 0.4$.

After reaching a statistically steady solution, first-order and second order statistics are computed. Denote $\langle\cdot\rangle$ the averaging operator over the two homogeneous

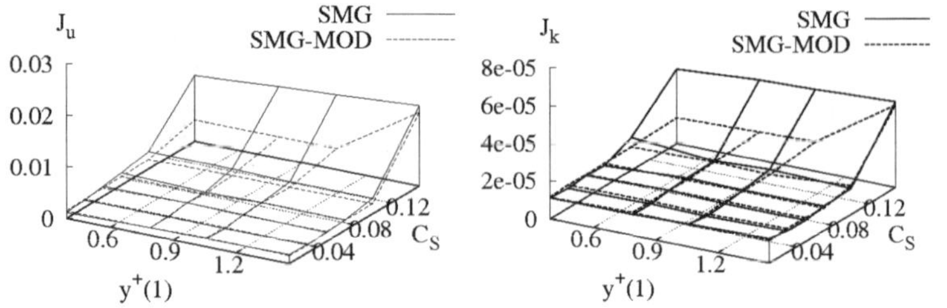

Fig. 2 Cost functionals for channel flow $Re_\tau = 395$, Left: mean velocity. Right: kinetic energy.

directions and in time. The quantities of interest are the mean velocity $U = \langle u \rangle$, the turbulent kinetic energy $k = \frac{1}{2}\langle (u - \langle u \rangle)^2 \rangle$ and its normalized variants $U^+ = \frac{U}{u_\tau}$ and $k^+ = \frac{k}{u_\tau^2}$. The L^2-error functional of the LES results compared to the DNS data is

$$J_u(y^+(1), C) = \left(\sum_{i=0}^{N_y} (U_i(y^+(1), C) - U_{i,\mathrm{DNS}})^2 \Delta y_i \right)^{1/2} \tag{15}$$

for the mean velocity (and similarly for kinetic energy J_k) with $\phi_i = \phi(y(i))$ and the spacing Δy_i in y-direction of cell i.

In Fig. 2, the dependence of the cost functionals J_u and J_k on C_S and $y^+(1)$ is shown for the Smagorinsky model. The result is robustness w.r.t. $C_S \in [0, 0.12]$ and $y^+(1) \in [0.5, 1.5]$. This means that a Newton-type approach to parameter calibration will not find local minima. In particular, the DIT-optimized value of C_S but also $C_S = 0$ (i.e., no turbulence model) are reasonable. The latter simulation can be seen as underresolved DNS on a layer-adapted mesh.

Reasonable results for the first and second order statistics are presented in Fig. 3 for the calibrated modified Smagorinsky model and the SA-DES model. The SA-DES model gives even better results and allows to avoid a damping of ν_t.

Channel flow at higher Re_τ: Now, the goal is to simulate turbulent channel flow at higher Reynolds number $Re_\tau = 4800$ using the calibrated model constants. A resolution of the wall layer regions (as for $Re_\tau = 395$) with a standard LES model is not feasible (on a single processor) due to the much finer mesh in all spatial directions and in time.

As DES-type approaches are still relatively expensive, the modified Smagorinsky LES model (WSMAG) and the SA-DES model (WSADES) are used with wall functions. This reduces the computing time by an order of magnitude due to the saving in grid points in wall-normal direction and due to the much larger time steps.

The results for the WSADES approach are given in Fig. 4. The original DES concept for coupling the RANS and LES regions gives two logarithmic layers, see [7]. The lower layer is the modeled log layer of the RANS model, while the upper layer is the resolved log-layer of the LES model. This causes a significant error in u_τ. This is subject to present and future research and will be presented elsewhere.

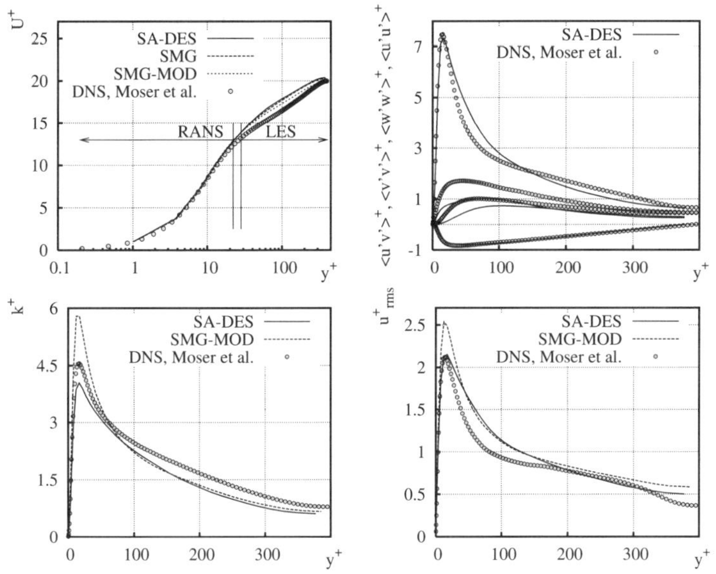

Fig. 3 Channel flow $Re_\tau = 395$ for modified Smagorinsky and SA-DES model: Upper left: Mean velocity U^+. Upper right: Fluctuations. Bottom left: Kinetic energy k^+. Bottom right: u^+_{rms}.

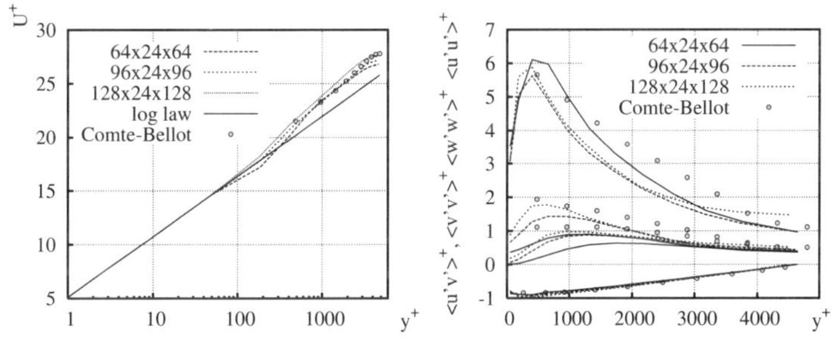

Fig. 4 SA-DES model with near-wall modeling (WSADES) for channel flow $Re_\tau = 4800$.

5 Summary and Conclusions

A strategy for calibration of model and discretization parameters of LES and DES within the framework of optimization techniques was presented. We use the DLR Theta code, which is an industrial RANS solver. Precurser studies on the benchmark problems of decaying isotropic turbulence and of turbulent channel flow at $Re_\tau = 395$ show that the central difference scheme (CDS) for the convective term is clearly superior to upwind schemes. Moreover it can be seen that second order accurate time discretization is necessary for proper calculation of second order statistics for turbulent channel flow.

A calibration of model and grid parameters was performed based on least-squares cost functionals for first and second order flow statistics. Best results for channel flow at $Re_\tau = 395$ are found for the calibrated SA-DES model which also avoids van Driest damping. Finally the optimized parameters are used for a simulation of turbulent channel flow at $Re_\tau = 4800$. A proper near-wall resolution is very expensive at such Reynolds numbers. Therefore LES and DES in combination with near-wall modeling based on wall functions are used and reasonable results are obtained.

Future work will be on turbulent channel flow at high Reynolds numbers with focus on more sophisticated methods for coupling hybrid wall-functions with LES. Another task will be on continuation of the wall-resolved LES for the flow over a backward facing step.

Acknowledgements The first author gratefully acknowledges financial support by the Deutsche Forschungsgemeinschaft (DFG) via GK 1023. This work benefited from the discussions of the second author within the DESider project (Detached Eddy Simulation for Industrial Aerodynamics) funded by the European Union and administrated by the CEC, Research Directorate-General, Growth Programme, under Contract No. AST3-CT-2003-502842. Finally we thank Dr. R. Kessler for valuable discussions and support with the DLR Theta code.

References

1. Becker, R., Braack, M., Vexler, B.: Parameter identification for chemical models in combustion problems. Applied Numerical Mathematics **54(3-4)**, 519–536 (2005)
2. Comte-Bellot, G., Corrsin, S.: Simple eulerian time correlation of full- and narrow-band velocity signals in grid-generated, 'isotropic' turbulence. Journal of Fluid Mechanics **48(2)**, 273–337 (1971)
3. Guermond, J., Minev, P., Shen, P.: An overview of projection methods for incompressible flows. Computer Methods in Applied Mechanics and Engineering **196**, 6011–6045 (2006)
4. Johansson, H., Runesson, K., Larsson, F.: Parameter identification with sensitivity assessment and error computation. GAMM-Mitteilungen **30(2)**, 430–457 (2007)
5. Morinishi, Y., Vasilyev, D.: A recommended modification to the dynamic two-parameter mixed subgrid scale model for large eddy simulation of wall bounded turbulent flow. Physics of Fluids **13**, 3400–3410 (2001)
6. Moser, R., Kim, J., Mansour, N.: Direct numerical simulation of turbulent channel flow up to $Re_\tau = 590$. Physics of Fluids **11**, 943–946 (1999)
7. Nikitin, N.V., Nicoud, F., Wasistho, B., Squires, K.D., Spalart, P.R.: An approach to wall modeling in large-eddy simulation. Physics of Fluids **12(7)**, 1629–1632 (2000)
8. Rhie, C.M., Chow, W.L..: Numerical study of the turbulent flow past an airfoil with trailing edge separation. AIAA Journal **21**, 1525–1532 (1983)
9. Spalart, P.R., Allmaras, S.R.: A one-equation turbulence model for aerodynamics flows. AIAA Paper 1992-0439 (1992)
10. Spalart, P.R., Jou, W.H., Strelets, M., Allmaras, S.R.: Comments on the feasibility of les for wings, and on a hybrid rans/les approach. August, 1997, In: Advances in DNS/LES, edited by C. Liu and Z. Liu (Greyden, Columbus, OH, 1997) (1997)
11. Travin, A., Shur, M., Spalart, P.R., Strelets, M.: Improvement of delayed detached-eddy simulation for les with wall modelling. In: Proceedings of the European Conference on Computational Fluid Dynamics (ECCOMAS CFD 2006) (2006)
12. Zhang, X.Q.: Identification of model and grid parameters for incompressible turbulent flows. Ph.D. thesis, University Göttingen (2007)

Fluid-Structure Interaction

Numerical Solution of Transonic and Supersonic 2D and 3D Fluid–Elastic Structure Interaction Problems

J. Dobeš, J. Fürst, H. Deconinck, and J. Fořt

Abstract We solve the system of Euler or Navier-Stokes equations in conservative ALE formulation, possibly supplemented by a suitable turbulence model. The structural dynamics is described by the equations of anisotropic elastic continuum with large or small displacements. The problem is closed by suitable interface conditions. We present a cell centered finite volume method with linear least square reconstruction with nonlinear WENO type weights, a cell centered finite volume method with linear reconstruction and Barth's limiter and a residual distribution scheme to solve the CFD sub-problem. To allow a large time step, all the considered methods use implicit time stepping formulated in dual time. The elastic problem and the mesh motion is solved by a simple finite element method. The whole problem is formulated in dual time and solved by a simple sub-iteration procedure. We present several numerical tests documenting the behavior of the methods. These include 2D transonic turbulent flow past a forced oscillatory pitching NACA0012 airfoil, 2D inviscid supersonic panel flutter and 3D inviscid flow past the elastic AGARD 445.6 wing.

1 Introduction

Consideration of problems of interaction between fluids and structural bodies is an important part of the design process. Despite a large progress in the improvement of

Jiří Dobeš, Jiří Fürst, and Jaroslav Fořt
Department of Technical Mathematics, Faculty of Mechanical Engineering, Czech Technical University, Karlovo Náměstí 13, Prague, CZ-121 35 Czech Republic, e-mail: {Jiri.Dobes, Jiri.Furst, Jaroslav.Fort}@fs.cvut.cz

Herman Deconinck
Von Karman Institute for Fluid Dynamics, Waterloosesteenweg 72, B-1640 Sint-Genesius-Rode, Belgium, e-mail: Deconinck@vki.ac.be

the computational methods [6], there is still a need for highly accurate and efficient numerical methods.

We will focus on a partitioned approach, i.e. we use distinct methods for computational fluid dynamics (CFD) and computational structural mechanics (CSM) coupled via interface conditions. This allows to use finly tuned methods for each subproblem what is especially important in transonic CFD. In this article, we present several variants of finite volume and residual distribution methods for solving CFD problems. For CSM, we use a simple finite element method. The coupled problem is solved by a sub-iteration approach.

2 Formulation of the Problem

Three sub-problems are considered: the fluid flow-field, the elastic dynamics and the motion of the fluid mesh. The sub-problems are connected via interface conditions.

First, we define a sufficiently smooth ALE mapping in d spatial dimensions

$$\mathscr{A}_t : \Omega_0 \subset \mathbb{R}^d \to \Omega_t \subset \mathbb{R}^d, \qquad \mathbf{x}(\mathbf{y},t) = \mathscr{A}_t(\mathbf{y}). \tag{1}$$

with Jacobian

$$J_{\mathscr{A}_t} = \det\left(\frac{\partial \mathbf{x}}{\partial \mathbf{y}}\right). \tag{2}$$

The domain velocity $\mathbf{w}$ is defined as

$$\mathbf{w}(\mathbf{x},t) = \left.\frac{\partial \mathbf{x}}{\partial t}\right|_{\mathbf{y}} (\mathbf{y}(\mathbf{x},t),t). \tag{3}$$

with the following equivalence

$$\nabla_x \cdot \mathbf{w} \equiv \left.\frac{1}{J_{\mathscr{A}_t}} \frac{\partial J_{\mathscr{A}_t}}{\partial t}\right|_{\mathbf{y}}. \tag{4}$$

We consider a system of compressible Navier-Stokes equations in ALE formulation

$$\left.\frac{1}{J_{\mathscr{A}_t}} \frac{\partial J_{\mathscr{A}_t} U}{\partial t}\right|_{\mathbf{y}} + \nabla_x \cdot [\mathbf{F}(U) - \mathbf{S}(U,\nabla U) - U\mathbf{w}] = 0, \tag{5}$$

where $U = (\rho, \rho\mathbf{u}, e) : \mathbb{R}^{d+1} \to \mathbb{R}^q$ are unknowns, $\mathbf{F} : \mathbb{R}^q \to \mathbb{R}^{d\times q}$ is the vector of inviscid flux functions and $\mathbf{S}$ is a vector of viscous flux functions, q being the number of unknowns. The system is supplemented by equations for turbulent variables in a similar form. Suitable initial and boundary conditions are prescribed. As a simplification, we also consider a system of Euler equations.

The elastic equations are formulated in Lagrange system of coordinates (denoted by $\tilde{\cdot}$) with unknown displacement $\tilde{u}_i = \tilde{x}_i' - \tilde{x}_i$. Dynamic equation for the continuum is given by

$$\rho \frac{\partial^2 \tilde{u}_i}{\partial t^2} = \frac{\partial \sigma_{ij}}{\partial \tilde{x}_j} + f_i, \tag{6}$$

where the stress tensor is given by generalized Hooke's law

$$\sigma_{ij} = c_{ijkl}\varepsilon_{kl}, \tag{7}$$

with c_{ijkl} a given elastic tensor: constant of material. Tensor of deformation is defined as

$$\varepsilon_{ij} = \frac{1}{2}\left(\frac{\partial \tilde{u}_i}{\partial \tilde{x}_j} + \frac{\partial \tilde{u}_j}{\partial \tilde{x}_i} + \frac{\partial \tilde{u}_k}{\partial \tilde{x}_i}\frac{\partial \tilde{u}_k}{\partial \tilde{x}_j}\right). \tag{8}$$

The quadratic term is insignificant for small displacements and it is some times neglected.

The tensor of forces acting to the elastic body is equal to the tensor resulting from the fluid flow. Displacement of the fluid boundary is equal to the displacement of elastic boundary, what gives a condition for the ALE mapping $\mathscr{A}_t$.

The full problem reads: find a solution of CFD problem $U : \Omega_{t,\mathrm{fluid}} \times [0, T_{\max}] \to \mathbb{R}^q$, a solution of CSM problem $\tilde{\mathbf{u}} : \Omega_{\mathrm{elastic}} \times [0, T_{\max}] \to \mathbb{R}^d$ and the solution of the fluid mesh motion problem $\mathscr{A}_t : \Omega_{0,\mathrm{fluid}} \times [0, T_{\max}] \to \Omega_{t,\mathrm{fluid}}$ such that all the sub–problems and the interface conditions are simultaneously satisfied.

3 Numerical Methods

3.1 Finite Volume Method

We have developed two variants of finite volume methods. Both work on general unstructured (hybrid) meshes, both are in cell centered settings, and use the three points backward implicit integration method (2BDF) formulated in dual time. The first method (Method I) uses AUSM type of splitting and the linear least square reconstruction with nonlinear weights (WLSQR), see e.g. [7]. The viscous terms are discretized centrally on the dual mesh. The second method (Method II) uses linear least square reconstruction with Barth's limiter [1]. The numerical flux is approximated by the Roe's Riemann solver. The moving mesh terms are treated for both methods following the approach of [8], the time derivative of Jacobian is approximated by the time derivative of the element volume and the velocity of the faces is approximated as

$$(\mathbf{w} \cdot \mathbf{n})^{h,n+1} \approx \frac{1}{\Delta t}\int_{t^n}^{t^{n+1}} (\dot{\mathbf{x}} \cdot \mathbf{n})\,\mathrm{d}t. \tag{9}$$

The integral is discretized in one Gauss point in two dimensions or two Gauss points three dimensions to ensure a satisfaction of the discrete geometric conservation law. It states that a constant solution is exactly preserved on a deforming grid.

3.2 Residual Distribution Method

The class of residual distribution (RD) schemes has emerged as an appealing alternative to better known finite volume and finite element schemes. A fist extension of the first order N RD scheme to the moving meshes has been published in [9]. An improved version of the N scheme satisfying the discrete maximum principle for scalar case is given in [3]. The extension of the second order LDA scheme was published in [4]. Both the schemes can be combined into nonlinear Bx scheme. See [4] and references therein.

For the LDA scheme we use equivalency of the scalar linearity preserving RD schemes with the Petrov-Galerkin FEM formulation [2]. The solution and the mesh velocity are approximated by the linear Galerkin shape functions ψ_i from the current domain configuration. The Petrov–Galerkin test function is given on each element E by $\varphi_i^E = \psi_i + \beta_i - 1/(d+1)$, where β_i is the RD distribution coefficient (matrix) for the node i. For the mesh velocity term we use $\nabla_x \cdot (U\mathbf{w}) = U\nabla_x \cdot \mathbf{w} + \nabla_x U \cdot \mathbf{w}$ and the identity (4). This terms have to be treated carefully to retain conservativity of the scheme [9]. This formulation gives us the (semi-discrete) element contribution

$$\phi_i^{E,\mathrm{LDA}} = \frac{1}{J^h_{\mathscr{A}_t}} \sum_{j\in E} \left.\frac{\partial J^h_{\mathscr{A}_t} u_j}{\partial t}\right|_{\mathbf{y}} m_{ij}^E + \phi_i^{E,\mathrm{sLDA}} - \frac{1}{J^h_{\mathscr{A}_t}} \left.\frac{\partial J^h_{\mathscr{A}_t}}{\partial t}\right|_{\mathbf{y}} \left(\sum_{j\in E} u_j m_{ij}^E\right), \tag{10}$$

where $m_{ij}^E = \int_E \varphi_i \psi_j \,\mathrm{d}\mathbf{x}$ is the element contribution to the mass matrix and $\phi_i^{E,\mathrm{sLDA}}$ is the well known nodal contribution from the *steady* version of the LDA scheme [11]. The 2BDF time discretization formula is used, i.e. the ALE time derivative is approximated by

$$\frac{1}{J^h_{\mathscr{A}_t}} \left.\frac{\partial J^h_{\mathscr{A}_t} u_j}{\partial t}\right|_{\mathbf{y}} = \frac{\alpha^{n+1}\mu(E^{n+1})u_j^{n+1} + \alpha^n \mu(E^n) u_j^n + \alpha^{n-1}\mu(E^{n-1})u_j^{n-1}}{t^{n+1} - t^n} \tag{11}$$

with coefficients

$$\alpha^{n+1} = \frac{1+2\tau}{1+\tau}, \quad \alpha^n = -1-\tau, \quad \alpha^{n-1} = \frac{\tau^2}{1+\tau}, \quad \tau = \frac{t^{n+1}-t^n}{t^n - t^{n-1}}. \tag{12}$$

The measure (volume) of the element is denoted by $\mu(E)$. All the terms in (10) are evaluated at time level $n+1$.

The N scheme is formulated with diagonally lumped mass matrix, the geometric source term is divided into the convective part and the velocity divergence term, the latter treated by the point-vise discretization on dual grid [3]. The 2BDF time discretization is again used.

$$\phi_i^{E,\mathrm{N}} = \frac{\alpha^{n+1}\mu(E^{n+1})u_i^{n+1} + \alpha^n \mu(E^n) u_i^n + \alpha^{n-1}\mu(E^{n-1})u_i^{n-1}}{d+1} + \phi_i^{E,\mathrm{sN}}$$

$$-\frac{\sum_{j\in E} u_j^{n+1}}{d+1}\frac{\alpha^{n+1}\mu(E^{n+1})+\alpha^{n}\mu(E^{n})+\alpha^{n-1}\mu(E^{n-1})}{d+1}. \quad (13)$$

Here $\phi_i^{E,\mathrm{sN}}$ is the well known nodal contribution from the *steady* version of the N scheme [11].

Both the schemes are blended using blending coefficient $\theta = \min(1, sc^2 h)$, $sc = (\frac{\partial p}{\partial t} + \nabla_x p \cdot \mathbf{v})^+/\delta_{pv}$, where $\mathbf{v}$ is the velocity vector of the flow, p is the static pressure, h the diameter of the element and δ_{pv} is a product of the characteristic pressure and velocity in the domain. Finally, sum of the element contributions to each node is driven towards zero using a dual-time approach. The Bx scheme respects the discrete geometric conservation law.

3.3 *Numerical Method for Structural Dynamics*

The governing equations are solved by means of a standard finite element method with bi-quadratic Lagrangian elements. Newmark method is used for the time discretization. The similar method is used for the mesh movement and a simple subiteration approach couples the problems together. The interface conditions are derived from the virtual works considerations [2].

4 Numerical Results

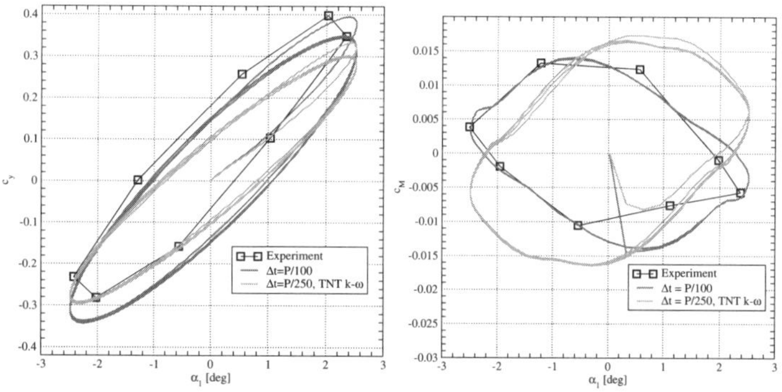

Fig. 1 NACA 0012. Method I. Lift and moment coefficients, inviscid and turbulent calculations.

The first test case involves the NACA 0012 airfoil, which is sinusoidally pitching around its a quarter chord (test case AGARD CT 5, see e.g. [3]). The free stream Mach number is 0.755. The airfoil performs a pitching motion

$$\alpha = 2.51^\circ \sin(2kt) + 0.016^\circ, \qquad k = \frac{\omega c}{2u_\infty} = 0.0814, \tag{14}$$

where c is the chord, u_∞ is the free-stream velocity and ω the frequency. The Reynolds number is $Re = 5.5 \cdot 10^6$. Results given by the inviscid calculations are present in Fig. 1. Relatively good agreement with experimental data can be observed. We present also results of turbulent computations. Two equation Kok's k–ω turbulence model was used. Unfortunately, overall agreement with the experimental data is worse than for the inviscid calculation. This is especially true for the moment coefficient. We expect that the differences are given mainly by the inappropriate choice of the turbulence model for unsteady calculation. Future investigation is needed.

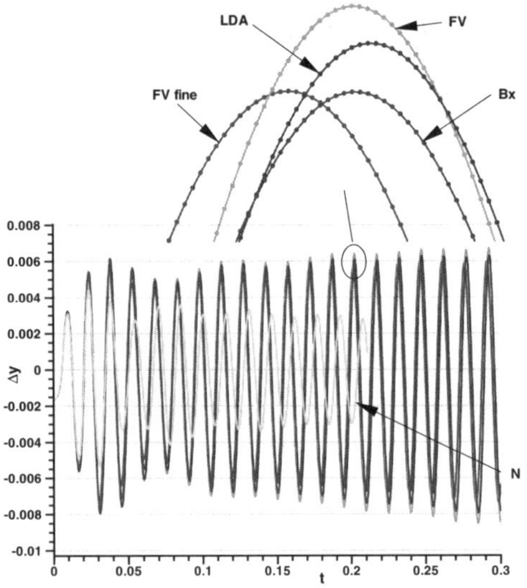

Fig. 2 Panel flutter problem. Dependence of the integral of deflection on time for different numerical schemes. $Ma_\infty = 2.2$.

As the second test, the supersonic panel flutter is considered [10]. An elastic panel with infinite aspect ratio is clamped on both edges. Its upper side is exposed to the supersonic air-stream, while the lower side resides in the still air with the same pressure as in the upper side. The panel has length $L = 0.5$ m, a uniform thickness $h = 1.25 \cdot 10^{-3}$ m, Young modulus $E = 7.728 \cdot 10^{10}\,\mathrm{N/m^2}$, Poisson ration $\nu = 0.33$ and the density $\rho_s = 2710\,\mathrm{kg/m^3}$. The plane strain assumption was used. The flow conditions are given by $p_\infty = 25714\,\mathrm{Pa}$ and $\rho_\infty = 0.4\,\mathrm{kg/m^3}$. The critical Mach number Ma_∞^{cr} that is, the lowest free stream Mach number for which an unstable aeroelastic mode of the panel appears, is given in the reference [10]. Using theoretical method the authors get $Ma_\infty^{\mathrm{cr}} \approx 2.27$ and using their numerical scheme $Ma_\infty^{\mathrm{cr}} \approx 2.23$, what they consider an "excellent agreement".

The elastic panel is discretized with 60×2 elements. The computational domain is formed by a half-circle of diameter $R = 5$. The mesh consists of 3451 nodes and 6722 triangular elements, giving 50 elements along the panel. One more computation is performed on a regular quadrilateral mesh of 300×100 elements with 100 elements along the panel (referred as "fine"). The integral of the deflection of the panel for $Ma = 2.2$ is plotted in Figure 2. The neutral response was correctly reproduced. Although the LDA scheme is only linearly stable, it is able to capture weak shock waves in the non-oscillating manner. The nonlinear Bx scheme gives similar results as the LDA scheme, which are very different from the first order N scheme.

Finally, the transonic flutter of the AGARD 455.6 wing ("solid model") is considered [5]. The elastic wing was discretized using 350 tri-quadratic elements. The elastic constants are uncertain and we have set it as for the green Honduras mahogany [12]. The CFD mesh consists of 22k nodes and 118k tetrahedral elements. The neutral response regime was chosen, which is characterized by the flutter speed index of 0.5214 and the free stream Mach number $Ma_\infty = 0.92$ with flow medium Freon-12. One period was divided to 120 time-steps. The integral of the wing velocity is plotted in Fig. 3. For the FV scheme, a small negative damping is observed, while by the LDA scheme the neutral response was correctly reproduced. The difference between the measured oscillation period and the computed period is less than 3 %, see Tab. 1. Considering uncertainty of the elastic constants we judge this result more accurate than one can expect.

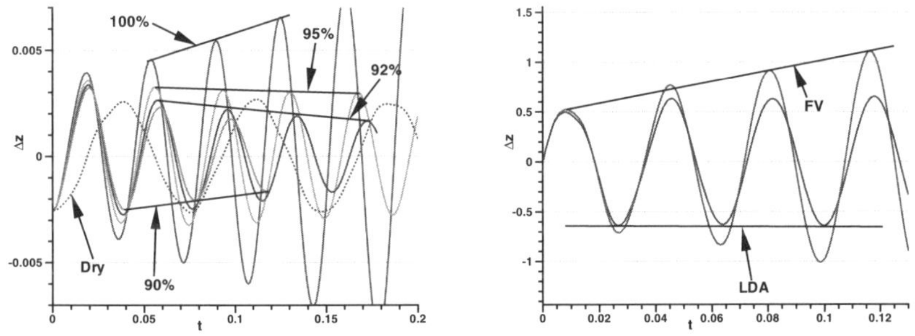

Fig. 3 AGARD 445.6 wing. Time dependence of the volume integral of the wing velocity for the measured neutral response regime.

Method	Period T	Error
Measured	0.036508	
LDA (BC1)	0.036265	0.66 %
FV	0.035746	2.08 %
LDA (BC2)	0.035436	2.93 %

Table 1 AGARD 445.6 wing. Oscillation period computed from fist 3 cycles

5 Conclusions

We have presented a method for calculation of the fluid/structure interaction in transonic and supersonic flow regions. We use three field formulation. We have especially focused on the development of the numerical method for fluid flow. We present three alternative methods, two finite volume and one of residual distribution type. The results documenting the behavior of the methods are included.

Acknowledgements This work was partially supported by Research plan of Ministry of Education, Youth and Sports of the Czech Republic No. 6840770003.

References

1. Barth, T.J., Jesperson, D.C.: The design and application of upwind schemes on unstructured meshes. AIAA Paper 89–0366, AIAA (1989)
2. Dobeš, J.: Numerical algorithms for the computation of unsteady compressible flow over moving geometries – application to fluid-structure interaction. Ph.D. thesis, ČVUT Praha, Czech Republic/Von Karman Institute for Fluid Dynamics, Belgium/Université Libre de Bruxelles, Belgium (2007). To appear.
3. Dobeš, J., Deconinck, H.: A second order space-time residual distribution method for solving compressible flow on moving meshes. AIAA Paper 2005-0493, AIAA (2005). Presented on 43rd AIAA Aerospace Sciences Meeting and Exhibit 10–13 January 2005, Reno, Nevada
4. Dobeš, J., Deconinck, H.: An ALE formulation of the multidimensional RDS for computations on moving meshes. In: H. Deconinck, E. Dick (eds.) The Fourth International Conference on Computational Fluid Dynamics, July 10–14. Ghent, Belgium (2006)
5. E. Carson Yates, J.: AGARD standard aeroelastic configurations for dynamic response. Candidate configuration I.-Wing 445.6. Technical Memorandum 100492, NASA (1987)
6. Farhat, C.: Encyclopedia of Computational Mechanics, chap. CFD-Based Nonlinear Computational Aeroelasticity. John Wiley & Sons, Ltd. (2004)
7. Fürst, J.: A weighted least square scheme for compressible flows. Flow, Turbulence and Combustion **76**(4), 331–342 (2006)
8. Koobus, B., Farhat, C.: Second-order time-accurate and geometrically conservative implicit schemes for flow computations on unstructured dynamic meshes. Comput. Methods Appl. Mech. Engrg. **170**(1–2), 103–129 (1999)
9. Michler, C., Sterck, H.D., Deconinck, H.: An arbitrary Lagrangian Eulerian formulation for residual distribution schemes on moving grids. Computers and Fluids **32**(1), 59–71 (2003)
10. Piperno, S., Farhat, C.: Partitioned procedures for the transient solution of coupled aeroelastic problems part II: Energy transfer analysis and three-dimensional applications. Comput. Meths. Appl. Mech. Engrg. **190**(24), 3147–3170 (2001)
11. Ricchiuto, M.: Construction and analysis of compact residual discretizations for conservation laws on unstructured meshes. Ph.D. thesis, Université Libre de Bruxelles, Von Karman Institute for Fluid Dynamics (2005)
12. Yosibash, Z., Kirby, R.M., Myers, K., Szabó, B., Karniadakis, G.: High-order finite elements for fluid-structure interaction problems. In: 44th AIAA/ASME/ASCE/AHS Structures, Structural Dynamics, and Materials Conference, 7-10 April, Norfolk, Virginia, 2003-1729. AIAA (2003). URL http://www.dam.brown.edu/people/yosibash/home_files/2003_1729.pdf

Numerical Simulations of Flow Induced Vibrations of a Profile

R. Honzátko, J. Horáček, and K. Kozel

Abstract The work deals with a numerical solution of the interaction of two-dimensional inviscid incompressible flow and a vibrating profile with two degrees of freedom. The profile can oscillate around an elastic axis and in the vertical direction. The mathematical model is represented by the system of incompressible unsteady Euler equations. Numerical schemes in the form of finite volume method are applied on a structured quadrilateral C-mesh. Two strategies, an artificial compressibility approach and a dual-time stepping method, are employed for numerical solution of governing equations. The motion of the profile is described by a system of two linear ordinary differential equations that are transformed to the system of first order ordinary differential equations and solved numerically using multistage four-order Runge-Kutta method. Deformations of the computational domain due to the profile motion are treated using the Arbitrary Lagrangian-Eulerian method. Numerical schemes used satisfy the geometric conservation law. Numerical simulations of flow-induced vibrations are performed for different upstream flow velocities past the profile NACA 0012 and the results are presented for translation and rotation of the profile in time domain. Moreover, pressure and velocity fields around the profile are shown at several time levels. The two numerical strategies are compared respectively.

Radek Honzátko
Department of Technical Mathematics, Faculty of Mechanical Engineering, Czech Technical University in Prague, Karlovo náměstí 13, 121 35 Praha 2 - Nové Město, Czech Republic, e-mail: honzatko.radek@seznam.cz

Jaromír Horáček
Institute of Thermomechanics AS CR, Dolejškova 5, 182 00 Praha 8, Czech Republic, e-mail: jaromirh@it.cas.cz

Karel Kozel
Department of Technical Mathematics, Faculty of Mechanical Engineering, Czech Technical University in Prague, Karlovo náměstí 13, 121 35 Praha 2 - Nové Město, Czech Republic, e-mail: karel.kozel@fs.cvut.cz

1 Mathematical Model

The governing equations are two-dimensional incompressible Euler equations in dimensionless conservative form:

$$\mathbb{D}W_t + F_{x_1} + G_{x_2} = 0, \tag{1}$$

where $W = (p, u_1, u_2)^T$, $F = (u_1, u_1^2 + p, u_1 u_2)^T$, $G = (u_2, u_1 u_2, u_2^2 + p)^T$ and $\mathbb{D} = \text{diag}(0, 1, 1)$. Here, W is the vector of conservative variables, F, G are inviscid physical fluxes and $\mathbb{D}$ is a diagonal matrix. Space coordinates are denoted x_1, x_2. Symbols $\mathbf{u} = (u_1, u_2)$ and p stand for velocity vector and pressure, respectively.

The method of artificial compressibility [3] and a time marching method are used for steady state computations. The artificial compressibility method consist in modifying governing equations by adding of the time derivative of pressure to the continuity equation. It is represented by substitution of the matrix $\mathbb{D}$ by the diagonal matrix

$$\mathbb{D}_\beta = \text{diag}\left(\tfrac{1}{\beta^2}, 1, 1\right) \tag{2}$$

in Eq. (1), where $\beta \in \mathbb{R}^+$ is a parameter.

Upstream conditions are prescribed values of the vector $\mathbf{u}$, i. e. $\mathbf{u} = \mathbf{u}_\infty$, pressure is extrapolated from the flow field. The downstream condition is a given value of p at the outlet boundary, i. e. $p = p_{\text{out}}$. The other values of the vector of conservative variables at the outlet boundary W_{out} are extrapolated. Wall conditions are nonpermeability conditions, i. e. $(u, v)_n = 0$ (normal component of velocity vector is equal to zero).

In the case of an unsteady flow two approaches, an artificial compressibility and a dual-time stepping method, are applied. In the first case, the mathematical model is modified according to the original one by the use of the identical matrix $\mathbb{D}_\beta$ instead of $\mathbb{D}$ in Eq. (1) as for the artificial compressibility method in steady state simulations. To eliminate the distinction of this mathematical model from the original one, the parameter β has to be a big positive number. Ideally, $\beta \to \infty$. The profit of this approach is that it allows to apply time-marching algorithms. In the second case, the mathematical model is reformulated [5, 2, 1] to be handled by a time-marching steady-state solver. This approach requires the addition of derivatives with respect to a fictious pseudo time τ to each of the three equations to give

$$\mathbb{D}_\beta W_\tau + \mathbb{D}W_t + F_{x_1} + G_{x_2} = 0. \tag{3}$$

A steady-state solution in pseudo time $(\partial p/\partial \tau,\ \partial u_1/\partial \tau,\ \partial u_2/\partial \tau \to 0)$ corresponds to an instantaneous unsteady solution in real time t. The unsteady flow calculation has thus been transformed into a series of steady-state calculations in pseudo time τ. It provides a possibility to develop a time-accurate time-marching scheme for unsteady incompressible flows.

2 The Arbitrary Lagrangian-Eulerian Method

Since, unsteady flow simulations are performed in a deforming domain due to the profile motion the Arbitrary Lagrangian-Eulerian (ALE) method is employed to the simulations. It is based on an ALE mapping

$$\mathscr{A}_t : \Omega_{\text{ref}} \rightarrow \Omega_t \,, \quad \xi \rightarrow \mathbf{x}(\xi,t) = \mathscr{A}_t(\xi) \tag{4}$$

of the reference configuration $\Omega_{\text{ref}} \equiv \Omega_0$ onto the current configuration Ω_t, with the ALE velocity $\mathbf{w} = \partial \mathscr{A}_t / \partial t(\xi)$, where $\xi = \mathscr{A}_t^{-1}(\mathbf{x})$. Then, Eq. (1) can be written in the ALE form as

$$(JW)_t|_\xi + J\left(\tilde{F}_{x_1}(W,w_1) + \tilde{G}_{x_2}(W,w_2)\right) = 0\,, \tag{5}$$

where $J = \det(d\mathbf{x}/d\xi)$ is the Jacobian of the ALE mapping $\xi \rightarrow \mathbf{x}$, $(JW)_t|_\xi$ represents the ALE time derivative [4] of JW, $\tilde{F} = F(W) - w_1 \mathbb{D}W$, $\tilde{G} = G(W) - w_2 \mathbb{D}W$ and $\mathbf{w} = (w_1, w_2)$.

Wall conditions on moving boundaries are given by $(u,v)_n = \mathbf{w}_n$, where $\mathbf{w}_n$ is a normal component of the ALE velocity to the boundary. It is assumed that the ALE velocity at each point on the surface of the moving boundary is equal to the velocity of its motion.

2.1 The Finite Volume Method in ALE Formulation

The finite volume method relies on a discretization of a computational domain into control volumes, usually called cells, C_i with boundaries denoted by ∂C_i. In the ALE formulation, these cells move and deform in time. Eq. (5) can be integrated over these control cells. At first, the integration is carried out over a reference cell $C_i(0)$ in the ξ space as follows

$$\int_{C_i(0)} (JW)_t|_\xi \, d\Omega_\xi + \int_{C_i(0)} J\left(\tilde{F}_{x_1}(W,w_1) + \tilde{G}_{x_2}(W,w_2)\right) d\Omega_\xi = 0. \tag{6}$$

Since the partial time derivative in Eq. (6) is evaluated at a constant ξ, the order of the derivation and integration can be changed. Hence

$$\frac{d}{dt}\int_{C_i(0)} W J \, d\Omega_\xi + \int_{C_i(0)} \left(\tilde{F}_{x_1}(W,w_1) + \tilde{G}_{x_2}(W,w_2)\right) J \, d\Omega_\xi = 0. \tag{7}$$

The inverse transformation to the time varying cells in Eq. (7) results in

$$\frac{d}{dt}\int_{C_i(t)} W \, d\Omega_\mathbf{x} + \int_{C_i(t)} \left(\tilde{F}_{x_1}(W,w_1) + \tilde{G}_{x_2}(W,w_2)\right) d\Omega_\mathbf{x} = 0. \tag{8}$$

Finally, integrating by parts the second integral yields the governing integral equation

$$\frac{d}{dt}\int_{C_i(t)} W\, d\Omega_{\mathbf{x}} + \oint_{\partial C_i(t)} \left(\tilde{F}(W,w_1)dx_2 - \tilde{G}(W,w_2)dx_1\right) = 0\,. \tag{9}$$

Note that for the artificial compressibility approach, the matrix $\mathbb{D}$ in $\tilde{F}$ and $\tilde{G}$ is substituted by the matrix $\mathbb{D}_\beta$.

3 Description of the Profile Motion

It is supposed that the vibrating profile has two degrees of freedom. The profile can oscillate in the vertical direction and in the angular direction around a so-called elastic axis EA (see Fig. 1). The vertical and torsional motion with small vibration amplitudes is described by the system of two ordinary differential equations:

$$\begin{aligned} m\ddot{h} + S_\varphi \ddot{\varphi} + k_{hh}h + d_{hh}\dot{h} &= -L(t)\,,\\ S_\varphi \ddot{h} + I_\varphi \ddot{\varphi} + k_{\varphi\varphi}\varphi + d_{\varphi\varphi}\dot{\varphi} &= M(t)\,, \end{aligned} \tag{10}$$

where, h is vertical displacement of the elastic axis (downwards positive) [m], φ is rotation angle around the elastic axis (clockwise positive) [rad], m is mass of the profile [kg], S_φ is static moment around the elastic axis [kg m], k_{hh} is bending stiffness [N/m], I_φ is inertia moment around the elastic axis [kg m^2] and $k_{\varphi\varphi}$ is torsional stiffness [N m/rad]. The coefficients of the proportional damping are considered in the form $d_{hh} = \varepsilon k_{hh}$ and $d_{\varphi\varphi} = \varepsilon k_{\varphi\varphi}$, where $\varepsilon \in \mathbb{R}$ is a small parameter. For more details on equations of airfoil motion, see, e.g., [9].

The aerodynamic lift force L [N] acting in the vertical direction (upwards positive) and the torsional moment M [N m] (clockwise positive) in the case of inviscid flow are defined as

$$L(t) = d \oint_{\Gamma_w(t)} \mathrm{p} n_2\, dl, \tag{11}$$

$$M(t) = d \oint_{\Gamma_w(t)} \mathrm{p}(-n_2 r_1 + n_1 r_2)\, dl, \tag{12}$$

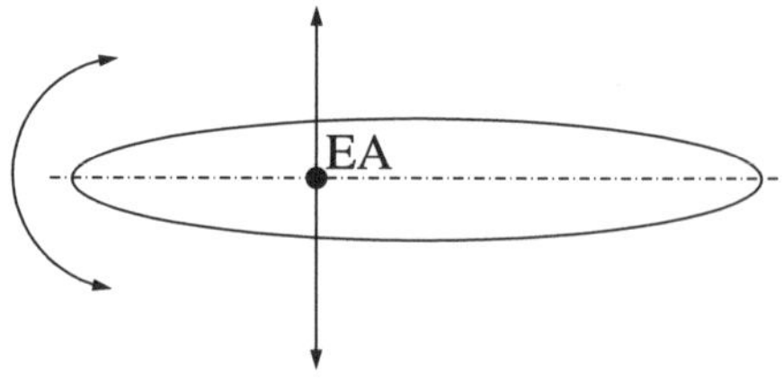

Fig. 1 Vertical and rotational motion of a profile.

Fig. 2 Airfoil segment.

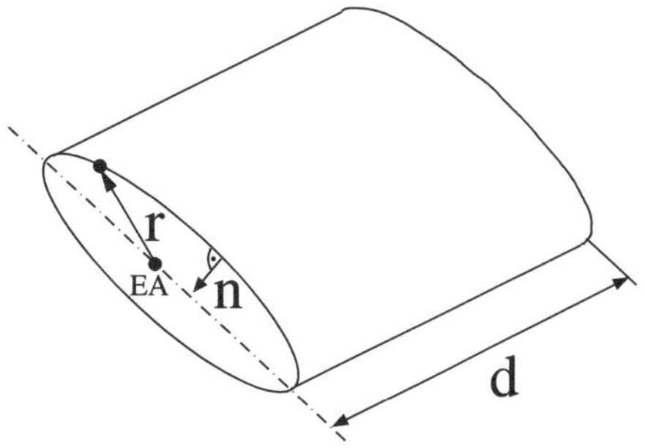

where p is pressure [Pa], d is airfoil depth [m] (see Fig. 2), $\mathbf{n} = (n_1, n_2)$ is unit inner normal to the profile surface $\Gamma_w(t)$, $\mathbf{r} = (r_1, r_2) = (x_1 - x_1^{\mathrm{EA}}, x_2 - x_2^{\mathrm{EA}})$ [m], $\mathbf{dl} = (-n_2, n_1)\,dl$ [m], (x_1, x_2) is a point on the profile surface and $(x_1^{\mathrm{EA}}, x_2^{\mathrm{EA}})$ are coordinates of the elastic axis (inside of the profile). Eqs. (11) and (12) together with the boundary conditions for velocity on moving boundaries represent the coupling of the fluid with the structure.

The system of Eqs. (10) is completed with the initial conditions prescribing values $h(0)$, $\varphi(0)$, $\dot{h}(0)$, $\dot{\varphi}(0)$. Furthermore, it is transformed to the system of first-order ordinary differential equations and solved numerically by the fourth-order Runge-Kutta method.

4 Numerical Scheme

The artificial compressibility approach is applied to the cell-centered explicit Lax-Wendroff (Richtmyer form) scheme (LWR) in a form of predictor-corrector with an added artificial viscosity (see, e.g., [6]). In the case of the dual-time stepping method (DTSM), the derivatives with respect to the real time are discretized using a three-point backward formula, which results in an implicit scheme of second-order of accuracy in time

$$W_\tau = -\frac{3W^{n+1} - 4W^n + W^{n-1}}{2\Delta t} - R\left(W^{n+1}\right) = -\tilde{R}\left(W^{n+1}\right), \tag{13}$$

where

$$R(W) = \mathbb{D}_\beta^{-1} F_{x_1} + \mathbb{D}_\beta^{-1} G_{x_2}. \tag{14}$$

The fluxes in Eq. (14) are discretized centrally in the finite volume method and an artificial dissipation term is added. The four-stage Runge-Kutta scheme is used for marching in pseudo time τ.

In the ALE formulation the inviscid fluxes are evaluated on the mesh configuration and with grid velocities such that a geometric conservation law (see, e.g., [8, 7]):

$$S_i(\mathbf{x}^{n+1}) - S_i(\mathbf{x}^n) = \int_{t^n}^{t^{n+1}} \oint_{\partial C_i(\mathbf{x})} \mathbf{wn}\, dl\, dt$$

is verified. Here, $S_i(\mathbf{x}^n) = \int_{C_i(t^n)} d\Omega_{\mathbf{x}}$ and $\mathbf{n}$ denotes an outer normal vector to the cell boundaries. The superscript n is associated with the real time and $\mathbf{x}$ represents the domain configuration at time t^n.

5 Numerical Results

Numerical results of flow induced vibrations are presented for the profile NACA 0012. A structured quadrilateral C-mesh is used in numerical simulations. The following input quantities were considered [9]: $m = 0.086622$ kg, $S_\varphi = -0.000779673$ kg m, $I_\varphi = 0.000487291$ kg m^2, $k_{hh} = 105.109$ N/m, $k_{\varphi\varphi} = 3.695582$ N m/rad, $d = 0.05$ m, $\rho = 1.225$ kg/m^3, profile chord $c = 0.3$ m. The position of the elastic axis of the profile measured along the chord from the leading edge is $x_{\mathrm{EA}} = 0.4c = 0.12$ m. The far-field flow velocities $U_\infty = \|\mathbf{u}_\infty\| = 5, 10, 15, 20, 25, 30, 35, 40, 41$ m/s were considered.

At time $t = 0$, the profile is released with the initial values $h(0) = -0.05$ m, $\dot{h}(0) = 0$, $\varphi(0) = 6°$, $\dot{\varphi}(0) = 0$.

First, the results obtained using the dual-time stepping method are presented. Fig. 3 shows the angle of rotation φ [°] (left panel) and the vertical displacement h [mm] (right panel) of the profile in dependence on time t [s]. The results refer to the upstream flow velocities $U_\infty = 5, 25$ and 41 m/s. For the velocity $U_\infty = 5$ m/s and $U_\infty = 25$ m/s, the system is stable and the free vibrations are damped by aerodynamic forces. For $U_\infty = 41$ m/s an unstable behaviour can be seen and a divergence type of instability is observed.

Further, numerical results achieved by the two strategies used for unsteady flows are compared in Fig. 4 showing comparison of the angle of rotation φ [°] and vertical displacement h [mm] of the profile in dependence on time t [s] for the upstream velocity $U_\infty = 10$ m/s. Full lines concern the artificial compressibility approach, while dash-dot lines concern the dual-time stepping method. Good agreement of the results can be observed.

6 Conclusion

The numerical solver using finite volume method for simulations of interaction of inviscid incompressible flows and a vibrating profile with two degrees of freedom has been developed. The numerical calculations performed with the smaller far-field velocities give expected results for angle and vertical displacements of the profile in time domain, i.e., the increasing damping with increasing the flow velocity. The instability of the system was observed for the far-field velocities above the critical

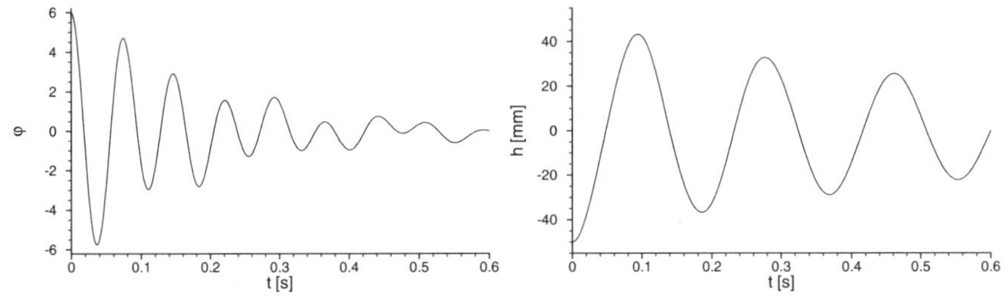

(a) $U_\infty = 5$ m/s

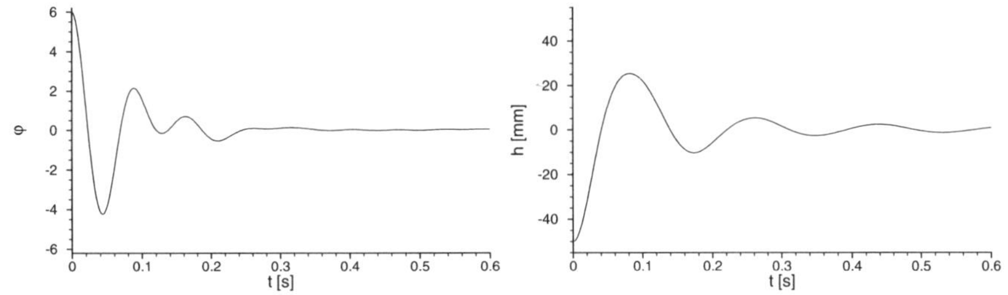

(b) $U_\infty = 25$ m/s

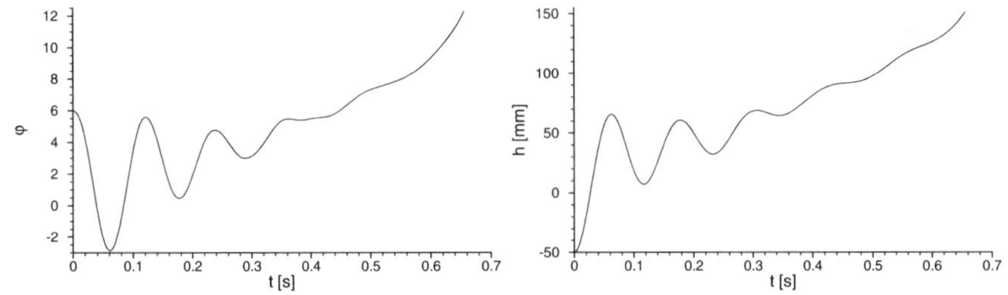

(c) $U_\infty = 41$ m/s

Fig. 3 Angle of rotation φ [°] and vertical displacement h [mm] in dependence on time t [s] for upstream flow velocities $U_\infty = 5, 25, 41$ m/s.

(here for $U_\infty = 41$ m/s). The solutions are in agreement with the numerical results presented in [9].

Two distinct approaches for numerical simulations of unsteady governing equations have been applied and the results for the smaller flow velocities ($U_\infty \leq 10$ m/s) are in a good agreement. For higher flow velocities, the DTSM scheme is faster then the LWR one as far as the computational time is concerned. It is mainly due to the time step limitation in the case of the LWR scheme.

Acknowledgements This work has partly been supported by the Research Plan of the Ministry of Education of the Czech Republic No 6840770010 and project of the Grant Agency of the Academy of Science of the Czech Republic No IAA200760613 "Computer modelling of aeroelastic phenomena for real fluid flowing past vibrating airfoils particularly after the loss of system stability".

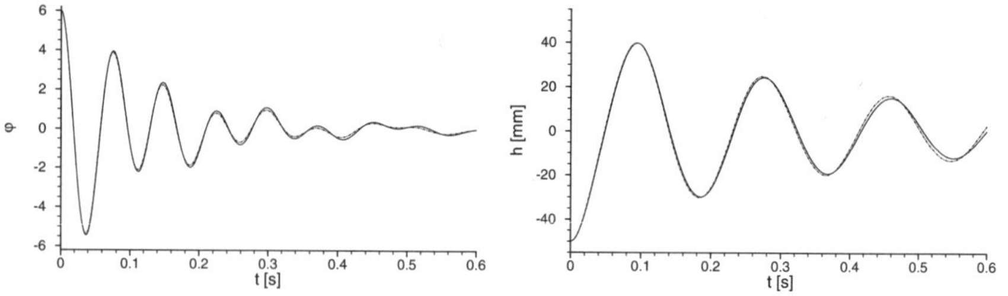

(a) Angle of rotation in dependence on time. (b) Vertical displacement in dependence on time.

Fig. 4 Comparison of the results for angle of rotation φ [°] and vertical displacement h [mm] in dependence on time t [s] obtained by the two different methods ($U_\infty = 10$ m/s): ——, data computed using LWR scheme; – · –, data computed using DTSM scheme.

References

1. Arnone, A., Liou, M.S., Povinelli, A.L.: Multigrid time-accurate integration of Navier-Stokes equations. In: AIAA Paper 93-3361 (1993)
2. Arnone, A., Marconcini, M., Pacciani, R.: On the use of dual time stepping in unsteady turbomachinery flow calculations. In: ERCOFTAC Bulletin, 42, pp. 37–42 (1999)
3. Chorin, A.J.: A numerical method for solving incompressible viscous flow problems. Journal of Computational Physics **2**(1), 12–26 (1967)
4. Formaggia, L., Nobile, F.: Stability analysis of second-order time accurate schemes for ALE-FEM. Computer methods in applied mechanics and engineering **193**, 4097–4116 (2004)
5. Gaitonde, A.L.: A dual-time method for two-dimensional unsteady incompreessible flow calculations. International journal for numerical methods in engineering **41**, 1153–1166 (1998)
6. Honzátko, R., Kozel, K., Horáček, J.: Numerical solution of 2D incompressible flow over a vibrating airfoil. In: CMFF'06 Conference Proceedings, pp. 233–240. Budapest University of Technology and Economics, Department of Fluid Mechanics, Budapest (2006)
7. Koobus, B., Farhat, C.: Second-order time-accurate and geometrically conservative implicit schemes for flow computations on unstructured dynamic meshes. Computer methods in applied mechanics and engineering **170**, 103–129 (1999)
8. Lesoinne, M., Farhat, C.: Geometric conservation laws for flow problems with moving boundaries and deformable meshes, and their impact on aeroelastic computations. Computer methods in applied mechanics and engineering **134**, 71–90 (1996)
9. Sváček, P., Feistauer, M., Horáček, J.: Numerical simulation of flow induced airfoil vibrations with large amplitudes. Journal of Fluids and Structures (2006). DOI 10.1016/j.jfluidstructs.2006.10.005

A Semi-Implicit Algorithm Based on the Augmented Lagrangian Method for Fluid-Structure Interaction

C.M. Murea

Abstract The paper presents a semi-implicit algorithm based on the Augmented Lagrangian Method for solving an unsteady fluid-structure interaction. At each time step, the position of the interface is determined in an explicit way. Then, the Augmented Lagrangian Method is employed for solving a fluid-structure coupled problem, such that the continuity of the velocity as well as the continuity of the stress hold on the interface. During the augmented lagrangian iterations, the fluid mesh does not move, which reduces the computational effort. Numerical results are presented.

1 Governing Equations

Let us denote by Ω^S the undeformed structure domain. We shall assume that its boundary admits decomposition $\partial\Omega^S = \Gamma_D \cup \Gamma_N \cup \Gamma_0$. On Γ_D the displacement will be prescribed and on Γ_N the stress is known. The initial fluid domain Ω_0^F is bounded by: Σ_1 the inflow section, Σ_2 the bottom boundary, Σ_3 the outflow section and Γ_0 the top boundary. The boundary Γ_0 is common of both domains and it represents the initial position of the fluid-structure interface. An example of initial configuration is represented in Figure 1 at the left, where $\Gamma_D = [AB] \cup [CD]$, $\Gamma_N = [DA]$ and $\Gamma_0 = [BC]$.

Under the action of the fluid stress, the structure will be deformed. At the time instant t, the fluid occupies the domain Ω_t^F bounded by the moving interface Γ_t and by the rigid boundary $\Sigma = \Sigma_1 \cup \Sigma_2 \cup \Sigma_3$ (see Figure 1 at the right).

We have assumed that the structure is governed by the linear elasticity equations, while Navier-Stokes model have been employed for the fluid flow.

Cornel Marius Murea
Laboratoire de Mathématiques, Informatique et Applications, Université de Haute-Alsace, 4, rue des Frères Lumière, 68093 Mulhouse, France,
e-mail: cornel.murea@uha.fr http://www.edp.lmia.uha.fr/murea/

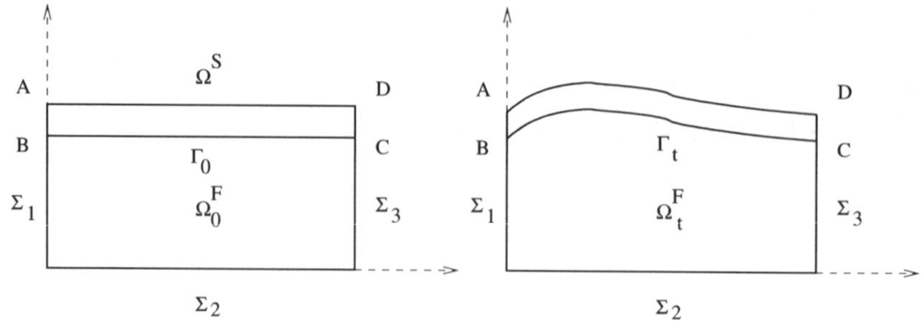

Fig. 1 Initial (left) and intermediate (right) geometrical configurations

Linear elasticity equations

Find the displacement $\mathbf{u} = (u_1, u_2)^T : \Omega^S \times [0,T] \to \mathbb{R}^2$ of the structure such that

$$\rho^S \frac{\partial^2 \mathbf{u}}{\partial t^2} - \nabla \cdot \sigma^S = \mathbf{f}^S, \quad \text{in } \Omega^S \times (0,T) \tag{1}$$

$$\sigma^S = \lambda^S (\nabla \cdot \mathbf{u}) \mathbb{I}_2 + 2\mu^S \varepsilon(\mathbf{u}) \tag{2}$$

$$\varepsilon(\mathbf{u}) = \frac{1}{2}\left(\nabla \mathbf{u} + (\nabla \mathbf{u})^T\right) \tag{3}$$

$$\mathbf{u} = 0, \quad \text{on } \Gamma_D \times (0,T) \tag{4}$$

$$\sigma^S \mathbf{n}^S = 0, \quad \text{on } \Gamma_N \times (0,T) \tag{5}$$

where $\rho^S > 0$ is the mass density of the structure, $\lambda^S > 0$ and $\mu^S > 0$ are the Lamé parameters, $\mathbf{f}^S$ is the applied volume force, $\mathbf{n}^S$ is the unit outer normal vector along the boundary $\partial\Omega^S$.

Navier-Stokes equations

Find the velocity $\mathbf{v}$ and the pressure p of the fluid such that

$$\rho^F \left(\frac{\partial \mathbf{v}}{\partial t} + (\mathbf{v} \cdot \nabla)\mathbf{v}\right) - \nabla \cdot \sigma^F = \mathbf{f}^F, \quad \forall t \in (0,T), \forall \mathbf{x} \in \Omega_t^F \tag{6}$$

$$\nabla \cdot \mathbf{v} = 0, \quad \forall t \in (0,T), \forall \mathbf{x} \in \Omega_t^F \tag{7}$$

$$\sigma^F = -p\mathbb{I}_2 + 2\mu^F \varepsilon(\mathbf{v}) \tag{8}$$

$$\sigma^F \mathbf{n}^F = \mathbf{h}_{in}, \quad \text{on } \Sigma_1 \times (0,T) \tag{9}$$

$$\sigma^F \mathbf{n}^F = \mathbf{h}_{out}, \quad \text{on } \Sigma_3 \times (0,T) \tag{10}$$

$$\mathbf{v} \cdot \mathbf{n}^F = 0, \quad \text{on } \Sigma_2 \times (0,T) \tag{11}$$

$$\boldsymbol{\tau}^F \cdot (\sigma^F \mathbf{n}^F) = 0, \quad \text{on } \Sigma_2 \times (0,T) \tag{12}$$

where $\rho^F > 0$ and $\mu^F > 0$ are the mass density and the viscosity of the fluid, $\mathbf{f}^F$ are the applied volume forces, h_{in} and h_{out} are prescribed boundary stress, $\mathbf{n}^F$ and $\boldsymbol{\tau}^F$ are the unit outer normal and tangential vectors along the boundary $\partial\Omega_t^F$.

The interface Γ_t is the image of the boundary Γ_0 by the map

$$\mathbb{T}(\mathbf{X}) = \mathbf{X} + \mathbf{u}(\mathbf{X},t).$$

Interface conditions

$$\mathbf{v}(\mathbf{X}+\mathbf{u}(\mathbf{X},t),t) = \frac{\partial \mathbf{u}}{\partial t}(\mathbf{X},t), \ \forall (\mathbf{X},t) \in \Gamma_0 \times (0,T) \tag{13}$$

$$\left(\sigma^F \mathbf{n}^F\right)_{(\mathbf{X}+\mathbf{u}(\mathbf{X},t),t)} \omega(\mathbf{X},t) = -\left(\sigma^S \mathbf{n}^S\right)_{(\mathbf{X},t)}, \ \forall (\mathbf{X},t) \in \Gamma_0 \times (0,T) \tag{14}$$

where $\omega(\mathbf{X},t) = \left\|\mathrm{cof}(\nabla\mathbb{T})\,\mathbf{n}^S\right\|_{\mathbb{R}^2}$. The equations (13) and (14) represent the continuity of velocity and of stress at the interface, respectively.
Initial conditions

$$\mathbf{u}(\mathbf{X},t=0) = \mathbf{u}^0(\mathbf{X}), \text{ in } \Omega^S \tag{15}$$

$$\frac{\partial \mathbf{u}}{\partial t}(\mathbf{X},t=0) = \dot{\mathbf{u}}^0(\mathbf{X}), \text{ in } \Omega^S \tag{16}$$

$$\mathbf{v}(\mathbf{X},t=0) = \mathbf{v}^0(\mathbf{X}), \text{ in } \Omega_0^F \tag{17}$$

The governing equations for fluid-structure interaction are (1)–(17).

2 Weak Formulation with Lagrangian Multiplier

Multiplying equation (1) by $\mathbf{w}^S = 0$ on Γ_D and from the Green formula, we obtain

$$\int_{\Omega^S} \rho^S \frac{\partial^2 \mathbf{u}}{\partial t^2} \cdot \mathbf{w}^S + a_S\left(\mathbf{u},\mathbf{w}^S\right) = \int_{\Omega^S} \mathbf{f}^S \cdot \mathbf{w}^S + \int_{\Gamma_0} \left(\sigma^S \mathbf{n}^S\right) \cdot \mathbf{w}^S, \quad \forall \mathbf{w}^S = 0 \text{ on } \Gamma_D \tag{18}$$

where

$$a_S\left(\mathbf{u},\mathbf{w}^S\right) = \int_{\Omega^S} \lambda^S (\nabla\cdot\mathbf{u})\left(\nabla\cdot\mathbf{w}^S\right) + \int_{\Omega^S} 2\mu^S \varepsilon(\mathbf{u}) : \varepsilon\left(\mathbf{w}^S\right).$$

Arbitrary Lagrangian Eulerian (ALE) framework for fluid equations

Let $\widehat{\Omega}^F$ be a reference fixed domain. Let $\mathscr{A}_t$, $t \in [0,T]$ be a family of transformations such that $\mathscr{A}_t(\widehat{\mathbf{x}}) = \widehat{\mathbf{x}}$, $\forall \widehat{\mathbf{x}} \in \Sigma_1 \cup \Sigma_2 \cup \Sigma_3$, $\mathscr{A}_t(\Gamma_0) = \Gamma_t$, $\mathscr{A}_t\left(\widehat{\Omega}^F\right) = \Omega_t^F$, where $\widehat{\mathbf{x}} = (\widehat{x}_1,\widehat{x}_2)^T \in \widehat{\Omega}^F$ represent the ALE coordinates and $\mathbf{x} = (x_1,x_2)^T = \mathscr{A}_t(\widehat{\mathbf{x}})$ the Eulerian coordinates. Let $\mathbf{v}$ be the velocity of the fluid in the Eulerian coordinates. The corresponding function in the ALE framework $\widehat{\mathbf{v}} : \widehat{\Omega}^F \times [0,T] \to \mathbb{R}^2$ is defined by $\widehat{\mathbf{v}}(\widehat{\mathbf{x}},t) = \mathbf{v}(\mathscr{A}_t(\widehat{\mathbf{x}}),t) = \mathbf{v}(\mathbf{x},t)$. We denote the ALE time derivative by $\left.\frac{\partial \mathbf{v}}{\partial t}\right|_{\widehat{\mathbf{x}}}(\mathbf{x},t) = \frac{\partial \widehat{\mathbf{v}}}{\partial t}(\widehat{\mathbf{x}},t)$ and the domain velocity by $\boldsymbol{\vartheta}(\mathbf{x},t) = \frac{\partial \mathscr{A}_t}{\partial t}(\widehat{\mathbf{x}})$.

The Navier-Stokes equations in the ALE framework give: find the velocity $\mathbf{v}$ and the pressure p of the fluid such that

$$\rho^F\left(\left.\frac{\partial \mathbf{v}}{\partial t}\right|_{\widehat{\mathbf{x}}}+((\mathbf{v}-\boldsymbol{\vartheta})\cdot\nabla)\mathbf{v}\right)-2\mu^F\nabla\cdot\varepsilon(\mathbf{v})+\nabla p=\mathbf{f}^F,\quad \forall t\in(0,T),\forall\mathbf{x}\in\Omega_t^F,$$

$$\nabla\cdot\mathbf{v}=0,\quad \forall t\in(0,T),\forall\mathbf{x}\in\Omega_t^F.$$

Multiplying the above equations by $\mathbf{w}^F$ and q respectively and using the Green formula, we have

$$\int_{\Omega_t^F}\rho^F\left.\frac{\partial \mathbf{v}}{\partial t}\right|_{\widehat{\mathbf{x}}}\cdot\mathbf{w}^F+\int_{\Omega_t^F}\rho^F(((\mathbf{v}-\boldsymbol{\vartheta})\cdot\nabla)\mathbf{v})\cdot\mathbf{w}^F+a_F\left(\mathbf{v},\mathbf{w}^F\right)+b_F\left(\mathbf{w}^F,p\right)$$
$$=\int_{\Omega_t^F}\mathbf{f}^F\cdot\mathbf{w}^F+\int_{\Sigma_1}\mathbf{h}_{in}\cdot\mathbf{w}^F+\int_{\Sigma_3}\mathbf{h}_{out}\cdot\mathbf{w}^F+\int_{\Gamma_t}\left(\sigma^F\mathbf{n}^F\right)\cdot\mathbf{w}^F,\forall\mathbf{w}^F\cdot\mathbf{n}^F=0\text{ on }\Sigma_2,$$
$$b_F(\mathbf{v},q)=0,\quad\forall q,$$

where

$$a_F\left(\mathbf{v},\mathbf{w}^F\right)=\int_{\Omega_t^F}2\mu^F\varepsilon(\mathbf{v}):\varepsilon\left(\mathbf{w}^F\right)\ \text{ and }\ b_F\left(\mathbf{w}^F,q\right)=-\int_{\Omega_t^F}\left(\nabla\cdot\mathbf{w}^F\right)q.$$

Now, we focus on the treatment of the interface conditions. The continuity of the velocity on the interface (13) will be replaced by

$$\int_{\Gamma_0}\boldsymbol{\zeta}\cdot\left(\widehat{\mathbf{v}}-\frac{\partial\mathbf{u}}{\partial t}\right)\omega=0,\quad\forall\boldsymbol{\zeta}:\Gamma_0\to\mathbb{R}^2.$$

Denoting by $\boldsymbol{\eta}=\sigma^F\mathbf{n}^F:\Gamma_t\to\mathbb{R}^2$, we have

$$\int_{\Gamma_t}\left(\sigma^F\mathbf{n}^F\right)\cdot\mathbf{w}^F=\int_{\Gamma_t}\boldsymbol{\eta}\cdot\mathbf{w}^F=\int_{\Gamma_0}\widehat{\boldsymbol{\eta}}\cdot\widehat{\mathbf{w}}^F\omega.$$

If $\widehat{\mathbf{w}}^F=\mathbf{w}^S$ on Γ_0 and the continuity of the stress on the interface holds (14), then

$$\int_{\Gamma_0}\widehat{\boldsymbol{\eta}}\cdot\widehat{\mathbf{w}}^F\omega=\int_{\Gamma_0}\widehat{\boldsymbol{\eta}}\cdot\mathbf{w}^S\omega=-\int_{\Gamma_0}\left(\sigma^S\mathbf{n}^S\right)\cdot\mathbf{w}^S.$$

Partitioned procedures by lagrange multiplier

Find $\mathbf{u}=0$ on Γ_D, $\mathbf{v}\cdot\mathbf{n}^F=0$ on Σ_2, p and $\boldsymbol{\eta}$ such that

$$\int_{\Omega^S}\rho^S\frac{\partial^2\mathbf{u}}{\partial t^2}\cdot\mathbf{w}^S+a_S\left(\mathbf{u},\mathbf{w}^S\right)-\int_{\Gamma_0}\widehat{\boldsymbol{\eta}}\cdot\mathbf{w}^S\omega=\int_{\Omega^S}\mathbf{f}^S\cdot\mathbf{w}^S,\quad\forall\mathbf{w}^S=0\text{ on }\Gamma_D,\tag{19}$$

$$\int_{\Omega_t^F}\rho^F\left.\frac{\partial \mathbf{v}}{\partial t}\right|_{\widehat{x}}\cdot\mathbf{w}^F+\int_{\Omega_t^F}\rho^F(((\mathbf{v}-\boldsymbol{\vartheta})\cdot\nabla)\mathbf{v})\cdot\mathbf{w}^F+a_F\left(\mathbf{v},\mathbf{w}^F\right)+b_F\left(\mathbf{w}^F,p\right)$$
$$+\int_{\Gamma_t}\boldsymbol{\eta}\cdot\mathbf{w}^F=\int_{\Omega_t^F}\mathbf{f}^F\cdot\mathbf{w}^F+\int_{\Sigma_1}\mathbf{h}_{in}\cdot\mathbf{w}^F+\int_{\Sigma_3}\mathbf{h}_{out}\cdot\mathbf{w}^F,\forall\mathbf{w}^F\cdot\mathbf{n}^F=0\text{ on }\Sigma_2,\tag{20}$$

$$b_F(\mathbf{v},q)=0,\quad\forall q,\tag{21}$$

$$\int_{\Gamma_0} \boldsymbol{\zeta} \cdot \left(\widehat{\mathbf{v}} - \frac{\partial \mathbf{u}}{\partial t} \right) \omega = 0, \quad \forall \boldsymbol{\zeta}. \tag{22}$$

The solution of the problem (19)–(22) verifies the boundary condition (14) weakly.

3 Time Integration Schema

Let us denote by $\Delta t > 0$ the time step, $t_{n+1} = (n+1)\Delta t$, and $\mathbf{u}^{n+1}$, $\dot{\mathbf{u}}^{n+1}$, $\ddot{\mathbf{u}}^{n+1}$ are approximations of $\mathbf{u}(\cdot, t_{n+1})$, $\frac{\partial \mathbf{u}}{\partial t}(\cdot, t_{n+1})$ $\frac{\partial^2 \mathbf{u}}{\partial t^2}(\cdot, t_{n+1})$ respectively.

The Newmark algorithm for structure equations gives: find $\mathbf{u}^{n+1}$, $\dot{\mathbf{u}}^{n+1}$, $\ddot{\mathbf{u}}^{n+1}$ such that

$$\int_{\Omega^S} \rho^S \ddot{\mathbf{u}}^{n+1} \cdot \mathbf{w}^S + a_S\left(\mathbf{u}^{n+1}, \mathbf{w}^S\right) - \int_{\Gamma_0} \widehat{\boldsymbol{\eta}}^{n+1} \cdot \mathbf{w}^S \omega = \int_{\Omega^S} \mathbf{f}^S \cdot \mathbf{w}^S, \ \forall \mathbf{w}^S = 0 \text{ on } \Gamma_D, \tag{23}$$

$$\dot{\mathbf{u}}^{n+1} = \dot{\mathbf{u}}^n + \Delta t \left((1-\delta)\ddot{\mathbf{u}}^n + \delta \ddot{\mathbf{u}}^{n+1} \right), \tag{24}$$

$$\mathbf{u}^{n+1} = \mathbf{u}^n + \Delta t \dot{\mathbf{u}}^n + (\Delta t)^2 \left(\left(\frac{1}{2} - \theta \right) \ddot{\mathbf{u}}^n + \theta \ddot{\mathbf{u}}^{n+1} \right), \tag{25}$$

where $\delta \in [0,1]$ and $\theta \in [0, \frac{1}{2}]$ are two parameters. This schema is of first order if $\delta \neq \frac{1}{2}$ and of second order if $\delta = \frac{1}{2}$. If $\delta \geq \frac{1}{2}$ and $2\theta \geq \delta$, the Newmark algorithm is unconditional stable.

Implementation of Newmark algorithm: the v-form

If $\delta \neq 0$, from the equation (24), it follows

$$\ddot{\mathbf{u}}^{n+1} = \frac{1}{\Delta t\, \delta} \left(\dot{\mathbf{u}}^{n+1} - \dot{\mathbf{u}}^n - \Delta t (1-\delta) \ddot{\mathbf{u}}^n \right).$$

We replace the above expression in (25) and we get

$$\begin{aligned} \mathbf{u}^{n+1} &= \mathbf{u}^n + \Delta t \dot{\mathbf{u}}^n + (\Delta t)^2 \left(\left(\frac{1}{2} - \theta \right) \ddot{\mathbf{u}}^n + \frac{\theta}{\Delta t\, \delta} \left(\dot{\mathbf{u}}^{n+1} - \dot{\mathbf{u}}^n - \Delta t (1-\delta) \ddot{\mathbf{u}}^n \right) \right) \\ &= \mathbf{u}^n + \Delta t \left(1 - \frac{\theta}{\delta} \right) \dot{\mathbf{u}}^n + (\Delta t)^2 \left(\frac{1}{2} - \frac{\theta}{\delta} \right) \ddot{\mathbf{u}}^n + \Delta t \frac{\theta}{\delta} \dot{\mathbf{u}}^{n+1}. \end{aligned}$$

Next, we inject $\ddot{\mathbf{u}}^{n+1}$ and $\mathbf{u}^{n+1}$ into the equation (23) and we obtain the problem: find $\dot{\mathbf{u}}^{n+1}$ such that

$$\mathscr{A}_S\left(\dot{\mathbf{u}}^{n+1}, \mathbf{w}^S\right) = \mathscr{L}_S\left(\mathbf{w}^S\right) + \int_{\Gamma_0} \widehat{\boldsymbol{\eta}}^{n+1} \cdot \mathbf{w}^S \omega, \quad \forall \mathbf{w}^S = 0 \text{ on } \Gamma_D$$

where

$$\mathscr{A}_S\left(\dot{\mathbf{u}}^{n+1},\mathbf{w}^S\right)=\int_{\Omega^S}\rho^S\frac{1}{(\Delta t)\,\delta}\dot{\mathbf{u}}^{n+1}\cdot\mathbf{w}^S+a_S\left(\Delta t\frac{\theta}{\delta}\dot{\mathbf{u}}^{n+1},\mathbf{w}^S\right),$$

$$\mathscr{L}_S\left(\mathbf{w}^S\right)=\int_{\Omega^S}\mathbf{f}^S\cdot\mathbf{w}^S+\int_{\Omega^S}\rho^S\left(\frac{1}{(\Delta t)\,\delta}\dot{\mathbf{u}}^n+\frac{1-\delta}{\delta}\ddot{\mathbf{u}}^n\right)\cdot\mathbf{w}^S$$
$$-a_S\left(\mathbf{u}^n+\Delta t\left(1-\frac{\theta}{\delta}\right)\dot{\mathbf{u}}^n+(\Delta t)^2\left(\frac{1}{2}-\frac{\theta}{\delta}\right)\ddot{\mathbf{u}}^n,\mathbf{w}^S\right).$$

For approximation of fluid equations, we employ a time integration algorithm based on the backward Euler schema and a semi-implicit treatment of the convection term: find $\mathbf{v}^{n+1}$ and p^{n+1} such that

$$\int_{\Omega^F_{t_{n+1}}}\rho^F\left(\frac{\mathbf{v}^{n+1}-\mathbf{V}^n}{\Delta t}\right)\cdot\mathbf{w}^F+\int_{\Omega^F_{t_{n+1}}}\rho^F\left(\left(\left(\mathbf{V}^n-\boldsymbol{\vartheta}^{n+1}\right)\cdot\nabla\right)\mathbf{v}^{n+1}\right)\cdot\mathbf{w}^F$$
$$+a_F\left(\mathbf{v}^{n+1},\mathbf{w}^F\right)+b_F\left(\mathbf{w}^F,p^{n+1}\right)=\int_{\Omega^F_{t_{n+1}}}\mathbf{f}^F\cdot\mathbf{w}^F$$
$$+\int_{\Sigma_1}\mathbf{h}_{in}^{n+1}\cdot\mathbf{w}^F+\int_{\Sigma_3}\mathbf{h}_{out}^{n+1}\cdot\mathbf{w}^F-\int_{\Gamma_{t_{n+1}}}\boldsymbol{\eta}^{n+1}\cdot\mathbf{w}^F,\quad\forall\mathbf{w}^F\cdot\mathbf{n}^F=0\text{ on }\Sigma_2$$
$$b_F\left(\mathbf{v}^{n+1},q\right)=0,\quad\forall q$$

where $\mathbf{V}^n(\mathbf{x})=\mathbf{v}^n\left(\mathscr{A}_{t_n}\circ\mathscr{A}_{t_{n+1}}^{-1}(\mathbf{x})\right)$ and $\boldsymbol{\vartheta}^{n+1}(\mathbf{x})=\frac{\mathscr{A}_{t_{n+1}}(\widehat{\mathbf{x}})-\mathscr{A}_{t_n}(\widehat{\mathbf{x}})}{\Delta t}=\frac{\mathbf{x}-\mathscr{A}_{t_n}\circ\mathscr{A}_{t_{n+1}}^{-1}(\mathbf{x})}{\Delta t}$.

Equivalently, we can write the before equations in a concise form: find $\mathbf{v}^{n+1}$ and p^{n+1} such that

$$\mathscr{A}_F\left(\mathbf{v}^{n+1},\mathbf{w}^F\right)+b_F\left(\mathbf{w}^F,p^{n+1}\right)=\mathscr{L}_F\left(\mathbf{w}^F\right)-\int_{\Gamma_{t_{n+1}}}\boldsymbol{\eta}^{n+1}\cdot\mathbf{w}^F,\forall\mathbf{w}^F\cdot\mathbf{n}^F=0\text{ on }\Sigma_2,$$
$$b_F\left(\mathbf{v}^{n+1},q\right)=0,\quad\forall q.$$

Semi-implicit time advancing scheme

The term "semi-implicit" used for the fully algorithm means that the interface position is computed explicitly, while the displacement of the structure, velocity and the pressure of the fluid are computed implicitly. This kind of algorithm was introduced in [1]. Our algorithm propose a different strategy based on the Augmented Lagrangian Method in order to get the continuity of the velocity and of the stress on the fluid-structure interface.

Let ρ and r be two positive real parameters.

From n to $n+1$ do

Step 1. Explicit prediction $\widetilde{\mathbf{u}}^{n+1}=\mathbf{u}^n+\Delta t\,\dot{\mathbf{u}}^n+\frac{(\Delta t)^2}{2}\ddot{\mathbf{u}}^n$

Step 2. Harmonic extension $\Delta\widetilde{\mathbf{d}}^{n+1}=0$, $\widetilde{\mathbf{d}}^{n+1}=\widetilde{\mathbf{u}}^{n+1}$ on Γ_0, $\widetilde{\mathbf{d}}^{n+1}=0$ on $\Sigma_1\cup\Sigma_2\cup\Sigma_3$

Step 3. Build mesh $\widetilde{\mathscr{T}}_h^{n+1}=\mathbb{T}\left(\widehat{\mathscr{T}_h}\right)$, where $\mathbb{T}(\widehat{\mathbf{x}})=\widehat{\mathbf{x}}+\widetilde{\mathbf{d}}^{n+1}(\widehat{\mathbf{x}})$

Step 4. Mesh velocity $\widetilde{\boldsymbol{\vartheta}}^{n+1}(\mathbf{x})=\frac{\widetilde{\mathbf{d}}^{n+1}(\widehat{\mathbf{x}})-\widetilde{\mathbf{d}}^n(\widehat{\mathbf{x}})}{\Delta t}$

Step 5. $\mathbf{v}^S_{old} = \dot{\mathbf{u}}^n$, $\widehat{\boldsymbol{\lambda}} = \widehat{\boldsymbol{\eta}}^n$

Step 6. Solve fluid-structure problem by the Augmented Lagrangian Method in the fixed mesh $\widetilde{\mathscr{T}_h^{n+1}}$ as follows

Step 6.1. Solve fluid problem: find $\mathbf{v}^F$ and p^F such that

$$\mathscr{A}_F\left(\mathbf{v}^F,\mathbf{w}^F\right)+b_F\left(\mathbf{w}^F,p^F\right) = \mathscr{L}_F\left(\mathbf{w}^F\right) - \int_{\Gamma_{\tilde{\mathbf{u}}}} \boldsymbol{\lambda}\cdot\mathbf{w}^F - r\int_{\Gamma_{\tilde{\mathbf{u}}}}\left(\mathbf{v}^F-\mathbf{v}^S_{old}\right)\cdot\mathbf{w}^F,$$
$$b_F\left(\mathbf{v}^F,q\right) = 0,$$

Step 6.2. Solve structure problem: find $\widehat{\mathbf{v}}^S$ such that

$$\mathscr{A}_S\left(\widehat{\mathbf{v}}^S,\mathbf{w}^S\right) = \mathscr{L}_S\left(\mathbf{w}^S\right) + \int_{\Gamma_0}\widehat{\boldsymbol{\lambda}}\cdot\mathbf{w}^S\omega + r\int_{\Gamma_0}\left(\widehat{\mathbf{v}}^F-\widehat{\mathbf{v}}^S\right)\cdot\mathbf{w}^S\omega$$

Step 6.3. If $\left\|\widehat{\mathbf{v}}^F-\widehat{\mathbf{v}}^S\right\|_{\Gamma_0} < tol$ then break
else $\widehat{\boldsymbol{\lambda}} = \widehat{\boldsymbol{\lambda}} + \rho\left(\widehat{\mathbf{v}}^F-\widehat{\mathbf{v}}^S\right)$; $\mathbf{v}^S_{old} = \widehat{\mathbf{v}}^S$; goto Step 6.1.

Step 7. Update
$\mathbf{v}^{n+1} = \mathbf{v}^F$, $p^{n+1} = p^F$, $\widehat{\boldsymbol{\eta}}^{n+1} = \widehat{\boldsymbol{\lambda}}$, $\dot{\mathbf{u}}^{n+1} = \widehat{\mathbf{v}}^S$,
$\ddot{\mathbf{u}}^{n+1} = \ddot{\mathbf{u}}^{n+1} = \frac{1}{\Delta t\,\delta}\left(\dot{\mathbf{u}}^{n+1}-\dot{\mathbf{u}}^n-\Delta t\,(1-\delta)\,\ddot{\mathbf{u}}^n\right)$,
$\mathbf{u}^{n+1} = \mathbf{u}^n + \Delta t\left(1-\frac{\theta}{\delta}\right)\dot{\mathbf{u}}^n + (\Delta t)^2\left(\frac{1}{2}-\frac{\theta}{\delta}\right)\ddot{\mathbf{u}}^n + \Delta t\frac{\theta}{\delta}\dot{\mathbf{u}}^{n+1}$.

The major advantage of this implementation consists in using a fixed mesh during the iterations **6.1**, **6.2**, **6.3**. The continuity of the velocity on the interface holds in the sence $\left\|\widehat{\mathbf{v}}^F-\widehat{\mathbf{v}}^S\right\|_{\Gamma_0} < tol$.

4 Numerical Results

The physical and numerical parameters used for the numerical simulations are introduced bellow. The length of the fluid domain is $L = 6\ cm$, the inflow Σ_1 and outflow Σ_3 sections are segments of length 1. The viscosity of the fluid is $\mu = 0.035\ \frac{g}{cm\cdot s}$, the volume force in fluid is $\mathbf{f}^F = (0,0)^T$, the outflow traction $\mathbf{h}_{out} = (0,0)^T$ and the inflow traction $\mathbf{h}_{in}(t) = \left(2000\left(1-cos(\frac{2\pi t}{0.025})\right),0\right)^T$ if $0 \le t \le 0.025$ and $\mathbf{h}_{in}(t) = (0,0)^T$ if $t > 0.025$.

We set the thickness of elastic wall $h = 0.1\ cm$, the Young's modulus $E = 0.75\cdot 10^6\ \frac{g}{cm\cdot s^2}$, the Poisson's ratio $\nu = 0.3$ and the mass density $\rho^S = 1.1\ \frac{g}{cm^3}$.

We have employed in Newmark algorithm the parameters: $\delta = 0.5$, $\theta = 0.25$. The time step is $\Delta t = 5\cdot 10^{-5}$, the number of time steps is $N = 1000$ which gives the final time $T = 0.05$. In augmented lagrangian algorithm, the penalization parameters $r = \rho = 10^4$ have been used.

Triangular finite elements have been employed: $\mathbb{P}_1$+*bubble* for fluid velocity and $\mathbb{P}_1$ for fluid pressure. For the structure, triangular $\mathbb{P}_1$ was used. The fluid and structure meshes are not necessary compatible on the interface.

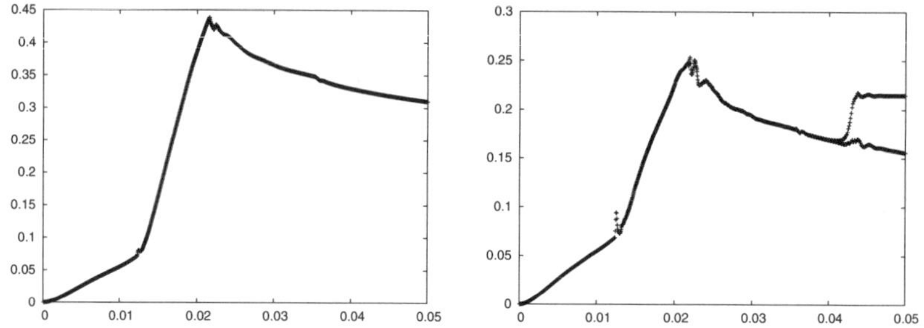

Fig. 2 Time history of the computed $\|\widehat{\mathbf{v}}^F - \widehat{\mathbf{v}}^S\|_{\Gamma_0}$ when $r = \rho = 10^4$ (left) and $r = \rho = 2 \cdot 10^4$ (right)

At each time step, 20 iterations are performed during the augmented lagrangian algorithm (steps **6.1**, **6.2**, **6.3**). We recall that the fluid and structure matrices do not change during these iterations.

When the penalization parameters $r = \rho$ increase, the difference between fluid and structure velocity on the interface diminishes (see Figure 2). Unfortunately, in the case $r = \rho = 2 \cdot 10^4$, oscillations appear after the time instant $t = 0.04$. Consequently, we can not use very large penalization parameters.

5 Concluding Remarks

A numerical procedure for solving a fluid-structure interaction problem was presented. At each time step, the position of the interface fluid-structure is determined in an explicit way. Then, the Augmented Lagrangian Method is employed for solving a fluid-structure coupled problem, such that the continuity of the velocity as well as the continuity of the stress hold on the interface. During the augmented lagrangian iterations, the fluid mesh does not move, which reduces the computational effort.

References

1. Fernández, M.A., Gerbeau, J.F., Grandmont, C.: A projection semi-implicit scheme for the coupling of an elastic structure with an incompressible fluid. Internat. J. Numer. Methods Engrg. **69**(4), 794–821 (2007)

Automated Multi-Level Substructuring for a Fluid-Solid Vibration Problem

M. Stammberger and H. Voss

Abstract The Automated Multi-Level Substructuring (AMLS) method has been developed to reduce the computational demands of frequency response analysis and has recently been proposed as an alternative to iterative projection methods like Lanczos or Jacobi–Davidson for computing a large number of eigenvalues for huge symmetric eigenvalue problems. Based on Schur complements and modal approximations of submatrices on several levels AMLS constructs a projected eigenproblem which yields good approximations of eigenvalues at the lower end of the spectrum. In this paper we discuss a structure preserving AMLS variant for nonsymmetric eigenproblems governing free vibrations of fluid–solid structures.

1 Introduction

Over the last few years, a new method for huge linear eigenvalue problems

$$Kx = \lambda Mx \tag{1}$$

where $K \in \mathbb{C}^{n\times n}$ and $M \in \mathbb{C}^{n\times n}$ are Hermitian and positive definite, known as *Automated Multi–Level Substructuring (AMLS)*, has been developed by Bennighof and co-authors, and has been applied to frequency response analysis of complex structures [1, 2]. Here the large finite element model is recursively divided into very many substructures on several levels based on the sparsity structure of the system matrices. A relatively small number of eigenmodes associated with the resulting subdomains

Markus Stammberger
Institute of Numerical Simulation, Hamburg University of Technology, D-21071 Hamburg, e-mail: markus.stammberger@tuhh.de

Heinrich Voss
Institute of Numerical Simulation, Hamburg University of Technology, D-21071 Hamburg, e-mail: voss@tuhh.de

and separating interfaces are used to represent approximately the global modes of the entire structure, thereby reducing the size of the finite element model substantially yet yielding satisfactory accuracy over a wide frequency range of interest.

Recent studies ([7, 10], e.g.) in vibro-acoustic analysis of passenger car bodies, where very large FE models with more than six million degrees of freedom appear and several hundreds of eigenfrequencies and eigenmodes are needed, have shown that for this type of problems AMLS is considerably faster than Lanczos type approaches.

On each level of the hierarchical substructuring AMLS consists of two steps. First for every substructure of the current level a congruence transformation is applied to the matrix pencil to decouple in the stiffness matrix the substructure from the degrees of freedom of higher levels. Secondly, the dimension of the problem is reduced by modal truncation of the corresponding diagonal blocks discarding eigenmodes according to eigenfrequencies which exceed a predetermined cut-off frequency. Hence, AMLS is nothing else but a projection method where the large problem under consideration is projected to a search space spanned by a smaller number of eigenmodes of clamped substructures on several levels.

Nonsymmetric eigenproblems governing free vibrations of fluid-solid structures are covered in the following way [7, 10]: one first solves the symmetric eigenproblems governing free vibrations of the fluid and the structure independently, and the original problem is then projected to the space spanned by these eigenmodes. So, the coupling is not considered when constructing the search space, but only in the projected problem. In this paper we propose an AMLS variant which incorporates the coupling already into the reduction process.

The paper is organized as follows. Section 2 summarizes the automated multi-level substructuring method for linear eigenvalue problems. Section 3 introduces the nonsymmetric eigenvalue problem governing free vibrations of a fluid-solid structure and presents the usual approach for solving it, and Section 4 proposes the structure preserving variant of AMLS. The paper closes with a numerical example demonstrating the improvement by the new approach.

2 Automated Multi-Level Substructuring

In this section we summarize the *Automated Multi-Level Substructuring* (AMLS) method for the linear eigenvalue problem

$$Kx = \lambda Mx \tag{2}$$

which was developed by Bennighof and co-workers [1, 2] over the last few years, who applied it to solve frequency response problems involving large and complex models. Here, K is the stiffness matrix and M the mass matrix of a finite element model of a structure. Both matrices are assumed to be large and sparse, and are symmetric and positive definite.

We first consider the component mode synthesis method (CMS method) which is the essential building block of the AMLS method. Assume that the graph of the matrix $|K|+|M|$ is partitioned into substructures. We distinguish only between local (i.e. interior) and interface degrees of freedom. Then K and M (after reordering) have the following form:

$$K = \begin{pmatrix} K_{\ell\ell} & K_{\ell i} \\ K_{i\ell} & K_{ii} \end{pmatrix} \quad \text{and} \quad M = \begin{pmatrix} M_{\ell\ell} & M_{\ell i} \\ M_{i\ell} & M_{ii} \end{pmatrix} \tag{3}$$

where $K_{\ell\ell}$ and $M_{\ell\ell}$ are block diagonal.

Annihilating $K_{\ell i}$ by block Gaussian elimination and transforming the local coordinates to modal degrees of freedom one obtains the equivalent pencil

$$(P^TKP, P^TMP) = \left(\begin{pmatrix} \Omega & 0 \\ 0 & \tilde{K}_{ii} \end{pmatrix}, \begin{pmatrix} I & \hat{M}_{\ell i} \\ \hat{M}_{i\ell} & \tilde{M}_{ii} \end{pmatrix} \right) \quad \text{with } P = \begin{pmatrix} \Phi & -K_{\ell\ell}^{-1}K_{\ell i} \\ 0 & I \end{pmatrix}. \tag{4}$$

Here Ω is a diagonal matrix containing the substructure eigenvalues, i.e. $K_{\ell\ell}\Phi = M_{\ell\ell}\Phi\Omega$, $\Phi^T M_{\ell\ell}\Phi = I$, and Φ contains in its columns the corresponding eigenvectors. Notice that $K_{\ell\ell}$ and $M_{\ell\ell}$ are block diagonal, and therefore it is quite inexpensive to eliminate $K_{\ell i}$ and to solve the interior eigenproblems.

In structural dynamics (4) is called Craig–Bampton form of the eigenvalue problem (2) corresponding to the partitioning (3).

Selecting some eigenmodes of problem (4) and dropping the rows and columns in (4) corresponding to the other modes one arrives at the component mode synthesis method (CMS) introduced by Hurty [6] and Craig and Bampton [4]. Usually, the modes associated with eigenvalues not exceeding a cut off threshold are kept. However, in a recent paper Lia, Bai and Gao [8] suggested a different choice based on a moment–matching analysis.

If the diagonal matrix $\tilde{\Omega}$ contains on its diagonal the kept eigenvalues, and $\tilde{M}_{\ell i}$ and $\tilde{M}_{i\ell}$ the retained rows and columns of $\hat{M}_{\ell i}$ and $\hat{M}_{i\ell}$, respectively, then the CMS approximations to the eigenpairs of (2) are obtained from the reduced eigenvalue problem

$$\begin{pmatrix} \ddot{\Omega} & 0 \\ 0 & \tilde{K}_{ii} \end{pmatrix} y = \lambda \begin{pmatrix} I & \ddot{M}_{\ell i} \\ \tilde{M}_{i\ell} & \tilde{M}_{ii} \end{pmatrix} y \tag{5}$$

AMLS generalizes CMS in the following way. Again the graph of $|K|+|M|$ is partitioned into a small number of subgraphs, but more generally than in CMS these subgraphs in turn are substructured on a number p of levels. This induces the following partitioning of the index set $I = \{1,\dots,n\}$ of degrees of freedom. I_1 is the set of indices corresponding to interface degrees of freedom on the coarsest level, and for $j = 2,\dots,p$ define I_j to be the set of indices of interface degrees of freedom on the j-th level which are not contained in I_{j-1}. Finally, let I_{p+1} be the set of interior degrees of freedom on the finest level.

With these notations the first step of AMLS is CMS with cut-off frequency γ applied to the finest substructuring. After j steps, $1 \le j \le p-1$, one derives a reduced pencil

$$\left(\begin{pmatrix} \Omega_p & O & O \\ O & K_{\ell\ell}^{(j)} & K_{\ell i}^{(j)} \\ O & K_{i\ell}^{(j)} & K_{ii}^{(j)} \end{pmatrix}, \begin{pmatrix} M_{pp}^{(j)} & M_{p\ell}^{(j)} & M_{pi}^{(j)} \\ M_{\ell p}^{(j)} & M_{\ell\ell}^{(j)} & M_{\ell i}^{(j)} \\ M_{ip}^{(j)} & M_{i\ell}^{(j)} & M_{ii}^{(j)} \end{pmatrix}\right). \tag{6}$$

where p denotes the degrees of freedom obtained in the spectral reduction in the previous steps, ℓ collects the indices in I_{p+1-j}, and i corresponds to the index set $\cup_{k=1}^{p-j} I_k$ of interface degrees of freedom on levels which are not yet treated. Applying the CMS method to the south–east 2×2 blocks of the matrices, i.e. annihilating the off–diagonal block $K_{\ell i}^{(j)}$ by block Gaussian elimination, and reducing the set of ℓ–indices by spectral truncation with cut-off frequency γ one arrives at the next level. After p CMS steps and a final spectral truncation of the lower–right blocks one obtains the reduction of problem (2) by AMLS.

3 Fluid-Solid Vibrations

Vibrations of fluid-solid structures are governed by the linear eigenvalue problem

$$\begin{pmatrix} K_s & C \\ 0 & K_f \end{pmatrix} \begin{pmatrix} u_s \\ p_f \end{pmatrix} = \lambda \begin{pmatrix} M_s & 0 \\ -\rho C^T & M_f \end{pmatrix} \begin{pmatrix} u_s \\ p_f \end{pmatrix} \tag{7}$$

where K_s and K_f are the stiffness matrices, and M_s and M_f are the mass matrices of the structure and the fluid, respectively. u_s is the structure displacement vector, p_f the fluid pressure vector, C is the coupling matrix between fluid and structure, and ρ is the fluid density.

Problem (7) is known to have real eigenvalues [9], but since it is not symmetric AMLS as introduced in Section 2 is not applicable.

Kropp et al. [7, 10] suggested to solve the eigenvalue problems $K_s\phi_s = \omega_s M_s \phi_s$ for the structure and $K_f\phi_f = \omega_f M_f \phi_f$ for the fluid by symmetric AMLS independently, and to project problem (7) to the subspace spanned by $\begin{pmatrix} \Phi_s \\ 0 \end{pmatrix}$ and $\begin{pmatrix} 0 \\ \Phi_f \end{pmatrix}$, where the columns of Φ_s and Φ_f are the eigenmodes of the structure and fluid eigenproblem, respectively, which do not exceed a given cut off frequency. Thus they obtain a projected eigenproblem

$$\begin{pmatrix} \Omega_s & \Phi_s^T C \Phi_f \\ 0 & \Omega_f \end{pmatrix} \begin{pmatrix} \tilde{u}_s \\ \tilde{p}_f \end{pmatrix} = \lambda \begin{pmatrix} \Phi_s^T M_s \Phi_s & 0 \\ -\rho \Phi_f^T C^T \Phi_s & \Phi_f^T M_f \Phi_f \end{pmatrix} \begin{pmatrix} \tilde{u}_s \\ \tilde{p}_f \end{pmatrix} \tag{8}$$

of the same structure as the original problem (7), but of much smaller dimension. Note, that the coupling is not taken into consideration in the reduction process but only in the projection of the eigenproblem.

A different approach was considered in [5]. Eliminating p_f in (7) one obtains the rational eigenvalue problem

$$K_s u_s = \lambda M_s u_s + \lambda \rho C (K_f - \lambda M_f)^{-1} C^T u_s. \tag{9}$$

Applying AMLS to the symmetric matrix pencil (K_s, M_s) and applying all transformations and projections to the coupling matrix C as well one obtains a rational eigenvalue problem of the same structure as (9) of much smaller dimension. This approach suffers the same weakness as the approach above that the coupling is not included into the reduction process.

4 A Structure Preserving Version of AMLS

In this section we propose a modified AMLS algorithm for solving the fluid-solid vibration problem (7) in order to capture the interaction of fluid and solid in a more appropriate way.

Similar to the AMLS method for symmetric problem the joint graph of $K := \begin{pmatrix} K_s & C \\ 0 & K_f \end{pmatrix}$ and $M := \begin{pmatrix} M_s & 0 \\ -\rho C^T & M_f \end{pmatrix}$ is substructured recursively on several levels, but differently from the approach of Kropp and Heiserer the coupling matrix is incorporated into the substructuring process. Hence, any substructure may consist solely of degrees of freedom from the fluid or from the solid, or caused by the coupling matrix it may contain degrees of freedom of both types.

The crucial point is to modify the AMLS algorithm such that the structure of (7) is preserved for the reduced problem, and all eigenvalues are still real.

Pure fluid or solid substructures obviously can be treated in the same way as in the symmetric AMLS method, i.e. they can be decoupled in the stiffness matrix by Gaussian elimination and reduced by modal truncation. We now describe a typical general reduction step.

After a couple of reduction steps for problem (7) one arrives at the following matrices where the unknowns have been reordered appropriately, and ρ has been set to 1 to save some space

$$\begin{pmatrix} K_{pp} & 0 & K^f_{p\ell} & 0 & K^f_{pi} & 0 & K^f_{pj} \\ K^s_{\ell p} & K^s_{\ell\ell} & C_\ell & K^s_{\ell i} & C_{\ell i} & K^s_{\ell j} & C_{\ell j} \\ 0 & 0 & K^f_{\ell\ell} & 0 & K^f_{\ell i} & 0 & K^f_{\ell j} \\ K^s_{ip} & K^s_{i\ell} & C_{i\ell} & K^s_{ii} & C_i & K^s_{ij} & C_{ij} \\ 0 & 0 & K^f_{i\ell} & 0 & K^f_{ii} & 0 & K^f_{ij} \\ K^s_{jp} & K^s_{j\ell} & C_{j\ell} & K^s_{ji} & C_{ji} & K^s_{jj} & C_j \\ 0 & 0 & K^f_{j\ell} & 0 & K^f_{ji} & 0 & K^f_{jj} \end{pmatrix}, \begin{pmatrix} K_{pp} & M^s_{p\ell} & M^f_{p\ell} & M^s_{pi} & M^f_{pi} & M^s_{pj} & M^f_{pj} \\ M^s_{\ell p} & M^s_{\ell\ell} & 0 & M^s_{\ell i} & 0 & M^s_{\ell j} & 0 \\ M^f_{\ell p} & -C^T_\ell & M^f_{\ell\ell} & -C^T_{i\ell} & M^f_{\ell i} & -C^T_{j\ell} & M^f_{\ell j} \\ M^s_{ip} & M^s_{i\ell} & 0 & M^s_{ii} & 0 & M^s_{ij} & 0 \\ M^f_{ip} & -C^T_{\ell i} & M^f_{i\ell} & -C^T_i & M^f_{ii} & -C^T_{ji} & M^f_{ij} \\ M^s_{jp} & M^s_{j\ell} & 0 & M^s_{ji} & 0 & M^s_{jj} & 0 \\ M^f_{jp} & -C^T_{\ell j} & M^f_{j\ell} & -C^T_{ij} & M^f_{ji} & -C^T_j & M^f_{jj} \end{pmatrix}. \tag{10}$$

Here p denotes the degrees of freedom obtained in the reduction steps on previous levels, ℓ collects the degrees of freedom to be handled in the current step, i corresponds to the index set of parent substructures, and j denotes interface variables of even coarser levels. A superscript s denotes the structure part, and f the fluid part of the model. Notice that the K and M part of the lower-right 6×6 block (i.e. the part which is obtained if the C blocks are replaced by 0) are symmetric and positive definite.

Annihilating $K^s_{\ell k}$ and $K^f_{\ell k}$, $k \in \{i, j\}$ by symmetric block Gaussian elimination one obtains

$$\begin{pmatrix} K_{pp} & 0 & K^f_{p\ell} & 0 & \tilde{K}^f_{pi} & 0 & \tilde{K}^f_{pj} \\ K^s_{\ell p} & K^s_{\ell\ell} & C_\ell & 0 & \tilde{C}_{\ell i} & 0 & \tilde{C}_{\ell j} \\ 0 & 0 & K^f_{\ell\ell} & 0 & 0 & 0 & 0 \\ \tilde{K}^s_{ip} & 0 & \tilde{C}_{i\ell} & \tilde{K}^s_{ii} & \tilde{C}_i & \tilde{K}^s_{ij} & \tilde{C}_{ij} \\ 0 & 0 & 0 & 0 & \tilde{K}^f_{ii} & 0 & \tilde{K}^f_{ij} \\ \tilde{K}^s_{jp} & 0 & \tilde{C}_{j\ell} & \tilde{K}^s_{ji} & \tilde{C}_{ji} & \tilde{K}^s_{jj} & \tilde{C}_j \\ 0 & 0 & 0 & 0 & \tilde{K}^f_{ji} & 0 & \tilde{K}^f_{jj} \end{pmatrix}, \begin{pmatrix} K_{pp} & M^s_{p\ell} & M^f_{p\ell} & \tilde{M}^s_{pi} & \tilde{M}^f_{pi} & \tilde{M}^s_{pj} & \tilde{M}^f_{pj} \\ M^s_{\ell p} & M^s_{\ell\ell} & 0 & \tilde{M}^s_{\ell i} & 0 & \tilde{M}^s_{\ell j} & 0 \\ M^f_{\ell p} & -C^T_\ell & M^f_{\ell\ell} & -\tilde{C}^T_{i\ell} & \tilde{M}^f_{\ell i} & -\tilde{C}^T_{j\ell} & \tilde{M}^f_{\ell j} \\ \tilde{M}^s_{ip} & \tilde{M}^s_{i\ell} & 0 & \tilde{M}^s_{ii} & 0 & \tilde{M}^s_{ij} & 0 \\ \tilde{M}^f_{ip} & -\tilde{C}^T_{\ell i} & \tilde{M}^f_{i\ell} & -\tilde{C}^T_i & \tilde{M}^f_{ii} & -\tilde{C}^T_{ji} & \tilde{M}^f_{ij} \\ \tilde{M}^s_{jp} & \tilde{M}^s_{j\ell} & 0 & \tilde{M}^s_{ji} & 0 & \tilde{M}^s_{jj} & 0 \\ \tilde{M}^f_{jp} & -\tilde{C}^T_{\ell j} & \tilde{M}^f_{j\ell} & -\tilde{C}^T_{ij} & \tilde{M}^f_{ji} & -\tilde{C}^T_j & \tilde{M}^f_{jj} \end{pmatrix} \tag{11}$$

where a tilde indicates that the associated matrix that has been modified in the elimination.

The next step is to solve the substructure eigenvalue problem

$$\begin{pmatrix} K^s_{\ell\ell} & C_\ell \\ 0 & K^f_{\ell\ell} \end{pmatrix} \begin{pmatrix} \phi \\ \psi \end{pmatrix} = \omega \begin{pmatrix} M^s_{\ell\ell} & 0 \\ -C^T_\ell & M^f_{\ell\ell} \end{pmatrix} \begin{pmatrix} \phi \\ \psi \end{pmatrix}. \tag{12}$$

which has real eigenvalues and eigenvectors due to the matrix structure described above.

If $(\phi^T, \psi^T)^T$ is a right eigenvector of (12) corresponding to the positive eigenvalue ω, then it is easily seen that $(\phi^T, \frac{1}{\omega}\psi^T)^T$ is a left eigenvector corresponding to ω.

We assume that the eigenvectors are normalized such that

$$\left(\phi_i^T, \frac{1}{\omega_i}\psi_i^T\right) \begin{pmatrix} M^s_{\ell\ell} & 0 \\ -C^T_\ell & M^f_{\ell\ell} \end{pmatrix} \begin{pmatrix} \phi_j \\ \psi_j \end{pmatrix} = \delta_{ij} \tag{13}$$

As in standard AMLS for symmetric problems we neglect eigenvectors corresponding to eigenvalues exceeding a given cut-off-frequency. Let the columns of Φ_ℓ and Ψ_ℓ be the structure and fluid part of the kept eigenvectors, respectively, and let the diagonal matrix Ω contain the according eigenvalues. Multiplying the matrices in (11) by

$$\begin{pmatrix} I & 0 & 0 & 0 & 0 \\ 0 & \Phi_\ell^T & \Omega^{-1}\Psi_\ell^T & 0 & 0 \\ 0 & 0 & 0 & I & 0 \\ 0 & 0 & 0 & 0 & I \end{pmatrix} \quad \text{and} \quad \begin{pmatrix} I & 0 & 0 & 0 \\ 0 & \Phi_\ell & 0 & 0 \\ 0 & \Psi_\ell & 0 & 0 \\ 0 & 0 & I & 0 \\ 0 & 0 & 0 & I \end{pmatrix} \tag{14}$$

from the left and right, respectively, one finally ends up with the reduced matrices in step ℓ

$$\begin{pmatrix} K_{pp} & \tilde{K}_{p\ell} & 0 & \tilde{K}^f_{pi} & 0 & \tilde{K}^f_{pj} \\ \tilde{K}_{\ell p} & \Omega & 0 & \tilde{C}_{\ell i} & 0 & \tilde{C}_{\ell j} \\ \tilde{K}^s_{ip} & \tilde{K}_{i\ell} & \tilde{K}^s_{ii} & \tilde{C}_i & \tilde{K}^s_{ij} & \tilde{C}_{ij} \\ 0 & 0 & 0 & \tilde{K}^f_{ii} & 0 & \tilde{K}^f_{ij} \\ \tilde{K}^s_{jp} & \tilde{K}_{j\ell} & \tilde{K}^s_{ji} & \tilde{C}_{ji} & \tilde{K}^s_{jj} & \tilde{C}_j \\ 0 & 0 & 0 & \tilde{K}^f_{ji} & 0 & \tilde{K}^f_{jj} \end{pmatrix}, \begin{pmatrix} M_{pp} & \tilde{M}_{p\ell} & \tilde{M}^s_{pi} & \tilde{M}^f_{pi} & \tilde{M}^s_{pj} & \tilde{M}^f_{pj} \\ \tilde{M}_{\ell p} & I & \tilde{M}^s_{\ell i} & \tilde{M}^f_{\ell i} & \tilde{M}^s_{\ell j} & \tilde{M}^f_{\ell j} \\ \tilde{M}^s_{ip} & \tilde{M}^s_{i\ell} & \tilde{M}^s_{ii} & 0 & \tilde{M}^s_{ij} & 0 \\ \tilde{M}^f_{ip} & \tilde{M}^f_{i\ell} & -\tilde{C}^T_i & \tilde{M}^f_{ii} & -\tilde{C}^T_{ji} & \tilde{M}^f_{ij} \\ \tilde{M}^s_{jp} & \tilde{M}^s_{j\ell} & \tilde{M}^s_{ji} & 0 & \tilde{M}^s_{jj} & 0 \\ \tilde{M}^f_{jp} & \tilde{M}^f_{j\ell} & -\tilde{C}^T_{ij} & \tilde{M}^f_{ji} & -\tilde{C}^T_j & \tilde{M}^f_{jj} \end{pmatrix} \tag{15}$$

Obviously the lower right 4×4 block of these matrices has the same structure as the lower right 6×6 block of the matrices in (10) demonstrating that the structure of the matrices is preserved in the AMLS reduction. Decoupling and reducing all coarser substructures in the same way one finally arrives at the projected eigenvalue problem.

5 Numerical Results

To evaluate the performance of the AMLS method above we consider the free vibrations of a tube bundle immersed in a fluid (cf. [3]). Discretizing by linear Lagrangian elements one obtains an eigenvalue problem (7) with 143082 degrees of freedom which has already been examined with a variant of AMLS for the rational eigenproblem (9) in [5]. The problem was partitioned into 2045 substructures on 11 levels.

In this problem the structure and the fluid are coupled very strongly, i.e. the resonance frequencies of the uncoupled problems do not approximate the resonance frequencies of the coupled problem well. For instance, there are 12 eigenfrequencies of the fluid which are less than the smallest structure eigenvalue ω_s whereas the coupled system has 18 eigenvalues not exceeding ω_s.

Applying AMLS for symmetric definite problems, it is usually sufficient to choose the cut-off-frequency as a small multiple of the maximum desired eigenvalue. For coupled problems, this cut-off-frequency has to be increased to get acceptable relative errors of the eigenvalue approximation. In our example we determined all eigenvalues less than 3.5, and performed the calculation with a cut-off-frequency of $\omega_c = 100$.

Figure 1 shows the relative errors of the eigenvalues obtained by AMLS incorporating the coupling into the reduction process (crosses) as compared to relative errors achieved by the AMLS approach in [7] where the coupling is included only into the projected problem (circles). Taking into account the coupling of the substructures in the reduction process diminishes the relative errors for all considered eigenvalues. In particular the ones with large relative errors in the ordinary AMLS method are improved considerably.

Differently from the original AMLS method the new method requires to solve non-symmetric eigenproblems, but this concerns only a small number of substructures on the finest level containing degrees of freedom on the interface between the fluid and the structure and there dimension is very small. Hence, the overall cost is increased only marginally.

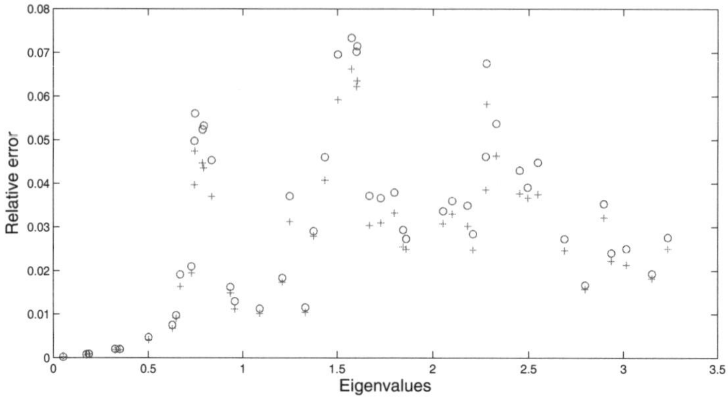

Fig. 1 Relative error for the eigenvalue approximations received from the method described above (crosses) and the method described in [7] (circles)

References

1. Bennighof, J.: Vibroacoustic frequency sweep analysis using automated multi-level substructuring. In: Proceedings of the AIAA 40th SDM Conference. St. Louis, Missouri (1999)
2. Bennighof, J., Lehoucq, R.: An automated multilevel substructuring method for the eigenspace computation in linear elastodynamics. SIAM J. Sci. Comput. **25**, 2084 – 2106 (2004)
3. Conca, C., Planchard, J., Vanninathan, M.: Fluid and Periodic Structures, *Research in Applied Mathematics*, vol. 24. Masson, Paris (1995)
4. Craig Jr., R., Bampton, M.: Coupling of substructures for dynamic analysis. AIAA J. **6**, 1313–1319 (1968)
5. Elssel, K., Voss, H.: Reducing sparse nonlinear eigenproblems by Automated Multi-Level Substructuring. Tech. rep., Institute of Mathematics, Hamburg University of Technology (2005). To appear in Adv. Engrg. Software
6. Hurty, W.: Vibration of structure systems by component-mode synthesis. J.Engrg.Mech.Div., ASCE **86**, 51–69 (1960)
7. Kropp, A., Heiserer, D.: Efficient broadband vibro–acoustic analysis of passenger car bodies using an FE–based component mode synthesis approach. J.Comput.Acoustics **11**, 139 – 157 (2003)
8. Liao, B.S., Bai, Z., Gao, W.: The important modes of subsystems: A moment-matching approach. Internat. J. Numer. Meth. Engrg. **70**, 1581 – 1597 (2007)
9. Ma, Z.D., Hagiwara, I.: Improved mode–superposition technique for modal frequency response analysis of coupled acoustic–structural systems. AIAA J. **29**, 1720 – 1726 (1991)
10. Stryczek, R., Kropp, A., Wegner, S., Ihlenburg, F.: Vibro-acoustic computations in the mid-frequency range: efficiency,evaluation, and validation. In: Proceedings of the International Conference on Noise & Vibration Engineering, ISMA 2004. KU Leuven (2004). CD-ROM ISBN 90-73802-82-2

On Numerical Approximation of Fluid-Structure Interaction Problems

P. Sváček

Abstract In this paper the problem of mutual interaction fluid flow over an airfoil with control section is addressed. The numerical approximation of turbulent incompressible viscous flow modelled by Reynolds Averaged Navier-Stokes equations is described. The application of the method on an aeroelastic problem is shown.

1 Introduction

Nonlinear fluid/structure interaction problems arise in many engineering and scientific applications. During the last years, significant advances have been made in the development and use of computational methods for fluid flows with structural interactions. The more efficient computational techniques were reached with the increasing computational power, see for example [1]. As the valuable information coming from fluid-structure interaction analysis need to be performed in many fields of industry (automobile, airplane) as well as in biomedicine, the analysis of fluid-structure interaction problems become more effective and more general. Since there is a need for effective fluid-structure interaction analysis procedures, various approaches have been proposed. In current simulations, arbitrary Lagrangian-Eulerian (ALE) formulations are now widely used. The ALE method is straightforward; however there is a number of important computational issues, cf. [5].

In the present study, attention is paid to the following aspects: second order time discretization and space finite element discretization of the Reynolds Averaged Navier-Stokes equations, Galerkin Least Squares (GLS) stabilization of the FEM, the choice of stabilization parameters, discretization of the structural model, numerical realization of the nonlinear discrete problem including the coupling of the fluid flow and airfoil motion. The developed sufficiently accurate and robust method is applied to a technically relevant case of flow-induced airfoil vibrations.

P. Sváček
Czech Technical University Prague, Faculty of Mechanical Engineering, Karlovo nám. 13, 121 35 Praha 2, Czech Republic, e-mail: Petr.Svacek@fs.cvut.cz

2 Fluid Model

In order to describe the mathematical model of the relevant technical problem we start with the two-dimensional Reynolds Averaged Navier-Stokes equations, where the Reynolds stresses are approximated with the aid of a Boussinesq approximation, see, e.g. , [10]. The moving domain is taken into account by the Arbitrary Lagrangian-Eulerian method.

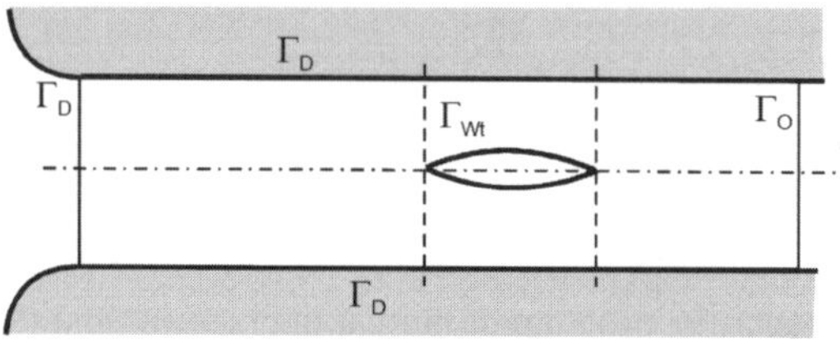

Fig. 1 Computational domain for the channel flow over DCA profile. Figure shows the Dirichlet part of boundary Γ_D, the moving part of boundary Γ_{Wt} and the outlet Γ_O.

2.1 Reynolds Equations

In order to take into account the turbulent flow we consider the fluid flow to be modelled by the system of Reynolds equations

$$\frac{D^{\mathscr{A}}\mathbf{v}}{Dt} - \nabla\cdot\left((\nu+\nu_T)\left(\nabla\mathbf{v}+(\nabla\mathbf{v})^T\right)\right) + ((\mathbf{v}-\mathbf{w}_D)\cdot\nabla)\mathbf{v} + \nabla p = 0, \quad \text{in } \Omega_t \tag{1}$$
$$\nabla\cdot\mathbf{v} = 0,$$

where $\mathbf{v}$ denotes the vector of mean part of the velocity, p denotes the mean part of the kinematic pressure, ν denotes the kinematic viscosity, and ν_T denotes the turbulent viscosity, [10]. In equation (1) we denote by $\frac{D^{\mathscr{A}}}{Dt}$ the ALE derivative and by $\mathbf{w}_D$ the domain velocity, cf. [9], [4]. The system is equipped with an initial condition and suitable boundary conditions, see Fig. 1.

2.2 Time Discretization

We consider a partition $0 = t_0 < t_1 < \cdots < T$, $t_k = k\tau$, with a time step $\tau > 0$, of the time interval $[0,T]$ and approximate the solution $\mathbf{v}(\cdot,t_n)$ and $p(\cdot,t_n)$ (defined in Ω_{t_n}) at time t_n by $\mathbf{v}^n$ and p^n, respectively. For the time discretization we employ a second-order two-step scheme using the computed approximate solution $\mathbf{v}^{n-1}$ in $\Omega_{t_{n-1}}$ and $\mathbf{v}^n$ in Ω_{t_n} for the calculation of $\mathbf{v}^{n+1}$ in the domain $\Omega_{t_{n+1}} = \Omega_{n+1}$.

We define for a fixed time $t = t_{n+1}$ the finite element spaces $\mathscr{W}$, $\mathscr{X}$ by

$$\mathscr{W} = \mathbf{H}^1(\Omega_{t_{n+1}}), \quad \mathscr{X} = \left\{\mathbf{z} \in \mathscr{W} : \mathbf{z} = 0 \text{ on } \Gamma_D \cup \Gamma_{W t_{n+1}} \right\}, \quad \mathscr{Q} = L^2(\Omega_{t_{n+1}}).$$

We approximate the ALE velocity $\mathbf{w}(t_{n+1})$ by $\mathbf{w}^{n+1}$ and set $\widehat{\mathbf{v}}^i = \mathbf{v}^i \circ \mathscr{A}_{t_i} \circ \mathscr{A}_{t_{n+1}}^{-1}$ (the symbol $\circ$ denotes the composite function). The vector-valued functions $\widehat{\mathbf{v}}^i$ are defined in the domain $\Omega_{t_{n+1}}$.

Then, on each time level t_{n+1}, the second-order two-step ALE time discretization yields the problem of finding unknown functions $\mathbf{v}^{n+1} : \Omega_{t_{n+1}} \to \mathbb{R}^2$ and $p^{n+1} : \Omega_{t_{n+1}} \to \mathbb{R}$ satisfying the equations

$$\begin{aligned}
&\frac{3\mathbf{v}^{n+1} - 4\widehat{\mathbf{v}}^n + \widehat{\mathbf{v}}^{n-1}}{2\tau} + (\tilde{\mathbf{w}} \cdot \nabla)\mathbf{v}^{n+1} - \nabla \cdot \left((\nu + \nu_T)\left(\nabla \mathbf{v} + (\nabla \mathbf{v})^T\right)\right) + \nabla p^{n+1} = 0,\\
&\nabla \cdot \mathbf{v}^{n+1} = 0,
\end{aligned} \tag{2}$$

in $\Omega_{t_{n+1}}$ and the boundary conditions. Here, $\tilde{\mathbf{w}}$ stands for the local transport velocity $\tilde{\mathbf{w}} = \mathbf{v}^{n+1} - \mathbf{w}^{n+1}$. The problem (2) is then weakly formulated in the standard form.

Problem 1 (Weak formulation of Navier-Stokes in ALE form). Find $U = (\mathbf{v}, p)$ such that

$$a(U;U,V) = f(V), \qquad \text{for all } V = (\mathbf{z}, q) \in \mathscr{X} \times \mathscr{Q}, \tag{3}$$

and appropriate Dirichlet boundary conditions are satisfied. The forms are defined by

$$\begin{aligned}
a(U^*;U,V) = {} & \left(\frac{3\mathbf{v}}{2\tau}, \mathbf{z}\right)_\Omega + c(\tilde{\mathbf{w}}; \mathbf{v}, \mathbf{z}) + \int_{\Gamma_O} \frac{1}{2}(\mathbf{v}\cdot\mathbf{n})\mathbf{v}\cdot\mathbf{z}\, dS, + \int_\Omega \frac{1}{2}(\nabla \cdot \mathbf{w}^{n+1})\mathbf{v}\cdot\mathbf{z}\, dx\\
+ {} & \left((\nu + \nu_T)\left(\nabla \mathbf{v} + (\nabla \mathbf{v})^T\right), \nabla \mathbf{z}\right)_\Omega - (p, \nabla \cdot \mathbf{z})_\Omega + (\nabla \cdot \mathbf{v}, q)_\Omega\\
f(V) = {} & \int_\Omega \frac{4\widehat{\mathbf{v}}^n - \widehat{\mathbf{v}}^{n-1}}{2\tau} \cdot \mathbf{z}\, dx - \int_{\Gamma_O} p_{\text{ref}} \mathbf{z} \cdot \mathbf{n}\, dS,
\end{aligned}$$

where $\Omega = \Omega_{n+1}$, $\tilde{\mathbf{w}} = \mathbf{v}^* - \mathbf{w}^{n+1}$ and where the trilinear skew-symmetric form c is defined by the relation $c(\mathbf{u}; \mathbf{v}, \mathbf{w}) = \int_{\Omega_{n+1}} \left[\frac{1}{2}(\mathbf{u}\cdot\nabla)\mathbf{v}\cdot\mathbf{w} - \frac{1}{2}(\mathbf{u}\cdot\nabla)\mathbf{w}\cdot\mathbf{v}\right] dx$.

2.3 Stabilized Finite Element Method

In order to apply the Galerkin FEM, we approximate the spaces $\mathscr{W}$, $\mathscr{X}$, $\mathscr{Q}$ from the weak formulation by finite dimensional subspaces $\mathscr{W}_\Delta$, $\mathscr{X}_\Delta$, $\mathscr{Q}_\Delta$, $\Delta \in (0, \Delta_0)$, $\Delta_0 > 0$, $\mathscr{X}_\Delta = \{\mathbf{v}_\Delta \in \mathscr{W}_\Delta; \mathbf{v}_\Delta|_{\Gamma_D \cap \Gamma_{W_t}} = 0\}$. In practical computations we assume that the domain Ω_{n+1} is a polygonal approximation of the region occupied by the fluid at time t_{n+1} and the spaces W_Δ, $\mathscr{X}_\Delta$, $\mathscr{Q}_\Delta$ are defined over a triangulation $\mathscr{T}_\Delta$ of the domain Ω_{n+1}, formed by a finite number of closed triangles $K \in \mathscr{T}_\Delta$. We use the standard assumptions on the system of triangulation. Here Δ denotes the size of the mesh $\mathscr{T}_\Delta$. The spaces W_Δ, $\mathscr{X}_\Delta$ and $\mathscr{Q}_\Delta$ are formed by piecewise polynomial functions, i.e. non-conforming equal order finite elements are used.

The standard Galerkin approximation of the weak formulation may suffer from two sources of instabilities. One instability is caused by the possible incompatibility of the pressure and velocity pairs of finite elements. It can be overcome by the use of pressure stabilizing terms. Further, the dominating convection requires to introduce some stabilization of the finite element scheme, as, e.g. upwinding or streamline-diffusion method. In order to overcome both difficulties, a modified Galerkin Least Squares method is applied, cf. ([6]).

We start with the definition of two parts $\mathscr{R}_K^a$ and $\mathscr{R}_K^f$ of the *local element residual* on the element $K \in \mathscr{T}_\Delta$

$$\mathscr{R}_K^a(\tilde{\mathbf{w}};\mathbf{v},p) = \frac{3\mathbf{v}}{2\tau} - \nabla\cdot\left((\nu+\nu_T)\left(\nabla\mathbf{v}+(\nabla\mathbf{v})^T\right)\right) + (\tilde{\mathbf{w}}\cdot\nabla)\mathbf{v} + \nabla p, \tag{4}$$
$$\mathscr{R}_K^f(\hat{\mathbf{v}}_n,\hat{\mathbf{v}}_{n-1}) = \frac{1}{2\tau}(4\hat{\mathbf{v}}_n - \hat{\mathbf{v}}_{n-1}).$$

The function $\widetilde{\mathbf{w}} = \mathbf{v}^* - \mathbf{w}^{n+1}$ stands for the transport velocity. Further, the GLS stabilizing terms are introduced:

$$\mathscr{L}(U_\Delta^*;U_\Delta,V_\Delta) = \sum_{K\in T_\Delta} \delta_K\left(\mathscr{R}_K^a(\tilde{\mathbf{w}};\mathbf{v},p),(\tilde{\mathbf{w}}\cdot\nabla)\mathbf{z}+\nabla q\right)_K,$$
$$\mathscr{F}(V_\Delta) = \sum_{K\in T_\Delta} \delta_K\left(\mathscr{R}_K^f(\hat{\mathbf{v}}_n,\hat{\mathbf{v}}_{n-1}),(\tilde{\mathbf{w}}\cdot\nabla)\mathbf{z}+\nabla q\right)_K, \tag{5}$$

and the grad-div stabilization terms are defined by

$$\mathscr{P}_\Delta(U,V) = \sum_{K\in\mathscr{T}_\Delta} \tau_K(\nabla\cdot\mathbf{v},\nabla\cdot\mathbf{z})_K. \tag{6}$$

Problem 2. GLS stabilized problem Find $U_\Delta = (\mathbf{v},p) \in \mathscr{W}_\Delta \times \mathscr{Q}_\Delta$ such that $\mathbf{v}$ satisfies approximately the Dirichlet boundary conditions and the equation

$$a(U_\Delta;U_\Delta,V_\Delta) + \mathscr{L}(U_\Delta;U_\Delta,V_\Delta) + \mathscr{P}_\Delta(U_\Delta,V_\Delta) = f(V_\Delta) + \mathscr{F}(V_\Delta), \tag{7}$$

holds for all $V_\Delta = (\mathbf{z},q) \in \mathscr{X}_\Delta \times \mathscr{Q}_\Delta$.

The following choice of parameters is used

$$\tau_K = \nu_K\left(1 + Re^{loc} + \frac{h_K^2}{\nu_K\,\tau}\right), \qquad \delta_K = \frac{h_K^2}{\tau_K},$$

where $\nu_K = |\nu + \nu_T|_{0,2,K}$, h_K denotes the local element size and the local Reynolds number Re^{loc} is defined as $Re^{loc} = \frac{h\|\mathbf{v}\|_K}{2\nu_K}$.

3 Spalart-Allmaras Turbulence Model

In the addressed technical problem the high Reynolds numbers means, that the flow becomes turbulent. In order to capture the turbulence phenomena, a turbulence model shall be employed. The system of equations (1) is coupled with a nonlinear partial differential equation for an additional quantity $\widetilde{\nu}$. The Spalart-Allmaras one equation turbulence model is employed:

$$\frac{\partial \widetilde{\nu}}{\partial t} + \mathbf{v} \cdot \nabla \widetilde{\nu} = \frac{1}{\beta} \left[\sum_{i=1}^{2} \frac{\partial}{\partial x_i} \left((\nu + \widetilde{\nu}) \frac{\partial \widetilde{\nu}}{\partial x_i} \right) + c_{b_2} (\nabla \widetilde{\nu})^2 \right] + G(\widetilde{\nu}) - Y(\widetilde{\nu}), \tag{8}$$

where $G(\widetilde{\nu})$ and $Y(\widetilde{\nu})$ are functions of the tensor of rotation of the mean velocity $(\omega_{ij})_{ij}$ and of the wall distance y. Here, the components of the rotation tensor are defined by $\omega_{ij} = \frac{1}{2}\left(\frac{\partial v_i}{\partial x_j} - \frac{\partial v_j}{\partial x_i}\right)$. The turbulent viscosity ν_T is defined by

$$\nu_T = \widetilde{\nu} \frac{\chi^3}{\chi^3 + c_v^3}, \qquad \chi = \frac{\widetilde{\nu}}{\nu}. \tag{9}$$

Furthermore we use the following relations (see also [11])

$$G(\widetilde{\nu}) = c_{b_1} \widetilde{S} \widetilde{\nu}, \qquad Y(\widetilde{\nu}) = c_{w_1} \frac{\widetilde{\nu}^2}{y^2} \left(\frac{1 + c_{w_3}^6}{1 + c_{w_3}^6 / g^6} \right)^{\frac{1}{6}}, \qquad \widetilde{S} = \left(S + \frac{\widetilde{\nu}}{\kappa^2 y^2} f_{\nu_2} \right),$$

$$f_{\nu_2} = 1 - \frac{\chi}{1 + \chi f_{\nu_1}}, \qquad g = r + c_{w_2}(r^6 - r), \qquad r = \frac{\widetilde{\nu}}{\widetilde{S} \kappa^2 y^2}, \qquad S = \sqrt{2 \sum_{i,j} \omega_{ij}^2},$$

and y denotes the distance from a wall. The following choice of constants is used $c_{b_1} = 0.1355$, $c_{b_2} = 0.622$, $\beta = \frac{2}{3}$, $c_v = 7.1$, $c_{w_2} = 0.3$, $c_{w_3} = 2.0$, $\kappa = 0.41$, $c_{w_1} = c_{b_1}/\kappa^2 + (1 + c_{b_2})/\beta$.

The equation (8) is time discretized with the aid of the backward Euler scheme. The numerical solution of the Spalart-Allmaras problem is performed with the aid of finite element method. The Galerkin approximations are known to be unstable for large mesh Péclet numbers. The convection-diffusion character of the problem requires to apply a stabilization procedure as streamline upwind/Petrov-Galerkin (SUPG). The use of the standard SUPG stabilization does not avoid local oscillations near sharp layers. In the FEM context the discontinuity capturing techniques (or shock capturing techniques) are usually employed. These stabilization techniques introduce additional dissipation in *crosswind* direction, cf. [8], [2]. In practical computations we follow the stabilization procedure from cf. [7].

4 Structure Model

As a structure model we consider an airfoil within the control section as shown in Fig. 2. The elastic axis is denoted by EO, the trailing edge flap is hinged at point EF at distance $\tilde{\Delta}$ after the elastic axis. By h, α and β the plunging of the elastic axis, pitching of the airfoil and rotation of the flap is denoted, respectively (see Fig. 2).

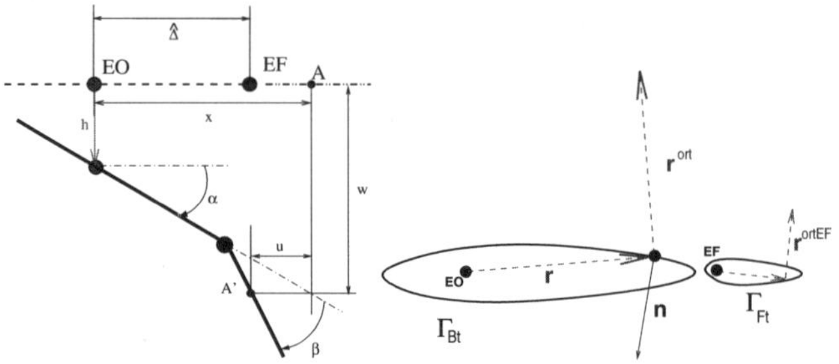

Fig. 2 The airfoil consisting from the front section with the surface Γ_{Bt} and the control section with the surface Γ_{Ft}.

The fluid motion generates an aerodynamic lift $L = L(t)$, an aerodynamic moment $M = M(t)$ and an hinge moment $M_\beta = M_\beta(t)$. By k_h, k_α and k_β the spring constant of the wing bending, the wing torsional stiffness and the flap hinge moment are denoted, respectively. The mass matrix $\mathbb{M}$ of the structural system is defined with the aid of the entire airfoil mass m, the moment of inertia I_α of the airfoil around the elastic axis and the flap moment of inertia I_β of the flap around the hinge. The equations of the motion for a flexible supported rigid airfoil with flap, cf. [3], read

$$\mathbb{M}\ddot{\eta} + \mathbb{B}\dot{\eta} + \mathbb{K}\eta = \mathbf{f}, \tag{10}$$

where

$$\mathbb{M} = \begin{pmatrix} m & S_\alpha & S_\beta \\ S_\alpha & I_\alpha & (c_\beta - e)bS_\beta + I_\beta \\ S_\beta & (c_\beta - e)bS_\beta + I_\beta & I_\beta \end{pmatrix}, \mathbb{K} = \operatorname{diag}(k_{hh}, k_{\alpha\alpha}, k_{\beta\beta})$$

and $\mathbb{D} = \operatorname{diag}(d_{hh}, d_{\alpha\alpha}, d_{\beta\beta})$, $\eta = (h, \alpha, \beta)^T$, $\mathbf{f} = \left(-L, M, M_\beta\right)^T$.

4.1 Aerodynamical Forces

In order to evaluate the aerodynamical forces we define the airfoil boundary Γ_{Wt} (the airfoil boundary moves in time) divided into the control section part $\Gamma_{Ft} \subset \Gamma_{Wt}$ and into the front part Γ_{Bt}, where $\Gamma_{Wt} = \Gamma_{Ft} \cup \Gamma_{Bt}$ and $\Gamma_{Ft} \cap \Gamma_{Bt} = \emptyset$, see Fig. 2.

Then, the aerodynamical lift force L acting in the vertical direction, the torsional moment M, the drag force D and the aerodynamical moment M_β acting on the control section part are defined by

$$L = -l \int_{\Gamma_{Wt}} \sum_{j=1}^{2} \tau_{2j} n_j \, \mathrm{dS}, \; M = l \int_{\Gamma_{Wt}} \sum_{i,j=1}^{2} \tau_{ij} n_j r_i^{\mathrm{ort}} \, \mathrm{dS}, \quad M_\beta(t) = l \int_{\Gamma_{Ft}} \sum_{i,j=1}^{2} \tau_{ij} n_j r_i^{\mathrm{ortEF}} \, \mathrm{dS}, \tag{11}$$

where

$$\tau_{ij} = \rho \left[-p\delta_{ij} + \nu \left(\frac{\partial u_i}{\partial x_j} + \frac{\partial u_j}{\partial x_i} \right) \right], \tag{12}$$

and r^{ort}, r^{ortEF}, $\mathbf{n}$, x_{EO}, x_{EF} are shown in Fig. 2.

5 Numerical Results

The numerical method was applied on a problem of a channel flow over a DCA profile. The comparison of the pressure distribution over the surface is presented in Fig. 3.

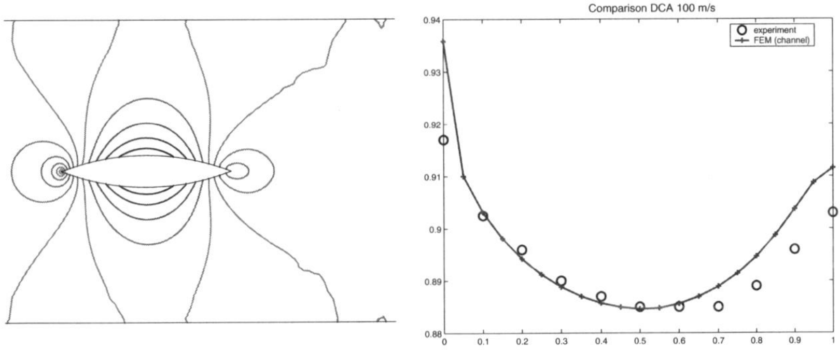

Fig. 3 Numerical approximation pressure around and on the DCA profile. The comparison with the experimental data is shown on the right-hand side.

The aeroelastic simulation for the airfoil with the flap section is shown in Fig. 4, where we set $m = 0.08662\,\mathrm{kg}$, $S_\alpha = -0.7796 \cdot 10^{-3}\,\mathrm{kg\ m}$ $S_\beta = 0\,\mathrm{kg\ m}$ $I_\alpha = 4.87291 \cdot 10^{-4}\,\mathrm{kg\ m^2}$ $\tilde{\Delta} = 0.12\,\mathrm{m}$, $I_\beta = 10^{-6}\,\mathrm{kg\ m^2}$ $k_{hh} = 105.109\,\mathrm{N\,m^{-1}}$, $k_{\alpha\alpha} = 3.695582\mathrm{N\,m\,rad^{-1}}$ and $k_\beta = 0.025\mathrm{N\,m\,rad^{-1}}$.

Acknowledgements This research was supported under grant No. 201/05/P142 of the Grant Agency of the Czech Republic and under the Research Plan MSM 6840770003 of the Ministry of Education of the Czech Republic.

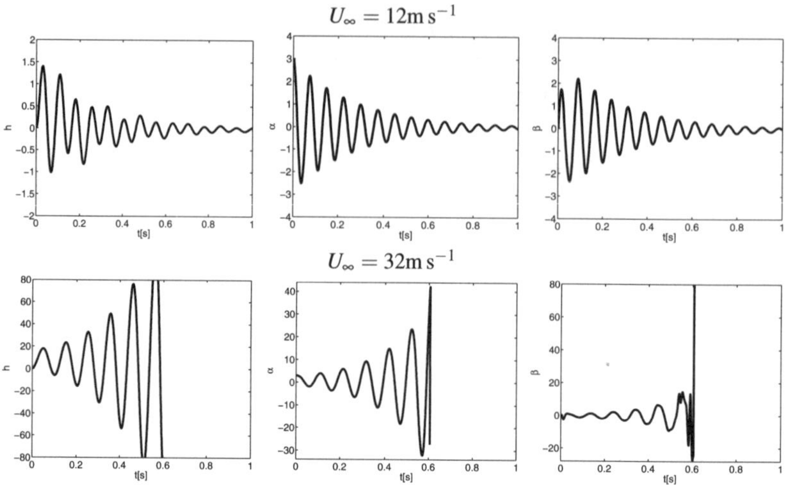

Fig. 4 The aeroelastic response for the airfoil with three degrees of freedom for far field velocity 12 m / s and 32 m / s

References

1. Bathe, K.J. (ed.): Computational Fluid and Solid Mechanics 2007. Elsevier (2007)
2. Codina, R.: A discontinuity capturing crosswind-dissipation for the finite element solution of the convection diffusion equation. Computational Methods in Applied Mechanical Engineering **110**, 325–342 (1993)
3. Dowell, E.H.: A Modern Course in Aeroelasticity. Kluwer Academic Publishers, Dodrecht (1995)
4. Feistauer, M., Horáček, J., Sváček, P.: Numerical simulation of flow induced airfoil vibrations with large amplitudes. Journal of Fluids and Structure **23**(3), 391–411 (2007). ISSN 0889-9746
5. Förster, C., Wall, W.A., Ramm, E.: On the geometric conservation law in transient flow calculations on deforming domains. International Journal For Numerical Methods In Fluids **50**, 1369–1379 (2006)
6. Gelhard, T., Lube, G., Olshanskii, M.A., Starcke, J.H.: Stabilized finite element schemes with LBB-stable elements for incompressible flows. Journal of Computational and Applied Mathematics **177**, 243–267 (2005)
7. Houzeaux, G.: A geometrical domain decomposition method in computational fluid dynamics. Ph.D. thesis, UPC (2003)
8. Hughes, T.J.R., Franca, L.P., Balestra, M.: A new finite element formulation for computational fluid dynamics: V. circumventing the Babuška-Breezzi condition: a stable Petrov-Galerkin formulation of the Stokes problem accomodating equal order interpolation. Computer Methods in Applied Mechanical Engineering **59**, 85–89 (1986)
9. Nomura, T., Hughes, T.J.R.: An arbitrary Lagrangian-Eulerian finite element method for interaction of fluid and a rigid body. Computer Methods in Applied Mechanics and Engineering **95**, 115–138 (1992)
10. Pope, S.B.: Turbulent Flows. Cambridge University Press (2000)
11. Wilcox, D.C.: Turbulence Modeling for CFD. DCW Industries (1993)

Optimal Control Problems

Fishways Design: An Application of the Optimal Control Theory

L.J. Alvarez-Vázquez, A. Martínez, M.E. Vázquez-Méndez, and M.A. Vilar

Abstract The main objective of this work is to present an application of mathematical modelling and optimal control theory to an ecological engineering problem related to preserve and enhance natural stocks of salmon and other fish which migrate between saltwater to freshwater. Particularly, we study the design (first) and the management (second) of a hydraulic structure (*fishway*) that enable fish to overcome stream obstructions as a dam or a weir. The problems are formulated within the framework of the optimal control of partial differential equations. They are approximated by discrete unconstrained optimization problems and then, solved by using a gradient free method (the Nelder-Mead algorithm). Finally, numerical results are showed for a standard real-world situation.

1 Introduction

Many types of fish undertake migrations on a regular basis, on time scales ranging from daily to annual, and with distances ranging from a few meters to thousands of kilometers. In this work we take attention on *diadromous fish* which migrate between salt and fresh water. The best known diadromous fish is salmon, which is capable of going hundreds of kilometers upriver. When man makes a barrier in a river (for example, a dam or a weir) he must install a *fishway* to allow salmon (and other fish) to overcome it.

Fishways are hydraulic structures placed on or around man-made barriers to assist the natural migration of diadromous fish. There are several types of fishways,

L.J. Alvarez-Vázquez and A. Martínez
Departamento de Matemática Aplicada II. ETSI Telecomunicación. Universidad de Vigo. 36310 Vigo. Spain, e-mail: lino@dma.uvigo.es, aurea@dma.uvigo.es

M.E. Vázquez-Méndez and M.A. Vilar
Departamento de Matemática Aplicada. Escola Politécnica Superior. Universidad de Santiago de Compostela. 27002 Lugo. Spain, e-mail: ernesto@usc.es, miguel@usc.es

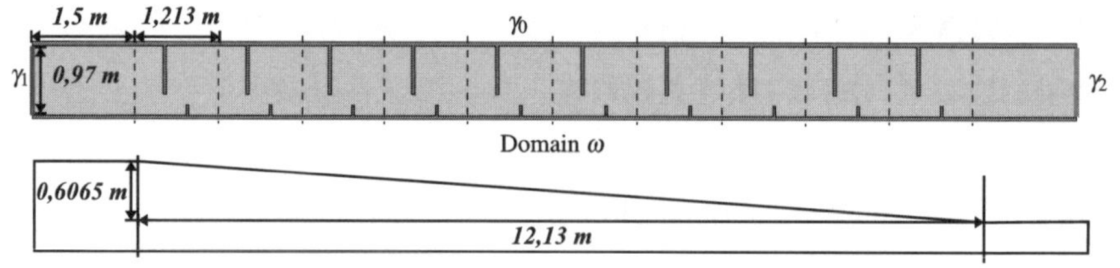

Fig. 1 Ground plant (domain ω) and elevation of the fishway under study.

but the more generally adopted for upstream passage of fish in stream obstructions is the *vertical slot fishway*. It consists of a rectangular channel with a sloping floor that is divided into a number of pools (see Fig. 1). Water runs downstream in this channel, through a series of vertical slots from one pool to the next one below. The water flow forms a jet at the slot, and the energy is dissipated by mixing in the pool. Fish ascends, using its burst speed, to get past the slot, then it rests in the pool till the next slot is tried. As said above, the objective of a fishway is enabling fish to overcome obstructions. In order to get it, water velocity in the fishway must be controlled. In the zone of the channel near the slots, the velocity must be close to a desired velocity suitable for fish leaping and swimming capabilities. In the remaining of the fishway, the velocity must be close to zero for making possible the rest of the fish. Moreover, in all the channel, flow turbulence must be minimized.

If a new fishway is going to be built, water velocity can be controlled through the location and length of the baffles separating the pools. On the opposite, if the fishway is already built, it only can be controlled by determining the flux of inflow water. In this work we are going to use mathematical modelling and optimal control theory to study these two situations: first related to the optimal design of a new fishway, and second related to the optimal management of a fishway already built. So, in (section 2) we present a mathematical model (shallow water equations) to simulate the water velocity in a fishway, and give a mathematical expression to evaluate the quality of that velocity field. Section 3 studies the problem related to the optimal design of a fishway to be built: we describe the problem, formulate it as a shape optimization problem, and approximate it by a discrete unconstrained optimization problem. Section 4 is devoted to the second problem, related to the management of a fishway already built. Finally, in section 5 we give numerical results for a standard fishway.

2 Numerical Simulation of Water Velocity in a Fishway

Let $\omega \subset \mathbb{R}^2$ be the ground plant of a standard fishway consisting of a rectangular channel divided into ten pools with baffles and sloping floor, and two transition pools, one at the beginning and other at the end of the channel, without baffles and flat floor. A scheme of the fishway can be seen in Fig. 1: water enters by the left side and runs downstream to the right side, and fish ascend in the opposite direction.

Water flow in the domain ω along the time interval $(0,T)$ is governed by:

$$\left.\begin{array}{ll} \dfrac{\partial H}{\partial t} + \nabla.\mathbf{Q} = 0 & \text{in } \omega \times (0,T) \\ \dfrac{\partial \mathbf{Q}}{\partial t} + \nabla.(\dfrac{\mathbf{Q}}{H} \otimes \mathbf{Q}) + gH\nabla(H-\eta) = \mathbf{f} & \text{in } \omega \times (0,T) \end{array}\right\} \quad (1)$$

where

- $H(x,y,t)$ is the height of water at point $(x,y) \in \omega$ at time $t \in (0,T)$,
- $\mathbf{u}(x,y,t) = (u,v)$ is the averaged horizontal velocity of water,
- $\mathbf{Q}(x,y,t) = \mathbf{u}H$ is the areal flow per unit depth,
- g is the gravity acceleration,
- $\eta(x,y)$ represents the bottom geometry of the fishway,
- $\mathbf{f}$ collects all the effects of bottom friction, atmospheric pressure, and so on.

Shallow water equations (1) must be completed with a set of initial and boundary conditions. We define three different parts in the boundary of ω: the lateral boundary of the channel, γ_0, the inflow boundary, γ_1, and the outflow boundary, γ_2. We consider $\mathbf{n}$ the unit outer normal vector to boundary $\gamma_0 \cup \gamma_1 \cup \gamma_2$. Thus, we assume the normal flux to be null on the lateral walls of the fishway, we impose an inflow flux in the normal direction, and we fix the height of water on the outflow boundary, i.e.

$$\left.\begin{array}{ll} H(0) = H_0,\ \mathbf{Q}(0) = \mathbf{Q}_0 & \text{in } \omega \\ \mathbf{Q}.\mathbf{n} = 0 & \text{on } \gamma_0 \times (0,T) \\ \mathbf{Q} = q\mathbf{n} & \text{on } \gamma_1 \times (0,T) \\ H = H_2 & \text{on } \gamma_2 \times (0,T) \end{array}\right\} \quad (2)$$

By using this notation we can give a mathematical expression to evaluate the quality of water velocity in the fishway. We have two objectives:

1. In the zone of the channel near the slots (say the lower third) the velocity must be as close as possible to a typical horizontal velocity c suitable for fish leaping and swimming capabilities, and in the remaining of the fishway the velocity must be close to zero for making possible the rest of the fish. That is, water velocity $\mathbf{u} = \frac{\mathbf{Q}}{H}$ must be close to the following target velocity:

$$\mathbf{v}(x_1,x_2) = \begin{cases} (c,0), & \text{if } x_2 \leq \frac{1}{3}W, \\ (0,0), & \text{otherwise,} \end{cases} \quad (3)$$

   where W is the width of the channel (in our case, $W = 0.97\,m$).
2. Flow turbulence must be minimized in all the channel.

According to this, for a weight parameter $\xi \geq 0$, we define the objective function

$$J = \frac{1}{2}\int_0^T \int_\omega \|\frac{\mathbf{Q}}{H} - \mathbf{v}\|^2 + \frac{\xi}{2}\int_0^T \int_\omega |curl(\frac{\mathbf{Q}}{H})|^2 \quad (4)$$

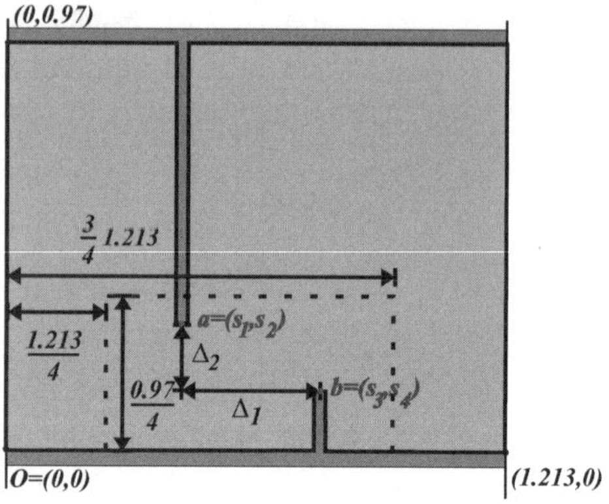

Fig. 2 Scheme of the first pool.

In order to evaluate J, firstly we have to solve the shallow water equations (1) with initial and boundary conditions (2). In this work we use an implicit discretization in time, upwinding the convective term by the method of characteristics, and Raviart-Thomas finite elements for the space discretization (the whole details of the numerical scheme can be seen in [3]). So, for the time interval $(0,T)$ we choose a natural number N, consider the time step $\Delta t = \frac{T}{N} > 0$ and define the discrete times $t_k = k\Delta t$ for $k = 0,\ldots,N$. We also consider a Lagrange-Galerkin finite element triangulation τ_h of the domain ω. Thus, the numerical scheme provides us, for each discrete time t_k, with an approximated flux $\mathbf{Q}_h^k$ and an approximated height H_h^k, which are piecewise-linear polynomials and discontinuous piecewise-constant functions, respectively. With these approximated fields we can compute the approximated velocity $\mathbf{u}_h^k = \frac{\mathbf{Q}_h^k}{H_h^k}$, and approach the value of J by

$$J_h^{\Delta t} = \frac{\Delta t}{2} \sum_{k=1}^{N} \sum_{E\in\tau_h} \left[\int_E \|\mathbf{u}_h^k - \mathbf{v}\|^2 + \xi \int_E |curl(\mathbf{u}_h^k)|^2\right] \tag{5}$$

3 First Problem: Design of a Fishway to Be Built

In this section we are going to study the optimal design of a new fishway: we will control the water velocity through the location and length of the baffles in the pools. For the channel described in previous section, we assume that the structure of the ten pools is the same (the shape of the complete fishway is determined by the shape of the first pool), and we take the two midpoints corresponding to the end of the baffles in the first pool (points $a = (s_1, s_2)$ and $b = (s_3, s_4)$ in Fig. 2) as design variables. We are looking for points a and b which provide the best velocity for fish (minimizing the function J given by (4)), but, previously, we must impose several design constraints on these points: first, we assume that points a and b are inside the dashed rectangle of Fig. 2, that is, the following eight relations must be satisfied:

$$\frac{1}{4}1.213 \leq s_1, s_3 \leq \frac{3}{4}1.213, \qquad 0 \leq s_2, s_4 \leq \frac{1}{4}0.97. \tag{6}$$

The second type of constraints are related to the fact that the vertical slot must be large enough so that fish can pass comfortably through it. This translates into:

$$\Delta_1 = s_3 - s_1 \geq 0.1, \qquad \Delta_2 = s_2 - s_4 \geq 0.05. \tag{7}$$

Then, the first optimization problem can be formulated as follows:
Problem ($\mathscr{P}_1$): Find ω, that is, $s = (a,b) = (s_1, s_2, s_3, s_4) \in \mathbb{R}^4$ verifying constraints (6) and (7), in such a way that $\mathbf{Q}$ and H, given by the solution of the state system (1)-(2) on the fishway $\omega = \omega(s)$, minimize the cost function $J \equiv J(s)$ given by (4).

A mathematical analysis of problem ($\mathscr{P}_1$) can be found in [1]. For its numerical resolution we propose a penalty method. Particularly, for a large enough parameter $\alpha_1 > 0$, we approximate ($\mathscr{P}_1$) by the discrete unconstrained optimization problem

$$\min \Phi_1(s) \tag{8}$$

where, for $s \in \mathbb{R}^4$, the value of $\Phi_1(s)$ can be computed from the following algorithm:
Step 1. Consider the corresponding domain $\omega(s)$ and generate its new mesh $\tau_h(s)$.
Step 2. Solve the state system (1)-(2) on $\omega(s)$ as proposed in section 2, and compute $J_h^{\Delta t} \equiv J_h^{\Delta t}(s)$ given by (5).
Step 3. Define $\tilde{\Phi}_1(s)$ in such a way that $\tilde{\Phi}_1(s) \leq 0 \Leftrightarrow s$ verifies (6) and (7), that is,

$$\begin{aligned}\tilde{\Phi}_1(s) = \max\{&\tfrac{1}{4}1.213 - s_1, \tfrac{1}{4}1.213 - s_3, s_1 - \tfrac{3}{4}1.213, s_3 - \tfrac{3}{4}1.213, -s_2, \\ &-s_4, s_2 - \tfrac{1}{4}0.97, s_4 - \tfrac{1}{4}0.97, 0.1 - s_3 + s_1, 0.05 - s_2 + s_4\}\end{aligned} \tag{9}$$

Step 4. Compute the value of the discrete penalty function

$$\Phi_1(s) = J_h^{\Delta t}(s) + \alpha_1 \max\{\tilde{\Phi}_1(s), 0\}. \tag{10}$$

To solve the problem (8) we use a gradient free method, the Nelder-Mead algorithm [5], which merely compares function values. Although the Nelder-Mead algorithm is not guaranteed to converge in the general case, it has good convergence properties in low dimensions. Moreover, to prevent stagnation at a non-optimal point, we use the oriented restarting recently proposed by Kelley [4].

4 Second Problem: Management of a Fishway Already Built

The second problem consists of the optimal management of a fishway already built. We consider an existing fishway (the domain $\omega \in \mathbb{R}^2$ is known and fixed) and we look for the flux across the inflow boundary providing a suitable water velocity in the fishway. Here, in the state system (1)-(2) the domain ω is a datum, and the control variable is the function $q(t)$, inflow flux for the time interval $(0,T)$. For this problem, we also have constraints on the control: since we need to inject water through γ_1,

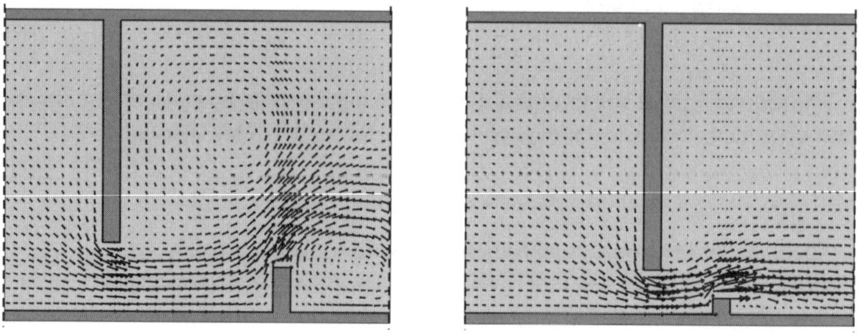

Fig. 3 Problem $(\mathscr{P}_1)$: Initial (left) and optimal (right) velocities in the central pool at $t = 300s$.

q must be negative and - due technological reasons - bounded. Thus, we are led to consider only the admissible fluxes in the set $U_{ad} = \{l \in L^2(0,T) \ : \ -B \leq l \leq 0\}$, with $B > 0$ a technological bound for water inflow.

Then, the optimal management of the fishway is formulated as:

Problem $(\mathscr{P}_2)$: Find the inflow flux $q \in U_{ad}$ such that, verifying the state system $(1)-(2)$, minimizes the cost function $J \equiv J(q)$ given by (4).

From a mathematical point of view, $(\mathscr{P}_2)$ is very different from $(\mathscr{P}_1)$ (a mathematical analysis of $(\mathscr{P}_2)$ can be seen in [2]), however its numerical resolution can be done in a similar way. In effect, due to technological reasons (flow control mechanisms cannot act upon water flow in a continuous way, but discontinuously at short time periods) we seek the control among the piecewise-constant functions. So, for the time interval $[0,T]$ we choose a number $M \in \mathbb{N}$, we consider the time step $\Delta\tau = \frac{T}{M} > 0$ and we define the discrete times $\tau_m = m\Delta\tau$ for $m = 0,1,\ldots,M$. Thus, a function $q \in L^2(0,T)$ which is constant at each subinterval determined by the grid $\{\tau_0, \tau_1, \ldots, \tau_M\}$ is completely fixed by the set of values $q^{\Delta\tau} = (q^0, q^1, \ldots, q^{M-1}) \in \mathbb{R}^M$, where $q^m = q(\tau_m)$, $m = 0,\ldots,M-1$. For a given $q^{\Delta\tau}$, the shallow water equations can be solved as proposed in section 2, and we can compute the value of $J_h^{\Delta t} \equiv J_h^{\Delta t}(q^{\Delta\tau})$ given by (5). Then, for a large enough penalty parameter $\alpha_2 > 0$, the problem $(\mathscr{P}_2)$ can be approximated by

$$\min \Phi_2(q^{\Delta\tau}) \tag{11}$$

where, for $q^{\Delta\tau} \in \mathbb{R}^M$, the value of $\Phi_2(q^{\Delta\tau})$ is computed from the following steps:

Step 1. Solve the state system (1)-(2) with a boundary condition $q^{\Delta\tau}$ on γ_1, and compute $J_h^{\Delta t} \equiv J_h^{\Delta t}(q^{\Delta\tau})$ given by (5).

Step 2. Compute the value of the discrete penalty function

$$\Phi_2(q^{\Delta\tau}) = J_h^{\Delta t}(q^{\Delta\tau}) + \alpha_2 \sum_{m=0}^{M-1} \max\{q^m, -B - q^m, 0\} \tag{12}$$

For small M, problem (11) can be also solved by using the Nelder-Mead method.

5 Numerical Results

We present here numerical results for a fishway as given in Fig. 1. Both initial and boundary conditions have been taken as constant: $\mathbf{Q}_0 = (0,0)\,m^2s^{-1}$, $H_0 = H_2 = 0.5\,m$. The time interval was $T = 300\,s$. For second member $\mathbf{f}$ consider the bottom friction stress with a Chezy coefficient of $57.36\,m^{0.5}s^{-1}$. For the cost function we have taken a target velocity $c = 0.8\,ms^{-1}$ and $\xi = 0$. Finally, for the time discretization we have taken $N = 3000$. Although we have developed many numerical experiences, we only present here one example for each problem.

5.1 *Optimal Design: Problem* $(\mathscr{P}_1)$

We have taken a constant inflow flux $q = -\frac{0.065}{0.97}m^2s^{-1}$, $\alpha_1 = 500$ and for the different space discretizations we have tried several regular triangulations of about 9500 elements. Thus, applying the Nelder-Mead algorithm, we have passed, after 76 function evaluations, from an initial cost $\Phi_1 = 1046.74$ for a random shape, to the minimum cost $\Phi_1 = 239.44$, corresponding to the optimal design variables $a = (0.577, 0.147)$, $b = (0.818, 0.054)$. Fig. 3 shows the water velocity at final time of the simulation in the central pool, corresponding to the initial random configuration (left), and to the optimal configuration given by a and b (right). It can be seen how, in the controlled case, the optimal velocity is close to the target velocity $\mathbf{v}$, and the two large recirculation regions at both sides of the slot are highly reduced.

5.2 *Optimal Management: Problem* $(\mathscr{P}_2)$

In this case we have taken a fixed fishway given by $a = (0.525, 0.121)$, $b = (0.660, 0610)$, a regular triangulation of 10492 elements, $\alpha_2 = 10^4$, $B = 0.12m^2s^{-1}$ and $M = 4$. We have passed, after 66 function evaluations, from an initial cost $\Phi_2 = 612.37$ to the minimum cost $\Phi_2 = 431.27$, corresponding to the optimal flux $q^0 = -0.114$, $q^1 = -0.085$, $q^2 = -0.066$, $q^3 = -0.116$. Fig. 4 shows water velocities at $t = 200\,s$ and $t = 300\,s$ in the middle of the fishway, corresponding to the initial random flux (left), and to the optimal flux (right).

6 Conclusions

Optimization and optimal control, joined to mathematical modelling, have shown to be a powerful tool for hydraulic engineering problems. Particularly, two realistic problems have been formulated and solved: a shape optimization problem arising in the building of a new fishway, and a boundary optimal control problem related to

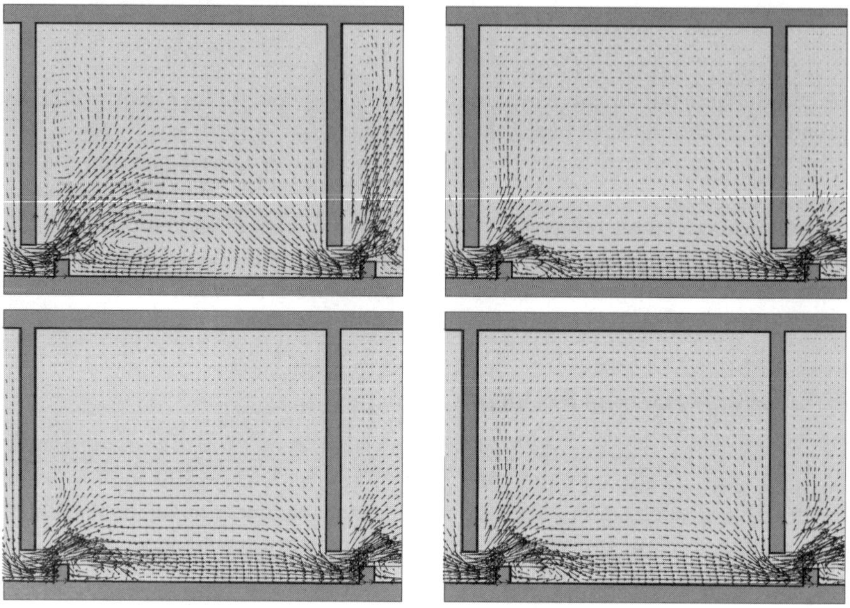

Fig. 4 Problem ($\mathscr{P}_2$): Initial (left) and optimal (right) velocities in the middle of the fishway at times $t = 200s$ and $t = 300s$.

the management of a fishway already built. From numerical experiences we observe that (a) controlling the water velocity in fishways is necessary, if we want them to fulfil their task, and (b) controlling the flux of inflow water in a vertical slot fishway can be useful, but a good shape design is vital to guarantee a correct performance.

Acknowledgements The research contained in this work was supported by Project MTM2006-01177 of Ministerio de Educación y Ciencia (Spain).

References

1. Alvarez-Vázquez, L., Martínez, A., Vázquez-Méndez, M., Vilar, M.: An optimal shape problem related to the realistic design of river fishways. Ecol. Eng. **in press**
2. Alvarez-Vázquez, L., Martínez, A., Vázquez-Méndez, M., Vilar, M.: Vertical slot fishways: Mathematical modeling and optimal management. Journal of Computational and Applied Mathematics (2007). DOI 10.1016/j.cam.2007.04.016
3. Bermúdez, A., Rodríguez, C., Vilar, M.A.: Solving shallow water equations by a mixed implicit finite element method. IMA J. Numer. Anal. **11**(1), 79–97 (1991)
4. Kelley, C.T.: Detection and remediation of stagnation in the Nelder-Mead algorithm using a sufficient decrease condition. SIAM J. Optim. **10**(1), 43–55 (electronic) (1999)
5. Nelder, J., Mead, R.: A simplex method for function minimization. Computer J. **7**, 308–313 (1965)

Moving Domain by Galerkin-Level Set Strategy: Application to Shape Geodesics

L. Blanchard and J.P. Zolésio

Abstract In this paper, we use the concept of connecting tubes introduced in [7][5] and we consider the geodesic tube construction between two sets according to the tube shape metric. The first section is devoted to the tube analysis and formulation associated to the Galerkin-Level Set strategy. This new variational formulation consists in parameterizing the level set function in a finite vector space. Consequently, the aim of a Galerkin-Level Set method is more focused on topology that on high accuracy for the boundary approximation. However, the main advantage of this method, over the traditional level set formulation, concerns the standard partial differential equation (PDE) evolution for the level set function that, in the Galerkin-Level Set method turns into a system of ordinary differential equations and we avoid any "usual" instability. The second section concerns a shape identification problem associated to an Hilbert space metric using the Galerkin-Level Set method. In the last section, the geodesic tube construction is made by a optimization process based on a shape gradient calculus. Finally, a geodesic tube construction between two sets is validated by numerical experiments in 3D.

1 Tube Connection by Galerkin-Level Set Strategy

Large evolutions of smooth domains and geometries are nicely modeled by level-sets in the form : $\Omega(t) = \{x \in \mathbb{D} \mid \Phi(t,x) > 0\}$, where $\mathbb{D}$ is a smooth bounded domain in $\mathbb{R}^n$, and where the feasible set for the level set function is : $\Phi \in \{\Phi \in C^1(\bar{I} \times \bar{\mathbb{D}}), \forall x \in \partial\mathbb{D}, \Phi(t,x) = -1\}$. Consequently $\Omega(t)$ is an open subset in $\mathbb{D}$ verifying $\bar{\Omega}(t) \subset \mathbb{D}$.

L. Blanchard
INRIA, Sophia-Antipolis,France e-mail: Louis.Blanchard@sophia.inria.fr

J.P. Zolésio
CNRS and INRIA, Sophia-Antipolis,France e-mail: Jean-Paul.Zolesio@sophia.inria.fr

The shape evolution analysis consists in the modeling of dynamical systems for the vector field $\mathbf{V}(t,x)$ see [3], so that the flow mapping $T_t(\mathbf{V})$ which maps Ω_0 onto $\Omega(t)$ furnishes the dynamic of the boundaries $\Gamma_t = \partial\Omega(t)$ through the equation $\Omega(t) = T_t(\mathbf{V})(\Omega_0)$. According to the velocity (speed) method from [3], we know that the domain $\Omega(t)$ evolves with the speed

$$\mathbf{V}(t,x) = \frac{-\partial_t \Phi(t,x)}{\|\nabla\Phi(t,x)\|}\nabla\Phi(t,x). \tag{1}$$

1.1 Tube Metric and Geodesic Tube Formulation

We introduce the concept of tubes by the product space $I \times \mathbb{D}$ where $I = [0,1]$ is the time interval. We consider the set $Q = \bigcup_{0\leq t\leq 1}\{t\} \times \Omega(t)$ defined up to a $n+1$ dimensional measure set and we denote ζ to be the characteristic function of the tube: $\zeta(t,x) = \chi_{\Omega(t)}(x)$ with $\zeta^2 = \zeta$. From now, let us consider two open sets Ω_0 and Ω_1, we focus on the construction of an admissible tube connecting the initial domain Ω_0 to the final domain Ω_1. Following [6], we consider the set of tubes connecting Ω_0 and Ω_1, two given domains in $\mathbb{D}$:

$$\mathscr{T}(\Omega_0,\Omega_1) = \left\{ \begin{array}{l} \zeta \in L^\infty(I\times\mathbb{D}) \text{ and piecewise } C^1, \\ \zeta^2 = \zeta, \end{array} \left[\begin{array}{l} \zeta(0) = \chi_{\Omega_0}, \\ \zeta(1) = \chi_{\Omega_1}. \end{array}\right] \right\} \tag{2}$$

Then, we focus on the construction of an optimal tube between the given domains Ω_0 and Ω_1, that is to say a geodesic tube. The classification of admissible tubes between Ω_0 and Ω_1 (see Fig. 1) is done by the choice of a criteria called *tube metric function* defined as follows:

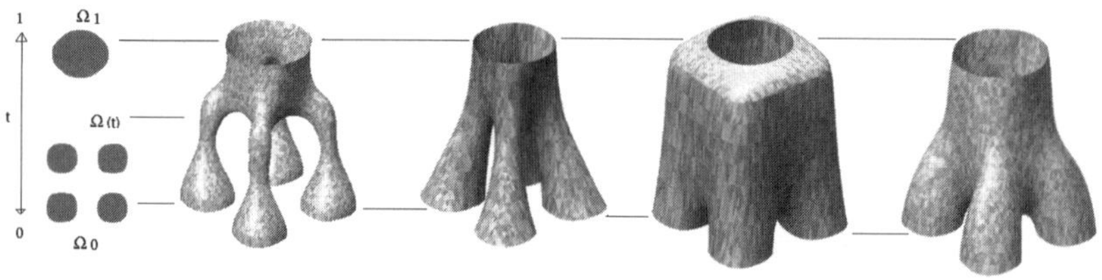

Fig. 1 Example of different connecting tubes between Ω_0 and Ω_1.

Definition 1. In order to evaluate the minimal tubes path between two open domains Ω_0 and Ω_1, we define the *tube metric function* d_c described as follows :

$$d_c(\Omega_0,\Omega_1) = \inf_{\zeta\in\mathscr{T}(\Omega_0,\Omega_1)} \|\partial_t\zeta\|_{L^1\left(I,M^1(\mathbb{D})\right)} \tag{3}$$

where $\mathscr{T}(\Omega_0,\Omega_1)$ is defined by (2) and $M^1(\mathbb{D})$ is the bounded Banach space with measure.

Proposition 1. *The tube metric function d_c is equivalent to:*

$$d_c(\Omega_0,\Omega_1) = \inf_{\zeta\in\mathscr{T}(\Omega_0,\Omega_1)} \int_0^1 \int_{\Gamma_t} |v(t,x)|\, d\Gamma\, dt \tag{4}$$

where v is the normal component of the velocity on the boundary $\Gamma_t = \partial\Omega(t)$.

Proof. We consider the geodesic tube characterization through the convection of a measurable set by a vector field $\mathbf{V}$. Using [6], the bounded measure norm of ζ turns into

$$\|\partial_t\zeta\|_{M^1(\mathbb{D})} = \int_{\Gamma_t} |v(t,x)|\, d\Gamma. \qquad \square \tag{5}$$

Proposition 2. *In the case of smooth domains the function $d_c(\Omega_0,\Omega_1)$ is a metric and this metric is conjectured to be the "Courant metric" developped in [3].*

1.2 Connecting Tubes Using a Galerkin-Level Set Strategy

The Galerkin approach consists in limiting the Level-Set analysis to a finite dimensional subspace : $\mathscr{E}\subset C^1(\bar{\mathbb{D}})$, where $\mathscr{E}$ is spanned by linear independent functions E_l defined over $\mathbb{D}$. Thus, according to a parametrization of the level set function Φ in the finite basis of functions E_l, the moving domain $\Omega(t)$ is defined by

$$\Omega(t) = \left\{x\in\mathbb{D} \;\mid\; \Phi(t,x) = \sum_{l=1}^{m} \lambda_l(t)\, E_l(x) > 0\right\}. \tag{6}$$

The metric d_c turns to be the infimum of

$$\int_0^1 \int_{\Gamma_t=\Phi_t^{-1}(0)} \frac{|\partial_t\Phi(t,x)|}{\|\nabla\Phi(t,x)\|}\, d\Gamma\, dt. \tag{7}$$

The infimum is taken over the family of the connecting admissible continuous vectors $\Lambda(t) = \big(\lambda_1(t),\dots,\lambda_m(t)\big)$. The admissibility means the connecting property $\Omega(0)=\Omega_0$ and $\Omega(1)=\Omega_1$. This admissibility condition turns in fact into a condition on the vector $\Lambda(t)$. The determination of admissible Λ_0 and Λ_1 will be described in the next section. Finally, using the Galerkin-Level Set formulation the function d_c defined by (4), as an infimum over characteristic functions $\zeta\in\mathscr{T}(\Omega_0,\Omega_1)$, can be reformulated as an infimum over parameters $\Lambda\in\mathscr{T}_\Lambda(\Omega_0,\Omega_1)$ where the set of connecting tubes between Ω_0 and Ω_1 becomes

$$\mathscr{T}_\Lambda(\Omega_0,\Omega_1) = \left\{\Lambda(t) = \big(\lambda_1(t),\dots,\lambda_m(t)\big)\in\big(C^0(I)\big)^m\,,\ \begin{bmatrix}\Lambda(0)=\Lambda_0,\\ \Lambda(1)=\Lambda_1.\end{bmatrix}\right\}$$

Consequently, the construction of a geodesic tube between Ω_0 and Ω_1 can be done through an iterative process, starting from an admissible set, which has to reduce the criterion d_c using an optimization method based on shape gradient calculus. In general, the initialization of an admissible set is difficult but in our case, the initial tube can be choosen as a convex combination of λ_0 and $\lambda_1 : \Lambda(t) = \Lambda_0(1-t) + \lambda_1 t$ (see Fig 2).

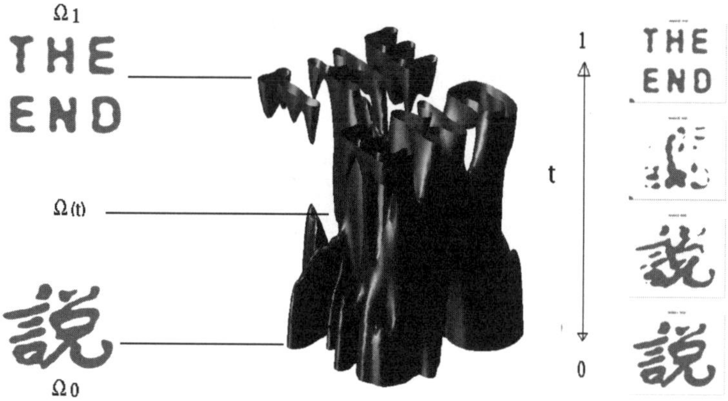

Fig. 2 Example of a connecting tube between Ω_0 and Ω_1 with parameters $\Lambda(t) = \Lambda_0(1-t) + \lambda_1 t$.

2 Determination of Parameters Λ_0 and Λ_1

This parameter identification (done through the calculus of the parameters in the Galerkin-Level Set expansion of the Level-Set function Φ) is performed by minimising the distance between the given domain Ω_i and the domain $\Omega(i)$, for $i \in \{0,1\}$. Indeed, we propose a metric associated to the Hilbert space H^s. Then, for a smooth domain Ω, we have $\chi_\Omega \in H^s(\mathbb{D})$, $0 \leq s < \frac{1}{2}$. Thus we consider the metric associated to that regularity noted $\delta_s(\Omega_i, \Omega(i))$ and defined as follows (see [2]):

$$\forall s \in [0, \frac{1}{2}[, \quad \delta_s(\Omega_i, \Omega(i)) = \| \chi_{\Omega_i} - \chi_{\Omega(i)} \|_{L^2(\mathbb{D})} + \| \chi_{\Omega_i} - \chi_{\Omega(i)} \|_s \tag{8}$$

where

$$\|\chi_\Omega\|_s^2 = \int_{\mathbb{D}} \int_{\mathbb{D}} \frac{|\chi_\Omega(x) - \chi_\Omega(y)|^2}{|x-y|^{-(n+2s)}} \, dxdy.$$

The parameter identification problem is: $\forall i \in \{0,1\} \quad \min_{\Lambda(i) \in \left(C^0(I)\right)^m} \delta_s(\Omega_i, \Omega(i))^2$.

2.1 Gradient Method for Parameters Identification

Proposition 3. *The funtional* $J(\Omega) = \delta_s\big(\Omega_i, \Omega\big)^2$ *is shape differentiable. For perturbation vector fields* $\mathbf{V} \in C_0^1(\mathbb{D}, \mathbb{R}^n)$*, the Eulerian derivative is expressed as follows*

$$dJ\big(\Omega;\mathbf{V}\big) = \int_{\Gamma} F(x)\,\langle \mathbf{V}(x), \mathbf{n}(x)\rangle_{\mathbb{R}^n}\, d\Gamma \tag{9}$$

where : $F(x) = 1 - 2\chi_{\Omega_i}(x) + 2\displaystyle\int_{\mathbb{D}} \frac{1 - 2\chi_{\Omega}(y) + 2\left[\chi_{\Omega_i}(y) - \chi_{\Omega_i}(x)\right]}{|x-y|^{-(n+2s)}}\, dy,$
and where $\langle .,. \rangle_{\mathbb{R}^n}$ *denotes the inner product on* $\mathbb{R}^n$, $\mathbf{n}$ *is the external normal vector on the boundary* Γ *and* $d\Gamma$ *is the arclength measure on* Γ.

Proof. The proof uses the structure theorem [3] (see [2] for details) . □

For $i \in \{0,1\}$, we search $\Lambda(i)$ by an iterative gradient method. The initialization is done by the choice of a starting estimate $\Lambda^0(i) = \big(\lambda_1^0(i), \ldots, \lambda_m^0(i)\big)$. Then, the update is performed by the iteration step : $\Lambda^{k+1}(i) = \Lambda^k(i) - \rho\, g^k(i)$, $\rho > 0$, where $g^k(i)$ is the gradient of the functional $j\big(\Lambda^k(i)\big) = J\big(\Omega_{\Lambda^k(i)}\big)$ with respect to $\Lambda^k(i)$. Then, to calculate the gradient g, we introduce the following notation:

$$\forall \gamma = (\gamma_1, \ldots, \gamma_m) \in \mathbb{R}^m \quad , \quad \Omega_\gamma = \Big\{ x \in \mathbb{D} \mid \Phi_\gamma(x) = \sum_{l=1}^{m} \gamma_l E_l(x) > 0 \Big\}.$$

Proposition 4. *The functional* $j(\Lambda) = J(\Omega_\lambda)$ *is Gateaux differentiable. Moreover, by chain rule the Gateaux derivative is given by* $j'(\Lambda, \mu) = dJ(\Omega_\Lambda; \mathbf{V}_{\Lambda,\mu})$ *where the vector fields* $\mathbf{V}_{\Lambda,\mu} \in C_0^1(\mathbb{D}, \mathbb{R}^n)$ *satisfy*

$$\mathbf{V}_{\Lambda,\mu}(x) = \frac{-\Phi_\mu(x)}{\|\nabla \Phi_\Lambda(x)\|^2} \nabla \Phi_\Lambda(x). \tag{10}$$

Using equations (9), (10) and the following equation : $\mathbf{n}(x) = \frac{-\nabla\Phi_\Lambda(x)}{\|\nabla\Phi_\Lambda(x)\|}$, *we get:*

$$j'(\Lambda, \mu) = \frac{\partial j(\Lambda + \varepsilon\mu)}{\partial \varepsilon}\Big|_{\varepsilon=0} = \int_{\Gamma_\Lambda = \partial\Omega_\Lambda} F(x) \frac{\Phi_\mu(x)}{\|\nabla \Phi_\Lambda(x)\|} d\Gamma = \langle \mu\, , g \rangle_{\mathbb{R}^m} \tag{11}$$

where g*, the gradient of the functional* $j(\Lambda)$ *with respect to* Λ*, verify:*

$$g = \Big(\int_{\Gamma_\Lambda} F(x) \frac{E_1(x)}{\|\nabla \Phi_\Lambda(x)\|} d\Gamma, \ldots, \int_{\Gamma_\Lambda} F(x) \frac{E_m(x)}{\|\nabla \Phi_\Lambda(x)\|} d\Gamma \Big). \tag{12}$$

2.2 Numerical Experiments

We present a 2D shape identification problem using the Galerkin-Level Set strategy. The given domain Ω_0 which we want to identify is a Chinese sinogram (*shuō*). We consider a Galerkin-Level Set expansion of the function Φ in Fourier series of dimension $m = 400$. We start with a smooth initial domain $\Omega_{t=0}$ corresponding to the lower frequency of the Fourier series : $\Lambda^0(0) = \big(1, 0, \ldots, 0\big)$ (left picture in Fig. 3 and Fig 4). In the right graph of Fig. 3 we show the evolution of $\Omega(t)$. The algorithm detects the contour of the given domain Ω_0 after 15 iterations. That illustrates the efficiency of the method.

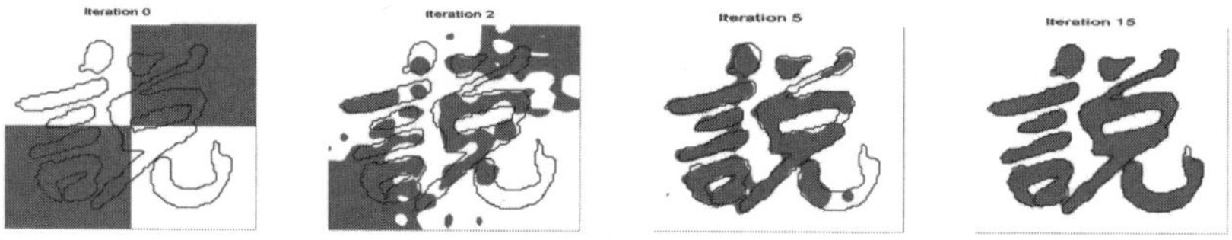

Fig. 3 Evolution of the moving domain $\Omega(t)$ using a Galerkin-Level Set strategy.

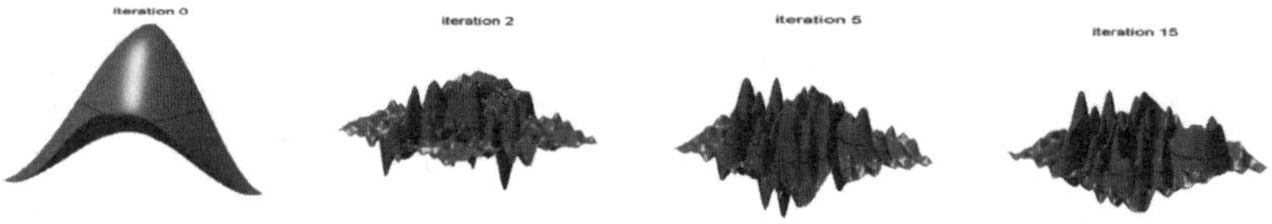

Fig. 4 Evolution of the Level Set function $\Phi(t,.)$ using a Galerkin-Level Set strategy.

3 Geodesic Tube Construction Using the Federer Theorem

In this section we describe an optimization method based on a shape gradient calculus to construct a geodesic tube. For numerical calculus, the *tube metric function* d_c defined by (4) is approximated by an integral on the domain $\Omega(t)$ using the following Federer measure decomposition theorem (see [4]):

Theorem 1. Federer measure decomposition theorem: *Let us consider a regular function $g \in L^1(\mathbb{D})$ and $\forall h > 0$ the domain $U_h(\Gamma_t) = \Big\{ x \in \mathbb{D} \mid \|\Phi(t,x)\| < h \Big\}$, we have:*

$$\int_{U_h(\Gamma)} g(x)\,dx = \int_{-h}^{+h} \left(\int_{\Phi^{-1}(z)} \frac{g(x)}{\|\nabla\Phi(x)\|}\, d\Gamma \right) dz. \qquad (13)$$

Corollary 1. *Assuming the mapping :* $z \in [-h, +h] \to \int_{\Phi^{-1}(z)} \frac{F(x)}{\|\nabla\Phi(x)\|}\, d\Gamma$ *to be continuous, we obtain :* $\int_{\Gamma} \frac{F(x)}{\|\nabla\Phi(x)\|}\, d\Gamma = \frac{1}{2h}\int_{U_h(\Gamma)} F(x)\, dx + o(1)$ *for* $h \to 0$. *In this approximation, the main advantage is that the denominator term* $\|\nabla\Phi(x)\|$ *has been eliminated.*

Moreover, to reduce the complexity of the *tube metric function* (3), we consider a polynomial decomposition of the parameters $\Lambda(t) = P_\alpha(t)\Lambda_1 + \left(1 - P_\alpha(t)\right)\Lambda_0$ with $P_\alpha(t) = \sum_{i=1}^{M} \alpha_i e_i(t)$ where $\{e_i(t)\}_{1\le i\le M}$ is a polynomial basis of degree M, and where the new parameters $\alpha = (\alpha_1, \dots, \alpha_M) \in \mathbb{R}^M$ in the optimization process do not depend on time. Then, using the Federer theorem (13) and the Galerkin-Level Set strategy in the determination of parameters Λ_0 and Λ_1, we consider an approximation of the integral in time of the measure norm of ζ defined by (5)

$$\tilde{J}_h(\alpha) = \int_0^1 \frac{1}{2h}\int_{U_h(\Gamma_t)} |\partial_t\Phi(t,x)|\, dx\, dt = \int_0^1 \frac{|\dot{P}_\alpha(t)|}{2h}\int_{U_h(\Gamma_t)} |\Phi_1(x) - \Phi_0(x)| dx\, dt \tag{14}$$

where $\forall i \in \{0,1\}$, $\Lambda_i = (\lambda_{i1}, \dots, \lambda_{im})$ and $\Phi_i(x) = \sum_{l=1}^{m} \lambda_{il} E_l(x)$.

Consequently, we consider the problem $(\tilde{\mathscr{P}})$: $\min_{\alpha\in K(\Omega_0,\Omega_1)} \tilde{J}_h(\alpha)$.

where the set of connecting tubes is $K(\Omega_0, \Omega_1) = \left\{ \alpha \in \mathbb{R}^M \;,\; \begin{bmatrix} P_\alpha(0) = 0 \\ P_\alpha(1) = 1 \end{bmatrix} \right\}$.

The geodesic tube construction is done by an iterative gradient method. The initialization consists in the construction of an initial tube through the choice of starting parameters $\alpha^0 = (\alpha_1^0, \dots, \alpha_m^0)$ such that $\alpha^0 \in K(\Omega_0, \Omega_1)$. Then, the update is performed by the iteration step : $\alpha^{k+1} = \alpha^k - \rho\, \tilde{g}^k$, $\rho > 0$, where $\tilde{g}^k$ is the gradient of the functional $\tilde{J}_h(\alpha^k)$ with respect to α^k. Then, to calculate the gradient $\tilde{g}$, we introduce the following notation: $\forall\gamma = (\gamma_1, \dots, \gamma_M) \in \mathbb{R}^M$

$$\Phi_\gamma(t,x) = P_\gamma(t)\,\Phi_1(x) + \left(1 - P_\gamma(t)\right)\Phi_0(x) \;,\; \Psi_\gamma(t,x) = P_\gamma(t)\left[\Phi_1(x) - \Phi_0(x)\right]$$

Proposition 5. *The functional* $\tilde{J}_h$ *is Gateaux differentiable. There exists a vector field* $W \in C_0^1(\mathbb{D}, \mathbb{R}^n)$ *such that*

$$\tilde{J}_h'(\alpha,\beta) = \frac{\partial \tilde{J}_h(\alpha + \varepsilon\beta)}{\partial\varepsilon}\Big|_{\varepsilon=0}$$

$$= \frac{1}{2h}\int_0^1 \int_{U_h(\Gamma_t)} \left[\partial_\varepsilon\left(|\partial_t\Phi_{\alpha+\varepsilon\beta}(t,x)|\right)\Big|_{\varepsilon=0} + div\left(|\partial_t\Phi_\alpha(t,x)|\,\mathbf{W}_{\alpha,\beta}(t,x)\right)\right] dx\, dt \tag{15}$$

where $\mathbf{W}$ *depends linearly on* β *as follows :* $\mathbf{W}_{\alpha,\beta}(t,x) = \frac{-\Psi_\beta(t,x)}{\|\nabla\Phi_\alpha(t,x)\|^2}\nabla\Phi_\alpha(t,x)$.
Consequently, we get $\tilde{J}_h'(\alpha,\beta) = \langle\beta\,,\tilde{g}\rangle_{\mathbb{R}^M}$ *where* $\tilde{g}$ *is the gradient of the functional* $\tilde{J}_h(\alpha)$ *with respect to* α *(see [1] for more details).*

3.1 Numerical Experiment for Geodesic Tube Construction

We present a 3D tube optimization corresponding to the problem $(\tilde{\mathscr{P}})$. The bottom left figure in Fig 5 shows the geodesic tube obtained after 7 iterations in the gradient method described previously. Comparing the distribution on time of the functional $\tilde{J}_h$, for different iterations using the bottom right figure in Fig. 5, we conclude that the geodesic tube is a smoother tube than the initial tube with a more homogeneous distribution.

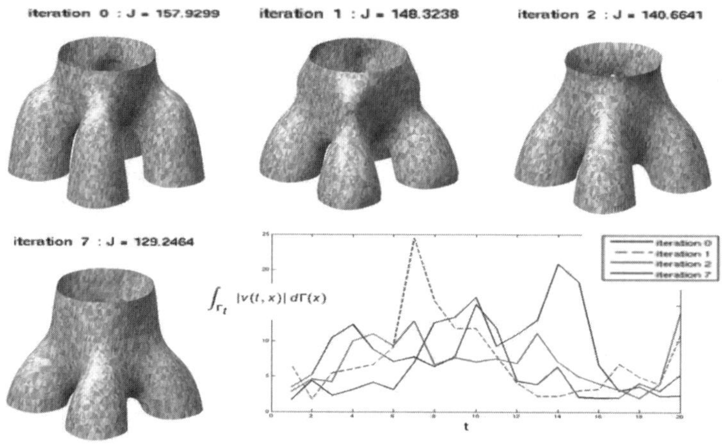

Fig. 5 Tube evolution and evolution of the distribution on time of the functional : $\tilde{J}_h$.

References

1. Blanchard, L., Zolésio, J.P.: Galerkin strategy for level set shape analysis : Application to geodesic tube. Proceeding of the 23nd IFIP TC7, Cracovia (2007)
2. Blanchard, L., Zolésio, J.P.: Morphing by moving shape modeling with galerkin approximation. International Conference on Nonlinear Problems in Aviation and Aerospace.Seenith Sivasundaram (Editor) (to be appear)
3. Delfour, M.C., Zolésio, J.P.: Shapes and geometries. Analysis, differential calculus, and optimization. SIAM (2001)
4. Federer, H.: Curvature measures. Trans. Amer. Math. Soc. pp. 418–491 (1959)
5. Zolésio, J.: Optimal tubes: Geodesic metric, euler flow, moving domain. Free and Moving Boundaries: Analysis, Simulation and Control **volume**(252), 203–213 (2005). Taylor & Francis CRC Press
6. Zolésio, J.P.: Shape topology by tube geodesic. In Information Processing: Recent Mathematical Advances in Optimization and Control **Presse de l'Ecole des Mines de Paris**, 185–204 (2004)
7. Zolésio, J.P.: Control of moving domains, shape stabilization and variational tube formulations. Control of Coupled Partial Differential Equations **Volume 155**, 329–382 (2007)

Numerical Analysis of a Control and State Constrained Elliptic Control Problem with Piecewise Constant Control Approximations

K. Deckelnick and M. Hinze

Abstract We consider an elliptic optimal control problem with control and pointwise state constraints. The cost functional is approximated by a sequence of functionals which are obtained by discretizing the state equation with the help of linear finite elements and enforcing the state constraints in the nodes of the triangulation. Controls are discretized piecewise constant on every simplex of the triangulation. Error bounds for control and state are obtained both in two and three space dimensions.

1 Introduction

Let $\Omega \subset \mathbb{R}^d$ $(d = 2,3)$ be a bounded domain with a smooth boundary $\partial\Omega$ and consider the differential operator

$$Ay := -\sum_{i,j=1}^{d} \partial_{x_j}\left(a_{ij}y_{x_i}\right) + \sum_{i=1}^{d} b_i y_{x_i} + cy,$$

where for simplicity the coefficients a_{ij}, b_i and c are supposed to be smooth functions on $\bar{\Omega}$. Furthermore, the operator A is assumed to be uniformly elliptic. We associate with A the bilinear form

$$a(y,z) := \int_\Omega \Big(\sum_{i,j=1}^{d} a_{ij}(x) y_{x_i} z_{x_j} + \sum_{i=1}^{d} b_i(x) y_{x_i} z + c(x) yz \Big) dx, \qquad y,z \in H^1(\Omega)$$

Klaus Deckelnick
Institut für Analysis und Numerik, Otto–von–Guericke–Universität Magdeburg, Universitätsplatz 2, 39106 Magdeburg, Germany e-mail: Klaus.Deckelnick@mathematik.uni-magdeburg.de

Michael Hinze
Schwerpunkt Optimierung und Approximation, Universität Hamburg, Bundesstraße 55, 20146 Hamburg, Germany, e-mail: michael.hinze@uni-hamburg.de

and suppose that this form is coercive on $H^1(\Omega)$ with constant $c_1 > 0$. From the above assumptions it follows that for a given $f \in (H^1(\Omega))'$ the elliptic boundary value problem

$$\begin{aligned} Ay &= f \text{ in } \Omega \\ \Sigma_{i,j=1}^d a_{ij} y_{x_i} \mu_j &= 0 \text{ on } \partial\Omega \end{aligned} \tag{1}$$

has a unique weak solution $y \in H^1(\Omega)$ which we denote by $y = \mathscr{G}(f)$. Here, μ is the unit outward normal to $\partial\Omega$. Furthermore, if $f \in L^2(\Omega)$, then the solution y belongs to $H^2(\Omega)$ and satisfies

$$\|y\|_{H^2} \leq C\|f\|. \tag{2}$$

In the above, $\|\cdot\| = \|\cdot\|_{L^2}$ and $\|\cdot\|_{H^m} = \|\cdot\|_{W^{m,2}}$, where $\|\cdot\|_{L^p}$ and $\|\cdot\|_{W^{m,p}}$ denote the norms in $L^p(\Omega)$ and $W^{m,p}(\Omega)$ respectively.

We are interested in the following control problem

$$\begin{aligned} &\min_{u \in U_{ad}} J(u) = \frac{1}{2}\int_\Omega |y - y_0|^2 + \frac{\alpha}{2}\int_\Omega |u|^2 \\ &\text{subject to } y = \mathscr{G}(u) \text{ and } y(x) \leq b(x) \text{ in } \Omega. \end{aligned} \tag{3}$$

Here, $U_{ad} := \{v \in L^2(\Omega) \,|\, a_l \leq v \leq a_u \text{ a.e. in } \Omega\} \subseteq L^2(\Omega)$ denotes the set of admissible controls, where $\alpha > 0$ and $a_l < a_u$ are given constants. Furthermore, we suppose that $y_0 \in H^1(\Omega)$ and $b \in W^{2,\infty}(\Omega)$ are given functions.

For the case without control constraints the finite element analysis of problem (3) is carried out in [6]. In the present work we extend the analysis to the case of control and pointwise state constraints using techniques which are applicable to a wider class of control problems.

From here onwards we impose the following assumption which is frequently referred to as *Slater condition* or interior point condition.

Assumption 1 $\exists \tilde{u} \in U_{ad}$ *with* $\mathscr{G}(\tilde{u}) < b$ *in* $\bar{\Omega}$.

Since the state constraints form a convex set and the set of admissible controls is closed and convex it is not difficult to establish the existence of a unique solution $u \in U_{ad}$ to this problem. In order to characterize this solution we introduce the space $\mathscr{M}(\bar{\Omega})$ of Radon measures which is defined as the dual space of $C^0(\bar{\Omega})$ and endowed with the norm $\|\lambda\|_{\mathscr{M}(\bar{\Omega})} = \sup_{f \in C^0(\bar{\Omega}), |f| \leq 1} \int_{\bar{\Omega}} f d\lambda$. Using [3, Theorem 5.2] we then infer (compare also [2, Theorem 2]):

Theorem 1. *Let* $u \in U_{ad}$ *denote the unique solution of* (3)*. Then there exist* $\lambda \in \mathscr{M}(\bar{\Omega})$ *and* $p \in L^2(\Omega)$ *such that with* $y = \mathscr{G}(Bu)$ *there holds*

$$\int_\Omega pAv = \int_\Omega (y - y_0)v + \int_{\bar{\Omega}} v d\lambda \;\; \forall v \in H^2(\Omega) \text{ with } \sum_{i,j=1}^d a_{ij} v_{x_i} \mu_j = 0 \text{ on } \partial\Omega, \tag{4}$$

$$\int_\Omega (p + \alpha u)(v - u) \geq 0 \qquad \forall v \in U_{ad}, \tag{5}$$

$$\lambda \geq 0,\; y(x) \leq b(x) \text{ in } \Omega \text{ and } \int_{\bar{\Omega}} (b - y) d\lambda = 0. \tag{6}$$

Our aim is to develop and analyze a finite element approximation of problem (3). We start by approximating the cost functional J by a sequence of functionals J_h where h is a mesh parameter related to a sequence of triangulations. The definition of J_h involves the approximation of the state equation by linear finite elements. The controls are discretized by piecewise constant functions which satisfy the constraints elementwise. Denoting by u_h the corresponding minimum of J_h with associate state y_h we shall prove the following error bounds,

$$\|u-u_h\|_{L^2},\ \|y-y_h\|_{H^1} \leq \begin{cases} Ch|\log h|, & \text{if } d=2, \\ C\sqrt{h}, & \text{if } d=3. \end{cases}$$

For a detailed discussion of the related contributions [4, 5, 9] we refer the reader to [6, 7, 8], where also numerical examples are presented, and also constraints on the gradient of the state are considered..

2 Finite Element Discretization

Let $\mathscr{T}_h$ be a triangulation of Ω with maximum mesh size $h := \max_{T\in\mathscr{T}_h} \operatorname{diam}(T)$ and vertices $x_1,\ldots,x_m$. We suppose that $\bar{\Omega}$ is the union of the elements of $\mathscr{T}_h$ so that element edges lying on the boundary are curved. In addition, we assume that the triangulation is quasi-uniform in the sense that there exists a constant $\kappa > 0$ (independent of h) such that each $T \in \mathscr{T}_h$ is contained in a ball of radius $\kappa^{-1}h$ and contains a ball of radius κh. Let us define the space of linear finite elements

$$X_h := \{v_h \in C^0(\bar{\Omega}) \,|\, v_h \text{ is a linear polynomial on each } T \in \mathscr{T}_h\}$$

as well as the space of piecewise constant functions

$$Y_h := \{v_h \in L^2(\Omega) \,|\, v_h \text{ is constant on each } T \in \mathscr{T}_h\}.$$

Let $Q_h : L^2(\Omega) \to Y_h$ be the orthogonal projection onto Y_h so that

$$(Q_h v)(x) := \fint_T v, \quad x \in T, T \in \mathscr{T}_h,$$

where $\fint_T v$ denotes the average of v over T. In what follows it is convenient to introduce a discrete approximation of the operator $\mathscr{G}$. For a given function $v \in L^2(\Omega)$ we denote by $z_h = \mathscr{G}_h(v) \in X_h$ the solution of the discrete Neumann problem

$$a(z_h, v_h) = \int_\Omega v v_h \ \text{ for all } v_h \in X_h.$$

It is well–known that for all $v \in L^2(\Omega)$

$$\|\mathscr{G}(v) - \mathscr{G}_h(v)\| \leq Ch^2\|v\|. \tag{7}$$

The corresponding estimate in L^∞ will be crucial for our analysis.

Lemma 1. *There exists a constant C which only depends on the data such that*

$$\|\mathscr{G}(v)-\mathscr{G}_h(v)\|_{L^\infty} \le Ch^2|\log h|^2 \qquad \textit{for all } v \in U_{ad}.$$

Proof. Let $v \in U_{ad}, z = \mathscr{G}(v), z_h = \mathscr{G}_h(v)$. Since $U_{ad} \subset L^\infty(\Omega)$ with $\|v\|_{L^\infty} \le \max(|a_l|,|a_u|)$, elliptic regularity theory implies that $z \in W^{2,q}(\Omega)$ for all $1 < q < \infty$. In addition, it is well–known that one has

$$\|z\|_{W^{2,q}} \le Cq\|v\|_{L^q} \qquad (C \text{ independent of } q)$$

by tracking the constants in the analysis. As a result we have

$$\|z\|_{W^{2,q}} \le Cq \qquad \text{for all } v \in U_{ad}. \tag{8}$$

Using Theorem 2.2 and the following Remark in [10] we have

$$\|z-z_h\|_{L^\infty} \le C|\log h| \inf_{\chi \in X_h} \|z-\chi\|_{L^\infty}, \tag{9}$$

which, combined with a well–known interpolation estimate and (8), yields

$$\|z-z_h\|_{L^\infty} \le Ch^{2-\frac{d}{q}}|\log h|\|z\|_{W^{2,q}} \le Cqh^{2-\frac{d}{q}}|\log h|$$

for all $v \in U_{ad}$. Choosing $q = |\log h|$ gives the result. □

In order to approximate (3) we introduce a discrete counterpart of U_{ad},

$$U_{ad}^h := \{v_h \in Y_h \,|\, a_l \le v_h \le a_u \text{ in } \Omega\}.$$

Note that $U_{ad}^h \subset U_{ad}$ and that $Q_h v \in U_{ad}^h$ for $v \in U_{ad}$. Since $Q_h v \to v$ in $L^2(\Omega)$ as $h \to 0$ we infer from (2), the continuous embedding $H^2(\Omega) \hookrightarrow C^0(\bar{\Omega})$ and Lemma 1 that

$$\mathscr{G}_h(Q_h v) \to \mathscr{G}(v) \text{ in } L^\infty(\Omega) \text{ for all } v \in U_{ad}. \tag{10}$$

Problem (3) is approximated by the following sequence of control problems depending on the mesh parameter h:

$$\begin{aligned} &\min_{u \in U_{ad}^h} J_h(u) := \frac{1}{2}\int_\Omega |y_h - y_0|^2 + \frac{\alpha}{2}\int_\Omega |u|^2 \\ &\text{subject to } y_h = \mathscr{G}_h(u) \text{ and } y_h(x_j) \le b(x_j) \text{ for } j = 1,\dots,m. \end{aligned} \tag{11}$$

Problem (11) represents a convex finite-dimensional optimization problem of similar structure as problem (3), but with only finitely many equality and inequality constraints for state and control, which form a convex admissible set. Again we can apply [3, Theorem 5.2] which together with [2, Corollary 1] yields (compare also the analysis of problem (P) in [4])

Lemma 2. *Problem (11) has a unique solution $u_h \in U_{ad}^h$. There exist $\lambda_1,\dots,\lambda_m \in \mathbb{R}$ and $p_h \in X_h$ such that with $y_h = \mathscr{G}_h(u_h)$ and $\lambda_h = \sum_{j=1}^m \lambda_j \delta_{x_j}$ we have*

$$a(v_h, p_h) = \int_\Omega (y_h - y_0)v_h + \int_{\bar\Omega} v_h d\lambda_h \qquad \forall v_h \in X_h, \tag{12}$$

$$\int_\Omega (p_h + \alpha u_h)(v_h - u_h) \ge 0 \qquad \forall v_h \in U_{ad}^h, \tag{13}$$

$$\lambda_j \ge 0,\, y_h(x_j) \le b(x_j),\, j = 1,\dots,m \text{ and } \int_{\bar\Omega} (I_h b - y_h) d\lambda_h = 0. \tag{14}$$

Here, δ_x denotes the Dirac measure concentrated at x and I_h is the usual Lagrange interpolation operator.

As a first result for (11) we prove bounds on the discrete states and the discrete multipliers.

Lemma 3. *Let $u_h \in U_{ad}^h$ be the optimal solution of (11) with corresponding state $y_h \in X_h$ and adjoint variables $p_h \in X_h$ and $\lambda_h \in \mathcal{M}(\bar\Omega)$. Then there exists $\bar h > 0$ such that*

$$\|y_h\|, \|\lambda_h\|_{\mathcal{M}(\bar\Omega)} \le C, \quad \|p_h\|_{H^1} \le C\gamma(d,h) \qquad \textit{for all } 0 < h \le \bar h,$$

where $\gamma(2,h) = \sqrt{|\log h|}$ and $\gamma(3,h) = h^{-\frac{1}{2}}$.

Proof. Since $\mathcal{G}(\tilde u) \in C^0(\bar\Omega)$, Assumption 1 implies that there exists $\delta > 0$ such that

$$\mathcal{G}(\tilde u) \le b - \delta \qquad \text{in } \bar\Omega. \tag{15}$$

It follows from (10) that there is $\bar h > 0$ with

$$\mathcal{G}_h(Q_h \tilde u) \le b - \frac{\delta}{2} \qquad \text{in } \bar\Omega \text{ for all } 0 < h \le \bar h. \tag{16}$$

Since $Q_h \tilde u \in U_{ad}^h$, (13), (12) and (16) imply

$$\begin{aligned}
0 &\le \int_\Omega (p_h + \alpha u_h)(Q_h\tilde u - u_h) = \int_\Omega p_h(Q_h\tilde u - u_h) + \alpha \int_\Omega u_h(Q_h\tilde u - u_h)\\
&= a(\mathcal{G}_h(Q_h\tilde u) - y_h, p_h) + \alpha \int_\Omega u_h(Q_h\tilde u - u_h)\\
&= \int_\Omega (\mathcal{G}_h(Q_h\tilde u) - y_h)(y_h - y_0) + \int_{\bar\Omega} (\mathcal{G}_h(Q_h\tilde u) - y_h) d\lambda_h + \alpha \int_\Omega u_h(Q_h\tilde u - u_h)\\
&\le C - \frac{1}{2}\|y_h\|^2 + \sum_{j=1}^m \lambda_j \big(b(x_j) - \frac{\delta}{2} - y_h(x_j)\big) = C - \frac{1}{2}\|y_h\|^2 - \frac{\delta}{2}\sum_{j=1}^m \lambda_j
\end{aligned}$$

where the last equality is a consequence of (14). It follows that $\|y_h\|, \|\lambda_h\|_{\mathcal{M}(\bar\Omega)} \le C$. In order to bound $\|p_h\|_{H^1}$ we insert $v_h = p_h$ into (12) and deduce with the help of the coercivity of A, a well–known inverse estimate and the bounds we have already obtained that

$$\begin{aligned}
c_1\|p_h\|_{H^1}^2 &\le a(p_h, p_h) = \int_\Omega (y_h - y_0)p_h + \int_{\bar\Omega} p_h d\lambda_h\\
&\le \|y_h - y_0\|\,\|p_h\| + \|p_h\|_{L^\infty}\|\lambda_h\|_{\mathcal{M}(\bar\Omega)} \le C\|p_h\| + C\gamma(d,h)\|p_h\|_{H^1}.
\end{aligned}$$

Hence $\|p_h\|_{H^1} \le C\gamma(d,h)$ and the lemma is proved. □

3 Error Analysis

An important ingredient in our analysis is an error bound for a solution of a Neumann problem with a measure valued right hand side. Let A be as above and consider

$$\begin{aligned} A^* q &= \tilde{\lambda}\llcorner\Omega \quad \text{in } \Omega \\ \textstyle\sum_{i=1}^d \left(\sum_{j=1}^d a_{ij} q_{x_j} + b_i q\right)\mu_i &= \tilde{\lambda}\llcorner\partial\Omega \text{ on } \partial\Omega. \end{aligned} \tag{17}$$

Theorem 2. Let $\tilde{\lambda} \in \mathscr{M}(\bar{\Omega})$. Then there exists a unique weak solution $q \in L^2(\Omega)$ of (17), i.e.

$$\int_\Omega qAv = \int_{\bar{\Omega}} v\, d\tilde{\lambda} \qquad \forall v \in H^2(\Omega) \text{ with } \sum_{i,j=1}^d a_{ij} v_{x_i} \mu_j = 0 \text{ on } \partial\Omega.$$

Furthermore, q belongs to $W^{1,s}(\Omega)$ for all $s \in (1, \frac{d}{d-1})$. For the finite element approximation $q_h \in X_h$ of q defined by

$$a(v_h, q_h) = \int_{\bar{\Omega}} v_h\, d\tilde{\lambda} \qquad \text{for all } v_h \in X_h$$

the following error estimate holds:

$$\|q - q_h\| \le C h^{2-\frac{d}{2}} \|\tilde{\lambda}\|_{\mathscr{M}(\bar{\Omega})}. \tag{18}$$

Proof. A corresponding result is proved in [1] for the case of an operator A without transport term subject to Dirichlet conditions, but the arguments can be adapted to our situation. We omit the details. □

We are now prepared to prove our main result.

Theorem 3. *Let u and u_h be the solutions of (3) and (11) respectively. Then we have for $0 < h \le \bar{h}$*

$$\|u - u_h\| + \|y - y_h\|_{H^1} \le \begin{cases} Ch|\log h|, & \text{if } d = 2, \\ C\sqrt{h}, & \text{if } d = 3. \end{cases}$$

Proof. We test (5) with u_h, (13) with $Q_h u$ and add the resulting inequalities. Keeping in mind that $u - Q_h u \perp Y_h$ we obtain

$$\begin{aligned} &\int_\Omega \left(p - p_h + \alpha(u - u_h)\right)(u_h - u) \\ &\ge \int_\Omega (p_h + \alpha u_h)(u - Q_h u) = \int_\Omega (p_h - Q_h p_h)(u - Q_h u). \end{aligned}$$

As a consequence,

$$\alpha\|u-u_h\|^2 \leq \int_\Omega (u_h-u)(p-p_h) - \int_\Omega (p_h-Q_hp_h)(u-Q_hu) \equiv I+II. \quad (19)$$

Let $y^h := \mathscr{G}_h(u) \in X_h$ and denote by $p^h \in X_h$ the unique solution of

$$a(w_h,p^h) = \int_\Omega (y-y_0)w_h + \int_{\bar\Omega} w_h d\lambda \qquad \text{for all } w_h \in X_h.$$

Applying Theorem 2 with $\tilde\lambda = (y-y_0)dx+\lambda$ we infer

$$\|p-p^h\| \leq Ch^{2-\frac{d}{2}}\left(\|y-y_0\| + \|\lambda\|_{\mathscr{M}(\bar\Omega)}\right). \quad (20)$$

Recalling that $y_h = \mathscr{G}_h(u_h), y^h = \mathscr{G}_h(u)$ and observing (12) as well as the definition of p^h we can rewrite the first term in (19)

$$I = \int_\Omega (u_h-u)(p-p^h) - \|y-y_h\|^2 + \int_\Omega (y-y_h)(y-y^h) + \\ + \int_{\bar\Omega} (y_h-y^h)d\lambda + \int_{\bar\Omega} (y^h-y_h)d\lambda_h.$$

Applying Young's inequality we deduce

$$|I| \leq \frac{\alpha}{4}\|u-u_h\|^2 - \frac{1}{2}\|y-y_h\|^2 + C\left(\|p-p^h\|^2 + \|y-y^h\|^2\right) \\ + \int_{\bar\Omega} (y_h-y^h)d\lambda + \int_{\bar\Omega} (y^h-y_h)d\lambda_h. \quad (21)$$

Let us estimate the integrals involving the measures λ and λ_h. Since $y_h - y^h \leq (I_hb-b)+(b-y)+(y-y^h)$ in $\bar\Omega$ we deduce with the help of (6), Lemma 1 and an interpolation estimate

$$\int_{\bar\Omega} (y_h-y^h)d\lambda \leq \|\lambda\|_{\mathscr{M}(\bar\Omega)}\left(\|I_hb-b\|_\infty + \|y-y^h\|_\infty\right) \leq Ch^2|\log h|^2.$$

On the other hand $y^h - y_h \leq (y^h-y)+(b-I_hb)+(I_hb-y_h)$, so that (14), Lemma 1 and Lemma 3 yield

$$\int_{\bar\Omega} (y^h-y_h)d\lambda_h \leq \|\lambda_h\|_{\mathscr{M}(\bar\Omega)}\left(\|b-I_hb\|_\infty + \|y-y^h\|_\infty\right) \leq Ch^2|\log h|^2.$$

Inserting these estimates into (21) and recalling (7) as well as (18) we obtain

$$|I| \leq \frac{\alpha}{4}\|u-u_h\|^2 - \frac{1}{2}\|y-y_h\|^2 + Ch^{4-d} + Ch^2|\log h|^2. \quad (22)$$

Let us next examine the second term in (19). Since $u_h = Q_hu_h$ and Q_h is stable in $L^2(\Omega)$ we have

$$|II| \leq 2\|u-u_h\| \, \|p_h - Q_h p_h\| \leq \frac{\alpha}{4}\|u-u_h\|^2 + Ch^2\|p_h\|_{H^1}^2$$

using an interpolation estimate for Q_h. Combining this estimate with (22), Lemma 3 and (19) we finally obtain

$$\|u-u_h\|^2 + \|y-y_h\|^2 \leq Ch^{4-d} + Ch^2|\log h|^2 + Ch^2\gamma(d,h)^2$$

which implies the estimate on $\|u-u_h\|$. The bound on $\|y-y_h\|_{H^1}$ is derived with standard finite element techniques and the bound on $\|u-u_h\|$. □

Acknowledgements The authors gratefully acknowledge the support of the DFG Priority Program 1253 entitled Optimization With Partial Differential Equations.

References

1. Casas, E.: L^2 estimates for the finite element method for the Dirichlet problem with singular data. Numer. Math. **47**, 627–632 (1985)
2. Casas, E.: Control of an elliptic problem with pointwise state constraints. SIAM J. Cont. Optim. **24**, 1309–1322 (1986)
3. Casas, E.: Boundary control of semilinear elliptic equations with pointwise state constraints. SIAM J. Cont. Optim. **31**, 993–1006 (1993)
4. Casas, E.: Error Estimates for the Numerical Approximation of Semilinear Elliptic Control Problems with Finitely Many State Constraints. ESAIM, Control Optim. Calc. Var. **8**, 345–374 (2002)
5. Casas, E., Mateos, M.: Uniform convergence of the FEM. Applications to state constrained control problems. Comp. Appl. Math. **21** (2002)
6. Deckelnick, K., Hinze, M.: Convergence of a finite element approximation to a state constrained elliptic control problem. SIAM J. Numer. Anal. **45**, 1937–1953 (2007)
7. Deckelnick, K., Hinze, M.: A finite element approximation to elliptic control problems in the presence of control and state constraints. Preprint Reihe der Hamburger Angewandten Mathematik, HBAM 2007-01 (2007)
8. Deckelnick, K., Günther, A., Hinze, M.: Finite element approximation of elliptic control problems with constraints on the gradient. Priority Program 1253, German Research Foundation, Preprint-Number SPP1253-08-02 (2007)
9. Meyer, C.: Error estimates for the finite element approximation of an elliptic control problem with pointwise constraints on the state and the control. WIAS Preprint 1159 (2006)
10. Schatz, A.H.: Pointwise error estimates and asymptotic error expansion inequalities for the finite element method on irregular grids. I: Global estimates. Math. Comput. **67**(223), 877–899 (1998)

Globalization of Nonsmooth Newton Methods for Optimal Control Problems

C. Gräser

Abstract We present a new approach for the globalization of the primal-dual active set or equivalently the nonsmooth Newton method applied to an optimal control problem. The basic result is the equivalence of this method to a nonsmooth Newton method applied to the nonlinear Schur complement of the optimality system. Our approach does not require the construction of an additional merit function or additional descent direction. The nonsmooth Newton directions are naturally appropriate descent directions for a smooth dual energy and guarantee global convergence if standard damping methods are applied.

1 Introduction

We consider the optimal control problem of minimizing the functional

$$\mathscr{J}(y,u) = \frac{1}{2}\|y - y_d\|_0^2 + \frac{\alpha}{2}\|u\|_0^2$$

subject to the constraints

$$u \le \psi \qquad \text{and} \qquad -\Delta y = u$$

where the state y and the control u are from suitable function spaces on the domain Ω and $\|\cdot\|_0$ denotes the $L^2(\Omega)$-norm.

It turned out that semismooth or nonsmooth Newton methods and interior point methods are amongst the most efficient techniques to deal with this kind of problem. Interior point methods (e.g., [2, 9]) regularize the problem by a sequence of barrier functions to overcome the nonsmoothness inherited by the inequality constraints.

Carsten Gräser
Freie Universität Berlin, Institut für Mathematik, Arnimallee 6, D - 14195 Berlin, Germany, e-mail: graeser@math.fu-berlin.de

Nonsmooth Newton methods introduced in [7, 8] solve such problems directly by certain linearizations of nonsmooth operators. These methods have the advantage that no regularization parameter has to be controlled. They differ in the nonsmooth operator used to incorporate the inequality constraints and in the concept of differentiability and semismoothness used to obtain linearizations and convergence results. There are finite dimensional (e.g., [1, 5, 6]) as well as infinite dimensional (e.g., [6, 10]) approaches. The convergence results are in general only local and globalization is tackled by the construction of merit functions and descent directions for these functions if the Newton directions fail to be suitable.

The approach used in [6] is shown to be equivalent to the primal-dual active set method and a global convergence result is obtained. However, this only holds under restrictive assumptions on α and if the linear subproblems are solved exactly. In [5] the nonsmooth Newton idea is applied to a discrete nonlinear Schur complement such that the nonsmooth Newton directions are natural descent directions for a dual minimization problem. Global convergence is achieved by damping based on the dual energy.

The present paper shows that the primal-dual active set method and the Schur complement nonsmooth Newton method basically coincide. Hence the dual approach of [5] offers a natural way to globalize the method of [6].

The paper is organized as follows. In Section 2 various reformulations of the problem are presented. Section 3 recalls the methods in [6] and [5]. Finally the equivalence of both methods is shown in Section 4.

2 An Optimal Control Problem

Using Green's formula and appropriate Sobolev spaces the above problem can be formulated in weak form as

$$(\mathscr{M})\quad \begin{cases} \text{Find } (y,u) \in H_0^1(\Omega) \times L^2(\Omega) \text{ such that} & \\ \qquad\qquad \mathscr{J}(y,u) \le \mathscr{J}(w,v) & \\ \text{s.t.} \quad u \in \mathscr{K}, \quad (\nabla y, \nabla v) = (u,v) \quad \forall v \in H_0^1(\Omega) \end{cases}$$

with the convex set $\mathscr{K} = \{v \in L^2(\Omega) : v \le \psi \text{ a.e in } \Omega\}$. Since the presented results only hold in the finite dimensional case we restrict our considerations to a corresponding discrete problem. Only the solution of the arising algebraic problem and not the discretization itself will be discussed. Therefore this problem is formulated in terms of vectors and matrices instead of discrete function spaces. Furthermore, to simplify notation we use $y \in \mathbb{R}^m$ and $u, \psi \in \mathbb{R}^n$ as well for the discrete approximations of the state, the control and the obstacle respectively. The discrete analogue of $(\mathscr{M})$ then reads:

$$
(\mathrm{M}) \begin{cases} \text{Find } (y,u) \in \mathbb{R}^m \times \mathbb{R}^n \text{ such that} \\ \qquad\qquad J(y,u) \le J(w,v) \\ \text{s.t.} \qquad u \in K, \qquad Ly = Iu \end{cases}
$$

with the discrete convex set $K = \{v \in \mathbb{R}^n : v \le \psi\}$ and the discrete convex energy

$$
J(y,u) = \frac{1}{2}\langle D_1 y, y\rangle + \frac{\alpha}{2}\langle D_2 u, u\rangle - \langle b, y\rangle.
$$

$b \in \mathbb{R}^m$ is a discrete approximation of the linear functional $(y_d, \cdot)$. The vectors y and u may have different dimensions since they will in general come from different discrete spaces. In the following we assume that D_1 and D_2 are symmetric and positive definite and L is assumed to be invertible. This is the case e.g. if $(\mathscr{M})$ is discretized by piecewise linear finite elements. Notice that D_1, D_2 and I are discrete analogues of the identity operator. For a finite element discretization they represent discrete L^2 inner products coupling functions from possibly different finite element spaces. Hence the matrices might differ in general.

Using the subdifferential $\partial\chi_K$ of the indicator functional χ_K of K the optimality system of (M) is given by

$$
(\mathrm{S}) \begin{cases} \text{Find } (y,u,w) \in \mathbb{R}^m \times \mathbb{R}^n \times \mathbb{R}^m \text{ such that} \\ \begin{pmatrix} D_1 & 0 & L^T \\ 0 & \alpha D_2 + \partial\chi_K & -I^T \\ L & -I & 0 \end{pmatrix} \begin{pmatrix} y \\ u \\ w \end{pmatrix} \ni \begin{pmatrix} b \\ 0 \\ 0 \end{pmatrix}. \end{cases}
$$

The elimination of the state $y(u) = L^{-1}Iu$ leads to a reduced problem

$$
u \in K: \qquad \tilde{J}(u) \le \tilde{J}(v) \qquad \forall v \in K
$$

with the energy $\tilde{J}(u) := J(L^{-1}Iu, u)$. Its optimality system is given by

$$
(\mathrm{PD}) \begin{cases} \text{Find } (u,\lambda) \in \mathbb{R}^n \times \mathbb{R}^n \text{ such that} \\ \qquad Au + \lambda = f \\ \qquad u \le \psi, \lambda \ge 0, \lambda(u-\psi) = 0 \end{cases}
$$

with $A = (L^{-1}I)^T D_1 (L^{-1}I) + \alpha D_2$ and $f = (L^{-1}I)^T b$. Notice that analogue formulations can also be derived in the infinite dimensional case.

3 Primal-Dual Active Set and Schur Nonsmooth Newton Method

The primal-dual active set method for the discrete problem is based on the primal-dual formulation (PD). For given $u^0, \lambda^0 \in \mathbb{R}^n$ and fixed parameter $c > 0$ it reads:

Algorithm 3.1 (Primal-Dual Active Set Method)

1. *Set* $\mathscr{A}_k = \{i : \lambda_i^k + c(u_i^k - \psi_i) > 0\}$ *and* $\mathscr{I}_k = \{1,\dots,n\} \setminus \mathscr{A}_k$
2. *Solve*

$$Au^{k+1} + \lambda^{k+1} = f,$$

$$u_i^{k+1} = \psi_i \text{ for } i \in \mathscr{A}_k, \quad \lambda_i^{k+1} = 0 \text{ for } i \in \mathscr{I}_k.$$

In [6] it is shown that the method can be interpreted as semismooth Newton method and that the following convergence results hold.

Theorem 1 (cf. [6]). *The sequence* (u^k, λ^k) *generated by Alg. 3.1 converges superlinearly to the solution* (u^*, λ^*) *of* (PD) *if* $\|(u^0, \lambda^0) - (u, \lambda)\|$ *is sufficiently small. Furthermore, it converges for arbitrary* (u^0, λ^0) *if A is the sum of an M-matrix and a sufficiently small perturbation matrix. For the case of a discretized control problem the latter is the case if* α *is small enough.*

Unfortunately global convergence is in general not preserved if α is too small or if the linear systems are solved inexactly. A similar result also holds in the infinite dimensional case (cf. [6]).

Another approach for the solution of the discrete algebraic problem, introduced in [5], is based on the elimination of the primal unknowns in (S) by

$$\begin{pmatrix} y(w) \\ u(w) \end{pmatrix} := \begin{pmatrix} D_1 & 0 \\ 0 & \alpha D_2 + \partial \chi_K \end{pmatrix}^{-1} \begin{pmatrix} b - L^T w \\ I^T w \end{pmatrix} \tag{1}$$

where $(\alpha D_2 + \partial \chi_K)^{-1}(z)$ is the single-valued preimage of z. Similar to linear saddle point problems (S) can be equivalently formulated as an unconstrained minimization problem

$$(\mathrm{M}^*) \begin{cases} \text{Find } w \in \mathbb{R}^n \text{ such that} \\ \qquad\qquad h(w) \le h(v) \qquad \forall v \in \mathbb{R}^n. \end{cases}$$

with the energy

$$h(w) = -\mathscr{L}(y(w), u(w), w) \tag{2}$$

where $\mathscr{L} : \mathbb{R}^m \times \mathbb{R}^n \times \mathbb{R}^m$ is the Lagrange functional associated with the saddle point problem (S) given by

$$\mathscr{L}(y, u, w) = J(y, u) + \chi_K(u) + \langle Ly - Iu, w \rangle .$$

Proposition 1 (cf. [5]). *The energy h defined by* (2) *is strictly convex, coercive, and continuously differentiable. Its gradient is given by the Lipschitz continuous operator*

$$\nabla h(w) = -(Ly(w) - Iu(w)).$$

This operator is also denoted the nonlinear Schur complement operator of (S).

Using these properties descent algorithms of the form

$$w^{k+1} = w^k + \rho^k d^k \tag{3}$$

with appropriate descent directions d^k and step sizes ρ^k can be applied to solve (M*). Choosing $d^k = -\nabla h(w^k)$ leads to the gradient method which is equivalent to a nonlinear Uzawa method (cf., e.g., [3, 4]) applied to (S). Since ∇h is Lipschitz continuous a nonsmooth Newton approach is used to obtain linearizations S_k of ∇h at w^k leading to the (damped) nonsmooth Newton method:

Algorithm 3.2 (Schur Nonsmooth Newton Method)

1. *Solve* $d^k = -S_k^{-1} \nabla h(w^k)$
2. *Set* $w^{k+1} = w^k + \rho^k d^k$

The linearizations S_k of the Schur complement ∇h at w^k are given by

$$S_k = LD_1^{-1}L^T + I\widehat{D}_2(w^k)I^T$$

where $\widehat{D}_2(w^k)$ is the linearization of $(\alpha D_2 + \partial \chi_K)^{-1}$ constructed by

$$\widehat{D}_2(w^k) = T_{\overline{\mathscr{A}}_k} \left(I - T_{\overline{\mathscr{A}}_k} + T_{\overline{\mathscr{A}}_k} \alpha D_2 T_{\overline{\mathscr{A}}_k} \right)^{-1} T_{\overline{\mathscr{A}}_k}$$

using the projection $T_{\overline{\mathscr{A}}_k}$ associated with the active set $\overline{\mathscr{A}}_k$ given by

$$(T_{\overline{\mathscr{A}}_k})_{ij} = \begin{cases} 1 & \text{if } i = j \text{ and } i \notin \overline{\mathscr{A}}_k \\ 0 & \text{else} \end{cases}, \qquad \overline{\mathscr{A}}_k = \{ i : u(w_k)_i = \psi_i \}.$$

It is easy to see that these directions are descent directions, i.e. $\langle d^k, \nabla h(w^k) \rangle < 0$ if $\nabla h(w^k) \neq 0$. The selected step sizes should guarantee sufficient descent. This can be achieved by selecting them using the Armijo rule or bisection such that they are *efficient*, i.e. they satisfy

$$\exists C > 0 \, \forall k: \qquad h(w^k + \rho^k d^k) \leq h(w^k) - C \left(\frac{\langle \nabla h(w^k), d^k \rangle}{\|d^k\|} \right)^2. \tag{4}$$

Having the descent property the following convergence result can be obtained.

Theorem 2 (cf. [5]). *Assume that the stepsizes ρ^k are efficient. Then the sequence generated by Alg. 3.2 converges to the solution w^* of* (M*) *for arbitrary w^0.*

This convergence result is preserved if inexact evaluation of S_k^{-1} is considered:

Algorithm 3.3 (Inexact Schur Nonsmooth Newton Method)

1. *Solve* $\tilde{d}^k = -S_k^{-1} \nabla h(w^k) + \varepsilon^k$
2. *Set* $w^{k+1} = w^k + \rho^k \tilde{d}^k$

Theorem 3 (cf. [5]). *Assume that the step sizes ρ^k are efficient, that the vectors $\tilde{d}^k$ are descent directions, and that $\|\varepsilon^k\| / \|\tilde{d}^k\| \to 0$. Then the sequence generated by Alg. 3.3 converges to the solution w^* of* (M*) *for arbitrary initial iterates w^0.*

It is easy to see that these convergence results also hold if the S_k are replaced by preconditioners $\overline{S}_k$ defined analogously using the smaller active sets

$$\overline{\overline{\mathscr{A}}}_k = \{i : (\alpha D_2 u(w_k))_i \neq (I^T w_k)_i\} \subset \overline{\mathscr{A}}_k.$$

Remark 1. Each damping strategy requires possibly multiple evaluation of either h or ∇h. These evaluations incorporate the solution of the nonlinear convex problem in (1). However, solving this problem is in general very cheap since D_2 represents the L^2 inner product or the identity and not a differential operator. Thus a nonlinear Gauß-Seidel method will converge to machine accuracy in a few steps.

Remark 2. It is easy to see that Alg. 3.2 terminates with the exact solution in one step in a sufficiently small neighborhood of the solution. Thus a superlinear convergence result within an unknown neighborhood would not provide further insight.

4 Globalization of the Primal-Dual Active Set Method

Now we analyze the relation between the presented methods. In the following let (u^k, λ^k) be the sequence generated by Alg. 3.1. Defining the sequences

$$y^k = L^{-1} I u^k, \qquad w^k = L^{-T} b - L^{-T} D_1 L^{-1} I u^k$$

the multiplier is given by

$$\lambda^{k+1} = I^T w^{k+1} - \alpha D_2 u^{k+1} \tag{5}$$

and the linear system in Alg. 3.1 is equivalent to

$$\begin{pmatrix} D_1 & 0 & L^T \\ 0 & T_{\mathscr{I}_k} + T_{\mathscr{A}_k} \alpha D_2 & -T_{\mathscr{A}_k} I^T \\ L & -I & 0 \end{pmatrix} \begin{pmatrix} y^{k+1} \\ u^{k+1} \\ w^{k+1} \end{pmatrix} = \begin{pmatrix} b \\ T_{\mathscr{I}_k} \psi \\ 0 \end{pmatrix}. \tag{6}$$

A simple computation shows that the constant c in the definition of the active set drops out after the first iteration. More precisely using (5) we have

Lemma 1. *Let $k > 0$ then the sets $\mathscr{A}_k$ can for arbitrary $c_i > 0$ be written as*

$$\mathscr{A}_k = \{i : (I^T w^k - \alpha D_2 u^k)_i + c_i(u_i^k - \psi_i) > 0\}.$$

Proof. Since $k > 0$ each index i is either in $\mathscr{A}_{k-1}$ or in $\mathscr{I}_{k-1}$. Hence $(u_i^k - \psi_i)$ or $(I^T w^k - \alpha D_2 u^k)_i = \lambda_i^k$ must be zero. Therefore the sum is positive iff one of these expressions is positive. Having $c, c_i > 0$ it is clear that

$$c(u_i^k - \psi_i) > 0 \Leftrightarrow (u_i^k - \psi_i) > 0 \Leftrightarrow c_i(u_i^k - \psi_i) > 0. \qquad \square$$

The definition of $\mathscr{A}_k$ is closely related to the Euclidean projection which differs from the projection with respect to D_2 if it is not diagonal. However, we need this projection in the following since it is the proper discrete analogue of the continuous L^2-projection. Therefore we assume

(A) D_2 is a diagonal matrix.

This is not very restrictive since (A) holds if u is discretized by finite differences, by piecewise constant finite elements, or by piecewise linear finite elements using mass lumping for D_2.

Using (A) and lemma 1 with $c_i = (\alpha D_2)_{ii}$ we get

Lemma 2. *Let $k > 0$ then*

$$\mathscr{A}_k = \{i : (\alpha D_2 u(w_k))_i \neq (I^T w_k)_i\}. \qquad \square$$

By this representation $u(w^k)$ can be expressed by a linear equation depending on the active set $\mathscr{A}_k$. Thus from (6) we get

$$\begin{pmatrix} D_1 & 0 & L^T \\ 0 & T_{\mathscr{I}_k} + T_{\mathscr{A}_k} \alpha D_2 T_{\mathscr{A}_k} & -T_{\mathscr{A}_k} I^T \\ L & -I T_{\mathscr{A}_k} & 0 \end{pmatrix} \begin{pmatrix} y^{k+1} - y(w^k) \\ u^{k+1} - u(w^k) \\ w^{k+1} - w^k \end{pmatrix} = \begin{pmatrix} 0 \\ 0 \\ \nabla h(w^k) \end{pmatrix}. \tag{7}$$

Here we used the fact that $u^{k+1} - u(w^k) = T_{\mathscr{A}_k}(u^{k+1} - u(w^k))$. By this representation we instantly get the main result:

Theorem 4. *Assume that assumption (A) holds. For $k > 0$ Alg. 3.1 is equivalent to the iteration*

$$w^{k+1} = w^k \underbrace{- \overline{S}_k^{-1} \nabla h(w^k)}_{d^k} \tag{8}$$

with the symmetric, positive definite preconditioner

$$\overline{S}_k = L D_1^{-1} L^T + I T_{\mathscr{A}_k} \left(T_{\mathscr{I}_k} + T_{\mathscr{A}_k} \alpha D_2 T_{\mathscr{A}_k} \right)^{-1} T_{\mathscr{A}_k} I^T$$

in the sense that (y^k, u^k) is computed from w^k using (6). *Thus global convergence of this descent method can be achieved by introducing appropriate damping parameters ρ^k in* (8) *even in the case of inexact evaluation in the sense of theorem 3.*

Proof. Equivalence follows from (7) by elimination of the state and control variables. The convergence results are direct consequences of theorems 2 and 3 since we have $T_{\mathscr{I}_k} = I - T_{\mathscr{A}_k}$ and $\mathscr{A}_k = \overline{\overline{\mathscr{A}}}_k$. $\square$

Remark 3. Theorem 4 shows that the primal-dual active set method and the (undamped) Schur complement nonsmooth Newton method applied to control problems basically coincide. The interpretation as descent method provides a natural way to globalize the method using damping even in the case of inexact solution of the linear systems. Notice that no artificial merit function and descent directions have to be constructed if this approach is used.

Remark 4. The above result does not carry over to the continuous problem. Since the natural embedding $\mathscr{I} : H_0^1(\Omega) \to H_0^1(\Omega)'$ is not invertible the elimination of the state used in (1) is in general only possible in a suitable subspace. In this subspace Alg. 3.3 can also be defined in the infinite dimensional case. However, the convergence theory given in [5] is no longer applicable.

This might suggest that the convergence gets slower for larger discrete problems. The numerical results in [5] seem to contradict this even in the case that damping is applied. Fig. 1 (taken from [5]) shows the number of interation steps for the solution to round-off errors over refinement levels for bilateral constraints, picewise constant y_d and $\alpha = 10^{-8}$ where damping is essential. The extension of the ideas to the infinite dimensional case is the topic of current research.

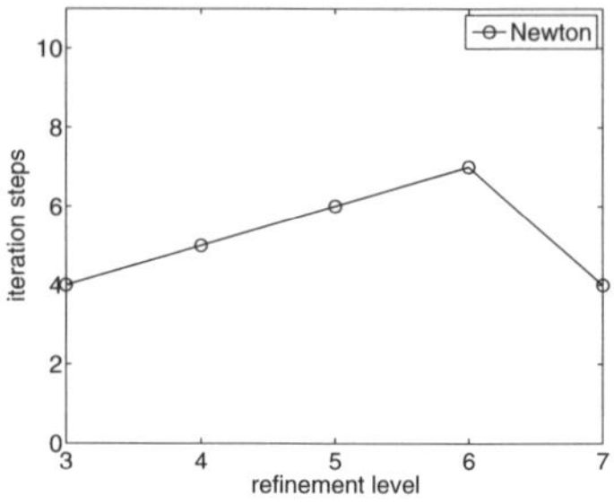

Fig. 1 Mesh dependence for $\alpha = 10^{-8}$

Acknowledgements This work has been funded in part by the Deutsche Forschungsgemeinschaft under contract Ko 1806/3-2.

References

1. Bergounioux, M., Ito, K., Kunisch, K.: Primal-dual strategy for constrained optimal control problems. SIAM J. Control Optim. **37**, 1176–1194 (1999)
2. Deuflhard, P., Weiser, M.: The central path towards the numerical solution of optimal control problems. Tech. rep., Zuse Institute Berlin (2001)
3. Ekeland, I., Temam, R.: Convex Analysis. No. 1 in Studies in Mathematics and its Applications. North-Holland, Amsterdam New York Oxford (1976)
4. Glowinski, R., Lions, J., Trémolières, R.: Numerical Analysis of Variational Inequalities. No. 8 in Studies in Mathematics and its Applications. North-Holland Publishing Company, Amsterdam New York Oxford (1981)
5. Gräser, C., Kornhuber, R.: Nonsmooth Newton methods for set-valued saddle point problems. Preprint A/04/2006, FU Berlin (2006)
6. Hintermüller, M., Ito, K., Kunisch, K.: The primal-dual active set strategy as a semismooth Newton method. SIAM J. Optim. **13**(3), 865–888 (2003)
7. Kummer, B.: Newton's method based on generalized derivatives for nonsmooth functions: Convergence analysis. In: W. Oettli, D. Pallaschke (eds.) Advances in optimization (Lambrecht, 1991), pp. 171–194. Springer, Berlin (1992)
8. Qi, L., Sun, J.: A nonsmooth version of Newtons's method. Mathematical Programming **58**, 353–367 (1993)
9. Schiela, A., Weiser, M.: Superlinear convergence of the control reduced interior point method for pde contrained optimization. Tech. rep., Zuse Institute Berlin (2005). To appear in COAP
10. Ulbrich, M.: Nonsmooth Newton-like methods for variational inequalities and constrained optimization problems in function spaces. Habilitationsschrift, TU München (2002)

An Inexact Trust-Region SQP Method with Applications to PDE-Constrained Optimization

M. Heinkenschloss and D. Ridzal

Abstract We present a trust-region sequential quadratic programming (SQP) method with iterative linear system solves, for the solution of smooth nonlinear equality-constrained optimization problems. Stopping criteria for iterative solvers are selected by the optimization algorithm, based on its overall progress. Global convergence is ensured and unnecessary oversolving of linear systems is eliminated. The algorithm is applied to several PDE-constrained optimization problems.

1 Introduction

Sequential quadratic programming (SQP) methods are used successfully for the solution of smooth nonlinear programming problems (NLPs). SQP methods compute an approximate solution of the NLP by solving a sequence of quadratic subproblems, which are built from a quadratic model of the Lagrangian and a linear Taylor approximation of the constraints. The solution of these subproblems requires the solution of linear systems in which the system operator involves the constraint Jacobian or its adjoint. For PDE-constrained problems, the solution of such linear systems is often performed using iterative solvers, which require carefully selected stopping criteria. These must be chosen based on the overall progress of the optimization algorithm, in order to ensure global convergence and avoid unnecessary

Matthias Heinkenschloss
Department of Computational and Applied Mathematics, MS-134, Rice University, 6100 Main Street, Houston, TX 77005-1892, USA. e-mail: heinken@rice.edu

Denis Ridzal
Optimization and Uncertainty Quantification, MS-1320, Sandia National Laboratories[†], P.O. Box 5800, Albuquerque, NM 87185-1320, USA. e-mail: dridzal@sandia.gov
[†]Sandia is a multiprogram laboratory operated by Sandia Corporation, a Lockheed Martin Company, for the United States Department of Energy's National Nuclear Security Administration under Contract DE-AC04-94-AL85000.

oversolving of linear systems. This paper gives a brief sketch of a trust-region SQP algorithm with iterative (i.e. inexact) linear system solves, for the solution of equality-constrained NLPs of the type

$$\min \; f(x) \tag{1a}$$

$$\text{s.t.} \;\; c(x) = 0, \tag{1b}$$

where $f : \mathscr{X} \to \mathbb{R}$ and $c : \mathscr{X} \to \mathscr{C}$ for some Hilbert spaces $\mathscr{X}$ and $\mathscr{C}$. We identify the dual $\mathscr{X}^*$ of $\mathscr{X}$ with $\mathscr{X}$. The description of the algorithm is followed by its application to a collection of PDE-constrained optimization problems.

For a detailed overview of the treatment of inexact linear system solves in constrained optimization see [7, 8, 10]. Trust-region SQP methods with inexact linear system solves are presented in [8, 11]. The methods in [8, 11] are formulated in the reduced space, i.e. in part rely on a decomposition of optimization variables into a basic and a nonbasic set. This splitting is not always convenient and prohibits the use of many efficient solvers for linear systems of the KKT type.

An inexact line-search SQP method for convex problems was recently introduced in [4]. Nonconvex problems are addressed in [3]; it is required that the Hessian in the quadratic subproblem is modified to exhibit certain positive definiteness properties.

Our trust-region SQP method builds on [8]. The control of inexactness in linear system solves is guided by the progress of the optimization algorithm. The derived stopping conditions for iterative solvers can be easily implemented and require little computational overhead. In contrast to [8, 11], our algorithm is formulated in the full space and can take advantage of preconditioners for KKT systems. In contrast to [3], our trust-region approach does not require a modification of the Hessian in the quadratic subproblem, even in the presence of nonconvexity.

2 Composite-Step Trust-Region SQP Algorithm with Iterative Augmented System Solves

Only a basic outline of the overall algorithm is included, with an emphasis on the choice of stopping conditions for linear system solves. For more details, see [7, 10].

Let $\mathscr{L} : \mathscr{X} \times \mathscr{C} \to \mathbb{R}$, $\mathscr{L}(x,\lambda) = f(x) + \langle \lambda, c(x) \rangle_{\mathscr{C}}$ be the Lagrangian functional for (1) and let $H_k = H(x_k, \lambda_k)$ be the Hessian of the Lagrangian $\nabla_{xx}\mathscr{L}(x_k,\lambda_k)$ or a replacement thereof. Trust-region SQP methods compute an approximate solution of (1) by approximately solving a sequence of subproblems derived from

$$\min \; \frac{1}{2}\langle H_k s_k, s_k \rangle_{\mathscr{X}} + \langle \nabla_x \mathscr{L}(x_k,\lambda_k), s_k \rangle_{\mathscr{X}} + \mathscr{L}(x_k,\lambda_k) \tag{2a}$$

$$\text{s.t.} \;\; c_x(x_k)s_k + c(x_k) = 0, \qquad \|s_k\|_{\mathscr{X}} \le \Delta_k. \tag{2b}$$

To deal with the possible incompatibility of the constraints (2b) we apply a Byrd-Omojokun-like composite step approach. See [5, Sec. 15.4.2] for an overview.

The trial step s_k is split into a quasi-normal step n_k and a tangential step t_k. The role of the quasi-normal step n_k is to improve feasibility, by approximately solving

$$\min \quad \|c_x(x_k)n + c(x_k)\|_{\mathscr{C}}^2 \tag{3a}$$
$$\text{s.t} \quad \|n\|_{\mathscr{X}} \le \zeta\Delta_k, \tag{3b}$$

where $\zeta \in (0,1)$ is a fixed constant. Once the quasi-normal step n_k is computed, the tangential step t_k is computed as an approximate solution of the subproblem

$$\min \quad \frac{1}{2}\langle H_k(t+n_k), t+n_k\rangle_{\mathscr{X}} + \langle\nabla_x\mathscr{L}(x_k,\lambda_k), t+n_k\rangle_{\mathscr{X}} + \mathscr{L}(x_k,\lambda_k) \tag{4a}$$
$$\text{s.t.} \quad c_x(x_k)t = 0, \qquad \|t+n_k\|_{\mathscr{X}} \le \Delta_k. \tag{4b}$$

The trial step $s_k = n_k + t_k$ is accepted or rejected based on the augmented Lagrangian merit function $\phi(x,\lambda;\rho) = \mathscr{L}(x,\lambda) + \rho\|c(x)\|_{\mathscr{C}}^2$ and the related quantities *ared* and *pred*, the actual and predicted reduction, see [7, 10]. One step of the composite-step trust-region SQP algorithm involves the following tasks.

Algorithm 1 *(One step of the Composite-Step Trust-Region SQP Algorithm)*

1. *Compute quasi–normal step* n_k.
2. *Compute tangential step* t_k.
3. *Compute new Lagrange multiplier estimate* λ_{k+1}.
4. *Update penalty parameter* ρ_k*; based on* $ared_k/pred_k$*, accept or reject new iterate* $x_{k+1} = x_k + s_k$ *and update* Δ_{k+1} *from* Δ_k.

Linear systems are solved in Steps 1–3 of Alg. 1. If iterative solvers are used, the stopping criteria must be carefully selected to ensure global convergence of the optimization algorithm and to avoid unnecessary oversolving of linear systems. We discuss the selection of stopping criteria below. One difficulty in devising them is that the (approximate) solution of one linear system feeds into subsequent linear systems. A global convergence theory for Alg. 1 with the choices of iterative linear system solves specified below is given in [7, 10].

Computation of the Quasi-Normal Step We use a dogleg method to compute an approximate solution of (3), see, e.g., [5, Sec. 7.5.3]. Let n_k^{cp} be the Cauchy point, i.e., the solution of $\min\{\|c_x(x_k)n + c(x_k)\|_{\mathscr{C}}^2 : n = -\alpha c_x(x_k)^* c(x_k), \alpha \ge 0\}$. The minimizer α^{cp} satisfies $\alpha^{cp} = \|c_x(x_k)^*c(x_k)\|_{\mathscr{X}}^2 / \|c_x(x_k)c_x(x_k)^*c(x_k)\|_{\mathscr{C}}^2$, i.e. the computation of n_k^{cp} requires no linear system solves. If $\|n_k^{cp}\|_{\mathscr{X}} \ge \zeta\Delta_k$, we set the quasi-normal step to $n_k = \zeta\Delta_k n_k^{cp}/\|n_k^{cp}\|_{\mathscr{X}}$.

If $\|n_k^{cp}\|_{\mathscr{X}} < \zeta\Delta_k$, we compute an approximate minimum norm solution n_k^N of $\min\|c_x(x_k)n + c(x_k)\|_{\mathscr{C}}^2$ and compute the quasi-normal step by moving from n_k^{cp} as far as possible toward n_k^N while staying within the trust region of radius $\zeta\Delta_k$. The minimum norm solution n_k^N can be computed by solving the *augmented system*

$$\begin{pmatrix} I & c_x(x_k)^* \\ c_x(x_k) & 0 \end{pmatrix}\begin{pmatrix} n_k^N \\ y \end{pmatrix} = \begin{pmatrix} 0 \\ -c(x_k) \end{pmatrix}, \tag{5}$$

see [1, p. 7]. If this system is to be solved iteratively, we solve for $\delta n_k = n_k^N - n_k^{cp}$ rather than n_k^N. In particular, we solve

$$\begin{pmatrix} I & c_x(x_k)^* \\ c_x(x_k) & 0 \end{pmatrix} \begin{pmatrix} \delta n_k \\ y \end{pmatrix} = \begin{pmatrix} -n_k^{cp} + e_k^1 \\ -c_x(x_k) n_k^{cp} - c(x_k) + e_k^2 \end{pmatrix}, \tag{6}$$

where we allow nonzero residuals e_k^1 and e_k^2. Given a fixed constant $C^{qn} \in (0,1)$, the residual size is restricted via the stopping criterion

$$\|e_k^1\|_{\mathscr{X}} + \|e_k^2\|_{\mathscr{C}} \leq C^{qn} \, \|c_x(x_k) n_k^{cp} + c(x_k)\|_{\mathscr{C}}. \tag{7}$$

Computation of the Tangential Step We approximately solve (4) using a variant of the Steihaug-Toint (ST) truncated conjugate gradient (CG) method [5, Sec. 7.5.1], where $c_x(x_k)t = 0$ is enforced using null-space projections. Every iteration i of the ST-CG algorithm requires a projection of the type $z^i = Pr^i$, computed by solving

$$\begin{pmatrix} I & c_x(x_k)^* \\ c_x(x_k) & 0 \end{pmatrix} \begin{pmatrix} z^i \\ y^i \end{pmatrix} = \begin{pmatrix} r^i \\ 0 \end{pmatrix}. \tag{8}$$

If the augmented system (8) is solved inexactly, several subtle but important modifications to the ST-CG algorithm are necessary, see [7, 10] for details. Here we only address the stopping criterion for the iterative solution of the augmented system.

Let k be the current SQP iteration, let $g_k = H_k n_k + \nabla_x \mathscr{L}(x_k, \lambda_k)$, let $i = 0, 1, \ldots$ be the ST-CG iteration counter, let z^0 be the first projection [7, 10], and let $j = 0, 1, \ldots$ be the running index of the linear solver for (8). At every solver iteration j, we have the relationship

$$\begin{pmatrix} I & c_x(x_k)^* \\ c_x(x_k) & 0 \end{pmatrix} \begin{pmatrix} \mathsf{z}_j \\ \mathsf{y}_j \end{pmatrix} = \begin{pmatrix} r^i + e_j^1 \\ 0 + e_j^2 \end{pmatrix}, \tag{9}$$

where e_j^1, e_j^2 are nonzero residuals. Given a fixed constant $C^t > 0$, the linear solver is stopped with $(z^i \;\; y^i) = (\mathsf{z}_j \;\; \mathsf{y}_j)$, for $i \geq 1$, when

$$\|e_j^1\|_{\mathscr{X}} + \|e_j^2\|_{\mathscr{C}} \leq \min\left\{ \frac{\|z^0\|_{\mathscr{X}}}{\|g_k\|_{\mathscr{X}}}, \frac{\Delta_k}{\max\{1, \|g_k\|_{\mathscr{X}}\}}, C^t \right\} \|\mathsf{z}_j\|_{\mathscr{X}}. \tag{10}$$

Lagrange Multiplier Update The Lagrange multiplier estimate λ_{k+1} is typically computed as an approximate solution of $\min \|\nabla_x f(x_k) + c_x(x_k)^* \lambda\|_{\mathscr{X}}$. The corresponding least-squares estimate can be obtained by solving the augmented system

$$\begin{pmatrix} I & c_x(x_k)^* \\ c_x(x_k) & 0 \end{pmatrix} \begin{pmatrix} z \\ \lambda_{k+1} \end{pmatrix} = \begin{pmatrix} -\nabla_x f(x_k) \\ 0 \end{pmatrix}. \tag{11}$$

If iterative solvers are used, it is advantageous to solve for $\delta\lambda_k = \lambda_{k+1} - \lambda_k$ rather than λ_{k+1} directly. In particular, we solve

$$\begin{pmatrix} I & c_x(x_k)^* \\ c_x(x_k) & 0 \end{pmatrix} \begin{pmatrix} z \\ \delta\lambda_k \end{pmatrix} = \begin{pmatrix} -\nabla_x f(x_k) - c_x(x_k)^* \lambda_k + e_k^1 \\ e_k^2 \end{pmatrix}. \tag{12}$$

Given constants $C^{\lambda,abs} > 0$ and $C^{\lambda,rel} \in (0,1)$, we stop the linear solver iteration if

$$\|e_k^1\|_{\mathscr{X}} + \|e_k^2\|_{\mathscr{C}} \leq \min\left\{C^{\lambda,rel}\|\nabla_x f(x_k) + c_x(x_k)^*\lambda_k\|_{\mathscr{X}},\, C^{\lambda,abs}\right\}. \tag{13}$$

Remark 1. Stopping criteria (7), (13) are easily imposed in practice, as all quantities on the right-hand sides of (7), (13) are known prior to calling the linear solver. Stopping criterion (10), on the other hand, is tied to the size of the linear solver iterate z_j, and thus requires a more sophisticated implementation.

3 Numerical Results

We report on the application of the inexact trust-region SQP algorithm to three optimal control problems. In all cases the algorithm is applied to the discretized problem. In particular, the spaces $\mathscr{X}$ and $\mathscr{C}$ are chosen to be $\mathbb{R}^n$ and $\mathbb{R}^m$. In all examples, augmented systems (6), (9), (12) are solved using a preconditioned GMRES algorithm, with the stopping criteria given by (7), (10), (13) respectively. To analyze the benefits of inexactness control, we compare the inexact SQP algorithm to a similar trust-region SQP algorithm in which the direct solves of the augmented systems (5), (8), (11) are replaced by preconditioned GMRES solves with a *fixed* relative tolerance τ. This is what one may do if one is forced to use iterative linear system solvers in an existing NLP code. We refer to this algorithm as the *fixed-tolerance SQP algorithm*. In order to make the comparison more balanced, we add iterative refinement [6] to null-space projections in the ST-CG component of the fixed-tolerance SQP algorithm. This correction procedure helps to reduce the number of ST-CG iterations and aids convergence of the optimization algorithm. Iterative refinement is not needed in the inexact SQP algorithm.

Optimal Control of the Poisson–Boltzmann Equation We consider a distributed control problem related to the design of semiconductor devices. In such applications, a common objective is to match the current measured at an Ohmic contact Γ_o of the device to a prescribed value, by modifying a reference doping profile $\widehat{u}$. For a mathematical treatment of the semiconductor equations we refer to [9]. Related optimization problems have been studied in, e.g. [2].

We consider a simplified model problem, governed by the Poisson–Boltzmann potential equation. We solve

$$\min\ \frac{\alpha}{2}\int_{\Gamma_o}(\nabla y(x)\cdot\nu - \nabla\widehat{y}(x)\cdot\nu)^2 ds + \frac{\beta}{2}\int_{\Omega}(u(x)-\widehat{u}(x))^2 dx + \frac{\gamma}{2}\int_{\Omega}|\nabla u(x)|^2 dx \tag{14a}$$

$$\text{s.t.}\quad -\nabla\cdot(k(x)\nabla y(x)) = \exp(y(x)) - \exp(-y(x)) - u(x) \qquad \text{in } \Omega, \tag{14b}$$

$$y(x) = 0 \qquad \text{on } \Gamma_D, \tag{14c}$$

$$(k(x)\nabla y(x))\cdot\nu = 0 \qquad \text{on } \Gamma_N, \tag{14d}$$

where $\alpha, \beta, \gamma > 0$, $k \in L^\infty(\Omega)$ with $k \geq \kappa > 0$ a.e. in Ω, ν is the outward unit normal, y is the potential, and u is the doping.

Our computational domain is $\Omega = (0,12) \times (0,6)$. We let $\Gamma_D = ((0,2) \cup (5,7) \cup (10,12)) \times \{6\}$ (source, gate, drain, see Fig. 1), $\Gamma_N = \partial\Omega \setminus \Gamma_D$, and $\Gamma_o = (10,12) \times \{6\}$ (drain). The permittivity is set to $k = 1$. We attempt to match the flux $\nabla \widehat{y} \cdot \nu = 0$ on Γ_o. It is shown in, e.g., [9, L. 3.2.1,Thm. 3.3.1] that for every $u \in L^\infty(\Omega)$, the differential equation (14b-d) has a unique solution $y \in H^2(\Omega)$. Standard techniques can be used to show the well-posedness of the infinite-dimensional problem (14) with states $y \in H^2(\Omega)$ and controls $u \in H^1(\Omega)$.

The problem is discretized using piecewise linear finite elements on a regular mesh with 4096 triangles. We solve (14) for several target doping profiles $\widehat{u}$, characterized by two positive numbers a and b. A target doping (a,b) means that $\widehat{u}(x) = a$ for $x \in ([0,2] \times [3,6]) \cup ([10,12] \times [3,6])$, and $\widehat{u}(x) = b$ elsewhere, see Fig. 1, resulting in optimization problems with varied difficulty levels. Cost functional parameters are $\alpha = b$, $\beta = b \cdot 10$, and $\gamma = b \cdot 10^{-5}$.

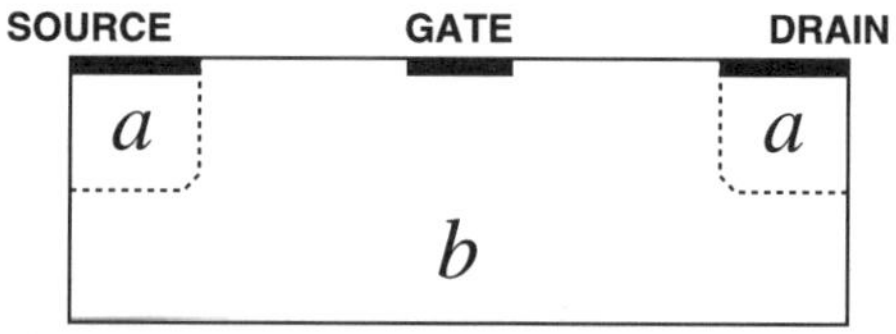

Fig. 1 Sketch of a MESFET semiconductor device and the corresponding doping profile (a,b).

The performance of the inexact SQP algorithm is compared to that of the fixed-tolerance SQP algorithm. We terminate both algorithms when $\|c(x_k)\|_{\mathscr{C}} < 10^{-6}$ and $\|\nabla_x \mathscr{L}(x_k, \lambda_k)\|_{\mathscr{X}} < 10^{-6}$. The ST-CG iteration is stopped with a relative residual tolerance of 10^{-2}. The GMRES solver is preconditioned with an ILU factorization. In the inexact SQP algorithm we set $C^{\lambda,abs} = 10^4$ and use three sets of parameters $(C^t, C^{qn}, C^{\lambda,rel})$, denoted by inxS for $(1, 10^{-1}, 10^{-2})$, inxM for $(10^{-1}, 10^{-2}, 10^{-3})$, and inxF for $(10^{-2}, 10^{-3}, 10^{-4})$, generating different local convergence behaviors of the SQP algorithm. We run the fixed-tolerance SQP algorithm with several choices of a fixed relative GMRES tolerance τ for all augmented system solves.

Table 1 Number of *SQP* / **GMRES** iterations. Inexact (inxF) vs. fixed-tolerance SQP algorithm.

doping (a,b)	inxF	$\tau = 10^{-10}$	$\tau = 10^{-8}$	$\tau = 10^{-6}$	$\tau = 10^{-4}$	$\tau = 10^{-2}$
$(1,1)$	*4* / **683**	*4* / **3105**	*4* / **2564**	F	F	F
$(10^1, 10^{-1})$	*8* / **534**	*8* / **3058**	*8* / **2417**	F	F	F
$(10^2, 10^{-2})$	*10* / **845**	*11* / **3574**	*11* / **2753**	F	F	F
$(10^3, 10^{-3})$	*19* / **1384**	*20* / **7847**	*20* / **6177**	F	F	F
$(10^4, 10^{-4})$	*23* / **2687**	*24* / **9107**	*24* / **6988**	F	F	F
$(10^5, 10^{-5})$	*26* / **2113**	*29* / **9542**	*29* / **7564**	F	F	F

F indicates failure: exceeded 50 SQP iterations *or* violated minimal trust–region radius of 10^{-8}.

We make several observations. First, from Table 1 it is evident that the fixed-tolerance SQP algorithm fails to converge for a relative GMRES stopping tolerance of and above 10^{-6}, regardless of the difficulty of the optimization problem. The inexact SQP algorithm, on the other hand, shows robust behavior, confirming the theoretical results of [7, 10]. Second, the inexact SQP algorithm uses on average four times less GMRES iterations compared to the best fixed-tolerance guess. This is due to the fact that the inexact SQP algorithm allows very loose GMRES stopping tolerances, based on its overall progress, see Fig. 2. Tolerances of 10^{-1} are not uncommon. Third, Table 2 indicates that our inexact SQP algorithm provides an effective mechanism for reducing the average number of linear solver iterations (per linear solver call) at the expense of a slower rate of convergence. This feature is, for example, desired in applications for which efficient iterative linear system solvers are not available.

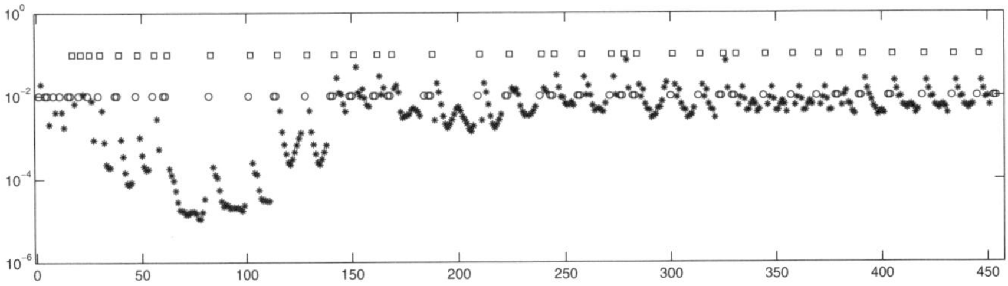

Fig. 2 Relative GMRES stopping tolerances for the inexact SQP algorithm with parameters inxS. Here $\square, *, \circ$ denote GMRES stopping tolerances for augmented solves in the quasi-normal step, tangential step, and Lagrange multiplier computation, respectively.

Table 2 Number of SQP iterations, total number of GMRES iterations, and avg. number of GMRES iterations per call to GMRES. The target doping profile is $(10^3, 10^{-3})$.

	inxS	inxM	inxF	$\tau = 10^{-10}$	$\tau = 10^{-8}$	$\tau = 10^{-6}$	$\tau = 10^{-4}$
sqp iter's	42	24	19	20	20	F	F
gmres iter's	1605	1633	1384	7847	6177	F	F
avg./call	3.5	5.7	6.7	16.6	13.1	F	F

Optimal Control of Navier–Stokes and Burgers' Equations We apply the inexact SQP algorithm to a boundary control problem governed by the Navier–Stokes equations, and to a distributed control problem governed by Burgers' equation. Detailed problem descriptions and mathematical formulations are given in [10, Ch.5]. Numerical experiments are set up as for the Poisson-Boltzmann problem. Tables 3 and 4 fully confirm our previous observations. The inexact algorithm yields significant savings in the total number of GMRES iterations. In addition, SQP iterations can be traded for cheaper linear system solves.

Table 3 Number of SQP iterations, total number of GMRES iterations, and avg. number of GMRES iterations per call to GMRES. Navier-Stokes problem.

	inxS	inxM	inxF	$\tau = 10^{-10}$	$\tau = 10^{-8}$	$\tau = 10^{-6}$	$\tau = 10^{-4}$
sqp iter's	22	9	8	8	8	F	F
gmres iter's	4424	2891	3449	12627	11706	F	F
avg./call	10.1	13.7	18.5	27.2	25.2	F	F

Table 4 Number of SQP iterations, total number of GMRES iterations, and avg. number of GMRES iterations per call to GMRES. Burgers' problem.

	inxS	inxM	inxF	$\tau = 10^{-11}$	$\tau = 10^{-9}$	$\tau = 10^{-7}$	$\tau = 10^{-5}$
sqp iter's	61	26	17	15	15	15	15
gmres iter's	6473	5761	2574	13682	13523	15340	23774
avg./call	12.2	14.1	14.7	19.2	17.7	16.4	14.4

Acknowledgements The work of M. Heinkenschloss was supported in part by ACI-0121360 and DMS-0511624. The work of D. Ridzal was supported by the DOE-SC ASCR Office through the John von Neumann Research Fellowship.

References

1. Björck, Å.: Numerical Methods for Least Squares Problems. SIAM, Philadelphia (1996)
2. Burger, M., Pinnau, R.: Fast optimal design of semiconductor devices. SIAM J. Appl. Math. **64**(1), 108–126 (2003)
3. Byrd, R.H., Curtis, F., Nocedal, J.: An inexact Newton method for nonconvex equality constrained optimization. Tech. rep., Department of Electrical Engineering and Computer Science, Northwestern University (2007)
4. Byrd, R.H., Curtis, F., Nocedal, J.: An inexact SQP method for equality constrained optimization. Tech. rep., Department of Electrical Engineering and Computer Science, Northwestern University (2007)
5. Conn, A.R., Gould, N.I.M., Toint, P.L.: Trust–Region Methods. SIAM, Philadelphia (2000)
6. Gould, N.I.M., Hribar, M.E., Nocedal, J.: On the solution of equality constrained quadratic programming problems arising in optimization. SIAM J. Sci. Comput. **23**(4), 1376–1395 (electronic) (2001)
7. Heinkenschloss, M., Ridzal, D.: A Matrix-Free Trust-Region SQP Method for Equality–Constrained Optimization. Tech. rep., Sandia National Laboratories, Albuquerque, NM, USA (2008)
8. Heinkenschloss, M., Vicente, L.N.: Analysis of inexact trust–region SQP algorithms. SIAM J. Optim. **12**, 283–302 (2001)
9. Markowich, P.A.: The Stationary Semiconductor Device Equations. Computational Microelectronics. Springer Verlag (1986)
10. Ridzal, D.: Trust–Region SQP Methods with Inexact Linear System Solves for Large–Scale Optimization. Ph.D. thesis, Department of Computational and Applied Mathematics, Rice University, Houston, TX (2006)
11. Ulbrich, S.: Generalized SQP-methods with "parareal" time-domain decomposition for time-dependent PDE-constrained optimization. In: Real-Time PDE-Constrained Optimization, *Computational Science and Engineering*, vol. 3, pp. 145–168. SIAM, Philadelphia (2007)

Semi–Monotonic Augmented Lagrangians for Optimal Control and Parameter Identification

D. Lukáš and Z. Dostál

Abstract Optimization and inverse problems governed by partial differential equations are often formulated as constrained nonlinear programming problems via the Lagrange formalism. The nonlinearity is treated using the sequential quadratic programming. A numerical solution then hinges on an efficient iterative method for the resulting saddle–point systems. In this paper we apply a semi–monotonic augmented Lagrangians method, recently proposed and analyzed by the second author, for equality and simple–bound constrained quadratic programming subproblems arising from optimal control and parameter identification. Provided multigrid preconditioning of primal and dual space inner products and of the Hessian the algorithm converges at $O(1)$ matrix–vector multiplications. Numerical results are given for applications in image segmentation and 2–dimensional magnetostatics discretized using lowest–order Lagrange finite elements.

1 Introduction

Many engineering problems involve a solution to partial differential equations (PDE), which describe a physical field under consideration, and a design of some parameters that influence data of the PDE so that the solution has required properties. A typical example is optimal control, where the design parameters control forcing terms in the PDE. Another example is parameter identification, where the design parameters are material coefficients of the PDE operator such that the corresponing solution fits best a given (measured) field. The latter problem is rather close

Dalibor Lukáš
VŠB–Technical University of Ostrava, 17. listopadu 15, 708 33 Ostrava–Poruba, Czech Republic, e-mail: dalibor.lukas@vsb.cz

Zdeněk Dostál
VŠB–Technical University of Ostrava, 17. listopadu 15, 708 33 Ostrava–Poruba, Czech Republic, e-mail: zdenek.dostal@vsb.cz

to an optimal topology design, in which we additionally require the coefficients to be discrete–valued.

In either case we consider the following optimization problem:

$$\min_{u\in\mathscr{U}_{\rm ad}} \mathscr{I}(u,y) \quad \text{s.t.} \quad \mathscr{B}(u,y)=\mathscr{G} \text{ on } \mathscr{Y}', \tag{1}$$

where $\mathscr{U}_{\rm ad} := \{u \in \mathscr{U} \,:\, \underline{u} \le u \le \overline{u}\}$ for $\underline{u},\overline{u} \in \mathscr{U}$, $\underline{u} < \overline{u}$, where $\mathscr{U}$ denotes a Hilbert space, typically $\mathscr{U} \subset L^2(\Omega)$ as in Section 3, of design parameters, $y \in \mathscr{Y}$ with $\mathscr{Y}$ being another Hilbert space of state PDE solutions, where $\mathscr{I} : \mathscr{U} \times \mathscr{Y} \to \mathrm{I\!R}$ denotes a twice continuously differentiable objective functional, $\mathscr{U}'$ and $\mathscr{Y}'$ denote the related dual spaces, where further $\mathscr{B} : \mathscr{U} \times \mathscr{Y} \to \mathscr{Y}'$ is an elliptic linear or bilinear PDE operator in case of optimal control or parameter identification, respectively, and where $\mathscr{G} \in \mathscr{Y}'$ in either case. Finally, we consider the quadratic approximations

$$\min_{v\in V_{\rm ad}} h^i(v) \quad \text{s.t.} \quad B^i(v)=g^i, \quad i=0,1,2,\ldots \tag{2}$$

at iteration points $v^i := (u^i,y^i) \in V_{\rm ad}$, where $h^i(v) := (1/2)\langle A^i v, v\rangle_V - \langle f^i, v\rangle_V$, $A^i := \mathscr{I}''(v^i)$, $f^i := -\mathscr{I}'(v^i)$, $B^i := \mathscr{B}'(v^i)$, $g^i := \mathscr{G} - \mathscr{B}(v^i)$, $V := \mathscr{U} \times \mathscr{Y}$ and $V_{\rm ad} := \mathscr{U}_{\rm ad} \times \mathscr{Y}$. For details and analysis of some SQP schemes we refer to [1].

The concern of our paper is an efficient solution to problems (2). We base our exposition on a Semi–Monotonic Augmented Lagrangian method for Bound and Equality constrained qp–problems (SMALBE), which has been recently proposed and analyzed by the second author in [2]. Note that we have recently applied a similar method to the Stokes problem, see [5]. The method relies on uniformly bounded spectra of Hessians A^i, which we can often assure via a geometric multigrid preconditioning. Then, the algorithm is proven to converge at $O(1)$ matrix–vector multiplications provided we have an optimal convergence method for the inner simple–bound constrained quadratic programming subproblems. Such a method was proposed and analyzed in [3], it is based on conjugate gradients and we call it Modified Proportioning with Reduced Gradient Projections (MPRGP). The rest of the paper is organized as follows: In Section 2 we describe the algorithms SMALBE and MPRGP preconditioned with multigrid and we refer to the main theoretical result on the optimal convergence. Finally, In Section 3 we present two benchmarks, namely an optimal control for image segmentation and a parameter identification for 2–dimensional magnetostatics, and give numerical results in terms of convergence.

2 The Algorithm SMALBE

Let us consider the problem (2) and from now on skip the index i. Denote by $Q := \mathscr{Y}$ the space of Lagrange multipliers. Let I_V and I_Q denote the inner product (Riesz isomorphism) operators on the Hilbert spaces V and Q, respectively, let $g \in \mathrm{Range}(B)$, $\mathrm{Range}(B)$ be closed, and let $V_{\rm BE} := \{v \in V_{\rm ad} \,:\, Bv = g\}$ be nonempty. Denote by $B^T : Q \to V'$ the adjoint operator to B. We assume that there exists $\underline{\rho} > 0$ such that the operator $A+\rho B^T I_Q^{-1} B$ is elliptic with the constant $\underline{\lambda} > 0$. For arbitrary $v \in V_{\text{ad}}$, $q \in Q$ and $\rho \geq \underline{\rho}$ the augmented Lagrange functional related to (2) reads as follows: $L(v,q,\rho) := h(v) + \langle Bv - g, q\rangle_Q + (\rho/2)\|Bv-g\|_{Q'}^2$, the related Fréchet derivative is $F(v,q,\rho) := L_v'(v,q,\rho) = Av - f + B^T q + \rho B^T I_Q^{-1}(Bv-g)$. Note that evaluations of dual norms are due to the Riesz theorem, e.g. $\|\varphi\|_{V'} = \sqrt{\langle \varphi, I_V^{-1}\varphi\rangle_V}$. Then, the problem (2) is equivalent to the saddle–point problem: $\min_{v\in V_{\text{ad}}} \max_{q\in Q} L(v,q,\rho)$ and it has a unique and stable solution $v^* \in V_{\text{ad}}$, while a related Lagrange multiplier $q^* \in Q$ need not be unique.

For numerical solution we will make use of the classical augmented Lagrangian algorithm, where in outer iterations we maximize over Q, increase the penalty ρ and solve the following simple–bound constrained inner qp–subproblems:

$$\min_{v\in V_{\text{ad}}} L(v,q,\rho), \tag{3}$$

where $q \in Q$ and $\rho \geq \underline{\rho}$ are fixed. We present Algorithm 1, which is a modification of the classical augmented Lagrangian algorithm such that we additionally employ an adaptive precision control for solution to the simple–bound constrained subproblems (5), and a special update rule for ρ assuring a monotonic increase of L. The evaluation of $\|F^P\|_{V(v^{(k+1)})^*}$ is described in Section 2.1. For details we refer to [2].

Algorithm 1 Semi–monotonic augmented Lagrangians with adaptive prec. control

Given $\eta > 0$, $\beta > 1$, $\nu > 0$, $\rho^{(0)} \geq \underline{\rho}$, $q^{(0)} \in Q$, precision $\varepsilon > 0$, feasibility precision $\varepsilon_{\text{feas}} > 0$
for $k := 0,1,2,\ldots$ **do**
  Find $v^{(k+1)} \in V_{\text{ad}} : \|F^P(v^{(k+1)}, q^{(k)}, \rho^{(k)})\|_{V(v^{(k+1)})^*} \leq \min\left\{\nu\|Bv^{(k+1)} - g\|_{Q'}, \eta\right\}$
  if $\|F^P(v^{(k+1)}, q^{(k)}, \rho^{(k)})\|_{V(v^{(k+1)})^*} \leq \varepsilon$ and $\|Bv^{(k+1)} - g\|_{Q'} \leq \varepsilon_{\text{feas}}$ **then**
    break
  end if
  $q^{(k+1)} := q^{(k)} + \rho^{(k)} I_Q^{-1} B v^{(k+1)}$
  if $k > 0$ and $L(v^{(k+1)}, q^{(k+1)}, \rho^{(k)}) < L(v^{(k)}, q^{(k)}, \rho^{(k-1)}) + \frac{\rho^{(k)}}{2}\|Bv^{(k+1)} - g\|_{Q'}^2$ **then**
    $\rho^{(k+1)} := \beta \rho^{(k)}$
  else
    $\rho^{(k+1)} := \rho^{(k)}$
  end if
end for
$v^{(k+1)}$, $q^{(k+1)}$ is the solution.

2.1 Inner Iterations: MPRGP

Let us give a precise meaning of F^P. Let V_{ad} be convex nonempty and closed. Then (3) is equivalent to the following variational inequality:

$$\text{Find } v^* \in V_{\text{ad}} \quad \text{such that} \quad \langle F(v^*, q, \rho), v - v^* \rangle_V \geq 0 \quad \forall v \in V_{\text{ad}}. \tag{4}$$

For $v \in V_{\text{ad}}$ by $M(v) := \{w \in V \mid \exists \bar{t} > 0\, \forall t \in [-\bar{t}, \bar{t}] \,:\, v + tw \in V_{\text{ad}}\}$ we denote a vector subspace of feasible full–directions, by $N(v) := M(v)^{\perp}$ its orthogonal complement, by $N^+(v) := \{w \in N(v) \mid \exists \bar{t} > 0\, \forall t \in (0, \bar{t}] \,:\, v + tw \in V_{\text{ad}}\}$ a half–space of feasible half–directions, and by $N^-(v) := -N^+(v)$ the other half–space of nonfeasible half–directions. Then, we can equivalently translate the variational inequality (4) into the following nonsmooth equality:

$$[F^P(v^*, q, \rho)](w) = 0 \quad \forall w \in V \qquad \text{or} \qquad \|F^P(v^*, q, \rho)\|_{V(v)^*} = 0, \tag{5}$$

where for any $v \in V_{\text{ad}}$, $w \in V$, decomposed into $w = w_M + w_N^+ + w_N^-$ with $w_M \in M(v)$, $w_N^+ \in N^+(v)$, and $w_N^- \in N^-(v)$, we define the following projection of $F(v, q, \rho)$:

$$[F^P(v, q, \rho)](w) = \langle F(v, q, \rho), w_M \rangle_V + \min\left\{\langle F(v, q, \rho), w_N^+ \rangle_V, 0\right\},$$

where the additive terms are applications of the so–called free and chopped gradient, respectively, and where

$$\begin{aligned} \|F^P(v, q, \rho)\|^2_{V(v)^*} &:= \|F(v, z, \rho)\|^2_{M(v)'} + \|F^P(v, z, \rho)\|^2_{N(v)^*}, \\ \|F^P(v, z, \rho)\|_{N(v)^*} &:= \sup_{w_N^+ \in N^+(v)} \left|[F^P(v, q, \rho)](w_N^+)\right| / \|w_N^+\|_V. \end{aligned}$$

In the inner iterations, i.e. the first line in the for–cycle of Algorithm 1, we shall approximately solve the auxiliary subproblems (5), for which we recommend to use Algorithm 2. It is based on the conjugate gradient method with proportioning and reduced gradient projections. We denote by $\|A\|_V := \sup_{\|w\|_V = 1} \|Aw\|_{V'}$ the norm of the linear continuous mapping. For more details we refer to [3].

2.2 Analysis: Linear Complexity, Multigrid Preconditioning

Now we refer to the main theoretical result given in [2, Theorem 5]. It indicates that Algorithm 1 works optimally provided a uniformly bounded spectra of all the Hessians independently of the discretization level. Thus, we can construct a multigrid preconditioner for A denoted by $\widehat{A}$, and substitute each occurence of v by $\widehat{A}^{-1/2}\widehat{v}$, which will guarantee boundeness of the Hessian spectra as well as linear complexity of number of the inner CG–iterations. However, under this substitution we would change the simple–bound constraint to a linear inequality constraint, which might be more tricky to handle. Therefore, we recommend to use a diagonal preconditioner for A only. Note that the optimal convergence theory was proven only in finite dimension, but we believe that a generalization to Sobolev spaces will be as straightforward as it has recently turned up in case of equality constrained quadratic programming, see [5].

Algorithm 2 Modified proportioning with reduced gradient projections

Given $\Gamma > 0$, $\overline{\alpha} \in (0, 2\|A\|_V^{-1}]$, $\eta > 0$, $\nu > 0$, $v^{(0)} \in V_{\mathrm{ad}}$, $q \in Q$, $\rho \geq \underline{\rho}$, prec. $\varepsilon > 0$, $\varepsilon_{\mathrm{feas}} > 0$
for $l := 0, 1, 2, \ldots$ **do**
  if $\|F^P(v^{(l)}, q, \rho)\|_{V(v^{(l)})^*} \leq \min\left\{\nu\|Bv^{(l)} - g\|_{Q'}, \eta\right\}$ or $\left(\|F^P(v^{(l)}, q, \rho)\|_{V(v^{(l)})^*} \leq \varepsilon\right.$ and $\left.\|Bv^{(l)} - g\|_{Q'} \leq \varepsilon_{\mathrm{feas}}\right)$ **then**
    break
  end if
  if $\|F^P(v^{(l)}, q, \rho)\|_{N(v)^*} < \Gamma\|F(v^{(l)}, q, \rho)\|_{M(v)'}$ **then**
    Generate $v^{(l+1)}$ by the conjugate gradient step
    if $v^{(l+1)} \notin V_{\mathrm{ad}}$ **then**
      Generate $v^{(l+1/2)}$ by the maximal feasible conjugate gradient step
      Generate $v^{(l+1)}$ as a feasible $(\overline{\alpha})$ addition of the free gradient and project onto V_{ad}
      Restart conjugate gradients with the free gradient
    end if
  else
    Restart conjugate gradients with the chopped gradient
    Generate $v^{(l+1)}$ by the conjugate gradient step
  end if
end for
$v^{(l+1)}$ is the solution.

Additionally, for proper measurements of dual norms we need applications of inverses of I_V and I_Q to be of the linear complexity too. Thus, we can replace applications of I_V^{-1} and I_Q^{-1} by approximate inverse applications $\widehat{I_V}^{-1}$ and $\widehat{I_Q}^{-1}$ using multigrid again.

3 Numerical Results

We present numerical results for two benchmark problems. First, we consider an optimal control problem for image segmentation, see [4] for an introduction of a similar formulation. Given $\Omega := (0,1)^2$, a noisy colour (red, green, blue components) image data $y^{\mathrm{d}} \in \left[L^2(\Omega)\right]^3$ and a regularization parameter $\mu > 0$, we look for sources $u \in \left[L^2(\Omega)\right]^3$ that produced homogeneous colour segments in the image. This leads to the following optimal control problem:

$$\min_{(u,y) \in \mathscr{U}_{\mathrm{ad}} \times \mathscr{Y}} \left\{ \frac{1}{2}\|y(x) - y^{\mathrm{d}}(x)\|^2_{[L^2(\Omega)]^3} + \frac{\mu}{2}\|u(x)\|^2_{[L^2(\Omega)]^3} \right\} \text{ s.t. } -\triangle y = u \text{ on } \mathscr{Y}',$$

where $\mathscr{U}_{\mathrm{ad}} := \{u \in [L^2(\Omega)]^3 \mid 0 \leq u(x) \leq 1 \text{ a.e. in } \Omega\}$ and $\mathscr{Y} := \left[H_0^1(\Omega)\right]^3$.

By numerical experiments we realized that it is enough to proceed with an SQP method, where we neglect the simple–bound constraint and solve the following unconstrained saddle–point system and then project the resulting u^i onto $\mathscr{U}_{\mathrm{ad}}$:

$$\begin{pmatrix} I_{L^2} & 0 & \text{sym.} \\ 0 & \mu I_{L^2} & \text{sym.} \\ -\triangle & -I_{L^2\to H^1} & 0 \end{pmatrix} \begin{pmatrix} y^i_j \\ u^i_j \\ q^i_j \end{pmatrix} = \begin{pmatrix} I_{L^2} y^{\mathrm{d}}_j \\ 0 \\ 0 \end{pmatrix} \quad \text{for} \quad j = 1,2,3,$$

where I_{L^2} stands for the identity (inner product) operator on $L^2(\Omega)$ and $I_{L^2\to H^1}$ stands for the orthogonal projection from $L^2(\Omega)$ to $H^1(\Omega)$. We use a finite element method and employ linear nodal Lagrange elements for both $\mathscr{U}$ and $Q = \mathscr{Y}$. We construct geometric multigrid preconditioners for $I_Q = -\triangle$ and $I_V = I_{[L^2(\Omega)]^3}$ so that a point diagonal smoother with 3 symmetric smoothing iterations is applied for the latter. For $I_Q = -\triangle$ we test a point additive smoother as well as a block Gauss–Seidel smoother with 3 symmetric smoothing iterations again. Since we have neglected the bound constraint, we use the preconditioned conjugate gradients (PCG) method instead of MPRGP. Numerical results for the first SQP iteration and for the red component with $\mu := 10^{-4}$ and relative precisions $\varepsilon = \varepsilon_{\text{feas}} = 10^{-5}$ are depicted in Fig. 1 and Table 1. The results are similar for the other colour components. Note that the number of iterations holds about a constant. Yet we have to improve our implementation in Matlab in order to get large–scale simulations in a reasonable time. Note that a similar problem was solved in [6] by an indefinite multigrid preconditioner within the conjugate gradients method with less computational effort.

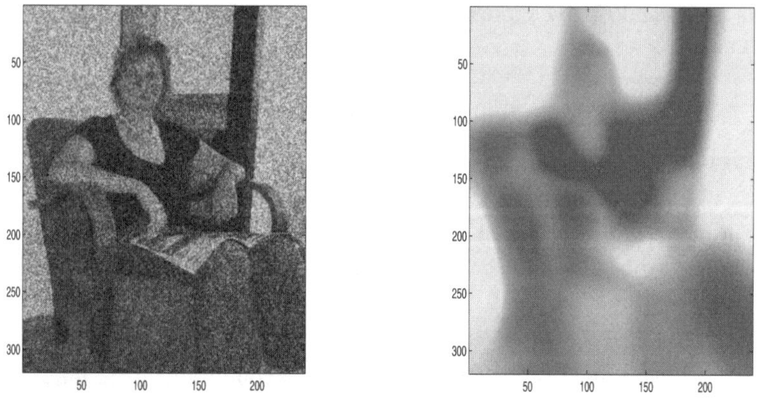

Fig. 1 Image segmentation: original noisy image y^{d} and the reconstructed smooth segments u^1

Second, we consider a parameter identification problem for 2–dimensional magnetostatics. Given a rectangular domain $\Omega \subset \mathbb{R}^2$, a measured magnetic field distribution $B^{\mathrm{g}} = \mathrm{curl}(y^{\mathrm{g}})$, where $y^{\mathrm{g}} \in H^1_0(\Omega)$, a forcing electric current term $g \in L^2(\Omega)$ such that $\mathrm{div}(g) = 0$, reluctivity of the air $\nu_0 := 4\pi 10^{-7}$, a minimal reluctivity of ferromagnetic components $\nu_1 = \nu_0/5000$ and a regularization parameter $\mu > 0$, we search for a distribution $u \in L^2(\Omega)$ of the magnetic reluctivity that has caused the

Table 1 Numerical experiments for optimal control in image segmentation

level l	$\dim(Q_l)$	point additive smoother		block Gauss–Seidel smoother	
		SMALBE/PCG iterations	**total PCG iterations**	**SMALE/PCG iterations**	**total PCG iterations**
1	336	5/7,6,9,7,9	**38**	5/7,6,9,7,9	**38**
2	1271	5/9,10,10,11,13	**53**	5/7,6,9,9,9	**40**
3	4941	5/9,11,11,12,12	**55**	5/7,9,8,6,9	**39**
4	19481	5/8,11,12,13,13	**57**	out of time	
5	77361	5/9,11,11,12,13	**56**	out of time	

measured magnetic field B^g. This leads to the following problem:

$$\min_{(u,y)\in\mathscr{U}_{\text{ad}}\times\mathscr{Y}} \left\{ \frac{1}{2}\|\nabla y(x) - \nabla y^g(x)\|^2_{L^2(\Omega)} + \frac{\mu}{2}\|u(x)\|^2_{L^2(\Omega)} \right\}$$

$$\text{s.t.} \quad -\operatorname{div}\left((\nu_0 + (\nu_1 - \nu_0)u)\,\nabla y\right) = g \text{ on } \mathscr{Y}',$$

where $\mathscr{U}_{\text{ad}} := \{u \in L^2(\Omega) \mid 0 \le u(x) \le 1 \text{ a.e. in } \Omega\}$ and $\mathscr{Y} := H^1_0(\Omega)^3$.

SQP approximations now lead to simple–bound and equality constrained qp–subproblems with the following Hessian:

$$\begin{pmatrix} -\triangle & 0 & \text{sym.} \\ 0 & \mu I_{L^2} & \text{sym.} \\ -\operatorname{div}(q^i\nabla\cdot) & -\operatorname{div}(\cdot\nabla y^i) & 0 \end{pmatrix}.$$

However, it turned out by numerical experiments that at some SQP iterations the design search set V^i_{BE} is empty. As a remedy we relax the upper bound constraint such that $0 \le u(x) \le \gamma^i$, where $\gamma^i \to 1_+$. For approximation we use linear Lagrange finite elements (fe) for $\mathscr{Y} = Q$ and elementwise constant basis for $\mathscr{U}$. We build a geometric multigrid preconditioner for $-\triangle$ with 3 symmetric Gauss–Seidel smoothing steps. The inner product on $\mathscr{U}$ can be inverted directly, since the fe–approximations of $I_{\mathscr{U}}$ are diagonal matrices. Therefore, we can also use the tensor–product preconditioner for the Hessians without changing the simple–bound constraint into a linear constraint, and we can make use of MPRGP. Numerical results for the first and second SQP iteration with $\mu := 10^{-4}$ and relative precisions $\varepsilon = \varepsilon_{\text{feas}} = 10^{-4}$ are depicted in Fig. 2 and Table 2. While the number of SMALBE iterations seems to be about constant, yet we have to improve preconditioning of $A + \rho B^T I_Q^{-1} B$, see the increasing numbers of CG–iterations as well as expansion steps in Table 2.

Acknowledgements This work has been supported by the Czech Grant Agency under the grant GAČR 201/05/P008, by the Czech Ministry of Education under the project MSM6198910027, and by the Czech Academy of Science under the project AVČR 1ET400300415.

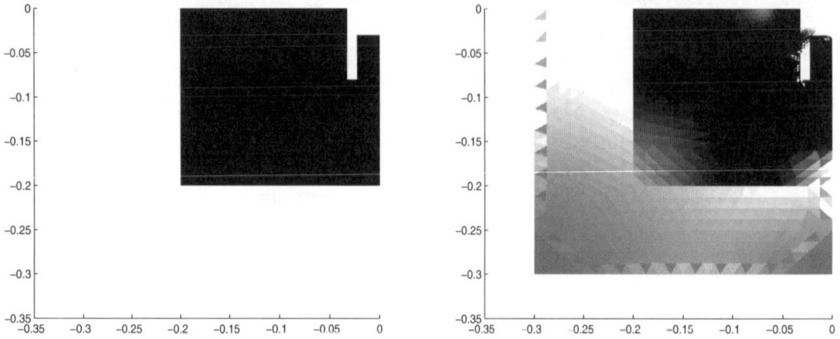

Fig. 2 Parameter identification in 2–dimensional magnetostatics: original and reconstructed ferromagnetics distribution u

Table 2 Numerical experiments for parameter identification in 2–dimensional magnetostatics

level primal / dual DOFs (y's + u's) / (q's)	1st SQP iteration **SMALBE / total inner steps** PCG + exp. + prop. steps	2nd SQP iteration **SMALBE / total inner steps** PCG + expansion + proportioning steps
1 91+187 / 91	**7 / 15+0+0** 1, 3, 2, 2, 3, 2, 2	**5 / 78+16+1** 15+13+0, 26+2+1, 12+1+0, 16+0+0, 9+0+0
2 373+784 / 373	**7 / 17+0+0** 2, 3, 2, 3, 3, 2, 2	**5 / 152+18+3** 20+11+0, 27+6+1, 35+1+1, 37+0+0, 33+0+1
3 1574+2992 / 1574	**8 / 22+0+0** 2, 3, 3, 3, 3, 3, 3, 2	**4 / 185+52+3** 19+30+0, 73+19+2, 46+3+0, 47+0+1
4 6292+11968 / 6292	**8 / 30+0+0** 3, 4, 4, 4, 4, 4, 4, 3	**5 / 377+176+5** 26+80+0, 160+87+4, 76+7+1, 73+2+0, 42+0+0

References

1. Burger, M., Mühlhuber, W.: Numerical approximation of an SQP-type method for parameter identification. SIAM J. Numer. Anal. **40**(5), 1775–1797 (2002)
2. Dostál, Z.: An optimal algorithm for bound and equality constrained quadratic programming problems with bounded spectrum. Computing **78**(4), 311–328 (2006)
3. Dostál, Z., Schöberl, J.: Minimizing quadratic functions over non–negative cone with the rate of convergence and finite termination. Comput. Optim. Appl. **30**(1), 23–44 (2005)
4. Hintermüller, M.: A combined shape Newton and topology optimization technique in real-time image segmentation. In: L. Biegler, O.Ghattas, M. Heinkenschloss, D. Keyes, B. van Bloemen Waanders (eds.) Real-Time PDE-Constrained Optimization. SIAM, Philadelphia (2006)
5. Lukáš, D., Dostál, Z.: Optimal multigrid preconditioned semi–monotonic augmented Lagrangians applied to the Stokes problem. Numer. Linear Algebra Appl. **14**(9), 741–750 (2007)
6. Schöberl, J., Zulehner, W.: Symmetric indefinite preconditioners for saddle point problems with applications to PDE–constrained optimization. SIAM J. Matrix Anal. **29**(3), 752–773 (2007)

An Optimal Control Problem for Stochastic Linear PDE's Driven by a Gaussian White Noise

H. Manouzi and S. Hou

Abstract A computationally efficient technique for the numerical solution of constrained optimal control problems governed by linear stochastic partial differential equations (SPDEs) is considered in this paper. Using the Wiener-Itô chaos expansion of the solution and the control, the stochastic problem is reformulated to a set of deterministic equations. To obtain these chaos coefficients, we use the usual Galerkin finite element method using standard techniques. Once this representation is computed, the statistics of the numerical solution can be easily evaluated. To illustrate our ideas we consider an optimal control problem of a linear elliptic equation with a quadratic cost functional and a distributed stochastic control which lies in the Hida distribution spaces.

1 Introduction

In the recent years, there has been an increased interest in applications of stochastic partial differential equations. Stochastic partial differential equation (SPDEs) are known to be an effective tool in modeling complex physical and engineering phenomena.

The mathematical treatment of SPDEs is more involved than deterministic PDEs. In the literature, various stochastic Galerkin methods have been applied to various linear and nonlinear problems, e.g. [1, 3, 8].

We shall use the approach of white noise analysis where both the Hida space and the more general Kondratiev space of distributions are to our disposal and where

H. Manouzi
Laval University, Department of Mathematics and Statistic, Quebec, Canada,
e-mail: hm@mat.ulaval.ca

S. Hou
Iowa State University, Department of Mathematics, Ames, IA 50011, U.S.A ,
e-mail: hou@math.iastate.edu

generalized solutions say $u(x,\omega)$ are treated in the sense that $\omega \longrightarrow u(x,\omega)$ is a stochastic distribution for a.a. x, e.g. [5, 6, 7]. An advantage of this approach is that one can establish a theory of nonlinear operations on distributions in order to handle a wide class of optimal control for nonlinear SPDEs .

In this paper we consider optimal control problems for linear stochastic partial differential equations (SPDEs) with quadratic cost functionals and distributed stochastic controls. We shall use the Wiener-Itô chaos expansion which represents a stochastic function $u(x,\omega) = \sum_{\alpha\in I} u_\alpha(x)H_\alpha(\omega)$ as formal series with respect to an orthonormal basis H_α. This decomposition separates the deterministic effects, described by the coefficients u_α, from the randomness that is covered by the basis functions H_α. The orthogonality of H_α enable us to reduce linear SPDEs to a system of decoupled deterministic equations for the coefficients $u_\alpha(x)$. Standard deterministic numerical methods can be applied to solve it sufficiently accurately. The main statistics, such as mean, covariance and higher order statistical moments can be calculated by simple formulas involving only these deterministic coefficients. Moreover, in the procedure described above, there is no randomness directly involved in the simulations. One does not have to deal with the selection of random number generators, and there is no need to solve the SPDE equations realization by realization. Instead, coefficient equations are solved once and for all.

An outline of the paper is as follows. In section 2 we review notation and introduce some function spaces. In section 3, we shall first give the existence and uniqueness of stochastic elliptic PDEs and then show that there is a unique optimal solution for a minimization problem constrained by stochastic elliptic equations. Finally we will derive a system of deterministic equations that the optimal solutions must satisfy and show how one can reconstruct particular realizations of the solution directly from Wiener-Itô chaos expansions once the coefficients are available.

2 A Review of Some Stochastic Sobolev Spaces

2.1 White Noise Space

Let $\mathscr{S}(\mathbb{R}^d)$ be the Schwartz space of smooth, rapidly decreasing functions on $\mathbb{R}^d$, and let $\mathscr{S}'(\mathbb{R}^d)$ be the dual space of tempered distributions. By the Bochner-Minlos theorem [7], there exists a unique probability measure μ, called the white noise probability measure, on the Borel σ-algebra on $\mathscr{S}'$ with characteristic functional

$$E_\mu[e^{i\langle\cdot,\eta\rangle}] := \int_{\mathscr{S}'} e^{i\langle\omega,\eta\rangle} d\mu(\omega) = e^{-\frac{1}{2}\|\eta\|^2_{L^2(\mathbb{R}^d)}}, \ \eta \in \mathscr{S}(\mathbb{R}^d). \tag{1}$$

The random variable $\langle\cdot,\eta\rangle_{\mathscr{S}',\mathscr{S}}$ defined on the probability space $\Omega = (\mathscr{S}', \mathscr{B}(\mathscr{S}'), \mu)$ thus follows a Gaussian distribution with mean zero and variance $\|\eta\|^2_{L^2(\mathbb{R}^d)}$, and can be interpreted as the stochastic integral w.r.t a Brownian motion B_x defined on $\mathbb{R}^d$, i.e.

$$\langle \omega, \eta \rangle_{\mathscr{S}',\mathscr{S}} = \int_{\mathbb{R}^d} \eta(t,x) dB_x(\omega), \ \omega \in \mathscr{S}', \eta \in \mathscr{S}$$

2.2 Chaos Decomposition

A chaos decomposition is an orthonormal expansion in the Hilbert space $L^2(\mathscr{S}')$ of quadratic integrable functions defined on $(\mathscr{S}', \mathscr{B}(\mathscr{S}'), \mu)$.

For $n \in \mathbb{N}_0, x \in \mathbb{R}$, define the Hermite polynomial $h_n(x) = (-1)^n e^{x^2/2} \frac{d^n}{dx^n}(e^{-x^2/2})$ and for $n \in \mathbb{N}$ define the Hermite functions

$$\xi_n(x) = \pi^{-1/4}((n-1)!)^{-1/2} e^{-x^2/2} h_{n-1}(\sqrt{2}x)$$

It is well-known that $\xi_n \in \mathscr{S}(\mathbb{R})$, $||\xi_n||_\infty \leq 1$ $(n \in \mathbb{N})$, and that $\{\xi_n : n \in \mathbb{N}\}$ constitutes an orthonormal basis in $L^2(\mathbb{R}, dx)$. We let $\{\eta_j\}_{j \in \mathbb{N}} \subset \mathscr{S}(\mathbb{R}^d)$ denote the orthonormal basis for $L^2(\mathbb{R}^d, dx)$ constructed by taking tensor-products of Hermite functions: $\eta_j(x) = \xi_{\delta_1^{(j)}}(x_1) \xi_{\delta_1^{(j)}}(x_2) \cdots \xi_{\delta_d^{(j)}}(x_d)$, $j = 1, 2, \cdots$, where $\delta^{(j)} = (\delta_1^{(j)}, \delta_2^{(j)}, \cdots, \delta_d^{(j)})$ is the jth multi-index number in some fixed ordering of all d-dimensional multi-indices $\delta = (\delta_1, \cdots, \delta_d)$. It follows that $\{\eta_j\}$ is an orthonormal basis for $L^2(\mathbb{R}^d)$. Let I denote the set of multi-indices $\alpha = (\alpha_1, \alpha_2, \cdots)$ where all $\alpha_i \in \mathbb{N}$ and only finitely many $\alpha_i \neq 0$ and let H_α denote the stochastic variables $H_\alpha(\omega) = \Pi_{j=1}^\infty h_{\alpha_j}(\langle \omega, \eta_j \rangle)$.

The family $\{H_\alpha : \alpha \in I\}$ constitutes an orthogonal basis for $L^2(\mathscr{S}', \mathscr{B}(\mathscr{S}'), \mu)$ and it holds $E[H_\alpha H_\beta] = \alpha! \delta_{\alpha\beta}$ [4].

Thus, any f in $L^2(\mathscr{S}')$ has a unique representation $f = \sum_{\alpha \in I} f_\alpha H_\alpha$ where $f_\alpha \in \mathbb{R}$ and $||f||^2_{L^2(\mu)} = \sum_{\alpha \in I} f_\alpha^2 \alpha!$.

Next, we introduce a family of stochastic Banach spaces needed for variational problems. This type of spaces are often used for the Hilbert space treatment of SPDEs of Wick type, other references include [2, 9, 10, 11, 13, 14].

2.3 Stochastic Sobolev Spaces

We shall use the notation $(2\mathbb{N})^\alpha := \prod_{j=1}^\infty (2^d \, \delta_1^{(j)} \cdots \delta_d^{(j)})^{\alpha_j}$ where $(\delta_1^{(j)} \cdots \delta_d^{(j)})$ is related to the basis $\{\eta_j\}$ by

$$\eta_j(x) = \xi_{\delta_1^{(j)}}(x_1) \xi_{\delta_1^{(j)}}(x_2) \cdots \xi_{\delta_d^{(j)}}(x_d) \, , \ j = 1, 2, \cdots$$

We have the following result [12]:

Lemma 1. *We have that*

$$\sum_{\alpha\in I} e^{p(2\mathbb{N})^{\alpha}} < \infty$$

if and only if $p < 0$.

Definition 1. Let $-1 \leq \rho \leq 1$ and $k \in \mathbb{R}$ and let V be a Banach space. We define the stochastic Banach spaces $(\mathscr{S})^{\rho,k,V}$ as the set of all formal sums

$$(\mathscr{S})^{\rho,k,V} := \left\{ v = \sum_{\alpha\in I} v_\alpha H_\alpha \; : \; v_\alpha \in V \;\text{ and }\; \|v\|_{\rho,k,V} < \infty \right\}$$

where $\|\cdot\|_{\rho,k,V}$ denote the norm

$$\|u\|_{\rho,k,V} := \left(\sum_{\alpha\in I} (\alpha!)^{1+\rho} \|u_\alpha\|_V^2 \, e^{k(2\mathbb{N})^{\alpha}} \right)^{\frac{1}{2}}$$

Lemma 2. *If V is a separable Hilbert space, then the space $(\mathscr{S})^{\rho,k,V}$ with the inner product*

$$(u,v)_{\rho,k,V} := \sum_{\alpha\in I} (u_\alpha, v_\alpha)_V (\alpha!)^{1+\rho} e^{k(2\mathbb{N})^{\alpha}}$$

is a separable Hilbert space. Moreover, if $k' \leq k$ then $\mathscr{S}^{\rho,k,V} \hookrightarrow \mathscr{S}^{\rho,k',V}$.

3 An Elliptic Distributed Stochastic Control Problem

Let $\mathscr{D}$ denote an open, bounded spatial domain in $\mathbb{R}^d$ with a boundary $\partial\mathscr{D}$ and $k,l \in \mathbb{R}$ such that $k < l$ and let $\lambda > 0$. As a prototype example of boundary control problems for SPDEs we will consider in the paper the minimization of the quadratic cost functional

$$\mathscr{J}(u,g) = \frac{1}{2}\|u-\overline{u}\|^2_{-1,k,H_0^1(\mathscr{D})} + \frac{\lambda}{2}\|g\|^2_{-1,l,L^2(\mathscr{D})} \tag{2}$$

subject to the elliptic SPDE for the state u and the stochastic control g:

$$-\Delta u(x,\omega) = f(x,\omega) + g(x,\omega), \qquad \text{in } \mathscr{D}\times\Omega \tag{3}$$

$$u(x,\omega) = 0, \qquad \text{on } \partial\mathscr{D}\times\Omega \tag{4}$$

where $\overline{u}(x,\omega) = \sum_{\alpha\in I} \overline{u}_\alpha(x) H_\alpha(\omega) \in (\mathscr{S})^{-1,k,H_0^1(\mathscr{D})}$ is a desired state.

Here, $f(x,\omega) = \sum_{\alpha\in I} f_\alpha(x) H_\alpha(\omega) \in (\mathscr{S})^{-1,k,L^2(\mathscr{D})}$ is the stochastic forcing term, $g = \sum_{\alpha\in I} g_\alpha(x) H_\alpha(\omega) \in (\mathscr{S})^{-1,l,L^2(\mathscr{D})}$ is the control term.

For this control problem we will give the existence and uniqueness of an optimal solution. We will derive an optimality system of equations that the optimal solutions must satisfy and discuss the chain of optimality systems.

3.1 Existence and Uniqueness of an Optimal Solution

We give a variational formulation of (3)–(4) and show that a solution is unique. Formally, assuming enough regularity on the data and the solution, we may write

$$(-\Delta u, v)_{-1,k,L^2(\mathscr{D})} = (f, v)_{-1,k,L^2(\mathscr{D})} + (g, v)_{-1,k,L^2(\mathscr{D})}, \forall v \in (\mathscr{S})^{-1,k,H_0^1(\mathscr{D})} \quad (5)$$

Integrating the left-hand side of (5) gives the bilinear form

$$(-\Delta u, v)_{-1,k,L^2(\mathscr{D})} = (\nabla u, \nabla v)_{-1,k,L^2(\mathscr{D})}$$

For a given pair $(k,l) \in \mathbb{R}^2$ with $k < l$ the variational formulation of (3)-(4) is: Find $u \in (\mathscr{S})^{-1,k,H_0^1(\mathscr{D})}$ such that

$$(\nabla u, \nabla v)_{-1,k,L^2(\mathscr{D})} = (f, v)_{-1,k,L^2(\mathscr{D})} + (g, v)_{-1,k,L^2(\mathscr{D})}, \forall v \in (\mathscr{S})^{-1,k,H_0^1(\mathscr{D})} \quad (6)$$

Theorem 1. *For every $k, l \in \mathbb{R}$ with $k < l$, $f \in (\mathscr{S})^{-1,k,L^2(\mathscr{D})}$ and $g \in (\mathscr{S})^{-1,l,L^2(\mathscr{D})}$ the problem (6) has a unique solution $u \in (\mathscr{S})^{-1,k,H_0^1(\mathscr{D})}$ and it holds*

$$\|u\|_{-1,k,H_0^1(\mathscr{D})} \leq C\{\|f\|_{-1,k,L^2(\mathscr{D})} + \|g\|_{-1,l,L^2(\mathscr{D})}\} \quad (7)$$

for a suitable positive constant C.

Proof. It clear that the bilinear form $(\nabla u, \nabla v)_{-1,k,H_0^1(\mathscr{D})}$ is symmetric, continuous and coercive in $(\mathscr{S})^{-1,k,H_0^1(\mathscr{D})}$. To apply the Lax-Milgram Theorem it suffices to show that the linear form $(f, v)_{-1,k,L^2(\mathscr{D})} + (g, v)_{-1,k,L^2(\mathscr{D})}$ is continuous. Using the definition of the inner product in $(\mathscr{S})^{-1,k,L^2(\mathscr{D})}$ we may write

$$\begin{aligned}
(g, v)_{-1,k,L^2(\mathscr{D})} &= \Sigma_{\alpha \in I} (g_\alpha, v_\alpha)_{L^2(\mathscr{D})} e^{k(2\mathbb{N})^\alpha} \\
&\leq \Sigma_{\alpha \in I} \|g_\alpha\|_{L^2(\mathscr{D})} e^{\frac{l}{2}(2\mathbb{N})^\alpha} \|v_\alpha\|_{L^2(\mathscr{D})} e^{\frac{k}{2}(2\mathbb{N})^\alpha} e^{(\frac{k}{2} - \frac{l}{2})(2\mathbb{N})^\alpha} \\
&\leq \|g\|_{-1,l,L^2(\mathscr{D})} \|v\|_{-1,k,L^2(\mathscr{D})} \sum_{\alpha \in I} e^{(\frac{k}{2} - \frac{l}{2})(2\mathbb{N})^\alpha} \\
&\leq C \|g\|_{-1,l,L^2(\mathscr{D})} \|v\|_{-1,k,H_0^1(\mathscr{D})}
\end{aligned}$$

where $\sum_{\alpha \in I} e^{(\frac{k}{2} - \frac{l}{2})(2\mathbb{N})^\alpha} < \infty$, since $\frac{k-l}{2} < 0$. □

Theorem 2. *The optimal control problem (2)-(4) admits a unique solution $(u, g) \in (\mathscr{S})^{-1,k,H_0^1(\mathscr{D})} \times (\mathscr{S})^{-1,l,L^2(\mathscr{D})}$.*

Proof. The proof follows standard arguments for proving existence of an optimal solution for deterministic optimal control problems. We provide the proof here for completeness. Set

$$\mathscr{U}_{\text{ad}} \equiv \{(u, g) \in (\mathscr{S})^{-1,k,H_0^1(\mathscr{D})} \times (\mathscr{S})^{-1,l,L^2(\mathscr{D})} : (u, g) \text{ satisfies (3)-(4)}, \mathscr{J}(u, g) < \infty\}$$

$\mathscr{U}_{\mathrm{ad}}$ is obviously nonempty. Let $\{(u^{(n)}, g^{(n)})\} \subset \mathscr{U}_{\mathrm{ad}}$ be a minimizing sequence of $\mathscr{J}(u,g)$, i.e., $\{(u^{(n)}, g^{(n)})\}$ satisfy

$$\lim_{n\to\infty} \mathscr{J}(u^{(n)}, g^{(n)}) = \inf_{(u,g)\in\mathscr{U}_{\mathrm{ad}}} \mathscr{J}(u,g) \tag{8}$$

A convergent sequence is bounded and this implies

$$\|g^{(n)}\|_{(\mathscr{S})^{-1,l,L^2(\mathscr{D})}} \le C_1 .$$

This estimate and Theorem 1 in turn yields

$$\|u^{(n)}\|_{(\mathscr{S})^{-1,k,H_0^1(\mathscr{D})}} \le C\{\|f\|_{-1,k,L^2(\mathscr{D})} + C_1\} .$$

Hence we have the weak convergence

$$u^{(n)} \rightharpoonup \widehat{u} \text{ in } (\mathscr{S})^{-1,k,H_0^1(\mathscr{D})} \quad \text{and} \quad g^{(n)} \rightharpoonup \widehat{g} \text{ in } (\mathscr{S})^{-1,l,L^2(\mathscr{D})}$$

for some $\widehat{u} \in (\mathscr{S})^{-1,k,H_0^1(\mathscr{D})}$ and $\widehat{g} \in (\mathscr{S})^{-1,l,L^2(\mathscr{D})}$. Passing to the limit in the weak form of (3)-(4) we can see that $(\widehat{u}, \widehat{g})$ satisfies (3)-(4), i.e., $(\widehat{u}, \widehat{g}) \in \mathscr{U}_{\mathrm{ad}}$.

Using the sequential weak lower semicontinuity of the functional $\mathscr{J}$ we have

$$\mathscr{J}(\widehat{u}, \widehat{g}) \le \liminf_{n\to 0} \mathscr{J}(u^{(n)}, g^{(n)}) = \inf_{(u,g)\in\mathscr{U}_{\mathrm{ad}}} \mathscr{J}(u,g)$$

Hence, we conclude that $\mathscr{J}(\widehat{u}, \widehat{g}) = \inf_{(u,g)\in\mathscr{U}_{\mathrm{ad}}} \mathscr{J}(u,g)$ so that $\mathscr{J}(\widehat{u}, \widehat{g})$ is an optimal solution.

The uniqueness follows from the abstract uniqueness result for the minimization of quadratic functionals with linear constraints. □

3.2 The Chain of Optimality Systems

Let (u,g) be the optimal solution of the elliptic optimal control problem (2)-(4). Defining the functional

$$F(g) = \mathscr{J}(u(g), g), \text{ where } u(g) \text{is the solution of the state equation}$$

Using classical necessary optimality conditions we can prove:

Theorem 3. *If $(\tilde{u}, \tilde{g})$ is the solution of (2)-(4), then $\tilde{g} = -\frac{1}{\lambda} p$ where p is the solution of the problem*

$$\begin{cases} -\Delta p = \tilde{u} - \overline{u}, & \text{in } \mathscr{D}\times\Omega \\ p \quad = 0, & \text{on } \partial\mathscr{D}\times\Omega \end{cases}$$

Conversely, if a pair $(\tilde{u}, \tilde{p}))$ obeys the systems

$$\begin{cases} -\Delta \tilde{u} = f - \frac{1}{\lambda}\tilde{p}, & in\ \mathscr{D} \times \Omega \\ -\Delta \tilde{p} = \tilde{u} - \overline{u}, & in\ \mathscr{D} \times \Omega \\ \tilde{u}, \tilde{p} \ = 0, & on\ \partial\mathscr{D} \times \Omega \end{cases}$$

then the pair $(\tilde{u}, -\frac{1}{\lambda}\tilde{p})$ *is the solution of (2)-(4).*

Using the orthogonality of H_α and if $\tilde{u} = \sum_{\alpha\in I} u_\alpha H_\alpha$, $\tilde{g} = \sum_{\alpha\in I} g_\alpha H_\alpha$ solves (2)-(4), then the chaos coefficients of the solution must solve the following set of deterministic problems: for each $\alpha \in I$

$$\begin{cases} -\Delta u_\alpha = f_\alpha - \frac{1}{\lambda} p_\alpha, & \text{in } \mathscr{D} \\ u_\alpha \ = 0, & \text{on } \partial\mathscr{D} \end{cases}$$

$$\begin{cases} -\Delta p_\alpha = u_\alpha - \overline{u}_\alpha, & \text{in } \mathscr{D} \\ p_\alpha \ = 0, & \text{on } \partial\mathscr{D} \end{cases}$$

$$g_\alpha = -\frac{1}{\lambda} p_\alpha$$

3.3 The Finite Element Approximation

Since the stochastic control problem are recasted in deterministic manner, their numerical solutions can be obtained using standard method widely used for PDEs. A finite element discretization of the optimality system can be defined in the usual manner.

For $N, K \in \mathbb{N}$, we define the cutting $I_{N,K} \subset I$ by

$$I_{N,K} = \{0\} \cup \bigcup_{n=1}^{N} \bigcup_{k=1}^{K} \left\{ \alpha \in \mathbb{N}_0^k : \quad |\alpha| = n \quad \text{and} \quad \alpha_k \neq 0 \right\}$$

Assume that the approximated chaos coefficients $(u_{h,\alpha}, p_{h,\alpha})$, $\forall \alpha \in I_{N,K}$ are computed by solving the deterministic optimality system, then the first two statistical moments of u are given by

$$E_\mu[u_h(x,y,\omega)] = u_{h,0}(x,y), \qquad E_\mu[u_h^2(x,y,\omega)] = \sum_{\alpha\in I_{N,K}} |\, u_{h,\alpha}(x,y)\,|^2 .$$

In practice, stochastic simulations of the solution can be carried out as follows: First, generate K independent standard Gaussian variables $X(\omega) = (X_i(\omega))$ $(i = 1, \ldots, K)$ using some random number generator, and then form the sums

$$u_h(x,y,\omega) = \sum_{\alpha\in I_{N,K}} u_{h,\alpha}(x,y) H_\alpha(X(\omega)), \quad p_h(x,y,\omega) = \sum_{\alpha\in I_{N,K}} p_{h,\alpha}(x,y) H_\alpha(X(\omega)), \tag{9}$$

where

$$H_\alpha(X(\omega)) := \prod_{j=1}^{K} h_{\alpha_j}(X_j(\omega)).$$

The advantage of this approach is that it enables us to generate random samples easily and fast. For example, in situations where one is interested in repeated simulations, one may compute the chaos coefficients in advance, store them, and produce the simulations whenever they are needed.

References

1. Babuška, I., Tempone, R., Zouraris, G.E.: Galerkin finite element approximations of stochastic elliptic partial differential equations. SIAM J. Numer. Anal. **42**(2), 800–825 (electronic) (2004)
2. Benth, F.E., Gjerde, J.: Convergence rates for finite element approximations of stochastic partial differential equations. Stochastics Stochastics Rep. **63**(3-4), 313–326 (1998)
3. Ghanem, R.: Ingredients for a general purpose stochastic finite elements implementation. Comput. Methods Appl. Mech. Engrg. **168**(1-4), 19–34 (1999)
4. Hida, T., Kuo, H.H., Potthoff, J., Streit, L.: White noise, *Mathematics and its Applications*, vol. 253. Kluwer Academic Publishers Group, Dordrecht (1993). An infinite-dimensional calculus
5. Holden, H., Lindstrøm, T., Øksendal, B., Ubøe, J., Zhang, T.S.: The Burgers equation with a noisy force and the stochastic heat equation. Comm. Partial Differential Equations **19**(1-2), 119–141 (1994)
6. Holden, H., Lindstrøm, T., Øksendal, B., Ubøe, J., Zhang, T.S.: The pressure equation for fluid flow in a stochastic medium. Potential Anal. **4**(6), 655–674 (1995)
7. Holden, H., Øksendal, B., Ubøe, J., Zhang, T.: Stochastic partial differential equations. Probability and its Applications. Birkhäuser Boston Inc., Boston, MA (1996). A modeling, white noise functional approach
8. Jardak, M., Su, C.H., Karniadakis, G.E.: Spectral polynomial chaos solutions of the stochastic advection equation. In: Proceedings of the Fifth International Conference on Spectral and High Order Methods (ICOSAHOM-01) (Uppsala), vol. 17, pp. 319–338 (2002)
9. Manouzi, H., Seaïd, M., Zahri, M.: Wick-stochastic finite element solution of reaction-diffusion problems. J. Comput. Appl. Math. **203**(2), 516–532 (2007)
10. Manouzi, H., Theting, T.G.: Mixed finite element approximation for the stochastic pressure equation of Wick type. IMA J. Numer. Anal. **24**(4), 605–634 (2004)
11. Manouzi, H., Theting, T.G.: Numerical analysis of the stochastic Stokes equations of Wick type. Numer. Methods Partial Differential Equations **23**(1), 73–92 (2007)
12. Pilipović, S., Seleši, D.: Expansion theorems for generalized random processes, Wick products and applications to stochastic differential equations. Infin. Dimens. Anal. Quantum Probab. Relat. Top. **10**(1), 79–110 (2007)
13. Theting, T.G.: Solving Wick-stochastic boundary value problems using a finite element method. Stochastics Stochastics Rep. **70**(3-4), 241–270 (2000)
14. Våge, G.: Variational methods for PDEs applied to stochastic partial differential equations. Math. Scand. **82**(1), 113–137 (1998)

A Priori Error Analysis for the Finite Element Approximation of Elliptic Dirichlet Boundary Control Problems

S. May, R. Rannacher, and B. Vexler

Abstract This article presents recent results of an a priori error analysis for the finite element approximation of Dirichlet boundary control problems governed by elliptic partial differential equations. For a standard model problem error estimates are proven for the primal variable, the control, as well as the associated adjoint variable. These estimates are of optimal order with respect to the solution's regularity to be expected on polygonal domains. The proofs rely on the Euler-Lagrange formulation of the optimal control problem and employ standard duality techniques and optimal-order L^p error estimates for the finite element Ritz projection. These estimates improve corresponding results in the literature and are supported by computational experiments. The details are contained in [9].

1 Dirichlet Boundary Control

On a convex polygonal domain $\Omega \subset \mathbb{R}^2$, we consider the following elliptic Dirichlet boundary control problem:

$$J(u,q) := \tfrac{1}{2}\|u-u_d\|^2_{L^2(\Omega)} + \tfrac{\alpha}{2}\|q\|^2_{L^2(\partial\Omega)} \ \rightarrow \ \min! \tag{1}$$

$$\text{such that} \quad -\Delta u = f \ \text{ in } \Omega, \quad u = q \ \text{ on } \partial\Omega. \tag{2}$$

Here, u_d and f are given sufficiently smooth functions, and $\alpha > 0$. The difficulty with this problem is that the natural "control space" $Q := L^2(\partial\Omega)$ does not fit the

Sandra May, Rolf Rannacher
Institute of Applied Mathematics, University of Heidelberg, INF 293/294, D-69120 Heidelberg, Germany, e-mail: sandra.may@iwr.uni-heidelberg.de, rannacher@iwr.uni-heidelberg.de

Boris Vexler
Lehrstuhl für Mathematische Optimierung, Technische Universität München, Fakultät für Mathematik, Boltzmannstraße 3, Garching b. München, Germany, e-mail: vexler@ma.tum.de

"state space" $H^1(\Omega)$ since the trace mapping $\gamma: H^1(\Omega) \to L^2(\partial\Omega)$ is not surjective. We therefore use the following "very weak" formulation of the state equation (see Lions/Magenes [8], and also Berggren [1] for the corresponding finite element analysis): *For given* $q \in L^2(\partial\Omega)$ *find* $u \in L^2(\Omega)$ *such that*

$$-(u,\Delta\phi) + \langle q, \partial_n\phi\rangle = (f,\phi) \quad \forall \phi \in H_0^1(\Omega) \cap H^2(\Omega). \tag{3}$$

Here, $(\cdot,\cdot) = (\cdot,\cdot)_{L^2(\Omega)}$ denotes the L^2 inner product on the domain Ω and $\langle\cdot,\cdot\rangle = \langle\cdot,\cdot\rangle_{L^2(\partial\Omega)}$ that on its boundary $\partial\Omega$. The corresponding norms will be denoted by $\|\cdot\| = \|\cdot\|_{L^2(\Omega)}$ and $|\cdot| = |\cdot|_{L^2(\partial\Omega)}$, respectively. By the results in [1] (see also [9]) there exists a uniquely determined solution to problem (3). For an overview of formulations of Dirichlet boundary control problems, see Kunisch/Vexler [7].

With the "solution operator" $S: L^2(\partial\Omega) \to L^2(\Omega)$ defined by

$$-(Sq,\Delta\phi) + \langle q, \partial_n\phi\rangle = (f,\phi) \quad \forall \phi \in H_0^1(\Omega) \cap H^2(\Omega), \tag{4}$$

the optimal control problem (1), (2) can be written in the reduced form

$$j(q) := J(Sq,q) \to \min! \quad q \in L^2(\partial\Omega). \tag{5}$$

The first directional derivative of $j(\cdot)$ at some point $q \in L^2(\partial\Omega)$ is given by

$$j'(q)(\chi) = \alpha\langle q,\chi\rangle - \langle \partial_n z, \chi\rangle, \quad \chi \in L^2(\partial\Omega), \tag{6}$$

where $z = z(q) \in H_0^1(\Omega) \cap H^2(\Omega)$ is the associated "adjoint state" determined by the equation

$$-(\psi,\Delta z) = (Sq - u_d, \psi) \quad \forall \psi \in L^2(\Omega). \tag{7}$$

The resulting first-order necessary optimality condition for the optimal solution $\hat{q}$ and the corresponding adjoint state $\hat{z}$,

$$j'(\hat{q})(\chi) = \alpha\langle \hat{q},\chi\rangle - \langle \partial_n \hat{z}, \chi\rangle = 0 \quad \forall \chi \in L^2(\partial\Omega), \tag{8}$$

implies a coupling between the regularity of $\hat{z}$ and that of $\hat{q}$:

$$\hat{z} \in W^{2,p}(\Omega) \quad \Rightarrow \quad \hat{q} \in W^{1-\frac{1}{p},p}(\partial\Omega).$$

This in turn involves $\hat{u} \in H^1(\Omega)$ because of (3). On the other hand equation (8) states that unavoidable corner singularities in the adjoint state $\hat{z}$ carry over to the optimal control $\hat{q}$. This problem-inherent difficulty causes a reduction in the orders of convergence to be expected for the finite element approximation of problem (1), (2).

2 Finite Element Discretization and Statement of Results

The finite element discretization of the optimization problem (1), (2) uses a standard weak formulation of the state equation which is possible due to higher regularity of the actual solution pair $\{\hat{u},\hat{q}\}$:

$$J(u_h,q_h) \rightarrow \min! \quad \text{on} \quad V_h \times V_h^{\partial}, \tag{9}$$

$$\text{such that} \quad (\nabla u_h, \nabla \phi_h) = (f,\phi_h) \quad \forall \phi_h \in V_{h,0}, \quad u_h|_{\partial\Omega} = q_h. \tag{10}$$

Here, $V_h \subset H^1(\Omega)$ is the standard space of "linear" finite elements, $V_{h,0} \subset V_h$ its subspace of functions vanishing along $\partial\Omega$, and V_h^{∂} the corresponding space of traces at $\partial\Omega$ of elements from V_h. For the solutions $\{\hat{u}_h,\hat{q}_h\} \in V_h \times V_h^{\partial}$ of these approximate problems the suboptimal error estimate

$$|\hat{q}-\hat{q}_h| + \|\hat{u}-\hat{u}_h\| = \mathcal{O}(h^{1-\frac{1}{p}}) \tag{11}$$

is among the results in Casas/Raymond [3] for a problem with (possible) additional control constraints. Our contributions are the L^2 error estimates

$$\|\hat{u}-\hat{u}_h\| = \mathcal{O}(h^{\frac{3}{2}-\frac{1}{p}}), \quad \|\hat{z}-\hat{z}_h\| = \mathcal{O}(h^{2-\frac{2}{p}}), \tag{12}$$

and the negative-norm estimate

$$|\hat{q}-\hat{q}_h|_{H^{-1}(\partial\Omega)} + \|\hat{u}-\hat{u}_h\|_{H^{-\frac{1}{2}}(\Omega)} = \mathcal{O}(h^{2-\frac{2}{p}}), \tag{13}$$

in the absence of control constraints. Here, the value of $p \in [2,p_*)$ is limited by $p_* = \omega_{\max}(\omega_{\max}-\frac{\pi}{2})^{-1}$, where $\omega_{\max} \in (\frac{\pi}{2},\pi)$ is the maximum inner angle of the polygonal domain Ω (see Grisvard [5] and Jakovlev [6]).

3 Numerical Results

The orders of convergence stated in (11), (12), and (13) are partially confirmed by computational tests. Here, two test configurations are considered:

- regular domain (unit square) and known analytic solution: The results shown in Table 1 and Table 2 indicate optimal-order convergence with respect to sufficiently weak negative Sobolev norms.
- more general domain with $\omega_{\max} = \frac{5}{6}\pi$ and unknown solution: The results shown in Table 3 indicate the sharpness of the estimates (11) and (12) with respect to the critical parameter $p_* = \frac{5}{2}$.

For details and other test settings we refer to [9]. The computations have been done using the software packages GASCOIGNE [4] and RoDoBo [10].

Remark. In the case of the unit square with known analytical solution and uniform cartesian meshes, a "superapproximation" effect occurs. This leads to full second-order convergence in L^2 of all three quantities. Further, the order of convergence observed in Table 3 for the adjoint state $\hat{z}$ seems to correspond rather to $\mathcal{O}(h^{2-1/p})$ (rate ≈ 1.6) than to $\mathcal{O}(h^{2-2/p})$ (rate $\approx$ 1.2), which we are currently able to prove.

Table 1 Test case with known analytic solution: convergence rates for a sequence of tensor-product meshes with 10% random shift of interior nodal points after each uniform refinement step

	$\|\hat{q}-\hat{q}_h\|$		$\|\|\hat{u}-\hat{u}_h\|\|$		$\|\|\hat{z}-\hat{z}_h\|\|$	
# cells	error	rate	error	rate	error	rate
256	1.30e-01	1.92	1.30e-02	1.82	8.19e-05	2.18
1 024	4.53e-02	1.52	4.46e-03	1.54	2.00e-05	2.03
4 096	2.69e-02	0.75	1.98e-03	1.17	4.98e-06	2.01
16 384	1.74e-02	0.63	8.89e-04	1.16	1.24e-06	2.01
65 536	9.65e-03	0.85	3.48e-04	1.35	3.11e-07	1.99
expected		1.00		1.50		2.00

Table 2 Test case with known analytical solution: convergence rates for a sequence of perturbed tensor-product meshes with 10% random shift of interior nodal points after each uniform refinement step

	$\langle\hat{q}-\hat{q}_h,1\rangle$		$(\hat{u}-\hat{u}_h,1)$	
# cells	error	rate	error	rate
256	-6.54e-01	2.02	6.54e-03	2.02
1 024	-1.65e-01	1.99	1.65e-03	1.99
4 096	-4.10e-02	2.01	4.11e-04	2.01
16 384	-1.02e-02	2.00	1.02e-04	2.01
65 536	-2.51e-03	2.03	2.46e-05	2.05
expected		2.00		2.00

Table 3 Test case on nonrectangular domain with angle $\omega_{\max} = \frac{5}{6}\pi$ and unknown solution:

	$\|\hat{q}-\hat{q}_h\|$		$\|\|\hat{u}-\hat{u}_h\|\|$		$\|\|\hat{z}-\hat{z}_h\|\|$	
# cells	error	rate	error	rate	error	rate
256	2.42e-02	0.87	4.04e-03	1.31	6.95e-04	1.75
1 024	1.47e-02	0.72	1.72e-03	1.23	2.43e-04	1.52
4 096	8.43e-03	0.80	7.25e-04	1.25	7.21e-05	1.75
16 384	5.35e-03	0.65	3.26e-04	1.15	2.21e-05	1.71
65 536	3.40e-03	0.65	1.47e-04	1.15	6.76e-06	1.71
expected		0.60		1.10		1.20

4 Ideas of Proof

The usual approach to error estimation for the approximation (9), (10) is based on the coercivity property of the second derivative of the reduced functional $j(\cdot)$:

$$j''(q)(\chi,\xi) = \alpha\langle\chi,\xi\rangle + (\tilde{B}\chi,\tilde{B}\xi), \quad \chi,\xi \in L^2(\partial\Omega), \tag{14}$$

where $\tilde{B}: L^2(\partial\Omega) \to L^2(\Omega)$ is an arbitrary extension operator. Then, the estimate

$$j''(\hat{q})(\xi,\xi) \geq \alpha\langle\xi,\xi\rangle, \quad \xi \in L^2(\partial\Omega), \tag{15}$$

and some argument using Galerkin orthogonality leads to the basic error estimate

$$|\hat{q}-\hat{q}_h| \leq \underbrace{c(1+\alpha^{-1})}_{=:\,c_\alpha} h^{1-\frac{1}{p}} \underbrace{\{ \|\hat{q}\|_{H^{1-\frac{1}{p}}(\partial\Omega)} + \|f\| + \|\hat{z}\|_{W^{2,p}(\Omega)} \}}_{=:\,\Sigma_p}. \tag{16}$$

The quantity Σ_p will be used in the following estimates to specify the regularity requirement for the optimal solution. For later use, we introduce the harmonic extension operator $B: H^{\frac{1}{2}}(\partial\Omega) \to H^1(\Omega)$, which is defined by

$$(\nabla Bq, \nabla\psi) = 0 \quad \forall\psi \in H_0^1(\Omega), \quad Bq_{|\partial\Omega} = q, \tag{17}$$

and its discrete analogue $B_h: V_h^\partial \to V_h$ defined by

$$(\nabla B_h q_h, \nabla\psi_h) = 0 \quad \forall\psi_h \in V_{h,0}, \quad B_h q_{h|\partial\Omega} = q_h. \tag{18}$$

Since $\hat{v}_h := \hat{u}_h - B_h\hat{q}_h$ is the Ritz projection of $\hat{v} := \hat{u} - B\hat{q}$, we have $\|\hat{v}-\hat{v}_h\| \leq ch^2\|f\|$. Consequently,

$$\begin{aligned}\|\hat{u}-\hat{u}_h\| &\leq \|\hat{v}-\hat{v}_h\| + \|B(\hat{q}-\hat{q}_h)\| + \|(B-B_h)\hat{q}_h\| \\ &\leq \underbrace{\|\hat{v}-\hat{v}_h\|}_{=\mathcal{O}(h^2)} + \underbrace{\|\hat{q}-\hat{q}_h\|_{H^{-\frac{1}{2}}(\partial\Omega)}}_{=\mathcal{O}(h^{1-\frac{1}{p}+?})} + \underbrace{\|(B-B_h)\hat{q}_h\|}_{=\mathcal{O}(h^{\frac{3}{2}-\frac{1}{p}})}.\end{aligned} \tag{19}$$

Hence to obtain optimal-order error estimates for $\hat{u}-\hat{u}_h$, we have to derive error estimates in "negative" Sobolev norms such as

$$\|\hat{q}-\hat{q}_h\|_{H^{-1}(\partial\Omega)} = \sup_{\xi\in H^1(\partial\Omega)} \frac{\langle\hat{q}-\hat{q}_h,\xi\rangle}{\|\xi\|_{H^1(\partial\Omega)}}. \tag{20}$$

As usual, this is done via "duality arguments". This requires comparable variational formulations for characterizing the optimal solutions $\{\hat{u},\hat{q},\hat{z}\}$ and $\{\hat{u}_h,\hat{q}_h,\hat{z}_h\}$ and the use of "Galerkin orthogonality".

5 The Karush-Kuhn-Tucker (KKT) Systems and the Duality Argument

The solution of the optimization problem is characterized by the set of equations

$$-(\hat{u},\Delta\phi)+\langle\hat{q},\partial_n\phi\rangle=(f,\phi) \qquad \forall\phi\in H_0^1(\Omega)\cap H^2(\Omega), \tag{21}$$
$$\alpha\langle\hat{q},\chi\rangle-\langle\partial_n\hat{z},\chi\rangle=0 \qquad \forall\chi\in L^2(\partial\Omega), \tag{22}$$
$$-(\Delta\hat{z},\psi)-(\hat{u},\psi)=-(u_d,\psi) \quad \forall\psi\in L^2(\Omega). \tag{23}$$

Due to the higher regularity of the optimal solution, we can rewrite the optimality system. With the function $\hat{v}:=\hat{u}-B\hat{q}\in H_0^1(\Omega)$, the triplet $\{\hat{v},\hat{q},\hat{z}\}\in H_0^1(\Omega)\times H^{\frac{1}{2}}(\partial\Omega)\times H_0^1(\Omega)$ satisfies the equations

$$(\nabla\hat{v},\nabla\phi)=(f,\phi) \qquad \forall\phi\in H_0^1(\Omega), \tag{24}$$
$$\alpha\langle\hat{q},\chi\rangle+(\hat{v}+B\hat{q},B\chi)=(u_d,B\chi) \quad \forall\chi\in H^{\frac{1}{2}}(\partial\Omega), \tag{25}$$
$$(\nabla\hat{z},\nabla\psi)-(\hat{v}+B\hat{q},\psi)=-(u_d,\psi) \quad \forall\psi\in H_0^1(\Omega). \tag{26}$$

Analogously the solution of the discrete optimization problem is characterized by the set of equations

$$\hat{u}_{h|\partial\Omega}=\hat{q}_h, \quad (\nabla\hat{u}_h,\nabla\phi_h)=(f,\phi_h) \qquad \forall\phi_h\in V_{h,0}, \tag{27}$$
$$\alpha\langle\hat{q}_h,\chi_h\rangle+(\hat{u}_h,B_h\chi_h)=(u_d,B_h\chi_h) \quad \forall\chi\in V_h^{\partial}, \tag{28}$$
$$(\nabla\hat{z}_h,\nabla\psi_h)-(\hat{u}_h,\psi_h)=-(u_d,\psi_h) \quad \forall\psi_h\in V_{h,0}. \tag{29}$$

With the function $\hat{v}_h:=\hat{u}_h-B_h\hat{q}_h\in V_{h,0}$ the triplet $\{\hat{v}_h,\hat{q}_h,\hat{z}_h\}\in V_{h,0}\times V_h^{\partial}\times V_{h,0}$ satisfies the equations

$$(\nabla\hat{v}_h,\nabla\phi_h)=(f,\phi_h) \qquad \forall\phi_h\in V_{h,0}, \tag{30}$$
$$\alpha\langle\hat{q}_h,\chi_h\rangle+(\hat{v}_h+B_h\hat{q}_h,B_h\chi_h)=(u_d,B_h\chi_h) \quad \forall\chi_h\in V_h^{\partial}, \tag{31}$$
$$(\nabla\hat{z}_h,\nabla\psi_h)-(\hat{v}_h+B_h\hat{q}_h,\psi_h)=-(u_d,\psi_h) \qquad \forall\psi_h\in V_{h,0}. \tag{32}$$

Combining the continuous and discrete optimality conditions, we obtain the following perturbed Galerkin orthogonality relation for the errors $\{e_v,e_q,e_z\}$:

$$(\nabla e_v,\nabla\phi_h)=0 \qquad \forall\phi_h\in V_{h,0}, \tag{33}$$
$$\alpha\langle e_q,\chi_h\rangle=-(\hat{v}+B\hat{q}-u_d,B\chi_h)+(\hat{v}_h+B_h\hat{q}_h-u_d,B_h\chi_h) \quad \forall\chi_h\in V_h^{\partial}, \tag{34}$$
$$(\nabla e_z,\nabla\psi_h)=(e_v+B\hat{q}-B_h\hat{q}_h,\psi_h) \qquad \forall\psi_h\in V_{h,0}. \tag{35}$$

Now, we have provided the bases for the duality argument announced above for estimating the control error in negative Sobolev norms such as (20).

For given $\xi\in H^1(\partial\Omega)$, we introduce the reduced dual problem: *Find* $\{w^z,w^q\}\in H_0^1(\Omega)\times H^{\frac{1}{2}}(\partial\Omega)$, *such that*

$$(\nabla\psi,\nabla w^z)+(\psi,Bw^q)=0 \quad \forall\psi\in H_0^1(\Omega), \tag{36}$$

$$\alpha\langle\chi,w^q\rangle+(B\chi,Bw^q)=\langle\chi,\xi\rangle \quad \forall\chi\in H^{\frac{1}{2}}(\partial\Omega). \tag{37}$$

Taking $\chi := e_q$ as test function and using (33) - (35) gives us

$$\begin{aligned}\langle e_q,\xi\rangle &= \alpha\langle e_q,w^q-P_h^\partial w^q\rangle+(Be_q,B(w^q-P_h^\partial w^q))\\ &\quad+(Be_q,(B-B_h)P_h^\partial w^q)-(e_v,B_hP_h^\partial w^q)\\ &\quad-((B-B_h)\hat{q}_h,B_hP_h^\partial w^q)+\underbrace{(\nabla\hat{z},\nabla(B-B_h)P_h^\partial w^q)}_{\text{difficult term}},\end{aligned}$$

where $P_h^\partial : L^2(\partial\Omega)\to V_h^\partial$ is the L^2 projection onto V_h^∂. The first terms on the right-hand side can be bounded by $ch^{2-\frac{2}{p}}\Sigma_p$, which requires lengthy but standard estimates mainly for the error of the L^2 projection, $I-P_h^\partial$, and the harmonic extensions, $B-B_h$.

For estimating the difficult last term, we set $q_h := P_h^\partial w^q$ and use the "Neumann projection" $R_h^N : H^1(\Omega)\to V_h$ to get rid of $B_hq_h \notin V_{h,0}$:

$$(\nabla\hat{z},\nabla(B-B_h)q_h) = \underbrace{(\nabla(\hat{z}-R_h^N\hat{z}),\nabla Bq_h)}_{=:\Lambda_1\text{(estimated in a standard way)}} + \underbrace{(\nabla R_h^N\hat{z},\nabla(B-B_h)q_h)}_{=:\Lambda_2\text{ (difficult term)}}.$$

Let a_i denote the nodal points of the finite element mesh and $\phi_h^i\in V_h$ the corresponding "nodal basis functions" satisfying $\|\nabla\phi_h^i\|\le c$. The critical term Λ_2 is then treated as follows:

$$\begin{aligned}\Lambda_2 &= (\nabla R_h^N\hat{z},\nabla(B-B_h)q_h)\\ &= \sum_{a_i\in\bar{\Omega}} R_h^N\hat{z}(a_i)(\nabla\phi_h^i,\nabla(B-B_h)q_h)\\ &= \sum_{a_i\subset\partial\Omega} R_h^N\hat{z}(a_i)(\nabla\phi_h^i,\nabla(B-B_h)q_h)\\ &\le \sum_{a_i\in\partial\Omega} |R_h^N\hat{z}(a_i)|\|\nabla\phi_h^i\|_{L^2(\mathrm{supp}(\phi_h^i))}\|\nabla(B-B_h)q_h\|_{L^2(\mathrm{supp}(\phi_h^i))}\\ &\le ch^{-\frac{1}{2}}\Big(\sum_{a_i\in\partial\Omega} h|R_h^N\hat{z}(a_i)|^2\Big)^{\frac{1}{2}}\|\nabla(B-B_h)q_h\|\\ &\le ch^{-\frac{1}{2}}|R_h^N\hat{z}-\hat{z}|\|\nabla(B-B_h)q_h\|\\ &\le ch^{2-\frac{1}{p}}\|\hat{z}\|_{W^{2,p}(\Omega)}|q_h|_{H^1(\partial\Omega)}\\ &\le ch^{2-\frac{1}{p}}\|\hat{z}\|_{W^{2,p}(\Omega)}|w^q|_{H^{1-\frac{1}{p}}(\partial\Omega)}\\ &\le ch^{2-\frac{2}{p}}\Sigma_p|\xi|_{L^p(\partial\Omega)}.\end{aligned}$$

Here, the trace estimate

$$\|v\|_{L^p(\partial\Omega)} \leq \varepsilon\|\nabla v\|_{L^p(\Omega)} + cp\varepsilon^{-\frac{1}{p-1}}\|v\|_{L^p(\Omega)}, \quad 0 < \varepsilon \leq 1, \tag{38}$$

and the optimal order L^p error estimate

$$\|R_h^N \hat{z} - \hat{z}\|_{L^p(\Omega)} + h\|\nabla(R_h^N \hat{z} - \hat{z})\|_{L^p(\Omega)} \leq ch^2\|\hat{z}\|_{W^{2,p}(\Omega)} \tag{39}$$

for the Neumann projection are used. The latter can be derived, for $2 \leq p \leq p_* < \infty$, by the techniques developed in [11]. This gives us

$$\|\hat{q} - \hat{q}_h\|_{H^{-1}(\partial\Omega)} \leq c_\alpha^2 h^{2-\frac{2}{p}} \Sigma_p, \tag{40}$$

where $c_\alpha \approx 1 + \alpha^{-1}$, and then also the estimate

$$\|\hat{u} - \hat{u}_h\|_{H^{-\frac{1}{2}}(\Omega)} + \|\hat{z} - \hat{z}_h\| \leq c_\alpha^2 h^{2-\frac{2}{p}} \Sigma_p. \tag{41}$$

For the details of the lengthy proof, we refer to [9].

References

1. Berggren, M.: *Approximations of very weak solutions to boundary-value problems*, SIAM J. Numer. Anal. 42, 860–877 (2004).
2. Brenner, S. and L. R. Scott: *The Mathematical Theory of Finite Element Methods*, Springer, Berlin-Heidelberg-New York, 1996.
3. Casas, E. and J.-P. Raymond: *Error estimates for the numerical approximation of Dirichlet boundary control for semilinear elliptic equations*, SIAM J. Contr. Optim. 45, 1586–1611 (2006).
4. GASCOIGNE: *A Finite Element Software Library*, http://www.gascoigne.uni-hd.de, 2006.
5. Grisvard, P.: *Elliptic Problems in Nonsmooth Domains*, Pitman, London, 1985.
6. Jakovlev, G. N.: *Boundary properties of functions of the class $W_p^{(1)}$ in regions with corners*, Dokl. Akad. Nauk. SSSR 140, 73–76 (1961).
7. Kunisch, K. and B. Vexler: *Constrained Dirichlet boundary control in L^2 for a class of evolution equations*, SIAM J. Contr. Optim. 46, 1726–1753 (2007).
8. Lions, J. L. and E. Magenes: *Non-Homogeneous Boundary Value Problems and Applications*, Springer, Berlin, 1972.
9. May, S., R. Rannacher, and B. Vexler: *Error analysis for a finite element approximation of elliptic Dirichlet boundary control problems*, Preprint, Institute of Applied Mathematics, University of Heidelberg, 2008.
10. RoDoBo: *A C++ Optimization Software for stationary and instationary PDE with interfaces to Gascoigne*, http://rodobo.uni-hd.de, 2006.
11. Rannacher, R. and L. R. Scott: *Some optimal error estimates for piecewise linear finite element approximations*, Math. Comp. 38, 437–445 (1982).

A Priori Error Analysis for Space-Time Finite Element Discretization of Parabolic Optimal Control Problems

D. Meidner and B. Vexler

Abstract In this article we discuss a priori error estimates for Galerkin finite element discretizations of optimal control problems governed by linear parabolic equations and subject to inequality control constraints. The space discretization of the state variable is done using usual conforming finite elements, whereas the time discretization is based on discontinuous Galerkin methods. For different types of control discretizations we provide error estimates of optimal order with respect to both space and time discretization parameters taking into account the spatial and the temporal regularity of the optimal solution. For the treatment of the control discretization we discuss different approaches extending techniques known from the elliptic case. For detailed proofs and numerical results we refer to [18, 19].

1 Introduction

A priori error analysis for finite element discretizations of optimization problems governed by partial differential equations is an active area of research. While many publications are concerned with elliptic problems, see, e.g., [1, 7, 11, 12, 13, 20], there are only few published results on this topic for parabolic problems, see [14, 16, 17, 21, 23]. In this paper we consider an optimal control problem governed by the heat equation. For the discretization of the state equation we employ a space-time finite element discretization and discuss different approaches for the discretization of the control variable. Extending techniques for treating inequality constraints on the control variable known for elliptic problems, we derive a priori error estimates

Dominik Meidner
Lehrstuhl für Mathematische Optimierung, Technische Universität München, Fakultät für Mathematik, Boltzmannstraße 3, Garching b. München, Germany, e-mail: meidner@ma.tum.de

Boris Vexler
Lehrstuhl für Mathematische Optimierung, Technische Universität München, Fakultät für Mathematik, Boltzmannstraße 3, Garching b. München, Germany, e-mail: vexler@ma.tum.de

taking into account spatial and temporal regularity of the solutions. The detailed proofs and numerical results illustrating our considerations are presented in [18, 19].

For a convex, polygonal spatial domain $\Omega \subset \mathbb{R}^n$ ($n = 2,3$) and a time interval $I = (0,T)$ we consider the state space X and the control space Q given as

$$X := \left\{ v \mid v \in L^2(I,V) \text{ and } \partial_t v \in L^2(I,V^*) \right\}, \quad V = H_0^1(\Omega), \quad Q = L^2(I,L^2(\Omega)).$$

The inner product of $L^2(\Omega)$ and the corresponding norm are denoted by $(\cdot,\cdot)$ and $\|\cdot\|$. The inner product and the norm of Q are denoted by $(\cdot,\cdot)_I$ and $\|\cdot\|_I$. The state equation in a weak form for the state variable $u = u(q) \in X$, the control variable $q \in Q$ and the initial condition $u_0 \in V$ reads:

$$(\partial_t u, \varphi)_I + (\nabla u, \nabla \varphi)_I + (u(0), \varphi(0)) = (f + q, \varphi)_I + (u_0, \varphi(0)) \quad \forall \varphi \in X. \qquad (1)$$

The optimal control problem under consideration is formulated as follows:

$$\text{Minimize } J(q,u) := \frac{1}{2}\|u - \hat{u}\|_I^2 + \frac{\alpha}{2}\|q\|_I^2 \text{ subject to (1) and } (q,u) \in Q_{\text{ad}} \times X, \quad (2)$$

where $\hat{u} \in L^2(I,L^2(\Omega))$ is a given desired state and $\alpha > 0$ is the regularization parameter. The admissible set Q_{ad} describes box constraints on the control variable:

$$Q_{\text{ad}} := \left\{ q \in Q \mid q_a \le q(t,x) \le q_b \quad \text{a.e. in } I \times \Omega \right\}.$$

Throughout we consider two situations: the problem without control constraints, i.e., $q_a = -\infty$, $q_b = +\infty$, $Q_{\text{ad}} = Q$, and the problem with control constraints, i.e., $q_a, q_b \in \mathbb{R}$ with $q_a < q_b$. In both cases the optimal control problem (2) is known to possess a unique solution $(\bar{q},\bar{u})$, see, e.g., [15]. This solution is characterized by the optimality system involving an adjoint equation for the adjoint variable $z = z(q) \in X$ satisfying

$$-(\varphi, \partial_t z)_I + (\nabla \varphi, \nabla z)_I + (z(T), \varphi(T)) = (\varphi, u(q) - \hat{u})_I \quad \forall \varphi \in X. \qquad (3)$$

The optimal solution $(\bar{q},\bar{u})$ and the corresponding adjoint state $\bar{z} = z(\bar{q})$ exhibit for any $p < \infty$ when $n = 2$ and $p \le 6$ when $n = 3$ the following regularity properties (see [19, Proposition 2.3]):

$$\bar{u}, \bar{z} \in L^2(I, H^2(\Omega) \cap H^1(I, L^2(\Omega)),$$
$$\bar{q} \in L^2(I, W^{1,p}(\Omega)) \cap H^1(I, L^2(\Omega)).$$

For the time discretization we use the lowest order discontinuous Galerkin method dG(0), see [9, 10], which is a variant of the implicit Euler scheme. We exploit a time partitioning

$$0 = t_0 < t_1 < \cdots < t_{M-1} < t_M = T$$

with corresponding time intervals $I_m := (t_{m-1}, t_m]$ of length k_m and a semidiscrete space

$$X_k^0 := \left\{ v_k \in L^2(I,V) \;\middle|\; v_k\big|_{I_m} \in \mathscr{P}_0(I_m,V),\; m = 1,2,\ldots,M \right\}.$$

Here, the discretization parameter k is defined as the maximum of all step sizes k_m. The dG(0) semidiscretization of the state equation (1) for given control $q \in Q$ reads: Find a state $u_k = u_k(q) \in X_k^0$ such that

$$B(u_k, \varphi) = (f + q, \varphi)_I + (u_0, \varphi_0^+) \quad \forall \varphi \in X_k^0, \tag{4}$$

where the bilinear form $B(\cdot,\cdot)$ is defined as

$$B(u_k,\varphi) := \sum_{m=1}^{M} (\partial_t u_k, \varphi)_{I_m} + (\nabla u_k, \nabla \varphi)_I + \sum_{m=2}^{M} ([u_k]_{m-1}, \varphi_{m-1}^+) + (u_{k,0}^+, \varphi_0^+)$$

with $[u_k]_m$ denoting the jump of the function u_k at t_m. The semidiscrete optimization problem for the dG(0) time discretization has the form:

$$\text{Minimize } J(q_k, u_k) \text{ subject to (4) and } (q_k, u_k) \in Q_{\text{ad}} \times X_k^0. \tag{5}$$

The unique solution of this problem is denoted by $(\bar{q}_k, \bar{u}_k)$ and is characterized using a semidiscrete adjoint solution $z_k = z_k(q) \in X_k^0$ determined by:

$$B(\varphi, z_k) = (\varphi, u_k(q) - \hat{u})_I \quad \forall \varphi \in X_k^0. \tag{6}$$

By stability estimates (see [18, Theorems 4.1, 4.3, Corollaries 4.2, 4.5]) we deduce that the semidiscrete optimal state $\bar{u}_k$ and adjoint state $\bar{z}_k = z_k(\bar{q})$ have the regularity

$$\bar{u}_k\big|_{I_m}, \bar{z}_k\big|_{I_m} \in L^2(I_m, H^2(\Omega)) \cap H^1(I_m, L^2(\Omega)),\; m = 1,2,\ldots,M$$

uniformly in k.

For the spatial discretization we use a usual conforming finite element space $V_h \subset V$ consisting of cellwise (bi-/tri-)linear functions over a quasi-uniform mesh $\mathscr{T}_h$, see, e.g., [3] for standard definitions. Then, the so called cG(1)dG(0) discretization of the state equation for given control $q \in Q$ has the form: Find a state $u_{kh} = u_{kh}(q) \in X_{k,h}^{0,1}$ such that

$$B(u_{kh}, \varphi) = (f + q, \varphi)_I + (u_0, \varphi_0^+) \quad \forall \varphi \in X_{k,h}^{0,1}, \tag{7}$$

where $X_{k,h}^{0,1}$ is defined similar to the semidiscrete space X_k^0 as

$$X_{k,h}^{0,1} := \left\{ v_{kh} \in L^2(I,V_h) \;\middle|\; v_{kh}\big|_{I_m} \in \mathscr{P}_0(I_m,V_h),\; m = 1,2,\ldots,M \right\} \subset X_k^0.$$

The corresponding optimal control problem is given as

$$\text{Minimize } J(q_{kh}, u_{kh}) \text{ subject to (7) and } (q_{kh}, u_{kh}) \in Q_{\text{ad}} \times X_{k,h}^{0,1}. \tag{8}$$

As above, the unique solution of this problem is denoted by $(\bar{q}_{kh}, \bar{u}_{kh})$ and can be characterized using the corresponding adjoint solution $z_{kh} = z_{kh}(q) \in X^{0,1}_{k,h}$ determined by

$$B(\varphi, z_{kh}) = (\varphi, u_{kh}(q) - \hat{u})_I \quad \forall \varphi \in X^{0,1}_{k,h}. \tag{9}$$

To obtain a fully discrete optimal control problem we consider a subspace $Q_d \subset Q$ of the control space and come up with the following problem:

$$\text{Minimize } J(q_\sigma, u_\sigma) \text{ subject to (7) and } (q_\sigma, u_\sigma) \in Q_{d,\text{ad}} \times X^{0,1}_{k,h}, \tag{10}$$

where $Q_{d,\text{ad}} = Q_d \cap Q_{\text{ad}}$. The solution of this problem is denoted by $(\bar{q}_\sigma, \bar{u}_\sigma)$ where the discretization parameter σ consists of all discretization parameters, i.e., $\sigma = (k,h,d)$. In the following sections we will consider different choices of the discrete control space Q and provide error estimates for the error $\|\bar{q} - \bar{q}_\sigma\|_I$.

2 Error Analysis for Problems without Control Constraints

In this section, we investigate the optimal control problem (2) in the situation when no constraints are imposed on the control, i.e., in the case of $Q_{\text{ad}} = Q$. The control space Q is discretized in time as the state space X, i.e., by piecewise constant polynomials on the subintervals I_m. The space discretization of Q is done either by piecewise (bi-/tri-)linear finite elements (as employed also for discretizing the state space) or by cellwise constant polynomials.

For this combination of state and control discretizations, the following estimate for the error in the control variable holds (cf. [18, Theorem 6.1]):

Theorem 1. *The error between the the solution $\bar{q} \in Q$ of the continuous optimization problem* (2) *and the solution $\bar{q}_\sigma \in Q_d$ of the discrete optimization problem* (10) *can be estimated as*

$$\|\bar{q} - \bar{q}_\sigma\|_I \leq \frac{C}{\alpha} k \{ \|\partial_t u(\bar{q})\|_I + \|\partial_t z(\bar{q})\|_I \}$$
$$+ \frac{C}{\alpha} h^2 \{ \|\nabla^2 u_k(\bar{q})\|_I + \|\nabla^2 z_k(\bar{q})\|_I \} + \left(2 + \frac{C}{\alpha}\right) \inf_{p_d \in Q_d} \|\bar{q} - p_d\|_I.$$

The constants C are independent of the mesh size h, the size of the time steps k, and the choice of the discrete control space $Q_d \subset Q$.

By applying standard interpolation estimates to the infimum term we obtain the optimal asymptotic orders of convergence of $\|\bar{q} - \bar{q}_\sigma\|_I = \mathcal{O}(k + h^2)$ for the piecewise (bi-/tri-)linear space discretization of the control space Q and $\|\bar{q} - \bar{q}_\sigma\|_I = \mathcal{O}(k + h)$ for the cellwise constant space discretization of the controls.

3 Error Analysis for Problems with Control Constraints

In this section, we provide estimates of the error in terms of the control variable for the constrained optimal control problem (2) with different choices of the discrete control space Q_d.

3.1 Cellwise Constant Discretization

Like in the unconstrained case, the controls are discretized with respect to time by piecewise constant polynomials. The space discretization is done here by cellwise constant polynomials. For proving the desired error estimate, we extend the techniques presented in [8] for elliptic problems to the case of parabolic optimal control problems. This demands the introduction of the solution $\bar{q}_d$ of the purely control discretized problem:

$$\text{Minimize } J(q_d, u_d) \text{ subject to (1) and } (q_d, u_d) \in Q_{d,\text{ad}} \times X. \tag{11}$$

By means of this problem, the following estimate for the error in the control variable can be proven (cf. [19, Corollary 5.3]):

Theorem 2. *Let $\bar{q} \in Q_{ad}$ be the solution of the optimal control problem (2), $\bar{q}_\sigma \in Q_{d,ad}$ be the solution of the discretized problem (10), where the cellwise constant discretization for the control variable is employed. Let moreover $\bar{q}_d \in Q_{d,ad}$ be the solution of the purely control discretized problem (11). Then the following estimate holds:*

$$\|\bar{q} - \bar{q}_\sigma\|_I \leq \frac{C}{\alpha} k \{\|\partial_t \bar{q}\|_I + \|\partial_t u(\bar{q}_d)\|_I + \|\partial_t z(\bar{q}_d)\|_I\}$$
$$+ \frac{C}{\alpha} h \{\|\nabla \bar{q}\|_I + \|\nabla z(\bar{q}_d)\|_I + h \left(\|\nabla^2 u_k(\bar{q}_d)\|_I + \|\nabla^2 z_k(\bar{q}_d)\|_I\right)\} = \mathcal{O}(k+h).$$

We note, that all terms in the above estimate which depend on discretization parameters k and d are uniformly bounded with respect to these parameters. Therefore the constant in $\mathcal{O}(k+h)$ is independent of all discretization parameters and depends only on problem data.

3.2 Cellwise Linear Discretization

A better convergence result for the error can be achieved by discretizing the controls with piecewise (bi-/tri-)linear finite elements in space instead of using only piecewise constant trial functions. In this section, we treat this discretization combined with the already known time discretization by piecewise constant polynomials.

The desired estimate is obtained by adapting the techniques described in [4, 6] to parabolic problems.

The analysis for this configuration is based on an assumption on the structure of the active sets. For each time interval I_m we group the cells K of the mesh $\mathscr{T}_h$ depending on the value of $\bar{q}_k$ on K into three sets $\mathscr{T}_h = \mathscr{T}_{h,m}^1 \cup \mathscr{T}_{h,m}^2 \cup \mathscr{T}_{h,m}^3$ with $\mathscr{T}_{h,m}^i \cap \mathscr{T}_{h,m}^j = \emptyset$ for $i \neq j$. The sets are chosen as follows:

$$\mathscr{T}_{h,m}^1 := \{\, K \in \mathscr{T}_h \mid \bar{q}_k(t_m,x) = q_a \text{ or } \bar{q}_k(t_m,x) = q_b \text{ for all } x \in K \,\}$$
$$\mathscr{T}_{h,m}^2 := \{\, K \in \mathscr{T}_h \mid q_a < \bar{q}_k(t_m,x) < q_b \text{ for all } x \in K \,\}$$
$$\mathscr{T}_{h,m}^3 := \mathscr{T}_h \setminus (\mathscr{T}_{h,m}^1 \cup \mathscr{T}_{h,m}^2)$$

Hence, the set $\mathscr{T}_{h,m}^3$ consists of the cells which contain the free boundary between the active and the inactive sets for the time interval I_m.

Assumption 1. *We assume that there exists a positive constant C independent of k, h, and m such that*

$$\sum_{K \in \mathscr{T}_{h,m}^3} |K| \leq Ch.$$

We note, that this assumption is fulfilled, if the boundary of active sets consists of a finite number of rectifiable curves. Similar assumptions are used in [2, 20].

Under this assumption, the following estimate for the error in the control variable can be proven (cf. [19, Corollary 5.8]):

Theorem 3. *Let $\bar{q} \in Q_{ad}$ be the solution of the optimal control problem* (2) *and $\bar{q}_\sigma \in Q_{d,ad}$ be the solution of the discrete problem* (10)*, where the cellwise (bi-/tri-)linear discretization for the control variable is employed. Then, if Assumption 1 is fulfilled, the following estimate holds for $n < p \leq \infty$ provided $z_k(\bar{q}_k) \in L^2(I, W^{1,p}(\Omega))$:*

$$\|\bar{q} - \bar{q}_\sigma\|_I \leq \frac{C}{\alpha} k \{\|\partial_t u(\bar{q})\|_I + \|\partial_t z(\bar{q})\|_I\} + \frac{C}{\alpha}\left(1 + \frac{1}{\alpha}\right)\{h^2 \|\nabla^2 u_k(\bar{q}_k)\|_I$$
$$+ h^2 \|\nabla^2 z_k(\bar{q}_k)\|_I + h^{\frac{3}{2}-\frac{1}{p}} \|\nabla z_k(\bar{q}_k)\|_{L^2(I,L^p(\Omega))}\} = \mathscr{O}(k + h^{\frac{3}{2}-\frac{1}{p}}).$$

For elliptic optimal control problems and cellwise (bi-/tri-)linear discretization of the control space, the convergence order $\mathscr{O}(h^{\frac{3}{2}})$ can be obtained, see [2, 5, 22]. Especially in two space dimensions a corresponding estimate follows from the above theorem for the parabolic problem. In this case a uniform bound for $z_k(\bar{q}_k) \in L^2(I, W^{1,p}(\Omega))$ for all $p < \infty$ can be obtained leading to $\mathscr{O}(k + h^{\frac{3}{2}-\varepsilon})$ for each $\varepsilon > 0$.

3.3 Variational Approach

In this section we prove an estimate for the error $\|\bar{q} - \bar{q}_\sigma\|_I$ in the case of no control discretization, cf. [13]. In this case we choose $Q_d = Q$ and thus, $Q_{d,\text{ad}} = Q_{\text{ad}}$. This implies $\bar{q}_\sigma = \bar{q}_{kh}$. However, $\bar{q}_{kh}$ is in general not a finite element function corresponding to the spatial mesh $\mathscr{T}_h$. This fact requires more care for the computation of $\bar{q}_{kh}$, see [13] for details. On the other hand, this approach provides the optimal order of convergence (cf. [19, Corollary 5.11]):

Theorem 4. *Let $\bar{q} \in Q_{ad}$ be the solution of optimization problem* (2) *and $\bar{q}_{kh} \in Q_{ad}$ be the solution of the discretized problem* (8). *Then the following estimate holds:*

$$\|\bar{q} - \bar{q}_{kh}\|_I \le \frac{C}{\alpha} k\{\|\partial_t u(\bar{q})\|_I + \|\partial_t z(\bar{q})\|_I\} + \frac{C}{\alpha} h^2\{\|\nabla^2 u_k(\bar{q})\|_I + \|\nabla^2 z_k(\bar{q})\|_I\} = \mathscr{O}(k+h^2).$$

3.4 Post-Processing Strategy

In this section, we extend the post-processing techniques initially proposed in [20] to the parabolic case. We discretize the control by piecewise constants in time and space as in Subsection 3.1. To improve the quality of the approximation, we additionally employ the post-processing step

$$\tilde{q}_\sigma := P_{Q_{\text{ad}}}\left(-\frac{1}{\alpha} z_{kh}(\bar{q}_\sigma)\right), \tag{12}$$

which makes use of the projection

$$P_{Q_{\text{ad}}} \colon Q \to Q_{\text{ad}}, \quad P_{Q_{\text{ad}}}(r)(t,x) = \max(q_a, \min(q_b, r(t,x))).$$

We obtain the following error estimate (cf. [19, Corollary 5.17]):

Theorem 5. *Let $\bar{q} \in Q_{ad}$ be the solution of the optimal control problem* (2) *and $\tilde{q}_\sigma \in Q_{ad}$ be given by means of* (12) *employing the adjoint state $z_{kh}(\bar{q}_\sigma)$ related to the solution $\bar{q}_\sigma$ of the discrete problem* (10), *where the cellwise constant discretization for the control variable is employed. Let, moreover, Assumption 1 be fulfilled and $n < p \le \infty$. Then, it holds*

$$\begin{aligned}\|\bar{q} - \tilde{q}_\sigma\|_I \le \frac{C}{\alpha}\left(1+\frac{1}{\alpha}\right) k\{\|\partial_t u(\bar{q})\|_I + \|\partial_t z(\bar{q})\|_I\} \\ + \frac{C}{\alpha}\left(1+\frac{1}{\alpha}\right) h^2\{\|\nabla^2 u_k(\bar{q}_k)\|_I + \frac{1}{\alpha}\|\nabla z_k(\bar{q}_k)\|_I + \left(1+\frac{1}{\alpha}\right)\|\nabla^2 z_k(\bar{q}_k)\|_I\} \\ + \frac{C}{\alpha^2}\left(1+\frac{1}{\alpha}\right) h^{2-\frac{1}{p}} \|\nabla z_k(\bar{q}_k)\|_{L^2(I,L^p(\Omega))} = \mathscr{O}\left(k + h^{2-\frac{1}{p}}\right)\end{aligned}$$

provided that $z_k(\bar{q}_k) \in L^2(I, W^{1,p}(\Omega))$.

Since in two space dimensions the adjoint solution $z_k(\bar{q}_k)$ is uniformly bounded in $L^2(I, W^{1,p}(\Omega))$ with respect to k for all $p < \infty$, the presented estimate leads to the almost optimal order of convergence $\mathscr{O}(k + h^{2-\varepsilon})$ for each $\varepsilon > 0$.

References

1. Arada, N., Casas, E., Tröltzsch, F.: Error estimates for a semilinear elliptic optimal control problem. Comput. Optim. Appl. **23**, 201–229 (2002)
2. Becker, R., Vexler, B.: Optimal control of the convection-diffusion equation using stabilized finite element methods. Numer. Math. **106**(3), 349–367 (2007)
3. Brenner, S., Scott, R.L.: The mathematical theory of finite element methods. Springer Verlag, Berlin Heidelberg New York (2002)
4. Casas, E.: Using piecewise linear functions in the numerical approximation of semilinear elliptic control problems. Adv. Comput. Math. **26**(1-3), 137–153 (2007)
5. Casas, E., Mateos, M.: Error estimates for the numerical approximation of boundary semilinear elliptic control problems. Continuous piecewise linear approximations. In: Systems, control, modeling and optimization, pp. 91–101. Springer, New York (2006). IFIP Int. Fed. Inf. Process., 202
6. Casas, E., Mateos, M.: Error estimates for the numerical approximation of neumann control problems. Comput. Optim. Appl. (2007). DOI 10.1007/s10589-007-9056-6. Published electronically
7. Casas, E., Mateos, M., Tröltzsch, F.: Error estimates for the numerical approximation of boundary semilinear elliptic control problems. Comput. Optim. Appl. **31**(2), 193–220 (2005)
8. Casas, E., Tröltzsch, F.: Error estimates for linear-quadratic elliptic control problems. In: V. Barbu, I. Lasiecka, D. Tiba, C. Varsan (eds.) Analysis and Optimization of Differential Systems, pp. 89–100. Kluwer Academic Publishers, Boston (2003). Proceeding of International Working Conference on Analysis and Optimization of Differential Systems
9. Eriksson, K., Estep, D., Hansbo, P., Johnson, C.: Computational Differential Equations. Cambridge University Press, Cambridge (1996)
10. Eriksson, K., Johnson, C., Thomée, V.: Time discretization of parabolic problems by the discontinuous Galerkin method. M2AN Math. Model. Numer. Anal. **19**, 611–643 (1985)
11. Falk, R.: Approximation of a class of optimal control problems with order of convergence estimates. J. Math. Anal. Appl. **44**, 28–47 (1973)
12. Geveci, T.: On the approximation of the solution of an optimal control problem governed by an elliptic equation. M2AN Math. Model. Numer. Anal. **13**, 313–328 (1979)
13. Hinze, M.: A variational discretization concept in control constrained optimization: The linear-quadratic case. Comput. Optim. and Appl. **30**(1), 45–61 (2005)
14. Lasiecka, I., Malanowski, K.: On discrete-time Ritz-Galerkin approximation of control constrained optimal control problems for parabolic systems. Control Cybern. **7**(1), 21–36 (1978)
15. Lions, J.L.: Optimal Control of Systems Governed by Partial Differential Equations, *Grundlehren Math. Wiss.*, vol. 170. Springer, Berlin (1971)
16. Malanowski, K.: Convergence of approximations vs. regularity of solutions for convex, control-constrained optimal-control problems. Appl. Math. Optim. **8**, 69–95 (1981)
17. McNight, R.S., Bosarge jr., W.E.: The Ritz-Galerkin procedure for parabolic control problems. SIAM J. Control Optim. **11**(3), 510–524 (1973)
18. Meidner, D., Vexler, B.: A priori error estimates for space-time finite element approximation of parabolic optimal control problems. Part I: Problems without control constraints. SIAM J. Control Optim. **47**(3), 1150–1177 (2008)
19. Meidner, D., Vexler, B.: A priori error estimates for space-time finite element approximation of parabolic optimal control problems. Part II: Problems with control constraints. SIAM J. Control Optim. **47**(3), 1301–1329 (2008)
20. Meyer, C., Rösch, A.: Superconvergence properties of optimal control problems. SIAM J. Control Optim. **43**(3), 970–985 (2004)
21. Rösch, A.: Error estimates for parabolic optimal control problems with control constraints. Zeitschrift für Analysis und ihre Anwendungen ZAA **23**(2), 353–376 (2004)
22. Rösch, A.: Error estimates for linear-quadratic control problems with control constraints. Optim. Methods Softw. **21**(1), 121–134 (2006)
23. Winther, R.: Error estimates for a Galerkin approximation of a parabolic control problem. Ann. Math. Pura Appl. (4) **117**, 173–206 (1978)

Multigrid Methods for Linear Elliptic Optimal Control Problems

M. Vallejos and A. Borzì

Abstract Multigrid optimization schemes that solve linear elliptic optimal control problems are discussed. For the solution of these problems, a comparison is made between the multigrid for optimization (MGOPT) method and the collective smoothing multigrid (CSMG) method.

Examples are given to illustrate and validate both techniques.

1 Introduction

Although the multigrid strategy was first introduced to design solvers for elliptic boundary value problems, it is now considered as one of the most promising approaches for the development of efficient optimization schemes. Some recent developments include the application of one-shot multigrid schemes to optimality systems [2, 5], to unconstrained optimization problems [7, 8, 9], and to inverse problems [10, 11]. Moreover, finite element multigrid methods applied to optimal control problems use different approaches on the smoothing procedure [1, 12, 13, 14].

The purpose of this paper is to investigate two representative multigrid methods for optimization: the collective smoothing multigrid method (CSMG) and the multigrid for optimization method (MGOPT). In our investigation we consider the application of these methods for solving linear elliptic optimal control problems. While both schemes are based on the well known full approximation storage (FAS) scheme

---

Michelle Vallejos
Institut für Mathematik und Wissenschaftliches Rechnen, Karl-Franzens-Universität Graz, Heinrichstr. 36, 8010 Graz, Austria, and University of the Philippines, Diliman, Quezon City, Philippines, e-mail: michelle.vallejos@uni-graz.at

Alfio Borzì
Dipartimento e Facoltà di Ingegneria, Università degli Studi del Sannio, 82100 Benevento, Italia, and Institut für Mathematik und Wissenschaftliches Rechnen, Karl-Franzens-Universität Graz, Heinrichstr. 36, 8010 Graz, Austria, e-mail: alfio.borzi@unisannio.it

[6], they represent different approaches to the solution of optimization problems. The CSMG scheme solves optimal control problems by solving the corresponding PDE optimality system in one shot treating all optimization variables collectively. As typical in multigrid development, this approach needs to customize the collective smoothing and intergrid transfer operators for each individual problem, i.e. the CSMG cannot be used as a black-box solver for all optimization problems. On the other hand, an appropriate design of the CSMG multigrid components results in a robust algorithm with typical multigrid efficiency. This fact is proved in [5] for linear control problems.

The motivation for investigating the MGOPT scheme [8, 9] is that it can be formulated in a way that is not problem specific and therefore it appears to have much larger applicability. In the MGOPT scheme the multigrid solution process represents the outer loop where the control function is considered as the unique dependent variable. The inner loop in this scheme consists of a classical one-grid optimization scheme and the other MGOPT components are chosen as those typical of a geometric multigrid approach. The essential condition for a 'successful' application of the MGOPT scheme is that the reduced Hessian, that is the Hessian of the optimization problem in the space of the control function, be positive definite. This is a much less restrictive requirement than ellipticity of the optimality system and the related smoothing and coarsening properties, which are required in the CSMG method. In this paper, we consider the MGOPT method for linear optimal control problems, comparing its numerical performance with that of the CSMG method.

2 Optimal Control Framework

An optimal control problem is formulated as follows

$$\begin{aligned} &\min_{u \in U} J(y,u) \\ &c(y,u) = 0 \end{aligned} \tag{1}$$

where $c(y,u) = 0$ is a partial differential equation (PDE) that represents the equality constraint. This equation is defined in $\Omega \subset \Re^d$. The state and control variables of the constraint c are denoted by y and u, respectively. We consider a cost functional of the tracking type given by $J(y,u) = h(y) + \nu g(u)$, where $\nu > 0$ is the weight of the cost of the control. The functions g and h are required to be twice continuously differentiable, bounded from below and $g(u) \to \infty$ as $\|u\| \to \infty$.

Given an optimization problem, the optimality system represents the first-order necessary conditions for a minimum. In order to derive these conditions, let $c: Y \times U \to Z$ for appropriate Hilbert spaces Y, U and Z, and consider the following Lagrange functional $L(y,u,p) = J(y,u) + \langle c(y,u), p \rangle_{Z,Z^*}$, where p is the Lagrange multiplier, also known as the adjoint variable. By equating to zero the Fréchet derivatives of L with respect to the triple (y,u,p), we have the optimality system

$$\begin{aligned} c(y,u) &= 0, \\ h'(y) + c_y(y,u)^* p &= 0, \\ \nu g'(u) + c_u(y,u)^* p &= 0. \end{aligned} \tag{2}$$

(* means adjoint.) The first differential equation in (2) is called the state equation and the second one is the adjoint equation. The last equation yields the optimality condition.

To better understand the importance of the last equation we introduce the reduced cost functional $\hat{J}(u) = J(y(u),u)$, where $y(u)$ denotes the unique solution of the state equation for a given u. One can show that the gradient of $\hat{J}(u)$ with respect to u is given by

$$\nabla \hat{J}(u) = \nu g'(u) + c_u(y,u)^* p(u), \tag{3}$$

where $p(u)$ solves the adjoint equation for a given u.

In a convex setting where the optimal control solution is unique, solving the optimality system is equivalent to solving the optimal control problem. However, in general, $c(y,u) = 0$ may be locally non convex wherein problem (1) may have multiple extremals. Therefore additional conditions must be satisfied to guarantee that the solution is a minimizer. For the second-order optimality conditions, we assume that (y,u,p) satisfy the optimality system (2) and the following

$$L_{zz}(y,u,p)(v,v) \geq c_1 \|v\|^2, \qquad c_1 > 0 \quad \forall\, v \in \mathscr{N}(c'(y,u)), \tag{4}$$

where $z = (y,u)$ and $c'(y,u)$ represents the linearized constraint. We assume that the null space $\mathscr{N}(c'(y,u))$ can be represented by $\mathscr{N}(c'(y,u)) = T(y,u)U$, where U is the space where the control is defined and $T(y,u) = \begin{bmatrix} -c_y^{-1} c_u \\ I_u \end{bmatrix}$, such that c_y and c_u are evaluated at $(y(u),u)$. Therefore, we can write condition (4) as

$$\nabla^2 \hat{J}(u)(w,w) \geq c_2 \|w\|^2 \qquad c_2 > 0 \quad \forall\, w \in U. \tag{5}$$

The operator $\nabla^2 \hat{J}(u) = T^*(y,u) L_{zz}(y,u,p) T(y,u)$ is the reduced Hessian where y and p solve the state and the adjoint equations for a given u. Hence, $\nabla^2 \hat{J}(u)$ is given by

$$\nabla^2 \hat{J} = L_{uu} + C^* L_{yy} C - L_{uy} C - C^* L_{yu} \tag{6}$$

where $C = C(y,u) = c_y^{-1}(y,u) c_u(y,u)$. The reduced Hessian matrix $\nabla^2 \hat{J}$ is symmetric. Thus, condition (5) requires that the smallest eigenvalue of the reduced Hessian be positive.

3 Linear Elliptic Optimal Control Problems

In this section, we discuss linear elliptic optimal control problems. The corresponding optimality systems are presented and the multigrid solutions will be detailed in the next section. We assume that Ω is an open, bounded and convex subset of $\Re^2$ or its boundary $\partial\Omega$ is $C^{1,1}$ smooth.

We focus on the following linear elliptic optimal control problem

$$\begin{aligned} \min_{u\in U} J(y,u) :&= \tfrac{1}{2}\|y-z\|^2_{L^2(\Omega)} + \tfrac{\nu}{2}\|u\|^2_{L^2(\Omega)}, \\ -\Delta y - u &= f \quad \text{in} \quad \Omega, \\ y &= 0 \quad \text{on} \quad \partial\Omega, \end{aligned} \tag{7}$$

where $z \in L^2(\Omega)$ is the target function, $f \in L^2(\Omega)$ and $U = L^2(\Omega)$. This choice of cost functional J corresponds to $h(y) = \frac{1}{2}\|y-z\|^2_{L^2(\Omega)}$ and $g(u) = \frac{1}{2}\|u\|^2_{L^2(\Omega)}$. In this case, u is a distributed control over Ω. The solution of problem (7) is characterized by the following optimality system

$$\begin{aligned} -\Delta y - u &= f \quad \text{in} \quad \Omega, \quad y = 0 \quad \text{on} \quad \partial\Omega, \\ -\Delta p + y &= z \quad \text{in} \quad \Omega, \quad p = 0 \quad \text{on} \quad \partial\Omega, \\ \nu u - p &= 0 \quad \text{in} \quad \Omega. \end{aligned} \tag{8}$$

From Eqs. (3) and (6), the reduced gradient and the reduced Hessian are given by $\nabla\hat{J}(u) = \nu u - p$ and $\nabla^2\hat{J}(u) = \nu \mathrm{I} + \Delta^{-2}$. Since $\nabla^2\hat{J}(u)$ is strictly positive, the solution to the optimality system (8) is guaranteed to be a minimizer. However, smaller ν correspond to more flat minima and more stiff optimality systems.

4 Discretization Scheme and Collective Smoothing

Our discussion on multigrid methods for optimization requires to define a hierarchy of problems $A_k u_k = f_k$ in Ω_k, indexed by $k = 1,2,\ldots,L$. Here Ω_k denotes the set of grid points with uniform grid spacing h_k for the finite difference discretization in Ω taken as a square domain. We have $h_1 > h_2 > \cdots > h_L > 0$, so that $h_{k-1} = 2h_k$. The number of interior grid points will be n_k, A_k is a $n_k \times n_k$ matrix, and any function in Ω_k is a vector of size n_k. We denote this vector space with V_k. In the space V_k, we introduce the inner product $(\cdot,\cdot)_k$ with the corresponding norm $\|u\|_k = \sqrt{(u,u)_k}$.

For multigrid purpose we define a restriction operator $I_k^{k-1} : V_k \to V_{k-1}$ and a prolongation operator $I_{k-1}^k : V_{k-1} \to V_k$. We require that they satisfy $(I_k^{k-1}u,v)_{k-1} = (u,I_{k-1}^k v)_k$ for all $u \in V_k$ and $v \in V_{k-1}$. That is, the restriction operator is the adjoint of the interpolation operator.

Now we consider the discrete version of the optimality system (8). We have

$$\begin{aligned} -\Delta_k y_k - u_k &= f_k, \\ -\Delta_k p_k + y_k &= z_k, \\ \nu u_k - p_k &= 0. \end{aligned} \tag{9}$$

Let $x \in \Omega_k$ where $x = (ih_k, jh_k)$ and i,j are the index of the grid points arranged lexicographically. We use the standard five point stencil for the Laplacian. We first set $A_{i,j} = -(y_{i-1,j} + y_{i+1,j} + y_{i,j-1} + y_{i,j+1}) - h^2 f_{i,j}$, and $B_{i,j} = -(p_{i-1,j} + p_{i+1,j} + p_{i,j-1} + p_{i,j+1}) - h^2 z_{i,j}$. The values $A_{i,j}$ and $B_{i,j}$ are considered constant during the update of the variables at ij. Hence, we have the following system of equations of three variables $y_{i,j}$, $u_{i,j}$ and $p_{i,j}$

$$\begin{aligned} A_{i,j} + 4y_{i,j} - h^2 u_{i,j} &= 0, \\ B_{i,j} + 4p_{i,j} + h^2 y_{i,j} &= 0, \\ \nu u_{i,j} - p_{i,j} &= 0. \end{aligned} \tag{10}$$

Let $w_k = (y_k, u_k, p_k)$. A collective smoothing step on w updates the values $y_{i,j}$, $u_{i,j}$, and $p_{i,j}$ such that the resulting residuals of the state and adjoint equations at that point are zero. Since (10) is a linear system, we can compute the updates for the variables $y_{i,j}$, and $p_{i,j}$ in the following way

$$\begin{aligned} y_{i,j}(u_{i,j}) &= \tfrac{1}{4}(h^2 u_{i,j} - A_{i,j}), \\ p_{i,j}(u_{i,j}) &= \tfrac{1}{16}(-h^4 u_{i,j} + h^2 A_{i,j} - 4B_{i,j}). \end{aligned} \tag{11}$$

To obtain an update $u_{i,j}$, we require that it satisfies the optimality condition $\nabla \hat{J}(u) = \nu u - p(u) = 0$. Hence, we have $u_{i,j} = \frac{1}{16\nu + h^4}(h^2 A_{i,j} - 4B_{i,j})$. With this $u_{i,j}$ we use (11) to update the values of the state and adjoint variable at the i, j grid point. A sweep of this smoothing scheme consists in an ordered sequential update of $y_{i,j}$, $u_{i,j}$ and $p_{i,j}$ on all grid points.

5 The Multigrid Method

In this section we present the two multigrid schemes for solving elliptic optimal control problems, the CSMG and the MGOPT method.

The CSMG scheme is based on the nonlinear multigrid full approximation storage (FAS) method applied to the optimality system with a collective smoothing. This method shows mesh independence due to its robustness with respect to the value of the weight of the control. Some recent applications of the CSMG method to linear control problems with state and control constraints are presented in [3, 4].

The multigrid for optimization (MGOPT) method, is very similar to the CSMG scheme. The MGOPT method was first introduced by Nash [9] and Lewis and Nash [7, 8] as an extension of the multigrid scheme to optimization problems.

To illustrate both methods we consider a discrete problem

$$A_k w_k = f_k \tag{12}$$

where A_k represents a discrete linear operator on Ω_k. The MGOPT method is applied to solve $\min_{u_k} \left(\hat{J}_k(u_k) - (f_k, u_k)_k\right)$. Hence in this case, $w := u$ and $A_k u_k = \nabla \hat{J}_k(u_k)$. In the CSMG case, we solve (9) and define $w := (y, u, p)$.

Let the smoothing iteration be given by S_k such that we get an update $w_k^l = S_k(w_k^{l-1}, f_k)$. Starting with an initial approximation w_k^0, we apply γ_1 times the smoothing scheme and obtain $w_k^{\gamma_1}$. Now, the desired solution w_k can be written as $w_k = w_k^{\gamma_1} + e_k$ for some error e_k. Therefore, (12) can be written as $A_k(w_k^{\gamma_1} + e_k) = f_k$ or equivalently as

$$A_k(w_k^{\gamma_1} + e_k) - A_k w_k^{\gamma_1} = f_k - A_k w_k^{\gamma_1}. \tag{13}$$

Next, we will represent (13) on a coarser grid Ω_{k-1}. Define $w_{k-1} := I_k^{k-1} w_k^{\gamma_1} + e_{k-1}$. Here, w_{k-1} represents a coarse-grid approximation to w_k. On the left hand side of (13), we represent A_k by A_{k-1} and $w_k^{\gamma_1}$ by $I_k^{k-1} w_k^{\gamma_1}$. On the other side, we apply the restriction operator I_k^{k-1} and we get $I_k^{k-1}(f_k - A_k w_k^{\gamma_1})$. Hence, we have the following equation

$$A_{k-1} w_{k-1} = I_k^{k-1}(f_k - A_k w_k^{\gamma_1}) + A_{k-1}(I_k^{k-1} w_k^{\gamma_1}). \tag{14}$$

We define $\tau_{k-1} = A_{k-1}(I_k^{k-1} w_k^{\gamma_1}) - I_k^{k-1} A_k w_k^{\gamma_1}$ then (14) can be written as

$$A_{k-1} w_{k-1} = I_k^{k-1} f_k + \tau_{k-1}. \tag{15}$$

The term τ_{k-1} is called the fine-to-coarse residual/gradient correction. The solution of (14) gives the error $e_{k-1} := w_{k-1} - I_k^{k-1} w_k^{\gamma_1}$. Therefore, we have a correction to the fine grid approximation as $w_k^{\gamma_1+1} = w_k^{\gamma_1} + \alpha_k I_{k-1}^k (w_{k-1} - I_k^{k-1} w_k^{\gamma_1})$. For CSMG, $\alpha_k = 1$, and for MGOPT, α_k is the step length obtained after a line search procedure in the direction given by $I_{k-1}^k (w_{k-1} - I_k^{k-1} w_k^{\gamma_1})$. Finally, we apply γ_2 iterations of the smoothing algorithm to damp possible high frequency errors that may arise from the coarse grid correction process. The following algorithm presents the method described above.

Algorithm (MG Algorithm)
Initialize w_k^0 to be an initial approximation at resolution k. If $k = 1$, solve $A_k w_k = f_k$ and return. Else if $k > 1$,

1. Apply γ_1 iterations of a smoothing algorithm. $w_k^l = S_k(w_k^{l-1}, f_k), \quad l = 1, 2, \ldots, \gamma_1$
2. Restrict the solution of (1). $w_{k-1}^{\gamma_1} = I_k^{k-1} w_k^{\gamma_1}$
3. Compute the fine-to-coarse correction. $\tau_{k-1} = A_{k-1} w_{k-1}^{\gamma_1} - I_k^{k-1} A_k w_k^{\gamma_1}$
4. Compute the right hand side. $f_{k-1} = I_k^{k-1} f_k + \tau_{k-1}$
5. Apply γ cycles of MG (γ_1, γ_2) to the coarse grid problem $A_{k-1} w_{k-1} = f_{k-1}$.
6. Prolongate the error. $e_k = I_{k-1}^k (w_{k-1} - w_{k-1}^{\gamma_1})$
7. Coarse grid correction. $w_k^{\gamma_1+1} = w_k^{\gamma_1} + \alpha_k e_k$
8. Apply γ_2 iterations of a smoothing algorithm. $w_k^l = S_k(w_k^{l-1}, f_k), \quad l = \gamma_1 + 2, \ldots, \gamma_1 + \gamma_2 + 1$

6 Numerical Results

In this section, we present the results of the experiments on the computational performance of the proposed multigrid schemes as solvers for distributed elliptic optimal control problems. Using different values of the cost of the control, we gathered the CPU time (in seconds) until a stopping tolerance of 10^{-8} is satisfied. For all computations, we use $\gamma_1 = \gamma_2 = 2$ pre and post smoothing steps. This means that one iteration of the CSMG and the MGOPT method uses $\gamma_1 + \gamma_2 = 4$ iterations of the gradient method (GM) on the finest grid. We consider problem (7) with the zero function as an initial guess, $\Omega = (0,1) \times (0,1)$ and $f, z \in L^2(\Omega)$ given by

$$f(x,y) = 1,$$
$$z(x,y) = \begin{cases} 2, & \text{on} \quad (0.25,0.75) \times (0.25,0.75) \\ 1, & \text{otherwise} \end{cases}.$$

Notice that z is not attainable because of the boundary condition. The numerical results are shown in Table 1. We can see that for CSMG using different parameters ν, the method converges within fifteen seconds. We obtain independence on the parameter ν and on the mesh size. For the one grid optimization scheme using gradient method (GM) and nonlinear conjugate gradient (NCG), we have a constant increase in the number of iterations as the parameter ν decreases. This property is also inherited by the MGOPT method using both GM and NCG as smoothing algorithms. Note that the step length α in this problem is inversely proportional to the weight of the control ν, i.e., $\alpha \propto \frac{1}{\nu}$. The collective smoothing discussed in Sect. 5 is robust with respect to the changes of the value of the weight of the control ν. This fact makes the CSMG algorithm a useful tool in solving problems in the limit case of bang-bang control [3, 5].

While it is shown that in this case, the CSMG scheme is superior to the MGOPT scheme, we see that the MGOPT scheme provides a framework to accelerate classical optimization schemes and because of its modularity it can be easily applied to a large class of PDE-based optimization problems.

7 Conclusion

We presented two multigrid schemes for solving linear elliptic optimal control problems, the CSMG and the MGOPT methods. The numerical results show that CSMG is faster compared to the MGOPT method. It is also mesh independent and parameter independent. MGOPT on the other hand does not exhibit similar properties as CSMG. However, they are more easier to implement since any optimization algorithm can be used as a smoothing iteration, unlike the smoothing iteration for CSMG which is only useful for particular problems being solved. Faster optimization algorithm for the MGOPT method results to a faster convergence of the MGOPT scheme.

Table 1 Results using CSMG, onegrid gradient method (GM) and nonlinear conjugate gradient (NCG), and MGOPT with GM and NCG

ν	mesh	$\hat{J}$	CSMG	GM	MGOPT[1]	NCG	MGOPT[2]
	65 × 65	0.846	1	1	1	1	1
10^{-1}	129 × 129	0.827	3	5	4	4	3
	257 × 257	0.817	15	23	17	18	17
	65 × 65	0.718	1	2	2	1	1
10^{-2}	129 × 129	0.701	3	11	9	5	4
	257 × 257	0.692	15	57	40	23	21
	65 × 65	0.348	1	12	8	1	1
10^{-3}	129 × 129	0.337	3	63	22	7	5
	257 × 257	0.331	15	312	97	34	26
	65 × 65	0.163	1	28	23	5	4
10^{-4}	129 × 129	0.154	3	144	74	26	23
	257 × 257	0.151	15	723	345	132	89

[1] MGOPT with GM
[2] MGOPT with NCG

References

1. Apel, T., Schöberl, J.: Multigrid methods for anisotropic edge refinement. SIAM J. Numer. Anal. **40**(5), 1993–2006 (2002)
2. Borzì, A.: Multigrid methods for parabolic distributed optimal control problems. J. Comput. Appl. Math. **157**(2), 365–382 (2003)
3. Borzì, A.: High-order discretization and multigrid solution of elliptic nonlinear constrained optimal control problems. J. Comput. Appl. Math. **200**(1), 67–85 (2007)
4. Borzì, A., Kunisch, K.: A multigrid scheme for elliptic constrained optimal control problems. Comput. Optim. Appl. **31**(3), 309–333 (2005)
5. Borzì, A., Kunisch, K., Kwak, D.Y.: Accuracy and convergence properties of the finite difference multigrid solution of an optimal control optimality system. SIAM J. Control Optim. **41**(5), 1477–1497 (2002)
6. Brandt, A.: Multi-level adaptive solutions to boundary-value problems. Math. Comp. **31**(138), 333–390 (1977)
7. Lewis, R., Nash, S.: A multigrid approach to the optimization of systems governed by differential equations. AIAA-2000-4890 (2000)
8. Lewis, R., Nash, S.: Model problems for the multigrid optimization of systems governed by differential equations. SIAM J. Sci. Comput. **26**(6), 1811–1837 (2005)
9. Nash, S.: A multigrid approach to discretized optimization problems. Optim. Methods Softw. **14**(1-2), 99–116 (2000)
10. Oh, S., Bouman, C., Webb, K.: Multigrid tomographic inversion with variable resolution data and image spaces. IEEE Trans. Image Process. **15**(9), 2805–2819 (2006)
11. Oh, S., Milstein, A., Bouman, C., Webb, K.: A general framework for nonlinear multigrid inversion. IEEE Trans. Image Process. **14**(1), 125–140 (2005)
12. Schöberl, J., Zulehner, W.: On Schwarz-type smoothers for saddle point problems. Numer. Math. **95**(2), 377–399 (2003)
13. Schöberl, J., Zulehner, W.: Symmetric indefinite preconditioners for saddle point problems with applications to PDE-constrained optimization problems. SIAM J. Matrix Anal. Appl. **29**(3), 752–773 (2007)
14. Zulehner, W.: A class of smoothers for saddle point problems. J. Computing **65**, 227–246 (2000)

Optimization, Inverse Problems

An Active Curve Approach for Tomographic Reconstruction of Binary Radially Symmetric Objects

I. Abraham, R. Abraham, and M. Bergounioux

Abstract This paper deals with tomographic reconstruction of radially symmetric objects from a single radiograph, in order to study the behavior of shocked material. Usual tomographic reconstruction algorithms (such as generalized inverse or filtered back-projection) cannot be applied here. To improve the reconstruction, we assume that the object is binary so that it may be described by curves that separate the two materials. We present a BV-model that leads to a non local Hamilton-Jacobi equation, via a level set strategy.

1 Introduction

We are interested here in a very specific application of tomographic reconstruction for a physical experiment which goal is to study the behavior of a material under a shock. During the deformation of the object, we obtain an X-ray radiography by high speed image capture. We suppose that the object is radially symmetric, so that one radiograph is enough to reconstruct the 3D object.

Physicists are looking for the shape of the interior at some fixed interest time. At that time, the interior may be composed of several holes which also may be very irregular. We deal here with a synthetic object that contains all the standard difficulties that may appear (see Fig. 1). These difficulties are characterized by:

- Several disconnected holes.
- A small hole located on the symmetry axis (which is the area where the details are difficult to recover).

Isabelle Abraham
CEA Ile de France- BP 12, 91680 Bruyères le Châtel, France e-mail: isabelle.abraham@cea.fr

Romain Abraham · Maïtine Bergounioux
Laboratoire MAPMO- Fédération Denis Poisson, CNRS, Université d'Orléans, BP 6759 Orléans cedex 02, France e-mail: romain.abraham@univ-orleans.fr, maitine.bergounioux@univ-orleans.fr

- Smaller and smaller details on the boundary of the top hole in order to determine a lower bound detection.

Our framework is completely different from the usual tomographic point of view, and usual techniques (such as filtered back-projection) are not adapted to our case.

2 A Continuous Model in BV Space

Let us explicit the projection operator involved in the tomography process. This operator, denoted by H, is given, for every function $f \in L^\infty(\mathbb{R}_+ \times \mathbb{R})$ with compact support, by

$$\forall (u,v) \in \mathbb{R} \times \mathbb{R}, \qquad Hf(u,v) = 2\int_{|u|}^{+\infty} f(r,v)\frac{r}{\sqrt{r^2-u^2}}dr. \tag{1}$$

For more details one can refer to [1]. Similarly the adjoint operator H^* of H is

$$\forall (r,z) \in \mathbb{R} \times \mathbb{R}, \qquad H^*g(r,z) = 2\int_{0}^{|r|} g(u,z)\frac{|r|}{\sqrt{r^2-u^2}}du. \tag{2}$$

Thanks to the symmetry, this operator characterizes the Radon transform of the object and so is invertible. However, the operator H^{-1} is not continuous with respect to the suitable topologies. Consequently, a small variation on the measure g leads to significant errors on the reconstruction. As radiographs are strongly perturbed, applying H^{-1} to data leads to a poor reconstruction. Due to the experimental setup there are two main perturbations:

- A blur, due to the detector response and the X-ray source spot size. We denote by F the effect of blurs and consider the simplified case where F is supposed to be linear.
- A noise which is supposed for simplicity to be an additive Gaussian white noise of mean 0, denoted by ε.

Consequently, the projection of the object f will be $g = F(Hf) + \varepsilon$. The comparison between the theoretical projection Hf and the perturbed one is shown in Fig. 1. The reconstruction using the inverse operator H^{-1} applied to g is given by Fig. 2. It is clear from Fig. 2 that the use of the inverse operator is not suitable. In what follows, we will call "experimental data " the image which corresponds to the blurred projection of a "fictive" object of density 0 with some holes of known "density" $\lambda > 0$. Consequently, the space of admissible objects will be the set $\mathscr{F}$ of functions f that take values in $\{0, \lambda\}$. Such functions $f \in \mathscr{F}$ are defined on $\mathbb{R}^2$ and have compact support included in some open bounded subset of $\mathbb{R}^2$, say Ω. Note that $\mathscr{F}$ is a subspace of the bounded variation functions space

$$BV(\Omega) = \{u \in L^1(\Omega) \mid J(u) < +\infty\}$$

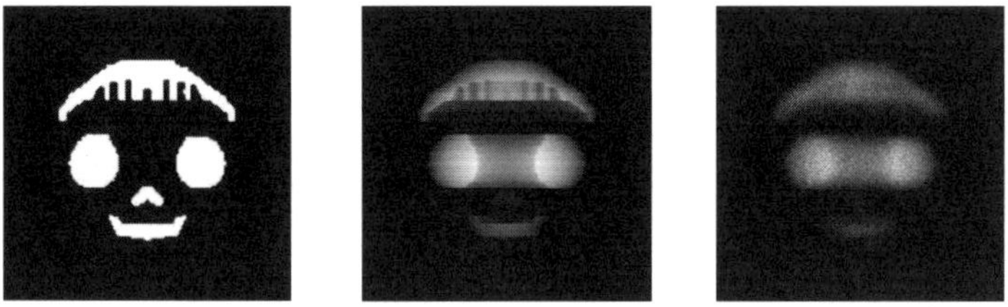

Fig. 1 Synthetic object, theoretical projection Hf, real projection with noise and blur.

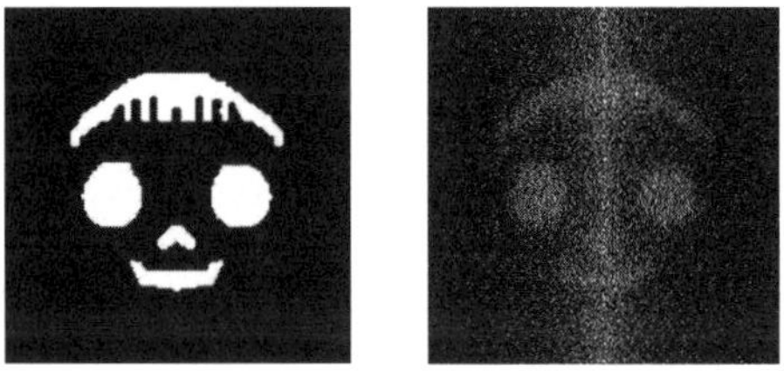

Fig. 2 Comparison between the real object on the left-hand side and the reconstruction computed with H^{-1} applied to the real projection on the right-hand side.

where $J(u)$ stands for the total variation of u (see [3] for example). Furthermore, a function f of $\mathscr{F}$ is characterized by the knowledge of the curves that limit the two areas where f is equal to λ and to 0. Indeed, as the support of the function f is bounded, these curves are disjoint Jordan curves and the inside density λ whereas the outside one is 0. We use here a standard technique for image reconstruction: we introduce an energy functional that is a function of γ_f where γ_f is a set of disjoint Jordan curves. We may split this energy in two terms:

- The first one is a matching term. It is the usual L^2-norm between $F(Hf)$ and the data g (where H is given by (1)), so that $E_1(\gamma_f) = \frac{1}{2}\|F(Hf) - g\|_2^2$. Here $\|\cdot\|_2$ stands for thc $L^2(\Omega)$- norm and $g \subset L^2(\Omega)$.
- The second term is a regularization term: $E_2(\gamma_f) = \ell(\gamma_f)$ where $\ell(\gamma_f)$ denotes the length of the curve γ_f.This penalization term may be also viewed as the total variation $J(f)$ (up to a multiplicative constant) of the function f because of the binarity.

Hence the total energy functional is

$$E(\gamma_f) = \frac{1}{2}\|F(Hf) - g\|_2^2 + \alpha \ell(\gamma_f) \tag{3}$$

which is an adaptation of the well-known Mumford-Shah energy functional introduced in [4]. The "optimal" value of $\alpha > 0$ may depend on the data.
Therefore, we consider the following minimization problem

$$(\mathscr{P}) \qquad \min \ \{ E(\gamma_f) \mid f \in BV(\Omega),\ f(x) \in \{0, \lambda\} \text{ a.e. on } \Omega \},$$

and we first get an existence result:

Theorem 1. *Problem* ($\mathscr{P}$) *admits at least one solution.*

Proof. See Theorem 1 of [2] □

3 Computation of the Energy Derivative

Now, we are interested in optimality conditions. In what follows, for mathematical reasons, we have to add an extra assumption: the curves γ_f are $\mathscr{C}^1$ so that the normal vector of the curves is well-defined (as an orthogonal vector to the tangent one). Unfortunately, with this assumption, we cannot compute easily the derivative of the energy in the $BV(\Omega)$ framework. Indeed we need regular curves and we do not know if the $BV(\Omega)$ minimizer provides a curve with the required regularity. Moreover, the set of constraints is not convex and it is not easy to compute the Gateaux derivative (no admissible test functions).

So we have few hope to get classical optimality conditions and we rather compute minimizing sequences. We focus on those that are given via the gradient descent method inspired by [4]. Formally, we look for a family of curves $(\gamma_t)_t \geq 0$ such that $\dfrac{\partial E}{\partial \gamma}(\gamma_t) \leq 0$ that is: $E(\gamma_t)$ decreases as $t \to +\infty$. Let us compute the energy variation when we operate a small deformation on the curves γ. In other words, we will compute the energy Gâteau derivative for a small deformation $\delta\gamma$:

$$\frac{\partial E}{\partial \gamma}(\gamma) \cdot \delta\gamma = \lim_{t \to 0} \frac{E(\gamma + t\delta\gamma) - E(\gamma)}{t}.$$

We first focus on local deformations $\delta\gamma$. Let (r_0, z_0) be a point P of γ. We consider a local reference system which center is P and axis are given by the tangent and normal vectors at P. We denote (ξ, η) the new generic coordinates in this reference system and still denote $f(\xi, \eta) = f(r, z)$ for convenience. We apply the implicit functions theorem to parametrize the curve: there exist a neighborhood U of P and a $\mathscr{C}^1$ function h such that, for every $(\xi, \eta) \in U$,

$$(\xi, \eta) \in \gamma \iff \eta = h(\xi).$$

Eventually, we get a neighborhood U of P, a neighborhood I of ξ_0 and a $\mathscr{C}^1$ function h such that

$$\gamma \cap U = \left\{ (\xi, \eta) \in \mathbb{R}^2 \mid \eta = h(\xi),\ \xi \in I \right\}.$$

The local parametrization is oriented along the outward normal $\mathbf{n}$ to the curve γ at point P. More precisely, we define the local coordinate system $(\tau, \mathbf{n})$ where τ is the usual tangent vector, $\mathbf{n}$ is the direct orthonormal vector; we set the curve orientation so that $\mathbf{n}$ is the outward normal. The function f if then defined on U by

$$f(\xi,\eta) = \begin{cases} \lambda \text{ if } \eta < h(\xi)\,, \\ 0 \text{ if } \eta \geq h(\xi)\,. \end{cases}$$

We then consider a local (limited to U) deformation $\delta\gamma$ along the normal vector. This is equivalent to handling a $\mathscr{C}^1$ function δh whose support is included in I. The new curve γ_t obtained after the deformation $t\delta\gamma$ is then parametrized by

$$\eta = \begin{cases} h(\xi) + t\delta h(\xi) & \text{for } (\xi,\eta) \in U\,, \\ \gamma & \text{otherwise.} \end{cases}$$

This defines a new function f_t:

$$f_t(\xi,\eta) = \begin{cases} f(\xi,\eta) \text{ if } (\xi,\eta) \notin U\,, \\ \lambda \quad \text{if } (\xi,\eta) \in U \cap \{\eta < h(\xi) + t\delta h(\xi)\}\,, \\ 0 \quad \text{if } (\xi,\eta) \in U \cap \{\eta \geq h(\xi) + t\delta h(\xi)\}\,. \end{cases} \tag{4}$$

We will also set $\delta f_t = f_t - f$. This deformation is described in Fig. 3.

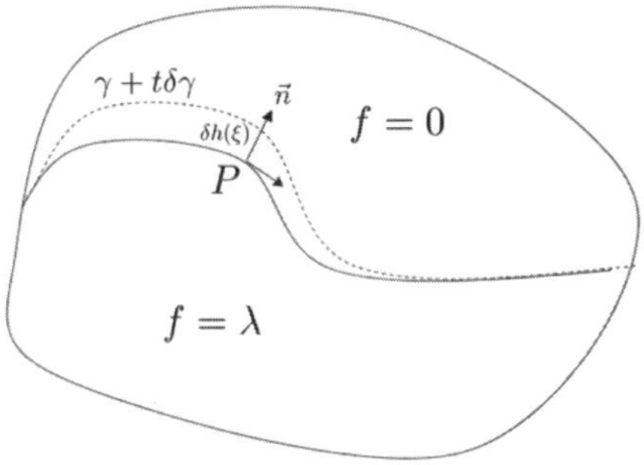

Fig. 3 Description of a local deformation of the initial curve γ. P is the current point, U is the neighborhood of P in which the deformation is restricted to and $\gamma + t\delta\gamma$ is the new curve after deformation. The interior of the curve is the set where $f = \lambda$.

The energy E_2 Gâteau derivative has already been computed in [4] and is

$$\frac{\partial E_2}{\partial \gamma}(\gamma)\delta\gamma = -\int_\gamma \text{curv}(\gamma)\big(\xi, h(\xi)\big)\delta h(\xi)d\xi\,,$$

where curv denotes the curvature of the curve and δh is the parametrization of $\delta\gamma$. It remains to compute the derivative of E_1 . We first estimate δf_t: a simple computation shows that

$$\delta f_t(\xi,\eta) = \begin{cases} 0 \text{ if } \eta \geq h(\xi) + t\delta h(\xi) \text{ or } \eta(\xi) \leq h(\xi)\,, \\ \lambda \text{ if } h(\xi) \leq \eta \leq h(\xi) + t\delta h(\xi)\,, \end{cases} \quad \text{in case } \delta h \geq 0\,,$$

and

$$\delta f_t(\xi,\eta) = \begin{cases} 0 \text{ if } \eta \le h(\xi) + t\delta h(\xi) \text{ or } \eta(\xi) \ge h(\xi)\,, \\ -\lambda \text{ if } h(\xi) \ge \eta \ge h(\xi) + t\delta h(\xi)\,, \end{cases} \quad \text{in case } \delta h \le 0\,.$$

Now we compute $E_1(\gamma_t) - E_1(\gamma)$ where γ (resp. γ_t) is the curve associated to the function f (resp. f_t):

$$\begin{aligned} &E_1(\gamma_t) - E_1(\gamma) \\ &\quad = \frac{1}{2}\int_{\mathbb{R}}\int_{\mathbb{R}} \left((g - FHf_t)^2 - (g - FHf)^2\right)(u,v)\,du\,dv \\ &\quad = \frac{1}{2}\int_{\mathbb{R}}\int_{\mathbb{R}} \left((g - FHf - FH\delta f_t)^2 - (g - FHf)^2\right)(u,v)\,du\,dv \\ &\quad = -\int_{\mathbb{R}}\int_{\mathbb{R}} (g - FHf)(u,v)\,FH\delta f_t(u,v)\,du\,dv + \underbrace{\frac{1}{2}\int_{\mathbb{R}}\int_{\mathbb{R}} (FH\delta f_t)^2(u,v)\,du\,dv}_{=o(t)} \\ &\quad = -\langle (g - FHf), FH\delta f_t\rangle_{L^2} + o(t) \\ &\quad = -\langle H^*F^*(g - FHf), \delta f_t\rangle_{L^2} + o(t). \end{aligned}$$

To simplify the notations, we denote by $\mathscr{A} f := (H^*Fg - H^*FFHf)$ so that we need to compute

$$\lim_{t\to 0} \frac{1}{t} \langle \mathscr{A} f, \delta f_t\rangle_{L^2}.$$

As δf_t is zero out of the neighbourhood U, we have

$$\langle \mathscr{A} f, \delta f_t\rangle_{L^2} = \int_U \mathscr{A} f(\xi,\eta)\,\delta f_t(\xi,\eta)\,d\xi\,d\eta.$$

In the case $\delta h \ge 0$, we have,

$$\langle \mathscr{A} f, \delta f_t\rangle_{L^2} = \lambda \int_{\xi\in\gamma}\int_{\eta=h(\xi)}^{\eta=h(\xi)+t\delta h(\xi)} \mathscr{A} f(\xi,\eta)\,d\xi\,d\eta.$$

As the function $\mathscr{A} f$ is continuous (and thus bounded on U), we may pass to the limit by dominated convergence and get

$$\lim_{t\to 0} \frac{1}{t} \langle \mathscr{A} f, \delta f_t\rangle_{L^2} = \lambda \int_\gamma \mathscr{A} f(\xi, h(\xi))\,\delta h(\xi)\,d\xi.$$

In the case $\delta h < 0$, we have

$$\langle \mathscr{A} f, \delta f_t\rangle_{L^2} = \int (-\lambda) \int_{\xi\in\gamma}\int_{\eta=h(\xi)+t\delta h(\xi)}^{\eta=h(\xi)} \mathscr{A} f(\xi,\eta)\,d\xi\,d\eta$$

and we obtain the same limit as in the non negative case. Finally, the energy derivative is

$$\frac{\partial E}{\partial \gamma}(\gamma_t)\cdot\delta\gamma_t = -\int_\gamma \left(\lambda \mathscr{A} f + \alpha\,\mathrm{curv}(\gamma_t)(\xi, h(\xi))\right)\delta h(\xi)\,d\xi\,. \tag{5}$$

As $\delta h = \langle \delta \gamma_t, \mathbf{n} \rangle$ formula (5) may be written

$$\frac{\partial E}{\partial \gamma}(\gamma_t) \cdot \delta \gamma_t = -\int_\gamma \lambda \left(\lambda \mathscr{A} f + \alpha \mathrm{curv}(\gamma_t)(s)\right) \langle \delta \gamma_t, \mathbf{n} \rangle \, ds \tag{6}$$

where $\mathbf{n}$ denotes the outward pointing normal unit vector of the curve γ, $\langle \cdot, \cdot \rangle$ denotes the usual scalar product in $\mathbb{R}^2$ and $c(s)$ is a positive coefficient that depends on the curvilinear abscissa s.

The latter expression is linear and continuous in $\delta\gamma$, this formula is also true for a non local deformation (which can be achieved by summing local deformations).

4 Front Propagation and Level Set Method

Now we consider a family of curves $(\gamma_t)_{t \geq 0}$ that converges toward a local minimum of the energy functional. From equation (6), it is clear that if the curves (γ_t) evolve according to the differential equation

$$\frac{\partial \gamma}{\partial t} = (\lambda \mathscr{A} f + \alpha \mathrm{curv}(\gamma_f))\mathbf{n}, \tag{7}$$

the total energy will decrease.

The level set method consists in viewing the curves γ as the 0-level set of a smooth real function ϕ defined on $\mathbb{R}^2$ (see [5]). The function f that we are looking for is then given by

$$f(x) = \lambda \mathbb{1}_{\phi(x)>0}.$$

Let us write the evolution PDE for functions $\phi_t = \phi(t, \cdot)$ that correspond to the curves γ_t. Let $x(t)$ be a point of the curve γ_t and let us follow that point during the evolution. We know that this point evolves according to equation (7)

$$x'(t) = \left(\lambda \mathscr{A} f + \alpha \ \mathrm{curv}(\gamma_f)\right)(x(t))\mathbf{n}.$$

We can rewrite this equation in terms of the function ϕ recognizing that

$$\mathbf{n} = \frac{\nabla \phi}{|\nabla \phi|} \quad \text{and} \quad \mathrm{curv}(\gamma) = \mathrm{div}\left(\frac{\nabla \phi}{|\nabla \phi|}\right)$$

where ∇ stands for the gradient of ϕ with respect to x, $|\cdot|$ denotes the Euclidean norm. The evolution equation becomes

$$x'(t) = \left(\lambda^2 \mathscr{A}\left(\mathbb{1}_{\phi(t,\cdot)>0}\right) + \alpha \ \mathrm{div}\left(\frac{\nabla \phi}{|\nabla \phi|}\right)\right) \frac{\nabla \phi}{|\nabla \phi|}(t, x(t)).$$

Then, as the point $x(t)$ remains on the curve γ_t, it satisfies $\phi_t(x(t)) = \phi(t, x(t)) = 0$. By differentiating this expression, we obtain

$$\frac{\partial \phi}{\partial t} + \langle \nabla \phi, x'(t) \rangle = 0$$

which leads to the following evolution equation for ϕ:

$$\frac{\partial \phi}{\partial t} + |\nabla_x \phi| \left(\lambda^2 \mathscr{A} \left(1_{\phi(t,\cdot)>0} \right) + \alpha \operatorname{div} \left(\frac{\nabla \phi}{|\nabla \phi|} \right) \right) = 0 \, ,$$

that is

$$\frac{\partial \phi}{\partial t} = |\nabla \phi| \left(\lambda^2 H^* F F H \left(1_{\phi(t,\cdot)>0} \right) - \alpha \operatorname{div} \left(\frac{\nabla \phi}{|\nabla \phi|} \right) - \lambda H^* F g \right) . \qquad (8)$$

The above equation is an Hamilton-Jacobi equation which involves a non local term (through H and F). Such equations are difficult to handle especially when it is not monotone (which is the case here). In particular, existence and/or uniqueness of solutions (even in the viscosity sense) are not clear. Nevertheless, some numerical experiments have been carried out using excplicit schemes based on Sethian's techniques for level-set methods and give interesting results.

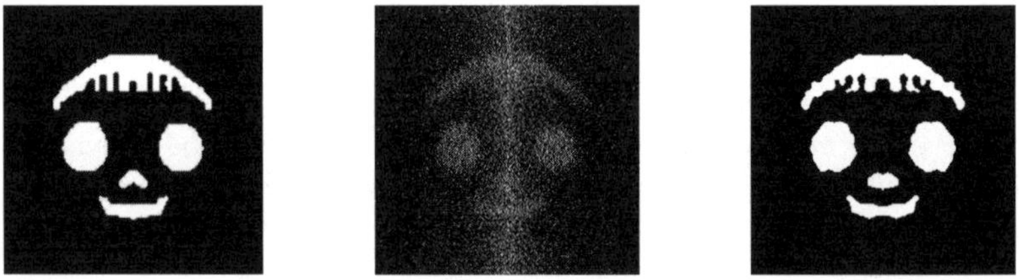

Fig. 4 Synthetic object, H^{-1} applied to the real projection , computed solution.

References

1. Abraham, I., Abraham, R., Bergounioux, M.: An active curve approach for tomographic reconstruction of binary radially symmetric objects (2006). URL http://hal.archives-ouvertes.fr/. Preprint, HAL-084855
2. Abraham, R., Bergounioux, M., Trélat, E.: A penalization approach for tomographic reconstruction of binary radially symmetric objects (2007). URL http://hal.archives-ouvertes.fr/. Preprint, HAL-00139494
3. Ambrosio, L., Fusco, N., Pallara, D.: Functions of bounded variation and free discontinuity problems. Oxford mathematical monographs. Oxford University Press (2000)
4. Mumford, D., Shah, J.: Optimal approximations by piecewise smooth functions and associated variational problem. Comm. Pure and Appl. Math **42**, 577–685 (1989)
5. Sethian, J.: Theory, algorithm and applications of level set method for propagating interface. Acta Numerica **5**, 309–395 (1996)

A Single-Pass Scheme for the Mean Curvature Motion of Convex Curves

E. Carlini

Abstract A new algorithm to solve the stationary level set equation describing the Mean Curvature Motion of convex curve is presented. This algorithm is a single pass scheme obtained by coupling the min-max type discrete operator for the MCM equation with the fast marching level set for the Eikonal equation. An adaption of the step, approximating the characteristics, is required to obtain an accurate method. Here an adaptive strategy is proposed.

1 Introduction

The scope of the paper is to develop a fast scheme for the stationary equation modelling the Mean Curvature Motion (MCM):

$$\begin{cases} \operatorname{div}\left(\frac{DT(x)}{|DT(x)|}\right)|DT(x)|+1=0 & x\in\Omega \\ T(x)=0 & x\in\partial\Omega, \end{cases} \tag{1}$$

where Ω is a bounded convex set of $\mathbf{R}^2$. The level set $\Gamma_t=\{x\in\mathbf{R}^2\ :\ T(x)=t\}$ represents a curve propagating by the mean curvature flow, see [4].
The present framework follows a single step scheme, known as the Fast Marching Method (FMM), proposed by Tsitsiklis and Sethian in [8, 6], for first order Eikonal equations modelling front propagation. Starting from these pioneering papers there has been quite a big effort to generalize the method to handle more complicated motions. Interesting extensions have been obtained for positive time-dependent anisotropic speed, see [7], and for time-dependent changing sign speed, see [2]. But so far, to our knowledge, no fast marching scheme has been developed for the mean curvature flow. The explanation is quite simple, to do that one has to

Dipartimento di Matematica, Università Sapienza di Roma, P.le Aldo Moro 2, Roma, e-mail: carlini@mat.uniroma1.it

face the degeneracy of the second order operator, which implies that the generalized characteristics are orthogonal to the gradient. This is the main difficulty of developing a single-pass scheme for (1).
In the paper [5], the authors look at (1) as a deterministic control game and they prove that the minimum exit time T^ε satisfies the following dynamic programming principle

$$T^\varepsilon(x) = \min_{a\in S^1}\max\left\{T^\varepsilon\left(x-\sqrt{2}\,\varepsilon a\right), T^\varepsilon\left(x+\sqrt{2}\,\varepsilon a\right)\right\}+\varepsilon^2, \tag{2}$$

with $S^1 = \{a\in \mathbf{R}^2, |a|=1\}$, they also prove that T^ε is uniformly approximating the solution of (1), when Ω is strictly convex bounded and smooth.
In [3], we have analyzed the fully discrete version of (2), developing an iterative min-max type scheme. It is then natural to develop a method that can solve the min-max scheme in a single pass. The key idea is to build the FMM for the Eikonal equation such that the upwind difference structure of the scheme, used to solve the equation, makes the information propagating one way, namely the direction orthogonal to the curve. We borrow this idea, even if the present situation is rather more complicated, since the informations are propagating in direction tangential to the curve. This means, that only in a very few cases one can choose a feasible step Δ to make the scheme work for all the evolutions. In more general cases we have to adapt the step Δ.
In Sect. 2 we present the scheme for Δ fixed, in Sect. 3 the adapted version and finally in Sect. 4 some numerical tests.

The presentation of the algorithm and the numerical tests are in two dimensions for simplicity of the presentation, but the scheme can be generalised to any number of dimensions.

2 The Algorithm

We will describe the algorithm to built the single-pass scheme approximating (1) modelling a convex domain in $\mathbf{R}^2$ collapsing by MCM.
We introduce a space step Δx and consider a uniform lattice $L=\{(x_{i,j})=(i\Delta x, j\Delta x), (i,j)\in \mathbf{Z}^2\}$.
We extend the solution at zero outside the domain Ω and we define χ as the following discrete function :

$$\chi_{i,j} = \begin{cases} 1 & \text{if } x_{i,j}\in \mathbf{Z}^2\setminus\Omega \\ -1 & \text{if } x_{i,j}\in\Omega. \end{cases} \tag{3}$$

The nodes (i,j) such that $\chi_{i,j}=1$ are points where the solution is known, this means that its value will not be computed. We call the set $\{x\in\mathbf{R}^2 : I[\chi](x)=1\}$ the *accepted domain*, where $I(\cdot)$ denotes the linear interpolation of a discrete function defined on the lattice L.
The idea of the scheme is the same as the standard FMM : the curve is propagated

considering a narrow band around the front containing the candidate times. Then as in the standard FMM we call the neighborhood of the node $(i,j) \in \mathbf{Z}^2$ the set

$$V(i,j) \equiv \{(k,m) \in \mathbf{Z}^2 : |(k,m)-(i,j)| \leq 1\}$$

and Narrow Band the following set:

$$NB(\chi) \equiv \{(i,j) \in \mathbf{Z}^2 : \chi_{i,j} = -1 \text{ and } \chi_{k,m} = 1 \ \exists (k,m) \in V(i,j)\}.$$

The values are computed using the operator described in [3]:

$$T^{n+1}_{i,j} = \min_{a \in S^1} \max_{b=-1,1} H_{i,j}[T^n](a,b) \tag{4}$$

with

$$H_{i,j}[T^n] = \begin{cases} I[T^n](x_{i,j} + \sqrt{2\Delta}ab) + \Delta & \text{if } x_{i,j} + \sqrt{2\Delta}ab \in \Omega \\ \tilde{\Delta} & \text{else} \end{cases} \tag{5}$$

where $\tilde{\Delta}$ is such that $x_{i,j} + \sqrt{2\tilde{\Delta}}ab \in \partial\Omega$.

At each iteration of the algorithm the nodes with the minimun value are removed from the Narrow Band and moved to the accepted points (then their values will not be recomputed again).

Initialization

1. *Initialization of the matrix χ and of the time T :*
   $\chi^0_{i,j} = \chi_{i,j} \ \ (i,j) \in \mathbf{Z}^2$
   $$T^0_{i,j} = \begin{cases} 0 & \text{if } x_{i,j} \in \mathbf{Z}^2 \setminus \Omega \\ \infty & \text{if } x_{i,j} \in \Omega \end{cases}$$

2. *Initialization of the values T on the Narrow Band:*
   compute $T^1_{i,j}$ using (4) on each $(i,j) \in NB(\chi^0)$

3. $n = 1$

Main cycle

4. $\hat{T}^n = \min\left\{T^n_{i,j},\ (i,j) \in NB(\chi^{n-1})\right\}$

5. *Definition of the new Accepted Points:* $NA^n = \{(i,j) \in NB(\chi^{n-1}) \ : \ T^n_{i,j} = \hat{T}^n\}$

6. *Update of χ^n:* $\chi^n_{i,j} = 1 \quad \text{if } (i,j) \in NA^n$

7. if $NB(\chi^n)$ is empty stop

8. compute $T^{n+1}_{i,j}$ using (4), for all $(i,j) \in V(l,m)$ such that $\chi^n_{l,m} = -1$ and $(l,m) \in NA^n$

9. set $T^{n+1} = T^n, \quad \chi^{n+1} = \chi^n$ and go to step 4

Boundary conditions
To avoid spurious oscillations coming from the use of interpolation of points near the boundary, we do not perform interpolation in a narrow band close to the boundary and we use special boundary conditions as it has been described in [3].
Complexity
Let us suppose M the number of nodes inside the domain Ω, considering that at each stage at least one node is accepted, the algorithm terminates at the most after M iterations. Let us assume that M is a constant and let us consider the cost of computing $T^n_{i,j}$. The optimization problem can be solved by several methods, but the cost is always independent of the dimension M, thus we can estimate it to be $O(1)$. Since at each iteration only the values of nodes in the neighbourhood of the accepted points are recomputed and once a point is accepted its value is not computed again, the cost of each stage is given by the minimization in step 4. We can implement the algorithm using an binary heap and then obtain each iteration in $O(logM)$ time. Finally the total complexity is estimate by $O(MlogM)$.
How to choose Δ
This is an important point: to make the algorithm work we have to guarantee that the characteristics directions a^*, solving the min-max operator in (4), are tracking information from the *accepted domain*.
We have proved in [3] that the consistency error for (4) using a linear interpolation is given by

$$L_{\Delta,\Delta x} = O(\Delta^{1/2}) + O(\Delta x^2/\Delta), \tag{6}$$

and the scheme results monotone independently of the choice of the ratio between step Δ and Δx. In principle we are allowed to choose even a very large Δ and still obtain convergence. This observation is what makes our algorithm work, since we can use a step Δ large enough, to be able to reach the part of the domain where the solution has been computed, and still be consistent.
Now, how to choose such a Δ? The easiest case is when the curvature is nearly the same for every point of the curve. In this case one can fix a Δ satisfying the requirement at the beginning of the algorithm and keep the same step for all the iterations. See the numerical test 1 in the last section.
For more general curves, the space Δ have to be correctly chosen, depending on the curve. This observation motivates the next section.

3 The Adaptive Algorithm

To be able to treat more general curves, i.e. non smooth and non strictly convex curves, we need to adapt the step Δ such that the scheme is always tracking the information in an *up-wind* fashion.
This is an important issue for accuracy reasons, as it has been remarked in [1]. The

technique we use here is quite different. Points on the curve with higher curvature need smaller steps to maintain accuracy, on the contrary one is allowed to increase the step Δ on points with low curvature, the latter case corresponds to the situation where the points need larger Δ to reach the *accepted domain*. We do not need error estimators to decide when to modify Δ, instead we check if the difference between the candidate time $\hat{T}^n_{i,j}$ and highest value accepted is greater than a fixed threshold. If so, we need to increase Δ.
We only allow the step Δ to increase, since once the points with higher curvature have been smoothed, these type of points will not appear anymore. This is a consequence of the fact, that convex domains become asymptotically circular and shrink to a point under the MCM.
The difference between the adapted algorithm and the previous is in the main cycle: the below step 4*b* should be added to the previous algorithm after step 4 and the new variable Min should be initialized at zero in the Initialization steps.

4b. *Adaption step* Δ

1. while $\hat{T}^n - Min > c_1 \Delta x$ do:

   a. $\sqrt{\Delta} = c_2 \sqrt{\Delta}$

   b. Compute $T^n_{i,j}$ using (4) on each $(i,j) \in NB(\chi^{n-1})$

   c. $\hat{T}^n = \min\left\{ T^n_{i,j},\ (i,j) \in NB(\chi^{n-1}) \right\}$

2. if $\hat{T}^n > Min$ then $Min = \hat{T}^n$

How to choose c_1, c_2*?*
We do not have a satisfactory answer to do this, obviously we want $c_1 > 0$ and $c_2 > 1$; in the experiments we have chosen $c_1 = 1$ and $c_2 = 1.5$.
Adaptive phase
The new step 4b is devoted to adapt the step if necessary. The necessary situations are when the selected candidate time has an increase too large with respect to the highest accepted value. This is the situation, when the characteristics have not reached the *accepted domain*. We then need to lengthen the step and recompute the candidate time on each node of the Narrow Band.
Since we are changing the step Δ it could happen, that the sequence of time $\hat{T}^n$ is not increasing. Therefore we need to update the value Min only if it is smaller than the last accepted value. Thereafter the condition to adapt Δ is checked on the greatest value accepted.
Complexity
This version of the algorithm is more expensive than the previous, since we need to recompute the solution on each node of the Narrow Band at the iterations where we need to change the step Δ. Supposing that this step is required only a constant number of times we again obtain the solution in $O(M \log M)$ operations.

4 Numerical Tests

We investigate the performance of the scheme in three typical situations: a smooth curve for which we know the exact solution and therefore can compute the numerical errors, a smooth curve and a non smooth and non strictly convex curve for which an adaptation of the step Δ is necessary to obtain an accurate approximation.

Collapse of a Circle
We consider as initial set Ω a circle of ray one centered in the origin of the numerical domain $[-1.5, 1.5] \times [-1.5, 1.5]$. We can compute the errors, comparing the approximated solution with the exact solution $T(x,y) = (1 - x^2 - y^2)/2$.
Fig.1 shows the approximated solution $(T^n_{i,j})_{i,j}$ obtained with $\Delta x = 0.015$ and its

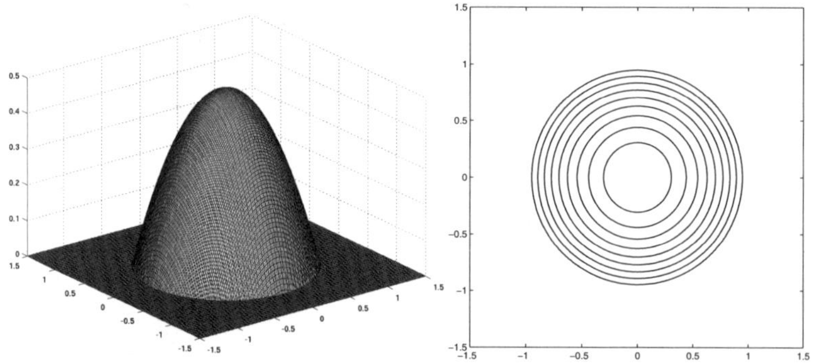

Fig. 1 Numerical solution T^n (on the left) and its level sets (on the right)

Table 1 L_∞ errors and rate for Test 1

Δx	Δ	E_∞	rate
0.12	0.1	2.33×10^{-2}	
0.06	0.05	1.23×10^{-2}	0.92
0.03	0.025	6.96×10^{-3}	0.82
0.015	0.0125	3.56×10^{-3}	0.96

level set Γ_{t_m} with $t_m = 0.05 * m$ and $m = 1, ..., 9$.
To compute the numerical solution T^n we have used the algorithm described in Sect. 2. The errors are computed in the discrete L_∞ norm:

$$E_\infty = \max_{i,j} |T(x_i, y_j) - T^n_{i,j}|.$$

In Table 1, we represent the errors computed running the code with the space step and the step Δ satisfying the ratio relation $\Delta x/\Delta = 1.2$. The obtained numerical order of convergence is always higher than the theoretical 1/2 given by (6).

Collapse of a Square
We consider as initial set Ω a square with side 2. This case is definitely more difficult than the previous one. The starting curve is not strictly convex and there are points where the curvature is infinite: the corners, and where it is zero: the sides of the square.
This is the typical case where we need to adapt the step Δ.
In Fig. 2 we compare the level set $0, 0.01, 0.05$ and $0.1 * m$ with $m = 1, \ldots 6$ of T^n obtained by the algorithm given in Sect.2 with the algorithm given in Sect.3. We have used $\Delta x = 0.03$ and $\Delta = \Delta x$ for the non adapted scheme and a starting step $\Delta = (2 * \Delta x)^2$ for the adapted scheme. The picture clearly show the advantage using the second one: the corners are the first point collapsing, these points are solved with very small Δ and good accuracy is achieved.

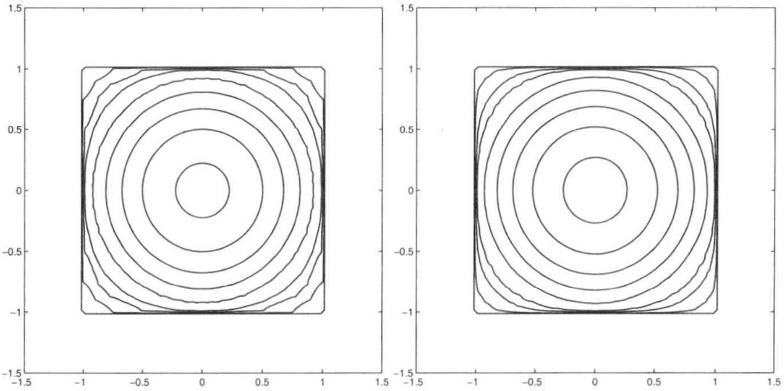

Fig. 2 Contour levels for test 2 obtained with algorithm of Sect.2 (on the left) and with algorithm of Sect.3 (on the right)

Collapse of an Ellipse
We consider as initial set Ω an ellipse with axis $a = \sqrt{0.5}$ and $b = \sqrt{2}$ centered in the origin . This is another case where we need to adapt the step Δ.
In Fig. 3 we show the level set Γ_{t_m} with $t_m = 0.1 * m$ and $m = 1, .., 5$ of T^n obtained by the adaptive algorithm in Sect.3. We have used $\Delta x = 0.025$ and a starting step $\Delta = (2 * \Delta x)^2$.

Conclusions and Observations

We have proposed a single-pass scheme for convex curves collapsing by the MCM and an adaption of the method to handle non strictly convex and non smooth curves. Here, we have not investigated an efficient method to evaluate the min-max operator,

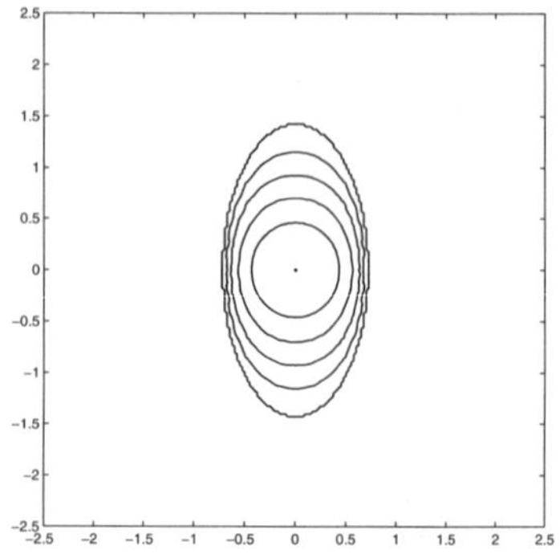

Fig. 3 Contour levels for test 3 obtained with algorithm of Sect.3

we are aware that to gain the denomination of fast we need to be able to propose a method that can solve the min-max operator with very few iterations. We defer the study of this problem as well of convergence and its generalization to non convex curves to a future paper.

Acknowledgements The author would like to thank Prof.R.Ferretti for fruitful discussions and valuable advise.

References

1. Carlini, E., Falcone, M., Ferretti, R.: A time–adaptive semi–lagrangian approximation to mean curvature motion. In: A. Bermúdez de Castro, D. Gómez, P. Quintela, P. Salgado (eds.) Numerical Mathematics and Advanced Applications , ENUMATH 2005, pp. 724–731. Springer, Berlin (2006)
2. Carlini, E., Falcone, M., Forcadel, N., Monneau, R.: Convergence of a generalized fast marching method for a non-convex eikonal equation. To appear in SIAM J.Num.Anal. (2008)
3. Carlini, E., Ferretti, R.: A semi-lagrangian approximation of min-max type for the stationary mean curvature equation. To appear in Proceeding ENUMATH 2007 (2007)
4. Evans, L.C., Spruck, J.: Motion of level sets by mean curvature. I. J. Differential Geom. **33**(3), 635–681 (1991)
5. Robert, K., Serfaty, S.: A deterministic-control based approach to motion by curvature. Comm. Pure Appl. Math **59**(59), 344–407 (2006)
6. Sethian, J.A.: Level Set Methods and Fast Marching Methods: Evolving Interfaces in Computational Geometry, Fluid Mechanics, Computer Vision and Materials Science. Cambridge University Press, Cambridge (1999)
7. Sethian, J.A., Vladimirsky, A.: Ordered upwind methods for static Hamilton-Jacobi equations: theory and algorithms. SIAM J. Numer. Anal. **41**(1), 325–363 (electronic) (2003)
8. Tsitsiklis, J.N.: Efficient algorithms for globally optimal trajectories. IEEE Trans. Automat. Control **40**(9), 1528–1538 (1995)

A Semi-Lagrangian Approximation of Min–Max Type for the Stationary Mean Curvature Equation

E. Carlini and R. Ferretti

Abstract We propose a technique to treat degenerate elliptic equations, focusing on the model problem of the stationary Mean Curvature Motion equation in two space dimensions. This technique may be interpreted as a stationary, fully discrete version of the schemes proposed in slightly different forms by Catté et al. and by Kohn and Serfaty. We study consistency and monotonicity of the scheme and a correct implementation of Dirichlet boundary conditions. Numerical tests are also presented.

1 Introduction

Given a convex open domain $\Omega \subset \mathbb{R}^2$, we consider the stationary Partial Differential Equation (PDE)

$$\begin{cases} \operatorname{div}\left(\frac{DT(x)}{|DT(x)|}\right)|DT(x)|+1=0 & x\in\Omega, \\ T(x)=0 & x\in\partial\Omega. \end{cases} \quad (1)$$

It has been shown in [4] that the evolutive PDE

$$\begin{cases} v_t(x,t)=\operatorname{div}\left(\frac{Dv(x,t)}{|Dv(x,t)|}\right)|Dv(x,t)| & (x,t)\in\mathbb{R}^2\times[0,\infty). \\ v(x,0)=v_0(x), & x\in\mathbb{R}^2. \end{cases} \quad (2)$$

E. Carlini
Dipartimento di Matematica, Universitá di Roma "La Sapienza", Roma, Italy,
e-mail: carlini@mat.uniroma1.it

R. Ferretti
Dipartimento di Matematica, Universitá di Roma Tre, Roma, Italy
e-mail: ferretti@mat.uniroma3.it

is related to (1) since $v(x,t) = T(x) - t$ for $x \in \Omega$ and $t > 0$ when $v(x,0) = T(x)$. In general $v_0(x)$ can be any continuous function such that $v_0(x) \equiv 0$ if $x \in \partial\Omega$, and it is positive outside Ω and negative inside. It is well known that (2) describes the Mean Curvature Motion (MCM) of the initial curve $\Gamma_0 = \partial\Omega$ in the framework of level set techniques. More precisely, the (generalized) MCM of Γ_0 at time t, denoted by Γ_t, is the zero level set of the solution $v(x,t)$. In some sense, solving (2) gives a redundant information since *any level set* of v moves by Mean Curvature. On the other hand, if Ω is convex, by solving (1) instead of (2), one can compute the Mean Curvature flow of Γ_0 until it collapses by

$$\Gamma_t = \left\{ x \in \mathbb{R}^2 : T(x) = t \right\}. \tag{3}$$

In other words, the function $T(x)$ represents the time at which the point x is reached by the front Γ_t.

A number of approximation schemes has been proposed for (2) and (1), but we will focus here on variable stencil (or large time-step) techniques. In addition to the space discretization step, in this class of schemes a second parameter adjusts the diameter of the stencil in the difference formula (this diameter is typically $O(\sqrt{\Delta t})$ in time-dependent situations). In order to have a consistent scheme, the stencil must be oriented orthogonally with respect to the gradient of the solution.

As fare as we are aware of, the first scheme in this streamline has been proposed by Catté, Dibos and Koepfler [3] for the case of (2). In [3] the correct orientation of the stencil is the result of a min–max operation. More recently, Falcone and Ferretti [5] have proposed a similar scheme for (2) in which the stencil is oriented according to some finite-difference estimate of the gradient.

In the stationary case (1), a recent paper by Kohn and Serfaty (see [6]) defines a semi-discrete approximation of $T(x)$ via a differential games approach by means of the following semi-continuous function:

$$T^\varepsilon(x) = \begin{cases} 0 & \text{if } x \in \mathbb{R}^2 \backslash \Omega \\ \varepsilon^2 & \text{if } \exists a \in S^1 : x \pm \sqrt{2}\,\varepsilon a \in \mathbb{R}^2 \backslash \Omega,\, x \in \Omega \\ k\varepsilon^2 & \text{if } \exists a \in S^1 : \max\left(T^\varepsilon\left(x + \sqrt{2}\,\varepsilon a\right), T^\varepsilon\left(x - \sqrt{2}\,\varepsilon a\right)\right) = (k-1)\varepsilon^2, \\ & x \in \Omega,\, \forall k \geq 1, \end{cases} \tag{4}$$

which in turn satisfies the dynamic programming principle:

$$T^\varepsilon(x) = \min_{a \in S^1} \max\left\{ T^\varepsilon\left(x - \sqrt{2}\,\varepsilon a\right), T^\varepsilon\left(x + \sqrt{2}\,\varepsilon a\right) \right\} + \varepsilon^2, \tag{5}$$

where $S^1 = \{a \in \mathbb{R}^2, |a| = 1\}$. They prove that, under the assumption that Ω is smooth, bounded and strictly convex, T^ε is an approximation of the solution of (1):

$$\|T^\varepsilon(x) - T(x)\|_{L^\infty(\Omega)} \leq C\varepsilon \tag{6}$$

where C depends only on the C^3 norm of T; in the paper they also prove that for convex domains, T is actually C^3.

The approximation scheme proposed here stems from this approach, but in obtaining a fully discrete scheme we will rather use continuous reconstructions, which may take advantage of the smoothness of the solution.

The outline of the paper is the following. In the next section we will present the construction of the scheme. Next the issues of consistency and monotonicity will be addressed, and in the last section we will present some numerical tests.

2 The Scheme

Using $\Delta = \sqrt{\varepsilon}$ in (5) as a first discretization parameter, but still keeping the variable x continuous (this would correspond to a time-discretization of (2) with time step $\Delta t = \Delta$), we start by giving a semi-discrete version of (1):

$$\begin{cases} U^\Delta(x) = \min_{a\in S^1} \max\left\{U^\Delta\left(x-\sqrt{2\Delta}\,a\right), U^\Delta\left(x+\sqrt{2\Delta}\,a\right)\right\} + \Delta & x\in\Omega \\ U^\Delta(x) = 0 & x\in\partial\Omega. \end{cases} \tag{7}$$

Then, we introduce an orthogonal space grid of step Δx: $\mathscr{G}_{\Delta x} = \{x_j = j\Delta x,\ j\in\mathbb{Z}^2\}$. We define by $\Omega^\Delta = \{j : x_j \in \Omega\cap\mathscr{G}_{\Delta x}\}$, denote by $\mathbf{u} = (u_j)_{j\in\mathbb{Z}^2}$, where u_j is the numerical approximation of $T(x)$at the node $x_j\in\Omega^\Delta$ and $u_j = 0$ if $j\in\mathbb{Z}^2\backslash\Omega^\Delta$. We also denote by $I[\mathbf{u}](x)$ an interpolation of $\mathbf{u}$ computed at $x\in\mathbb{R}^2$. In the sequel, we will assume that this interpolation is monotone (e.g. piecewise bilinear).
Since we have extended by zero the discrete approximation $\mathbf{u}$ outside Ω, performing a plain interpolation at points close to the boundary of Ω could lead to an undesired numerical dissipation. For this reason we define:

$$\begin{cases} \Delta_j(a) = \max\left\{h : 0\le h\le\Delta : x_j+\sqrt{2h}\,a\in\Omega\right\} & j\in\Omega^\Delta, a\in S^1 \\ \Delta_j^* = \min_{a\in S^1}\max\left\{\Delta_j(a),\Delta_j(-a)\right\} & j\in\Omega^\Delta, \end{cases} \tag{8}$$

and give the following "boundary conditions" at nodes close to $\partial\Omega$ where $\Delta_j^* < \Delta$:

$$\begin{cases} u_j = \Delta_j^* & \Delta_j^* < \Delta,\ j\in\Omega^\Delta \\ u_j = 0 & \mathbb{Z}^2\backslash\Omega^\Delta. \end{cases} \tag{9}$$

Note that Δ is adjusted in order to bring the points $x\pm\sqrt{2\Delta_j^*}\,a$ precisely on the boundary. This makes it useless to perform an interpolation since the boundary value is known.

Projecting the semi-discrete approximation (7) on the grid $\mathscr{G}_{\Delta x}$ and performing an interpolation whenever the points $x_j\pm\sqrt{2\Delta}\,a$ are not grid nodes, we obtain:

$$u_j = \min_{a\in S^1}\max\left\{I[\mathbf{u}]\left(x_j-\sqrt{2\Delta}\,a\right), I[\mathbf{u}]\left(x_j+\sqrt{2\Delta}\,a\right)\right\} + \Delta \qquad j\in\Omega^\Delta. \tag{10}$$

We solve (10) by the fixed point iteration:

$$u_j^{n+1} = \min_{a\in S^1}\max\left\{I[\mathbf{u}^n]\left(x_j-\sqrt{2\Delta}\,a\right), I[\mathbf{u}^n]\left(x_j+\sqrt{2\Delta}\,a\right)\right\}+\Delta \qquad j\in\Omega^\Delta. \tag{11}$$

As we will remark later, this amounts to seeking the solution of the stationary problem as the limit solution of a suitable evolutive problem. This allows to give an upper bound on the number of iterations.

Denoting the scheme (11) by the compact notation $u_j^{n+1} = S_j(\mathbf{u}^n)$, and using (8) into (11), we obtain

$$\begin{cases} u_j^{n+1} = S_j(\mathbf{u}^n) & \Delta_j^* > \Delta,\ j\in\Omega^\Delta \\ u_j = \Delta_j^* & \Delta_j^* \le \Delta,\ j\in\Omega^\Delta \\ u_j = 0 & j\in\mathbb{Z}^2\backslash\Omega^\Delta \end{cases} \tag{12}$$

where Δ_j^* is defined in (8) for any $j\in\Omega^\Delta$.

3 Convergence of the Scheme

We split this section in two subsections dealing with respectively consistency and monotonicity. Convergence of the scheme then follows by the Barles – Souganidis theorem (see [1]). In fact, (12) may be interpreted as a scheme for the evolutive equation

$$v_t = \mathrm{div}\left(\frac{Dv}{|Dv|}\right)|Dv|+1 \tag{13}$$

in which we are interested in the solution up to the extinction time of the curve (this is a consequence of the comparison principle, see [4]).

3.1 Consistency

First, we note that, as it has been proved in [7], we can write

$$F(Du,D^2u) = \mathrm{div}\left(\frac{Du(x)}{|Du(x)|}\right)|Du(x)| = \min_{a\in S^1, a\cdot Du=0}\left\{a^T D^2u(x)a\right\}, \tag{14}$$

and that in turn, the right-hand side of (14) can be rewritten as

$$\begin{aligned}\min_{a\in S^1, a\cdot Du=0}\left\{a^T D^2u(x)a\right\} &= \min_{a\in S^1}\max\left\{a^T D^2u(x)a - a\cdot Du, a^T D^2u(x)a + a\cdot Du\right\} \\ &= \tilde{a}(x)^T D^2u(x)\tilde{a}(x),\end{aligned} \tag{15}$$

where we have denoted by $\tilde{a}(x)$ the minimizer (note that since $a \in S^1$, the choice of the sign is irrelevant). Here and in the sequel, we make use of the fact that if two functions $f_1(a)$ and $f_2(a)$ depend continuously on a and span the same set, their max is minimized when they attain the same value. We assume that the interpolation error of $I[\cdot]$ is $O(\Delta x^r)$ on smooth functions; a typical situation in which interpolation is also monotone (as required in the following subsection) is $\mathbb{P}_1$ or $\mathbb{Q}_1$ interpolation for which $r = 2$.

Let $\mathbf{w}$ be a vector containing the samples of a smooth solution, so that $w_j = u(x_j)$. Once written the scheme as $u_j = S_j(\mathbf{w})$, the consistency error $L_{\Delta x,\Delta}$ of the scheme is defined by the equality

$$u(x_j) = S_j(\mathbf{w}) + \Delta\, L_{\Delta x,\Delta}(x_j). \tag{16}$$

Now, at internal nodes S_j is defined by:

$$S_j(\mathbf{w}) = \min_{a\in S^1} \max\left\{ I[\mathbf{w}]\left(x_j + \sqrt{2\Delta}\, a\right), I[\mathbf{w}]\left(x_j - \sqrt{2\Delta}\, a\right)\right\} + \Delta \tag{17}$$

and therefore, denoting by $\bar{a}_j$ the minimizer in (17),

$$S_j(\mathbf{w}) = I[\mathbf{w}]\left(x_j + \sqrt{2\Delta}\, \bar{a}_j\right) + \Delta. \tag{18}$$

The values within the max can be expressed as

$$\begin{aligned} I[\mathbf{w}]\left(x_j \pm \sqrt{2\Delta}\, a\right) &= u\left(x_j \pm \sqrt{2\Delta}\, a\right) + O(\Delta x^r) \\ &= u(x_j) \pm \sqrt{2\Delta}\, Du(x_j)\cdot a + \Delta a^T D^2 u(x_j) a \\ &\quad + O\left(\Delta^{3/2}\right) + O(\Delta x^r). \end{aligned} \tag{19}$$

The max of $I[\mathbf{w}](x_j \pm \sqrt{2\Delta}\, a)$ is minimized when they coincide, so that

$$\begin{aligned} &u(x_j) + \sqrt{2\Delta}\, Du(x_j)\cdot \bar{a}_j + \Delta \bar{a}_j^T D^2 u(x_j)\bar{a}_j + O\left(\Delta^{3/2}\right) + O(\Delta x^r) \\ &= u(x_j) - \sqrt{2\Delta}\, Du(x_j)\cdot \bar{a}_j + \Delta \bar{a}_j^T D^2 u(x_j)\bar{a}_j + O\left(\Delta^{3/2}\right) + O(\Delta x^r) \end{aligned} \tag{20}$$

and therefore

$$\sqrt{2\Delta}\, Du(x_j)\cdot \bar{a}_j = O\left(\Delta^{3/2}\right) + O(\Delta x^r). \tag{21}$$

Note that (21) shows that the min–max operation basically selects the direction orthogonal to the gradient. Using (18), (19) and (21), (17) gives

$$S_j(\mathbf{w}) = u(x_j) + \Delta \bar{a}_j^T D^2 u(x_j)\bar{a}_j + O\left(\Delta^{3/2}\right) + O(\Delta x^r) + \Delta. \tag{22}$$

Using now (14), (15) and (21), we obtain

$$
\begin{aligned}
F\left(Du(x_j),D^2u(x_j)\right) &= \tilde{a}(x_j)^T D^2u(x_j)\tilde{a}(x_j) \\
&\leq \max\left\{\bar{a}_j^T D^2u(x_j)\bar{a}_j - \bar{a}_j\cdot Du(x_j), \bar{a}_j^T D^2u(x_j)\bar{a}_j + \bar{a}_j\cdot Du(x_j)\right\} \\
&= \bar{a}_j^T D^2u(x_j)\bar{a}_j + O(\Delta) + O\left(\frac{\Delta x^r}{\Delta^{1/2}}\right). \qquad (23)
\end{aligned}
$$

Neglecting irrelevant error terms, equations (22) and (23) imply

$$
u(x_j) - S_j(\mathbf{w}) + \Delta\left(F\left(Du(x_j),D^2u(x_j)\right)+1\right) \leq O\left(\Delta^{3/2}\right) + O(\Delta x^r), \qquad (24)
$$

and taking into account that $u(x)$ satisfies (1), we can bound the truncation error from one side as

$$
L_{\Delta x,\Delta}(x_j) \leq O\left(\Delta^{1/2}\right) + O\left(\frac{\Delta x^r}{\Delta}\right). \qquad (25)
$$

By interchanging the roles of the minimizers $\tilde{a}$ and $\bar{a}$ we can obtain the reverse inequality, so that at last

$$
|L_{\Delta x,\Delta}(x_j)| \leq O(\Delta^{1/2}) + O\left(\frac{\Delta x^r}{\Delta}\right). \qquad (26)
$$

Lastly, note that this estimate has been obtained for internal points, but it also holds a fortiori at nodes where $\Delta_j^* < \Delta$, since in this case the term related to interpolation disappears.

3.2 Monotonicity

We start from two vectors $\mathbf{v}$, $\mathbf{w}$ such that $\mathbf{v} \geq \mathbf{w}$ and prove that $S(\mathbf{v}) \geq S(\mathbf{w})$ (the inequalities being intended component by component). Denoting by a_j^* the direction achieving the min max in $S_j(\mathbf{v})$, we can write $S_j(\mathbf{v})$ as

$$
\begin{aligned}
S_j(\mathbf{v}) &= \min_{a\in S^1}\left\{\max\left(I[\mathbf{v}]\left(x_j+\sqrt{2\Delta}\,a\right), I[\mathbf{v}]\left(x_j-\sqrt{2\Delta}\,a\right)\right)\right\} + \Delta \\
&= I[\mathbf{v}]\left(x_j+\sqrt{2\Delta}\,a_j^*\right) + \Delta. \qquad (27)
\end{aligned}
$$

An analogous computation for $\mathbf{w}$ gives

$$
\begin{aligned}
S_j(\mathbf{w}) &= \min_{a\in S^1}\left\{\max\left(I[\mathbf{w}]\left(x_j+\sqrt{2\Delta}\,a\right), I[\mathbf{w}]\left(x_j-\sqrt{2\Delta}\,a\right)\right)\right\} + \Delta \\
&= I[\mathbf{w}]\left(x_j+\sqrt{2\Delta}\,\bar{a}_j\right) + \Delta \\
&\leq \max\left\{I[\mathbf{w}]\left(x_j-\sqrt{2\Delta}\,a_j^*\right), I[\mathbf{w}]\left(x_j+\sqrt{2\Delta}\,a_j^*\right)\right\} + \Delta \\
&= I[\mathbf{w}]\left(x_j+\sqrt{2\Delta}\,a_j^*\right) + \Delta, \qquad (28)
\end{aligned}
$$

in which the inequality is implied by the optimality of $\bar{a}_j$ in $S_j(\mathbf{w})$, and the sign is arbitrarily chosen since both a^* and $-a^*$ are minimizers. Taking into account that $I[\mathbf{v}](x) \geq I[\mathbf{w}](x)$ by the monotonicity of $I[\cdot]$, we finally obtain

$$S_j(\mathbf{w}) \leq I[\mathbf{w}]\left(x_j + \sqrt{2\Delta}\, a_j^*\right) + \Delta \leq I[\mathbf{v}]\left(x_j + \sqrt{2\Delta}\, a_j^*\right) + \Delta = S_j(\mathbf{v}). \tag{29}$$

Again, it is easy to check that monotonicity also holds at nodes close to the boundary, at which $\Delta_j^* < \Delta$.

4 Numerical Tests

We show in this section the behaviour of the scheme on two classical tests of curve propagation, the circle and the square. In the first test the exact solution is smooth and its explicit expression is known so that numerical errors can be computed, in the second the initial curve is nonsmooth and we are interested in capturing the geometry of solutions near the corners. The exact collapse time is $T^* = 0.5$ for the first test and $0.5 < T^* < \sqrt{2}/2$ for the second test. Multiplying Δ by the number of iterations, we get a time of the same order of magnitude as T^*.

Test 1. In the case the initial curve is a circle with radius R centered in the origin, and the exact solution is $T(x) = (R^2 - |x|^2)/2$.
We compute the errors in the discrete norm,

$$|E|_{L^\infty} = \max_{j \in \Omega^\Delta} |u_j^n - T(x_j)|, \tag{30}$$

for a circle with radius $R = 1$. The results are shown in Table 1; for each test we report the space step Δx, the step Δ, the error, the number of iteration needed to reach the fixed point and the rate of convergence. The test shows a good performance with

Table 1 Errors and convergence rates for test 1

Δx	Δ	$\lvert E\rvert_{L^\infty}$	iterations	rate
0.12	0.5	$2.15 \cdot 10^{-2}$	2	
0.06	0.1	$1.30 \cdot 10^{-2}$	8	0.7
0.03	0.05	$7.08 \cdot 10^{-3}$	14	0.9
0.015	0.025	$3.11 \cdot 10^{-3}$	25	1.2

a CFL relationship $\Delta t/\Delta x = 1.7$ (typical of first order PDEs) and the convergence order goes beyond the theoretical value of $1/2$. This is partly due to the smoothness of the solution, as well as to the fact that in this case the min–max operator selects a direction *exactly* orthogonal to the gradient, so that the first (and lower order) term in the consistency error would disappear. In Fig. 1 (left) we plot the contour levels of $T(x)$ at 0, 0.1, 0.2, 0.3, 0.4, 0.5 representing the collapse of the circle.

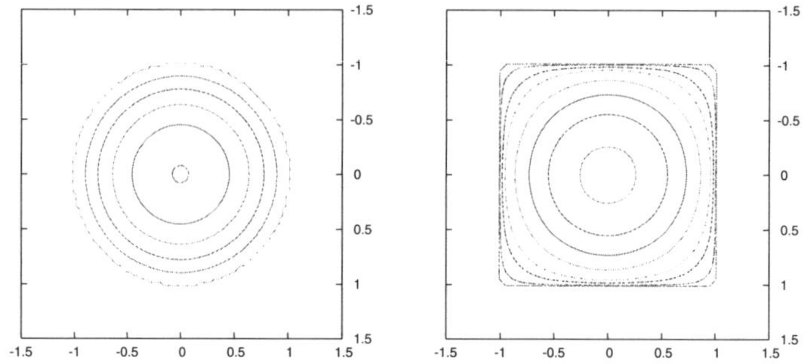

Fig. 1 Level curves of $T(x)$ at 0, 0.1, 0.2, 0.3, 0.4, 0.5 for the circle (left) and at 0, 0.01, 0.05, 0.1, 0.2, 0.3, 0.4, 0.5 for the square (right)

Test 2. In this test we follow the evolution of a nonsmooth initial boundary (a square). Here a parabolic CFL condition is generally required to follow properly the evolution of the corners. The ability of the scheme in resolving small structures is discussed in [2], where we also propose an adaptive (with respect to Δ) technique to overcome the problem. We show in Fig. 1 (right) the behaviour of the scheme on a suitable set of level curves, with $\Delta x = 0.06$ and $\Delta t = 0.0036$, i.e. with a parabolic relationship. It has taken 208 iterations for the scheme to reach the fixed point.

References

1. Barles, G., Souganidis, P.E.: Convergence of approximation schemes for fully nonlinear second order equations. Asymptotic Anal. **4**(3), 271–283 (1991)
2. Carlini, E., Falcone, M., Ferretti, R.: A time-adaptive semi-Lagrangian approximation to mean curvature motion. In: Numerical mathematics and advanced applications, pp. 732–739. Springer, Berlin (2006)
3. Catté, F., Dibos, F., Koepfler, G.: A morphological scheme for mean curvature motion and applications to anisotropic diffusion and motion of level sets. SIAM J. Numer. Anal. **32**(6), 1895–1909 (1995)
4. Evans, L.C., Spruck, J.: Motion of level sets by mean curvature. I. J. Differential Geom. **33**(3), 635–681 (1991)
5. Falcone, M., Ferretti, R.: Consistency of a large time-step scheme for mean curvature motion. In: Numerical mathematics and advanced applications, pp. 495–502. Springer, Milano (2003)
6. Kohn, R.V., Serfaty, S.: A deterministic-control-based approach to motion by curvature. Comm. Pure Appl. Math. **59**(3), 344–407 (2006)
7. Soner, H.M., Touzi, N.: A stochastic representation for the level set equations. Comm. Partial Differential Equations **27**(9-10), 2031–2053 (2002)

Topological Derivative Based Methods for Non–Destructive Testing

A. Carpio and M.–L. Rapún

Abstract We propose a numerical strategy to reconstruct scatterers buried in a medium when the incident radiation (electromagnetic, thermal, acoustic) is governed by Helmholtz transmission problems. The scattering problem is recast as a shape optimization problem with the Helmholtz equation as a constraint and the scatterer as a design variable. Our method is based on the (successive) computation of topological derivatives of the associated shape functional for updated guesses of the scatterer. We present an efficient scheme to compute the required topological derivatives at each step. The scheme combines explicit expressions for the topological derivatives in terms of the solutions of forward and adjoint transmission problems with BEM–FEM approximations. Our technique applies in either spatially homogeneous or inhomogeneous media. Finally, a two dimensional numerical test illustrates the ability of the method to reconstruct buried shapes.

1 The Inverse and Forward Scattering Problems

We consider a medium $\mathscr{R}$ where some unknown objects are buried. The union of all of them will be denoted by Ω. To simplify, the surrounding medium $\mathscr{R}$ is taken to be the whole plane or space.

The configuration is illuminated by an incident wave, which interacts with the medium and the obstacles, generating transmitted and reflected waves. The total wave field (formed by incident, transmitted and reflected waves) solves a transmission boundary problem, the so–called forward problem.

A. Carpio
Dpto. Matemática Aplicada, Fac. Matemáticas, Universidad Complutense de Madrid, 28040 Madrid, Spain, e-mail: ana_carpio@mat.ucm.es

M.–L. Rapún
Dpto. Fundamentos Matemáticos, E.T.S.I.Aeronáuticos, Universidad Politécnica de Madrid, 28040 Madrid, Spain, e-mail: marialuisa.rapun@upm.es

The total wave field is then measured on a sampling surface Γ_{meas}. In practice, the total field is known on a finite set of receptors located at Γ_{meas} for several incident waves. The inverse problem consists in finding the shape and structure of the obstacles such that the solutions of the forward problems agree with the measured values at the receptors.

A variational approach is used to find an approximate solution for the inverse problem. Instead of looking for the domain Ω such that $u = u_{meas}$ on Γ_{meas}, we seek a domain minimizing the functional

$$J(\mathscr{R} \setminus \overline{\Omega}) := \int_{\Gamma_{meas}} |u - u_{meas}|^2, \tag{1}$$

u being the solution of the corresponding forward problem in $\mathscr{R} \setminus \overline{\Omega}$ and u_{meas} the total field on Γ_{meas}.

In this work we assume that the incident radiation is a plane wave $u_{inc}(\mathbf{x}) = e^{\imath \kappa_e \mathbf{d} \cdot \mathbf{x}}$ in the direction $\mathbf{d}$ and the original forward problem for the total field is a Helmholtz transmission problem

$$\begin{cases} \Delta u + k_e^2(\mathbf{x}) u = 0, & \text{in } \Omega_e := \mathbb{R}^d \setminus \overline{\Omega}, \\ \Delta u + k_i^2(\mathbf{x}) u = 0, & \text{in } \Omega, \\ u^- = u^+, \quad \partial_n^- u = \partial_n^+ u, & \text{on } \Gamma, \\ r^{(d-1)/2} \left(\partial_r (u - u_{inc}) - \imath \kappa_e (u - u_{inc}) \right) \to 0, & \text{as } r := |\mathbf{x}| \to \infty. \end{cases} \tag{2}$$

The wave–numbers k_e, k_i are assumed to be known, depend smoothly on the position and satisfy $k_e(\mathbf{x}) \geq k_{e,0} > 0$, $k_i(\mathbf{x}) \geq k_{i,0} > 0$, $\forall \mathbf{x} \in \mathbb{R}^d$, and $k_e(\mathbf{x}) = \kappa_e$, $|\mathbf{x}| > R > 0$. The symbols $\pm$ are employed to denote traces and normal derivatives from each side of the boundary Γ. The normal vector points inside Ω. The transmission conditions represent the continuity of the solution and the normal derivative across the interface. They are used when the obstacle is 'penetrable' by the radiation. Neumann (sound–hard obstacles) or Dirichlet (sound–soft obstacles) conditions can be handled with similar techniques. The last condition in (2) is the standard Sommerfeld radiation condition, which implies that only outgoing waves are allowed.

From the physical point of view, assuming that the properties of the inclusion Ω are known beforehand may not be realistic. The tools developed in this paper are the basis of a method to reconstruct both the obstacles and their properties. This will be done in a forthcoming paper [3].

2 Topological Derivatives

Topological derivative methods are a powerful tool to solve inverse scattering problems associated with shape reconstruction. They have the ability of providing good initial guesses of the scatterers without any a priori information about their shape or location. Moreover, iterative schemes based on the computation of topological derivatives allow for topological changes in the updated guesses and seem to be faster than methods inspired in level–sets [11]. Recent work on topological derivatives focuses on problems with constant parameters [1, 5, 7, 10]. However, spatial inhomogeneities [4, 8] occur in many important applications such as tomography or photothermal reconstruction. Spatial variations are incorporated in [3].

The topological derivative (TD in the sequel) of a shape functional $J(\mathscr{R})$ is a way to measure its sensitivity when infinitesimal balls are removed from the region $\mathscr{R}$ at each point $\mathbf{x}$. The formal definition is [12]

$$D_T(\mathbf{x},\mathscr{R}) = \lim_{\varepsilon\to 0} \frac{J(\mathscr{R}) - J(\mathscr{R}\setminus\overline{B_\varepsilon(\mathbf{x})})}{\text{Volume}(B_\varepsilon(\mathbf{x}))}, \qquad \mathbf{x}\in\mathscr{R}.$$

Regions where the TD takes large negative values indicate locations where an object should be placed. The definition can be generalized to other shapes than balls, but this is not relevant in our study.

For our particular shape functional (1), we use a procedure that relates topological and shape derivatives, and allows for the computation of the TD as a limit of shape derivatives in domains with vanishing inclusions (see [5] for details). The shape derivatives involve the solutions of forward and adjoint problems in such domains. One can perform asymptotic expansions of the solutions of Helmholtz equations with vanishing inclusions to compute the limit. In problems with variable coefficients, these expansions are made directly on the partial differential equation (and not on integral expressions in terms of Green functions, as in [7]). For the sake of brevity, we write below the resulting expressions omitting the proofs. A rigorous justification will appear in [3].

When we do not have any a priori information about the location of the obstacles, we study the functional in the whole plane or space, that is, $\mathscr{R} = \mathbb{R}^d$, $d = 2,3$. In this case, we have the following result.

Theorem 1. *For any $\mathbf{x}\in\mathbb{R}^d$, the TD of the cost functional (1) is*

$$D_T(\mathbf{x},\mathbb{R}^d) = Re\left[(k_i^2(\mathbf{x}) - k_e^2(\mathbf{x}))u(\mathbf{x})\overline{w}(\mathbf{x})\right], \tag{3}$$

where u is the solution to the forward problem (2) with $\Omega = \emptyset$, i.e.,

$$\begin{cases} \Delta u + k_e^2(\mathbf{x})u = 0, & \textit{in } \mathbb{R}^d, \\ r^{(d-1)/2}\left(\partial_r(u-u_{inc}) - \imath\kappa_e(u-u_{inc})\right) \to 0, & \textit{as } r\to\infty, \end{cases} \tag{4}$$

and w solves the adjoint problem

$$\begin{cases} \Delta w + k_e^2(\mathbf{x})w = (u_{meas}-u)\delta_{\Gamma_{meas}}, & \textit{in } \mathbb{R}^d, \\ r^{(d-1)/2}\left(\partial_r w + \imath\kappa_e w\right) \to 0, & \textit{as } r\to\infty, \end{cases} \tag{5}$$

$\delta_{\Gamma_{meas}}$ being the Dirac delta function on Γ_{meas}.

Notice that the true scatterer enters the TD through the measured data u_{meas} influencing the adjoint field. For constant parameters, formula (3) was already obtained in [7].

When an initial guess Ω_{ap} is available, we compute the TD in $\mathscr{R} = \mathbb{R}^d \setminus \overline{\Omega}_{ap}$. Formally, the resulting expression for the TD is the same as before, but now u and w solve forward and adjoint problems with obstacle Ω_{ap}.

Theorem 2. *For any* $\mathbf{x} \in \mathbb{R}^d \setminus \overline{\Omega}_{ap}$, *the TD of the cost functional (1) is*

$$D_T(\mathbf{x}, \mathbb{R}^d \setminus \overline{\Omega}_{ap}) = Re\left[(k_i^2(\mathbf{x}) - k_e^2(\mathbf{x}))u(\mathbf{x})\overline{w}(\mathbf{x})\right],$$

where u is the solution to the forward problem (2) with $\Omega = \Omega_{ap}$ and w solves

$$\begin{cases} \Delta w + k_e^2(\mathbf{x})w = (u_{meas} - u)\delta_{\Gamma_{meas}}, & \text{in } \mathbb{R}^d \setminus \overline{\Omega}_{ap}, \\ \Delta w + k_i^2(\mathbf{x})w = 0, & \text{in } \Omega_{ap}, \\ w^- = w^+, \quad \partial_n w^- = \partial_n w^+, & \text{on } \partial\Omega_{ap} \\ r^{(d-1)/2}(\partial_r w + \imath\kappa_e w) \to 0, & \text{as } r := |\mathbf{x}| \to \infty. \end{cases} \tag{6}$$

In some applications, Helmholtz operators in the forward problem are replaced by $\text{div}(\alpha_s(\mathbf{x})\nabla u) + k_s^2(\mathbf{x})u$, $s = e, i$, and the continuity condition on the normal derivatives across the interface is replaced by $\alpha_i(\mathbf{x})\partial_n u^- = \alpha_e(\mathbf{x})\partial_n u^+$. The topological derivative of (1) is

$$\text{Re}\left[\frac{2\alpha_e(\mathbf{x})(\alpha_e(\mathbf{x}) - \alpha_i(\mathbf{x}))}{\alpha_e(\mathbf{x}) + \alpha_i(\mathbf{x})}\nabla u(\mathbf{x})\nabla\overline{w}(\mathbf{x}) + (k_i^2(\mathbf{x}) - k_e^2(\mathbf{x}))u(\mathbf{x})\overline{w}(\mathbf{x})\right], \tag{7}$$

where the forward and adjoint fields solve modified versions of (2) and (6) with second order operators in divergence form and generalized transmission conditions.

3 Iterative Scheme

Let us assume first that k_e is constant, that is, $k_e = \kappa_e$ and $k_i \equiv k_i(\mathbf{x})$. We describe afterwards how to tackle the case $k_e \equiv k_e(\mathbf{x})$.

Initialization

To compute a first approximation of the scatterer, we calculate the TD of (1) when $\Omega = \emptyset$ using Theorem 1. Since k_e is constant, the solution to the forward problem is the incident wave, $u(\mathbf{x}) = e^{\imath\kappa_e \mathbf{d}\cdot\mathbf{x}}$. The conjugate of the solution w to the adjoint problem is

$$\overline{w} = \int_{\Gamma_{meas}} \Phi_{\kappa_e}(|\cdot - \mathbf{y}|)(\overline{u_{meas} - u})(\mathbf{y})\, d\ell_{\mathbf{y}}, \tag{8}$$

Φ_{κ_e} being the outgoing fundamental solution of the operator $\Delta + \kappa_e^2$, i.e., $\Phi_{\kappa_e}(r) := (\imath/4)H_0^{(1)}(\kappa_e r)$ when $d = 2$ and $\Phi_{\kappa_e}(r) := \exp(\imath\kappa_e r)/(4\pi r)$ when $d = 3$.

We use these formulas to evaluate the TD on a grid. Then, we choose

$$\Omega_1 = \{\mathbf{x} \in \mathbb{R}^d,\ D_T(\mathbf{x}, \mathbb{R}^d) \leq -C_1\}, \quad C_1 > 0. \tag{9}$$

Iteration

A nested sequence $\Omega_j \subset \Omega_{j+1}$ is constructed in such a way that

$$\Omega_{j+1} = \Omega_j \cup \{\mathbf{x} \in \mathbb{R}^d \setminus \overline{\Omega}_j,\ D_T(\mathbf{x}, \mathbb{R}^d \setminus \overline{\Omega}_j) \leq -C_{j+1}\}, \quad C_{j+1} > 0.$$

A similar idea was suggested in [6], imposing a volume constraint on C_{j+1} and Ω_{j+1}. Our approach is more empirical. We select C_{j+1} examining the values of $D_T(\mathbf{x}, \mathbb{R}^d \setminus \overline{\Omega}_j)$. The proposed value is accepted if $D_T(\mathbf{x}, \mathbb{R}^d \setminus \overline{\Omega}_{j+1})$ does not present a sudden increase near $\partial\Omega_{j+1}$, which would indicate that spurious points have been included in the approximation.

At each step $D_T(\mathbf{x}, \mathbb{R}^d \setminus \overline{\Omega}_j)$ is computed using Theorem 2. When $\Omega = \Omega_j$, the adjoint and forward fields are calculated adapting the mixed FEM–BEM formulation of (2) proposed in [9]. Our strategy follows the classical treatment of mixed problems for Laplace or Helmholtz equations in which both the scalar field and the flux must be calculated, see [2]. We consider two unknowns defined in Ω, the solution itself $u \in L^2(\Omega)$ and its gradient $\mathbf{a} := \nabla u \in H(\mathrm{div}, \Omega) := \{\mathbf{b} \in (L^2(\Omega))^2 \,|\, \nabla \cdot \mathbf{b} \in L^2(\Omega)\}$, together with a couple of unknowns defined on Γ, the interior trace $\xi := u^- \in H^{1/2}(\Gamma)$ and a exterior density $\varphi \in H^{-1/2}(\Gamma)$ such that

$$u = u_{inc} + \int_\Gamma \Phi_{\kappa_e}(|\cdot - \mathbf{y}|)\varphi(\mathbf{y})d\ell_{\mathbf{y}}, \qquad \text{in } \Omega_e. \tag{10}$$

The regularity assumptions on u and ∇u imply that $u \in H^1(\Omega)$ and the traces make sense in these spaces. In terms of the new four unknowns, we write the forward problem (2) as

$$\begin{cases} \mathbf{a} - \nabla u = 0, \quad \nabla \cdot \mathbf{a} + k_i^2 u = 0, & \text{in } \Omega, \\ u^- - \xi = 0, \quad \xi - V^{\kappa_e}\varphi = u_{inc}, & \text{on } \Gamma, \\ \mathbf{a} \cdot \mathbf{n} + (\frac{1}{2}I - J^{\kappa_e})\varphi = \partial_{\mathbf{n}} u_{inc}, & \text{on } \Gamma, \end{cases} \tag{11}$$

where the boundary operators V^{κ_e} and J^{κ_e} are defined on Γ by

$$V^{\kappa_e}\varphi := \int_\Gamma \Phi_{\kappa_e}(|\cdot - \mathbf{y}|)\varphi(\mathbf{y})d\ell_y, \qquad J^{\kappa_e}\varphi := \int_\Gamma \partial_{n(\mathbf{y})}\Phi_{\kappa_e}(|\cdot - \mathbf{y}|)\varphi(\mathbf{y})d\ell_y.$$

If the solution u of the original forward problem (2) belongs to $H^1(\Omega) \cap H^1_{loc}(\Omega_e)$ (as it happens when Γ is $\mathscr{C}^2$), then we have a solution to (11). Reciprocally, if u solves (11), then $u \in H^1(\Omega)$, u defined by (10) belongs to $H^1_{loc}(\Omega_e)$ and together solve problem (2). Once we have established the equivalence of both formulations, we work with (11) and treat u and $\mathbf{a}$ as independent unkowns. In variational form, this problem is also equivalent to

$$\begin{cases} \int_\Omega \mathbf{a} \cdot \mathbf{b}\, d\mathbf{z} + \int_\Omega u(\nabla \cdot \mathbf{b})\, d\mathbf{z} - \int_\Gamma \xi(\mathbf{b} \cdot \mathbf{n})\, d\ell = 0, & \forall \mathbf{b} \in H(\mathrm{div}, \Omega), \\ \int_\Omega (\nabla \cdot \mathbf{a}) v\, d\mathbf{z} + \int_\Omega k_i^2 u v\, d\mathbf{z} = 0, & \forall v \in L^2(\Omega), \\ \int_\Gamma (\mathbf{a} \cdot \mathbf{n})\eta\, d\ell + \int_\Gamma (\frac{1}{2}I - J^{\kappa_e})\varphi\, \eta\, d\ell = \int_\Gamma \partial_{\mathbf{n}} u_{inc}\, \eta\, d\ell, & \forall \eta \in H^{1/2}(\Gamma), \\ \int_\Gamma \xi \psi\, d\ell - \int_\Gamma V^{\kappa_e}\varphi\, \psi\, d\ell = \int_\Gamma u_{inc}\, \psi\, d\ell, & \forall \psi \in H^{-1/2}(\Gamma). \end{cases}$$

This formulation makes it easier to select finite elements ensuring continuity of u and $\mathbf{a} \cdot \mathbf{n}$ across Γ. We can extend it to operators in divergence form, allowing for a fast evaluation of the gradients when computing the topological derivative (7). Another advantage is the simple far field representation (10), which in addition does not depend on how far is Γ_{meas} from the obstacles. We use then a FE approximation in Ω and a BE approximation in Ω_e.

For the numerical solution in 2D, we combine finite elements on a triangular grid (with spaces of piecewise constant functions for the scalar field and the lowest order Raviart–Thomas elements for the flux) with a conforming approximation of the boundary unknowns using periodic piecewise constant functions and continuous piecewise linear functions defined on uniform staggered grids. The choice of adequate spaces in 3D is also feasible. Coding difficulties arise due to the non–matching grids of the FEM and BEM routines. For further details we refer to [9].

If Γ_{meas} is far form the obstacles, pure FEM methods involve big computational domains, resulting in systems with huge matrices. Combining FEM and BEM, we reduce the number of unknowns. The systems to be solved are much smaller. Their matrices are mostly sparse (except for small blocks which are rather full) and have a useful block structure. Both facts can be exploited to store conveniently the matrix entries and choose adequate solvers, see [9].

To solve the adjoint problem, we split $\overline{w} = w_1 + w_2$. The first term w_1 is defined by (8) in Ω_e and vanishes in Ω. The second term w_2 solves

$$\begin{cases} \Delta w_2 + \kappa_e^2 w_2 = 0, & \text{in } \mathbb{R}^d \setminus \overline{\Omega}, \\ \Delta w_2 + k_i^2(\mathbf{x}) w_2 = 0, & \text{in } \Omega, \\ w_2^- - w_2^+ = w_1^+, \quad \partial_n^- w_2 - \partial_n^+ w_2 = \partial_n w_1^+, & \text{on } \Gamma, \\ r^{(d-1)/2}(\partial_r w_2 - \imath \kappa_e w_2) \to 0, & \text{as } r \to \infty. \end{cases}$$

Then, the BEM–FEM formulation proposed for the forward problem applies. The only difference with respect to the forward problem is the right hand side. To compute numerically the TD, we assemble one matrix and solve two problems with different right hand sides.

Spatially Dependent k_e

Let us consider now the general case $k_e = k_e(\mathbf{x})$. When $\Omega = \emptyset$, the solution of the forward problem is not the incident wave. However, $k_e(\mathbf{x}) = \kappa_e$ when $|\mathbf{x}| > R$. Then, we decompose $u = u_{inc} + v$ in $\mathbb{R}^2 \setminus \overline{B}_R$ and $u = v$ in B_R, v being the solution to

$$\begin{cases} \Delta v + \kappa_e^2 v = 0, & \text{in } \mathbb{R}^2 \setminus \overline{B}_R, \\ \Delta v + k_e^2(\mathbf{x}) v = 0, & \text{in } B_R, \\ v^- - v^+ = u_{inc}, \quad \partial_n v^- - \partial_n v^+ = \partial_n u_{inc}, & \text{on } \Gamma_R, \\ r^{(d-1)/2}(\partial_r v - \imath \kappa_e v) \to 0, & \text{as } r \to \infty. \end{cases} \tag{12}$$

We can use the BEM–FEM formulation proposed above (now B_R plays the role of Ω and k_e the role of k_i). The solution of the corresponding adjoint problem is again

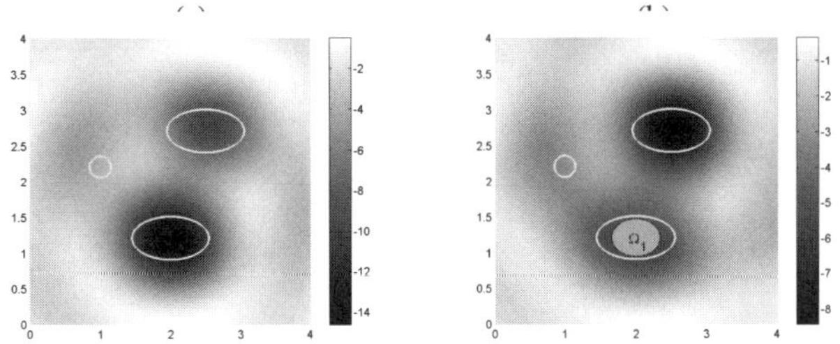

Fig. 1 Topological derivative: (a) $\Omega_{ap} = \emptyset$, (b) $\Omega_{ap} = \Omega_1$.

decomposed as $\overline{w} = w_1 + w_2$, where w_1 is defined as in (8) and w_2 solves a problem similar to (12) with w_1^+ and $\partial_n w_1^+$ in the right hand side instead of u_{inc} and $\partial_n u_{inc}$. Again, the BEM–FEM formulation can be applied here and both, the forward and the adjoint problems will share the same matrix. Finally, when $\Omega \neq \emptyset$, the same ideas are easily adapted, since in this case the conditions on Γ are directly imposed in the variational formulation.

4 Numerical Experiments

We present a test in a simple geometry with three objects. Their location is represented by white lines in all plots in Fig. 1 and 2. The surrounding medium is assumed to be homogeneous, with wave–number $k_e = 2$. The wave–number for the buried objects is the space dependent function $k_i(x,y) = 1 + y/4$. This choice does not correspond to any real physical problem, but illustrates the performance of the method (see also [8] for similar tests in elasticity problems).

The total wave–field was measured on a set of 30 uniformly distributed sampling points on the circle of radius 4 and center (2,2) for 25 planar incident waves $(\cos\theta_j, \sin\theta_j)$ with angles uniformly distributed in $[0, 2\pi)$.

Since we do not have any a priori information about the number, size or location of the obstacles, we have computed first the TD using Theorem 1. The computed values are shown in Fig. 1(a). The presence of the object at the bottom is clearly detected and the smallest one is completely ignored. The TD also seems to detect the third object. Notice that, although the inclusions at the top and at the bottom have exactly the same size and shape, there is a bigger difference between the wave number inside the object at the bottom and the surrounding media than between the object at the top and the exterior media. This justifies the better reconstruction of the object at the bottom. We have been very cautious when selecting the threshold C_1 in (9) to characterize the obstacles in order to avoid spurious points. Only the approximation of the object denoted by Ω_1 in Fig. 1(b) has been trusted.

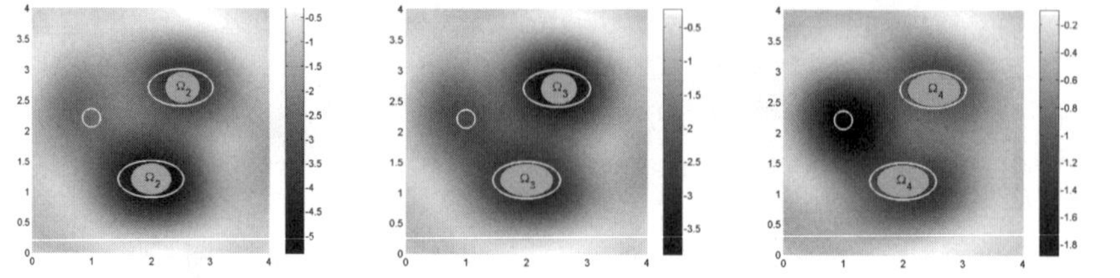

Fig. 2 Topological derivative when $\Omega_{ap} = \Omega_2, \Omega_3, \Omega_4$.

Once a first guess in known, we compute the TD using Theorem 2. The obtained values are shown in Fig. 1(b). The lowest values of the updated TD indicate the location and approximated size of a new object. The smallest one remains undetected.

Finally, in Fig. 2 we represent the next three iterations of the scheme. The iteration converges very fast. A few steps provide satisfactory reconstructions. Furthermore, the small object, ignored in the first trials, is captured.

Acknowledgements The authors are partially supported by MAT2005-05730-C02-02, MEC–FEDER MTM2007-63204, BSCH–UCM PR27/05-13939 and CM-910143 projects.

References

1. Bonnet, M., Constantinescu, A.: Inverse problems in elasticity. Inverse Problems **21**, R1–R50 (2005)
2. Brezzi, F., Fortin, M.: Mixed and hybrid finite element methods. Springer-Verlag, New York (1991)
3. Carpio, A., Rapún, M.L.: Solving inhomogeneous inverse problems by topological derivative methods. Submitted
4. Dorn, O., Lesselier, D.: Level set methods for inverse scattering. Inverse Problems **22**, R67–R131 (2006)
5. Feijoo, G.: A new method in inverse scattering based on the topological derivative. Inverse Problems **20**, 1819–1840 (2004)
6. Garreau, S., Guillaume, P., Masmoudi, M.: The topological asymptotic for pde systems: the elasticity case. SIAM J. Control Optim. **39**, 1756–1778 (2001)
7. Guzina, B., Bonnet, M.: Small-inclusion asymptotic of misfit functionals for inverse problems in acoustics. Inverse Problems **22**, 1761–1785 (2006)
8. Guzina, B., Chikinev, I.: From imaging to material identification: A generalized concept of topological sensitivity. Inverse Problems **55**, 245–279 (2007)
9. Rapún, M.L., Sayas, F.J.: A mixed–fem and bem coupling for the approximation of the scattering of thermal waves in locally non–homogeneous media. Math. Model. Numer. Anal. **40**, 871–896 (2006)
10. Samet, B., Amstutz, S., Masmoudi, M.: The topological asymptotic for the helmholtz equation. SIAM J. Control Optim. **42**, 1523–1544 (2003)
11. Santosa, F.: A level set approach for inverse problems involving obstacles. ESAIM Control, Optim. Calculus Variations **1**, 17–33 (1996)
12. Sokowloski, J., Zochowski, A.: On the topological derivative in shape optimization. SIAM J. Control Optim. **37**, 1251–1272 (1999)

A Characteristics Driven Fast Marching Method for the Eikonal Equation

E. Cristiani and M. Falcone

Abstract We introduce a new Fast Marching method for the eikonal equation. The method is based on the informations driven by characteristics and it is an improvement with respect to the standard Fast Marching method since it accepts more than one node at every iteration using a dynamic condition. We analyze the method and present several tests on fronts evolving in the normal direction with variable velocities including cases where a change in topology occurs.

1 Introduction

The success of the level set method in the analysis and in the simulation of complex interface problems is mainly due to its capability to handle the changes in topology of the interfaces. The price to pay is the fact that we look for a function $u : \mathbf{R}^n \times [0,T] \to \mathbf{R}$ and we locate the front Γ_t at time t just considering the 0-level set of $u(x,t)$, so we add one extra dimension to the original problem which lives in $\mathbf{R}^n$.

The Fast Marching (FM) method has been proposed to cut down the computational complexity of the level set method (see [7, 9] for the origin of the method and [1, 2, 8] for some recent developments). It was developed for the time-independent eikonal equation

$$\begin{cases} c(x)|\nabla T(x)| = 1, & x \in \mathbf{R}^n \backslash \Omega_0 \\ T(x) = 0, & x \in \Gamma_0 = \partial\Omega_0 \end{cases} \quad (1)$$

where Ω_0 is a closed bounded set and $c(x)$ is Lipschitz continuous and strictly positive. The front is recovered as the t-level set of the minimal time function $T(x)$

Emiliano Cristiani
ENSTA, 32, Boulevard Victor, Paris, France, e-mail: emiliano.cristiani@ensta.fr

Maurizio Falcone
Dipartimento di Matematica, SAPIENZA - Università di Roma, e-mail: falcone@mat.uniroma1.it

which is the unique viscosity solution of (1). Note that the solution u of the level set method can be written as $u(x,t) = T(x) - t$, so (1) is related both to the evolution of a front with velocity $c(x)$ in the normal direction and to the minimum time problem (see [3, 6]).

The FM method sets-up the computation in a narrow band near the front and updates the narrow band in order to follow the evolution of the front at every time. In this way it eliminates the extra dimension introduced by the model since at every iteration the computation is performed in a neighborhood of the front, *i.e.* we work on $O(\sqrt{N})$ nodes if the grid has N nodes. Once a node exits the narrow band it is accepted, *i.e.* it is no more computed in the following iterations. The FM method accepts only one node at every iteration (the node with the minimal value T) to guarantee that the evolution of the front is tracked correctly. However, in several situations one has the impression that the FM method can be improved and that it is not needed to accept just one point at every iteration. For example, this is the case when the initial configuration Ω_0 is convex so that in the normal evolution of the front there are neither a merging nor a crossing of characteristics. The main contribution here is to modify the FM method based on the semi-Lagrangian approximation (see [2, 3] for details) developing a new algorithm which accepts several points provided some local conditions are satisfied (see Section 2). In this way we drop the search for the minimum value *at every iteration* via a Min-Heap structure. We note that another Group Marching algorithm has been proposed by Kim in [5] to speed up the standard FM method based on the finite difference scheme. Our algorithm is different in two respects. The first is that we use a different local rule for the computation whereas the second is related to the condition which allows our method to accept more nodes.

2 The Characteristics Fast Marching Method

In this section we present the algorithm and some considerations about its computational cost. Note that the basic algorithm is developed in order to deal with the evolution of a *single* front although it works often for several merging fronts. An extension to the case of general merging front is also given. Via the Kružkov tranform $v(x) = 1 - e^{-T(x)}$ the equation (1) can be rewritten in the fixed point form $v(x) = F[v(x)] := \min_{a \in B(0,1)}\{c(x)a \cdot \nabla v(x)\} + 1$ where $B(0,1)$ is the unit ball centered in 0. Let us just recall the basic semi-Lagrangian scheme we will use as local rule for the computation. The value of v at the node x_i will be denoted by v_i.

$$\begin{cases} v_i = \min\limits_{a \in B(0,1)} \{e^{-h} v(x_i - hc(x_i)a) + 1 - e^{-h}, & x_i \in \mathbf{R}^n \backslash \Omega_0 \\ v_i = 0, & x_i \in \partial\Omega_0 \end{cases} \tag{2}$$

We chose a variable (fictitious) time step $h = h_i$ such that $c(x_i)h_i = \Delta x$ and we use a linear interpolation to compute $v(x_i - h_i c(x_i)a)$.

2.1 Main Idea

In several situations the accept-the-node-with-minimum-value condition of FM method appears to be too restrictive, particularly when $c(x) \equiv c_0$ and Ω_0 is a small ball. In fact, in this case one can accept almost all the nodes in the narrow band at the same time without loosing informations.

In the Characteristics Fast Marching (CFM) method the point of view is reversed with respect to the FM method. Instead of declaring accepted the node with the minimum value in the narrow band (*i.e.* the first node which will be reached by the front in the next iteration), we look for the node in the narrow band with the maximal velocity c_{max}, *i.e.* the node from which we can cover the distance Δx and enter the accepted zone in the minimum time. Once we have c_{max} we can compute the time step $\Delta t = \frac{\Delta x}{c_{max}}$ for that iteration which is the time needed by the fastest node to reach the accepted zone. While *all* the nodes with the maximal velocity c_{max} reach the accepted zone the other nodes in the narrow band come closer to the accepted zone without touching it. In order to take into account this displacement (which is smaller than Δx), we introduce the *local time* t_i^{loc}. The local time is set to 0 when the node i enters the narrow band and it is increased at each iteration by the (variable) increment Δt until the node is accepted. At each iteration we label as accepted all the nodes having a local time t_i^{loc} large enough to safisfy two conditions: they reach the accepted zone moving at speed $c(x_i)$ *and* they are computed by (2) just using nodes in the accepted region.

2.2 The CFM Algorithm for a Single Front

Let us introduce the algorithm. In the following, the set of the nodes belonging to the narrow band will be denoted by *NB*.

Initialization

1. The nodes belonging to the initial front Γ_0 are located and labeled as *accepted*. They form the set $\widetilde{\Gamma_0}$. The value of v of these nodes is set to 0.
2. *NB* is defined as the set of the neighbors of $\widetilde{\Gamma_0}$, external to Γ_0.
3. Set $t_i^{loc} := 0$ for any $i \in NB$.
4. The remaining nodes are labeled as *far*, their value is set to 1 (corresponding to $T = +\infty$).

Main Cycle

1. Compute $c_{max} = \max\{c_i \ : \ i \in NB\}$ and set $\Delta t := \Delta x / c_{max}$.
2. For any $i \in NB$:
   a. Update the local time: $t_i^{loc} := t_i^{loc} + \Delta t$.
   b. If $t_i^{loc} \cdot c_i \geq \Delta x$ then
      i. Compute v_i by the scheme (2).

ii. Check if v_i is computed using only *accepted* nodes. If yes, set `flag`=true, else set `flag`=false.
iii. If `flag`=true, then
- Label i as *accepted* and remove i from *NB*.
- Define *FN* as the set of the *far* neighbors of i. Include *FN* in *NB* and set $t_k^{loc} := 0$ for any $k \in FN$.

3. If not all nodes are *accepted* go back to 1.

2.3 Front Merging

Let us extend the previous algorithm to the more interesting situation when more fronts are merging together. This a delicate point because we risk to accept too many points before they reach the correct value. One of the main differences with respect to the FM method is that the narrow band does not follow exactly the level sets of the solution. For example, if $c(x)$ is constant, the narrow band of a rectangular front evolves without smoothing corners contrary to the real level sets of the exact solution. In computing the evolution of a single front this is not a major difficulty, since the correctness of the solution is guaranteed by the fact that we accept *only* those nodes computed by other already accepted nodes. However, the evolution of m fronts merging together can be difficult to follow in some cases (but not in all cases). In fact, a single node can be reached first by the narrow band corresponding to one on those fronts and it is accepted according to these informations whereas it should be accepted according to the informations driven by another front. In Fig. 1 we can

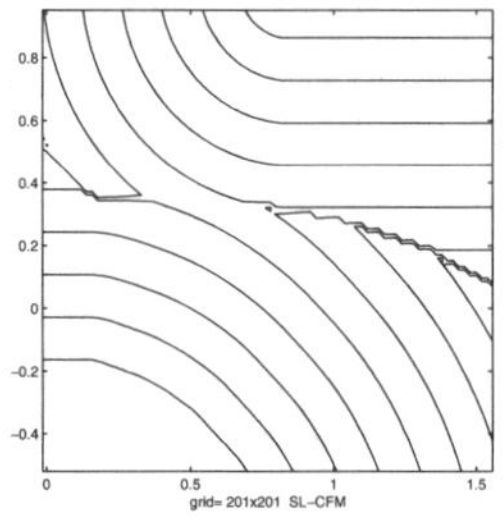

Fig. 1 The merging zone in the case the CFM method fails

see a front merging where this situation occurs and produces a wrong solution in the merging zone. Note that this difficulty can not be solved reducing the space step Δx and it seems that there is no way to solve the problem without making some important changes to the algorithm.

In order to deal with m merging fronts we start considering we can compute by CFM method the m value functions $v^{(k)}$, $k = 1, \dots, m$ corresponding to the evolutions of the m fronts on m copies of the domain and than take the minimum $v := \min\{v^{(1)}, \dots, v^{(m)}\}$ as final solution. Of course this choice leads to multiply by m the time required for computation (we do not consider here problems related to memory) and then it is not efficient. To overcome this problem we use the following strategy. When the k-th CFM method accepts the node i with value $v_i^{(k)}$, it checks if another CFM method has already computed a value $\tilde{v}_i < v_i^{(k)}$ for the same node. If this is the case, the k-th CFM method avoids to enlarge the narrow band from the node i. This procedure leads to a very small overlapping zone between fronts and a CPU time comparable with the CFM method for a single front (see Fig. 2).

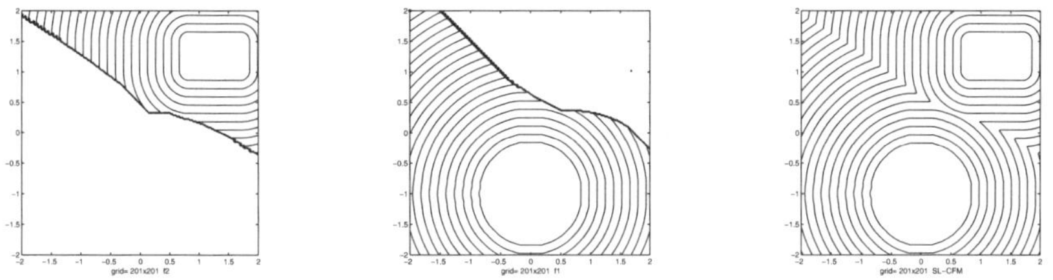

Fig. 2 Merging via two applications of CFM method

Computational cost

The computational cost of the CFM method is more complicated to determine with respect to the FM method. This is mainly due to the fact that it is very difficult to say *a priori* how many times the `flag` remains fixed to false for each node. By the experiments it seems that this happens when the velocity field $c(x)$ increases along characteristics.

Searching for the maximal velocity c_{max} in the narrow band costs $O(\ln N_{nb})$ where N_{nb} is the number of nodes in the narrow band (bounded by N but expected to be of order $\sqrt{N}$).

When the velocity is constant all the nodes of the narrow band are accepted at the same time, then the number of times we need to search for c_{max} in the narrow band is divided by a factor N_{nb} with respect to the FM method and this is surely the most powerful feature of the CFM method.

In conclusion, for a constant velocity $c(x) \equiv c_0$ we expect the computational cost to be of order $O((N \ln N_{nb})/N_{nb}) = O(\sqrt{N} \ln \sqrt{N})$.

3 Numerical Results

In this section we present some numerical tests in order to compare the CFM method with the standard iterative semi-Lagrangian (SL) scheme and the FM method based on the semi-Lagrangian scheme (FM-SL) introduced in [2]. The domain of computation is $[-2,2]^2$. We used MATLAB 7.0 on a Processor Intel dual core 2x2.80 GHz with 1 GB RAM.

Test 1: constant velocity. $\Gamma_0 = (0,0)$. $c(x,y) \equiv 1$. Solution: $T(x,y) = \sqrt{x^2+y^2}$.

Test 2: velocity depending on x**.** $\Gamma_0 = (0,0)$. $c(x,y) = x+3$.

Test 3: velocity depending on x **and** y**.** $\Gamma_0 = (0,0)$. $c(x,y) = |x+y|$.

Test 4: merging of two fronts. $\Gamma_0 =$ a rectangle and a circle. $c(x,y) \equiv 1$.

Table 1 Errors and CPU time for Test 1

method	Δx	L^∞ error	L^1 error	CPU time (sec)
CFM	0.08	0.0329	0.3757	0.27
FM-SL	0.08	0.0329	0.3757	0.58
SL (46 it)	0.08	0.0329	0.3757	9.7
CFM	0.04	0.0204	0.2340	1.14
FM-SL	0.04	0.0204	0.2340	2.44
SL (86 it)	0.04	0.0204	0.2340	70.95
CFM	0.02	0.0122	0.1406	4.9
FM-SL	0.02	0.0122	0.1406	10.56
SL (162 it)	0.02	0.0122	0.1406	530.56

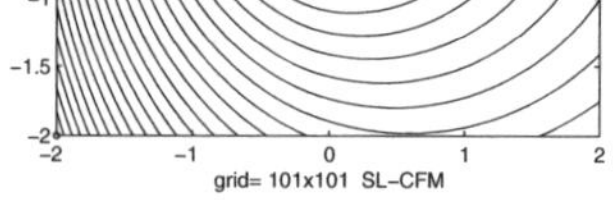

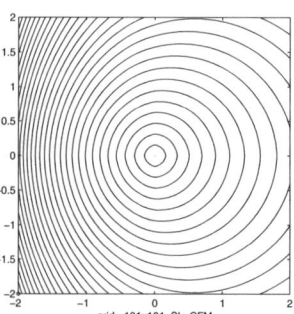

Fig. 3 Numerical result for Test 2

Comments

By Test 1 we can see that the semi-Lagrangian iterative method is much slower than both Fast Marching methods although all methods compute exactly the same approximation of the viscosity solution of equation (1).

Table 2 CPU time for Test 2

method	Δx	CPU time (sec)
CFM	0.04	1.19
FM-SL	0.04	2.34
CFM	0.02	5.20
FM-SL	0.02	10.44

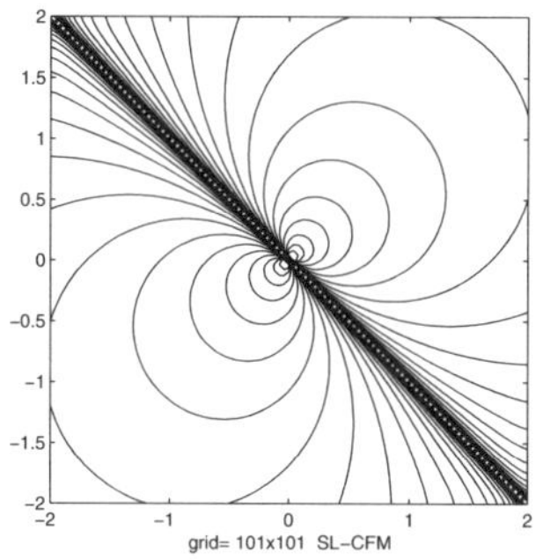

Fig. 4 Numerical result for Test 3

Table 3 CPU time for Test 3

method	Δx	CPU time (sec)
CFM	0.04	3.5
FM-SL	0.04	2.5
CFM	0.02	16.73
FM-SL	0.02	11.59

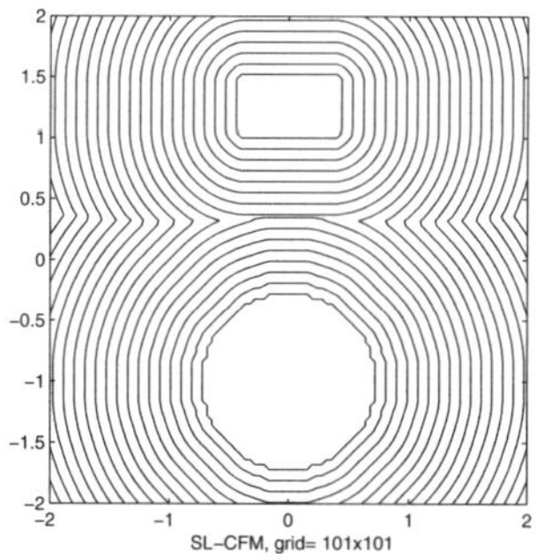

Fig. 5 Numerical result for Test 4

Table 4 CPU time for Test 4

method	Δx	CPU time (sec)
CFM	0.04	1.05
FM-SL	0.04	2.12
CFM	0.02	5.08
FM-SL	0.02	9.53

As expected, the CPU time for CFM method is smaller than that of FM method when the velocity field is constant or have relatively small variations in the domain (Test 1 and 2).

When the function $c(x)$ is increasing along characteristics (Test 3) the CFM method is slower than the FM method, this is due to the fact that the `flag` is often false after the step (ii) of the algorithm so that a lot of nodes in the narrow band are computed but only few of them are accepted. Note that in the FM method not all the nodes in the narrow band are computed at each iterations but only the new entries. In Test 4 the evolution of two merging fronts is computed without any modification of the basic algorithm and again the CPU time for the CFM method is about the half of the time needed by the FM method.

References

1. Cristiani, E.: Fast Marching and semi-Lagrangian methods for Hamilton-Jacobi equations with applications. Ph.D. thesis, Dipartimento di Metodi e Modelli Matematici per le Scienze Applicate, SAPIENZA - Università di Roma, Rome, Italy, February 2007
2. Cristiani, E., Falcone, M.: Fast semi-Lagrangian schemes for the Eikonal equation and applications. SIAM J. Numer. Anal., **45** (2007), pp. 1979–2011
3. Falcone, M.: The minimum time problem and its applications to front propagation. In A. Visintin e G. Buttazzo (eds.), "Motion by mean curvature and related topics", De Gruyter Verlag, Berlino, 1994
4. Falcone, M., Giorgi, T., Loreti, P.: Level sets of viscosity solution: some applications to fronts and rendez-vous problems. SIAM J. Appl. Math., **54** (1994), pp. 1335–1354
5. Kim, S.: An $O(N)$ level set method for eikonal equations. SIAM J. Sci. Comput., **22** (2001), pp. 2178–2193
6. Sethian, J. A.: Level set methods and fast marching methods. Evolving interfaces in computational geometry, fluid mechanics, computer vision, and materials science. Cambridge University Press, 1999
7. Sethian, J. A.: A fast marching level set method for monotonically advancing fronts. Proc. Natl. Acad. Sci. USA, **93** (1996), pp. 1591–1595
8. Sethian, J. A., Vladimirsky, A.: Ordered upwind methods for static Hamilton-Jacobi equations: theory and algorithms. SIAM J. Numer. Anal., **41** (2003), pp. 325–363
9. Tsitsiklis, J. N.: Efficient algorithms for globally optimal trajectories. IEEE Tran. Automatic. Control, **40** (1995), pp. 1528–1538

Hierarchical Model Reduction for Advection-Diffusion-Reaction Problems

A. Ern, S. Perotto, and A. Veneziani

Abstract Some engineering problems ranging from blood flow to river flow, from internal combustion engines to electronic devices have been recently modelled by coupling problems with different space dimensions (geometrical multiscale method). In this paper we focus on a new approch, where different levels of detail of the problem at hand stem from a different selection of the dimension of a suitable function space. The coarse and fine models are thus identified in a straightforward way. Moreover this approach lends itself to an automatic model adaptive strategy. The approach is addressed on a 2D linear advection-diffusion reaction problem.

1 Motivations

Many engineering problems of practical interest, even though formulated in 3D, exhibit a spatial dimension predominant over the others. This is the case, for instance, of river dynamics, blood flow problems or internal combustion engines. In these cases, it is sometimes possible to resort to downscaled models where only the dominant space dependence is considered (e.g., the Euler equations come from a 1D approximation of blood flows). Nevertheless the simplifying assumptions at the basis of these models can fail locally, essentially where "transversal" dynamics are relevant (e.g., a lake in a river network, an aneurysm in a blood vessel). Ideally, in correspondence with these configurations, one would like to locally enhance the 1D approximation via a proper higher-dimensional enrichment. In the so-called geometrical multiscale approach these enrichments consist of 2D or 3D models. Here

Alexandre Ern
CERMICS, Ecole Nationale des Ponts et Chaussées, ParisTech, 6 et 8, av. Blaise Pascal, 77455 Marne la Vallée cedex 2, France, e-mail: ern@cermics.enpc.fr

Simona Perotto and Alessandro Veneziani
MOX, Dipartimento di Matematica "F. Brioschi", Politecnico di Milano, Via Bonardi 9, I-20133 Milano, Italy, e-mail: {simona.perotto, alessandro.veneziani}@polimi.it

we follow a different strategy. We simplify the reference problem (the *full model*) by tackling in a different manner the dependence of the solution on the leading direction and on the transverse ones. The former is spanned by a classical piecewise polynomial basis. The latter are expanded into a *modal basis*. We end up with a real *hierarchy* of simplified models (the *reduced models*), distinguishing one another for the different number of modal transversal functions. From a computational viewpoint, independently of the dimension of the full problem, the reduced formulation leads us to a system of 1D problems (associated with the leading direction), coupled by the transversal information. In this work we present preliminary results of this approach applied to a 2D elliptic framework.

A similar approach can be found in [1, 2, 4, 5], though confined to a thin domain setting. Our proposal is potentially more effective than these approaches, as our reduced model is a system of 1D (rather than 2D) problems, also for a 3D full problem.

2 The Full Problem

Let us consider a linear advection diffusion reaction (ADR) problem. For the sake of simplicity we assume the computational domain Ω in $\mathbb{R}^2$ and homogeneous Dirichlet boundary conditions.

Let $\mu \in L^\infty(\Omega)$, with $\mu \geq \mu_0 > 0$, the diffusivity coefficient, $\mathbf{b} = (b_1, b_2)^T \in [L^\infty(\Omega)]^2$ the convective field and $\sigma \in L^\infty(\Omega)$ the reaction coefficient. We assume $\nabla \cdot \mathbf{b} \in L^\infty(\Omega)$. Moreover for the well-posedness of the problem we assume $-\frac{1}{2}\nabla \cdot \mathbf{b} + \sigma \geq 0$ a.e. in Ω. Finally, let $f \in L^2(\Omega)$ be the forcing term. Standard notation for the Sobolev spaces as well as for the spaces of functions bounded a.e. in Ω is adopted.

The weak formulation of the problem reads: find $u \in V = H_0^1(\Omega)$ s.t.

$$\int_\Omega \mu \nabla u \cdot \nabla v \, dxdy + \int_\Omega (\mathbf{b} \cdot \nabla u + \sigma u) v \, dxdy = \int_\Omega f v \, dxdy \quad \forall v \in V. \tag{1}$$

Furthermore we assume that the domain Ω can be represented as a 2D *fiber bundle*, i.e.

$$\Omega = \bigcup_{x \in \Omega_{1D}} \gamma_x, \tag{2}$$

where Ω_{1D} is a supporting one-dimensional domain, while $\gamma_x \subset \mathbb{R}$ represents the 1D (transversal) *fiber* associated with $x \in \Omega_{1D}$. In practice we distinguish in Ω a *leading direction*, associated with Ω_{1D}, and a secondary *transversal direction*, represented by the fibers γ_x. This choice finds its justification in the hydrodynamic as well as haemodynamic applications we are interested in, where the dominant direction is provided by the blood or the water main stream, respectively (see Fig. 1, left).

We map domain Ω into a reference domain $\widehat{\Omega}$, where the analysis is easier. For this purpose, for any $x \in \Omega_{1D}$, we introduce the map $\psi_x : \gamma_x \to \widehat{\gamma}_1$ between the generic fiber $\gamma_x \subset \mathbb{R}$ and a 1D reference fiber $\widehat{\gamma}_1$. The domain $\widehat{\Omega}$ is identified as the rectangle

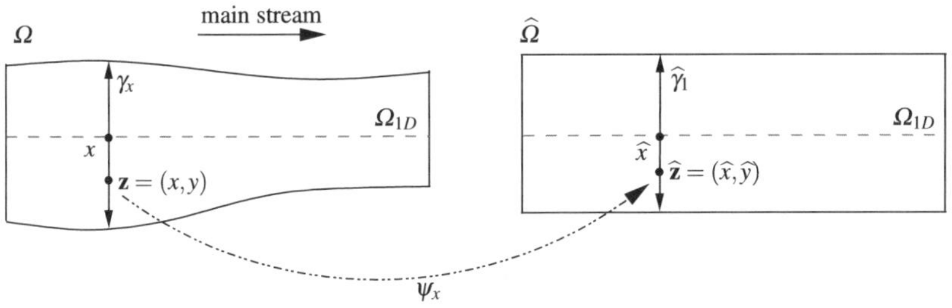

Fig. 1 The map ψ_x

with sides of length $|\Omega_{1D}|$ and $|\widehat{\gamma}_1|$. The map ψ_x thus simply acts on the fiber length (see Fig. 1). Throughout the paper we denote with $\mathbf{z} = (x,y)$ and $\widehat{\mathbf{z}} = (\widehat{x},\widehat{y})$ the generic point in Ω and the corresponding point in $\widehat{\Omega}$ via the map ψ_x, respectively, where $x \equiv \widehat{x} \in \Omega_{1D}$ while $\widehat{y} = \psi_x(y) \in \widehat{\gamma}_1$, with $y \in \gamma_x$.

A predominant role in the applications of our interest is played by the so-called *affine map*, given by $\widehat{y} = \psi_x(y) = L(x)^{-1}\left[y - g\right]$, where $L(x) = |\gamma_x|$ is the length fiber while g is a suitable shift factor. In particular when $L(x) = L =$ constant, the physical domain Ω itself coincides with a rectangle.

3 The Reduced Setting

The fiber structure introduced on the domain Ω is the starting point in defining the dimensional reduction. We resort to different function spaces along the supporting fiber Ω_{1D} rather than the transversal ones γ_x, in the spirit of a *model anisotropy*.
In more detail, we associate with Ω_{1D} the function space $V_{1D} \equiv H_0^1(\Omega_{1D})$, whose functions account for the homogeneous Dirichlet boundary conditions on $\partial\Omega_{1D}$. On the transversal reference fiber we introduce a *modal basis* $\{\varphi_k\}$, with $k \in \mathbb{N}$, where $\varphi_k : \widehat{\gamma}_1 \to \mathbb{R}$ and $\{\varphi_k\}$ is assumed $L^2(\widehat{\gamma}_1)$-orthonormal. The functions φ_k take into account the boundary conditions on $\partial\Omega_\gamma = \bigcup_{x\in\Omega_{1D}} \partial\gamma_x$. The transversal function space is therefore given by $V_{\widehat{\gamma}_1} = span\{\varphi_k\}$.

Different choices can be pursued for the modal functions φ_k (see [3], Remark 1). Here we adopt trigonometric functions, according to a classical Fourier expansion.

By properly combining the space V_{1D} with the modal basis $\{\varphi_k\}_k$, we define the *reduced space*, for any fixed a priori $m \in \mathbb{N}$,

$$V_m = \left\{ v_m(x,y) = \sum_{k=0}^{m} \varphi_k(\psi_x(y))\, \widetilde{v}_k(x), \text{ with } \widetilde{v}_k \in V_{1D} \right\}. \tag{3}$$

The $L^2(\widehat{\gamma}_1)$-orthogonality of the modal functions implies that the frequency coefficients $\widetilde{v}_k$ in (3) are

$$\widetilde{v}_k(x) = \int_{\widehat{\gamma}_1} v_m(x, \psi_x^{-1}(\widehat{y}))\varphi_k(\widehat{y})\,d\widehat{y}, \quad \text{with} \quad k = 0, \ldots, m. \tag{4}$$

Convergence of an approximation u_m to (1) stems essentially from the following assumptions: *i)* the *conformity* of the reduced space V_m in V, i.e. that $V_m \subset V, \forall m \in \mathbb{N}$; *ii)* the *spectral approximability* of V_m in V, namely that $\lim_{m \to +\infty} \left(\inf_{v_m \in V_m} \|v - v_m\|_V \right) = 0$, for any $v \in V$. These two requirements basically lead to proper regularity assumptions on the map ψ_x as well as on the spaces V_{1D} and $V_{\widehat{\gamma}_1}$ (for further details, see [3]). Throughout the paper we assume that these two hypotheses are fulfilled.

3.1 The ADR Reduced Form

The reduced formulation of the ADR equation (1) entails solving such a problem on the subspace V_m of V in (3).

Thus, for any $m \in \mathbb{N}$, we can state the so-called ADR *reduced problem*: find $u_m \in V_m$ such that

$$\int_\Omega \mu \nabla u_m \cdot \nabla v_m \,dxdy + \int_\Omega \left(\mathbf{b} \cdot \nabla u_m + \sigma u_m\right) v_m \,dxdy = \int_\Omega f v_m \,dxdy \quad \forall v_m \in V_m. \tag{5}$$

The well-posedness as well as the strong consistency of this formulation are guaranteed by assumption *i)* above.

Actually the reduced formulation (5) amounts to solving a system of $(m+1)$ coupled 1D problems, with coefficients computed on the reference fiber $\widehat{\gamma}_1$. For this purpose we introduce the Jacobian $\mathscr{J}(\widehat{y}) = \left(\partial \psi_x(y)/\partial y\right)|_{y=\psi_x^{-1}(\widehat{y})}$ associated with the map ψ_x. Moreover we define $\mathscr{D}(\widehat{y}) = \left(\partial \psi_x(y)/\partial x\right)|_{y=\psi_x^{-1}(\widehat{y})}$, representing a *deformation index* of the current domain Ω with respect to the reference one.

Let us exploit in (5) the representation $u_m(x,y) = \sum_{j=0}^m \widetilde{u}_j(x)\,\varphi_j(\psi_x(y))$ of u_m as a function of V_m and identify the test function v_m with $v_m(x,y) = \vartheta(x)\,\varphi_k(\psi_x(y))$, for any $\vartheta \in V_{1D}$ and any $k = 0, \ldots, m$, to get

$$\begin{aligned}
&\sum_{j=0}^m \Big[\int_\Omega \mu(x,y) \nabla\left(\widetilde{u}_j(x)\,\varphi_j(\psi_x(y))\right) \cdot \nabla\left(\vartheta(x)\,\varphi_k(\psi_x(y))\right) dxdy \\
&+ \int_\Omega \Big(\mathbf{b}(x,y) \cdot \nabla\left(\widetilde{u}_j(x)\,\varphi_j(\psi_x(y))\right) + \sigma(x,y)\widetilde{u}_j(x)\,\varphi_j(\psi_x(y))\Big)\,\vartheta(x)\,\varphi_k(\psi_x(y))\,dxdy \Big] \\
&= \int_\Omega f(x,y)\vartheta(x)\,\varphi_k(\psi_x(y))\,dxdy.
\end{aligned} \tag{6}$$

We analyze separately the different terms. Moving from the gradient expansion

$$\nabla\left(w(x)\,\varphi_s(\psi_x(y))\right) = \varphi_s(\psi_x(y)) \begin{bmatrix} \frac{dw(x)}{dx} \\ 0 \end{bmatrix} + w(x)\,\varphi_s'(\psi_x(y)) \begin{bmatrix} \frac{\partial \psi_x(y)}{\partial x} \\ \frac{\partial \psi_x(y)}{\partial y} \end{bmatrix}, \tag{7}$$

with $\varphi_s'(\psi_x(y)) = d\varphi_s(\psi_x(y))/d\psi_x(y)$, for $s = 0,\dots,m$, and $w \in V_{1D}$, we rewrite the diffusive contribution in (6) as the sum of 1D diffusive-, convective-, and reactive-terms with respect to the unknowns $\widetilde{u}_j$, since

$$\int_{\Omega_{1D}} \Big\{ \Big(\int_{\gamma_x} \mu(x,y)\,\varphi_j(\psi_x(y))\,\varphi_k(\psi_x(y))\,dy \Big) \frac{d\widetilde{u}_j(x)}{dx}\,\frac{d\vartheta(x)}{dx} \tag{8}$$

$$+\Big(\int_{\gamma_x} \mu(x,y)\,\varphi_j(\psi_x(y))\,\varphi_k'(\psi_x(y))\,\frac{\partial \psi_x(y)}{\partial x}\,dy \Big) \frac{d\widetilde{u}_j(x)}{dx}\,\vartheta(x)$$

$$+\Big(\int_{\gamma_x} \mu(x,y)\,\varphi_j'(\psi_x(y))\,\varphi_k(\psi_x(y))\,\frac{\partial \psi_x(y)}{\partial x}\,dy \Big) \widetilde{u}_j(x)\,\frac{d\vartheta(x)}{dx}$$

$$+\Big(\int_{\gamma_x} \mu(x,y)\,\varphi_j'(\psi_x(y))\,\varphi_k'(\psi_x(y)) \Big\{ \Big[\frac{\partial \psi_x(y)}{\partial x}\Big]^2 + \Big[\frac{\partial \psi_x(y)}{\partial y}\Big]^2 \Big\} dy \Big) \widetilde{u}_j(x)\,\vartheta(x) \Big\}\,dx.$$

Similarly, we recast the convective term in (6) as the sum of a 1D convective and a 1D reactive term:

$$\int_{\Omega_{1D}} \Big\{ \Big(\int_{\gamma_x} b_1(x,y)\,\varphi_j(\psi_x(y))\,\varphi_k(\psi_x(y))\,dy \Big) \frac{d\widetilde{u}_j(x)}{dx}\,\vartheta(x) \tag{9}$$

$$+\Big(\int_{\gamma_x} \varphi_j'(\psi_x(y))\varphi_k(\psi_x(y)) \Big[b_1(x,y)\frac{\partial \psi_x(y)}{\partial x} + b_2(x,y)\frac{\partial \psi_x(y)}{\partial y} \Big] dy \Big) \widetilde{u}_j(x)\,\vartheta(x) \Big\}\,dx.$$

Finally the reactive contribution in (6) leads to a reactive term with respect to the $\widetilde{u}_j$'s:

$$\int_{\Omega_{1D}} \Big(\int_{\gamma_x} \sigma(x,y)\varphi_j(\psi_x(y))\,\varphi_k(\psi_x(y))\,dy \Big) \widetilde{u}_j(x)\,\vartheta(x)\,dx. \tag{10}$$

In practice all the integrals above on γ_x, as well as the forcing term in (6), are computed on the reference fiber $\widehat{\gamma}_1$, by properly exploiting the map ψ_x (i.e. both the Jacobian $\mathscr{J}(\widehat{y})$ and the deformation index $\mathscr{D}(\widehat{y})$). A straightforward arrangement of the terms in (8), (9) and (10) allows us to reformulate problem (5) as follows: for $j = 0,\dots,m$, find $\widetilde{u}_j \in V_{1D}$ such that, $\forall \vartheta \in V_{1D}$,

$$\sum_{j=0}^{m} \Big\{ \int_{\Omega_{1D}} \Big[\underbrace{\widehat{r}_{kj}^{\,1,1}(x)\,\frac{d\widetilde{u}_j(x)}{dx}\,\frac{d\vartheta(x)}{dx}}_{\text{(I)}} + \underbrace{\widehat{r}_{kj}^{\,1,0}(x)\,\frac{d\widetilde{u}_j(x)}{dx}\,\vartheta(x) + \widehat{r}_{kj}^{\,0,1}(x)\,\widetilde{u}_j(x)\,\frac{d\vartheta(x)}{dx}}_{\text{(II)}}$$

$$+\underbrace{\widehat{r}_{kj}^{\,0,0}(x)\,\widetilde{u}_j(x)\,\vartheta(x)}_{\text{(III)}} \Big] dx \Big\} = \int_{\Omega_{1D}} \Big[\int_{\widehat{\gamma}_1} f(x,\psi_x^{-1}(\widehat{y}))\varphi_k(\widehat{y})\,|\mathscr{J}^{-1}(\widehat{y})|\,d\widehat{y} \Big]\,\vartheta(x)\,dx, \tag{11}$$

with $k = 0,\dots,m$, where

$$\widehat{r}_{kj}^{\,s,t}(x) = \int_{\widehat{\gamma}_1} r_{kj}^{s,t}(x,\widehat{y})\,|\mathscr{J}^{-1}(\widehat{y})|\,d\widehat{y}, \quad \text{for} \quad s,t = 0,1, \tag{12}$$

and

$$r_{kj}^{1,1}(x,\widehat{y}) = \mu\left(x,\psi_x^{-1}(\widehat{y})\right)\varphi_j(\widehat{y})\,\varphi_k(\widehat{y}),$$
$$r_{kj}^{0,1}(x,\widehat{y}) = \mu\left(x,\psi_x^{-1}(\widehat{y})\right)\varphi_j'(\widehat{y})\,\varphi_k(\widehat{y})\,\mathscr{D}(\widehat{y}),$$
$$r_{kj}^{1,0}(x,\widehat{y}) = \mu\left(x,\psi_x^{-1}(\widehat{y})\right)\varphi_j(\widehat{y})\,\varphi_k'(\widehat{y})\,\mathscr{D}(\widehat{y}) + b_1\left(x,\psi_x^{-1}(\widehat{y})\right)\varphi_j(\widehat{y})\,\varphi_k(\widehat{y}),$$
$$r_{kj}^{0,0}(x,\widehat{y}) = \mu\left(x,\psi_x^{-1}(\widehat{y})\right)\varphi_j'(\widehat{y})\varphi_k'(\widehat{y})\left\{\left[\mathscr{D}(\widehat{y})\right]^2 + \left[\mathscr{J}(\widehat{y})\right]^2\right\} +$$
$$\sigma\left(x,\psi_x^{-1}(\widehat{y})\right)\varphi_j(\widehat{y})\varphi_k(\widehat{y}) + \varphi_j'(\widehat{y})\varphi_k(\widehat{y})\left\{b_1\left(x,\psi_x^{-1}(\widehat{y})\right)\mathscr{D}(\widehat{y}) + b_2\left(x,\psi_x^{-1}(\widehat{y})\right)\mathscr{J}(\widehat{y})\right\}.$$

Notice that in (11) the dependence of the reduced solution u_m on the main stream and on the transversal directions is split: coefficients $\widehat{r}_{kj}^{s,t}$ essentially collect the transversal contribution to the domain Ω_{1D}. We still recognize in (11) an ADR problem, the terms (I), (II) and (III) representing the diffusive, convective and reactive contribution, respectively.

From (8), (9) and (10) it is easy to see that the conversion from the full to the reduced framework is not one to one. Indeed a purely diffusive (advective) full term also yields reduced advective-reactive (reactive) contributions. The possible self-adjointness of the full problem is thus usually lost in the reduced framework. This property can be preserved in a few cases by a proper choice of the map ψ_x and of the reduced space V_m (see [3]).

From a computational viewpoint, solving (11) requires dealing with a small number of coupled 1D problems, provided that the modal index m is small enough. This is likely more convenient than solving the full problem (1).

Finally we point out that the computation of the $\widehat{r}_{kj}^{s,t}$'s in (12) simplifies considerably under particular assumptions on the data, e.g. for constant coefficients μ, $\mathbf{b}$, σ, or when the map ψ_x is affine (see [3] for the details).

4 Finite Element Approximation of the Reduced Problem

Formulation (11) can be understood as a *model semidiscretization* of the full problem (1), the transversal direction being discretized via the modal basis $\{\varphi_k\}$.

With a view to a full discretization of (1), we introduce a partition $\mathscr{T}_h$ of Ω_{1D} into sub-intervals $K_j = (x_{j-1}, x_j)$ of width $h_j = x_j - x_{j-1}$, and set $h = \max_j h_j$. We associate with $\mathscr{T}_h$ a finite element space $V_{1D}^h \subset V_{1D}$, with $\dim(V_{1D}^h) = N_h$, such that a standard density hypothesis of V_{1D}^h in V_{1D} is guaranteed.

The *discrete reduced formulation* is thus represented by system (11) solved on the subspace V_{1D}^h of V_{1D}, the test function ϑ coinciding now with the generic basis function ϑ_l of the finite element space, for $l = 1, \ldots, N_h$. Moreover, by expanding the unknown coefficients $\widetilde{u}_j^h$ in terms of the basis $\{\vartheta_i\}_{i=1}^{N_h}$ itself and by properly varying the indices k and l, we get a linear system with a $(m+1)N_h \times (m+1)N_h$ block matrix A. All the $N_h \times N_h$-blocks share the sparsity pattern proper of the adopted 1D finite element approximation, with the consequent benefits both in storing and solving the associated algebraic system.

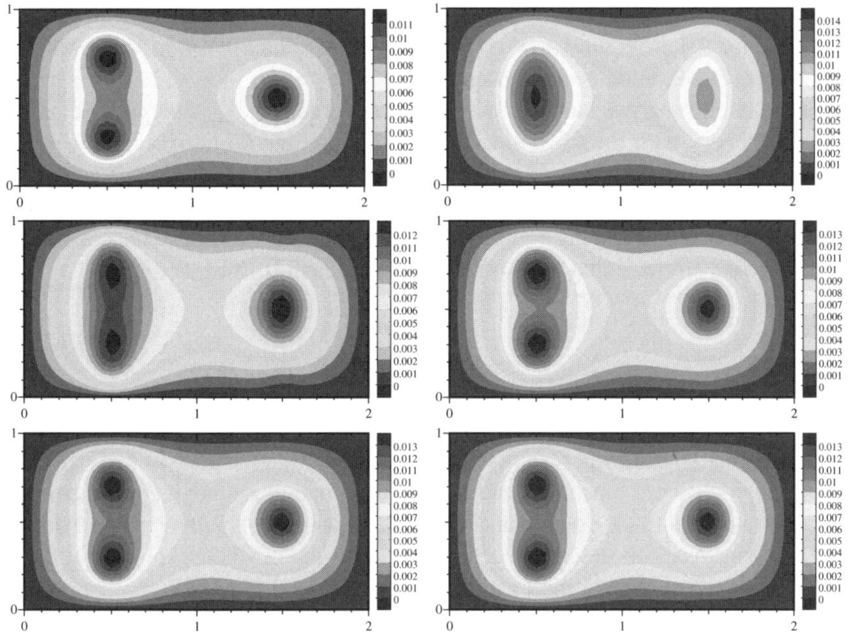

Fig. 2 Full solution u and reduced solutions u_2, u_4, u_6, u_8, u_9 (top-bottom, left-right)

5 Numerical Assessment

We look for a reliable and sufficiently accurate approximation of the full solution u to (1) by properly selecting the reduced space V_m in (3), namely the modal index m. The choice of m represents a crucial issue. It should be a trade-off between the needs to capture the main features of u and to contain the computational cost.

We adopt here a heuristic strategy where we first fix the index $m = 0$ and then we gradually increase such a value, while keeping it constant along the whole domain Ω_{1D}.

Let us focus on a purely diffusive differential problem exhibiting a heterogeneity in the corresponding source term. We solve the Poisson problem on the domain $\Omega = (0,2) \times (0,1)$, completed with homogeneous Dirichlet boundary conditions. The forcing term is localized in 3 circular regions of Ω, the function f in (1) coinciding with the characteristic function $\chi_{D_1 \cup D_2 \cup D_3}$, with $D_1 = \{(x,y) : (x-1.5)^2 + (y-0.5)^2 \leq 0.01\}$, $D_2 = \{(x,y) : (x-0.5)^2 + (y-0.25)^2 \leq 0.01\}$ and $D_3 = \{(x,y) : (x-0.5)^2 + (y-0.75)^2 \leq 0.01\}$. The associated full solution exhibits a peak in correspondence with each of the areas D_i, for $i = 1,2,3$ (see Fig. 2, top-left). Figure 2 gathers the reduced solutions corresponding to different choices for the modal index m. In particular it is evident the expected failure of the reduced solution u_2 (Fig. 2, top-right) to detect the two peaks of u along the straight line $x = 0.5$.

On the contrary u_2 already matches the exact value in correspondence with the peak in D_1 (notice the different scales). Nevertheless the reliability of u_m increases as m gets larger (Fig. 2, middle and bottom row).

These preliminary results confirm the convergence expected from classical Galerkin theory with the fulfillment of the assumptions *i)* and *ii)* in Sect. 3. It is worth pointing out that, even if a purely diffusive full model leads to an advective-diffusive-reactive reduced problem, the latter does not seem to suffer from convective or reactive numerical instabilities if $\mathscr{D}(\widehat{y})$ is small enough, since the convective-reactive terms are weighted by the diffusive coefficient μ itself (see [3]).

6 Conclusions and Future Developments

The preliminary numerical results in Sect. 5 suggest that the proposed dimensional reduction could be a reasonable approach for containing computational costs, in particular when both the domain and the problem at hand exhibit a "main stream direction". Many aspects deserve to be investigated. First of all the extension of the reduced approach to more complex problems (e.g., Oseen, Navier-Stokes equations). A second issue is the set-up of a mathematically sound procedure for selecting the proper modal index m. A possible approach could be based on the comparison between u_m and u_{m^+}, with $m^+ > m$. We investigate extensively this issue in [3]. An alternative solution is based on a domain decomposition approach, where different values of m are used in different parts of Ω ([3]): for instance, in the example of Fig. 2, a smaller value of m could suffice on the right half of the domain but not in the left half. In perspective this approach is suited to being coupled with a proper a posteriori modeling error analysis to get an automatic tool able to detect the most appropriate value m in the different parts of the domain in the spirit of a *model dimension adaptivity*.

References

1. Ainsworth, M.: A posteriori error estimation for fully discrete hierarchic models of elliptic boundary value problems on thin domains. Numer. Math. **80**, 325–362 (1998)
2. Babuška, I., Schwab, C.: A posteriori error estimation for hierarchic models of elliptic boundary value problems on thin domains. SIAM J. Numer. Anal. **33**, 221–246 (1996)
3. Ern, A., Perotto, S., Veneziani, A.: Hierarchical model reduction: a domain decomposition approach. In preparation (2008)
4. Vogelius, M., Babuška, I.: On a dimensional reduction method. I. The optimal selection of basis functions. Math. Comp. **37**, 31–46 (1981)
5. Vogelius, M., Babuška, I.: On a dimensional reduction method. II. Some approximation-theoretic results. Math. Comp. **37**, 47–68 (1981)

A Semi-Lagrangian Scheme for the Open Table Problem in Granular Matter Theory

M. Falcone and S. Finzi Vita

Abstract We introduce and analyze a new scheme for the approximation of the two-layer model proposed in [10] for growing sandpiles over an open flat table. The method is based on a semi-Lagrangian approximation of the nonlinear terms which allows to obtain more accurate results with respect to the standard finite difference approximation. We present several features of the scheme and give some hints on its implementation. Finally we show some tests where it is compared with a previously studied finite difference approach.

1 Introduction

Mathematical models for the dynamics of granular materials have recently been studied in several papers (see e.g. [1]), since a complete and realistic description of many phenomena in this field is far from being achieved. In this paper we restrict ourselves to the rather simple case of growing sandpiles over a flat bounded open table under the action of a given vertical source, neglecting wind effects and avalanches. For this phenomenon two main differential models have been proposed. The first one is the variational model of Prigozhin [11], where the surface flow of sand is supposed to exist only at critical slope, that is the maximal admissible slope α for any stationary configuration of sand (here for simplicity it is assumed $\alpha = 1$). For that model there are a sufficiently developed theory and efficient numerical schemes which use duality arguments (see [2]). The second approach has been introduced by Hadeler and Kuttler [10] as an extension of the known BCRE model [3] (see also [4]). In their model the pile is obtained summing two distinct

Maurizio Falcone
Dipartimento di Matematica, SAPIENZA - Università di Roma, e-mail: falcone@mat.uniroma1.it

Stefano Finzi Vita
Dipartimento di Matematica, SAPIENZA - Università di Roma, e-mail: finzi@mat.uniroma1.it

layers: the standing layer u and the rolling layer v. The two layers interact during the growth process, the exchange term between u and the v being described by the nonlinear term in the following system of partial differential equations:

$$\begin{cases} u_t = (1-|Du|)v & \text{in } \Omega \times (0,T) \\ v_t = \operatorname{div}(vDu) - (1-|Du|)v + f & \text{in } \Omega \times (0,T) \\ u(\cdot,0) = v(\cdot,0) = 0 \quad \text{in } \Omega, & u = 0 \quad \text{on } \Gamma \times (0,T)\,. \end{cases} \tag{1}$$

In (1) f represents the source, the table is assumed initially empty, and no boundary condition is needed on v (the characteristic lines for the second equation are all directed towards the interior of Ω). Note that this model allows for the movement of rolling grains even at sub-critical slope. It also seems to be better suited for the description of fast processes or small details over the surface (see [12] for a detailed comparison between the two models). However, the mathematical theory for the two-layer model is still incomplete (even a general result of existence for its solutions is lacking). It is interesting to note that the two distinct models have essentially the same set of admissible stationary solutions $(\overline{u},\overline{v})$ and that a characterization of such equilibria has been given, using the theory of viscosity solutions, in [5]: the maximal equilibrium $\overline{u}$ is given by the function $d(x)$ measuring at any point $x \in \Omega$ its distance to the boundary, and an integral representation is proved for $\overline{v}$ along of the transport rays of sand.

Anyway, as it was observed in [8], only in particular cases the asymptotic behavior is the same (and then $\overline{u}(x) = d(x)$). In general, the effective stationary standing solution $\overline{u}$ for the model (1) is not explicitly known. It is also for such a reason that in [8] we proposed a finite difference scheme which preserves the properties of the model at the discrete level, and which is able to give a numerical characterization of stationary solutions for every general source support in a square open table. A disadvantage of this approach is that finite difference schemes on a regular structured grid restrict the numerical discrete flow essentially to the axis directions, breaking the homogeneous character of the real phenomenon (cones of sand are in general approximated by pyramids oriented along the grid axes). This numerical anisotropy is typical of finite difference (FD) schemes and trying to eliminate it is particularly important in view of the extension of schemes to the so-called partially-open table problem, that is when high vertical walls bound portions of the table boundary. In fact, in such a problem an infinite number of transport rays can meet at certains points on the boundary, yielding a discontinuous and even unbounded rolling layer $v \in L^1(\Omega)$ at the equilibrium, and strong numerical difficulties in its description (see [6]). This has motivated the development of a semi-Lagrangian (SL) approach which will be discussed in this paper. By construction, the SL approach mimics the method of characteristics in order to follow in a more accurate way the real direction of sand flow. We introduce the scheme and its properties in Section 2, and in Section 3 we present first numerical tests in the regular case of the open table problem, showing a slight but clear improvement in the accuracy with respect to the standard FD approach. These results let us conjecture that in the partially-open table problem the SL approach could be of interest, despite of its time-consuming feature. Experiments in this direction are actually under consideration.

A detailed presentation of SL methods for evolutive nonlinear partial differential equations of the first order can be found in [7]; see also [9] for a convergence result of a first order SL scheme to the viscosity solution.

2 The Semi-Lagrangian Scheme and its Properties

Let us start from the first equation in (1) and observe that it can be equivalently written in one dimension as

$$u_t = v(1-|u_x|) = v\left(1+\min_{a\in B(0,1)} au_x\right) , \tag{2}$$

where $B(0,1)$ denotes the closed unit ball of $\mathbf{R}$, that is the interval $[-1,1]$. In this way the nonlinear term $|u_x|$ is written as the minimum over the directional derivatives of u and is actually computed comparing the values for $a=1$, $a=-1$ and $a=0$. Note that in two dimensions we can use exactly the same formula to compute $|\nabla u|$. At the discrete level, if we introduce in $\Omega=(\alpha,\beta)$ a uniform grid of points $G=\{x_i : x_i=\alpha+i\Delta x, i=0,..,N\}$, (2) can be written as

$$u_i^{n+1} = u_i^n + \Delta t v_i^n\left(1+\min_{a\in\{-1,0,1\}}\frac{u^n(x_i+a\Delta t)-u_i^n}{\Delta t}\right) , \tag{3}$$

where, as usual, u_i^n denotes the approximate value for $u(x_i,n\Delta t)$. If $\Delta t\neq\Delta x$ and $a\neq 0$, then the point $z_i(a)\equiv x_i+a\Delta t$ is not a node of the grid G and the value of u^n in it has to be computed, for example, by linear interpolation. For the equation in v, we have also second order terms. In fact,

$$v_t = v_x u_x + v u_{xx} - v(1-|u_x|) + f ; \tag{4}$$

let us use the standard second order central difference $D^2u_i^n$ to approximate $u_{xx}(x_i)$ whereas u_x at any node is replaced by the maximal (in absolute value) finite difference in the left and right directions, Du_i^n. A simple trick is to replace the nonlinear term at the node x_i by the previously computed difference $(u_i^{n+1}-u_i^n)$. Finally, for v_x we use the discrete directional derivative in the direction of u_x. After a simplification, we finally obtain the following discretization of (4)

$$v_i^{n+1} = v^n(x_i+\Delta t Du_i^n) - (u_i^{n+1}-u_i^n) + \Delta t(v_i^n D^2u_i^n + f_i) . \tag{5}$$

Of course, also the term $v^n(x_i+\Delta t Du_i^n)$ requires a local reconstruction by interpolation. In order to complete the scheme we need to add initial and boundary conditions

$$u_i^0 = v_i^0 = 0 \quad \forall i ; \quad u_0^n = u_N^n = 0 \quad \forall n . \tag{6}$$

The extension of this approach to $\mathbf{R}^2$ is straightforward.We assume for simplicity that $\Omega = (0,1)\times(0,1)$, that is the unit square table, and we introduce a uniform grid of points $G = \{x_{i,j} = (i\Delta x, j\Delta x), i,j = 0,..,N\}$, where $\Delta x = 1/N$ is again the space discretization step. Then the analogue of (3)-(5) and (6) is

$$\begin{cases} u_{i,j}^{n+1} = u_{i,j}^n + \Delta t v_{i,j}^n \left(1 + \dfrac{1}{\Delta t} \min\limits_{a\in B(0,1)} (u^n(x_{i,j}+a\Delta t) - u_{i,j}^n)\right) , \\ v_{i,j}^{n+1} = v^n(x_{i,j}+\Delta t Du_{i,j}^n) - (u_{i,j}^{n+1} - u_{i,j}^n) + \Delta t(v_{i,j}^n D^2 u_{i,j}^n + f_{i,j}) , \\ u_{i,j}^0 = v_{i,j}^0 = 0 \quad \forall i,j \,; \qquad u_{i,j}^n = 0 \quad \forall n, \text{ if } x_{i,j}\in\partial\Omega \, , \end{cases} \tag{7}$$

where $u_{i,j}^n$ denotes the approximate value for $u(x_{i,j}, n\Delta t)$. In order to compute the value of $u^n(x_{i,j}+a\Delta t)$ we use now a bilinear interpolation with respect to the four vertices of the cell containing $x_{i,j}+a\Delta t$. The same local reconstruction is required for the term $v^n(x_{i,j}+\Delta t Du_{i,j}^n)$. The minimum term in (7) is approximated by comparing the values of u^n on a finite set of directions (for example the eight directions $\theta_k = k\pi/4$, $k = 1,..,8$, plus the origin, that is the node under consideration). $D^2 u_{i,j}^n$ denotes the standard five-points second order difference which replaces $\Delta u(x_{i,j})$.

2.1 Properties of the SL Scheme

Here we present some features of the above scheme, showing that it preserves the physical properties of the continuous model. At the moment we are able to give a complete proof of that only in the one-dimensional case, but these properties are confirmed by all the 2D experiments. Further details and complete proofs will be presented in a forthcoming paper.

Theorem 1. *Let $f\geq 0$ in Ω and*

$$\frac{\Delta t}{\Delta x} \leq \min\left(\frac{1}{2}, \frac{1}{C\|f\|_\infty}\right), \tag{8}$$

where C is a positive constant. Then, the sequences u^n and v^n defined in (3)-(5)-(6) satisfy the following properties for any n:

1. *positivity and monotonicity in u:* $0 \leq u^n \leq u^{n+1}$;
2. *positivity in v:* $v^n \geq 0$;
3. *sub-critical slope in u:* $|Du^n| \leq 1$.

Proof. The structure of the proof is the same of that of Theorem 3.1 of [8]. The above properties are proved by induction. Here we only discuss some details which are characteristic of the semi-Lagrangian scheme. In particular, for the nonlinear term in (3) we remark that

$$-1 \leq \min_{a\in\{-1,0,1\}} \frac{u^n(x_i+a\Delta t) - u_i^n}{\Delta t} \leq 0 \, , \quad \forall n,i \, ; \tag{9}$$

moreover, since the intermediate values are computed in this case by linear interpolation and $\Delta t \leq \Delta x$, we have for example that

$$\frac{u^n(x_i + \Delta t) - u_i^n}{\Delta t} = \frac{u_{i+1}^n - u_i^n}{\Delta x} \,. \tag{10}$$

At the same time, if for example $Du_i^n > 0$, we can rewrite

$$v^n(x_i + \Delta t Du_i^n) = v_i^n + \frac{\Delta t}{\Delta x}(v_{i+1}^n - v_i^n)Du_i^n \,, \tag{11}$$

and the proof of positivity for v^n follows the same arguments used for the FD scheme in [8], yielding the bound $2\Delta t \leq \Delta x$.

The stronger stability bound (8) has to be assumed in order to achieve the gradient constraint in every interval $[x_i, x_{i+1}]$. An accurate discussion on admissible configurations allows to rewrite the different quotient in (9) always in terms of the slope of u^n inside the interval under consideration. Then, considering the difference $(u_{i+1}^n - u_i^n)$ as defined by (3), one gets sub-criticality if $(\Delta t \ v_i^n) \leq \Delta x$, which holds for the uniform boundedness of v^n in terms of the source f. □

3 Numerical Results

In this section we present some numerical tests in two dimensions to compare the SL method with the old FD method. For computation we used MATLAB 7.0 on a Processor Intel Pentium M740, 1.73GHz with 80Gb RAM.

Test 1: Constant Source on all of Ω

We assume $f \equiv 0.5$ in all the domain $\Omega = (0,1) \times (0,1)$. In this case the ridge set S (the set where d is singular) is given by the two diagonals of the square and, following [5], the stationary solutions are given by $\overline{u}(x) = d(x)$ and

$$\overline{v}(x) = 0 \ \text{ on } S, \quad \overline{v}(x) = \int_0^{\tau(x)} f(x + tD\overline{u}(x))\, dt \ \text{ in } \Omega \setminus S, \tag{12}$$

where $\tau(x)$ denotes the distance of x from the set S along the transport ray through x. In Fig. 1 we show the computed solutions $\overline{u}$ and $\overline{v}$ for $\Delta x = 0.02$, $\Delta t = 0.01$ (the stopping criterion for the equilibrium was: $\|u^{n+1} - u^n\|_\infty \leq 10^{-6}$).

We show in Table 1 the results of the two schemes at the same final time when $\Delta t/\Delta x$=0.4, comparing the L^∞ and L^1 norms of the error for u. It can be seen that the two methods are both approximately of first order. Nevertheless, the SL scheme is slightly more accurate in the L^∞ (+12%) and the L^1 (+30%) norms respectively. Naturally, it is also much more expensive due to the minimum term computation at any iteration.

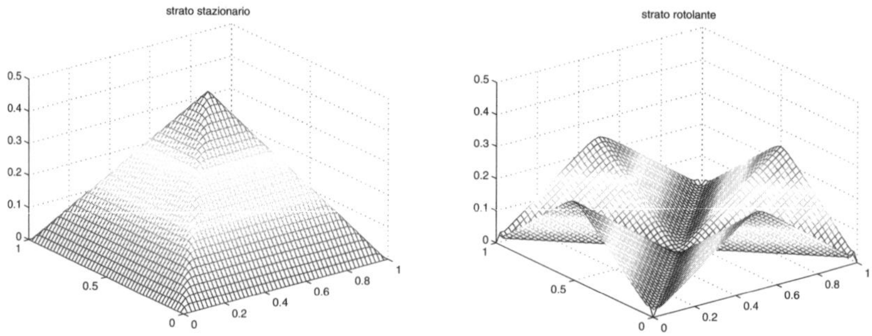

Fig. 1 Test 1: Numerical results for u and v

Table 1 Errors for Test 1 at CFL=0.4

method	Δx	L^∞ error	L^1 error
FD	0.05	0.01718	0.03812
SL	0.05	0.01499	0.02692
FD	0.025	0.00878	0.02027
SL	0.025	0.00770	0.01406
FD	0.0125	0.00457	0.01043
SL	0.0125	0.00404	0.00718

Test 2: Constant Source on a Connected Subdomain of Ω

Assume now that the constant source is concentrated in a small square inside Ω (see Fig. 2), and it is zero elsewhere. An explicit formula for u is not known in this case (with some efforts v could be still computed by formula (12)), and Fig. 2 shows the stationary solution u as computed by the SL algorithm. In Fig. 3 the level lines found for u are compared with those found by the FD scheme.

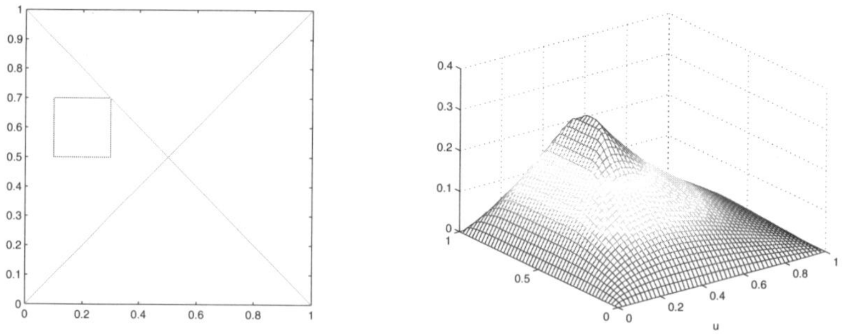

Fig. 2 Test 2: The source support and the computed standing layer u

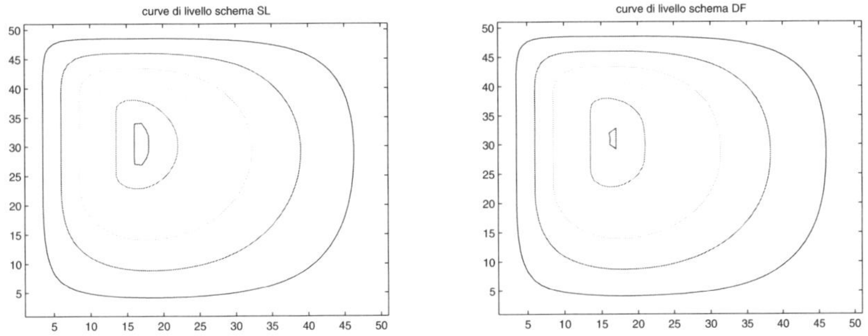

Fig. 3 Test 2: Level curves for u with the SL and the FD schemes

Test 3: Constant Source on a Disconnected Subdomain of Ω

Assume now that the constant source is concentrated in two distinct small squares inside Ω, and it is zero elsewhere. Fig. 4 shows this source support together with the stationary solution u as computed by the SL algorithm. In Fig. 5 the level lines found for u by the two algorithms are compared.

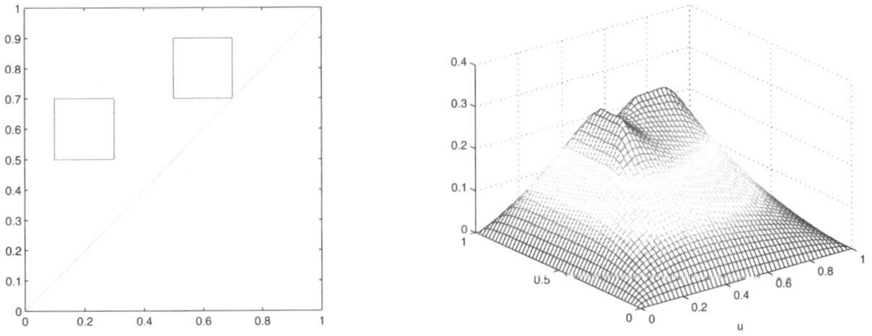

Fig. 4 Test 3: The source support and the computed standing layer u

Conclusions

The above results seem to show the SL scheme, although enough expensive, can be more accurate than the corresponding FD scheme in the description of singularity regions for the solution. This can be seen by comparing the higher level sets of the two numerical solutions, those corresponding to the crests of the piles. This feature

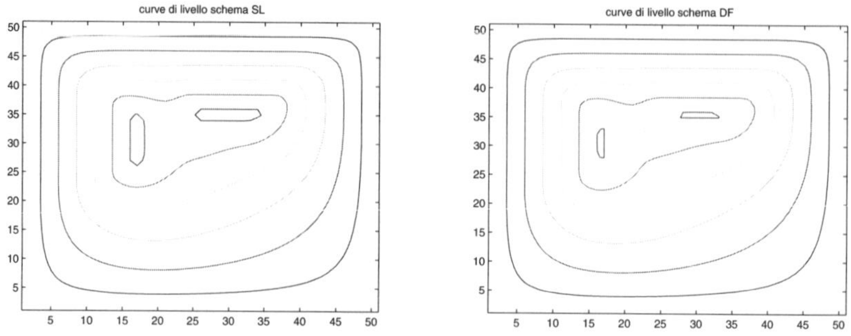

Fig. 5 Test 3: Level curves for u with the SL and the FD schemes

can be very important for more complicated problems and geometries, like for the wall and the silos problems, where hardest singularities can arise (see [6]).

Acknowledgements We thank Alessia Pacella for the numerical experiments.

References

1. Aranson, I.S., Tsimring, L.S.: Patterns and collective behavior in granular media: theoretical concepts. arXiv:cond-mat/0507419 (2005)
2. Barrett, J.W., Prigozhin, L.: Dual formulations in critical state problems. Interfaces and Free Boundaries **8**, 347–368 (2006)
3. Bouchaud, J.-P., Cates, M.E., Ravi Prakash, J., Edwards, S.F.: A model for the dynamics of sandpile surfaces. J. Phys. I France **4**, 1383–1410 (1994)
4. Boutreux, T., de Gennes, P.-G.: Surface flows of granular mixture, I. General principles and minimal model. J. Phys. I France **6**, 1295–1304 (1996)
5. Cannarsa, P., Cardaliaguet, P.: Representation of equilibrium solutions to the table problem for growing sandpiles. J. Eur. Math. Soc. (JEMS) **6**, 435–464 (2004)
6. Crasta, G., Finzi VIta, S.: An existence result for the sandpile problem on flat tables with walls. To appear; preprint available at http://cpde.iac.rm.cnr.it/ (2008)
7. Falcone, M., Ferretti, R.: Semi-Lagrangian schemes for Hamilton-Jacobi equations, discrete representation formulae and Godunov methods. J. Comp. Physics **175**, 559–575 (2002)
8. Falcone, M., Finzi Vita, S.: A finite difference approximation of a two-layer system for growing sandpiles. SIAM J. Sci. Comput. **28**, 1120–1132 (2006)
9. Falcone M., Giorgi T.: An approximation scheme for evolutive Hamilton-Jacobi equations. In: W.M. McEneaney, G. Yin, Q. Zhang (eds.), "Stochastic Analysis, Control, Optimization and Applications: A Volume in Honor of W.H. Fleming", Birkhäuser, 289–303 (1999)
10. Hadeler, K.P., Kuttler, C.: Dynamical models for granular matter. Granular Matter **2**, 9–18 (1999)
11. Prigozhin, L.: Variational model of sandpile growth. Euro. J. Appl. Math. **7**, 225–235 (1996)
12. Prigozhin, L., Zaltzman, B.: Two continuous models for the dynamics of sandpiles surface. Physical Review E **63**, 041505 (2001)

On a Variational Approximation of the Effective Hamiltonian

M. Falcone and M. Rorro

Abstract The approximation of the effective Hamiltonian is a challenging problem with a strong impact on many applications e.g. to the study of dynamical systems, weak KAM theory, homogenization, mass transfer problems. In this paper we present a numerical approximation of the variational approach proposed by C. Evans in [4], discuss its consistency and give some hints regarding its implementation. Finally, we compare this approach to the numerical implementation of the min-max formula proposed by Gomes and Oberman [6].

1 Introduction

The cell problem, as originally introduced circa in 1988 by [7], corresponds to the following equation

$$H(x, Du + P) = \lambda \text{ on } \mathbb{T}^N \tag{1}$$

where $\mathbb{T}^N$ is the unit flat torus, $P \in \mathbb{R}^N$ is a fixed vector and the unknown is the pair (u, λ). Let us assume that H is Lipschitz continuous, coercive and convex in the second variable. Then, it is well known (see [7], [3] for details) that for each fixed $P \in \mathbb{R}^N$ there exists a unique real number λ such that (1) has a periodic Lipschitz continuous viscosity solution. Such unique value, as a function of P, is the so called the *effective Hamiltonian* related to H and it is usually denoted by $\overline{H}(P)$. The major difficulty is to determine the function $\overline{H} : \mathbb{R}^N \to \mathbb{R}$, a rather difficult and heavy computation even for low dimensional problems.

Maurizio Falcone
Dipartimento di Matematica, SAPIENZA - Università di Roma, e-mail: falcone@mat.uniroma1.it

Marco Rorro
Dipartimento di Metodi e Modelli Matematici per le Scienze Applicate, SAPIENZA - Università di Roma, e-mail: rorro@caspur.it

Since the effective Hamiltonian verifies the following identity

$$\overline{H}(P) = \inf_{u \in C^1(\mathbb{T}^N)} \sup_{x \in \mathbb{T}^N} H(x, Du + P) \tag{2}$$

usually indicated as the *min-max formula* (see [2] and [6]) one can think that the above characterization can lead to an algorithm. In fact, a direct discretization has been proposed in [6] and some examples have been computed using that formula. The main idea in that approach is to discretize $C^1(\mathbb{T}^N)$ by piecewise linear functions (the P_1 approximation of finite elements) and then apply a min-max search to the discrete formula using MATLAB. Here we try to improve the performances of the above method using the `FFSQP` library [10] which has been conceived to solve nonlinear min-max problems. Although this code has a better performance with respect to the MATLAB `minimax` function used in [6], the method is still too expensive in terms of *CPU* times and seems to be inadequate to compute $\overline{H}(P)$ with a reasonable accuracy (see the last section for details). This experience has motivated further efforts to find new ways to compute the effective Hamiltonian. We note that one of the main difficulties in both problems (1) and (2) is that, even if the value of $\overline{H}(P)$ is unique for each fixed P, the solution of (1) or the minimizer of (2) is in general not unique. In [9] and [8] different regularizations of (1) are considered (see [1] for some a priori estimates). Here we follow a different variational approach based on a regularization of the min-max formula (2) that was proposed in [4]. This approach is in principle less accurate than the direct discretization of the min-max formula because we have replaced the original problem by a regularized problem introducing an additional error. However, as we will explain in the sequel, this approach is simpler and more efficient from a computational point of view. The numerical results presented in Section 3 show that the CPU times are drastically reduced by this method. We note that for this approach a strict convexity of the Hamiltonian in the second variable is required.

2 A Variational Approximation

As we said in the introduction, our starting point is the approximation of the effective Hamiltonian proposed by Evans in [4]. This is defined by

$$\overline{H}_k(P) := \frac{1}{k} \log\left(\int_{\mathbb{T}^n} e^{kH(x, Du_k + P)} \mathrm{d}x \right), \tag{3}$$

where $k \in \mathbb{N}$ and $u_k \in C^1(\mathbb{T}^N)$ is a minimizer of the functional

$$I_k[u_k] = \int_{\mathbb{T}^n} e^{kH(x, Du_k + P)} \mathrm{d}x$$

subject to the normalization

$$\int_{\mathbb{T}^N} u_k \mathrm{d}x = 0.$$

Theorem 1 ([4]). *Suppose $H(x,p)$ is strictly convex in p. Then:*

$$\overline{H}(P) = \lim_{k \to +\infty} \overline{H}_k(P). \tag{4}$$

Moreover, the following estimates hold true:

$$\overline{H}_k(P) \leq \overline{H}(P) \leq \overline{H}_k(P) + C\frac{\log k}{k} \tag{5}$$

for any $k \in \mathbb{N}$. Let us introduce the new variable $v_k = u_k + Px$ and observe that it satisfies the Euler–Lagrange equation

$$\operatorname{div}\left(e^{kH(x,Dv_k)} D_p H(x,Dv_k)\right) = 0. \tag{6}$$

The idea of the approximation is then to construct a finite difference approximation of (6) to compute v_k. For simplicity, let us fix the dimension to $N = 1$ and assume that the grid G is a standard lattice $G = \{x_i : x_i = i\Delta x\}$. Using a standard second order finite difference approximation for v_x and a central approximation for v_{xx}

$$v_x = \frac{v_{i+1} - v_{i-1}}{2\Delta x} \quad \text{and} \quad v_{xx} = \frac{v_{i+1} - 2v_i + v_{i-1}}{\Delta x^2} \tag{7}$$

we end up with a sparse nonlinear system of n equations in the n unknown $v_1, \ldots, v_n$ (n is the number of nodes). Since the term v_i is contained only in the discretization of the second derivative, it is easier to solve the i-th equation with respect to v_i, $v_i = F_i(v_{i+1}, v_{i-1})$, by the iterative scheme

$$v_i^{m+1} = F_i(v_{i+1}^m, v_{i-1}^m) \quad \text{for } i = 1, \ldots, n. \tag{8}$$

Once a minimizer is obtained, we compute $\overline{H}_k(P)$ by a normalization of formula (3), that we get by adding and subtracting $\max_{x \in \mathbb{T}^N} H(x,Dv_k)$, to obtain

$$\overline{H}_k(P) = \max_{x \in \mathbb{T}^N} H(x,Dv_k) + \frac{1}{k} \log\left(\int_{\mathbb{T}^n} e^{k\left(H(x,Dv_k) - \max_{x \in \mathbb{T}^N} H(x,Dv_k)\right)} dx\right). \tag{9}$$

It is easy to verify that these formulae are equivalent. Formula (9) is more convenient from a numerical point of view. In fact, by this formula we can avoid the overflow error due to the computation of the exponential under the integral. Let us explain this procedure by an example in one dimension. Consider the following Hamiltonian

$$H(x,Du) = \frac{1}{2}|Du|^2 - f(x).$$

The Euler–Lagrange equation for the functional $I_k[v_k]$ is

$$\operatorname{div}\left(e^{k\left(\frac{1}{2}|Dv_k|^2-f(x)\right)}Dv_k\right)=0.$$

We can write it in the equivalent form

$$k\left(v_x v_{xx}-f_x(x)\right)v_x+v_{xx}=-kf_x(x)v_x+\left(kv_x^2+1\right)v_{xx}=0,$$

which by finite differences (7) becomes

$$-kf_x(x)\left(\frac{v_{i+1}-v_{i-1}}{2\Delta x}\right)+\left(k\left(\frac{v_{i+1}-v_{i-1}}{2\Delta x}\right)^2+1\right)\left(\frac{v_{i+1}-2v_i+v_{i-1}}{\Delta x^2}\right)=0. \tag{10}$$

Finally, we can write (10) in a fixed point form

$$v_i=F_i(v_{i+1},v_{i-1}):=\frac{v_{i+1}+v_{i-1}}{2}-\Delta x^3\frac{kf_x(x)(v_{i+1}-v_{i-1})}{k(v_{i+1}-u_{i-1})^2+4\Delta x^2}.$$

In higher dimension, the Euler–Lagrange equation and its finite difference approximation have been computed and solved respect to v_i, using the symbolic computation software MAPLE. The fixed point scheme has been implemented thorough the software HJPACK [5], exploiting the structure and the parallelism already defined in that software.

3 The Gomes-Oberman Discretization Revisited

The approximated effective Hamiltonian, $\overline{H}^{\Delta x}$, proposed in [6], from a direct discretization of the min-max formula (2), is

$$\overline{H}^{\Delta x}(P)=\inf_{u\in W^1}\operatorname*{ess\,sup}_{x\in\mathbb{T}^N}H(x,Du+P)$$

where $W^1\equiv\left\{w:\mathbb{T}^N\to\mathbb{R}:w\in C(\mathbb{T}^N)\text{ and }Dw(x)=c_j\,\forall x\in T_j,\forall j\right\}$, T_j is a family of simplices such that $\mathbb{T}^N=\bigcup_j T_j$ and $\Delta x\equiv\max_j\operatorname{diam}(T_j)$.

Proposition 1 ([6]). *Suppose $H(x,p)$ is convex in p. Then $\overline{H}^{\Delta x}(P)$ is convex,*

$$\overline{H}(P)=\lim_{\Delta x\to 0}\overline{H}^{\Delta x}(P) \tag{11}$$

and

$$\overline{H}(P)\le\overline{H}^{\Delta x}(P) \tag{12}$$

It is interesting to note that some a priori error estimates are also available for this approximation scheme. More precisely, when u is Lipschitz continuous (which is the case when $H(x,p)$ is strictly convex in p), we have

$$\overline{H}^{\Delta x}(P)\le\overline{H}(P)+O(\Delta x^{1/2}). \tag{13}$$

and when $u \in C^2(\mathbb{T}^N)$, the above estimate can be improved to

$$\overline{H}^{\Delta x}(P) \leq \overline{H}(P) + O(\Delta x). \tag{14}$$

It is natural to discretize the spatial variable by computing the supremum only on the nodes of the triangulation x_i, $i = 1, \ldots, n$. So the fully discrete min-max problem is

$$\min_{u \in W^1} \max_{x_i} H(x_i, Du(x_i) + P). \tag{15}$$

The spatial approximation introduces an additional error of $O(\Delta x)$, which is proportional to the Lipschitz constant (in the x variable) of H. The min-max problem (15) is a finite dimensional nonlinear optimization problem

$$\min_{Du} F(Du) \tag{16}$$

where the map $F : \mathbb{R}^{Nn} \to \mathbb{R}^{Nn}$ with N the dimension and n the number of nodes is defined by $\max_i F_i(Du) = \max_i H(x_i, Du + P)$. We note that the map F is still convex as long as H is convex. In [6] a discretization of Du by finite difference is used setting $Du(x_i) = (u_{i+1} - u_i)/\Delta x$ which can be solved by using standard optimization routines. The periodicity of u is automatically verified imposing

$$\frac{u_{n+1} - u_n}{\Delta x} = \frac{u_1 - u_n}{\Delta x}. \tag{17}$$

Instead of introducing a discretization of $Du(x_i)$, we consider it as an independent variable, c_i, and so we consider the linear constrained optimization problem

$$\min_{c_i} \max_i H(x_i, c_i + P) \quad \text{subject to } \sum_i c_i = 0. \tag{18}$$

The linear constraint is equivalent to impose u to be periodic. Although the linear constraint makes the problem harder, it improves the accuracy of the solution as $H \notin C^1$. We note that the min-max problem can be rewritten, with the addition of a new variable y, as a nonlinear constrained optimization problem

$$\min_{c_i, y} \{y : H(x_i, c_i + P) \leq y\}. \tag{19}$$

In [6], the `fminimax` function, contained in the MATLAB Optimization Toolbox, is used to solve the problem. Here instead we use the optimization routine `ffsqp` [10]. Both the algorithms are based on Sequential Quadratic Programming. `ffsqp` tackles the problem directly, without using the equivalent form (19). It also provides two kinds of line search (monotone and nonmonotone). We use the nonmonotone line search, which forces a decrease of the objective function within at most four iterations, since the monotone line search (of Armijo type) does not work in some experiments where H is not strictly convex. We use `ffsqp` providing the gradient of the linear constraint and let it compute the gradient of the objective function. It uses a forward finite differences approximation.

4 Numerical Results

An implicit solution for $\overline{H}$ if $N=1$ and $H(x,p)=\frac{1}{2}(Du+P)^2-f(x)$ with f periodic of period one is (see [7] for details)

$$\overline{H}(P)=\begin{cases} -\min f & \text{if } |P|<\int_0^1\sqrt{2[f(x)-\min f]}\mathrm{d}x \\ c:|P|=\int_0^1\sqrt{2[f(x)+c]}\mathrm{d}x & \text{otherwise.}\end{cases} \tag{20}$$

We use this result in the following experiment to validate the schemes.

The one dimensional pendulum

The Hamiltonian of the one dimensional pendulum is $\mathscr{H}_1(x,Du)=\frac{1}{2}|Du|^2+\cos 2\pi x$ From (20) we get that $\overline{\mathscr{H}}_1(P)=1$ if $|P|<4\pi^{-1}$. Let us define

$$e(P)=\begin{cases} 1-\overline{\mathscr{H}}_1^{\Delta}(P) & \text{if } |P|<4\pi^{-1} \\ \\ |P|-\mathtt{quad}\left(\sqrt{2(\cos(2\pi x)+\overline{\mathscr{H}}_1^{\Delta}(P)}\right) & \text{otherwise.}\end{cases}$$

where $\overline{\mathscr{H}}_1^{\Delta}$ is the computed effective Hamiltonian and `quad` is a MATLAB function we use to compute the integral in (20) with $c=\overline{\mathscr{H}}_1^{\Delta}$ and with a tolerance of 10^{-12}. In the following tables we show the L^1 and L^∞ norm of e

$$\|e\|_1=\sum_i |e(P_i)| \qquad \|e\|_\infty=\max_i |e(P_i)| \tag{21}$$

and the numerical order of convergence, i.e.

$$\frac{\log(\|e_{\Delta x_1}\|/\|e_{\Delta x_2}\|)}{\log(\Delta x_1/\Delta x_2)}$$

where $\Delta x_1=n^{-1}$ refers to the previous approximation and $\mathscr{H}_1^{\Delta}$ has been computed over a grid on $[0,2\pi]$ with $\Delta P=(4\pi)^{-1}$. We note that the L^1 and L^∞ error for the min-max approximation (Table 1) are almost the same. In fact most of the error is concentrated at $P=4/\pi$ where $\overline{\mathscr{H}}_1$ is not differentiable.

In the variational approximation one would like to fix k as big as possible, or better, if $|\overline{H}_k-\overline{H}_k^{\Delta}|\le C(\Delta x)^p$ and $k=C(\Delta x)^{-q}$, since

Table 1 L^1 and L^∞ norm of e and deduced order by the min-max approximation

n	L^1	L^∞	L^1-order	L^∞-order
10	3.7×10^{-3}	3.2×10^{-3}		
50	1.0×10^{-4}	1.0×10^{-4}	2.2	2.1
100	2.4×10^{-5}	2.4×10^{-5}	2.1	2.1
200	5.5×10^{-6}	5.5×10^{-6}	2.1	2.1
400	1.3×10^{-6}	1.3×10^{-6}	2.0	2.0

$$\begin{aligned}|\overline{H}-\overline{H}_k^{\Delta}| &\le |\overline{H}-\overline{H}_k|+|\overline{H}_k-\overline{H}_k^{\Delta}| \le C\left(\frac{\log k}{k}+(\Delta x)^p\right)\\ &\le C\left((\Delta x)^q\log\left((\Delta x)^{-q}\right)+(\Delta x)^p\right)\\ &= C(\Delta x)^p\left((\Delta x)^{q-p}\log\left((\Delta x)^{-q}\right)+1\right)\end{aligned} \tag{22}$$

one would like to choose $q=p+1$. Unfortunately the scheme shows some instability if k is too big respect to n, above all if P belongs to the flat part of $\overline{H}$, in this case $(-4/\pi,4/\pi)$. So we fix $k=2n/5$ $(q=1)$ and the stopping criterion $\varepsilon=n^{-3}/10$. Hence the predominant part of the error is $\log n/n$, as confirmed also from the deduced order in Table 2. We note that most of the error is concentrated in flat part of $\overline{H}$, which causes the increase of the L^1 error.

Table 2 L^1 and L^∞ norm of e and deduced order by the variational approximation.

n	L^1	L^∞	L^1-order	L^∞-order
10	5.8	3.9×10^{-1}		
50	1.9	1.2×10^{-1}	0.7	0.7
100	1.1	6.9×10^{-2}	0.8	0.8
200	6.3×10^{-1}	3.9×10^{-2}	0.8	0.8
400	3.5×10^{-1}	2.2×10^{-2}	0.8	0.8

Two uncoupled penduli

Consider the Hamiltonian $\mathscr{H}_2(x,Du)=\frac{1}{2}|Du|^2+\cos 2\pi x+\cos 2\pi y$. In this case

$$\overline{\mathscr{H}}_2(P)=\overline{\mathscr{H}}_1(P_1)+\overline{\mathscr{H}}_1(P_2)$$

where $P=(P_1,P_2)$. The error behaves like in the one dimensional case, but the computational cost, that can be neglected in the one dimensional case, here begins to be relevant, as shown in Table 3 and 4 where it is shown the CPU time needed to compute just one value of $\overline{H}$ on an IBM power5 at 1.9GHz. Although the direct discretization of the min-max formula is more accurate, its computational cost is too high.

Table 3 $e(P)$ at $P = (4\pi^{-1}, 4\pi^{-1})$ and CPU time for the min-max method

n	$e(P)$	order	seconds
10	6.1×10^{-3}		6s
20	1.4×10^{-3}	2.1	1183s

Table 4 $e(P)$ at $P = (4\pi^{-1}, 4\pi^{-1})$ and CPU time for the variational method

n	$e(P)$	order	seconds
10	6.5×10^{-2}		0s
20	5.3×10^{-2}	0.3	0s
50	3.1×10^{-2}	0.6	0s
100	1.9×10^{-2}	0.7	4s
200	1.2×10^{-2}	0.7	67s
400	6.9×10^{-3}	0.8	591s
800	4.0×10^{-3}	0.8	23910s

Acknowledgements The second author wishes to thank Craig Evans for his advice and for the many enlightning discussions. The authors wish to thank also the CASPUR consortium for the technical support to the numerical tests on parallel architectures.

References

1. Camilli, F., Capuzzo Dolcetta, I., Gomes, D.A.: Error estimates for the approximation of the effective hamiltonian (2006). URL http://cpde.iac.rm.cnr.it/file_uploaded/FES30074.pdf. preprint
2. Contreras, G., Iturriaga, R., Paternain, G.P., Paternain, M.: Lagrangian graphs, minimizing measures and Mañé's critical values. Geom. Funct. Anal. **8**(5), 788–809 (1998)
3. Evans, L.C.: Periodic homogenisation of certain fully nonlinear partial differential equàtions. Proc. Roy. Soc. Edinburgh Sect. A **120**(3-4), 245–265 (1992)
4. Evans, L.C.: Some new PDE methods for weak KAM theory. Calc. Var. Partial Differential Equations **17**(2), 159–177 (2003)
5. Falcone, M., Lanucara, P., Rorro, M.: HJPACK Version 1.9 User's Guide (2006). URL http://www.caspur.it/hjpack/user_guide1.9.pdf
6. Gomes, D.A., Oberman, A.M.: Computing the effective Hamiltonian using a variational approach. SIAM J. Control Optim. **43**(3), 792–812 (electronic) (2004)
7. Lions, P.L., Papanicolau, G., Varadhan, S.: Homogenization of Hamilton–Jacobi equations. Unpublished
8. Qian, J.: Two approximations for effective Hamiltonians arising from homogenization of Hamilton-Jacobi equations (2003). UCLA, Math Dept., preprint
9. Rorro, M.: An approximation scheme for the effective Hamiltonian an applications. Appl. Numer. Math. **56**(9), 1238–1254 (2006)
10. Zhou, J.L., Tits, A.L., Lawrence, C.T.: User's Guide for FFSQP Version 3.7: A FORTRAN Code for Solving Constrained Nonlinear (Minimax) Optimization Problems, Generating Iterates Satisfying All Inequality and Linear Constraints (1997). URL http://www.aemdesign.com/download-ffsqp/ffsqp-manual.pdf

Estimation of Diffusion Coefficients in a Scalar Ginzburg-Landau Equation by Using Model Reduction

M. Kahlbacher and S. Volkwein

Abstract Proper orthogonal decomposition (POD) is a powerful technique for model reduction of linear and non-linear systems. It is based on a Galerkin type discretization with basis elements created from the system itself. In this work POD is applied to estimate scalar parameters in a scalar non-linear Ginzburg-Landau equation. The parameter estimation is formulated in terms of an optimal control problem that is solved by an augmented Lagrangian method combined with a sequential quadratic programming algorithm. A numerical example illustrates the efficiency of the proposed solution method.

1 Introduction

Proper orthogonal decomposition (POD) is a method to derive low order models for systems of differential equations. It is based on projecting the system onto subspaces consisting of basis elements that contain characteristics of the expected solution. This is in contrast to, e.g., finite element techniques, where the elements of the subspaces are uncorrelated to the physical properties of the system that they approximate. It is successfully used in different fields including signal analysis and pattern recognition (see, e.g., [4]), fluid dynamics and coherent structures (see, e.g., [7, 17]) and more recently in control theory (see, e.g., [13]). The relationship between POD and balancing is considered in [12, 16, 20]. In contrast to POD approximations, reduced-basis element methods for parameter dependent elliptic are investigated in [1, 14, 15], for instance.

Martin Kahlbacher
Institute for Mathematics and Scientific Computing, University of Graz, Heinrichstrasse 36, 8010-Graz, Austria, e-mail: martin.kahlbacher@uni-graz.at

Stefan Volkwein
Institute for Mathematics and Scientific Computing, University of Graz, Heinrichstrasse 36, 8010-Graz, Austria, e-mail: stefan.volkwein@uni-graz.at

In this work we continue our research in [10] and apply a POD Galerkin approximation to estimate diffusion coefficients in a scalar, non-linear Ginzburg-Landau equation. The corresponding parameter identification problem is formulated as an optimal control problem with inequality constraints for the parameters. To solve this optimization problem with a scalar inequality constraint we apply an augmented Lagrangian method (see, e.g., [2, 3]) combined with a globalized sequential quadratic programming (SQP) algorithm as described in [6]. In [9] error estimates for POD Galerkin schemes for linear and certain semi-linear elliptic, parameter dependent systems are proved. The resulting error bounds depend on the number of POD basis functions and on the parameter grid that is used to generate the snapshots and to compute the POD basis.

The paper is organized in the following manner: In Section 2 we introduce the parameter identification problem and review some pre-requisites. The POD approximation is explained shortly in Section 3 and numerical examples are carried out in Section 4.

2 Identification Problem

Let $\Omega \subset \mathbb{R}^r$, $r = 2$ or 3, be a bounded open domain and let $\Gamma = \partial\Omega$ denote the boundary of Ω. We suppose that Ω is split into m measurable disjunct subdomains Ω_i, i.e.,

$$\Omega = \bigcup_{i=1}^{m} \Omega_i \quad \text{and} \quad \Omega_i \cap \Omega_j = \emptyset \text{ for } i \neq j.$$

We consider a *scalar Ginzburg-Landau equation*

$$-\nabla \cdot (c\nabla u) + qu + u^3 = f \qquad \text{in } \Omega, \tag{1a}$$

$$c\frac{\partial u}{\partial n} + \sigma u = g \qquad \text{on } \Gamma, \tag{1b}$$

where $q \in L^\infty(\Omega)$ with $q(\mathbf{x}) \geq q_a > 0$ for almost all (f.a.a.) $\mathbf{x} \in \Omega$, $f \in L^2(\Omega)$, $\sigma \in L^\infty(\Gamma)$ with $\sigma(\mathbf{s}) \geq 0$ f.a.a. $\mathbf{s} \in \Gamma$ and $g \in L^2(\Gamma)$. Furthermore, c is supposed to be constant on the subdomains Ω_i:

$$c(\mathbf{x}) = c^i \quad \text{f.a.a. } \mathbf{x} \in \Omega_i \cup \left(\overline{\Omega}_i \cap \Gamma\right),\ 1 \leq i \leq m,$$

with positive c_i's. Problem (1) is a simplified model of the full Ginzburg-Landau equations of superconductivity valid in the absence of internal fields [18].

A function $u \in H^1(\Omega)$ is called a *weak solution* to (1) if

$$\sum_{i=1}^{m} \int_{\Omega_i} c^i \nabla u \cdot \nabla \varphi + \left(qu + u^3 - f\right)\varphi \,\mathrm{d}\mathbf{x} + \int_\Gamma (\sigma u - g)\varphi \,\mathrm{d}\mathbf{s} = 0 \quad \forall \varphi \in H^1(\Omega). \tag{2}$$

By standard Galerkin procedure it follows that (2) admits a unique solution $u \in H^1(\Omega)$. If Ω is convex with a Lipschitz-continuous boundary or if Ω has a boundary of class C^2 we derive $u \in H^2(\Omega)$ from regularity results for elliptic equations.

Remark 1. Note that for $c \equiv 1$, $q \equiv -1$, $f \equiv 0$, $g \equiv 0$, $\sigma \equiv 0$ problem (1) has the form

$$-\Delta u - u + u^3 = 0 \text{ in } \Omega \quad \text{and} \quad \frac{\partial u}{\partial n} = 0 \text{ on } \Gamma. \tag{3}$$

Then, the constant functions $u \equiv 0$, $u \equiv -1$, and $u \equiv 1$ solve (3). Hence, the assumption $q \geq q_a > 0$ almost everywhere in Ω is essential to prove uniqueness. ◇

The goal of the identification problem is to identify the diffusion coefficient c, i.e., the c_i's, from (perturbed) measurement u_d for the solution u to (1) on Γ. Therefore, we introduce the quadratic cost functional $J : H^1(\Omega) \times \mathbb{R}^m \to [0,\infty)$ by

$$J(u,c) = \frac{\alpha}{2} \int_\Gamma \left|u - u_d\right|^2 \mathrm{d}\mathbf{x} + \frac{1}{2} \sum_{i=1}^m \kappa_i \left|c^i\right|^2 \tag{4}$$

for $u \in H^1(\Omega)$ and $c = (c^1,\ldots,c^m) \in \mathbb{R}^m$. The optimal control problem is of the form

$$\min J(u,c) \quad \text{subject to (s.t.)} \quad (u,c) \in H^1(\Omega) \times C_{ad} \text{ solves (2)}, \tag{\textbf{P}}$$

where the set of admissible diffusion coefficients is given by

$$C_{ad} = \left\{c = (c^1,\ldots,c^m) \in \mathbb{R}^m \,|\, c^i \geq c_a \text{ for } i = 1,\ldots,m\right\}$$

with a positive scalar c_a.

Using standard arguments it can be proven that **(P)** possesses a (local) solution $x_* = (u_*, c_*) \in H^1(\Omega) \times C_{ad}$. To characterize an optimal solution of **(P)** we introduce the Lagrange function $L : H^1(\Omega) \times \mathbb{R}^m \times H^1(\Omega) \to \mathbb{R}$ by

$$L(u,c,p) = J(u,c) + \sum_{i=1}^m \int_{\Omega_i} c^i \nabla u \cdot \nabla p + \left(qu + u^3 - f\right) p \,\mathrm{d}\mathbf{x} + \int_\Gamma \left(\sigma u - g\right) p \,\mathrm{d}s$$

for $(u,c,p) \in H^1(\Omega) \times \mathbb{R}^m \times H^1(\Omega)$. Existence of a *Lagrange multiplier* (or *dual state*) p_* associated with $x_* = (u_*, c_*)$ is shown in [5, Theorem 3.3], where p_* satisfies the *dual system* (here written in its strong form)

$$-\nabla \cdot \left(c_* \nabla p_*\right) + q p_* + 3(u_*)^2 p_* = 0 \text{ in } \Omega, \quad c \frac{\partial p_*}{\partial n} + \sigma p_* = u_d - u_* \text{ on } \Gamma$$

and the *variational inequality*

$$\sum_{i=1}^m \left(\kappa_i c_*^i + \int_{\Omega_i} \nabla u_* \cdot \nabla p_* \,\mathrm{d}\mathbf{x}\right)\left(c^i - c_*^i\right) \geq 0 \quad \text{for all } c = (c^1,\ldots,c^m) \in C_{ad}.$$

To solve (3) we continue our earlier work [9, 10] and apply an augmented Lagrangian method combined with an globalized SQP algorithm. The discretization of the state and the dual equations is carried out by a POD Galerkin approximation. We refer the reader to [8, 10], where a scalar potential parameter is identified in a linear elliptic partial differential equation.

3 POD Approximation

In this section we introduce briefly the POD method. Suppose that for points $c_j = (c_j^1,\ldots,c_j^m) \in C_{ad}$, $j = 1,\ldots,n$, we know (at least approximately) the solution u_j to (1), e.g., by utilizing a finite element or finite difference discretization. We set

$$\mathscr{V} = \operatorname{span}\{u_1,\ldots,u_n\} \subset H^1(\Omega)$$

with $d = \dim \mathscr{V} \le n$. Then the *POD basis of rank* $\ell \le d$ is given by the solution to

$$\min_{\psi_1,\ldots,\psi_\ell} \sum_{j=1}^n \Big\| u_j - \sum_{i=1}^\ell \langle u_j, \psi_i\rangle_{H^1(\Omega)}\, \psi_i \Big\|^2_{H^1(\Omega)} \quad \text{s.t.} \quad \langle \psi_i, \psi_j\rangle_{H^1(\Omega)} = \delta_{ij}. \tag{5}$$

The solution to (5) is characterized by the eigenvalue problem

$$\mathscr{R}\psi_i = \lambda_i \psi_i, \quad 1 \le i \le \ell,$$

where $\lambda_1 \ge \lambda_2 \ge \ldots \ge \lambda_\ell \ge \ldots \ge \lambda_d > 0$ denote the eigenvalues of the linear, bounded, self-adjoint, and non-negative operator $\mathscr{R} : H^1(\Omega) \to \mathscr{V}$ defined by

$$\mathscr{R}z = \sum_{j=1}^n \langle u_j, z\rangle_{H^1(\Omega)}\, u_j \quad \text{for } z \in H^1(\Omega);$$

see [7, 11, 19]. Suppose that we have determined a POD basis $\{\psi_i\}_{i=1}^\ell$. We set

$$V^\ell = \operatorname{span}\{\psi_1,\ldots,\psi_\ell\} \subset \mathscr{V} \subset H^1(\Omega).$$

Then the following relation holds

$$\sum_{j=1}^n \Big\| u_j - \sum_{i=1}^\ell \langle u_j, \psi_i\rangle_{H^1(\Omega)}\, \psi_i \Big\|^2_{H^1(\Omega)} = \sum_{i=\ell+1}^d \lambda_i.$$

Next we introduce the *POD Galerkin scheme for* (2). The function $u^\ell = \sum_{i=1}^\ell u_i^\ell \psi_i \in V^\ell$ solves

$$\begin{aligned} &\sum_{j=1}^m \int_{\Omega_j} c_j \nabla u^\ell \cdot \nabla \psi \,\mathbf{dx} + \int_\Omega \big(q u^\ell + (u^\ell)^3\big)\psi \,\mathbf{dx} + \int_\Gamma \sigma u^\ell \psi \,\mathbf{ds} \\ &\quad = \int_\Omega f\psi \,\mathbf{dx} + \int_\Gamma g\psi \,\mathbf{ds} \quad \forall \psi \in V^\ell. \end{aligned} \tag{6}$$

Problem (6) is a non-linear system for the ℓ unknown modal coefficients $u_1^\ell, \ldots, u_\ell^\ell \in \mathbb{R}$. If

$$\mathscr{E}(\ell) = \frac{\sum_{i=1}^{\ell} \lambda_i}{\sum_{i=1}^{d} \lambda_i} \approx 1 \quad \text{for } \ell \ll d,$$

holds, (6) is called a *low-dimensional model* for (2).

4 Numerical Example

In this section we present a numerical example for the identification problem. The numerical test is executed on a standard 3.0 GHz desktop PC. We are using the MATLAB 7.1 package together with FEMLAB 3.1.

Run 1 Let $\Omega = \{\mathbf{x} = (x_1, x_2) \,|\, x_1^2 + x_2^2 < 1\}$ be the open unit cirle in $\mathbb{R}^2$ and the subdomains Ω_1, Ω_2 be given as

$$\Omega_1 = \Omega \setminus \Omega_2, \quad \Omega_2 = \left\{\mathbf{x} = (x_1, x_2) \in \Omega \,|\, \tfrac{x_1^2}{a^2} + \tfrac{(x_2 - 0.5)^2}{b^2} < 1\right\}$$

with $a = 0.75$ and $b = 0.45$; see Figure 1 (left plot). Thus, the diffusion coeffi-

Fig. 1 Domain Ω and subdomains Ω_1, Ω_2 (left plot); FE solution (right plot).

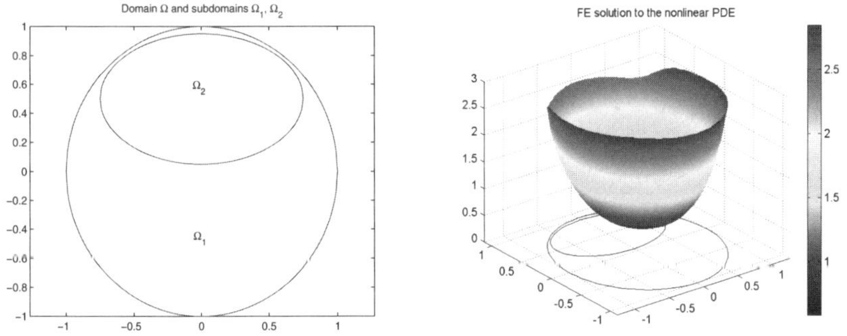

cient is given by $c = (c_1, c_2)$. In (1) we choose $q \equiv 10$, $f \equiv 3$, $\sigma \equiv 2$, $g(\mathbf{x}) = 10 + \cos(\pi x_1/2) \cdot \cos(\pi x_2/2)$ for $\mathbf{x} = (x_1, x_2) \in \Gamma$. For $\bar{c} = (0.7, 1.4)$ we calculate a finite element (FE) solution $\bar{u}^h = \bar{u}^h(\bar{c}) \in H^1(\Omega)$ using standard piecewise linear FE ansatz functions on a triangular mesh with 1147 degrees of freedom. The CPU time for the FE solve is 13.5 seconds. The FE solution is plotted in Figure 1 (right plot). To derive a POD basis we choose the diffusion values $c_j = (\eta_k, \eta_l) \in \mathbb{R}_+^2$, $1 \le j \le n$, with

$$j = 5(k-1) + l \text{ for } 1 \le k, l \le 5, \quad \eta_k = 0.5 + \frac{k-1}{4} \text{ for } k = 1, \ldots, 5$$

and compute the corresponding FE solutions $u_j^h = u^h(c_j) \in H^1(\Omega)$ to (1), i.e., we have $n = 25$ snapshots $\{u_j^h\}_{j=1}^n$. The computation of the snapshots requires 341 seconds. Next we compute the POD basis of rank ℓ as described in Section 3. The decay of the largest normalized eigenvalues $\lambda_i / \sum_{j=1}^d \lambda_j$, $1 \le i \le 9$, is plotted in Figure 2.

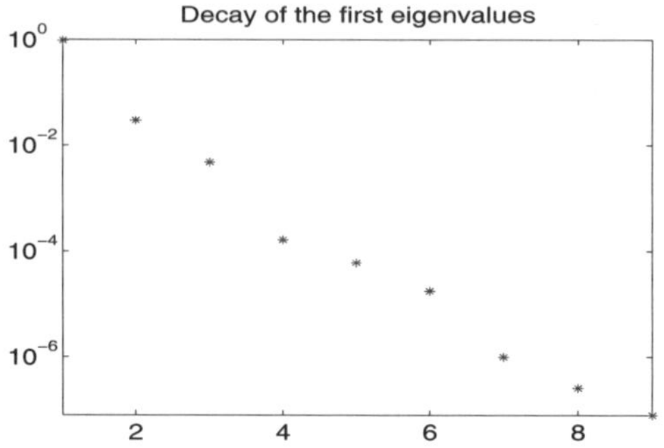

Fig. 2 Decay of the largest 9 normalized eigenvalues $\lambda_i / \sum_{j=1}^d \lambda_j$.

For the POD Galerkin approximation we choose $\ell = 7$ POD basis functions. The computation of the POD solution $\bar{u}^\ell = \bar{u}^\ell(\bar{c})$ for the diffusion parameter $\bar{c}$ requires 0.1 second. The relative error in the H^1-norm between the FE state $\bar{u}^h$ and the POD state $u^\ell(\bar{c})$ is 0.25%. Furthermore, the relative errors between the FE state and the POD state decreases with increasing number ℓ of POD basis functions (see Table 1). Let us mention that in [9] error estimates for POD Galerkin schemes are derived.

Table 1 Relative errors between the FE state and the POD state for different numbers ℓ of POD basis functions applying in the POD Galerkin approximation for (1).

	$\ell = 4$	$\ell = 5$	$\ell = 6$	$\ell = 7$	$\ell = 8$
$\frac{\|\bar{u}^h - \bar{u}^\ell\|_{H^1(\Omega)}}{\|\bar{u}^h\|_{H^1(\Omega)}}$	0.01160	0.00291	0.00289	0.00247	0.00246

Next turn to the identification problem. We set $c_a = 0.01$ in the definition of the admissible set C_{ad} of diffusion coefficients. Moreover, we choose the weights $\alpha = 100$ and $\kappa = 10^{-5}$ for the cost functional. For $\bar{c}$ we have already computed the FE solution $\bar{u}^h$ to (1). Let for any $\mathbf{x} \in \overline{\Omega}$ the term $\varepsilon(\mathbf{x}) \in [-1, 1]$ denote a random variable and let $\delta \ge 0$ be a given perturbation. In (4) we set $u_d = u_d^h|_\Gamma$ for the desired state, where $u_d^h = (1 + \delta\varepsilon)\bar{u}^h$. The goal of the identification problem is to recover $c_{\text{ideal}} = \bar{c}$ from the perturbed measurement u_d for $u_{\text{ideal}}^h = \bar{u}^h$ on the boundary Γ. We choose the perturbation $\delta = 0.05$ (i.e., 5% noise) and apply the augmented Lagrangian method combined with a globalized SQP method to determine a numerical solution (u_*^ℓ, c_*^ℓ) with $c_*^\ell = (0.7041, 1.4018)$. This gives the relative error

$$\frac{|c_*^\ell - c_{\text{ideal}}|_2}{|c_{\text{ideal}}|_2} \approx 0.0029 = 0.29\%,$$

in the diffusion parameter, where $|\cdot|_2$ denotes the Euclidean norm in $\mathbb{R}^2$. Moreover, the relative errors in the state variable to the ideal data u_{ideal} and to the noisy data $(1+\delta\varepsilon)\bar{u}^h$ are presented in Table 2.

Table 2 Relative errors of the POD state u_*^ℓ compared to the ideal data u_{ideal} and to the noisy data u_d with 5% noise.

	$\frac{\|u_*^\ell - u\|_{H^1(\Omega)}}{\|u\|_{H^1(\Omega)}}$	$\frac{\|u_*^\ell - u\|_{L^2(\Omega)}}{\|u\|_{L^2(\Omega)}}$	$\frac{\|u_*^\ell - u\|_{L^2(\Gamma)}}{\|u\|_{L^2(\Gamma)}}$
$u = u_{\text{ideal}}^h$	0.0061	0.0018	0.0014
$u = u_d$	0.3406	0.0215	0.0235

Notice that the optimization algorithm only needs 1.6 seconds CPU time, whereas the FE based augmented Lagrangian method combined with a globalized SQP solver stops after 286 seconds. However, the CPU time for the computation of the snapshots is larger that the CPU time for the FE optimization method. The advantage of POD regarding computing times appears significantly when the identification problem has to be solved several times (e.g., for different data u_d) so that the already computed POD basis can be utilized again. For instance, if we take the perturbation $\delta = 0.03$ for the measurement data u_d and solve the optimal control prolem again we obtain an optimal solution after 2.2 seconds with $c_*^\ell = (0.7019, 1.3996)$ that leads to the relative error

$$\frac{|c_*^\ell - c_{\text{ideal}}|_2}{|c_{\text{ideal}}|_2} \approx 0.0012 = 0.12\%.$$

The relative errors in the state variable are presented in Table 3.

Table 3 Relative errors of the POD state u_*^ℓ compared to the ideal data u_{ideal} and to the noisy data u_d with 3% noise.

	$\frac{\|u_*^\ell - u\|_{H^1(\Omega)}}{\|u\|_{H^1(\Omega)}}$	$\frac{\|u_*^\ell - u\|_{L^2(\Omega)}}{\|u\|_{L^2(\Omega)}}$	$\frac{\|u_*^\ell - u\|_{L^2(\Gamma)}}{\|u\|_{L^2(\Gamma)}}$
$u = u_{\text{ideal}}^h$	0.0041	0.0014	0.0007
$u = u_d$	0.2122	0.0129	0.0141

Acknowledgements The authors gratefully acknowledge support by the Austrian Science Fund FWF under grant no. P19588-N18.

References

1. M. Barrault, Y. Maday, N.C. Nguyen, and A.T. Patera. An empirical interpolation method: application to efficient reduced-basis discretization of partial differential equations. *Comptes Rendus de'l Académie des Sciences Paris*, Ser. I 339:667-672, 2004.
2. D.P. Bertsekas. *Constrained Optimization and Lagrange Multipliers*. Academic Press, Ney York, (1982).
3. D.P. Bertsekas. *Nonlinear Programming*. 2nd edition, Athena Scientific, Belmont, (1999).
4. K. Fukuda. *Introduction to Statistical Recognition*. Academic Press, New York, (1990).
5. M.D. Gunzburger, L. Hou, and T.P. Svobodny. Finite element approximations of an ptimal control problem associated with the scalar Ginzburg-Landau equation. *Computers Math. Applic.*, 21:123-131, 1991.
6. M. Hintermüller. A primal-dual active set algorithm for bilaterally control constrainted optimal control problems. *Quarterly of Applied Mathematics*, **61**, 131–160, (2003).
7. P. Holmes, J.L. Lumley, and G. Berkooz. *Turbulence, Coherent Structures, Dynamical Systems and Symmetry*. Cambridge Monographs on Mechanics, Cambridge University Press, 1996.
8. M. Kahlbacher. *POD for parameter estimation of bilinear elliptic problems*. Diploma thesis, Institute for Mathematics and Scientific Computing, University of Graz, October 2006.
9. M. Kahlbacher and S. Volkwein. Galerkin proper orthogonal decomposition methods for parameter dependent elliptic systems. *Discussiones Mathematicae: Differential Inclusions, Control and Optimization*, 27:95-117, 2007.
10. M. Kahlbacher and S. Volkwein. Model reduction by proper orthogonal decomposition for estimation of scalar parameters in elliptic PDEs. In Proceedings of *ECCOMAS CFD*, P. Wesseling, E. Onate, and J. Periaux (eds.), Egmont aan Zee, 2006.
11. K. Kunisch and S. Volkwein. *Galerkin proper orthogonal decomposition methods for a general equation in fluid dynamics. SIAM J. Numer. Anal.*, 40:492-515, 2002.
12. S. Lall, J.E. Marsden and S. Glavaski. Empirical model reduction of controlled nonlinear systems. In: *Proceedings of the IFAC Congress*, vol. F, 473-478, 1999.
13. H.V. Ly and H.T. Tran. Modelling and control of physical processes using proper orthogonal decomposition. *Mathematical and Computer Modeling*, 33:223-236, 2001.
14. L. Machiels, Y. Maday, and A.T. Patera. Output bounds for reduced-order approximations of elliptic partial differential equations. *Computer Methods in Applied Mechanics and Engineering*, 190:3413-3426, 2001.
15. Y. Maday and E.M. Rønquist. A reduced-basis element method. *Journal of Scientific Computing*, 17, 1-4, 2002.
16. C.W. Rowley. Model reduction for fluids, using balanced proper orthogonal decomposition. *International Journal of Bifurcation and Chaos*, 15:997-1013, 2005.
17. L. Sirovich. Turbulence and the dynamics of coherent structures, parts I-III. *Quarterly of Applied Mathematics*, XLV:561-590, 1987.
18. M. Tinkham. *Introduction to Superconductivity*. McGraw Hill, New York, 1975.
19. S. Volkwein. *Model Reduction using Proper Orthogonal Decomposition*. Lecture Notes, Institute of Mathematics and Scientific Computing, University of Graz. see `http://www.uni-graz.at/imawww/volkwein/POD.pdf`
20. K. Willcox and J. Peraire. Balanced model reduction via the proper orthogonal decomposition. *American Institute of Aeronautics and Astronautics (AIAA)*, 40, 2323-2330, 2002.

Reduced Order Models (POD) for Calibration Problems in Finance

E.W. Sachs and M. Schu

Abstract In this paper we consider the calibration of mathematical models for option pricing to observed data on the market. As a model for the underlying stock prices we use a jump diffusion process which results for the price of a call option in a partial integro-differential equation. We employ the dual - Dupire-type - version of it in order to improve the efficiency of the original calibration problem. To reduce the complexity of the problem even further, we use a reduced order model technique based on proper orthogonal decomposition techniques to obtain a model for the option price which is considerably smaller in size, but still copies the original model at a surprising accuracy. In the second half of the paper, we present numerical results which support these findings.

1 Introduction

The pricing of options in the financial markets has become a mathematically challenging problem in various areas of mathematics. Here we address some issues in the calibration of pricing models.

The basic model for the price of a call option was given by Black and Scholes [4] in form of a parabolic partial differential equation. If $S \in [0,\infty)$ denotes the stock price and $t \in [0,T]$ the time variable, then the price of a European call option with strike price K and expiration date T is given by $C(t,S)$, where S is the current stock price at current time t. $C(\cdot,\cdot)$ is a solution of

Ekkehard W. Sachs
Department of Mathematics, Virginia Tech, Blacksburg, VA 24060 and
Universität Trier, FB IV - Department of Mathematics, 54286 Trier, Germany,
e-mail: sachs@uni-trier.de

Matthias Schu
Universität Trier, FB IV - Department of Mathematics, 54286 Trier, Germany

$$-C_t = \frac{1}{2}\sigma^2 S^2 C_{SS} + rSC_S - rC \qquad (t,S) \in [0,T] \times [0,\infty) \tag{1}$$

with a final condition $C(T,S) = \max\{S-K,0\}$ and boundary condition $C(t,0) = 0$. The parameter r represents the riskfree interest rate and σ is the volatility of the underlying asset.

A major improvement of the original Black-Scholes model represents the local volatility model, where the volatility σ is no longer constant, but depends 'locally' on time and stock price, i.e. $\sigma(t,S)$.

Often, one can find many options traded on the market for the same asset with different expiration times T_i and strike prices K_i and we denote the observed market prices of these options by $C_i^{obs}, i = 1,...,n$. Given $\sigma(t,S)$ we have for each pair (T_i,K_i) a model price $C_i(t,S), i = 1,...,n$ defined through the solution of (1) with $T = T_i$ and $K = K_i$. The calibration problem at a time t_0 with observed stock price S_{t_0} is a nonlinear least squares problem

$$\min_{\sigma(\cdot,\cdot)} \quad \sum_{i=1}^{n} (C_i(t_0, S_{t_0}; \sigma) - C_i^{obs})^2. \tag{2}$$

However, note that the use of the local volatility model in calibration, causes a single function evaluation to require the solution of n partial differential equations.

For this problem there is an alternative approach, i.e. one can use a very efficient approach for the calibration problem by using the Dupire equation [6] and [3] instead. Consider a solution $D(T,K)$ depending on the expiration time T and the strike price K of the following partial differential equation, Dupire's equation,

$$D_T = \frac{1}{2}\sigma(T,K)^2 K^2 D_{KK} + rKD_K \qquad (T,K) \in [t_0,\infty] \times [0,\infty) \tag{3}$$

with $D(t_0,K) = \max\{S_{t_0} - K, 0\}$ as initial and $D(T,0) = S_{t_0}$ as boundary condition.

Note that we can consider the original call price $C(t,S)$ not only in dependence of time t and stock price S, but also in dependence on the 'initial conditions', the duration T and the strike price K by $C(t,S;T,K)$. Then it can be shown, that the following identity holds

$$C_i(t,S) = C(t,S;T_i,K_i) = D(T_i,K_i), \qquad i = 1,...,n. \tag{4}$$

Hence the nonlinear least squares problem (2) can be rewritten as

$$\min_{\sigma(\cdot,\cdot)} \quad \sum_{i=1}^{n} (D(T_i,K_i;\sigma) - C_i^{obs})^2. \tag{5}$$

This requires in contrast to the previous formulation for each function evaluation not n, but only one solution of a partial differential equation.

The goal of this contribution is to set up the calibration problem for option price models using Lévy processes and to apply Dupire's framework. Having formulated this problem as an optimization problem with **partial integro-differential**

equations (PIDE), we derive from the **proper orthogonal decomposition (POD)** framework a reduced order model. The main issue and original aspect in this paper is to apply the POD framework to a PIDE and present numerical results in this case. They support the statement that this is a reasonable approach to obtain reduced order models for this particular application in finance.

2 Partial Integro-Differential Equation

In the past decade several extensions of the Black-Scholes model for the pricing of options have been developed. In this paper we consider a jump diffusion model based on Lévy processes. Here f is the probability density for the distribution of jumps sizes and λ a parameter which regulates the frequency of the jumps. The pricing process is given by a PIDE, see [5]. Recently, [1] and [14] derived a Dupire-like PIDE for the original PIDE, similar to the Black-Scholes framework. The price $D(T,K)$ of an option with expiration time T and the strike price K is given by the following PIDE:

$$\begin{aligned}&\bar{D}_T + \tfrac{1}{2}\sigma^2(T,K)K^2\bar{D}_{KK} + (r-\lambda\zeta)K\bar{D}_K + \lambda(1+\zeta)\bar{D} \\ &\quad -\lambda\int_{-\infty}^{\infty}\bar{D}(T,Ke^{-y})e^y f(y)\,dy = 0, \qquad (T,K)\in[t_0,T_{max}]\times[0,\infty)\end{aligned} \tag{6}$$

with an initial condition $\bar{D}(t_0,K)=\max\{S_{t_0}-K,0\}$ depending on the current time t_0 and the current stock price S_{t_0} and boundary condition $\bar{D}(T,0)=S_{t_0}$.

For the density function we use Merton's model and define f as

$$f(y) = \frac{1}{\sqrt{2\pi}\sigma_J}e^{-\frac{(y-\mu_J)^2}{2\sigma_J^2}}$$

with μ_J the expected value and σ_J the standard deviation of the normally distributed jump sizes and $\zeta = exp\{\frac{\sigma_J^2}{2}+\mu_J\}-1$.

One can consider various calibration problems, here we look as in the Black-Scholes case at the volatility σ for calibration. In other papers, see e.g. [5], [8], [11], the distribution function f is parametrized and used for calibration. This will be the subject of future research in the POD context.

For numerical reasons we apply a variable transformation $x=\ln(K/S_{t_0})$. We obtain a PIDE defined on $[t_0,T_{max}]\times(-\infty,\infty)$:

$$\begin{aligned}&D_T + \tfrac{1}{2}\sigma^2(T,x)D_{xx} + \left(r+\tfrac{1}{2}\sigma^2(T,x)-\lambda\zeta\right)D_x + \lambda(1+\zeta)D \\ &\qquad\qquad -\lambda\int_{-\infty}^{+\infty} D(T,x-y)e^y f(y)\,dy = 0\end{aligned} \tag{7}$$

with initial condition $D(t_0,x) = \max\{S_{t_0} - S_{t_0}e^x, 0\}$. We restrict the solution interval for D to $[x_{min}, x_{max}]$ with boundary conditions $D(T,x) = S_{t_0}$ for $x \leq x_{min}$ and $D(T,x) = 0$ for $x \geq x_{max}$.

The weak formulation of the PIDE with a given set of basis functions $w_1, \ldots, w_n$ on $[x_{min}, x_{max}]$ is given by

$$\int_{x_{min}}^{x_{max}} D_T(T,x) w_j(x) dx = a(T; D(T,\cdot), w_j), \qquad j = 1, \ldots, n \tag{8}$$

with a bilinear form

$$\begin{aligned} a(T;v,w) = &- \int_{x_{min}}^{x_{max}} \frac{\sigma^2(T,x)}{2} v_x(x) w_x(x) dx \\ &- \int_{x_{min}}^{x_{max}} \left(r + \frac{\sigma^2(T,x)}{2} - \lambda\zeta + \sigma(T,x)\sigma_x(T,x) \right) v_x(x) w(x) dx \\ &- \lambda(1+\zeta) \int_{x_{min}}^{x_{max}} v(x) w(x) dx + \lambda \int_{x_{min}}^{x_{max}} \int_{-\infty}^{+\infty} v(x-y) w(x) f(y) e^y dy\, dx. \end{aligned} \tag{9}$$

If we approximate the solution $D(T,x)$ by $\sum_{i=1}^{n} \alpha_i(T) w_i(x)$ then the following system of ODEs needs to be solved

$$\sum_{i=1}^{n} \dot{\alpha}_i(T) \langle w_i, w_j \rangle_{L^2} = \sum_{i=1}^{n} \alpha_i(T) a(T; w_i, w_j), \quad j = 1, \ldots, n, \quad T \in [t_0, T_{max}]. \tag{10}$$

For the numerical solution, we use a fully implicit Euler scheme for $m = 0, \ldots, N$

$$\sum_{i=1}^{n} \alpha_i(T_{m+1}) \Big(\langle w_i, w_j \rangle_{L^2} - \Delta T\, a(T_{m+1}; w_i, w_j) \Big) = \sum_{i=1}^{n} \alpha_i(T_m) \langle w_i, w_j \rangle_{L^2}.$$

For an efficient numerical solution there are several approaches known, see e.g.[2], [12], [10], [15].

3 Reduced Order Model by Proper Orthogonal Decomposition

The calibration of the PIDE model requires many function evaluations during the optimization phase. Each function evaluation itself needs a new numerical solution of the PIDE. In order to save computing time, one could use a less complex model to approximate the PIDE solver.

There is an abundant amount of literature on POD, a recent reference is [9] and its citations.

In [13] an approach using the reduced basis approach is applied to a calibration problem of fitting option prices for a local volatility model.

The calibration problem using the PIDE (7) is given in the following form, where D_n is the solution of the Galerkin approximation (10).

$$\min_{\sigma(\cdot,\cdot)} \sum_{i=1}^{n}(D_n(T_i,x_i;\sigma)-C_i^{obs})^2 \quad \text{where} \quad D_n(T,x)=\sum_{i=1}^{n}\alpha_i(T;\sigma)w_i(x). \tag{11}$$

In the Proper Orthogonal Decomposition one computes for a given choice of σ a solution of the PIDE (10) and stores the solution values over time $D_n(T_m,x)$, so called snapshots, in a long matrix for the coefficients $\alpha_i(T_m)$

$$Y=[\alpha(T_0),...,\alpha(T_N)] \qquad \alpha(T_m)=(\alpha_1(T_m),...,\alpha_n(T_m))^T.$$

Then one applies a Singular Value Decomposition to the matrix $\bar{Y}=M^{1/2}YD^{1/2}$ (where M is the mass matrix of the basis functions w_i and D is a diagonal weighting matrix) and looks at the decay of the singular values. With $L \ll N$ only the first L vectors $\psi_0,...\psi_L$ are stored, which correspond to the L largest singular values of $\bar{Y}$. Based on these vectors one can compute a new set of basis function $v_1,...,v_L$ which can be used instead of $w_1,...,w_N$ in a Galerkin approximation. Hence the approximating solution of the PIDE by $D_n(T,x)=\sum_{i=1}^{n}\alpha_i(T;\sigma)w_i(x)$ is replaced by $D_{pod}(T,x)=\sum_{k=1}^{L}\beta_k(T;\sigma)v_k(x)$. This results, in contrast to (10) in a small system of ODEs of size L

$$\sum_{k=1}^{L}\dot{\beta}_k(T)\langle v_k,v_j\rangle_{L^2}=\sum_{k=1}^{L}\beta_k(T)a(T;v_k,v_j), \quad j=1,...,L, \;\; T\in[t_0,T_{max}] \tag{12}$$

which might not exhibit the sparsity pattern as for a finite element basis, but its dimension is small.

The calibration problem (11) is replaced by

$$\min_{\sigma(\cdot,\cdot)} \sum_{i=1}^{n}(D_{pod}(T_i,x_i;\sigma)-C_i^{obs})^2 \quad \text{where} \quad D_{pod}(T,x)=\sum_{k=1}^{L}\beta_k(T;\sigma)v_k(x). \tag{13}$$

The big advantage is that a function evaluation for this POD-function is by far faster than for the original function. During the course of an optimization iteration the parameter value σ might move away from the one, which the POD model is based on. In this case, one has to reboot the process, i.e. compute a new POD basis for the modified value of σ, see e.g. [7].

4 Numerical Results

We give some numerical results to illustrate the computational savings by using a reduced order model.

First, we demonstrate the accuracy of the solution of the POD model. Since the Black-Scholes model is a diffusion equation with a convective term, which in finance often is not too large, it is no surprise that a POD approximation works well also for this case. However, it is not clear, if the same holds also for the PIDE which also includes an integral term in addition to the convection term.

We use the parameters $\lambda = 0.5$, $\mu_J = 0$, $\sigma_J = 0.5$. Furthermore $T_{max} = 0.5$ and discretization parameters are 500 steps in x-direction and 150 steps in T-direction.

We compare the Finite Element solution and the POD solution of the PDE or PIDE. The deviation is measured as $\frac{1}{151}\sum_{i=0}^{150}\|D(T_i,\cdot)-D_{POD}(T_i,\cdot)\|_{L^2}^2$ with T_i the grid points in T-direction. The results are shown in Table 1.

Table 1 Accuracy of POD approximations, PIDE model

POD Basis El.	Deviation	smallest sing. val.
3	8.36e-001	1.63e+003
4	1.36e-001	1.76e+002
5	2.17e-002	2.28e+001
6	3.31e-003	3.11e+000
7	4.86e-004	4.25e-001
8	6.82e-005	5.71e-002
9	9.17e-006	7.46e-003
10	1.18e-006	9.46e-004

One can see that the results in Table 1 show a fast decay in the value of the smallest singular values. Hence it is expected that deviation of the POD model from the FEM model is decaying rapidly too.

In the following tables we use a matlab subroutine 'fminunc' to solve the calibration problem for σ in the Black-Scholes case and also for the PIDE.

Let L denote the number of POD basis elements used in D_{POD}, σ_{opt} the calculated solution, $f_{pod}(\sigma_{opt})$ the optimal value of the objective function in (13) and *time* is the time needed to solve the optimization problem.

We first look at the special case of a constant volatility. The optimal value is $\sigma = .30$, the starting value always $\sigma = .25$ and the POD basis is calculated only once for the starting value. Table 2 shows the results of the calibrating problem without jumps $\lambda = 0$, this means the Black-Scholes case in the first four columns.

Table 2 Optimal Parameters for POD Approximations

	Black-Scholes model			PIDE model		
L	σ_{opt}	$f_{pod}(\sigma_{opt})$	*time*	σ_{opt}	$f_{pod}(\sigma_{opt})$	*time*
4	0.3007	0.2610	0.6410	0.2806	0.1525	22.78
5	0.2991	0.0701	0.6870	0.2827	0.0904	22.67
6	0.2981	0.0338	0.6560	0.2845	0.0691	22.73
7	0.2977	0.0254	0.6250	0.2861	0.0549	22.58
8	0.2974	0.0238	0.6250	0.2874	0.0426	22.55
9	0.2974	0.0237	0.7340	0.2884	0.0333	22.91
10	0.2974	0.0239	0.9840	0.2892	0.0272	22.69
500	0.3000	0.0000	2.1090	0.3000	0.0000	340.58

The last line in Table 2 denotes the optimization with the full evaluation of D, in other words the optimization using the common finite element method with 500 discretization points in x-direction in every function evaluation. We see good approximations to the optimal solution and a remarkable time savings of the POD-method even in the simple case of only one parameter.

In the case of the PIDE model with $\lambda = 0.5$ the results are in the last three columns in Table 2. Here the computations are more expensive than in the Black-Scholes model due to the integral term. This causes a double-integral in the bilinear form which destroys the sparse-structure of the stiffness matrix and makes the solution of one finite element problem far more expensive.

Therefore, a reduced order model approach should work well in this case - and this can be confirmed from the quality of the results, which is as good as in the Black-Scholes case. The speedup for the PIDE case is significantly higher than in the Black-Scholes case - it ranges around a factor of 15.

In Table 3 we study the case of a time (maturity) dependent volatily $\sigma(T)$. The components of the solution vector $\sigma_{opt} = (0.30, 0.28, 0.32)$ represent the size of the piecewise linear modelled volatility at three different grid points. L again denotes the number of POD basis functions and $f_{pod}(\sigma^L_{opt})$ is the optimal value of the objective function. Start value always was $\sigma_{start} = (0.25, 0.25, 0.25)$.

Table 3 POD Approximation, 3 Parameters, Black-Scholes model

L	σ^L_{opt}	$f_{pod}(\sigma^L_{opt})$
4	(0.3085, 0.2594, 0.3289)	0.0486
5	(0.3014, 0.2716, 0.3210)	0.0320
6	(0.2995, 0.2753, 0.3180)	0.0248
7	(0.2989, 0.2768, 0.3168)	0.0240
8	(0.2986, 0.2775, 0.3164)	0.0236
500	(0.3000, 0.2800, 0.3200)	0.0000

Table 4 POD Approximation, 8 Parameters, PIDE model

L	$\|\sigma^L_{opt} - \sigma_{opt}\|$	$f_{pod}(\sigma^L_{opt})$	grid
4	0.1123	0.1399	250
	0.0428	0.0147	500
5	0.0958	0.0743	250
	0.0361	0.0096	500

The last Table 4 shows the PIDE case ($\lambda = 0.5$) with maturity dependent volatility with 8 discretization points. Here we use a kind of nested iteration. Starting with a constant volatility σ_{start} we try to get closer to the solution by optimizing first on a coarse grid and then use this solution to calculate a new POD basis when we optimize on the finer grid. The finest grid has 500 steps in K-direction.

5 Conclusion

Models for the evaluation of option prices reach higher levels of sophistication. If one uses jump processes in the stochastic description, this leads to parabolic differential equations which - in addition to the convection term - also include an integral term. The latter one is known to require special numerical techniques for an efficient solution of the partial integro-differential equation. We use this model to calibrate it to given data. We reformulate it using a Dupire-like approach to make it more amenable to calibration problems. After that we apply a model reduction technique, here proper orthogonal decomposition, to obtain a model of smaller size. The numerical results show that this is a promising approach in reducing the computing time substantially.

References

1. Achdou, Y.: An inverse problem for a parabolic variational inequality with an integro-differential operator. SIAM J. Control Optim. To appear
2. Almendral, A., Oosterlee, C.: Numerical valuation of options with jumps in the underlying. Applied Numerical Mathematics **53**, 1–18 (2005)
3. Andersen, L., Andreasen, J.: Jump-diffusion processes: Volatility smile fitting and numerical methods for option pricing. Rev. of Derivatives Res. **4**, 231–262 (2000)
4. Black, F., Scholes, M.: The pricing of options and corporate liabilities. Journal of Political Economy **81**(3), 637–654 (1973)
5. Cont, R., Tankov, P.: Financial Modelling with Jump Processes. Chapman and Hall (2004)
6. Dupire, B.: Pricing with a smile. Risk **7**(1), 18–20 (1994)
7. Fahl, M.: Trust region methods for flow control based on reduced order modelling. Ph.D. thesis, Universität Trier (2001)
8. He, C., Kennedy, J.S., Coleman, T.F., Forsyth, P.A., Li, Y., Vetzal, K.: Calibration and hedging under jump diffusion. Review of Derivative Research **9**, 1–35 (2006)
9. Hinze, M., Volkwein, S.: Error estimates for abstract linear-quadratic optimal control problems using proper orthogonal decomposition. Computational Optimization and Applications (2008). To appear
10. Ikonen, S., Toivanen, J.: Numerical valuation of European and American options with Kou's jump-diffusion model. Amamef Conference on Numerical Methods in Finance, Inria-Rocquencourt, France (2006)
11. Kindermann, S., Mayer, P., Albrecher, H., Engl, H.: Identification of the local speed function in a Levy model for option pricing. Journal of Integral Equations and Applications (2007). To appear
12. Matache, A.M., von Petersdorff, T., Schwab, C.: Fast deterministic pricing of options on Lévy driven assets. Mathematical Modelling and Numerical Analysis **38**(1), 37–71 (2004)
13. Pironneau, O.: Calibration of options on a reduced basis. Tech. rep., Laboratoire Jacques-Louis Lions, Université Paris VI (2006)
14. Pironneau, O.: Dupire identities for complex options. Tech. rep., Laboratoire Jacques-Louis Lions, Université Paris VI (2006)
15. Sachs, E.W., Strauss, A.: Efficient solution of a partial integro-differential equation in finance. Applied Numerical Mathematics (2008). To appear

Ordinary Differential Equations

A Collocation Method for Quadratic Control Problems Governed by Ordinary Elliptic Differential Equations

W. Alt, N. Bräutigam, and D. Karolewski

Abstract We investigate discretizations for a class of quadratic optimal control problems governed by one-dimensional elliptic differential equations. In contrast to the papers [3] dealing with finite element approximations and [2, 1] dealing with finite difference approximation, the dicretizations considered here are based on a collocation method using quadratic splines for the state equation. Under the assumption that the optimal control has bounded variation we prove discrete and continuous quadratic convergence of approximating controls.

1 Introduction

In this paper we consider the one-dimensional elliptic optimal control problem

$$
\begin{aligned}
\text{(CP1)} \quad & \min J(z,u) = \frac{1}{2}\int_0^T \left(|z(t)-z_d(t)|^2 + \nu\,|u(t)|^2\right) dt \\
\text{s.t.} \quad & \ddot{z}(t) + Az(t) = Bu(t) + c(t) \qquad \text{for a.a. } t \in [0,T], \\
& z(0) = z(T) = 0, \\
& a \le u(t) \le b \qquad \text{for a.a. } t \in [0,T],
\end{aligned}
$$

where $u \in L_2(0,T;\mathbb{R}^m)$, $z, z_d \in W_2^2(0,T;\mathbb{R}^n)$, $\dot{c} \in BV(0,T;\mathbb{R}^n)$, $A \in \mathbb{R}^{n\times n}$ is symmetric and positive semidefinite, $B \in \mathbb{R}^{n\times m}$ and $a,b \in \mathbb{R}^m$, $a < b$. The symbol T denotes the total length of the interval and $\nu > 0$ is a regularizing parameter. By $|\cdot|$ we denote the Euclidian norm in $\mathbb{R}^n$ resp. $\mathbb{R}^m$.

Walter Alt and Dominik Karolewski
Institut für Angewandte Mathematik, Friedrich-Schiller-Universität Jena, Ernst-Abbe-Platz 2, D–07743 Jena, Germany, e-mail: alt@minet.uni-jena.de

Nils Bräutigam
Institut für Angewandte Mathematik, Friedrich-Alexander-Universität Erlangen-Nürnberg, Martensstr. 3, D–91058 Erlangen, Germany, e-mail: Nils.Braeutigam@am.uni-erlangen.de

In our previous paper [3] we investigated finite element approximations of Problem (CP1), while the papers [2, 1] deal with finite difference approximation. In the present paper we derive error estimates for a discretization of (CP1) based on collocation splines. This requires a complete new analysis of the adjoint equation and their discretization. As a consequence, some changes in the analysis of the convergence of the discrete controls are also necessary. The analysis of this method is based on error estimates for elliptic differential equations in Sendov/Popov [13]. As in Alt/Bräutigam [1] we assume that the derivative of the optimal control has bounded variation and derive an error estimate of order h^2 in a discrete norm. Then, according to Meyer/Rösch [11] (see also Alt et al. [3], Alt/Bräutigam [1]) we construct a new control for which we prove a continuous error estimate of order h^2.

Related results can found in Grossmann/Roos [7], Tröltzsch [14], Dontchev et al. [6]), Arada et al. [4], Casas/Tröltzsch [5], Malanowski [10], Hinze [8], and Rösch [12]. For a more detailed discussion we refer to [3].

The following notations are used. By $X(0,T;\mathbb{R}^n)$ we denote a space of functions on $[0,T]$ with values in $\mathbb{R}^n$. We refer to $L_2(0,T;\mathbb{R}^n)$ as the Hilbert space of square-integrable functions with the usual scalar product $(\cdot,\cdot)$ and the corresponding norm $\|\cdot\|_2$. By $L_\infty(0,T;\mathbb{R}^n)$ we denote the space of essentially bounded functions with norm $\|\cdot\|_\infty$, and for $k \geq 1$, $p \in \{2,\infty\}$ by $W_p^k(0,T;\mathbb{R}^n)$ the usual Sobolev spaces. The linear space of functions $z \in W_p^k(0,T;\mathbb{R}^n)$ for $p > 1$ satisfying the boundary conditions $z(0) = z(T) = 0$ is denoted by $W_{p,0}^k(0,T;\mathbb{R}^n)$, and we refer to the linear space of functions with bounded total variation by $BV(0,T;\mathbb{R}^n)$. Finally, $\mathrm{V}_a^b w$ denotes the total variation of the function w on $[a,b]$. Throughout the paper c is a generic constant that has different values in different relations and which is independent of the mesh spacing h.

An outline of our paper follows. After a short short discussion of optimality conditions for the continuous problem in the subsequent section we introduce quadratic splines and the collocation method for ordinary elliptic differential equations in Section 3. The main result is the error estimate for the numerical solution of the state equation. In Section 4 we investigate the discrete control problem and derive an discrete error estimate of order two for the discrete optimal controls. Afterwards we construct a new discrete control for which we derive continuous error estimates of quadratic order. Finally, we discuss a numerical example.

2 Optimality Conditions

It is a well-known fact that the mapping $y \mapsto z$ where

$$-\ddot{z}(t) + Az(t) = y(t) \quad \text{for a.a. } t \in [0,T], \quad z(0) = z(T) = 0, \tag{1}$$

defines a continuous linear operator $S\colon L_2(0,T;\mathbb{R}^n) \to W_{2,0}^2(0,T;\mathbb{R}^n)$ which assigns to each $y \in L_2(0,T;\mathbb{R}^n)$ the unique solution $z = S(y)$. Making use of the solution operator S, Problem (CP1) can be equivalently written in the form

$$\text{(CP2)} \qquad \min F(u) = J(S(Bu+e),u), \qquad \text{s.t. } u \in U^{\mathrm{ad}},$$

where $U^{\mathrm{ad}} = \{u \in L_2(0,T;\mathbb{R}^m) \mid a \le u(t) \le b \text{ for a.a. } t \in [0,T]\}$. Problem (CP2) has a unique solution $\bar{u}$ which is characterized by the pointwise variational inequality

$$\left(B^{\mathsf{T}}\bar{p}(t) + \nu\bar{u}(t)\right)^{\mathsf{T}}(u - \bar{u}(t)) \ge 0 \quad \forall u \in U \tag{2}$$

for a.a. $t \in [0,T]$, where $U = \{u \in \mathbb{R}^m \mid a \le u \le b\}$ and $\bar{p} = S^*(S(Bu+e) - z_d)$ is the adjoint state, i.e., the unique solution of

$$-\ddot{p}(t) + Ap(t) = (S(B\bar{u}+e))(t) - z_d(t) \quad \text{for a.a. } t \in [0,T], \quad p(0) = p(T) = 0. \tag{3}$$

The optimality condition (2) implies that $\bar{u}(t)$ is the projection of $-\frac{1}{\nu}B^{\mathsf{T}}\bar{p}(t)$ onto $[a,b]$, i.e.,

$$\bar{u} = \Pi_{[a,b]}\left(-\frac{1}{\nu}B^{\mathsf{T}}\bar{p}\right), \tag{4}$$

(cf. Malanowski [10]). This further implies $\bar{u} \in W^1_\infty(0,T;\mathbb{R}^m)$, especially the optimal control $\bar{u}$ is a Lipschitz continuous function so that (2) holds true for all $t \in [0,T]$.

3 Discretization of the State Equation

Consider the uniform grid $G = \{t_i = ih \mid i = 0,\dots,N\}$ with mesh size $h = T/N$, $N \ge 2$. By $B_{2,i}$, $i = -1,\dots,n+1$, we denote the B-splines of order 2 (see e.g. Sendov [13], Section 4.3), and by $W_h(0,T;\mathbb{R}^n)$ we denote the finite-dimensional linear space of quadratic splines which is the span of $\{B_{2,-1},\dots,B_{2,N+1}\}$. The subspace of functions with vanishing boundary values $w_h(0) = w_h(T) = 0$ is denoted by $W_{h,0}(0,T;\mathbb{R}^n)$. Further, we denote by $V_h(0,T;\mathbb{R}^n)$ the finite-dimensional linear space of continuous and piecewise linear functions on the grid G. A function $v_h \in V_h(0,T;\mathbb{R}^n)$ is uniquely defined by its values $v_0,\dots,v_N$ at the grid points.

The state equation (1), resp. the operator S, is then discretized by the operator S_h, where $S_h(y) \in W_{h,0}(0,T;\mathbb{R}^n)$ is the unique quadratic spline satisfying the collocation and boundary conditions

$$\begin{aligned} -(\ddot{S}_h y)(t_i) + A(S_h y)(t_i) &= y(t_i),\ i = 0,\dots,N, \\ (S_h y)(0) = (S_h y)(T) &= 0. \end{aligned} \tag{5}$$

This discretization is stable as shown in the following theorem.

Theorem 1. *Let $y \in W^1_\infty(0,T;\mathbb{R}^n)$ with $\dot{y} \in BV(0,T;\mathbb{R}^n)$. Then for sufficiently small h (i.e., $h < \sqrt{8\|A\|_\infty}$) we have*

$$\|S_h y\|_\infty \le \tilde{c}\,\|y\|_\infty \tag{6}$$

with a constant $\tilde{c}$ independent of y and h.

Proof. The proof is similar to the proof in the single valued case. For this we refer to Theorem 7.3 in Sendov [13] and for more details to Karolewski [9]. □

Next we define on $V_h(0,T;\mathbb{R}^n)$ the scalar product

$$\langle f,g\rangle_h = h\sum_{i=0}^{N} f(t_i)^{\mathsf{T}} g(t_i)$$

with the associated norm $\|f\|_h = \sqrt{\langle f,f\rangle_h}$. By S_h^* we denote the discretization of the adjoint operator S^*. The two discrete operators have an adjoint like property with respect to this scalar product, i.e.,

$$\langle S_h v_h, w_h\rangle_h = \langle v_h, S_h^* w_h\rangle_h + 2h^2 g(v_h,w_h) \quad \forall v_h, w_h \in V_h(0,T;\mathbb{R}^n) \tag{7}$$

with $g(v_h,w_h) = |v_h(0)| + |v_h(T)| + |w_h(0)| + |w_h(T)|$. The proof of this equality uses the linear systems for the coefficients together with a transformation into a system with zero boundary coefficients. The correction term stems from this step and is unnecessary for vanishing boundary values of v_h, w_h. On the space $V_{h,0}(0,T;\mathbb{R}^n)$ the operator S_h is therefore selfadjoint, i.e., $S_h = S_h^*$.

Estimates for the discretization error $Sy - S_h y$ play a crucial role in the proof of error estimates for discretizations of the control problem (CP2).

Theorem 2. *Let $y \in W_\infty^1(0,T;\mathbb{R}^n)$ with $\dot y \in BV(0,T;\mathbb{R}^n)$. Then for sufficiently small h we have the error estimates*

$$\|Sy - S_h y\|_\infty \le \gamma(\|y\|_\infty + \mathrm{V}_0^T \dot y)\, h^2, \tag{8}$$

$$\|(S^*(Sy - z_d) - S_h^*(S_h y - z_d)\|_\infty \le \tilde\gamma(\|y\|_{1,\infty} + \mathrm{V}_0^T \dot y + \|z_d\|_{2,2})\, h^2, \tag{9}$$

with constants γ, $\tilde\gamma$ independent of y and h.

Proof. By Theorem 7.4 and Corollary 7.7 of Sendov [13] we obtain with $z = Sy$

$$\|Sy - S_h y\|_\infty \le \gamma_1 \left(\mathrm{V}_0^T \dddot z + \|z\|_\infty + \|y\|_\infty\right) h^2 \tag{10}$$

with a constant γ_1 independent of h. Since $z = Sy$ we have $\dddot z = A\dot z - \dot y$, and therefore

$$\mathrm{V}_0^T \dddot z \le \|A\| \mathrm{V}_0^T \dot z + \mathrm{V}_0^T \dot y \le T\|A\|\,\|\ddot z\|_\infty + \mathrm{V}_0^T \dot y \le T\|A\|^2\,\|\dot z\|_\infty + T\|A\|\,\|z\|_\infty + \mathrm{V}_0^T \dot y.$$

Using the fact that $\max\{\|z\|_\infty, \|\dot z\|_\infty\} = \|z\|_{1,\infty} \le \gamma_2 \|y\|_\infty$ with a constant γ_2 independent of y we get

$$\mathrm{V}_0^T \dddot z \le \gamma_3 \|y\|_\infty + \mathrm{V}_0^T \dot y$$

with a constant γ_3 independent of y. Together with (10) this shows (8). □

4 Discretization of the Control Problem

By $P_h : C(0,T;\mathbb{R}^n) \to V_h(0,T;\mathbb{R}^n)$ we denote the interpolation operator defined by $(P_h u)(t_i) = u(t_i)$ for $i = 0,\dots,N$. Using the operators S_h, P_h and the space $V_h(0,T;\mathbb{R}^m)$ we discretize problem (CP2) in the following way:

$$(\text{CP})_h \qquad \min \frac{1}{2}\|S_h(Bu_h+e) - P_h z_d\|_h^2 + \frac{\nu}{2}\|u_h\|_h^2$$
$$\text{s.t. } u_h \in U_h^{\text{ad}} = U^{\text{ad}} \cap V_h(0,T;\mathbb{R}^m).$$

Problem $(\text{CP})_h$ has a unique solution $\bar{u}_h \in V_h(0,T;\mathbb{R}^m)$ which is characterized by the optimality conditions

$$\langle B^\mathsf{T} p_h(\bar{u}_h) - \nu \bar{u}_h, \zeta_h - \bar{u}_h \rangle_h \geq 0 \quad \forall \zeta_h \in U_h^{\text{ad}}, \tag{11}$$

where $p_h(\bar{u}_h) = S_h^*(S_h(B\bar{u}_h + e) - P_h z_d)$ is the discrete adjoint state for $\bar{u}_h$.

At the end of Section 2 we have shown that the solution $\bar{u}$ of (CP2) belongs to the space $W_\infty^1(0,T;\mathbb{R}^m)$. Some of the results of the previous section, which are used here, are formulated in terms of the variation of $\dot{y}$. Therefore, in the sequel we require additional regularity of $\bar{u}$. As in Alt et al. [3] and Dontchev et al. [6] we assume that the derivative of $\bar{u}$ has bounded variation.

First we derive a result on discrete quadratic convergence for the solutions $\bar{u}_h \in V_h(0,T;\mathbb{R}^m)$ of the problems $(\text{CP})_h$.

Theorem 3. *Let $\bar{u}$ be the solution of* (CP2) *with the assumption $\dot{\bar{u}} \in BV(0,T;\mathbb{R}^m)$ and let $\bar{u}_h \in V_h(0,T;\mathbb{R}^m)$ be the solution of the discrete problem* $(\text{CP})_h$. *Then*

$$\|P_h\bar{u} - \bar{u}_h\|_h \leq c\,h^2 \tag{12}$$

holds true with a constant c independent of h.

Proof. Since $\bar{u}_h$ is feasible for problem (CP2), we have by (2)

$$\left(B^\mathsf{T}(P_h\bar{p})(t_i) + \nu\,(P_h\bar{u})(t_i)\right)^\mathsf{T}\left(\bar{u}_h(t_i) - (P_h\bar{u})(t_i)\right) \geq 0, \quad i = 0,\dots,N,$$

because $\bar{u}(t_i) = (P_h\bar{u})(t_i)$. Summing up all inequalities and adding (11) with $\zeta_h = P_h\bar{u}$

$$\langle B^\mathsf{T}(P_h\bar{p} - p_h(\bar{u}_h)) + \nu\,(P_h\bar{u} - \bar{u}_h), \bar{u}_h - P_h\bar{u}\rangle_h \geq 0,$$

which is equivalent to

$$\begin{aligned}\nu\,\|\bar{u}_h - P_h\bar{u}\|_h^2 &\leq \langle P_h\bar{p} - p_h(\bar{u}_h), B(\bar{u}_h - P_h\bar{u})\rangle_h \\ &= \langle P_h\bar{p} - p_h(\bar{u}), B(\bar{u}_h - P_h\bar{u})\rangle_h + \langle p_h(P_h\bar{u}) - p_h(\bar{u}_h), B(\bar{u}_h - P_h\bar{u})\rangle_h,\end{aligned}$$

where the last equality follows from $p_h(P_h\bar{u}) = p_h(\bar{u})$. For the second term we obtain by (7)

$$\begin{aligned}\langle p_h(P_h\bar{u}) - p_h(\bar{u}_h), B(\bar{u}_h - P_h\bar{u})\rangle_h &= \langle S_h^* S_h B(P_h\bar{u} - \bar{u}_h), B(\bar{u}_h - P_h\bar{u})\rangle_h \\ &= \langle S_h B(P_h\bar{u} - \bar{u}_h), S_h B(\bar{u}_h - P_h\bar{u})\rangle_h \\ &= -\|S_h B(P_h\bar{u} - \bar{u}_h)\|_h^2 \leq 0.\end{aligned}$$

Note that in case of $a < 0 < b$ the solution of (CP2) and its discretization vanish at the boundary, because the adjoint state is defined with this property and by virtue of (4). For the first term we have

$$\begin{aligned}\langle \bar{p} - p_h(P_h\bar{u}), B(\bar{u}_h - P_h\bar{u})\rangle_h &\leq \|P_h\bar{p} - p_h(\bar{u})\|_h \|B\| \, \|P_h\bar{u} - \bar{u}_h\|_h \\ &\leq c\,\|P_h\bar{p} - p_h(\bar{u})\|_\infty \, \|P_h\bar{u} - u_h\|_h \leq c h^2 \, \|P_h\bar{u} - \bar{u}_h\|_h.\end{aligned}$$

where the last inequality follows from (9). Dividing by $\|P_h\bar{u} - \bar{u}_h\|_h$, we obtain (12). □

Theorem 3 shows only *discrete* convergence for the solutions $\bar{u}_h \in V_h(0,T;\mathbb{R}^m)$ of the problems $(\mathrm{CP})_h$. Interpolation error estimates imply that the *continuous* error $\|\bar{u} - \bar{u}_h\|_2$ is only of order $3/2$. Therefore, we adopt the idea of Meyer/Rösch [11] to construct, based on (4), a new feasible control by

$$\tilde{u}_h = \Pi_{[a,b]}\left(B^{\mathsf{T}} p_h(\bar{u}_h)\right), \tag{13}$$

for which we can prove continuous convergence of order 2.

Theorem 4. *Let $\bar{u}$ be the solution of problem* (CP2) *with $\dot{\bar{u}} \in BV(0,T;\mathbb{R}^m)$ and $\bar{u}_h \in V_h(0,T;\mathbb{R}^m)$ being the solution of the discrete problem* $(\mathrm{CP})_h$. *Then for the control $\tilde{u}_h$ defined by* (13) *we have the continuous error estimate*

$$\|\bar{u} - \tilde{u}_h\|_\infty \leq c h^2$$

with a constant c independent of h.

Proof. The projection operator $\Pi_{[a,b]}$ is Lipschitz continuous. Therefore we have

$$\|\bar{u} - \tilde{u}_h\|_\infty \leq c\,\|p(\bar{u}) - p_h(\bar{u}_h)\|_\infty \leq c\,\|p(\bar{u}) - p_h(\bar{u})\|_\infty + c\,\|p_h(\bar{u}) - p_h(\bar{u}_h)\|_\infty.$$

For the first term we use (9) and by (6) and Theorem 3 we get for the second term

$$\|p_h(\bar{u}) - p_h(\bar{u}_h)\|_\infty = \|S_h^* S_h B(P_h\bar{u} - \bar{u}_h)\|_\infty \leq c\,\|P_h\bar{u} - \bar{u}_h\|_h \leq c h^2,$$

which implies the assertion. □

5 Numerical Example

In order to illustrate the result of Theorem 4, we consider the same problem as in Hinze [8], Section 4.2, for which the optimal solution is known. We choose the parameters $T = 1$, $z_d \equiv 2$, $\nu = 0.1$, $a = -\infty$, $b = 2.5(\sqrt{2}-1)$ and $e(t) = -2 + t^2 - t - \min\{-\frac{1}{\nu}(t^2 - t), b\}$. The optimal control is then given by

$$\bar{u}(t) = \min\{-\frac{1}{\nu}(t^2 - t), b\},$$

Figure 1 shows the discrete and the exact solution, which cannot be distinguished from the control $\tilde{u}_h$ in the picture.

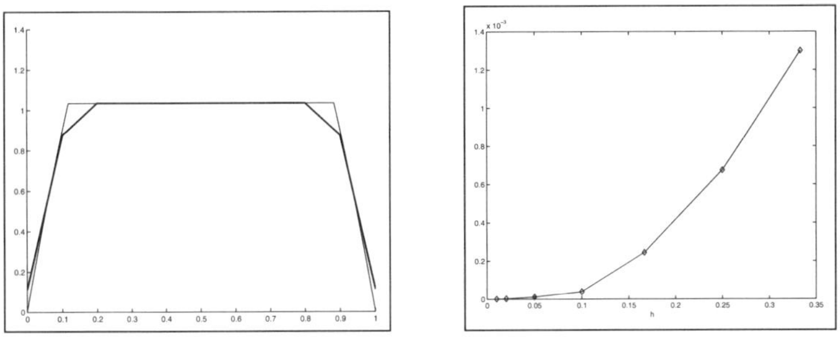

Fig. 1 In the left picture we see the discrete (thick line) and the exact control (thin line). In the right picture the error $\|\bar{u} - \tilde{u}_h\|_\infty$ is sketched against the mesh size h.

Table 1 shows the errors for the discrete optimal control $\bar{u}_h$ and the control $\tilde{u}_h$ obtained in the postprocessing step. For comparison the results of the finite element discretization from [3] and for the finite difference discretization from [2] are listed in Table 2. In all three cases the order of continuous convergence for the solutions of the discretized problems is only linear. After the post-processing step (13) the order is improved to quadratic order.

Table 1 Errors for collocation method in dependence of mesh size h.

h	$\|\bar{u} - \bar{u}_h\|_\infty$	$\|\bar{u} - \bar{u}_h\|_\infty / h$	$\|\bar{u} - \tilde{u}_h\|_\infty$	$\|\bar{u} - \tilde{u}_h\|_\infty / h^2$
1/3	0.4889	1.467	0.2447	2.202
1/4	0.4241	1.696	0.1502	2.401
1/6	0.2548	1.529	0.0674	2.428
1/10	0.1317	1.317	0.0245	2.447
1/20	0.0923	1.846	0.0062	2.472
1/50	0.0245	1.225	$9.953 \cdot 10^{-4}$	2.488
1/100	0.0150	1.500	$2.494 \cdot 10^{-4}$	2.494

Table 2 Errors from [3] (left columns) and [2] (right columns).

h	$\|\bar{u}-\bar{u}_h\|_\infty$	$\|\bar{u}-\tilde{u}\|_\infty$	$\|\bar{u}-\bar{u}_h\|_\infty$	$\|\bar{u}-\tilde{u}\|_\infty$
1/3	1.0355	0.2532	0.6696	0.2531
1/4	0.9382	0.1550	0.5485	0.1556
1/6	0.6952	0.0686	0.3062	0.0694
1/10	0.4502	0.0248	0.1111	0.0250
1/20	0.2375	0.0062	0.0870	0.0063
1/50	0.0980	$9.9840 \cdot 10^{-4}$	0.0179	0.0010
1/100	0.0495	$2.4985 \cdot 10^{-4}$	0.0145	$2.5 \cdot 10^{-4}$

References

1. Alt, W., Bräutigam, N.: Discretization of elliptic control problems with time dependent parameters. Accepted by Applied Numerical Mathematics
2. Alt, W., Bräutigam, N.: Finite-difference discretizations of quadratic control problems governed by ordinary elliptic differential equations. Accepted by Computational Optimization and Applications
3. Alt, W., Bräutigam, N., Rösch, A.: Error estimates for finite element approximations of elliptic control problems. Discussiones Mathematicae: Differential Inclusions, Control and Optimization **27**, 7–23 (2007)
4. Arada, N., Casas, E., Tröltzsch, F.: Error estimates for the numerical approximation of a semilinear elliptic control problem. Computational Optimization and Applications **23**, 201–229 (2002)
5. Casas, E., Tröltzsch, F.: Error estimates for linear-quadratic elliptic control problems. In: V. Barbu (ed.) Analysis and Optimization of Differential Systems, pp. 89–100. Kluwer Academic Press (2003)
6. Dontchev, A.L., Hager, W.W., Veliov, V.V.: Second-order Runge-Kutta approximations in control constrained optimal control. SIAM Journal on Numerical Analysis **38**, 202–226 (2000)
7. Grossmann, C., Roos, H.G.: Numerik partieller Differentialgleichungen. Teubner, Wiesbaden (2005)
8. Hinze, M.: A variational discretization concept in control constrained optimization: The linear-quadratic case. Computational Optimization and Applications **30**, 45–61 (2005)
9. Karolewski, D.: Diskretisierung Elliptischer Steuerungsprobleme mit quadratischen Splines. Master's thesis, Friedrich-Schiller-Universität Jena (2007)
10. Malanowski, K.: Convergence of approximations vs. regularity of solutions for convex, control-constrained optimal control problems. Applied Mathematics and Optimization **8**, 69–95 (1981)
11. Meyer, C., Rösch, A.: Superconvergence properties of optimal control problems. SIAM Journal on Control and Optimization **43**, 970–985 (2004)
12. Rösch, A.: Error estimates for linear-quadratic control problems with control constraints. Optimization Methods and Software **21**, 121–134 (2006)
13. Sendov, B., Popov, V.A.: The Averaged Moduli of Smoothness. Wiley-Interscience, New York (1988)
14. Tröltzsch, F.: Optimale Steuerung partieller Differentialgleichungen. Vieweg, Wiesbaden (2005)

A Road Traffic Model with Overtaking: Continuation of the Oscillatory Patterns

L. Buřič and V. Janovský

Abstract We investigate microscopic models of the road traffic. In particular, we consider *car-following* models for a single-line traffic flow on a circular road. The classical differentiable models break down at the time instant when two cars collide. Nevertheless, the natural action of a driver would be to overtake the slower car. In our previous work, we proposed a model which simulates an overtaking. The model implicitly defines a maneuver consisting of deceleration/acceleration just shortly before/after the overtaking. We observed a large variety of oscillatory solutions (*oscillatory patterns*) of the model. In case $N = 3$ (three cars on the route), we can supply a finite classification list of these patterns. In the present contribution, we stick to $N = 3$, and formulate our model as a particular *Filippov system* i.e., ODE with discontinuous righthand sides. We define oscillatory patterns as invariant objects of this Filippov system. We use the standard software (AUTO97) to continue these patterns with respect to a parameter.

1 Introduction

In order to understand dynamics of traffic flows, mathematical models are in current use, see e.g. [9] for a recent review. Microscopic modelling of vehicular traffic is usually based on the so called *Follow-the-Leader models*, see [8].

Consider the system

$$x_i' = y_i\,, \quad y_i' = \frac{1}{\tau}\left[V(x_{i+1} - x_i) - y_i\right]\,, \quad x_{N+1} = x_1 + L\,, \quad i = 1, \ldots, N\,. \qquad (1)$$

Lubor Buřič
Department of Mathematics, Institute of Chemical Technology, Technická 5, 16628 Prague 6, Czech Republic, e-mail: lubor.buric@vscht.cz

Vladimír Janovský
Charles University, Faculty of Mathematics and Physics, Sokolovská 83, 18675 Prague 8, Czech Republic, e-mail: janovsky@karlin.mff.cuni.cz

It models N cars on a circular road of the length L. The pairs (x_i, y_i) are interpreted as the position x_i and the velocity y_i of the car number i. The acceleration y_i' of each car depends on the difference between the car velocity y_i and the *optimal velocity function* $V = V(x_{i+1} - x_i)$. We assume that this function $r \mapsto V(r)$, with domain $r \geq 0$, satisfies the following assumptions: V is positive valued and monotonically increasing, $V(0) = 0$, $V(r) \to V^{max} > 0$ as $r \to \infty$, there exists a positive constant b such that $V''(r) > 0$ ($V''(r) < 0$) if $r < b$ ($r > b$), see [1, 2]. Reciprocal value of the parameter τ is called *sensitivity*, [1]. In all forthcoming computations we will consider the hyperbolic optimal velocity function defined as

$$V(r) = V^{max} \frac{\tanh(a(r-1)) + \tanh(a)}{1 + \tanh(a)}, \tag{2}$$

where V^{max} and a are positive constants, see [2]. The choice of V imposes a driving law and we assume that this law is the same for all N drivers. The difference

$$h_i \equiv x_{i+1} - x_i, \quad i = 1, \ldots, N, \tag{3}$$

is called *headway* (of the i-th car). Given an initial condition $[x^0, y^0] \in \mathbb{R}^N \times \mathbb{R}^N$, the system (1) defines a flow on $\mathbb{R}^N \times \mathbb{R}^N$. Without loss of generality, we assume

$$s \leq x_1^0 \leq x_2^0 \leq \cdots \leq x_{N-1}^0 \leq x_N^0 \leq L + s, \tag{4}$$

$$y^0 = (y_1^0, \ldots, y_N^0), \quad y_i^0 > 0, \quad i = 1, \ldots, N, \tag{5}$$

where $s \in \mathbb{R}$ is an arbitrary phase shift.

The smallest instant of time $t_E > 0$ is called an *event* provided that there exists $i \in \{1, \ldots, N\}$ such that the relevant headway component of the flow vanishes at t_E, i.e., $h_i(t_E) \equiv x_{i+1}(t_E) - x_i(t_E) = 0$. In the traditional interpretation, the cars No i and No $i+1$ *collide* at t_E and the classical model breaks down. In [3, 6] we proposed a different interpretation: At t_E, the car No i **overtakes** the car No $i+1$. We suggested an algorithm how to proceed for $t > t_E$. By induction, the Overtaking Algorithm generate a flow $[x(t), y(t)] = \Pi(t, [x^0, y^0])$ on any finite interval $0 \leq t \leq T_{\max}$. The phase curve $[x(t), y(t)]_{0 \leq t \leq T_{\max}}$ is continuous and piecewise smooth. The algorithm generates a finite set $t_E = \{t_E(s)\}_{s=1}^{Z}$ called *event sequence*. The time $t_E(s)$ is related to an event of the overtaking: a car number i_s overtakes a car number j_s. Symbolically, $[i_s \to j_s]$. Each event $t_E(s)$ is related to a transposition of cars on the route.

2 The Overtaking Model as a Filippov System

For the sake of simplicity, let us consider $N = 3$. There are only two possible car orderings along the route, "123" and "132". The ordering is related to the sign of permutation of the cars. In order to measure the distance between the cars No i and No j we introduce a new variable which we call the *gap*:

$$h_{i,j} = x_j - x_i\,, \quad i < j\,, \quad h_{i,j} = L - h_{j,i}\,, \quad i > j\,. \tag{6}$$

The idea is to define the flow Π in different state variables namely, the velocity $y \in \mathbb{R}^3$ and the gap components $h_{1,2}$, $h_{2,3}$ and $h_{3,1}$.

To that end we modify the optimal velocity function to be L-periodic and discontinuous: We define $\tilde{V} = \tilde{V}(r)$ via the formula (2) on the interval $[0, L)$, and extend the function $\tilde{V} = \tilde{V}(r)$ periodically on the whole $\mathbb{R}$.

The relevant system possesses the first integral. In fact, we can eliminate one gap variable, say $h_{3,1}$,

$$h_{3,1} = L - h_{1,2} - h_{2,3}\,. \tag{7}$$

For details, see [5]. The resulting system reads as follows:

$$h'_{1,2} = y_2 - y_1\,, \quad h'_{2,3} = y_3 - y_2\,, \tag{8a}$$

$$y'_1 = \frac{1}{\tau}\left[\tilde{V}(h_{1,2}) - y_1\right]\,, \quad y'_2 = \frac{1}{\tau}\left[\tilde{V}(h_{2,3}) - y_2\right]\,, \tag{8b}$$

$$y'_3 = \frac{1}{\tau}\left[\tilde{V}(L - h_{1,2} - h_{2,3}) - y_3\right]\,, \tag{8c}$$

for $lL < h_{1,2} < (1+l)L$, $mL < h_{2,3} < (1+m)L$, $nL < h_{3,1} < (1+n)L$, where $l, m, n \in \mathbb{Z}$ and $l + m + n$ is **even**, and

$$h'_{1,2} = y_2 - y_1\,, \quad h'_{2,3} = y_3 - y_2\,, \tag{9a}$$

$$y'_1 = \frac{1}{\tau}\left[\tilde{V}(h_{1,2} + h_{2,3}) - y_1\right]\,, \tag{9b}$$

$$y'_2 = \frac{1}{\tau}\left[\tilde{V}(L - h_{1,2}) - y_2\right]\,, \quad y'_3 = \frac{1}{\tau}\left[\tilde{V}(L - h_{2,3}) - y_3\right]\,, \tag{9c}$$

for $lL < h_{1,2} < (1+l)L$, $mL < h_{2,3} < (1+m)L$, $nL < h_{3,1} < (1+n)L$, where $l, m, n \in \mathbb{Z}$ and $l + m + n$ is **odd**.

The righthand sides of (8) and (9) are related to particular values $l, m, n \in \mathbb{Z}$. Domains of equations (8) and (9) are *voxels*, which are referred to by triples $\{l, m, n\}$. The voxels are "colored" as the **even/odd** ones. The complement of voxels are the hyperplanes

$$h_{i,j} = kL\,, \quad k \in \mathbb{Z}\,. \tag{10}$$

They play the role of switches between the systems (8) and (9).

Both (8)&(9) define a *Filippov system* i.e., the system of five ODE's with discontinuous righthand sides, see [7]. Given an initial condition $[y^0, h^0] \in \mathbb{R}^3 \times \mathbb{R}^3$, so that $y^0 > 0$, $h^0 = (h^0_{1,2}, h^0_{2,3}, h^0_{3,1})$ satisfies (7), and h^0 is inside of a voxel, then we solve the initial value problem for (8)&(9). Let $[h(t), y(t)] \equiv \Theta(t, [h^0, y^0])$ be its solution at $t \geq 0$. As an example, consider

$$h^0 = [1.7628, 0.5425, 0.1947]\,, \quad y^0 = [2.6305, 3.7393, 2.1614]\,, \tag{11}$$

$L = 2.5$, $a = 2$, $V_{max} = 7$, $\tau = 1$. The trajectory of flow Θ for $0 \leq t \leq 40$, projected to the scaled gap space, is shown on Fig. 1 (on the left). The points on the phase curve which are marked by the filled squares belong to the hyperplanes (10) and hence they are related to the overtaking. Recording of the relevant time generates the event sequence t_E. It is apparent that the phase curve evolves into an invariant pattern. It is related to an oscillatory solution of the system (8)&(9).

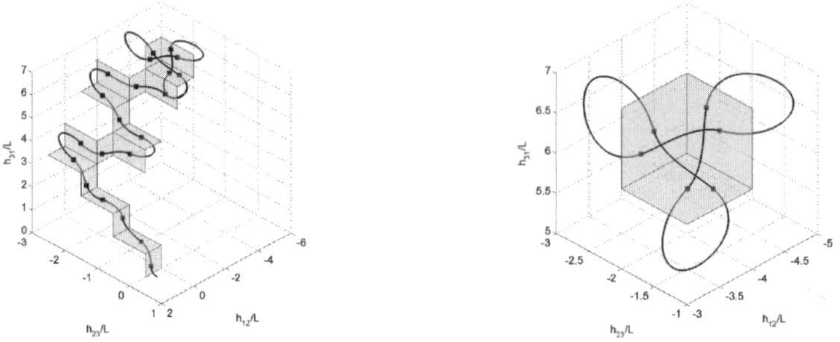

Fig. 1 On the left: Transition to an invariant object. On the right: Zoom of the invariant object called rotating wave of class 2, the period $T = 7.9481$

3 Continuation of the Oscillatory Patterns

In [6], we observed five classes of oscillatory patterns which are limit sets of the flow Π: two classes of rotating waves and three types of patterns, which are nicknamed as the three-legged dog patterns (essentially, phase-shifted reflectionally symmetric oscillations of two cars while the third car oscillates differently). We introduced the notion of a *periodic event map*. The particular patterns were classified by spatial and temporal symmetries of the relevant event maps. We will refer to these patterns as wave-1, wave-2, 3dog, 3dog-1 and 3dog-2. Hence, on Fig. 1 (on the right) there is wave-2 projected on the gap space. Remaining patterns projected on the gap space can be found in [4].

Wave-2 corresponds to the event map

$$G_E = \{[1 \to 2], [3 \to 2], [2 \to 3], [1 \to 3], [3 \to 1], [2 \to 1], \text{etc.}\} . \qquad (12)$$

Since there are six events in one period, the *event map period* p_E, see [6] for the notion, equals to $p_E = 6$. Similarly, the event map for wave-1 is

$$G_E = \{[1 \to 2], [3 \to 2], [3 \to 1], [2 \to 1], [2 \to 3], [1 \to 3], \text{etc.}\} . \qquad (13)$$

It means that $p_E = 6$. The 3dog pattern is related to the event map

$$G_E = \{[1 \to 2], [2 \to 1], \text{etc.}\} , \qquad (14)$$

while 3dog-1 corresponds to

$$G_E = \{[1 \to 3], [2 \to 3], \text{etc.}\} . \tag{15}$$

The event maps of both 3dog and 3dog-1 patterns are two-periodic, i.e. $p_E = 2$. Finally, 3dog-2 pattern produces the event map

$$G_E = \{[3 \to 1], [3 \to 2], [1 \to 2], [3 \to 2], [3 \to 1], [2 \to 1], \text{etc.}\} , \tag{16}$$

with $p_E = 6$.

The main advantage of the Filippov formulation (8)&(9) is that it allows to *continue* an invariant object with respect to a parameter. We are able to continue all patterns. Each pattern is defined as a boundary value problem on $\mathbb{R}^{5p_E}$ with special boundary value conditions. Defining equations are rather complicated and hence we have to give them elsewhere. The continuation is performed using the package AUTO97. We review selected numerical results.

We considered five branches of particular patterns, see Fig. 2. Continuation of each branch collapsed at two points referred to as L_{min} and L_{max}, according to the appropriate L coordinate. The continuation collapse refers to an event which is degenerated in the sense that transmission of the trajectory is not *transversal* at that point: the particular trajectory touches/crosses a hyperplane (10) in a tangent direction. We observe the former scenario in case of wave-2, see Fig. 3. Note that the trajectory touches the boundary $h_{1,2} = -L$. A similar observation can be made for both 3dog, see Fig. 5, and 3dog-2 patterns, see Fig. 7. As far as the latter scenario is concerned, in Fig. 4 the trajectory crosses the hyperplane $h_{1,2} = 0$ in a tangent direction. The case is manifested as a cusp in projection on the $(h_{1,2}, y_1)$-plane. Similar behavior can be observed for 3dog-1 pattern, see Fig. 6. Note that, in general, solid/dashed lines refer to trajectories related to the "123"/ "132" ordering. The analysis of the extremal cases L_{min} and L_{max} anticipates a qualitative change (a bifurcation) of the patterns.

Finally, we briefly mention stability of the branches on Fig. 2: Two folds on the branch wave-1 suggest that, say for $L = 3.1$, three wave-1 patterns coexist, see Fig. 8 (on the left). Dynamical simulation indicates that the solid cycle is stable while the dashed ones are unstable. For example, unstable cycle labeled as U1 evolves into 3dog pattern, see Fig. 8 (on the right).

Acknowledgements The research of the first author was supported by the grant MSM 6046137306 of the Ministry of Education, Youth and Sports, Czech Republic. The second author was supported by the Grant Agency of the Czech Republic (grant No. 201/06/0356) and also by the research project MSM 0021620839 of The Ministry of Education, Youth and Sports, Czech Republic.

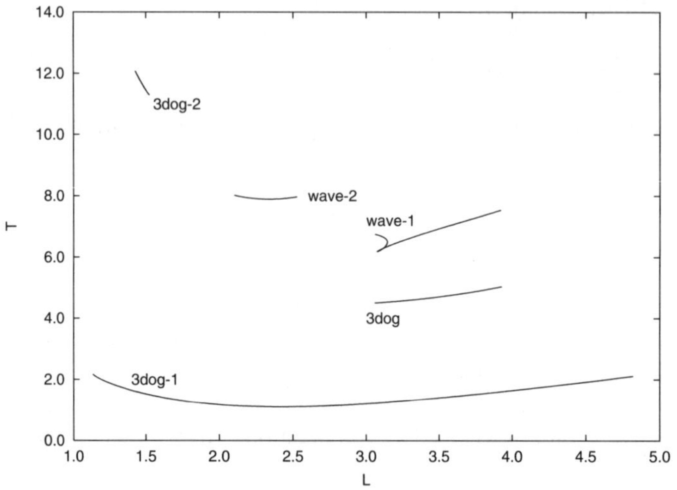

Fig. 2 Continuation of patterns: the length L vs. pattern period T

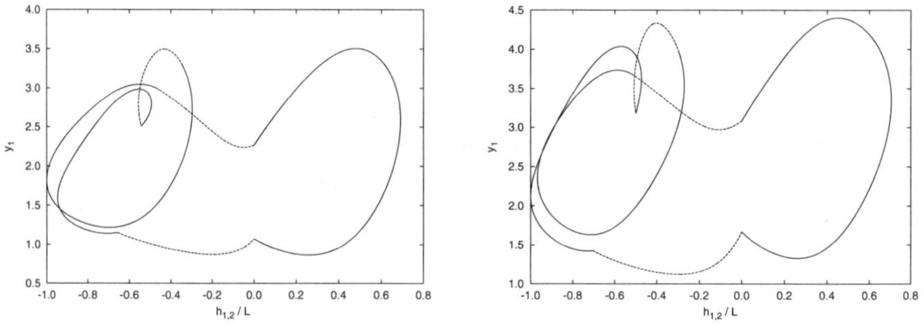

Fig. 3 Wave-2 for critical values of L: $L_{min} = 2.10218$ (on the left), $L_{max} = 2.52677$ (on the right)

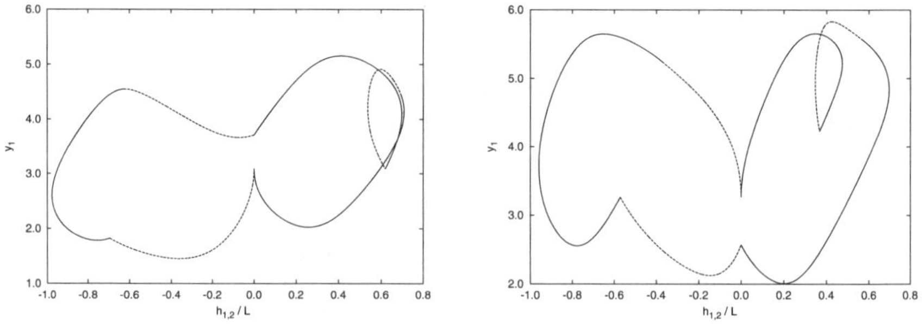

Fig. 4 Wave-1 for critical values of L: $L_{min} = 3.05990$ (on the left), $L_{max} = 3.92221$ (on the right)

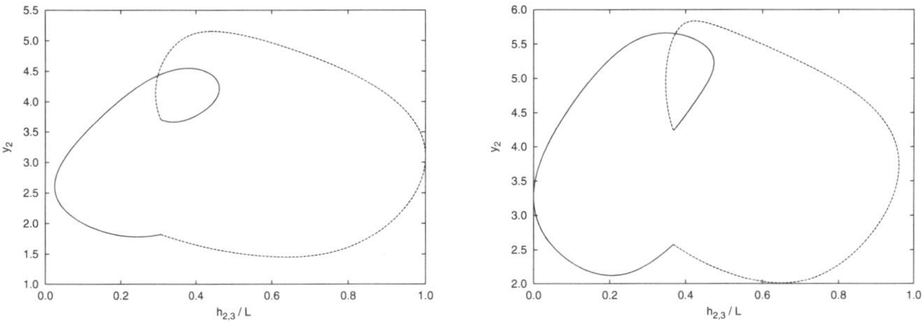

Fig. 5 3dog for critical values of L: $L_{min} = 3.06008$ (on the left), $L_{max} = 3.92738$ (on the right)

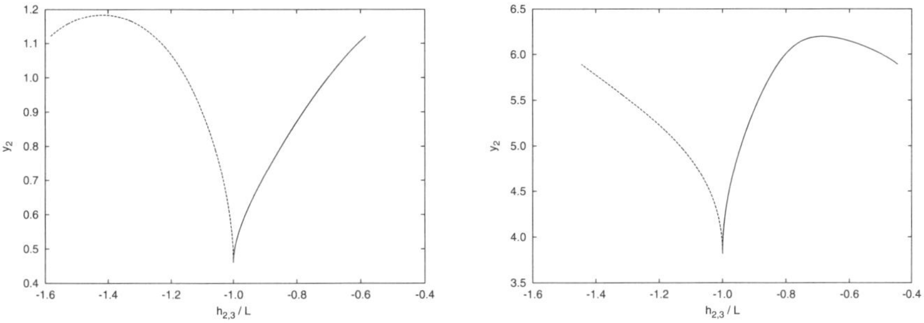

Fig. 6 3dog-1 for critical values of L: $L_{min} = 1.13705$ (on the left), $L_{max} = 4.82000$ (on the right)

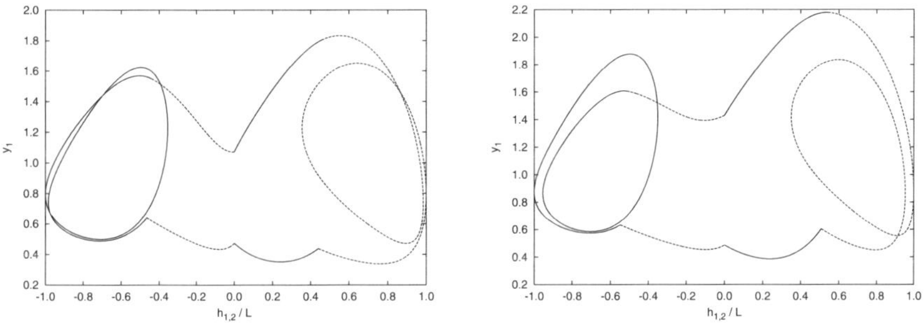

Fig. 7 3dog-2 for critical values of L: $L_{min} = 1.42784$ (on the left), $L_{max} = 1.52387$ (on the right)

References

1. Bando, M., Hasebe, K., Nakanishi, K., Nakayama, A., Shibata, A., Sugiyama, Y.: Phenomenological study of dynamical model of traffic flow. J. Phys. I France **5**, 1389–1399 (1995)
2. Bando, M., Hasebe, K., Nakayama, A., Shibata, A., Sugiyama, Y.: Dynamical model of traffic congestion and numerical simulation. Phys. Rev. E **51**, 1035–1042 (1995)
3. Buřič, L., Janovský, V.: The traffic problem: modelling of the overtaking. In: T.S. Simos (ed.) International Conference on Numerical Analysis and Applied Mathematics 2006, pp. 166–169. WILEY-VCH Verlag, Weinheim (2006)

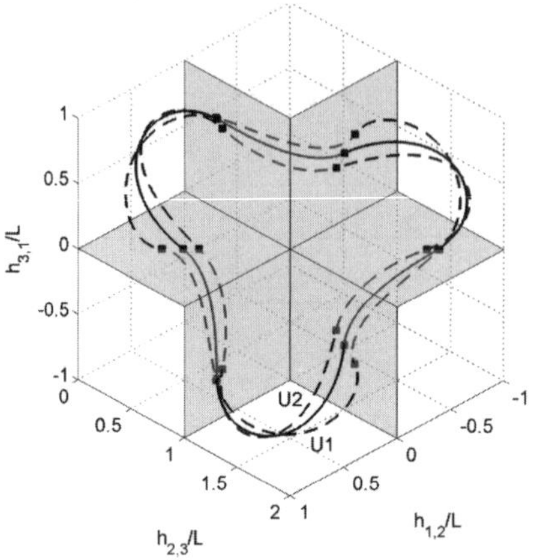

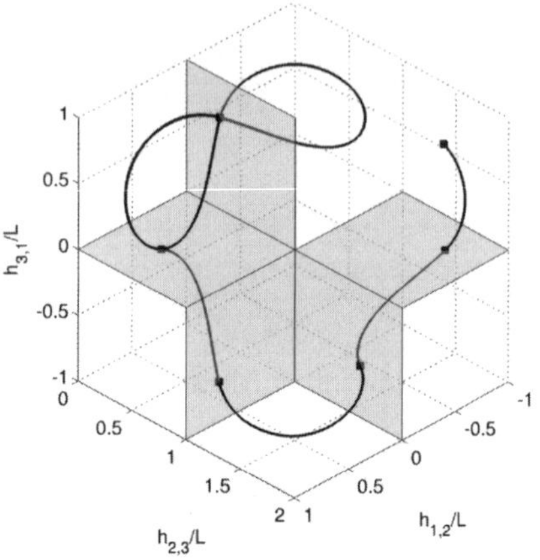

Fig. 8 Three coexisting waves-1 as $L = 3.1$ (on the left). The dashed waves-1 labeled as U1 and U2 are unstable. E.g., by a small perturbation, U1 evolves into 3dog pattern (on the right)

4. Buřič, L., Janovský, V.: The overtaking in a class of traffic models: A Filippov system formulation. In: T.S. Simos (ed.) International Conference on Numerical Analysis and Applied Mathematics 2007, pp. 97–100. AIP Conference Proceedings 936, American Institute of Physics, Melville, New York (2007)
5. Buřič, L., Janovský, V.: The traffic problem: modeling of the overtaking. Submitted to Math. Method Appl. Sci. (2007). URL http://web.vscht.cz/buricl/pub/icnaam06.pdf
6. Buřič, L., Janovský, V.: On pattern formation in a class of traffic models. Physica D **237**, 28–49 (2008)
7. Filippov, A.F.: Differential Equations with Discontinuous Righthand Sides. Kluwer Academic Publishers, Dordrecht (1988)
8. Gazis, D.C., Herman, R., Rothery, R.: Nonlinear follow-the-leader models of traffic flow. Operations Res. **9**, 545 (1961)
9. Helbing, D.: Traffic and related self-driven many-partical systems. Rev. Modern Phys. **73**, 1067–1141 (2001)

A Second Order Scheme for Solving Optimization-Constrained Differential Equations with Discontinuities

A. Caboussat and C. Landry

Abstract A numerical method for the resolution of a system of ordinary differential equations coupled with a mixed constrained minimization problem is presented. This coupling induces discontinuities of some time-dependent variables when inequality constraints are activated or deactivated. The ordinary differential equations are discretized in time and combined with the first order optimality conditions of the optimization problem. We use a second order multistep method based on a predictor-corrector Adams scheme to detect the discontinuities by extrapolation of the trajectories. Optimization features, namely a sensitivity analysis, are exploited to compute the derivatives of the optimization variables and track the discontinuity points. The main difficulty consists in the impossibility of defining an explicit event function to characterize the activation or deactivation of a constraint. The order of convergence of our method is proved when inequality constraints are activated and numerical results for atmospheric organic particles are presented.

1 Introduction

Dynamic optimization problems arise when coupling an optimization problem with ordinary differential equations. They appear for instance in computational chemistry. We present here a mathematical model and a numerical method for the simulation of dynamic phase transition for a single atmospheric aerosol particle that exchanges mass with the surrounding gas [1]. The mass transfer is described by ordinary differential equations while a mixed constrained minimization problem

Alexandre Caboussat
Department of Mathematics, University of Houston, 4800 Calhoun Rd, Houston, Texas 77204-3008, USA, e-mail: caboussat@math.uh.edu

Chantal Landry
Institute of Analysis and Scientific Computing, Ecole Polytechnique Fédérale de Lausanne, 1015 Lausanne, Switzerland, e-mail: chantal.landry@epfl.ch

determines the thermodynamic equilibrium of the particle, *i.e.* the partitioning of organics between different liquid phases.

Let $(0,T)$ be the interval of integration with $T > 0$. Let us denote by $\mathbf{b}(t)$ the concentration-vector of the s chemical components present in the particle at time $t \in (0,T)$. The dynamic optimization problem consists in finding $\mathbf{b}(t), \mathbf{x}_\alpha(t), y_\alpha(t)$ satisfying

$$\begin{aligned}
&\frac{\mathrm{d}}{\mathrm{d}t}\mathbf{b}(t) = \mathbf{f}(\mathbf{b}(t), \mathbf{x}_\alpha(t)), \qquad a.e.\ t \in (0,T), \qquad \mathbf{b}(0) = \mathbf{b}_0, \\
&(y_\alpha(t), \mathbf{x}_\alpha(t)) = \arg \min_{\tilde{y}_\alpha, \tilde{\mathbf{x}}_\alpha} \sum_{\alpha=1}^{p} \tilde{y}_\alpha g(\tilde{\mathbf{x}}_\alpha) \\
&\text{s.t. } \mathbf{e}^T \tilde{\mathbf{x}}_\alpha = 1,\ \tilde{\mathbf{x}}_\alpha > 0,\ \tilde{y}_\alpha \geq 0,\ \alpha = 1, \ldots, p, \quad \sum_{\alpha=1}^{p} \tilde{y}_\alpha \tilde{\mathbf{x}}_\alpha = \mathbf{b}(t),
\end{aligned} \tag{1}$$

where p is the number of possible liquid phases present in the aerosol, $\tilde{y}_\alpha$ is the total number of moles in phase α, $\tilde{\mathbf{x}}_\alpha$ is the mole-fraction concentration vector in phase α, $\mathbf{e}$ is the vector $(1, \ldots, 1)^T$, g is the Gibbs free energy and $\mathbf{f}$ is the mass flux between the aerosol and the surrounding media.

In the following sections we consider a model problem with linear equality constraints. Let $\mathbf{f} : \mathbb{R} \times \mathbb{R}^s \times \mathbb{R}^m \to \mathbb{R}^s$ be Lipschitz continuous and bounded, $g \in C^\infty(\mathbb{R}^m)$ and $A \in \mathbb{R}^{s \times m}$, with $s < m$. The problem reads: Find $\mathbf{b} : (0,T) \to \mathbb{R}^s$ and $\mathbf{z} : (0,T) \to \mathbb{R}^m$ satisfying

$$\begin{aligned}
\frac{\mathrm{d}}{\mathrm{d}t}\mathbf{b}(t) &= \mathbf{f}(t, \mathbf{b}(t), \mathbf{z}(t)), \qquad a.e.\ t \in (0,T), \qquad \mathbf{b}(0) = \mathbf{b}_0, \\
\mathbf{z}(t) &= \arg\min_{\tilde{\mathbf{z}}} g(\tilde{\mathbf{z}}) \qquad s.t. \quad A\tilde{\mathbf{z}} = \mathbf{b}(t), \quad \tilde{\mathbf{z}} \geq 0.
\end{aligned} \tag{2}$$

The loss of regularity of the variable $\mathbf{z}$ occurs when one of the inequality constraints $z_i(t) \geq 0$, $i = 1, \ldots, m$ is activated or deactivated. In this paper we only discuss the activation of constraints. The main difference between problems arising in control systems theory [10] and the present problem resides in the fact that the underlying energy g is minimized for a.e. $t \in (0,T)$ along the trajectory, and not only at the final time T.

The numerical scheme to solve (2) is introduced in the next section in the case without activation or deactivation of constraints. Then the algorithm for the detection of the discontinuities is presented. A theoretical result is given in a particular case and numerical results for the system (1) finally show the accuracy and efficiency of our method.

2 Numerical Algorithm without Tracking of Discontinuities

In order to solve the system (2), we opt for a splitting algorithm between differential and optimization operators (see [1] for another approach). Hence we fully exploit the characteristics of the minimization problem to ensure the admissibility of the solution.

The differential equations are solved with the Crank-Nicolson scheme. Let $h > 0$ be a fixed time step, $t^n = nh$, $n = 0,\dots,N$, the discretization of $(0,T)$ with $t^N = T$, and $\mathbf{b}^n$ and $\mathbf{z}^n$ denote respectively the approximations of $\mathbf{b}(t^n)$ and $\mathbf{z}(t^n)$. The differential equations discretized in time consist in finding $\mathbf{b}^{n+1} \in \mathbb{R}^s$ and $\mathbf{z}^{n+1} \in \mathbb{R}^m$ at each time step that satisfy

$$\frac{1}{h}(\mathbf{b}^{n+1} - \mathbf{b}^n) = \frac{1}{2}\mathbf{f}(t^n, \mathbf{b}^n, \mathbf{z}^n) + \frac{1}{2}\mathbf{f}(t^{n+1}, \mathbf{b}^{n+1}, \mathbf{z}^{n+1}). \tag{3}$$

We solve this equation with a fixed-point method. At each time step, a sequence of fixed-point iterates $(\mathbf{b}^{n+1,\ell}, \mathbf{z}^{n+1,\ell})$ is computed as follows:

(i) setting $\mathbf{b}^{n+1,0} = \mathbf{b}^n$ and $\mathbf{z}^{n+1,0} = \mathbf{z}^n$,
(ii) for $\ell = 0,\dots,r$

  (a) solve the equation for $\mathbf{b}^{n+1,\ell+1}$ with a Newton method

$$\frac{1}{h}(\mathbf{b}^{n+1,\ell+1} - \mathbf{b}^n) = \frac{1}{2}\mathbf{f}(t^n, \mathbf{b}^n, \mathbf{z}^n) + \frac{1}{2}\mathbf{f}(t^{n+1}, \mathbf{b}^{n+1,\ell+1}, \mathbf{z}^{n+1,\ell}),$$

  (b) solve the optimization problem in (2) with $\mathbf{b}^{n+1,\ell+1}$ to obtain $\mathbf{z}^{n+1,\ell+1}$,
  (c) if $\|\mathbf{b}^{n+1,\ell+1} - \mathbf{b}^{n+1,\ell}\|_2 \leq tol \cdot \|\mathbf{b}^{n+1,\ell+1}\|_2$ then return,

(iii) set $\mathbf{b}^{n+1} = \mathbf{b}^{n+1,\ell+1}$ and $\mathbf{z}^{n+1} = \mathbf{z}^{n+1,\ell+1}$,

where r is a given maximal number of iterations and tol is a given tolerance.

Hence, at each iteration of the fixed-point method we have to solve the optimization problem to determine $\mathbf{z}^{n+1,\ell+1}$. The resolution of the optimization problem is based on a primal-dual interior-point method detailed in [2]. The main principle consists in relaxing the inequality constraint $\tilde{\mathbf{z}} \geq 0$ by adding a slack variable $\tilde{\mathbf{w}}$ that is incorporated into a logarithmic barrier term in the objective function. Let $\nu > 0$ be a given parameter. The minimization problem written as in the system (2) becomes

$$\begin{aligned} &\min\ g(\tilde{\mathbf{z}}) - \nu \sum_{i=1}^{m} \ln(\tilde{w}_i) \\ &s.t.\ A\tilde{\mathbf{z}} = \mathbf{b}, \quad \tilde{z}_i - \tilde{w}_i = 0, \quad \tilde{w}_i > 0, \quad i = 1,\dots,m. \end{aligned} \tag{4}$$

The objective function and the constraints of (2) being continuous, the solution of (4) converges to the solution of the initial problem (2) as ν tends to zero [6].

We write the first order optimality conditions corresponding to (4) in the fixed-point algorithm. After elimination of the slack variables, we obtain

$$\begin{aligned} \nabla g(\mathbf{z}^{n+1,\ell+1}) + A^T \boldsymbol{\lambda}^{n+1,\ell+1} - \boldsymbol{\theta}^{n+1,\ell+1} &= \mathbf{0}, \\ A\mathbf{z}^{n+1,\ell+1} &= \mathbf{b}^{n+1,\ell+1}, \\ z_i^{n+1,\ell+1}\theta_i^{n+1,\ell+1} - \nu &= 0, \quad i = 1,\dots,m, \\ z_i^{n+1,\ell+1},\ \theta_i^{n+1,\ell+1} &> 0, \quad i = 1,\dots,m, \end{aligned} \tag{5}$$

where $\boldsymbol{\lambda}^{n+1,\ell+1}$ and $\boldsymbol{\theta}^{n+1,\ell+1}$ are dual (Kuhn-Tucker) multipliers.

Starting with an interior-point parameter ν^0, the above nonlinear system is solved by applying one Newton iteration, then decreasing the parameter $\nu^k = \xi \nu^{k-1}$, $\xi \in (0,1)$, and repeating the process until convergence is reached [2, 6].

The interior-point method does allow constraints to be deactivated, since the solution of the relaxed minimization problem converges to the one of the original problem when $\nu \to 0$ [6]. From the numerical viewpoint, at each step of the algorithm, if $j \in \{1,\dots,m\}$ such that $|z_j^{n+1,\ell+1}| < \varepsilon$ exists (where ε is a given bound), then the jth constraint is activated in the interval $[t^n, t^{n+1}[$.

3 Tracking of Discontinuities

When an inequality constraint is activated or deactivated, the variable $\mathbf{z}$ loses its regularity. In order to preserve the order of our numerical method, the discontinuity point has to be detected with enough accuracy [5, 7, 8]. An arising difficulty here is the absence of a function expressing explicitly the time when a constraint is activated, the variable $\mathbf{z}$ being the result of a minimization problem for given $\mathbf{b}$.

The principle of our tracking method is an extrapolation method inspired by [4]. Let us assume that the jth constraint $z_j(t) > 0$ activates and that the activation occurs during the interval $[t^n, t^{n+1}[$. We are looking for a fractional time step τ such that $z_j(t^n + \tau) = 0$. A Taylor expansion gives

$$0 = z_j(t^n + \tau) = z_j(t^n) + \tau \frac{\mathrm{d}z_j}{\mathrm{d}t}(t^n) + \mathcal{O}(\tau^2).$$

Hence, the time when the discontinuity occurs is estimated by

$$\tau \approx -z_j(t^n) / \frac{\mathrm{d}z_j}{\mathrm{d}t}(t^n). \tag{6}$$

The value $z_j(t^n)$ is already approximated by z_j^n, but the derivative $\frac{\mathrm{d}z_j}{\mathrm{d}t}(t^n)$ remains to be estimated. Starting from the chain rule

$$\frac{\mathrm{d}z_j}{\mathrm{d}t}(t^n) = \sum_{i=1}^{s} \frac{\partial z_j}{\partial b_i}(\mathbf{b}(t^n)) \cdot \frac{\mathrm{d}}{\mathrm{d}t} b_i(t^n), \tag{7}$$

the approximation of the derivatives $\frac{\partial z_j}{\partial b_i}(\mathbf{b}(t^n))$ is derived with a sensitivity analysis [6]. The differentiation of the first order optimality conditions (5) with $\nu = 0$, relative to $\mathrm{d}b_i$, $i = 1,\dots,s$, leads to the linear systems for the variations of the solutions $\mathbf{z}, \boldsymbol{\lambda}$ due to a variation of the data b_i.

$$\begin{pmatrix} \nabla^2 g(\mathbf{z}) & A^T \\ A & 0 \end{pmatrix} \begin{pmatrix} \frac{\mathrm{d}\mathbf{z}}{\mathrm{d}b_i} \\ \frac{\mathrm{d}\boldsymbol{\lambda}}{\mathrm{d}b_i} \end{pmatrix} = \begin{pmatrix} \mathbf{0} \\ \mathbf{e}_i \end{pmatrix}, \tag{8}$$

where $\mathbf{e}_i$ is the usual unit vector defined by $(e_i)_k = \delta_{ik}$ for $k = 1, \ldots, s$. The derivatives $\frac{\partial z_j}{\partial b_i}(\mathbf{b}(t^n))$ are approximated by

$$\frac{\partial z_j}{\partial b_i}(\mathbf{b}(t^n)) \approx \frac{\mathrm{d} z_j}{\mathrm{d} b_i}(\mathbf{b}^n), \qquad \text{for } i = 1, \ldots, m. \tag{9}$$

The derivatives db_i/dt in (7) are approximated with the 2-steps Adams-Bashforth method with non-constant time step:

$$\frac{\mathrm{d}}{\mathrm{d}t}\mathbf{b}(t^n) \approx \mathbf{f}(t^n, \mathbf{b}^n, \mathbf{z}^n) + \frac{\tau}{2h}\left(\mathbf{f}(t^n, \mathbf{b}^n, \mathbf{z}^n) - \mathbf{f}(t^{n-1}, \mathbf{b}^{n-1}, \mathbf{z}^{n-1})\right). \tag{10}$$

Combining (6), (9) and (10) we obtain the following equation of second order in τ

$$0 = z_j + \tau \sum_{i=1}^{s} \frac{\mathrm{d} z_j}{\mathrm{d} b_i} f_i(t^n, \mathbf{b}^n, \mathbf{z}^n) + \frac{\tau^2}{2h} \sum_{i=1}^{s} \frac{\mathrm{d} z_j}{\mathrm{d} b_i} \left(f_i(t^n, \mathbf{b}^n, \mathbf{z}^n) - f_i(t^{n-1}, \mathbf{b}^{n-1}, \mathbf{z}^{n-1})\right),$$

that admits a unique positive root τ for a time step h sufficiently small.

Once the fractional time step τ is computed, a predictor-corrector method based on two-steps Adams-Bashforth and Adams-Moulton schemes is used to approximate $\mathbf{b}$ at $t^n + \tau$, namely:

$$\begin{aligned}
\mathbf{b}^{n+1}_{pred} &= \mathbf{b}^n + \tau\left[\left(1 + \frac{\tau}{2h}\right)\mathbf{f}(t^n, \mathbf{b}^n, \mathbf{z}^n) - \frac{\tau}{2h}\mathbf{f}(t^{n-1}, \mathbf{b}^{n-1}, \mathbf{z}^{n-1})\right] && \text{(predictor)},\\
\mathbf{b}^{n+1} &= \mathbf{b}^n + \frac{\tau}{2}\left[\mathbf{f}(t^n + \tau, \mathbf{b}^{n+1}_{pred}, \mathbf{z}^{n+1}_{pred}) + \mathbf{f}(t^n, \mathbf{b}^n, \mathbf{z}^n)\right] && \text{(corrector)},
\end{aligned}$$

where $\mathbf{z}^{n+1}_{pred}$ is obtained by solving the optimization problem as in (2) with $\mathbf{b}^{n+1}_{pred}$. This method has a low computation cost, since all the terms in the above equations are already known before the tracking of the discontinuity except $\mathbf{z}^{n+1}_{pred}$ in the corrector's equation, that has to be computed in addition.

4 Theoretical Results

Error estimates for the approximations of the location and time when a discontinuity occurs are obtained in a simplified case by using nonlinear techniques presented in [3, 9]. We assume that (i) $\mathbf{z}(0) > 0$ and the jth inequality constraint is the first to be activated in the time interval $(0, T)$ and (ii) the optimization algorithm [2] gives an exact solution.

Let us denote by $t^\star$ the first time when the event $z_j(t) = 0$ occurs and consider the particular case when the event is geometrically defined by the intersection of the trajectory $\mathbf{b}(t)$ with a given hyperplane of $\mathbb{R}^s$. We describe this plane by the parametric equations $\mathbf{OC} + \sum_{i=1}^{s-1} \beta_i \mathbf{d}_i$, where $\mathbf{O}$ is the origin of our axes, $\mathbf{C}$ is a point in the hyperplane, $\mathbf{d}_i$, $i = 1, \ldots, s-1$, are direction vectors and β_i, $i = 1, \ldots, s-1$, are the unknown variables.

Let us define the function $\mathbf{F} : \mathbb{R} \times \mathbb{R}^{s-1} \rightarrow \mathbb{R}^s$ by $\mathbf{F}(t, \beta_1, \ldots, \beta_{s-1}) = \mathbf{b}(t) - \mathbf{OC} - \sum_{i=1}^{s-1} \beta_i \mathbf{d}_i$, which vanishes at the intersection point denoted by $(t^\star, \beta_1^\star, \ldots, \beta_{s-1}^\star)$.

Let $\mathbf{b}_h$ be the linear spline interpolation of $\mathbf{b}^n$, $n = 0, \ldots, N$. We define the numerical approximation $\mathbf{F}_h : \mathbb{R} \times \mathbb{R}^{s-1} \rightarrow \mathbb{R}^s$ by $\mathbf{F}_h(t, \beta_1, \ldots, \beta_{s-1}) = \mathbf{b}_h(t) - \mathbf{OC} - \sum_{i=1}^{s-1} \beta_i \mathbf{d}_i$. The function $\mathbf{F}_h$ vanishes at the approximated intersection point denoted by $(t_h^\star, \beta_{1,h}^\star, \ldots, \beta_{s-1,h}^\star)$.

As in the previous section we assume that there exists $n = n(h) \leq N - 1$ such that $t^\star \in [t^n, t^{n+1}[$ (note that, when h becomes smaller, the index $n = n(h)$ becomes larger). We can establish the theorem (*a priori* error estimates):

Theorem 1. *Assume that the functions* $\mathbf{F}$ *and* $\mathbf{F}_h$ *admit zeros in* $(0, T)$, *denoted by* $(t^\star, \beta_1^\star, \ldots, \beta_{s-1}^\star)$ *and* $(t_h^\star, \beta_{1,h}^\star, \ldots, \beta_{s-1,h}^\star)$ *resp. Furthermore assume that* $\mathbf{b}$ *can be extended in a* C^3 *manner in the neighborhood of the discontinuity point, and* $DF(t^\star, \beta_1^\star, \ldots, \beta_{s-1}^\star)$ *is regular. Then:*

(i) there exists $h_1 > 0$ *and a constant* $C_1 > 0$ *such that*

$$\|\mathbf{F}_h(t^\star, \beta_1^\star, \ldots, \beta_{s-1}^\star)\|_\infty \leq C_1 h^2, \quad \forall h < h_1; \tag{11}$$

(ii) there exist $h_2 > 0$, $\delta > 0$ *and a ball centered in* $(t^\star, \beta_1^\star, \ldots, \beta_{s-1}^\star)$ *with radius* δ, *denoted by* $\mathscr{B}((t^\star, \beta_1^\star, \ldots, \beta_{s-1}^\star), \delta)$, *such that for all* $h < h_2$ *there exists a unique* $(t_h^\star, \beta_{1,h}^\star, \ldots, \beta_{s-1,h}^\star) \in \mathscr{B}((t^\star, \beta_1^\star, \ldots, \beta_{s-1}^\star), \delta)$ *satisfying* $\mathbf{F}_h(t_h^\star, \beta_{1,h}^\star, \ldots, \beta_{s-1,h}^\star) = \mathbf{0}$. *Moreover there exists a constant* C *independent of* $h < h_2$ *such that the following a priori error estimates holds*

$$\|(t^\star, \beta_1^\star, \ldots, \beta_{s-1}^\star) - (t_h^\star, \beta_{1,h}^\star, \ldots, \beta_{s-1,h}^\star)\|_\infty \leq C \|\mathbf{F}_h(t^\star, \beta_1^\star, \ldots, \beta_{s-1}^\star)\|_\infty. \tag{12}$$

The proof follows [3, 9]. Relationships (11) and (12) allow to conclude:

$$\exists\, h_0 > 0,\, C_0 > 0 \text{ s.t. } |t^* - t_h^*| + \|\mathbf{b}(t^*) - \mathbf{b}_h(t_h^*)\|_2 \leq C_0 h^2, \quad \forall h < h_0.$$

5 Numerical Results

Numerical results for the detection of discontinuities are presented for the phase equilibrium problem (1) described in the introduction.

If the aerosol is a mixture of three chemical components, the solution $\mathbf{b}$ of (1) and its numerical approximation can be represented on a two-dimensional phase diagram [2]. The phase diagram for a system composed by hexacosanol, pinic acid and water is illustrated in Figure 1 (a) where the digits (1,2,3) represent the number of deactivated constraints in each area.

In the particular case when all $y_\alpha(t)$ are strictly positive, the exact solution $\mathbf{b}$ and the exact time of activation $t^\star$ when the third inequality constraint is activated are known. Starting from an initial aerosol composition of 20% of hexacosanol, 20% of pinic acid and 60% of water, and with a time step equals to $0.01s$, we apply the algorithm of resolution with the tracking of the discontinuity. The numerical

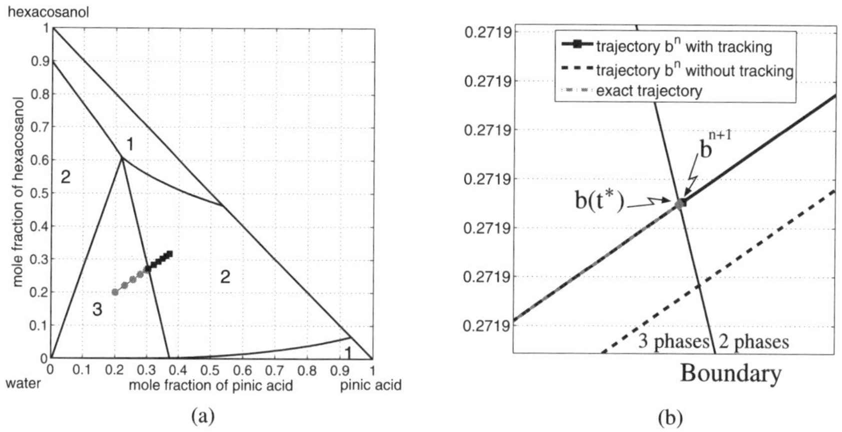

Fig. 1 (a) Numerical approximation of the trajectory of **b** on the phase diagram of hexacosanol-pinic acid-water. (b) Zoomed-in view near the discontinuity point.

approximation of **b** is depicted in Figure 1 (a). The grey circles refer to the approximation $\mathbf{b}^n$ when the three constraints are deactivated, whereas the black squares refer to $\mathbf{b}^n$ with only two deactivated constraints. The number of activated constraints, and the discontinuity point located on the boundary between the areas with 3 and 2 activated constraints are accurately computed.

At each time step, the fixed-point algorithm stops in less than 3 iterations (for $tol = 10^{-5}$), while the interior-point method requires less than 25 iterations (for a tolerance of 10^{-13} on the increments in the infinity-norm), leading to a CPU time of $0.0028\,s$ per time step with an Intel processor of $2.40\,GHz$.

Figure 1 (b) is a zoom of the trajectory of the numerical solution $\mathbf{b}^n$, $n = 0, \ldots, N$, near the discontinuity point. The exact solution **b** is represented, as well as the corresponding discontinuity point $\mathbf{b}(t^\star)$, and the approximated trajectory of the numerical solutions with and without the tracking of the discontinuity. The discontinuity point obtained numerically is very close to the one obtained analytically and the trajectory with tracking is nearly superimposed with the exact trajectory, as opposed to the one without tracking.

Figure 2 (a) shows the evolution of the variables $y_\alpha(t)$, $\alpha = 1, \ldots, 3$. The solid lines are the exact solutions y_1, y_2, and y_3, whereas the markers ■, ◆ and • represent respectively the numerical values y_1^n, y_2^n and y_3^n for $n = 0, \ldots, N$. The approximation of the time of the discontinuity is efficiently computed and markers are located at the point where the curves y_1, y_2 and y_3 are not continuously differentiable.

Finally, Figure 2 (b) shows the convergence order for the error $|t^\star - (t^n + \tau)|$ on the approximation of the time of the constraint activation $t^\star$. This result numerically confirms that the method is convergent to the second order.

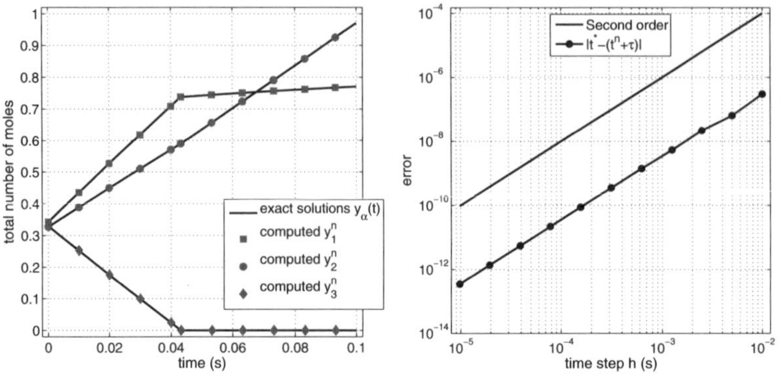

(a) (b)

Fig. 2 (a) Plot of the exact solutions y_1, y_2 and y_3. The markers ■, ♦ and • are respectively the computed values y_1^n, y_2^n and y_3^n for $n = 0, \ldots, N$ (b) Log-log convergence plot of the error $|t^\star - (t^n + \tau)|$.

References

1. Amundson, N., Caboussat, A., He, J., Landry, C., Seinfeld, J.H.: A dynamic optimization problem related to organic aerosols. C. R. Math. Acad. Sci. Paris **344**(8), 519–522 (2007)
2. Amundson, N., Caboussat, A., He, J.W., Seinfeld, J.H.: A primal-dual interior-point method for an optimization problem related to the modeling of atmospheric organic aerosols. Journal of Optimization Theory and Applications **130**(3), 375–407 (2006)
3. Caloz, G., Rappaz, J.: Numerical Analysis for Nonlinear and Bifurcation Problems, *Handbook of Numerical Analysis (P.G. Ciarlet, J.L. Lions eds)*, vol. 5, pp. 487–637. Elsevier, Amsterdam (1997)
4. Esposito, J.M., Kumar, V.: A state event detection algorithm for numerically simulating hybrid systems with model singularities. ACM Transactions on Modeling and Computer Simulation (TOMACS) **17**, 1–22 (2007)
5. Faugeras, B., Pousin, J., Fontvieille, F.: An efficient numerical scheme for precise time integration of a diffusion-dissolution/precipitation chemical system. Math. Comp. **75**(253), 209–222 (2005)
6. Fiacco, A.V., McCormick, G.P.: Nonlinear programming : sequential unconstrained minimization techniques. Wiley, New York (1968)
7. Gear, C.W., Østerby, O.: Solving ordinary differential equations with discontinuities. ACM Trans. Math. Software **10**(1), 23–44 (1984)
8. Hairer, E., Nørsett, S.P., Wanner, G.: Solving ordinary differential equations. I, *Springer Series in Computational Mathematics*, vol. 8, second edn. Springer-Verlag, Berlin (1993). Nonstiff problems
9. Rappaz, J.: Numerical approximation of PDEs and Clément's interpolation. In: Partial differential equations and functional analysis, *Oper. Theory Adv. Appl.*, vol. 168, pp. 237–250. Birkhäuser, Basel (2006)
10. Veliov, V.: On the time-discretization of control systems. SIAM J. Control Optim. **35**(5), 1470–1486 (1997)

Optimal Load Changes for a Molten Carbonate Fuel Cell Model

K. Chudej, K. Sternberg, and H.J. Pesch

Abstract Molten carbonate fuel cells are a promising technology for future stationary power plants. In order to enhance service life a more detailed understanding of the dynamical behavior is needed. This is enabled by a hierarchy of mathematical models based on chemical and physical laws. These mathematical models allow numerical simulation and optimal control of load changes. Mathematically speaking, we solve an optimal control problem subject to a degenerated partial differential equation system coupled with an integro differential-algebraic equation system. New numerical results are presented.

1 Introduction

Several types of fuel cells exist which are suited for different applications due to their different behavior [7]. Stacks of molten carbonate fuel cells are used for stationary power and heat supply [6]. Due to the high operation temperature an internal reforming, i.e. production of H_2 from CH_4 (or other fuel gases) in the fuel cell, is possible. Moreover clean exhaust gases are produced. In order to enhance service life, hot spots and high temperature gradients inside the fuel cell have to be avoided. Recently a hierarchy of mathematical models for a single (averaged) MCFC has been developed in order to describe the dynamical behavior of important physical and chemical variables [2, 6]. We will demonstrate, how one of these models can

Kurt Chudej
Universität Bayreuth, Lehrstuhl für Ingenieurmathematik, 95440 Bayreuth,
e-mail: kurt.chudej@uni-bayreuth.de

Kati Sternberg
Merz Pharmaceuticals GmbH, Frankfurt, formerly: Universität Bayreuth, Lehrstuhl für Ingenieurmathematik, 95440 Bayreuth

Hans Josef Pesch
Universität Bayreuth, Lehrstuhl für Ingenieurmathematik, 95440 Bayreuth

be used for simulation and optimal control of load changes during electrical power production.

Mathematically speaking, we have to compute optimal boundary control functions for an optimal control problem subject to a partial differential-algebraic equation system coupled with an integro differential-algebraic equation system. Due to the continual model updates during the project time, we chose the approach "first discretize, then optimize". We present recent numerical results for faster load changes [4].

2 Mathematical Model of MCFC

The following 2D crossflow model of a single molten carbonate fuel cell is based on [2, 3], see also [4]. The important mathematical variables are depicted in Fig. 1, which describe the gas flow through the anode gas channel, the catalytic combustor, the reversal chamber, the cathode gas channel and the cathode recycle.

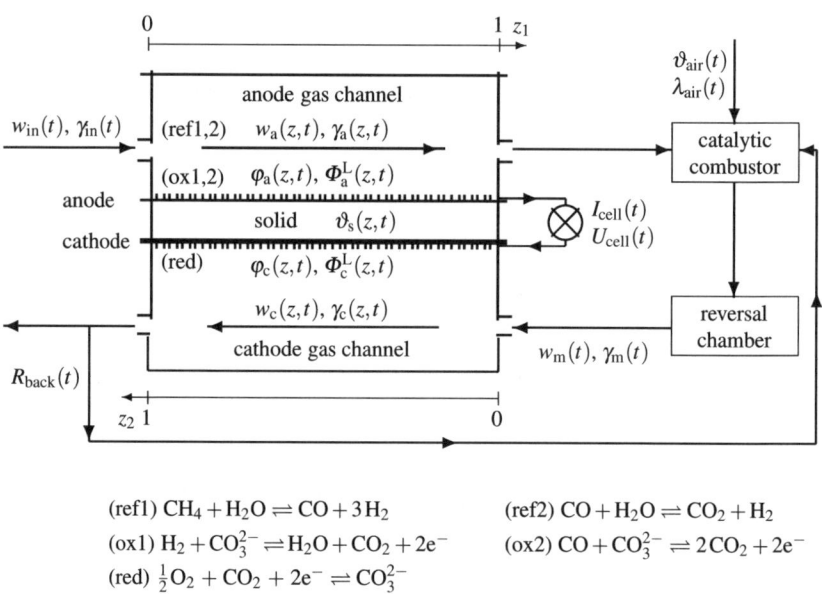

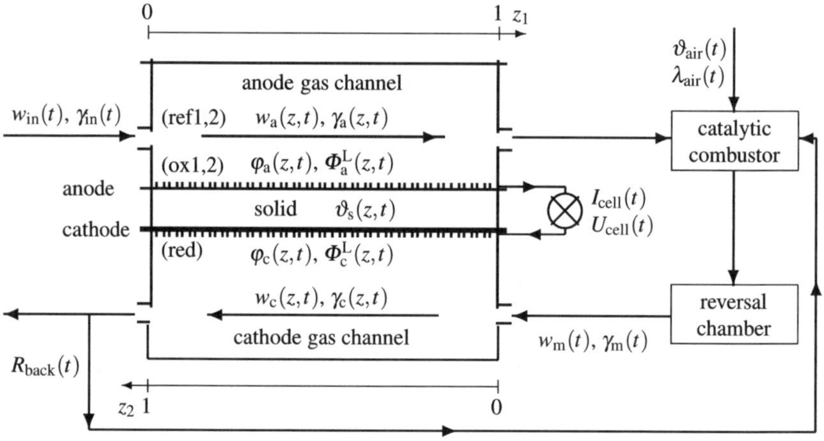

(ref1) $CH_4 + H_2O \rightleftharpoons CO + 3H_2$ (ref2) $CO + H_2O \rightleftharpoons CO_2 + H_2$

(ox1) $H_2 + CO_3^{2-} \rightleftharpoons H_2O + CO_2 + 2e^-$ (ox2) $CO + CO_3^{2-} \rightleftharpoons 2CO_2 + 2e^-$

(red) $\frac{1}{2}O_2 + CO_2 + 2e^- \rightleftharpoons CO_3^{2-}$

Fig. 1 2D crossflow model of a molten carbonate fuel cell with mathematical variables $w = (\chi, \vartheta)$, χ = vector of molar fractions, ϑ = temperature, φ = vector of partial pressures, Φ = electrical potential, U_{cell} = cell voltage, I_{cell} = cell current

The spatial domain and boundaries are given in Fig. 2. All variables are dimensionless. One unit of the dimensionless time t equals 12.5 seconds. The index set

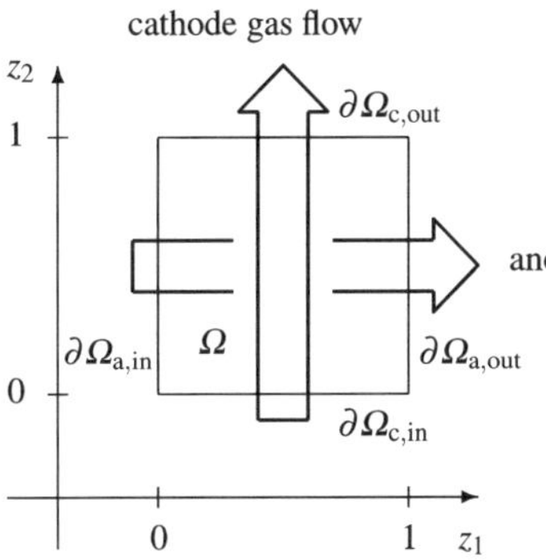

Fig. 2 2D cross flow model: (spatial) domain, boundaries ($\partial\Omega = \partial\Omega_{a,in} \cup \partial\Omega_{a,out} \cup \partial\Omega_{c,in} \cup \partial\Omega_{c,out}$), gas flow directions

$\mathscr{I} := \{CH_4, H_2O, H_2, CO, CO_2, O_2, N_2\}$ is used for molar fractions and partial pressures. In the following there always holds $i \in \mathscr{I}$, $j \in \{a,c\}$.

States, prescribed by PDEs, PDAEs and DAEs:

- In the anode gas channel (j =a) resp. the cathode gas channel (j =c): molar fractions $\chi_{i,j}(z,t)$, gas temperatures $\vartheta_j(z,t)$, and molar flow densities $\gamma_j(z,t)$. Near the electrodes: partial pressures $\varphi_{i,j}(z,t)$. Only a subset of chemical substances is needed during the numerical solution of the mathematical model.
  Abbreviation: $w_j(z,t) = ((\chi_{i,j})_{i\in\mathscr{I}}, \vartheta_j)$, $w_{a/c} = (w_a, w_c)$.
- In the solid (electrolyte): temperature $\vartheta_s(z,t)$.
- At the entry of the cathode gas channel: molar fractions $\chi_{i,m}(t)$, temperatures $\vartheta_m(t)$, and molar flow density $\gamma_m(t)$.
  Abbreviation: $w_m(t) = ((\chi_{i,m})_{i\in\mathscr{I}}, \vartheta_m)$
- Potentials $\Phi_a^L(z,t)$, $\Phi_c^L(z,t)$, cell voltage $U_{cell}(t)$.
  Abbreviation: $\Phi_{a/c}^L = (\Phi_a^L, \Phi_c^L)$.

Variables, which may serve as components of the boundary control $u(t)$:

- At the entry of the anode gas channel: molar fractions $\chi_{i,in}(t)$, gas temperature $\vartheta_{in}(t)$, and molar flow density $\gamma_{in}(t)$.
  Abbreviation: $w_{in}(t) = ((\chi_{i,in})_{i\in\mathscr{I}}, \vartheta_{in})$
- At the entry of the catalytic combustor: Gas temperature $\vartheta_{air}(t)$, air number $\lambda_{air}(t)$.
- Switch for cathode recycle: $R_{back}(t) \in [0,1]$.

Variables, which are prescribed:

- The cell current $I_{cell}(t)$ is prescribed, and usually either constant or a step function for a load change from one constant level to another constant level.

Partial differential-algebraic equations with boundary conditions:

$$\frac{\partial \vartheta_s}{\partial t} = \mu_1 \frac{\partial^2 \vartheta_s}{\partial z_1^2} + \mu_2 \frac{\partial^2 \vartheta_s}{\partial z_2^2} + \psi_1(\vartheta_s, w_{a/c}, \varphi_{a/c}, \Phi^L_{a/c}, U_{cell}), \quad \frac{\partial \vartheta_s}{\partial n}|_{\partial\Omega} = 0, \tag{1}$$

$$\frac{\partial w_a}{\partial t} = -\gamma_a \vartheta_a \frac{\partial w_a}{\partial z_1} + \psi_2(\vartheta_s, w_a, \varphi_a, \Phi^L_a), \quad w_a|_{\partial\Omega_{a,in}} = w_{in}(t), \tag{2}$$

$$\frac{\partial w_c}{\partial t} = -\gamma_c \vartheta_c \frac{\partial w_c}{\partial z_2} + \psi_3(\vartheta_s, w_c, \varphi_c, \Phi^L_c, U_{cell}), \quad w_c|_{\partial\Omega_{c,in}} = w_m(t), \tag{3}$$

$$0 = -\frac{\partial(\gamma_a \vartheta_a)}{\partial z_1} + \psi_4(\vartheta_s, w_a, \varphi_a, \Phi^L_a), \quad \gamma_a|_{\partial\Omega_{a,in}} = \gamma_{in}(t), \tag{4}$$

$$0 = -\frac{\partial(\gamma_c \vartheta_c)}{\partial z_2} + \psi_5(\vartheta_s, w_c, \varphi_c, \Phi^L_c, U_{cell}), \quad \gamma_c|_{\partial\Omega_{c,in}} = \gamma_m(t), \tag{5}$$

$$0 = \psi_6(\vartheta_s, \chi_a, \varphi_a, \Phi^L_a), \qquad 0 = \psi_7(\vartheta_s, \chi_c, \varphi_c, \Phi^L_c, U_{cell}), \tag{6}$$

$$\frac{\partial \Phi^L_{a/c}}{\partial t} = \psi_8(\vartheta_s, \varphi_{a/c}, \Phi^L_{a/c}, U_{cell}, I_{a/e/c}; I_{cell}). \tag{7}$$

Integro differential-algebraic equations:

$$\frac{dU_{cell}}{dt} = \psi_9(I_{a/e/c}; I_{cell}), \qquad I_a(t) = \int_\Omega i_a(\vartheta_s, w_a, \varphi_a, \Phi^L_a)\,dz, \tag{8}$$

$$I_c(t) = \int_\Omega i_c(\vartheta_s, w_c, \varphi_c, \Phi^L_c, U_{cell})\,dz, \qquad I_e(t) = \int_\Omega i_e(\Phi^L_{a/c})\,dz, \tag{9}$$

$$\frac{dw_m}{dt} = \psi_{10}\Big(w_m, \int_{\partial\Omega_{a,out}} w_a\,dz_2, \int_{\partial\Omega_{a,out}} \gamma_a\,dz_2, \int_{\partial\Omega_{c,out}} w_c\,dz_1, \int_{\partial\Omega_{c,out}} \gamma_c\,dz_1, \lambda_{air}, \vartheta_{air}, R_{back}\Big), \tag{10}$$

$$\gamma_m(t) = \psi_{11}\Big(w_m, \int_{\partial\Omega_{a,out}} w_a\,dz_2, \int_{\partial\Omega_{a,out}} \gamma_a\,dz_2, \int_{\partial\Omega_{c,out}} w_c\,dz_1, \int_{\partial\Omega_{c,out}} \gamma_c\,dz_1, \lambda_{air}, \vartheta_{air}, R_{back}\Big). \tag{11}$$

Initial conditions:

$$\begin{aligned} &\vartheta_s|_{t=0} = \vartheta_{0,s}(z),\ w_a|_{t=0} = w_{0,a}(z),\ w_c|_{t=0} = w_{0,c}(z),\ w_m|_{t=0} = w_{0,m}, \\ &\Phi^L_a|_{t=0} = \Phi^L_{0,a}(z),\ \Phi^L_c|_{t=0} = \Phi^L_{0,c}(z),\ U_{cell}|_{t=0} = U_{0,cell} \end{aligned} \tag{12}$$

The PDAE system consists of a parabolic heat equation (1), hyperbolic transport equations (2–3) with given wind direction (because it is known a priori that $\gamma_{a/c}, \vartheta_{a/c}$ are positive), and partial differential-algebraic equations (4–7). A detailed index analysis of (1–12) yields differential time index $\nu_t = 1$ (see [5], some small modifications have to be made in order to allow a switched on cathode recycle). Therefore *consistent* initial conditions are given by (12), no initial conditions can be prescribed for the algebraic variables $\gamma_{a/c/m}$, $\varphi_{a/c}$, $I_{a/c/e}$.

An obvious engineering approach is to use the method of lines (MOL) by discretizing the spatial partial derivatives. This is simplified by the a priori knowledge of the wind direction of the hyperbolic equations. A five-point star for the (scaled) Laplacian and suitable upwind formulas are used for the spatial partial derivatives and quadrature formulas for the spatial integrals. This yields a very large semi-explicit DAE of index $\nu = 1$.

$$M\dot{x}(t) = g(x(t), u(t)), \qquad M[x(0) - x_0] = 0, \qquad M = \mathrm{diag}(I, O) \tag{13}$$

3 Optimization Scenarios and Numerical Results

One drawback of molten carbonate fuel cells is the slow system reaction for a load change (i.e. a change in the cell current I_{cell} during the operation of the MCFC). The potentials $\Phi^{\mathrm{L}}_{\mathrm{a/c}}$, U_{cell} react very fast, whereas the molar quantities and especially the solid temperature ϑ_{s} react only slowly. Fast load changes induce temperature changes, which have to be compensated especially for increasing cell current.

The goal of the optimal control is therefore to control the system after a load change as fast as possible into the new stationary state. A good indicator for the stationary state is the cell voltage U_{cell}, which reacts very fast and significantly on a sudden load change of the cell current and moreover reacts also on changes of the slowest variable, the solid temperature ϑ_{s}.

The following technologically interesting scenario is analyzed [4]: The system input cell current is prescribed as a discontinuous step function

$$I_{\mathrm{cell}}(t) = \begin{cases} I_{\mathrm{cell},1} = 0.7 \text{ if } t \le t^\star \\ I_{\mathrm{cell},2} = 0.6 \text{ if } t > t^\star \end{cases}. \tag{14}$$

Initial conditions (12) at $t = 0 \le t^\star$ are the stationary solution for constant $I_{\mathrm{cell},1}$.

Find optimal boundary control functions $u : [0, t_{\mathrm{f}}] \to I\!R^6$, such that the functional

$$J[u] = \int_{t^\star}^{t_{\mathrm{f}}} L dt \text{ with } L = \left[U_{\mathrm{cell}}(t) - U_{\mathrm{cell,2,stat}}\right]^2, \; U_{\mathrm{cell,2,stat}} = 30.788 \tag{15}$$

is minimized s.t. PDAE/integro-DAE (1–12) and control constraints $u(t) \in U$.

In the discretized version the PDAE/integro-DAE constraint (1–12) is replaced by the semi-explicit DAE (13) and then solved by the software package NUDOCCCS (Büskens [1]) which transforms this problem into a nonlinear programming problem which is finally solved by a SQP algorithm.

Since this surrogate problem is still too time consuming, a slightly modified problem is solved. A sequence of optimal control problems

$$\left(\min \int_{t_k}^{t_{k+1}} L dt \;\; \text{s.t. (13) and } u(t) \in U\right)_{k=1,\ldots,5} \tag{16}$$

is solved. A logarithmic type grid $t_1 = t^\star = 0, t_2 = 0.1, t_3 = 1.1, t_4 = 11.1, t_5 = 111.1, t_6 = t_{\mathrm{f}} = 1111.1$ is used, due to the different time scales of the variables. Initial conditions for the first optimal control problem are the stationary solution for $I_{\mathrm{cell},1}$. Initial conditions for the k-th optimal control problem are the free final conditions of the $(k-1)$-th optimal control problem. In each time interval $[t_k, t_{k+1}]$ an equidistant control grid of 21 points is used. Spatial discretization is 3×3.

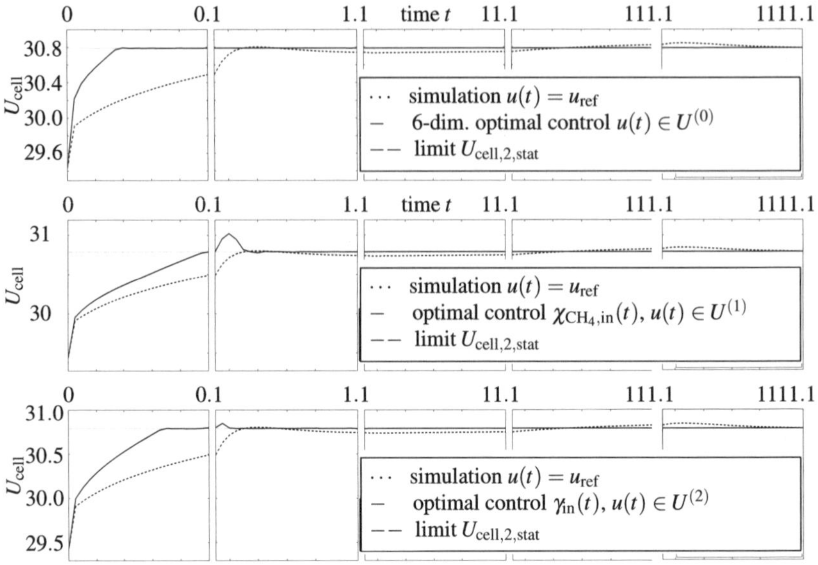

Fig. 3 Simulated and (sub)optimal controlled cell voltage U_{cell}

Numerical solutions are presented for seven different admissible control sets $U^{(0)} \stackrel{\text{def}}{=} \{u \in I\!R^6 \mid \underline{u}_i \leq u_i \leq \overline{u}_i, i = 1, \ldots, 6\}$, $U^{(p)} \stackrel{\text{def}}{=} U^{(0)} \cap \{u \in I\!R^6 \mid u_i = u_{i,\text{ref}}, i \neq p\}$, $p = 1, \ldots, 6$, see Table 1 and Figure 3 and 5. The other molar fractions at the anode inlet fulfill $\chi_{H_2O,\text{in}} = 1 - \chi_{CH_4,\text{in}}$, $\chi_{k,\text{in}} = 0$, $k \notin \{CH_4, H_2O\}$. Computational time for the (sub)optimal control with a 6-dimensional control vector $u(t) \in U^{(0)}$ is about 3 days, for the other scenarios with a scalar control about 2 hours. Fig. 3 and 5 present the cell voltage $U_{\text{cell}}(t)$ on the five time intervals $[t_k, t_{k+1}]$.

Table 1 Control constraints and reference values

	$\underline{u}_i$	$u_{i,\text{ref}}$	$\overline{u}_i$
$u_1 = \chi_{CH_4,\text{in}}$	0.25	1.0/3.5	0.4
$u_2 = \gamma_{\text{in}}$	0.85	1.0	1.5
$u_3 = \vartheta_{\text{in}}$	2.8	3.0	3.2
$u_4 = \lambda_{\text{air}}$	1.5	2.2	2.5
$u_5 = \vartheta_{\text{air}}$	1.0	1.5	2.5
$u_6 = R_{\text{back}}$	0.4	0.5	0.6

A fast increase of the cell voltage can be seen in the simulation until $t \approx 0.005$. This is the immediate consequence of the very fast change of the electrical variables. A moderate increase until $t \approx 0.015$ is due to the fast change of the molar quantities. The final stationary value is reached only after about $t \approx 1000$ (over 3 hours), due to the slow changes in the solid temperature ϑ_s. The oscillating behavior while reaching the new stationary cell voltage has undesirable effects on the cell power $P_{cell} = I_{cell} U_{cell}$ and should be avoided.

The new stationary cell voltage is reached significantly earlier in the optimal control scenarios. For the 6-dimensional control function the cell voltage is in a 0.1%-tube around the new stationary solution after $t \approx 0.03$.

The following scenarios examine, whether the technological expensive control of all input variables is really needed. Especially the scalar control of the molar fraction $\chi_{CH_4,in}$ or of the molar flow density γ_{in} give promising results, see Fig. 3. Most favorable seems the scalar control of the molar flow density γ_{in}: the excess cell voltage in the second time interval is lower. Moreover this approach can be easily realized in practice. The scalar optimal control γ_{in} is given in Fig. 4. Fig. 5 presents the less efficient other scalar optimal control scenarios for reference purposes.

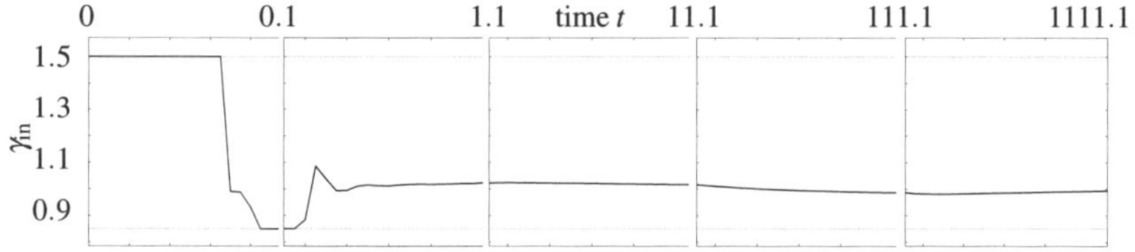

Fig. 4 Scalar (sub)optimal control γ_{in}

4 Conclusion and Outlook

Several suboptimal boundary control strategies were numerically computed and compared for a complicated partial differential algebraic equation system modelling realistically the dynamical behavior of a molten carbonate fuel cell. Although technologically relevant optimal control results could be computed, the huge computational time needed demands model reduction techniques, which are currently under development.

Acknowledgements This research was partly funded by the German Federal Ministry of Education and Research within the project "Optimierte Prozessführung von Brennstoffzellensystemen mit Methoden der Nichtlinearen Dynamik". The second author was funded by the *Promotionsabschlussstipendium des HWP-Programms "Chancengleichheit für Frauen in Forschung und Lehre"*.

We are indebted to Prof. Dr. Büskens from University of Bremen for providing us with the direct optimal control software package NUDOCCCS.

We are indebted to Dr.-Ing. Heidebrecht and Prof. Dr.-Ing. Sundmacher from the Max-Planck-Institut für Dynamik komplexer technischer Systeme, Magdeburg, for providing us with the realistic fuel cell model and to Dipl.-Ing. Berndt and Dipl.-Ing. Koch from the management of the IPF Heizkraftwerksbetriebsges. mbH, Magdeburg, for their continual support.

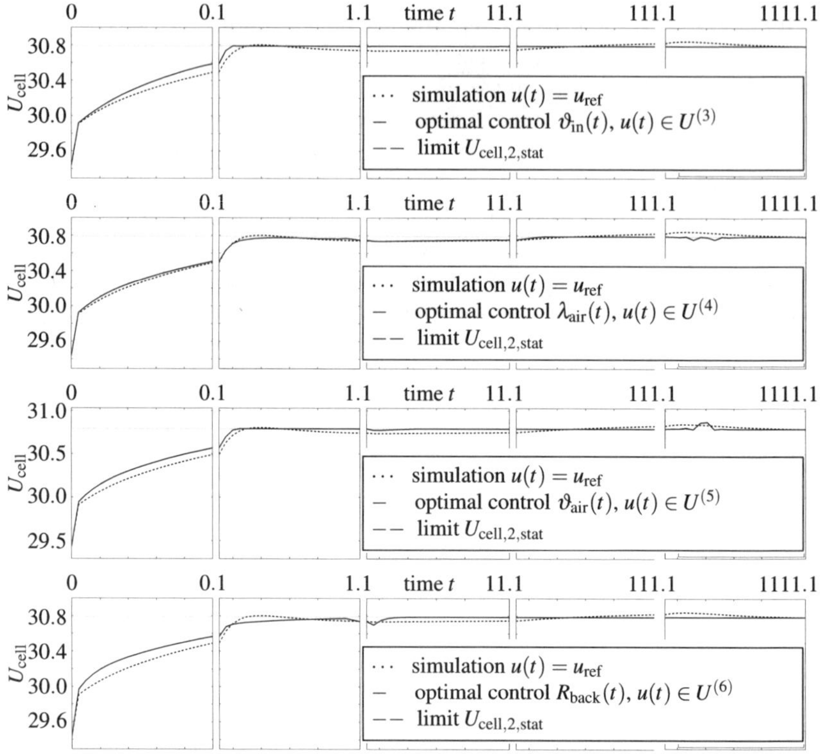

Fig. 5 Simulated and (sub)optimal controlled cell voltage U_{cell}

References

1. Büskens, C.: Optimierungsmethoden und Sensitivitätsanalyse für optimale Steuerprozesse mit Steuer- und Zustands-Beschränkungen. Ph.D. thesis, Universität Münster (1998)
2. Heidebrecht, P.: Modelling, analysis and optimisation of a molten carbonate fuel cell with direct internal reforming (DIR-MCFC). Ph.D. thesis, Universität Magdeburg (2004)
3. Heidebrecht, P., Sundmacher, K.: Dynamic model of a cross-flow molten carbonate fuel cell with direct internal reforming. Journal of the Electrochemical Society **152**, A2217–A2228 (2005)
4. Sternberg, K.: Simulation, Optimale Steuerung und numerische Sensitivitätsanalyse einer Schmelzkarbonat-Brennstoffzelle mithilfe eines partiell differential-algebraischen Gleichungssystems. Ph.D. thesis, Fakultät für Mathematik, Physik und Informatik, Universität Bayreuth (2007)
5. Sternberg, K., Chudej, K., Pesch, H.J.: Suboptimal control of a 2D molten carbonate fuel cell PDAE model. Mathematical and Computer Modelling of Dynamical Systems **13**(5), 471–485 (2007)
6. Sundmacher, K., Kienle, A., Pesch, H.J., Berndt, F., Huppmann, G.: Molten Carbonate Fuel Cells – Modeling, Analysis, Simulation, and Control. Wiley-VCH, Weinheim (2007)
7. Winkler, W.: Brennstoffzellenanlagen. Springer, Berlin (2002)

Differential DAE Index for Reactive Euler Equations

A. Hmaidi and P. Rentrop

Abstract For the numerical simulation of a reacting gas flow, we extend the Euler equations to take into account the interaction of the chemical species. Moreover we describe the thermodynamic properties of the gas such as pressure, internal energy and temperature using appropriate closure equations. The system obtained by the combination of the Euler equations and the closure equations is a differential-algebraic system. After modeling the chemical source terms, we focus on the closure relations that complete the Euler equations for an ideal gas and for a real gas mixture. By investigating the DAE-index of such systems for ideal gases, we show the DAE-index for the time integration is 1, whereas this must not be true for real gases.

1 Introduction

Due to environmental and economical reasons, the development and the improvement of gas turbines by increasing the efficiency and reducing fuel consumption and pollutant formation has become more essential than ever before. The numerical study of reactive flows is a very important step towards reaching these goals in modern power plants. The modeling of the gas flow is based on conservation principles. The three fundamental principles in fluid dynamics are the conservation of mass, momentum and energy. For one-dimensional inviscid flows we obtain the 1D-Euler gas equations

$$\begin{pmatrix} \rho \\ \rho u \\ \rho E \end{pmatrix}_t + \begin{pmatrix} \rho u \\ \rho u^2 + p \\ (\rho E + p)u \end{pmatrix}_x = \mathbf{0} \tag{1}$$

Ayoub Hmaidi and Peter Rentrop
Chair of Numerical Analysis M2, Munich University of Technology
Boltzmannstrasse 3, D-85748 Garching, e-mail: {hmaidi, toth-pinter}@ma.tum.de

which form a system of partial differential equations of first order for the four unknown states ρ, u, p and E.
The conservation principles are also valid in the case of chemically reacting gas flow. The closure conditions - relations between pressure, density and internal energy- complete the system. It is the initial resp. boundary conditions and the closure equations, i.e. pressure and internal energy laws, that are different from one application to the other. Although in many cases (like high pressure or low temperature) not appropriate [9], the ideal gas law is often preferred to the real gas law for its simplicity. In our investigation, we consider a reactive multispecies flow with ideal gas or real gas closure equations and study the influence, the analytical properties and the numerical consequences due to the DAE-index of such choices.

2 Euler Equations for Multispecies Reacting Gas

For a 1-d reactive gas flow [4] consisting of N_s species undergoing N_r reactions the Euler gas equations are extended by the conservation of mass for each chemical component, which yields a system of $N_s + 2$ partial differential equations.

$$\begin{pmatrix} \rho_1 \\ \rho_2 \\ \vdots \\ \rho_{N_s} \\ \rho u \\ \rho E \end{pmatrix}_t + \begin{pmatrix} \rho_1 u \\ \rho_2 u \\ \vdots \\ \rho_{N_s} u \\ \rho u^2 + p \\ (\rho E + p)u \end{pmatrix}_x = \begin{pmatrix} W_1 \dot{\omega}_1 \\ W_2 \dot{\omega}_2 \\ \vdots \\ W_{N_s} \dot{\omega}_{N_s} \\ 0 \\ \dot{Q}_c \end{pmatrix} \tag{2}$$

$\rho_s(t,x)$	density of species X_s	W_s	molecular weight of species X_s
$\rho(t,x)$	mixture density of gas	$p(t,x)$	mixture pressure
$u(t,x)$	mixture velocity	$E(t,x)$	total non-chemical energy
$\dot{\omega}_s$	production rate of species X_s	$\dot{Q}_c$	chemical heat release

Remark: For an N_{dim}-dimensional flow we get $(N_s + N_{dim} + 1)$ equations: one has to replace the scalar velocity u by the vector $\mathbf{u}$ and apply the adequate divergence operator $\nabla \cdot$ instead of the spatial derivative ∂_x.

In contrast to the standard flow equations, for a multispecies flow [4] we need further conditions: one for each partial pressure p_s and one involving the temperature T. These additional conditions are called closure equations and represent constraints which force the solution to lie on a manifold.

With the operator $\mathscr{F}$ standing for a spatial semidiscretization of the convection and the chemical source term and $\mathscr{G}$ representing the algebraic constraints due to the closure relations, we write the system of the Euler and the closure equations as

$$\dot{\mathbf{U}} = \mathscr{F}(\mathbf{U}, \mathbf{V}), \tag{3}$$

$$\mathbf{0} = \mathscr{G}(\mathbf{U}, \mathbf{V}). \tag{4}$$

$\mathbf{U} = (\rho_1, \rho_2, .., \rho_{N_s}, \rho u, \rho E)$ represents the vector of the differential variables and $\mathbf{V} = (p_1, p_2, .., p_{N_s}, T)$ the vector of algebraic variables (pressure and temperature). Taking the total time derivative of the algebraic constraints equation

$$\mathbf{0} = \frac{\partial \mathscr{G}}{\partial U}\dot{\mathbf{U}} + \frac{\partial \mathscr{G}}{\partial V}\dot{\mathbf{V}} \tag{5}$$

enables us to write the system in the form

$$\begin{pmatrix} I & 0 \\ K & J \end{pmatrix} \begin{pmatrix} \dot{\mathbf{U}} \\ \dot{\mathbf{V}} \end{pmatrix} = \begin{pmatrix} \mathscr{F}(\mathbf{U},\mathbf{V}) \\ \mathbf{0} \end{pmatrix} \tag{6}$$

with submatrices $K := \frac{\partial \mathscr{G}}{\partial U}$ and $J := \frac{\partial \mathscr{G}}{\partial V}$.

3 The Differential Index

After semidiscretization in space, the complete set of conservations laws and closure conditions form a large time-dependent system of differential algebraic equations (DAE system). DAEs arise in many contexts in science and engineering [5, 7]. They refer to combinations of differential equations and algebraic (eventually nonlinear) equations. This coupled system of equations forces the solution to lie on a manifold, which may cause new theoretical and numerical problems [3].
To classify these difficulties the index was introduced. Roughly speaking, the index is a measure for the additional effort required to solve a DAE in comparison to an ODE. The larger the index, the worse the analytical and numerical problems get. In the following we restrict to the standard differential index for the time integration. For analytical investigation it is sufficient to study the semi-explicit system

$$\dot{y} = f(y,z), \; y \in \mathbb{R}^{n_y}, \tag{7}$$
$$0 = g(y,z), \; z \in \mathbb{R}^{n_z}. \tag{8}$$

In order to remind of the index concept we introduce the linear algebra index.
Given the implicit linear system

$$A\dot{x} + Bx = f(t) \tag{9}$$

with A, B being $(n \times n)$ matrices, x, $f \in \mathbb{R}^n$, A is allowed to be singular. If the matrix B is not regular, then we require the matrix pencil $A - \sigma B$ to be regular (except for a finite number of generalized eigenvalues σ).

The index of (9) corresponds to the nilpotency of the eigenvalue $\lambda = 0$ of the matrix pencil $(A\lambda + B)$. Index-0 means that there exists a complete set of eigenvalues for $\lambda = 0$ and that no algebraic relation needs to be satisfied.

In the index-1 case we may solve (8) locally for z due to the implicit function theorem : $z = G(y)$. Inserting z into (7), we obtain $\dot{y} = f(y, G(y))$, the so-called state space form. The DAE system is called an index-1 problem, since only one differentiation of (8) is needed to achieve $\dot{z} = -g_z^{-1} g_y f$.

Index-k means that we have to apply k differentiations of the constraint (8) in order to get an explicit form for the variable z.

An inherent problem is that we have to satisfy the underlying manifolds resulting from each differentiation. Moreover we have to choose consistent initial values. Standard numerical schemes like Rosenbrock-Wanner, BDF's or Radau [6] can solve up to index-2 problems reliably. In the case of higher indices [1], projection techniques or a reformulation of the problem [2] are necessary.

4 Chemical Source Terms

We consider a multicomponent mixture [4] of N_s species in which N_r chemical reactions take place

$$\sum_{s=1}^{N_s} \nu_{sr}^f X_s \rightleftharpoons \sum_{s=1}^{N_s} \nu_{sr}^b X_s, \quad r = 1, 2, \cdots, N_r.$$

ν_{sr}^f and ν_{sr}^b are the stoichiometric coefficients for the r-th forward and backward reaction of species X_s. The net reaction rate $\dot{\Omega}_r$ of the r-th reaction is given by

$$\dot{\Omega}_r = k_f^r \prod_{s=1}^{N_s} [X_s]^{\nu_{sr}^f} - k_b^r \prod_{s=1}^{N_s} [X_s]^{\nu_{sr}^b} \tag{10}$$

where $[X_s]$ denotes the concentration of the species X_s and reads $[X_s] = \frac{\rho_s}{W_s}$. k_f and k_b are the Arrhenius rate coefficients for the forward and backward reactions. The chemical production rate of species X_i is obtained through

$$\dot{\omega}_i = \sum_{r=1}^{N_r} (\nu_{ir}^b - \nu_{ir}^f) \dot{\Omega}_r. \tag{11}$$

Further, the *chemical heat release* [8] of all reactions reads $\dot{Q}_c = -\sum_{s=1}^{N_s} W_s h_s^0 \dot{\omega}_s$, where h_s^0 denotes the heat of formation of species X_s. The chemical heat release couples the chemical reactions to the flow internal energy via the energy equation.

5 Closure Conditions for Ideal Gases

The ideal gas law provides a relationship between pressure, density and temperature

$$p_s = \rho_s \frac{\mathscr{R} T}{W_s}, \quad \text{for all species } X_s, \; s = 1, .., N_s. \tag{12}$$

Since the temperature is related to the internal properties of the system, we need equations which relate the internal energy or enthalpy to the other variables such as density, pressure and temperature [4, 8].

The internal energy of an ideal gas X_s varies according to

$$de_s = \frac{\partial e_s}{\partial T}|_v dT = c_{v,s} dT \tag{13}$$

where c_v denotes the specific heat at constant volume. With the specific heat at constant pressure c_p, we get an equivalent expression for the internal enthalpy h_s

$$dh_s = \frac{\partial h_s}{\partial T}|_p dT = c_{p,s} dT. \tag{14}$$

Lemma 1. *For ideal gases, the DAE system has index 1.*

Proof. According to Dalton's law, the total pressure of a mixture of ideal gases is

$$p = \sum_{s=1}^{N_s} p_s = \sum_{s=1}^{N_s} \rho_s \frac{\mathscr{R}T}{W_s} = \rho \frac{\mathscr{R}T}{W} \tag{15}$$

where the mixture molecular weight W is defined through

$$\frac{\rho}{W} = \sum_{s=1}^{N_s} \frac{\rho_s}{W_s}. \tag{16}$$

Furthermore, the following relations are valid for an ideal gas mixture

$$\rho d = \sum_s \rho_s d_s \text{ for } d \in \{e, c_v, h, c_p, \frac{1}{W}\}. \tag{17}$$

Using the closure equations for the partial pressures p_s and the internal energy e, we get the matrices K and J

$$K = \begin{pmatrix} \frac{-\mathscr{R}T}{W_1} & & & \\ & \ddots & & \\ & & \frac{-\mathscr{R}T}{W_{N_s}} & \\ \frac{p}{\rho^2} & \cdots & \frac{p}{\rho^2} \; \frac{u}{\rho} & \frac{-1}{\rho} \end{pmatrix}, \quad J = \begin{pmatrix} 1 & & & -\mathscr{R}\frac{\rho_1}{W_1} \\ & \ddots & & \vdots \\ & & 1 & -\mathscr{R}\frac{\rho_{N_s}}{W_{N_s}} \\ -\frac{1}{\rho} & \cdots & -\frac{1}{\rho} & c_p \end{pmatrix}. \tag{18}$$

With the thermodynamic relation

$$\frac{p}{\rho} = \frac{\mathscr{R}}{W} T = (c_p - c_v) T \tag{19}$$

we simplify the last line of the matrices and obtain

$$K = \begin{pmatrix} \frac{-\mathscr{R}T}{W_1} & & & & \\ & \ddots & & & \\ & & \frac{-\mathscr{R}T}{W_{N_s}} & & \\ 0 & .. & 0 & \frac{u}{\rho} & -\frac{1}{\rho} \end{pmatrix}, \quad J = \begin{pmatrix} 1 & & -\mathscr{R}\frac{\rho_1}{W_1} \\ & \ddots & \vdots \\ & & 1 \; -\mathscr{R}\frac{\rho_{N_s}}{W_{N_s}} \\ & & c_v \end{pmatrix}. \tag{20}$$

Since $c_v > 0$, J is always invertible and thus the system has index 1. □

6 Closure Conditions for Real Gases

The behaviour of real gas usually agrees with ideal gas properties within a 5% range at normal pressure and temperature. However at high pressure or low temperature, the thermodynamic properties of a real gas deviate significantly from those expected from an ideal gas. Therefore the closure conditions must be reformulated.
Since Dalton's law is not valid for real gases, the total pressure of the mixture is given through a general algebraic relation

$$p = \pi(\rho_1, .., \rho_{N_s}, p_1, .., p_{N_s}, T). \tag{21}$$

Furthermore the pressure law for species X_s is modified to take the form

$$p_s = \rho_s \frac{\mathscr{R}T}{W_s} Z_s. \tag{22}$$

Z_s is the compressibility factor of species X_s. It is obvious that for ideal gases Z_s equals 1. To take real gas effects into account, the compressibility factor is usually modelled as a multivariable function of pressure, molar volume, temperature, etc.
A class of these models [9] consists in a polynomial representation of Z_s in either the inverse of the molar volume v_s or the pressure p_s considering temperature-dependent coefficients. A standard model for the compressibility factor is:

$$Z(\frac{1}{v}, T) = 1 + \sum_i \frac{\alpha_i(T)}{v^i}, \tag{23}$$

where the molar volume v is defined as $\frac{1}{v} := \frac{\rho}{W}$.

We complete the system with closure equations for the internal energy : the internal energy equation of species X_s is extended with real gas effects and takes the form

$$de_s = \frac{\partial e_s}{\partial T}|_v dT + \frac{\partial e_s}{\partial v}|_T dv \tag{24}$$

$$= c_{v,s} dT + \left(T \frac{\partial p_s}{\partial T}|_v - p_s \right) dv \tag{25}$$

$$= c_{v,s} dT - \mathscr{R}T^2 \frac{\partial Z_s}{\partial T}|_\rho \frac{d\rho_s}{\rho_s}. \tag{26}$$

Finally we define the mixture internal energy by a general algebraic law

$$e = \varepsilon(\rho_1,..,\rho_{N_s},e_1,..,e_{N_s},T) \tag{27}$$

Lemma 2. *Depending on $\frac{\partial \varepsilon}{\partial T}$, the system has an index of 1 or higher.*

Proof. The vector of constraints G is given by the following closure equations for the internal pressures $p_{s\{s=1,..,N_s\}}$

$$0 = g_s(\rho_s, p_s, T) = p_s - \frac{\rho_s}{W_s}\mathscr{R}TZ_s = p_s - \frac{\rho_s}{W_s}\mathscr{R}T\left[1+\sum \alpha_{s,i}(T)\left(\frac{\rho_s}{W_s}\right)^i\right] \tag{28}$$

and the closure equation for the internal energy

$$0 = g_e(\rho_1,..,\rho_{N_s},\rho v,\rho E,T) = e+\frac{1}{2}u^2 - E = \varepsilon(\rho_1,..,\rho_{N_s},e_1,..,e_{N_s},T)+\frac{1}{2}u^2 - E. \tag{29}$$

We obtain the matrices K and J with the following structure

$$K = \begin{pmatrix} \frac{\partial g_1}{\partial \rho_1} & & & & \\ & \ddots & & & \\ & & \frac{\partial g_{N_s}}{\partial \rho_{N_s}} & & \\ \frac{\partial g_e}{\partial \rho_1} & \cdots & \frac{\partial g_e}{\partial \rho_{N_s}} & \frac{\partial g_e}{\partial(\rho u)} & \frac{\partial g_e}{\partial(\rho E)} \end{pmatrix}, \quad J = \begin{pmatrix} 1 & & & \frac{\partial g_1}{\partial T} \\ & \ddots & & \vdots \\ & & 1 & \frac{\partial g_{N_s}}{\partial T} \\ & & & \frac{\partial g_e}{\partial T} \end{pmatrix}. \tag{30}$$

It is then obvious that the index of the system depends on $\frac{\partial g_e}{\partial T}$, *i.e.* $\frac{\partial \varepsilon}{\partial T}$. □

7 Results of Numerical Simulations

To show the influence of ideal or real gas closure conditions we run numerical simulations. To separate effects of convection from those of the closure equations, we decided to consider a continuous stirred-tank reactor so that chemical reactions take place at spatially-constant temperature, pressure and density. With this assumption, the convective part of the Euler equations vanishes and we can model the reactor by an ODE system in addition to the appropriate set of closure equations.

As a benchmark, we consider a reactor containing four chemical species denoted A, B, C and D undergoing the following reactions

$$A+B \rightleftharpoons C+D \tag{31}$$

$$A+D \rightleftharpoons B+C. \tag{32}$$

For the same set of initial conditions, we study the reaction kinetics using either the ideal or the real gas closure equations. We used the multistep solver *ode*15*s* of MATLAB which can solve index-1 DAE. The qualitative evolution of the concentrations is quite similar. However, within a relatively short time span, the mass fractions obtained in both cases clearly deviate from one another as shown in Fig. 1.

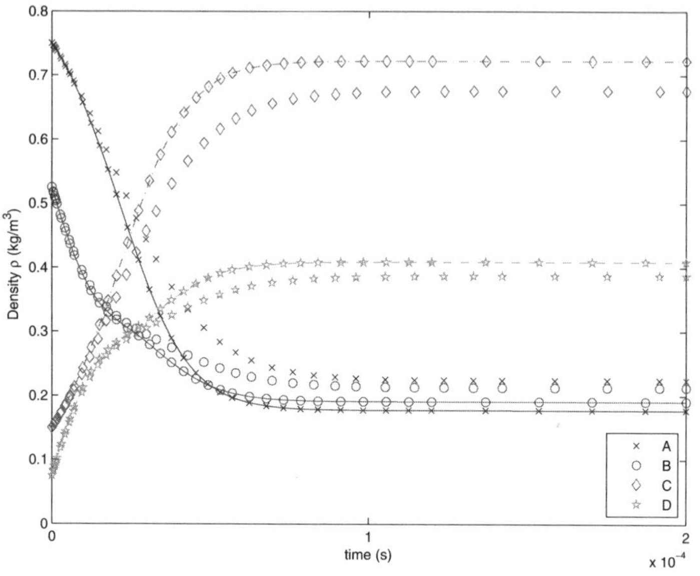

Fig. 1 Reaction kinetics for ideal (dots) and real gas (line) for the 4 reactants *A* to *D*

Acknowledgements This work has been supported within the Siemens doctoral program. The authors thank Prof.Dr. Albert Gilg and Dr. Utz Wever from Siemens Corporate Technology Munich for their support.

References

1. Campbell, S., März, R.: Direct transcription solution of high index optimal control problems. SIAM J. Comp. Appl. Math. **202**, 186–202 (2007)
2. Gear, C.: DAE index transformation. SIAM J. Sci. Stat. Comput. **9**, 39–47 (1988)
3. Günther, M., Wagner, Y.: Index concepts for linear mixed systems of differential-algebraic and hyperbolic-type equations. SIAM J. Sci. Comput. **22**, 1610–1629 (2000)
4. Hmaidi, A.: Mathematical modeling and numerical simulation of reactive 1-D Euler gas equations. Bachelor's thesis, Munich University of Technology (2005)
5. Hoschek, M., Rentrop, P., Wagner,Y.: Network approach and differential-algebraic systems in technical applications. Surv. Math. Ind. **9**, 49–76 (1999)
6. Kunkel, P., Mehrmann, V., Seidel, S.: A MATLAB Toolbox for the numerical solution of differential-algebraic equations. TU Berlin Preprint 16 (2005)
7. Neumeyer, T., Rentrop, P.: The DAE-aspect of the charge cycle in a combustion engine. Appl. Numer. Math. **25**, 287–295 (1997)
8. Poinsot, T., Veynante, D.: Theoretical and numerical combustion. R.T. Edwards (2005)
9. Rist, D.: Dynamik realer Gase. Springer (1996)

Application of a First Order Asymptotic Method for Modeling Singularly Perturbed BVPs

N. Parumasur, P. Singh, and V. Singh

Abstract We present a modified numerical algorithm based on a first order asymptotic procedure for solving singularly perturbed second-order boundary-value problems (BVPs). We outline the derivation of the asymptotic procedure and show how the method may be adapted for solving BVPs. The algorithm is similar to the shooting method. We present some test cases, including the Fisher equation of mathematical biology, and the results indicate that the method would be useful for solving BVPs.

1 Introduction

In this paper we are concerned with boundary value problems exhibiting one or more transition layers near the boundary. In regular perturbation theory there are methods that give solutions that are valid for finite intervals of time and also for all time. In this paper we study equations in which the order drops when the bifurcation parameter is set to zero and in these kinds of problems the solutions undergo rapid transition as one moves from one boundary point to the other and this makes it difficult to find solutions that are even valid in small intervals. We present a modified numerical algorithm based on a first order asymptotic procedure for solving singularly perturbed second-order boundary-value problems (BVPs). We follow the algorithm of the asymptotic expansion for solving singularly perturbed systems of ordinary differential equations [5, 6]. The present article deals with BVPs and serves as an extension of the work considered in the article presented in ENUMATH 2005 [2]. A similar approach for solving singularly perturbed initial value problems and evolution equations has been proposed in [1, 4, 8]. In that context the method is referred to as the diffusion approximation. The asymptotic procedure has also been

N. Parumasur, P. Singh, and V. Singh
University of KwaZulu-Natal, School of Mathematical Sciences, Private Bag X54001, Durban, 4000, South Africa, e-mail: [parumasurn1,singhp,singhvs]@ukzn.ac.za

applied in kinetic theory and the method referred to as the prompt-jump approximation [3].

We consider the boundary value problem

$$\begin{aligned} \varepsilon\frac{d^2x}{dt^2}+A\frac{dx}{dt}+f(x)&=0\\ x(0)&=\alpha\\ x(1)&=\beta \end{aligned} \tag{1}$$

where $x(t), f(x),\ \alpha,\beta \in \mathbf{R}$ and ε is a small perturbation parameter. The eigenvalues of the matrix A have all positive real parts, hence A is invertible.

We convert the boundary value problem (1) to the initial value problem by dropping the second boundary condition $x(1)=\beta$ and replacing it by the initial condition

$$x'(0)=s. \tag{2}$$

The initial-value problem has a uniquely determined solution $x(t)=x(t,s)$ which depends on the choice of the initial value s for $x'(0)$. To determine the value of s consistent with the right-hand boundary condition we must find a zero $s=\bar{s}$ of the function $F(s)=x(1,s)-\beta$. Assuming that $F(s)\in C^2(0,1)$, one can use the secant method to determine $\bar{s}$. Starting with an initial approximation s^0, one then has to iteratively compute values s^i according to

$$s^{i+1}=s^i-\frac{F(s^i)\triangle s^i}{F(s^i+\triangle s^i)-F(s^i)} \tag{3}$$

where

$$\frac{F(s^i+\triangle s^i)-F(s^i)}{\triangle s^i}\approx F'(s^i)$$

and $F(s^i)$ is determined by solving an initial-value problem

$$\begin{aligned} \varepsilon\frac{d^2x}{dt^2}+A\frac{dx}{dt}+f(x)&=0\\ x(0)&=\alpha\\ x'(0)&=s^{(i)} \end{aligned} \tag{4}$$

up to $t=1$.

In this paper instead of applying the "shooting method" to the stiff second order system of equations (1), we first apply the first order asymptotic procedure to the system and then adapt the "shooting method" to the resulting non-stiff first order system. We briefly discuss the procedure.

In section 2 we outline the derivation of the asymptotic procedure and show how the method may be adapted for solving BVPs. In section 3 we provide some test cases, including the Fisher equation of mathematical biology, and present the numerical results.

2 Outline of the Asymptotic Procedure

The second order system (1) is converted to the first order system

$$
\begin{aligned}
\varepsilon \frac{dz}{dt} &= -Az - f(x) \\
\frac{dx}{dt} &= z \\
x(0) &= \alpha \\
z(0) &= s
\end{aligned}
\tag{5}
$$

We truncate expansions to first order in ε in order to derive the first order version of the steady state approximation. The functions x and z in (5) are each decomposed into a bulk solution depending on t and an initial layer solution depending on $\tau = t/\varepsilon$. Hence

$$
\begin{aligned}
z(t) &= \bar{z}(t) + \tilde{z}(\tau) + O(\varepsilon^2) \\
x(t) &= w(t) + \tilde{x}(\tau) + O(\varepsilon^2)
\end{aligned}
\tag{6}
$$

where

$$
\begin{aligned}
\bar{z}(t) &= \bar{z}_0(t) + \varepsilon \bar{z}_1(t) \\
\tilde{z}(\tau) &= \tilde{z}_0(\tau) + \varepsilon \tilde{z}_1(\tau) \\
\tilde{x}(\tau) &= \tilde{x}_0(\tau) + \varepsilon \tilde{x}_1(\tau).
\end{aligned}
\tag{7}
$$

We note that the bulk solution w for the slow variable x remains unexpanded. Substituting (6) into (5) we obtain upon equating functions of t and τ separately

$$\varepsilon \frac{d\bar{z}}{dt} = -A\bar{z} - f(w), \tag{8}$$

$$\frac{dw}{dt} - \bar{z}, \tag{9}$$

$$\frac{d\tilde{x}}{d\tau} = \varepsilon \tilde{z}, \tag{10}$$

$$\frac{d\tilde{z}}{d\tau} = -A\tilde{z} + f(w(\varepsilon\tau)) - f(w(\varepsilon\tau) + \tilde{x}). \tag{11}$$

Since the original system is autonomous, we can replace $\bar{z}(t)$ by

$$\bar{z}(t) = \phi_0(w(t)) + \varepsilon \phi_1(w(t)) \tag{12}$$

where $\bar{z}_0(t) = \phi_0(w(t))$ and $\bar{z}_1(t) = \phi_1(w(t))$ are functions of $w(t)$. Substituting (12) into (8) we obtain

$$\varepsilon \left[\frac{d\phi_0}{dw}\frac{dw}{dt} + \varepsilon \frac{d\phi_1}{dw}\frac{dw}{dt} \right] = -A(\phi_0 + \varepsilon \phi_1) - f(w). \tag{13}$$

Substituting (12) into (13) because of (9) and equating coefficients of powers of ε we obtain

$$\phi_0(w) = -A^{-1}f(w), \tag{14}$$

$$\frac{d\phi_0}{dw}\phi_0(w) = -A\phi_1(w). \tag{15}$$

which is solved for $\phi_1(w)$, yielding

$$\phi_1(w) = -A^{-2}\frac{df(w)}{dw}A^{-1}f(w). \tag{16}$$

Hence, from (9) we obtain $w(t)$ as the solution of the first order system

$$\frac{dw}{dt} = -A^{-1}\left[I + \varepsilon A^{-1}\frac{df(w)}{dw}A^{-1}\right]f(w). \tag{17}$$

In order to obtain the initial condition $w(0)$ for (17) we substitute (7) into (10) and equate powers of ε to obtain the initial layer functions

$$\tilde{x}_0(\tau) = 0. \tag{18}$$

$$\tilde{x}_1(\tau) = -A^{-1}e^{-A\tau}(s + A^{-1}f(w(0))). \tag{19}$$

Using the original initial conditions together with (18) and (19) we obtain

$$w(0) := w(0,s) = \alpha + \varepsilon A^{-1}[s + A^{-1}f(\alpha)]. \tag{20}$$

Now taking $t = 1$ in equation (6) and (19) it can be shown that

$$w(1) := w(1,s) = \beta + \varepsilon A^{-1}e^{-A/\varepsilon}[s + A^{-1}f(w(0,s))]. \tag{21}$$

We solve the first order system

$$\begin{aligned}\frac{dw}{dt} &= -A^{-1}\left[I + \varepsilon A^{-1}\frac{df(w)}{dw}A^{-1}\right]f(w)\\ w(0) &= w(0,s).\end{aligned} \tag{22}$$

numerically and denote the solution at $t = 1$ by $\hat{w}(1,s)$.

It remains to provide a suitable value for s. This is obtained by defining $F(s) = \hat{w}(1,s) - w(1,s)$ and iteratively solving (3). A convenient starting value for the iteration is obtained by setting $\varepsilon = 0$ in equation (1) and we obtain $s^0 = -A^{-1}f(\alpha)$. We are essentially trying to find an optimal s such that both $w(0,s)$ and $w(1,s)$ lie on the solution curve $w(t)$ of equation (17).

Alternatively, the problem may be cast in the fixed point form

$$s^{i+1} = G(s^i) \tag{23}$$

where

$$G(s) = \frac{1}{\varepsilon} e^{A/\varepsilon} A[(\hat{w}(1,s) - \beta) - A^{-1} f(w(0,s))].$$

3 Numerical Modeling

We consider several examples to demonstrate the performance of the proposed method. The numerical results are obtained with Mathematica Version 6.0 on a Dell PC.

Example 1.

$$\varepsilon \frac{d^2x}{dt^2} + \frac{dx}{dt} - 1 = 0$$
$$x(0) = 0$$
$$x(1) = 0$$

We choose $\varepsilon = 0.01$. Figures 1 and 2 shows a plot of the solution profiles $x(t)$ (numerical solution) and $w(t)$ (asymptotic solution), corresponding to different values of s in (3). It is seen that the solution profiles are almost identical (in the bulk region) after the first iteration.

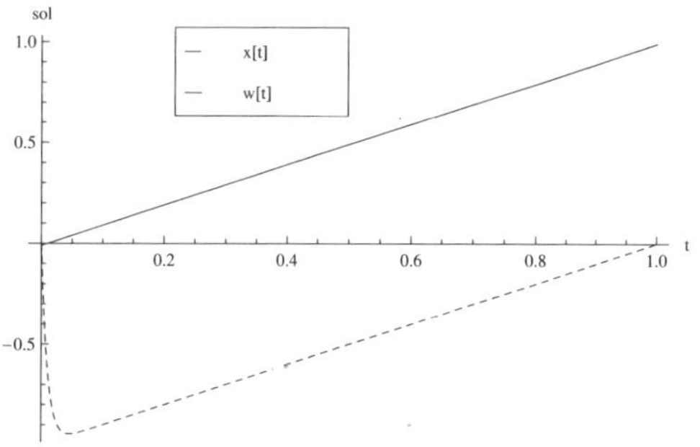

Fig. 1 Numerical solution $x(t)$ (dotted) versus first order asymptotic solution $w(t)$ (solid) $(s = s^0)$.

Example 2.

$$\varepsilon \frac{d^2x}{dt^2} + 2\frac{dx}{dt} + \exp(x) = 0$$
$$x(0) = 0$$
$$x(1) = 0$$

We choose $\varepsilon = 0.001$. Figures 3, 4 and 5 shows a plot of the solution profiles $x(t)$ (numerical solution) and $w(t)$ (asymptotic solution), corresponding to different

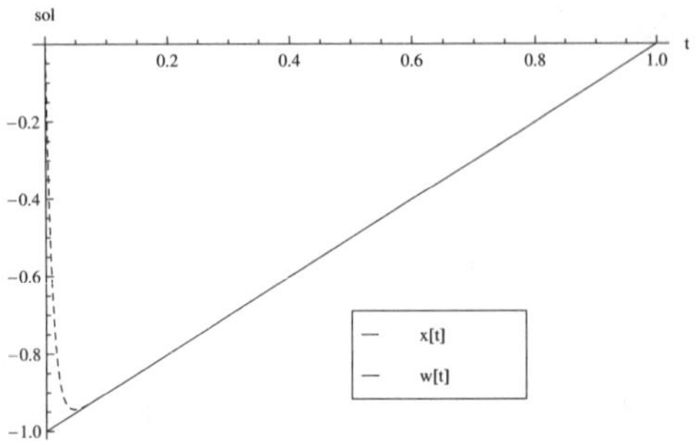

Fig. 2 Numerical solution $x(t)$ (dotted) versus first order asymptotic solution $w(t)$ (solid) after first Newton iteration ($s = s^1$).

values of s in (3). It is seen that the solution profiles are almost identical (in the bulk region) after the second iteration.

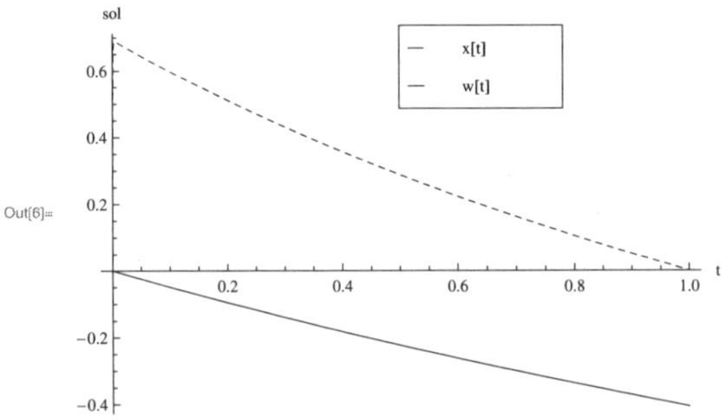

Fig. 3 Numerical solution $x(t)$ (dotted) versus first order asymptotic solution $w(t)$ (solid) ($s = s^0$).

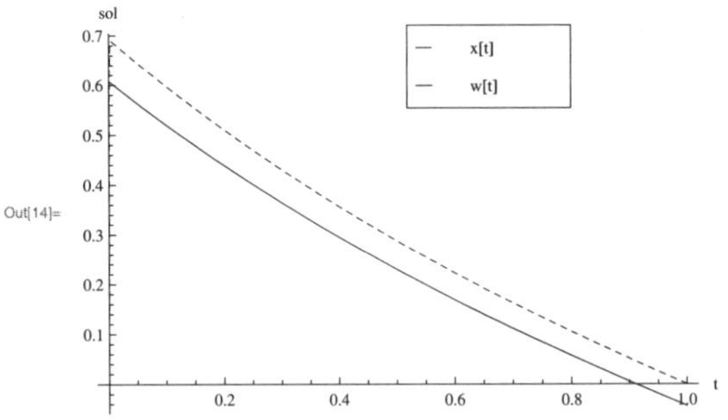

Fig. 4 Numerical solution $x(t)$ (dotted) versus first order asymptotic solution $w(t)$ (solid) after first Newton iteration ($s = s^1$).

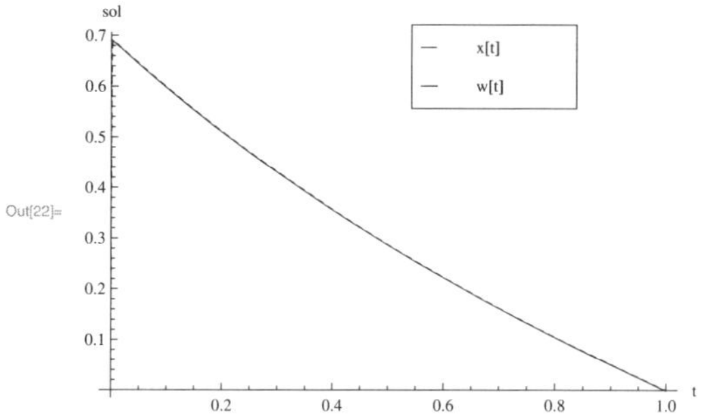

Fig. 5 Numerical solution $x(t)$ (dotted) versus first order asymptotic solution $w(t)$ (solid) after second Newton iteration ($s = s^2$).

Example 3. We consider the Fisher equation from mathematical biology[7]

$$\varepsilon \frac{d^2x}{dt^2} + \frac{dx}{dt} + x(1-x) = 0$$
$$x(-\infty) = 1, \quad x(\infty) = 0$$
$$0 < \varepsilon \leq 0.25$$

Using standard perturbation analysis it can be shown that the first order asymptotic solution is given by

$$x(t) = (1+e^t)^{-1} + \varepsilon e^t (1+e^t)^{-2} \ln \left[4e^t(1+e^t)^{-2}\right]. \tag{24}$$

Hence, we use $x(0) = \alpha$ and $x(1) = \beta$ from this solution and apply the present algorithm.

We use $\varepsilon = 0.1$. Figures 6 and 7 depicts the absolute error ($|x(t) - w(t)|$). It is seen that the error smoothes out uniformly after the first iteration.

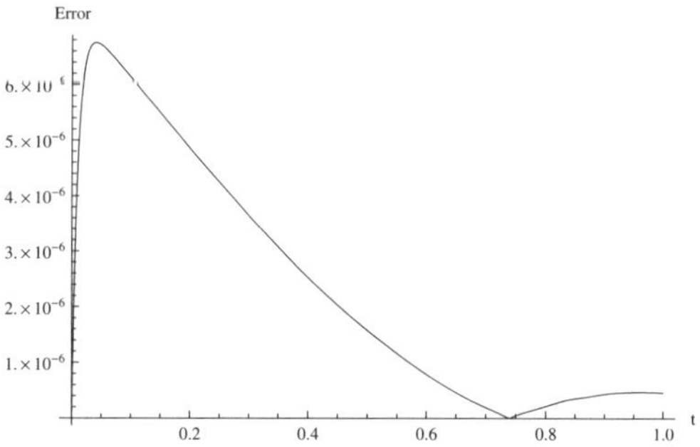

Fig. 6 Absolute Error: Numerical solution $x(t)$ and first order asymptotic solution $w(t)$ ($s = s^0$).

The perturbation parameter ε varies between 0.0001 and 0.1 representing both small and large values. The preceding numerical examples demonstrate fine convergence of the bulk solution $w(t)$. Furthermore the computation time to obtain the asymptotic solution $w(t)$ was superior to that of solving the original system for $x(t)$

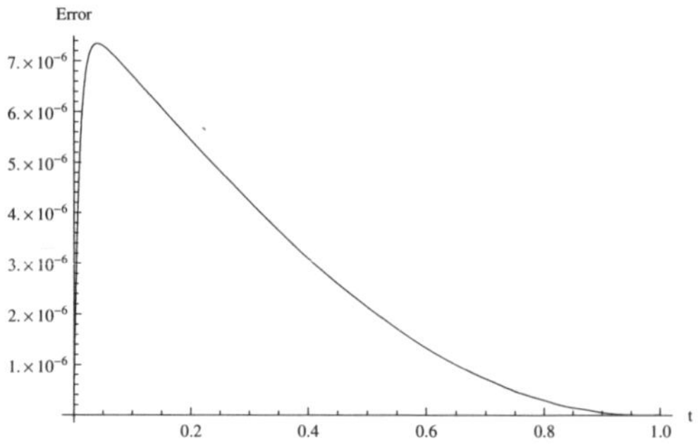

Fig. 7 Absolute Error: Numerical solution $x(t)$ and first order asymptotic solution $w(t)$ after first Newton iteration ($s = s^1$).

in Mathematica. In Example 2 the solution of $w(t)$ was obtained in under 1 second whilst the solution of $x(t)$ took roughly 14 seconds. The extension to multidimensional case is the subject of the expanded version of the present exposition [9].

Acknowledgements We would like to thank Professor Janusz Mika for suggesting applying the proposed method to BVPs.

References

1. Banasiak, J., Parumasur, N., Kozakiewicz, J.: Diffusion aproximation of linear kinetic equations with non-equilibrium data-computational experiments. Transport Theory Statist. Phys. **34**(6), 475–496 (2005)
2. Banasiak, J., Parumasur, N., Kozakiewicz, J.: Numerical modelling of kinetic equations. In: P.Q. Alfredo Bermúdez de Castro Dolores Gómez, P. Salgado (eds.) Numerical mathematics and advanced applications, pp. 618–625. Springer-Verlag, Berlin (2006)
3. Beauwens, R., Mika, J.: On the improved prompt jump approximation. In: 17th IMACS World Congress (2005). URL http://imacs2005.ec-lille.fr
4. Kozakiewicz, J.: Singularly perturbed evolution equations with applications to kinetic theory: Analytic methods and numerical investigations. PhD thesis, University of KwaZulu-Natal (2006)
5. Mika, J., Kozakiewicz, J.: First-order asymptotic expansion method for singularly perturbed systems of second-order ordinary differential equations. Comput. Math. Appl. **25**(3), 3–11 (1993)
6. Mika, J., Palczewski, A.: Asymptotic analysis of singularly perturbed systems of ordinary differential equations. Comput. Math. Appl. **21**(10), 13–32 (1991)
7. Murray, J.: Mathematical Biology. Springer-Verlag, Berlin (1989)
8. Parumasur, N., Banasiak, J., Kozakiewicz, J.: Challenges in the numerical solution for models in transport theory. Transport Theory Statist. Phys. **36**(1-3), 67–78 (2007)
9. Parumasur, N., Singh, P., Singh, V.: Application of an asymptotic procedure for modeling singurlalry perturbed boundary value problems. Nonlinear Analysis Real World Applications (submitted).

Solid Mechanics

Newton-Like Solver for Elastoplastic Problems with Hardening and its Local Super-Linear Convergence

P.G. Gruber and J. Valdman

Abstract We discuss a solution algorithm for quasi-static elastoplastic problems with hardening. Such problems can be described by a time dependent variational inequality, where the displacement and the plastic strain fields serve as primal variables. After discretization in time, one variational inequality of the second kind is obtained per time step and can be reformulated as each one minimization problem with a convex energy functional which depends smoothly on the displacement and non-smoothly on the plastic strain. There exists an explicit formula how to minimize the energy functional with respect to the plastic strain for a given displacement. By substitution, the energy functional can be written as a functional depending only on the displacement. The theorem of Moreau from convex analysis states that this energy functional is differentiable with an explicitly computable first derivative. The second derivative of the energy functional does not exist, hence the plastic strain minimizer is not differentiable on the elastoplastic interface, which separates the continuum in elastically and plastically deformed parts. A Newton-like method exploiting slanting functions of the energy functional's first derivative instead of the nonexistent second derivative is applied.

1 Introduction

We consider a quasi-static initial-boundary value problem for small strain elastoplasticity with hardening. Throughout the paper, the linear isotropic hardening law is considered. Several interesting computational techniques for solving the elastoplastic problem with various kinds of hardening can be found in [8, 10, 1].

After discretizing in time we consider only one time step, which is described as a minimization problem for an unknown displacement u and plastic strain field p,

Peter G. Gruber and Jan Valdman
SFB F013 'Numerical and Symbolic Scientific Computing', Johannes Kepler University Linz, 4040 Linz, Austria e-mail: {peter.gruber, jan.valdman}@sfb013.uni-linz.ac.at

see [2]. There, it has been already shown that a method of alternating minimization convergences globally and linearly. The minimization with respect to the plastic strain can be calculated locally by using an explicitly known dependence [1] of the plastic strain on the total strain, i.e., $p = \tilde{p}(\varepsilon(u))$. Thus, an equivalent energy minimization problem for the displacement u only,

$$J(u) := \bar{J}(u, \tilde{p}(\varepsilon(u))) \to \min,$$

can be defined. Since the dependencies of the energy functional on the second argument, and of the minimizer $\tilde{p}$ on the total strain $\varepsilon(u)$ are not smooth, the Fréchet derivative $\mathrm{D}J(u)$ seems not to exist. However, a damped generalized Newton method introduced in [1] converges globally and linearly. The super-linear convergence is discussed but not proved there.

The main theoretical contribution of this paper is the extension of the analysis done in [1]. We show that the structure of the energy functional $J(u)$ satisfies the assumptions of Moreau's theorem from convex analysis and therefore the energy functional $J(u)$ is Fréchet differentiable with an explicitly computable Fréchet derivative $\mathrm{D}J(u)$. The second derivative of the energy functional, $\mathrm{D}^2J(u)$, does not exist. This is due to the non-differentiability of the plastic strain minimizer $\tilde{p}$ on the elastoplastic interface, which separates the deformed continuum in elastically and plastically deformed parts.

By the concept of slant differentiability [3] we define a Newton-like method using slanting functions instead of the usual derivative. We call such method a slant Newton method for short. One of the main results in [3] is, that a slant Newton method converges locally super-linear under the same assumptions as the classical Newton method. The main task is to find slanting functions for the mapping $\max\{0,\cdot\}$, which occurs within the formula of the plastic minimizer $\tilde{p}$ and causes its non-differentiability. Such slanting functions are easy to find in the spatial discrete case (e.g. after the FEM discretization). This explains the local super-linear convergence, which was originally conjectured in Remark 7.5 of [1].

The spatially continuous case is more complicated and requires some extra regularity assumptions for the trial stress in each slant Newton step. To the best knowledge of the authors, there are no theoretical results yet known, which would guarantee the required regularity properties. Thus, our work may serve as a starting point for more regularity analysis concerning elastoplastic problems.

2 Problem Description

We consider an elastoplastic one time step minimization problem with isotropic hardening as described in [2]. Let t_k for $k \in \{1,\ldots,N\}$ denote the kth time step, $\Omega \subset \mathbb{R}^3$ be a bounded domain with Lipschitz continuous boundary $\partial\Omega$, split into two parts Γ_N and Γ_D. We define $V := \left[H^1(\Omega)\right]^3$, $V_0 := \{v \in V \mid v = 0 \text{ on } \Gamma_D\}$, $V_D := \{v \in V \mid v = u_D \text{ on } \Gamma_D\}$ for $u_D \in \left[H^{1/2}(\Gamma_D)\right]^3$, $Q := [L_2(\Omega)]^{3\times 3}$, and $\overline{\mathbb{R}} := \mathbb{R} \cup \{+\infty\}$.

Moreover, we consider the body force $f_k \in [L_2(\Omega)]^3$, the traction $g_k \in [L_2(\Gamma_N)]^3$, the previous time step approximation of the plastic strain $p_{k-1} \in Q$ and the hardening parameter $\alpha_{k-1} \in L_2(\Omega)$ to be given such that $\alpha_{k-1} \geq 0$ almost everywhere.

Then the elastoplastic energy functional $\bar{J}_k : V_D \times Q \to \overline{\mathbb{R}}$ is defined by

$$\bar{J}_k(v,q) := \frac{1}{2}\|\varepsilon(v) - q\|_{\mathbb{C}}^2 + \psi_k(q) - l_k(v), \tag{1}$$

where

$$\langle A,B\rangle_F := \sum_{i,j} a_{ij}b_{ij}, \qquad \|A\|_F := \langle A,A\rangle_F^{1/2}, \tag{2}$$

$$\langle q_1,q_2\rangle_{\mathbb{C}} := \int_\Omega \langle \mathbb{C}q_1(x), q_2(x)\rangle_F \,\mathrm{d}x, \qquad \|q\|_{\mathbb{C}} := \langle q,q\rangle_{\mathbb{C}}^{1/2}, \tag{3}$$

$$\tilde{\alpha}_k(q) := \alpha_{k-1} + \sigma_y H \|q - p_{k-1}\|_F, \tag{4}$$

$$\psi_k(q) := \begin{cases} \int_\Omega \left(\frac{1}{2}\tilde{\alpha}_k(q)^2 + \sigma_y\|q - p_{k-1}\|_F\right) \mathrm{d}x & \text{if } \operatorname{tr} q = \operatorname{tr} p_{k-1}, \\ +\infty & \text{else}, \end{cases} \tag{5}$$

$$l_k(v) := \int_\Omega f_k \cdot v \,\mathrm{d}x + \int_{\Gamma_N} g_k \cdot v \,\mathrm{d}s. \tag{6}$$

Here, $\sigma_y > 0$ and $H > 0$ denote the material constants yield stress and modulus of hardening, respectively, and $\mathbb{C}$ denotes the elastic stiffness tensor, which uses the Lamé constants $\lambda > 0$ and $\mu > 0$.

The convex functional $\bar{J}_k$ expresses the mechanical energy of the deformed system at the kth time step. Notice, that $\bar{J}_k$ is smooth with respect to the displacements v, but not with respect to the plastic strains q. The goal is to find a displacement u_k and a plastic strain p_k such that the energy $\bar{J}_k$ is minimized:

Problem 1. Find $(u_k, p_k) \in V_D \times Q$ which satisfy $\bar{J}_k(u_k,p_k) = \inf_{(v,q)\in V_D\times Q} \bar{J}_k(v,q)$.

A short summary on the mathematical modeling of this minimization problem starting from the classical formulation can be found in [7].

3 New Contribution

In [2] a method of an alternate minimization regarding the displacement v and the plastic strain q was investigated to solve Problem 1. The global linear convergence of the resulting method was shown and a local super-linear convergence was conjectured. Another interesting technique is to reduce Problem 1 to a minimization problem with respect to the displacements v only. This can be achieved by substituting the explicit minimizer of J_k with respect to the plastic strain field for some given displacement v, $q = \tilde{p}_k(\varepsilon(v))$. We will observe that such reduced minimization problem is smooth with respect to the displacements v and its derivative is explicitly computable.

The following theorem is formulated for functionals mapping from a Hilbert space $\mathbb{H}$ provided with the scalar product $\langle \circ, \diamond \rangle_{\mathbb{H}}$ and the norm $\|\cdot\|_{\mathbb{H}} := \langle \cdot, \cdot \rangle_{\mathbb{H}}^{1/2}$. If a function F is Fréchet differentiable, we shall denote its derivative in a point x by $\mathrm{D}F(x)$ and its Gâteaux differential in the direction y by $\mathrm{D}F(x;y)$. We refer to [4] concerning the definitions of convex, proper, lower semi continuous, and coercive.

Theorem 1. *Let the function $f : \mathbb{H} \times \mathbb{H} \to \overline{\mathbb{R}}$ be defined*

$$f(x,y) = \frac{1}{2}\|x-y\|_{\mathbb{H}}^2 + \psi(x) \tag{7}$$

where ψ is a convex, proper, lower semi continuous, and coercive function of $\mathbb{H}$ into $\overline{\mathbb{R}}$. Then $F(y) := \inf_{x\in\mathbb{H}} f(x,y)$ maps into $\mathbb{R}$, and there exists a unique function $\tilde{x} : \mathbb{H} \to \mathbb{H}$ such that $F(y) = f(\tilde{x}(y), y)$ for all $y \in \mathbb{H}$. Moreover, there holds:

1. *F is strictly convex and continuous in $\mathbb{H}$.*
2. *F is Fréchet differentiable with the Fréchet derivative*

$$\mathrm{D}F(y) = \langle y - \tilde{x}(y), \cdot \rangle_{\mathbb{H}} \quad \textit{for all } y \in \mathbb{H}. \tag{8}$$

Proof. See [9, 7.d. Proposition].

We apply Theorem 1 to Problem 1 and obtain the following proposition.

Proposition 1. *Let $k \in \{1,\ldots,N\}$ denote the time step, and let $\bar{J}_k$ be defined as in (1). Then there exists a unique mapping $\tilde{p}_k : Q \to Q$ satisfying*

$$\bar{J}_k(v, \tilde{p}_k(\varepsilon(v))) = \inf_{q\in Q} \bar{J}_k(v,q) \quad \forall v \in V_D. \tag{9}$$

Let J_k be a mapping of V_D into $\mathbb{R}$ defined as

$$J_k(v) := \bar{J}_k(v, \tilde{p}_k(\varepsilon(v))) \quad \forall v \in V_D. \tag{10}$$

Then, J_k is strictly convex and Fréchet differentiable. The associated Gâteaux differential reads

$$\mathrm{D}J_k(v;w) = \langle \varepsilon(v) - \tilde{p}_k(\varepsilon(v)), \varepsilon(w) \rangle_{\mathbb{C}} - l_k(w) \quad \forall w \in V_0 \tag{11}$$

with the scalar product $\langle \circ, \diamond \rangle_{\mathbb{C}}$ defined in (3) and l_k defined in (6).

Proof. Recall, that the functional $\bar{J}_k : V \times Q \to \overline{\mathbb{R}}$ defined in (1) using (3), (5), and (6) can be decomposed as $\bar{J}_k(v,q) = f_k(\varepsilon(v),q) - l_k(v)$, where the functional $f_k : Q \times Q \to \overline{\mathbb{R}}$ reads $f_k(s,q) := \frac{1}{2}\|q-s\|_{\mathbb{C}}^2 + \psi_k(q)$. Theorem 1 states an existence of a unique minimizer $\tilde{p}_k : Q \to Q$ which satisfies the condition $f_k(\varepsilon(v), \tilde{p}_k(\varepsilon(v))) = \inf_{q\in Q} f_k(\varepsilon(v),q)$, where the functional $F_k(\varepsilon(v)) := f_k(\varepsilon(v), \tilde{p}_k(\varepsilon(v)))$ is strictly convex and differentiable with respect to $\varepsilon(v) \in Q$. Since $\varepsilon : v \to \varepsilon(v)$ is a Fréchet differentiable, linear and injective mapping of V_D into Q, the compound functional $F_k(\varepsilon(v))$ is Fréchet differentiable and strictly convex with respect to $v \in V_D$. Considering the Fréchet differentiability and linearity of l_k with respect to $v \in V_D$, we conclude the strict convexity and Fréchet differentiability of the functional J_k defined in (10). The explicit form of the Gâteaux differential $\mathrm{D}J_k(v;w)$ in (11) results from the linearity of the two mappings l_k and ε, and the Fréchet derivative $DF_k(\varepsilon(v);\cdot) = \langle \varepsilon(v) - \tilde{p}_k(\varepsilon(v)), \cdot\rangle_{\mathbb{C}}$ as in (8), combined with the chain rule.

The minimizer $\tilde{p}_k$ can be calculated by hand (see [1, 5]) and it exactly recovers the classical return mapping algorithm [10]. Let the trial stress $\tilde{\sigma}_k : Q \to Q$ at the kth time step and the yield function $\phi_{k-1} : Q \to \mathbb{R}$ at the $k-1$st time step be defined by

$$\tilde{\sigma}_k(q) := \mathbb{C}(q - p_{k-1}) \quad \text{and} \quad \phi_{k-1}(\sigma) := \|\mathrm{dev}\sigma\|_F - \sigma_y(1 + H\alpha_{k-1}), \tag{12}$$

where $\mathrm{dev}A := A - \frac{A:I}{I:I}I$ denotes the deviator of a matrix A, and where I denotes the identity matrix. Then, the minimizer $\tilde{p}_k$ reads

$$\tilde{p}_k(\varepsilon(v)) = \frac{1}{2\mu + \sigma_y^2 H^2} \max\{0, \phi_{k-1}(\tilde{\sigma}_k(\varepsilon(v)))\} \frac{\mathrm{dev}\tilde{\sigma}_k(\varepsilon(v))}{\|\mathrm{dev}\tilde{\sigma}_k(\varepsilon(v))\|_F} + p_{k-1}. \tag{13}$$

We obtain a smooth minimization problem by using J_k as in (10) with $\tilde{p}_k$ as in (13):

Problem 2. Find $u_k \in V_D$ such that $J_k(u_k) = \inf_{v \in V_D} J_k(v)$.

To solve this smooth minimization problem one could apply Newton's Method $v^{j+1} = v^j - (\mathrm{D}^2 J_k(v^j))^{-1} \mathrm{D}J_k(v^j)$, but unfortunately the second derivative does not exist, since the max–function in (13) is not differentiable. Therefore, we apply a Newton-like method which uses slanting functions (see [3]) instead of the second derivative. We shall call such method a Slant Newton Method. Henceforth, let X and Y be Banach spaces, and $\mathscr{L}(X,Y)$ denote the set of all linear mappings of X into Y.

Definition 1. Let $U \subseteq X$ be an open subset and $x \in U$. A function $F : U \to Y$ is said to be *slantly differentiable at x* if there exists a mapping $F^o : U \to \mathscr{L}(X,Y)$ which is uniformly bounded in an open neighborhood of x, and a mapping $r : X \to Y$ with $\lim_{h\to 0} \frac{\|r(h)\|_Y}{\|h\|_X} = 0$ such, that $F(x+h) = F(x) + F^o(x+h)h + r(h)$ holds for all $h \in X$ satisfying $(x+h) \in U$. We say, $F^o(x)$ is a *slanting function for F at x*. F is called *slantly differentiable in U* if there exists $F^o : U \to \mathscr{L}(X,Y)$ such that F^o is a slanting function for F for all $x \in U$. F^o is then called a *slanting function for F in U*.

Theorem 2. *Let $U \subseteq X$ be an open subset, and $F : U \to Y$ be a slantly differentiable function with a slanting function $F^o : U \to \mathscr{L}(X,Y)$. We suppose, that $x^* \in U$ is a solution to the nonlinear problem $F(x) = 0$. If $F^o(x)$ is non-singular for all $x \in U$ and $\{\|F^o(x)^{-1}\|_{\mathscr{L}(Y,X)} : x \in U\}$ is bounded, then the Newton-like iteration*

$$x^{j+1} = x^j - F^o(x^j)^{-1} F(x^j) \tag{14}$$

converges super-linearly to x^*, provided that $\|x^0 - x^*\|_X$ is sufficiently small.

Proof. See [3, Theorem 3.4] or [6, Theorem 1.1].

We apply the Slant Newton Method (14) to elastoplasticity by choosing $F = \mathrm{D}J_k$ as in (11). Due to [6, Proposition 4.1], the max-function is slantly differentiable as a mapping of $L_p(\Omega)$ into $L_q(\Omega)$ if $p > q$ but not if $p \leq q$. Therefore, if there holds $\phi_{k-1}(\tilde{\sigma}_k(\varepsilon(v))) \in L_{2+\delta}(\Omega)$ for some $\delta > 0$, then $\mathrm{D}J_k$ (cf. (11),(13)) has a slanting function which reads

$$(\mathrm{D}J_k)^o(v;w,\bar{w}) := \langle \varepsilon(w) - \tilde{p}_k^o(\varepsilon(v);\varepsilon(w)), \varepsilon(\bar{w})\rangle_{\mathbb{C}} \tag{15}$$

with a slanting function for $\tilde{p}_k$, e. g.,

$$\tilde{p_k}^o(\varepsilon(v);q) := \begin{cases} 0 & \text{if } \beta_k \leq 0, \\ \xi\left(\beta_k \operatorname{dev} q + (1-\beta_k)\frac{\langle \operatorname{dev}\tilde{\sigma}_k, \operatorname{dev} q\rangle_F}{\|\operatorname{dev}\tilde{\sigma}_k\|_F^2}\operatorname{dev}\tilde{\sigma}_k\right) & \text{else}, \end{cases} \tag{16}$$

where the abbreviations $\xi := \frac{2\mu}{2\mu+\sigma_y^2 H^2}$, $\tilde{\sigma}_k := \tilde{\sigma}_k(\varepsilon(v))$ and $\beta_k := \frac{\phi_{k-1}(\tilde{\sigma}_k)}{\|\operatorname{dev}\tilde{\sigma}_k\|_F}$ with ϕ_{k-1} and $\tilde{\sigma}_k$ defined in (12) are used (see [5, Corollary 2]). $(\mathrm{D}J_k)^o$ in Equation (15) is commonly known as the *consistent tangent*, see [10]. For fixed $v \in V_D$, the bilinear form $(\mathrm{D}J_k)^o(v;\cdot,\cdot)$ in (15) is elliptic and bounded in V_0 (see [5, Lemma 2]).

Corollary 1. *Let $k \in \{1,\dots,N\}$ and $\delta > 0$ be fixed and t_k denote the kth time step. Let the mapping $\mathrm{D}J_k : V_D \to V_0{}^*$ be defined $\mathrm{D}J_k(v) := \mathrm{D}J_k(v;\circ)$ as in (11), and $(\mathrm{D}J_k)^o : V_D \to \mathscr{L}(V_0, V_0{}^*)$ be defined $(\mathrm{D}J_k)^o(v) := (\mathrm{D}J_k)^o(v;\diamond,\circ)$ as in (15). Then, the Slant Newton iteration*

$$v^{j+1} = v^j - \left[(\mathrm{D}J_k)^o(v^j)\right]^{-1}\mathrm{D}J_k(v^j)$$

converges super-linearly to the solution u_k of Problem 2, provided that $\|v^0 - u_k\|_V$ is sufficiently small, and that $\phi_{k-1}(\tilde{\sigma}_k(\varepsilon(v)))$ as in (12) is in $L_{2+\delta}(\Omega)$ for all $v \in V_D$.

Proof. We check the assumptions of Theorem 2 for the choice $F = \mathrm{D}J_k$. Let $v \in V_D$ be arbitrarily fixed. The mapping $(\mathrm{D}J_k)^o(v) : V_0 \to V_0{}^*$ serves as a slanting function for $\mathrm{D}J_k$ at v, since $\phi_{k-1}(\tilde{\sigma}_k(\varepsilon(v)))$ is in $L_{2+\delta}(\Omega)$ everywhere in V_D. Moreover, $(\mathrm{D}J_k)^o(v) : V_0 \to V_0{}^*$ is bijective if and only if there exists a unique element w in V_0 such, that for arbitrary but fixed $f \in V_0{}^*$ there holds

$$(\mathrm{D}J_k)^o(v;w,\bar{w}) = f(\bar{w}) \quad \forall \bar{w} \in V_0\,. \tag{17}$$

Since the bilinear form $(\mathrm{D}J_k)^o(v)$ is elliptic and bounded, we apply the Lax-Milgram Theorem to ensure the existence of a unique solution to (17). Finally, with κ_1 denoting the v-independent ellipticity constant for $(\mathrm{D}J_k)^o(v;\diamond,\circ)$, the uniform boundedness of $[(\mathrm{D}J_k)^o(\cdot)]^{-1} : V_D \to \mathscr{L}(V_0{}^*, V_0)$ follows from the estimate

$$\begin{aligned} \|[(\mathrm{D}J_k)^o(v)]^{-1}\| &= \sup_{w^* \in V_0{}^*} \frac{\|[(\mathrm{D}J_k)^o(v)]^{-1} w^*\|}{\|w^*\|_{V_0{}^*}} = \sup_{w \in V_0} \frac{\|w\|_V}{\|(\mathrm{D}J_k)^o(v;w,\cdot)\|_{V_0{}^*}} \\ &= \sup_{w \in V_0} \inf_{\bar{w} \in V_0} \frac{\|w\|_V\,\|\bar{w}\|_V}{|(\mathrm{D}J_k)^o(v;w,\bar{w})|} \leq \sup_{w \in V_0} \frac{\|w\|_V^2}{|(\mathrm{D}J_k)^o(v;w,w)|} \leq \frac{1}{\kappa_1}\,. \end{aligned}$$

Remark 1. Notice the required assumption on the integrability of ϕ_{k-1}. It is still an open question, under which extra conditions this property can be satisfied. The local super linear convergence in the spatially discrete case (after FE-discretization) can be shown without any additional assumption, see [5, Proposition 3].

4 Numerical Example

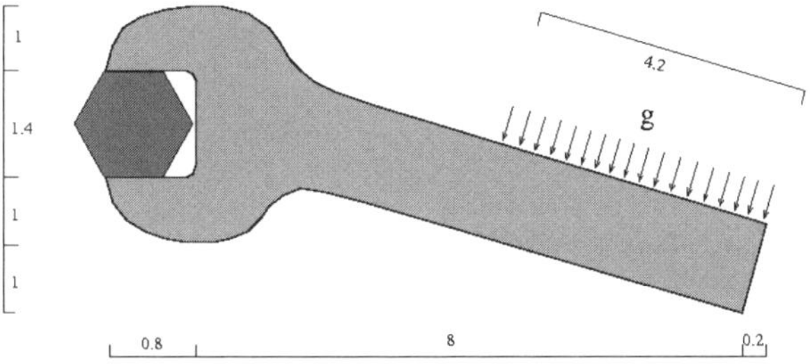

Fig. 1 Problem setup.

This example simulates the deformation of a screw-wrench under pressure. The problem geometry is shown in Figure 1: A screw-wrench sticks on a screw (homogeneous Dirichlet boundary condition) and a surface load g is applied to a part of the wrench's handhold in interior normal direction. The material parameters are set $E = 2e8$, $\nu = 0.3$, $\sigma_Y = 2e6$, $H = 0.001$, and the traction intensity amounts $|g| = 6e4$. Figure 2 shows the yield function (right) and the elastoplastic zones (left), where purely elastic zones are light, and plastic zones are dark. The displacement of the domain is multiplied by factor 10. Table 1 reports on the super linear convergence of the Newton-like method for graded uniform meshes.

Level	1	...	4	5	6	7	8
DOF	202	...	10590	41662	165246	658174	2627070
step 1	3.61e-03	...	1.09e-01	1.31e-01	1.48e-01	1.58e-01	1.63e-01
step 2	2.35e-06	...	3.70e-02	5.69e-02	6.93e-02	7.96e-02	8.83e-02
step 3	1.53e-11	...	4.38e-03	7.58e-03	1.32e-02	2.99e-02	4.16e-02
step 4	4.57e-15	...	1.10e-04	4.03e-04	2.43e-03	3.56e-03	4.97e-03
step 5		...	2.92e-08	5.96e-06	2.18e-04	1.20e-04	2.11e-04
step 6			4.16e-14	2.94e-10	1.50e-05	1.03e-05	2.06e-05
step 7				7.86e-14	3.89e-09	1.16e-09	1.39e-06
step 8					1.55e-13	2.99e-13	6.77e-09
step 9							5.93e-13

Table 1 The relative error in displacements $|v^j - v^{j-1}|_\varepsilon / \left(|v^j|_\varepsilon + |v^{j-1}|_\varepsilon\right)$ is displayed for graded uniform meshes, where $|v|_\varepsilon := \left(\int_\Omega \langle \varepsilon(v), \varepsilon(v)\rangle_F \, dx\right)^{1/2}$. DOF denotes degrees of freedom.

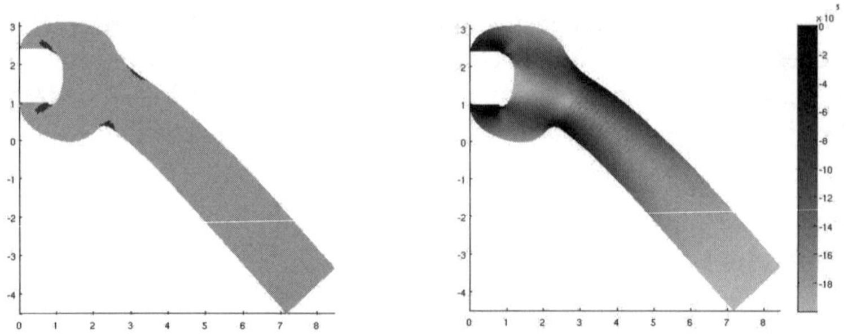

Fig. 2 Elastoplastic zones (left) and yield function (right) of the deformed wrench geometry. The displacement is magnified by a factor 10 for visualization reasons.

Acknowledgements The authors are pleased to acknowledge support by the Austrian Science Fund 'Fonds zur Förderung der wissenschaftlichen Forschung (FWF)' for their support under grant SFB F013/F1306 in Linz, Austria. The idea of looking at the elastoplastic formulation in terms of the Moreau Theorem came out during working progresses with H. Gfrerer, J. Kienesberger and U. Langer. We would also like to thank D. Knees and C. Wieners for an extended overview on literature about regularity results in elastoplasticity.

References

1. Alberty, J., Carstensen, C., Zarrabi, D.: Adaptive numerical analysis in primal elastoplasticity with hardening. Comput. Methods Appl. Mech. Eng. **171**(3-4), 175–204 (1999)
2. Carstensen, C.: Domain decomposition for a non-smooth convex minimalization problems and its application to plasticity. Numerical Linear Algebra with Applications **4**(3), 177–190 (1997)
3. Chen, X., Nashed, Z., Qi, L.: Smoothing methods and semismooth methods for nondifferentiable operator equations. SIAM J. Numer. Anal. **38**(4), 1200–1216 (2001)
4. Ekeland, I., Témam, R.: Convex Analysis and Variational Problems. SIAM (1999)
5. Gruber, P.G., Valdman, J.: Newton-like solver for elastoplastic problems with hardening and its local super-linear convergence. SFB Report 2007-06, Johannes Kepler University Linz, SFB F013 "Numerical and Symbolic Scientific Computing" (2007)
6. Hintermüller, M., Ito, K., Kunisch, K.: The primal-dual active set strategy as a semismooth newton method. SIAM J. Optim. **13**(3), 865–888 (2002)
7. Kienesberger, J., Valdman, J.: An efficient solution algorithm for elastoplasticity and its first implementation towards uniform h- and p- mesh refinements. In: A.B. de Castro, D. Gomez, P. Quintela, P. Salgado (eds.) Numerical Mathematics and Advanced Applications: Enumath 2005 (2006)
8. Korneev, V.G., Langer, U.: Approximate solution of plastic flow theory problems, *Teubner-Texte zur Mathematik*, vol. 69. Teubner-Verlag, Leipzig (1984)
9. Moreau, J.J.: Proximité et dualité dans un espace hilbertien. Bull. Soc. Math. France **93**, 273–299 (1965)
10. Simo, J.C., Hughes, T.J.R.: Computational Inelasticity. Springer-Verlag New York (1998)

On a Fictitious Domain Method for Unilateral Problems

J. Haslinger, T. Kozubek, and R. Kučera

Abstract Two variants of the fictitious domain method are compared. The first one enforces unilateral conditions by Langrange multipliers defined on the boundary γ of the original domain ω so that the computed solution has a singularity on γ that can result in an intrinsic error. The second one uses an auxiliary boundary Γ located outside of $\overline{\omega}$ on which a new control variable is introduced in order to satisfy the conditions on γ. Therefore the singularity is moved away from $\overline{\omega}$ so that the computed solution is smoother in ω. It is experimentally shown that the discretization error is significantly smaller in this case.

1 Introduction

This contribution deals with numerical realization of elliptic boundary value problems with unilateral boundary conditions using a fictitious domain method. Any fictitious domain formulation [2] extends the original problem defined in a domain ω to a new (fictitious) domain Ω with a simple geometry (e.g. a box) which contains $\overline{\omega}$. The main advantage consists in possibility to use a uniform mesh in Ω leading to a structured stiffness matrix. This enables us to apply highly efficient multiplying procedures [6].

Fictitious domain formulations of problems with the classical Dirichlet or Neumann boundary conditions lead after a finite element discretization typically to algebraic saddle-point systems. For their solution one can use the algorithm studied in [4] that combines the Schur complement reduction with the null-space method. The situation is not so easy for unilateral problems since their weak formulation contains a non-differentiable projection operator. Fortunately, a resulting algebraic

Jaroslav Haslinger
Charles University, Prague, Czech Republic, e-mail: hasling@karlin.mff.cuni.cz

Tomáš Kozubek and Radek Kučera
VŠB-TU Ostrava, Czech Republic e-mail: tomas.kozubek@vsb.cz, radek.kucera@vsb.cz

representation is described by a system that is semi-smooth in the sense of [1] so that a generalized Newton method can be applied. This method has been already used in [5] for solving complementarity problems. In our case each Newton step relates to a mixed Dirichlet-Neumann problem and therefore the algorithm from [4] can be used for solving inner linear systems. Due to the *superlinear* convergence rate of the Newton iteration [1], the computations are only slightly more expensive than the solution of pure Dirichlet or Neumann problems.

In this paper we compare two variants of the fictitious domain method. The first one enforces unilateral conditions by Langrange multipliers defined on the boundary γ of the original domain ω. Therefore the fictitious domain solution has a singularity on γ that can result in an intrinsic error of the computed solution. The second one uses an auxiliary boundary Γ located outside of $\overline{\omega}$ on which we introduce a new control variable in order to satisfy the conditions on γ. In the second approach the singularity is moved away from $\overline{\omega}$ so that the computed solution is smoother in ω. We shall experimentally show that the discretization error is significantly smaller in this case. Both approaches are theoretically justified on the level of continuous setting [3]. On the other hand some theoretical issues remain still open for the discrete problem. For more details we refer to [3, 4].

2 Setting of the Problem

We shall consider the following unilateral problem in a bounded domain $\omega \subset \mathbb{R}^2$ with the Lipschitz boundary γ:

$$-\Delta u + u = f \quad \text{in } \omega, u \geq g, \ \frac{\partial u}{\partial n_\gamma} \geq 0, \ \frac{\partial u}{\partial n_\gamma}(u-g) = 0 \quad \text{on } \gamma, \tag{1}$$

where $f \in L^2_{loc}(\mathbb{R}^2)$, $g \in H^{1/2}(\gamma)$ are given functions and $\frac{\partial}{\partial n_\gamma}$ stands for the normal derivative of a function on γ. We denote by $(\cdot,\cdot)_{k,S}$ the scalar product in $H^k(S)$, $k \geq 0$ integer ($H^0(S) := L^2(S)$).

The weak form of (1) reads as follows:

$$\left.\begin{array}{l} \textit{Find } u \in H^1(\omega) \textit{ such that} \\ (u,v)_{1,\omega} = (f,v)_{0,\omega} + \langle \frac{\partial u}{\partial n_\gamma}, v\rangle_\gamma \ \ \forall v \in H^1(\omega), \\ \frac{\partial u}{\partial n_\gamma} \in H_+^{-1/2}(\gamma), \quad \langle \mu - \frac{\partial u}{\partial n_\gamma}, u-g\rangle_\gamma \geq 0 \ \ \forall \mu \in H_+^{-1/2}(\gamma), \end{array}\right\} \tag{2}$$

where $\langle \cdot,\cdot \rangle_\gamma$ denotes the duality pairing between $H^{-1/2}(\gamma)$ and $H^{1/2}(\gamma)$. It is well-known that this problem has a unique solution.

Next, we shall suppose that $\frac{\partial u}{\partial n_\gamma} \in L^2_+(\gamma)$. Thus the duality pairing in (2) is represented by the $L^2(\gamma)$-scalar product and the inequality in (2) is equivalent to

$$\frac{\partial u}{\partial n_\gamma} = P(\frac{\partial u}{\partial n_\gamma} - \rho(u-g)), \tag{3}$$

where P denotes the projection of $L^2(\gamma)$ onto $L^2_+(\gamma)$ and $\rho > 0$ is arbitrary but fixed.

We shall present two variants of a fictitious domain formulation. To this end we choose a bounded domain Ω having a simple shape such that $\overline{\omega} \subset \Omega$ and construct a closed curve $\Gamma \subset \Omega$ surrounding ω. We shall distinguish two cases concerning the mutual positions of γ and Γ:

(i) $\gamma \equiv \Gamma$, (non-smooth variant); (ii) $\mathrm{dist}(\gamma, \Gamma) > 0$, (smooth variant).

Instead of (2), we propose to solve the extended problem in Ω called the *fictitious domain formulation* of (1):

$$\left.\begin{aligned} &\textit{Find } (\hat{u}, \lambda) \in H^1_0(\Omega) \times H^{-1/2}(\Gamma) \textit{ such that} \\ &(\hat{u}, v)_{1,\Omega} = (f, v)_{0,\Omega} + \langle \lambda, v \rangle_\Gamma \quad \forall v \in H^1_0(\Omega), \\ &\frac{\partial \hat{u}_{|\omega}}{\partial n_\gamma} = P(\frac{\partial \hat{u}_{|\omega}}{\partial n_\gamma} - \rho(\hat{u}_{|\omega} - g)), \end{aligned}\right\} \tag{4}$$

where $\langle \cdot, \cdot \rangle_\Gamma$ stands for the duality pairing between $H^{-1/2}(\Gamma)$ and $H^{1/2}(\Gamma)$. It is readily seen that $\hat{u}_{|\omega}$ solves (2), where $\hat{u}$ is the first component of the solution to (4). An existence analysis for this problem is discussed in [3].

3 Discretization

Let us consider finite dimensional subspaces $V_h \subset H^1_0(\Omega)$, $\Lambda_H(\gamma) \subset L^2(\gamma)$, $\Lambda_H(\Gamma) \subset L^2(\Gamma)$ such that $\dim V_h = n$, $\dim \Lambda_H(\gamma) = \dim \Lambda_H(\Gamma) = m$. By a discretization of (4) we mean the following problem:

$$\left.\begin{aligned} &\textit{Find } (\hat{u}_h, \lambda_H) \in V_h \times \Lambda_H(\Gamma) \textit{ such that} \\ &(\hat{u}_h, v_h)_{1,\Omega} = (f, v_h)_{0,\Omega} + (\lambda_H, v_h)_{0,\Gamma} \quad \forall v_h \in V_h, \\ &\delta_H \hat{u}_h = P(\delta_H \hat{u}_h - \rho(\tau_H \hat{u}_h - g_H)), \end{aligned}\right\} \tag{5}$$

where $\delta_H \hat{u}_h$, $\tau_H \hat{u}_h$ and g_H are appropriate approximations of $\frac{\partial \hat{u}_{h|\omega}}{\partial n_\gamma}$, $\hat{u}_{h|\gamma}$ and g, respectively, in $\Lambda_H(\gamma)$ [3]. The algebraic representation of (5) can be written in the form

$$F(\bar{y}) = 0 \tag{6}$$

with $F : \mathbb{R}^{n+m} \mapsto \mathbb{R}^{n+m}$ defined by

$$F(\bar{y}) := \begin{pmatrix} A\bar{u} - B_\Gamma^\top \bar{\lambda} - \bar{f} \\ G(\bar{u}) \end{pmatrix}, \quad \bar{y} := \begin{pmatrix} \bar{u} \\ \bar{\lambda} \end{pmatrix}, \tag{7}$$

where $G(\bar{u}) := C_\gamma \bar{u} - \max\{0, C_\gamma \bar{u} - \rho(B_\gamma \bar{u} - \bar{g})\}$ and the max-function is understood componentwisely. Here, $A \in \mathbb{R}^{n\times n}$ denotes the standard stiffness matrix, $B_\gamma, B_\Gamma \in \mathbb{R}^{m\times n}$ are the Dirichlet trace matrices related to γ, Γ, respectively, $C_\gamma \in \mathbb{R}^{m\times n}$ is the Neumann trace matrix on γ and $\bar{f} \in \mathbb{R}^n$, $\bar{g} \in \mathbb{R}^m$.

The equation (6) is nonsmooth due to the presence of the max-function. Fortunately, it is semismooth in the sense of [1] so that a semismooth variant of the Newton method can be used.

4 Algorithm

The concept of semismoothness uses slant differentiability of a function. Here, we recall basic results of [1] related to our problem.

Let Y, Z be Banach spaces and $\mathscr{L}(Y,Z)$ denote the set of all bounded linear mappings of Y into Z. Let $U \subseteq Y$ be an open subset and $F : U \mapsto Z$ a function.

Definition 1. (i) The function F is called slantly differentiable at $y \in U$ if there exists a mapping $F^o : U \mapsto \mathscr{L}(Y,Z)$ such that $\{F^o(y+h)\}$ are uniformly bounded for sufficiently small $h \in Y$ and

$$\lim_{h\to 0} \frac{1}{\|h\|} \|F(y+h) - F(y) - F^o(y+h)h\| = 0.$$

The function F^o is called a slanting function for F at y.
(ii) The function F is called slantly differentiable in U if there exists $F^o : U \mapsto \mathscr{L}(Y,Z)$ such that F^o is a slanting function for F at every point $y \in U$. The function F^o is called a slanting function for F in U.

Theorem 1. *Let F be slantly differentiable in U with a slanting function F^o. Suppose that $y^* \in U$ is a solution to the nonlinear equation $F(y) = 0$. If $F^o(y)$ is non-singular for all $y \in U$ and $\{\|F^o(y)^{-1}\| : y \in U\}$ is bounded, then the Newton method*

$$y^{k+1} = y^k - F^o(y^k)^{-1} F(y^k)$$

converges superlinearly to y^*, provided that $\|y^0 - y^*\|$ is sufficiently small.

Let us focus on the max-function $\psi(y) = \max\{0, y\}$ with $Y = Z = \mathbb{R}$. This function is slantly differentiable and

$$\psi^o(y) = \begin{cases} 1, & y > 0, \\ \sigma, & y = 0, \\ 0, & y < 0, \end{cases}$$

is the slanting function in $\mathbb{R}$ for an arbitrary (but fixed) real number σ. Since the convergence rate of the Newton method does not depend on the choice of a slanting function, we shall use $\psi^o(0) = 0$ below.

The function F defined by (7) is slantly differentiable in $\mathbb{R}^{n+m}$ with

$$F^o(\bar{y}) = \begin{pmatrix} A & -B_\Gamma^\top \\ G^o(\bar{u}) & 0 \end{pmatrix},$$

where $G^o(\bar{u}) = (G_1^o(\bar{u}), \dots, G_m^o(\bar{u}))^\top$ with

$$G_i^o(\bar{u}) = C_{\gamma,i} - \psi^o(C_{\gamma,i}\bar{u} - \rho(B_{\gamma,i}\bar{u} - g_i))(C_{\gamma,i} - \rho B_{\gamma,i}), \quad i = 1, \dots, m,$$

and the subscript i denotes the i-th row of the corresponding matrix. A more convenient setting of F^o uses an active set terminology.

Let $\mathscr{M} := \{1, 2, \dots, m\}$. We define the sets of *inactive* and *active* indices at $\bar{y} = (\bar{u}^\top, \bar{\lambda}^\top)^\top \in \mathbb{R}^{n+m}$ by

$$\mathscr{I}(\bar{u}) := \{i \in \mathscr{M} : C_{\gamma,i}\bar{u} - \rho(B_{\gamma,i}\bar{u} - g_i) \le 0\},$$

$$\mathscr{A}(\bar{u}) := \{i \in \mathscr{M} : C_{\gamma,i}\bar{u} - \rho(B_{\gamma,i}\bar{u} - g_i) > 0\}.$$

It is easily seen that

$$G_i^o(\bar{u}) = \begin{cases} C_{\gamma,i}, & i \in \mathscr{I}(\bar{u}), \\ \rho B_{\gamma,i}, & i \in \mathscr{A}(\bar{u}), \end{cases}$$

therefore

$$G^o(\bar{u}) = D(\mathscr{I}(\bar{u}))C_\gamma + \rho D(\mathscr{A}(\bar{u}))B_\gamma,$$

where $D(\mathscr{S})$ denotes the diagonal matrix for $\mathscr{S} \subseteq \mathscr{M}$ defined by

$$D(\mathscr{S}) = \operatorname{diag}(s_1, \dots, s_m) \quad \text{with} \quad s_i = \begin{cases} 1, & i \in \mathscr{S}, \\ 0, & i \notin \mathscr{S}. \end{cases}$$

The Newton method leads to the following active-set type algorithm.

Algorithm ASM (Active-Set Method)

(0) Set $k := 0$ and choose $\rho > 0$, $\varepsilon_u > 0$ ($\varepsilon_u = 10^{-5}$).
Initialize $\bar{u}^0 \in \mathbb{R}^n$ and $\bar{\lambda}^0 \in \mathbb{R}^m$.
(1) Define the inactive and active sets by: $\mathscr{I}^k := \mathscr{I}(\bar{u}^k)$, $\mathscr{A}^k := \mathscr{A}(\bar{u}^k)$.
(2) Solve:

$$\begin{pmatrix} A & -B_\Gamma^\top \\ D(\mathscr{I}^k)C_\gamma + \rho D(\mathscr{A}^k)B_\gamma & 0 \end{pmatrix} \begin{pmatrix} \bar{u}^{k+1} \\ \bar{\lambda}^{k+1} \end{pmatrix} = \begin{pmatrix} \bar{f} \\ \rho D(\mathscr{A}^k)\bar{g} \end{pmatrix}.$$

(3) Set $err(k) := \|\bar{u}^{k+1} - \bar{u}^k\| / \|\bar{u}^{k+1}\|$. If $err(k) \le \varepsilon_u$, return $\bar{u} := \bar{u}^{k+1}$.
(4) Set $k := k + 1$ and go to step (1).

Remark 1. The algorithm has the finite terminating property provided that all possible matrices in the step (2) are non-singular. This follows directly from the fact that the number of the active sets is finite so that the active set corresponding to the solution, and hence the solution itself, must be found in a finite number of the Newton iterations. Unfortunately the non-singularity of the matrices is guaranteed in some special cases only [3]. But in general, this question remains still open.

Remark 2. It is readily seen that ρ can be discarded from the linear systems in the step (2). Indeed, if $k \geq 1$ then ρ does not play any role in the definitions of $\mathscr{I}^k$ and $\mathscr{A}^k$ since always either $C_{\gamma,i}\bar{u}^k = 0$ or $B_{\gamma,i}\bar{u}^k - g_i = 0$. Moreover, an appropriate choice of the initial iterate $\bar{u}^0$ (e.g related to the Dirichlet problem) makes it possible to omit ρ completely from the algorithm.

The finite terminating property mentioned in Remark 1 assumes the exact solution of the linear systems in the step (2). Numerical experiments however show that the inexact implementation is more efficient. In order to maintain the finite terminating property, we drive the precision control in solving inner linear systems adaptively. Our main idea consists in respecting $err(k-1)$ achieved in the previous Newton iteration and, if the progress is not sufficiently large then the precision of the inner loop is increased independently of $err(k-1)$. Denoting $\delta(k)$ the upper bound for the relative residual terminating iterations of the inner solver [4] in the k-th Newton step, we can express our strategy by

$$\delta(k) := \min\{\varepsilon_{min} \times err(k-1), c_{fact} \times \delta(k-1)\}$$

with $0 < \varepsilon_{min} < 1$, $0 < c_{fact} < 1$, $err(0) = 2$ and $\delta(0) = \varepsilon_{min}/c_{fact}$ (typically $\varepsilon_{min} = 10^{-2}$ and $c_{fact} = 0.2$).

5 Numerical Experiments

We illustrate the efficiency of the presented method on the model problem (1), in which $\omega = \{(x,y) \in \mathbb{R}^2 | (x-0.5)^2/0.4^2 + (y-0.5)^2/0.2^2 < 1\}$, $f \equiv -10$ and $g(x,y) = 5\sin(2\varphi)(r^2 + r(\cos\varphi + \sin\varphi) + 0.5)^{1/2} - 1.5$ on γ, where (φ, r) is the polar coordinate of $(x-0.5, y-0.5)$. In the fictitious domain formulation (4) we take $\Omega = (0,1) \times (0,1)$. Moreover we replace $H_0^1(\Omega)$ by $H_{per}^1(\Omega)$ (periodic functions on Ω) that enables us to apply multiplying procedures based on circulant matrices [6]. In the discretized problem (5) we consider V_h formed by piecewise bilinear functions on a uniform rectangulation of Ω with a stepsize h and $\Lambda_H(\gamma)$, $\Lambda_H(\Gamma)$ defined by piecewise constant functions on partitions of polygonal approximations of γ, Γ, respectively. The curve Γ is constructed by shifting γ three h units in the direction of the outward normal vector n_γ and $H/h = 5$; see Figure 1. The definition of $\delta_H \hat{u}_h$ in (5) uses averaging of gradients.

Figure 2 shows the solution $\hat{u}_h$ for $h = 1/256$. In Tables 1, 2 we report the number of primal variables (n), the number of active ($m_{\mathscr{A}} = |\mathscr{A}|$) and inactive ($m_{\mathscr{I}} = |\mathscr{I}|$) control variables, the number of outer (Newton) iterations, the total number of inner (BiCGSTAB) iterations, the computational time and the errors of approximate solutions in the indicated norms (the comparisons are done with respect to the reference solution computed on the fine mesh with $h = 1/2048$). From the errors, we determine the convergence rate of fictitious domain approaches.

Fig. 1 Geometry of the problem.

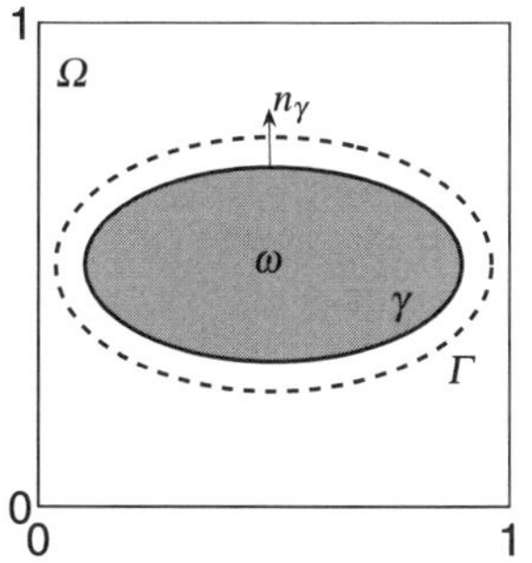

Fig. 2 Solution $\hat{u}_h$ and obstacle g.

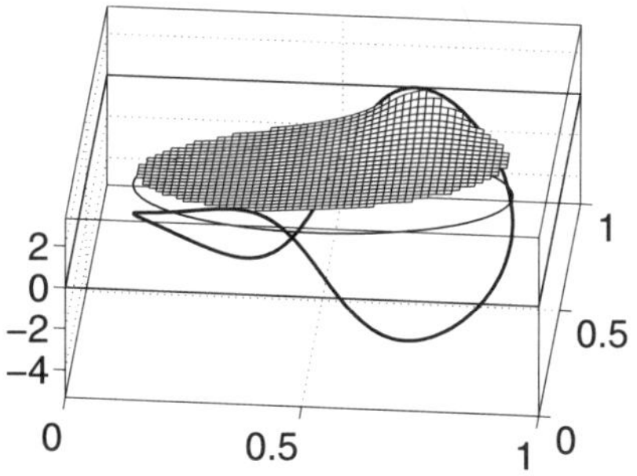

Table 1 Non-smooth fictitious domain formulation ($\gamma \equiv \Gamma$).

Step h	$n/m_{\mathscr{A}}/m_{\mathscr{I}}$	out./$\sum$inn.its.	C.time[s]	$\mathrm{Err}_{L^2(\omega)}$	$\mathrm{Err}_{H^1(\omega)}$	$\mathrm{Err}_{L^2(\gamma)}$
1/128	16641/6/44	11/48	0.81	4.0280e-003	1.7229e-001	1.5350e-002
1/256	66049/13/87	11/62	3.50	2.3784e-003	1.0700e-001	6.3671e-003
1/512	263169/23/175	12/76	31.31	1.9782e-003	1.1129e-001	4.5262e-003
1/1024	1050625/45/351	11/118	185.32	1.0554e-003	8.3205e-002	2.3919e-003
Convergence rates:				0.6063	0.3094	0.8538

Table 2 Smooth fictitious domain formulation ($\gamma \not\equiv \Gamma$).

Step h	$n/m_{\mathscr{A}}/m_{\mathscr{I}}$	out./$\sum$inn.its.	C.time[s]	$\mathrm{Err}_{L^2(\omega)}$	$\mathrm{Err}_{H^1(\omega)}$	$\mathrm{Err}_{L^2(\gamma)}$
1/128	16641/6/44	10/41	0.6875	5.6320e-003	2.6868e-001	2.2502e-002
1/256	66049/13/87	9/73	3.891	1.9606e-003	1.2138e-001	7.3177e-003
1/512	263169/23/175	9/90	34.11	2.8203e-004	2.4997e-002	1.2019e-003
1/1024	1050625/45/351	9/104	161	2.2655e-005	5.5767e-003	1.4466e-004
Convergence rates:				2.6670	1.9051	2.4450

In Table 3 we compare efficiency of the inexact and exact (with $\delta(k) \equiv 10^{-12}$) implementations of Algorithm ASM. When the active and inactive sets corresponding to the solution are recognized then the exact implementation finds immediately the solution. The inexact implementation divides computations of the solution into several Newton iterations, the total cost is however considerably smaller.

The last experiment in Table 4 documents a convergence property of Algorithm ASM. To this end we compute $ratio(k) := \|u^k - \bar{u}\| / \|u^{k-1} - \bar{u}\|$ which tends to zero proving the superlinear convergence. Let us note that the exact implementation is used in this test.

Acknowledgements This research is supported by the grant GAČR 201/07/0294 and by the research project MSM0021620839. The second author also acknowledges the support of the grant IAA1075402.

Table 3 Iteration history for $h = 1/256$ and various implementations.

	Exact		Inexact, $c_{fact} = 0.01$		Inexact, $c_{fact} = 0.5$	
k	inn.its.	$m_{\mathcal{A}^k}/m_{\mathcal{I}^k}$	inn.its.	$m_{\mathcal{A}^k}/m_{\mathcal{I}^k}$	inn.its.	$m_{\mathcal{A}^k}/m_{\mathcal{I}^k}$
0	65	33/67	1	33/67	2	33/67
1	84	26/74	2	25/75	1	21/79
2	70	20/80	13	16/84	5	16/84
3	69	16/84	23	14/86	8	19/81
4	54	14/86	34	13/87	10	16/84
5	51	13/87	13	13/87	9	13/87
6	0	13/87			4	13/87
7					6	13/87
8					6	13/87
9					6	13/87
10					3	13/87
11					6	13/87
12					4	13/87
$\sum$inn.its.	393		86		70	

Table 4 Convergence of Algorithm ASM for $\gamma \equiv \Gamma$ and $h = 1/512$.

k	$err(k)$	$ratio(k)$	$m_{\mathcal{A}^k}/m_{\mathcal{I}^k}$
0	1	-	66/132
1	0.2957	0.2632	49/149
2	0.1083	0.3412	39/159
3	0.0387	0.3387	33/165
4	0.0174	0.4019	29/169
5	0.0068	0.3802	26/172
6	0.0024	0.3321	24/174
7	0.0004	0.1556	23/175
8	0	0	23/175

References

1. Chen, X., Nashed, Z., Qi, L.: Smoothing methods and semismooth methods for nondifferentiable operator equation. SIAM J. Numer. Anal. **38**, 1200–1216 (2000)
2. Glowinski, R., Pan, T., Periaux, J.: A fictitious domain method for Dirichlet problem and applications. Comput. Meth. Appl. Mech. Eng. **111**, 283–303 (1994)
3. Haslinger, J., Kozubek, T., Kučera: Fictitious domain formulation of unilateral problems: analysis and algorithms. In preparing (2008)
4. Haslinger, J., Kozubek, T., Kučera, R., Peichel, G.: Projected Schur complement method for solving non-symmetric systems arising from a smooth fictitious domain approach. Num. Lin. Algebra Appl. **14**, 713–739 (2007)
5. Hintermüller, M., Ito, K., Kunisch, K.: The primal-dual active set strategy as a semismooth Newton method. SIAM J. Optim. **13**, 865–888 (2002)
6. Kučera, R.: Complexity of an algorithm for solving saddle-point systems with singular blocks arising in wavelet-Galerkin discretizations. Appl. Math. **50**(3), 291–308 (2005)

Bifurcations in Contact Problems with Local Coulomb Friction

J. Haslinger, R. Kučera, and O. Vlach

Abstract This contribution illustrates the bifurcation behaviour of solutions to contact problems with local Coulomb friction. The bifurcation character of solutions is well-known for models with a low number of degrees of freedom. Our aim is to show that a similar phenomen occurs when a finite element approximation with a high number of degrees of freedom is used. We experimentally find a critical value of the coefficient of friction in which one branch of solutions splits into two.

1 Introduction

Contact problems with local Coulomb friction belong to challenging mathematical problems which remained unsolved for a long time. Recent results on the existence of solutions to this class of problems can be found in [1]. On the other hand, a complete description of the structure of solutions is still missing in a general case. For discrete problems the situation is slightly better. Systems with a very small number of degrees of freedom can be solved "by hand" so that all solutions are available: see for ex. [5] where the system was parametrized by applied loads P and [4] where the parametrization by a coefficient of friction $\mathscr{F}$ is used. Nevertheless it is not still clear if and how these results can be extended to finite element models with a very high number of dof. which are already close to a continuous model. In this contribution we focus on the parametrization by $\mathscr{F}$. To our knowledge there are only few results valid for any finite number of dof., namely (a) the existence of locally lipschitz continuous branches of solutions (see [4]) (b) the existence of a solution for any coefficient of friction and uniqueness of the solution if $\mathscr{F}$ is below a critical value which (unfortunately) depends on a discretization parameter of

Jaroslav Haslinger
Charles University, Prague, Czech Republic, e-mail: hasling@karlin.mff.cuni.cz

Radek Kučera and Oldřich Vlach
VŠB-TU Ostrava, Czech Republic e-mail: radek.kucera@vsb.cz, oldrich.vlach2@vsb.cz

a finite element model (see [2]). In practice this means that for a given finite element partition one may have a different number of solutions depending on the value of $\mathscr{F}$. The aim of this paper is to document this phenomenon experimentally for "real" discretizations: one branch of the solutions splits into (at least) two ones for $\mathscr{F}$ passing a critical value. As a model of friction we use Coulomb's law with a coefficient which depends on a solution.

2 Setting of the Problem

Let us consider an elastic body represented by a bounded domain $\Omega \subset \mathbb{R}^d$ $(d=2,3)$ with the Lipschitz boundary $\partial\Omega = \overline{\Gamma}_u \cap \overline{\Gamma}_p \cap \overline{\Gamma}_c$ where Γ_u, Γ_p, Γ_c are non–empty, disjoint parts of $\partial\Omega$. On each part different boundary conditions are prescribed: Ω is fixed along Γ_u, while surface tractions of density P act on Γ_p. The body is unilaterally supported by a rigid foundation S along Γ_c. For the sake of simplicity we shall suppose that S is either a half-plane $(d=2)$ or a half-space $(d=3)$ and there is no gap between Ω and S in the undeformed state. Finally, Ω is subject to body forces of density F. Our aim is to find an equilibrium state of Ω taking into account friction between Ω and S which obeys the classical Coulomb law with a coefficient of friction $\mathscr{F}$ *depending* on a solution. An equilibrium state is characterized by a displacement vector $u : \Omega \mapsto \mathbb{R}^d$ which satisfies the equilibrium equations of linear elasticity in Ω, the classical boundary conditions on Γ_u and Γ_p and the following unilateral and friction conditions on Γ_c:

$$T_n := T(u)\cdot n \le 0, \;\; u_n := u\cdot n \le 0, \;\; T_n u_n = 0 \;\text{ on } \Gamma_c \tag{1}$$

$$\left.\begin{array}{l} \|T_t(u)\| \le -\mathscr{F}(\|u_t\|)T_n(u) \;\text{ on } \Gamma_c \\ u_t(x) \ne 0 \Rightarrow T_t(u)(x) = \mathscr{F}(\|u_t\|)T_n(u)\frac{u_t}{\|u_t\|}(x), \;\; x \in \Gamma_c \end{array}\right\} \tag{2}$$

where $T_n(u), T_t(u) := T(u) - T_n(u)n$ is the normal, tangential component of a stress vector $T(u)$, respectively which corresponds to u; u_n, $u_t := u - u_n n$ is the normal, tangential component of a displacement vector u, respectively. The symbol $\|\ \|$ in (2) stands for the absolute value of a scalar $(d=2)$ or the Euclidean norm of a vector $(d=3)$. Finally, $\mathscr{F}$ is a coefficient of friction whose value depends on the magnitude of u_t on Γ_c.

Assuming that Ω is made of a linear elastic material which obeys a linear Hooke law characterized by elasticity coefficients $c_{ijkl} \in L^\infty(\Omega)$, the weak form of our problem is given by the following *implicit* variational inequality:

$$\left.\begin{array}{l} \text{Find } u \in K \text{ such that} \\ a(u, v-u) + j(u,u,v) - j(u,u,u) \ge L(v-u) \quad \forall v \in K \end{array}\right\} \tag{$\mathscr{P}$}$$

The meaning of symbols is as follows (the summation convention is adopted):

$$\mathbb{V} = \{v \in (H^1(\Omega))^d |\ v = 0 \text{ on } \Gamma_u\}$$
$$K = \{v \in \mathbb{V} |\ v_n \le 0 \text{ on } \Gamma_c\}$$
$$a(u,v) := \int_\Omega c_{ijkl}\varepsilon_{kl}(u)\varepsilon_{ij}(v)\,dx\ ,\quad \varepsilon_{kl}(u) = \frac{1}{2}\Big(\frac{\partial u_k}{\partial x_l} + \frac{\partial u_l}{\partial x_k}\Big)$$
$$L(v) := \int_\Omega F_i v_i\,dx + \int_{\Gamma_p} P_i v_i\,ds\ ,\quad F \in (L^2(\Omega))^d,\ P \in (L^2(\Gamma_p))^d$$
$$j(u,v,w) := -\langle \mathscr{F}(||u_t||)T_n(v), ||w_t||\rangle\ ,$$

where $\langle\ ,\ \rangle$ is the duality pairing between X_n (the space of $v_{n|\Gamma_c}$, $v \in \mathbb{V}$) and its dual X_n'. The cone of non–negative elements from X_n' will be denoted by X_{n+}'. Finally, let $X_t^+ = \{\varphi \in L^2(\Gamma_c)\,|\,\exists v \in \mathbb{V}: \varphi = ||v_t|| \text{ on } \Gamma_c\}$.

The existence of solutions to $(\mathscr{P})$ under appropriate assumptions on data, and in particular on $\mathscr{F}$ has been established in [1]. The numerical realization of $(\mathscr{P})$ is based on an equivalent fixed–point formulation. For $(\varphi, g) \in X_t^+ \times X_{n+}'$ fixed let us consider the following contact problem with given friction and the coefficient $\mathscr{F}_\varphi := \mathscr{F}(\varphi)$:

$$\left.\begin{array}{l}\text{Find } u := u(\varphi,g) \in K \text{ such that}\\ a(u,v-u) + j(\varphi,g,v) - j(\varphi,g,u) \ge L(v-u) \quad \forall v \in K\end{array}\right\} \qquad (\mathscr{P}(\varphi,g))$$

and define the mapping $\Phi : X_t^+ \times X_{n+}' \mapsto X_t^+ \times X_{n+}'$ by

$$\Phi(\varphi,g) = (||u_{t|\Gamma_c}||, -T_n(u)) \tag{3}$$

where $u \in K$ is the unique solution of $(\mathscr{P}(\varphi,g))$. Comparing the definitions of $(\mathscr{P})$ and $(\mathscr{P}(\varphi,g))$ we see that $u \in K$ solves problem $(\mathscr{P})$ if and only if it solves problem $\mathscr{P}(||u_{t|\Gamma_c}||, -T_n(u))$ or equivalently, $(||u_{t|\Gamma_c}||, -T_n(u))$ is a fixed point of Φ.

3 Discretization of $(\mathscr{P})$, Properties of the Discrete Model

Let Ω be a polygonal $(d = 2)$ or a polyhedral $(d = 3)$ domain and $\mathscr{T}_h$ be a partition of $\overline{\Omega}$ into triangles $(d = 2)$ or tetrahedra $(d = 3)$ such that $diam\,T \le h\ \forall T \in \mathscr{T}_h$. With any $\mathscr{T}_h$ we associate the spaces V_h, $\mathbb{V}_h$:

$$V_h = \{v_h \in C(\overline{\Omega})|\ v_{h|T} \in P_1(T)\ \forall T \in \mathscr{T}_h,\ v_h = 0 \text{ on } \Gamma_u\}\ ,\quad \mathbb{V}_h = (V_h)^d\ .$$

By $\mathscr{V}_h = V_{h|\Gamma_c}$ we denote the space of restrictions on Γ_c of functions from V_h while $\mathscr{V}_h^+$ stands for the set of non–negative elements of $\mathscr{V}_h$. Further, let $\mathscr{T}_H$ be a partition of $\overline{\Gamma}_c$ into segments S_H, $diam\,S_H \le H\ \forall S_H \in \mathscr{T}_H$. On any $\mathscr{T}_H$ we construct the space L_H of piecewise constant functions:

$$L_H = \{\mu_H \in L^2(\Gamma_c)|\quad \mu_{H|S_H} \in P_0(S_H)\ \forall S_H \in \mathscr{T}_H\}$$

and its subset Λ_H of all non–negative functions. For any $(\varphi_h, g_H) \in \mathscr{V}_h^+ \times \Lambda_H$ given we define the following auxiliary problem:

$$\left.\begin{array}{l}\text{Find } (u_h,\lambda_H) \in \mathbb{V}_h \times \Lambda_H \text{ such that} \\ a(u_h, v_h - u_h) + j(\varphi_h, g_H, v_h) - j(\varphi_h, g_H, u_h) \geq \\ \qquad L(v_h - u_h) - (\lambda_H, v_{hn} - u_{hn})_{0,\Gamma_c} \quad \forall v_h \in \mathbb{V}_h \\ (\mu_H - \lambda_H, u_{hn})_{0,\Gamma_c} \leq 0 \quad \forall \mu_H \in \Lambda_H \end{array}\right\} \qquad (\mathscr{P}(\varphi_h, g_H))_h^H$$

$(\mathscr{P}(\varphi_h, g_H))_h^H$ is a mixed formulation of the contact problem with given friction and the coefficient $\mathscr{F}_{\varphi_h} := \mathscr{F} \circ \varphi_h$ which uses the dualization of the unilateral constraint $u_{hn} \leq 0$ on Γ_c. Next we shall suppose that $\mathbb{V}_h$ and Λ_H are such that the following condition guaranteeing the uniqueness of a solution to $(\mathscr{P}(\varphi_h, g_H))_h^H$ is satisfied:

$$(\mu_H, v_{hn})_{0,\Gamma_c} = 0 \;\; \forall v_h \in \mathscr{V}_h \quad \Rightarrow \quad \mu_H = 0\,. \tag{4}$$

This enables us to define the mapping $\Phi_{hH} : \mathscr{V}_h^+ \times \Lambda_H \mapsto \mathscr{V}_h^+ \times \Lambda_H$ by

$$\Phi_{hH}(\varphi_h, g_H) = (r_h||u_{ht|_{\Gamma_c}}||, \lambda_H)\,,$$

where (u_h, λ_H) is the solution of $(\mathscr{P}(\varphi_h, g_H))_h^H$ and $r_h : C(\overline{\Gamma}_c) \mapsto \mathscr{V}_h$ is a linear approximation operator preserving the monotonicity property: $v \geq 0$ on $\overline{\Gamma}_c \Rightarrow r_h v \in \mathscr{V}_h^+$ (the Lagrange interpolation operator, e.g.). Since $-\lambda_H$ can be interpreted as the discrete normal stress on Γ_c, the mapping Φ_{hH} can be viewed to be a discretization of Φ defined by (3).

Definition 1. By a discrete solution of the contact problem with Coulomb friction and the coefficient depending on a solution we call any function $u_h \in \mathbb{V}_h$ such that (u_h, λ_H) is a solution of $(\mathscr{P}(r_h||u_{ht|_{\Gamma_c}}||, \lambda_H))_h^H$, i.e. $(r_h||u_{ht|_{\Gamma_c}}||, \lambda_H)$ is a fixed point of Φ_{hH}.

Let us recall main results concerning the existence and uniqueness of the fixed point of Φ_{hH}. Proofs for 2D problems can be found in [3] but their adaptation to the 3D case is easy.

Theorem 1. *It holds:*

(a) if $\mathscr{F} \in C(\mathbb{R}_+^1)$, $0 \leq \mathscr{F}(t) \leq \mathscr{F}_{max}$ $\forall t \in \mathbb{R}_+^1$, where $\mathscr{F}_{max}$ is given then there exists at least one fixed point of Φ_{hH};
(b) if, in addition to (a), $\mathscr{F}$ is Lipschitz continuous in $\mathbb{R}_+^1$:

$$|\mathscr{F}(t_1) - \mathscr{F}(t_2)| \leq l|t_1 - t_2| \quad \forall t_1, t_2 \in \mathbb{R}_+^1$$

so Φ_{hH} is in $\mathscr{V}_h^+ \times \Lambda_H$: $\exists q > 0$ such that

$$||\Phi_{hH}(\varphi_h, g_H) - \Phi_{hH}(\overline{\varphi}_h, \overline{g}_H)|| \leq q||(\varphi_h, g_H) - (\overline{\varphi}_h, \overline{g}_H)|| \tag{5}$$

holds for every $(\varphi_h, g_H), (\overline{\varphi}_h, \overline{g}_H) \in \mathscr{V}_h^+ \times \Lambda_H$, where

$$||(\varphi_h, g_H)|| := ||\varphi_h||_{0,\Gamma_c} + ||g_H||_h, \quad ||g_H||_h := \sup_{\mathbb{V}_h} \frac{(g_h, v_{hn})_{0,\Gamma_c}}{||v_h||_{1,\Omega}}.$$

The constant q in (5) *depends on* $\Omega, h, H, \mathscr{F}_{max}$ *and* l *in such a way that for* Ω, h, H *fixed,* $q \to 0+$ *if* $\mathscr{F}_{max}, l \to 0+$.

Remark 1. There exist $\overline{\mathscr{F}} > 0, \bar{l} > 0$ both depending on Ω, h and H such that if $\mathscr{F}_{max} \leq \overline{\mathscr{F}}$ and $l \leq \bar{l}$ the mapping Φ_{hH} is contractive in $\mathscr{V}_h^+ \times \Lambda_H$ so that Φ_{hH} has a unique fixed point and the method of successive approximations converges.

Remark 2. If the following Babuška–Brezzi condition and the inverse inequality are satisfied, i.e.

$$||\mu_H||_h \geq \beta ||\mu_H||_{X_n'}, \quad ||\mu_H||_{0,\Gamma_c} \leq \overline{\beta} H^{-1/2} ||\mu_H||_{X_n'}, \quad \forall \mu_H \in L_H$$

where $\beta, \overline{\beta} > 0$ do not depend on $h, H > 0$ then the bounds $\overline{\mathscr{F}}, \bar{l}$ guaranteeing the uniqueness of the solution are bounded from above by $\sqrt{hH}$, i.e. are mesh–dependent ([3]).

Let us comment on the previous results. Unlike to the continuous setting in which the existence of a solution has been shown for $\mathscr{F}$ small enough, a solution to the discrete model exists for any $\mathscr{F}$ satisfying (a) of Theorem 1 regardless of the shape of $\Omega, \mathscr{F}_{max}, l$ and the applied forces F and P. Moreover, if $\mathscr{F}_{max}$ and l are small enough, the solution to the discrete model is unique. Unfortunately, this uniqueness result depends on the mesh norms h, H as follows from Remark 2. One of ways how a possible non–uniqueness comes to light is that the method of successive approximations used for finding fixed points of Φ_{hH} depends on the choice of initial approximations. In the next section we illustrate this phenomenon on model examples in 2D and 3D: starting from two different initial approximations we find two different fixed points for a particular coefficient of friction $\mathscr{F}$. Then taking the same examples (with the same $\mathscr{T}_h$ and $\mathscr{T}_H$) but replacing $\mathscr{F}$ by $\xi\mathscr{F}$, where $\xi \to 0+$ we find (accordingly to our theoretical results) a critical value $\overline{\xi} > 0$ for which originally two different fixed points will coincide for $\xi < \overline{\xi}$ using the same initial approximations as before.

4 Examples with Branching Solutions

We start with a 2D problem. The body represented by $\Omega = (0,10) \times (0,1)\,[m]$ is made of an elastic material characterized by the Young modulus $E = 21.19e10\,[Pa]$ and Poisson's ratio $\sigma = 0.277$. The partition of $\partial\Omega$ into Γ_u, $\Gamma_p = \Gamma_{p1} \cup \Gamma_{p2}$ and Γ_c is seen from Fig. 1. The surface tractions P are linearly distributed along Γ_{p1} and Γ_{p2} starting from the following values: $P_{|\Gamma_{p1}}(0,1) = (0, 1.e6)\,[N]$, $P_{|\Gamma_{p1}}(10,1) = (0, -8.e6)\,[N]$, $P_{|\Gamma_{p2}}(10,1) = (-10.e6, 10.e6)\,[N]$ and $P_{|\Gamma_{p2}}(10,0) = (-10.e6, -3.e6)\,[N]$. The body forces are neglected. The graph of the coefficient of friction $\mathscr{F}$ is shown in Fig. 2.

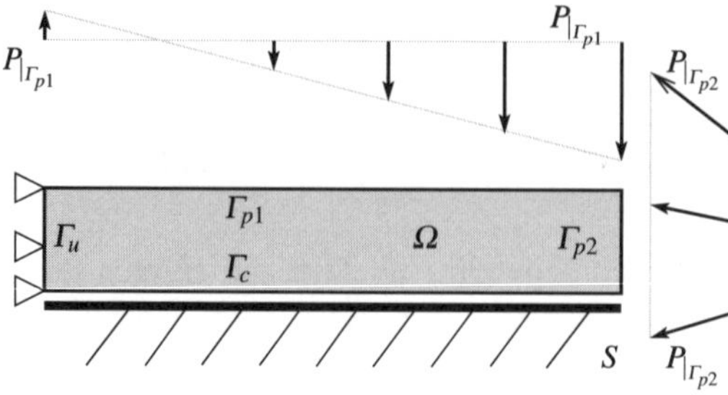

Fig. 1 Geometry of the problem

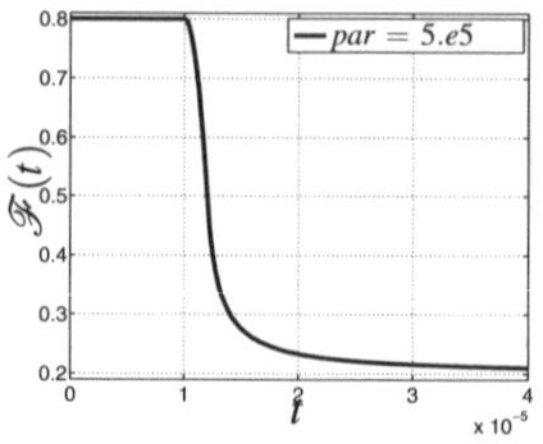

Fig. 2 The graph of $\mathscr{F}$

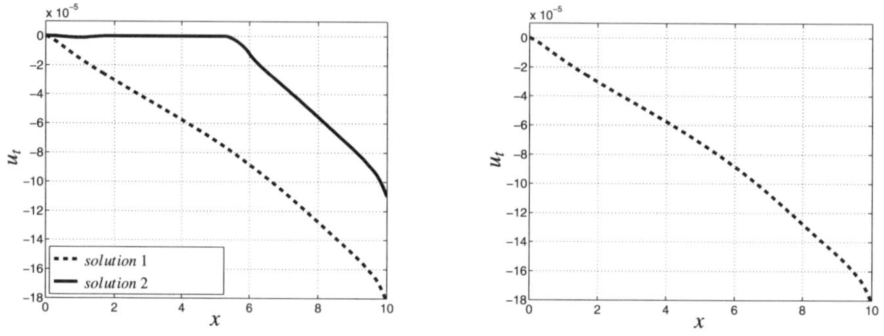

a) $\xi = 0.98581$ *b)* $\xi = 0.9858$

Fig. 3 Tangential displacements on Γ_c

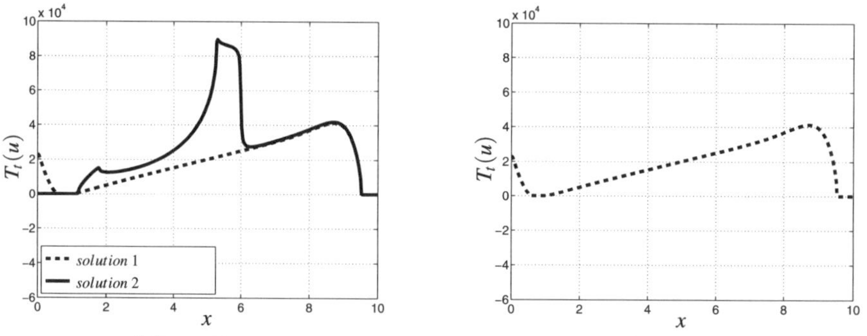

a) $\xi = 0.98581$ *b)* $\xi = 0.9858$

Fig. 4 Friction force $T_t(u)$ on Γ_c

The numerical realization of each iterative step of the method of successive approximations is based on its dual formulation (for details see [3]). The used partitions of $\overline{\Omega}$ and $\overline{\Gamma}_c$ give 26 640 primal variables and 720 dual variables (discrete contact stresses). Two different initial approximations were used, namely $(\varphi_h^{(0)}, g_H^{(0)}) = (0,0)$ corresponding to a contact problem without friction and $(\overline{\varphi}_h^{(0)}, \overline{g}_H^{(0)}) = (0, 1.e8)$

(a contact problem with a high slip bound). Starting from them, two different fixed points of Φ_{hH} were obtained. Since the most significant differences are in the tangential direction we focus on it. One of these solutions is such that a slip occurs along the whole Γ_c ($solution1$) while both stick and slip zone are present for the second one ($solution2$). Now instead of $\mathscr{F}$ we take $\xi\mathscr{F}$. If $\xi = 0.98581$ then again two solutions with the same character as for $\xi = 1$ appear. On the other hand if $\xi = 0.9858$ both solutions meet together and only one solution with a slip along the whole Γ_c is obtained (see Figs. 3 and 4).

Now we switch to 3D problems. Let the body be represented by $\Omega = (0,10) \times (0,1) \times (0,1)\,[m]$. The decomposition of the boundary $\partial\Omega$ into Γ_u, Γ_p and Γ_c, as well as the applied surface tractions P are seen from Fig. 5. The Young modulus E, Poisson's ratio σ and the coefficient of friction $\mathscr{F}$ are the same as in the 2D case. The body forces are neglected again. Discretizations of $\overline{\Omega}$ and $\overline{\Gamma}_c$ are such that the total number of the primal, dual variables is 30000 and 12700, respectively. The initial approximations for the method of successive approximations are the same as before. Denote $\mathscr{F}_\xi := \xi\mathscr{F}$. For $\xi = 1.37689$ we get two different solutions: the one sliding along the whole Γ_c, the other one with a stick and slip zone as shown in Fig. 6. The norm of $T_t(u)$ on Γ_c is depicted in Fig. 7 and the distribution of $T_n(u)$ and u_n on Γ_c are shown in Figs. 8 and 9 for ($solution$ 2). Setting $\xi = 1.37688$ both solutions joint together. The obtained solution slides along the whole Γ_c, i.e. it has the character of ($solution$ 1).

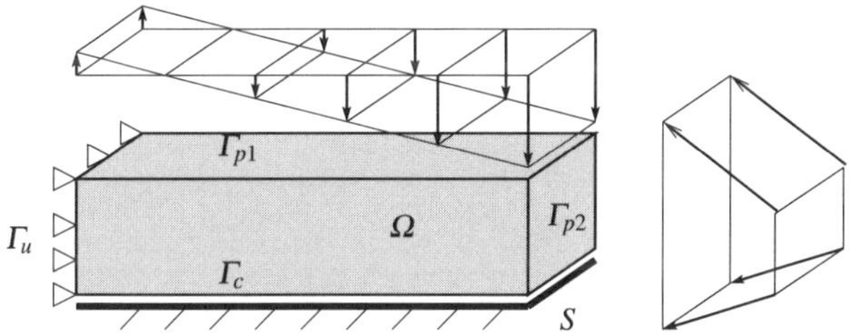

Fig. 5 Geometry of the problem

Acknowledgements This research was supported by the grants GAČR 201/07/0294 and MSM0021620839.

References

1. Eck, C., Jarušek, J., Krbec, M.: Unilateral contact problems. Variational methods and existence theorems, *Pure and Applied Mathematics (Boca Raton)*, vol. 270. Chapman & Hall/CRC, Boca Raton, FL (2005)
2. Haslinger, J.: Approximation of the Signorini problem with friction, obeying the Coulomb law. Math. Methods Appl. Sci. **5**(3), 422–437 (1983)

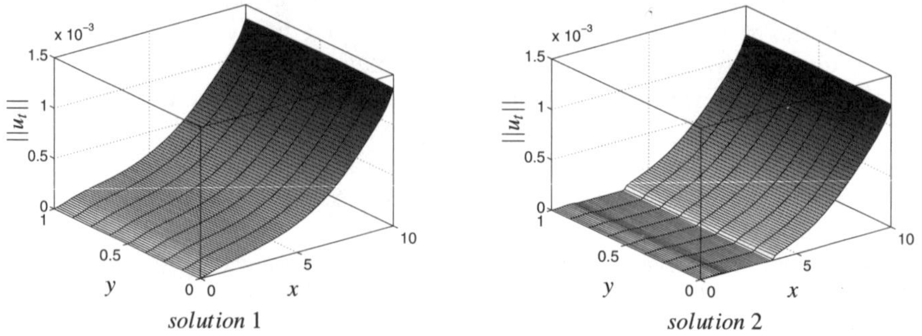

Fig. 6 The norm of the tangential displacements for $\xi = 1.37689$

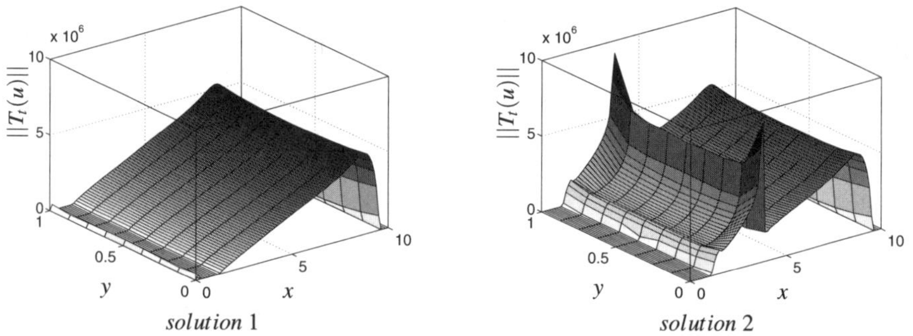

Fig. 7 The norm of the tangential stresses for $\xi = 1.37689$

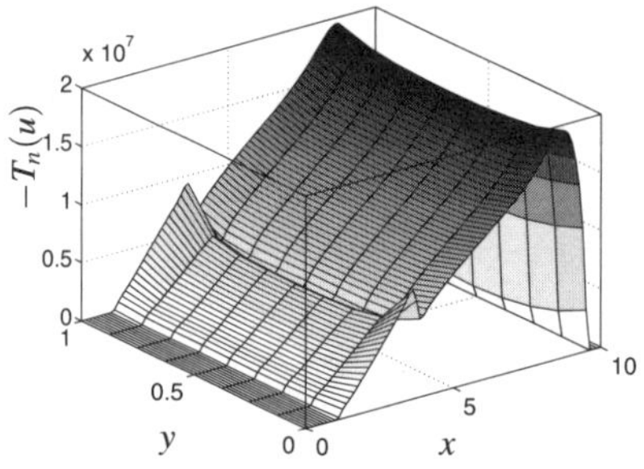

Fig. 8 The normal stresses for $\xi = 1.37689$

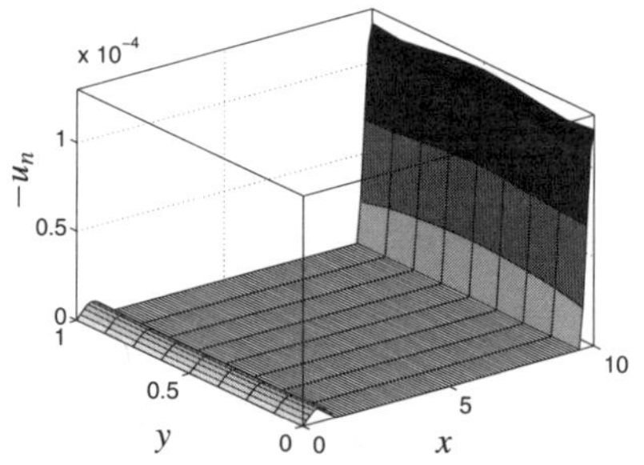

Fig. 9 The normal displacement for $\xi = 1.37689$

3. Haslinger, J., Vlach, O.: Approximation and numerical realization of 2D contact problems with Coulomb friction and a solution–dependent coefficient of friction. J. Comput. Appl. Math. **197**(2), 421–436 (2006)
4. Hild, P., Renard, Y.: Local uniqueness and continuation of solutions for the discrete Coulomb friction problem in elastostatics. Quart. Appl. Math. **63**(3), 553–573 (2005)
5. Janovský, J.: Catastrophic features of Coulomb friction model. In: Proceedings of the MAFELAP IV (J. Whiteman, ed.), pp. 259–264 (1982)

Point Load on a Shell

A.H. Niemi, H. Hakula, and J. Pitkäranta

Abstract We study the fundamental (normal point load) solution for shallow shells. The solution is expressed as a Fourier series and its properties are analyzed both at the asymptotic limit of zero shell thickness and when the thickness has a small positive value. Some results of benchmark computations using both high- and low-order finite elements are also presented.

1 Introduction

According to the two-dimensional models of linear shell theory, the deformation of the middle surface of a thin shell under a given load is obtained by minimizing a quadratic energy functional of the form

$$\mathscr{F}(\mathbf{u}) = \mathscr{A}_m(\mathbf{u},\mathbf{u}) + \mathscr{A}_s(\mathbf{u},\mathbf{u}) + t^2\mathscr{A}_b(\mathbf{u},\mathbf{u}) - 2\mathscr{Q}(\mathbf{u}), \tag{1}$$

where t is the thickness of the shell and $\mathscr{A}_m$, $\mathscr{A}_s$, $t^2\mathscr{A}_b$ and $\mathscr{Q}$ correspond to the deformation energy due to stretching, deformation energy due to transverse shearing, deformation energy due to bending and the external load functional, respectively. Further, $\mathbf{u} = (u, v, w, \theta, \psi)$ is a vector field on the middle surface Γ of the shell that defines the tangential displacements u, v and normal deflection w of the middle surface as well as the rotations θ, ψ of its normal.

We consider here the problem of shell deformation under a normal point load so that the load functional is assumed to have the form

$$\mathscr{Q}(\mathbf{u}) = F\langle\delta_P, w\rangle = Fw(P), \tag{2}$$

Antti H. Niemi, Harri Hakula, and Juhani Pitkäranta
Institute of Mathematics, Helsinki University of Technology, P.O.Box 1100, Espoo FI-02015 TKK, Finland, e-mail: antti.h.niemi@tkk.fi, harri.hakula@tkk.fi and juhani.pitkaranta@tkk.fi

where $P \in \Gamma$ is a point at the middle surface. The relevance of this problem is undeniable, as the solution is the fundamental solution, or *Green's function*, for normal loads and so it has been studied widely in classical shell theory, see e.g. [1] and the references therein. Anyway, it seems that closed form solutions have been obtained only for spherical shells and that the detailed behavior of the solution near point P is still an open problem when the thickness t is small — especially in hyperbolic and parabolic shell geometries.

Our aim here is to give some solutions to this problem and to find out how accurately these solutions can be approximated with finite elements. Our starting point is a 'shallow' version of the classical shell model where certain geometrical simplifications are assumed, see [6]. Within this simplified model, we analyze fundamental solutions that can be expressed as Fourier series and focus first on the asymptotic limit solution at $t = 0$. In model cases this can be expressed explicitly in the sense of distributions. We conclude that the transverse deflection w of the asymptotic solution has a term of the form $w \sim F\delta_P$ in all geometries. The remaining part of w is smooth when the shell is elliptic, but in hyperbolic and parabolic shell geometries there arises additional line δ-distributions along the characteristic lines of the middle surface.

Concerning the more realistic situation where the thickness t has a small positive value, we conclude as follows:

1. In all shell geometries the asymptotic term $w \sim F\delta_P$ is spread into a 'hot spot' of width $\sim \sqrt{Rt}$ around P, where R is the curvature length scale of the shell.
2. The line δ-distributions in the hyperbolic and parabolic cases are spread to 'ridges' of width $\sim \sqrt[n]{R^{n-1}t}$, where $n = 3$ in the hyperbolic case and $n = 4$ in the parabolic case.

We support these conclusions also by numerical experiments based on truncated Fourier series and finite element computations using both high- and low-order elements.

2 Classical Shell Theory

For a shell consisting of homogeneous isotropic material with Poisson ratio ν, the energy functionals in (1) are given by

$$\begin{aligned}
\mathscr{A}_m(\mathbf{u},\mathbf{u}) &= \int_\Gamma \left[\nu(\beta_{11}+\beta_{22})^2 + (1-\nu)(\beta_{11}^2 + 2\beta_{12}^2 + \beta_{22}^2)\right] \,\mathrm{d}\Gamma, \\
\mathscr{A}_s(\mathbf{u},\mathbf{u}) &= \frac{1-\nu}{2}\int_\Gamma \left(\rho_1^2 + \rho_2^2\right) \,\mathrm{d}\Gamma, \\
\mathscr{A}_b(\mathbf{u},\mathbf{u}) &= \frac{1}{12}\int_\Gamma \left[\nu(\kappa_{11}+\kappa_{22})^2 + (1-\nu)(\kappa_{11}^2 + 2\kappa_{12}^2 + \kappa_{22}^2)\right] \,\mathrm{d}\Gamma.
\end{aligned} \tag{3}$$

In a general geometry, the strains β_{ij}, ρ_i and κ_{ij}, $i,j=1,2$ are related linearly to the displacement components u,v,w,θ,ψ via variable coefficients that depend locally on the fundamental forms of Γ, see e.g. [1].

We assume here that the middle surface of the shell, in the neighborhood of point P, is represented in the form

$$z(x,y)=\frac{1}{2}ax^2+cxy+\frac{1}{2}by^2, \tag{4}$$

where x and y are Cartesian coordinates in the tangent plane Ω with origin P. The leading terms of the strain expressions at P may then be written as

$$\begin{aligned}
&\beta_{11}=\frac{\partial u}{\partial x}+aw, \quad \beta_{22}=\frac{\partial v}{\partial y}+bw, \quad \beta_{12}=\frac{1}{2}\left(\frac{\partial u}{\partial y}+\frac{\partial v}{\partial x}\right)+cw,\\
&\rho_1=\theta-\frac{\partial w}{\partial x}, \quad \rho_2=\psi-\frac{\partial w}{\partial y},\\
&\kappa_{11}=\frac{\partial \theta}{\partial x}, \quad \kappa_{22}=\frac{\partial \psi}{\partial y}, \quad \kappa_{12}=\frac{1}{2}\left(\frac{\partial \theta}{\partial y}+\frac{\partial \psi}{\partial x}\right).
\end{aligned} \tag{5}$$

The use of these expressions may be justified (formally, see [6]) also in a neighborhood of P, in which the middle surface Γ is shallow with respect to the tangent plane Ω, i.e. $r=\sqrt{x^2+y^2}$ is small compared with R. One may then as well set $\Gamma \hookrightarrow \Omega$ and $d\Gamma \hookrightarrow dxdy$ when evaluating the strain energy functionals (3).

The above model can be simplified further by neglecting transverse shear energy which usually is small. This can be accomplished by eliminating the rotations θ,ψ from the classical *Kirchhoff-Love constraints*

$$\rho_1=\theta-\frac{\partial w}{\partial x}=0, \quad \rho_2=\psi-\frac{\partial w}{\partial y}=0. \tag{6}$$

The strain energy takes then the form

$$\mathscr{F}(\mathbf{u})=\mathscr{A}_m(\mathbf{u},\mathbf{u})+t^2\mathscr{A}_b(\mathbf{u},\mathbf{u})-2\mathscr{Q}(\mathbf{u}), \tag{7}$$

where now $\mathbf{u}=(u,v,w)$ and the bending strains are given by

$$\kappa_{11}=\frac{\partial^2 w}{\partial x^2}, \quad \kappa_{22}=\frac{\partial^2 w}{\partial y^2}, \quad \kappa_{12}=\frac{\partial^2 w}{\partial x \partial y}.$$

The minimizer of (7) satisfies the Euler equations

$$0=-\frac{\partial\beta_{11}}{\partial x}-\nu\frac{\partial\beta_{22}}{\partial x}-(1-\nu)\frac{\partial\beta_{12}}{\partial y},$$

$$0=-\nu\frac{\partial\beta_{11}}{\partial y}-\frac{\partial\beta_{22}}{\partial y}-(1-\nu)\frac{\partial\beta_{12}}{\partial x}, \tag{8}$$

$$F\delta_P=(a+\nu b)\beta_{11}+(\nu a+b)\beta_{22}+2(1-\nu)c\beta_{12}+\frac{t^2}{12}\Delta^2 w,$$

where $\Delta=\frac{\partial^2}{\partial x^2}+\frac{\partial^2}{\partial y^2}$ is the usual two-dimensional Laplacian. These equations together constitute a system of total order eight and they can be expressed equivalently as the *fundamental shell equation*

$$\frac{t^2}{12}\Delta^4 w+(1-\nu^2)\Delta_m^2 w=F\Delta^2\delta_P, \tag{9}$$

where Δ_m is a second order partial differential operator which represents membrane forces and is defined as

$$\Delta_m=a\frac{\partial^2}{\partial y^2}+b\frac{\partial^2}{\partial x^2}-2c\frac{\partial^2}{\partial x\partial y}.$$

In view of (4) and the usual classification of differential operators, the operator Δ_m is called elliptic/hyperbolic/parabolic in accordance with the geometric nature of the middle surface at P. Note also that when $a=b=c=0$, Eq. (9) reduces to the well known biharmonic equation representing the bending of a flat plate under a concentrated load.

To get an understanding of the curvature effects that couple membrane and bending action in shell deformations, we analyze solutions of (9) that can be expanded as Fourier series of the form

$$w(x,y)=\sum_{m,n=1}^{\infty}W_{mn}\cos\left((m-\tfrac{1}{2})\pi x\right)\cos\left((n-\tfrac{1}{2})\pi y\right). \tag{10}$$

Actually, this form was used already in [5], where we introduced a set of benchmark problems for the numerical evaluation of finite element algorithms.

Assume now that $P=(0,0)$ so that

$$F\delta_P(x,y)=F\sum_{m,n=1}^{\infty}\cos\left((m-\tfrac{1}{2})\pi x\right)\cos\left((n-\tfrac{1}{2})\pi y\right).$$

By using the shorthand notation $M=(m-\frac{1}{2})\pi$ and $N=(n-\frac{1}{2})\pi$, we may write formally $\Delta=-M^2-N^2$ and $\Delta_m=-aN^2-bM^2-2cMN$ in (9); hence, the Fourier coefficients of the transverse deflection are given by

$$W_{mn}=\frac{12F(M^2+N^2)^2}{t^2(M^2+N^2)^4+12(1-\nu^2)(aN^2+bM^2+2cMN)^2}. \tag{11}$$

We consider three model cases where the curvature parameters a, b, c are chosen as follows

1. $a = b = 1/R$ & $c = 0$ (Elliptic shell)
2. $c = 1/R$ & $a = b = 0$ (Hyperbolic shell, characteristic lines $x = 0$ & $y = 0$)
3. $b = 1/R$ & $a = c = 0$ (Parabolic shell, characteristic line $x = 0$)

Let us study first the asymptotic limit solutions at $t = 0$. In the classical shell-membrane theory one usually sets $t = 0$ in (8) and then solves the system of equations by carefully relaxing the boundary and regularity conditions on $\mathbf{u}$. Here we do nicely by expanding the Fourier coefficients (11) at $t = 0$ as follows:

1. $W_{mn} = \frac{FR^2}{1-\nu^2}$
2. $W_{mn} = \frac{FR^2}{1-\nu^2}\left(\frac{1}{2} + \frac{1}{4}\frac{M^2}{N^2} + \frac{1}{4}\frac{N^2}{M^2}\right)$
3. $W_{mn} = \frac{FR^2}{1-\nu^2}\left(1 + 2\frac{N^2}{M^2} + \frac{N^4}{M^4}\right)$

In $\Omega = (-1,1) \times (-1,1)$ these correspond to explicit solutions of the form

1. $w(x,y) = \frac{FR^2}{1-\nu^2}\delta_P(x,y)$
2. $w(x,y) = \frac{FR^2}{1-\nu^2}\left(\frac{1}{2}\delta_P(x,y) + \frac{1}{16}(|y|-1)\delta''(y) + \frac{1}{16}(|x|-1)\delta''(x)\right)$
3. $w(x,y) = \frac{FR^2}{1-\nu^2}\left(\delta_P(x,y) + \frac{1}{2}(|x|-1)\delta''(y) + \frac{1}{12}(|x|^3 - 3x^2 + 2)\delta''''(y)\right)$

Assume next that $t = \frac{1}{1000}$. We show in Figs. 1 and 2 contour plots of the deflection w in the hyperbolic and parabolic cases with the parameter values set as $R = 1$, $\nu = \frac{1}{3}$ and $F = -1$. These results have been obtained by truncating the Fourier series (10), (11) at $m = n = 1000$. We observe that in different shell geometries the main features of the deformations are rather similar close to P, but highly different away from P.

This behavior can be anticipated also from the Fourier coefficients (11). Namely, the curvature effects do not interact significantly with Fourier modes that vary in length scales $\ll \sqrt{Rt} \simeq 0.03$, but come into play when $M^2 + N^2 \sim \frac{1}{Rt}$ basically in the same way in any shell geometry. Concerning longer length scales, i.e. Fourier modes with $M, N < \frac{1}{\sqrt{Rt}}$, we may reason as follows. In the hyperbolic case one finds that

$$W_{mn} \sim \frac{12FM^2}{t^2M^6 + 48(1-\nu^2)R^{-4}},$$

when $N \sim R^{-1}$ so that W_{mn} grows with M until $M \sim \sqrt[3]{R^{-2}t^{-1}}$ and the same holds when the roles of M and N are exchanged. These properties are reflected in Fig. 1 as line layers decaying in the length scale $L \sim \sqrt[3]{R^2t} \simeq 0.10$ from the characteristic lines. In the parabolic case we have

$$W_{mn} \sim \frac{12FN^4}{t^2N^8 + 12(1-\nu^2)R^{-6}}$$

when $M \sim R^{-1}$ so that here W_{mn} grows with N until $N \sim \sqrt[4]{R^{-3}t^{-1}}$ in accordance with the line layer decaying in the length scale $L \sim \sqrt[4]{R^3t} \simeq 0.18$ in Fig. 2.

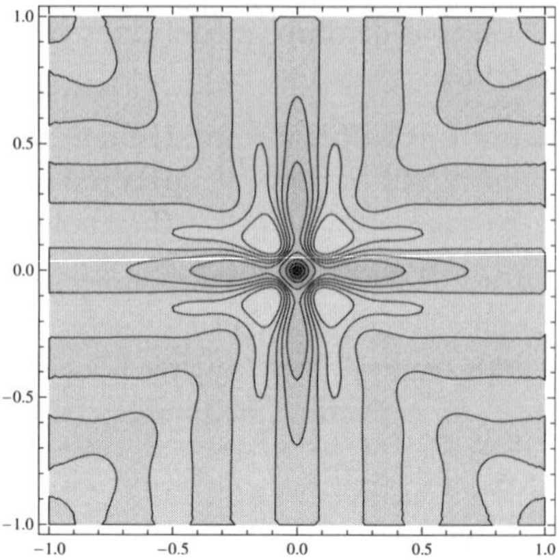

Fig. 1 Hyperbolic shell: Transverse deflection w at $t = \frac{1}{1000}$ due to the point load at $P = (0,0)$. Largest displacements take place within a 'hot spot' of width $\sim \sqrt{Rt} \simeq 0.03$ around P. In addition, the deformation features 'ridges' of width $\sim \sqrt[3]{R^2 t} \simeq 0.10$ along the characteristic lines $x = 0$ and $y = 0$.

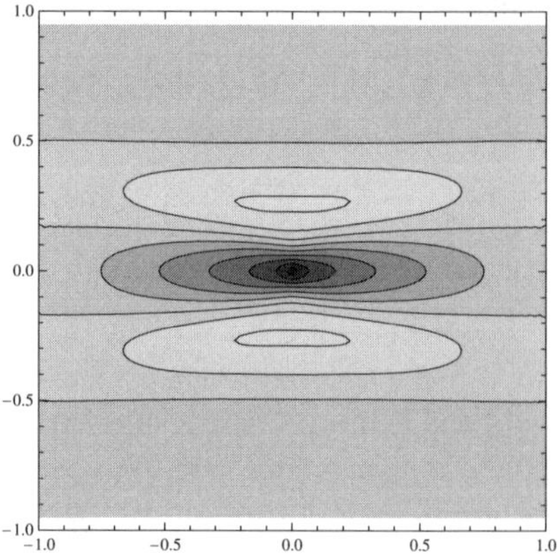

Fig. 2 Parabolic shell: Transverse deflection w at $t = \frac{1}{1000}$ due to the point load at $P = (0,0)$. Largest displacements take place within a 'hot spot' of width $\sim \sqrt{Rt} \simeq 0.03$ around P. In addition, the deformation features a 'ridge' of width $\sim \sqrt[4]{R^3 t} \simeq 0.18$ along the characteristic line $x = 0$.

3 Benchmark Computations: h–FEM *versus* p–FEM

In this section, we construct finite element approximations to the fundamental solutions in the hyperbolic and parabolic cases. Since the imposition of the Kirchhoff-Love constraints (6) in a finite element space is rather complicated, we take the 5-field model (1)–(5) as our starting point here. We approximate each displacement component separately in the same way by using a standard scalar finite element space $V_{h,p} \subset H^1(\Omega)$ associated to subdivision of $\Omega = (-1,1) \times (-1,1)$ into rectangular elements with side length at most h and shape functions spanning all polynomials of given degree p, $p \geq 1$. On the boundary $\partial\Omega$, we impose as kinematic constraints the symmetry/antisymmetry conditions corresponding to (10), cf. [5].

We start by setting up two rectangular 'macroelement' meshes on Ω based on the specific structure of the solution in hyperbolic and parabolic cases, see Fig. 3. Our goal is to find out which is more efficient way to increase the accuracy of the approximation: raising the polynomial degree p within each macroelement or decreasing h in the lowest-order case ($p = 1$) by refining the mesh.

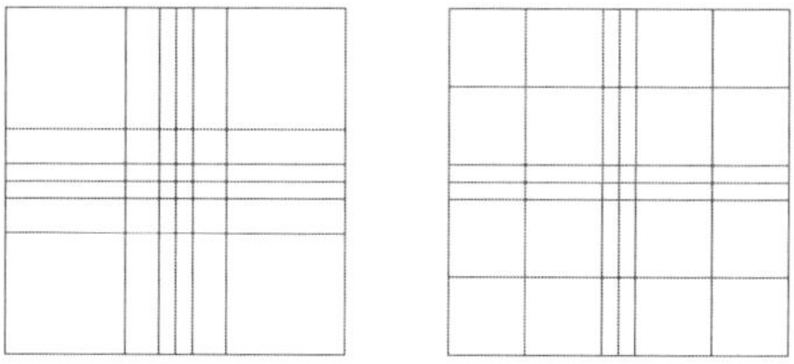

Fig. 3 Macroelement meshes for hyperbolic and parabolic shells at $t = \frac{1}{1000}$.

In the latter approach involving *bilinear* shape functions, we modify the transverse shear strains as

$$\rho_1 \hookrightarrow \Pi_x \rho_1, \quad \rho_2 \hookrightarrow \Pi_y \rho_2,$$

where Π_x and Π_y are defined elementwise as averaging operators in the coordinate direction indicated by the subscript so as to avoid *shear locking*. Among the possible numerical tricks aiming at avoiding *membrane locking*, we choose the one where the membrane strains are computed using the plane elastic strains on Ω:

$$\beta_{11} \hookrightarrow \Pi_x \beta_{11} + cR_y w, \quad \beta_{22} \hookrightarrow \Pi_y \beta_{22} + cR_x w,$$
$$\beta_{12} \hookrightarrow \frac{1}{2}\left(\frac{\partial u}{\partial y} + \frac{\partial v}{\partial x} + c\Pi_x w + c\Pi_y w + aR_x w + bR_y w\right).$$

Here R_x, R_y are certain difference operators, see [2, 3, 4] for more details on this formulation and its relation to current engineering practice.

We compare the above strategies by setting $p = 12$ in the 'p-version' and by subdividing each macroelement uniformly into 64 rectangles in the 'h-version' so that we have approximately 12000 degrees of freedom in both cases. The results of benchmark computations are reported in Figs. 4–6 showing the transverse deflection along the line $x = \frac{1}{2}$ as well as along the characteristic line $y = 0$ in hyperbolic and parabolic shell geometries. The results show that the 'h-version' is here clearly inferior to the 'p-version' — especially in resolving the line layer in hyperbolic geometry. In view of the theoretical predictions in [4], this is not so surprising.

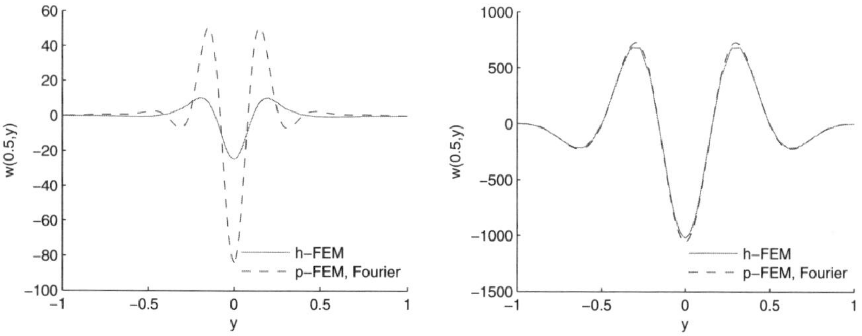

Fig. 4 Hyperbolic & Parabolic shells: Comparison of different methods along the line $x = \frac{1}{2}$.

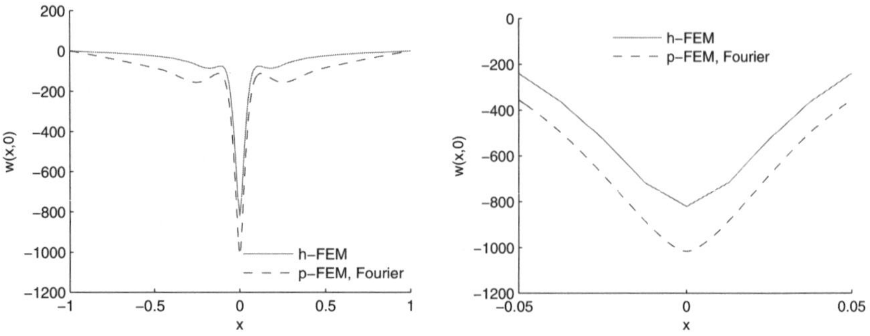

Fig. 5 Hyperbolic shell: Comparison of different methods along the characteristic line $y = 0$.

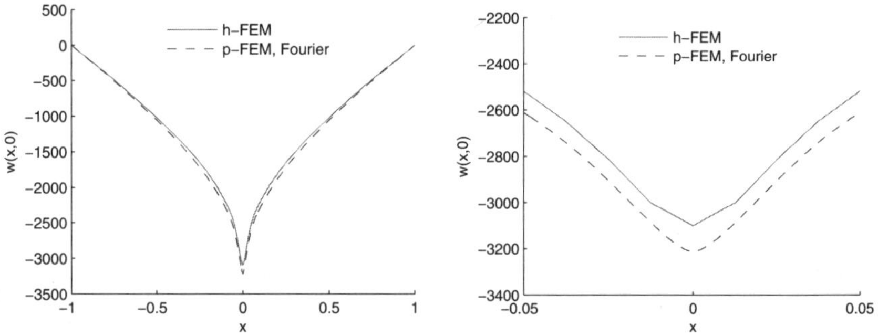

Fig. 6 Parabolic shell: Comparison of different methods along the characteristic line $y = 0$.

References

1. Lukasiewicz, S.: Local loads in plates and shells. Sijthoff & Noordhoff, The Netherlands (1979)
2. Malinen, M.: On the classical shell model underlying bilinear degenerated shell finite elements. Int. J. Numer. Meth. Engng **52**, 389–416 (2001)
3. Malinen, M.: On the classical shell model underlying bilinear degenerated shell finite elements: general shell geometry. Int. J. Numer. Meth. Engng **55**, 629–652 (2002)
4. Niemi, A.H.: Approximation of shell layers using bilinear elements on anisotropically refined rectangular meshes. Preprint submitted to Comput. Methods Appl. Mech. Engrg (2007)
5. Niemi, A.H., Pitkäranta, J., Hakula, H.: Benchmark computations on point-loaded shallow shells: Fourier vs. FEM. Comput. Methods Appl. Mech. Engrg. **196**, 894–907 (2007)
6. Pitkäranta, J., Matache, A.M., Schwab, C.: Fourier mode analysis of layers in shallow shell deformations. Comput. Methods Appl. Mech. Engrg. **190**, 2943–2975 (2001)

Printing: Krips bv, Meppel, The Netherlands
Binding: Stürtz, Würzburg, Germany